Recommended Dietary Allowances (RDA) and Adequate Intakes (AI) for Vitamins

Age (yr)	Thiamin RDA (mg/day)	Riboflavin RDA (mg/day)	Niacin RDA (mg/day)[a]	Biotin AI (μg/day)	Pantothenic acid AI (mg/day)	Vitamin B_6 RDA (mg/day)	Folate RDA (μg/day)[b]	Vitamin B_{12} RDA (μg/day)	Choline AI (mg/day)	Vitamin C RDA (mg/[illegible]	Vitamin A RDA (μg/[illegible]	Vitamin [illegible] AI (μg/d[illegible]	Vitamin [illegible] RDA (n[illegible]	Vitami[illegible] AI (μg/day)
Infants														
0–0.5	0.2	0.3	2	5	1.7	0.1	65	0.4	125	40	400	5	4	2.0
0.5–1	0.3	0.4	4	6	1.8	0.3	80	0.5	150	50	500	5	5	2.5
Children														
1–3	0.5	0.5	6	8	2	0.5	150	0.9	200	15	300	5	6	30
4–8	0.6	0.6	8	12	3	0.6	200	1.2	250	25	400	5	7	55
Males														
9–13	0.9	0.9	12	20	4	1.0	300	1.8	375	45	600	5	11	60
14–18	1.2	1.3	16	25	5	1.3	400	2.4	550	75	900	5	15	75
19–30	1.2	1.3	16	30	5	1.3	400	2.4	550	90	900	5	15	120
31–50	1.2	1.3	16	30	5	1.3	400	2.4	550	90	900	5	15	120
51–70	1.2	1.3	16	30	5	1.7	400	2.4	550	90	900	10	15	120
>70	1.2	1.3	16	30	5	1.7	400	2.4	550	90	900	15	15	120
Females														
9–13	0.9	0.9	12	20	4	1.0	300	1.8	375	45	600	5	11	60
14–18	1.0	1.0	14	25	5	1.2	400	2.4	400	65	700	5	15	75
19–30	1.1	1.1	14	30	5	1.3	400	2.4	425	75	700	5	15	90
31–50	1.1	1.1	14	30	5	1.3	400	2.4	425	75	700	5	15	90
51–70	1.1	1.1	14	30	5	1.5	400	2.4	425	75	700	10	15	90
>70	1.1	1.1	14	30	5	1.5	400	2.4	425	75	700	15	15	90
Pregnancy														
≤18	1.4	1.4	18	30	6	1.9	600	2.6	450	80	750	5	15	75
19–30	1.4	1.4	18	30	6	1.9	600	2.6	450	85	770	5	15	90
31–50	1.4	1.4	18	30	6	1.9	600	2.6	450	85	770	5	15	90
Lactation														
≤18	1.4	1.6	17	35	7	2.0	500	2.8	550	115	1200	5	19	75
19–30	1.4	1.6	17	35	7	2.0	500	2.8	550	120	1300	5	19	90
31–50	1.4	1.6	17	35	7	2.0	500	2.8	550	120	1300	5	19	90

NOTE: For all nutrients, values for infants are AI.

[a]Niacin recommendations are expressed as niacin equivalents (NE), except for recommendations for infants younger than 6 months, which are expressed as preformed niacin.

[b]Folate recommendations are expressed as dietary folate equivalents (DFE).

[c]Vitamin A recommendations are expressed as retinol activity equivalents (RAE).

[d]Vitamin D recommendations are expressed as cholecalciferol.

[e]Vitamin E recommendations are expressed as α-tocopherol.

Recommended Dietary Allowances (RDA) and Adequate Intakes (AI) for Minerals

Age (yr)	Sodium AI (mg/day)	Chloride AI (mg/day)	Potassium AI (mg/day)	Calcium AI (mg/day)	Phosphorus RDA (mg/day)	Magnesium RDA (mg/day)	Iron RDA (mg/day)	Zinc RDA (mg/day)	Iodine RDA (μg/day)	Selenium RDA (μg/day)	Copper RDA (μg/day)	Manganese AI (mg/day)	Fluoride AI (mg/day)	Chromium AI (μg/day)	Molybdenum RDA (μg/day)
Infants															
0–0.5	120	180	400	210	100	30	0.27	2	110	15	200	0.003	0.01	0.2	2
0.5–1	370	570	700	270	275	75	11	3	130	20	220	0.6	0.5	5.5	3
Children															
1–3	1000	1500	3000	500	460	80	7	3	90	20	340	1.2	0.7	11	17
4–8	1200	1900	3800	800	500	130	10	5	90	30	440	1.5	1.0	15	22
Males															
9–13	1500	2300	4500	1300	1250	240	8	8	120	40	700	1.9	2	25	34
14–18	1500	2300	4700	1300	1250	410	11	11	150	55	890	2.2	3	35	43
19–30	1500	2300	4700	1000	700	400	8	11	150	55	900	2.3	4	35	45
31–50	1500	2300	4700	1000	700	420	8	11	150	55	900	2.3	4	35	45
51–70	1300	2000	4700	1200	700	420	8	11	150	55	900	2.3	4	30	45
>70	1200	1800	4700	1200	700	420	8	11	150	55	900	2.3	4	30	45
Females															
9–13	1500	2300	4500	1300	1250	240	8	8	120	40	700	1.6	2	21	34
14–18	1500	2300	4700	1300	1250	360	15	9	150	55	890	1.6	3	24	43
19–30	1500	2300	4700	1000	700	310	18	8	150	55	900	1.8	3	25	45
31–50	1500	2300	4700	1000	700	320	18	8	150	55	900	1.8	3	25	45
51–70	1300	2000	4700	1200	700	320	8	8	150	55	900	1.8	3	20	45
>70	1200	1800	4700	1200	700	320	8	8	150	55	900	1.8	3	20	45
Pregnancy															
≤18	1500	2300	4700	1300	1250	400	27	12	220	60	1000	2.0	3	29	50
19–30	1500	2300	4700	1000	700	350	27	11	220	60	1000	2.0	3	30	50
31–50	1500	2300	4700	1000	700	360	27	11	220	60	1000	2.0	3	30	50
Lactation															
≤18	1500	2300	5100	1300	1250	360	10	13	290	70	1300	2.6	3	44	50
19–30	1500	2300	5100	1000	700	310	9	12	290	70	1300	2.6	3	45	50
31–50	1500	2300	5100	1000	700	320	9	12	290	70	1300	2.6	3	45	50

Tolerable Upper Intake Levels (UL) for Vitamins

Age (yr)	Niacin (mg/day)[a]	Vitamin B_6 (mg/day)	Folate (μg/day)[a]	Choline (mg/day)	Vitamin C (mg/day)	Vitamin A (μg/day)[b]	Vitamin D (μg/day)	Vitamin E (mg/day)[c]
Infants								
0–0.5	—	—	—	—	—	600	25	—
0.5–1	—	—	—	—	—	600	25	—
Children								
1–3	10	30	300	1000	400	600	50	200
4–8	15	40	400	1000	650	900	50	300
Adolescents								
9–13	20	60	600	2000	1200	1700	50	600
14–18	30	80	800	3000	1800	2800	50	800
Adults								
19–70	35	100	1000	3500	2000	3000	50	1000
>70	35	100	1000	3500	2000	3000	50	1000
Pregnancy								
≤18	30	80	800	3000	1800	2800	50	800
19–50	35	100	1000	3500	2000	3000	50	1000
Lactation								
≤18	30	80	800	3000	1800	2800	50	800
19–50	35	100	1000	3500	2000	3000	50	1000

[a]The UL for niacin and folate apply to synthetic forms obtained from supplements, fortified foods, or a combination of the two.

[b]The UL for vitamin A applies to the preformed vitamin only.

[c]The UL for vitamin E applies to any form of supplemental α-tocopherol, fortified foods, or a combination of the two.

Tolerable Upper Intake Levels (UL) for Minerals

Age (yr)	Sodium (mg/day)	Chloride (mg/day)	Calcium (mg/day)	Phosphorus (mg/day)	Magnesium (mg/day)[d]	Iron (mg/day)[b]	Zinc (mg/day)	Iodine (μg/day)	Selenium (μg/day)	Copper (μg/day)	Manganese (mg/day)	Fluoride (mg/day)	Molybdenum (μg/day)	Boron (mg/day)	Nickel (mg/day)
Infants															
0–0.5	—[e]	—[e]	—	—	—	40	4	—	45	—	—	0.7	—	—	—
0.5–1	—[e]	—[e]	—	—	—	40	5	—	60	—	—	0.9	—	—	—
Children															
1–3	1500	2300	2500	3000	65	40	7	200	90	1000	2	1.3	300	3	0.2
4–8	1900	2900	2500	3000	110	40	12	300	150	3000	3	2.2	600	6	0.3
Adolescents															
9–13	2200	3400	2500	4000	350	40	23	600	280	5000	6	10	1100	11	0.6
14–18	2300	3600	2500	4000	350	45	34	900	400	8000	9	10	1700	17	1.0
Adults															
19–70	2300	3600	2500	4000	350	45	40	1100	400	10,000	11	10	2000	20	1.0
>70	2300	3600	2500	3000	350	45	40	1100	400	10,000	11	10	2000	20	1.0
Pregnancy															
≤18	2300	3600	2500	3500	350	45	34	900	400	8000	9	10	1700	17	1.0
19–50	2300	3600	2500	3500	350	45	40	1100	400	10,000	11	10	2000	20	1.0
Lactation															
≤18	2300	3600	2500	4000	350	45	34	900	400	8000	9	10	1700	17	1.0
19–50	2300	3600	2500	4000	350	45	40	1100	400	10,000	11	10	2000	20	1.0

[d]The UL for magnesium applies to synthetic forms obtained from supplements or drugs only.

[e]Source of intake should be from human milk (or formula) and food only.

NOTE: An Upper Limit was not established for vitamins and minerals not listed and for those age groups listed with a dash (—) because of a lack of data, not because these nutrients are safe to consume at any level of intake. All nutrients can have adverse effects when intakes are excessive.

Personal Nutrition

SIXTH EDITION

Personal Nutrition

SIXTH EDITION

Marie A. Boyle
SOUTHERN MAINE COMMUNITY COLLEGE

Sara Long
SOUTHERN ILLINOIS UNIVERSITY

Australia • Brazil • Canada • Mexico • Singapore • Spain
United Kingdom • United States

Personal Nutrition, Sixth Edition
Marie A. Boyle and Sara Long

Publisher: Peter Marshall
Development Editor: Elizabeth Howe
Assistant Editor: Elesha Feldman
Editorial Assistant: Lauren Vogelbaum
Technology Project Manager: Donna Kelley
Marketing Manager: Jennifer Somerville
Marketing Assistant: Catie Ronquillo
Marketing Communications Manager: Jessica Perry
Project Manager, Editorial Production: Sandra Craig
Creative Director: Rob Hugel
Art Director: Lee Friedman
Print Buyer: Doreen Suzuki
Permissions Editor: Roberta Broyer
Production: Martha Emry
Text Designer: Diane Beasley
Photo Researcher: Stephen Forsling
Copy Editor: Mary Douglas
Illustrations: Jim Atherton
Cover Designer: Charlie Wayne Denison
Cover Image: ©Lee Frost/Robert Harding World Imagery/ Getty Images
Compositor: Lachina Publishing Services
Printer: CTPS

Printed in China
3 4 5 6 7 10 09 08 07

For more information about our products, contact us at:
Thomson Learning Academic Resource Center
1-800-423-0563
For permission to use material from this text or product, submit a request online at **http://www.thomsonrights.com.**
Any additional questions about permissions can be submitted by e-mail to **thomsonrights@thomson.com.**

Library of Congress Control Number: 2006904581

ISBN-13: 978-0-495-01934-3
ISBN-10: 0-495-01934-8

Thomson Higher Education
10 Davis Drive
Belmont, CA 94002-3098
USA

DEDICATION

In memory of my father,
David M. Boyle,
and his love of everything Irish,
especially his daughters,
and my mother,
Marie T. Boyle,
a lifelong educator,
cherished mother,
and treasured friend.

—Marie A. Boyle Struble

This is dedicated to my husband,
the love of my life: Kevin.
I cannot comprehend
what life would be without you!

—Sara Long Roth

About the Authors

MARIE BOYLE STRUBLE, PhD, RD, received her BA in psychology from the University of Southern Maine and her MS and PhD in nutrition from Florida State University in Tallahassee, Florida. She is coauthor of the community nutrition textbook *Community Nutrition in Action: An Entrepreneurial Approach* and Chair of the Dietetics Program at Southern Maine Community College in South Portland, Maine. She also teaches online distance courses in such areas as Community Nutrition, Nutrition and Aging, and Nutrition Applications of Psychological and Sociological Issues for the Graduate Program in Nutrition at the College of Saint Elizabeth in Morristown, New Jersey. Her other professional activities include serving as an author and reviewer for the American Dietetic Association and Society for Nutrition Education. She coauthored the current Position Paper of the American Dietetic Association on *Addressing World Hunger, Malnutrition, and Food Insecurity*. She also serves as Editor of the *Journal of Hunger and Environmental Nutrition* by Haworth Press. She is a member of the American Dietetic Association, the American Public Health Association, and the Society for Nutrition Education.

SARA LONG ROTH, PhD, RD, is Professor in the Department of Animal Science, Food and Nutrition and Director, Didactic Program in Dietetics at Southern Illinois University. Prior to obtaining her PhD in health education, she practiced as a clinical dietitian for 11 years. Her specialty areas are medical nutrition therapy, nutrition education, and food and nutrition assessment. She is an active leader in national, state, and district dietetic associations where she has served in numerous elected and appointed positions, including the Commission on Accreditation of Dietetics Education, and the Commission on Dietetic Registration. Dr. Long is coauthor of *Understanding Nutrition Therapy and Pathophysiology, Medical Nutrition Therapy: A Case Study Approach, Foundations and Clinical Applications of Nutrition: A Nursing Approach*, and *Essentials of Nutrition and Diet Therapy*. Dr. Long has received various awards and honors for teaching, including Outstanding Dietetic Educator (ADA) and Outstanding Educator for the College of Agricultural Sciences.

Contents in Brief

Contents

11 The Life Cycle: Conception through the Later Years 345

12 Food Safety and the Global Food Supply 385

Appendixes

Preface

This Sixth Edition of *Personal Nutrition* reflects the same vision we had in writing the first edition of this book almost 20 years ago—that is, to apply basic nutrition concepts to personal everyday life. The text is designed to support the many one- to four-credit introductory nutrition courses available to students today from a variety of majors. This edition incorporates the many changes that have taken place in the field of nutrition in recent years, and offers all readers the opportunity to develop practical skills in making decisions regarding their personal nutrition and health. Our challenge has been to teach facts about nutrition, to nurture critical thinking skills, and to motivate readers to apply what they learn in daily life.

Nutrition is a subject that is forever changing. Since the last edition was published, the USDA published new *Dietary Guidelines for Americans*, as well as the MyPyramid Food Guidance System. We have witnessed the emergence of new research findings regarding the phytochemicals and the pervasive marketing of functional foods for disease prevention. Additionally, we have been challenged by the increasing cultural, ethnic, and generational diversity of our society, recent advances in biotechnology, and the parallel trends toward supersized food portions and obesity. Nutrition claims bombard us frequently in advertising and articles about diet and nutrition on television, radio, and the Internet, and in newspapers and magazines. It is important that consumers have the knowledge to evaluate the nutrition issues and controversies. This Sixth Edition of *Personal Nutrition* continues to provide a sieve through which to separate the valid nutrition information from the rest.

Chapter Content

Chapter 1 introduces the basic nutrients the body needs and provides a personal invitation to eat well for optimum health. It assists the reader in becoming a sophisticated consumer of new information about nutrition, and explores the factors that affect food choices, including the media, advertising, and cultural factors. Chapter 2 describes the new MyPyramid food guide and related nutrition tools, and the most recent dietary guidelines needed to help make sound food choices. It provides sample food labels for understanding the nutrition information, terminology, and health claims found on labels. Chapter 2 also includes a section on various international and ethnic cuisines that highlights the multicultural heritage of our country. Chapters 3 through 7 present the nutrients and show how they all work together to nourish the body. The chapters on vitamins and minerals spotlight the emerging importance of the antioxidant nutrients and phytonutrients and also feature new colorful food photos depicting excellent food sources for individual vitamins and minerals. Alcohol is covered in depth in Chapter 8 and provides students with important information on alcohol's relationship to nutrition and health, helping them make informed and responsible decisions. Chapter 9 discusses weight management issues and compares major weight-loss programs. Chapter 10 addresses the relationships between nutrition and personal fitness. Chapter 11 describes the special nutrition needs and concerns that arise during the various stages of the life cycle from conception through old age. The new MyPyramid for Kids Food Guide is included. Chapter 12 addresses consumer concerns about the safety of our food supply and provides a glimpse at some of the problems and advantages of the newer food technologies.

Features

The new *Eat Well Be Well* features throughout the text motivate readers to make good health a priority and provide suggestions for making the best food and lifestyle choices for healthy living and disease prevention. This feature includes practical tips for today's student that offer health benefits for a lifetime. Topics include "Whole Grains for Health," "Nourish the Heart," "Never Say 'Diet'," "An Eating Pattern for Longevity," and "Color Your Plate for Health."

The *Savvy Diner* feature in every chapter provides practical suggestions for healthy eating and reinforces the recommendations made in the *Dietary Guidelines for Americans*. The *Savvy Diners* include tips for supermarketing, choosing healthy portion sizes, consuming heart-healthy diets, eating more beans, preserving vitamins in foods, seasoning foods without excess salt, dining out defensively, eating for peak performance, creating tasteful meals for one, and practicing home food safety.

The *Ask Yourself* sections at the beginning of each chapter contain a set of true/false questions designed to provide readers with a preview of the chapter's contents. Answers to the questions appear on the following page. New to this edition is *In Review* end-of-chapter study questions that follow every chapter, providing a consistent resource for students to review key concepts from the text. Answers are included in Appendix E. The *Nutrition on the Web* feature is located at the end of each chapter and contains an annotated list of World Wide Web addresses that provides links to reliable sources of nutrition and health information from the Internet related to the chapter's topics. Moreover, you can link with the Internet addresses presented in this book through the publisher's Nutrition Resource Center online at **www.thomsonedu.com/nutrition/boyle**. A new feature—*The Menu of Online Study Tools*—appears at the end of every chapter. The Menu contains a list of helpful study tools (practice tests, flash cards, glossary, web links, and animations) and video clips available on the student website.

The *Nutrition Action* features that appear in every chapter are magazine-style essays that keep you abreast of current topics important to the nutrition-conscious consumer. The *Nutrition Action* features address topics such as fast food, smart snacking, the Mediterranean Diet, food allergies, medicinal herbs, diet and blood pressure, and aging well with exercise. Several *Nutrition Action* sections have been updated to reflect the latest issues in the field. For example, "Carbohydrates—Friend or Foe?" helps consumers choose healthful carbohydrates while making sense of the carbohydrate debate; "Diet Confusion: Weighing the Evidence" helps readers make sense of the current weight-loss scene; and "Eat Fresh Eat Local" includes the earth-friendly benefits of eating fresh and locally grown organic foods.

Scorecards are hands-on features included in every chapter. *Scorecards* allow readers to evaluate their own nutrition behaviors and knowledge in many areas. Some of the *Scorecards* assist readers in assessing their longevity, overall diet, fruit and vegetable consumption, calcium intake, weight status, exercise habits, and food safety know-how.

The final special feature of each chapter is the *Spotlight*—which have been thoroughly updated. Each addresses a common concern people have about nutrition. *Spotlight* topics include nutrition and the media, ethnic cuisines, alternative sweeteners, diet and heart disease, the benefits derived from soy foods, nutrition and cancer prevention, osteoporosis, fetal alcohol syndrome, eating disorders, popular fitness aids and supplements, and childhood obesity. Chapter 12 *Spotlight* covers the many factors that influence nutrition and food insecurity among the people of the world, and underscores that the practical suggestions offered throughout this book for attaining the ideals of personal nutrition are the very suggestions that best support the health of the whole earth as well. The *Spotlights* continue in their question and answer format to encourage the reader to ask further questions about nutrition issues. We encourage you to ask us questions, too, in care of the publisher.

The Appendixes have been updated. Appendix A provides a colorfully illustrated introduction to the workings of the human body; Appendix B presents aids to calculations, including how to calculate the percentage of calories from fat in one's diet. Appendix B also provides a series of photos depicting the U.S. Food Exchange System. Appendix C includes the Canadian Dietary Guidelines and recommendations for physical activity; Appendix D includes the chapter reference notes; Appendix E includes answers to the end-of-chapter review questions, and Appendix F includes our Table of Food Composition. The Glossary of terms that follows the Appendixes provides a quick reference to the nutrition terminology defined in the margins of the text and can be used as a review tool.

We welcome you to the fascinating subject of nutrition. We hope that the book speaks to you personally and that you find it practical for your everyday use. We hope, too, that by reading it you may enhance your own personal nutrition and health.

Acknowledgments

We are grateful to the many individuals who have made contributions to the development of this Sixth Edition of *Personal Nutrition*. We thank our family and friends for their continued support and encouragement throughout this endeavor and countless others. We appreciate the insights provided by our colleagues—especially to all those individuals who have contributed their expertise to previous editions of this text—including Eleanor Whitney, Diane Morris, Gail Zyla, Kathleen Shimomura Morgan, and Kathy Roberts. Their insights are reflected in this new edition, still. We are indebted to Brian K. Jones. Without his assistance, we would never have been able to meet deadlines. Thank you, Brian—you will make SIUC proud and be an asset to the profession of dietetics! We appreciate the hard work and expertise of the team of authors preparing ancillary material for this edition. Thanks to Art Gilbert and Elizabeth Morton for their work on the instructor's manual, Charalee Allen for her excellent electronic slide presentations, Judy Kaufman for creating test bank questions, and Melissa Wdowik for preparing website content. Thanks to Shazia Nathoo at Axxya Systems for creating the food composition table found in Appendix F and the computerized diet analysis program that accompanies this book. Special thanks go to the editorial team and their staff: Peter Marshall, Publisher; Elizabeth Howe, Developmental Editor; Martha Emry, Production Service; and Sandra Craig, Project Manager. Their guidance ensured the highest quality of work throughout all facets of this production. We are especially grateful to Martha Emry, because this text would not appear as it does today without her tireless commitment to excellence. We appreciate the work of Donna Kelley in creating and managing the technology products that accompany this text, and Elesha Feldman for her work on the student and instructor resources. As always, our gratitude goes to Jennifer Somerville, Marketing Manager, and her team for their fine efforts in marketing this book. Our thanks to the many sales representatives who will introduce this new book to its readers. Our appreciation goes to other members of the production team:

Jim Atherton, artist; Mary Douglas, copy editor; Martha Ghent, proofreader; and Stephen Forsling, photo researcher. We are also indebted to everyone at Lachina Publishing Services for their hard work and diligence in producing a text to be proud of. Last, but not least, we owe much to our colleagues who provided expert reviews of the manuscript, not only for their ideas and suggestions, many of which made their way into the text, but also for their continued enthusiasm, support, and interest in *Personal Nutrition*. Thanks to all of you:

Debra Boardley, *University of Toledo*
Art Gilbert, *University of California, Santa Barbara*
Joanne Gould, *Kean University*
Judy Kaufman, *Monroe Community College*
Zaheer Ali Kirmani, *Sam Houston State University*
Lori Miller Kohler, *Minneapolis Community and Technical College*
Rosa A. Mo, *University of New Haven*
Elizabeth K. Morton, *University of South Carolina*
Robin S. Schenk, *Buffalo State, SUNY*
Tricia Steffen, *Nova Southeastern University*
Melissa Wdowik, *University of North Carolina, Charlotte*
Sharman Willmore, *Cincinnati State Technical and Community College*
Linda O. Young, *University of Nebraska, Lincoln*

MARIE BOYLE STRUBLE
SARA LONG ROTH
JUNE 2006

Personal Nutrition

SIXTH EDITION

The Art of Understanding Nutrition

1

ASK YOURSELF . . .

Which of the following statements about nutrition are true, and which are false? For each false statement, what is true?

1. It is possible to have an appetite without being hungry.
2. Most people obtain information about nutrition from health professionals.
3. The way people choose to live and eat can affect their health and quality of life as they age.
4. Vitamins and minerals supply calories.
5. You can order a low-fat, balanced meal at a fast-food outlet.
6. Healthful diets cost more than relatively unhealthful diets.
7. When a person suffers from malnutrition, it means that he or she is taking in too few nutrients.
8. A nutritionist is a professional who is certified to advise people on nutrition.
9. The notion of eating insects repels people around the world.
10. The more current a dietary claim, the more you can trust its accuracy and reliability.

Answers found on the following page.

Tell me what you eat, and I will tell you what you are.

Anthelme Brillat-Savarin (1755–1826, French politician and gourmet; author of *Physiology of Taste*)

CONTENTS

Stroll down the aisle of any supermarket, and you'll see all manner of foods touting such claims as "low-fat," "low-carb," "low-calorie," and "fat-free." Flip through the pages of just about any magazine, and you're likely to find advice on how to lose weight. Walk into any gym, and you'll probably hear members discussing the merits of one performance-enhancing food or another. All this boils down to the fact that nutrition has become part and parcel of the American lifestyle.

It wasn't always that way, however. The field of **nutrition** is a relative newcomer on the scientific block. Although Hippocrates recognized diet as a component of health back in 400 B.C., only in the past one hundred years or so have researchers begun to understand that carbohydrates, fats, and proteins are needed for normal growth. The next nutrition breakthrough—the discovery of the first vitamin—occurred in the early 1900s. It wasn't until 1928, when an organization called the American Institute of Nutrition was formed, that nutrition was officially looked upon as a distinct field of study.[1]* It took several more decades before nutrition achieved its current status as one of the most talked about scientific disciplines.

*Reference notes for each chapter are in Appendix D.

Photo © Daryl Solomon/Envision

nutrition the study of foods, their nutrients and other chemical components, their actions and interactions in the body, and their influence on health and disease.

Today we spend billions of dollars each year to investigate the many aspects of nutrition, a science that encompasses not only the study of vitamins, minerals, and other nutrients, but also such diverse subjects as alcohol, caffeine, and pesticides. In addition, nutrition scientists continually expand our understanding of the impact food has on our bodies by examining research in chemistry, physics, biology, biochemistry, genetics, immunology, and other nutrition-related fields. A number of other disciplines also make valuable contributions to the study of nutrition. These related fields include psychology, anthropology, epidemiology, geography, agriculture, ethics, economics, sociology, and philosophy.

Even though science has shown us that to some extent we really are what we eat, many consumers have become more confused than ever about how to translate the steady stream of new findings about nutrition into a lifestyle of healthful eating. Reportedly, Americans spend $42 billion a year on diet and health books. As Table 1-1 illustrates, people of all ages make food purchasing decisions based on claims regarding nutrition and health.[2] Each additional nugget of nutrition news that comes along raises new concerns: Is caffeine bad for me? Should I take

TABLE 1-1

Proportion of Food Shoppers Who Have Purchased a Product Based on a Specified Health-Related Claim on the Package

		% by generation*			
Claim	Total %	Gen Y	Gen X	Boomers	Matures
Low-fat	63	52	61	64	66
Whole-grain	62	49	55	68	61
Low in saturated fats	55	44	49	55	60
Low calorie	52	42	52	54	52
High in calcium	51	58	55	47	48
High in vitamin C	51	57	59	49	46
Low sodium	48	31	46	47	56
Vitamin-rich or vitamin-fortified	47	51	52	47	41
Sugar-free	46	33	46	46	49
Reduces risk of heart disease	42	31	36	43	49
Low-carb	40	41	39	43	36
Reduces risk of cancer	26	19	19	24	35
No. of Interviews	**1,001**	**75**	**221**	**382**	**292**

*Generation birth years: Generation Y = 1981–1994; Generation X = 1965–1980; Baby Boomers = 1946–1964; Matures = Pre-1946.

SOURCE: A. E. Sloan, "Top Ten Global Food Trends," *Food Technology* 59 (2005): 31. Data from the Food Marketing Institute, *Trends in the United States: Consumer Attitudes and the Supermarket, 2005 Edition* (Washington, D.C.: The Research Department, Food Marketing Institute).

Ask Yourself Answers: **1.** True. **2.** False. Most Americans look first to television for nutrition information, then to magazines, and then to newspapers. **3.** True. **4.** False. Only protein, carbohydrate, and fat supply calories. **5.** True. **6.** False. People can save money when they switch from a typical high-fat diet to the grain-based, produce-rich diet recommended by health experts. **7.** False. Malnutrition can be caused either by taking in too few nutrients or by consuming excess nutrients. **8.** False. A nutritionist is a person who claims to specialize in the study of nutrition, but some are self-described experts whose training is questionable. A registered dietitian (RD), however, is recognized as a nutrition expert—with training in nutrition, food science, and diet planning. **9.** False. It's true that most people in the United States and Canada find the idea of eating insects repulsive because the practice is not part of the American culture. But people in many other countries, because they have been brought up in cultures in which insects have long been a traditional food, consider dishes prepared with insects a delicacy. **10.** False. If a nutrition claim is too new, it may not have been adequately tested. Findings must be confirmed many times over by experiments and evaluated in light of other knowledge before they can be translated into recommendations for the public.

vitamin supplements? Do diet pills work? Can a sports drink improve my performance? Do pesticides really pose a hazard?

Some manufacturers and media outlets feed into the confusion by offering health-conscious consumers unreliable products and misleading dietary advice, often making unsubstantiated claims for a number of nutritional products, including supplements touted as fat melters, muscle builders, and energy boosters. Unfortunately, misinformation runs rampant in the marketplace. Americans spend more than $30 billion annually on medical and nutritional **health fraud** and **quackery,** up from only $1 billion to $2 billion in the early 1960s.[3] Consider, for example, that college athletes alone may spend hundreds of dollars a month on nutritional supplements, even though most of the products pitched to serious exercisers are useless and, in some cases, potentially harmful. At the same time, the sale of weight-loss foods, products, and services—not all of them sound—has become a $40 billion industry.[4]

To be sure, the widespread interest in nutrition has generated some positive changes in the marketplace. Whereas the sale of fresh fruit, salads, and other low-fat items such as frozen yogurt in fast-food chains was virtually unheard of at one time, those eateries couldn't survive in the current nutrition-conscious environment without offering such healthful fare.[5] (See the Nutrition Action feature later in the chapter for tips on eating healthfully at fast-food outlets.) By the same token, food manufacturers have responded to consumer concerns about diet by developing new technologies, such as the creation of fat substitutes, to provide shoppers with an unprecedented number of choices at the supermarket.

With the amount of nutrition information and the number of food alternatives always on the rise, choosing a healthful diet can seem like a daunting task. Fortunately, you don't need a degree in nutrition to put the principles of the science to use in your own life. A basic understanding of nutrition can go a long way in helping you protect your health (and your wallet). This book lays the foundation you need to take nutrition science out of the laboratory and move it into your kitchen, both today and tomorrow. The first steps are understanding the nature of the nutrients themselves and exploring the current thrust of the field of nutrition.

health fraud conscious deceit practiced for profit, such as the promotion of a false or an unproven product or therapy.

quackery fraud. A quack is a person who practices health fraud.

quack = to boast loudly

nutrients substances obtained from food and used in the body to promote growth, maintenance, and repair. The nutrients include carbohydrate, fat, protein, vitamins, minerals, and water.

essential nutrients nutrients that must be obtained from food because the body cannot make them for itself.

© Najiah Feanny/CORBIS SABA

▲ *Some stores sell pills and potions touted as fat melters, energy boosters, and muscle builders.*

The Nutrients in Foods

Almost any food you eat is mostly water, and some foods are as high as 99 percent water. The bulk of the solid materials consists of carbohydrate, fat, and protein. If you could remove these materials, you would detect a tiny residue of minerals, vitamins, and other materials. Water, carbohydrate, fat, protein, vitamins, and some of the minerals are **nutrients.** Some of the other materials are not nutrients. The six classes of nutrients are carbohydrate, fat, protein, vitamins, minerals, and water.

A complete chemical analysis of your body would show that it is made of similar materials in roughly the same proportions as most foods. For example, if you weigh 150 pounds (and if that is a desirable weight for you), your body contains about 90 pounds of water and about 30 pounds of fat. The other 30 pounds consist of mostly protein, carbohydrate, and the major minerals of your bones—calcium and phosphorus. Vitamins, other minerals, and incidental extras constitute a fraction of a pound.

Scientists use the term **essential nutrient** to describe the nutrients that the body must obtain from food. About 40 nutrients are known to be essential; that is, they are compounds that the body cannot make for itself but are indispensable to life processes. How can you be sure you're getting all the nutrients you need? The rest of this chapter, along with the diet planning tools presented in Chapter 2, will help you design a diet that covers all of your body's needs.

The six classes of nutrients:

- carbohydrate
- fat
- protein
- vitamins
- minerals
- water

The Energy-Yielding Nutrients

The energy-yielding nutrients:

- carbohydrate
- fat
- protein

On being broken down in the body, or digested, three of the nutrients—carbohydrate, protein, and fat—yield the **energy** that the body uses to fuel its various activities. In contrast, vitamins, minerals, and water, once broken down in the body, do not yield energy but perform other tasks, such as maintenance and repair. Each gram of carbohydrate and protein consumed supplies your body with 4 calories, and each gram of fat provides 9 calories (see Figure 1-1). Only one other compound that people consume provides calories, and that is alcohol, which provides 7 calories per gram. Alcohol is not considered a nutrient, however, because it does not help maintain or repair body tissues the way nutrients do.

The body uses energy from carbohydrate, fat, and protein to do work or generate heat. This energy is measured in **calories**—familiar to most everyone as markers of how "fattening" foods are. If your body doesn't "release" the energy you obtained from a food soon after you've eaten it, it stores it, usually as body fat, for use later. If excess amounts of protein, fat, or carbohydrate are eaten fairly regularly, the stored fat builds up over time and leads to obesity. Too much of any food, whether lean meat (a protein-rich food), potatoes (a high-carbohydrate food), or butter (a fatty food), can contribute excess calories that result in overweight.

Vitamins, Minerals, and Water

Unlike carbohydrate, fat, and protein, **vitamins** and **minerals** do not supply energy, or calories. Instead, they regulate the release of energy and other aspects of **metabolism.** As Table 1-2 shows, there are 13 vitamins, each with its special role to play. Vitamins are divided into two classes: water-soluble (the B vitamins and vitamin C) and fat-soluble (vitamins A, D, E, and K). This distinction has many implications for the kinds of foods that provide the different vitamins and how the body uses them, as you will see in Chapter 6.

The minerals also perform important functions. Some, such as calcium, make up the structure of bones and teeth. Others, including sodium, float about in the body's fluids, where they help regulate crucial bodily functions, such as heartbeat and muscle contractions.

Often neglected but equally vital, **water** is the medium in which all the body's processes take place. Some 60 percent of your body's weight is water, which carries materials to and from cells and provides the warm, nutrient-rich bath in which cells thrive. Water also transports hormonal messages from place to place. When energy-yielding nutrients release energy, they break

energy the capacity to do work, such as moving or heating something.

calorie the unit used to measure energy.

vitamins organic, or carbon-containing, essential nutrients that are vital to life and needed in minute amounts.

vita = life
amine = containing nitrogen

minerals inorganic compounds, some of which are essential nutrients.

metabolism encompasses all of the chemical and physical reactions occurring in living cells, including the reactions by which the body obtains and uses energy from foods.

water provides the medium for life processes.

TABLE 1-2
The Vitamins and Minerals

The Vitamins			The Minerals			
The water-soluble vitamins:		**The fat-soluble vitamins:**	**The major minerals:**		**The trace minerals:**	
B vitamins		Vitamin A	Calcium	Potassium	Chromium	Manganese
Thiamin	Vitamin B_{12}	Vitamin D	Chloride	Sodium	Copper	Molybdenum
Riboflavin	Folate	Vitamin E	Magnesium	Sulfur	Fluoride	Selenium
Niacin	Biotin	Vitamin K	Phosphorus		Iodine	Zinc
Vitamin B_6	Pantothenic acid				Iron	
Vitamin C						

NOTE: A number of trace minerals are currently under study to determine possible dietary requirements for humans. These include arsenic, boron, cadmium, cobalt, lead, lithium, nickel, silicon, tin, and vanadium.

© John Kelly/Stone/Getty Images

© Felicia Martinez/PhotoEdit

▲ *Remember that 1 gram is a very small amount. For instance, 1 teaspoon of sugar weighs roughly 5 grams.*

Calorie Value of Carbohydrate, Fat, and Protein

If you know the number of grams of carbohydrate, fat, and protein in a food, you can calculate the number of calories in it. Simply multiply the carbohydrate grams by 4, the fat grams by 9, and the protein grams by 4. Add the totals together to obtain the number of calories. For example, a deluxe fast-food hamburger contains about 45 grams of carbohydrate, 27 grams of protein, and 39 grams of fat:

45 grams of carbohydrate × 4 calories	=	180 calories
39 grams of fat × 9 calories	=	351 calories
27 grams of protein × 4 calories	=	108 calories
Total:		639 calories

The percentage of your total energy intake from carbohydrate, fat, and protein can then be determined by dividing the number of calories from each energy nutrient by the total calories, and then multiplying your answer by 100 to get the percentage:

$$\text{calories from carbohydrate} = \frac{45 \times 4 \text{ cal/g}}{639} = 0.281 \times 100 = 28\%$$

$$\text{calories from fat} = \frac{39 \times 9 \text{ cal/g}}{639} = 0.548 \times 100 = 55\%$$

$$\text{calories from protein} = \frac{27 \times 4 \text{ cal/g}}{639} = 0.168 \times 100 = 17\%$$

See Appendix B for help with figuring percentages and other calculations.

FIGURE 1-1
CALORIC VALUES OF CARBOHYDRATE, PROTEIN, FAT, AND ALCOHOL

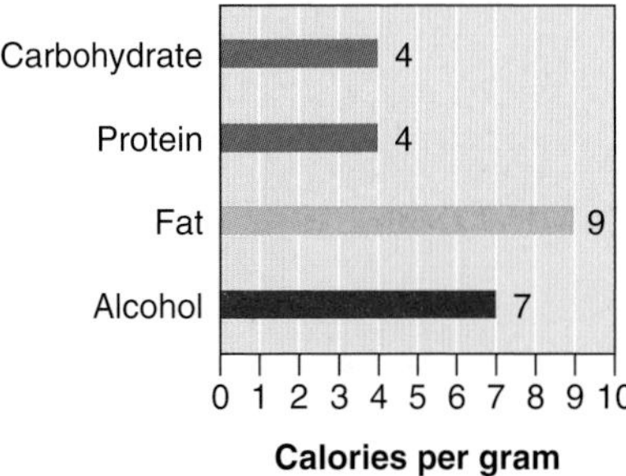

down into water and other simple compounds. Without water, you could live only a few days.

Each day your body loses water in the form of sweat and urine. Therefore, you must replace large amounts of it—on the order of two to three quarts a day. To be sure, you don't need to *drink* that much water daily, because the foods and other beverages you consume do supply some of the water you need.

Nutrition and Health Promotion

In the past, scientists investigating the role that diet plays in health focused on the consequences of getting too little of one nutrient or another. Until the end of World War II, in fact, nutrition researchers concentrated on eliminating deficiency diseases such as **goiter,** a condition in which the thyroid gland swells from lack of the mineral iodine, and **pellagra,** inflammation of the skin caused by deficiency of the B vitamin niacin.

These days, the focus is just the opposite. Deficiency diseases have been virtually eliminated in America because of our country's abundant food supply and the practice of fortifying food with essential nutrients (adding iodine to salt, for example). Yet diseases related to **malnutrition** in the form of dietary excess and imbalance run rampant. Five of the leading causes of death—heart disease, cancer, stroke, diabetes, and hypertension—have been linked to diet (see Table 1-3). Another three are associated with excessive alcohol consumption—accidents, suicide, and liver disease.[6] **Overnutrition** contributes to other ills as well, including obesity and dental disease. Because obesity and a sedentary lifestyle are linked with chronic diseases, such as diabetes, heart disease, and certain cancers, it can be projected that the increased rates of obesity will lead to increased deaths each year, not to mention hospitalizations, disability, lost time on the job, and poor quality of life for many Americans.[7]

goiter (GOY-ter) enlargement of the thyroid gland caused by iodine deficiency.

pellagra (pell-AY-gra) niacin deficiency characterized by diarrhea, inflammation of the skin, and, in severe cases, mental disorders.

malnutrition any condition caused by an excess, deficiency, or imbalance of calories or nutrients.

overnutrition calorie or nutrient overconsumption severe enough to cause disease or increased risk of disease; a form of malnutrition.

TABLE 1-3

Leading Causes of Death in the United States, 2003

Many of the major killers, such as heart disease, some types of cancer, stroke, and diabetes, are influenced by a number of factors, including a person's genetic makeup, eating and exercise habits, exposure to tobacco, and other lifestyle practices.[a] Blue shading indicates a cause of death in which diet can play a part; yellow shading indicates a cause of death in which alcohol use can play a part.

Rank	Cause of Death	Percentage of Total Deaths
1	Heart disease	28.0%
2	Cancer	22.7%
3	Stroke	6.5%
4	Chronic lower-respiratory diseases	5.2%
5[b]	Motor vehicle and other accidents	4.3%
6	Diabetes mellitus	3.0%
7	Influenza and pneumonia	2.7%
8	Alzheimer's disease	2.6%
9	Kidney disease	1.7%
10	Infections of the blood (septicemia)	1.4%
11	Suicide	1.3%
12	Chronic liver disease and cirrhosis	1.1%
13	Hypertension (high blood pressure)	0.9%
14	Parkinson's disease	0.7%
15	Pneumonitis (pneumonia caused by aspiration of solid or liquid materials into the lungs)	0.7%

[a]Lifestyle related (68%); Diet related (61%); Alcohol related (7%).

[b]The leading cause of death for persons aged 15–24 is motor vehicle and other accidents, followed by homicide, suicide, cancer, and heart disease. About half of all accident fatalities are alcohol related.

SOURCE: Centers for Disease Control and Prevention, *National Vital Statistics Report, 2005*; available at www.cdc.gov/nchs.

This is not to say that diet is the sole culprit causing these diseases. A number of environmental, behavioral, social, and genetic factors work together to determine a person's likelihood of suffering from a **degenerative disease.** For example, diet notwithstanding, someone who smokes, doesn't exercise regularly, and has a parent who suffered a heart attack is more likely to end up with heart disease than a nonsmoker who works out regularly and does not have a close relative with heart disease. The way to alter disease risk is to concentrate on changing the daily habits that can be controlled. The results can be significant.

Table 1-4 summarizes ways to improve health with sound nutrition practices. As you read Table 1-4, keep in mind that although everyone can benefit from eating a healthful diet that complies with the guidelines, some people stand to gain more than others. Those who have high blood cholesterol levels, for instance, are already at risk for heart disease, thereby making it especially important for them to eat a healthy diet and maintain a healthful weight. By the same token, those who have close relatives with, say, diabetes, would do well to keep their weight down and pay particular attention to the other nutrition guidelines that help stave off the condition. (The chapters that follow explain the link between diet and chronic diseases in more detail and offer advice on how to follow each dietary recommendation.)

The exact proportion that dietary factors contribute to each health problem can only be estimated, but some experts speculate that they account for a third or more of all cases of both cancer and heart disease.[8] Moreover, some elements appear to play a more integral role than others in determining disease risk. A high-fat diet, for instance, raises the risk of some types of cancer, heart disease,

degenerative disease chronic disease characterized by deterioration of body organs as a result of misuse and neglect. Poor eating habits, smoking, lack of exercise, and other lifestyle habits often contribute to degenerative diseases, including heart disease, cancer, osteoporosis, and diabetes.

TABLE 1-4
Eating to Beat the Odds

Dietary Recommendation	To Help Reduce the Risk of . . .
Fat: Reduce total fat intake to 20%–35% of total calories. Reduce saturated fat intake to less than 10% of calories and the intake of cholesterol to less than 300 mg daily.[a,b]	Some types of cancer, obesity, heart disease, and possibly gallbladder disease.
Weight: Achieve and maintain a desirable weight.[a]	Diabetes, high blood pressure, stroke, cancers (especially breast and uterine), osteoarthritis, and gallbladder disease.
Carbohydrates and fiber: Increase consumption of fruits, vegetables, legumes, and whole grains.[a,c]	Diabetes, heart disease, and some types of cancer.
Sodium: Limit daily intake of salt (sodium chloride).[d]	High blood pressure and stroke.
Alcohol: Avoid completely or drink only in moderation.[e]	Heart disease, high blood pressure, liver disease, stroke, some forms of cancer, and malformations in babies born to mothers who drink alcohol during pregnancy.
Sugar: Choose and prepare foods and beverages with little added sugars or caloric sweeteners. Aim for less than 10% of total calories from sugars.[f]	Tooth decay and gum disease.
Calcium: Maintain adequate intake (1,000–1,300 mg a day).	Osteoporosis (adult bone loss), bone fractures, and possibly colon cancer.

[a]Pay particular attention to this guideline if you have glucose intolerance, high blood cholesterol, high triglyceride levels, or high blood pressure.
[b]The intake of fat and cholesterol can be reduced by substituting fish, poultry without skin, lean meats, and low-fat or fat-free dairy products for fatty meats and whole-milk products; by choosing more vegetables, fruits, cereals, and legumes; and by limiting oils, fats, egg yolks, and fried or other fatty foods.
[c]Every day eat five or more servings of a combination of vegetables and fruits, especially green and orange vegetables and citrus fruits, and six or more servings of a combination of breads, cereals, and legumes. Specific recommendations are provided in Figure 2-5 on page 44.
[d]Limit the use of salt in cooking and avoid adding it to foods at the table. Highly processed, salty, salt-preserved, and salt-pickled foods should be consumed sparingly.
[e]Moderate drinking is defined as no more than one drink a day for the average-sized woman and no more than two drinks a day for the average-sized man. A drink is any alcoholic beverage that delivers ½ ounce of pure ethanol: 5 ounces of wine, 12 ounces of beer, or 1½ ounces of hard liquor (whiskey, scotch, etc.).
[f]Added sugars are those incorporated into foods and beverages during production. This does not include naturally occurring sugars such as fructose in fruits.

SOURCES: Adapted from U.S. Department of Health and Human Services and U.S. Department of Agriculture, *Dietary Guidelines for Americans, 2005*, 6th ed. (Washington, D.C.: U.S. Government Printing Office, January 2005); and Institute of Medicine, *Dietary Reference Intakes for Energy, Carbohydrate, Fiber, Fat, Fatty Acids, Cholesterol, Protein, and Amino Acids* (Washington, D.C.: National Academy Press, 2002).

and obesity, which in turn may contribute to a number of other problems, including diabetes and high blood pressure.

A National Agenda for Improving Nutrition and Health

Some people do things that are not good for their health. They overeat, smoke, refuse to wear a helmet when riding a bicycle, never wear seat belts when driving, fail to take their blood pressure medication—the list is endless. These behaviors reflect personal choices, habits, and customs that are influenced and modified by social forces. We call these lifestyle behaviors, and they can be changed if the individual is so motivated. **Health promotion** focuses on changing human behavior—getting people to eat healthy diets, be active, get regular rest, develop leisure-time hobbies for relaxation, strengthen social networks with family and friends, and achieve a balance among family, work, and play.[9]

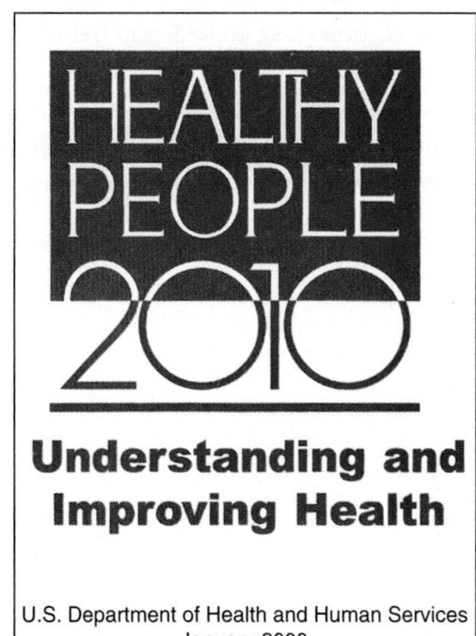

health promotion helping people achieve their maximum potential for good health.

The relative importance of certain dietary recommendations was underscored by their appearance in the U.S. Department of Health and Human Services' official health promotion strategy laid out in the year 2000 for improving the nation's health. Called *Healthy People 2010: Understanding and Improving Health*, the plan of action presents a national health promotion and disease prevention agenda for the first decade of the twenty-first century.[10] The report set health objectives for the year 2010 with two broad goals designed to help all Americans

Healthy Aging—Three Powerful Steps: Eat Smart, Move More, Start Early

Every week, it seems, we hear about new research findings showing how our lifestyle habits influence our health. Consider that researchers who monitored the habits and health of a group of some 7,000 Californians for nearly two decades were able to pinpoint seven common lifestyle elements associated with optimal quality of life and longevity: Avoiding excess alcohol, not smoking, maintaining a healthy weight, exercising regularly, sleeping 7 to 8 hours a night, eating breakfast, and eating nutritious, regular meals. In fact, after 20 years, those who had adhered to these healthful habits were only half as likely to have died as those who hadn't. They were also half as likely to have suffered the types of disabilities that interfere with day-to-day living. Granted, the researchers speculated that some of the habits—for example, sleeping 7 to 8 hours a night—are not necessarily as beneficial as, say, the habit of exercising regularly. Rather, regular eating and sleeping habits are most likely to be signs that people make the time and have enough control of their lives to take care of their health.[11]

1 Eat Smart.

Nutrition shares responsibility with other lifestyle factors for maintaining good health. By the time you are 65 years old, you will have eaten about 100,000 pounds of food. Each bite may or may not have brought with it the nutrients you needed. The impact of the food you have eaten, together with your lifestyle habits, accumulates over a lifetime, and people who have lived and eaten differently all their lives are in widely different states of health by the time they reach 65. *Make healthful eating a priority right now.* Eat a wide variety of nutritious foods every day, to help you get all the nutrients you need for good health. Researchers repeatedly report that people who regularly consume a variety of plant foods, such as fruits, vegetables, legumes, nuts, and whole grains, have reduced risks of heart disease, stroke, diabetes, certain cancers, and other chronic diseases.[12] The key to disease prevention and optimal health is not in eating or avoiding a certain food but, rather, in creating a lifestyle that includes time for preparing nutritious meals and enjoying regular physical activity.

2 Move More.

The U.S. government's 2005 *Dietary Guidelines for Americans* raised the goals for exercise.[13] A minimum of 30 minutes of moderately intense exercise (brisk walking between classes, bicycling from your dorm to classes, stair climbing instead of taking the elevators, etc.) on most or all days of the week can reduce the risk of chronic diseases such as diabetes, heart disease, and some types of cancer. However, 60 minutes a day may be needed for weight loss or to prevent weight gain with age, and 60 to 90 minutes of exercise per day may be required to sustain a weight loss. It doesn't matter whether the 60 minutes is accumulated in small increments (such as walking between classes or taking the stairs instead of the elevator several times a day) or at one time.

3 Start Early.

A person who practices good health habits can expect to delay the onset of even minimal disability by several years, compared with a person who practices few or none of them.[14] If you believe in accepting the things you can't control and controlling the things you can, you are in luck. Your nutritional health can be controlled, and it deserves your conscientious attention.

The findings in Figure 1-2 illustrate that you can change the probable length and quality of your life. Because nutrition is involved in at least half of the preceding lifestyle recommendations, it clearly plays a key role in maintaining good health. This chapter's Scorecard feature, The Longevity Game, further demonstrates this point.

achieve their full potential by increasing the quality and years of healthy life and by eliminating health disparities. Table 1-5 lists the nutrition-related objectives considered to be top health priorities for the present decade.

Several *Healthy People* goals focus on risk reduction and specify targets for the intake of nutrients such as fat, saturated fat, and calcium and of foods such as fruits, vegetables, and whole-grain products. Other risk-reduction goals set targets to reduce the prevalence of obesity, to increase the proportion of people who adopt sound dietary practices, and to increase the proportion of people who reduce their use of salt and sodium in foods and at the table. Each *Healthy People 2010* objective has a target for specific improvements to be achieved by the year 2010.

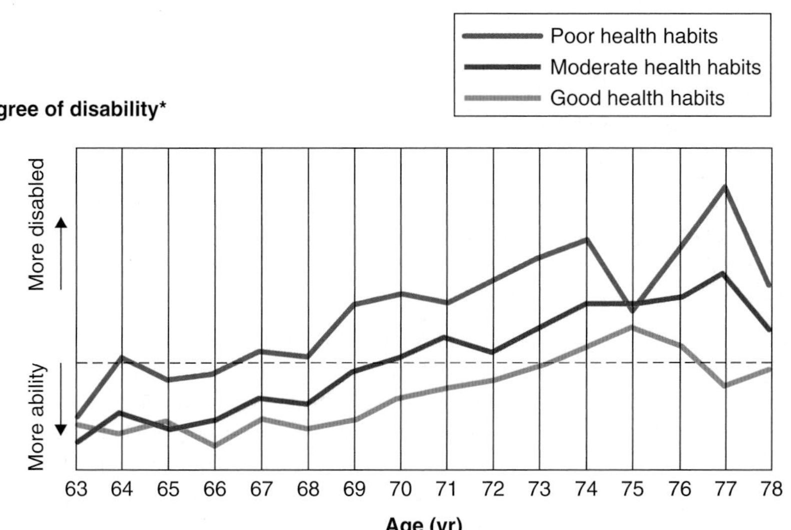

FIGURE 1-2
HEALTHY AGING

In a study of more than 1,700 people, those who smoked the least, maintained a healthy weight, and exercised regularly not only lived longer but postponed disability. As shown here, people with the best health habits delayed the onset of even minimal disability to about age 73, compared with age 66 for those with the poorest health habits.

*The dashed line represents minimal disability, defined as having some difficulty performing the everyday tasks of daily living (such as bathing, dressing, eating, walking, toileting, and getting outside).

SOURCE: *New England Journal of Medicine* (April 9, 1998): pp. 1035–1041.

How are Americans doing in terms of meeting the *Healthy People 2010* goals? When *Healthy People 2010* was released in 2000, life expectancy was 76 years. Today, the average life expectancy at birth is 77 years, and death rates for heart disease, stroke, and certain types of cancer have declined.[15] Since the 1980s, however, the prevalence of overweight has soared, and the trend toward obesity

TABLE 1-5
Healthy People 2010 **Nutrition-Related Objectives for the Nation**[a]

Disease-Related Objectives

Reduce rates of heart disease and stroke, hypertension, cancer, diabetes, osteoporosis, and tooth decay.

Nutrition Objectives

Increase prevalence of healthy weight and decrease the prevalence of obesity.

Increase rates of safe and effective weight-loss practices (such as healthy eating and exercise).

Increase proportion of worksites that offer nutrition or weight-management classes or counseling.

Increase proportion of people who meet the recommended dietary intakes for fat, saturated fat, sodium, and calcium in the diet.[b]

Increase dietary intakes of fruits, vegetables, and grain products (especially whole grains).[c]

Increase proportion of mothers who breastfeed, children whose intake of meals and snacks at school contribute to overall dietary quality, and schools teaching essential nutrition topics.

Reduce rates of growth retardation among low-income children and rates of iron deficiency in young children, women of childbearing age, and low-income pregnant women.

Increase food security among U.S. households and, in so doing, reduce hunger.

Food Safety Objectives[d]

Reduce deaths from food allergy.

Increase the proportion of consumers who practice four essential food safety behaviors when handling food: washing hands, avoiding cross contamination, cooking meats thoroughly, and chilling foods promptly.

Reduce occurrences of improper food-safety techniques in retail food establishments.

[a]The complete list of *Healthy People 2010* objectives may be viewed online at www.health.gov/healthypeople/.
[b]Recommendations are: 30% of calories or less from fat; 10% of calories or less from saturated fat; 2,300 mg or less of sodium and 1,300 mg of calcium for children (ages 9 to 18); 1,000 mg of calcium for adults (ages 19 to 50); and 1,200 mg of calcium for adults over 50.
[c]See Table 2-3 on page 42 and Figure 2-5 on page 44 for specific recommendations.
[d]See Chapter 12 for more on the topic of food safety.

SOURCE: *Healthy People 2010: Understanding and Improving Health* (Washington, D.C.: U.S. Department of Health and Human Services, 2000).

Scorecard The Longevity Game

You can't look into a crystal ball to find out how long you will live. But you can get a rough idea of the number of years you're likely to survive based largely on your lifestyle today as well as certain givens, such as your family history. To do so, play the Longevity Game.

Start at the top line—age 77, the average life expectancy for adults in the United States today. For each of the 11 lifestyle areas, add or subtract years as instructed. If an area doesn't apply to you, go on to the next one. If you are not sure of the exact number to add or subtract, make a guess. Don't take the score too seriously, but do pay attention to those areas where you lose years; they could point to habits you might want to change.

START WITH	77
1. Exercise	______
2. Relaxation	______
3. Driving	______
4. Blood pressure	______
5. 65 and working	______
6. Family history	______
7. Smoking	______
8. Drinking	______
9. Gender	______
10. Weight	______
11. Age	______
YOUR FINAL SCORE:	______

1. Exercise. If your job requires regular, vigorous activity, or if you work out each day, add 3 years. If you don't get much exercise at home, on the job, or at play, subtract 3 years.
2. Relaxation. If you have a laid-back approach to life (you roll with the punches), add 3 years. If you're aggressive, hard-driving, or anxious (suffer from sleepless nights, bite your nails, etc.), subtract 3 years. If you consider yourself unhappy, subtract another year.
3. Driving. Drivers under age 30 who have received traffic tickets in the past year or who have been involved in an accident should subtract 4 years. For other violations, subtract one. If you always wear seatbelts, add a year.
4. Blood pressure. Although high blood pressure is a major contributor to common killers (heart attacks and strokes) it can be lowered effectively through drugs and changes in lifestyle. The problem is that rises in blood pressure can't be felt, so many victims don't know they have it and therefore never receive lifesaving treatment. If you *know* your blood pressure, add 1 year.
5. 65 and working. If you are at the traditional retirement age or older and still working, add 3.
6. Family history. If any grandparent has reached age 85, add 2; if all grandparents have reached age 80, add 6. If a parent died of a stroke or heart attack before age 50, minus 4. If a parent or brother or sister has (or had) diabetes since childhood, minus 3.
7. Smoking. Cigarette smokers who finish more than two packs a day, minus 8; one or two packs a day, minus 6; one-half to one pack, minus 3.
8. Drinking. If you drink two cocktails (or beers or glasses of wine) a day, subtract 1 year. For each additional daily libation, subtract 2.
9. Gender. Women live longer than men. Females add 3 years; males subtract 3 years.
10. Weight. If you avoid eating fatty foods and don't add salt to your meals, your heart will probably remain healthy longer, entitling you to add 2 years.

    Now, weigh in: overweight by 50 pounds or more, minus 8; 30 to 40 pounds, minus 4; 10 to 29 pounds, minus 2.
11. Age. How long you have already lived can help predict how much longer you'll survive. If you're under 30, the jury is still out. But if your age is 30 to 39, plus 2; 40 to 49, plus 3; 50 to 69, plus 4; 70 or over, plus 5.

SOURCE: From "The Longevity Game," by Northwestern Mutual Life Insurance Company, with permission.

continues (see Figure 1-3). In fact, overweight has increased among all ethnic and age subgroups of the population. One contributing factor is that people seem to be taking fewer steps to control their weight by adopting sound dietary patterns and being physically active. Some 38 percent of adults engage in no leisure-time physical activity. Little or no improvements are noted for dietary fat intake and consumption of fruits, vegetables, and whole grains.

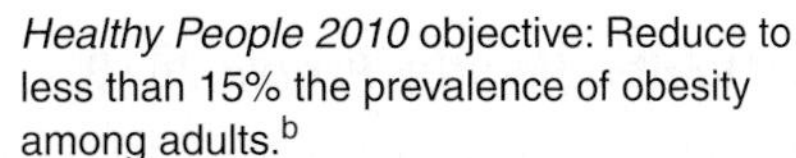
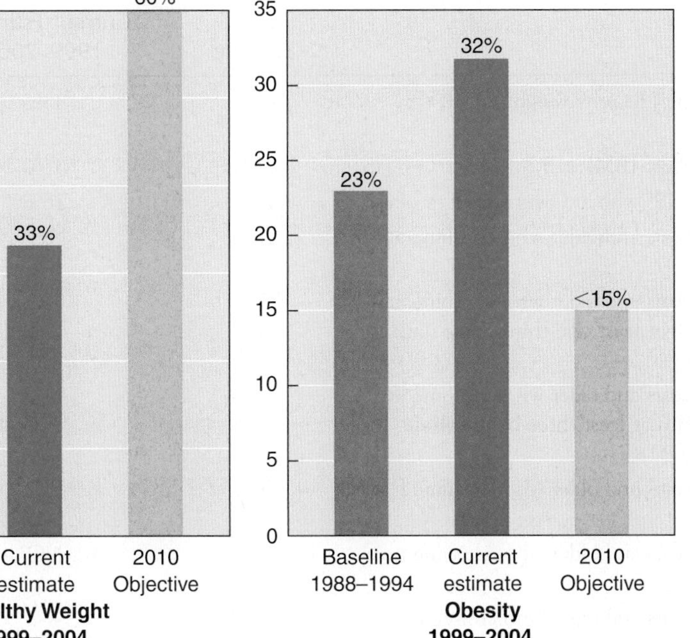

FIGURE 1-3
HEALTHY WEIGHT GOALS

During the baseline period (1988–1994), only 42 percent of adults were at a healthy weight, whereas 23 percent of adults were identified as obese. Current estimates (1999–2004) suggest that even fewer people (33 percent) are at a healthy weight and that some 32 percent of adults are obese.

[a] Healthy weight is defined as having a body mass index (BMI) from 18.5–24.9.
[b] Obesity is defined as having a BMI at or above 30.

SOURCE: Baseline data for healthy weight and obesity among adults is from National Health and Nutrition Examination Survey (NHANES III), 1988–1994. Current estimates are from NHANES 1999–2004. Public Health Service, U.S. Department of Health and Human Services, *Healthy People 2010 Progress Review: Nutrition and Overweight* (January 21, 2004); C. L. Ogden and coauthors, Prevalence of overweight and obesity in the United States, 1999–2004, *Journal of the American Medical Association* 295 (2006): 1549–1555.

Table 1-6 (see page 14) shows the current status of the U.S. population for various nutrition and health-related lifestyle habits and compares it with the nation's goals for 2010. Although these are just a few of the goals for nutrition and health spelled out in *Healthy People 2010*, they represent some of the priorities for maintaining good health. Much of the practical information presented later in this chapter and in those that follow is aimed at guiding you toward developing eating and lifestyle habits that will help you achieve the goals.

Understanding Our Food Choices

The choices you make about what to eat can have a profound impact on your health, both now and in your later years. The healthful eater resists disease and other stresses better than a person with poor dietary habits and is more likely to enjoy an active, vigorous lifestyle for a greater number of years. Even so, the nutritional profile of various foods ranks as only one of many factors that influence your eating habits. Whether you realize it or not, each time you sit down to a meal you bring to the table such factors as your own personal preferences, cultural traditions, and economic considerations. These influences exert as great an impact on your eating habits as does **hunger**—the physiological need for food—and **appetite**—the psychological desire for food, which may arise in response to the sight, smell, or thought of food even when you're not hungry. The following sections examine some of the most influential factors in making food choices.

Availability

Our diets are limited by the types and amounts of food available through the food supply, which, in turn, is influenced by many forces. Because we have the geographic area, climate, soil conditions, labor, and capital necessary to maintain a large agricultural industry, Americans enjoy what is arguably the most abundant food supply in the world. In addition, unlike many other less wealthy countries,

hunger the physiological need for food.

appetite the psychological desire to eat, which is often but not always accompanied by hunger.

TABLE 1-6
Status Report on Healthy Lifestyle Habits: *Healthy People 2010*

This table shows the percentage of adults practicing certain healthy lifestyle habits compared to the *Healthy People 2010* goals for the nation.

Healthy Lifestyle Objective	Baseline	Current Estimate 1999–2004	Target
Increase the proportion of adults who are at a healthy weight (BMI[a] 18.5–24.9).	42%	33%	60%
Reduce the proportion of adults who are obese (BMI ≥ 30).	23%	32%	15%
Reduce the proportion of children (ages 6–19) who are overweight or obese.	11%	16%	5%
Increase the proportion of persons ages 2 years and older who consume at least two daily servings of fruit.	28%	NC[b]	75%
Increase the proportion of persons ages 2 years and older who consume at least three daily servings of vegetables, with at least one-third being dark green or deep yellow vegetables.	3%	NC	50%
Increase the proportion of persons ages 2 years and older who consume at least six daily servings of grain products, with at least three being whole grains (e.g., whole wheat bread and oatmeal).	7%	NC	50%
Increase the proportion of persons ages 2 years and older who consume less than 10% of calories from saturated fat.	36%	NC	75%
Increase the proportion of persons ages 2 years and older who consume no more than 30% of calories from fat.	33%	NC	75%
Increase the proportion of persons ages 2 years and older who consume 2,400 mg or less of sodium daily.	21%	—[c]	65%
Increase the proportion of persons ages 2 years and older who meet dietary recommendations for calcium.	46%	—	75%
Increase the proportion of adults who engage regularly, preferably daily, in moderate physical activity for at least 30 minutes per day.	32%	33%	50%

[a]BMI (Body Mass Index) is calculated as weight in kilograms (kg) divided by the square of height in meters $(m)^2$. [BMI = weight (kg)/height (m^2)]. To estimate BMI using pounds (lb) and inches (in.), divide weight in pounds by the square of height in inches. Then multiply the resulting number by 703 [BMI = weight (lb)/height $(in.)^2 \times 703$].
[b]NC = little or no change.
[c]— = cannot assess; limited data.

SOURCE: Public Health Service, U.S. Department of Health and Human Services, *HP 2010 Progress Review* (2004).

the United States and Canada have the resources needed to import and distribute a wide variety of foods from other countries—everything from kiwi from New Zealand to mangoes from the tropics.

History has shown, however, that when it comes to health, an abundant food supply can be a double-edged sword. Access to many types of foods allows people to choose high-fat diets that are rich in meats and other fatty foods, which can contribute to increased rates of heart disease and other problems. That's one of the reasons why degenerative diseases are sometimes referred to as diseases of affluence.

Income, Food Prices, and Convenience

As most college students know firsthand, the amount of money available to spend on food can mean the difference between ordering pizza every night and resigning yourself to a steady diet of peanut butter and jelly sandwiches. Extremely low incomes can make it difficult for people to buy enough food to meet their minimum nutritional needs, thereby putting them at risk for **undernutrition.**

undernutrition severe underconsumption of calories or nutrients leading to disease or increased susceptibility to disease; a form of malnutrition.

We know that a balanced diet of healthy foods is associated with reduced risk of chronic diseases and death, particularly from heart disease and cancer. One

(Text continues on page 18.)

Nutrition Action

Good and Fast—A Guide to Eating on the Run, or Has Your Waistline Been Supersized?

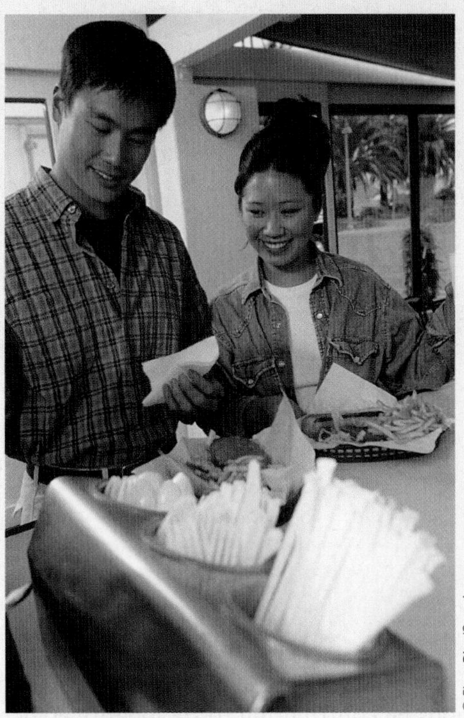
© PhotoDisc/Getty Images

Contrary to popular belief, your professors do not conspire to plan due dates for assignments, quizzes, and exams on the same day. As a student, your schedule can get hectic, especially at midterm and the end of a semester. In the working world, much like college, you will also have deadlines and scheduling conflicts that may leave little time for everyday activities like sitting down to a meal with friends or family. Even if your life is not hectic, chances are you've stood in line for a burger and fries, a slice of pizza, a taco, or a muffin and coffee at least once this week. Almost 60 percent of Americans eat out for lunch at least once a week.[16] And nearly 25 percent eat lunch out five or more times a week.[17]

Even though a meal eaten on the run fits easily into a busy schedule, it's not necessarily so simple to work it into the dietary guidelines recommended by major health organizations. Eating away from home is not an invitation to forget about good nutrition, eat more ("supersize"), or eat differently than you would at home.[18] Can you eat away from home and still eat healthfully? It is possible with some planning.

Fast food does not have to be an unhealthy option, as long as you don't supersize. Fast foods commonly contain more fat, including more saturated fat, less fiber, more cholesterol, and more calories than meals made at home.

Portion sizes of foods have been on the increase since the early 1970s.[19] In the 1950s, a single-serve bottle of a soft drink was 6 ounces. Today, a single-serve bottle is 20 ounces. A typical bagel purchased today at 4–7 ounces is almost twice the size bagels used to be at 2–3 ounces.[20]

Cheeseburger
Small fries
Small soft drink
690 calories
24 g fat
8 g saturated fat

Cheeseburger
Supersize fries
Supersize soft drink
1,350 calories
43 g fat
13 g saturated fat

Along with the larger portions come more calories. Could these larger portions be contributing to the obesity epidemic in the United States? Yes! When people are served more food, they eat more.[21] Is it any coincidence, then, that as portion sizes have increased over the past two decades, the commonness of overweight and obesity among adults and children also has increased?[22] We don't think so.

The enlarging size of American food portions is linked to a practice used by the food industry called "value" marketing. Consumers are prompted by point-of-purchase displays and verbally from employees to spend a little extra money to "upgrade" to larger portions, leaving the customer with the feeling of "What a deal I got!"[23] For the food sellers, costs to increase portion sizes are small, but the profit margin is huge.

So how do you, the typical harried college student, get fast meals without increasing the calories and fat in your daily diets? Remember these strategies next time you find yourself in a time crunch and in line at the closest fast-food restaurant.

Strategy 1: Don't supersize. It's a great marketing ploy to get you to buy more, but you will also be eating more fat, cholesterol, and salt—and weighing more as a result. Think small. Stick with smaller burgers, sliced meat, or grilled chicken sandwiches with mustard, catsup, lettuce, onion, pickles, and tomatoes.

Strategy 2: Think grilled, not fried. Frying foods adds about 50 percent more fat and/or calories compared with items that have not been fried. (At Wendy's, a grilled chicken sandwich has 8 grams of fat versus 18 in a deep-fried, breaded chicken sandwich.[24]) Order a baked potato that you can dress with low-fat margarine and/or low-fat or fat-free sour cream. If you can't live without French fries, get them on occasion, but most of the time, try to substitute a salad (with low-fat, fat-free, or regular dressing on the side) or a baked potato with low-fat toppings.

Strategy 3: Hold the mayo. Each spoonful of mayonnaise adds about 100 calories, nearly all of which are fat. Most fast-food sandwiches contain more than just one spoonful of mayo in their toppings and special sauces. The same goes for tartar sauce. Order a fish sandwich without it, and you'll trim at least 70 fat-laden calories (the amount in just 1 tablespoon) from your meal. Ask for lots of lettuce and tomatoes and less sour cream and guacamole on your nachos and tacos. A tablespoon of either sour cream or guacamole adds about 25 calories to Mexican fare. A few extra chunks of tomato, on the other hand, supply a negligible number of calories, no fat, and a good deal of vitamin C.

Strategy 4: Avoid all-you-can-eat restaurants. No explanation necessary. Remember moderation and variety.

Strategy 5: "Just say no." Did you know that a 16-ounce soft drink adds 200 calories to a meal and nothing else? A medium chocolate shake can add 350 calories to a meal, and a large shake can add 770 calories. Wash your meal down with low-fat milk instead of a milk shake, and you'll cut the fat and calorie count at least in half. Or, better yet, in the continuing effort to get enough fluids, drink water instead.

Strategy 6: Balance fast-food meals with other food choices during the day.[25] If you can't avoid fast food, then adjust your portion sizes and food choices at other meals. Increasing physical activity will help counterbalance extra calories.

Strategy 7: Split your order—share with a friend. Or split dessert. Frequently a few bites can satisfy a sweet tooth.

Strategy 8: Bring your lunch. You'll save money and time in addition to planning a healthy lunch in less time than you would probably spend in line or in the drive-through line. Try taking leftovers in a microwavable container. Make double batches when you cook, and put some in the freezer in single-serving containers for lunch.

Strategy 9: Choose grab-and-go foods. Keep these foods handy for a fast and healthy snack: breadsticks, baby carrots, fig bars, graham crackers, pretzels, fresh fruit, dried fruit, fruit juices (not fruit drinks), low-fat yogurt, string cheese, and low-fat popcorn.

Strategy 10: If all else fails, go for the obvious low-calorie choices. For example, Subway sells subs with 6 grams of fat or less. Don't let the word *chicken* or *fish* fool you. Many health-conscious consumers have heard the advice to choose skinless poultry and fish instead of relatively high-fat red meat. But when it comes to chicken nuggets and fish patties coated with batter and deep fried, it is a different story. Six chicken nuggets, for example, typically contain as many calories (about 300) as an entire burger. What's more, many chicken and fish sandwiches chalk up as much fat as a pint and a half of ice cream. Even rotisserie-style chicken contains a large amount of fat and calories if you don't remove the skin before eating it. When ordering a pizza, hold the sausage and pepperoni, and ask for mushrooms, green peppers, and onions instead. Pizza is an excellent source of calcium—the bone-building mineral that many Americans don't get enough of—as well as protein, carbohydrate, and a number of vitamins and minerals. But two slices of *pepperoni* pizza can easily contain 100 more calories and twice as much fat as the same amount topped with onions, green peppers, and mushrooms.

© PhotoDisc/Getty Images

▲ *Numerous factors influence your food choices, including these:*
- *Hunger, appetite, and food habits*
- *Nutrition knowledge, health beliefs/concerns, and practices*
- *Availability, convenience, and economy*
- *Advertising and the media*
- *Early experiences, social interactions, and cultural traditions*
- *Personal preference, taste, and psychological needs*
- *Values, such as political views, environmental concerns, and religious beliefs*

reason people may not choose healthy foods over foods that are low in nutrients and high in fat is cost. If you are like most Americans, you think it's "more expensive" to eat healthy foods like fruits and vegetables. But is the cost of a healthy diet really significantly higher than the standard American diet of "convenience" foods, snacks, bakery items, soft drinks, and other less nutritious foods? Compare cost of convenience foods with the same foods made from scratch (most cost more).

A consumer's *perception* of the cost of various foods can also play a role in his or her choices. For example, two barriers that prevent people from adopting healthful eating habits are the beliefs that it would be too expensive and inconvenient. See the margin list for the most frequently cited roadblocks to healthful eating.[26] In fact, 40 percent of consumers who answered one survey said that fruits, vegetables, seafood, and other elements of a low-fat, nutrient-rich diet would strain their budgets. But some research has shown that switching from a high-fat diet to one that is lower in fat can reduce food costs.[27] Just cutting back on the amount of meat and poultry—the source of much of the fat in the American diet and where many of our food dollars are spent—goes a long way in trimming food budgets.

Research reported in the *Journal of the American Dietetic Association* found that not only is a diet of healthy foods less costly, but it also promotes weight loss.[28] In one study, overweight children and their parents were encouraged to consume more "nutrient-dense" foods like fruits and vegetables (high in nutrients and low in fat and calories) and less "empty-calorie" foods (low in nutrients and high in fat and/or calories). After one year, food costs decreased, and the overweight children and parents were 5 to 8 percent less overweight than they were at the beginning of the study![29]

Perceived Barriers to Healthful Eating

- Healthy foods are not always available from fast-food and take-out restaurants.
- It costs more to eat healthy foods.
- I'm too busy to take the time to eat healthfully.
- I hear too much conflicting information about which foods are healthy and which foods are not.
- Healthy foods don't taste as good.
- The people I usually eat with do not eat healthy foods.

Advertising and the Media

Television and radio commercials as well as magazines and newspapers play an extremely powerful role in influencing our food choices and our knowledge of nutrition. Given today's health-conscious environment, food manufacturers promote the nutritional merits of their products more than ever before. In fact, in some of the most popular women's magazines, the number of food and beverage ads containing nutrition claims increased by nearly 100 percent in the last two decades.[30] Unfortunately, advertising is not always created with the consumer's best interest in mind. Much of the food advertising that we're exposed to from the earliest ages is aimed at selling products that aren't the optimal choices to regularly include in a healthful diet. For example, the great majority of television commercials geared to children and aired on Saturday mornings promote high-fat, sugary foods such as candy and sugar-coated cereals.[31] In fact, the food industry spends more than $11 billion overall for food, beverage, and fast-food advertising. This amount dwarfs the almost $5 million spent by the government in promoting its 5 A Day campaign to encourage people of all ages to increase their consumption of fruits and vegetables.[32] As a result, commercials promoting good nutrition are relatively few and far between.

Along with advertising, the media rank among the most influential sources of diet and nutrition information, which in turn affects our food choices. Consider that most Americans look first to television as a source of nutrition information; then to magazines, newspapers, radio, family or friends, books, the Internet; and finally to doctors (see Figure 1-4).[33] Unfortunately, the reliability of information delivered by the media varies considerably. The Spotlight at the end of the chapter will help you learn how to evaluate the nutrition information you receive via the media.

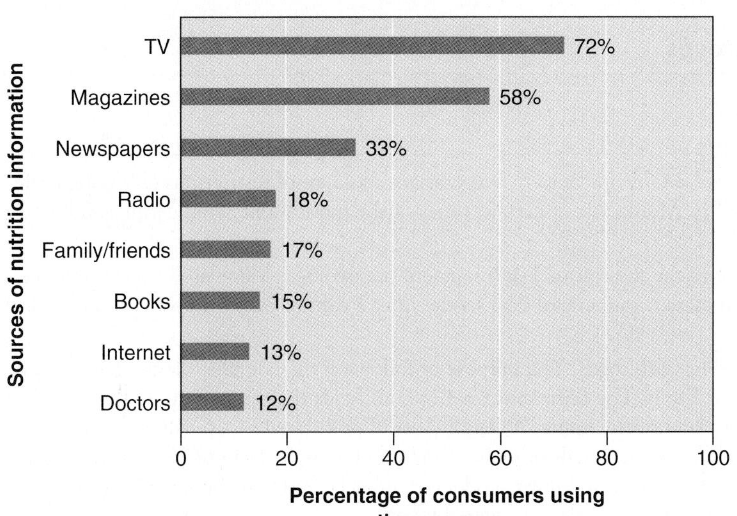

FIGURE 1-4
CONSUMER SOURCES OF NUTRITION INFORMATION*

*NOTE: Consumers could cite more than one source.

SOURCE: Reprinted with permission from the American Dietetic Association, *Nutrition and You: Trends 2002*, American Dietetic Association. © Copyright 2002.

Social and Cultural Factors

A person's social and cultural groups have a significant impact on his or her food choices. **Social groups** such as families, friends, and coworkers tend to exert the most influence. The family, particularly the wife/mother, plays one of the most powerful roles in determining our food choices. That makes sense because the family is the first social group a person encounters as well as the one to which he or she typically belongs for the longest period of time. The values, attitudes, and traditions of our family can have a lasting effect on our food choices. Think of the holiday food traditions in your own family. Treasured recipes or rituals surrounding holiday meals are often passed from one generation to the next.

Friends, coworkers, and members of other social networks also influence our food choices and eating behavior. For instance, many weight-loss programs feature group sessions made up of people who are in the same boat. Thus, they can support one another in their efforts to lose weight. Social pressure can also push us to eat meals we might not choose on our own. For example, if you are a guest in another country or in a friend's home, choosing not to partake of the food and drink that's offered might be considered rude. By the same token, it's natural to join your friends on a spontaneous trip for ice cream or pizza even when you're not hungry.

Culture also determines our food choices to a large extent. Many of our eating habits arise from the traditions, belief systems, technologies, values, and norms of the culture in which we live. For example, in the United States and Canada, the idea of eating insects is generally considered repulsive. But many people throughout the rest of the world relish dishes prepared with various bugs, including locust dumplings (northern Africa), red-ant chutney (India), water beetles in shrimp sauce (Laos), and fried caterpillars (South Africa).[34]

One of the ways people of different cultures often come together to share their heritage is by sampling each other's traditional foods. American consumers are particularly fortunate because they typically don't have to travel far to taste the food of different cultures. **Ethnic cuisine,** ranging from Chinese to Mexican to Italian to Indian food, has become embedded in American culture. (The Spotlight in Chapter 2 discusses the eating patterns of different cultures in detail.)

Religion is one aspect of culture that affects the food choices of millions of people worldwide. The practice of giving and abstaining from food has long been used by many cultures as a way to show devotion, respect, and love to a supreme being or power. Also, dietary customs play an important role in the practice of many of the world's major religions. As Table 1-7 shows, many religions specify which foods their followers may eat and how those foods must be prepared.

▲ *The act of eating is complex. We derive many benefits—both physical and emotional—when we eat foods.*

social group a group of people, such as a family, who depend on one another and share a set of norms, beliefs, values, and behaviors.

culture knowledge, beliefs, customs, laws, morals, art, and literature acquired by members of a society and passed along to succeeding generations.

ethnic cuisine the traditional foods eaten by the people of a particular culture.

TABLE 1-7
Dietary Practices of Selected Religious Groups

Religious Group	Dietary Practices*
Buddhist	Dietary customs vary depending on sect. Many are lacto-ovovegetarians, because of restrictions on taking a life. Some eat fish, and most eat no beef or poultry. Monks fast at certain times of the month and avoid eating solid food after the noon hour.
Hindu	All foods thought to interfere with physical and spiritual development are avoided. Many are lacto-vegetarians and/or avoid alcohol. The cow is considered sacred—an animal dear to the Lord Krishna. Beef is never consumed, and often pork is avoided.
Jewish	*Kashrut* is the body of Jewish law dealing with foods. The purpose of following the complex dietary laws is to conform to the Divine Will as expressed in the Torah. The term *kosher* denotes all foods that are permitted for consumption. To "keep kosher" means to follow dietary laws in the home. A lengthy list of prohibited foods, called *treyf,* includes pork and shellfish. The laws define how birds and mammals must be slaughtered, how foods must be prepared, and when they may be consumed. For example, dairy foods and meat products cannot be eaten at the same meal. During Passover, special laws are observed, such as the elimination of any leavened foods.
Mormon	Alcoholic beverages and coffee, tea, and other beverages containing caffeine are avoided. Mormons are encouraged to limit meat intake and emphasize grains in the diet.
Muslim	Overeating is discouraged, and consuming only two-thirds of capacity is suggested. Dietary laws are called *halal.* Prohibited foods are called *haram,* and they include pork and birds of prey. Laws define how animals must be slaughtered. Alcoholic drinks are not allowed. Fasting is required from sunup to sundown during the month of Ramadan.
Roman Catholic	Meat is not consumed on Fridays during Lent (40 days before Easter). No food or beverages (except water) are to be consumed 1 hour before taking communion.
Seventh Day Adventist	Most are lacto-ovovegetarians. If meat is consumed, pork is avoided. Tea, coffee, and alcoholic beverages are not allowed. Water is not consumed with meals, but is drunk before and after meals. Followers refrain from using seasonings and condiments. Overeating and snacking are discouraged.

*Many of the religious guidelines regarding food have practical applications for the society. For example, the Hindu prohibition against killing cattle respects the needs for Indian farmers to use cattle for power and cattle dung for fuel. Cows also supply milk to make dairy products.

SOURCE: Adapted from Religious Food Practices, accessed at www.eatethnic.com; and M. Boyle and D. Holben, *Community Nutrition in Action: An Entrepreneurial Approach*, 4th ed. (Belmont, CA: Wadsworth, 2006), 510.

Personal Values or Beliefs

See the Nutrition Action feature in Chapter 12 for more about the benefits of purchasing locally produced, seasonally available, and organically grown foods.

Some people adopt a certain way of eating or making choices based on a larger worldview. For instance, many environmentally conscious people, believing that raising animals for human consumption strains the world's supply of land and water, choose to abstain from meat as much as possible in an effort to preserve the Earth's resources. Others may choose to boycott certain manufacturers' items for political reasons, perhaps because they disagree with a company's advertising practices.

Over the past decade, many people have begun to make food choices based on sustainability issues and the benefits of purchasing locally produced, seasonally available, and organically grown food.[35] Sustainability in the food system is defined as society's ability to shape its economic and social systems to maintain both natural resources and human life, and it involves building locally based, self-reliant food systems.[36] When consumers purchase foods that have been produced locally, a greater proportion of the profits remain with local farmers, providing them with a livable income while supporting local economies.[37] Additionally, the purchase of locally produced foods protects the environment by reducing the use of fossil fuel (for transporting food) and packaging materials.[38]

Other Factors That Affect Our Food Choices

One of the main reasons you choose to eat certain foods is your preference for certain tastes. Just about everyone enjoys sweet foods, for example, because

The Savvy Diner: You CAN Afford to Eat Nutritious Foods—Tips for Supermarketing

Want to save money and consume less fat and calories? Healthy eating starts with healthy food shopping. With preparation and some label-reading skills, you can leave the grocery store with a supply of lower fat, lower calorie basics like these:[39]

- Fat-free or low-fat milk, yogurt, cheeses, and cottage cheese
- Light or reduced-fat margarines
- Eggs/egg substitutes
- Whole-grain breads, bagels, and English muffins
- Low-fat flour tortillas, soft corn tortillas
- Plain cereals (dry or cooked)
- Rice, pasta (remember, it's what you put on these grain products that increases the fat/calorie content)
- Fresh, frozen, or canned fruits packed in juice
- Fresh, frozen, or canned vegetables
- Dry beans and peas
- Skinless, white-meat chicken or turkey
- Fish and shellfish (not battered and fried)
- Beef: round, sirloin, chuck, loin, and extra lean ground beef
- Pork: leg, shoulder, tenderloin
- Low-fat or nonfat salad dressings, salsa, herbs, and spices

The next time you shop for groceries, keep in mind these smart shopping tips:[40]

1. Buy local foods and fresh foods in season. Use the local newspaper to find the best seasonal buys and special sale items.
2. Shop from a list to help avoid buying unnecessary items. Keep a running list in your kitchen, and note items you need to replace.
3. Read the ingredients list and Nutrition Facts label on packaged foods; compare amounts of fat, sodium, calories, and nutrients in similar products. Ingredients are listed in order of quantity.
4. Use "sell by" and "best if used by" dates to ensure quality and freshness. Buy only the amount you or your family will eat before the food spoils.
5. Shop the perimeter of the grocery store to find many fresh whole foods: fresh produce; low-fat dairy products; lean meats, poultry, and fish; and whole-grain breads. Maneuver down the aisles only for specific items on your list such as canned tomato products, spices, and canned or dry beans.

humans are born with an affinity for sugar.[41] In addition, we usually prefer foods that have happy associations for us—foods prepared for special occasions, those given to us by a loved one when we were children, or those eaten by an admired role model. By the same token, intense aversion to certain foods—perhaps foods you were given when you were sick or foods you were forced to eat as a child—can be strong enough to last a lifetime. Your parents may have taught you to prefer certain foods and pass up others for reasons of their own, without even being aware they were doing so.

Food habits are also intimately tied to deep psychological needs, such as an infant's association of food with a parent's love. Yearnings, cravings, and addictions with profound meaning and significance sometimes surface as food behavior. Some people respond to stress—positive or negative—by eating; others use food to fill a void, such as lack of satisfying personal relationships or fulfilling work.

The influences on people's eating habits are as many and varied as the individuals themselves. Our food choices reflect our own unique cultural legacies, philosophies, and beliefs. To think of food as nothing more than a source of nutrients would deny food's rich symbolism and meaning and take away much of the pleasure of breaking bread with friends and family. As you read this book and consider ways to improve your own eating habits, take time to reflect on your unique background and think about how you can integrate your knowledge of nutrition into your cultural heritage and philosophies.

▲ *Even at many American-style restaurants, you can experience other cultures by sampling from the various ethnic cuisines found on the menu.*

Spotlight

How Do You Tell If It's Nutrition Fact or Nutrition Fiction?

Most people want to eat a healthy diet and lower their chances of cancer and other chronic diseases. Frequently, however, information they read or hear about healthy eating from one source is contradicted by another. It's not always easy to recognize accurate information from misinformation. Do you know which of the following statements are true or false? Check your answers with the correct answers at the end of this section.*

1. It is difficult for busy people to eat a balanced diet.
2. People who graduate from college are smart enough not to be victimized by nutrition misinformation.
3. Sugar is a major cause of hyperactivity in children.
4. No special training is legally required to offer nutrition information to the public.
5. Most health food retailers have been educated about the research done regarding the products they sell.
6. Protein and/or amino acid supplements help bodybuilders, recreational weight lifters, and other athletes improve their performance by increasing muscle size.
7. Most nutrition-related books and magazine articles undergo prepublication review by experts.
8. Cigarette smoking is the leading cause of preventable death in the United States.
9. Herbal products are as safe and effective as many drugs prescribed by physicians.
10. It is illegal for manufacturers of dietary supplements to print false or misleading information about their products.

*Adapted from Quackwatch. Consumer Health I.Q. Test. Available at www.quackwatch.org/04ConsumerEducation/iq.html; accessed December 14, 2002.

You've just watched a television commercial for a vitamin supplement that is guaranteed to produce a laundry list of benefits, including fewer colds, a better complexion, and a decreased risk of cancer. Should you buy it? You've just read a magazine article with a plan for quick weight loss. Should you believe it? Someone who plays the same sport as you do says that improving your diet will help your game. Where do you go for help? A friend just sent you an e-mail saying that bananas imported from Costa Rica are infected with necrotizing fasciitis. Should you pass this on to all of your friends?

We all find ourselves faced with such decisions at one time or another. It's crucial to know how to protect ourselves from nutrition misinformation. Health fraud costs consumers more than $30 billion each year, and money down the drain is just one problem stemming from misleading nutritional information. Although some fraudulent claims about nutrition are harmless and may make for a good laugh, others can be harmful. False claims about nutritional products have been known to bring about malnutrition, birth defects, mental retardation, and, in extreme cases, even death.

Negative effects due to following false nutritional claims can happen in two ways. First, the product in question may cause direct harm. Even a seemingly innocuous substance such as vitamin A can cause severe liver damage over time if taken in large enough amounts. Second, using bogus nutritional remedies can cause problems because such remedies can build false hope and might keep a consumer from obtaining sound, scientifically tested medical treatment. A person who relies on a so-called anticancer diet as a cure for the disease, for example, might forego possible lifesaving interventions such as surgery or chemotherapy.

The following questions and answers will help you learn to evaluate the nutrition information that you see in the media or on the Internet. They will help you develop the necessary skills to view nutrition claims with a skeptic's eye or, at the very least, to help you decide when to find a qualified professional to help you evaluate the information.

Judging by what I've read on television, it seems as though nutritionists are always changing their minds. One week the headlines say to take vitamin E to help prevent heart disease, and the next week they say that vitamin E may not, in fact, prevent the disease. Why is there so much controversy?

Part of the confusion stems from the way the media interpret findings of scientific research. A good case in point is the controversy over whether a high-fiber diet protects against colon and rectal cancers—diseases that affect some 148,000 Americans each year. (A detailed discussion of nutrition and cancer appears in the Chapter 6 Spotlight.)

The fiber and colon cancer connection dates back to the early 1970s when scientists observed that colon cancer was extremely uncommon in areas of the world where the diet consisted largely of unrefined foods and little meat. Researchers theorized that dietary fiber may protect against colon cancer by binding bile (a chemical substance needed in fat digestion) and speeding the passage of wastes and potentially harmful compounds through the colon. Since then, other studies have also suggested that those who eat a high-fiber diet have a lower risk of colon and rectal cancers.[42]

However, in 1998, a flurry of headlines threatened to pull the pedestal out from under the popular fiber theory, asking, "Fiber: Is It Still the Right Choice?" A Harvard-based study published in the *New England Journal of Medicine*, one of the most prestigious medical journals, suggested that fiber did nothing to prevent cancer.[43] The 16-year trial of almost 90,000 nurses—called the Nurses' Health Study—found that nurses who ate low-fiber diets (less than 10 grams daily) were no more likely to develop colon cancer than those eating higher levels of fiber (about 25 grams daily). As a result, the researchers concluded that the study provided no support for the theory that fiber could reduce risk of colon cancer.

The news was surprising and reinforces the need for research studies to be duplicated. All studies have some limitations, and a number of questions can be raised regarding conclusions of the Nurses' Health Study. For example, the study relied on participants to recall their eating habits accurately.

Is this type of self-reported dietary information reliable? Back in 1980, the nurses were asked for information about their intakes of "dark" breads. However, food labels at that time did not list the fiber content of breads, and some wheat breads on the shelf had similar amounts of fiber as is found in white bread. Did the nurses mistakenly consider "dark" bread the same as 100 percent whole-wheat bread?

Another question is "What are the optimal levels of fiber intake for colon cancer protection?" Some experts believe it may take more than 25 grams of fiber a day to show cancer-protective effects, which might explain the lack of effect noted in the Nurses' Health Study.

This fiber story illustrates how news reports based on only one study can leave the public with the impression that scientists can't make up their minds. It seems as if one week scientists are saying fiber is good, and the next week the word is that fiber doesn't do any good at all. The truth is that health experts and major health organizations continue to urge adults to get 21 to 38 grams of fiber a day. (The average fiber intake today is about 15 grams.)

Contrary to what some headlines imply, reputable scientists do not base their dietary recommendations for the public on findings of one or two studies. Scientists are still conducting research to determine whether fiber does in fact help prevent colon cancer, and, if so, what types of fiber and in what amount. Scientists design their research to test theories, such as the notion that eating a high-fiber diet is associated with lower risk of cancer. Other factors, however, often complicate the matter at hand. The study of fiber and colon cancer is complex due to many other factors linked with colon cancer development, including inactivity, obesity, saturated fat intake, low calcium or folate intakes, and others.

So, should you hang on to your high-fiber cereals and vegetables? Yes, according to the American Institute for Cancer Research. Its analysis of more than 4,500 research studies provides evidence that increased intake of fiber may be associated with decreased incidence of colon cancer.[44] Most important, however, the health benefits of fruits, vegetables, legumes, and whole grains go beyond the possible protection from colon cancer.[45] Diets rich in fiber from these foods are also strongly associated with reduced risks of heart disease, high blood pressure, type 2 diabetes, and diverticular disease, a condition that can lead to painful inflammation of the large intestine. Fiber also promotes a feeling of fullness after you eat, which can help with weight control.

How can I tell if a nutrition news story is noteworthy and a source of credible nutrition information?

You can critique nutrition news you read by asking a series of questions (see Table 1-8). Consider the following points as a checklist for separating the bogus news stories from those worth your attention:

- The study described in the news story should be published in a journal that uses experts in the field to review research results (called *peer review*). These reviewers serve to point out flaws in research design and can challenge researchers' conclusions before the study is published.
- Be sure that the report is about recent research. The science of nutrition continues to develop from the results of new studies that employ state-of-the-art methods and technology and benefit from scrutiny of experts current in the particular field of study.
- Are the reported results from an **epidemiological study** or an **intervention study?** Epidemiological studies examine populations to determine food patterns and health status over time. These population studies are useful in uncovering **correlations** between two factors (for example, whether a high calcium intake early in life reduces the incidence of bone fractures later in life). However, they are not considered as conclusive as intervention studies. A correlation between two factors may suggest a cause-and-effect relationship between the factors *but does not prove it*.

  Intervention studies examine the effects of a specific treatment or intervention on a particular group of subjects and compare the results to a similar group of people not receiving the treatment. An example is a cholesterol-monitoring study in which half the subjects follow dietary advice to lower their blood cholesterol and half do not.

  Ideally, intervention studies should be randomized and controlled—that is, subjects are assigned

TABLE 1-8

Questions to Ask about a Research Report

- Was the research done by a credible institution? A qualified researcher?
- Is this a preliminary study? Have other studies reached the same conclusions?
- Was the study done with animals or humans?
- Was the research population large enough? Was the study long enough?
- Who paid for the study? Might that affect the findings? Is the science valid despite the funding source?
- Was the report reviewed by peers?
- Does the report avoid absolutes, such as "proves" or "causes"?
- Does the report reflect appropriate context? For example, how does the research fit into a broader picture of scientific evidence and consumer lifestyles?
- Do the results apply to a certain group of people? Do they apply to someone of your age, gender, and health condition?
- What do follow-up reports from qualified nutrition experts say?

SOURCE: Position of the American Dietetic Association: Food and Nutrition Misinformation, *Journal of the American Dietetic Association* 106 (2006): 605. Reprinted with permission from Elsevier.

to either an **experimental group** or a **control group** based on a random selection process. Each subject has an equal chance of being assigned to either group. The experimental group receives the "treatment" being tested; the control group receives a **placebo** or neutral substance. If possible, neither the researcher nor participants should know which subjects have been assigned to which group until the end of the experiment. A randomized, controlled study helps ensure that the study's conclusions are a result of the treatment and minimizes chances that results are due to a placebo effect or bias on the part of the researcher.

- To achieve validity—accuracy in results—studies must generally include a sufficiently large number of people (for example, intervention studies of 50 or more persons). This reduces the chances that the results are simply a coincidence and helps generalize the conclusions of the study to a wider audience.
- Look for similarities between the subjects in the study and yourself. The more you have in common with the participants (age, diet pattern, gender, and so forth), the more pertinent the study results may be for you. It's wise to be leery of media stories based solely on animal experiments. Although different species may respond in some way for better or for worse to a treatment, the same response will not necessarily occur in humans.

  For example, suppose rats became sick when injected with the amount of substance X that a person would normally consume in a year's time. A researcher might correctly conclude that substance X could cause illness. But in a case like this, remember that a rat is a tiny animal with a different physical makeup than a human being. Moreover, the rodent received a huge dose of substance X all at once. Thus, the experiment does not indicate whether small amounts of substance X taken by a person over the course of a year would produce the same harmful effect. A news story that implies otherwise misses the mark.
- Even if an experiment is carefully designed and carried out perfectly, however, its findings cannot be considered definitive until they have been confirmed by other research. Testing and retesting reduce the possibility that the outcome was simply the result of chance, error, or oversight on the part of the experimenter. Every study should be viewed as preliminary until it becomes just one addition to a significant body of evidence pointing in the same direction.

When making dietary recommendations for the public, experts pool the results of different types of studies, such as analyses of food patterns for groups of people and carefully controlled studies of people in hospitals or clinics. Before drawing any conclusions, they then consider the evidence from all of the research. In fact, the dietary guidelines spelled out in the 1,300-page *Diet and Health* report are based on the results of hundreds of studies. The bottom line is that if you read a report in the newspaper or watch one on television that advises making a dramatic change in your diet or lifestyle based on the results of one study, don't take it to heart.

The findings may make for a good story, but they're not worth taking too seriously.

Why doesn't the government do something to prevent the media from delivering misleading nutrition information?

The **First Amendment** guarantees freedom of the press. Accordingly, it is possible for people to express whatever views they like in the media, whether sound, unsound, or even dangerous. This freedom is a cornerstone of the U.S. Constitution, and to deny it would be to deny democracy. Writers cannot be punished by law for publishing misinformation unless it can be proved in court that the information has caused a reader bodily harm.

Fortunately, most professional health groups maintain committees to combat the spread of health and nutrition misinformation. Many professional organizations have banded together to form the National Council against Health Fraud (NCAHF). The NCAHF monitors radio, television, and other advertising, and it investigates complaints. For more information on separating nutrition fact from nutrition fiction, visit these websites:

- www.ncahf.org
- www.quackwatch.org
- www.cdc.gov

Is the Internet a reliable source of nutrition and health information?

Information is rampant on the Internet.[46] In a sense, the Internet *is* information, and the information is continually being revised and *created.* Internet information exists in many forms (facts, statistics, stories, opinions), is created for many purposes (to entertain, to inform, to persuade, to sell, to influence), and varies in quality from good to bad.

The information you find on the Internet is only as good as its source. One method for determining whether information found on the Internet is reliable and of good quality is the CARS Checklist. The acronym CARS stands for *credibility, accuracy, reasonableness*, and *support*.[47]

- *Credibility*. Check credentials of the author (if there is one!) or sponsoring organization. Is the author or organization respected and well known as a source of sound, *scientific* information? Evidence of a lack of credibility includes no posted author and even the presence of misspelled words or bad grammar. A credible sponsor will use a professional approach to designing the website.
- *Accuracy*. Check to ensure that the information is current, factual, and comprehensive. If important facts, consequences, or other information are missing, the website may not be presenting a complete story. Evidence of a lack of accuracy includes no date on the document, the use of sweeping generalizations, and the presence of outdated information. How often is the site updated? Reliable sites are updated regularly with a date posted on the site. Watch out for testimonials masquerading as scientific evidence—a common method for promoting questionable products on the Internet.
- *Reasonableness*. Evaluate the information for fairness, balance, and consistency. Does the author present a fair, balanced argument supporting his or her ideas? Are the author's arguments rational? Has he or she maintained objectivity in discussing the topic? Does the author have an obvious—or somewhat hidden—conflict of interest? Evidence of a lack of reasonableness includes gross generalizations ("Foods not grown organically are all toxic and shouldn't be eaten") and outlandish claims ("Kombucha tea will cure cancer and diabetes").
- *Support*. Check to see whether supporting documentation is cited for scientific statements. Does the website refer to legitimate scientific journals and publications? Is the source of information clearly presented? An Internet document that fails to show its sources is suspect.

Many reputable health and professional organizations and government agencies host websites on the Internet. To help you explore the World Wide Web, the publishers of this book maintain a website with links to reliable nutrition and health-related information at **www.thomsonedu.com/nutrition.**

Does the First Amendment make it legal for companies to say whatever they want about the products they sell?

No. Unlike journalists, purveyors of products are bound by law to make only true statements about their wares. The Food and Drug Administration (FDA) holds the authority to prosecute companies that display false nutrition information on product labels or enclosures, and the Federal Trade Commission (FTC) can take to task manufacturers who make fraudulent or misleading statements in their advertisements. Nevertheless, combating health fraud is an overwhelming job requiring enormous amounts of time and money. As one FDA official has put it, "Quack promoters have learned to stay one step ahead of the laws either by moving from state to state or by changing their corporate names."[48]

How can I tell whether a product is bogus?

It's not always easy. Given that many misleading claims are supposedly backed by scientific-sounding statements, it is difficult for even informed consumers to separate fact from fiction. The general rule of thumb is: If it sounds too good to be true, it probably is. The following red flags can help you spot a quack:

- *The promoter claims that the medical establishment is against him or her and that the government won't accept this new "alternative" treatment.* If the government or medical community doesn't accept a treatment, it's because the treatment hasn't been proven to work. Reputable professionals don't suppress knowledge about fighting disease. On the contrary, they welcome new remedies for illness, provided the treatments have been carefully tested.

- *The promoter uses testimonials and anecdotes from satisfied customers to support claims.* Valid nutrition information comes from careful experimental research, not from random tales. A few people's reports that the product in question "works every time" are never acceptable as sound scientific evidence.
- *The promoter uses a computer-scored questionnaire for diagnosing "nutrient deficiencies."* Those computers are programmed to suggest that just about everyone has a deficiency that can be reversed with supplements the promoter just happens to be selling, regardless of the consumer's symptoms or health.
- *The promoter claims that the product will make weight loss easy.* Unfortunately, there is no simple way to lose weight. Again, if a claim sounds too good to be true, it probably is.
- *The promoter promises that the product is made with a "secret formula" available only from this one company.* Legitimate health professionals share their knowledge of proven treatments so that others can benefit from it.
- *The treatment is available only through the back pages of magazines, over the phone, or by mail-order ads in the form of news stories or 30-minute commercials (known as infomercials) in talk-show format.* Results of studies on credible treatments are reported first in medical journals and then administered by a doctor or other health professional. If information about a treatment appears only elsewhere, it probably can't withstand scrutiny.*

If I do buy a product, say, to help me lose weight, but I still need some advice about dieting, should I check with a nutritionist?

To answer that question, first consider the following. About 25 years ago, Charlie Herbert became a professional of the International Academy of Nutrition Consultants. Another member of the household, Sassafras Herbert, met all the requirements for membership in the American Association of Nutrition and Dietary Consultants, a "professional association dedicated to maintaining ethical standards in nutritional and dietary consulting." The only qualification for membership is a $50 fee, regardless of your background (or even your species). Charlie Herbert is a cat, and Sassafras is a poodle. The two obtained their "credentials" with the help of the late Victor Herbert, M.D., former professor of medicine, at Mount Sinai School of Medicine, New York City, and a leader in combating nutrition fraud. Dr. Herbert had his pets added to the membership rosters of

MINIGLOSSARY

accreditation approval; in the case of hospitals or university departments, approval by a professional organization of the educational program offered. There are phony accrediting agencies; the genuine ones are listed in a directory called *Accredited Institutions of Postsecondary Education.*

control group a group of individuals with characteristics that match the group being treated in an intervention study but who receive a sham treatment or no treatment at all.

correlation a simultaneous change in two factors, such as a decrease in blood pressure with regular aerobic activity (a direct or positive correlation) or the decrease in incidence of bone fractures with increasing calcium intakes (an inverse or negative correlation).

correspondence school a school from which courses can be taken and degrees granted by mail. Schools that are accredited offer respectable courses and degrees.

diploma mill a correspondence school that grinds out degrees—sometimes worth no more than the cost of the paper they are printed on—the way a grain mill grinds out flour.

epidemiological study a study of a population that searches for possible correlations between nutrition factors and health patterns over time.

experimental group the participants in a study who receive the real treatment or intervention under investigation.

First Amendment the amendment to the U.S. Constitution that guarantees freedom of the press.

intervention study a population study examining the effects of a treatment on experimental subjects compared to a control group.

nutritionist a person who claims to be capable of advising people about their diets. Some nutritionists are registered dietitians, whereas others are self-described experts whose training is questionable.

placebo a sham or neutral treatment given to a control group; an inert, harmless "treatment" that the group's members cannot recognize as different from the real thing.

registered dietitian (RD) a professional who has graduated from a program of dietetics accredited by the Commission on Accreditation for Dietetics Education (CADE) of the American Dietetic Association (ADA), has completed an internship program or the equivalent to gain practical skills, has passed a registration examination, and maintains competencies through continuing education. Some states require licensing for dietitians, thereby requiring anyone who wants to use the title "dietitian" to receive permission by passing a state examination.

those organizations to demonstrate how easy it is for anyone to get fake nutrition credentials. This is because in some states, the term ***nutritionist*** is not legally defined at present.

Before you pay a fee or follow a nutritionist's advice, inquire about the person's credentials. Some "nutritionists" obtain their diplomas and titles without undergoing the rigorous training required to obtain a legitimate degree in nutrition. Lax state laws make it possible for irresponsible **correspondence schools**—also called **diploma mills**—to grant degrees to unqualified individuals for nothing more than a fee.

How can I check a nutritionist's credentials?

You can call the institution the person claims has awarded the degree. To find out about the existence or reputation of an institution of higher learning, you can go to any library and ask for a directory of colleges and universities called *Accredited Institutions of Postsecondary Education,* published by the American Council on Education (www.acenet.edu). Be suspicious of diplomas or degrees issued by institutions that cannot prove that they have **accreditation** from the Council on Education.

Another option is to find out whether the person is a special type of nutritionist, known as a **registered dietitian (RD).** The RD, an especially meaningful credential, has a standard definition—a professional who has fulfilled coursework required by the American Dietetic Association (ADA), including courses in psychology, business, chemistry, anatomy, physiology, advanced nutrition, medical nutrition therapy, food science, and food service administration.

In addition, the RD has completed an internship that includes on-the-job training for counseling people about diet and has passed a national registration exam. All registered dietitians must keep their credentials current by completing regular continuing education requirements.

You can check on any RD by asking for that person's registration number and calling the Commission on Dietetic Registration (312-899-0040). If you'd like to contact an RD but don't know where to find one, call the ADA's Consumer Nutrition Hot Line at (800) 366-1655 for a referral or visit the ADA website **(www.eatright.org).**

© Courtesy of Marilyn Herbert

▲ *Charlie and Sassafras display their professional credentials.*

*If you think you've been duped by a quack, write the FDA, Office of Consumer Affairs and Information, HFC-110, 5600 Fishers Lane, Rockville, MD 29857; your state attorney general's office; the Federal Trade Commission, Correspondence Branch, Sixth and Pennsylvania Avenues, N.W., Washington, D.C. 20580; and/or the media responsible for running the ad. If you ordered the product by mail, call the U.S. Postal Service and ask for a fraud packet. If you suspect that a website is promoting misinformation or marketing bogus goods or services, e-mail the FDA at otcfraud@cder.fda.gov.

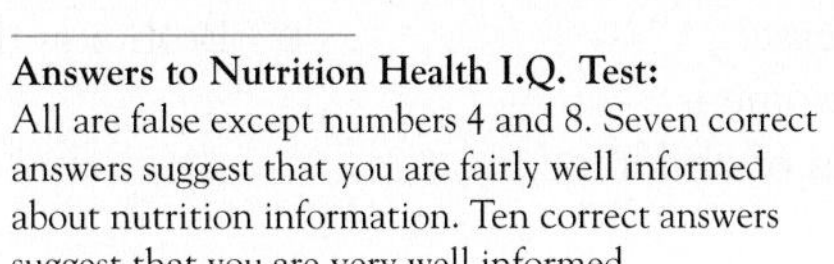
Answers to Nutrition Health I.Q. Test:
All are false except numbers 4 and 8. Seven correct answers suggest that you are fairly well informed about nutrition information. Ten correct answers suggest that you are very well informed.

In Review

1. Experts believe that dietary factors may account for __________ of all cancer and heart disease.
   a. 10%
   b. 50%
   c. 20%
   d. 30%

2. Nutrition information can come from a variety of sources. Where does the average American obtain the majority of his or her information?
   a. television
   b. Internet
   c. dietitians
   d. doctors

3. What is the leading cause of death in the United States?
   a. cancer
   b. heart disease
   c. liver disease
   d. diabetes

4. What is the calorie value of a meal that supplies 110 g of carbohydrates, 25 g of protein, 20 g of fat, and 5 g of alcohol?
   a. 160
   b. 345
   c. 560
   d. 755

5. One of the two broad goals of the *Healthy People 2010* initiative includes:
   a. reducing a person's risk for heart disease.
   b. increasing the intake of calcium of women.
   c. increasing fruit and vegetable intakes of children.
   d. eliminating health disparities.

6. Because claims made in advertisements about the benefits of nutrition products are required by law to be true, they can be believed.
   a. true
   b. false

7. Which of the following is a professional in the field of nutrition and has been trained in sound evidence-based nutrition?
   a. Registered Dietitian, RD
   b. Board Certified Nutritionist, BCN
   c. Fellow of the Academy of Nutrition, FAN
   d. World Weight Federation, WWF

8. Eating a fast-food meal is really not consistent with a healthy lifestyle.
   a. true
   b. false

9. By chemical analysis, what nutrient is present in highest amounts in most foods?
   a. fat
   b. water
   c. protein
   d. carbohydrate

10. The term *essential* describes a nutrient that the body:
    a. cannot make for itself in sufficient quantity.
    b. must obtain from food.
    c. cannot be found in food.
    d. Both a and b are correct.
    e. Both a and c are correct.

Menu of Online Study Tools

A variety of study tools for this chapter are available at our website to deepen your understanding of chapter concepts. Go to

www.thomsonedu.com/nutrition/boyle

to find

- Practice tests
- Flashcards
- Glossary
- Web links
- Animations
- Chapter summaries, learning objectives, and crossword puzzles

Nutrition on the Web

www.thomsonedu.com/nutrition/boyle
Go to the *Personal Nutrition* site to check for the latest updates to chapter topics or to access links to related websites.

www.healthfinder.gov
Scroll down to Smart Choices for online screening checklists and information on healthy lifestyles; this site offers many links to other reliable health-related sites.

www.eatright.org
The American Dietetic Association home page includes frequently asked questions, nutrition resources, and many reliable links to other food and nutrition sites.

www.5aday.com
Promotes the 5 A Day for Better Health campaign.

www.hc-sc.gc.ca/
A Canadian health and nutrition information home page.

www.healthypeople.gov
Updates on *Healthy People 2010* initiative.

www.nutrition.gov
A government resource that provides easy access to all online government information on nutrition and dietary guidance.

www.nal.usda.gov/fnic
The Food and Nutrition Information Center.

www.pueblo.gsa.gov
A site for consumer food and nutrition information.

www.health.nih.gov
A search engine from the National Institutes of Health with access to Medline and PubMed databases.

www.nlm.nih.gov/medlineplus/healthfraud.html
Health fraud resources and tips on where to report fraud cases.

The Pursuit of an Ideal Diet

2

ASK YOURSELF . . .

Which of the following statements about nutrition are true, and which are false? For each false statement, what is true?

1. It is wise to eat the same foods every day.
2. Milk is such a perfect food that it alone can provide all the nutrients a person needs.
3. Cookies cannot be included in a healthful diet.
4. When it comes to nutrients, more is always better.
5. A person's energy needs are based on his or her age, gender, and physical activity levels.
6. From a nutritional standpoint, there is nothing wrong with grazing on snacks all day, provided the snacks meet nutrient needs without supplying too many calories.
7. If you don't meet your recommended intake for a nutrient every day, you will end up with a deficiency of that nutrient.
8. If a food label claims that a product is low fat, you can believe it.
9. Most dietitians encourage people to think of their diets in terms of the four basic food groups.
10. According to the government, people should try to eat at least 2 cups of fruit and 2½ cups of vegetables—totaling nine servings—a day.

Answers found on the following page.

If the doctors of today will not become the nutritionists of tomorrow, the nutritionists of today will become the doctors of tomorrow.

Thomas Edison (1847–1931, American inventor)

CONTENTS

For most people, eating is so habitual that they give hardly any thought to the foods they choose to eat. Yet, as Chapter 1 emphasized, the foods you select can have a profound effect on the quality, and possibly even the length, of your life. Given all the statistics and government mandates presented so far, however, designing a healthful diet may seem like a complicated matter involving a rigid regimen that excludes certain foods from the diet. Fortunately, that's not the case. The government, as well as many major health organizations, has devised dietary guidelines and tools (such as food labels) to help you choose the most healthful diet. This chapter provides an overview of some of the best guides and tools and shows you how to use them.

As you read the following pages, keep in mind that one of the biggest misconceptions about planning a healthful diet is believing that some foods, say, carrots and celery sticks, are "good," whereas others, like cookies and candy, are "bad." People who categorize foods this way often feel guilty every time they "splurge" on a so-called bad food.

The overall diet is what really counts. A diet consisting of nothing but carrot sticks is just as unhealthful as one made up of only candy bars. The trick is choosing a healthful balance of foods. The ideal diet contains primarily foods that supply adequate nutrients, fiber, and calories without an excess of fat, sugar, sodium, or alcohol.

adequacy characterizes a diet that provides all of the essential nutrients, fiber, and energy (calories) in amounts sufficient to maintain health.

balance a feature of a diet that provides a number of types of foods in balance with one another, such that foods rich in one nutrient do not crowd out of the diet foods that are rich in another nutrient.

calorie control control of consumption of energy (calories); a feature of a sound diet plan.

The ABCs of Eating for Health

When you plan a diet for yourself, try to make sure it has the following characteristics:

- **adequacy** (to provide enough of the essential nutrients, fiber, and energy—in the form of calories)
- **balance** (to avoid overemphasis on any food type or nutrient at the expense of another)
- **calorie control** (to supply the amount of energy you need to maintain desirable weight—not more, not less)
- **moderation** (to avoid excess amounts of unwanted constituents, such as fat, salt, or sugar)
- **variety** (to consume different foods rather than eating the same meals day after day)

Equally important, be sure that your diet suits you—that it consists of foods that fit your personality, family and cultural traditions, lifestyle, and budget. At best, your diet can be a source of both pleasure and good health.

Adequacy Any nutrient can be used to demonstrate the importance of dietary *adequacy*. For example, iron is an essential nutrient that your body loses daily and must be replaced continually by iron-rich foods. If your diet does not provide adequate iron—that is, it lacks food sources of the mineral—you can develop a condition known as iron-deficiency anemia. If you add iron-rich foods such as meat, fish, poultry, and legumes to your diet, the condition is likely to disappear soon. (More information about iron appears in Chapter 7.)

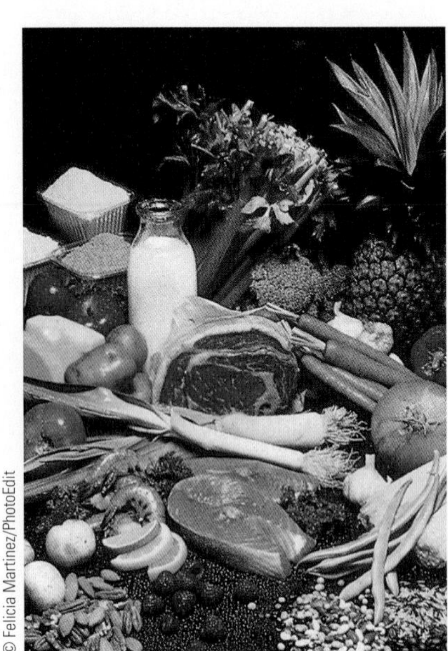

▲ *Variety fosters good nutrition.*

Balance To appreciate the importance of *balance*, consider a second essential nutrient. Calcium plays a vital role in building a strong frame that can withstand the gradual loss of bone that occurs with age. Thus, adults are advised to consume three cups of milk or milk products (the best sources of this bone-building mineral) daily to meet their calcium needs. Foods that are rich in calcium typically lack iron, however, and vice versa, so you have to balance the two in your diet.

Balancing the whole diet is a juggling act that, if successful, provides enough, but not too much, of each of the 40-odd nutrients the body needs for good health. As you will see later in the chapter, you can design a diet that is both adequate and balanced by using food group plans that help you choose from various groups the specific amounts of foods that should be eaten each day.

Calorie Control To maintain a desirable weight, energy intakes should not exceed energy needs. *Calorie control* helps ensure a balance between energy we take in from food and energy we expend in activity. A cup and a half of ice cream, for example, has about the same amount of calcium as a cup of milk, but the ice cream may contain more than 500 calories, whereas the milk may supply only 90. When it comes to iron, a 3-ounce serving of beef pot roast provides

Ask Yourself Answers: **1.** False. It is unwise to eat the same foods day in and day out; your diet will lack variety and probably will not supply all the nutrients your body needs. **2.** False. Milk rates as an excellent source of nutrients such as calcium and protein, but it contains only very small amounts of iron and several other essential nutrients. **3.** False. Any food can fit into a healthful eating plan. It's the total diet, not individual foods, that can be either good or bad for health. **4.** False. Too much or too little of a nutrient is often equally harmful. **5.** True. **6.** True. **7.** False. Even if you don't meet your recommended intake for a nutrient, you still may be consuming a sufficient amount of the nutrient. **8.** True. **9.** False. Dietitians today encourage people to think about eating from the five food groups in the MyPyramid Food Guide. **10.** True.

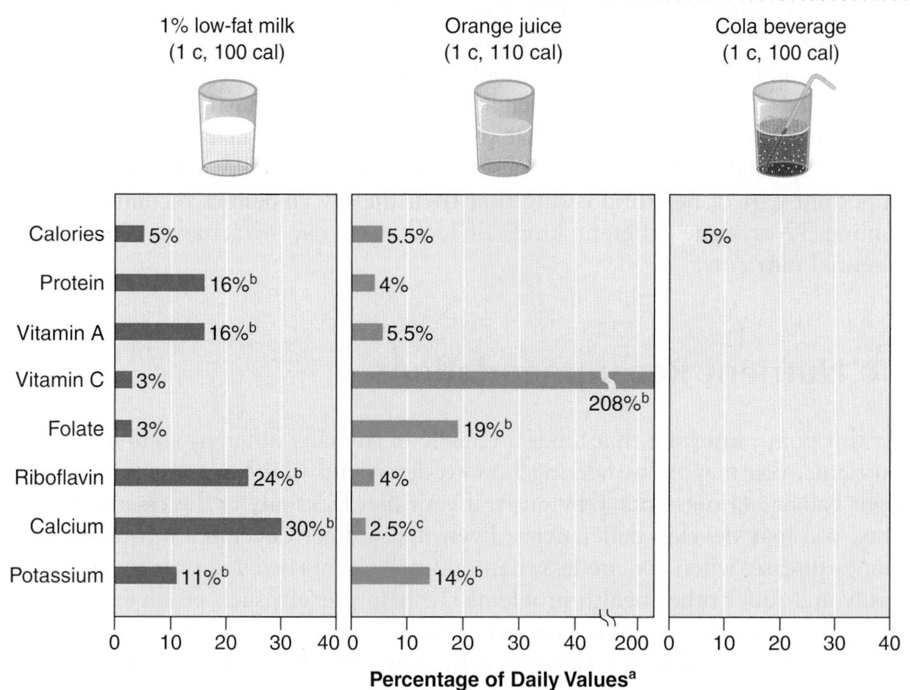

FIGURE 2-1
NUTRIENT DENSITY OF SELECTED BEVERAGES

Understanding the concept of nutrient density will help you locate foods that are high in the amount of nutrients provided per calorie. Compare the nutrient contributions made by the three beverages shown here. The figures show that orange juice and milk are good sources of several essential nutrients, whereas the cola beverage simply provides calories (from sugar).

[a] The Daily Values, used on food labels, are based on 2,000 calories a day for adults and children over 4 years old and are useful for making comparisons among foods in terms of their nutrient composition.
[b] A good source of this nutrient (more than 10 percent of the recommended intake).
[c] Certain brands of orange juice are fortified with calcium and provide 30 percent of the recommended value.

about twice the amount of the mineral as a 3-ounce serving of canned tuna. But whereas the beef contains 240 calories, the tuna adds only about 100 calories to the diet. The choice of which one to eat depends on personal preferences as well as the nutrient and calorie content of the other foods in the diet.

Those who are trying to ensure optimal intakes of nutrients without excess calories should be sure to include foods that are rich in nutrients (protein, vitamins, and minerals) but relatively low in calories and fat. Such foods are referred to as **nutrient dense.** A baked potato, for example, contains more iron and vitamin C for its calories than French fries. Hence, it is more nutrient dense. Figure 2-1 compares the nutrient density of selected beverages.

Moderation Another characteristic of a healthy diet is *moderation*. In other words, try to eat meals that do not contain excessive amounts of any one nutrient, particularly dietary fat—the culprit linked to a number of chronic diseases. That's not to say that you should choose only foods that supply little or no fat. Such an approach is unrealistic and will only lead to frustration. A more moderate philosophy to adopt is the 80/20 rule: If you eat low-fat, nutrient-dense foods (and remember to exercise) at least 80 percent of the time, you probably won't reverse the healthful benefits of these foods if you splurge occasionally the remaining 20 percent of the time.

Variety Aside from avoiding the monotony of eating the same foods day after day, we need *variety* in our diet for two reasons: (1) Some foods are better sources of nutrients needed in such small amounts that we don't consciously plan diets around them, and (2) a limited diet can supply excess amounts of undesirable substances such as chemical contaminants. Eating many different foods, on the other hand, greatly reduces the likelihood that large amounts of a potential toxin will be consumed.

Research underscores that variety is one of the hallmarks of a healthful diet. Consider that a team of U.S. scientists examined the eating habits of more than 10,000 people in the early 1970s and found that those whose food choices were the least varied were most likely to have died 20 years later. In addition, they

moderation the attribute of a diet that provides no unwanted constituent in excess.

variety a feature of a diet in which different foods are used for the same purposes on different occasions—the opposite of *monotony.*

nutrient dense refers to a food that supplies large amounts of nutrients relative to the number of calories it contains. The higher the level of nutrients and the fewer the number of calories, the more nutrient dense the food is.

found that many of the people surveyed failed to regularly include items from major food groups in their meals. About 25 percent left out calcium-rich dairy products, and 17 percent went without fruit while 46 percent did not eat vegetables—both of which contain fiber and many essential nutrients.[1] If your diet lacks variety, chances are good that you're missing out on many nutrients necessary for optimal health. The Japanese, incidentally, recognize variety as such an important part of healthful eating that their dietary guidelines recommend consuming 30 or more different kinds of food *every* day to achieve a balance of essential nutrients.[2]

Nutrient Recommendations

At this point, knowing that foods are made of so many different combinations of nutrients, you may be wondering how to determine whether you are eating the right balance of nutrients. Obviously, if your diet lacks any of the essential nutrients, you may develop deficiencies. Even if you don't develop a full-blown deficiency disease, when you are less than optimally nourished you may get sick more easily and suffer other health problems. To help prevent such problems and provide a benchmark for people's nutrient needs, experts in Canada and the United States devised the **Dietary Reference Intakes (DRI)** for planning and assessing diets of healthy people in both countries.

The Dietary Reference Intakes (DRI)

A committee of nutrition experts selected by the National Academy of Sciences (NAS) sets forth the DRI—a set of daily nutrient standards based on the latest scientific evidence regarding diet and health. (DRI for all age groups are listed on the inside front cover.) The first set, called the Recommended Dietary Allowances (RDA), was published in 1941 and has been revised ten times since then. Since 1997, the NAS has developed a series of reports on the DRI (see margin list) that expands and updates the RDA in the United States and Recommended Nutrient Intakes (RNI) in Canada.[3]

The DRI Reports:

- Calcium, Vitamin D, Phosphorus, Magnesium, and Fluoride, 1997
- Folate, Vitamin B_{12}, other B Vitamins, and Choline, 1998
- Vitamins C and E, Selenium, and Carotenoids, 2000
- Vitamins A and K and Trace Minerals, 2002
- Energy, Macronutrients, and Physical Activity, 2002
- Water, Potassium, Sodium, Chloride, and Sulfate, 2004

Full texts of the reports are available at www.nap.edu.

Although the DRI are widely used, many people have misconceptions about their meaning and intent. To get a proper perspective about the DRI, consider the following facts:

- The DRI estimate the energy and nutrient needs of *healthy* people. People with certain medical problems often have different nutritional needs.
- Separate recommendations are made for different groups of people. For instance, the DRI committee issues one set of recommendations for children ages 4 through 8, another set for adult men, another for pregnant women, and so on.
- The DRI are recommendations that apply to *average* daily intakes. They are not intended to be nutrient requirements that individuals must meet day in and day out. The DRI take into account differences among individuals and establish a range within which the nutritional needs of virtually all healthy people in a particular age and gender group will be covered (see Figure 2-2).
- The DRI are illustrated in Figure 2-2 and include: Estimated Average Requirement (EAR), Recommended Dietary Allowances (RDA), Adequate Intakes (AI), Tolerable Upper Intake Levels (UL), Estimated Energy Requirement (EER), and Acceptable Macronutrient Distribution Range (AMDR).
- The DRI may evolve over time because of new scientific evidence indicating a need for reevaluation of current concepts and recommended intakes.[4]

Dietary Reference Intakes (DRI) a set of reference values for energy and nutrients that can be used for planning and assessing diets for healthy people.

FIGURE 2-2
THE CORRECT VIEW OF THE DRI

This figure shows that people with intakes below the EAR are likely to have dietary adequacy of 50 percent or less. People with intakes between the EAR and RDA are likely to have adequacy between 50 and 97–98 percent. Intakes between the RDA or AI and the UL are likely to be adequate. At intakes above the UL, the risk of harm increases.

People often think that more is better when it comes to nutrients. Too much of a good thing can be dangerous, however, so the RDA, AI, and AMDR fall within an optimal margin of safety, and the UL help people avoid harmful excesses of nutrients.

SOURCE: Adapted from Institute of Medicine, *Dietary Intakes for Energy, Carbohydrate, Fiber, Fat, Fatty Acids, Cholesterol, Protein, and Amino Acids* (Washington, D.C.: National Academy Press, 2002).

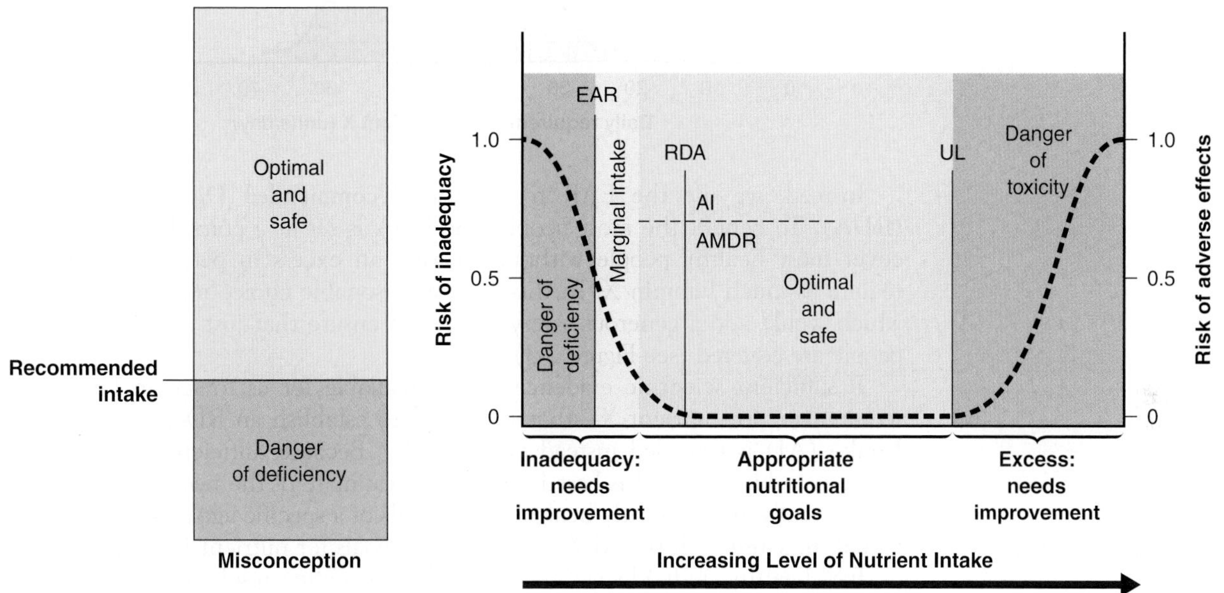

The DRI for Nutrients

The DRI aim to prevent nutrient deficiencies in a population as well as reduce risk for chronic diseases such as heart disease, cancer, and osteoporosis. To understand the process of developing the new DRI, consider the following discussion. Suppose we were the committee members, and we were called upon to set the DRI for nutrient X. First, we would determine the **requirement**—how much of that nutrient the average healthy person needs to prevent a deficiency. To do so, we would review scientific research and explore how the body stores the nutrient, what the consequence of a deficiency might be, what causes depletion of the nutrient, and what other factors affect a person's need for nutrient X. We would also consider current concepts regarding the amount of the nutrient needed for reducing risk of chronic disease (for example, optimal vitamin E intake and reduced risk of heart disease, optimal calcium intake and reduced risk of bone fractures later in life, and so on). However, we know that people vary in the amount of a given nutrient they need. Mr. A might need an average of 40 units of nutrient X daily for optimal health; Ms. B might need 35; Mr. C, 60; and so on.

Our next task would be to determine what amount of nutrient X to recommend for the general public. In other words, what amount of nutrient X would we consider optimal for covering most people's needs without posing a hazard? One option would be to set it at the average requirement for nutrient X (shown in Figure 2-3 at 45 units). This is the **Estimated Average Requirement (EAR)**—the amount of a nutrient that is estimated to meet the requirement for the nutrient in half of the people of a specific age and gender. But if we did so and people took us literally, half the population—Mr. C and everyone else whose requirement is greater than 45—would eventually develop deficiencies.

requirement the minimum amount of a nutrient that will prevent the development of deficiency symptoms. Requirements differ from the RDA and AI, which include a substantial margin of safety to cover the requirements of different individuals.

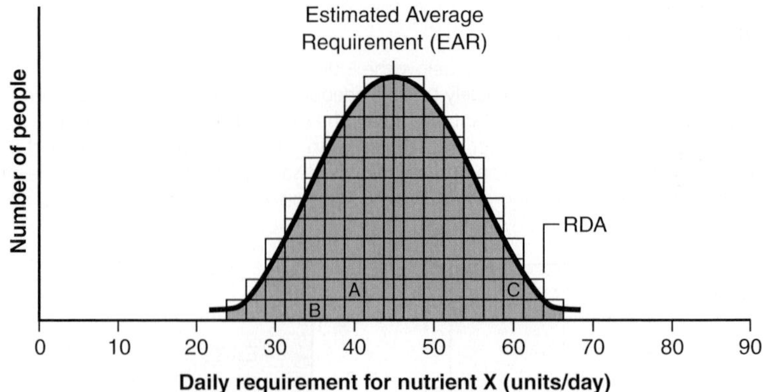

FIGURE 2-3
NUTRIENT REQUIREMENTS VARY FROM PERSON TO PERSON

Each square represents a person with different nutrient needs. For example, the squares labeled A, B, and C represent Mr. A, Ms. B, and Mr. C. The EAR is the amount of a nutrient estimated to meet the needs of half of the people in a particular age and gender group. The RDA for the nutrient is the amount that covers about 98 percent of the population.

Instead, we use the EAR to set the **Recommended Dietary Allowance (RDA).** To benefit the most people, the RDA is set at a point high enough to cover most healthy people without creating an excess in people who do not require as much vitamin X. In this case, a reasonable choice might be 63 units, which would add a generous safety margin to ensure that just about all healthy people are covered (see Figure 2-3).

If sufficient scientific evidence is not available for us to set an EAR—the requirement for nutrient X—that is needed to establish an RDA, an **Adequate Intake (AI)** is provided instead of an RDA.* Because sufficient scientific evidence is lacking, the AI is based on our best estimate of the need for nutrient X in practically all apparently healthy individuals of a specific age and gender. Individuals may use both the RDA and the AI as goals for nutrient intake.

For healthy individuals, no established benefits result from consuming amounts of a nutrient that exceed the recommended intake (the RDA or AI). However, the DRI include a **Tolerable Upper Intake Level (UL),** which is the maximum daily intake of a nutrient that is unlikely to pose risk of *adverse* effects in healthy people. The UL exceeds the RDA and is not intended to be a recommended level of intake (refer to Figure 2-2). The need for setting ULs is the

MINIGLOSSARY OF DRI TERMS

Dietary Reference Intakes (DRI) a set of reference values for energy and nutrients that can be used for planning and assessing diets for healthy people.

Estimated Average Requirement (EAR) the amount of a nutrient that is estimated to meet the requirement for the nutrient in half of the people of a specific age and gender. The EAR is used in setting the RDA.

Recommended Dietary Allowance (RDA) the average daily amount of a nutrient that is sufficient to meet the nutrient needs of nearly all (97–98 percent) healthy individuals of a specific age and gender.

Adequate Intake (AI) the average amount of a nutrient that appears to be adequate for individuals when there is not sufficient scientific research to calculate an RDA. The AI exceeds the EAR and possibly the RDA.

Tolerable Upper Intake Level (UL) the maximum amount of a nutrient that is unlikely to pose any risk of adverse health effects to most healthy people. The UL is not intended to be a recommended level of intake.

Estimated Energy Requirement (EER) the average calorie intake that is predicted to maintain energy balance in a healthy adult of a defined age, gender, weight, height, and level of physical activity, consistent with good health.

Acceptable Macronutrient Distribution Range (AMDR) a range of intakes for a particular energy source (carbohydrates, fat, protein) that is associated with a reduced risk of chronic disease while providing adequate intakes of essential nutrients.

*This text refers to the nutrient recommendations (for example, RDA and AI) as DRI recommended intakes.

result of more and more people using large doses of nutrient supplements and the increasing availability of **fortified foods.** Individuals can use the UL to determine whether their levels of nutrient intakes may pose risk of adverse effects over time.

The DRI represent a major shift in thinking about nutrient requirements for humans, from prevention of nutrient deficiencies to prevention of chronic disease. They also herald a new thinking about the role of dietary supplements in achieving good health—such that, for some individuals at higher risk, use of nutrient supplements may be desirable in order to meet recommended intakes. For example, older women who are at risk of osteoporosis may benefit from dietary calcium supplements to help maintain bone mineral mass. In addition, the development of the UL signals the widespread recognition that high intakes of nutrients can create a degree of risk.

fortified foods foods to which nutrients have been added, either because they were not already present or present in insignificant amounts. Examples: margarine with added vitamin A, milk with added vitamin D, certain brands of orange juice with added calcium, and breakfast cereals with added nutrients and nonnutrients.

The DRI for Energy and the Energy Nutrients

The DRI report on energy and the energy nutrients provides guidelines for the United States and Canada regarding the consumption of energy, carbohydrates, fiber, fat, fatty acids, cholesterol, and protein, and it includes guidelines for physical activity as well. To meet the body's daily energy and nutritional needs while minimizing risk for chronic disease, consumption of the energy nutrients should resemble the pattern shown in Figure 2-4.

To reduce the risk of chronic disease, adults and children should spend at least 1 hour every day doing a moderately intense physical activity (such as brisk

FIGURE 2-4
RECOMMENDED DIETARY INTAKE RANGES FOR ENERGY NUTRIENTS

A balanced diet is composed of approximately 10–35 percent protein, 45–65 percent carbohydrates (with no more than 10 percent of this amount from added sugars or caloric sweeteners), and 20–35 percent from fat, including the fats from meat, poultry, fish, eggs, milk and milk products, oils, and nuts. Currently, most Americans consume about 15 percent of their calories from protein, 49 percent from carbohydrates (with 16 percent of this amount coming from simple sugars), and 32–34 percent of their total calories from fat.

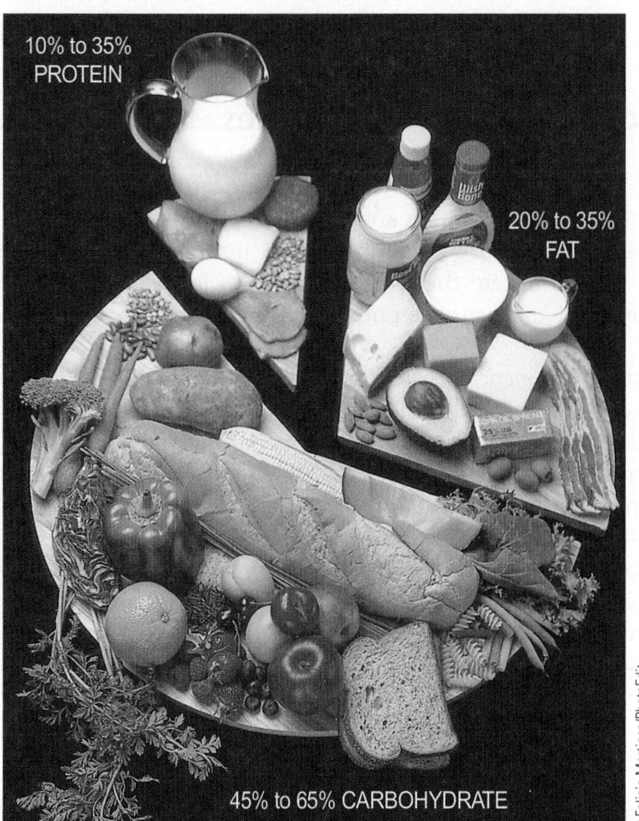

Dietary Reference Intakes (DRI) for Carbohydrates, Fats, and Protein	
Carbohydrates	45 to 65 percent of total calories
Fats	20 to 35 percent of total calories
Protein	10 to 35 percent of total calories

© Bob Thomas/Getty Images

▲ *People vary in the amount of a given nutrient they need. The challenge of the DRI is to determine the best amount to recommend for everybody.*

walking) or 20 to 30 minutes 4 to 7 days per week in a high-intensity activity (such as running or cycling). The DRI report stresses the need to balance caloric intake with physical activity, recommending total calories to be consumed by individuals of given heights, weights, and genders for each of four different levels of physical activity (from sedentary to very active). For more about weight management and the new exercise guidelines, see Chapters 9 and 10 as well as the inside back cover of this text.

The DRI Committee bases its "1 hour a day total" exercise goal on several studies of daily energy expenditure by individuals who maintain a healthy weight. Daily energy expenditure can be cumulative, so the 60 minutes does not need to be done all at once. Daily activities such as climbing the stairs at school or the office, housecleaning, or walking to your destination from a distant parking spot can be combined with moderate physical activities such as brisk walking.

Nutrition Recommendations from WHO

- Energy: Sufficient to support normal growth, physical activity, and body weight (BMI = 20–22)
- Total fat: 15 to 30 percent of total energy
  - Saturated fat: 0 to 10 percent of total energy
  - Dietary cholesterol: 0 to 300 milligrams per day
- Total carbohydrate: 55 to 75 percent of total energy
  - Dietary fiber: 27 to 40 grams per day
  - Added sugars: 0 to 10 percent of total energy
- Protein: 10 to 15 percent of total energy
- Salt: Less than 5 grams per day, preferably iodized
- Fruits and vegetables: At least 400 grams (about 1 pound) daily
- Physical activity: One hour per day of moderate intensity on most days of the week

Other Recommendations

Different nations and international groups have set forth different sets of standards similar to the DRI. Another widely used set of recommendations comes from two international organizations: the Food and Agriculture Organization (FAO) and the World Health Organization (WHO).[5] The FAO/WHO recommendations listed in the margin are considered sufficient to address the needs of nearly all people around the world. They differ from the DRI in that they are devised with slightly different priorities and purposes in mind. They assume, for example, that most people's diets contain protein of a lower quality than the protein in the diets of people in the United States. As a result, they recommend higher amounts of that nutrient (Chapter 5 discusses protein quality). The FAO/WHO recommendations also take into consideration the fact that worldwide, people are generally smaller and more physically active than people in the United States.

The Challenge of Dietary Guidelines

As Chapter 1 pointed out, health authorities are as concerned today about widespread nutrient excesses among Americans as they used to be about nutrient deficiencies. This is where dietary guidelines come into play. Among the most widely used of the guidelines are the *Dietary Guidelines for Americans*. The *Nutrition Recommendations for Canadians* presented in Appendix C are also commonly used.

The *Dietary Guidelines for Americans 2005* provide science-based advice to promote health and to reduce risk for chronic diseases through diet and physical activity. They emphasize variety, calorie control, moderation, and food safety. The guidelines also emphasize physical activity because it increases energy expenditure and helps in weight control.

Table 2-1 presents the *Dietary Guidelines* grouped into nine general topics along with their key recommendations. The first three topics focus on choosing nutrient-dense foods based on energy needs, maintaining a healthful body weight, and engaging in regular physical activity. The fourth topic, "Food Groups to Encourage," focuses on selecting a variety of fruits, vegetables, whole grains, and milk. The next three topics advise people to choose sensibly in their use of fats, carbohydrates, and salt. Topic eight provides advice for drinking alcoholic beverages in moderation, if at all. The final topic addresses basic food-safety skills.

TABLE 2-1

Key Recommendations of the *Dietary Guidelines for Americans 2005*

Adequate Nutrients within Energy Needs

- Consume a variety of nutrient-dense foods and beverages within and among the basic food groups; limit the intake of saturated and *trans* fats, cholesterol, added sugars, salt, and alcohol.
- Meet recommended intakes within energy needs by adopting a balanced eating pattern, such as the USDA MyPyramid Food Guidance System.

Weight Management

- To maintain body weight in a healthy range, balance calories from foods and beverages with calories expended.
- To prevent gradual weight gain over time, make small decreases in food and beverage calories and increase physical activity.

Physical Activity

- Engage in regular physical activity and reduce sedentary activities to promote health, psychological well-being, and a healthy body weight.
    - To reduce the risk of chronic disease in adulthood, engage in at least 30 minutes of moderate-intensity physical activity, above usual activity, at work or home on most days of the week.
    - For most people, greater health benefits can be obtained by engaging in physical activity of more vigorous intensity or longer duration.
    - To help manage body weight and prevent gradual, unhealthy body weight gain in adulthood, engage in approximately 60 minutes of moderate- to vigorous-intensity activity on most days of the week while not exceeding caloric intake requirements.
    - To sustain weight loss in adulthood, participate in at least 60 to 90 minutes of daily moderate-intensity physical activity while not exceeding caloric intake requirements. Some people may need to consult with a healthcare provider before participating in this level of activity.
- Achieve physical fitness by including cardiovascular conditioning, stretching exercises for flexibility, and resistance exercises or calisthenics for muscle strength and endurance.

Food Groups to Encourage

- Consume a sufficient amount of fruits and vegetables while staying within energy needs. Nine servings including 2 cups of fruit and 2½ cups of vegetables per day are recommended for a reference 2,000-calorie intake, with higher or lower amounts depending on individual calorie needs.
- Choose a variety of fruits and vegetables each day, including selections from all five vegetable subgroups (dark green, orange, legumes, starchy vegetables, and other vegetables) several times a week.
- Consume 3 or more ounce-equivalents of whole-grain products per day, with the rest of the recommended grains coming from enriched or whole-grain products. In general, at least half the grains should come from whole grains.
- Consume 3 cups per day of fat-free or low-fat milk or equivalent milk products.

Fats

- Consume less than 10 percent of calories from saturated fats and less than 300 milligrams of cholesterol per day, and keep *trans* fat consumption as low as possible.
- Keep total fat intake between 20 to 35 percent of calories; choose from mostly polyunsaturated and monounsaturated fat sources, such as fish, nuts, and vegetable oils.
- When selecting and preparing meat, poultry, dry beans, and milk or milk products, make choices that are lean, low fat, or fat free.
- Limit intake of fats and oils high in saturated and/or *trans* fatty acids, and choose products low in such fats and oils.

Carbohydrates

- Choose fiber-rich fruits, vegetables, and whole grains often.
- Choose and prepare foods and beverages with little added sugars.
- Reduce the incidence of dental caries by practicing good oral hygiene and consuming sugar- and starch-containing foods and beverages less frequently.

Sodium and Potassium

- Choose and prepare foods with little salt (less than 2,300 milligrams of sodium per day, or approximately 1 teaspoon of salt).
- At the same time, consume potassium-rich foods, such as fruits and vegetables.

Alcoholic Beverages

- Those who choose to drink alcoholic beverages should do so sensibly and in moderation—defined as the consumption of up to one drink per day for women and up to two drinks per day for men.
- Some individuals should not consume alcoholic beverages, including those who cannot restrict their alcohol intake, women of childbearing age who may become pregnant, pregnant and lactating women, children and adolescents, individuals taking medications that can interact with alcohol, and those with specific medical conditions.
- Individuals should avoid alcoholic beverages when engaging in activities that require attention, skill, or coordination, such as driving or operating machinery.

Food Safety

- To avoid microbial foodborne illness:
    - Clean hands, food contact surfaces, and fruits and vegetables.
    - Meat and poultry should not be washed or rinsed, to avoid spreading bacteria to other foods.
    - Separate raw, cooked, and ready-to-eat foods while shopping, preparing, or storing foods.
    - Cook foods to a safe internal temperature.
    - Chill perishable food promptly and defrost foods properly.
    - Avoid unpasteurized milk or any products made from it; avoid raw or undercooked eggs or foods containing raw eggs, raw or undercooked meat and poultry, unpasteurized juices, and raw sprouts.

NOTE: These guidelines are intended for adults and healthy children ages 2 and older. The *Dietary Guidelines for Americans 2005* contains additional recommendations for specific populations. The full document is available at www.healthierus.gov/dietaryguidelines.

(Text continues on page 43.)

Nutrition Action Grazer's Guide to Smart Snacking

© PhotoDisc/Getty Images

It's 3:30 in the afternoon, and that sound you just heard is coming from your stomach. Physiologically speaking, the human digestive system is customized for us to eat about every 4 hours to maintain our energy level. So, it's not unreasonable to get a little hungry between meals or before we go to bed. In fact, it is now clear that healthy snacking can fit into any eating plan and is important to everyone's health.[6]

But, although **grazing** is "in," many people feel a twinge of guilt now and then about between-meal munching. Perhaps parental warnings from childhood that snacks can spoil a meal linger in the back of many minds. Nevertheless, nutritious nibbling can make it easier for many people to eat healthfully.

Snacks can supply essential vitamins, minerals, and calories to diets of young children, who often cannot eat large portions at mealtime because of their small stomachs and appetites. Teenagers typically don't seem to have the time (or inclination) to eat regularly, due to busy school schedules, basketball or volleyball practice, music lessons, or other activities. Snacks account for approximately 25 percent of the calories they eat and are often more energy dense than their meals.[7]

Snacks also contribute to nutritional needs of adults, who may find that fitting meals into a busy schedule is difficult. Even senior citizens, whose lifestyles tend to be less hectic, can benefit from grazing. That's because lack of activity, certain medications, and isolation can blunt a formerly hearty appetite, making frequent, small meals more desirable than large breakfasts, lunches, and dinners. And, of course, a college student's busy life makes it difficult, if not impossible, to plan meals and snacks in advance, so snacking is a factor at any age.

The key to healthful snacking for everyone is to choose low-fat, high-fiber, nutrient-rich foods instead of snacks that add fat and calories to the diet and little else in the way of nutrients. A snack with a balance of carbohydrate, some fat, and some protein will satisfy hunger for a longer period of time than food with only carbohydrate or sugars (such as candy or soft drinks). How do you achieve this balance? Choosing snacks that contain at least two food groups can provide yummy and healthy snack possibilities.

grazing eating small amounts of food at intervals throughout the day rather than—or in addition to—eating regular meals.

Some snacks that appear nutritious may be deceiving. For example, fruit drinks, mixes, and punches are loaded with sugar (usually in the form of high fructose corn syrup) and are more similar to soft drinks than to fruit juice. Fruit rolls and bars are nutritionally similar to jams and jelly, not fresh fruit. For these snacks, sugar is added and most of the nutrients are lost when the fruit is processed with heat. Energy and protein bars can also be deceiving because, like candy bars, many are loaded with sugar and fat. Even some varieties of microwave popcorn don't rate well as snacks because they contain so much oil and salt. (You can easily make your own popcorn in the microwave oven without adding excessive extras.) When you feel like snacking, try some of the alternatives offered in Table 2-2. These snacks provide at least two food groups and 100–250 calories per serving.[8] Use Table 2-3 as a guide for choosing healthy snacks instead of calorie-dense snacks with little nutritional value. In addition, consider the following tips next time you're in the mood to grab a snack:

- Stock your refrigerator and kitchen cupboards with healthy foods like fruit juices, low-fat yogurt, fresh fruits and vegetables, plain popcorn, pretzels, whole-grain crackers, and low-fat cheeses so that they are close at hand. Nibblers often reach for a snack just to have something to munch on rather than because of a desire for the food itself. If nutritious choices are easy to get to, chances are that's what you'll eat.
- Carry fresh fruit and crackers and cheese, or even half a sandwich, in your backpack or book bag so you won't have to resort to buying candy from a vending machine when you get the urge to munch.
- Create your own snacks. Mix together one cup each pretzels, peanuts, raisins, and sunflower seeds to take with you on your next bicycle trip or hike. As a substitute for cream cheese, blend ½ cup drained low-fat yogurt with ½ cup low-fat cottage cheese. Then, add a bit of chopped pineapple, strawberries, or other fruit, and spread this mixture on crackers, bagels, English muffins, or rice cakes.
- Make new versions of old favorites, such as *Frozen Bananas*, *Chili Popcorn*, and *Mexican Snack Pizzas*.[9]
- Snack with a friend. If you're craving a candy bar or chips and nothing else will do, try splitting a bar or a bag with a friend. That way, you'll satisfy your craving without going too far overboard on calories, fat, or salt.
- Try to brush your teeth—or at least rinse your mouth thoroughly—after snacking to prevent tooth decay. (See the Nutrition Action feature in Chapter 3 for a detailed explanation of the role of diet in dental health.)

TABLE 2-2
Smart Snacking

- Low-fat yogurt with ½ c cereal
- Fruit and yogurt smoothies
- Half a bagel with 1 tablespoon peanut butter
- 1-oz serving trail mix (nuts, dried fruits)
- 1 c cereal with low-fat or skim milk or soy milk
- A *small* handful of tortilla chips with salsa or low-fat bean dip
- Fruit and cheese: apples or grapes with low-fat American, cheddar, or provolone cheese
- 1–2 tbsp peanut butter on an apple, celery, or carrots
- One slice pizza
- One piece of whole fruit (apple, banana, orange)
- ½ c fruit salad
- Small bag of baked potato chips
- Ice cream sandwich

Frozen Bananas

Mix 1 tablespoon peanut butter with ¼ cup evaporated skim milk. Cut a banana in half, and roll the halves in the peanut butter mixture. Then roll the coated banana halves in bran cereal. Place in the freezer until frozen. Makes two banana halves, each of which supplies about 165 calories, 4 grams of fat, and 149 milligrams of sodium.

Chili Popcorn

Mix 1 quart popped popcorn with 1 tablespoon melted margarine. In a separate bowl, mix 1¼ teaspoons chili powder, ¼ teaspoon cumin, and a dash of garlic powder. Sprinkle seasonings over popcorn and mix well. Makes about four 1-cup servings, each of which contains approximately 50 calories, 3 grams of fat, and 42 milligrams of sodium.

Mexican Snack Pizzas

Split two English muffins and toast lightly. Mix ¼ cup tomato paste; ¼ cup canned, drained, chopped kidney beans; 1 tablespoon each chopped onion and chopped green pepper; and ½ teaspoon oregano. Spread mixture on muffin halves. Top with ¼ cup shredded part-skim mozzarella cheese and broil until cheese is bubbly (about 2 minutes). Garnish with ¼ cup shredded lettuce. Makes four servings, each of which contains 95 calories, 2 grams of fat, and 300 milligrams of sodium.

TABLE 2-3
What's in a Muncher's Healthy Snacking Menu?

100 Calories of Grains				
1½ slices whole wheat bread	or	⅔ Lender's Original® wheat bagel	or	2⁄7 Dunkin' Donuts® plain bagel
1 c Kellogg's Corn Flakes®	or	scant ½ c Kellogg's Just Right® with fruit and nuts	or	¼ c Kellogg's Healthy Choice Low Fat Granola® (without raisins)
100 Calories of Fruit				
1¼ fresh apple	or	1 c unsweetened applesauce	or	½ c sweetened applesauce
2 heaping cups fresh strawberries	or	½ c frozen, sweetened strawberries	or	2 tbsp strawberry jam
100 Calories of Dairy				
⅗ c 1% cottage cheese	or	4⅓ tbsp grated Parmesan cheese	or	2 tbsp cream cheese
100 Calories of Potato				
½ plain baked potato w/skin	or	½ c mashed potato	or	10 French fries
100 Calories of Protein Foods (meats, poultry, fish, beans, nuts)				
2 oz roasted tip round beef	or	1.4 oz extra lean broiled ground beef	or	1.2 oz regular broiled ground beef
2.1 oz roasted chicken breast without skin	or	1.4 oz KFC Original Recipe® chicken breast with skin		
3 oz baked/broiled haddock	or	⅛ oz Mrs. Paul's Select Cut® frozen haddock fillets (breaded)		
⅜ c chickpeas	or	¼ c hummus (made with chickpeas)		
29 pistachios	or	15 almonds	or	6 macadamia nuts
100 Calories of "Eat Sparingly" Foods				
Scant ½ cup soft-serve vanilla frozen yogurt	or	⅜ c regular vanilla ice cream	or	3⁄16 c Haagen-Dazs® vanilla ice cream
2⅓ reduced fat Oreos®	or	1⅖ regular Oreos®	or	10⁄11 fudge-covered Oreo
⅘ Reese's Peanut Butter Cup®	or	½ Chunky® bar	or	5⁄14 Snickers® bar

SOURCE: Adapted from "For 100 Calories, You Can Have . . . ," *Tufts University Health & Nutrition Letter* (July 2000): 8.

As you can see, these guidelines recommend that people choose diets that:

- Emphasize fruits, vegetables (especially dark green and orange vegetables and legumes), whole grains, and fat-free or low-fat milk and milk products
- Include lean meats, poultry, fish, beans, eggs, and nuts
- Are low in saturated fats, *trans* fats, cholesterol, salt (sodium), added sugars, and alcohol

Healthy eating and regular physical activity enable people of all ages to work productively, enjoy life, and feel their best. They also help children grow, develop, and do well in school. The goal of the *Dietary Guidelines* is to help people decrease their risk of some forms of cancer, heart disease, obesity, diabetes, high blood pressure, stroke, osteoporosis, and liver disease—the so-called **lifestyle diseases.** Following such recommendations certainly makes sense, given the considerable potential health benefits they confer.

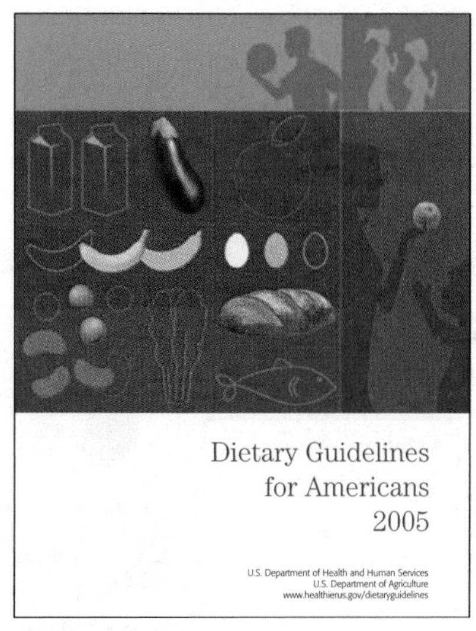

Dietary Guidelines
for Americans
2005
U.S. Department of Health and Human Services
U.S. Department of Agriculture
www.healthierus.gov/dietaryguidelines

Introducing the MyPyramid Diet-Planning Tool

Although the DRI and the dietary guidelines provide good frameworks for healthful eating, planning daily menus requires use of other, more specific tools. Dietitians and other nutrition experts often rely on a number of tools that you, too, can use to assess and plan your own diet.

One of the most helpful, easy-to-use diet-planning tools is the **food group plan,** which separates foods into specific groups and then specifies the number of **servings** from each group to eat each day. One example of such a plan is the Four Food Group Plan, devised in the 1950s and taught to consumers for nearly four decades. Several years ago, however, scientists updated this plan. The revised version, shown in the margin, was called the Food Guide Pyramid. It contained five food groups rather than four, and the tip of the pyramid was not considered to be a major food group because the foods found there provided extra calories and little else in the way of nutrients.

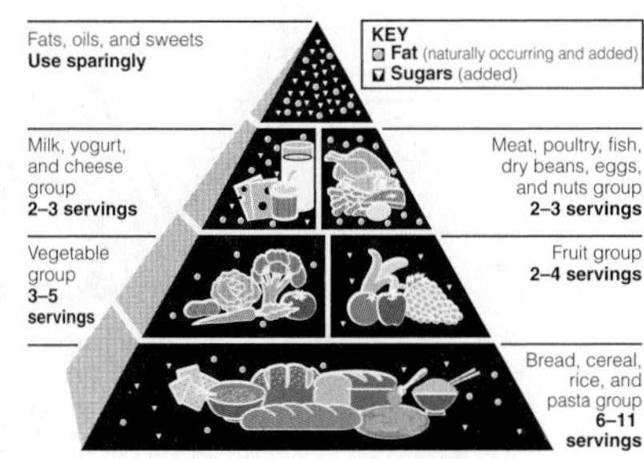

In 2005 USDA released a new symbol and interactive food guidance system, taking into consideration new nutrition knowledge gained over the years and the current consumption patterns of Americans. The MyPyramid food guide replaces the Food Guide Pyramid and is presented in detail in Figure 2-5. This new tool includes five food groups and presents tips for choosing foods from within each group.

The new MyPyramid Food Guidance System provides food-based guidance to help implement the recommendations of the *Dietary Guidelines for Americans 2005* and provides a visual aid to assist in improving diet and lifestyle. MyPyramid is designed to help you:

- Make smart choices from every food group
- Find your balance between food and physical activity
- Get the most nutrition out of your calories by focusing on nutrient-rich foods in sensible portion sizes

Use the Power of the Pyramid to Achieve a Healthy Lifestyle

MyPyramid is designed to help consumers choose foods that supply a good balance of nutrients, and it aims to moderate or limit dietary components often consumed in excess, such as saturated fat, *trans* fat, sugar, sodium, and alcohol, in keeping with the U.S. government's *Dietary Guidelines for Americans*. The new

lifestyle diseases conditions that may be aggravated by modern lifestyles that include too little exercise, poor diets, and excessive drinking and smoking. Lifestyle diseases are also referred to as *diseases of affluence*.

food group plan a diet-planning tool, such as MyPyramid, that groups foods according to similar origin and nutrient content and then specifies the amount of food a person should eat from each group.

serving the standard amount of food used as a reference to give advice regarding how much to eat (such as a 1 cup serving of milk).

FIGURE 2-5
MyPYRAMID: STEPS TO A HEALTHIER YOU

SOURCE: U.S. Department of Agriculture, 2005.

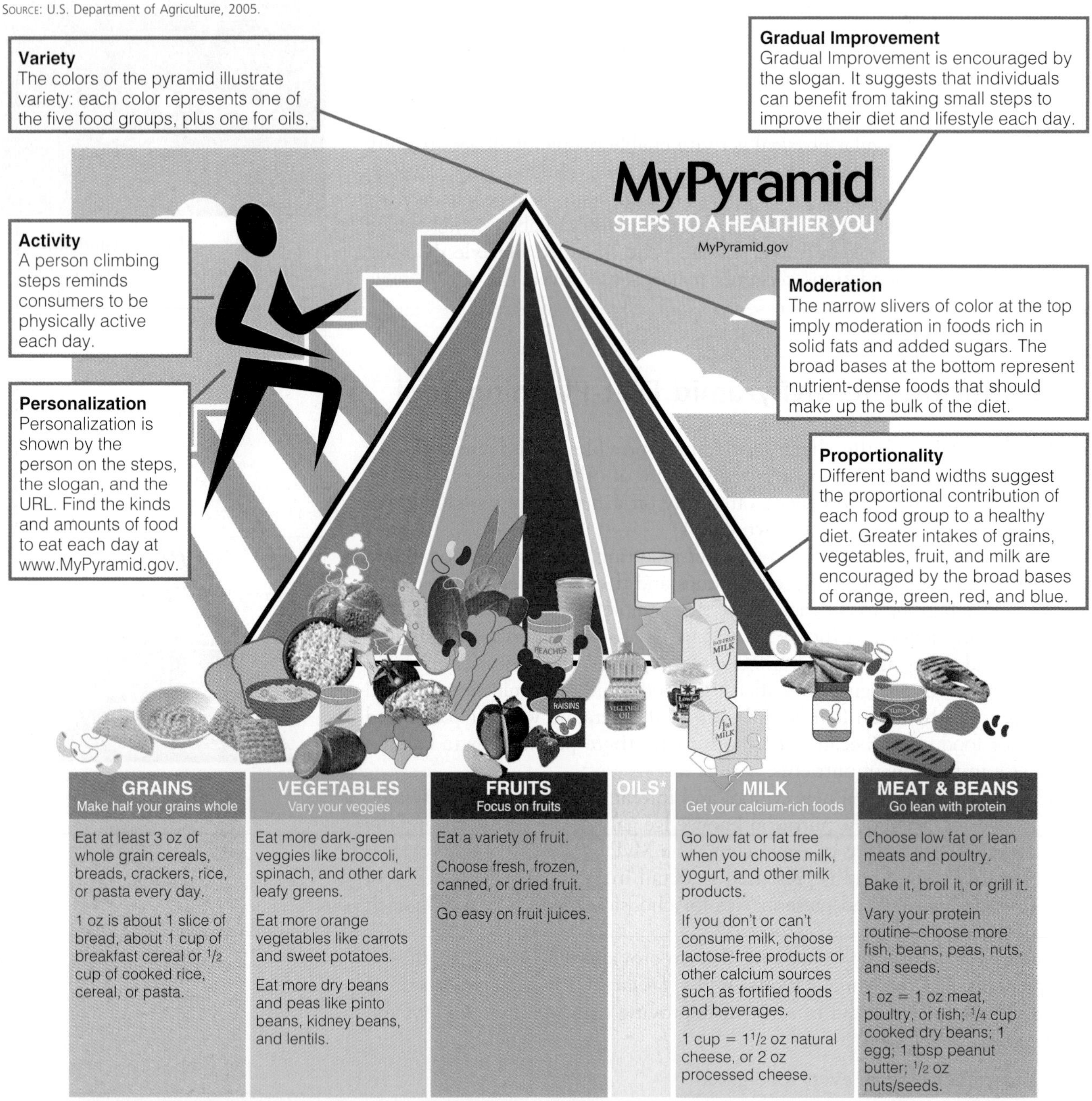

Variety
The colors of the pyramid illustrate variety: each color represents one of the five food groups, plus one for oils.

Gradual Improvement
Gradual Improvement is encouraged by the slogan. It suggests that individuals can benefit from taking small steps to improve their diet and lifestyle each day.

MyPyramid
STEPS TO A HEALTHIER YOU
MyPyramid.gov

Activity
A person climbing steps reminds consumers to be physically active each day.

Moderation
The narrow slivers of color at the top imply moderation in foods rich in solid fats and added sugars. The broad bases at the bottom represent nutrient-dense foods that should make up the bulk of the diet.

Personalization
Personalization is shown by the person on the steps, the slogan, and the URL. Find the kinds and amounts of food to eat each day at www.MyPyramid.gov.

Proportionality
Different band widths suggest the proportional contribution of each food group to a healthy diet. Greater intakes of grains, vegetables, fruit, and milk are encouraged by the broad bases of orange, green, red, and blue.

GRAINS Make half your grains whole	VEGETABLES Vary your veggies	FRUITS Focus on fruits	OILS*	MILK Get your calcium-rich foods	MEAT & BEANS Go lean with protein
Eat at least 3 oz of whole grain cereals, breads, crackers, rice, or pasta every day. 1 oz is about 1 slice of bread, about 1 cup of breakfast cereal or 1/2 cup of cooked rice, cereal, or pasta.	Eat more dark-green veggies like broccoli, spinach, and other dark leafy greens. Eat more orange vegetables like carrots and sweet potatoes. Eat more dry beans and peas like pinto beans, kidney beans, and lentils.	Eat a variety of fruit. Choose fresh, frozen, canned, or dried fruit. Go easy on fruit juices.		Go low fat or fat free when you choose milk, yogurt, and other milk products. If you don't or can't consume milk, choose lactose-free products or other calcium sources such as fortified foods and beverages. 1 cup = 1 1/2 oz natural cheese, or 2 oz processed cheese.	Choose low fat or lean meats and poultry. Bake it, broil it, or grill it. Vary your protein routine–choose more fish, beans, peas, nuts, and seeds. 1 oz = 1 oz meat, poultry, or fish; 1/4 cup cooked dry beans; 1 egg; 1 tbsp peanut butter; 1/2 oz nuts/seeds.
For a 2,000 calorie diet, you need the amounts below from each food group. To find the amounts that are right for you, go to www.MyPyramid.gov.					
Eat 6 oz every day.	Eat 2 1/2 cups every day.	Eat 2 cups every day.		Get 3 cups every day; for kids aged 2 to 8, It's 2.	Eat 5 1/2 oz every day.

*Make most of your fat choices from fish, nuts, and vegetable oils. Find your allowance for oils at www.MyPyramid.gov.

pyramid's anatomy includes six key components that can help you achieve a healthy lifestyle, including:

Some examples of moderate physical activity are walking briskly, mowing the lawn, dancing, swimming, and bicycling on level terrain. Some examples of vigorous physical activity are jogging, high-impact aerobic dancing, swimming continuous laps, and bicycling uphill.

- **Activity.** The pyramid promotes regular physical activity and reduced sedentary activities. As part of an overall healthy diet, balancing caloric intake with energy needs can help reduce the risk of chronic disease, help prevent weight gain, and help sustain weight loss. Like the Dietary Guidelines, the pyramid recommends engaging in at least 30 minutes of moderate-to-vigorous physical activity on most days of the week to reduce the risk of chronic dis-

© (Left) Polara Studios; (Right) Sunstar/Photo Researchers, Inc.

◀ *The MyPyramid Food Guidance System calls for eating a variety of foods to get the nutrients you need and at the same time the right amount of calories to maintain a healthy weight. Remember to balance the energy consumed with the energy expended in play.*

ease; up to 60 minutes of activity to manage body weight; and at least 60 to 90 minutes of physical activity to sustain a weight loss while not exceeding caloric intake requirements.

- **Variety.** Eat foods from all food groups and subgroups. Different foods contain different nutrients and other substances known to be protective against chronic diseases. No one food or no single food group provides all essential nutrients in amounts necessary for good health. The pyramid creates a foundation for good nutrition and health by guiding us to make food selections that can supply all the necessary nutrients and possibly even reduce risk of certain chronic diseases.
- **Proportionality.** The pyramid shows the proportions of foods that should make up a healthful diet. The widths of the different bands suggest the proportional contribution of the five food groups to a healthy diet (refer to Figure 2-5). Eat more of some foods (fruits, vegetables, whole grains, and fat-free or low-fat milk products have the widest bands on the pyramid), and eat less of others (such as foods high in sugars).
- **Moderation.** The broad bases at the bottom of the pyramid represent the nutrient-rich foods that should make up the bulk of your diet. The narrow slivers at the top represent foods that are rich in **solid fats** and **added sugars.** Choose forms of foods from the bottom of the pyramid to limit intake of saturated or *trans* fats, added sugars, cholesterol, and salt.
- **Personalization.** One size doesn't fit all. MyPyramid can help you choose the foods and amounts of food that are right for you. For a quick estimate of what and how much you need to eat, enter your age, gender, and physical activity level in the MyPyramid Plan box at www.MyPyramid.gov. You can also print out a worksheet based on your personal pyramid to help you track your progress.
- **Gradual improvement.** The slogan "MyPyramid—Steps to a Healthier You" suggests that individuals can benefit from taking small steps to improve their diet and lifestyle every day (for example, taking the stairs instead of the elevator or escalator, or ordering a green salad instead of fries).

Anatomy of MyPyramid
- Activity
- Variety
- Moderation
- Proportionality
- Personalization
- Gradual improvement

For a hundred more tips, visit www.smallstep.gov.

Use the Power of the Pyramid to Build a Healthy Diet

You can determine the right amount of foods to eat to meet your personal energy needs and promote a healthy weight in three easy steps:

Step 1: **Estimate Your Daily Energy Needs.** Use Table 2-4 to find your estimated daily energy needs, or go to www.MyPyramid.gov and enter your

The pyramid isn't the only food group plan. Canada has one of its own—Canada's Food Guide (shown in Appendix C).

TABLE 2-4
Estimated Daily Calorie Needs for Adults[a]

Find your gender and age. Then select the activity level that best describes your lifestyle (sedentary, moderately active, or active).

Activity Level:	Sedentary[b]	Moderate[c]	Active[d]
Females			
19–30	2,000	2,200	2,400
31–50	1,800	2,000	2,200
51–60	1,600	1,800	2,200
61+	1,600	1,800	2,000
Males			
19–30	2,400	2,600	3,000
31–50	2,200	2,400	2,800
51+	2,000	2,200	2,600

[a] The calorie levels in each gender and age group are based on persons of average height and at a healthy weight. If you are overweight, your calorie needs may be higher to maintain your weight. To lose weight, you can follow the calorie level in the chart, depending on your body weight.

[b] **Sedentary:** less than 30 minutes a day of moderate physical activity in addition to daily activities.

[c] **Moderately active:** at least 30 minutes and up to 60 minutes a day of moderate physical activity in addition to daily activities. Examples of moderate physical activity include walking briskly (about 3½ mph), mowing the lawn, dancing, swimming, or bicycling on level terrain. A person should feel some exertion but should be able to carry on a conversation comfortably during the activity.

[d] **Active:** 60 or more minutes a day of moderate physical activity in addition to daily activities.

NOTE: For more information, visit www.MyPyramid.gov.

age, gender, and usual activity level. If you are moderately physically active, you need fewer calories than those who are more active. The calorie levels for the food intake patterns in MyPyramid were matched to age and gender groups using the **Estimated Energy Requirement (EER)** for a person of average height, healthy weight, and sedentary activity level in each age and gender group. The sedentary level was selected as a way to help people avoid overestimating their calorie needs.

Step 2: **Build Your Daily Eating Plan.** Use your estimated number of calories to build a daily eating plan that incorporates all five food groups. How much food do you need to eat each day? MyPyramid provides guidance by indicating an appropriate amount of food to eat from each food group that will supply not only necessary nutrients but also the adequate amounts of calories. Table 2-5 specifies the amounts of foods from each group that are needed daily to create a healthful diet at several calorie levels.

Step 3: **Let the Pyramid Guide Your Food Choices.** Eating is one of life's great pleasures. Different people like different foods and like to prepare the same foods in different ways. Because you can build a healthy menu using many foods in many ways, you have lots of room for choice. The pyramid can provide a starting point to develop healthful eating patterns while still allowing for personal preferences in food choices. Make choices from each major group in the pyramid and combine them however you like. Remember to choose a variety of foods that you enjoy.

The important thing to remember is that you can enjoy all foods as part of a healthy diet as long as you don't overdo it on fat (especially saturated fat and *trans* fat), added sugars, salt, and alcohol. Read labels to identify (and avoid) foods that are higher in saturated fats, *trans* fats, added sugars, and salt (sodium). Certain items in each food group, particularly the high-fat choices, may be difficult to fit into a healthful eating plan on a regular basis.

Estimated Energy Requirement (EER) The EER represents the average dietary energy intake that will maintain energy balance in a healthy person of a given gender, age, weight, height, and physical activity level.

TABLE 2-5

Build Your Eating Plan with Recommended Daily Amounts from Each Food Group

Find your calorie level at the top of the chart. Follow the column below your estimated calorie level to see how much food to eat from each of the food groups. (Estimated daily energy needs are shown in Table 2-4.) Recommended amounts for fruits, vegetables, and milk are measured in cups and those for grains and meats, in ounces.

Calorie Level:	1,600	1,800	2,000	2,200	2,400	2,600	2,800	3,000
Grains	5 oz	6 oz	6 oz	7 oz	8 oz	9 oz	10 oz	10 oz
Vegetables	2 cups	2½ cups	2½ cups	3 cups	3 cups	3½ cups	3½ cups	4 cups
Fruits	1½ cups	1½ cups	2 cups	2 cups	2 cups	2 cups	2½ cups	2½ cups
Oils	5 tsp	5 tsp	6 tsp	6 tsp	7 tsp	8 tsp	8 tsp	10 tsp
Milk	3 cups	3 cups	3 cups	3 cups	3 cups	3 cups	3 cups	3 cups
Meat & Beans	5 oz	5 oz	5½ oz	6 oz	6½ oz	6½ oz	7 oz	7 oz
Discretionary Calorie Allowance*	132 cal	195 cal	267 cal	290 cal	362 cal	410 cal	426 cal	512 cal

*Discretionary calories are the balance of calories remaining in a person's estimated energy allowance after accounting for the number of calories needed to meet recommended nutrient intakes through consumption of foods in low-fat or no added sugar forms. The calories assigned to discretionary calories may be used to increase intake from the basic food groups; to select foods from these groups that are higher in fat or with added sugars; to add oils, solid fats, or sugars to foods or beverages; or to consume alcohol.

MyPyramid can help you get the nutrients your body needs each day. Using the pyramid to assess and plan your own diet requires an understanding of how much food to consume from the various food groups and how much of a food counts as a serving. For instance, one slice of bread, one half a bagel, and ½ cup of pasta each counts as 1 ounce from the Grains group. When it comes to vegetables, ½ cup of raw or cooked vegetables or 1 cup of leafy raw vegetables chalks up a ½-cup serving. Table 2-6 shows the portions that count as a serving from each of the various food groups. Besides showing what foods are the equivalent of one serving from each food group, Table 2-6 also lists the nutrients supplied by each of the five food groups. To get an idea of how your current diet compares with the recommendations of MyPyramid, try the Rate Your Plate Scorecard on page 56.

For more information on diet planning with the MyPyramid tool, go to "Inside the Pyramid" at www.MyPyramid.gov. See the Food Gallery photos for examples of sensible-size portions of the foods you eat.

The Savvy Way to Meeting Nutrient Needs

The grains, fruits, vegetables, and milk groups serve as the foundation of a healthy diet because they supply the vitamins, minerals, and fiber many people's diets lack. When it comes to the milk group, low-fat and fat-free dairy products make the best choices. Although foods from the Meats and Beans group provide protein, iron, zinc, and other nutrients, they can often contain large amounts of fat as well as saturated fat. Thus, choosing wisely from this group goes a long way in limiting the fat content of your diet. Because lean cuts of meat, skinless poultry, and fish rank lower in fat than ground beef and chicken with skin, the leaner items should be chosen more often than the high-fat selections. Dry beans, which contain only a trace of fat, are good choices to add to the diet even more often.

In addition to the five food groups, we need a small amount of oil in our diet because oils provide vitamin E and essential fats. The Oils category includes vegetable oils and soft margarines containing no *trans* fat. Heart-healthy oils like olive, canola, and peanut oil are also good choices. (For more about fats, see Chapter 4.)

Making Use of MyPyramid
- Determine calorie needs.
- Note amount of food to eat from each food group.
- Plan meals and snacks.
- Balance food intake with physical activity.

The Savvy Way to Moderate Energy Intake

You can follow the pyramid guidelines to get enough nutrients without overdoing calories. The way to balance your energy needs is to make your food choices count. By incorporating the tips and strategies for healthful eating found in

TABLE 2-6
Are You Savvy about Recommended Amounts of Food to Eat?

Food Group	Goal[b]	Nutrients Supplied	Equivalent Amounts
Grain Group Make half your grains whole	6 oz	Protein, complex carbohydrate and fiber, folate, niacin, riboflavin, thiamin, magnesium, and iron	**1 ounce is equivalent to:** 1 slice bread, 1 mini bagel, 6" tortilla, ½ English muffin 1 cup dry cereal ½ cup cooked rice, pasta, cereal 1 packet instant oatmeal 3 cups popcorn
Whole grains	3 oz		
Other grains	3 oz		
Vegetable Group Vary your veggies	2½ cups	Vitamin C, beta-carotene, fiber, potassium, folate, calcium, magnesium	**½ cup is equivalent to:** ½ cup cut-up raw or cooked vegetables 1 cup raw leafy vegetable ½ cup vegetable juice
Dark green vegetables	3 cups/week		
Orange vegetables	2 cups/week		
Legumes (dry beans)	3 cups/week		
Starchy vegetables	3 cups/week		
Other vegetables	6½ cups/week		
Fruit Group Focus on fruits	2 cups	Vitamin C, beta-carotene, fiber, potassium, folate, magnesium	**½ cup is equivalent to:** ½ cup fresh, frozen, or canned fruit 1 medium fruit ¼ cup dried fruit ½ cup fruit juice
Milk Group Get your calcium-rich foods	3 cups	Protein, calcium, riboflavin, vitamin B_{12}, vitamins A and D	**1 cup is equivalent to:** 1 cup low-fat/fat-free milk, yogurt 1½ oz low-fat or fat-free natural cheese 2 oz low-fat or fat-free processed cheese ⅓ cup shredded cheese

Table 2-7 into your eating plan, you will get the nutrients you need for good health. Choose the most nutrient-rich foods from each of the food groups. This means choosing more whole foods that are naturally nutrient-rich, such as whole grains, fruits and vegetables, including legumes, and fat-free and low-fat milk and milk products. This can help you maintain a healthy weight and may lower your risk for developing chronic health problems such as diabetes, heart disease, or cancer. Be flexible and adventurous! Try new food choices in place of some of the less nutritious or higher calorie foods you usually eat. A sample eating pattern is given in Table 2-8 (page 54), and Table 2-9 (page 55) translates that eating pattern into a nutritious menu. The Eat Well Be Well feature on page 57 provides additional tips for including a variety of fruits and vegetables in your diet for maximum health benefits.

discretionary calorie allowance the balance of calories remaining in a person's energy allowance, after accounting for the number of calories needed to meet recommended nutrient intakes through consumption of nutrient-dense foods in low-fat or no added sugar forms. The calories assigned to discretionary calories may be used to increase intake from the basic food groups; to select foods from these groups that are higher in fat or with added sugars; to add oils, solid fats, or sugars to foods or beverages; or to consume alcohol.

Gaining Calorie Control: The Discretionary Calorie Allowance

If you consistently build your diet by choosing mostly nutrient-dense foods that are low in solid fat and added sugars, you may be able to meet your nutrient needs without using your full calorie allowance. If so, you may have what is called a **discretionary calorie allowance** for use in meeting the rest of your calorie needs (see Figure 2-6).

Most discretionary calorie allowances are very small, between 100 and 300 calories, especially for those who are not physically active. How do we track these

TABLE 2-6

Are You Savvy about Recommended Amounts of Food to Eat? (continued)

Food Group	Goal[b]	Nutrients Supplied	Equivalent Amounts
Meat & Bean Group Go lean with protein	5½ oz	Protein, zinc, iron, niacin, thiamin, phosphorus, vitamin B_6, vitamin B_{12}, vitamin E, fiber (legumes)	**1 ounce is equivalent to:** 1 oz cooked lean meats, poultry, fish 1 egg ¼ cup cooked dry beans or tofu 1 tbsp peanut butter ½ oz nuts or seeds ½ cup split pea soup
Oils Oils[c]	6 tsp (27 grams)	Vitamin E, essential fatty acids (see Chapter 4)	**1 tsp equivalent is:** 1 tsp soft margarine 1 tbsp low-fat mayonnaise 2 tbsp light salad dressing 1 tsp vegetable oil 1 oz peanuts, nuts, or sunflower seeds has 3 teaspoons of oil
Discretionary calorie allowance[d] **Example of distribution:** **Solid fat** **Added sugars[e]**	267 calories 18 grams 8 tsp		**1 tbsp added sugar equivalent is:** 1 tbsp jelly or jam ½ oz jelly beans 8 oz lemonade

[a] All servings are per day unless otherwise noted. Vegetable subgroup amounts are per week. To follow this eating pattern, food choices over time should provide these amounts of food from each group on average.

[b] Recommended amounts to eat are based on a 2,000-calorie diet. See Table 2-4 for information about gender/age/activity levels and appropriate calorie intakes. See Table 2-5 for more information on the food groups, amounts, and food intake patterns at other calorie levels, or go to www.MyPyramid.gov.

[c] The oils listed in this table are not considered to be part of discretionary calories because they are a major source of the vitamin E and polyunsaturated fatty acids, including the essential fatty acids, in the food pattern. In contrast, solid fats (i.e., saturated and *trans* fats) are listed separately as sources of discretionary calories.

[d] The discretionary calorie allowance is the remaining amount of calories in each calorie level after nutrient-dense forms of foods in each food group are selected. Discretionary calories may be used to increase the amount of food selected from each food group; to consume foods that are not in the lowest fat form (such as 2% milk or medium-fat meat) or that contain added sugars; to add oil, fat, or sugars to foods; or to consume alcohol. The table shows an example of how these calories may be divided between solid fats and added sugars.

[e] Added sugars are the sugars and syrups added to foods and beverages during processing or preparation, not the naturally occurring sugars in fruits or milk.

extra calories? One example is that a regular 12-ounce soda has 155 calories, but all 155 of these calories are from added sugars and, thus, are considered "discretionary" calories. Keep in mind that, for many people, the discretionary calorie allowance is totally used up in the foods they choose in each food group, such as higher fat meats, higher fat cheeses, whole milk, and sweetened bakery products.

Your discretionary calories can be used to:

- Eat additional nutrient-dense foods from each of the food groups, such as an extra container of low-fat yogurt or an extra piece of fruit

(Text continues on page 52.)

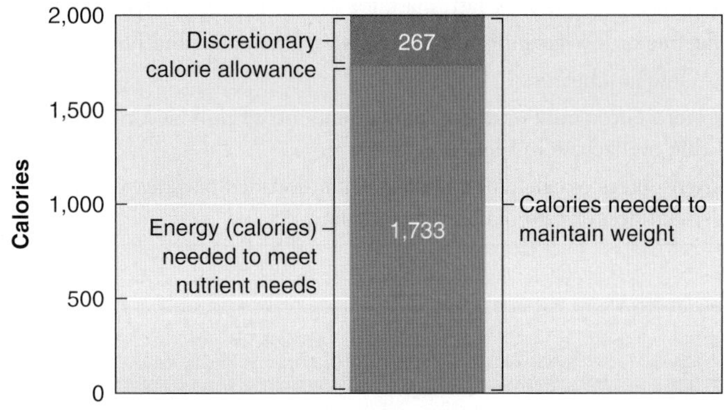

FIGURE 2-6
THE DISCRETIONARY CALORIE ALLOWANCE

TABLE 2-7
Making the Most of Your Own Personal Pyramid

Food Group	Strategies
 Grain Group	Check the ingredient list on grain product labels. For many whole-grain products, the words "whole" or "whole grain" will appear before the grain ingredient's name. Check the Nutrition Facts label for the fiber content of food products. Fiber content is a good clue to the amount of whole grain in the product. Choose 100% whole-grain breads, preferably, or mixed whole and white flour breads such as multi-grain or cracked wheat. To eat more whole grains, substitute a whole-grain product for a refined product—such as eating whole-wheat bread instead of white bread or brown rice instead of white rice. Add whole grains to mixed dishes such as soups, stews, and casseroles.
 Vegetable Group	Buy fresh vegetables in season. They cost less and are likely to be at their peak flavor. Stock up on frozen vegetables for quick and easy cooking in the microwave. Prepare main dishes, side dishes, and salads that include vegetables. Keep a bowl of cut-up vegetables in a see-through container in the refrigerator. Carrot and celery sticks are traditional, but consider broccoli, cucumber slices, or red or green pepper strips. Add dark-green or orange vegetables to soups, stews, casseroles, and stir-fries. Use romaine, spinach, or other dark leafy greens as salad greens, and eat green salads often. Choose main dishes, side dishes, and salads that include cooked dry beans or peas.
 Fruit Group	Use fruit in salads, toppings, desserts, and/or snacks regularly. Use fruit as a topping on cereal, pancakes, and other foods rather than sugars, syrups, or other sweet toppings. Buy fresh fruits in season when they may be less expensive and at their peak flavor. Buy fruits that are dried, frozen, and canned (in water or juice) as well as fresh, so that you always have a supply on hand. Choose whole or cut-up fruits more often as snacks or with meals, instead of juice.
 Milk Group	Drink fat-free (skim) or low-fat (1%) milk as a beverage. Use fat-free or low-fat milk or yogurt on cereal. Eat fat-free or low-fat yogurt as a snack. If you drink cappuccinos or lattes, ask for them with fat-free (skim) milk. Add fat-free or low-fat milk instead of water to oatmeal and hot cereals. Choose low-fat cheeses. Drink lactose-free milk or drink smaller amounts of milk at a time as options for those that are lactose intolerant. For other sources of calcium, include calcium-fortified beverages, fortified breakfast cereals, sardines, or tofu made with calcium.

SOURCE: Adapted from *Inside the Pyramid*, available at www.mypyramid.gov.

Tips	Nutrient-Dense Choices (Choose Most Often)	Less Nutrient-Dense Choices (Limit Choices)
Make at least half of your grain selections whole grains	*Whole grains:* Barley, brown and wild rice, bulgur, millet, oats, popcorn, quinoa, rye, wheat; whole-grain low-fat breads, cereals, crackers, and pastas *Enriched grain products:* breads, bagels, cereals, grits, pastas, rice, rolls, and tortillas	Biscuits, cakes, cookies, crackers, cornbread, croissants, Danish, doughnuts, fried rice, French toast, granola, muffins, pancakes, pies, presweetened cereals, taco shells, waffles
Choose from all five vegetable subgroups several times a week. Eat more dark-green vegetables, orange vegetables, and dry beans and peas.	*Dark green vegetables:* Broccoli and leafy greens such as arugula, kale, green-leaf and romaine lettuce; spinach, beet, collard, mustard, and turnip greens *Orange vegetables:* Carrots, pumpkin, sweet potatoes, winter squash *Legumes:* Black beans, black-eyed peas, chickpeas (garbanzo beans), kidney beans, pinto beans, soybeans, split peas, and lentils *Starchy vegetables:* Cassava, corn, green peas, and white potatoes *Other vegetables:* Artichokes, asparagus, bamboo shoots, beets, bok choy, Brussels sprouts, cabbage, cauliflower, celery, cucumbers, eggplant, green beans, iceberg lettuce, mushrooms, okra, onions, peppers, snow peas, tomatoes, vegetable juices, zucchini	Baked beans, candied sweet potatoes, coleslaw, French fries, potato salad, refried beans, scalloped potatoes
Choose a variety of fruits each day and consume no more than one-third of the recommended intake as fruit juice. Keep the amounts of fruit juice consumed to less than half of total fruit intake.	Apples, apricots, avocados, bananas, blueberries, cantaloupe, grapefruit, grapes, guava, kiwi, mango, oranges, papaya, peaches, pears, pineapple, plums, raspberries, strawberries, watermelon, dried fruit, unsweetened fruit juices	Canned or frozen fruit in syrup, juices, punches, and fruit drinks with added sugars
Make most milk group choices fat free or low fat (1%). Consume other calcium-rich foods if milk and milk products are not consumed.	Fat-free milk and fat-free milk products, buttermilk, cheeses, cottage cheese, yogurt, fat-free fortified soy milk	Reduced-fat milk, whole milk, reduced-fat milk and whole milk products such as cheeses, cottage cheese, yogurt, chocolate milk, fortified soy milk, custards and puddings, ice cream, ice milk, frozen yogurt, and sherbet

Continues

TABLE 2-7
Making the Most of Your Own Personal Pyramid (continued)

Food Group		Strategies
 © Polara Studios, Inc.	Meat & Bean Group	Select meat cuts that are low in fat and ground beef that is extra lean (at least 90% lean). Trim fat from meat and remove poultry skin before cooking or eating. Drain fat from ground meats after cooking. Use preparation methods that do not add fat, such as grilling, broiling, poaching, or roasting. Choose lean turkey, roast beef, ham, or low-fat luncheon meats for sandwiches instead of fatty luncheon meats such as regular bologna or salami. Select fish as a choice from this group more often, especially fish rich in omega-3 fatty acids, such as salmon, trout, and herring.* Choose dry beans or peas as a main dish often. Choose nuts as a snack, on salads, or in main dishes, to replace meat or poultry, not in addition to these.
 © Matthew Farruggio	Oils	Substitute vegetable oils for solid fats like butter, stick margarine, shortening, or lard. Substitute nuts for meat or cheese as a snack or as part of a meal.
 © Matthew Farruggio	Discretionary Calories	Limit products containing saturated fats, such as ground and processed meats, full-fat cheese, cream, ice cream, and fried foods. Limit foods containing partially hydrogenated vegetable oils, which contain *trans* fats, such as some commercially fried foods and some bakery goods. (Partially hydrogenated vegetable oils are listed on ingredient labels of food products.) Select baked, steamed, or broiled rather than fried foods most often. Select lean or low-fat foods most often. Solid fats that are added to foods are considered discretionary calories. Choose water, fat-free milk, or unsweetened tea or coffee as a beverage most often. Limit sweet snacks and desserts. Choose canned fruits in 100% fruit juice or water rather than syrup. Select unsweetened cereals; then, if desired, add sugar or other sweeteners only to taste.

*Women who may become pregnant, pregnant women, and nursing mothers should avoid some types of fish and eat types lower in mercury. For more information: www.cfsan.fda.gov/~dms/admehg3.html.

- Select limited amounts of foods that are not in their most nutrient-dense form and/or contain **solid fats** or **added sugars**, such as whole milk, full-fat cheese, sausage, biscuits, sweetened cereal, and sweetened yogurt
- Add fats or sweeteners to foods, such as sauces, gravies, sugar, syrup, butter, and jelly
- Eat or drink items that contain only fats, caloric sweeteners, and/or alcohol, such as candy, soda, wine, and beer

Added fats and sugars are always counted as discretionary calories, as in the following examples:

solid fats fats that are solid at room temperature, such as butter, lard, and shortening. These fats may be visible or may be a constituent of foods such as milk, cheese, meats, or baked products.

added sugars sugars and other caloric sweeteners that are added to foods during processing or preparation. Added sugars do not include naturally occurring sugars such as those that occur in milk and fruits.

Tips	Nutrient-Dense Choices (Choose Most Often)	Less Nutrient-Dense Choices (Limit Choices)
Make most choices lean or low fat. Choose a variety of different types of foods from this group each week. Include fish, dry beans and peas, nuts, and seeds, as well as meats, poultry, and eggs. Consider dry beans and peas as an alternative to meat or poultry as well as a vegetable choice.	Lean meat, fish, shellfish, poultry (no skin), eggs, legumes, tofu, tempeh, peanut butter, nuts, and seeds	Luncheon meats, ground beef, hot dogs, fried meats, fried fish, fried poultry, or fried eggs, poultry with skin, refried beans, sausages
Choose most fats from sources of monounsaturated and polyunsaturated fatty acids, such as fish, nuts, seeds, and vegetable oils.	Liquid vegetable oils, such as canola, corn, flaxseed, olive, peanut, safflower, sesame, soybean oil, and sunflower oils Foods naturally high in oils, such as avocados, nuts, olives, fatty fish, and shellfish Foods that are mainly oil, such as mayonnaise, oil-based salad dressings, and soft (tub or squeeze) margarine with no *trans* fats.	
Limit intakes of foods and beverages with solid fats or added sugars. Intakes should not exceed the discretionary calorie allowance. Choose and prepare foods and beverages with little added sugars or caloric sweeteners.	Most discretionary calorie allowances are very small, between 100 and 300 calories. For many people, the discretionary calorie allowance is totally used by the foods they choose in each food group, such as higher fat meats, cheeses, whole milk, or sweetened bakery products. Many people overspend their discretionary calorie allowance, choosing more added fats, sugars, and alcohol than their calorie budget allows.	Butter, bacon, cream, cream cheese, hard margarine, lard, pork rinds, sour cream, shortening, whipping cream Candy, gelatin, honey, jam, jelly, marshmallows, molasses, Popsicles, soft drinks, sugar, syrup Calories from alcoholic beverages (beer, wine, liquor) are counted among discretionary calories.

- The fat in reduced-fat or whole milk or milk products and the sugar and fat in chocolate milk, ice cream, and pudding
- The fat in higher fat meats (e.g., poultry with skin, higher fat luncheon meats, sausages)
- The sugars added to fruits, fruit juices, and fruits canned in syrup
- The fat in vegetables prepared with added fat
- The added fats and/or sugars in grain products such as sweetened cereals, higher fat crackers, pies, cakes, and cookies

TABLE 2-8
A Sample Eating Pattern for 2,000 Calories

Food Group	Recommended Amounts	Breakfast	Lunch	Snack	Dinner	Snack
Fruits	2 c	½ c		1 c	½ c	
Vegetables	2½ c		1 c		1½ c	
Grains	6 oz	1½ oz	2 oz		2 oz	½ oz
Milk	3 c	1 c			1 c	1 c
Meats & Beans	5½ oz		2 oz	1 oz	2½ oz	
Oils	6 tsp		2		4 tsp	
Discretionary Calories	267 cal					

Note that alcohol is not included in any portion of the pyramid, but like the added sugars, alcohol provides calories and no nutrients to speak of. The *Dietary Guidelines* recommend no more than one or two alcoholic drinks a day. A standard drink is a 12-ounce can or bottle of beer, a 5-ounce glass of wine, or a 1½-ounce shot of liquor.

The Savvy Diner

Rules of Thumb for Portion Sizes—It's All in Your Hands

In the "more must be better" viewpoint of eating in America today, it can be difficult to balance energy we take in through the food we eat with the energy we expend on a daily basis. This is evidenced by the larger and larger portion sizes we find in restaurants, on the grocery shelves, and in almost every aspect of eating in the United States today.

What's the difference between a serving and a portion? A *serving* is a standard amount of food used as a reference to give advice regarding how much food to eat (such as 1 cup of milk or its equivalent). The amount of a food that is considered a serving usually stays the same. A *portion,* on the other hand, is the amount of food *you* choose to eat, and it may vary from one meal or snack to the next based on your appetite or hunger. How do you determine how much food is in your portion? You can use the following images as a visual reference to "right-size" your portions and avoid consuming extra calories.

© Felicia Martinez/PhotoEdit
One fist clenched = 8 fl oz

© Felicia Martinez/PhotoEdit
Two hands, cupped = 1 cup

© Felicia Martinez/PhotoEdit
One hand, cupped = ½ c

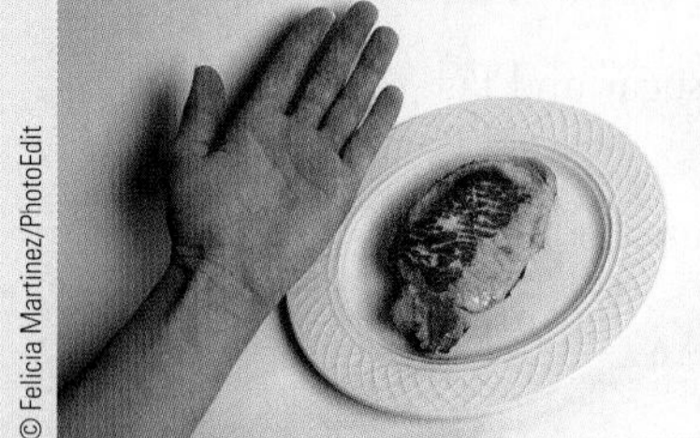
© Felicia Martinez/PhotoEdit
Palm of hand = 3 oz

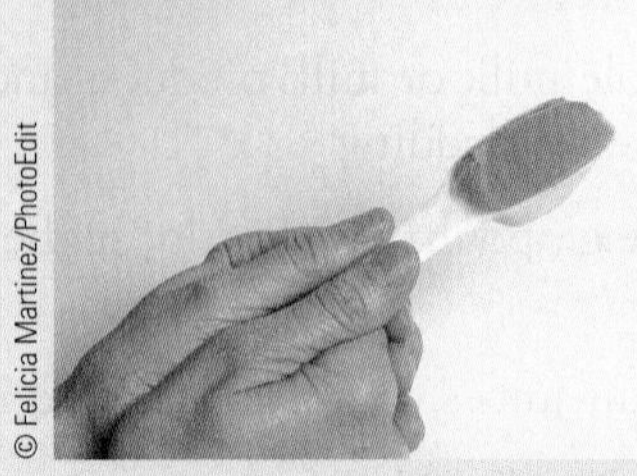
© Felicia Martinez/PhotoEdit
Two thumbs together = 1 tbsp

SOURCE: Adapted from USDA, *Dietary Guidelines for Americans,* 5th ed., 2000.

TABLE 2-9
A Sample Menu for 2,000 Calories

The menu below provides about 1,830 calories, leaving about 170 discretionary calories to spend on additional nutrient-dense foods, or on foods with added sugars or fats.

Amounts	Menu Item	Calories
	Breakfast	
1½ oz whole grains	1½ c whole-grain flakes cereal	162
1 c milk	1 c fat-free milk	90
1 c fruit	1 c sliced banana	105
	Lunch	
2 oz whole grains	2 slices rye bread	160
2 oz meat and beans	1½ oz tuna, packed in water, drained	55
1 c vegetables	2 tbsp kidney beans	25
	6 baby carrots, shredded	25
	¼ c shredded Romaine lettuce	2
	2 slices tomato	10
2 tsp oils	1 tbsp low-fat mayonnaise	35
	1 tsp Dijon mustard	5
8 oz unsweetened beverage		
	Snack	
½ c fruit	¼ c dried apricots	80
1 oz meat and beans	½ oz almonds (12 almonds)	80
	Dinner	
	Spinach salad:	
½ c vegetables	1 c raw baby spinach leaves	8
½ oz meat and beans	2 tbsp chickpeas	34
½ c fruit	½ c tangerine slices	30
2 tsp oils	2 tsp oil and vinegar dressing	60
	Pasta with meat sauce:	
2 oz enriched grains	1 c rigatoni pasta	197
1 c vegetables	½ c tomato sauce with	25
	½ c mixed frozen vegetables	25
2 oz meat and beans	2 oz cooked ground turkey breast,	130
2 tsp oils	sautéed in 2 tsp oil	90
1 c milk	1½ oz Parmesan cheese	190
	Snack	
1 c milk	1 c fat-free fruited yogurt	120
½ oz grains	3 graham crackers	90

More Tools for Diet Planning

In 1990, Congress passed one of the most important pieces of legislation of the twentieth century. Known as the Nutrition Labeling and Education Act, the law called for sweeping changes in the way foods are labeled in the United States. Officials at the U.S. Food and Drug Administration (FDA), the nation's food industry watchdog, spent several years devising regulations aimed at revamping the food label. By May 1994, food manufacturers had to relabel some 300,000 packaged foods sold in American supermarkets.[10] For consumers, the law ensures that food companies provide the kind of nutrition information that best allows people to select foods that fit into a healthful eating plan.

Scorecard: Rate Your Plate

To see how your diet measures up to the recommendations in the MyPyramid plan, follow these steps.

Step 1: Write down everything you ate yesterday, including meals *and* snacks. Make note of portion sizes as well. (Refer to Table 2-6 for help with conversions.)

Step 2: Identify the food group for each item you ate. (Refer to Figure 2-5 for help.)

Step 3: Using the five food groups, determine the amounts that are right for you. Go to www.mypyramid.gov and type in your age, sex, and activity level under the "MyPyramid Plan" and click on "submit." This will give you a plan with a good estimate of the amount of servings you need from each food group. In addition, it will give you the approximate number of calories you require.

Step 4: Circle the estimated amounts you should eat from the middle column. In the right column, write down the amounts you ate yesterday. (Refer to Table 2-6 and the *Savvy Diner* feature in this chapter for help with conversions.) Compare the two columns to see how your diet rates.

	AMOUNT YOU SHOULD EAT	AMOUNT YOU ATE
Grain group servings (ounces)	5 6 7 8 9 10	______
Vegetable group servings (cups)	2 2½ 3 4	______
Fruit group servings (cups)	2 2½ 3	______
Milk group servings (cups)	3	______
Meat & Bean group (ounces)	5 5½ 6 7	______
Solid Fats and Added Sugars	Use Discretionary Calories	______

Step 5: Decide what changes in your eating habits will make your diet more healthful. If your diet is "un-proportional," with too many foods coming from one food group, make gradual changes to develop a diet with more variety. The chapters that follow offer tips on how to do so.

Food Labels

Considering the great variety of packaged foods available, using the food label to understand the nutrients a food supplies or lacks is essential (see Figure 2-7). The label is one of the most important tools you can use to eat healthfully.

By law, all labels must contain the following:

- The name of the food, also known as the statement of identity.
- The name of the manufacturer, packer, or distributor, as well as the firm's city, state, and ZIP code.
- The net quantity, which tells you how much food is in the container so that you can compare prices. Net quantity has to be stated in both inch or pound units and metric units.
- The **ingredients list,** with items listed in descending order by weight. The first ingredient listed makes up the largest proportion of all the ingredients in the food, the second, the second largest amount, and so on. If the first ingredient in the list is sugar, for example, you know the food contains more sugar than anything else. The list is especially useful in helping people identify ingredients they avoid for health, religious, or other reasons (see Figure 2-8).
- The **Nutrition Facts panel,** unless the packge is small—no larger than 12 square inches of surface area (about the size of a small candy bar or a roll of breath mints). Small packages must carry a telephone number or address that consumers can use to obtain nutrition information (see Figure 2-9).

ingredients list a listing of the ingredients in a food, with items listed in descending order of predominance by weight. All food labels are required to bear an ingredients list.

Nutrition Facts panel a detailed breakdown of the nutritional content of a serving of a food that must appear on virtually all packaged foods sold in the United States.

Eat Well Be Well — Color Your Plate for Health with a Variety of Fruits and Vegetables

The message is simple: *Eat plenty of fruits and vegetables every day for better health.*[11] Most Americans, however, find this simple bit of nutritional advice challenging to put into practice. Overwhelming evidence points to the health benefits available from diets rich in fruits and vegetables because of the vitamins, minerals, and **phytochemicals** (*phyto* is the Greek word for plant) found in plant foods.

In the classic sense, the naturally occurring phytochemicals—for example, lycopene, a pigment that makes tomatoes red and watermelon pink—are not vitamins, minerals, or nutrients because they do not provide energy or building materials. However, scientists who study them say that phytochemicals may have the potential to slow the aging process; boost immune function; decrease blood pressure and cholesterol; prevent cataracts; prevent, slow, or even reverse certain cancers; and strengthen our hearts and circulatory systems. (More about the food sources and beneficial effects of phytochemicals appears in Chapter 6.)

How can you incorporate fruits and vegetables into your diet to achieve the maximum health benefits? Aim for the amounts of fruits and vegetables recommended in MyPyramid (about 2 cups of fruits and 2½ cups of vegetables on a 2,000 calorie diet).* Go to www.MyPyramid.gov to determine your own recommended intake of fruits and vegetables. Also consider the following three steps in your daily food planning:

*Low-fat diets rich in fruits and vegetables and low in saturated fat and cholesterol may reduce the risk of heart disease and some types of cancer—diseases associated with many factors.

SOURCE: © Produce for Better Health Foundation.

1 Color Your Plate with Health-Protective Foods.

Consume many differently colored fruits and vegetables. Select at least three differently colored fruits and vegetables a day. The red pigment in tomatoes has different bioactive properties than the orange pigments in carrots, sweet potatoes, melon, or winter squash. Add dark green leafy vegetables such as kale, mustard greens, collards, and spinach, along with broccoli and Brussels sprouts. Fruits are a vital powerhouse of the **antioxidant nutrients** that act to squelch cell-damaging molecules in the body. Blueberries top the list for being one of the highest in antioxidants. Be sure to include citrus fruits such as oranges and grapefruit regularly, as they contain many compounds that have antioxidant and other vital health benefits.

2 Be Adventurous: Select from as Wide a Variety of Fruits and Vegetables as Possible.

Foods within each group of the MyPyramid Food Guidance System have similar nutrient content. As part of a healthy diet, choosing foods from each group offers a variety of the nutrients necessary for health. But variety within food groups is valuable, too. Even fruits that may seem like comparable foods often provide a number of different nutrients—no food has them all. For example, oranges are an excellent source of vitamin C and folate but do not supply beta-carotene (a precursor of vitamin A); cantaloupe is high in beta-carotene.

3 Make It Easy on Yourself!

Keep fruit and sliced vegetables in easy-to-reach places, such as sliced vegetables in the refrigerator and fresh fruit on the table. For convenience, keep a variety of fresh, frozen, and canned fruits and vegetables on hand to add to soups, salads, rice dishes, and other menu items. Frozen fruits and vegetables are processed right after harvesting and contain nutrient profiles similar to fresh produce. If using canned items, choose fruits packed in fruit juice (not in syrup) and remove excess salt from vegetables by rinsing before using them.

*In general, 1 cup of fruit or 100% fruit juice, or ½ cup of dried fruit is considered 1 cup from the fruit group; 1 cup of raw or cooked vegetables or vegetable juice, or 2 cups of raw leafy vegetables are considered as 1 cup from the vegetable group.

Nutrition Facts Panel The Nutrition Facts panel must indicate the amount of certain mandatory nutrients that one serving of the food contains. When you consider the nutrition information, keep serving sizes in mind. Based on the amount of food most people eat at one time, the FDA has set forth a list of serving sizes for more than 100 food categories. Manufacturers must use these recommended serving sizes on food labels. For instance, the FDA might say that the serving size for product X must always be 8 ounces. This procedure ensures that consumers can easily compare one brand of a product to another without having to make difficult calculations to compare nutrient quantities for different serving

phytochemicals nonnutritive substances in plants that possess health-protective benefits.

antioxidant nutrients vitamins and minerals that protect other compounds from damaging reactions involving oxygen by themselves reacting with oxygen. The antioxidant nutrients are vitamin C, vitamin E, and beta-carotene. The mineral selenium also has a role in antioxidant reactions in the body.

FIGURE 2-7

HOW TO READ A FOOD LABEL

You can use food labels to help you make informed food choices for healthy eating practices. The food label allows you to compare similar products, determine the nutritional value of the foods you choose, and can increase your awareness of the links between good nutrition and reduced risk of chronic, diet-related diseases.

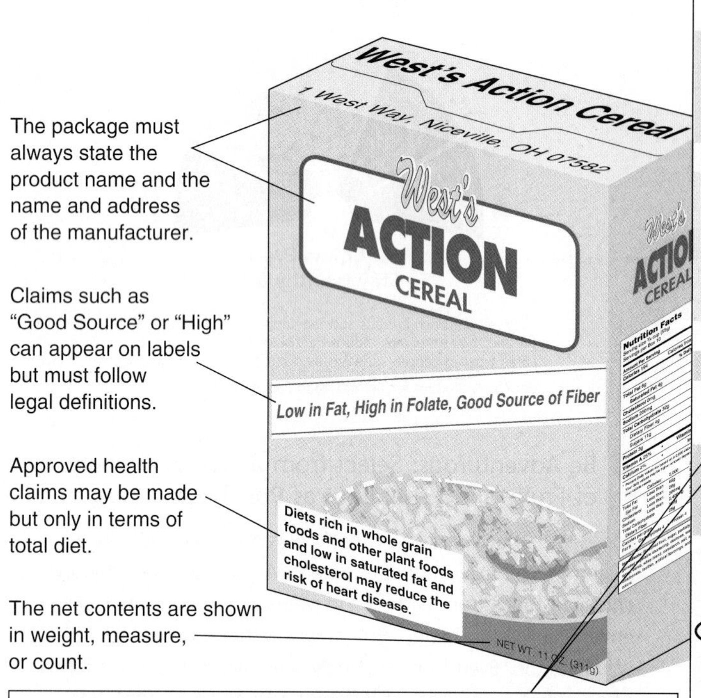
West's Action Cereal
1 West Way, Niceville, OH 07582
West's
ACTION
CEREAL
Low in Fat, High in Folate, Good Source of Fiber
Diets rich in whole grain foods and other plant foods and low in saturated fat and cholesterol may reduce the risk of heart disease.
NET WT. 11 OZ. (311g)

The package must always state the product name and the name and address of the manufacturer.

Claims such as "Good Source" or "High" can appear on labels but must follow legal definitions.

Approved health claims may be made but only in terms of total diet.

The net contents are shown in weight, measure, or count.

The only vitamins required to appear on the Nutrition Facts panel are vitamins A and C. If a manufacturer makes a nutrition claim about another vitamin, however, the amount of that nutrient in a serving of the product must also be stated on the panel. For instance, the cereal shown is touted as "High in Folate," so the percentage of the recommended intake for folate in a serving of the cereal (25%) is stated on the label. The manufacturer has the option of listing any other nutrients as well.

Nutrition Facts

Serving size 3/4 cup (55g)
Servings per Box 5

Amount Per Serving	
Calories 167	Calories from Fat 27
	% Daily Value*
Total Fat 3g	**5%**
Saturated Fat 1g	**5%**
***Trans* Fat** 0g	
Cholesterol 0mg	**0%**
Sodium 250mg	**10%**
Total Carbohydrate 32g	**11%**
Dietary Fiber 4g	**16%**
Sugars 11g	
Protein 3g	
Vitamin A	**0%**
Vitamin C	**15%**
Calcium	**10%**
Iron	**25%**
Vitamin D	**0%**
Thiamin	**25%**
Riboflavin	**25%**
Niacin	**25%**
Folate	**25%**
Phosphorus	**15%**
Magnesium	**10%**
Zinc	**6%**
Copper	**8%**

*Percent Daily values are based on a 2,000 calorie diet. Your daily values may be higher or lower depending on your calorie needs:

	Calories:	2,000	2,500
Total Fat	Less than	65g	80g
Sat Fat	Less than	20g	25g
Cholesterol	Less than	300mg	300mg
Sodium	Less than	2,400mg	2,400mg
Total Carbohydrate		300g	375g
Dietary Fiber		25g	30g

Calories per gram
Fat 9 • Carbohydrate 4 • Protein 4

Ingredients: Whole oats, milled corn, enriched wheat flour (contains niacin, reduced iron, thiamin mononitrate, riboflavin, folic acid), dextrose, maltose, high fructose corn syrup, brown sugar, partially hydrogenated cottonseed oil, coconut oil, walnuts, vitamin C (sodium ascorbate), vitamin A (palmitate), iron.

Calorie/gram reminder

Ingredients in descending order of predominance by weight

Start here

Serving size, number of servings per container, and calorie information

Limit these nutrients.

Information on sodium is required on food labels.

Get enough of these nutrients.

Guide to the % Daily Value: Quantities of nutrients per serving and percentage of Daily Value for nutrients based on a 2,000 calorie energy intake.
5% or less is low.
10% or more is good.
20% or more is high.

Calcium and iron are required on the label. Look for foods that provide 10% or more of these minerals.

Reference values

This allows comparison of some values for nutrients in a serving of the food with the needs of a person requiring 2,000 or 2,500 calories per day.

sizes—say, if one brand's label listed ingredients for a 6-ounce serving size and another listed ingredients for an 8-ounce serving size. The nutrient information that must appear on the Nutrition Facts panel is calories, calories from fat, total fat, saturated fat, *trans* fat, cholesterol, sodium, total carbohydrate, dietary fiber, sugars, protein, vitamin A, vitamin C, calcium, and iron (in that order).

These nutrients were chosen to appear on the Nutrition Facts panel because they address today's health concerns. Today, many people need to be concerned about getting an excess of certain nutrients, such as fat, rather than too few vitamins and minerals.

The FDA spelled out the ranking of the required nutrients to ensure that the label reflects the government's dietary priorities for the public. For example, fat falls near the top of the list because most consumers need to pay closer attention to the amount of fat in their diet. Most Americans eat too much fat, which raises the risk of developing heart disease, obesity, and cancer—chronic problems suffered by millions. Protein, on the other hand, appears near the bottom of the label because the amount of protein most Americans eat does not rate as a major health concern.

FIGURE 2-8
USING THE INGREDIENTS LIST ON FOOD LABELS

Oats 'N' More

Ingredients: whole grain oats, (includes the oat bran), modified corn starch, wheat starch, sugar, salt, oat fiber, trisodium phosphate, calcium carbonate, vitamin E (mixed tocopherols) added to preserve freshness. **Vitamins and Minerals:** iron and zinc (mineral nutrients), vitamin C (sodium ascorbate), vitamin B_6 (pyridoxine hydrochloride), riboflavin, thiamin mononitrate, niacinamide, folic acid, vitamin A (palmitate), vitamin B_{12}, vitamin D.

Morning Krisps

Ingredients: Sugar, wheat, corn syrup, honey, hydrogenated soybean oil, salt, caramel color, soy lecithin. **Vitamins and Iron:** sodium ascorbate (vitamin C), ferric phosphate (iron), niacinamide, pyridoxine hydrochloride (vitamin B_6), riboflavin, vitamin A palmitate, thiamin hydrochloride, BHT (preservative), folic acid, vitamin B_{12}, and vitamin D.

The cereals shown contain sugars. Learn to read the ingredients list. Labels list ingredients in order of amount by weight with the greatest amount of an ingredient present in the food listed first. Check labels for sugar terms, in addition to sugar, such as: brown sugar, corn syrup, dextrose, fructose, glucose, high-fructose corn syrup, honey, invert sugar, levulose, mannitol, molasses, sorbitol, and sucrose.

Note that only vitamins A and C, iron, and calcium appear on the nutrition panel. Those are the only vitamins and minerals (except for sodium) required to be on food labels, unless a manufacturer chooses to make a nutrition claim about another one. For instance, if a manufacturer says that a cereal is *high* in folate or **fortified** with niacin, the amount of folate or niacin in the product must appear on the label (see Figure 2-7).[12]

Daily Values The **Daily Values** for fats, sodium, carbohydrates, and fiber are calculated according to what experts deem a healthful diet for adults (see Table 2-10).

fortified food a food to which manufacturers have added 10 percent or more of the Daily Value for a particular nutrient.

Daily Value the amount of fat, sodium, fiber, and other nutrients health experts say should make up a healthful diet. The % Daily Values that appear on food labels tell you the percentage of a nutrient that a serving of the food contributes to a healthful diet.

FIGURE 2-9
TYPES OF FOOD LABELS

A container with less than 40 square inches of surface area for nutrition labeling is allowed to present fewer facts as shown in the format on the can of tuna. A more simplified format, as shown on the can of cola, is allowed on foods that do not contain significant amounts of nutrients. Packages with less than 12 square inches of surface area, such as small candy bars, do not need to carry nutrition information, although they must provide a telephone number or address for contacting the company to obtain nutrition information.

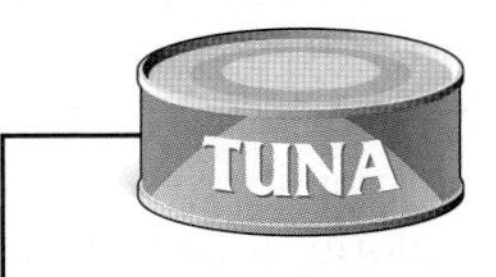

Nutrition Facts
Serving Size 1/4 cup (56g)
Servings about 2.5
Calories 80
Fat Cal. 25
*Percent Daily Values (DV) are based on a 2,000 calorie diet.

Amount/serving	% DV*	Amount/serving	% DV*
Total Fat 3g	5%	**Total Carb.** 0g	0%
Sat. Fat 1g	5%	Fiber 0g	0%
Trans Fat 0g		Sugars 0g	
Cholest. 35mg	12%	**Protein** 12g	22%
Sodium 250mg	10%		

Vitamin A 0% • Vitamin C 0% • Calcium 0% • Iron 3%

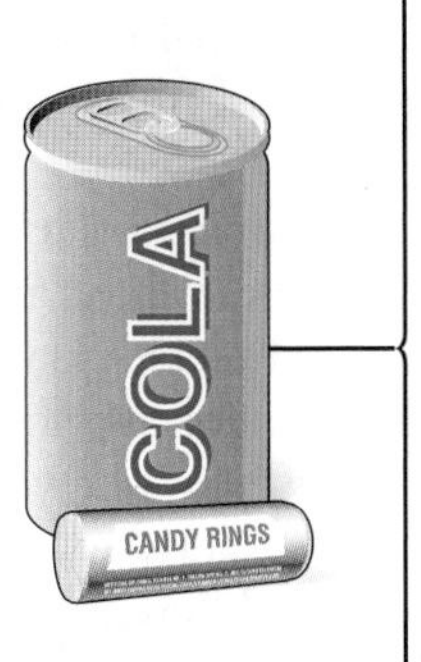

Nutrition Facts
Serving Size 1 can (360 ml)

Amount Per Serving	
Calories 140	
	% Daily Value*
Total Fat 0g	0%
Sodium 20mg	1%
Total Carbohydrate 36g	12%
Sugars 36g	
Protein 0g	0%

*Percent Daily Values are based on a 2,000 calorie diet.

For instance, since the Daily Value for fat recommends that no more than 30 percent of total calories should come from fat, the % Daily Value tells you the percentage of fat that a serving of the food contributes to a 2,000-calorie eater's fat "allowance." A 2,000-calorie diet was chosen as a good point of reference because that's about the amount recommended for most moderately active women, teenage girls, and sedentary men. Of course, more calories may be appropriate for many men, teenage boys, and active women. This is why the nutrition panel also shows Daily Values for a 2,500-calorie diet (refer to the bottom of the Nutrition Facts panel in Figure 2-7).

Courtesy: *Modern Maturity Magazine.*

"Henry likes nothing more than to curl up with a good label."

To understand how the Daily Values for fats, sodium, carbohydrates, and fiber are calculated, let's go through an example. First, look for the grams of total fat and the % Daily Value for the cereal shown in Figure 2-7. The label shows that a serving supplies 3 grams of fat, with a Daily Value of 5 percent. This means that a serving of the cereal contributes 5 percent of the total fat that a person eating 2,000 calories a day should consume.

Now look at the bottom of the Nutrition Facts panel, which indicates that someone eating 2,000 calories a day should take in no more than 65 grams of fat a day. Divide 3 (the number of fat grams in a serving of the cereal) by 65. Multiply that number by 100 to obtain a percentage. The answer is 5—that is, 5 percent of the total fat.

You can use the % Daily Values to get a good idea of how various foods fit into a healthful diet, regardless of the number of calories you eat. Consider a student who eats only 1,800 calories a day. If she snacks on two servings of potato chips with a Daily Value of 15 percent fat per serving, she's already taken in 30 percent of the fat someone eating 2,000 calories should have in an entire day. Because she eats less than 2,000 calories, the potato chips contribute slightly more than 30 percent. Thus, the 30 percent Daily Value shows that potato chips chalk up to a lot of fat for one snack. If she checks the % Daily Value for fat on a label of pretzels, on the other hand, she would see that one serving supplies only about 3 percent fat per serving. Thus, if she eats two servings, she's eaten less than 10 percent of the fat she can have, leaving her much less likely to go overboard on fat throughout the rest of the day. In other words, the % Daily Value column can give you a good idea of how different foods fit into your overall diet.

Some people find it easiest to bypass the % Daily Values and simply check the grams of total fat a serving of food supplies to see how much it adds to a daily fat tally. Let's say a man eats 2,000 calories a day and therefore should consume no more than 65 grams of fat a day. If he eats a muffin (15 grams of fat) and coffee with cream (10 grams) in the morning, he's already up to 25 grams of fat. That means he can have about 40 more grams during the rest of the day to stay within his fat "budget."

Once you determine the maximum number of fat grams you should have in a day, you can use the food label to get a good idea of how many grams of total fat the items you buy add to your daily tally. To figure out your fat allowance, check Table 4-5 on page 122.

You can also use the Daily Values to comparison shop. For example, if you're looking for a high-fiber cereal to increase the amount of fiber in your diet, you can check the % Daily Value for fiber on the labels of several brands of cereal. If a serving of Brand X's cereal has a Daily Value of 20 percent for fiber and Brand Y supplies only 5 percent, Brand X is an excellent source of fiber, whereas Brand Y is low in fiber.

© Zigy Kaluzny/Getty Images

▲ *Learn to read food labels to help you achieve a healthful diet.*

The % Daily Values for vitamins and minerals are calculated using standard values designed specifically for use on food labels. These values are shown in Table 2-10 and on the inside back cover of this book. These standard values for nutrients were created to help manufacturers avoid a stumbling block they face as they label foods. Because manufacturers don't know whether you're an 18-year-old woman or a 30-year-old man, they don't know exactly what your nutritional needs are. You may recall that the DRI include a different set of vitamin and mineral recommendations for each gender and age group.

TABLE 2-10
Daily Value Amounts

Fat	65 g	(30% of calories)
Saturated fat	20 g	(10% of calories)
Cholesterol	300 mg	—
Carbohydrate (total)	300 g	(60% of calories)
Fiber	25 g	(11.5 g per 1,000 calories)
Protein	50 g	(10% of calories)
Sodium	2,400 mg	
Vitamin C	60 mg	
Vitamin A	900 μg	
Calcium	1,000 mg	
Iron	18 mg	

NOTE: The values for energy-yielding nutrients are based on 2,000 calories a day.

To help get around the problem, the nutrient recommendations used for vitamins and minerals on labels represent the highest of all the values to ensure that virtually everyone in the population is covered. For most nutrients, the highest recommendation is for an adult man. When it comes to iron, however, the DRI for women is the highest (women require more iron than men), so the women's DRI is used as the standard value on labels.

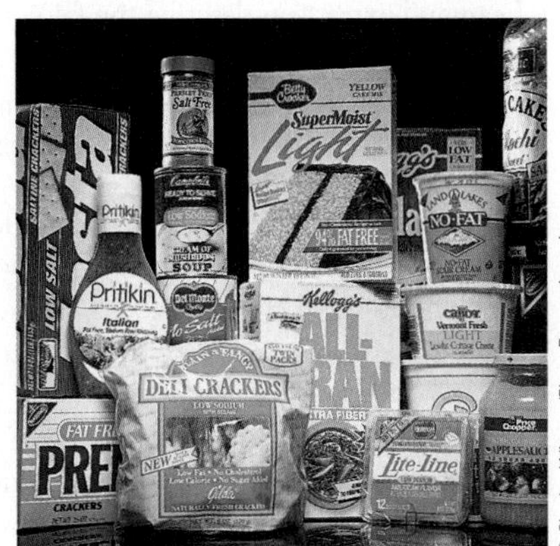

▲ *Nutrient content claims are strictly defined by the FDA.*

Nutrient Content Claims By law, foods carrying terms called **nutrient content claims**—low-fat, low-calorie, light, and so forth—must adhere to specific definitions spelled out by the Food and Drug Administration. For instance, a serving of a food dubbed low-fat must contain no more than 3 grams of fat. An item touted as low-calorie may provide no more than 40 calories per serving. Table 2-11 lists the claims commonly used on food labels and their legal definitions.

Health Claims A statement linking the nutritional profile of a food to a reduced risk of a particular disease is known as a **health claim.** The FDA has set forth very strict rules governing the use of such health claims.[13] For example, if a food's label bears health claims regarding calcium, a serving of the product must contain at least 20 percent of the Daily Value for calcium, among other restrictions. What's more, the manufacturers are allowed to imply only that the food "may" or "might" reduce risk of disease. They must also note the other factors, such as exercise, that play a role in prevention of the disease. Finally, they must phrase the claim so that the consumer can understand the relationship between the nutrient and the disease. For example, a health claim on a food low in fat, saturated fat, and cholesterol might read, "Whereas many factors affect heart disease, diets low in saturated fat and cholesterol may reduce the risk of this disease." The nutrient–disease relationships about which health claims can be made are listed in Table 2-12.[14]

Exchange Lists

Although food group plans provide sufficient detail to help most healthy people plan a good diet, **exchange lists** take meal planning a step further. As their name implies, exchange lists are simply lists of categories of foods, such as fruit, with portions specified in a way that allows the foods to be mixed and matched or exchanged with one another in the diet. For instance, you might strive to eat two servings of fruit each day, and the exchange list shows you that ½ cup of orange juice, a small banana, or a small apple each counts as a fruit.

Portion sizes within groups are determined by considering the calorie, protein, carbohydrate, and fat content of the food. For example, one fruit contains about 60 calories and 15 grams of carbohydrate. One starch, on the other hand,

nutrient content claims claims such as "low-fat" and "low-calorie" used on food labels to help consumers who don't want to scrutinize the Nutrition Facts panel get an idea of a food's nutritional profile. These claims must adhere to specific definitions set forth by the Food and Drug Administration.

health claim a statement on the food label linking the nutritional profile of a food to a reduced risk of a particular disease, such as osteoporosis or cancer. Manufacturers must adhere to strict government guidelines when making such claims.

exchange lists lists of foods with portion sizes specified. The foods on a single list are similar with respect to nutrient and calorie content and thus can be mixed and matched in the diet.

TABLE 2-11
Definitions of Nutrient Content Claims

Free means that a product contains none or only negligible amounts of fat, saturated fat, *trans* fat, cholesterol, sodium, sugar, and/or calories. For instance, "calorie free" means fewer than 5 calories per serving.

Low indicates that the food can be eaten frequently without exceeding dietary guidelines for fat, saturated fat, cholesterol, sodium, and/or calories. More specifically:

low-fat: 3 grams or fewer per serving*
low saturated fat: no more than 1 gram per serving
low sodium: no more than 140 milligrams per serving*
very low sodium: no more than 35 milligrams per serving
low cholesterol: no more than 20 milligrams and no more than 2 grams of saturated fat per serving*
low-calorie: no more than 40 calories per serving*

Lean and **extra lean** describe the fat content of meat, poultry, seafood, and game meats:

lean: fewer than 10 grams of fat, no more than 4.5 grams of saturated fat, and fewer than 95 milligrams of cholesterol per serving (or 100 grams)
extra lean: fewer than 5 grams of fat, fewer than 2 grams of saturated fat, and fewer than 95 milligrams of cholesterol per serving (or 100 grams)

High is used when a serving of a food contains 20 percent or more of the Daily Value for a particular nutrient.

Good source indicates that a serving of the food supplies 10 to 19 percent of the Daily Value for a particular nutrient.

Reduced denotes a product that has been nutritionally altered and contains 25 percent less of a nutrient such as fat or calories than the regular, unaltered product. A product cannot be dubbed "reduced," however, if the regular version of the food already meets the requirements for a "low" claim. That is, if a food is "low-fat" to begin with, it cannot be called "reduced" if manufacturers take even more fat out of it.

Less means that a food contains 25 percent less of a nutrient or calories than a comparable food. For example, pretzels containing 25 percent less fat than potato chips carry a "less fat" claim. *Fewer* can be used in the same way.

Light carries several meanings: First, a nutritionally altered food contains one-third fewer calories than or half the fat of the regular product. If fat supplies 50 percent or more of the calories to begin with, it must be reduced by half to be called "light." Second, if the sodium content of a low-fat, low-calorie food has been reduced by 50 percent but the food is not low in fat and calories, it must be labeled "light in sodium." Third, "light" can be used to describe a food's color and/or texture, as long as the label explains the intent. For example, "light brown sugar."

More means that a serving of the food contains at least 10 percent more of the Daily Value of a particular nutrient than the regular food. The label on calcium-fortified bread can state that the product contains "more calcium" than regular bread.

Percent fat-free is an indication of the amount of a food's weight that is fat-free, which can be used only on foods that are low-fat or fat-free to begin with. For instance, a food that weighs 100 grams with 3 grams from fat can be labeled "97 percent fat-free." Note that this term refers to the amount of the food that is fat-free by weight, not calories. If that same food supplies 100 calories, the 3 grams of fat contribute 27 of them (1 gram of fat contains 9 calories). This means that 27 of the 100 calories, or 27 percent of the total calories, come from fat.

Made with oat bran, no tropical oils claims, known as implied claims, are prohibited if they mislead consumers into believing a product supplies (or lacks) significant levels of nutrients. For example, a manufacturer can say a product is "made with oat bran" only if it contains enough oat bran to meet the definition for "good source" of fiber.

Healthy indicates that a food is low in fat and saturated fat; contains no more than 60 milligrams of cholesterol per serving; and provides at least 10 percent of the Daily Value for vitamin A, vitamin C, protein, calcium, iron, or fiber (main dishes must supply at least two of the six nutrients). In addition, the food must meet sodium requirements—no more than 360 milligrams of sodium per serving of individual foods and no more than 480 milligrams per main dish meal.

*On meals and main dish products such as frozen dinners, *low-calorie* can be used if the dish contains no more than 120 calories in 100 grams, or about 3.5 ounces. Specifically: *low sodium*, the dish supplies no more than 140 milligrams per 100 grams; *low cholesterol*, a maximum of 20 milligrams and 2 grams saturated fat per 100 grams; *light*, the dish or meal is low fat or low calorie.

provides about 80 calories, 15 grams of carbohydrate, 3 grams of protein, and a trace of fat. This breakdown makes exchange lists especially useful tools for people who follow carefully planned diets as a result of a health problem such as diabetes.

Exchange lists are also useful for people who are following calorie-controlled diets to lose weight. Dietitians sometimes give clients tailor-made diets centered on the exchange lists—say, a 1,500-calorie daily diet that might include eight starches, three vegetables, two fruits, and so forth. A person can take such a framework and use the exchange lists to choose a wide variety of foods that fit into the basic eating plan. Table 2-13 shows the seven common exchange lists. Typical portions used in the list are as follows:

Starch: 1 small potato/1 slice bread—80 calories
Fruit: 1 small orange—60 calories
Milk: 1 cup fat-free milk—90 calories

TABLE 2-12
Health Claims on Food Labels

Approved health claims with significant scientific agreement:[a]

- Calcium-rich foods and reduced risk of osteoporosis
- Low-sodium foods and reduced risk of high blood pressure
- Low-fat diet and reduced risk of cancer
- A diet low in saturated fat and cholesterol and reduced risk of heart disease
- Fiber-containing grain products, fruits, and vegetables and reduced risk of cancer
- Soluble fiber in fruits, vegetables, and grains and reduced risk of heart disease
- Soluble fiber in whole oats and psyllium seed husk and reduced risk of heart disease
- Fruit- and vegetable-rich diet and reduced risk of cancer
- Folate or folic acid and reduced risk of neural tube defects (i.e., newborn malformations such as spina bifida)
- Sugar alcohols and reduced risk of tooth decay
- Soy protein and reduced risk of heart disease
- Whole-grain foods and reduced risk of heart disease and certain cancers
- Plant sterol and plant stanol esters and reduced risk of heart disease
- Potassium and reduced risk of high blood pressure and stroke

A sampling of *qualified* health claims:[b]

- Omega-3 fatty acids and reduced risk of heart disease
- Nuts such as walnuts and reduced risk of heart disease
- Calcium and reduced risk of high blood pressure
- Calcium and reduced risk of colorectal cancer
- Monounsaturated fat from olive oil and reduced risk of heart disease
- B vitamins and reduced risk of heart disease

[a] The FDA does not require these claims to carry disclaimers.
[b] FDA requires label disclaimers when these claims are used, such as—Health claim: Although there is scientific evidence supporting the claim, the evidence is not conclusive.

Other carbohydrates: 3 gingersnaps—80 calories (but calories in this list vary)
Vegetable: ½ cup green beans—25 calories
Meat and meat substitutes: 1 ounce lean meat or low-fat cheese—55 calories
Fat: 1 teaspoon butter—45 calories

Food Composition Tables

You now have an understanding of the tools you need to set up a healthful eating plan for yourself. You may find yet another tool to be useful as well: **Food composition tables,** which list the exact number of calories, grams of fat, milligrams of sodium, and other nutrients found in commonly eaten foods.

Appendix F provides a food composition table that profiles the nutrient content of a wide variety of foods. Such tables offer the health-conscious eater a wealth of useful information—everything from the amount of vitamin C in an orange to the number of calories in an order of French fries to the amount of calcium in various cheeses. To be sure, the nutritional content of foods listed in food composition tables varies depending on cooking methods and other factors, and not every table lists every single nutrient. Still, food composition tables give fairly precise estimates of the nutrients in the foods you eat.

Computer buffs can take advantage of one of the many software packages containing a database that is, in essence, a food composition table. Simply plug in the foods you eat, and the computer will generate a profile of your daily diet. Dietitians often use such software to analyze people's diets as well as their recipes. More and more reasonably priced, reliable nutrition-analysis software is becoming widely available, and free diet analysis software programs are available online (see Nutrition on the Web at the end of this chapter).

TABLE 2-13
The Exchange Lists

Carbohydrate Group
- Starch
- Fruit
- Milk
  - Skim (fat-free)
  - Low-fat
  - Whole
- Other carbohydrates
- Vegetable

Meat and Meat Substitute Group
- Very lean
- Lean
- Medium-fat
- High-fat

Fat Group

food composition tables tables that list the nutrient profile of commonly eaten foods.

Spotlight A Tapestry of Cultures and Cuisines

American eating habits have become as diverse as the various ethnic and cultural groups that make up America's people. Throughout American history, immigrant groups—from Poles to Jews to Italians to Irish to Germans to Hispanics to African Americans to Asians—have had and continue to have profound effects on the collective American palate. As food writer and critic John Mariani has pointed out:

> *[T]he United States—a stewpot of cultures—has developed a gastronomy more varied . . . than that of any other country in the world. . . . In any major American city one will find restaurants representing a dozen national cuisines, including northern Italian trattorias, bourgeois French bistros, Portuguese seafood houses, Vietnamese and Thai eateries, Chinese dim sum parlors, Japanese sushi bars, and German rathskellers.*[15]

This Spotlight examines some of the more prevalent ethnic and regional food practices to see how they originated and how they fit into a healthful eating plan.

I love to eat Mexican food. Is it true that it is loaded with fat? Do the Mexican people eat a lot of high-fat food?

Although it's true that many menu selections in Mexican restaurants are loaded with high-fat ingredients such as cheese, ground beef, sour cream, guacamole, and fried tortillas, most dishes eaten regularly in Mexican American homes are much simpler. Breakfast, for example, might be tortillas (flat, thin corn or wheat pancakes) served with fried beans, eggs, or cereal, and a beverage. Lunches and dinners often consist of beans and rice, bread or tortillas, meat/sausage (often as part of a stew), a vegetable or lettuce and tomato, and a beverage.

© Michael Newman/PhotoEdit

The traditional Mexican diet drawing largely from Spanish and Indian influences was even simpler, containing mostly vegetables, including beans, squash, and maize (corn). In addition, it often included cactus parts, agave (a plant with spiny-margined leaves and flowers), chili peppers, **amaranth** (a grain), avocado, and **guava** (a sweet, juicy fruit with green or yellow skin and red or yellow flesh).

This type of diet is high in complex carbohydrates such as rice and vitamin A- and C-rich fruits and vegetables, making it a particularly healthful aspect of the traditional Mexican way of eating. Consider that no Mexican meal is complete without salsa, a low-fat condiment consisting of vitamin-rich tomatoes, chilis, and onions. Another plus of the Mexican diet is frequent use of beans, mostly the pinto variety, which rank as a particularly good source of fiber.

There are downsides of the Mexican diet, however. Most foods, even beans and rice, are typically fried rather than baked or broiled. For example, *frijoles refritos* (refried beans) are usually fried in lard and contain about 270 calories and 3 grams of fat per cup. Most flour tortillas are also made with lard, sometimes 1 or 2 teaspoons per tortilla. (Corn tortillas, on the other hand, typically contain very little fat.) Another drawback of the Mexican diet is the frequent consumption of high-fat meats, such as **chorizo** (spicy pork or beef sausage) and eggs, which are used in dishes such as **chiles rellenos** (roasted mild green chili pepper stuffed with cheese, dipped in egg batter, and fried), **burritos** (warm flour tortillas stuffed with a mixture of egg, meat, beans, and/or avocado), and **chilaquiles** (tortilla casserole often made with eggs or meat).

Fortunately, health-conscious Mexican-food lovers can take advantage of the popular cuisine with a little know-how. Instead of the usual American versions of Mexican fare, such as fried tortillas packed with ground beef and smothered in cheese and sour cream, opt for corn tortillas filled with, say, regular, unfried pinto beans mixed with chopped onion and topped with a sprinkle of shredded cheese and a generous portion of lettuce, salsa, a dollop of fat-free plain yogurt (a low-fat alternative to sour cream), and a garnish of sliced avocado. Or, instead of serving high-fat commercial tortilla chips with salsa, try making "no-fry" chips: Immerse several tortillas in warm water, drain quickly, cut into six to eight wedges, and place on a nonstick pan; bake in a 500-degree Fahrenheit oven for 3 to 4 minutes, flip, and continue to bake for another minute or two until golden brown.

Adventurous eaters who have access to a wide variety of exotic produce might want to try some of the different fruits and vegetables the Mexican people have enjoyed for decades, such as **jicama** (HE-cahmah) (yam bean root—a vegetable that is tan outside and white inside, has a mild chestnut flavor, and is always eaten raw).[16] Figure 2-10 shows produce and other foods commonly eaten by Mexican Americans.

FIGURE 2-10
MEXICAN AMERICAN FOODS AND MyPYRAMID[a]

[a] Discretionary calories can be used for bacon, butter, candy, cream cheese, fried pork rinds, lard, hard margarine, soft drinks, and sour cream.
[b] These foods are described in the Miniglossary.

SOURCE: Adapted from www.mypyramid.gov, 2005 and the *Pyramid Packet* © 1993, Penn State Nutrition Center, 417 East Calder Way, University Park, PA 16801.

GRAINS	VEGETABLES	FRUITS	OILS	MILK	MEAT & BEANS
bolillo[b], bread, cake, cereal, corn tortilla, crackers, flour tortilla, fried flour tortilla, graham crackers, macaroni, oatmeal, pastry, rice, sopa[b], spaghetti, sweet bread, taco shell	agave[b], beets, cabbage, carrots, cassava[b], chilis, corn, green tomatoes, iceberg lettuce, jicama[b], onion, peas, potato, prickly pear cactus leaves, purslane[b], squash, sweet potatoes, tomato, turnips	apple, avocado, banana, cherimoya[b], guava[b], mango, orange, papaya, pineapple, plantain[b], zapote[b]		cheddar cheese, custard, evaporated milk, ice cream, jack cheese, powdered milk, queso blanco, fresco, or mexicano[b]	beef, black beans, chicken, chorizo[b], eggs, fish, garbanzo beans, kidney beans, lamb, nuts, peanut butter, pinto beans, pork, refried beans,

Are foods served in Chinese restaurants in the United States, such as chop suey, egg rolls, and fortune cookies, traditional Chinese foods? A Chinese American friend of mine says they aren't. Is she right?

© Tony Freeman/PhotoEdit

Yes, it's true that chop suey and the like are American inventions. The traditional Chinese diet consists of much simpler, lower fat dishes.

Overall, about 80 percent of the calories in traditional Chinese fare comes from grains, legumes, and vegetables, while the other 20 percent comes from animal meats, fruits, and fat. In southern China, rice can be easily produced and provides the bulk of the complex carbohydrate in the diet. In northern areas, where wheat grows readily, noodles, dumplings, and steamed buns are staples. In all regions of the country, fruits and vegetables are eaten in abundance. Typically, they are not eaten raw but rather are steamed, added to soups, or stir-fried in peanut or corn oil or (less frequently) lard. In addition, vegetables are often salted, pickled, and dried, a practice resulting from lack of the facilities needed to refrigerate and transport fresh produce across the country. As for meat and other protein foods, pork is considered the staple meat, though poultry, eggs, lamb, and fish are eaten when available. Tofu, or soybean curd, is another staple often added to stir-fries or fermented into sauces. Dairy foods, such as milk and cheese, have never been part of the Chinese diet.

Traditional Chinese meals follow basically the same pattern. Breakfast might be rice congee (rice gruel containing bits of meat), a salty side dish such as pickles, and tea. Lunch typically consists of soup, rice, and mixed dishes made with vegetables and fish, meat, or poultry; and dinner is usually a larger version of lunch, occasionally followed by fresh fruit.

Four schools of cooking have developed within China as a result of differences in climate, food production, religion, and custom: Peking, Shanghai, Szechwan or Hunan, and Cantonese. Peking cooking comes from the north and northeast part of China and is distinguished by use of garlic, leeks, and scallions. It has given rise to familiar dishes such as Peking duck and spring rolls. Shanghai cuisine, from the east and coastal areas, is characterized by "red-cooking"—braising foods with large amounts of soy sauce and sugar. Pickled and salted vegetables are also frequently served with meat. In the west and central part of China, people favor the Szechwan or Hunan cooking style, in which chili peppers and hot pepper sauces are added liberally to dishes, and the food tends to be spicy and oily. The cuisine of southern China, known as Cantonese cooking, is the style Americans know best. Because foods are usually steamed or stir fried and chicken broth is often used as a cooking medium, Cantonese food tends to be the least fatty. One popular example of Cantonese food is **dim sum,** steamed or fried dumplings stuffed with pork, shrimp, beef, sweet paste, or preserves and steamed or fried.[17]

The regional differences in cooking styles and eating habits among people living in rural China make the country fertile ground for scientific research into effects of diet on disease. Another reason rural China is such an ideal place to conduct research is because most people residing in rural China today spend their entire lives within the vicinity of the community in which they were born. In addition, eating habits are dictated by climate and environment because the country has little or no means of transporting foods from one region to another. Thus, scientists have been able to carefully study eating patterns and disease rates in an attempt to see how the two are related. One of the largest such studies was led by T. Colin Campbell, Ph.D., of Cornell University. Begun in 1983, the research was carried out in 130 villages located in 65 counties of rural China and included 6,500 adults. Findings suggest the traditional, plant-based Chinese diet is associated with low rates of many of the chronic diseases that plague Americans, such as heart disease and some types of cancer.[18]

To be sure, Chinese food served in American Chinese restaurants is a far cry from the type eaten daily by rural Chinese people. Many Chinese restaurant meals are swimming in oil and contain much more meat and poultry and fewer vegetables than "real" Chinese food. Consider that a typical American Chinese meal might include won ton soup, barbecued spareribs, chicken lo mein, and fried rice—chalking up some 1,400 calories and more than 80 grams of fat. A typical meal eaten in rural China, on the other hand, would likely contain a heaping portion of rice, along with fiber- and nutrient-rich vegetables and less than an ounce of meat and fish and would thereby contain only a fraction of the fat.[19] Americans who want to enjoy both the flavor and health benefits of traditional Chinese cuisine can "stretch" one of the many delicious vegetable-based dishes with relatively large portions of rice and go easy on deep-fried appetizers such as egg rolls. People who enjoy cooking can also follow the Chinese people's lead and make low-fat, high-carbohydrate stir-fries with lots of vegetables, little oil, and small amounts of meat, poultry, or seafood. People who are sodium conscious may also want to go easy on soy sauce, which contains large amounts of the mineral, or try some of the reduced-sodium soy sauces on the market. Figure 2-11 shows Chinese foods placed on a food pyramid.

How does Italian food rate? I've heard that the olive oil in Italian pasta dishes is good for health.

Much of the Italian food served in the United States runs very high in fat and calories. Rich cream sauces in dishes such as fettuccine alfredo, an abundance of cheese and sausage in entrees, including meat-topped pizza and lasagna, and liberal use of olive oil to prepare all manner of pasta and other Italian fare can chalk up extraordinary amounts of fat and calories.

With a little modification, however, the Italian food so many Americans love can easily fit into a high-carbohydrate, low-fat diet. For example, substituting vegetables for sausage and pepperoni on pizza and in pasta sauces and lasagna reduces fat content considerably. Using reduced-fat cheeses, such as part-skim mozzarella and ricotta and low-fat cottage cheese, as substitutes for high-fat versions called for in many Italian recipes also helps skim some of the fat. When it comes to olive oil, a staple in traditional Italian cuisine, it's true that using it instead of butter is preferable from a health standpoint.

For reasons that are explained in detail in Chapter 4, olive oil and other vegetable oils are less likely than certain other types of fat, such as butter and lard, to boost blood cholesterol levels. Some scientists even believe that people can eat large amounts of fat—in the neighborhood of 35 to 40 percent of total calories—without raising their risk of heart disease, as long as the pre-

FIGURE 2-11
A CHINESE AMERICAN MyPYRAMID FOOD GUIDE[a]

[a] Discretionary calories can be used for bacon fat, butter, coconut milk, duck sauce, honey, lard, maltose syrup, sesame paste, suet, and sugar.
[b] These foods are described in the Miniglossary.

SOURCE: Adapted from www.mypyramid.gov, 2005 and the *Pyramid Packet* © 1993, Penn State Nutrition Center, 417 East Calder Way, University Park, PA 16801.

GRAINS	VEGETABLES	FRUITS	OILS	MILK	MEAT & BEANS
amaranth[b], barley, bing[b], dumplings, fried rice, glutinous rice[b], mantou[b], noodles (including cellophane noodles[b], rice sticks[b], rice vermicelli[b]), rice congee, rice flour, steamed rice, sorghum, wonton wrappers	bamboo shoots, black mushroom, bok choy[b], cabbage, celery, chiles, Chinese broccoli[b], choy sum[b], eggplant, garlic, ginger, green beans, leek, mustard greens, okra, onions, Oriental radish[b], peas, pickled cucumber, potatoes, scallions, spinach, sprouts, taro[b], tomatoes, turnip, water chestnut, watercress, yard-long beans[b]	carambola, guava, jujube[b], litchi[b], longan[b], mango, orange, papaya, persimmon, pummelo, watermelon		buffalo milk, cow's milk, fish bones, soybean milk, yogurt	bean paste, beef, chestnuts, chicken, duck, eggs, fish (e.g., carp, catfish) lamb, legumes (e.g., mung beans, soybeans), shellfish and other seafood (e.g., shrimp, squid), squab

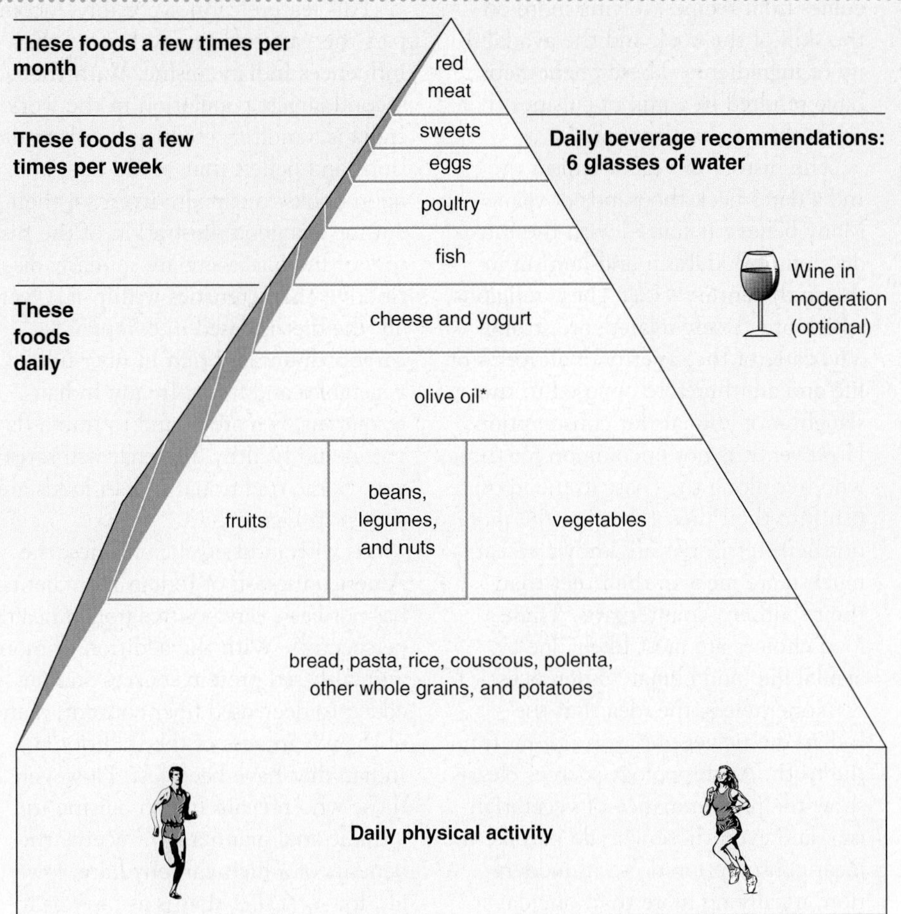

FIGURE 2-12
MEDITERRANEAN FOOD GUIDE PYRAMID

*Other oils rich in monounsaturated fats such as canola or peanut oil can be substituted for olive oil. People who are watching their weight should limit their oil consumption.

SOURCE: © 1994 Oldways Preservation & Exchange Trust, http://oldwayspt.org.

dominant fat is olive oil. This line of thinking stems from research conducted in southern Italy and the Greek island of Crete as well as southern regions of other European nations—including France, North African areas such as Morocco and Tunisia, and Middle Eastern countries such as Israel and Syria. Collectively known as the Mediterranean region, these countries share an overall dietary pattern that includes an abundance of fruits and vegetables, breads, and other grains, beans, nuts, and seeds; low to moderate amounts of cheese, yogurt, fish, and poultry; small amounts of red meat; moderate consumption of wine; and liberal use of olive oil. Scientists are fascinated with this region because people living there historically have enjoyed long lives and low rates of chronic diseases. That was particularly true in Crete during the 1950s and 1960s. At that time, researchers found that residents of the island shared one of the lowest rates of heart disease and cancer ever recorded and one of the longest life expectancies. That held true despite the fact that their diets contained nearly 40 percent of calories as fat—well above the 30 percent limit recommended by major U.S. health organizations.[20]

With this historical perspective in mind, some scientists recommend that Americans adopt a "Mediterranean diet" that includes lots of fruits, vegetables, and grains; little meat and other flesh food; moderate amounts of wine; and generous amounts of olive oil (see Figure 2-12). This advice has sparked a good deal of controversy within the scientific community, however. Many nutrition experts question the wisdom of recommending such a diet in America, where excess fat from olive oil and other sources might contribute to the epidemic of obesity in this country. In addition, many experts point out that diet is not the only lifestyle factor that may have promoted the long, healthy lives of the people of Crete and Italy several decades ago. For example, the residents of this region farmed the land and thereby engaged in far more physical activity than the average American, a factor

that probably played a role in their health and longevity. In addition, unlike many modern-day Americans, they enjoyed social support of extended networks of family and friends, another part of their lifestyle that probably contributed to their well-being. Given these caveats, major U.S. health organizations, including the American Dietetic Association, the American Heart Association, and the U.S. Department of Health and Human Services, maintain that Americans should keep the amount of total fat—whether in the form of olive oil or anything else—to no more than 20 to 35 percent of total calories.

Our advice is that you consider adopting aspects of the Mediterranean diet virtually all nutrition experts advocate: Eat an abundance of produce, grains, and legumes combined with moderate amounts of dairy products and relatively smaller amounts of meats, poultry, and fish.

I have heard that the traditional Indian diet is vegetarian; is that true?

"Traditional Indian Diet" can mean many different things. Other than the heavy use of vegetables and spices throughout India, Indian cuisine has few other shared characteristics. Geographic location, religious beliefs, and availability all contribute to the varied nature of the Indian diet. Furthering the diversity, Indian cooking rarely comes from recipes, relying more on the skill of the cook and the availability of ingredients. These phenomena have resulted in a mix of cuisines that are both tasty and various.

The history of vegetarianism in India dates back thousands of years. Many believe it started with the introduction of Buddhism and Jainism in the sixth century B.C.E. These religions, which are closely related, are strong advocates of the sanctity of all forms of life and are therefore opposed to the slaughter of animals for consumption. However, it is not uncommon for those who live along the coast to incorporate fish into their diet. Likewise, the more northern territories are known to eat much more meat in their diet than their southern counterparts. These food choices are most likely due to availability and climate differences.

Nonetheless, the idea that the Indian diet is vegetarian is not far from the truth. Anthropology sources clearly show the predominance of vegetarianism, and even those who do partake in meat consumption do so in moderation, usually no more than one day a week. More often, meat consumption is reserved for special occasions like weddings and celebrations, averaging about four times a year. Another interesting fact is that women are less likely to eat meat than men. This practice stems from the belief that meat causes masculine traits such as aggressiveness.

This feature cannot possibly encompass the vast culture and history that influences Indian cuisine. With the second largest population in the world, India is a multifaceted land with traditions and beliefs that would take books upon books to explain. Even so, their cuisine is a good illustration of the history of India as there are so many distinctive characteristics within it. Overall, the diet is based in complex carbohydrates and rich in fiber-full vegetables and fruits. In any Indian restaurant, you are bound to find a flavorful and healthy dish that will satisfy you! Some traditional Indian foods are shown in Figure 2-13.

As with most ethnic cuisines, the Americanization of Indian ingredients has not been very positive from a health perspective. With the addition of more animal-based protein sources and fat added to decreased fiber content, many of the advantages of the traditional Indian diet have been lost. However, those who prepare Indian cuisine in a traditional manner will receive the benefits of a high-carbohydrate, low-fat, low-cost diet that is as tasty as it is healthy.

What about "soul food" or "southern" cooking? Do most African Americans eat lots of it?

The term *soul food* was coined in the mid-1960s to promote ethnic pride and solidarity among African Americans,

FIGURE 2-13
FITTING INDIAN FOODS INTO THE MyPYRAMID FOOD GUIDE[a]

[a] Discretionary calories can be used for ghee,[b] coconut oil, and similar items.
[b] These foods are described in the Miniglossary.

GRAINS	VEGETABLES	FRUITS	OILS	MILK	MEAT & BEANS
basmati rice, bulgur wheat, chapati[b], millet, rice, roti[b]	broccoli, cabbage, carrots, cauliflower, cucumber, eggplant, onions, plantains, potato, squash, tomatoes, turnips	bananas, chutney, dates, figs, grapes, mango, melon, papaya		lassi[b], milk, yogurt	almonds, beef, cashews, chicken, chickpeas, lamb, lentils, shrimp

FIGURE 2-14
TRADITIONAL AFRICAN AMERICAN FOODS AND THE MyPYRAMID FOOD GUIDE[a]

[a] Discretionary calories can be used for butter, candy, fruit drinks, lard, meat drippings, chitterlings,[b] soft drinks, and vegetable shortening.
[b] These foods are described in the Miniglossary.

SOURCE: Adapted from www.mypyramid.gov, 2005 and the *Pyramid Packet* © 1993, Penn State Nutrition Center, 417 East Calder Way, University Park, PA 16801.

GRAINS	VEGETABLES	FRUITS	OILS	MILK	MEAT & BEANS
biscuits, cookies, cornbread, grits[b], pasta, rice	beets, broccoli, cabbage, corn, green peas, greens, hominy[b], okra, potatoes, spinach, squash, sweet potatoes, tomatoes, yams	apples, bananas, berries, fruit juice, peaches, watermelon		buttermilk, cheese, milk	blackeyed peas, beef, catfish, chicken, crab, crayfish, eggs, kidney beans, peanuts, perch, pinto beans, pork, red beans, red snapper, salmon, sardines, shrimp, tuna, turkey

© Bonnie Kamin/PhotoEdit

but origins of soul food date back to a much earlier time in history. Black-eyed peas, **grits** (coarsely ground cornmeal), collard greens, okra, and other soul foods evolved from the traditional diet of West African slaves living in the South. When West Africans were brought to the United States to work the fields, their dietary habits revolved around foods provided by slave owners. Corn was commonly given as a staple, and it was prepared in many forms, such as grits, cornmeal pudding, and **hominy** (hulled, dried corn kernels with certain parts removed). Also, because salt pork was frequently a staple supplied to slaves, pork fat was used to fry and flavor greens, breads, stews, and other foods. In addition, some owners allowed their slaves to grow vegetables in small plots. Slaves who farmed such plots grew American vegetables, including cabbage, collard and mustard greens, sweet potatoes, and turnips, and the slaves introduced okra and black-eyed peas, two West African favorites, to the United States. The slaves who worked as cooks in the homes of slave owners made popular other southern favorites, such as fried chicken and fried catfish.

After emancipation, eating patterns of African Americans did not change significantly, and they represent much of what we think of as southern cuisine today. The underpinnings of this eating style, which is very high in fat, remain corn-based dishes, greens, pork, and pork products such as **chitterlings** (chitlins—pig intestines) and ham hocks. Food habits of African Americans around the country tend to be influenced more by economic status and geographic location than by heritage, although soul food remains a symbol of identity and heritage for many African Americans.[21] Figure 2-14 shows traditional African American foods in pyramid form.

I see many foods marked "kosher" in the supermarket, especially during Jewish holidays such as Passover. Are kosher foods better for health than regular food items? Do other religious groups eat special foods?

As pointed out in Chapter 1 (refer to Table 1-7 on page 18), food is part of the symbolism and traditions of many major religions. In the predominantly Christian United States, however, most people's eating habits are not dictated by religion to a large extent.

Nevertheless, during the past few decades, "Jewish" foods have been growing in popularity among Americans of various religious backgrounds. For instance, bagels are now staple breakfast menu items in the United States.

Most traditional Jewish foods eaten in America come from a particular group known as Ashkenazic Jews—Jews from central and eastern European countries such as Russia, Germany,

© Bill Aron/PhotoEdit

FIGURE 2-15
FITTING JEWISH AMERICAN FOODS INTO THE MyPYRAMID FOOD GUIDE[a]

[a] Discretionary calories can be used for cream cheese, honey, jelly, hard margarine, marmalade, preserves, schmaltz,[b] sherbet, sour cream, and sugar.
[b] These foods are described in the Miniglossary.

SOURCE: Adapted from www.mypyramid.gov, 2005 and the *Pyramid Packet* © 1993, Penn State Nutrition Center, 417 East Calder Way, University Park, PA 16801.

GRAINS	VEGETABLES	FRUITS	OILS	MILK	MEAT & BEANS
bagel, barley, bialy[b], blintz, bulgur, challah[b], crepe, dumplings, hard rolls, honey cake, kasha[b], matzoh[b], noodle pudding, pastry, pita bread, pumpernickel bread, rye bread	artichokes, asparagus, beets/borscht, broccoli, Brussels sprouts, cabbage, carrot, cauliflower, corn, garlic, green beans, greens, leeks, olives, onion, peas, peppers, pickles, potatoes, sorrel, spinach, squash, sweet potatoes, tomatoes, turnips, yams	bananas, citrus fruits, dates, dried apples, dried apricots, dried pears, figs, grapes, melons, prunes, raisins		cottage cheese, edam cheese, farmer's cheese, gouda cheese, milk, Swiss cheese, yogurt	almonds, beef, beef tongue, brisket, chickpeas, chopped liver, corned beef, dry beans, eggs, gefilte fish[b], herring, lentils, lox[b], pastrami, poultry, salmon, sardines, smelt, smoked fish, split peas, tripe, veal

Poland, and Romania. (The other major group is Sephardic Jews, who come mainly from Spain and Portugal.) Although few Jewish people in the United States strictly abide by all dietary laws Judaism prescribes, many adhere to at least some rules of *kashrut*, biblical ordinances specifying which foods are **kosher,** or fit to eat.

Most people assume laws of kashrut were set forth to protect the health of the Jewish people. For example, a popular misconception is that kashrut forbids consumption of pork products because eating undercooked pork can cause serious illness. In truth, however, Jewish dietary laws are considered divine commandments set forth to maintain spiritual, not physical, health. Foods labeled as kosher are not necessarily more healthful than their unmarked counterparts. Instead, the designation kosher indicates that a food has been prepared in accordance with the basic tenets of kashrut. For instance, one principle of kashrut is separation of milk and meat products, meaning that an item containing, say, both ground beef and cheese would not be kosher.

Another tenet is selection of appropriate meat, poultry, and seafood items. Only animals with cloven hooves that chew their cud are allowed—cattle, sheep, goats, and deer. Chicken, turkey, goose, pheasant, and duck can be kosher, but birds of prey cannot. Seafood with both fins and scales can be kosher, whereas shellfish is forbidden.

Finally, as a result of the salting process used to prepare kosher animal foods, many traditional Jewish foods are high in sodium. Herring, smoked fish, canned beef, tongue, corned beef, and other deli-style meats are examples. Other traditional Jewish foods, many of which are high in fat, include **schmaltz** (chicken fat), **knishes** (potato pastry filled with ground meat or potato), cream cheese, and chopped liver.[22]

Figure 2-15 shows how some traditional Jewish foods fit into the MyPyramid Food Guide.

MINIGLOSSARY OF FOODS

fusion cuisine a term used to describe food that combines the elements of two or more cuisines—say, European and Oriental—to create a new one

MEXICAN AMERICAN

amaranth a golden-colored grain

agave a plant with spiny-margined leaves and flower

bolillo a roll-like bread often used instead of tortillas or to make sandwiches

burritos warm flour tortillas stuffed with a mixture of egg, meat, beans, and/or avocado

cassava a starchy root that is never eaten raw because it must be cooked to eliminate its bitter smell

cherimoya a fruit with a rough green outer skin and sherbetlike flesh

chilaquiles tortilla casserole often made with eggs or meat

chiles relleños roasted mild green chili pepper stuffed with cheese, dipped in egg batter, and fried

chorizo spicy beef or pork sausage

guava a sweet, juicy fruit with green or yellow skin and red or yellow flesh

jicama a crisp, bean root vegetable that is tan outside and white inside and is always eaten raw; as popular in Mexico as the potato is in the United States

plantain a greenish, starchy banana; because it is starchy even when ripe, it is never eaten raw and is usually pan fried

queso blanco, fresco, or Mexicano soft white cheese made of part-skim milk

sopa rice or pasta that is fried and cooked in consommé

purslane a leafy vegetable that can be used in salads or cooked like spinach

zapote an apple-size fruit with green skin and black flesh

CHINESE AMERICAN

bing thin pancakes

bok choy a vegetable with broad, white or greenish-white stalks and dark green leaves; also called *Chinese chard*

cellophane noodles thin, translucent noodles made from mung beans

Chinese broccoli a green leafy vegetable often stir fried; also called *Chinese kale*

choy sum a bright green vegetable commonly stir fried; also called *field mustard* or *Chinese flowering cabbage*

dim sum steamed or fried dumplings stuffed with pork, shrimp, beef, sweet paste, or preserves and steamed or fried

glutinous rice short-grained, opaque white rice that turns sticky when cooked.

jujube Chinese date

litchi small, round fruits with orange-red skin and opaque white flesh; also called *litchee* or *lychee*

longan a small round fruit with smooth brown skin and clear pulp

mantou steamed bread

oriental radish a large, cylindrically shaped vegetable with smooth skin; also called *daikon*

rice sticks flat, opaque, wide noodles made from rice flour

rice vermicelli thin white noodles made from rice flour

taro a starchy vegetable with brown hairy skin and a pink-purple interior

yard-long beans thin, tender string beans that grow to as long as 18 inches

INDIAN

chapatti a flat unleavened bread made with finely milled wholemeal flour

ghee a clarified butter without any milk solids or water

lassi a milkshake type of drink made from yogurt

roti an unleavened flatbread, resembling a tortilla, usually made from whole-wheat flour

AFRICAN AMERICAN

grits coarsely ground cornmeal

hominy hulled, dried corn kernels

chitterlings (chitlins) pig intestine

JEWISH AMERICAN

bialy a flat breakfast roll that is softer than a bagel

challah an egg-containing yeast bread, often braided, typically served on the Sabbath and holidays

gefilte fish a chopped fish mixture often made with pike and whitefish as well as matzoh crumbs, eggs, and seasonings

kasha cracked buckwheat, barley, millet, or wheat that is served as a cooked cereal or potato substitute

knish a potato pastry filled with ground meat, potato, or kasha

kosher fit, proper, or in accordance with religious law

lox smoked salmon

matzoh a cracker-like bread eaten most often at Passover

schmaltz chicken fat

In Review

1. Separate the following foods into two categories, those that are more nutrient dense and those that are more calorie dense.
   a. orange juice
   b. V8 juice
   c. orange soda
   d. Snickers Bar
   e. apple
   f. peach pie
   g. broccoli
   h. Gummi Bears
   i. oatmeal
   j. cheesecake

2. According to the *Dietary Guidelines for Americans 2005*, adults are advised to exercise at least 30 minutes a day to reduce the risk of chronic disease; 60 minutes daily to prevent weight gain; and up to 90 minutes daily if they have lost weight and want to keep it off.
   a. true
   b. false

3. According to U.S. nutrition labeling regulations, strong scientific evidence has been found for all of the following health claims regarding nutrition and disease except:
   a. sugar and diabetes.
   b. sodium and high blood pressure.
   c. calcium and osteoporosis.
   d. fats and cancer and heart disease.

4. Which of the following is true about Dietary Reference Intakes (DRI)?
   a. They provide a benchmark for people's nutrient needs.
   b. They are the same for everyone regardless of age, gender, or health status.
   c. They have not changed since 1941.
   d. They are the same as Recommended Dietary Allowances.

5. List and briefly explain the ABCs of healthy eating.

6. Reading food labels is a good way to make informed choices about food. Which of the following is true about food labeling?
   a. The words "low," "good," and "high" that appear on food labels really don't mean anything and can be used however the manufacturer sees fit.
   b. Every vitamin and mineral that a food item contains must be listed on the label.
   c. Food labels list their ingredients in order of amount by weight with the ingredient of the greatest amount listed first.
   d. Daily Values listed on food labels refer to the amount of that food you should eat in a day.

7. A statement linking the nutritional profile of a food to a reduced risk of a particular disease is known as a:
   a. nutrient content claim.
   b. health claim.
   c. disease reducer.
   d. composition claim.

8. The Exchange Lists:
   a. are an effective tool in meal planning.
   b. are used by dietitians to help clients follow specific diets.
   c. include Carbohydrate, Meat and Meat Substitute, and Fat groups.
   d. all of the above

9. Nutrient-dense foods are foods that:
   a. are iron-rich.
   b. contain a mixture of carbohydrate, fat, and protein.
   c. carry the USDA nutrition labeling.
   d. are rich in nutrients but relatively low in calories.

10. The MyPyramid Food Guidance System makes which of the following recommendations for a 2,000 calorie diet?
    a. Eat 2 cups of fruit and 2½ cups of vegetables a day.
    b. Eat 3 ounces of whole-grain foods a day.
    c. Drink 3 cups of fat-free or lowfat milk a day.
    d. all of the above

Menu of Online Study Tools

A variety of study tools for this chapter are available at our website to deepen your understanding of chapter concepts. Go to www.thomsonedu.com/nutrition/boyle to find

- Practice tests
- Flashcards
- Glossary
- Web links
- Animations
- Chapter summaries, learning objectives, and crossword puzzles

Nutrition on the Web

www.thomsonedu.com/nutrition/boyle
Go to the *Personal Nutrition* site to check for the latest updates to chapter topics or to access links to related websites.

www.nap.edu
Look here for updates about the new Dietary Reference Intakes (DRIs).

www.nal.usda.gov/fnic/dga/index.html
Find practical tips for following the Dietary Guidelines.

www.MyPyramid.gov
Tips and resources for using MyPyramid in planning a healthy diet.

www.nal.usda.gov/fnic/Fpyr/pyramid.html
Scroll down and click on the line for even more Food Pyramids to view a variety of Ethnic/Cultural/Special Audience Pyramids.

www.ars.usda.gov/ba/bhnrc/ndl
Click on "Search" for free food analyses. Just type in the food you want to analyze, and get a breakdown of its calories, fat, fiber, protein, vitamins, and minerals.

www.fao.org/ag/AGN/nutrition/education_guidelines_country_en.stm
Dietary guidelines and food guides from around the world.

www.health.gov
A portal to the websites of a number of multi-agency health initiatives and activities of the U.S. government.

www.ag.uiuc.edu/~food-lab/nat
A free diet analysis program developed at the University of Illinois–Urbana/Champaign. This site allows anyone to analyze the various nutrients in the foods they eat.

www.nal.usda.gov/fnic
Click on "Food Composition" for numerous food composition resources.

www.cfsan.fda.gov/label.html
Useful facts about food labels; updates on label health claims.

www.usda.gov/cnpp
The Interactive Healthy Eating Index—the USDA Center for Nutrition Policy and Promotion's online dietary assessment tool. After you provide a day's worth of dietary information, you will receive a "score" on the overall quality of your diet as compared to current recommendations for total fat, saturated fat, cholesterol, and sodium intakes.

The Carbohydrates: Sugar, Starch, and Fiber

3

ASK YOURSELF . . .

Which of the following statements about nutrition are true, and which are false? For each false statement, what is true?

1. Fruit sugar (fructose) is less fattening than table sugar (sucrose).
2. Foods high in complex carbohydrate (starch and fiber) are good choices when you are trying to lose weight.
3. People with diabetes should never eat sugar.
4. The primary role of dietary fiber is to provide energy.
5. The brain demands the sugar glucose to fuel its activities.
6. Honey and refined sugar are the same as far as the body is concerned.
7. Of all the components of foods that increase one's risk of diseases, sugars are probably the biggest troublemakers.
8. Breads that are brown in color have more fiber than white bread.
9. Some foods labeled sugar-free actually contain calorie-bearing sugars.
10. Artificial sweeteners are safe to use in moderation.

Answers found on the following page.

Rabbit said, "Honey or condensed milk with your bread?" [Pooh] was so excited that he said, "Both," and then, so as not to seem greedy, he added, "But don't bother about the bread, please."

from *Winnie the Pooh,* A. A. Milne (1882–1956, children's book author)

CONTENTS

Once upon a time, bread, potatoes, pasta, and other starchy foods were placed on the dieter's list of most-fattening or "illegal" foods. Still today, carbohydrate bashing is a popular pastime. This unattractive image undoubtedly comes from the practice of serving many carbohydrate-rich foods laden with fat—potatoes with sour cream and butter, vegetables or pasta with rich cream sauces, toast with butter, and salads with fat-rich dressings. People who need to lose weight must limit high-calorie foods, but they are ill advised to try to avoid all **carbohydrates.** Excess calories *are* fattening, and the fat, not the carbohydrate, raises the calorie count the most.

This chapter invites you to learn to distinguish between certain carbohydrates, such as starch and fiber, and others, such as concentrated sugars. You will learn to choose your carbohydrates by the company they keep (see Table 3-1).

The Body's Need for Carbohydrates

The primary role of carbohydrates is to provide the body with energy (calories), and for certain body systems (for example, the brain and the nervous system), carbohydrates are the preferred energy source. Carbohydrates are the ideal fuel for the body. There are only two alternative calorie sources: protein and fat. Protein-rich foods are usually expensive and provide no advantage over carbohydrates when used to provide fuel for the body. Fat-rich foods might be less expensive, but fat cannot be used efficiently as fuel by the brain and nerves, and diets

It's not the potatoes that are fattening; it's the butter, sour cream, or gravy that they put on us!
BUTTER
SOUR CREAM

TABLE 3-1
Choosing Carbohydrates by the Company They Keep

Food	Calories	Grams of Fat
Toast (2 slices)		
with margarine (2 tsp)	188	9
with low-sugar jelly/fruit spread (2 tsp)	139	2
Potato (medium)		
with margarine (1 tsp) and sour cream (1 tbsp)	287	10
with fat-free sour cream (2 tbsp)	223	0
Bagel (medium)		
with regular cream cheese (2 tbsp)	263	11
with fat-free cream cheese (2 tbsp)	188	1
Pasta (1 c)		
with Alfredo sauce (⅓ c)	390	20
with tomato and mushroom sauce (⅓ c)	232	4

SOURCE: Adapted from *Environmental Nutrition* 17 (February 1994): 2.

high in fat are associated with many chronic diseases. Thus, of all alternative food-energy sources, carbohydrates are preferred; they provide most of the day's energy for most of the world's people.

The brain and nervous system are sensitive to the concentration of glucose in the blood. Normal blood glucose levels are important for a feeling of well-being. When your blood glucose level becomes too high, you get sleepy; when the concentration falls too low, you get weak and shaky. Only when blood glucose is within the normal range can you feel energetic and alert. The person who wants to feel energetic and alert all day should make the effort to eat so as to maintain blood glucose levels in the normal range to fuel the critical work of the brain and nervous system.

Carbohydrate Basics

Carbohydrate-rich foods are obtained almost exclusively from plants. Milk is the only animal-derived food that contains significant amounts of carbohydrate. Carbohydrates are divided into two categories: complex carbohydrates and simple carbohydrates. **Complex carbohydrates** include starch and fiber. Starches make up a large part of the world's food supply—mostly as grains. For example, such staples as wheat, rice, and corn are rich sources of starch. Fiber is found abundantly in plants, especially in the outer portions of cereal grains, and in fruits, legumes, and most vegetables. **Simple carbohydrates** include naturally occurring sugars in fresh fruits, in some vegetables, in milk and milk products, and as added sugars in concentrated form, such as in honey, corn syrup, or sugar in the sugar bowl. All of these carbohydrates have characteristics in common, but they are of different merit nutritionally. Table 3-2 introduces the different types of carbohydrates.

carbohydrates compounds made of single sugars or multiple sugars and composed of carbon, hydrogen, and oxygen atoms.

carbo = carbon (C)
hydrate = water (H_2O)

complex carbohydrates long chains of sugars (glucose) arranged as starch or fiber; also called polysaccharides.

poly = many
saccharides = sugar unit

simple carbohydrates (sugars) the single sugars (monosaccharides) and the pairs of sugars (disaccharides) linked together.

Ask Yourself Answers: **1.** False. Fructose and sucrose are equally fattening because they have the same number of calories per gram. **2.** True. **3.** False. People with diabetes need to watch the total carbohydrate in their diets, but they can choose foods with sugar as a small portion of that total. **4.** False. Although certain fibers may provide negligible calories to the diet, fiber's primary role is to provide bulk for the digestive tract. **5.** True. **6.** True. **7.** False. Of all the things in foods associated with risk of diseases, fat is by far the biggest troublemaker. **8.** Not always. A brown color does not always mean the bread is high in fiber. The color can come from molasses or caramel. To be high in fiber, the label must say *whole-grain* or *whole-wheat*, not simply *wheat* bread. *Whole-grain* flour should be listed first in the ingredients list. **9.** True. **10.** True.

TABLE 3-2
Categories and Sources of Carbohydrate

Carbohydrate Type	Common Names	Examples of Food Sources
Monosaccharides		
Glucose ●	Dextrose, blood sugar	Fruits, sweeteners
Fructose ■	Fruit sugar, levulose	Fruits, honey, high-fructose corn syrup
Galactose ▲	—	Part of lactose, found in milk
Disaccharides		
Sucrose (glucose + fructose) ●—■ = ● + ■	Table sugar	Beet and cane sugar, fruit, most sweets
Lactose (glucose + galactose) ●—▲ = ● + ▲	Milk sugar	Milk and milk products
Maltose (glucose + glucose) ●—● = ● + ●	Malt sugar	Sprouted seeds
Polysaccharides		
Starches ●-●-●-●-●-●-●	Dextrins*	Potatoes, legumes, corn, wheat, rye, and other grains
Dietary fiber	Roughage, bulk:	Whole grains, legumes, fruits, vegetables
Insoluble fibers	cellulose, hemicellulose	Wheat products, brown rice, vegetables, legumes, seeds
Soluble fibers	pectins, gums, mucilages, some hemicelluloses	Oat products, barley, legumes, fruits, vegetables, seeds

*Starch can be broken down during food processing to shorter chains of glucose units known as *dextrins*. The word *dextrins* sometimes appears on food labels, because dextrins can be used as thickening agents in foods.

The Simple Carbohydrates: Monosaccharides and Disaccharides

All carbohydrates are composed of single sugars—known as **monosaccharides**—alone or in various combinations, and all carbohydrates but fiber can quickly be converted to **glucose** in the body. Green plants make glucose from carbon dioxide and water through a process known as *photosynthesis* in the presence of chlorophyll and sunlight, as illustrated in the margin.

Glucose is not a very sweet sugar, but plants can rearrange its atoms to form another sugar, **fructose,** which is sweet to the taste. Fructose is found mostly in fruits, in honey, and as part of another sugar—table sugar. Glucose and fructose are the most common single sugars in nature.

Some sugars are double sugars—known as **disaccharides**—made by bonding two single sugars together. When glucose and fructose are bonded together, they form **sucrose,** or table sugar, the product most people refer to when they use the term *sugar*. The sweet taste of sucrose comes primarily from the fructose in its structure. It occurs naturally in many fruits and vegetables. Sugar cane and sugar beets are two sources from which sucrose is purified

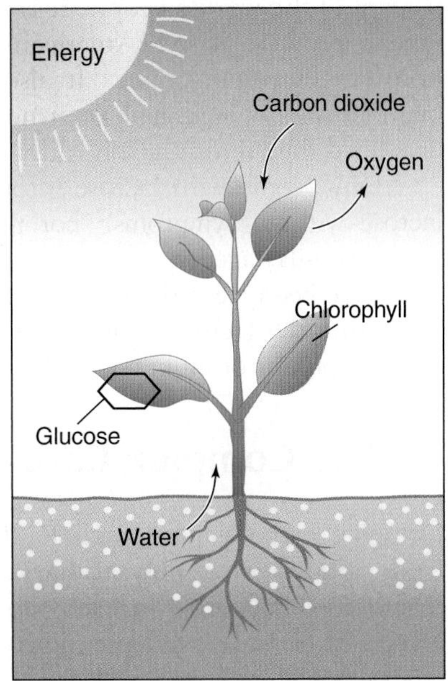

▲ ***Photosynthesis:*** *In the presence of chlorophyll and the energy of the sun, plants make glucose. Water, absorbed by the plant's roots, and carbon dioxide, absorbed through the plant's leaves, combine to form a molecule of glucose.*

glucose (GLOO-koce) the building block of carbohydrate; a single sugar used in both plant and animal tissues as quick energy. A single sugar is known as a **monosaccharide.**

mono = *one*

fructose (FROOK-toce) fruit sugar—the sweetest of the single sugars.

Another single sugar, *galactose* (ga-LACK-toce), occurs bonded to glucose in the sugar of milk.

sucrose (SOO-crose) a double sugar composed of glucose and fructose. A double sugar is known as a **disaccharide.**

di = *two*

TABLE 3-3
Lactose in Selected Dairy Products

Dairy Product	Lactose (grams)
Cheese (1 oz)	
Parmesan or Cream	0.7–0.8
Cheddar or American	0.5
Whole milk (1 c)	11.0
2% low-fat milk (1 c)	9.0–13.0
Fat-free milk (1 c)	12.0–14.0
Chocolate milk (1 c)	10.0–12.0
Lactose-reduced, low-fat milk (1 c)	3.3
Buttermilk (1 c)	9.0–11.0
Low-fat yogurt (1 c)*	11.0–15.0
Cottage cheese, low-fat (1 c)	7.0–8.0
Ice cream/ice milk (1 c)	9.0–10.0

*Choose yogurt with "active cultures"—they help to digest the lactose found in yogurt.

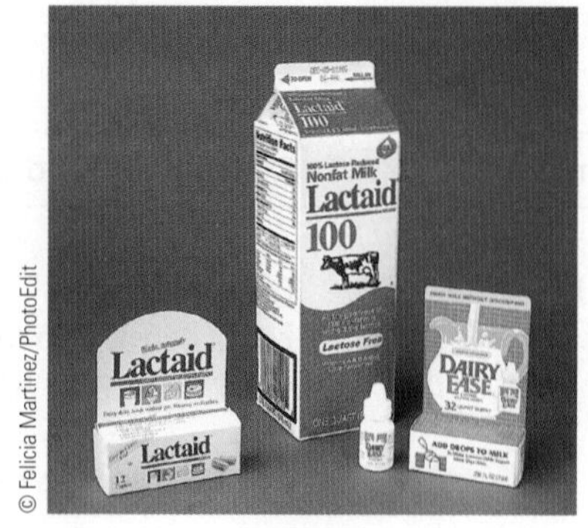

© Felicia Martinez/PhotoEdit

▲ *Lactose-reduced milk, enzyme solutions containing lactase for treating dairy products, and lactase tablets can help reduce the symptoms of lactose intolerance.*

and granulated to various extents to provide the brown, white, and powdered sugars available in the supermarket. Sucrose is one of the two most caloric ingredients of candy, cakes, pastries, frostings, cookies, presweetened ready-to-eat cereals, and other concentrated sweets. (The other major calorie contributor is fat, discussed in Chapter 4.)

Another double sugar, **maltose,** consists of two glucose units. It occurs in sprouting seeds and arises during the digestion of starch in the human body. The malt found in beer contains maltose. Enzymes used in the brewing process break down the long chains of starch in barley and wheat into maltose units.

Finally, **lactose,** the major sugar in milk, is a double sugar made by mammals from galactose and glucose units. A human baby is born with the digestive **enzymes** necessary to split lactose into its two simple sugars—glucose and galactose—so that they can be absorbed by the body. Lactose also facilitates the absorption of calcium and promotes the growth of beneficial bacteria in the intestines. Breast milk and infant formula, which contain lactose, are ideal foods for babies because they provide a simple, easily digested carbohydrate to meet an infant's energy needs.

When you eat a food containing lactose, the enzyme *lactase* in your small intestine first splits the double sugar into single sugars so that they can enter your bloodstream. Many people can lose the ability to digest lactose during or after childhood. Thereafter, these people may experience nausea, bloating, abdominal pain or cramping, diarrhea, or excessive gas after drinking milk or eating lactose-containing products. These conditions are created because when the intestinal bacteria use the lactose for energy, it produces gas and other products that irritate the intestine. This condition, called **lactose intolerance,** is inherited by about 70 percent of the world's people. It is most common in African Americans, Mediterranean peoples, Native Americans, and Asians, and less common in people of northern European origin.[1] It also can develop temporarily in people who are malnourished or sick, making it necessary for them to avoid milk and milk products until they return to a healthy state.

Many people with lactose intolerance are able to consume small amounts of lactose without symptoms.[2] For them, lower-lactose foods such as yogurt, acidophilus milk, aged cheeses, cottage cheese, or specially prepared milk products that have been treated with an enzyme to reduce lactose may be tolerated. Table 3-3 shows the lactose contents of fermented dairy products compared with the lactose content of milk.

maltose a double sugar composed of two glucose units.

lactose a double sugar composed of glucose and galactose; commonly known as milk sugar.

enzymes protein catalysts. A catalyst facilitates a chemical reaction without itself being altered in the process.

lactose intolerance inability to digest lactose as a result of a lack of the necessary enzyme lactase. Symptoms include nausea, abdominal pain, diarrhea, or excessive gas that occurs anywhere from 15 minutes to a couple of hours after consuming milk or milk products.

starch a plant polysaccharide composed of hundreds of glucose molecules, digestible by human beings.

polysaccharide a long chain of 10 or more glucose molecules linked together in straight or branched chains; another term for complex carbohydrates.

staple grain a grain used frequently or daily in the diet—for example, corn (in Mexico) or rice (in Asia).

The Complex Carbohydrates: Starch

Complex carbohydrates include starch and fiber. All starchy foods are plant foods. **Starch** is a **polysaccharide** made up of many glucose units bonded together—3,000 or so in each molecule of starch. Shorter carbohydrate chains composed of 3 to 10 glucose molecules are called *oligosaccharides*.

Seeds such as grains, peas, and beans are the richest starch sources. Most societies have a primary or **staple grain** that provides most of the people's food energy. In many Asian nations, the staple grain is rice. In Canada, the United States, and Europe, the staple grain is wheat. If you consider all the food products made from wheat—bread (and other baked goods made from wheat flour), cereals, and pasta—you will realize how pervasive this grain is in the food supply. The staple grains of other peoples include corn, millet, rye, barley, and oats.

TABLE 3-4
Types of Carbohydrates Found in Selected Foods in the MyPyramid Food Guide

	Sugar	Starch	Fiber
Grains Group			
Bread, cooked grains	✔	✔	✔
Vegetable Group			
Corn, peas	✔	✔	✔
Green beans	✔		✔
Potato	✔	✔	✔
Tomato	✔		✔
Fruit Group			
Apple, banana	✔ (mostly fructose)		✔
Orange juice	✔ (mostly fructose)		
Milk Group			
Milk	✔ (lactose)		
Meat & Beans Group			
Meat, poultry, fish, eggs			
Legumes	✔	✔	✔

▲ *Foods such as potatoes, dried beans and peas, rice and whole-grain breads, cereals, and pastas are especially nutritious because of their starch, fiber, vitamin, and mineral content—and because they are virtually fat and cholesterol free.*

A second important source of starch is the legume family, including dried beans and peas such as butter beans, kidney beans, pinto beans, navy beans, black-eyed peas, chickpeas, lentils, and soybeans. These vegetables are about 40 percent starch by weight and contain abundant protein. Root vegetables (such as yams) and tubers (such as potatoes) are other sources of starch that are important in many societies. Table 3-4 shows the types of carbohydrates found in foods from the USDA MyPyramid food guidance system.

The Bread Box: Refined, Enriched, and Whole-Grain Breads

For many people, grains supply much of the carbohydrate, or at least most of the starch, in a day's meals. Because grains have such a primary place in the diet, be sure that the grains you choose—wheat, rice, oats, or corn—contribute the nutrients you need. Also, be sure to learn the meanings of the words associated with the flours that make up the grain products you use—**refined, enriched, fortified,** and **whole grain.** This discussion of the nutritional differences between different breads provides an example of an important principle of nutrition: Foods that are far removed from their original state of wholeness may lack significant nutrients.

The part of the wheat plant that is made into flour and then into bread and other baked goods is the kernel. The wheat kernel (a whole grain) has four main parts. The **germ** is the part that grows into a wheat plant, and it contains concentrated food to support the new life. It is especially rich in protein, vitamins, and minerals. The **endosperm** is the soft, white inside portion of the kernel containing starch and protein. The **bran,** a protective coating around the kernel (which is similar in function to the shell of a nut), is also rich in nutrients and fiber. The **husk,** commonly called *chaff*, is unusable for most purposes except for animal feed.

In earlier times, people milled wheat by grinding it between two stones, sifting out the inedible chaff, but retaining the nutrient-rich bran and germ as well as the endosperm. Improved milling machinery made it possible to remove the dark, heavy germ and bran as well, leaving a whiter, smoother-textured flour. People came to look on this flour as more desirable than the crunchy, dark brown, "old-fashioned" flour but at first were unaware of the nutrition implications.

refined refers to the process by which the coarse parts of food products are removed. For example, refining wheat into flour involves removing three of the four parts of the kernel—the chaff, the bran, and the germ—leaving only the endosperm.

enriched refers to the process by which the B vitamins thiamin, riboflavin, niacin, folic acid, and the mineral iron are added to refined grains and grain products at levels specified by law.

fortified foods foods to which nutrients have been added. Typically, commonly eaten foods are chosen for fortification with added nutrients to help prevent a deficiency (iodized salt, milk with vitamin D) or to reduce the risk of chronic disease (juices with added calcium).

whole grain refers to a grain that is milled in its entirety (all but the husk), not refined. Whole grains include wheat, corn, rice, rye, oats, amaranth, barley, buckwheat, sorghum, and millet. Two others—bulgur and couscous—are processed from wheat grains.

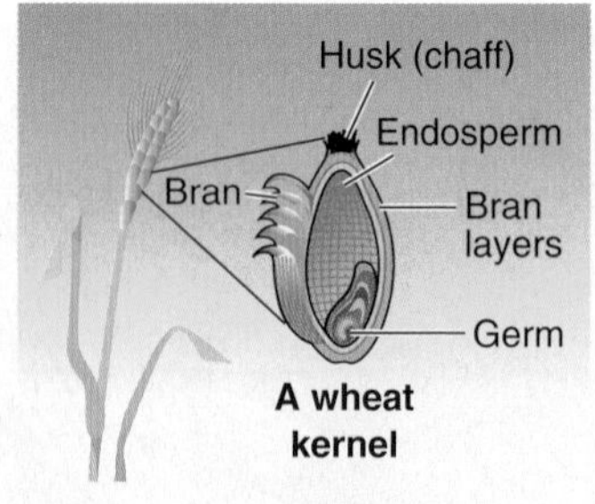

germ the nutrient-rich and fat-dense inner part of a whole grain.

endosperm the soft, white inside portion of a grain or kernel that contains starch and protein and provides energy.

bran the fibrous protective covering of a whole grain and source of fiber, B vitamins, and trace minerals.

husk the outer, inedible covering of a grain.

Bread eaters suffered a tragic loss of needed nutrients when they turned to white bread. A 1936 U.S. survey revealed that many people were suffering from deficiencies of the nutrients iron, thiamin, riboflavin, and niacin, which they had formerly received from bread. The Enrichment Act of 1942 standardized the return of these four lost nutrients to commercial flour. This legislation was amended in 1996 to include folic acid, a form of the B vitamin folate, which is considered important in the prevention of certain birth defects. Today, you can assume that almost all breads, grains such as rice, wheat products such as macaroni and spaghetti, and cereals, both cooked and ready to eat, have been enriched. Figure 3-1 shows that although enrichment makes refined bread comparable to whole-grain bread with respect to the enrichment nutrients, it does not do so with respect to other important nutrients. Therefore, although the enrichment of flour and other cereal products does improve them, it doesn't improve them enough: Whole-grain products are still preferred over enriched products. If bread is a staple food in your diet—that is, if you eat it every day—you are well advised to learn to like the hearty flavor of whole-grain bread.

The Complex Carbohydrates: Fiber

fibers the indigestible residues of food, composed mostly of polysaccharides. The best known of the fibers are cellulose, hemicellulose, pectin, and gums.

The **fibers** of a plant form the supporting structures of its leaves, stems, and seeds. Most fibers are polysaccharides, just as starch is, but with different bonds between the glucose units—bonds that cannot be broken by human digestive enzymes. The term *fiber* is used by almost everyone as though it represents a single entity. Fiber was known generations ago as "roughage." However, many com-

FIGURE 3-1
NUTRIENTS IN WHOLE-GRAIN, ENRICHED WHITE, AND UNENRICHED WHITE BREADS

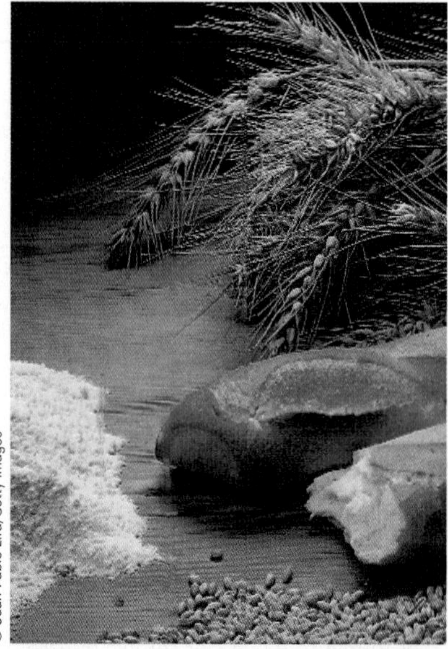

▲ *The nutrients present in the wheat plant at harvest are not always present in the wheat products you eat.*

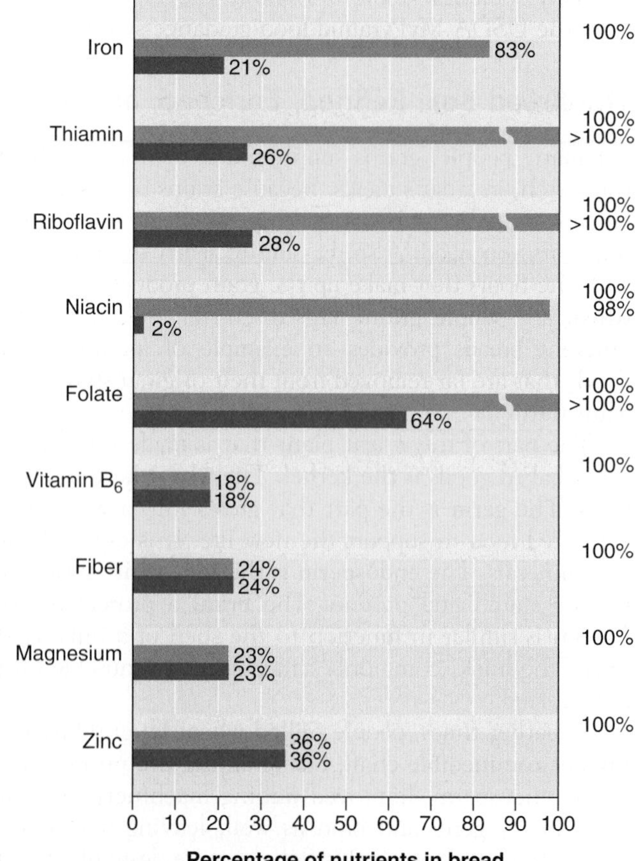

pounds, mostly carbohydrates, make up fiber.* Such compounds are familiar to us as the strings of celery, the skins of corn kernels, and the membranes separating the segments in citrus fruits. When isolated from plants, fiber may be used to thicken jelly (citrus pectin), to keep salad dressing from separating (guar gum), to provide bulk (wheat and other brans), and to create other effects on food texture and consistency.

The bonds that hold the units of fiber together cannot be broken by human digestive enzymes, but some can be broken by the bacteria that reside in the human digestive tract. However, very few (if any) calories are absorbed during this process. Fiber has two forms, **insoluble fiber** and **soluble fiber,** each of which exerts important effects on people's health.

insoluble fiber includes the fiber types called cellulose, hemicellulose, and lignin. Insoluble fibers do not dissolve in water.

soluble fiber includes the fiber types called pectin, gums, mucilages, some hemicelluloses, and algal substances (for example, carageenan). Soluble fibers either dissolve or swell when placed in water.

The Health Effects of Fiber

Many health experts encourage consumers to eat more fiber. According to recent evidence, inadequate levels of fiber in the diet are associated with several diseases (see the Miniglossary), whereas the consumption of recommended levels of fiber offers many health benefits.[3] Recall that dietary fiber is found only in plant foods, such as fruits, vegetables, legumes, and whole grains, and that it is the part of plant foods that human enzymes cannot digest.

Table 3-5 shows the potential health benefits from consuming these two types of fiber. As you can see from the table, not all fibers are created equal. Because insoluble and soluble fibers have different effects in the body, it is important to eat a variety of high-fiber foods to get both types.

Both types of fiber—soluble and insoluble—can help with weight control. In the stomach, they convey a feeling of fullness because they absorb water, and some of them delay the emptying of the stomach so that you feel full longer. Also, if you eat many high-fiber foods, you are likely to eat fewer empty-calorie foods such as concentrated fats and sweets. In fact, producers of some diet aids base the success of their products on the ability of certain fibers in the products to provide bulk and make you feel full.

TABLE 3-5
Health Benefits of Dietary Fiber

Health Problems	Fiber Type	Possible Health Benefits
Obesity	Insoluble/soluble	Replaces calories from fat, provides satiety, and prolongs eating time because of chewiness of food
Digestive tract disorders: Constipation Diverticulosis Hemorrhoids	Insoluble	Provides bulk and aids intestinal motility, binds bile acids
Colon cancer	Insoluble	Speeds transit time through intestines and may protect against prolonged exposure to carcinogens.*
Diabetes	Soluble	May improve blood sugar tolerance by delaying glucose absorption
Heart disease	Soluble	May lower blood cholesterol by slowing absorption of cholesterol and binding bile

*This effect is based on epidemiologic studies and is usually observed along with a reduced-fat intake. It is unclear whether the protection comes from fiber or from other components that accompany fiber in foods—such as vitamins (for example, folate), minerals (for instance, calcium), or phytochemicals.

Foods Rich in Insoluble Fiber

bran
brown rice
green beans
green peas
many vegetables
nuts
rice
seeds
skins/peels of fruits and vegetables
wheat bran
whole-grain products

Foods Rich in Soluble Fiber

barley
broccoli
carrots
corn
fruits (especially citrus)
legumes
oat bran
oats
potatoes
rye

*The woody material of heavy stems and bark is the noncarbohydrate lignin, which is classed by some as a fiber.

MINIGLOSSARY OF DISEASES AND CONDITIONS ASSOCIATED WITH LACK OF FIBER IN THE DIET

appendicitis inflammation and/or infection of the appendix, a sac protruding from the large intestine.

atherosclerosis (ATH-er-oh-scler-OH-sis) a type of cardiovascular disease characterized by the formation of fatty deposits, or plaques, in the inner walls of the arteries.

colon cancer cancer of the large intestine (colon), the terminal portion of the digestive tract (see Appendix A).

constipation hardness and dryness of bowel movements associated with discomfort in passing them.

diabetes a disease characterized by insufficiency or relative ineffectiveness of insulin that renders a person unable to regulate the blood glucose level normally.

diverticulosis (dye-ver-tic-you-LOCE-iss) outpocketings of weakened areas of the intestinal wall, like blowouts in a tire, that can rupture, causing dangerous infections.

hemorrhoids (HEM-or-oids) swollen, hardened (varicose) veins in the rectum, usually caused by the pressure resulting from constipation.

obesity body weight that is high enough above normal weight to constitute a health hazard (see also Chapter 9).

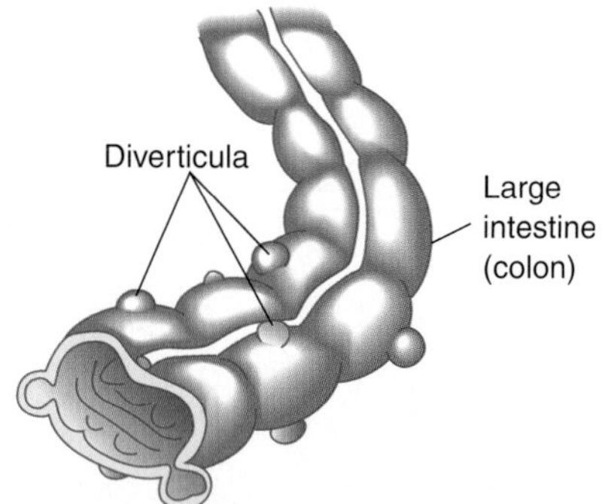

▲ ***Diverticulosis:*** *The outpocketings of intestinal linings that balloon through the weakened intestinal wall muscles are known as diverticula.*

Insoluble fibers, the type in wheat bran, hold water in the colon, thereby increasing bulk and stimulating the muscles of the digestive tract so that they retain their health and tone. The toned muscles can more easily move waste products through the colon for excretion. This prevents **constipation, hemorrhoids** (in which veins in the rectum swell, bulge out, become weak, and bleed), and **diverticulosis** (in which the intestinal walls become weak and bulge out in places in response to pressure needed to excrete waste when bulk is inadequate). These fibers may also speed up the passage of food through the digestive tract, thus shortening the time of exposure of the tissue to agents in food that might cause certain cancers, such as **colon cancer.**[4]

Soluble fibers, the type in beans and oats, are credited with reducing the risks of heart and artery disease—**atherosclerosis**—by lowering the level of cholesterol in the blood. It appears that the products of bacterial digestion of soluble fiber in the colon are absorbed into the body and may inhibit the body's production of cholesterol, as well as enhance the clearance of cholesterol from the blood.[5] Cholesterol levels may also decrease if food sources of soluble fiber (for example, barley, lentils, peas, beans, oat bran, or **psyllium**-enriched cereal) are used as part of a heart-healthy, low-cholesterol diet.[6] Certain soluble fibers also bind to cholesterol compounds and carry them out of the body with the feces, thus lowering the body's total cholesterol level.

Soluble fibers also improve the body's handling of glucose, even for people with diabetes, perhaps by slowing the digestion or absorption rate of carbohydrates.[7] Blood glucose levels therefore stay moderate, helping to prevent symptoms of **diabetes** or hypoglycemia. The list of fiber's contributions to human health, therefore, is impressive.

When people choose high-fiber foods in hope of receiving some of these benefits, they must choose with care. Wheat bran, which is composed mostly of cellulose, has no cholesterol-lowering effect, whereas oat bran and the fibers of legumes, carrots, apples, and grapefruits do lower blood cholesterol. On the other hand, the fiber of wheat bran in whole-wheat bread is one of the most effective stool-softening fibers that can help to prevent constipation and hemorrhoids. If one practical conclusion were to be drawn from what we know about fiber, it would have to be that all whole-plant foods seem to contain many kinds of fibers and thus can be expected to have the entire range of these beneficial effects. To obtain the greatest benefits from fiber, therefore, you must eat a variety of fiber-rich foods rather than taking doses of purified fiber, such as bran, from a single source.

psyllium seed husk, an ingredient in certain cereals and bulk-forming laxatives; contains both soluble and insoluble properties.

▲ *We are advised to increase our intakes of complex carbohydrates. Choose plenty of whole foods like these . . .*

. . . and fewer foods like these—foods that no longer resemble their original farm-grown products.

■ Guidelines for Choosing Carbohydrates

As it happens, most of the recent nutrition goals and guidelines have had one point in common. They tend to favor a return to a more **whole-food,** plant-based diet. Generally speaking, the more a food resembles the original, farm-grown product, the more nutritious it is likely to be. During processing, some nutrients may be lost, and food producers often add nutrient-poor ingredients such as sugar, salt, and fat. For example, a potato contains 20 milligrams of vitamin C, but the same number of calories in French fries contains only about 7 milligrams. Also, the same number of calories in potato chips contains only 2 milligrams of vitamin C. People who change their diet and begin to eat fewer processed foods and more whole-food, plant-based products can expect many health benefits.

Large whole apple with peel: 4 g fiber

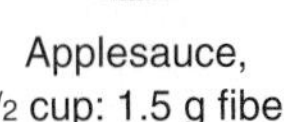

Applesauce, 1/2 cup: 1.5 g fiber

Apple juice, 3/4 cup: 0.2 g fiber

▲ *A whole food has the nutritional advantage over its processed forms.*

Complex Carbohydrates in the Diet

Complex carbohydrates are thought to be our most valuable energy nutrient. The *Dietary Guidelines for Americans* includes the following recommendation concerning intakes of complex carbohydrates in the diet: Choose fiber-rich fruits, vegetables, and whole grains often.[8] Likewise, the new USDA MyPyramid food guide illustrates the goal of healthy eating as a shift away from a diet based on high-protein, higher-fat foods to one that uses more of the beneficial complex carbohydrates. Thus, the new pyramid graphic reflects this shift visually with wider vertical bands, which encourage you to eat proportionally more servings of grain products, vegetables, and fruit (see Figure 2-5 on page 44). Table 3-6 lists the current recommendations for choosing carbohydrates for a healthful diet.

whole food a food that is altered as little as possible from the plant or animal tissue from which it was taken—such as milk, oats, potatoes, or apples.

TABLE 3-6
Recommendations for Carbohydrates in the Diet

	Carbohydrate	Added Sugars	Fiber
Dietary Guidelines	45%–65% of total calories	< 10% of total calories	14 g/1,000 calories consumed
DRI Committee	45%–65% of total calories	Maximum upper limit: 25% or less of total calories	Adequate Intake (AI) Men / Women Adults under 50 years: 38 g / 25 g Adults over 50 years: 30 g / 21 g
World Health Organization	55%–75% of total calories	0%–10% of total calories	Lower limit: 27 g/day Upper limit: 40 g/day
Current Intake	49%	≥ 16%	14–15 g

FIGURE 3-2
WHOLE GRAIN CONSUMPTION IN THE AVERAGE AMERICAN DIET

SOURCE: *Healthy People 2010 Progress Review: Nutrition and Overweight,* January 21, 2004.

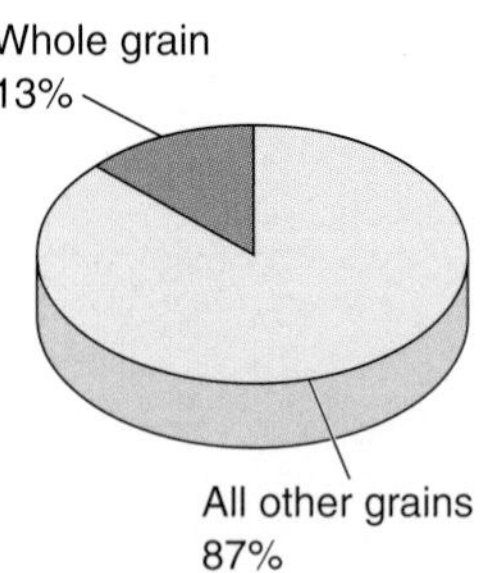

According to the Dietary Guidelines, at least half of the recommended grain servings consumed should be whole grains, particularly because of their fiber content. Today, whole grains make up only 13 percent of the grains eaten by Americans (see Figure 3-2).[9] See the Eat Well Be Well and Savvy Diner features that follow for more benefits and tips for making half your grains whole.

Grain products are low in fat, unless fat is added in processing, in food preparation, or at the table. For maximum benefit from grains, choose most often foods that contain little fat (less than 3 grams of fat per serving). Good choices include bagels, breadsticks, low-fat crackers, tortillas (not fried), couscous, English muffins, enriched breads, farina, grits, pancakes, pastas, popcorn (air-popped), pretzels, ready-to-eat cereals, rice, rice cakes, and taco shells. Remember to make at least half of your selections from whole grains (whole wheat, barley, oats, bulgur, quinoa, millet, rye) and whole-grain products.

Fiber in the Diet

Undoubtedly, including fiber in a daily meal plan has benefits—but how much is enough? How much is too much? Even fiber has potential to cause harm if taken in excess.

Because fiber carries water out of the body, too much fiber can cause dehydration and intestinal discomfort. Also, iron is mainly absorbed early during digestion, thus, because fiber speeds the movement of foods through the digestive system, it may limit the opportunity for the absorption of iron and other nutrients. Binders in some fibers link chemically with minerals such as calcium and zinc, making them unavailable for absorption by carrying them out of the body (more about this in Chapter 7). Too much bulk from the diet could reduce the total amount of food consumed and cause deficiencies of both nutrients and energy. The malnourished, the elderly, and children, because they eat small amounts of food anyway, are especially vulnerable to these concerns.

The Dietary Guidelines recommend 14 grams of fiber per 1,000 calories consumed.[10] Major health organizations agree that about 21 to 40 grams of dietary fiber daily—depending on energy needs—is a desirable intake (see Table 3-6).[11] The diet can supply that amount if it is high in fruits, vegetables, legumes, and whole grains with limited added fats and sugars—the same recommendations made in the Dietary Guidelines.

The wholesale addition of purified fiber to foods is ill advised because it can be taken so easily to extremes. People whose fiber intake consists of only one isolated type of fiber are deprived of the benefits the other types of fiber provide. On the other hand, if you include a variety of whole grains, legumes, nuts, fruits, and vegetables in your diet, you get the various types of fiber you need, together with a package of benefits—water, minerals, vitamins, the energy nutrients, and other beneficial substances.

Added Sugars: Use Discretion

The *Dietary Guidelines for Americans* includes the following recommendation concerning intakes of **added sugars** in the diet:

- Choose and prepare foods and beverages with little added sugars.
- Reduce the incidence of dental caries by practicing good oral hygiene and consuming sugar- and starch-containing foods and beverages less frequently.

The MyPyramid food guide distinguishes between naturally occurring sugars as found in fruits and concentrated added sugars as found in soft drinks or other calorically sweetened foods. As discussed in Chapter 2, if you consistently build your diet by choosing mostly nutrient-dense foods that are low in solid fat and added sugars, you may be able to meet your nutrient needs without using your

added sugars sugars and other caloric sweeteners that are added to foods during processing or preparation. Added sugars do not include naturally occurring sugars such as those that occur in milk and fruits.

full calorie allowance. If so, you may have what is called a "discretionary calorie allowance" for use in meeting your calorie needs (see Figure 2-6 on page 49).

However, most discretionary calorie allowances are very small, between 100 and 300 calories, especially for those who are not physically active. For many people, the discretionary calorie allowance is used by the foods they choose in each food group, such as higher-fat meats, higher-fat cheeses, whole milk, or higher-fat or sweetened snack and bakery products (such as biscuits, brownies, cakes, cheese curls, chow mein noodles, cookies, corn bread, corn chips, croissants, croutons, cupcakes, doughnuts, fried rice, granola, pastry, pizza crust, commercially popped popcorn, presweetened cereals, stuffing, toaster pastry, fried tortillas, and waffles).

The World Health Organization recommends that we limit our intake of added sugars to 10 percent of calories.[12] The Dietary Reference Intakes (DRI) Committee recommends that we choose most often the naturally occurring sugars present in nutrient-rich dairy products and fruits to minimize our intake of added sugars.[13] For people who otherwise meet their nutrient needs, maintain a healthy body weight, and still need additional calories to meet energy needs (such as athletes in training), the DRI Committee sets the *maximum* intake for added sugars at 25 percent of total calories or less because diets high in added sugars tend to displace intakes of vitamin A, calcium, iron, zinc, and other essential vitamins and minerals. For most people, 3 to 12 teaspoons per day are suggested as part of one's available discretionary calories. Table 3-7 shows the amounts of sugar in some common products.

Small amounts of added sugars can be included as part of a nutrient-dense diet and be within your MyPyramid calorie allowance:
- 3 tsp for 1,600 calories
- 5 tsp for 1,800 calories
- 8 tsp for 2,000 calories
- 9 tsp for 2,200 calories
- 12 tsp for 2,400 calories

▲ *A sampling of foods providing added sugars to the diet.*

TABLE 3-7
Sugar in Selected Foods

Food	Teaspoons of Sugar per Serving
Fruit drink, ade (12 oz)	12
Chocolate shake (10 oz)	9
Chocolate (2 oz)	8
Cola (12 oz)	8
Jellybeans (10)	7
Yogurt, fruit flavored (1 c)	7
Apple pie (⅙ of pie)	6
Cake, frosted (1/16 of cake)	6
Angel food cake (1/12 of cake)	5
Applesauce, sweetened (½ c)	5
Fig bars (1)	5
Fudge (1 oz)	5
Sherbet (½ c)	5
Cereal, Sugar Pops (¾ c)	4
Shredded Wheat (¾ c)	0
Doughnuts, glazed (1)	4
Fruit, canned in heavy syrup (½ c)	4
Gelatin dessert (½ c)	4
Chocolate milk, 2% (1 c)	3
Corn, canned (½ c)	3
Ice cream, ice milk, or frozen yogurt (½ c)	3
Syrup or honey (1 tbsp)	3
Dairy creamer (1 tbsp)	2
Doughnuts, plain (1)	2
Catsup (1 tbsp)	1
Chewing gum (2 sticks)	1
Cookie, Oreo type (1)	1
Sugar, jam, or jelly (1 tsp)	1

Eat Well Be Well Whole Grains for Health

Fruits, vegetables, and foods made from whole grains should form the foundation of a nutritious diet. Why the emphasis on whole-grain foods? Vitamins, minerals, fiber, and other protective substances including hundreds of phytochemicals contribute to the health benefits of whole-grain foods. Refined grains are low in fiber and in the protective substances that accompany the fiber in whole grains. A diet rich in whole grains may help protect you against many chronic diseases including diabetes, heart disease, and certain cancers, and it can also be an effective tool for managing your weight.

Evidence suggests that people consuming diets rich in whole-grain foods, including fiber-rich cereals, have improved insulin sensitivity and are less likely to develop **metabolic syndrome,** a clustering of risk factors associated with the development of type 2 diabetes and heart disease.[14] These risk factors include excess body weight (especially around the abdomen), insulin resistance (reduced ability of the hormone insulin to regulate blood glucose), high levels of fat particles in the blood (known as triglycerides), low-levels of the body's good form of cholesterol (known as HDL-cholesterol), and increased blood glucose levels. Recent data show that an estimated 24 percent of adults in the United States have metabolic syndrome. Metabolic syndrome can also be found among children, especially in those who are overweight.[15]

How can you incorporate whole grains into your diet to achieve the maximum health benefits and protection from chronic diseases? Very easily! Consider the following three tips:

1 Count to Three.

Consume three or more ounce-equivalents of whole-grain products per day, with the rest of the recommended grains coming from enriched or whole-grain products.* Go to www.MyPyramid.gov to determine your own recommended intake of whole grains. In general, at least half the grains you eat should come from whole grains.

2 Keep It Varied.

Whole grains differ from refined grains in the amount of fiber and nutrients they provide, and different whole-grain foods differ in the amount and type of fiber and nutrients they contain. Therefore, choose a variety of whole grains. Consume whole-grain breads and cereals (especially oatmeal), brown rice, and whole-grain pastas. Grind flaxseed in a coffee grinder and add it to your cereal. Experiment with new or unfamiliar grains like quinoa, bulgur (cracked wheat), barley, buckwheat, and amaranth.

3 Check the Label.

Choose whole grains and whole-grain foods over highly processed grains and cereals by doing some investigative shopping in the grocery store. Compare products by examining the ingredients list and Nutrition Facts panel (refer to Figures 2-7 and 2-8 in Chapter 2). Buy products that list a whole-grain or whole-wheat or other whole-grain flour *first* on the label's ingredient list. Look for brown rice, oatmeal, whole oats, bulgur, popcorn, whole rye, graham flour, pearl barley, whole wheat, whole-grain corn. If the label states "wheat flour," "enriched flour," or "degerminated cornmeal," the product is not whole grain. Believe it or not, food products labeled with the words *multigrain, stone-ground, 100% wheat, seven-grain,* or *bran* are usually *not* whole-grain products! Read the ingredient list to see if it is a whole grain.

Check the Daily Value listed in the Nutrition Facts panel to find out if a food is high or low in fiber. The "% Daily Value" for fiber provides a good clue about the amount of whole grain in the product. A percent Daily Value (% DV) of 5 percent for fiber is low, whereas 20 percent or more is high. Use the Nutrition Facts panel to help you choose products with a higher % Daily Value.

The level of whole grains in ready-to-eat cereal also depends on the sugar content. The more sugar, the less grain and fiber. For example, an ounce of unsweetened oat cereal may contain 3 grams of fiber, but the same size serving of its sweetened counterpart may contain 13 grams of sugar, and only 1 gram of fiber. Read the food label's ingredient list, looking for terms that indicate added sugars (sucrose, high-fructose corn syrup, honey, and molasses) that add extra calories. Choose grain products with fewer added sugars, fats, or oils. The Savvy Diner feature that follows presents additional pointers for easily incorporating whole grains into your diet.

*One serving of whole grains from the Grains Group equals the following: 1 slice whole-grain bread; ½ whole-grain English muffin, bun, or bagel; 1 ounce whole-grain dinner roll; 1 cup (1 ounce) ready-to-eat whole-grain cereal; ½ cup cooked cereal (oatmeal); ½ cup cooked brown rice or whole-wheat pasta; 5–6 whole-grain crackers; 3 cups popped popcorn.

metabolic syndrome a cluster of interrelated symptoms, including obesity, high blood pressure, abnormal blood lipids, and insulin resistance; highly associated with development of type 2 diabetes and heart disease.

Currently, adults in the United States consume an average of 21 teaspoons of sugar a day, or about 16 percent of total calories from sugar.[16] Added sugar intakes are highest in young adults, particularly men aged 19 to 30, with some individuals having added sugar intakes as high as 55 teaspoons of sugar a day. The major sources of added sugars in the American diet include (in decreasing order): regu-

The Savvy Diner Make Half Your Grains Whole

The message is simple: *Make at least half of your grain selections* ***whole*** *grains.*[17] Many consumers find this simple bit of nutritional advice challenging, but a few pointers can help you put this advice into practice. The following tips can get you started on building a healthy foundation for your diet by selecting a variety of whole grains and whole-grain products daily.

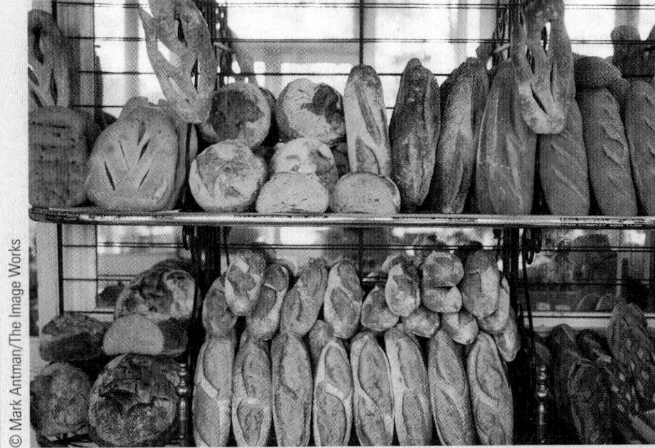

© Mark Antman/The Image Works

▲ *Enjoy the hearty flavor of whole grains.*

- Start your day with a high-fiber selection—a warm bowl of oatmeal with fresh fruit, bran cereals or shredded wheat with sliced fruit, low-fat apple bran muffin, buckwheat pancakes, or whole-grain English muffin.
- Whole grains are naturally low in fat and added sugars. Keep them that way by going easy on the fats and sugars you add as spreads, seasonings, or toppings—such as butter, margarine, oil, gravy, sauces, sugar, jams, and jellies.
- Whenever possible, substitute whole-grain flour for at least one-fourth of the all-purpose flour called for in baking.
- Make a fiber-rich snack mix from ready-to-eat whole-grain cereals, fat-free popcorn, and nuts, or enjoy air-popped or light microwaved popcorn.
- Try whole-wheat pasta or brown rice, and order sandwiches on whole-grain breads or whole-wheat tortillas for a change.
- Combine whole grains with other tasty, nutritious foods in mixed dishes. Try brown rice stuffing (cooked brown rice, onion, celery, and seasonings) in baked green peppers or tomatoes; pearl barley in vegetable soup; bulgur in vegetarian chili or salads.

lar soft drinks; sugars and candy; cakes, cookies, pies; fruit drinks; dairy products (such as ice cream, sweetened yogurt, and sweetened milk); and sweetened grains (such as waffles and cinnamon toast).[18] A later section of this chapter offers pointers on using discretion in "Keeping Sweetness in the Diet."

How the Body Handles Carbohydrates

Just as glucose is the original unit from which the many different carbohydrate foods are made, so is glucose the basic carbohydrate unit that each cell of the body uses for energy. Cells cannot use lactose, sucrose, or starch—they require glucose. The task of the **digestive system,** then, is to disassemble the double sugars and starch into single sugars so these monosaccharides can be absorbed into the blood. Following absorption, the liver converts to glucose any carbohydrates that were not absorbed in the form of glucose so they can be used by the cells. The cells can either store the glucose, use it for current energy needs, or convert it to fat.

The first digestive enzymes to work on starch are those in the saliva; they begin taking the starch apart, and the enzymes in the stomach and intestines continue digestive action (see Figure 3-3). The enzymes release the individual glucose units, which are absorbed across the intestinal wall into the blood (see Appendix A). Cooking facilitates the digestive process by spreading out the tightly packaged chains of glucose so that during **digestion** the digestive enzymes can break the chains down into glucose units for **absorption.** One to four hours after a meal, all the starch has been digested and absorbed and is circulating to the cells as glucose.

digestive system the body system composed of organs and glands associated with the ingestion and processing of food for absorption of nutrients into the body.

digestion the process by which foods are broken down into smaller absorbable products.

absorption the passage of nutrients or substances into cells or tissues; nutrients pass into intestinal cells after digestion and then into the circulatory system (for example, into the bloodstream).

FIGURE 3-3
A SUMMARY OF CARBOHYDRATE DIGESTION AND ABSORPTION

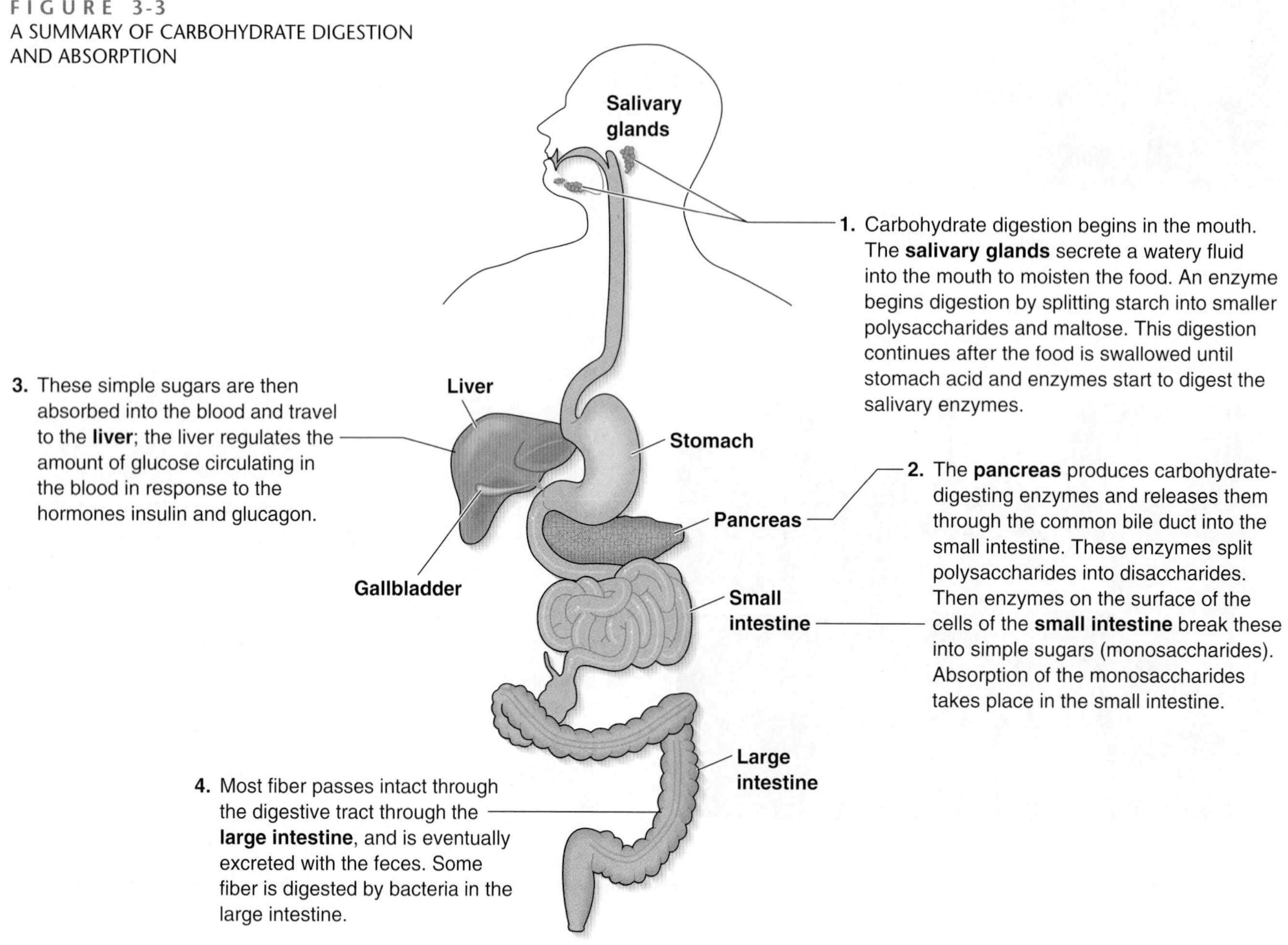

Salivary glands
1. Carbohydrate digestion begins in the mouth. The **salivary glands** secrete a watery fluid into the mouth to moisten the food. An enzyme begins digestion by splitting starch into smaller polysaccharides and maltose. This digestion continues after the food is swallowed until stomach acid and enzymes start to digest the salivary enzymes.
Liver
3. These simple sugars are then absorbed into the blood and travel to the **liver**; the liver regulates the amount of glucose circulating in the blood in response to the hormones insulin and glucagon.
Stomach
2. The **pancreas** produces carbohydrate-digesting enzymes and releases them through the common bile duct into the small intestine. These enzymes split polysaccharides into disaccharides. Then enzymes on the surface of the cells of the **small intestine** break these into simple sugars (monosaccharides). Absorption of the monosaccharides takes place in the small intestine.
Pancreas
Gallbladder
Small intestine
Large intestine
4. Most fiber passes intact through the digestive tract through the **large intestine**, and is eventually excreted with the feces. Some fiber is digested by bacteria in the large intestine.

To review the structure and function of the many body organs discussed in this section, turn to Appendix A.

If the blood delivers more glucose than the cells need, the liver and muscles take up the surplus to build the polysaccharide **glycogen.** The muscles hold two-thirds of the body's total store of this carbohydrate and use it during exercise. (Chapter 10 explores the relationship between glycogen and exercise and offers tips on how to make the most of glycogen stores.) The liver stores the remaining one-third of the body's glycogen, making it available as needed to help maintain the blood glucose level.

After the glycogen stores are full and the cells' immediate energy needs are met, the body takes a third path for using carbohydrates. For example, perhaps you have eaten a dinner that includes enough carbohydrates to fill your glycogen stores. Now, you watch a movie, eat popcorn, and drink a cola. Your digestive tract delivers glucose from the popcorn and soda to your liver; your liver breaks these extra energy compounds into small fragments and puts them together into the more permanent energy-storage compound—fat. The fat is then released from the liver, carried to the fatty tissues of the body, and deposited there. (Fat cells, too, can utilize excess glucose to make fat for storage.) Unlike the liver cells, however, which can store only about half a day's glucose needs as glycogen, the fat cells can store unlimited quantities of fat.

glycogen (GLY-co-gen) a polysaccharide composed of chains of glucose, manufactured in the body and stored in liver and muscle. As a storage form of glucose, liver glycogen can be broken down by the liver to maintain a constant blood glucose level when carbohydrate intake is inadequate.

Maintaining the Blood Glucose Level

Maintaining a normal blood glucose level depends on two safeguards, as shown in Figure 3-4. When it gets too high (for example, immediately after a meal), the blood glucose level can be corrected by siphoning off the excess glucose and con-

FIGURE 3-4
BLOOD GLUCOSE REGULATION

A. When a person eats, blood glucose rises. High blood glucose stimulates the pancreas to release insulin. Insulin serves as a key for entrance of blood glucose into cells. Liver and muscle cells store the glucose as glycogen. Excess glucose can also be stored as fat.

B. Later, when blood glucose is low, the pancreas releases glucagon, which serves as the key for the liver to break down stored glycogen into glucose and release it into the blood to raise blood glucose levels.

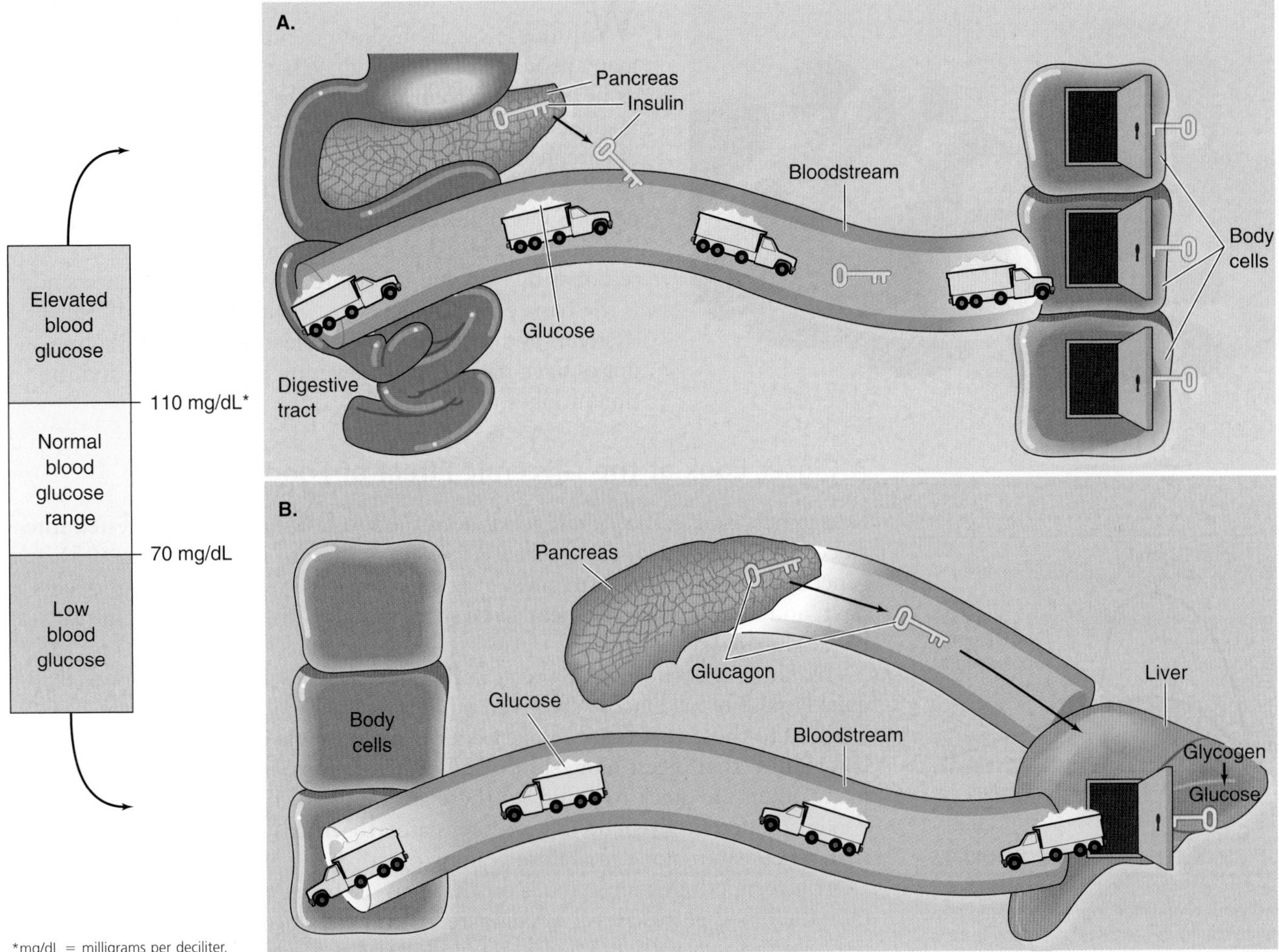

*mg/dL = milligrams per deciliter.

verting it into liver glycogen, muscle glycogen, or body fat. When it gets too low (for example, following an overnight fast), it can be replenished by drawing on liver glycogen stores.

When the blood glucose level rises, the body adjusts by storing the excess. The first organ to detect the excess glucose is the pancreas, which releases the hormone **insulin** in response. Most of the body's cells respond to insulin by taking up glucose from the blood to make glycogen or fat. Thus, the blood glucose level is quickly brought back down to normal as the body stores the excess. Insulin's opposing hormone, released by the pancreas when blood glucose is too low, is **glucagon,** which draws forth glucose from storage, making it available to supply energy. Insulin and glucagon both work to maintain the concentration of glucose in the blood within the normal range—neither too high nor too low.

Obviously, when the blood glucose level falls and stores are depleted, a meal or a snack can replenish the supply. An appropriate choice is to eat a balanced meal containing foods that offer carbohydrates (including fiber), protein, and fat for the following reasons.

insulin a hormone secreted by the pancreas in response to high blood glucose levels; it assists cells in drawing glucose from the blood.

glucagon (GLUE-cuh-gon) a hormone released by the pancreas that signals the liver to release glucose into the bloodstream.

Nutrition Action Carbohydrates—Friend or Foe?

© Buntrock/StockFood Munich/Stockfood America

With the low-carbohydrate trend now on its way out, many people are wondering "What are they going to tell us next about carbohydrates?" The answer to that question might lie in a decade-old theory, called the **glycemic effect** of carbohydrates, which is making its way back into the spotlight. The glycemic effect is the effect of food on a person's blood glucose and insulin response—how fast and how high the blood glucose rises and how quickly the body responds by bringing it back to normal. This phenomenon is the basis of many diets like "South Beach" and "Nutri System," and it relates to the effect particular carbohydrates have on raising or lowering blood glucose and the time it takes to do so.

A Closer Look at the Glycemic Effect of Foods

Researchers rank carbohydrate foods according to how fast each is digested into glucose and how much the food causes blood glucose to rise (see Figure 3-5). In general, legumes produce the most even blood glucose response, dairy products next, and fruits and cereals next. Pure sugar produces the greatest rise in the blood glucose level.[19]

By rating this effect with a scale called the **glycemic index (GI),** foods that quickly raise blood glucose are given a higher score, and lower scores are assigned to those that don't. Some believe this can help people pick "good carbohydrates" that digest more slowly and therefore increase **satiety.** Nevertheless many others are quick to point out there are well-known flaws in this system. Foods of different glycemic index scores eaten together behave much differently than when eaten alone, therefore voiding practical use of these scores.

A glycemic index ranks foods on the basis of the extent to which the foods raise the blood glucose level as compared with pure glucose. The blood glucose response to foods with a *high* glycemic index (such as white bread, cornflakes, waffles, and jelly beans) is fast and high. As shown in the margin on page 91, foods with a *low* glycemic index *tend* to be wholesome fiber-rich foods (such as fruits, legumes, whole oats, bran cereals) and dairy products. Because these foods are more slowly digested, they release glucose gradually into the blood.

FIGURE 3-5
GLYCEMIC EFFECT OF CARBOHYDRATES
Researchers rank foods according to how fast each is digested into glucose and how much the food causes blood glucose to rise following ingestion. In general, legumes produce the most even blood glucose response, whereas pure sugar produces the greatest rise in the blood glucose level.

- The carbohydrates in the meal provide a quick source of glucose.
- The protein in the meal stimulates glucagon secretion, which opposes insulin and prevents it from storing glucose too quickly.
- The soluble fibers and fat in the meal slow down digestion so that a steady stream of glucose is received rather than a sudden flood.[20]

By these standards, a bowl of whole-grain cereal and fresh fruit with low-fat milk for breakfast (offering complex carbohydrate, protein, and fat) is a good choice. Eating well-spaced, carefully chosen meals that provide the balance of protein, carbohydrates, and fat recommended in the *Dietary Guidelines* can prevent rapid rises and falls in the blood glucose level.

glycemic effect (also called *glycemic response*) the effect of food on a person's blood glucose and insulin response—how fast and how high the blood glucose rises and how quickly the body responds by bringing it back to normal.

glycemic index (GI) a ranking of foods based on their potential to raise blood glucose levels.

satiety (sah-TIE-eh-tee) the feeling of fullness and satisfaction that occurs after a meal.

The effect of a food on the blood glucose level is important to people with abnormalities of blood glucose regulation, notably diabetes and hypoglycemia. Rapid swings in blood glucose can also affect the performance of an athlete during an endurance event or an office employee following lunch.[21] However, a problem with using the glycemic index as a tool for food selection is that the actual glycemic index of a food can vary from person to person, as well as from meal to meal—depending on the overall composition of the meal. Due to the high variability of the index, it may take some experimenting on an individual's part to learn what combinations of foods yield the most satisfying effects.

However, the theory of so-called "good carbohydrates" is something that is agreed upon by many nutrition professionals. These "good carbohydrates" are usually thought of as containing little added sugar and having adequate amounts of complex carbohydrates and fiber—which often create a better GI score. In fact, the new dietary guidelines heavily reference this belief and recommend increasing intakes of fruits, vegetables, and whole grains.

Here is a comparison of glycemic index, carbohydrate, fiber, fat, and calorie content of two similar meals.

Meal 1	Meal 2
White bread	Wheat bread
Carrots	Black-eyed peas
Baked white potato	Sweet potato
Soda	Skim milk
Chicken breast	Chicken breast
Watermelon	Peaches
Total: (GI) 492, (fat) 4g, (fiber) 9g, (carb.) 77g, (calories) 620	**Total: (GI) 341, (fat) 3g, (fiber) 11g, (carb.) 84g, (calories) 663**

In this case, the lower glycemic index of the second meal does not necessarily mean the meal will be lower in calories and/or higher in fiber. This illustrates why use of the GI as a "dieting" tool might be up for debate.

Overall, people are wise to choose diets rich in fruits, vegetables, legumes, whole grains, low-fat dairy, nuts, fish, and lean meats. Quite often, such a diet will have a low glycemic index, plus an array of other healthful attributes. Here are some nutrition "do's" that ensure a healthy diet:

- Choose foods minimally processed and with fiber intact.
- Forgo items with sugar added.
- Look for "whole grain" as the main ingredient.

Whether it is the glycemic index or any other way of dealing with carbohydrates, these principles will not steer you and your diet wrong no matter what the current fad happens to be!

High Glycemic Index Foods

French, white, and other soft-textured breads or bagels
rice (medium-grain white or brown)
certain cereals (Cheerios, Corn Flakes, Rice Krispies)
waffles
mashed potatoes
watermelon
honey, regular soft drinks, jelly beans
pretzels

Intermediate Glycemic Index Foods

Cream of Wheat, instant oatmeal, Shredded Wheat
sourdough and rye breads
banana, pineapple, orange juice
ice cream
popcorn
raisins

Low Glycemic Index Foods

whole-grain, heavy-textured breads
long-grain brown or white rice
bran cereals, toasted Muesli cereal, whole oats
apples, oranges, peaches
baked beans, lentils, other legumes
carrots
milk, yogurt
sweet potatoes
tomato soup

Sample Ingredients List for a Whole-grain Food

Ingredients: Whole-wheat flour, water, wheat gluten, soybean and/or canola oil, yeast, salt, honey

Hypoglycemia and Diabetes

In some people, conditions develop in which the body is unable to maintain normal blood glucose levels without careful medical and dietary interventions. The following discussions help you appreciate the difficulties of the person with **hypoglycemia** whose insulin response is excessive or the person with **diabetes** whose insulin response is slow or ineffective.

Hypoglycemia

Suppose that your blood glucose falls, your glycogen reserves are exhausted, but you do *not* eat. Gradually, your body will shift into a fasting state, breaking down

hypoglycemia (HIGH-po-gligh-SEEM-eeuh) an abnormally low blood glucose concentration—below 60 to 70 mg/100 ml.

diabetes (dye-uh-BEET-eez) a disorder (technically termed *diabetes mellitus*) characterized by insufficiency or relative ineffectiveness of insulin, which renders a person unable to regulate the blood glucose level normally.

ketosis abnormal amounts of ketone bodies in the blood and urine; ketone bodies are produced from the incomplete breakdown of fat when glucose is unavailable for the brain and nerve cells.

hyperglycemia an abnormally high blood glucose concentration, often a symptom of diabetes.

its muscle to provide amino acids to the liver. In this state, the liver converts some of these amino acids into glucose to fuel the brain, and the fat released from the muscle cells is used to fuel other cells. (Chapter 9 describes this state, **ketosis,** in more detail.) Most times when this happens, the transition is smooth and not noticeable. But at other times, when your blood glucose level falls rapidly or below what is normal for you, you may experience symptoms of glucose deprivation to the brain—irritability, weakness, and dizziness. Your muscles may become weak, shaky, and trembling, and your heart may race in an attempt to speed more fuel to your brain. These symptoms disappear once calories are consumed. While *true* hypoglycemia is rare, the symptoms of low blood sugar may occur in people who are not eating often enough to keep their blood sugar levels from dropping. The treatment consists of eating balanced meals and learning to eat more frequently throughout the day (see the sample menu on page 94).

People who suspect they might have *true* hypoglycemia should consult a physician for diagnosis and treatment. It is a serious condition in which abnormal amounts of insulin are secreted, perhaps because of a pancreatic tumor or other health problem. As a result, the person's blood glucose is constantly too low. This type of hypoglycemia causes symptoms such as headache, confusion, fatigue, nervousness, sweating, or unconsciousness. People with hypoglycemia are told to avoid alcohol and snacks high in simple sugars. They usually benefit from eating six small, evenly spaced meals a day that include a variety of complex carbohydrates and lean protein foods. These foods will increase blood sugar levels slowly and more gradually than highly refined carbohydrates.

Diabetes

Because you know how the blood glucose level is maintained, you can appreciate the problem of people with diabetes whose insulin response is slow or ineffective. Most adults with diabetes fall into this category (see Table 3-8).[22] Even if the blood glucose level rises too high **(hyperglycemia),** glucose still fails to get into cells and the blood glucose level stays too high for an abnormally long time. The kidneys may respond by shifting some glucose into the urine so that it can be excreted. Short-term effects of hyperglycemia may include thirst, frequent urination, weakness, lack of ability to concentrate, hunger, and blurred vision. People with diabetes must be careful to eat regularly scheduled, balanced meals—providing a constant, steady, moderate flow of glucose to the bloodstream—so the body's insulin response can keep up with the need to process the glucose.

See Chapter 11 for a discussion of *gestational diabetes*—a form of diabetes seen only in pregnancy.

Research indicates that many people with diabetes actually do best on a diet that is high in complex carbohydrate-rich foods—as high as is recommended for

TABLE 3-8
Distinctions Between the Two Major Types of Diabetes

	Type 1 Diabetes	Type 2 Diabetes
Incidence	5%–10% of cases	90%–95% of cases
Age of Onset	<20	>45
Insulin Deficiency	Yes, pancreas unable to make insulin to meet needs	In some cases there may be insufficient insulin, or cells may be unresponsive to insulin.
Risk Factors	Genetic predisposition plus environmental factor (for example, viral infection)	Genetic predisposition plus obesity (especially central-type obesity), family history
Treatment	Insulin injections, diet, and exercise	Weight loss, diet, and exercise; oral drugs or insulin sometimes needed

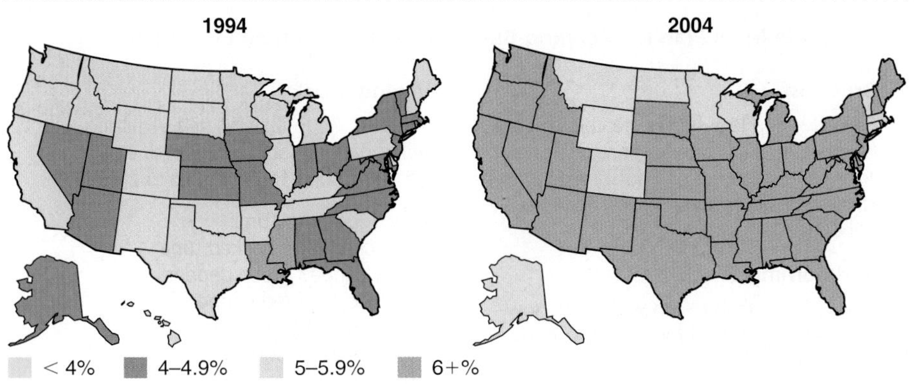

FIGURE 3-6
PREVALENCE OF DIAGNOSED DIABETES BY STATE, UNITED STATES, 1994 AND 2004

This series of maps displays the increasing prevalence (existing cases) of diagnosed diabetes among the adult populations of states between 1994 and 2004. In 1994, 14 states had a prevalence of less than 4% and only two states had a prevalence of 6% or greater. However, by 2004, no state had a prevalence of diagnosed diabetes of less than 4%, and 39 states had a prevalence of 6% or greater. In 2004, the age-adjusted prevalence of diagnosed diabetes ranged from a high of 10.8% in Puerto Rico to a low of 4.8% in Colorado.

SOURCE: Centers for Disease Control and Prevention

any healthy person. As previously mentioned, the starch and protein in these foods help to regulate the blood glucose level.

The most common type of diabetes—type 2 diabetes—is usually diagnosed in people over 45 and results from genetic factors, too little insulin, or cellular resistance to insulin.[23] As noted in Figure 3-6, the incidence of type 2 diabetes is rising, mostly because the U.S. population is aging, sedentary, and gaining excess weight.[24] Obesity is considered a major risk factor for this type of diabetes, especially when the excess fat is carried around the abdomen. Usually, eating a healthful diet, exercising, and losing weight can normalize blood glucose levels for people with type 2 diabetes. For others, oral medications and sometimes insulin shots are necessary.

Recent studies show that type 2 diabetes, which normally occurs only in adults, is now affecting an increasing number of children and adolescents. Obesity in children and teens seems to play a major role in the early development of type 2 diabetes.[25]

People with type 2 diabetes—those who make insulin but whose cells resist responding to it—tend to become obese, storing more fat than normal and being constantly hungry. This happens because the liver takes the glucose from the blood that the cells cannot take up, converts it to fat, and then ships it out to the fat cells for storage. The fat cells respond—slowly, to be sure—but ultimately they store all the fat that is sent to them. Due to the insulin resistance of body cells, insulin does not move glucose into cells effectively, and the message that energy fuels are coming in from food is delayed, causing the person with diabetes to be constantly hungry. Unfortunately, the larger the fat cells become, the more resistant they may be to insulin, thus making the diabetes worse. People with diabetes in their family are urged not to gain excess weight, for it is likely to precipitate the onset of the disease.

Persons with the less common type of diabetes—type 1—produce no (or very little) insulin and require insulin injections to control their blood sugar levels. Type 1 diabetes is more likely to be diagnosed during childhood and has been linked to a possible viral or allergic reaction in susceptible persons that causes the destruction of pancreatic cells that produce insulin. The person who does not produce any insulin is likely to experience a sudden onset of the disease. Such a person may lose weight rapidly because, without insulin, the cells cannot store glucose or fat. Thus, the two types of diabetes have opposite effects on body weight—one making the person fat and the other making the person thin. To determine your own risk for diabetes, take the test in the box on page 94.

The American Diabetes Association eased its restriction on sugar use in the diets of people with diabetes because sucrose-containing foods do not seem to exert any higher glycemic response than do some starchy foods.[26] However, according to their guidelines, any foods containing sucrose must be used in place of other carbohydrate-containing foods and should not exceed 5 to 10 percent of the total carbohydrate calories.

A Sample Meal Plan for Keeping Blood Sugar Levels on an Even Keel

Breakfast

Spread ¼ cup low-fat ricotta cheese on 2 halves of a small cinnamon raisin bagel. Top with 2 tbsp raisins and sprinkle with cinnamon. Heat under broiler for 1 to 2 minutes.

Midmorning Snack

Mix 1 sliced fresh peach with 1 cup low-fat plain yogurt and top with ¼ cup low-fat granola.

Lunch

Stuff a whole-wheat pita with lettuce, ¾ cup Zesty Vegetable Tuna (see recipe), and top with shredded lettuce.
1 cup fresh melon cubes.

Midafternoon Snack

Spread 2 tbsp of hummus over 4 slices of whole-grain melba toast.

Dinner

¾ cup Quick Kidney Bean Salad (see recipe)
3 oz grilled chicken breast
1 cup steamed broccoli florets
1 medium baked sweet potato
2 tsp butter

Evening Snack

Combine ¾ cup fat-free milk, 1 frozen peeled banana, and 1 tbsp peanut butter and blend for 1 to 2 minutes until smooth.

Zesty Vegetable Tuna
1 6-oz can water-packed tuna
½ cup chopped green pepper
½ cup chopped red pepper
½ cup shredded carrots
3 tbsp light mayonnaise
2 tbsp dijon mustard
3 tbsp balsamic vinegar
Combine last three ingredients, then add remaining ingredients.
Makes 3 servings.

Quick Kidney Bean Salad
1 16-oz can kidney beans, rinsed and drained
½ cup chopped celery
½ cup chopped onion
⅓ cup orange juice
2 tbsp cider vinegar
Mix all ingredients in a bowl and refrigerate until chilled.
Makes 4 servings.

SOURCE: Adapted from "Getting Even with Low Blood Sugar," *Health Wise* 5 (1999): 6.

What Is Your Risk for Diabetes?

1. I have given birth to a baby weighing more than nine pounds.
   Yes = 1 No = 0 ________
2. I have a sister or a brother with diabetes.
   Yes = 1 No = 0 ________
3. I have a parent with diabetes.
   Yes = 1 No = 0 ________
4. My weight is equal to or above that listed in the At-Risk Weight Chart shown at right.
   Yes = 5 No = 0 ________
5. I am under 65 years of age and I get little or no exercise during a typical day.
   Yes = 5 No = 0 ________
6. I am between 45 and 64 years of age.
   Yes = 5 No = 0 ________
7. I am 65 years or older.
   Yes = 9 No = 0 ________

Total: ________

At-Risk Weight Chart

If you weigh the same or more than the weight listed here for your height (BMI >27), you may be at risk for diabetes.

Height, without shoes (feet/inches)	Weight, without clothing (pounds)
4′10″	128
4′11″	133
5′0″	138
5′1″	143
5′2″	147
5′3″	152
5′4″	157
5′5″	162
5′6″	167
5′7″	172
5′8″	177
5′9″	183
5′10″	188
5′11″	193
6′0″	198
6′1″	204
6′2″	210
6′3″	218
6′4″	221

SCORING

3–9 points: Chances are you are at low risk for diabetes now. Maintain a healthy weight, eat nutritious low-fat meals, and exercise regularly to keep risk low.

10 or more points: You are at high risk for diabetes. See your doctor to be tested.

SOURCE: Reprinted with permission, © 2000, American Diabetes Association. For more information about diabetes, call 1-800-DIABETES.

Sugar and Health

Sugar is often in the headlines and has been accused of contributing to a host of human ills such as tooth decay, obesity, diabetes, heart disease, hyperactive behavior in children, and even criminal behavior. Nevertheless, research studies have not shown a direct link between sugar and any of these conditions, except tooth decay.[27] However, eating too many high-sugar foods could also mean that you are eating inadequate amounts of foods containing essential nutrients. On the other hand, if you eat a lot of high-sugar foods without eating less of other foods, you might be getting too many calories. Excess calories from any energy nutrient, even protein, are stored as body fat. Evidence from population studies in many countries shows that obesity rates increase as sugar consumption increases. One reason may be that many sugary foods, such as candy bars, are also high in fat.

The dietary guideline to limit sugar intake does not apply to *all* sugars in the diet. The diluted *naturally occurring* sugars found in milk and fruits should not be confused with concentrated refined sugars, such as table sugar, honey, and corn syrup. These concentrated sweets should be used in moderation, so they do not displace needed nutrients.

When people learn that fruit's energy comes from simple sugars, they may think that eating fruit is the same as consuming concentrated sweets such as candy or soft drinks. However, fruits differ from candy and soft drinks in important ways. Their sugars are diluted in large volumes of water, packaged in fiber, and mixed with many vitamins and minerals needed by the body (see Table 3-9). In contrast, concentrated sweets such as honey and table sugar are merely—as the popular phrase calls them—**empty-calorie foods.** How much sugar *do* you eat? Try the Carbohydrate Consumption Scorecard, which gives you the opportunity to check your diet for sugar and fiber and find out.

▲ *The honey and table sugar (sucrose) are* concentrated sweeteners, *each containing the simple sugars fructose and glucose. In the orange, these sugars are* naturally occurring sugars. *Compared with honey or sugar, the orange is more nutrient dense—providing vitamins, minerals, and fiber along with its sugars.*

Keeping Sweetness in the Diet

The taste of sweetness is a pleasure; the liking for it is innate. The *Dietary Guidelines for Americans* recommend that you "choose and prepare foods and beverages

empty-calorie foods a phrase used to indicate that a food supplies calories but negligible nutrients.

TABLE 3-9
Sample Nutrients in Fruits and Sugars

	Size of 100-calorie portion	Carbohydrate (g)	Fiber (g)	Vitamin A (µg)	Vitamin C (mg)	Folate (µg)	Potassium (mg)
Fruits							
Apricots	6	24	6	137	20	18	622
Apple	1 large	25	5	4	12	6	244
Cantaloupe	½ 5" melon	23	2	444	116	47	853
Orange	1¼ c sections	26	5	24	120	68	408
Pineapple	1¼ c chunks	24	2.5	2.5	30	20	219
Strawberries	2 c	20	6	4	164	50	478
Watermelon	1¼ c chunks	22	2	56	30	6	352
Sugars							
Cola beverage	8 oz	26	0	0	0	0	3
Honey	1½ tbsp	26	0	0	0	<1	22
Jelly	2 tbsp	26	0	0	<1	0	24
Sugar, white	2 tbsp	24	0	0	0	0	0
Sugar, brown	3 tbsp	24	0	0	0	0	90
Adult Dietary Reference Intake (DRI)		**≥130**	**21–38**	**700–900**	**75–90**	**400**	**4,700**

Scorecard

Carbohydrate Consumption
Rating Your Diet: How Sweet Is It?

Now that you are aware of some of the sources of added sugars, let's take a look at *your* diet. Check the box that most closely describes your eating habits to see how the foods you choose affect the amount of added sugars in your diet.

How often do you	SELDOM OR NEVER	1 OR 2 TIMES A WEEK	3 TO 5 TIMES A WEEK	ALMOST DAILY
1. Drink soft drinks, sweetened fruit drinks, or punches?	☐	☐	☐	☐
2. Choose sweet desserts and snacks, such as cakes, pies, cookies, and ice cream?	☐	☐	☐	☐
3. Use canned or frozen fruits packed in heavy syrup or add sugar to fresh fruit?	☐	☐	☐	☐
4. Eat candy?	☐	☐	☐	☐
5. Add sugar to coffee or tea?	☐	☐	☐	☐
6. Use jam, jelly, or honey on bread or rolls?	☐	☐	☐	☐

HOW DID YOU DO?

The more often you choose the items listed above, the higher your diet is likely to be in sugars. You may need to cut back on sugar-containing foods, especially those you checked as "3 to 5 times a week" or more. This does not mean eliminating these foods from your diet. You can moderate your intake of sugars by choosing foods that are high in sugar less often and by eating smaller portions.

Check Your Diet for Fiber

To reap the benefits of fiber, healthy adults need between 21 and 40 grams of fiber a day. Most Americans get about 15 grams a day. To figure how much fiber you consume in a day, write down the number of servings you eat in a typical day for each of the food categories below. Next, multiply by the factor shown. (This number represents the average amount of fiber from a serving in that food group.) Then add up the total and determine how you compare to the recommended goal.

FOOD GROUP	NUMBER OF SERVINGS	AMOUNT (g)
Vegetables (½ c cooked; 1 c raw)	______ × 2 g	= ______
Fruits (1 medium; ½ c cut; ¼ c dried)	______ × 2 g	= ______
Dried beans, lentils, split peas (½ c cooked)	______ × 8 g	= ______
Nuts, seeds (¼ c; 2 tbsp peanut butter)	______ × 2 g	= ______
Whole grains (1 slice bread; ½ c rice, pasta; ½ bun, bagel, muffin)	______ × 2 g	= ______
Refined grains[a] (1 slice bread; ½ c rice, pasta; ½ bun, bagel, muffin)	______ × 1 g	= ______
[b]Breakfast cereals (1 oz)	______ × g	= ______
	Total	= ______

[a] Refined grains refer to products made with white or wheat flour (not whole-wheat flour); white rice, or other processed grains.
[b] Check the Nutrition Facts panel on the cereal you eat to determine the number of grams of fiber in a 1-ounce serving.

SCORING

21–40 Good News! Your fiber intake meets the recommended goal.

15–20 You consume more fiber than the average American. Another serving from the fruit, vegetable, and whole-grain categories would help bring your fiber intake into the recommended goal area.

10–15 Your intake is similar to the average American intake. Consider adding a serving of legumes and additional servings of fruits, vegetables, and whole grains to your daily diet.

0–10 Your fiber intake is too low. Try getting at least five servings of fruits and vegetables, along with three servings of whole-grain foods each day. Add a high-fiber cereal to your morning routine, and consider adding legumes to some of your weekly meals.

with little added sugars or caloric sweeteners."[28] To help with this task while still catering to the sweet tooth, consider the following pointers:

- Use less of all sugars, including white sugar, brown sugar, honey, jelly, and syrups.
- Choose sensibly to limit your intake of beverages and foods that are high in added sugars, such as soft drinks, fruit drinks, candy, ice cream, cakes, cookies, and pies.
- Select fresh fruits or fruits that are canned without sugar or in fruit juice (rather than heavy syrup) to satisfy your urge for sweets.
- Learn to read the ingredients list. Check food labels for clues about added sugar content. If any of the following sugars appears first or second in the ingredients list, or if several names are listed, the food is likely to be high in sugar (see Figure 2-8 on page 59). Examples of added sugars include brown sugar, confectioner's sugar, corn syrup, dextrose, fructose, high-fructose corn syrup, fruit juice concentrate, honey, malt syrup, molasses, pancake syrup, powdered sugar, raw sugar, and white sugar.
- For dental health, remember that how frequently you eat sugar is as important as, and perhaps more important than, how much sugar you eat at one time (see the section that follows).
- Alternatives to sweet desserts might be whole-grain crackers, low-fat cheese, and yogurt. Snacks for children could include fruits, vegetables, string cheese, popcorn, homemade fruit juice pops, and other wholesome foods.
- Substitute fruit *juices* or water for fruit drinks, regular soft drinks, and punches that contain considerable amounts of sugar.
- Buy *unsweetened* cereals so that you can control the amount of sugar added. Many cereals are presweetened. Check the Nutrition Facts panel for the grams of sugar present. Many cereals list sugar first or second among their ingredients, indicating a high amount of added sugar.
- Experiment with reducing the sugar in your favorite recipes. Some recipes taste just the same even after a 50 percent reduction in sugar content.
- The sweet spices—allspice, anise, cardamom, cinnamon, ginger, and nutmeg—can replace substantial sugar in recipes. Use half as much sugar and increase one and a half times the amount of spice the recipe calls for. Increasing the amount of extracts like vanilla can enhance sweetness, too. Experiment with other extracts like maple, coconut, banana, and chocolate, or try adding dried fruit to baked goods for extra sweetness and nutrients as you decrease sugar.

▲ *Naturally sweet foods such as fruit can satisfy your sweet tooth.*

Still another alternative for reducing sugar intake is to use sugary foods that convey nutrients as well as calories. For example, rather than using table sugar as a topping, add raisins and banana slices, which are really very sweet. The person with nutrition sense and a taste for sweets can artfully combine the two by using sugar with creative imagination to enhance the flavors of nutritious foods.

You may wonder whether using artificial sweeteners to reduce some of the total sugar in your diet is a safe and recommended strategy. The Spotlight feature "Sweet Talk" at the end of this chapter will help you decide.

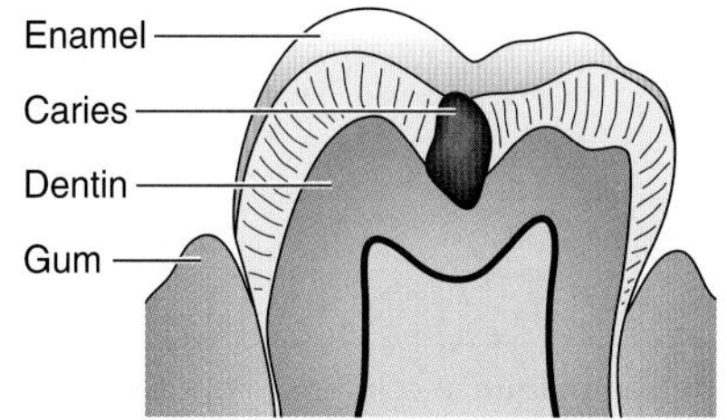

▲ *Bacteria living in the mouth feed on sugar found in foods and release an acid that can eat away at tooth enamel and result in a cavity.*

Keeping a Healthy Smile

For years, conventional wisdom has held that staying away from candy, cookies, sugary soda pop, and the like is the most important line of dietary defense against cavities. Even Aristotle asked in about 350 B.C., "Why do figs when they are sweet, produce damage to the teeth?"[29]

These days, dentists advise that cutting down on the amount of sugar eaten is not the only way to prevent **dental caries.** Every bit as important to consider (if not more) is whether a food clings to the teeth and lingers in the mouth as well as when and how often it's eaten. It has to do with the mouth's level of acid.

dental caries decay of the teeth, or cavities.
caries = rottenness

"Just pull my sweet tooth."

Each time you bite into a food, the bacteria that live in your mouth feed on the sugar in it and release an acid that eats away at tooth enamel. When a large number of bacteria living in a film referred to as **dental plaque** produce enough acid to dissolve a "hole" in the enamel over a period of time, the result is a cavity.

Although consuming sugary items such as soft drinks boosts acid production in the mouth, so does munching on starchy foods such as crackers or pretzels. This happens because enzymes in the saliva can break the carbohydrate in the cracker or pretzel into the simple sugars that bacteria feast on. If a crumb or two from a starchy food gets caught between the teeth, it might provide enough carbohydrate for bacteria to feed on for hours, thereby prolonging the teeth's exposure to acid. In fact, some research indicates that even high-sugar foods such as chocolate bars and hot fudge sundaes are less likely to contribute to cavity formation than stickier, starchier items that are likely to linger in the mouth, such as potato chips and crackers.[30]

To be sure, the carbohydrate content and stickiness of a food are only two of the many factors that influence the food's effect on teeth. Another is how often you eat the food. Each time you eat a carbohydrate-containing food, your teeth are bathed in acids for about 20 minutes. Thus, the more often you eat, the longer your teeth are exposed to harmful acids.[31]

Yet another factor to consider is the combination of foods that you eat. Starchy, sugary foods tend to be less harmful to the teeth when eaten with a meal than when consumed alone. One reason may be that the mouth makes more saliva during a full meal. Saliva production is crucial, because even though it helps make the sugar that encourages bacteria to thrive, it also clears food particles from the mouth and neutralizes destructive acids before they can dissolve the teeth. One study at the University of Iowa's College of Dentistry studied people who nibbled on foods that have a strong tendency to produce acid, such as raisins and chocolate bars, and then chewed sugarless gum. They found that within 10 minutes the gum helped stimulate the release of enough saliva to neutralize the acid flow that the sweets had caused.[33] In addition, some research suggests that, like sugarless chewing gum, foods such as cheese and peanuts help fight acid attacks that are stimulated by eating carbohydrate-rich foods.[34]

Of course, although saliva helps fight cavities, the best way to ensure good dental health is to brush your teeth as soon as possible after eating, or at least swish the mouth with water after a meal to help rinse the teeth and dislodge stuck particles. Flossing daily is also important because flossing rids the mouth not only of food but also of plaque—before it becomes so widespread that it produces tooth-threatening levels of acid. Regular visits to the dentist also play a role in keeping dental health up to snuff. In addition, drinking water containing fluoride, a mineral that helps strengthen tooth enamel, can prevent cavities. People whose drinking water does not supply fluoride should check with a dentist about the need for fluoride supplements.

Along with practicing good dental hygiene, eating a balanced diet helps keep the mouth healthy. One reason is that if the diet lacks essential nutrients, mouth tissues can become compromised, leaving them particularly vulnerable to infection. In fact, some experts believe that **periodontal disease** is especially severe among people who have poor diets.[35] Another reason to eat an adequate diet, particularly for children, is that it helps to ensure proper tooth development.[36] Similarly, eating well should rank as a high priority for a pregnant woman, whose unborn baby's teeth, among other things, start forming after just six weeks and begin to harden between the first and second trimester of pregnancy.

Incidentally, parents should never allow an infant to sleep with a bottle filled with sweetened liquids, fruit juices, milk, or formula. Little ones allowed to do so often develop what is known as **nursing bottle syndrome.** As a baby sucks on a bottle, the tongue pushes outward slightly and covers the lower teeth. If the infant falls asleep with the bottle in the mouth, the liquid bathes the teeth, particularly the upper teeth not protected by the tongue, thereby literally soaking them in cavity-causing carbohydrates for hours at a time.

For optimal dental health, the American Dental Association (ADA) recommends the following:[32]

- Eat a balanced diet.
- Keep snacking to a minimum, if possible. The ADA recognizes that some people, such as people with diabetes, may require snacks. For others, however, the ADA suggests limiting the number of snacks if brushing the teeth is not possible shortly after eating them.
- Eat sweets with meals rather than between them.
- Brush and floss thoroughly each day to remove dental plaque.
- Use an ADA-accepted fluoride toothpaste and mouth rinse and talk to your dentist about the need for supplemental fluoride.
- Visit a dentist regularly.
- Do not allow infants to sleep with bottles in their mouth that contain sweetened liquids, fruit juices, milk, or formula.

dental plaque a colorless film, consisting of bacteria and their by-products, that is constantly forming on the teeth.

periodontal disease inflammation or degeneration of the tissues that surround and support the teeth.

nursing bottle syndrome (also called *baby bottle tooth decay*) decay of all the upper and sometimes the back lower teeth that occurs in infants given carbohydrate-containing liquids when they sleep. The syndrome can also develop in babies given bottles of liquid to carry around and sip all day.

Spotlight Sweet Talk—Alternatives to Sugar

It all started back in 1879—the year that a substance called saccharin was first discovered and found to be able to sweeten foods without adding calories. Since that time, the sale of **artificial sweeteners** for use as tabletop sweeteners, such as Sweet 'N Low and Equal, and in artificially sweetened soft drinks, yogurt, and numerous other products has soared.[37] The five artificial sweeteners approved for use in the United States today are **acesulfame-K, aspartame, neotame, saccharin,** and **sucralose.*** But despite the ever-growing popularity of such sweeteners, doubts about their safety have stirred a good deal of controversy over the years. In addition, confusion about the characteristics of the various alternatives to sugar (see Table 3-10) and their role in managing problems such as obesity, diabetes, and tooth decay prevails among the millions of Americans who use them. The following discussion about sugar substitutes will help put matters into perspective.

Is it true that small amounts of saccharin cause cancer?

This assumption has never been proven. It's true that some studies of saccharin, the sweetener in Sweet 'N Low, have found that it can cause bladder cancer in laboratory rats. A human, however, would have to drink 850 cans of soft drinks a day to take in a dose equivalent to what the rats in those studies were given. Moreover, some research suggests that although very high doses of saccharin may promote bladder cancer in rats, the sweetener does not have the same effect on mice, hamsters, monkeys, or humans.[38] In addition, investigations of large groups of people have yet to establish any clear-cut link between saccharin consumption and the risk of cancer. One study conducted by the National Cancer Institute and involving more than 9,000 men and women showed no association between saccharin use and bladder cancer.[39] Thus, it appears that the health risks, if any, posed by saccharin are minuscule at most. People who fear getting cancer would be better off quitting smoking or reducing the amount of fat in their diet than worrying about putting Sweet 'N Low in their coffee now and then. As the American Medical Association's Council on Scientific Affairs has stated, "In humans, available evidence indicates that the use of artificial sweeteners, including saccharin, is not associated with an increased risk of bladder cancer."[40] The government removed saccharin from a list of potential cancer-causing agents, and in December 2000, federal legislation was signed to remove the saccharin warning label that had been required on saccharin-sweetened foods and beverages in the United States since 1977.[41]

Does aspartame cause headaches?

Not in most people. Although the U.S. Food and Drug Administration has received numerous complaints from consumers who claim to suffer from headaches, nausea, anxiety, and other symptoms after consuming aspartame-containing foods or beverages, scientific studies have never confirmed that the sweetener is truly the culprit. Scientists at Duke University, who conducted carefully controlled research into the matter, concluded that tablets containing the amount of aspartame in about 4 liters of diet soft drinks were

© Polara Studios

▲ *Aspartame, sucralose, and other sugar substitutes are used to sweeten a wide variety of products.*

no more likely to prompt headaches in people than **placebo** pills administered for the sake of comparison.[42] Furthermore, numerous careful scientific investigations carried out both before and after aspartame first appeared on the market have indicated that the product brings about adverse health effects only in a small group of people with a rare metabolic disorder known as **phenylketonuria,** or PKU. People born with PKU must carefully control their consumption of phenylalanine (one of the two amino acids that make up aspartame) to prevent health problems as severe as mental retardation. The American Medical Association's Council on Scientific Affairs, the Centers for Disease Control and Prevention, the Food and Drug Administration, the American Dietetic Association, and numerous other health organizations all consider the sweetener safe for use except by people with PKU.[43] Table 3-11 shows how to determine the amount of aspartame in your diet.

I have heard a lot about Splenda as a sweetener. Why is it so popular all of a sudden?

Splenda (sucralose) is billed as the sweetener that is "made from sugar so it tastes like sugar." Although taste is

*The herbal sweetener *stevia* is sold in the United States as a dietary supplement, but it has not received approval from FDA for use as a food additive. Derived from the leaves of a South American perennial plant, stevia has long been used in South America as a tabletop sweetener and in beverages, candy, gums, baked goods, dairy products, and frozen desserts.

TABLE 3-10
How Sweet It Is

Name (Trade Name)	Sweeteners Compared with Sucrose	Typical Uses	Characteristics[a]
Sweeteners approved for use in the United States:			
Acesulfame-K (Sunette/Sweet One)	200 times sweeter	Tabletop sweetener, puddings, gelatins, chewing gum, candies, baked goods, desserts, diet drink mixes	Stable in high temperatures; soluble in water. A new blend of Ace-K and aspartame synergizes their sweetening power.
Aspartame[b] (Equal)	200 times sweeter	General purpose sweetener in foods, beverages, chewing gum; tabletop sweetener	Loses sweetness at high temperatures and may lose sweetness over time
Neotame	7,000 times sweeter	General purpose sweetener in beverages, dairy products, frozen desserts, baked goods, and gums	Stable at high temperatures
Saccharin (Sweet 'N Low/ Sugar Twin)	300 times sweeter	Soft drinks, tabletop sweetener, baked goods, candy	Stable at high temperatures
Sucralose (Splenda)	600 times sweeter	Soft drinks, jams, frozen desserts, dairy products, baked goods, chewing gum, salad dressings, syrups, tabletop sweetener	Stable at various temperatures

[a] In addition to the characteristics listed, all of the sweeteners have a synergistic effect when combined with other sweeteners. In other words, together they enhance each other's sweetness, yielding a combined sweetness greater than the sum of all the substances' sweetness.
[b] The NutraSweet Company held the patent on aspartame from 1981 to 1992. Equal is still the most common trade name for aspartame, but others may become common in the future. The NutraSweet Company also produces neotame, approved for use in July 2002.

SOURCE: "Position of the American Dietetic Association: Use of Nutritive and Nonnutritive Sweeteners," *Journal of the American Dietetic Association* 98 (1998): 580–587.

purely subjective, it is true Splenda is a derivative of sugar (sucrose). Splenda is the result of a multi-step process that selectively substitutes three atoms of chlorine for three hydroxyl groups (OH groups) on the sugar molecule, resulting in a sweetener 600 times sweeter than regular sugar. Splenda provides no calories. Because Splenda is stable, even when used at extremely high or low temperatures, it can be used in a variety of frozen and cooked foods that some of its competitor sweeteners cannot.

Splenda's marketing has gone well beyond simply claiming sugar-like qualities. Its manufacturer has produced "Splenda Granular" that mimics regular sugar's characteristics and quality when cooking and specifically when baking. Because a molecule of sucralose is 600 times sweeter than a molecule of sucrose, the actual amount of Splenda you would need is much smaller. Therefore, bulking agents are added to allow use of Splenda cup for cup in place of sugar when cooking. A similar product that combines half regular sugar with half "Splenda Granular" is available for baking. Although this product contains calories, the baking results are claimed to be superior, and it also can be measured cup for cup like regular sugar.

Products that use Splenda tout that the taste is superior to other artificial sweeteners, but that is for the consumer to decide. The real advantage to using Splenda is its packaging and its quality of being highly interchangeable with regular sugar.

Can the use of artificial sweeteners bring about weight loss?

If only it were that simple. Obviously, because artificially sweetened foods typically contain fewer calories than their sugar-sweetened counterparts, substituting a low-calorie alternative for a sugar-laden food can allow a person to enjoy sweet-tasting foods and still save calories. But simply *adding* foods sweetened with sugar substitutes to the diet will not do the trick. Moreover, eating artificially sweetened foods to create an excuse to splurge on a different high-calorie food defeats the purpose. A person who drinks diet soda to justify eating an ice cream sundae later is reaping little, if any, benefit. In other words, it's *how* you fit artificial sweeteners into the rest of your diet that really counts. The American

TABLE 3-11

Checking the Aspartame in Your Diet

The Food and Drug Administration has set forth an "acceptable daily intake" of 50 milligrams of aspartame per kilogram (2.2 pounds) of body weight. Most people, however, take in much less—on the order of fewer than 5 milligrams per kilogram daily. To reach the FDA's limit, consider that a 150-pound adult would have to consume about 19 12-ounce cans of diet soda or 97 packets of Equal, and a 40-pound child would have to consume four 12-ounce cans of diet soft drinks or 24 packets of Equal. To see how much aspartame you're getting in your diet, check the following numbers for the average amount of aspartame found in typical artificially sweetened foods.

Aspartame-Sweetened Food	Milligrams of Aspartame
1 packet Equal tabletop sweetener	35
12 oz diet soda	170
8 oz sugar-free fruit yogurt	125
8 oz powdered drink	100
4 oz gelatin dessert	80
4 oz pudding	25

SOURCES: "Position of the American Dietetic Association: Use of Nutritive and Nonnutritive Sweeteners," *Journal of the American Dietetic Association* 98 (1998): 584; "Approximate Aspartame Content in Food and Drug Administration–Approved Categories," The NutraSweet Company.

Dietetic Association takes the position that individuals who desire to lose weight may choose to use artificial sweeteners, but they should do so as part of a sensible weight management program that includes a nutritious diet and regular physical activity.[44] (See Chapter 9 for healthy weight-management guidelines.)

On the flip side, some consumers may be under the impression that artificial sweeteners bolster the appetite. A report that aspartame sends mixed signals to the brain and thereby increases appetite as well as food consumption spawned the notion that artificial sweeteners can interfere with weight loss.[45] That report was based on comparisons of the effects of drinking plain water and drinking aspartame- and sugar-sweetened water, however, rather than consuming actual foods or beverages that are typically sweetened with aspartame. Moreover, the researchers did not measure how much food people ate after drinking the liquids. Thus, the study did not show how consuming aspartame-sweetened foods influences appetite and eating behavior in the real world.

Since that time, about a dozen investigations into the relationship between appetite and foods and beverages sweetened with the substance have indicated that aspartame either lessens or doesn't affect feelings of hunger or the amount of food ultimately eaten.[46] So for now, at least, it appears that aspartame's impact on appetite need not concern people who are trying to lose weight.

I have heard that foods sweetened with sugar substitutes do not contribute to tooth decay. Is this true?

Not necessarily. As pointed out earlier, any carbohydrate-containing food, be it sugar sweetened or otherwise, can promote tooth decay. When you eat a sugary food, the millions of bacteria lurking on the surfaces of your teeth and between them feast on the sugar and, in the process, release an acid that eats away at tooth enamel. Once inside the mouth, carbohydrates can be devoured by cavity-causing bacteria because enzymes in the saliva break the complex carbohydrates into simple sugars—on which bacteria thrive.

Should people with diabetes eat foods sweetened only with sugar substitutes?

That depends. It is often assumed that because one of the hallmarks of diabetes is high blood sugar (glucose), sugar from foods is the major culprit behind high blood sugar and should therefore be off limits for people with diabetes. But it's not that simple. The total amount of carbohydrate, including both simple and complex, exerts the most influence on blood glucose levels. What's more, many other factors, including the amount of fat and fiber in a food, affect the body's blood glucose response to it. That's why people with diabetes do not have to limit themselves to only artificially sweetened foods. Sugar-containing items can be incorporated into a carefully designed eating plan.

Another reason that blanket statements about the use of sugar substitutes are not made for people with diabetes is that the various types of sweeteners behave differently in the body. A group of sugar substitutes known as **alternative sweeteners** (fructose, sorbitol, mannitol, and xylitol), for instance, contain calories but were formerly recommended for people with diabetes because the body generally absorbs them more slowly than it does table sugar. Most of these alternative sweeteners, however, have side effects that detract from their desirability. Some research suggests, for example, that large amounts of fructose may contribute to increases in blood cholesterol levels, making fructose a poor choice for the many people whose diabetes goes hand in hand with heart disease. By the same token, some people experience diarrhea after consuming large amounts of sorbitol or mannitol, and, according to the American Diabetes Association, those sweeteners' overall effect on blood glucose control is insignificant. Products with sorbitol and mannitol may carry the following warning on the label because high intakes increase the risk of malabsorption: "Excess consumption may have a laxative effect." Thus, the role sugar substitutes play in the overall diet is a decision that should be made individually with the help of a dietitian and a physician.[47]

Isn't it true that sugar-free chewing gum doesn't contain any calories?

Not so—some sugar-free chewing gum is sweetened with certain alternative sweeteners, such as xylitol and sorbitol, also known as **sugar alcohols.** Although sugar alcohols impart a sweet taste and supply calories (about 8 per stick), unlike sucrose, they do not promote tooth decay. Xylitol may actually inhibit the production of tooth-damaging acid by the caries-producing bacteria in the mouth and prevent them from adhering to the teeth. For this reason, the FDA authorizes use of the health claim on food labels that sugar alcohols do not promote tooth decay.

Some manufacturers sweeten chewing gum with artificial sweeteners or a combination of sweeteners (for example, both xylitol and aspartame). Chewing gum sweetened with artificial sweeteners contains negligible calories and also does not promote tooth decay.

MINIGLOSSARY OF SWEETENERS

acesulfame-K (AY-see-sul-fame) a derivative of acetoacetic acid approved for use in the United States in 1988. Because it is not metabolized by the body, acesulfame-K does not contribute calories and is excreted from the body unchanged. It is currently approved for use in more than 70 countries and is found in more than 100 international products, including chewing gum, gelatins, nondairy creamers, powdered drink mixes, and puddings.

alternative sweeteners nutritive (calorie-containing) sweeteners such as fructose, sorbitol, mannitol, and xylitol.

artificial sweeteners nonnutritive sugar replacements such as acesulfame-K, aspartame, neotame, saccharin, and sucralose.

aspartame a dipeptide (see Chapter 5) containing the amino acids aspartic acid and phenylalanine and used in the United States and Canada since 1981. Although it is digested as protein and supplies calories, it is so sweet that only small amounts, which contribute negligible calories, are needed to sweeten foods. Thus, it is classified as a nonnutritive sweetener. Aspartame is often sold under the trade name NutraSweet, and it is also blended with lactose and an anticaking agent and sold commercially as Equal.

neotame a zero-calorie, heat-stable sweetener approved in 2002 that is a derivative of the dipeptide composed of aspartic acid and phenylalanine. It is 7,000 to 13,000 times as sweet as sugar depending on how it is used. Unlike aspartame, it is not metabolized to phenylalanine, and thus no special labeling for people with PKU is required.

phenylketonuria (FEN-il-KEY-toe-NU-ree-ah) **or PKU** an inborn error of metabolism, detectable at birth, in which the body lacks the enzyme needed to convert the amino acid phenylalanine to the amino acid tyrosine. If not detected and treated, derivatives of phenylalanine accumulate in the blood and tissues, where they can cause severe damage, including mental retardation.

placebo (plah-SEE-bo) a sham treatment given to a control group; an inert, harmless "treatment" that the group's members cannot recognize as different from the real thing. Using a placebo minimizes the chance that an effect of the treatment will appear to have occurred due to the healing effect of the *belief* in the treatment (known as the *placebo effect*), rather than the treatment itself.

saccharin a zero-calorie sweetener discovered in 1879 and used in the United States since the turn of the century. A possible link to bladder cancer led to saccharin being banned as a food additive in Canada, although it is available there as a tabletop sweetener. Saccharin is the sweetening agent in Sweet 'N Low and Sugar Twin.

sucralose a nonnutritive sweetener approved in 1998 as a tabletop sweetener and for use in a variety of desserts, confections, and nonalcoholic beverages. Sucralose is a noncaloric, heat-stable sweetener derived from a chlorinated form of sugar. Although sucralose is made from sugar, the body does not recognize it as sugar, and the sucralose molecule is excreted in the urine essentially unchanged.

sugar alcohols (mannitol, sorbitol, isomalt, xylitol) can be derived from fruits or commercially produced from dextrose and are absorbed more slowly and metabolized differently than other sugars in the human body. The sugar alcohols are not readily used by ordinary mouth bacteria and therefore are associated with less cavity formation. Although the sugar alcohols are used as sugar substitutes, they do add calories (about 1.5 to 3 calories per gram) to a food product. They are found in a wide variety of chewing gums, candies, and dietetic foods. Sorbitol and mannitol can have a laxative effect in some people.

In Review

1. Carbohydrates are:
   a. the same as fat.
   b. the body's preferred source of energy.
   c. found mainly in foods of animal origin.
   d. usually not part of a balanced diet.
2. Carbohydrates are separated into two distinct categories—simple and complex. Of the simple carbohydrates, name the single sugars (monosaccharides) and show how they combine to produce double sugars (disaccharides).
3. What is the only health concern currently linked to sugar?
   a. diabetes
   b. heart disease
   c. cancer
   d. tooth decay
4. According to the World Health Organization, what maximum percentage of daily calories should come from simple sugars?
   a. 35%
   b. 25%
   c. 10%
   d. 5%
5. What are the main sources of complex carbohydrates in the diet?
   a. leafy green vegetables
   b. dairy foods
   c. sugary sweets
   d. seeds, beans, and grains
6. Briefly explain where and in what form extra glucose is stored in the body.
7. Match the following terms with their definitions.
   a. refined
   b. enriched
   c. fortified
      1. the process by which the coarse parts of food products are removed
      2. foods to which nutrients have been added
      3. a process by which the B vitamins thiamin, riboflavin, niacin, folic acid, and the mineral iron are added to refined grains and grain products at levels specified by law
8. Which of the following is *not* true about fiber?
   a. It cannot be digested by humans.
   b. It provides four calories per gram.
   c. It is found only in plant sources.
   d. It can be soluble or insoluble.
9. ______ is how fast and how high the blood glucose rises and how quickly the body responds by bringing it back to normal.
   a. Diabetic effect
   b. Ketosis
   c. Glycemic effect
   d. Glycemic timing
10. Name three differences between type 1 and type 2 diabetes.

Menu of Online Study Tools

A variety of study tools for this chapter are available at our website to deepen your understanding of chapter concepts. Go to

www.thomsonedu.com/nutrition/boyle

to find

- Practice tests
- Flashcards
- Glossary
- Web links
- Animations
- Chapter summaries, learning objectives, and crossword puzzles

Nutrition on the Web

www.thomsonedu.com/nutrition/boyle
Go to the *Personal Nutrition* site to check for the latest updates to chapter topics or to access links to related websites.

www.healthfinder.gov
Search for information on lactose intolerance, tooth decay, diabetes, and artificial sweeteners.

http://ific.org
Search for information on sugars, sweeteners, and fiber.

www.cdc.gov/health/diabetes.htm
Provides guidelines and updates of information for people with diabetes.

www.ada.org
Information about dental caries and gum disease from the American Dental Association.

www.diabetes.org
The American Diabetes Association's site with consumer information for people with diabetes.

www.nlm.nih.gov
Free access to National Library of Medicine's Medline for information searches on a variety of health-related topics.

www.wheatfoods.org
Information about the benefits of eating a high-carbohydrate diet.

www.eatright.org
The American Dietetic Association's site with position papers and resources on carbohydrate topics.

www.flaxcouncil.ca
Look for information about the benefits of fiber from flaxseed.

The Lipids: Fats and Oils

4

ASK YOURSELF . . .

Which of the following statements about nutrition are true, and which are false? For each false statement, what is true?

1. The body can store fat in virtually unlimited amounts.
2. Dietary cholesterol is found only in animal foods.
3. A person's blood level of cholesterol is a predictor of that person's risk of having a heart attack.
4. For the health of your heart, the fat you should avoid eating, most of all, is cholesterol.
5. The more monounsaturated fats you consume, the better it is for your health.
6. Fruits are essentially fat free.
7. In general, the softest margarines are the most polyunsaturated.
8. Polyunsaturated fat has the same number of calories as saturated fat.
9. All the foods you eat should contain less than 10 percent of calories from saturated fat.
10. No one is free of atherosclerosis.

Answers found on the following page.

The ultimate reality of nutrition rests with the chemistry of the food we eat and its effects on the processes of life.

R. M. Deutsch (1928–1988, nutrition author and educator)

CONTENTS

While you and your friend are visiting a health fair at a local shopping mall, you learn that your blood cholesterol level is high. Your friend, a registered dietitian, urges you to see your health care provider to request a blood test to determine the level of triglycerides and the ratio of "good" to "bad" cholesterol in your blood. As you leave the mall, you notice a bookstore display with one diet book that urges you to cut your fat intake to virtually nothing to reduce cancer risk and another book touting the benefits of a high-fat diet for weight loss. Then you stop at the window of a health-food store. Your friend points to the freshly ground peanut butter, mentioning to you that it is not hydrogenated as you stare at the bottles of antioxidant and fish oil supplements that line the wall. Later, during the evening news, you notice a television commercial featuring a margarine that promises to reduce your blood cholesterol level. "What does all of this mean?" you ask.

Actually you may know more than you think you know about **lipids,** more commonly called **fats** and **oils.** The most obvious dietary sources of fat are oil, butter, margarine, and shortening. Other food sources that provide fat to the diet are meats, nuts, mayonnaise, salad dressings, eggs, bacon, gravy, cheese, ice cream, and whole milk. You may also know that egg yolk and liver are high in cholesterol, and you probably know that the cholesterol in the body is in some way related to heart disease. However, you may be confused by the many terms related to the fat in your diet. This chapter explores the terminology of fat and describes how fats can both contribute to health and detract from it.

lipids a family of compounds that includes triglycerides (fats and oils), phospholipids (lecithin), and sterols (cholesterol).

fats lipids that are solid at normal room temperature.

oils lipids that are liquid at normal room temperature.

satiety the feeling of fullness or satisfaction that people feel after meals.

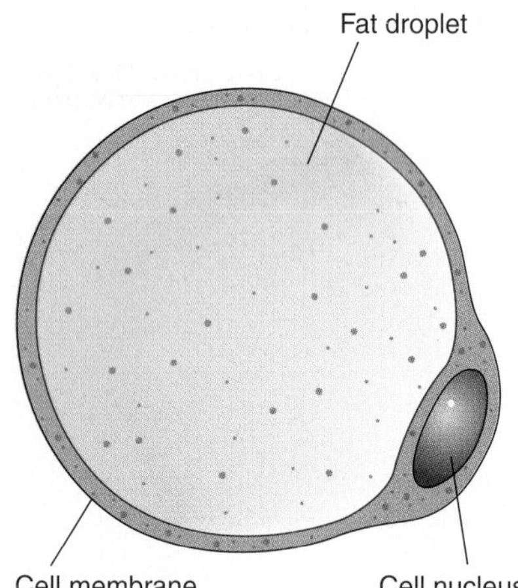

▲ A Fat Cell
Within the fat cell, lipid is stored in a droplet. This droplet can enlarge, and the cell membrane will grow to accommodate its swollen contents.

A Primer on Fats

Because most people have the impression that fat is bad for them, it may come as a surprise to learn that fats are valuable. More than valuable, some are absolutely essential, and some fats *must* be present in the diet for you to maintain good health. Even if you wanted to, it would be impossible to remove all the fat from your diet, because at least a trace of fat is found in almost all foods.

The Functions of Fats in the Body

Fat is the body's chief storage form for the energy (or calories) from food eaten in excess of immediate need. Fats provide about 60 percent of the energy needed to perform much of the body's work during rest and slightly more during extended bouts of light to moderately intense exercise.

Fat serves as an energy reserve. Whenever you eat, you store some fat, and within a few hours after a meal, you take the fat out of storage and use it for energy until the next meal. Both glucose and fat are stored after meals, and both are released later when needed as energy to fuel the cells' work. However, whereas excess carbohydrate and protein can be converted to fat, the process does not work in reverse. Fat cannot be converted back into protein and carbohydrate. Fat can serve only as an energy fuel for cells equipped to use it.

The body has scanty reserves of carbohydrate and virtually no protein to spare, but it can store fat in practically unlimited amounts. A pound of body fat is worth 3,500 calories, and a person's body can easily carry 30 to 50 pounds of fat without appearing fat at all.

Both fats and oils are found in your body, and both help to keep your body healthy. Table 4-1 summarizes the major functions of fats in the body. Fat is important to all your body's cells as a major component of cell membranes. Natural oils in the skin provide a radiant complexion; in the scalp they help to nourish the hair and make it glossy. Fat insulates the body and cushions the vital organs. It serves as a shock absorber. The fat under the skin also provides insulation from extremes of temperature, thus achieving internal climate control.

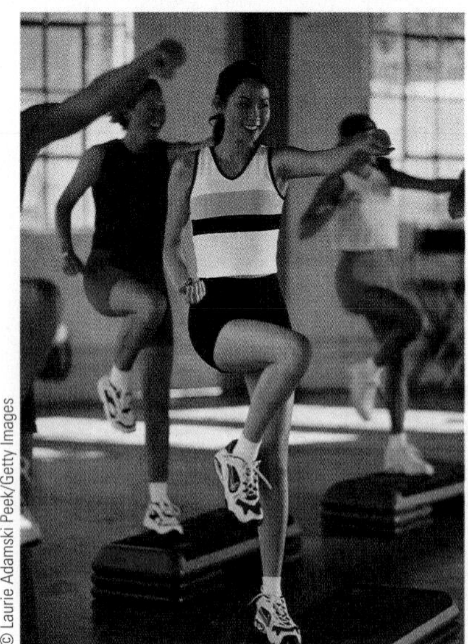

▲ *Muscles derive fuel from fat.*

The Functions of Fats in Foods

Fat is a nutrient found in many foods. As the most concentrated source of calories, fat contains more than twice as many calories, ounce for ounce, as protein or carbohydrate. High-fat foods may therefore deliver many *unneeded* calories in only a few bites to the person who is not expending much physical energy. Fats in foods also provide **satiety** by slowing the rate at which the stomach empties. This is the reason you feel fuller longer after eating meals that include fat.

Fat is important for another reason. Some essential nutrients are soluble in fat and therefore are found mainly in foods that contain it. These nutrients are the essen-

TABLE 4-1
The Functions of Fats

Fats in the Body
• Provide a concentrated source of energy • Serve as an energy reserve • Form the major components of cell membranes • Nourish skin and hair • Insulate the body from extremes of temperature • Cushion the vital organs to protect them from shock
Fats in Foods
• Provide calories • Provide satiety • Carry fat-soluble vitamins and essential fatty acids • Contribute aroma and flavor

Ask Yourself Answers: **1.** True. **2.** True. **3.** True. **4.** False. For the health of your heart, the fat you should avoid eating, most of all, is saturated fat. **5.** False. Monounsaturated fats such as olive or canola oil are good for the health of your heart, but like all fats, they should be consumed in moderation. **6.** True. **7.** True. **8.** True. **9.** False. Your *total* diet, not individual foods, should contain less than 10 percent of calories from saturated fat. **10.** True.

tial fatty acids (to be described shortly) and the fat-soluble vitamins—A, D, E, and K (described in Chapter 6). Fat also carries many dissolved compounds that give foods their aroma and flavor. This accounts for the aromatic smells associated with foods that are being fried, such as onions or French fries. It also helps explain why a plain doughnut is more flavorful than a plain roll—it is higher in fat. Table 4-1 summarizes the functions of fats in foods.

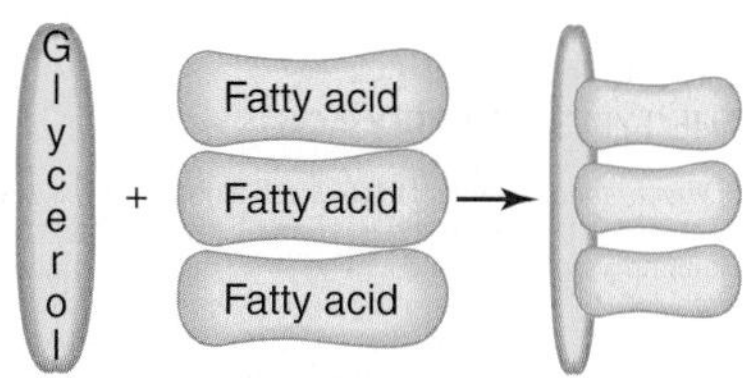

Glycerol + 3 Fatty acids → **Triglyceride**

▲ *Triglyceride*

A Closer View of Fats

When the energy from any of the energy-yielding nutrients is to be stored as fat, it is first broken into fragments—small molecules made of carbon, hydrogen, and oxygen. These fragments are then linked together into chains known as fatty acids—the major building blocks of **triglycerides,** which are the chief form of fat found in the body. Triglycerides are composed of three **fatty acids** that are attached to the compound **glycerol,** which is why glycerol is known as the "backbone" of triglycerides.

About 95 percent of the lipids in foods and in the human body are triglycerides. Other members of the lipid family are the **phospholipids** (of which **lecithin** is one) and the **sterols** (**cholesterol** is the best known of these).

Saturated versus Unsaturated Fats

Fatty acids differ from one another in two ways—in chain length and in degree of saturation. *Chain length* refers to the number of carbon atoms that are hooked together in the fatty acid.* Chain length is significant because it affects the solubility of fat in water—the short-chain fatty acids are somewhat soluble in water. Milk, butter, and cheese are rich in the short-chain fatty acids; vegetable oils and red meat contain triglycerides with long-chain fatty acids, which are insoluble in water.

More significant than chain length, the other way fatty acids differ from one another is in their degree of *saturation*. Saturation refers to the chemical structure—specifically to the number of hydrogen atoms the fatty acid chain is holding. If every available bond from the carbons is holding (or attached to) a hydrogen atom, we say the chain is a **saturated fatty acid**—filled to capacity, or saturated, with hydrogen (see Figure 4-1).

Sometimes, especially in the fatty acids in plants and fish, there is a place in the chain where hydrogens are missing and can easily be added. This is known as an "empty spot," or point of unsaturation (see Figure 4-1). A chain that possesses a point of unsaturation is an **unsaturated fatty acid.** If there is one point of unsaturation, it is a **monounsaturated fatty acid.** If there are two or more points of unsaturation, it is a **polyunsaturated fatty acid.**

The Essential Fatty Acids

The human body can synthesize all the fatty acids it needs from carbohydrate, fat, or protein, except two—**linoleic acid** and **linolenic acid.** These two cannot be made from other substances in the body or from each other, and they must be supplied by the diet; they are, therefore, **essential fatty acids.** Linoleic acid and linolenic acid are polyunsaturated fatty acids, widely distributed in the food supply, especially in plant and fish oils. Fortunately, they are readily stored in the adult body, making deficiencies unlikely. Still, deficiency symptoms can appear in the person deprived of these acids—a characteristic skin rash and, in children, poor growth.

*Short-chain fatty acids contain 4 to 6 carbons, medium-chain fatty acids contain 8 to 10 carbons, and long-chain fatty acids contain 12 or more carbons.

triglycerides (try-GLISS-er-ides) the major class of dietary lipids, including fats and oils. A triglyceride is made up of three units known as *fatty acids* and one unit called *glycerol.*

fatty acids basic units of fat composed of chains of carbon atoms with an acid group at one end and hydrogen atoms attached all along their length.

glycerol (GLISS-er-all) an organic compound that serves as the backbone for triglycerides.

phospholipid (FOSS-foh-LIP-ids) a lipid similar to a triglyceride but containing phosphorus; one of the three main classes of lipids.

lecithin (LESS-ih-thin) a phospholipid, a major constituent of cell membranes, manufactured by the liver, and also found in many foods.

sterols (STEER-alls) lipids with a structure similar to that of cholesterol; one of the three main classes of lipids.

cholesterol (koh-LESS-ter-all) one of the sterols, manufactured in the body for a variety of purposes.

saturated fatty acid a fatty acid carrying the maximum possible number of hydrogen atoms (having no points of unsaturation). Saturated fats are found in animal foods like meat, poultry, and full-fat dairy products, and in tropical oils such as palm and coconut.

unsaturated fatty acid a fatty acid with one or more points of unsaturation. Unsaturated fats are found in foods from both plant and animal sources. Unsaturated fatty acids are further divided into monounsaturated fatty acids and polyunsaturated fatty acids.

monounsaturated fatty acid a fatty acid containing one point of unsaturation, found mostly in vegetable oils such as olive, canola, and peanut.

polyunsaturated fatty acid (sometimes abbreviated PUFA) a fatty acid in which two or more points of unsaturation occur, found in nuts and vegetable oils such as safflower, sunflower, and soybean, and in fatty fish.

linoleic (lin-oh-LAY-ic) **acid, linolenic** (lin-oh-LEN-ic) **acid** polyunsaturated fatty acids, essential for human beings.

essential fatty acid a fatty acid that cannot be synthesized in the body in amounts sufficient to meet physiological need.

FIGURE 4-1
THE TYPES OF FATTY ACIDS

Omega, the last letter of the Greek alphabet (ω), is used by chemists to refer to the position of the end-most double bond in a fatty acid counting from the far end (the methyl [CH3] end). The omega-6 fatty acids have their end-most double bonds after the sixth carbon in the chain; the omega-3 acids, after the third. Linoleic acid and linolenic acid are *essential* fatty acids.

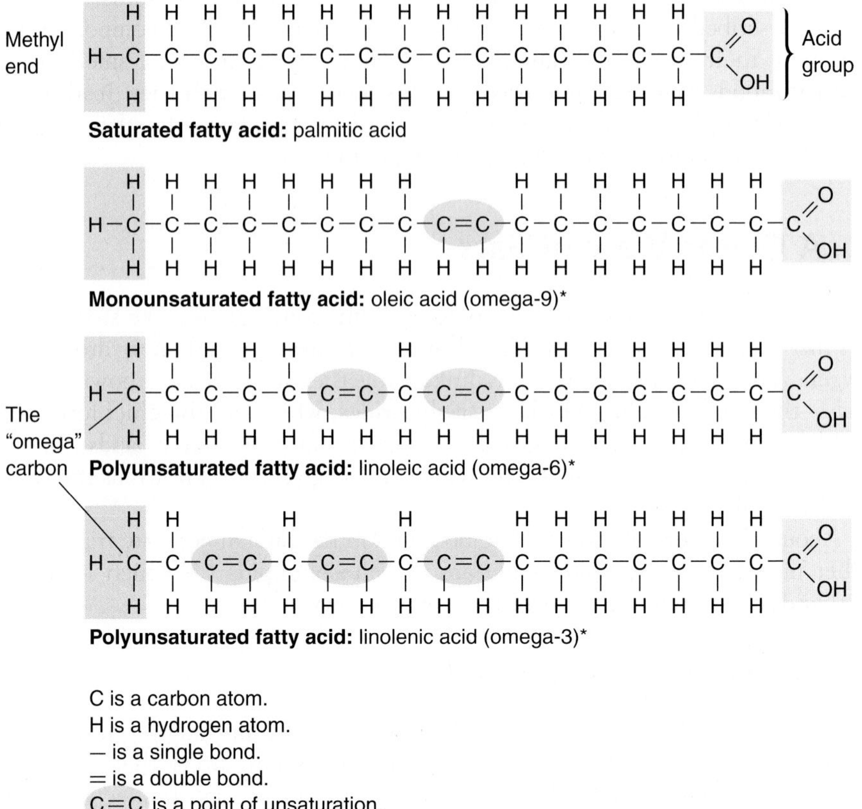

Omega-6 versus Omega-3 Fatty Acids

One further classification system for unsaturated fatty acids classifies the fatty acids as either **omega-6 fatty acids** or as **omega-3 fatty acids** (see Figure 4-1). Of the two essential fatty acids, linoleic acid is an omega-6 fatty acid, related to a whole series of others. Linolenic acid is an omega-3 fatty acid, with a similar family of its own.

Of interest in relation to dietary fat are findings about the omega-3 fatty acids found in fish oils, which offer a protective effect on health. Interest in fish oils was first kindled when someone thought to ask why the Eskimos of Greenland, who eat a diet very high in fat, have such a low rate of heart disease. The trail led to the abundance of fish they eat, then to the oils in those fish, and finally to the omega-3 fatty acids—EPA and DHA—in the oils. Now scientists are unraveling the mystery of what those fatty acids do.[1]

People who eat two to three meals of oily coldwater fish (such as salmon or bluefish) a week tend to have lower blood cholesterol and triglyceride levels and slower clot-forming rates.[2] More than 70 studies have now documented other connections as well. Diets rich in omega-3 fatty acids seem to bring about enhanced defenses against cancer (via the immune response) and reduced inflammation in arthritis and asthma sufferers.[3] The Eskimos apparently did well using fish for food. See the Nutrition Action feature on page 126 for more about the benefits of omega-3 fats and fish in the diet.

The Essential Fatty Acids
(Polyunsaturated)

Omega-6
Linoleic acid
Arachidonic acid

Omega-3
Linolenic acid
(DHA) (EPA)

▲ ***The Essential Fatty Acids***
The fatty acids in fish oils include eicosapentaenoic (EYE-kossa-PENTA-ee-NOH-ic) acid (EPA) and docosahexaenoic (DOE-cosa-HEXA-ee-NOH-ic) acid (DHA), both of which are omega-3 fatty acids.

Characteristics of Fats in Foods

The amount of unsaturated fatty acids in a fat affects the temperature at which the fat melts. The more unsaturated a fat, the more liquid it is at room temperature. In contrast, the more saturated a fat (the more hydrogen it has), the firmer it is. Thus, of three fats—beef fat, chicken fat, and corn oil—beef fat is the most

omega-6 fatty acids polyunsaturated fatty acids that have their end-most double bonds after the sixth carbon in the chain.

omega-3 fatty acids polyunsaturated fatty acids that have their end-most double bonds after the third carbon in the chain.

FIGURE 4-2
A COMPARISON OF SATURATED AND UNSATURATED FATTY ACIDS IN DIETARY FATS AND OILS

All fats are a mixture of saturated, monounsaturated, and polyunsaturated fatty acids (though we usually call them by the name of the fatty acid they have the most of). Look for the least saturated (red), and a good mixture of everything else (for example, canola oil). Polys (yellow and green) and monos (blue) lower cholesterol if you eat them in place of saturated fats. Alpha-linolenic acid (green) is an omega-3 polyunsaturated fat that may protect the heart. Canola, soy, and flaxseed oil are good sources.

SOURCE: USDA Nutrient Database for Standard Reference (Release 14), the National Sunflower Association, and the Flax Council of Canada. Reprinted with permission from *Nutrition Action Healthletter* (July/August 2002), p. 7.

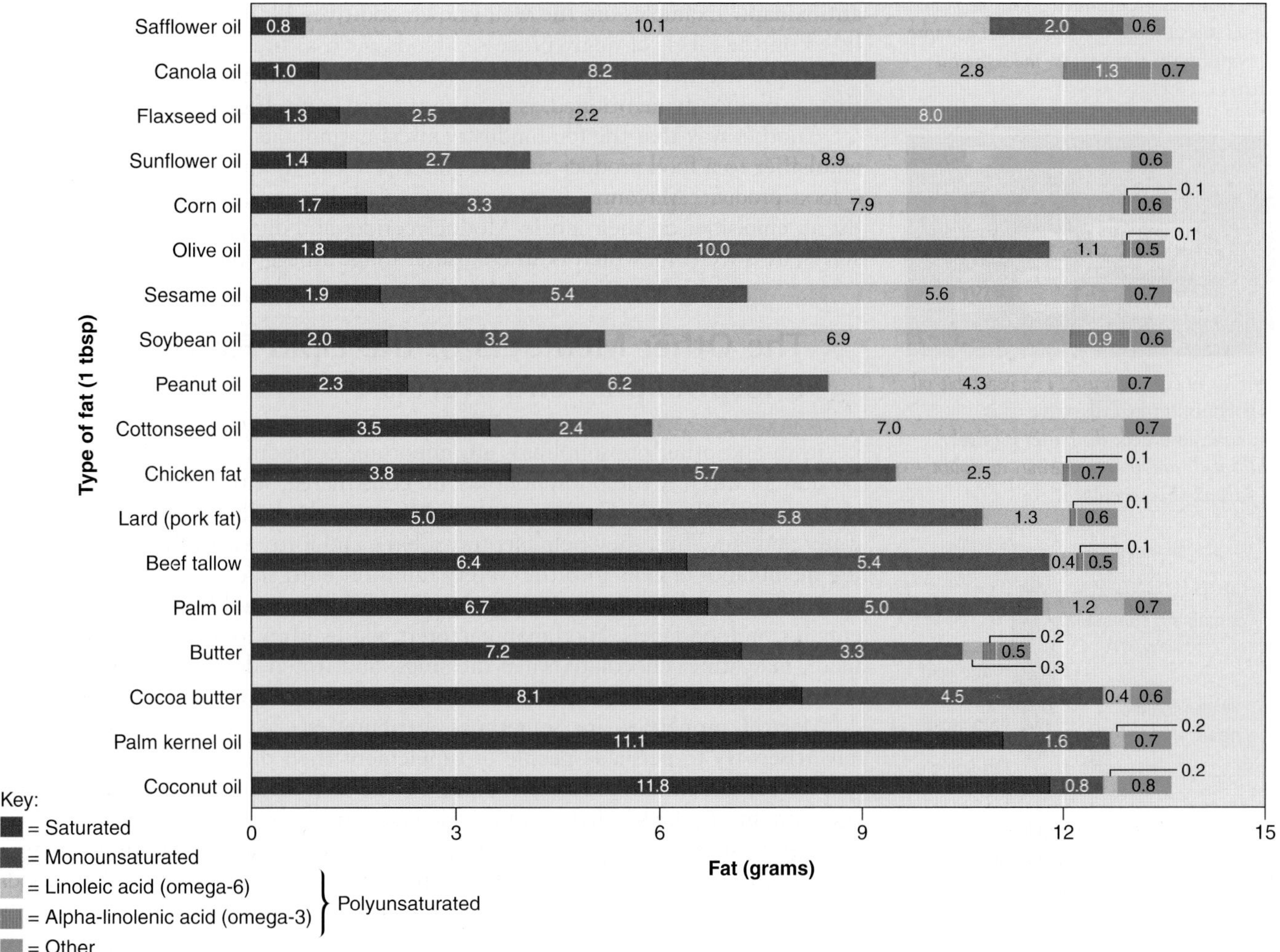

saturated and the hardest; chicken fat is less saturated and somewhat soft; and corn oil, which is the most unsaturated, is a liquid at room temperature. If your health care provider tells you to use monounsaturated fats or polyunsaturated fats, you can usually judge which ones to choose by their hardness at room temperature. Fats and oils contain mixtures of saturated, monounsaturated, and polyunsaturated fatty acids; the fatty acids that predominate determine whether the fat is solid or liquid at room temperature. Figure 4-2 compares the most common fats and oils with regard to their various types of fat.

Because fats differ chemically, they behave differently in foods. To control their characteristics, food manufacturers sometimes alter them—and sometimes the alterations have health consequences, as described later in the chapter.

Points of unsaturation in fatty acids are like weak spots in that they are vulnerable to attack by oxygen. When the unsaturated points react with oxygen, the oils become rancid. This is why unprocessed oils should be stored in tightly covered containers. If stored for long periods, they need refrigeration to prevent spoilage.

One way to prevent spoilage of oils containing unsaturated fatty acids is to change them chemically by **hydrogenation,** but this causes them to lose their

hydrogenation (high-droh-gen-AY-shun) the process of adding hydrogen to unsaturated fat to make it more solid and more resistant to chemical change.

© Quest Photographic, Inc.

▲ *The more unsaturated a fat, the more liquid it is at room temperature. The more polyunsaturated the fat is, the sooner it melts.*

© Michael Newman/PhotoEdit

▲ *Fats in common use. The vegetable oil is polyunsaturated; the olive oil is monounsaturated; the butter is largely saturated; and the margarine, no doubt, is partially hydrogenated.*

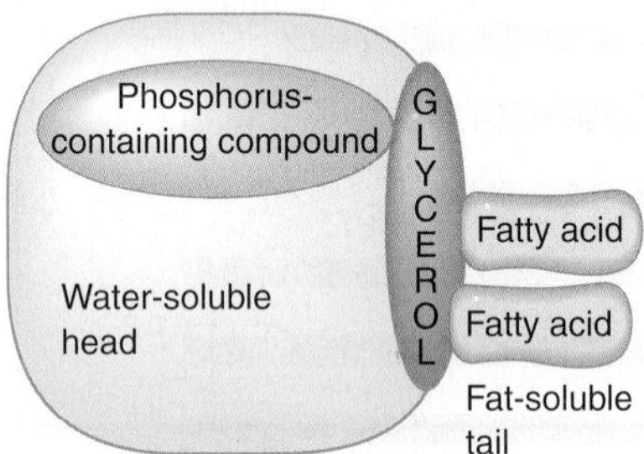

▲ ***A Phospholipid: Lecithin***
This phospholipid (lecithin) consists of a water-soluble head and a fat-soluble tail. Thus, lecithin can travel back and forth across lipid-containing cell membranes to the watery fluids on both sides. A phospholipid is a phosphorus-containing fat.

antioxidant (anti-OX-ih-dant) a compound that protects other compounds from oxygen by itself reacting with oxygen.

emulsifier a substance that mixes with both fat and water and can break fat globules into small droplets, thereby suspending fat in water.

bile a mixture of compounds, including cholesterol, made by the liver, stored in the gallbladder, and secreted into the small intestine. Bile emulsifies lipids to ready them for enzymatic digestion and helps transport them into the intestinal wall cells.

monoglyceride (mon-oh-GLISS-er-ide) a glycerol molecule with one fatty acid attached to it. A *diglyceride* is a glycerol molecule with two fatty acids attached to it.

unsaturated character and the health benefits that go with it. When food producers want to use a polyunsaturated oil such as corn oil to make a spreadable margarine, they hydrogenate the oil. Hydrogen is forced into the oil, some of the unsaturated fatty acids accept the hydrogen, and the oil becomes harder. The spreadable margarine that results is more saturated than the original oil but not as saturated as butter.

A second way to prevent spoilage of oils is to add a chemical that will compete for the oxygen and thus protect the oil. Such an additive is called an **antioxidant.** Examples are the well-known additives BHA and BHT,* which are listed on the labels of many processed foods, such as breakfast cereals and snack foods, and the natural antioxidants vitamin C and vitamin E. A third alternative, as previously mentioned, is to keep the product refrigerated.

Another way that the food industry alters natural fats and oils is by adding an **emulsifier** to a food product to allow fats and water to mix and remain mixed in a food product. Mayonnaise, margarines, salad dressings, and cake mixes often list an emulsifier on their labels. Monoglycerides and diglycerides are good emulsifiers, as is lecithin, which is an emulsifier found in egg yolk.

The Other Members of the Lipid Family: Phospholipids and Sterols

Lecithin and other phospholipids are important components of cell membranes. Because of the way the phospholipids are constructed, with a water-soluble head and a fat-soluble tail, they serve as emulsifiers in the body by joining with both water and fat. Thus, they help fats travel back and forth across the lipid-containing membranes of cells into the watery fluids on both sides. Because it functions as an emulsifier, lecithin is often used as a food additive and listed in the ingredients list on food labels. Food manufacturers add lecithin to foods such as mayonnaise, margarine, chocolate, salad dressings, and frozen desserts to keep the fats dispersed with the other ingredients.

Magical properties are sometimes attributed to lecithin, and health-food advertisers try to persuade people to supplement their diets with it. But lecithin is widespread in food and is also made by the liver in abundant quantities and therefore most people's diets contain adequate amounts.

Sterols, such as cholesterol, are large molecules with a multiple-ring structure as shown in the margin on page 113. Cholesterol is found only in animal foods and is also made in the body, where it is an important compound with many functions. It is a part of **bile,** which is necessary in the digestion of fats, and it is the starting material from which the sex hormones and many other hormones are made. In the skin, one of cholesterol's derivatives is made into vitamin D with the help of sunlight. It is also an important lipid in the structure of brain and nerve cells. In fact, cholesterol is a part of every cell. But, although it is widespread in the body and necessary to the body's function, it also is the major component of the plaque that narrows the arteries in the killer disease atherosclerosis. (See the Spotlight at the end of the chapter for more on heart disease.) Table 4-2 lists the cholesterol content of selected foods.

How the Body Handles Fat

A summary of lipid digestion is shown in Figure 4-3. After digestion in the upper small intestine, the products of fat digestion—fatty acids, glycerol, and **monoglycerides**—must enter the bloodstream to be of use to the body's cells. The shortest free fatty acids pass into the cells that line the intestine by simple diffu-

*BHA and BHT are butylated hydroxyanisole and butylated hydroxytoluene.

TABLE 4-2
Cholesterol in Foods[a]

(mg[b])	Sources
408	Liver, 3 oz
213	Egg, 1
107	Shrimp, 3 oz
85	Ground beef (lean), 3 oz
72	Prime rib, 3 oz
70	Pork chop (lean), 3 oz
58	Turkey (skinless), 3 oz
55	Chicken breast (skinless), 3 oz
47	Fish (baked cod), 3 oz
46	Sausage (1 link), 3 oz
35	Tuna (canned, in water), 3 oz
30	Milk (whole), 1 c
30	Ice cream (10% fat), ½ c
29	Cheddar cheese, 1 oz
26	Bologna (beef), 2 slices
17	Milk (reduced fat, 2%), 1 c
17	Yogurt (plain, low-fat), 1 c
11	Butter, 1 tsp
10	Cottage cheese (reduced fat), ½ c
7	Milk (1%), 1 c
0	Orange, 1 med
0	Broccoli, 1 c
0	Pinto beans, ½ c
0	Tofu, 4 oz

[a]Five foods contribute up to 70% of the cholesterol in the U.S. diet: eggs (30%), beef (16%), poultry (12%), cheese (6%), and milk (5%).
[b]Daily Value is 300 mg.

Cholesterol

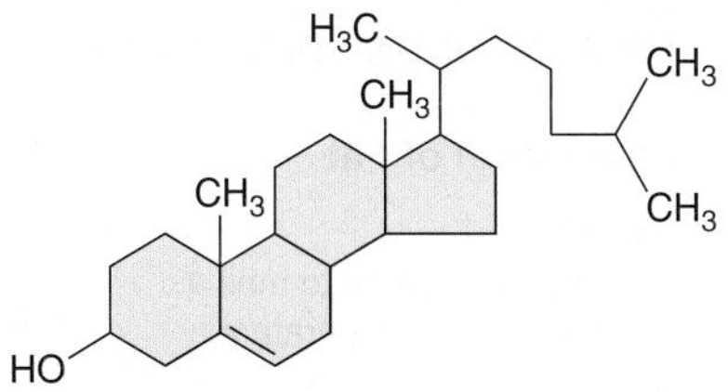

Sterols such as cholesterol have a multiple-ring structure. Cholesterol can be:

- Incorporated as an integral part of the structure of cell membranes
- Used to make bile for digestion
- Used to make sex hormones (estrogen and testosterone)
- Made into vitamin D
- Deposited in the artery walls, leading to plaque buildup and heart disease

sion. Because these short-chain fatty acids are somewhat water soluble, they can, without any further processing, enter the body's capillaries. Like the products of carbohydrate digestion, the short-chain fatty acids are transported from these capillaries through collecting veins to the capillaries of the liver. The liver cells pick them up and convert them to other substances the body needs. The glycerol follows the same path as the short-chain fatty acids because it, too, is water soluble.

The larger products of fat digestion (long-chain fatty acids, cholesterol, and phospholipids) are insoluble in water, a difficulty that must be overcome. The body's fluids—**lymph** and blood—are watery and will not accept these larger molecules as they are. The longer-chain fatty acids do pass into the intestinal cells, but there they reconnect with glycerol or with monoglycerides, forming new triglycerides. Then the cells package them for transport before releasing them into the lymph system.

The cells allow triglycerides and other lipids to form and combine with special proteins to make **chylomicrons,** one of the four types of **lipoproteins** found in the blood (see Figure 4-3, Part C). Within the body, the larger fats always travel in the form of lipoproteins. In this ingenious configuration, the water-soluble proteins enable the fats to travel in the watery body fluids. That way, when the tissues of the body need energy from fat, they can extract the energy they need from the lipoproteins in the blood and lymph. The remnants that remain are picked up by the liver, which dismantles them and reuses their parts. The characteristics of the four types of lipoproteins circulating in the blood are shown in Figure 4-4.

Lipoproteins are very much in the news these days. In fact, the health care provider who measures your blood lipid profile is interested not only in the types

lymph (LIMF) the body fluid that transports the products of fat digestion toward the heart and eventually drains back into the bloodstream; lymph consists of the same components as blood with the exception of red blood cells.

lipoproteins (LIP-oh-PRO-teens) clusters of lipids associated with protein that serve as transport vehicles for lipids in blood and lymph. The four main types of lipoproteins are chylomicrons, VLDL, LDL, and HDL.

FIGURE 4-3
A SUMMARY OF LIPID DIGESTION AND ABSORPTION

A. Digestion of Fat

1. Mouth
Some hard fats begin to melt as they reach body temperature.

2. Stomach
The stomach's churning action mixes fat with water and acid. A stomach enzyme accesses and breaks apart a small amount of fat. Fat is last to leave the stomach.

3. Liver, Gallbladder, and Small Intestine
Once in the small intestine, fat encounters **bile**, an emulsifier made in the liver (see Part B). The gallbladder, a storage organ, squirts bile into the contents of the small intestine to blend the fat with the watery digestive secretions.

4. Pancreas
Fat-digesting enzymes from the pancreas (pancreatic lipase) enter the small intestine. The enzymes can attack fat only after emulsification by bile. They break down the triglycerides to fatty acids, glycerol, and monoglycerides.

5. Large intestine
Some fat and cholesterol, trapped in fiber, are carried out of the body with other wastes.

B. Emulsification of Fat by Bile

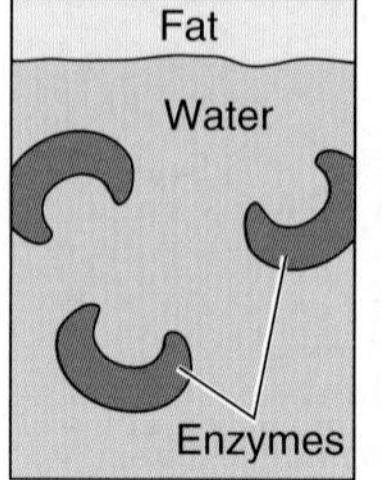

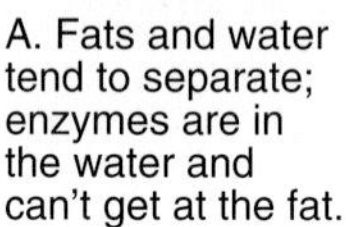
A. Fats and water tend to separate; enzymes are in the water and can't get at the fat.

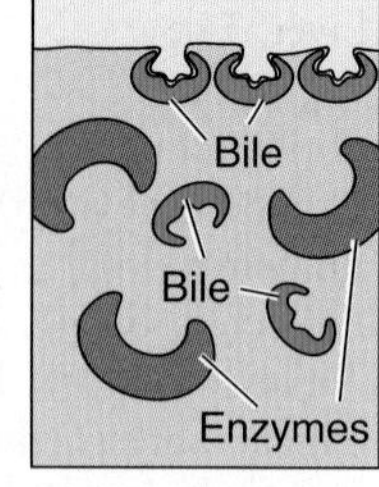

B. Bile (emulsifier) has affinity for fats and for water so it can bring them together.

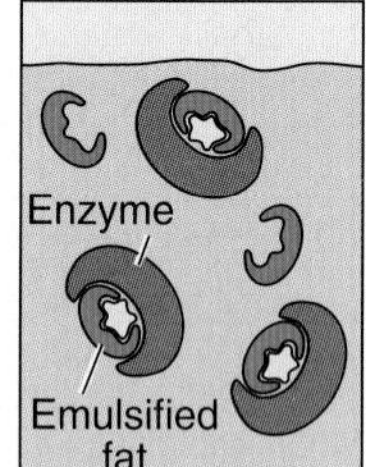

C. Small droplets of emulsified fat. The enzymes now have access to the fat, which is mixed in the water solution.

C. Absorption of Fat: The Chylomicron

Most of the newly digested fats are absorbed into lymph as part of a special package—the chylomicron. A chylomicron (lipoprotein) contains an interior of triglycerides and cholesterol surrounded by phospholipids. Proteins cover the structure. Such an arrangement of hydrophobic (water-fearing) molecules (the fatty acids) on the inside and hydrophilic (water-loving) molecules (proteins) on the outside allows lipids to travel through the watery fluids of the body.

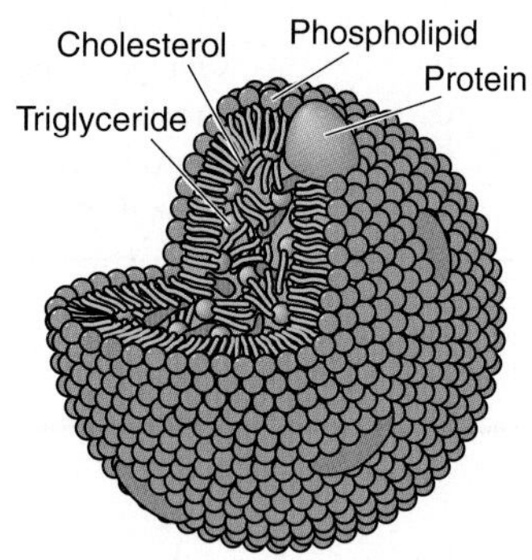

of fats in your blood (triglycerides and cholesterol) but also in the lipoproteins that carry them. One distinction among types of lipoproteins is very important because it has implications for the health of the heart and blood vessels—the distinction between **low-density lipoproteins (LDL)** and **high-density lipoproteins (HDL).** The more protein in the lipoprotein molecule, the higher its density.

Of course, the intestine also processes the few other dietary fats, such as phospholipids and cholesterol, which may have entered the body in food. These fats enter the circulation the same way the triglycerides do, and they also travel packaged in chylomicrons. After transport, they end up in the liver as part of the chylomicron remnants.

"Good" versus "Bad" Cholesterol

A silent, symptomless risk factor for heart disease is much talked about but little understood: elevated blood cholesterol levels. Blood cholesterol levels may be high for any of a number of reasons. Some people inherit tendencies to make too much cholesterol or to fail to destroy it on schedule. Others have high blood cholesterol for any or all of the following lifestyle reasons: eating too much saturated fat and ***trans* fat,** exercising too little, or carrying too much weight.[4] The blood lipid profile mentioned earlier can give you an idea of your standing as to

***trans* fatty acid** a type of fatty acid created when an unsaturated fat is hydrogenated. Found primarily in margarines, shortenings, commercial frying fats, and baked goods, *trans* fatty acids have been implicated in research as culprits in heart disease.

FIGURE 4-4
THE LIPOPROTEINS

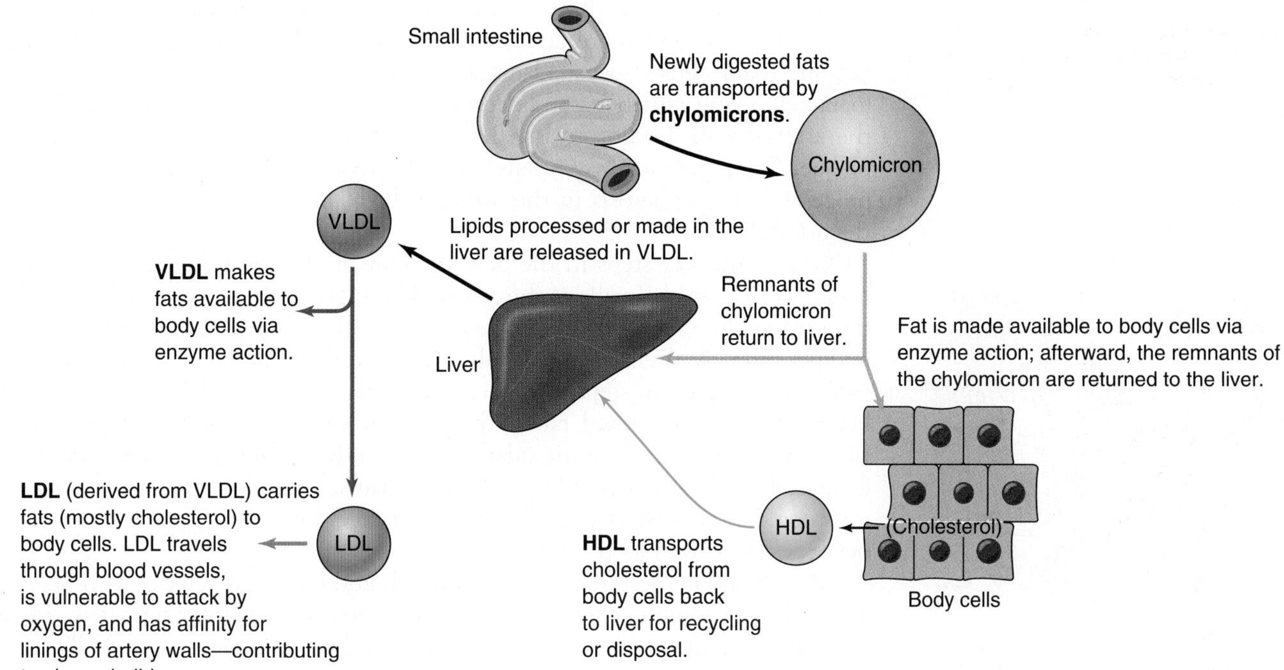
A. Functions and Interactions of Lipoproteins
Small intestine
Newly digested fats are transported by **chylomicrons**.
Chylomicron
VLDL
Lipids processed or made in the liver are released in VLDL.
VLDL makes fats available to body cells via enzyme action.
Remnants of chylomicron return to liver.
Liver
Fat is made available to body cells via enzyme action; afterward, the remnants of the chylomicron are returned to the liver.
LDL (derived from VLDL) carries fats (mostly cholesterol) to body cells. LDL travels through blood vessels, is vulnerable to attack by oxygen, and has affinity for linings of artery walls—contributing to plaque buildup.
LDL
HDL
(Cholesterol)
HDL transports cholesterol from body cells back to liver for recycling or disposal.
Body cells

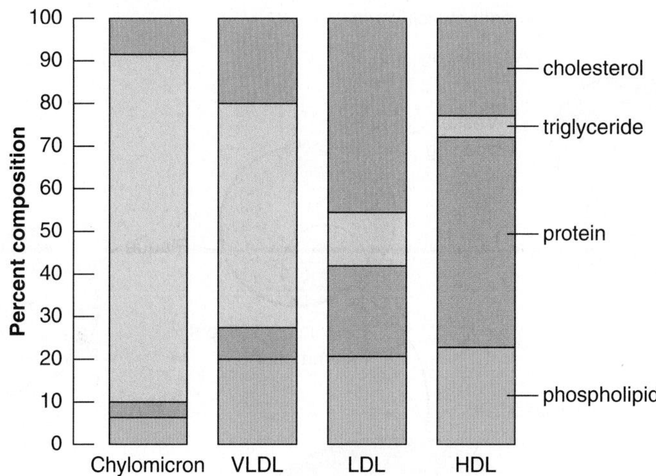
B. The Composition of Lipoproteins
100
90
80
70
60
50
40
30
20
10
0
Percent composition
cholesterol
triglyceride
protein
phospholipid
Chylomicron
VLDL
LDL
HDL

chylomicron (KIGH-loh-MY-cron) a type of lipoprotein that transports newly digested fat—mostly triglyceride—from the intestine through lymph and blood.

VLDL (very-low-density lipoprotein) carries fats packaged or made by the liver to various tissues in the body.

LDL (low-density lipoprotein) carries cholesterol (much of it synthesized in the liver) to body cells. A high blood cholesterol level usually reflects high LDL.

HDL (high-density lipoprotein) carries cholesterol in the blood back to the liver for recycling or disposal.

this risk factor. (Table 4-10 on page 135 in the Spotlight at the end of this chapter shows how to interpret your blood lipid profile.)

The underlying cause of most heart disease is atherosclerosis—the narrowing of the arteries caused by a buildup of cholesterol-containing plaque in the arterial walls (this chapter's Spotlight further discusses atherosclerosis and heart disease). The initiating step in the process of atherosclerosis is some form of injury or inflammation in the artery wall.[5] High blood pressure, high blood cholesterol levels, and cigarette smoke are potential sources of injury, as are the other causes listed in the margin (see also Table 4-9 on page 134). Raised LDL concentrations in the blood are a sign of high heart attack risk because LDLs in the blood tend to deposit cholesterol in the arteries.

Leading Risk Factors for Heart Disease

High LDL blood cholesterol level
Low HDL blood cholesterol level
High blood pressure
Cigarette smoking
Obesity
Physical inactivity
Diabetes
An "atherogenic" diet (high in saturated and *trans* fats, and low in fruits, vegetables, legumes, and whole grains)
Other risk factors cannot be changed: advanced age, male gender, and family history

oxidized LDL-cholesterol (o-LDL) the cholesterol in LDLs that is attacked by reactive oxygen molecules inside the walls of the arteries; o-LDL is taken up by scavenger cells and deposited in plaque.

foam cells cells from the immune system containing scavenged oxidized LDL cholesterol that are thought to initiate arterial plaque formation.

Researchers now theorize that LDL-cholesterol is damaging to the artery walls once it has been oxidized. Circulating LDL-cholesterol is more likely to settle along the linings of the artery walls after it first reacts with an unstable form of oxygen to become **oxidized LDL-cholesterol (o-LDL)** (see Figure 4-5).

Researchers believe that scavenger cells from the immune system known as macrophages ingest more and more of the o-LDL particles and eventually become **foam cells**—so called because of their resemblance to sea foam. These foam cells eventually burst and deposit their accumulated cholesterol as debris on the arterial wall, leading to the development of fatty streaks—the precursors of plaque. Thus, scientists speculate that the oxidized form of LDL catalyzes the process of atherosclerosis in the artery walls by attracting these macrophages to the arterial area.

"Excuse me, but which is it that practically kills you—polysaturated or polyunsaturated!?"

One of the key steps in the development of atherosclerosis is the accumulation of o-LDL in the walls of the artery. The more LDLs in the circulating blood, the greater the chance for oxidation to occur. This leaves more o-LDL available for ingestion by the scavenger cells and more debris left behind in the arterial wall. In theory then, steps to reduce the total amount of LDL circulating in the blood (reducing saturated fat in the diet) or steps to prevent the oxidation of LDL-cholesterol (ample antioxidants in the body) would reduce the formation of foam cells and cause less injury to arterial walls. Thus, the process of atherosclerosis could be slowed or possibly prevented. See Table 4-3 for the effects of various types of fat on blood lipids.

Most blood cholesterol is carried in LDL and correlates *directly* with heart disease risk, but some is carried in HDL and correlates *inversely* with risk. In fact, the most potent single predictor of heart attack risk may be the HDL level.

FIGURE 4-5
ATHEROSCLEROSIS

As LDL particles penetrate the walls of the arteries, they become oxidized-LDL and next are scavenged by the body's white blood cells. These foam cells are then deposited into the lining of the artery wall. This process, known as *atherosclerosis,* causes plaque deposits to enlarge, artery walls to lose elasticity, and the passage through the artery to narrow.

*Early injury may be the result of smoking, high blood pressure, elevated blood cholesterol, elevated homocysteine level, diabetes, genetics, or possibly a bacterial or viral infection.

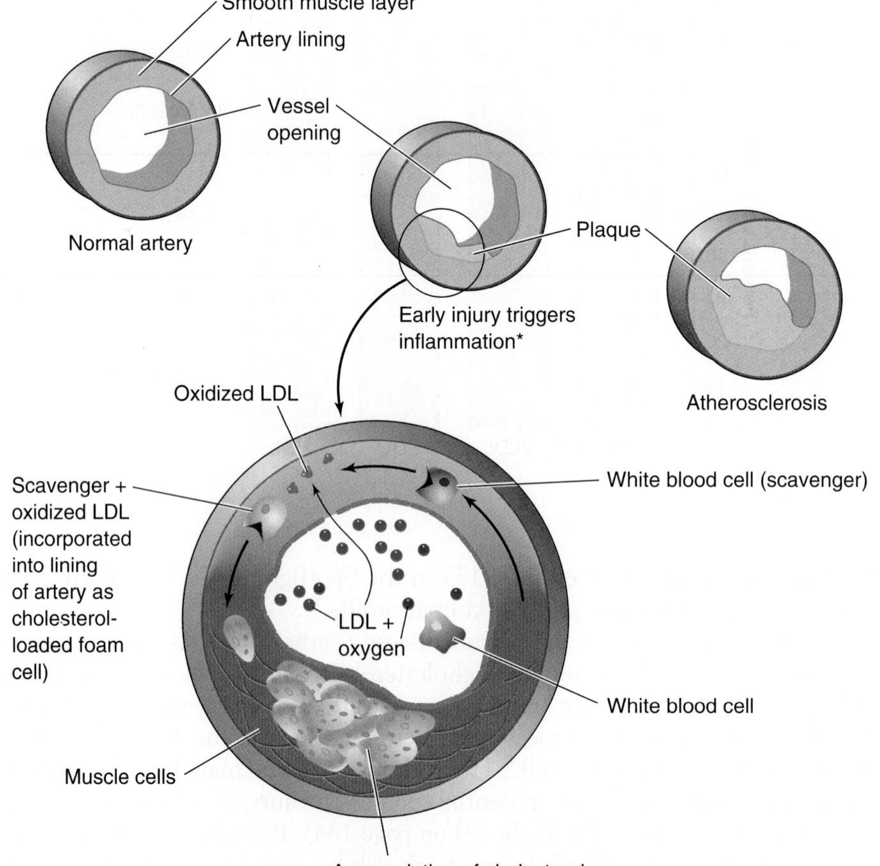

TABLE 4-3
The Effects of Various Kinds of Fat on Blood Lipids

Type of Fat[a]	Dietary Sources	Effects on Blood Lipids
Saturated Fat	All animal meats, beef tallow, butter, cheese, chocolate, coconut, cocoa butter, coconut oil, cream, hydrogenated oils, lard, palm oil, stick margarine, shortening, whole milk, and ice cream	Increases total cholesterol Increases LDL-cholesterol
Polyunsaturated Fat	Almonds, corn oil, cottonseed oil, filberts, fish, liquid/soft margarine, mayonnaise, pecans, safflower oil, sesame oil, soybean oil, sunflower oil, walnuts	If used to *replace* saturated fat in the diet, polyunsaturated fat may: Decrease total cholesterol Decrease LDL-cholesterol Decrease HDL-cholesterol[b]
Monounsaturated Fat	Almonds, avocados, canola oil, cashews, olive oil, olives, peanut butter, peanut oil, peanuts, poultry	If used to *replace* saturated fat in the diet, monounsaturated fat may: Decrease total cholesterol Decrease LDL-cholesterol without decreasing HDL-cholesterol
Omega-3 Fat	Canola and soybean oils, flaxseed, oily coldwater fish (salmon, mackerel, tuna), shellfish, soyfoods, walnuts, wheat germ	If used to *replace* saturated fat in the diet, omega-3 fat may: Decrease total cholesterol Decrease LDL-cholesterol Increase HDL-cholesterol Decrease triglycerides
***Trans* Fat**	Margarine (hard stick), cake, cookies, doughnuts, crackers, chips, meat and dairy products, hydrogenated peanut butter, shortening, many fast foods	Increases total cholesterol Increases LDL-cholesterol

[a]All fats, whether classified as mainly saturated fat, monounsaturated fat, or polyunsaturated fat, contain mixtures of saturated and unsaturated fats and provide the same number of calories: 9 calories per gram.
[b]If consumed in large amounts (>10% of total calories).

Research indicates that an acceptable total blood cholesterol reading of 200 mg/dl or below may not be protective against heart disease if the HDL level is low. Raised HDL concentrations relative to LDL represent cholesterol on its way out of the arteries back to the liver—and a reduced risk of heart attack.

Some cases of elevated blood cholesterol do not respond to changes in lifestyle. In such cases, cholesterol-lowering drugs might be prescribed. However, for many people, a few simple changes in diet can improve cholesterol readings, as discussed in the Eat Well Be Well feature that follows.

LDL, the "bad" cholesterol

HDL, the "good" cholesterol

Fat in the Diet

The question of what kind of fat to include in the diet can be puzzling. Research has potentially linked the fat in the diet to several diseases, including certain types of cancer, heart disease, arthritis, and gallbladder disease. The remainder of this chapter helps you apply what you have learned about fats—that is, how to choose foods that supply enough but not too much of the right kinds of fat to support optimal health and provide pleasure in eating. The current recommendations for dietary fat intake are listed in Table 4-4 on page 121.

Today, the average American diet includes about 34 percent of its calories from fat, about 12 percent of calories from saturated fat, and about 2.6 percent of calories from *trans* fat.[6] Although the percentage of calories from fat has

(Text continues on page 121.)

Eat Well Be Well Nourish the Heart

Heart disease rarely develops from a single risk factor. Clusters of risk factors usually occur, and a small increase in one risk factor, such as blood pressure, becomes more critical when combined with other risk factors, such as obesity. Luckily, a similar pattern happens in reverse. Even moderate changes in one risk factor can decrease several others at the same time. For example, a weight loss of just 5 to 10 pounds can reduce blood pressure in overweight people. Likewise, becoming physically active can lower blood pressure, increase HDL ("good" cholesterol), and help control weight.

In the past few years, attention has focused on reducing heart disease risk by both *reducing* contributory factors in the diet—notably, saturated fat and *trans* fat—and by *increasing* the intake of protective factors, such as fiber and the antioxidant vitamins. How does this translate into a heart-healthy lifestyle? What foods should you eat, for example, to achieve the goal of reducing your saturated fat intake to less than 10 percent of total calories, while also increasing your intake of the protective fibers and antioxidant nutrients? This feature offers some general tips that will help you make the heart-healthy food choices needed to achieve these goals. See this chapter's Spotlight for more tips on how to raise your HDL ("good" cholesterol) level and lower your LDL ("bad" cholesterol) level.

© Polara Studios

▲ *Enjoy a variety of nutritious foods for the health of your heart.*

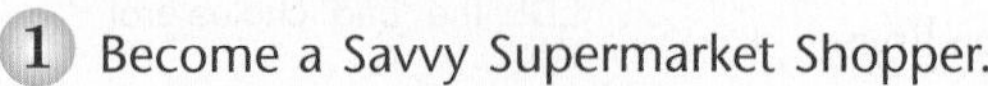

1 Become a Savvy Supermarket Shopper.

Learn to read food labels to help you choose nutritious, heart-healthy food products. Read manufacturers' labels to determine both the amounts and the types of fat contained in foods. When comparing foods, look at the Nutrition Facts panel, and choose the food with the lower amounts of saturated fat, *trans* fat, and cholesterol (see Figure 4-6). For saturated fat and cholesterol, keep in mind that 5 percent of the daily value (% DV) or less is low and 20 percent or more is high. There is no % DV for *trans* fat, but we are advised to keep our intake of *trans* fat as low as possible while consuming a nutritionally adequate diet.

2 Keep Blood Cholesterol at or Below the Recommended Levels.

Among the most influential diet-related factors that raise blood cholesterol levels are saturated fat and *trans* fat intakes. For every 1 percent reduction in a high blood cholesterol level, there is a 2 percent to 3 percent reduction in the risk of heart attack. As it turns out, the changes in diet that reduce blood cholesterol concentrations mostly do so by reducing LDL-cholesterol. Dietary modifications that help lower LDL-cholesterol include:

- *Substitute* highly monounsaturated fats (canola or olive oils) and highly polyunsaturated fats (vegetable oils and fish oils; refer to Table 4-3) for saturated fats in the diet because both lower LDL-cholesterol levels in the blood. Choose fats and oils with 2 grams or less of saturated fat per tablespoon, such as liquid and tub margarines and olive or canola oils.[7] Shop for margarine with liquid vegetable oil as the first ingredient. Look for the canola- or olive-oil-based margarines containing zero *trans* fat.
- *Replace* saturated fats and *trans* fats with unsaturated fats from vegetables, legumes (beans), and nuts. Choose healthful portions of skinless poultry, lean meat, and fish, especially omega-3-fatty-acid-rich fish such as salmon. Limit foods high in saturated fat, *trans* fat, and/or cholesterol, such as full-fat milk products, fatty meats, tropical oils (coconut or palm), partially hydrogenated vegetable oils, and egg yolks. Eat fewer high-fat desserts (such as cheesecake, ice cream, brownies, pies, pastries, and butter-and-cream-frosted cakes). Choose fruits as desserts most often. Use the tips in the box on page 120 to make your own favorite recipes more healthful.
- *Add bulk.* People on high-fiber diets have been shown to excrete more cholesterol and fat than those on low-fiber

FIGURE 4-6
CHECKING OUT THE FOOD LABEL FOR FAT INFORMATION

Papa Solo's French Bread Pizza
French Bread Pizza with Tomato Sauce and Mozzarella Cheese

Nutrition Facts
Serving Size 1 pizza (158 g)
Servings per Container 1

Amount Per Serving	
Calories 310	Calories from Fat 45
	% Daily Value*
Total Fat 5g	**8%**
Saturated Fat 2g	**10%**
Trans Fat 1.5g	
Polyunsaturated Fat 0.5g	
Monounsaturated Fat 1g	
Cholesterol 10mg	**3%**
Sodium 470mg	**20%**
Total Carbohydrate 49g	**16%**
Dietary Fiber 6g	**24%**
Sugars 3g	
Protein 20g	
Vitamin A 4% •	**Vitamin C** 0%
Calcium 30% •	**Iron** 15%

* Percent Daily Values are based on a 2,000 calorie diet. Your Daily Values may be higher or lower depending on your calorie needs:

		Calories: 2,000	2,500
Total Fat	Less than	65g	80g
Sat Fat	Less than	20g	25g
Cholesterol	Less than	300mg	300mg
Sodium	Less than	2,400mg	2,400mg
Total Carbohydrate		300g	375g
Dietary Fiber		25g	30g

Calories per gram
Fat 9 • Carbohydrate 4 • Protein 4

Ingredients: French bread (enriched unbleached wheat flour), water, wheat gluten, oat fiber, sugar, soybean oil, nonfat dry milk, mozzarella cheese, tomatoes. Contains 2% or less of each of the following: pizza spice mix (spices, salt, onion and garlic), sugar, coloring (dextrose, cabbage extract, annatto).

MAMA MIA's
Pepperoni Pizza-for-One

Nutrition Facts
Serving Size 1 pizza (191g)
Servings per Container 1

Amount Per Serving	
Calories 520	Calories from Fat 243
	% Daily Value*
Total Fat 27g	**42%**
Trans Fat 8g	
Saturated Fat 10g	**50%**
Cholesterol 25mg	**8%**
Sodium 1,280mg	**53%**
Total Carbohydrate 53g	**18%**
Dietary Fiber 4g	**16%**
Sugars 6g	
Protein 19g	
Vitamin A 45% •	**Vitamin C** 0%
Calcium 30% •	**Iron** 10%

* Percent Daily Values are based on a 2,000 calorie diet. Your Daily Values may be higher or lower depending on your calorie needs:

		Calories: 2,000	2,500
Total Fat	Less than	65g	80g
Sat Fat	Less than	20g	25g
Cholesterol	Less than	300mg	300mg
Sodium	Less than	2,400mg	2,400mg
Total Carbohydrate		300g	375g
Dietary Fiber		25g	30g

Calories per gram
Fat 9 • Carbohydrate 4 • Protein 4

Ingredients: CRUST: Wheat flour with malted barley flour, water, partially hydrogenated vegetable oil (soybean and/or cottonseed oil) with soy lecithin, soybean oil, yeast, high fructose corn syrup, salt; TOPPING: Part-skim mozzarella cheese substitute, pepperoni (pork and beef, salt, water, dextrose, spices, sodium nitrite), part-skim mozzarella cheese; SAUCE: Tomato puree, water, green peppers, salt, lactose and flavoring, spices, corn oil, xanthum gum.

Total fat refers to all the fat in the food: saturated, monounsaturated, polyunsaturated, and *trans* fat. Total fat, saturated fat, *trans* fat, and cholesterol information is required on the label. Listing the amount of monounsaturated and polyunsaturated fats in the food is voluntary.

Ingredients list—For sources of *total* fat, look for terms such as: coconut oil, diglycerides, lard, monoglycerides, oil, palmitate, palm oil, stearate, triglycerides, and vegetable shortening.

Also check ingredients lists for *hydrogenated* or *partially hydrogenated oil.* The higher on the list it appears, the more of both total fat and *trans* fatty acids the product probably contains, and the more difficult it is to fit into a healthful diet. As an alternative, look for similar products made with no hydrogenated oils and smaller amounts of total fat.

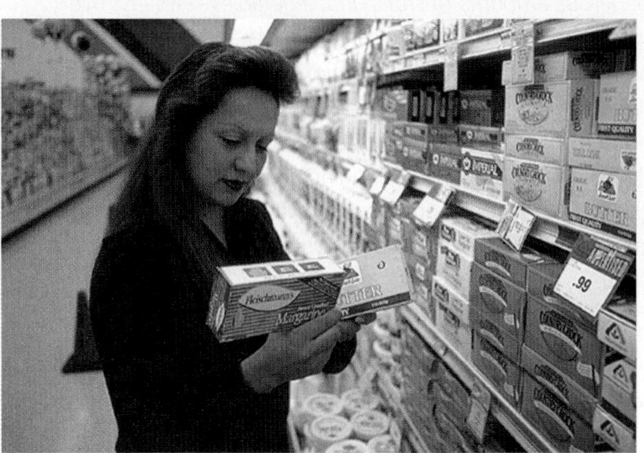

▸ *Be a savvy supermarket shopper. Read food labels to compare the total fat, saturated fat,* trans *fat, and cholesterol content of similar foods, in addition to fat calories per serving.*

Recipe Modification

Use this substitution information to modify your own favorite recipes. They'll be more healthful but will still look and taste as good as your originals!

Instead of:	**Substitute**:
1 whole egg	2 whipped egg whites or 1¼ cup egg substitute
Whole milk	Fat-free or 1 percent milk
Evaporated milk	Evaporated nonfat or low-fat milk
Heavy cream	Evaporated nonfat or low-fat milk
Sour cream	Reduced-fat or fat-free sour cream or plain yogurt
Solid shortening (baking)	Vegetable oil, margarine
Solid shortening (stir-frying)	Peanut or olive oil
Mayonnaise	Reduced-fat or fat-free mayonnaise
Cream cheese	Reduced-fat or fat-free cream cheese
Regular cheese	Cheese made from part-skim milk (mozzarella, Swiss lace, farmer) or reduced-fat cheeses
Salad dressing	Fat-free salad dressings or reduced-calorie dressing
White sauce (made with cream and butter)	Use low-fat milk and cornstarch or flour by blending the starch into cold liquid to eliminate the need for fat.

diets. A 5 percent decrease in total cholesterol can be achieved with a 5- to 10-gram increase in soluble fiber intake.[8] One reason for this is that the high-fiber diet decreases food's transit time through the digestive tract, allowing less time for cholesterol to be absorbed. When cholesterol from the diet is thus reduced, the body must turn to its own supply for making necessary body compounds. Diets high in fiber are typically low in fat and cholesterol—another advantage to emphasizing fiber. The various dietary fibers have varying effects on blood cholesterol. Soluble fibers such as that found in rolled oats, oat bran, and psyllium-fortified cereals have favorable effects on blood cholesterol, whereas wheat bran does not appear to be effective.[9] Apples, pears, peaches, oranges, and grapes are good sources of pectin, another type of cholesterol-lowering fiber.

- *Eat generous amounts of fruits and vegetables*—at least five servings a day—as a source of antioxidants and other protective compounds in the diet. Antioxidants, such as beta-carotene, vitamins C and E, and the mineral selenium, strengthen the body's natural defenses against cell damage by blocking the potentially damaging **free radicals** that arise as a part of numerous normal cell activities. Free radicals, the chemical compounds that oxidize LDL-cholesterol, become a problem when they are too plentiful. Antioxidants in the body may reduce the amount of LDL-cholesterol that becomes oxidized because the antioxidants (particularly vitamin E) can neutralize the highly reactive oxygen before it gets a chance to oxidize the LDL particle, thereby lessening the buildup of plaque in the artery walls.[10] Scientists are now exploring the potential abilities of all the antioxidant substances found in foods to protect against heart disease.[11]

3 Balance Energy Intake with Energy Needs.

For heart health, maintain a healthy body weight and a level of physical activity that keeps you fit and matches the number of calories you eat.[12] Walk, or do other activities for at least 30 minutes on most days. Moderate exercise, such as brisk walking, may both lower LDL levels and raise HDL levels if the activity is consistently pursued for long enough periods.

It seems that the factors affecting the health of the heart are all tangled together, but all evidence points to the same general recommendations. For good health and to avoid heart disease: do not smoke; reduce blood pressure and weight, if necessary; exercise regularly; and eat a heart-healthy diet.

free radicals highly toxic compounds created in the body as a result of chemical reactions that involve oxygen. Environmental pollutants such as cigarette smoke and ozone also prompt the formation of free radicals.

TABLE 4-4
Dietary Recommendations for Fats in the Diet*

From the *Dietary Guidelines for Americans*:
- Consume less than 10 percent of calories from saturated fatty acids and less than 300 mg/day of cholesterol, and keep *trans* fatty acid consumption as low as possible.
- Keep total fat intake between 20 to 35 percent of calories, with most fats coming from sources of polyunsaturated and monounsaturated fatty acids, such as fish, nuts, and vegetable oils.
- When selecting and preparing meat, poultry, dry beans, and milk or milk products, make choices that are lean, low fat, or fat free.
- Limit intake of fats and oils that are high in saturated and/or *trans* fatty acids, and choose products low in such fats and oils.

From the American Heart Association:
- Total fat: 30% or less of total calories
- Saturated fat and *trans* fat: less than 10% of calories
- Polyunsaturated fat: up to 10% of total calories
- Monounsaturated fat: up to 20% of total calories
- Dietary cholesterol: less than 300 mg/day on average

From the DRI Committee:
An *Acceptable Macronutrient Distribution Range (AMDR)* for:
- Total fat: 20%–35% of total calorie intake
- Polyunsaturated fats:
  - Omega-6 fats: 5%–10% of total calorie intake
    Adequate intake of linoleic acid (essential omega-6 fatty acid):

Men	Women
17 g/day	12 g/day

  - Omega-3 fats: 0.6%–1.2% of total calorie intake
    Adequate intake of linolenic acid (essential omega-3 fatty acid):

Men	Women
1.6 g/day	1.1 g/day

On Food Labels: Daily Values (DV) for a 2,000-calorie Diet:

Total Fat	Saturated Fat	Cholesterol
65 g	20 g	300 mg

*Refer to Table 4-3 for dietary sources of various types of fat.

decreased from 42 percent in 1970, the apparent downward trend is misleading because average calorie consumption has increased by 15 percent since 1970 and actual fat grams consumed has increased as well.[13] The *Dietary Guidelines for Americans* recommend that *total* fat does not exceed 20 to 35 percent of the day's total calories and comes from mostly polyunsaturated and monounsaturated fat sources such as fish, nuts, and vegetable oils. Saturated fat should be kept low and contribute less than 10 percent of calories. Table 4-5 shows you how to determine your daily fat allowance.

Scientists continue to debate which type of fat makes the best replacement for saturated fat in the diet. Researchers have shown that the substitution of monounsaturated for saturated fatty acids in the diet brings about a decrease in blood levels of LDL-cholesterol, without the decrease in HDL-cholesterol seen when polyunsaturated fats (at intakes greater than 10 percent of calories) are substituted for saturated fats.[14] Some researchers warn, however, that diets rich in any type of fat can be calorie dense and could worsen the problem of obesity in the United States and Canada.

"Fat on the plate" includes visible fats and oils, such as butter, the oil in salad dressing, and the fat you trim from a steak. It also refers to some you cannot see, such as the fat that marbles a steak or that is hidden in such foods as nuts,

TABLE 4-5

How to Determine Daily Fat Allowances

We are advised to choose a diet that is low in saturated fat and cholesterol and moderate in total fat. The DRI Committee has set an acceptable range of from 20 percent to 35 percent of total calories from fat. Recommended intake for saturated fat is less than 10 percent of total daily calories. Every single food you eat does not need to conform to these allowances if you balance higher fat items with lower fat foods throughout the day.

Use this table to find your total daily fat allowance, based on your daily calorie intake, to keep your fat grams within the acceptable range. To find a reasonable calorie intake for yourself, see the DRI table on the inside front cover of this book. Or, use the following quick method to determine your calorie allowance: Multiply your body weight in pounds by 15 (if you're active). If you weigh 150 pounds, you expend about 2,250 calories (150 × 15) in a typical day. If you're sedentary, multiply your weight by 13.

	Total Fat			Saturated Fat
Total Calories per Day	20% of Total Calories from Fat[a] (g/day)	30% of Total Calories from Fat[b] (g/day)	35% of Total Calories from Fat[a] (g/day)	Less than 10% of Total Calories from Saturated Fat (g/day)
1,200	27	40	47	13 or less
1,500	33	50	58	17 or less
1,800	40	60	70	20 or less
2,000[c,d]	40	65	75	20 or less
2,400[e]	53	80	93	27 or less
2,500[d]	55	80	95	25 or less
2,800	62	93	109	31 or less
3,000[f]	67	100	117	33 or less

[a]The DRI Committee set an Acceptable Macronutrient Distribution Range for fat intake at 20% to 35% of total calories. For more discussion about AMDR, see Chapter 2.

[b]The American Heart Association determines fat allowances based on a diet with no more than 30% of total calorie intake from fat.

[c]Percent Daily Values on Nutrition Facts Labels are based on a 2,000-calorie diet.

[d]Values for 2,000 and 2,500 calories are rounded to the nearest 5 grams to be consistent with the Nutrition Facts Label.

[e]Estimated energy requirement (EER) for 19-year-old women; subtract 7 calories per day for females for each year of age above 19. EER values are determined at four physical activity levels; the values above are for the "active" female.

[f]Estimated energy requirement (EER) for 19-year-old men; subtract 10 calories per day for males for each year of age above 19. EER values are determined at four physical activity levels; the values above are for the "active" male.

Example: If you eat 1,800 calories per day, multiply 1,800 by 10 percent *to determine the maximum number of calories that should come from saturated fat* in one day (1,800 × 0.1 = 180 calories from saturated fat). Since 1 gram of fat provides 9 calories, divide the calories from saturated fat by 9 to see the maximum amount of saturated fat you can have per day (180 ÷ 9 = 20 grams of saturated fat per day):

Total daily calories × 0.10 = total saturated fat calories
Total saturated fat calories ÷ 9 = total saturated fat grams

Likewise, *to determine an acceptable range (20% to 35% of calories) for total fat intake:*

Total daily calories × 0.20 = total fat calories
Total fat calories ÷ 9 = total fat grams

To determine maximum amount of calories from fat:

Total daily calories × 0.35 = total fat calories
Total fat calories ÷ 9 = total fat grams

cheese, biscuits, crackers, doughnuts, cookies, muffins, avocados, olives, fried foods, and chocolate. Most of the fat, especially saturated fat, in our diet comes from animal products (see Figure 4-7). Meats probably conceal most of the fat that people unwittingly consume. Many people, when choosing a serving of meat, don't realize that they are electing to eat a large amount of fat. When selecting beef or pork, look for the words *loin* or *round* on the label—these words

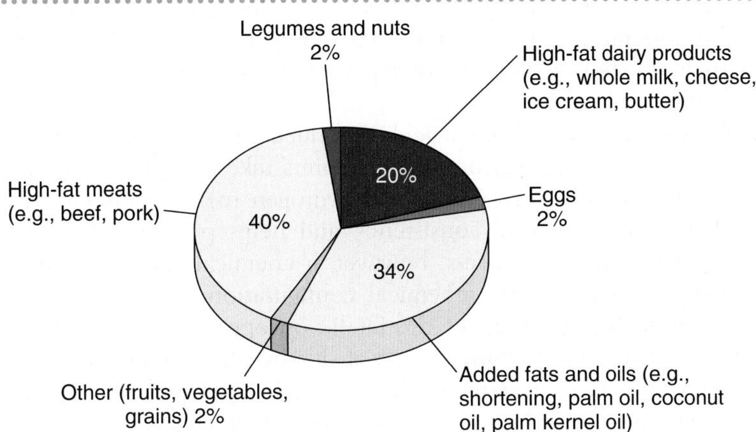

FIGURE 4-7
MAJOR SOURCES OF SATURATED FAT IN THE U.S. DIET

The current American diet contains about 12 percent of total calories as saturated fat.

represent lean cuts from which the fat can be trimmed. Table 4-6 shows a few examples of the differences in the saturated fat content of different forms of commonly eaten foods. Note that lower saturated fat choices can be made within the same food group (for example, low-fat milk versus whole milk).

▲ *A 3-ounce portion of lean beef, chicken, or fish is roughly the size of a deck of playing cards or the palm of the average woman's hand.*

The *Trans* Fatty Acid Controversy—Is Butter Better?

Conventional nutrition wisdom has long held that margarine is better than butter to use on food, including bagels and other breads. But in recent years, rumors have been spreading that margarine may not be the most healthful choice after all. The heart of the controversy is a growing body of research suggesting that a

TABLE 4-6
Compare the Saturated Fat and Calorie Content of Selected Foods

Food Category	Portion	Saturated Fat (grams)	Calories
Cheese			
• Regular cheddar cheese	1 oz	6.0	114
• Low-fat cheddar cheese	1 oz	1.2	49
Milk			
• Whole milk	1 cup	4.6	146
• Low-fat (1%) milk	1 cup	1.5	102
• Skim (fat-free) milk	1 cup	0	90
Frozen desserts			
• Regular ice cream	½ cup	4.9	145
• Low-fat frozen yogurt	½ cup	2.0	110
Ground beef			
• Regular ground beef (25% fat)	3 oz (cooked)	6.1	236
• Extra lean ground beef (5% fat)	3 oz (cooked)	2.6	148
Pork			
• Spareribs	3 oz (cooked)	9.5	337
• Ham, extra lean	3 oz (cooked)	1.5	123
Chicken			
• Fried chicken (leg with skin)	3 oz (cooked)	3.3	212
• Roasted chicken (breast no skin)	3 oz (cooked)	0.9	140

SOURCE: Adapted from Dietary Guidelines for Americans, 2005; ARS Nutrient Database for Standard Reference, Release 17.

FIGURE 4-8
TYPES OF UNSATURATED FATTY ACIDS: *CIS* VERSUS *TRANS*

Unsaturated fatty acids are either in *cis* form or in *trans* form, depending on the way in which the hydrogen atoms are attached to the points of unsaturation in the carbon chain. If the hydrogen atoms are attached to the same side of the points of unsaturation, the arrangement is called *cis*. If the hydrogen atoms are attached to different sides, the arrangement is called *trans*.

```
 H  H  H  H           H  H        H
 |  |  |  |           |  |        |
-C--C==C--C-         -C--C==C--C-
 |        |           |     |  |
 H        H           H     H  H

     cis                  trans
```

cis = same
trans = across

certain type of fat found in margarine may be as likely to boost blood levels of "bad" LDL-cholesterol as the saturated fat found in butter.[15] What's more, the research suggests that in large amounts, *trans* fatty acids lower "good" HDL-cholesterol in the blood.[16]

The alleged culprit, *trans* fatty acid, is formed when margarine is processed. Consider that to make margarine, manufacturers take a highly unsaturated vegetable oil and partially hydrogenate (add hydrogen to) it. Hydrogenation gives the spread its relatively solid consistency and helps protect against rancidity. During the hydrogenation process, however, a chemical "fluke" occurs. Hydrogenating the oil creates a new chemical configuration, known as a *trans* fatty acid, in which hydrogen atoms of the fat lie on opposite sides of the point of unsaturation in the carbon chain instead of side by side (see Figure 4-8). (Recall that the body of a fatty acid molecule consists of a chain of carbon atoms to which hydrogen is attached, as shown in Figure 4-1 on page 110.)

Because *trans* fatty acids are formed whenever oils are hydrogenated, margarine is only one of the foods on the market that contain *trans* fatty acids (see Figure 4-9). Shortenings, baked goods, certain brands of peanut butter, the commercial frying fats used in many fast-food outlets to cook French fries and other fried items, and many other foods that list hydrogenated or partially hydrogenated vegetable oil on their label are among the foods containing *trans* fatty acids. Fueling the controversy even further, an analysis by researchers at Harvard University suggested that more than 30,000 American deaths each year are attributable to *trans* fatty acids. The authors of the report went so far as to call for government legislation mandating that manufacturers include *trans* fatty acid amounts on food labels and phase out the use of partially hydrogenated oils in the United States.[17]

In January 2006, the Food and Drug Administration began requiring that manufacturers list the *trans* fat content on a separate line within the Nutrition Facts panel on all food products (as shown in Figure 4-10). Manufacturers can label a product "*trans* fat free" if the food contains less than 0.5 gram of *trans* fat and less than 0.5 gram of saturated fat per serving.[18] Canada also requires labeling of *trans* fat content of foods.

Many major health organizations point out that the *trans* fatty acid research does not indicate that consumers should switch back from eating margarine to eating butter, which contains an excess of saturated fat. That's because *trans* fatty acids account for only about 2.6 percent of the total calories in a typical American diet, whereas saturated fat comprises some 12 percent of calories.[19] And although the evidence that has come to light doesn't warrant going back to butter, it does underscore the value of keeping the intake of *total* saturated and *trans* fat to a minimum. The foods that contain large amounts of *trans* fatty acids—French fries, potato chips, deep-fat fried doughnuts—are also high in total fat. With the addition of *trans* fat to the Nutrition Facts panel, you can now evaluate the total fat, saturated fat, and *trans* fat content per serving for various food products (see Figure 4-10). Thus, health-conscious eaters who want to hedge their

FIGURE 4-9
MAJOR SOURCES OF *TRANS* FATTY ACIDS IN THE U.S. DIET

Trans fatty acids occur naturally in meat, poultry, and dairy products. They show up in packaged and fast foods to which hydrogenated vegetable oils have been added. Note that high levels of *trans* fatty acids go hand in hand with high-fat foods. Thus, if you eat a low-fat diet, you will consume minimal amounts of *trans* fatty acids.

SOURCE: *FDA Consumer*, 2003.

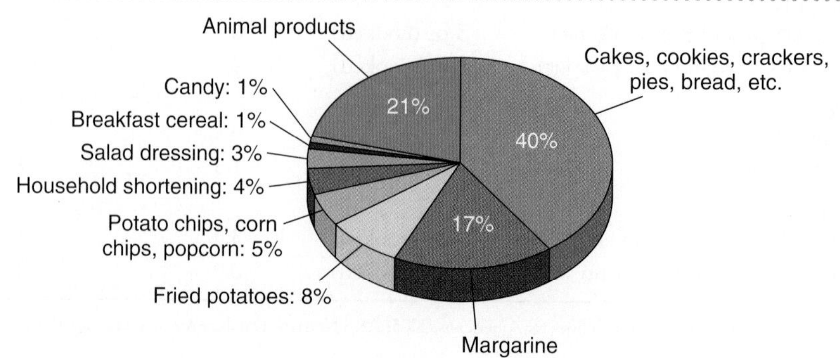

FIGURE 4-10
COMPARE SPREADS

Read manufacturers' labels to determine both the amounts and the types of fat contained in foods. Shop for margarine containing no more than 2 grams of saturated fat per tablespoon and with liquid vegetable oil as the first ingredient. Choose soft tub or liquid forms over stick; "hard" stick margarines typically contain the most *trans* fatty acids. Look for the canola- or olive-oil-based margarines containing zero *trans* fat.

SOURCE: Adapted from Center for Food Safety and Applied Nutrition, Office of Nutritional Products, Labeling, and Dietary Supplements, U.S. Food and Drug Administration, January 16, 2004.

	Butter	Margarine, stick	Margarine, tub
Nutrition Facts			
Serving Size	1 Tbsp (14 g)	1 Tbsp (14 g)	1 Tbsp (14 g)
Servings per Container	32	32	32
Amount Per Serving			
Calories	100	100	60
Calories from Fat	100	100	60
	% Daily Value	**% Daily Value**	**% Daily Value**
Total Fat	11g — **17%**	11g — **17%**	7g — **11%**
Saturated Fat	7g ← **35%**	2g ← **10%**	1g ← **5%**
Trans Fat	0g ←	3g ←	0.5g ←
Cholesterol	30mg → **10%**	0mg → **0%**	0mg → **0%**
Saturated Fat:	**7g**	**2g**	**1g**
+ *Trans* Fat:	**0g**	**3g**	**0.5g**
Combined Amt.:	**7g**	**5g**	**1.5g**

bets against developing high blood cholesterol levels should stick with the principles of heart-healthy eating outlined in the Savvy Diner feature that follows. The Scorecard on page 130 will help you rate your own dietary selections for fat.

The Savvy Diner Choose Fats Sensibly

It is not the food but how you prepare it that often determines the total fat (and calories) in a food. Compare the fat in 3 ounces of broiled chicken (3 grams fat, 141 calories) versus 3 ounces of fried chicken (18 grams fat, 364 calories). What makes the difference in calories? Fat. This feature suggests a variety of ways to reduce added fats in your diet when preparing home-cooked meals.

- Use nonstick sprays rather than fat to coat pans.
- Try reducing the fat in recipes a little at a time, and notice that you can do so and still have a tasty meal. If you reduce fat by one tablespoon of butter or oil, you lower the total fat in the product by about 12 grams and at least 100 calories. Or, to reduce the fat even more, try substituting unsweetened applesauce or fruit purees for the oil called for in your favorite cake, quick bread, or cookie recipe.

© Thomas Del Brase/Getty Images

▲ *Bake, broil, poach, or steam.*

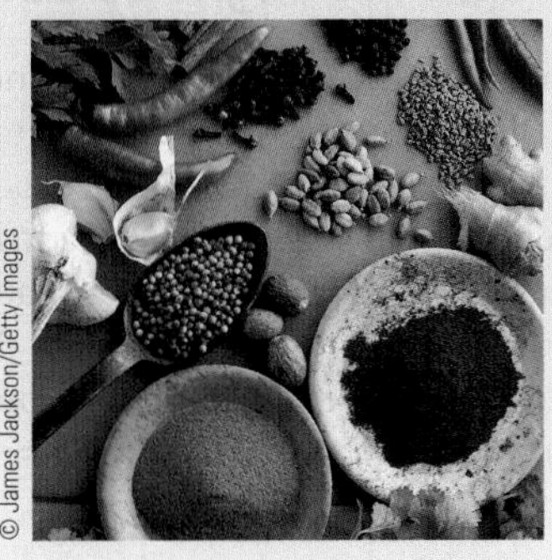
© James Jackson/Getty Images

▲ *Season with herbs and spices.*

- When you sauté a vegetable such as onion in butter, margarine, or oil, try reducing the amount used and substituting water, fat-free vegetable or chicken broth, or wine in its place.
- Prepare broths, soups, and stews ahead of time, refrigerate them, and then skim off hardened fat from the surface.
- Prepare lean cuts of meat, remove visible fat from meat, remove skin from poultry, and cook meats on a rack so that the invisible fat can drain off.
- For flavor in sauces and dressings, experiment with herbs and spices, onions or garlic, salsa, ginger, lemon juice, plain fat-free yogurt with lemon juice, mustard, or butter-flavored granules as replacements for butter, margarine, or oil.
- Try using fruit jam instead of butter on your bagel or toast. Most fruit jams contain half the calories per teaspoon of butter. Or, try the whipped varieties of butter, margarine, or cream cheese, which contain about half the calories of the regular types.

Nutrition Action Oh, Nuts! You Mean Fat Can Be Healthy?

© Quest Photographic, Inc.

Think that a low-fat diet is tasteless? Think again. All fat is not evil! Researchers are finding that people living in Greece, France, Italy, and other countries around the Mediterranean Sea have only a fraction of the heart disease and related deaths experienced by people living in Western, industrial countries like the United States.

What causes this difference? Genetic differences aside, other factors suggest that nutritional differences may play a role because dietary patterns in countries around the Mediterranean Sea vary significantly from those in industrialized countries like the United States, Canada, and northern European countries. What do people who live around the Mediterranean Sea eat that protects them from coronary heart disease? Let's see—the "Mediterranean diet" is rich in fruits, vegetables, grains, fish, and beans, whereas most American diets include much more meat and other animal products known to be high in saturated fat. Thus, the Mediterranean diet is high in the vitamins, minerals, fiber, and **phytochemicals** that keep bodies healthy and low in saturated fat because Mediterranean people consume so little meat and butter (see Figure 4-11). Most Americans don't eat enough whole-grain products, fruits, and vegetables to get the necessary vitamins, minerals, fiber, and phytochemicals that will protect their bodies from coronary heart disease and other chronic diseases.

When More Fat Might Be Better

Study the pyramid in Figure 4-11, and you will see that the Mediterranean diet is not fat free. About 30 percent of total calories are provided by fat. The key is the *type* of fat eaten. Only about 8 percent of total calories come from saturated fats, whereas average Americans consume about 12 percent of their calories from saturated fat. And, as you already know, saturated fat is the major dietary risk factor for heart disease.[20] What do these Mediterranean people use instead of animal fat (butter, meats, ice cream)? They use olive oil and other fats from plant sources that are higher in monounsaturated and polyunsaturated fats.[21]

phytochemicals (FIGH-toe-CHEM-icals) physiologically active compounds found in plants that appear to help promote health and reduce risk for cancer, heart disease, and other conditions. Also called *phytonutrients.*

phyto = plant

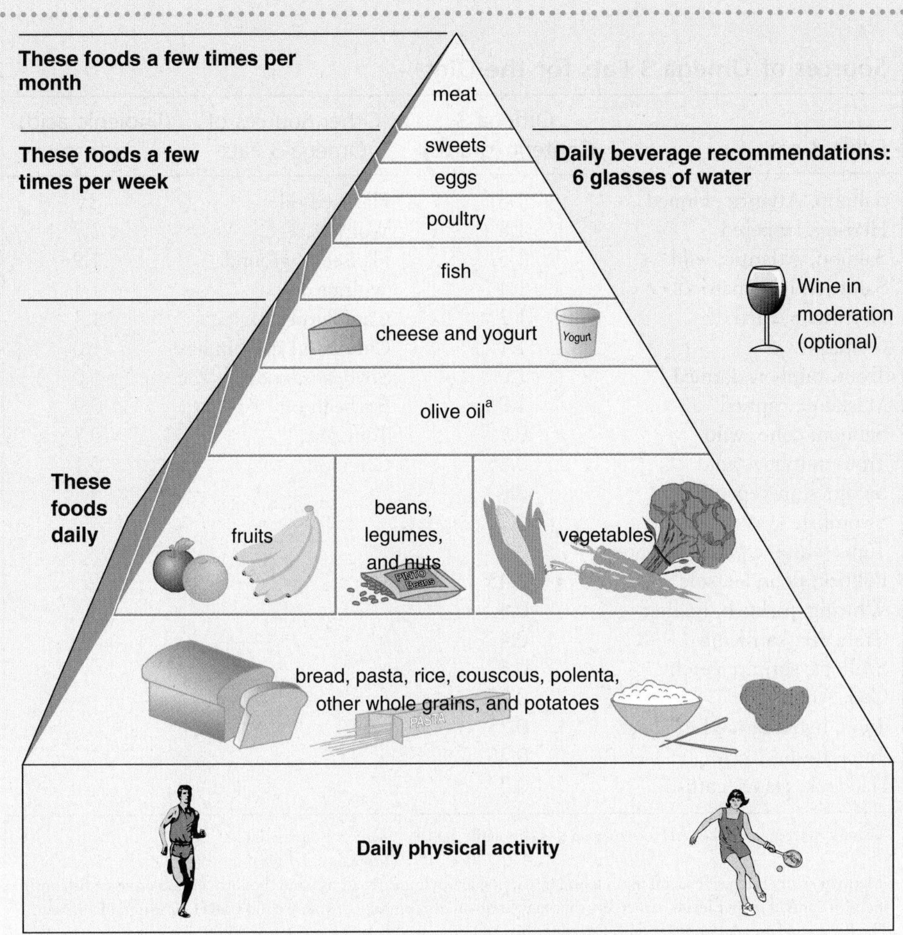

FIGURE 4-11
THE MEDITERRANEAN FOOD PYRAMID

*Other oils rich in monounsaturated fats, such as canola or peanut oil, can be substituted for olive oil. People who are watching their weight should limit oil consumption.

Adapted with permission. © 2000 Oldways Preservation & Exchange Trust, http://oldwayspt.org.

What's So Special about Olive Oil?

Olive oil is high in oleic acid, a monounsaturated fatty acid that seems to keep the heart healthy. Oleic acid helps keep "good" HDL-cholesterol (the artery cleaner) high while lowering levels of "bad" LDL-cholesterol (the artery clogger). It also contains a good ratio of omega-3 fatty acids to omega-6 fatty acids. This is a good thing. But keep in mind olive oil is still a fat. It provides as many calories as any other pure fat. Therefore, olive oil can promote overweight and obesity if too much is consumed. Other vegetable oils high in monounsaturated fats are peanut and canola oils.

Something's Fishy Here!

Another place to find omega-3 fats is fish, especially oily coldwater fish like salmon, cod, farmed catfish, lake trout, herring, bluefish, sardines, albacore tuna, mackerel, and shellfish (see Table 4-7). Eating these fish at least two times a week as part of a balanced diet can help reduce blood clot formation (which can lead to heart attack and stroke), decrease risk of arrhythmias (which can lead to sudden cardiac death), decrease triglyceride levels, decrease growth rate of atherosclerotic plaque, improve the health of the arteries, and slightly lower blood pressure (another risk factor for heart disease).[22]

▲ *Another place to find omega-3 fats is fish, especially fatty fish like salmon.*

TABLE 4-7
Sources of Omega-3 Fats for the Diet[a]

Fish[b] (3 oz)	Omega-3 Content (grams)[c]	Other Sources of Omega-3 Fats	(linolenic acid) g/tbsp
Salmon, Atlantic, farmed	1.8	Flaxseed oil	8.0
Herring, kippered	1.8	Walnuts, ¼ c	2.7
Salmon, Atlantic, wild	1.5	Flaxseeds, ground	1.9
Sardines, in tomato sauce	1.4	Walnut oil	1.4
Herring, pickled	1.2	Canola oil	1.3
Oysters	1.1	Canola oil mayonnaise	1.0
Trout, rainbow, farmed	1.0	Soybeans, cooked, ½ c	1.0
Mackerel, canned	1.0	Soybean oil	0.9
Salmon, coho, wild	0.9	Tofu, ½ c	0.7
Trout, rainbow, wild	0.85	Olive oil	0.1
Sardines, in vegetable oil	0.8		
Swordfish	0.7		
Tuna, white, canned in water	0.7		
Pollock, flounder, sole	0.45		
Whiting, rockfish, halibut	0.4		
Crab, Alaskan King	0.4		
Scallops, shrimp, perch	0.3		
Cod, Atlantic	0.25		
Tuna, light, canned in water	0.25		
Tuna, fresh	0.25		
Haddock, clams, catfish	0.2		

[a]Dietary Reference Intake (AI) for omega-3 fats (linolenic acid): Men: 1.6 g/day
Women: 1.1 g/day

[b]Methylmercury is a heavy metal toxin found in varying levels in nearly all fish and shellfish. However, some fish and shellfish contain higher levels of mercury that may harm an unborn baby or young child's developing nervous system. The Food and Drug Administration (FDA) and the Environmental Protection Agency (EPA) advise women who may become pregnant, pregnant women, lactating women, and young children to avoid some types of fish, including shark, swordfish, king mackerel, or tilefish, because they contain high levels of mercury. Instead, choose up to 12 ounces a week of a variety of fish and shellfish that are lower in mercury, such as shrimp, canned light tuna, salmon, flounder, pollock, or catfish. For updates on the mercury status of fish and shellfish, go to the Center for Food Safety at www.cfsan.fda.gov.

[c]Includes eicosapentaenoic acid (EPA) and docosahexaenoic acid (DHA).

SOURCE: USDA Nutrient Data Laboratory and P. M. Kris-Etherton, W. S. Harris, and L. J. Appel, Fish consumption, fish oil, omega-3 fatty acids, and cardiovascular disease, *Circulation* 106 (2002): 2747–2757.

Recently, the FDA approved a claim for a reduced risk of heart disease for foods, such as fish, that contain the EPA and DHA omega-3 fatty acids: "Supportive, but not conclusive research shows that consumption of EPA and DHA omega-3 fatty acids may reduce the risk of coronary heart disease."[23] So you don't like fish and think you can get the healthy fats you need from fish oil capsules? Think again. The American Heart Association recommends more research to confirm the health benefits of omega-3 fatty acid supplements.[24]

Nuts to You!

Other places to find omega-3 fatty acids are soybeans (actually a legume), pecans, walnuts, and flaxseed. The type of omega-3 fatty acid in these plant foods is less potent than that found in fish oils, but they are good sources,

© PhotoDisc/Getty Images

◀ *Nuts are rich in many nutrients and other beneficial substances but are also high in fat. Two whole walnuts or ten large peanuts contain the same amount of fat as is found in a teaspoon of butter or margarine (5 grams of fat and 45 calories).*

nonetheless.[25] Research has demonstrated that people who eat nuts tend to eat more dietary fat, but the fat is primarily monounsaturated. Also, nuts contain insoluble fiber, another dietary component that protects us from coronary heart disease.[26]

Current Western intakes of omega-3 fats are well below the levels that experts consider optimal.[27] Health experts now advise us to decrease consumption of foods rich in omega-6 fatty acids, such as vegetable oils (corn, safflower, sesame, sunflower), and increase the intake of omega-3 fatty acids (those found in fish, canola and soybean oils, flaxseed, walnuts, other nuts, and soy foods) to achieve a more healthful balance between the two types of fats.

What's Your Meal Mentality?

Another difference between how Americans and Mediterranean people eat is the *way* meals are eaten. The Mediterranean diet involves a wide variety of foods served in several small courses at a slower pace that tends to prevent overeating. Note that one of the dietary guidelines in Greece advises everyone to "eat slowly, preferably at regular times of the day, and in a pleasant environment." Overall, the Mediterranean diet emphasizes olive oil (but no other vegetable oils, hydrogenated oils, or tropical oils); less animal protein (particularly red meats); poultry, eggs, and sweets only a few times a week; and daily use of cheese, yogurt, fruits, vegetables, and whole-grain products. Doesn't sound too hard, does it? Still, other factors undoubtedly contribute to the lower incidence of heart disease in the Mediterranean region.

In understanding the paradox of how a moderate-fat diet (about 30 to 35 percent of calories from mostly monounsaturated fats) can coexist with a low incidence of heart disease, it is important to note that people from the Mediterranean region in general are more physically active, have the social and emotional support found in extended networks of family and friends, and consume more of each day's calories earlier in the day than people in the United States and Canada. Any one of these factors may also be significant in maintaining lower incidence of heart disease in the region. In addition to these possible lifestyle changes, a healthful change in eating patterns (with a Mediterranean twist) for most people in the United States and Canada would be to adopt a diet that is lower in saturated fat and higher in complex carbohydrates and fiber (fruits, vegetables, whole-grain breads and other grains, and legumes).

Scorecard: Rate Your Fats and Health IQ

DO YOU?	OFTEN	SOMETIMES	RARELY
Trim or drain the fat from meats, remove the skin from chicken, and serve fish-based meals?	10	5	1
Eat a variety of fresh, frozen, or canned fruits and vegetables?	10	5	1
Eat high-fat foods such as bacon, sausage, regular franks, and luncheon meat several times a week?	1	5	10
Limit whole eggs or egg yolks to four per week?	10	5	1
Read food labels when shopping?	10	5	1
Choose low-fat or fat-free milk, yogurt, cheese, and sour cream?	10	5	1
Bake, rather than fry, foods?	10	5	1
Maintain a healthy weight?	10	5	1
Think "eating right" and make trade-offs when eating out?	10	5	1
Choose doughnuts, croissants, or sweet rolls for breakfast?	1	5	10
Choose reduced-fat or fat-free products when available?	10	5	1
Routinely add margarine, butter, salad dressing, and sauces to foods?	1	5	10
Balance a high-fat dinner by choosing low-fat foods for breakfast and lunch?	10	5	1
Plan exercise (walking, running, swimming, bike riding) into your schedule on most days of the week?	10	5	1
TOTAL	____	____	____

SCORING

113–150 You practice heart-healthy habits. Keep up the good work.

75–112 Not a bad score, but there's room for improvement.

5–74 Too low a score! Learn more about the relationships between fats and health.

▲ *Large potato with 1 tablespoon butter and 1 tablespoon sour cream (14 grams fat, 350 calories).*

▲ *Large potato with 2 tablespoons fat-free sour cream or yogurt seasoned with chives (less than 1 gram fat, 235 calories).*

SOURCE: Adapted from American Dietetic Association and Mosby Great Performance, *Healthy Eating for the Whole Family* (St. Louis, MO: Mosby–Year Book, 1995), 5.

Understanding Fat Substitutes

Fat-free ice cream, cookies, cakes, and salad dressings have long been the stuff that dieters' dreams are made of. The food industry has now made such fatless fare a reality. With more and more low-fat and fat-free items introduced daily into supermarkets across the country, sales of fat-free products are soaring.

The boom in both low-fat and fat-free products can be accounted for by the country's expanding health consciousness. Although most Americans have heard the warnings that fat can contribute to heart disease, cancer, and obesity, many people find low-fat diets particularly unpalatable because fat adds a desirable flavor and texture—known as "mouth feel"—to foods. With innovations in food chemistry, however, the food industry can now concoct new recipes or ingredients that yield low-fat or nonfat products that retain the characteristic flavor and mouth feel of fat. The types of fat-reduction ingredients currently in use are listed in the margin.

Some manufacturers of fat-free cakes and cookies, for example, make those products by substituting egg whites for whole eggs and fat-free milk for whole milk as well as by removing the butter from their recipes. The end products are sweets that contain fewer calories and fewer grams of fat per serving than traditional desserts.[28] The sample menu shown in Table 4-8 illustrates the role of fat-replacers in the reduction of fat and calories in the diet.

Another technique manufacturers use to replace fat is to add starches, gums, and gels to their products. For example, Kraft uses cellulose gel, a complex carbohydrate, as a filler to make fat-free salad dressings and Sealtest nonfat dairy dessert.[29] Other companies add starches and gums—also complex carbohydrates—to items such as sauces and yogurts to skim fat from those foods. That's

Types and Uses of Fat-Replacer Ingredients

Carbohydrate based:
Carrageenan (a seaweed derivative), fruit purees, gelatin, gels derived from cellulose or starch, guar gum, xanthum gum, maltodextrins made from corn, corn starch (Stellar), polydextrose, Oatrim (made from oat fiber), and Z-trim (a modified form of insoluble fiber)
Uses: Dairy products, frozen desserts, salad dressings, baked goods, cake and cookie mixes, frostings.

Protein based:
Whey protein concentrate (Dairy-Lo), microparticulated protein products (Simplesse®, K-Blazer) made from whey, or milk and egg white protein
Uses: Cheese, butter, salad dressings, mayonnaise, sour cream, dairy products, ice cream, baked goods.

Fat based:
Mono- and diglycerides; Caprenin (a substitute for cocoa butter in candy) and Salatrim (found in reduced-fat baking chips) both contain long chain fatty acids, which are partially absorbed, and short-chain fatty acids (providing 5 calories per gram); Olestra (noncaloric artificial fat made from fatty acids and sucrose)
Uses: Savory snacks, chocolate confections, bakery products.

TABLE 4-8
The Role of Fat Replacers in Fat and Calorie Reduction

This sample menu shows the difference (in fat and calories) that eating foods containing fat replacers can make.

Regular Lunch

	Calories	Fat (grams)
Bread, 2 slices	130	2
American cheese, 1 oz	105	9
Bologna, 2 oz	180	17
Mayonnaise, 1 tbsp	100	11
Banana	105	0
Chocolate cookies, 2	140	6
	760	45

Fat-Replaced Lunch

	Calories	Fat (grams)
Bread, 2 slices	130	2
Reduced-fat cheese product, 1 oz	75	4
Fat-free bologna, 2 oz	40	0
Low-fat mayonnaise/dressing, 1 tbsp	25	1
Banana	105	0
Reduced-fat chocolate cookies, 2	120	3
	495	10

SOURCE: Adapted from P. Kurtzweil, "Taking the Fat Out of Food," *FDA Consumer* (July–August 1996).

Simplesse® the trade name for a protein-based, low-calorie artificial fat, approved by the FDA for use in foods such as frozen desserts; cannot be used for frying or baking.

olestra an artificial fat derived from vegetable oils and sugar combined in such a way that the body cannot break them down. Sold under the brand name Olean®, olestra does not contribute calories to food. It can, however, prevent absorption of some nutrients. Thus, the FDA requires all products made with it to bear this warning: "This Product Contains Olestra. Olestra may cause abdominal cramping and loose stools. Olestra inhibits the absorption of some vitamins and nutrients. Vitamins A, D, E, and K have been added."

because starches and gums hold water and impart a smooth creamy texture similar to that of fat, and they add form and structure to foods. These substitutes cannot, however, replace the fat used for cooking and frying.[30]

Yet another more innovative approach the food industry has taken to provide fat-free fare is the development of fat substitutes. In 1990, a substance called **Simplesse®** became the first such product to gain the approval of the Food and Drug Administration. Six years in the making, Simplesse is a mixture of food proteins such as egg white, whey, and milk protein that are cooked and blended to form tiny round particles that trap water. Inside the mouth, the particles roll over one another, and the tongue perceives them as a creamy, smooth liquid similar to fat.

The FDA allows use of Simplesse in foods including cheese, baked goods, ice creams, frozen desserts, mayonnaise, salad dressings, yogurts, sour cream, and butter. Foods made with this fat substitute contain considerably less fat and fewer calories than their traditional fat-containing counterparts.[31] The reason is that whereas fat contains 9 calories per gram, Simplesse supplies only 1 to 2. Simplesse cannot be used for frying or baking, because heat causes it to gel and lose its creaminess.

Another artificial fat approved for use in salty snacks and crackers is **olestra** (sold under the brand name Olean®). Created by Procter & Gamble, the olestra molecule resembles a triglyceride but is structured in a way that prevents its breakdown by digestive enzymes in the body, thereby allowing it to pass through the digestive tract completely unabsorbed (see Figure 4-12). Olestra interferes with the absorption of the fat-soluble vitamins (A, D, E, and K) as well as beta-carotene. To compensate for this effect, olestra must be fortified with the fat-soluble vitamins. Because twenty years of research consider olestra to be safe for use by consumers, FDA no longer requires products containing olestra to bear a warning label.[32] However, if large amounts of olestra are consumed, it can cause nutrient losses, abdominal cramping, and loose stools.

Certainly, the growing number of fat-free foods and fat substitutes provides consumers with viable alternatives to fattier fare. Nevertheless, many experts view the fat-free boom with skepticism. Although reduced-fat foods can help lower the overall fat content of the diet, they *do* contain calories, and they are *not* a replacement for a healthful diet rich in whole grains, fresh fruits, and vegetables. Nor are they likely to become the panacea that will prevent problems such as heart disease and obesity. When included in a low-fat diet, however, they can help consumers reach and adhere to the goal of taking in no more than 20 to 35 percent of total calories as fat.[33]

FIGURE 4-12
OLESTRA: A FAT-FREE FAT

Compare the structure of olestra to that of a triglyceride. Whereas triglycerides have three fatty acids attached to a glycerol core, olestra has up to six, seven, or eight fatty acids on a sucrose core. The extra fatty acids make it too large to be digested or absorbed by the body, which is why it adds no fat or calories to the diet. Therefore, a 1-ounce bag of potato chips, which typically has 10 grams of fat and 150 calories, has 0 grams of fat and 60 calories when Olean® is added in place of fat.

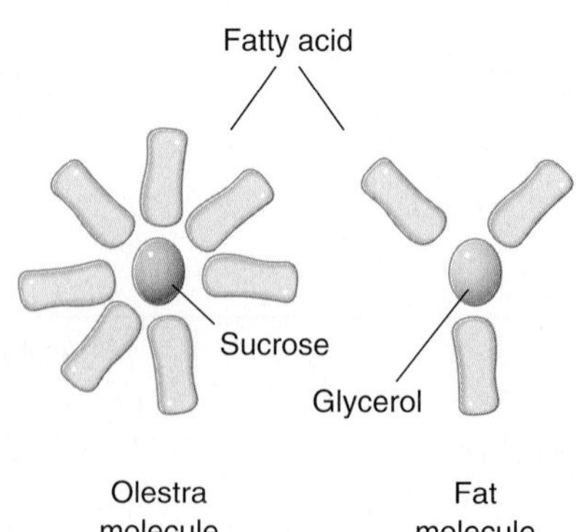

Spotlight

Diet and Heart Disease

More than half the people who die in the United States each year die of heart and blood vessel disease. The underlying condition that contributes to most of these deaths is atherosclerosis, which leads to closure of the arteries that feed the heart and brain and thus to heart attacks and strokes. In terms of direct health care costs, lost wages, and lost productivity, heart disease costs the United States more than $60 billion a year. There is little wonder, then, that much effort has been focused on preventing it.

The twin demons that lead to most forms of heart disease are atherosclerosis and hypertension. Atherosclerosis, the subject of this Spotlight, is the narrowing of the arteries caused by a buildup of cholesterol-containing plaque in the arterial walls. Hypertension (which is discussed in the Nutrition Action feature of Chapter 7) is high blood pressure, and each of these conditions aggravates the other.

How can I know whether I have atherosclerosis?

No one is free of atherosclerosis. The question is not whether you have it but how far advanced it is and what you can do to retard or reverse it. As already mentioned, atherosclerosis usually begins with the accumulation of soft mounds of lipid, known as *plaques*, along the inner walls of the arteries, especially at the branch points. These plaques gradually enlarge, making the artery walls lose their elasticity and narrowing the passage through them. Most people have well-developed plaques by the time they are 30.

Normally, the arteries expand with each heartbeat to accommodate the pulses of blood that flow through them. Because arteries that are hardened and narrowed by plaques cannot expand, the blood pressure rises. The increased pressure puts a strain on the heart and further damages the artery wall. At damaged points, plaques are especially likely to form; thus, the development of atherosclerosis is a self-accelerating process.

Hypertension makes atherosclerosis worse. A stiffened artery, already strained by each pulse of blood surging through it, is stressed even more if the internal pressure is high. Hypertension causes injured places within the blood vessels to develop more frequently, and plaques grow faster. Hardened arteries also fail to let blood flow freely through the body's blood pressure-sensing organs, the kidneys, which respond as if the blood pressure were too low and raise it further.

How can I slow the process down?

Learn your risk factors, and control the ones you can. Among the many factors linked to heart disease are smoking, gender (being male), age (men older than 45 and women older than 55), postmenopausal status in women, heredity, diabetes, high blood pressure, lack of exercise, obesity, high blood cholesterol level, and low HDL-cholesterol level (see Table 4-9). Some of the risk factors are also powerful *predictors* of heart disease. If you have none of the risk factors, the statistical likelihood of your developing heart disease may be only 1 in 100. If you have three major risk factors, your chance may rise to over 1 in 20. Some of the factors that have emerged as powerful predictors of risk are high LDL cholesterol, low HDL-cholesterol, high blood pressure, smoking, obesity, and physical inactivity.[34]

What can I do to reduce my risk?

Obviously, some risk factors cannot be altered. Being born male or inheriting a predisposition to develop high blood cholesterol or high blood pressure are factors beyond your control. Even so, you can make conscious choices that may reduce your risk of developing heart disease. Let's examine one of the two major risk factors related to diet—high blood cholesterol—with an eye toward learning which dietary and lifestyle changes will help reduce your heart disease risk.

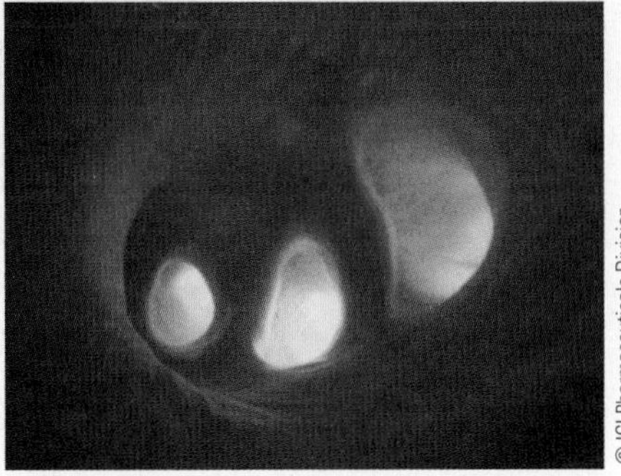

▲ *A normal artery provides open passage for blood to circulate.*

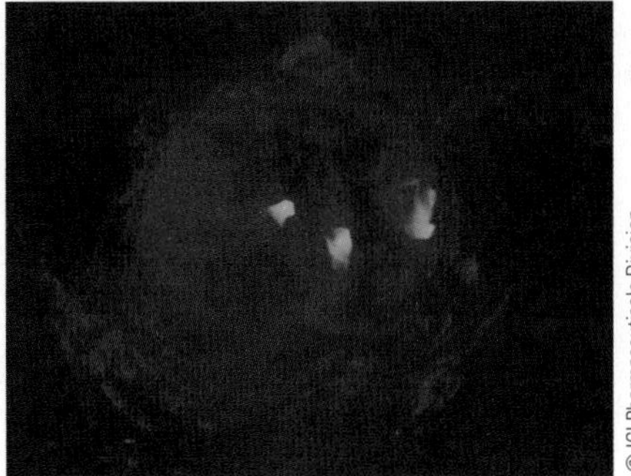

▲ *Plaques along an artery narrow the passage and obstruct blood flow.*

TABLE 4-9
Leading Risk Factors for Heart Disease

Heart disease rarely develops from a single risk factor. Clusters of risk factors usually occur, and a small increase in one risk factor, such as blood pressure, becomes more critical when combined with other risk factors. Luckily, a similar pattern happens in reverse. Even moderate changes in one risk factor can decrease several others at the same time. For example, a weight loss of just 5 to 10 pounds can reduce blood pressure in overweight people. Or, becoming physically active can lower blood pressure, increase HDL ("good" cholesterol), and help control weight.

Risk Factor	How to Minimize the Risk
Risk Factors You Can Change	
High LDL-cholesterol; Low HDL-cholesterol	Limit intake of cholesterol, saturated fat, and *trans* fat. Increase your intake of soluble fiber, soy foods, and omega-3 fats. Maintain a physically active lifestyle.
High blood pressure	Control high blood pressure with medication and a heart-healthy diet. Maintain a healthy weight. Losing just 5 to 10 pounds may lower your blood pressure.
Cigarette smoking	Stop smoking. Nicotine constricts blood vessels and forces your heart to work harder. Carbon monoxide reduces oxygen in blood and damages the lining of blood vessels.
Diabetes	Maintain proper weight. Losing excess weight helps control blood sugar level. Eat high-fiber foods. Limit saturated fat and sugar. Get regular exercise.
Physical inactivity	Get at least 60 minutes of moderate-paced physical activity on most days of the week. See Chapter 10 for more about exercise.
Obesity	Maintain a healthy weight and exercise. Being only 10 percent overweight increases heart disease risk. See Chapter 9 for more about weight management.
"Atherogenic" diet	Keep saturated fat to under 10 percent of daily calories. Substitute olive and canola oils for saturated fat. Increase fiber intake by eating cereal grains, legumes, fruits, and vegetables. Eat five to nine servings of fruits and vegetables a day to receive the beneficial antioxidants and phytochemicals they contain.
Stress	Get regular exercise. Avoid excessive caffeine and alcohol. Practice relaxation techniques. Maintain good social relationships.
Risk Factors You Can't Change	
Age	Men over age 45 and women over age 55 are at increased risk.
Gender	Men are at higher risk. Estrogen may protect women before menopause.
Genetics	Increased risk if you have a father or brother under age 55 or a mother or sister under 65 who had heart disease.

To what extent does a high blood cholesterol level raise the risk of developing heart disease?

The likelihood of a person developing or dying from heart disease increases as total blood cholesterol level rises. Figure 4-13 presents this relationship graphically, showing that the number of deaths from heart disease increases steadily among those with elevated blood cholesterol levels, particularly when the level rises above 200 milligrams per deciliter. Individuals with a blood cholesterol level in the neighborhood of 300 milligrams per deciliter run four times the risk of dying from heart disease compared with those whose cholesterol level is lower than 200 milligrams per deciliter. Having a low HDL-cholesterol value is now also recognized as being a risk factor for heart disease.[35] A low HDL value is defined as one below 40 milligrams per deciliter. A combined effort to lower LDL-cholesterol and raise HDL-cholesterol delivers a double punch in the fight against heart disease.

Reducing high blood cholesterol levels, particularly the "bad" LDL-cholesterol level, is thus an important strategy toward lowering the risk of heart disease. This is especially true for persons having one or more of the other major heart disease risk factors, such as smoking. Table 4-10 provides information for determining what to do if your blood cholesterol level is high and whether you might need to seek treatment.

What sorts of dietary changes should I make to reduce my total blood cholesterol and LDL-cholesterol levels?

Probably the most significant dietary changes you can make are to reduce the amounts of saturated fat and *trans* fat that you eat since high intakes are related to high blood levels of LDL-cholesterol.[36] The goals to work toward are to reduce your intake of saturated fat to less than 10 percent of total calories, and keep *trans* fatty acid consumption as low as possible. People with elevated levels of LDL-cholesterol,

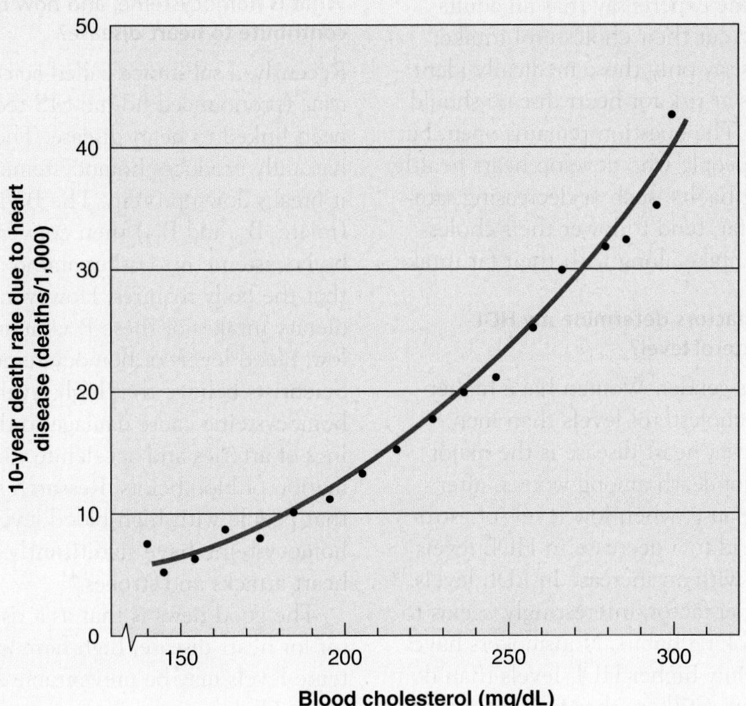

FIGURE 4-13
RELATIONSHIP BETWEEN BLOOD CHOLESTEROL LEVEL AND DEATH RATE FROM HEART DISEASE

Persons with blood cholesterol levels of 300 mg/dL are four times as likely to die from heart disease as those with blood cholesterol levels below 200 mg/dL.

SOURCE: Data from 361,662 men screened for the Multiple Risk Factor Intervention Trial (MRFIT). Adapted from National Cholesterol Education Program, *Report of the Expert Panel on Population Strategies for Blood Cholesterol Reduction* (Bethesda, MD: U.S. Department of Health and Human Services, Public Health Service, National Institutes of Health, National Heart, Lung, and Blood Institute, NIH Publication No. 90-3046, November 1990), 7.

TABLE 4-10

Standards for Blood Cholesterol Levels and Risk of Heart Disease

Evaluating Blood Lipid Levels

	Desirable	Borderline High	High Risk
Total blood cholesterol (mg/dL)	<200	200–239	≥240
LDL-cholesterol (mg/dL)	<100[a]	130–159	160–189[b]
HDL-cholesterol (mg/dL)	≥60	40–59	<40
Triglycerides, fasting (mg/dL)	<150	150–199	200–499[c]

Treating LDL-Cholesterol Levels

Risk Status	Target/Recommendations for LDL-Cholesterol
Low-risk individuals[d]	Less than 160 mg/dL Begin dietary therapy at 160 mg/dL; consider drug therapy at 190 mg/dL.
High-risk individuals[e]	Less than 130 mg/dL Begin dietary therapy at 130 mg/dL; consider drug therapy at 160 mg/dL.
Very high-risk individuals[f]	Less than 100 mg/dL Begin dietary therapy at 100 mg/dL; consider drug therapy at 130 mg/dL.

[a]100–129 mg/dL LDL indicates a near or above optimal level.
[b]≥190 mg/dL LDL indicates a very high risk.
[c]≥500 mg/dL triglycerides indicates very high risk.
[d]Low-risk individuals are those with one or no risk factors for heart disease. Major risk factors (other than LDL cholesterol) include cigarette smoking, hypertension or on antihypertensive medication, low HDL cholesterol (< 40 mg/dL), family history of premature heart disease, age (men ≥ 45 years; women ≥ 55 years).
[e]High-risk individuals are those who have at least two risk factors for CHD. (Men over age 45 and postmenopausal women are considered to have one risk factor.)
[f]Very high-risk individuals are those with known heart disease (history of heart attack or previous bypass surgery or angioplasty) and diabetes.

existing heart disease, diabetes, or combinations of risk factors are advised to reduce their intake of saturated fat even further, to less than 7 percent of total calories and their cholesterol intakes should be less than 200 milligrams per day.[37] See this Chapter's Eat Well Be Well feature for more specifics on reducing your LDL-cholesterol level.

Can the addition of plant sterols to the diet help reduce blood cholesterol levels?

Certain new brands of margarine contain phytosterols (plant stanols or plant sterols), which are compounds derived from plant oils that may actually reduce blood cholesterol when consumed as part of a low-fat diet.[38] A phytosterol is a phytochemical that is structurally similar to the steroid hormones (for example, estrogen). Examples of margarines containing phytosterols include Benecol and Take Control. Benecol uses a stanol ester derived from pine trees and Take Control uses one made from soy. These compounds block the absorption of dietary cholesterol from the intestine. The FDA has given the manufacturers approval for their label to state that the products "promote healthy cholesterol levels." Margarines containing these compounds, however, typically cost three times or more than regular margarine, and the safety of their long-term daily use is not known.

Should I also reduce my cholesterol intake?

Dietary cholesterol can raise LDL levels, depending on the amount consumed and on the body's ability to compensate by making less. The *Dietary Guidelines* recommend that we limit the cholesterol in our daily diets to 300 milligrams. On average, men consume 345 milligrams of cholesterol and women consume about 210 milligrams daily. The average intake for American men is greater than that consumed in countries with little heart disease.

Some experts say that all adults should cut their cholesterol intake; others say only those medically identified as at risk for heart disease should do so. The question remains open, but most people who develop heart-healthy eating habits, such as decreasing saturated fat, tend to lower their cholesterol intake along with their fat intake.

What factors determine my HDL-cholesterol level?

One is gender. Women have higher HDL-cholesterol levels than men. However, heart disease is the major cause of death among women after menopause, when low levels of estrogen lead to a decrease in HDL levels along with an increase in LDL levels.[39] Another factor, interestingly, seems to be smoking habits. Nonsmokers have uniformly higher HDL levels than do smokers. Still another factor is weight reduction for those who are overweight.[40]

If any dietary factors are significant, one important one may be to include some fish in the diet.[41] Another may be including soy foods and foods containing soluble fibers that lower LDL levels.* By far the most powerful influence on HDL levels is not a dietary factor at all—it is regular exercise. The discovery that exercise raises HDL levels has given great impetus to the physical fitness movement, and especially to the popularity of running and walking. The earliest reports indicated increased HDL levels in long-distance runners, and the continuing study of this elite group has repeatedly demonstrated that running does indeed elevate HDL. People do not, however, need to become competitive athletes to raise their HDL. Moderate exercise, such as walking, may both lower LDL levels and raise HDL levels if the activity is consistently pursued for long enough periods.

*For a discussion of the benefits of soy in prevention of heart disease, see the Spotlight in Chapter 5.

What is homocysteine, and how does it contribute to heart disease?

Recently, a substance called *homocysteine* (pronounced ho-mo-SIS-teen) has been linked to heart disease. The body naturally produces homocysteine when it breaks down protein. The B vitamins (folate, B_6, and B_{12}) then convert homocysteine into other amino acids that the body requires. However, when dietary intakes of these B vitamins are low, blood levels of homocysteine rise. Scientists believe that high levels of homocysteine cause damage to the linings of arteries and accelerate the formation of blood clots. Research shows that people with high blood levels of homocysteine have significantly more heart attacks and strokes.[42]

The good news is that as a risk factor for heart disease, high homocysteine levels may be preventable and reversible by consuming generous amounts of the B vitamins. Good sources of folate include orange juice; enriched breads, cereals, and other grain products; green leafy vegetables; and legumes. Vitamin B_6 is found in meat, chicken, fish, whole grains, and legumes. Vitamin B_{12} is found in fortified cereals, fortified soy milk, and all animal foods. (See Chapter 6 for more information about these B vitamins.)

What about alcohol? I've heard that moderate alcohol intake can be beneficial to heart disease.

The evidence with regards to alcohol and blood lipids points to a possible protective mechanism between moderate alcohol consumption and heart disease risk factors.[43] Investigators have reported that the consumption of moderate amounts of alcohol appears to raise HDL levels.[44] Moderate drinking is defined as no more than one drink a day for most women, and no more than two drinks a day for most men. A strong association between moderate alcohol consumption and low rates of heart disease was first observed in France and then in other wine-drinking areas of the Mediterranean. Researchers have since identified antioxidants and other compounds in red wine that decrease blood clotting, and this may help explain part of the "French paradox," or how a region with high fat intakes could have such low rates of heart disease. Researchers are quick to point out, however, that other factors certainly may contribute to the paradox. For example, the French consume about 57 percent of their day's calories before 2 P.M., whereas most Americans have only consumed about 38 percent of their calories by that time.[45] Also, the French mostly consume their wine with meals and eat more fruits and vegetables and leaner cuts of meat.

Scientists warn that caution is needed in recommending moderate alcohol consumption to the public. Many people need to refrain from alcohol consumption altogether (pregnant women, recovering alcoholics, people under age 21, persons on certain medications, and those intending to drive a vehicle).[46] Keep in mind that alcohol can have profoundly negative effects on the body and is associated with many disease states (see Chapter 8).

What about garlic? I've heard that garlic pills and whole cloves help lower blood cholesterol.

A sizeable body of evidence from around the world suggests that garlic may help lower blood cholesterol, reduce blood pressure, and help arteries remain elastic.[47] Although the research is promising, some studies have shown conflicting results regarding the effect of garlic on blood cholesterol. Further studies are needed before scientists can come to any conclusions about the usefulness of fresh garlic or garlic pills, which vary a great deal in composition, in lowering blood cholesterol levels. See Chapter 6 for more about the phytochemical benefits derived from garlic and its relatives (leeks, onions, chives, scallions, and shallots).

How do I translate these recommendations into a heart-healthy lifestyle?

Most people have a difficult time translating the dietary recommendations into actual meal patterns. This Chapter's Eat Well Be Well feature provided many tips for reducing your LDL-cholesterol level. The American Heart Association makes the following suggestions as well:[48]

- Eat a variety of antioxidant-rich fruits and vegetables. Choose five or more servings per day.
- Choose fat-free or low-fat dairy products, such as fat-free milk or low-fat or fat-free yogurt.
- Consume abundant legumes of many varieties, including soybeans, kidney beans, and lentils.
- Eat a variety of grain products, including whole grains.
- Choose skinless poultry, lean meat, and fish, especially omega-3-fatty-acid-rich fish such as salmon.
- Limit foods high in saturated fat, *trans* fat and/or cholesterol, such as full-fat milk products, fatty meats, tropical oils, partially hydrogenated vegetable oils, and egg yolks.
- Limit your intake of foods high in calories or low in nutrition, including foods like soft drinks and candy that have a lot of sugars.
- Adopt low-fat cooking methods, such as broiling, baking, steaming, and stir-frying.
- Consume alcohol only in moderation, if at all.
- Eat less than 6 grams of salt (sodium chloride) per day (2,400 milligrams of sodium).
- Have no more than one alcoholic drink per day if you're a woman and no more than two if you're a man. "One drink" means it has no more than ½ ounce of pure alcohol. Examples of one drink are 12 ounces of beer, 5 ounces of wine, 1½ ounce of 80-proof spirits, or 1 ounce of 100-proof spirits.

Attention to emotional health is also important in reducing the risk of heart disease. Both love and affection seem to affect the heart. People with many social ties appear to develop less heart disease than people with few or none. Similarly, pet owners, even owners of pet fish, have lower blood pressure than do people without pets. Clearly, the mystery of heart disease, like all the great human mysteries, involves the mind and spirit as well as the body. So nourish yourself in all ways—not just physically.

Should children, like their parents, eat low-fat diets for heart health?

The current Dietary Guidelines and experts representing 42 major U.S. health and professional organizations recommend that children aged 2 and older eat diets containing no more than 30 percent of total calories from fat and no more than 300 milligrams of cholesterol daily. The reason is that hardening of the arteries often begins in childhood. However, from birth to 2 years of age, a child's fat consumption should not be restricted, because fat is a concentrated source of the calories needed to ensure proper physical development.

The panel also advised that children or teens should get their blood cholesterol measured if they have one parent with a high blood cholesterol level. For children and adolescents, a total of 200 milligrams per deciliter or more is considered high, 170 to 199 is borderline high, and less than 170 is acceptable. In addition, children born to families with a history of premature heart disease should have both total blood cholesterol and HDL-cholesterol checked.[49]

In Review

1. All of the following statements about dietary fat are correct *except*:
   a. Fat carries water-soluble vitamins in food.
   b. Fat is a component of all cell membranes.
   c. Fat contributes 9 calories per gram.
   d. Fat can act as an insulator in the body.

2. Which of the following groups of foods contain cholesterol?
   a. cheese and eggs
   b. canola oil, peanut oil, and corn oil
   c. avocados, prunes, and bananas
   d. All of the above foods contain cholesterol.

3. Americans are urged to eat more fish due to its high content of:
   a. vitamin D.
   b. HDL-cholesterol.
   c. monounsaturated fat.
   d. omega-3-fatty acids.

4. The building blocks of the chief form of fat found in the body are:
   a. triglycerides.
   b. monoglycerides.
   c. antioxidants.
   d. fatty acids and glycerol.

5. Briefly explain why fats are classified as saturated, monounsaturated, or polyunsaturated fats.

6. A fat that is very unsaturated may __________ at room temperature.
   a. explode
   b. melt (or liquefy)
   c. solidify
   d. be a liquid
   e. Both b and d are correct.

7. Hydrogenation is a huge buzzword right now. What is it and why is it used?

8. What does it mean that linoleic acid and linolenic acid are "essential" to the human body?

9. Which of the following is true of *trans* fats?
   a. *Trans* fats lower LDL-cholesterol.
   b. *Trans* fats lower total cholesterol.
   c. *Trans* fats raise LDL-cholesterol.
   d. Both b and c are correct.

10. In general, individuals should try to keep:
    a. HDL and LDL cholesterol low.
    b. HDL cholesterol high and LDL cholesterol low.
    c. both high-density lipoprotein (HDL) and low-density lipoprotein (LDL) cholesterol high.
    d. None of the above; HDL and LDL are uncontrollable factors.

Menu of Online Study Tools

A variety of study tools for this chapter are available at our website to deepen your understanding of chapter concepts. Go to
www.thomsonedu.com/nutrition/boyle
to find

- Practice tests
- Flashcards
- Glossary
- Web links
- Animations
- Chapter summaries, learning objectives, and crossword puzzles

Nutrition on the Web

www.thomsonedu.com/nutrition/boyle
Go to the *Personal Nutrition site* to check for the latest updates to chapter topics or to access links to related websites.

www.eatright.org
Search for information on fats and oils, and review the ADA position paper on fat replacers.

www.mayoclinic.com
Provides practical health and nutrition information to consumers, current articles on a variety of nutrition topics, recipes with lower-fat versions of family-favorite recipes, and a quiz to test your cooking skills.

www.fda.gov
Click on Food and search for information on Olestra and other fat substitutes and labeling of *trans* fatty acids in foods.

www.cnn.com/HEALTH
The CNN Health page posts research updates; search for fat and cholesterol information for a healthy diet.

www.healthfinder.gov
This site presents a wide assortment of health and nutrition information from government agencies.

www.nhlbi.nih.gov/chd
The National Heart, Lung, and Blood Institute provides consumers and health professionals with practical information for cardiovascular health. This site explains how heart disease develops and provides tips for reducing your cholesterol level and risk for heart disease. Visit the Cyber Kitchen and Cyber Cafe to determine how well you estimate portion sizes, and go to Create a Diet to see how your diet compares to the current dietary recommendations for a healthy heart.

www.americanheart.org
The American Heart Association provides practical advice about following a healthy lifestyle, a risk assessment tool, a quiz to test your knowledge about diet, exercise, and heart disease, recipes, and useful links to related sites.

www.heartinfo.org
Provides information on healthy shopping and cooking, eating healthy away from home, general nutrition tips, heart disease, high blood pressure, supplements and heart disease, and weight management.

www.caloriecontrol.org
Search for information on fat replacers and reduced-fat products, low-fat recipes, and note the discussion on "Calories Still Count."

www.flaxcouncil.ca
The Flax Council of Canada gives useful information on the health benefits of flaxseed, soluble fiber, and omega-3 fatty acids. The site posts research updates, fact sheets, and recipes.

www.heartandstroke.ca
The Heart and Stroke Foundation of Canada offers a tutorial on heart-healthy food choices, an assessment of your risk for heart disease, and a quiz to test your heart health knowledge.

www.healthyfridge.org
This site provides information on heart disease, practical shopping tips for stocking a healthy refrigerator, heart-healthy recipes, and a quiz to test your saturated fat IQ.

www.ncbi.nlm.nih.gov/PubMed
A search engine to help you locate information from current scientific articles on any topic related to fat.

The Proteins and Amino Acids

5

ASK YOURSELF . . .

Which of the following statements about nutrition are true, and which are false? For each false statement, what is true?

1. When more protein is eaten than the body needs, it is stored intact in the body (the way fat is stored), so that it can be used when a person's diet falls short of supplying the day's need for essential proteins.
2. No new living tissue can be built without protein.
3. Whenever cells are lost, protein is lost.
4. All enzymes and hormones are made of protein.
5. When antibodies enter the body, they produce illness.
6. When a person doesn't eat enough food to meet the body's energy needs, the body devours its own protein tissue.
7. Once the body has assembled its proteins into body structures, it never lets go of them.
8. Milk protein is the standard against which the quality of other proteins is usually measured.
9. It is impossible to consume too much protein.
10. People who eat no meat need to eat a lot of special foods to get enough protein.

Answers found on the following page.

The amino acids of proteins are the raw materials of heredity, the keys to life chemistry, handed from generation to generation.

R. M. Deutsch (1928–1988, nutrition author and educator)

CONTENTS

The **proteins** are perhaps the most highly respected of the three energy nutrients, and the roles they play in the body are far more varied than those of carbohydrate or fat. First named 150 years ago after the Greek word *proteios* ("of prime importance"), proteins have revealed countless secrets about how living processes take place, and they account for many nutrition concerns. How do we grow? How do our bodies replace the materials they lose? How does blood clot? What makes us able to become immune to diseases we have been exposed to? To a great extent, the answers to these and many other such questions are found in an understanding of the nature of the proteins.

What Proteins Are Made Of

To appreciate the many vital functions of proteins, we must understand their structure. One key difference from carbohydrate and fat, which contain only carbon, hydrogen, and oxygen atoms, is that proteins contain nitrogen atoms. These nitrogen atoms give the name *amino* ("nitrogen containing") to the **amino acids** of which protein is made. Another key difference is that in contrast to the carbohydrates—whose repeating units, glucose molecules, are identical—the amino acids in a strand of protein are different from one another.

An Amino Acid: glycine

Amine group — H–N–C–C–OH (H above and below C, =O on acid carbon) — Acid group

Side group

An Amino Acid: phenylalanine

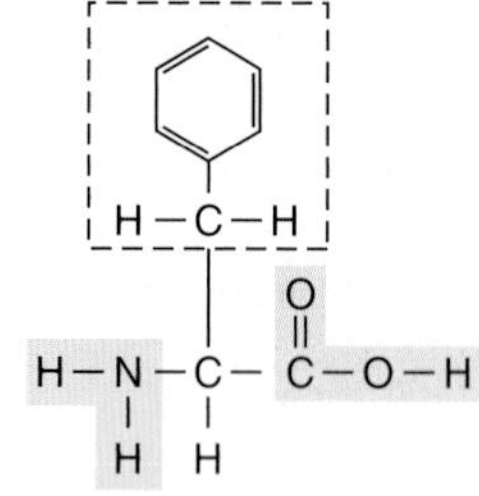

All amino acids have the same, simple chemical backbone with an **amine group** (the nitrogen-containing part) at one end and an acid group at the other end. The differences among the various amino acids are due to the varying structures of the chemical side chains that are attached to the backbone. Twenty amino acids with 20 different side chains make up most of the proteins of living tissue.

The side chains vary in complexity from a single hydrogen atom like that on glycine to a complex ring structure like that on phenylalanine. Not only do these structures differ in composition, size, and shape, but they also differ in electrical charge. Some are negative, some are positive, and some have no charge. These side chains help to determine the shapes and behaviors of the larger protein molecules that the amino acids make up.

Essential and Nonessential Amino Acids

The body can make about half of the amino acids (known as nonessential amino acids) for itself, when it has the needed parts—nitrogen to form the amine group and backbone fragments, which are derived from carbohydrate or fat. But even the healthy body cannot make some amino acids. These are known as **essential amino acids.*** If the diet does not supply them, the body cannot make the proteins it needs to do its work. The indispensability of the essential amino acids makes it necessary for people to eat protein food sources every day.

The nine essential amino acids for human adults that must be obtained from the diet:

histidine	phenylalanine
isoleucine	threonine
leucine	tryptophan
lysine	valine
methionine	

The nonessential amino acids—also important in nutrition:

alanine	glutamine
arginine	glycine
asparagine	proline
aspartic acid	serine
cysteine	tyrosine
glutamic acid	

Proteins as the Source of Life's Variety

In the first step of **protein synthesis,** each amino acid is hooked to the next. A bond, called a **peptide bond,** is formed between the amino end of one and the acid end of the next. Proteins are made of many amino acid units, from several dozen to many hundred.

A strand of protein is not straight; it is more like a tangled chain. The amino acids at different places along the strand are attracted to one another, and this attraction causes the strand to coil into a shape similar to that of a metal spring. Not only does the strand of amino acids form a long coil, but the coil tangles, forming a globular structure.

The charged amino acids are attracted to water, and in the body fluids they orient themselves on the outside of the globular structure. The neutral amino acids are repelled by water and are attracted to one another; they tuck themselves into the center, away from the body fluid. All these interactions among the amino acids and the surrounding fluid result in the unique architecture of each type of protein. Additional steps may be needed for the protein to become functional. A mineral or a vitamin may be needed to complete the unit and activate it, or several proteins may gather to form a functioning group.

The differing shapes of proteins enable them to perform different tasks in the body. In proteins that give strength and elasticity to body parts, several springs of amino acids coil together and form ropelike fibers. Other proteins, like those in the blood, do not have such structural strength but are water soluble, with a globular shape like a ball of steel wool. Some are shaped like hollow balls that

proteins compounds composed of atoms of carbon, hydrogen, oxygen, and nitrogen and arranged as strands of amino acids. Some amino acids also contain atoms of sulfur.

amino (a-MEEN-o) **acids** building blocks of protein; each is a compound with an amine group at one end, an acid group at the other, and a distinctive side chain.

amine (a-MEEN) **group** the nitrogen-containing portion of an amino acid.

essential amino acids amino acids that cannot be synthesized by the body or that cannot be synthesized in amounts sufficient to meet physiological need.

protein synthesis the process by which cells assemble amino acids into proteins. All individuals are unique because of minute differences in the ways their body proteins are made. The instructions for making all the proteins in our bodies are transmitted in the genetic information we receive at conception.

peptide bond a bond that connects one amino acid with another.

Ask Yourself Answers: **1.** False. When more protein is eaten than the body needs, it is *not* stored in the body (the way fat is stored), so it must be eaten every day to avoid protein depletion. **2.** True. **3.** True. **4.** False. All enzymes, but not all hormones, are made of protein. **5.** False. Antibodies protect the body from illness caused by antigens. **6.** True. **7.** False. Your body loses protein every day. **8.** False. Egg white protein, not milk protein, is the standard against which the quality of other proteins is usually measured. **9.** False. It is possible to consume too much protein. **10.** False. People who eat no meat can easily get enough protein without eating a lot of special foods.

*The distinction between essential and nonessential amino acids is not quite as clear-cut as the list in the margin makes it appear. For example, cysteine and tyrosine normally are not essential because the body makes them from methionine and phenylalanine. However, if there are not enough of these precursors from which to make cysteine and tyrosine, they must be supplied in the diet.

can carry and store minerals in their interiors. Still others provide support to tissues. Some—the enzymes—act on other substances to change them chemically.

Denaturation of Proteins

Proteins can undergo **denaturation,** resulting in distortion of shape by heat, alcohol, acids, bases, or the salts of heavy metals. The denaturation of a protein is the first step in the protein's breakdown. Denaturation is useful to the body in digestion. During the digestion of a food protein, an early step is denaturation by the stomach acid, which opens up the protein's structure, permitting digestive enzymes to break the peptide bonds (see Figure 5-1 on page 147). Denaturation can also occur during food preparation. For example, cooking an egg denatures the proteins of the egg and makes the egg firmer. Perhaps more important, cooking denatures two raw-egg proteins that bind the B vitamin biotin and the mineral iron, as well as a protein that slows the digestion of other proteins. Thus, cooking eggs liberates biotin and iron and aids in protein digestion.

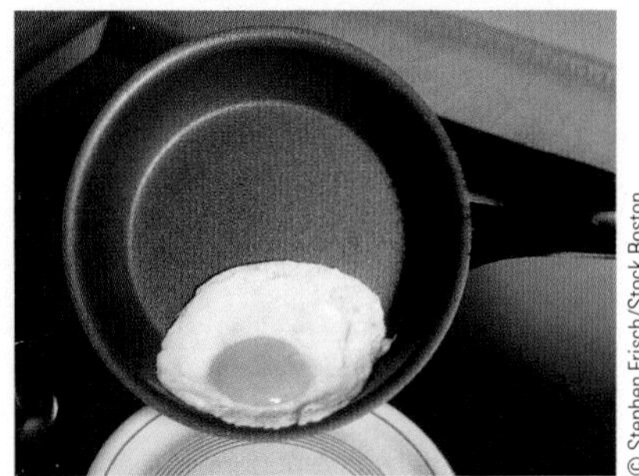

▲ *Cooking an egg denatures its proteins.*

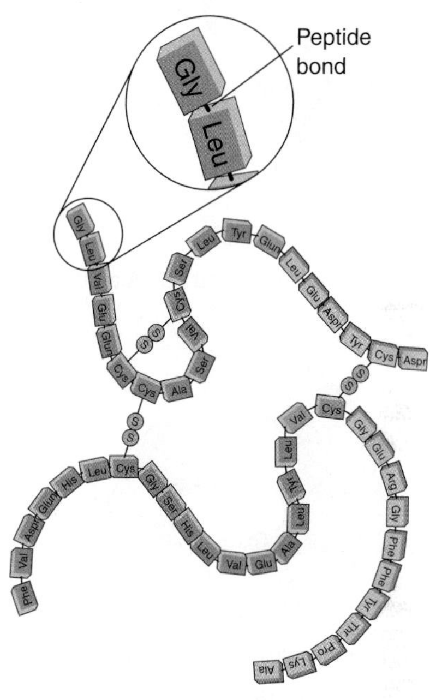

▲ **A Molecule of Insulin**
Amino acids are linked together with peptide bonds to form strands of protein. The sulfur groups (S) on two cysteine (cys) molecules can bond together, creating a "sulfur bridge" between the two protein strands.

The Functions of Body Proteins

No new living tissue can be built without protein, for protein is part of every cell. About 20 percent of our total body weight is protein. Proteins come in many forms: enzymes, antibodies, hormones, transport vehicles, oxygen carriers, tendons and ligaments, scars, the cores of bones and teeth, the filaments of hair, the materials of nails, and more (see Table 5-1). A few of the many vital functions of proteins are described here to show why they have rightfully earned their position of importance in nutrition.

Growth and Maintenance

One function of dietary protein is to ensure the availability of amino acids for building the proteins of new tissue. New tissue is needed in an embryo; in a growing child; in the blood that replaces that which has been lost in burns, hemorrhage, or surgery; in the scar tissue that heals wounds; or in new hair and nails. Not so obvious, but equally important, is the protein that helps replace worn-out

TABLE 5-1
The Functions of Body Proteins

1. **Growth and Maintenance**
   - Proteins provide *building materials*—amino acids—for growth and repair of body tissues.
   - *Body structures.* Proteins form vital parts of most body structures such as skin, nails, hair, membranes, muscles, teeth, bones, organs, ligaments, and tendons.
2. **Regulatory Roles**
   - *Enzymes.* Proteins facilitate numerous chemical reactions in the body; all enzymes are proteins.
   - *Hormones.* Some proteins act as chemical messengers, regulating body processes; not all hormones are proteins.
   - *Antibodies.* Proteins assist the body in maintaining its resistance to disease by acting against foreign disease-causing substances.
   - *Fluid balance.* Proteins help regulate the quantity of fluids in body compartments.
   - *Acid–base balance.* Proteins act as buffers to maintain the normal acid and base concentrations in body fluids.
   - *Transportation.* Proteins move needed nutrients and other substances into and out of cells and around the body.
3. **Energy Production**
   - *Energy.* Protein can be used to provide calories (4 calories per gram) to help meet the body's energy needs.

denaturation the change in shape of a protein brought about by heat, alcohol, acids, bases, salts of heavy metals, or other agents.

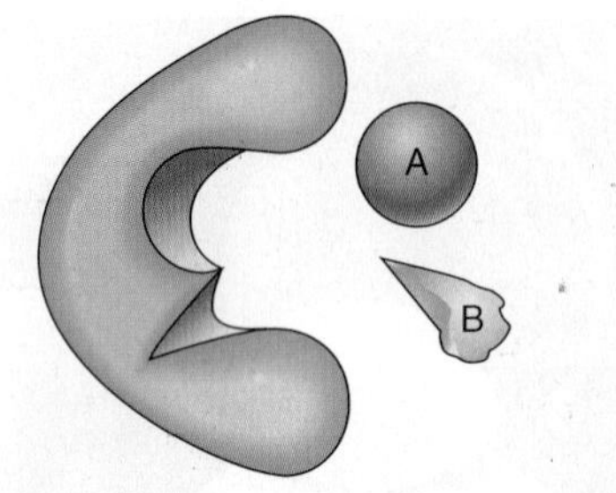

Enzyme plus two compounds, A and B

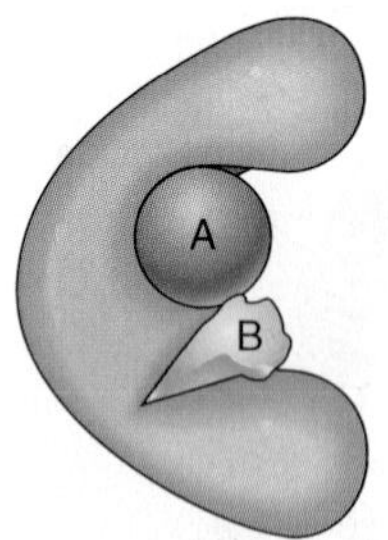

Enzyme complexed with A and B

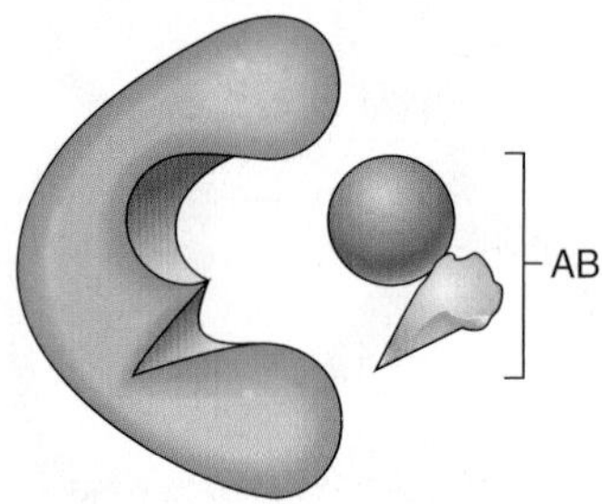

Enzyme plus new compound AB

▲ ***Enzyme Action***
Each enzyme facilitates a specific chemical reaction.

cells. The cells that line the digestive tract only live for about 3 days and are constantly being shed and excreted. You have probably observed that the cells of your skin die, rub off, and are replaced from underneath. For this new growth, amino acids must constantly be resupplied by food.

Enzymes

All enzymes are proteins, and they are among the most important proteins formed in living cells. Enzymes are catalysts—biological spark plugs—that help chemical reactions take place. There are thousands of enzymes inside a single cell, each type facilitating a specific chemical reaction. Enzymes are involved in such processes as the digestion of food, the release of energy from the body's stored energy supplies, and the growth and repair of tissue.

The question about how an enzyme can be specific for a particular reaction is only a partially understood mystery. Scientists believe that the surface of the enzyme is contoured so that it can recognize only the substances it works on and ignore others. The surface provides a site that attracts one or more specific chemical compounds and promotes a specific chemical reaction. For example, two substances might first become attached to the enzyme and then to each other. The newly formed product is then expelled by the enzyme into the fluid of the cell. Enzymes are the hands-on workers in the production and processing of all substances needed by the body.

Hormones

Hormones are similar to enzymes in creating profound effects. However, hormone molecules differ from enzyme molecules. For one thing, not all of them are made of protein. For another, hormones don't catalyze chemical reactions directly but, rather, are messengers that elicit the appropriate responses to maintain a normal environment in the body. Hormones regulate overall body conditions, such as the blood glucose level (the hormones insulin and glucagon) and the metabolic rate (thyroid hormone).

Antibodies

Of all the great variety of proteins in living organisms, the **antibodies** best demonstrate that proteins are specific to individual organisms. Antibodies form in response to the presence of antigens (foreign proteins or other large molecules) that invade the body. The foreign protein may be part of a bacterium, a virus, or a toxin, or it may be something present in food that causes a reaction we call an allergy. The body, after recognizing that it has been invaded, manufactures antibodies that deactivate the foreign substance. Without sufficient protein to make antibodies, the body cannot maintain its resistance to disease.

One of the most fascinating aspects of how antibodies respond to foreign substances is that each antibody is uniquely designed to destroy a specific foreign substance. An antibody that has been manufactured to combat one strain of flu virus is of no help in protecting a person against another strain. Once the body has learned to make a particular antibody, it never forgets, and the next time it encounters that same foreign substance, it will be equipped

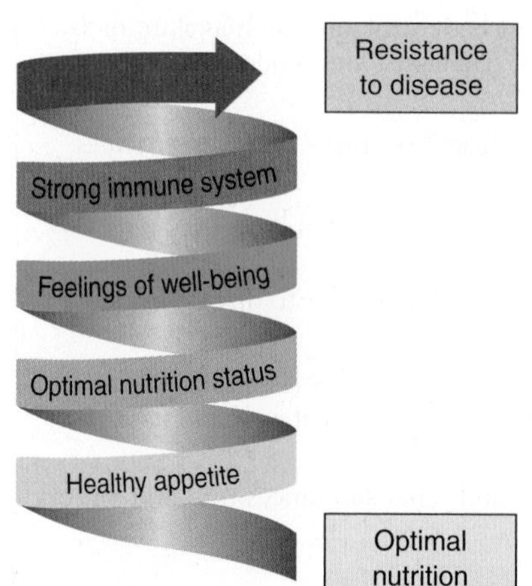

▲ *An optimal diet helps to provide strength and support to the body's immune system.*

hormones chemical messengers. Hormones are secreted by a variety of glands in the body in response to altered conditions. Each affects one or more target tissues or organs and elicits specific responses to restore normal conditions.

antibodies large proteins of the blood and body fluids that are produced by one type of immune cell in response to invasion of the body by unfamiliar molecules (mostly foreign proteins). Antibodies inactivate the foreign substances and so protect the body. The foreign substances are called *antigens*.

pieces—some single amino acids and many strands of two amino acids, **dipeptides,** some strands of three amino acids, **tripeptides,** and some longer chains. Digestion continues until almost all pieces of protein are broken into dipeptides, tripeptides, and free amino acids. Absorption of amino acids takes place all along the small intestine. As for dipeptides and tripeptides, the cells that line the small intestine capture them on their surfaces, split them into amino acids on the cell surfaces, absorb them, and then release them into the bloodstream.

Once they are circulating in the bloodstream, the amino acids are available to be taken up by any cell of the body. The cells can use them to make proteins, either for their own use or for secretion into the circulatory system for other uses.

If a *non*essential amino acid (that is, one the body can make for itself) is unavailable for a growing protein strand, the cell will make one and will continue attaching amino acids to the strand. If, however, an essential amino acid (one the body cannot make) is missing, the building of the protein will halt. The cell cannot hold partially completed proteins to complete them later, for example, the next day. Instead, it must dismantle the partial structures and return surplus amino acids to the circulation, making them available to other cells. If other cells do not soon pick up these amino acids and insert them into protein, the liver will remove their amine groups for the kidney to excrete. Other cells will then use the remaining fragments for other purposes. The nutritional need calling for the production of that particular protein will not be met.

dipeptides (dye-PEP-tides) protein fragments two amino acids long. A peptide is a strand of amino acids.

Phe
Val

tripeptides (try-PEP-tides) protein fragments three amino acids long.

Gly
Leu
Val

FIGURE 5-1
A SUMMARY OF PROTEIN DIGESTION AND ABSORPTION

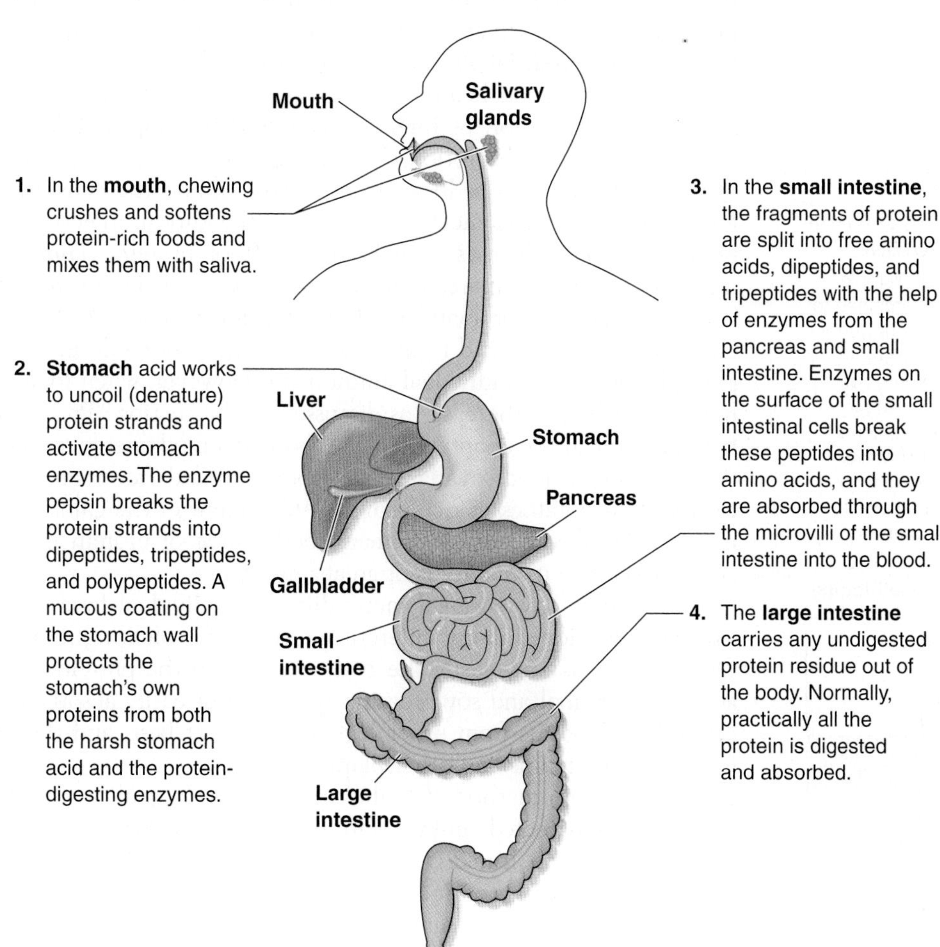
Mouth
Salivary glands
1. In the **mouth**, chewing crushes and softens protein-rich foods and mixes them with saliva.
2. **Stomach** acid works to uncoil (denature) protein strands and activate stomach enzymes. The enzyme pepsin breaks the protein strands into dipeptides, tripeptides, and polypeptides. A mucous coating on the stomach wall protects the stomach's own proteins from both the harsh stomach acid and the protein-digesting enzymes.
Liver
Stomach
Pancreas
Gallbladder
Small intestine
Large intestine
3. In the **small intestine**, the fragments of protein are split into free amino acids, dipeptides, and tripeptides with the help of enzymes from the pancreas and small intestine. Enzymes on the surface of the small intestinal cells break these peptides into amino acids, and they are absorbed through the microvilli of the small intestine into the blood.
4. The **large intestine** carries any undigested protein residue out of the body. Normally, practically all the protein is digested and absorbed.

complete proteins proteins containing all the essential amino acids in the right proportion relative to need. The *quality* of a food protein is judged by the proportions of essential amino acids that it contains relative to our needs. Animal and soy proteins are the highest in quality.

incomplete protein a protein lacking or low in one or more of the essential amino acids.

limiting amino acid a term given to the essential amino acid in shortest supply (relative to the body's need) in a food protein; it therefore *limits* the body's ability to make its own proteins.

complementary proteins two or more food proteins whose amino acid assortments complement each other in such a way that the essential amino acids limited in or missing from each are supplied by the others.

protein quality a measure of the essential amino acid content of a protein relative to the essential amino acid needs of the body.

protein digestibility–corrected amino acid score (PDCAAS) a measuring tool for determining protein quality. PDCAAS reflects both a protein's digestibility and its proportion of amino acids relative to human needs.

Protein Quality of Foods

The role of protein in food, as already mentioned, is not to provide body proteins directly but to supply the amino acids from which the body can make its own proteins. Because body cells cannot store amino acids for future use, it follows that all the essential amino acids must be eaten as part of a balanced diet. To manufacture body proteins, then, all the needed amino acids must be available to the cells. Three important characteristics of dietary protein, therefore, are (1) that it should supply at least the nine essential amino acids, (2) that it should supply enough other amino acids to make nitrogen available for the synthesis of whatever nonessential amino acids the cell may need to make, and (3) that it should be accompanied by enough food energy (preferably from carbohydrate and fat) to prevent sacrifice of its own amino acids for energy. This presents no problem to people who regularly eat **complete proteins,** such as those of meat, fish, poultry, cheese, eggs, milk, or many soybean products, as part of balanced meals.[2] The proteins of these foods contain ample amounts of all the essential amino acids relative to our bodies' need for them, and the rest of the diet provides protein-sparing energy and needed vitamins and minerals. An equally sound diet choice is to eat two or more **incomplete protein** foods from plants, each of which supplies the **limiting amino acid** in the other—also, of course, as part of a balanced diet. The *quality* of plant proteins (legumes, grains, and vegetables) having different *limiting* amino acids can therefore be balanced by combining different sources of plant proteins, either during a meal or over the course of a day, making sufficient amounts of all the essential amino acids available for protein synthesis. This strategy—using **complementary proteins**—is shown in Figure 5-2. Note that by combining a grain (whole-wheat bread) that is low in lysine but high in methionine with a legume (peanut butter) that is low in methionine but high in lysine, all the essential amino acids are provided.

A person in good health can be expected to use dietary protein efficiently. However, malnutrition or infection can seriously impair digestion (by reducing enzyme secretion), absorption (by causing degeneration of the absorptive surface of the small intestine or losses from diarrhea), and the cells' use of protein (by forcing amino acids to meet other needs). In addition, infections cause increased production of antibodies, which are made of protein. Thus, malnutrition or infection can greatly increase protein needs while making it hard to meet them.[3]

People usually eat many foods containing protein. Each food has its own characteristic amino acid balance, and a mixture of foods almost invariably supplies plenty of each individual amino acid. However, when food energy intake is limited, this is not the case (as discussed in the section "Protein-Energy Malnutrition," later in the chapter). Also, even when food energy intake is abundant, if the selection of foods available is severely limited (for example, when a single food such as potatoes or rice provides 90 percent of the calories), protein intake may not be adequate. The primary food source of protein must be taken into account because its quality is of great importance.

Researchers have studied many different individual foods as protein sources and have developed many different methods of evaluating their **protein quality.** One category is how easily the body can absorb the protein. In general, amino acids from animal and soy proteins are the most easily absorbed (90 to 100 percent). Amino acids from other legumes are next best, and the absorption rates for those from grains and other plant foods vary.

An important method of evaluating the protein quality of foods is the **protein digestibility–corrected amino acid score** or **PDCAAS,** which is used by the Dietary Reference Intake (DRI) Committee to evaluate protein needs. This method takes into account both the proportion of amino acids that a food provides relative to meeting human needs and the digestibility of the protein. Foods earning a perfect score of "100" (on a scale of 0 to 100) include egg white, fat-

FIGURE 5-2
HOW TWO PLANT PROTEINS COMBINE TO YIELD A COMPLETE PROTEIN

Two incomplete proteins (for example, legumes plus grains) can be combined to equal a complete protein (peanut butter sandwich). In this example, the peanut butter provides adequate amounts of the amino acid lysine but is lacking in methionine. The bread "complements" the peanut butter because it contains adequate methionine but is lacking in lysine. When combined as a sandwich, all essential amino acids are present.

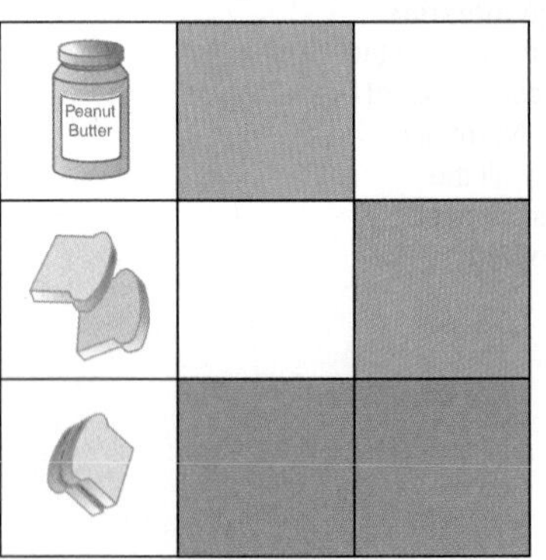

free milk, tuna, beef, and chicken; soy merits a score of 94; and most legumes have scores of 50 to 60. Manufacturers use the PDCAAS to determine the protein values listed on food labels, and it reflects the protein quality of food products.

When amino acids are wasted, their amine groups (which contain their nitrogen) cannot be stored. Therefore, the efficiency of a protein can be assessed experimentally by measuring the net loss of nitrogen from the body. The higher the amount of nitrogen retained, the higher the quality of the protein. This is the basis for determining the **biological value (BV)** of proteins. A high-quality protein by this standard is egg white protein, which has been designated the **reference protein** and given a score of 100. Other proteins are compared with it. The best guarantee of amino acid adequacy is to eat a variety of foods containing protein in the presence of adequate amounts of vitamins, minerals, and energy from carbohydrate and fat.

To calculate the percentage of calories you derive from protein:

1. Use your total calories as the denominator (example: 1,900 cal).
2. Multiply your total protein intake in *grams* by 4 cal/g to obtain calories from protein as the numerator (example: 70 g protein × 4 cal/g = 280 cal).
3. Divide to obtain a decimal, multiply by 100, and round off (example: 280/1,900 × 100 = 15% cal from protein).

Recommended Protein Intakes

Recommended protein intakes can be stated in two ways—as a percentage of total calories or as an absolute number (grams per day). The DRI Committee recommends that protein provide 10 percent to 35 percent of total caloric intake.[4] The recommended protein allowance for a healthy adult is 0.8 gram per kilogram (or 2.2 pounds) of desirable body weight per day.

The recommendation for protein uses the desirable, not the actual, weight for a given height because the desirable weight is proportional to the *lean* body mass of the average person. Lean body mass, not total weight, determines protein need. This is because fat tissue is composed largely of fat, which does not require much protein for maintenance.

The recommendations for protein intake are based on the assumption that the protein source will be a combination of plant and animal proteins, that it will be consumed with adequate calories from carbohydrate and fat, and that other nutrients in the diet will be adequate. These protein recommendations apply only to healthy individuals with no unusual metabolic need for protein.

To calculate your recommended protein intake (RDA):

1. Find the desirable weight for a person your height (see Appendix B). Assume this weight is appropriate for you.
2. Change pounds to kilograms (divide pounds by 2.2; 1 kilogram = 2.2 pounds).
3. Multiply kilograms by 0.8 g/kg.

Example (for a 5′8″ male):

1. Desirable weight: about 150 lb.
2. 150 lb ÷ 2.2 lb = 68 kg (rounded off).
3. 68 kg × 0.8 g/kg = 54 g protein (rounded off).

Protein and Health

With all the attention that has been paid to the health effects of starch, sugars, fibers, fats, oils, and cholesterol, protein has been slighted. Protein deficiency effects are well known because, together with energy deficiency, they are the world's main form of malnutrition. But the health effects of too much protein—and particularly the effects of proteins of different kinds—are far less well known. The following sections discuss protein deficiency, excess protein, and types of protein. The Nutrition Action feature in this chapter discusses the problem of protein-related food allergies in children and adults.

Chapter 10 discusses the protein needs of athletes, and the Spotlight in Chapter 10 presents the pros and cons of using popular amino acid and other supplements.

Protein-Energy Malnutrition

Protein deficiency and energy deficiency go hand in hand so often that public health officials have given a nickname to the pair: **protein-energy malnutrition (PEM).** The two diseases and their symptoms overlap all along the spectrum, but the extremes have names of their own. Protein deficiency is **kwashiorkor,** and energy deficiency is **marasmus.**[5]

Kwashiorkor is the Ghanaian name for "the evil spirit that infects the first child when the second child is born." In countries where kwashiorkor is prevalent, parents customarily give their newly weaned children watery cereal rather than the food eaten by the rest of the family. The child has been receiving the mother's breast milk, which contains high-quality protein designed to support

biological value (BV) a measure of protein quality, assessed by determining how well a given food or food mixture supports nitrogen retention.

reference protein egg white protein, the standard with which other proteins are compared to determine protein quality.

protein-energy malnutrition (PEM) also called protein-calorie malnutrition (PCM), the world's most widespread malnutrition problem, includes both kwashiorkor and marasmus as well as the states in which they overlap.

kwashiorkor (kwash-ee-OR-core) a deficiency disease caused by inadequate protein in the presence of adequate food energy.

marasmus (ma-RAZ-mus) an energy deficiency disease; starvation.

edema (eh-DEEM-uh) swelling of body tissue caused by leakage of fluid from the blood vessels, seen in (among other conditions) protein deficiency.

dysentery (DISS-en-terry) an infection of the digestive tract that causes diarrhea.

acquired immune deficiency syndrome (AIDS) an immune system disorder caused by the human immunodeficiency virus (HIV).

growth. However, when a new baby is born, the child is weaned and suddenly is fed only a weak drink with scant protein of very low quality. It is not surprising that the just-weaned child becomes sick when the new baby arrives.

The child who has been banished from its mother's breast faces this threat to life by engaging in as little activity as possible. Apathy is one of the earliest signs of protein deprivation. The body is collecting all its forces to meet the crisis and so cuts down on any expenditure of protein not needed for the heart, lungs, and brain. As the apathy increases, the child doesn't even cry for food. All growth ceases, and the child is no larger at age 4 than at age 2. New hair grows without the protein pigment that gives hair its color. The skin also loses its color, and open sores fail to heal. Digestive enzymes are in short supply, the digestive tract lining deteriorates, and absorption fails. The child can't assimilate what little food is eaten. Proteins and hormones that previously kept the fluids correctly distributed among the compartments of the body now are diminished, so that fluid leaks out of the blood (**edema**) and accumulates in the belly and legs. Blood proteins, including hemoglobin, are not synthesized, so the child becomes anemic, which increases the child's weakness and apathy. The kwashiorkor victim often develops a fatty liver, caused by a lack of the protein carriers that transport fat out of the liver. Antibodies to fight off invading bacteria are degraded to provide amino acids for other uses; the child becomes an easy target for any infection. Then **dysentery,** an infection of the digestive tract that causes diarrhea, further depletes the body of nutrients, especially minerals. Measles, which might make a healthy child sick for a week or two, kills the kwashiorkor child within two or three days. If the condition is caught in time, the starving child's life may be saved by careful nutrition therapy.

Children with marasmus suffer symptoms similar to those of children with kwashiorkor because both conditions cause loss of body protein tissue. However, there are differences between the two conditions. A marasmic child looks like a wizened little old person—just skin and bones. The child is often sick because his or her resistance to disease is low. All the muscles are wasted, including the heart muscle, and the heart is weak. Metabolism is so slow that body temperature is subnormal. There is little or no fat under the skin to insulate against cold. The experience of hospital workers with victims of this disease is that the victims' primary need is to be wrapped up and kept warm. The disease occurs most commonly in children from 6 months to 18 months of age. Because the brain normally grows to almost its full adult size within the first two years of life, marasmus impairs brain development and thus may have a permanent effect on a child's learning ability.

PEM is prevalent in Africa, Central America, South America, and Asia. Cases have also been reported on American Indian reservations and in the inner cities and impoverished rural areas of the United States.[6] PEM has also been recognized in many undernourished hospital patients, including those with anorexia nervosa, **AIDS,** cancer, and other wasting conditions. The extent and severity of malnutrition worldwide is a political and economic problem. It is discussed further in the Spotlight feature in Chapter 12.

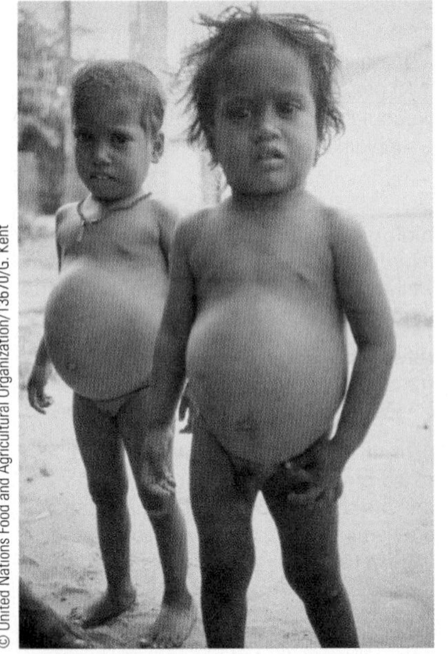

▲ *Kwashiorkor. These children have the characteristic edema and swollen belly often seen with kwashiorkor.*

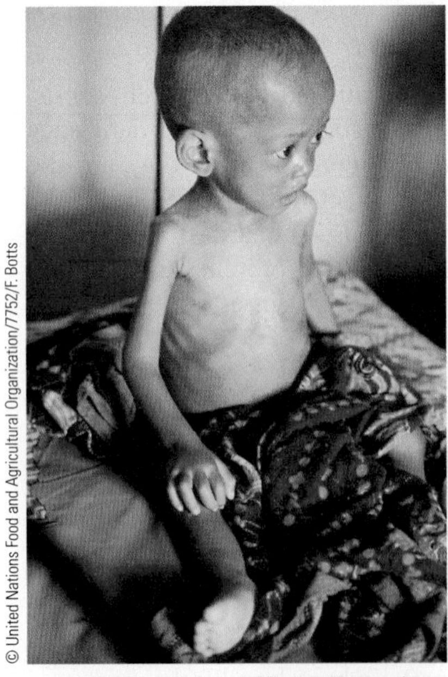

▲ *Marasmus. This child is suffering from the extreme emaciation of marasmus.*

Too Much Protein

Many of the world's people struggle to obtain enough food and enough protein to keep themselves alive, but in the developed countries, where protein is abundant, the problems of protein excess can be seen. Animals fed high-protein diets experience a protein overload effect, seen in the enlargement of their livers and kidneys. In human beings, diets high in animal protein necessitate higher intakes of calcium as well, because such diets promote calcium excretion.[7] Excess protein may also create an increased demand for vitamin B_6 in the diet, which the body requires to utilize the protein. The higher a person's intake of animal-protein sources such as meat, the more likely it is that fruits, vegetables, and grains will be crowded out of the diet, creating deficiences in other nutrients.

Although protein is essential to health, the body converts extra protein to energy (glucose), which is stored as body fat when energy needs are met. Despite the flood of new protein-packed snack bars and other products in the marketplace, there are no known benefits from consuming excess protein. The recommended upper limit for protein intake applies *when calorie intake is adequate.* Note the qualification "when calorie intake is adequate" in the preceding statement. Remember that your recommended protein intake can be stated as a percentage of calories in the diet or as a specific number of grams of dietary protein. The recommended protein intake for a 150-pound person is roughly 55 grams, or about 12 percent of their daily caloric intake. Fifty-five grams of protein is equal to 220 calories and equals 11 percent of a 2,000-calorie intake, which is reasonable for a 150-pound active person. If this person were to drastically reduce his or her caloric intake to, say, 800 calories a day, then 220 calories from protein is suddenly 28 percent of the total. However, it is still this person's recommended intake for protein, and a reasonable intake. It is the caloric intake that is unreasonable in this example. Similarly, if the person eats too many calories, say 4,000, this protein intake represents only 6 percent of the total caloric intake, yet it is still a reasonable intake. It is the caloric intake that may be unreasonable.

Thus, it is important to be careful when judging protein intakes as a percentage of calories. Always ask what the absolute number of grams is, too, and compare it with the recommended protein intake in grams. Recommendations stated as a percentage of calories are useful only when food energy intakes (calories) are within reason.

Protein in the Diet

Misconceived notions abound regarding protein in the diet; the most obvious of these is that more is better. American women eat about 60 to 65 grams of protein a day, notably higher than the recommended 46 grams a day. Young men average about 100 grams a day and drop to about 75 to 85 grams as older adults, still considerably higher than their recommended intake of 52 to 56 grams. Moreover, more than 65 percent of this protein comes from animal and dairy products (see Figure 5-3). Saturated fats supply half or more of the calories in some animal protein foods. You can better balance your food choices by selecting one-third or less of your protein from animal sources and the rest from plants. (See the Eat Well Be Well feature on page 154). To limit your intake of saturated fat, consider the following tips when shopping for protein foods:

- Dairy foods are excellent sources of protein, calcium, and other important nutrients. Look for fat-free and low-fat varieties for the recommended three servings each day. Look for low-fat cheeses that have less than 5 grams of fat per ounce, such as part-skim or fat-free ricotta or mozzarella, farmer's cheese, feta cheese, string cheese, or other reduced-calorie cheeses. These choices are lower in saturated fat and calories than their full-fat counterparts.

FIGURE 5-3
PROTEIN CONTRIBUTED BY FOOD GROUPS IN THE AVERAGE AMERICAN DIET

These foods provide 70% of the protein in the U.S. diet. The following foods also contribute at least 1% (in descending order): cold cuts (excluding ham), ready-to-eat cereal, white potatoes, sausage, flour and baking ingredients, ice cream, sherbet and frozen yogurt, nuts and seeds, cooked rice and other grains, and canned tuna.

SOURCE: Adapted from P. A. Cotton and coauthors, Dietary sources of nutrients among U.S. adults, 1994–1996, *Journal of the American Dietetic Association* 104 (2004): 921–930.

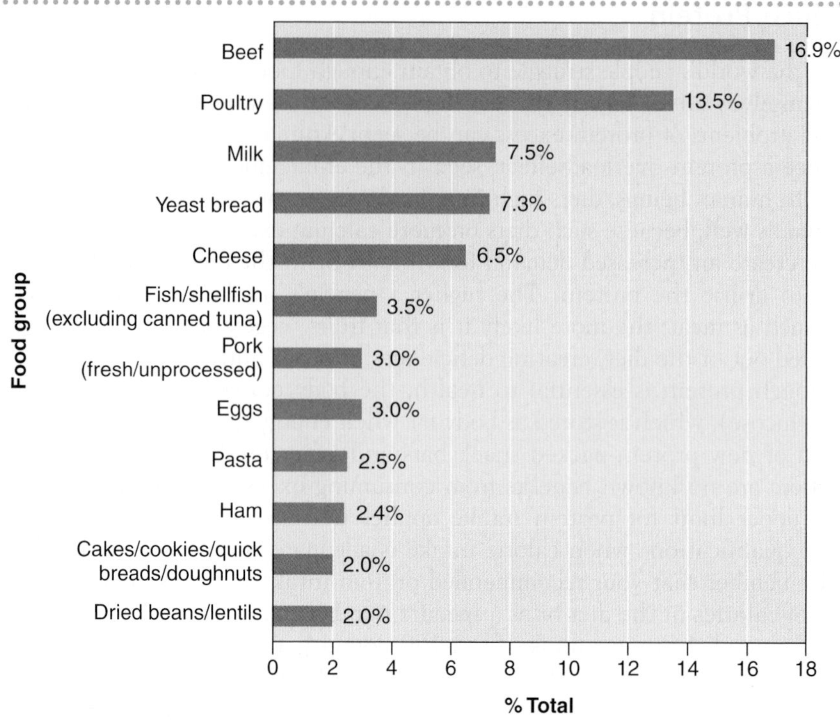

© Polara Studios

© Polara Studios

© Polara Studios

© Polara Studios

▲ *Foods that supply protein in abundance are shown here in the Milk Group and the Meat & Beans Group of the MyPyramid Food Guide (top two photos). Servings of foods from the Vegetable Group and the Grains Group can also contribute protein to the diet (bottom two photos).*

legumes (leg-GYOOMS) plants of the bean and pea family having roots with nodules that contain bacteria that can trap nitrogen from the air in the soil and make it into compounds that become part of the seed. The seeds are rich in high-quality protein compared with those of most other plant foods.

- Fish and shellfish—fresh, frozen, or canned in water—make excellent protein choices and are low in fat. Experts recommend you eat at least two fish meals per week.
- Meat, chicken, and fish all provide excellent protein, as well as iron, zinc, and vitamin B_{12}. We are advised to choose low-fat varieties. Look for the leanest meats:

  Flank steak, round steak, sirloin, tenderloin, or extra-lean ground beef

  Lean ham, Canadian bacon, pork tenderloin, and center-loin pork chops

  Chicken, turkey, or game hens without the skin; fresh ground turkey breast or chicken breast meat
- At the deli counter, select items with less than 1 gram of fat per ounce, such as lean ham, turkey or chicken breast, and lean roast beef.
- Nuts and nut butters are good sources of protein but are also high in fat. Look for nonhydrogenated varieties. A 1-tablespoon serving of peanut butter counts as 1 ounce of meat in the MyPyramid food guide.
- Eggs are another excellent source of protein. Also look for egg substitutes and substitute ¼ cup for one whole egg in recipes. One egg counts as 1 ounce of meat.
- Many interesting sources of protein are available. Try adding **legumes** to your meals using the tips in the Savvy Diner feature that follows. A ¼-cup serving counts as 1 ounce of meat in the MyPyramid food guide.

As the vegetarian knows, one can easily design a perfectly acceptable diet around plant foods alone by choosing an appropriate variety. This chapter's Spotlight reviews the benefits of soy products, an excellent source of protein in the diet.

Adequate protein in the diet is easy to obtain. A breakfast of one egg, two slices of wheat bread, and a glass of milk provide nearly 20 grams of protein. This meets about 35 percent of an average man's recommended protein intake of 56 grams and about 43 percent of a woman's recommended protein intake of 46 grams. The Scorecard on page 155 shows you how to estimate your own protein intake.

The Savvy Diner Eat More Beans

Legumes—dried beans, peas, and lentils—have been very highly praised in recent years.[8] The 2005 Dietary Guidelines recommend that Americans should eat more beans. On the MyPyramid Food Guide, legumes are the only food featured in two different categories, the Vegetables Group and the Meats & Beans Group. Legumes are rich in the B vitamins and fiber. They are good sources of protein, iron, and zinc and are naturally low in fat and cholesterol free. In January 2005, the Food and Drug Administration approved a new health claim for bean packages and cans: "Diets including beans may reduce your risk of heart disease and certain cancers."[9] However, Americans typically only eat about one-third the recommended amount of legumes—one cup instead of the recommended three cups per week for a 2,000 calorie diet (see bean recommendations for other calorie levels at MyPyramid.gov).

Here are three simple ways to increase your intake of this low cost, easy-to-include group of foods:

1. Enjoy adding more legumes to your weekly meals. Include legumes in entrees (tacos), side dishes (baked beans), soups (split pea or lentil), and salads (chickpeas from the salad bar, or three-bean combos). Recipes can be prepared from dried beans (soaked, rinsed, and cooked) or more quickly from canned beans (rinsed to remove excess sodium). Thorough rinsing is recommended to remove the gas-causing sugar known as raffinose found in beans.
2. Explore the many varieties of legumes used in cooking—as noted in the Miniglossary below.
3. Enjoy learning more about these nutritional powerhouses online. The following three sites can help:
   - *American Dry Bean Board:* www.americanbean.org/. Includes bean basics, photos, health tips, and 100+ recipes.
   - *Beans for Health Alliance:* www.beansforhealth.org/. A more global focus, with many links and a useful bibliography.
   - *USA Dry Peas, Lentils, and Chickpeas:* www.pea-lentil.com/. Nutrition information, cooking instructions, and recipes.

© Tom McCarthy Photography

Legumes include such plant foods as the soybean, kidney bean, garbanzo bean, black bean, lentil, garden pea, black-eyed pea, and lima bean.

MINIGLOSSARY OF LEGUMES

black, cuban, or **turtle beans** These medium-size, black-skinned ovals have a rich, sweet taste. They are best served in Mexican and Latin American dishes or thick soups and stews.

black-eyed peas These peas are small, oval shaped, and creamy white with a black spot. They have a vegetable flavor with mealy texture. Use in salads with rice and greens.

garbanzo beans or **chickpeas** These legumes are large, round, and tan colored. They have a nutty flavor and crunchy texture. Use in soups and stews and puréed for dips.

great northern beans This variety is medium white and kidney shaped. Enjoy the delicate flavor and firm texture in salads, soups, and main dishes.

kidney beans These familiar beans are large, red, and kidney shaped (the white variety is called *cannellini*).They have a bland taste and soft texture but tough skins. Use in chili, bean stews, and Mexican dishes (for red) or Italian dishes (for white).

lentils These legumes are small, flat, and round. Usually brown colored, lentils also can be green, pink, or red. They have a mild taste with firm texture. Best used when combined with grains or vegetables in salads, soups, or stews.

lima or **butter beans** Limas are soft and mealy in texture. They are flat, oval shaped, and white tinged with green. The smaller variety has a milder taste. Use in soups and stews.

pinto beans These medium ovals are mottled beige and brown with an earthy flavor. They are most often used in Mexican dishes, such as refried beans, stews, or dips.

red beans This versatile bean is a medium-size, dark-red oval. The taste and texture are similar to kidney beans. Use in soups and stews, and serve with rice.

soybeans You can find these creamy white ovals in numerous food products, such as tofu, flour, grits, and milk. They have a firm texture and bland flavor. See the Spotlight feature later in this chapter for a discussion of the possible health benefits derived from soy foods.

split peas Green or yellow, these small, halved peas supply an earthy flavor with a mealy texture. They are best used in soups and with rice or grains.

white navy beans These beans are small, white ovals and are best used in soups and stews and as baked beans.

SOURCE: Adapted from K. Mangum, *Life's Simple Pleasures: Fine Vegetarian Cooking for Sharing and Celebration* (Boise, ID: Pacific Press, 1990), 149.

The New American Plate—Reshape Your Protein Choices for Health

© PhotoDisc/Getty Images

The ancient inhabitants of South America liked to eat a kind of paste made from peanuts. But modern peanut butter came into being around 1890 as the bright idea of a St. Louis physician, who thought it would be a good health food for elderly people. It was not linked with jelly until the 1920s.[10]

Many health organizations now recommend a diet that emphasizes vegetables, fruits, legumes, and whole grains to protect against cancer, heart disease, stroke, diabetes, and obesity.[11] The key to getting enough, but not too much, protein seems to be to use a variety of plant-based foods and to de-emphasize meats.[12]

1 In the Kitchen

- Small meat portions tend to work best when mixed into dishes with lots of vegetables and grains; try stir-fries, pastas, soups and stews, burritos, and main-dish salads. For example, cook a large pot of soup, stew, or chili. Minimize the amount of meat you use, and load it up with vegetables (fresh or frozen) and cooked beans.
- Go meatless one or more days each week. Experiment with recipes from health-minded cookbooks.
- Take a fresh look at your favorite recipes. Try to use less meat and add more vegetables. Instead of chicken and broccoli stir-fry, try stir-fried veggies with a little chicken.
- For quick, colorful, meals rich in nutrients and flavor, try various combinations of stir-fried vegetables on beds of steamed brown rice, whole-grain bulgur, or couscous.

2 In the Lunch Box

- Get out of the peanut butter and jelly rut by filling sandwiches with water-packed tuna mixed with mandarin oranges, bean sprouts, and a bit of plain low-fat yogurt; chopped, cooked, skinless chicken combined with raw sliced vegetables and a little French dressing; cooked, mashed dried beans seasoned with chopped onion, garlic powder, rosemary, thyme, and pepper; or low-fat cottage cheese flavored with drained, chopped pineapple.
- Take a thermos filled with chili, vegetable soup, or a milk-based soup, such as cream of tomato, prepared with nonfat milk instead of a sandwich. Try cold lunches such as low-fat yogurt and fruit, brown rice with cubes of skinless poultry, or cooked pasta tossed with raw vegetables, low-fat cheese, and a bit of Italian dressing.

3 At the Table

- When dining out, choose an ethnic restaurant with plant-based entrées on the menu. Consider Spanish paella, Asian stir-fries, Moroccan stew, Indian curries, or French ratatouille as your entrée. Or try Chinese, Vietnamese, or Thai take-out with lots of rice and vegetables.
- Make whole grains, vegetables, and legumes the main event of your meals. At least two-thirds of your meal should come from these plant-based foods and one-third or less from lean meat, poultry, fish, or low-fat dairy products.

The Vegetarian Diet

More and more people are following vegetarian diets. The reasons for becoming vegetarian vary widely.[13] Some people have health reasons whereas others have religious or ethical reasons. Some believe that vegetarianism is ecologically sound, and others believe that it is less costly than the meat-eating alternative. In addition to the traditional types of vegetarians (see Table 5-2 on page 157), some people eat seafood but not other meats, and some include chicken and other poultry but not red meat. Whatever the particular reasons for choosing a vegetarian diet, the vegetarian needs to be aware of its implications for nutrition and health.[14]

Important goals for *any* diet planner include the following:

Scorecard Estimate Your Protein Intake

The average American consumes much more than his or her recommended protein intake. How do you compare? First, figure your recommended protein intake (divide your weight in pounds by 2.2 and then multiply by 0.8). Next, write down everything you ate and drank yesterday. Using the values given below, estimate the grams of protein you ate from both animal and plant sources. If an item is not listed here, use Appendix F to determine the amount of protein it contains. How close are you to your recommended protein intake? What percentage of your protein comes from animal versus plant sources?

Recommended Protein Intake: ________ **grams**

PROTEIN FOODS	AMOUNT	GRAMS OF PROTEIN IN 1 SERVING	GRAMS OF PROTEIN IN YOUR TYPICAL DIET
Animal Sources			
Hard cheese (e.g., cheddar)	1 oz	7	________
Cottage cheese	½ c	14	________
Milk	1 c	9	________
Yogurt	1 c	12	________
Egg	1 large	7	________
Poultry	3 oz	21	________
Ground beef, lean	3 oz	24	________
Beef steak, lean	3 oz	26	________
Pork chop, lean	3 oz	20	________
Other	____	____	________
		Animal Proteins Subtotal (grams):	________
Plant Sources			
Vegetables	½ c	2	________
Legumes, cooked	½ c	8	________
Tofu	4 oz	9	________
Cereals	1 c	2–6	________
Bread	1 slice	2	________
Tortilla	1	2	________
Rice	½ c	3	________
Pasta	½ c	3	________
Peanut butter	1 tbsp	4	________
Nuts	2 tbsp	3	________
Seeds	2 tbsp	3	________
Other	____	____	________
		Plant Proteins Subtotal (grams):	________
		Day's Total Protein:	________ **grams**

- To obtain neither too few nor too many calories—that is, to maintain a healthful weight.
- To obtain adequate quantities of complete protein.
- To obtain the needed vitamins and minerals.

In addition to these guidelines, vegetarians can use the special *Vegetarian Food Pyramid* (see Figure 5-4) to balance their diets.

Proteins

The vegetarian needs adequate amounts of all the essential amino acids. Because proteins from animals contain ample amounts of the essential amino acids, the lacto-ovovegetarian can get a head start on meeting protein needs by drinking recommended amounts of milk daily or by consuming the equivalent in milk products in the day's diet.

FIGURE 5-4
A VEGETARIAN FOOD PYRAMID AND MEAL PLANNING TIPS

SOURCE: V. Messina, V. Melina, and A. R. Mangels, A new food guide for North American vegetarians, *Journal of the American Dietetic Association* 103 (2003).

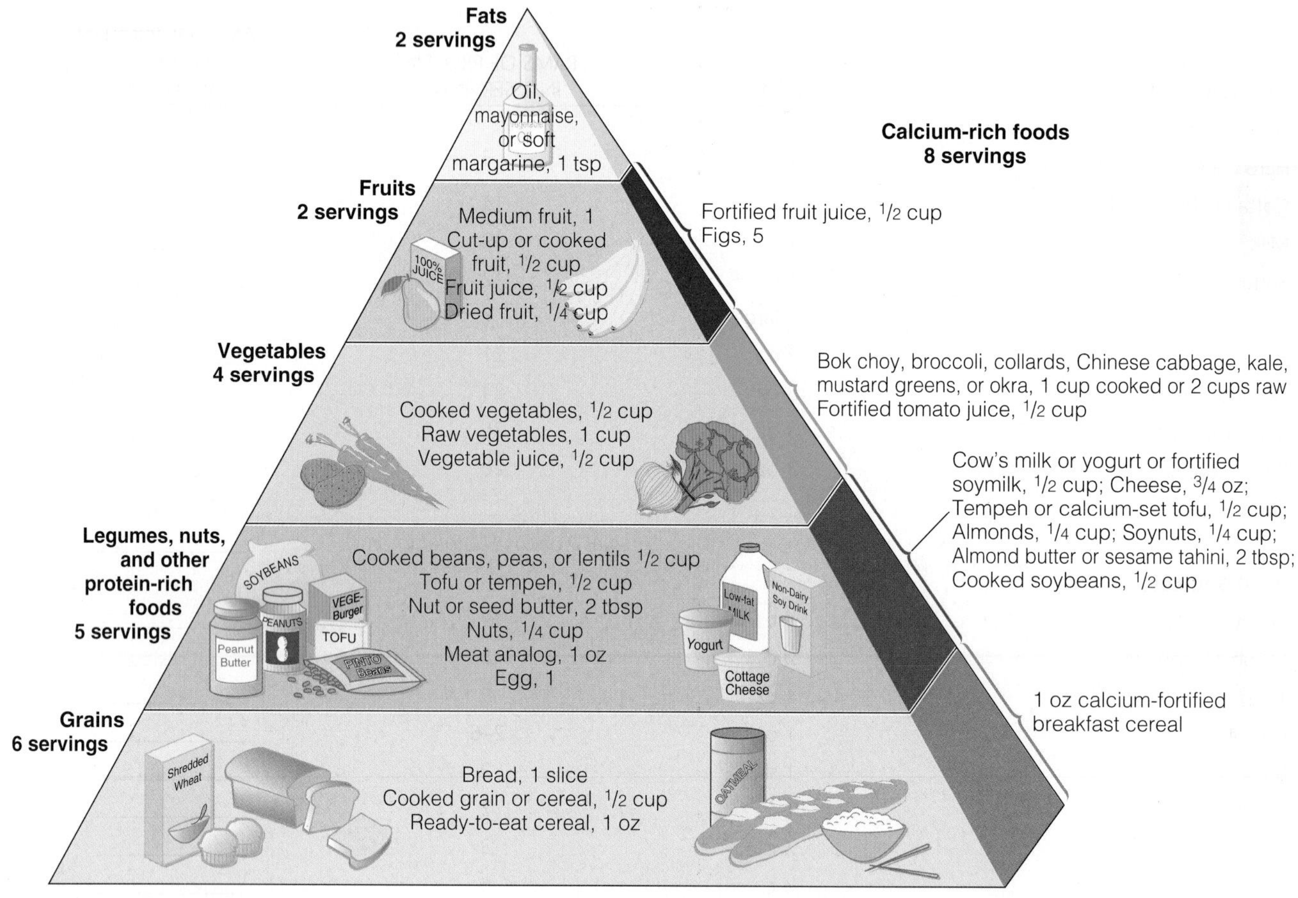

Meal Planning Tips

1. Choose a variety of foods.
2. The number of servings in each group is for minimum daily intakes. Choose more foods from any of the groups to meet energy needs.
3. A serving from the calcium-rich food group provides approximately 10% of adult daily requirements. Choose 8 or more servings per day. These also count towards servings from the other food groups in the guide. For example, 1/2 cup of fortified fruit juice counts as a calcium-rich food and also counts towards servings from the fruit group.
4. Include 2 servings every day of foods that supply omega-3 fats. Foods rich in omega-3 fat are found in the legumes/nuts group and in the fats group. A serving is 1 teaspoon of flaxseed oil, 3 teaspoons of canola or soybean oil, 1 tablespoon of ground flaxseed, or 1/4 cup walnuts. For the best balance of fats in your diet, olive and canola oils are the best choices for cooking.
5. Servings of nuts and seeds may be used in place of servings from the fats group.
6. Be sure to get adequate vitamin D from daily sun exposure or through fortified foods or supplements. Cow's milk and some brands of soy milk and breakfast cereals are fortified with vitamin D.
7. Include at least 3 good food sources of vitamin B_{12} in your diet every day. These include 1 tablespoon of *Red Star* Vegetarian Support Formula nutritional yeast, 1 cup fortified soy milk, 1/2 cup cow's milk, 3/4 cup yogurt, 1 large egg, 1 ounce fortified breakfast cereal, 1 1/2 oz fortified meat analog. If you don't eat these foods regularly (at least 3 servings per day), take a daily vitamin B_{12} supplement of 5 to 10 μg or a weekly B_{12} supplement of 2000 μg.
8. If you include sweets or alcohol in your diet, consume these foods in moderation. Get most of your daily calories from the foods in the Vegetarian Food Guide.

TABLE 5-2
Types of Vegetarians

Semivegetarian Some but not all groups of animal-derived products, such as meat, poultry, fish, seafood, eggs, milk, and milk products, are included in this diet.

Lactovegetarian Milk and milk products are included in this diet, but meat, poultry, fish, seafood, and eggs are excluded.
possible limiting nutrient: iron

Lacto-ovovegetarian Milk and milk products and eggs are included in this diet, but meat, poultry, fish, and seafood are excluded.
possible limiting nutrient: iron

Ovovegetarian Eggs are included in this diet, but milk and milk products, meat, poultry, fish, and seafood are excluded.
possible limiting nutrients: iron, vitamin D, calcium, riboflavin

Strict vegetarian/vegan All animal-derived foods, including meat, poultry, fish, seafood, eggs, milk, and milk products are excluded from this diet.
possible limiting nutrients: iron, vitamin D, calcium, riboflavin, vitamin B_{12}, high-quality protein

macrobiotic diet Extremely restrictive diet based on metaphysical beliefs and consisting mostly of legumes, whole grains, and certain vegetables.
when taken to extremes, includes only brown rice and water or herbal teas and can cause malnutrition and death

Adequate amounts of amino acids can be obtained from a plant-based diet when a varied diet is routinely consumed on a daily basis. Mixtures of proteins from unrefined grains, vegetables, legumes, seeds, and nuts eaten over the course of a day complement one another in their amino acid profiles so that deficits in one are made up by the assets of another.[15] Table 5-3 gives examples of how such mixtures of foods can be combined to form complete proteins.

© Polara Studios

▲ *Well-planned, plant-based meals consisting of a variety of whole grains, legumes, nuts, vegetables, fruits, and for some vegetarians, eggs and dairy products, can offer sound nutrition and health benefits to vegetarians and nonvegetarians alike.*

Vitamins

The lacto-ovovegetarian diet can be adequate in all vitamins, but several vitamins may be a problem for the vegan. One such vitamin is vitamin B_{12}, which doesn't occur naturally in plant foods but is available in fortified foods, such as breakfast cereals or **nutritional yeast** that is grown in a vitamin B_{12}-enriched environment. The vegan needs a reliable B_{12} source, such as vitamin B_{12}-fortified soy milk, breakfast cereals, or **meat replacements.** Some vegetarians use seaweeds, fermented soy, and other products in the belief that they provide vitamin B_{12} in adequate amounts, but these products are not currently recommended as reliable sources. A pregnant or lactating woman who is eating a vegan diet should be aware that her infant may develop a vitamin B_{12} deficiency that can damage the baby's nervous system, even if the mother remains healthy. Because large amounts of vitamin B_{12} are stored in the body, it may take years for a deficiency to develop. Vegan diets are not generally recommended for infants and young children.

Another vitamin of concern to the vegan is vitamin D.[16] The milk drinker is protected if the milk is fortified with vitamin D, but there is no practical source of vitamin D in plant foods. Fortified margarines, soy milk, and breakfast cereals can supply some vitamin D. Regular exposure to the sun can help prevent a deficiency, too.

Riboflavin, another B vitamin obtained from milk, is present in the diet of the vegan who eats ample servings of dark greens, whole and enriched grains, mushrooms, legumes, nuts, and seeds. The vegan who doesn't eat these foods, however, may not meet the body's riboflavin needs.

nutritional yeast a fortified food supplement containing B vitamins, iron, and protein that can be used to improve the quality of a vegetarian diet.

meat replacements textured vegetable protein products formulated to look and taste like meat, fish, or poultry. Many of these are designed to match the known nutrient contents of animal protein foods.

TABLE 5-3
Complementary Protein Combinations That Provide High-Quality Protein

Combine			Examples
Cereal Grains	+	**Legumes**	
Barley		Dried beans	Bean taco
Bulgur		Dried lentils	Chili and cornbread
Oats		Dried peas	Lentils or beans and rice
Rice		Peanuts	Peanut butter sandwich
Whole-grain breads			
Pasta			
Cornmeal			
Legumes (or grains)	+	**Seeds and Nuts**	
Dried beans		Sesame seeds	Hummus (chickpea and sesame paste)
Dried lentils		Sunflower seeds	Split pea soup and sesame crackers
Dried peas		Walnuts	Noodles with sesame seeds
Peanuts		Cashews	
		Nut butters	

Examples

© Amy Etra/PhotoEdit
Hummus and bread

© Michael Newman/PhotoEdit
Corn and black-eyed peas

© Tony Freeman/PhotoEdit
Peanut butter and wheat bread

© FoodPix/Jupiter Images
Tofu and rice

Minerals

Iron and zinc need special attention in the diets of all vegetarians.[17] Whole-grain products, soy foods, other legumes, dried fruit, nuts, and seeds are important sources of iron in the vegetarian diet. The iron in these foods, however, is not as easily absorbed by the body as that in meat. Because the vitamin C in fruits and vegetables can triple the absorption of the iron provided by other foods eaten at the same meal, vegetarian meals should be rich in foods offering vitamin C.

Zinc is widespread in plant foods, but its availability may be hindered by the fibers and other binders found in fruits, vegetables, and whole grains. Vegetarians are advised to eat varied diets that include wheat germ, legumes, nuts, seeds, and whole-grain products. Milk, yogurt, and cheese provide zinc to the lactovegetarian as well.

Special efforts are necessary to meet the vegan's calcium needs.[18] Whereas the milk-drinking vegetarian is protected from calcium deficiency, the vegan must find other sources of calcium. Some good sources of calcium are *regular* servings of calcium-fortified breakfast cereals and juices; legumes; firm-style tofu; other soy foods, including calcium-fortified soy milk; dried figs, some nuts, such as almonds; certain seeds, such as sesame seeds; and some vegetables, such as broccoli, collard greens, kale, mustard greens, turnip greens, okra, rutabaga, and Chinese cabbage (bok choy). The choices should be varied, because the absorption of the calcium provided by some of these foods is hindered by binders in them. The strict vegetarian is urged to use *calcium-fortified* soy milk, juices, or cereals in *ample quantities, regularly.*[19]

Health Benefits

Vegetarian protein foods are higher in fiber, richer in certain vitamins and minerals, and lower in fat as compared to meats.[20] Vegetarians can enjoy a nutritious diet that is very low in fat, provided that they eat high-fat foods such as margarine, oil, cheese, sour cream, and nuts in moderation. Table 5-4 offers tips for nutritious, easy-to-fix vegetarian meals and snacks.

Studies have found that people with vegetarian or near-vegetarian traditions, such as the Seventh-Day Adventists and the Chinese, have lower rates of heart disease, cancer, diabetes, and obesity than those consuming the typical North American diet.[21] Informed vegetarians are more likely to be at the desired weights for their heights and to have lower blood cholesterol levels, lower blood pressure, lower rates of certain types of cancer, better digestive function, and better health in other ways.[22] Even compared with people who are health conscious, vegetarians experience fewer deaths from cardiovascular disease. Often vegetarianism goes with a healthful lifestyle (no smoking, lower alcohol intakes, emphasis on supportive family life, and so forth), so it is unlikely that dietary practices *alone* account for all the aspects of improved health. However, they may contribute to it.

TABLE 5-4
Easy-to-Prepare Vegetarian Meals and Snacks

Breakfast
- Cold cereal (preferably iron enriched, as noted on the label); eat with fat-free or low-fat milk, yogurt, or soy milk.
- Hot cereals: add fresh fruit slices and yogurt and sprinkle with cinnamon.
- Toast, bagels: top with low-fat cheese, low-fat cottage cheese, or 1 to 2 tbsp of peanut butter.

Snacks
- Assorted fresh fruits and vegetables with yogurt dip.
- Low-fat cheese or peanut butter on rice cakes or crackers.
- Fat-free or low-fat yogurt.
- English muffin pizza with part-skim mozzarella cheese.
- Hummus with pita bread wedges and crisp vegetables.*

Lunch and Dinner
- Salads: add tofu, chickpeas, three-bean salad, kidney beans, low-fat cottage cheese, sunflower seeds, and hard-cooked egg.
- Salad dressings: add salad seasonings to plain yogurt or blenderized tofu.
- Pasta: add diced tofu and/or canned kidney beans to tomato sauce; top with grated, part-skim mozzarella.
- Baked potato: top it with canned beans, steamed vegetables, or low-fat cheese.
- Hearty soups: enjoy lentil, split pea, bean, and minestrone soups, either homemade or canned.
- Vegetarian pizza: top with nonfat or low-fat cheeses and lots of vegetables.

*To make hummus: Blend 1 cup cooked chickpeas, 2 tablespoons tahini (sesame seed paste), 2 tablespoons lemon juice, 1 clove minced garlic, and ¼ cup chopped fresh parsley in food processor until smooth. Chill and serve with pita bread wedges, crackers, or crisp vegetables. Makes two servings.

Nutrition Action Food Allergy—Nothing to Sneeze At

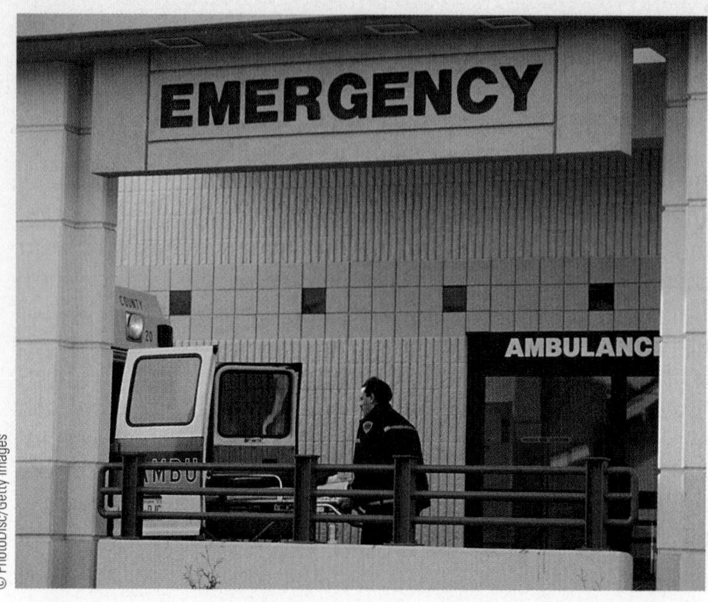

© PhotoDisc/Getty Images

In December 1995, a 33-year-old woman with peanut allergy read the contents of the label of a container of split pea soup. Peanuts were not listed. She ate only part of the soup and within minutes experienced a life-threatening reaction and was rushed to the emergency room for treatment. Following the incident, the split pea soup was analyzed and found to contain peanut flour as a component of the "flavoring" ingredients. The soup manufacturer subsequently discontinued using peanut flour in the product.[23]

The woman in the story belongs to the estimated 1 percent to 2 percent of the adult population in the United States suffering from food allergies. Interestingly, although a relatively small percentage actually have food allergies, nearly 25 percent of people "think" they do and develop **food aversions.**[24] What is the explanation? The answer may be in understanding the difference between **food allergy** and **food intolerance.** Although the physical response to a food allergy and food intolerance may be very similar, the difference between the two is whether or not the immune system is involved in the reaction.

food aversion a strong desire to avoid a particular food.

food allergy an adverse reaction to an otherwise harmless substance that involves the body's immune system.

food intolerance a general term for any adverse reaction to a food or food component that does not involve the body's immune system.

adverse reaction an unusual response to food, including food allergies and food intolerances.

Food Intolerance versus Food Allergy

Food intolerance is far more common than food allergy. A food intolerance is an **adverse reaction** to food that *does not* involve the immune system. Lactose intolerance is one example of a food intolerance. A person with lactose intolerance lacks the enzyme needed to digest lactose. When that person eats milk products, symptoms such as gas, cramps, and bloating can occur. Gluten intolerance is caused by an intolerance to gluten, a protein found in wheat, oats, barley, and rye. People with gluten intolerance must avoid products prepared with these grains. Some people may develop a sensitivity to various other agents in a food. Tyramine, found in cheese or red wine, can induce a headache in some people. Others may have a sensitivity to certain food additives such as monosodium glutamate (MSG), sulfites, or coloring agents. The physical reaction to these agents can include hives, rashes, nasal congestion, or asthma.

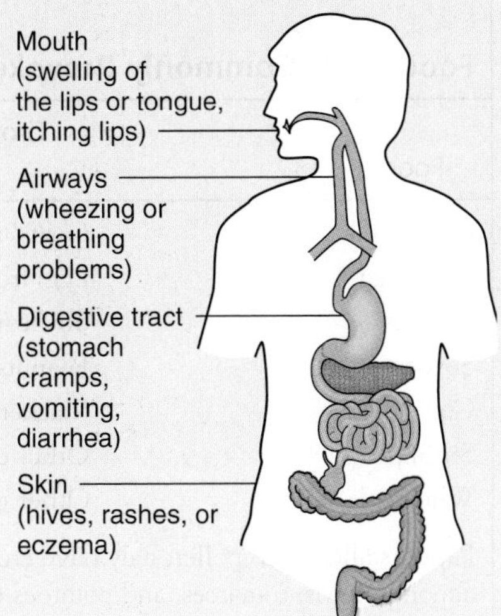

FIGURE 5-5
COMMON SITES FOR ALLERGIC REACTIONS

SOURCE: U.S. Food and Drug Administration, *FDA Courier* (May 1994).

A food allergy, on the other hand, is an abnormal response to a food triggered by the immune system. The allergic reaction involves three main components: **food allergens,** immunoglobulin E (IgE), and mast cells. Food allergens are the fragments of food that are responsible for the allergic reaction. They consist of proteins from the food that are not broken down during the digestive process, which then cross the gastrointestinal lining to enter the bloodstream. IgE is a type of protein called an antibody that circulates through the blood. When allergic people eat certain foods, their immune system reacts to the food allergen by making IgE that is specific to that food. Once released, the IgE antibody attaches to a cell found in all body tissues called the mast cell. The mast cells are specialized cells of the immune system that serve as the storehouse for various chemical substances, including **histamine.** Mast cells are found in all tissues, but they are especially common in the areas of the body that are typical sites of allergic reaction—the nose and throat, lungs, skin, and gastrointestinal tract. When an allergic reaction occurs, the food allergen interacts with the IgE on the surface of the mast cells, which triggers those cells to release histamine. Depending on the tissue in which the histamine is released, these chemicals will cause a person to have various symptoms of a food allergy (see Figure 5-5). The most severe allergic reaction is **anaphylaxis.** This potentially fatal condition occurs when several parts of the body experience food allergy reactions at the same time. Signs of anaphylaxis include difficulty breathing, swelling of the mouth and throat, a drop in blood pressure, and loss of consciousness. The reaction can occur in a few seconds or minutes, and without immediate medical attention, death may result. The foods most associated with anaphylactic reactions include peanuts, tree nuts (for example, walnuts, cashews), eggs, and shellfish.

According to the Food Allergy Network, eight foods cause 90 percent of all allergic reactions: egg, fish, milk, peanuts, shellfish, soy, tree nuts, and wheat.[25] In adults, the most common foods to cause allergic reactions include

food allergen a substance in food—usually a protein—that is seen by the body as harmful and causes the immune system to mount an allergic reaction.

histamine a substance released by cells of the immune system during an allergic reaction to an antigen, causing inflammation, itching, hives, dilation of blood vessels, and a drop in blood pressure.

anaphylaxis (an-ah-fa-LAX-is) a potentially fatal reaction to a food allergen causing reduced oxygen supply to the heart and other body tissues. Symptoms include difficulty breathing, low blood pressure, pale skin, a weak, rapid pulse, and loss of consciousness.

▲ *The top eight foods causing adverse reactions in some individuals: milk, eggs, peanuts, nuts, fish, shellfish, soy, and wheat.*

TABLE 5-5
Foods that Commonly Provoke Food Allergy

Food	Cross-Reacting Foods
Cows' milk*	Goats' milk, ewes' milk
Hens' eggs*	Eggs from other birds
Peanuts*	Soybeans, green beans, green peas
Soybeans	Peanuts, green beans, green peas
Cod	Mackerel, herring
Shrimp	Other crustaceans
Wheat	Other grains, mostly rye

Patients allergic to pollen may have cross reactions with hazelnuts, green apples, peaches, almonds, kiwis, tomatoes, and potatoes (birch pollen) or wheat, rye, and corn (grass pollen).

*Most common allergic foods for children.

SOURCE: Adapted from C. Bindslev-Jensen, "ABCs of Allergies: Food Allergy," *British Medical Journal* 316 (1998): 7140.

cross-reaction the reaction of one antigen with antibodies developed against another antigen.

shellfish, peanuts, tree nuts, fish, and egg. In children, the problem foods are typically egg, milk, and peanuts. **Cross-reaction** is a concern for someone diagnosed with a food allergy. For instance, if someone has a history of allergic reaction to shrimp, testing may show that person is also allergic to other shellfish, such as crab and lobster (see Table 5-5).

Children and Food Allergy

The prevalence of food allergy is greatest in the first few years of life, with up to 6 percent of children younger than 3 years experiencing allergic responses.[26] Increased susceptibility of infants to food allergic reactions is believed to be the result of their immature immune system. The immature digestive system may also allow more intact allergen proteins to enter the bloodstream. Cow's milk and soy are the most common allergens for infants, producing reactions such as hives, bloating, and diarrhea. Breastfeeding is recommended for the first 12 months to avoid early exposure to cow's milk or soy, thus avoiding allergy. Breast milk contains agents that stimulate the development and maintenance of the digestive tract, which further reduces the risk of food allergy. Early exposure to certain foods can also play a role in the development of food allergy in children. Delaying introduction of common allergenic foods may delay the onset of food allergies by allowing the digestive system and immune system to further develop before exposure to these known allergens. Young children often lose their sensitivity to most of the common allergenic foods in a few years. Unfortunately, sensitivity to certain foods, such as peanuts, tree nuts, fish, and shellfish, is rarely lost, and sensitivity persists into adulthood.

Diagnosis and Management of Food Allergy

Because both food intolerance and food allergy present very similar symptoms, diagnosing a food allergy involves determining if the reaction is mediated by the immune system. A complete physical examination can rule out the possibility that an underlying physical condition may be causing the symptoms. The physician takes a detailed case history and considers the type and timing of symptoms as well as the suspected offending food. Individuals may be asked to keep a food diary and record all physical symptoms. Once food allergy is identified as a likely cause of symptoms, confirmation of the diagnosis and positive identification of the allergen involves various tests. A prick skin test (PST) is a simple test done in the doctor's office. The doctor puts a drop of the substance being tested on the forearm of the patient and pricks it with a needle, which allows a tiny amount of the substance to enter the skin. If the patient is allergic, a small bump will occur at the site in approximately 15 minutes. The radioallergosorbent test (RAST) requires a blood sample. Laboratory tests are conducted using specific foods to find out if the patient has IgE antibodies to those foods. Also, the physician may recommend an elimination diet, in which the patient does not eat the food suspected of causing the allergy for a period of about two to four weeks. If the allergic symptoms improve, a diagnosis can be made.

The only treatment for food allergy is to avoid the offending food. People with food allergies must develop a skill for reading food labels. Effective January 2006, the Food and Drug Administration requires food labels to clearly state if food products contain any ingredients that contain protein derived from the eight major allergenic foods.[27] Manufacturers are required to identify in plain English the presence of ingredients that contain protein derived from milk, eggs, fish, crustacean shellfish, tree nuts, peanuts, wheat, or soybeans in the list of ingredients or to say "contains" followed by name of the source of the food allergen after or adjacent to the list of ingredients. The possibility of cross-contamination is also a problem for people with food allergies. When one food comes in contact with another food, trace amounts of each food mix with the other. Cross-contamination can occur in processing plants when different foods are processed on the same equipment. It can also occur at home—a knife inserted into the peanut butter jar and then used in the jelly will contaminate the jelly with peanut protein. The emerging field of biotechnology presents an additional challenge through genetic engineering (see Chapter 12). In genetic engineering, a gene from one species can be "spliced" into another species. The FDA requires that any food product of biotechnology that contains protein from a food that is known as a common allergen must be properly labeled.

Currently, no drugs are available to cure food allergy. However, some medications, such as antihistamines and corticosteroids, are used to treat the symptoms. Individuals with a history of anaphylactic reactions to food allergens typically carry self-injecting syringes of epinephrine to use in case of accidental exposure.

Many questions remain to be answered in the field of food allergy. As scientists gain insight into the reactions of the immune system to various components of allergens, we can expect advances in the diagnosis and treatment of this disorder.[28] As the mystery of DNA is unraveled, it is conceivable that food allergy can be prevented. Until that time, however, many organizations offer additional information on food allergy and food intolerance, as listed in the margin.

The Food Allergy Network
800-929-4040
www.foodallergy.org
American Academy of Allergy, Asthma, and Immunology
414-272-6071
www.aaaai.org
American Dietetic Association
800-877-1600
www.eatright.org
Asthma and Allergy Foundation of America
800-7-ASTHMA
www.aafa.org

Spotlight

Wonder Bean—The Benefits of Soy

Soybeans have a long, rich history in the Eastern world cuisine. The early Chinese recognized the importance of this food, and they called it *Ta Tau,* which means "greater bean." According to Chinese tradition, soybeans were named as one of the five most sacred crops by the emperor who reigned 5,000 years ago.

Not too long ago in the Western world, soybeans were fodder for livestock. During the 1970s, many people turned to a vegetarian style of eating, often as a form of protest, but more frequently as a way to adopt a healthier lifestyle. Soy protein became the meat substitute of choice. Twenty-five years later, soybeans are recognized as the ideal functional food—a food that has the potential to reduce the risk of disease.

Soy foods are currently a hot area of research. Recent findings show that substances such as phytoestrogens and isoflavones, which are found in soybeans, can lower cholesterol and help prevent disease. Numerous studies attest to the role soy foods may play in reducing risk for certain forms of cancer, heart disease, and osteoporosis as well as in controlling diabetes and easing a woman's transition through menopause. Is it any wonder that over 5,000 years ago the soybean was called the "greater bean"?

What is soy?

Soybeans are legumes, members of the same plant family that includes other beans, peas, and lentils. Among edible legumes, however, the soybean is somewhat unusual because it is relatively low in carbohydrates. Soybeans are, however, high in fiber.

Among plant foods, legumes are high in protein. What distinguishes the soybean from its cousins, however, is the nature of the protein: soybeans supply all of the essential amino acids needed for health. The amino acid pattern of soy protein is essentially the same quality as that found in meat, milk, and egg protein. Soybeans are the only vegetable food that contains complete protein. What emerges from this nutritional analysis of the soybean is the image of "balance." Soybeans as a food can basically stand alone, give or take a few vitamins and minerals. Add some vegetables to the beans, and you have created a high-quality, nutrient-dense meal!

© Scott Hirko/Hespenheide Design

You have mentioned isoflavones. What are they?

Isoflavones are a type of *phytoestrogen*—compounds that have a weak, estrogenic activity. Many types of phytoestrogenic compounds are available in edible plants. Foods made from soybeans have varying amounts of the isoflavones, depending on how they are processed (see Table 5-6). Foods such as tofu, soy milk, soy flour, and soy nuts have higher isoflavone concentrations than foods made with a combination of soy and grains. Soy sauce and soybean oil have virtually no isoflavones.

Research in several areas of health care has shown that consumption of soy foods may play a role in lowering

TABLE 5-6
Protein and Isoflavone Content of Selected Soy Foods

	Serving Size	Protein[a] (g)	Isoflavones[b] (mg)
Green soy beans, edamame	4 oz	14	50
Soy nuts	3 oz	20	90
Miso	2 tbsp	4	15
Soy milk	1 c	10	24
Soy flour, roasted	¼ c	7	42
Tempeh	4 oz	19	36
Tofu, firm	4 oz	13	24
Texturized soy protein (TVP), dry	¼ c	6	29
Soy burger (check label)	1	10–12	38–55
Soy protein isolate, dry	1 oz	23	28
Soy protein concentrate (alcohol extracted)	1 oz	17	4[c]

[a] Soy protein data (rounded to whole numbers) from USDA nutrient Database for Standard Reference, Release 15.
[b] Isoflavone data (rounded to whole numbers) from USDA–Iowa State University database on the Isoflavone Content of Foods, 1999.
[c] To isolate soy protein from defatted soybeans, the carbohydrates must be removed by using a solvent, which can remove some or all of the isoflavone content. Because isoflavones are soluble in alcohol, much of the isoflavone content is lost if alcohol or repeated water washings are used in the extraction process. Soy milk, soy flour, tofu, tempeh, and soy protein isolate are not prepared with alcohol or repeated water extraction and therefore have a higher isoflavone content than soy protein concentrate.

MINIGLOSSARY WHAT FOODS CONTAIN SOY?[29]

green soybeans (Edamame) These large soybeans are harvested when the beans are still green and sweet and can be served as a snack or a main vegetable dish, after boiling in water for 15 to 20 minutes. They are high in protein and fiber and contain no cholesterol.

hydrolyzed vegetable protein (HVP) Hydrolyzed vegetable protein (HVP) is a protein obtained from any vegetable, including soybeans. The protein is broken down into amino acids by a chemical process called acid hydrolysis. HVP is a flavor enhancer that can be used in soups, broths, sauces, gravies, flavoring and spice blends, canned and frozen vegetables, and meats and poultry.

meat alternatives Meat alternatives made from soybeans contain soy protein or tofu and other ingredients mixed together to simulate various kinds of meat. These meat alternatives are sold as frozen, canned, or dried foods.

miso Miso is a rich, salty condiment that characterizes the essence of Japanese cooking. The Japanese make miso soup and use miso to flavor a variety of foods. A smooth paste, miso is made from soybeans and a grain such as rice, plus salt and a mold culture, and then aged in cedar vats for one to three years. Miso should be refrigerated. Use miso to flavor soups, sauces, dressings, marinades, and pâtés.

nondairy soy frozen dessert Nondairy frozen desserts are made from soy milk or soy yogurt. Soy ice cream is one of the most popular desserts made from soybeans and can be found in many grocery stores.

soy cheese Soy cheese is made from soy milk. Its creamy texture makes it an easy substitute for sour cream or cream cheese. It can be found in a variety of flavors in many food stores.

soy flour Soy flour is made from roasted soybeans ground into a fine powder. Soy flour gives a protein boost to recipes. Soy flour is gluten-free so yeast-raised breads made with soy flour are more dense in texture. Replace one-quarter to one-third wheat flour with soy flour in recipes for muffins, cakes, cookies, pancakes, and quick breads.

soy protein, texturized Texturized soy protein usually refers to products made from texturized soy flour. Texturized soy flour is made by running defatted soy flour through an extrusion cooker, which allows for many different forms and sizes. When hydrated, it has a chewy texture. It is widely used as a meat extender.

soy yogurt Soy yogurt is made from soy milk. Its creamy texture makes it an easy substitute for sour cream or cream cheese. Soy yogurt can be found in a variety of flavors in many food stores.

soybeans As soybeans mature in the pod they ripen into a hard, dry bean. Most soybeans are yellow. However, brown and black varieties are also available. Whole soybeans can be cooked and used in sauces, stews, and soups.

soy milk, soy beverages Soybeans that are soaked, ground fine, and strained produce a fluid called soybean milk. Soy milk is an excellent source of high-quality protein and B vitamins. Look for calcium-fortified varieties.

soy-nut butter Made from roasted, whole soy nuts, which are then crushed and blended with soy oil and other ingredients, soy-nut butter has a slightly nutty taste, significantly less fat than peanut butter, and provides many other nutritional benefits.

soy nuts Roasted soy nuts are whole soybeans that have been soaked in water and then baked until browned. Soy nuts can be found in a variety of flavors, including chocolate covered. High in protein and isoflavones, soy nuts are similar in texture and flavor to peanuts. Try sprinkling some on salads.

tempeh Tempeh, a traditional Indonesian food, is a chunky, tender soybean cake. Whole soybeans, sometimes mixed with another grain such as rice or millet, are fermented into a rich cake of soybeans with a smoky nutty flavor. Tempeh can be marinated and grilled and added to soups, casseroles, or chili.

tofu and tofu products Tofu, also known as soybean curd, is a soft cheese-like food made by curdling fresh hot soy milk with a coagulant. Tofu is a bland product that easily absorbs the flavors of other ingredients with which it is cooked. Tofu is rich in high-quality protein and B vitamins and is low in sodium. Firm tofu is dense and solid and can be cubed and served in soups, stir fried, or grilled. Firm tofu is higher in protein, fat, and calcium than other forms of tofu. Silken tofu is a creamy product and can be used as a replacement for sour cream in many dip recipes.

risk for disease.[30] Soy isoflavones are being studied intensively to clarify their physiological effects. In some cases, research has shown that the isoflavones may be a key factor in the disease-fighting potential of soybeans.

It is important to keep in mind that our knowledge of the long-term effects of isoflavones is based on their content in soy foods. These foods have been consumed for hundreds of years, and are known to be safe. It is still best to obtain isoflavones by enjoying a variety of soy foods.

The new information about soy seems promising. What are the potential health benefits of adding soy foods to my diet?

SOY AND HEART DISEASE

High blood cholesterol is a major risk factor for heart disease. A great deal of evidence indicates that soy protein helps lower blood cholesterol levels. Replacing animal protein with soy protein in the diet lowers total and LDL cholesterol levels in people with high cholesterol.[31] A meta-analysis of 38 research studies concluded that soy protein lowers total and LDL cholesterol and triglycerides, without lowering HDL cholesterol in people with high cholesterol.[32] In these studies, the average consumption of soy protein was 47 grams per day. The greatest decreases in blood cholesterol were seen in those with the highest starting levels. Even adding soy protein to an omnivorous diet has been shown to produce this effect.[33] The newly approved health claim for food labels states that as little as 25 grams of soy protein per day may be enough to lower cholesterol levels. Refer to Table 5-6 for the protein content of selected soy foods.

SOY AND OSTEOPOROSIS

Soybeans and soy foods may help prevent and treat osteoporosis, a disease that weakens bones and often results in bone fractures. As women age, it becomes more important than ever to maintain adequate levels of calcium in bones. Soyfoods such as fortified soy milk, texturized soy protein, and tofu made with calcium salt are all good calcium sources.

Isoflavones found in soy protein may also play an important role in protecting bones.[34] A breakthrough study at the University of Illinois at Urbana concluded that consuming soybean isoflavones can increase bone mineral content and bone density.

As little as 40 grams of soy protein, consumed each day for 6 months, led to positive results in a test group of postmenopausal women.[35] Forty grams of soy protein can be found in 2 ounces of soy protein isolate (see Table 5-6).

SOY AND MENOPAUSE

The hormonal changes that occur during menopause can cause a variety of symptoms and increase risk for heart disease and osteoporosis.[36] Soy foods that contain phytoestrogens are being studied for their possible efficacy in decreasing the negative effects of menopause. Fluctuating levels of estrogen can cause hot flashes, night sweats, insomnia, vaginal dryness, or headaches. Hormone replacement therapy (HRT) had been commonly prescribed to help prevent the negative health effects of menopause. However, recent findings from the Women's Health Initiative indicate possible increased risk for breast cancer, stroke, and heart disease from HRT.[37]* Scientists are now investigating the question: Can soy foods provide the same kinds of health benefits as HRT, without the risks?

In women who are producing little estrogen, phytoestrogens may produce enough estrogenic activity to relieve symptoms such as hot flashes. A recent study found that women who were fed 45 grams of soy flour per day had a 40 percent reduction in the incidence of hot flashes.[38] From an epidemiological point of view, it is interesting to note that in Japan, where soy consumption is high, menopause symptoms of any kind are rarely reported. In addition, bones tend to be stronger in Asia, and broken hips and spinal fractures are less common.

Soy contains phytoestrogens in the form of isoflavones, genistein, and daidzein. These are known to have weak estrogenic effects when consumed by animals and humans.[39] Researchers continue to study the physiological effects of the isoflavones to find out whether they can serve some of the same functions as estrogen, and thereby decrease the health risks associated with menopause.

SOY AND CANCER

One out of every four deaths in the United States is due to cancer. Epidemiological studies show that populations that consume a typical Asian diet have lower incidences of breast, prostate, and colon cancers than those consuming a Western diet.[40] The Asian diet includes mostly plant foods, including legumes, fruits, and vegetables, and is low in fat. The Japanese have the highest consumption of soy foods. Japan has a very low incidence of hormone-dependent cancers. The mortality rate from breast and prostate cancers in Japan is about one-fourth that of the United States.[41] Some evidence suggests that the difference in cancer rates is not due to genetics, but rather to diet. Migration studies have shown that when Asians move to the United States and adopt a Western diet, they ultimately have the same cancer incidence as Americans.[42] Other long-term studies have noted an inverse association between regular consumption of miso soup and breast cancer risk in premenopausal women.[43] In Hawaii, a long-term study of 8,000 men of Japanese ancestry showed that men who ate tofu daily were only one-third as likely to get prostate cancer as those who ate tofu once a week or less.[44]

Soybeans contain five classes of compounds that have been identified as anticarcinogens.[45] Most of these compounds can be found in many different plant foods, but soy is the only significant dietary source of isoflavones. Soy isoflavones, especially genistein, have been the subject of a tremendous amount of cancer research.

Conflicting data abounds in almost all nutritional research, and soy is no exception. Some studies have actually shown a correlation between soy consumption and the development of breast cancer, and still others associate the time in life that soy is actually consumed with the extent of its effects.[46] Whatever the case, it could be years before scientists reach definite conclusions. The field of soy research is still relatively new with only a few researchers working on this area for more than two decades.[47] As this area of nutritional research evolves, more conflicting information will no doubt be published. Thus far, however, the potential benefits of soy seem to outweigh the risks.

SOY AND DIABETES

Another interesting benefit of soy foods is their effect on glucose control. Recently, scientists have become interested in the role of soy foods in regulating diabetes.[48] Emerging evidence indicates that the phytoestrogens, soy protein, and soy polysaccharides found in soy foods might play a beneficial role in diabetes. Although further long-term investigation is needed, this emerging area of research will be important in years to come. In addition, because soybeans are a complex carbohydrate and also have a low

*The National Institutes of Health (NIH) established the Women's Health Initiative in 1991 to study strategies for preventing the major causes of death and disability in postmenopausal women, including heart disease, breast and colon cancer, and osteoporosis.

glycemic index, they are an ideal food to help regulate blood glucose levels in people with diabetes. Overall it seems that soy foods can have a positive synergistic effect on diabetes.

How much soy should I eat on a daily basis to receive these health benefits?

Science has not yet established a recommended daily amount of soy consumption to achieve all of the health benefits mentioned here. In many cases, just one serving of soy per day may help improve your health. Although the new health claim for food labels states that 25 grams of soy protein each day may lower risk of heart disease, no recommendation for daily isoflavone intake has yet been made. In Asian countries, where people typically consume 25 to 40 milligrams of isoflavones a day, the incidence of osteoporosis, heart disease, and certain cancers is low. Still, many questions remain about how soy acts in the body and how much is needed for benefits at various stages of the life cycle. A balanced diet that includes soy is recommended, but loading up on one food, nutrient, or phytochemical is not advisable. Here are some quick tips for adding some soy to your diet:

- Pour soy milk over breakfast cereal once or twice a week.
- Cook oatmeal or cream of wheat in soy milk. You might try vanilla flavored soy milk.
- Make pancakes or French toast with vanilla soy milk.
- Use soy milk to make hot chocolate.
- Add firm tofu chunks in place of meat or poultry in stir-fries, fajitas, or shish kebabs.
- Try a "garden" burger or soy burger. Check food labels; not all are made with soy.
- Replace ground beef on nachos, pizzas, or in spaghetti sauce with crumbled soy burgers.
- Add tofu to the blender with your favorite seasoning packet, such as ranch, onion soup mix, or taco seasoning, and serve with tortilla chips or fresh vegetables.
- Add soy flour to quick bread recipes. It adds moisture and a soy protein boost. Just replace one-fourth of the total flour with soy flour in recipes for quick breads, muffins, or cakes.
- Enjoy soy-nut butter on bagels, breads, English muffins, or carrot and celery sticks.

© Polara Studios

Quick and Easy Soy Recipes

Strawberry–Banana Frosty
3 c plain or vanilla soy milk
1 c strawberries
1 ripe banana
Blend in blender until smooth.
Makes six servings.

Multigrain Apple Pancakes
1¼ c yellow corn meal
1½ c rolled oats
1½ c unbleached flour
1½ c soy flour
1 tsp cinnamon
1 tbsp baking powder
1½ c plain soy milk
1½ c applesauce

In a large bowl, combine the rolled oats, corn meal, unbleached flour, soy flour, cinnamon, and baking powder. Add the soy milk, and blend with a few swift strokes. Fold in the applesauce. Pour 1¼ cup of the batter on a hot nonstick griddle or pan. Cook for about 2 minutes or until bubbles appear on the surface. Flip the pancake and cook for another minute or until heated through. Serve the pancakes with maple syrup, fruit spread, or applesauce.
Makes 12 pancakes.

Soy Nut Trail Mix
2 c roasted soy nuts
1 c raisins
1 c oat-ring cereal
1 c dried cranberries
2 c mini-wheat squares cereal
½ c dried cherries

Mix all ingredients in large bowl or container. Keep tightly closed in container or zippered plastic bag.
Yield: 7 cups

SOURCE: *Soyfoods Cookbook,* www.soyfoods.com/recipes.

In Review

1. Which of the following is *not* true about protein?
   a. The primary role of protein is to provide building material for body tissues.
   b. The primary role of protein is to provide calories.
   c. Most Americans eat well above the RDA for protein each day.
   d. High protein intakes increase the body's excretion of calcium.
   e. We need to consume foods that supply a sufficient amount of all essential amino acids every day.

2. Why are some amino acids referred to as essential?
   a. They must be supplied by the diet.
   b. They are found in fortified breakfast cereals.
   c. They are more important than nonessential amino acids.
   d. They are needed in larger amounts than nonessential amino acids.
   e. Only b and c are correct.

3. All of the following food groups provide protein except:
   a. milk.
   b. grains.
   c. fruits.
   d. vegetables.

4. If an essential amino acid required for formation of a certain enzyme is missing in the diet:
   a. another amino acid will be substituted in its place so the enzyme can be made.
   b. synthesis of the enzyme will be halted.
   c. the partially synthesized enzyme will be stored in the body until the missing amino acid is supplied by the diet.
   d. the amino acid will be made from glucose.

5. List three functions of protein in the body.

6. ______ are the building blocks of proteins.
   a. Amino acids
   b. Lipids
   c. Calories
   d. Amino bases

7. Where do lacto-ovovegetarians get their protein?
   a. poultry meat and bovine meat
   b. seafood
   c. milk, cheese, and eggs
   d. vegetables only

8. What is the protein RDA for a person whose appropriate weight is 150 pounds?
   a. 55 grams
   b. 64 grams
   c. 80 grams
   d. 150 grams

9. Which of the following is *not* a benefit of choosing plant-based over animal-based protein?
   a. Plant-based proteins are higher in fiber.
   b. Plant-based proteins are often lower in fat.
   c. Plant-based proteins provide essential nutrients.
   d. Plant-based proteins actually burn calories.

10. Briefly describe a "complete protein" and a "high-quality protein."

Menu of Online Study Tools

A variety of study tools for this chapter are available at our website to deepen your understanding of chapter concepts. Go to

www.thomsonedu.com/nutrition/boyle

to find

- Practice tests
- Flashcards
- Glossary
- Web links
- Animations
- Chapter summaries, learning objectives, and crossword puzzles

■ Nutrition on the Web

www.thomsonedu.com/nutrition/boyle
Go to the *Personal Nutrition* site to check for the latest updates to chapter topics or to access links to related websites.

www.eatright.org
Search for information about protein in foods; view the ADA Position paper on Vegetarian Diets.

www.fda.gov
Go to Foods and search for information on vegetarian diets.

www.who.int/nut
Search this site for more information about protein-energy malnutrition worldwide.

www.nal.usda.gov/fnic/Fpyr/pyramid.html
View the Food Pyramid for Vegetarians and a variety of pyramids from other countries at this site.

www.vrg.org
The Vegetarian Resource Group provides information on vegetarianism, vegetarian books and recipes, and links to related sites; it hosts a fun Vegetarian Quiz to test your knowledge.

www.ncbi.nlm.nih.gov/PubMed
This search engine helps you locate information from current scientific articles on any topic related to protein.

www.soyfoods.com
This U.S. Soy Foods Directory website is an essential resource for anyone interested in learning more about soy foods. The site includes a searchable database, recipes, and research information about the health benefits of soy foods.

www.talksoy.com
The United Soybean Board provides information about soy foods and answers questions about soy foods.

www.nal.usda.gov/fnic/etext/macronut.html
Access general information about protein functions, requirements, and food sources. Includes many useful links to related information.

The Vitamins

6

ASK YOURSELF . . .

Which of the following statements about nutrition are true, and which are false? For each false statement, what is true?

1. The most important role that vitamins play is providing energy.
2. You can't overdose on vitamins because the body excretes them in the urine.
3. Several major public health associations recommend that all adults take antioxidant supplements.
4. Serving for serving, fruits and vegetables tend to be the richest sources of vitamins.
5. In general, nutrients are absorbed equally as well from foods as from supplements.
6. Vitamin C supplements prevent colds.
7. Oatmeal is an example of a functional food.
8. Fresh vegetables contain more vitamins than frozen vegetables.
9. Phytochemicals are beneficial nonnutrient substances found in fruits, vegetables, and whole grains.
10. Large doses of niacin can cause flushing, rash, and fatigue.

Answers found on the following page.

In France old Crainquebille sold leeks from a cart, leeks called "the asparagus of the poor." Now asparagus sells for the asking, almost, in California markets, and broccoli, that strong age-old green, leaps from its lowly pot to the Ritz's copper saucepan.

Who determines, and for what strange reasons, the social status of a vegetable?

M. F. K. Fisher (1908–1992, U.S. food writer)

CONTENTS

About a century ago, scientists ushered in a new era in the science of nutrition—the discovery of vitamins. They quickly realized that these substances, found in minute amounts in foods, were just as essential to health as fats, carbohydrates, and proteins. A diet lacking in one vitamin could cause a barrage of symptoms and, ultimately, death. Knowledge of the vital roles played by vitamins quickly advanced, and today life-threatening vitamin deficiencies are rare in developed countries such as the United States and Canada.

Still, the vitamin research that has been conducted during the past decade or so has marked the beginning of yet another chapter in the annals of nutrition. Throughout the past ten years, more and more scientists have investigated the possibility that large doses of certain vitamins will help stave off chronic diseases such as cancer and heart disease, problems that rank as major killers today. In fact, the study of vitamins, particularly a class known as the antioxidant vitamins, is one of the hottest, most widely publicized areas in nutrition research today. In addition, the pros and cons of taking vitamin supplements are the subject of heated debate among the scientific community. To help you sort through the steady stream of controversy regarding vitamins, this chapter explores the history, roles, and current thrust of research for the various vitamins and offers practical advice on how to incorporate the information into decisions about your own lifestyle.

scurvy the vitamin C deficiency disease characterized by bleeding gums, tooth loss, and, in severe cases, death.

rickets a disease that occurs in children as a result of vitamin D deficiency and that is characterized by abnormal growth of bone, which in turn leads to bowed legs and an outward-bowed chest.

pellagra (pell-AY-gra) niacin deficiency characterized by diarrhea, inflammation of the skin, and, in severe cases, mental disorders and death.

Turning Back the Clock

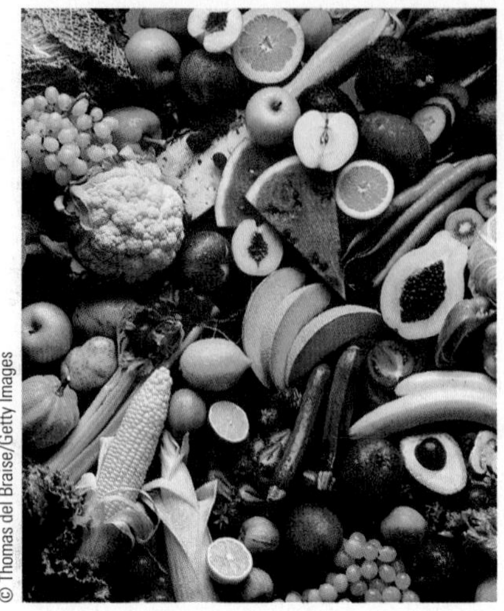

© Thomas del Braise/Getty Images

▲ *The more varied the kinds of fruits and vegetables you eat, the better nourished you are likely to be.*

Many of the vitamin deficiency diseases that have been virtually eliminated today were first recognized in Greek and Roman times and ultimately led to the discovery of vitamins centuries later. One of the most prevalent was **scurvy,** a disease characterized by bleeding gums, tooth loss, and even death, due to lack of vitamin C. The scourge of armies, sailors, and other travelers forced to do without vitamin C-rich foods for weeks on end, scurvy was recognized by Hippocrates, a Greek physician heralded today as the father of medicine.[1]

A cure for scurvy was not recorded until the 16th century, however, when a beverage made of spruce needles or oranges and lemons was recommended. In 1753, a British physician named James Lind published a famous report recommending consumption of herbs, lettuce, endive, watercress, and summer fruits to prevent scurvy. By the early 1800s, sailors in the British navy had been dubbed "limeys" because they were required to drink lemon or lime juice daily.[2] Although they still didn't know that vitamin C was the actual antidote, they did recognize that certain foods prevented and cured the illness.

Similarly, a deficiency disease called **rickets** dates back to Roman times, when children frequently suffered skeletal deformities as a result of a lack of vitamin D. By the 1600s, rickets was known as the English disease because it afflicted so many English children. Some 200 years later, cod liver oil was finally recognized as a cure for the disease; no one knew at the time, however, that the "magic" ingredient in the oil was vitamin D.

See the photo depicting rickets on page 188.

Another deficiency disease, called **pellagra,** was not recognized until 1730, when a Spanish physician named Gaspar Casal first described the crusty, dry, scabby, blackish patches of skin symptomatic of the disease. In Italy, the disease was named pellagra, from the Italian *pelle agra*, meaning "sour skin." Called *mal de la rosa* in Spanish, pellagra was thought to be incurable until Dr. Casal noticed that the people who developed the disease were typically poor and had inadequate diets made up of mostly corn and little meat.

By the 19th century, physicians had recognized that certain foods prevented or cured pellagra and other deficiency diseases. But they still hadn't determined exactly what it was in the various foods that worked as a remedy. By the middle of the 19th century, however, the science of chemistry had advanced to a point at which foods could be analyzed. Chemists had determined that foods consisted of fats, proteins, and carbohydrates along with minerals and water, and they assumed that they had identified all the nutritionally significant compounds.

Then, in the early 20th century, scientists detected minute amounts of other substances that they found to be essential in preventing disease and maintaining health. The substances were dubbed *vitamines*, a term coined in 1912 by a scientist named Dr. Casimir Funk to indicate that these substances were *vital* for sur-

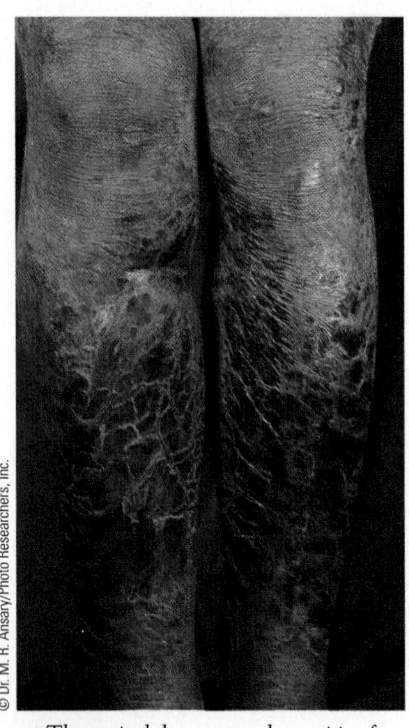

© Dr. M. R. Ansary/Photo Researchers, Inc.

▲ *The typical dry, crusty dermatitis of pellagra is found only on areas of the body exposed to sunlight.*

Ask Yourself Answers: **1.** False. Vitamins do not provide energy, though they do play roles in energy-yielding reactions in the body. **2.** False. Excess doses of all of the vitamins can be toxic. **3.** False. No major health organization recommends that all adults take antioxidant pills. **4.** True. **5.** False. In general, nutrients are absorbed best from foods because they are accompanied by other ingredients that facilitate their absorption. **6.** False. Vitamin C has never been proved to prevent colds; at best, it may reduce the severity of cold symptoms. **7.** True. Oatmeal, oat bran, and whole-oat products contain a soluble fiber shown to reduce cholesterol levels when eaten as part of a heart-healthy diet. **8.** False. Fresh vegetables do not necessarily contain more vitamins than their frozen counterparts, depending on such factors as how the fresh vegetable has been stored and how long since it has been harvested. **9.** True. **10.** True.

vival and that they contained nitrogen—that is, they were *amines*. (The *e* was later dropped when scientists discovered that some of the vitamins were not amines.)

Over the next few decades, scientists identified the various **vitamins,** established their chemical formulas, and determined their functions in the body. They also measured the amount of vitamins in various foods and determined human and animal requirements for the compounds. Knowledge of vitamins constantly evolves as scientists continue to study their actions in the human body.

Today, scientists recognize vitamins as potent compounds that perform many tasks in the body to promote growth and reproduction and maintain health and life. Vitamins constantly work to keep your nerves and skin healthy; build bone, teeth, and blood; and heal wounds, among many other things. Although they do not provide calories, vitamins are essential in helping the body make use of the calories consumed via foods.

vitamin a potent, indispensable compound that performs various bodily functions that promote growth and reproduction and maintain health. Vitamins are *essential* nutrients required in minute amounts and *organic,* meaning that they contain or are related to carbon compounds. Contrary to popular belief, vitamins do not supply calories.

coenzymes enzyme helpers; small molecules that interact with enzymes and enable them to do their work. Many coenzymes are made from water-soluble vitamins.

The Two Classifications of Vitamins

Vitamins fall into two categories: those that dissolve in water (water-soluble vitamins) and those that dissolve in fat (fat-soluble vitamins). To date, scientists have identified 13 vitamins, each with its own special roles to play (see Table 6-1). As Figure 6-1 shows, each of the food groups in the MyPyramid Food Guide supplies a number of vitamins. Eating plans that exclude entire food groups or fail to include the minimum number of servings from each group may lead to vitamin deficiencies over time.

The nine water-soluble vitamins—eight B vitamins and vitamin C—are found in the watery compartments of foods, such as the juice of an orange. These vitamins are distributed into water-filled compartments of the body, including the fluid that surrounds the spinal cord. The body excretes excess water-soluble vitamins if blood levels rise too high. As a result, these vitamins rarely reach toxic levels in the body. This is not to say, however, that excess levels cannot cause problems, at least in some people.

In contrast, the four fat-soluble vitamins—A, D, E, and K—are generally found in the fats and oils of foods. Because they are stored in the liver and in body fat, it is possible for megadoses of fat-soluble vitamins to build up to toxic levels in the body and cause undesirable side effects. Table 6-2 (page 176) shows the contrasts between the two classes of vitamins.

Water-soluble vitamins include:

- B vitamins
  - Thiamin
  - Riboflavin
  - Niacin
  - Vitamin B_6
  - Folate
  - Vitamin B_{12}
  - Biotin
  - Pantothenic acid
- Vitamin C

Fat-soluble vitamins include:

- Vitamin A
- Vitamin D
- Vitamin E
- Vitamin K

Water-Soluble Vitamins

In the body, water-soluble vitamins act as **coenzymes**—that is, they assist enzymes in doing their metabolic work within the body (see Figure 6-2 on page 176). You may recall from Chapter 5 that enzymes are proteins that act as catalysts to help boost chemical reactions in the body, as described on page 144.

FIGURE 6-1
GOOD SOURCES OF VITAMINS IN THE USDA MyPYRAMID FOOD GUIDE*

*Serving sizes shown here are based on a 2,000 calorie diet. Go to www.mypyramid.gov for a quick estimate of how much food you should eat from the different food groups based on your age, gender, and activity level.

GRAINS	VEGETABLES	FRUITS	OILS	MILK	MEAT & BEANS
6 oz every day	2½ cups every day	2 cups every day	6 tsp	3 cups every day	5½ oz every day
Niacin Riboflavin Thiamin Folate	Vitamin A (as beta-carotene) Vitamin C Vitamin K Riboflavin Folate	Vitamin A (as beta-carotene) Vitamin C Folate	Vitamin E	Riboflavin Vitamin A Vitamin D Vitamin B_{12}	Vitamin B_6 Vitamin B_{12} Niacin, Biotin Thiamin Vitamin D

TABLE 6-1
A Guide to the Vitamins

Vitamin (Chemical Name)	Best Sources[a]	Chief Roles	Deficiency Symptoms	Toxicity Symptoms[b]
Water-Soluble Vitamins				
Thiamin	Meat, pork, liver, fish, poultry, whole-grain and enriched breads, cereals and grain products, nuts, legumes	Helps enzymes release energy from carbohydrate; supports normal appetite and nervous system function	Beriberi: edema, heart irregularity, mental confusion, muscle weakness, apathy, impaired growth	None reported
Riboflavin	Milk, leafy green vegetables, yogurt, cottage cheese, liver, meat, whole-grain or enriched breads, cereals and grain products	Helps enzymes release energy from carbohydrate, fat, and protein; promotes healthy skin and normal vision	Eye problems, skin disorders around nose and mouth, magenta tongue, hypersensitivity to light	None reported
Niacin	Meat, eggs, poultry, fish, milk, whole-grain and enriched breads, cereals and grain products, nuts, legumes, peanuts	Helps enzymes release energy from energy nutrients; promotes health of skin, nerves, and digestive system	Pellagra: flaky skin rash on parts exposed to sun, loss of appetite, dizziness, weakness, irritability, fatigue, mental confusion, indigestion, delirium	Flushing, nausea, headaches, cramps, ulcer irritation, heartburn, abnormal liver function, rapid heartbeat with doses above 500 mg per day
Vitamin B_6 (pyridoxine)	Meat, poultry, fish, shellfish, legumes, fruits, soy products, whole-grain products, green leafy vegetables	Protein and fat metabolism; formation of antibodies and red blood cells; helps convert tryptophan to niacin	Nervous disorders, skin rash, muscle weakness, anemia, convulsions, kidney stones	Depression, fatigue, irritability, headaches, numbness, damage to nerves, difficulty walking
Folate (folacin, folic acid)	Green leafy vegetables, liver, legumes, seeds, citrus fruits, melons, enriched breads and grain products	Red blood cell formation; protein metabolism; new cell division	Anemia, heartburn, diarrhea, smooth red tongue, depression, poor growth, neural tube defects, increased risk of heart disease, stroke, and certain cancers	Diarrhea, insomnia, irritability, may mask a vitamin B_{12} deficiency
Vitamin B_{12} (cobalamin)	Animal products: meat, fish, poultry, shellfish, milk, cheese, eggs; fortified cereals	Helps maintain nerve cells; red blood cell formation; synthesis of genetic material	Anemia, smooth red tongue, fatigue, nerve degeneration progressing to paralysis	None reported
Pantothenic acid	Widespread in foods	Coenzyme in energy metabolism	Rare; sleep disturbances, nausea, fatigue	None reported
Biotin	Widespread in foods	Coenzyme in energy metabolism; fat synthesis; glycogen formation	Loss of appetite, nausea, depression, muscle pain, weakness, fatigue, rash	None reported

[a] The recommended intakes for the vitamins are listed on the inside front cover.
[b] The Tolerable Upper Intake Levels (UL) for the vitamins are listed on the inside front cover.

TABLE 6-1
A Guide to the Vitamins (continued)

Vitamin (Chemical Name)	Best Sources[c]	Chief Roles	Deficiency Symptoms	Toxicity Symptoms[d]
Water-Soluble Vitamins				
Vitamin C (ascorbic acid)	Citrus fruits, cabbage-type vegetables, tomatoes, potatoes, dark green vegetables, peppers, lettuce, cantaloupe, strawberries, mangoes, papayas	Synthesis of collagen (helps heal wounds, maintains bone and teeth, strengthens blood vessel walls); antioxidant; strengthens resistance to infection; helps body absorb iron	Scurvy: anemia, depression, frequent infections, bleeding gums, loosened teeth, pinpoint hemorrhages, muscle degeneration, rough skin, bone fragility, poor wound healing, hysteria	Intakes of more than 1 g per day may cause nausea, abdominal cramps, diarrhea, and increased risk for kidney stones
Fat-Soluble Vitamins				
Vitamin A	*Retinol:* fortified milk and margarine, cream, cheese, butter, eggs, liver *Beta-carotene:* Spinach and other dark leafy greens, broccoli, deep orange fruits (apricots, peaches, cantaloupe), and vegetables (squash, carrots, sweet potatoes, pumpkin)	Vision; growth and repair of body tissues; maintenance of mucous membranes; reproduction; bone and tooth formation; immunity; hormone synthesis; antioxidant (in the form of beta-carotene only)	Night blindness, rough skin, susceptibility to infection, impaired bone growth, abnormal tooth and jaw alignment, eye problems leading to blindness, impaired growth	Red blood cell breakage, nosebleeds, abdominal cramps, nausea, diarrhea, weight loss, blurred vision, irritability, loss of appetite, bone pain, dry skin, rashes, hair loss, cessation of menstruation, liver disease, birth defects
Vitamin D (cholecalciferol)	Self-synthesis with sunlight; fortified milk, fortified margarine, eggs, liver, fish	Calcium and phosphorus metabolism (bone and tooth formation); aids body's absorption of calcium	Rickets in children; osteomalacia in adults; abnormal growth, joint pain, soft bones	Deposits of calcium in organs such as the kidneys, liver, or heart, mental retardation, abnormal bone growth
Vitamin E	Vegetable oils, green leafy vegetables, wheat germ, whole-grain products, liver, egg yolk, salad dressings, mayonnaise, margarine, nuts, seeds	Protects red blood cells; antioxidant (protects fat-soluble vitamins); stabilization of cell membranes	Muscle wasting, weakness, red blood cell breakage, anemia, hemorrhaging	Doses over 800 IU/day may increase bleeding (blood clotting time)
Vitamin K	Bacterial synthesis in digestive tract, liver, green leafy and cabbage-type vegetables, soybeans, milk, vegetable oils	Synthesis of blood-clotting proteins and a blood protein that regulates blood calcium	Hemorrhaging, decreased calcium in bones	Interference with anticlotting medication; synthetic forms may cause jaundice

[c] The recommended intakes for the vitamins are listed on the inside front cover.
[d] The Tolerable Upper Intake Levels (UL) for the vitamins are listed on the inside front cover.

beriberi the thiamin deficiency disease, characterized by irregular heartbeat, paralysis, and extreme wasting of muscle tissue.

TABLE 6-2
General Characteristics of Water-Soluble and Fat-Soluble Vitamins

Characteristic	Water-Soluble B Vitamins and Vitamin C	Fat-Soluble Vitamins A, D, E, and K
Dietary intake	Excess intake usually detected and excreted by the kidneys.	Excess intake tends to be stored in fat-storage sites.
Body stores	Only a short-term storage supply available; daily intake recommended.	Long-term storage available in body tissues; regular intake recommended.
Deficiency	Deficiency symptoms appear relatively quickly.	Deficiency symptoms are slow to develop.
Toxicity	Lower risk of toxicity.	Higher risk of toxicity.
Absorption and transport	Easily absorbed into blood; travel freely in blood.	Like lipids, absorbed into lymph; many require protein carriers to travel in the blood.
Solubility	Dissolves in water	Dissolves in lipid
Stability in food handling and processing	Less stable	Stable
Role(s) in the body	Most B vitamins share similar roles; vitamin C serves many different roles in the body.	Each has unique roles in the body.

TABLE 6-3
Thiamin in Foods

(mg)	Sources
0.97	Pork chop (3 oz)
0.41	Sunflower seeds (2 tbsp)
≥0.35	Fortified cereal (1 c)
0.24	Salmon (4 oz)
0.23	Watermelon (1 slice)
0.23	Green peas (½ c)
0.22	Baked potato (1)
0.22	Enriched pasta (½ c)
0.21	Black beans (½ c)
0.21	Peanuts (⅓ c)
0.17	Black-eyed peas (½ c)
0.13	Oatmeal, cooked (½ c)
0.11	Sirloin steak (3 oz)
0.11	Orange (1)
0.11	Wheat bread (1 slice)
0.09	Milk, fat-free (1 c)

In foods, the water-soluble vitamins are relatively fragile. Although large amounts of them are naturally present in many foods, they can be washed out or destroyed during food storage, processing, and preparation. These effects are described in detail in the Savvy Diner feature later in this chapter.

Thiamin

One of the B vitamins, thiamin, acts primarily as a coenzyme in reactions that release energy from carbohydrates. Thiamin also plays a crucial role in processes involving the nerves. Thiamin is so vital to the functioning of the entire body that a deficiency affects the nerves, muscles, heart, and other organs. A severe deficiency, called **beriberi,** causes extreme wasting and loss of muscle tissue, swelling all over the body, enlargement of the heart, irregular heartbeat, and paralysis. Ultimately, the victim dies from heart failure. A mild thiamin deficiency, on the other hand, often mimics other conditions and typically manifests itself as vague, general symptoms such as stomachaches, headaches, fatigue, rest-

FIGURE 6-2
HOW A COENZYME WORKS

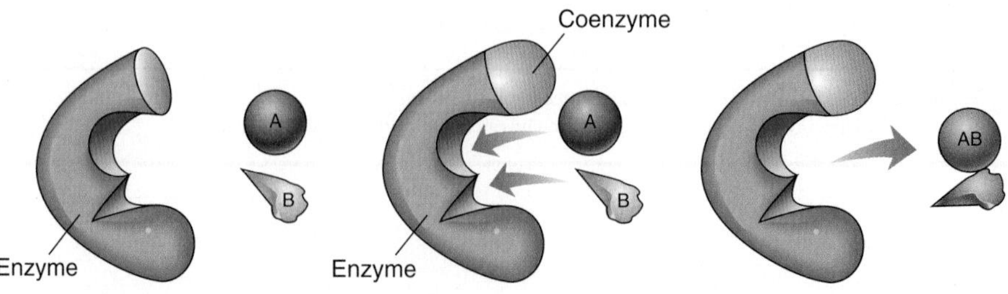
Coenzyme
A
B
Enzyme
A
B
Enzyme
AB
Without the coenzyme, compounds A and B don't respond to the enzyme.
With the coenzyme in place, A and B are attracted to the active side on the enzyme, and they react.
The reaction is completed. A new product, AB, has been formed.

SOURCES OF THIAMIN
Adult DRI (RDA) is 1.1 to 1.2 mg.*

*Dietary Reference Intakes (DRI) for all age groups are listed on the inside front cover.

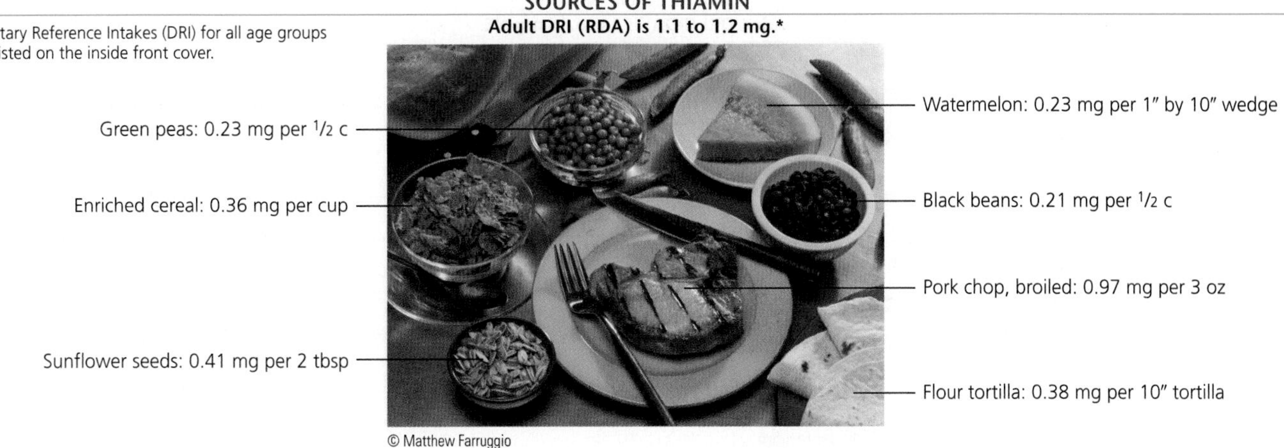

© Matthew Farruggio

lessness, sleep disturbances, chest pains, fevers, personality changes (aggressiveness and hostility), and neurosis.

Thiamin is found in a wide variety of foods, and virtually no single food will supply your daily needs in a single serving the same way that, say, an orange provides a plentiful supply of vitamin C. However, people who eat a balanced diet following the framework of the MyPyramid Food Guide typically consume plenty of thiamin. As Table 6-3 shows, thiamin is found in a variety of meats, legumes, fruits, and vegetables, as well as in all enriched and whole-grain products.

Riboflavin

Like thiamin, the B vitamin riboflavin acts as a coenzyme in energy-releasing reactions in the body. In addition, riboflavin helps to prepare fatty acids and amino acids for breakdown. Deficiencies of the vitamin, which are rare, are characterized by severe skin problems, including painful cracks at the corners of the mouth; a red, swollen tongue; and teary or bloodshot eyes.

Table 6-4 shows the riboflavin content of foods. Milk and dairy products contribute a good deal of the riboflavin in many people's diets. Meats are another good source, as are dark green vegetables such as broccoli. Leafy green vegetables and whole-grain or enriched bread and cereal products also supply a generous amount of riboflavin in most people's diets. Because riboflavin can be destroyed by the ultraviolet rays of the sun or by fluorescent lamps, milk is usually sold in protective cardboard or opaque plastic containers rather than in transparent glass bottles.

TABLE 6-4
Riboflavin in Foods

(mg)	Sources
0.60	Low-fat yogurt (1 c)
0.45	Fat-free milk (1 c)
0.37	Almonds (⅓ c)
≥0.35	Fortified cereal (1 c)
0.24	Pork chop (3 oz)
0.23	Ricotta cheese (½ c)
0.23	Sirloin steak (3 oz)
0.21	Beet greens, cooked (½ c)
0.21	Egg, cooked (1)
0.21	Spinach, cooked (½ c)
0.20	Ground beef (3 oz)
0.17	Cheddar cheese (1.5 oz)
0.16	Turkey (3 oz)
0.11	Asparagus, cooked (½ c)
0.10	Strawberries (1 c)
0.08	Wheat bread (1 slice)

SOURCES OF RIBOFLAVIN
Adult DRI (RDA) is 1.1 to 1.3 mg.

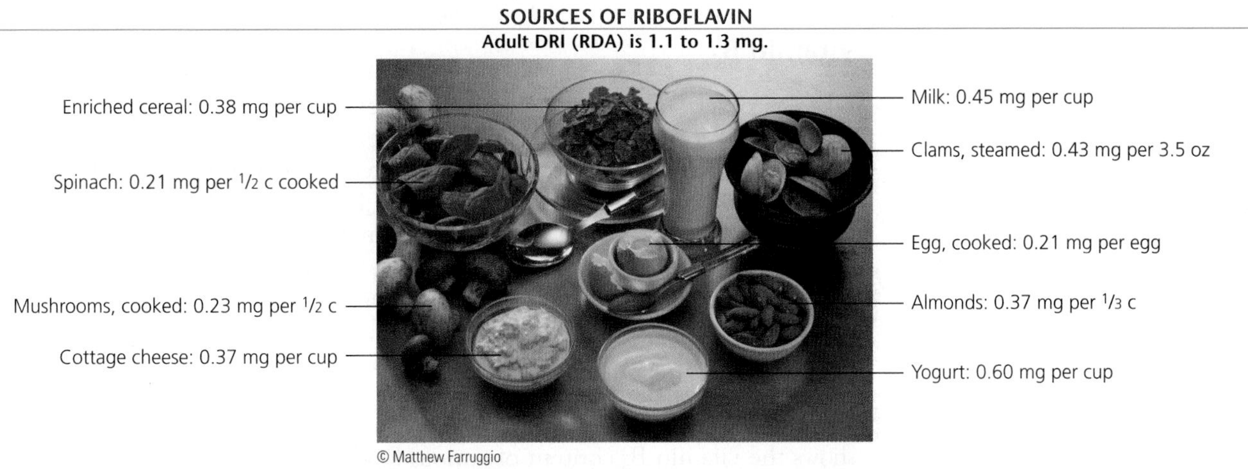

© Matthew Farruggio

SOURCES OF NIACIN
Adult DRI (RDA) is 14 to 16 mg NE.

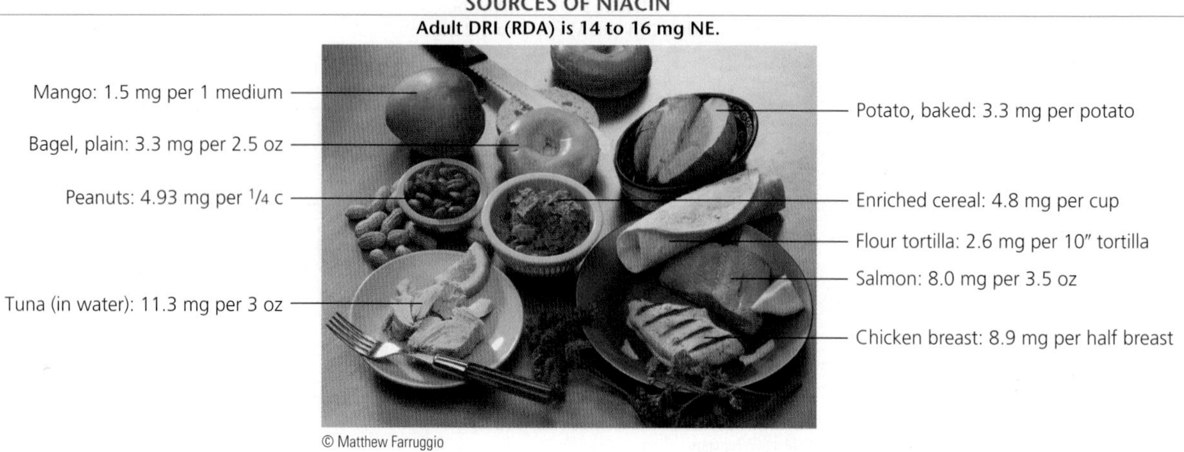

© Matthew Farruggio

TABLE 6-5
Niacin in Foods

(mg NE)	Sources
11.30	Tuna (3 oz)
8.90	Chicken breast (½)
6.05	Halibut (3 oz)
5.08	Ground beef (3 oz)
4.63	Turkey (3 oz)
4.29	Peanut butter (2 tbsp)
≥4.00	Fortified cereal (1 c)
3.31	Baked potato (1)
3.29	Sirloin steak (3 oz)
1.85	Flounder/sole (3 oz)
1.53	Cantaloupe (½)
1.49	Brown rice, cooked (½ c)
1.13	Wheat bread (1 slice)
0.97	Asparagus, cooked (½ c)
0.89	Broccoli, cooked (½ c)
0.86	Peach (1)

Niacin

Like thiamin and riboflavin, the B vitamin niacin is part of a coenzyme that is vital to producing energy. Without niacin to form this coenzyme, energy-yielding reactions come to a halt. Over time, a deficiency of niacin leads to the disease pellagra, characterized by diarrhea, dermatitis, and, in severe cases, dementia—a progressive mental deterioration resulting in delirium, mania or depression, and eventually death.

Although we can prevent niacin deficiency by eating a diet rich in niacin itself, consuming plenty of protein also staves off the problem. That's because the essential amino acid tryptophan, which is a component of protein, can be converted to niacin in the body. In fact, 60 milligrams of tryptophan yield 1 milligram of niacin. Thus, the DRI for niacin is expressed in niacin equivalents (NEs)—that is, the amount of niacin present in food, including the amount that can be theoretically made from the tryptophan in the food.

Milk, eggs, meat, poultry, and fish contribute the bulk of the niacin equivalents consumed by most people, followed by enriched breads and cereals. Table 6-5 shows the niacin content of some common foods.

Diet aside, in recent years, niacin has been increasingly used as a drug-like supplement to help lower cholesterol. Doses ranging from 10 to 15 times the RDA have been shown to reduce "bad" LDL-cholesterol and raise "good" HDL-cholesterol.[3] The hitch, however, is that such high doses of niacin can lead to side effects such as nausea, flushing of the skin, rash, fatigue, and liver damage. Because of the side effects, many experts argue that niacin pills should be sold not as over-the-counter dietary supplements but rather as drugs prescribed and taken only while under a physician's supervision.

Vitamin B_6

Like the other B vitamins, vitamin B_6 functions as a coenzyme and is an indispensable cog in the body's machinery. For example, vitamin B_6 plays many roles in protein metabolism. In fact, a person's requirement for vitamin B_6 is proportional to protein intakes. Because vitamin B_6 performs this and so many other tasks, a deficiency causes a multitude of symptoms, including weakness, irritability, and insomnia. Low levels of vitamin B_6 may also weaken the body's immune response and increase a person's risk for heart disease. Vitamin B_6 is found in meats, vegetables, and whole-grain cereals, and true vitamin B_6 deficiencies are rare—occurring in some people who eat inadequate diets and whose nutrient needs are higher than usual because of pregnancy, alcohol abuse, some diseases, use of certain prescription drugs, and other unusual circumstances. Table 6-6 shows the vitamin B_6 content of various foods.

SOURCES OF VITAMIN B_6
Adult DRI (RDA) is 1.3 mg.

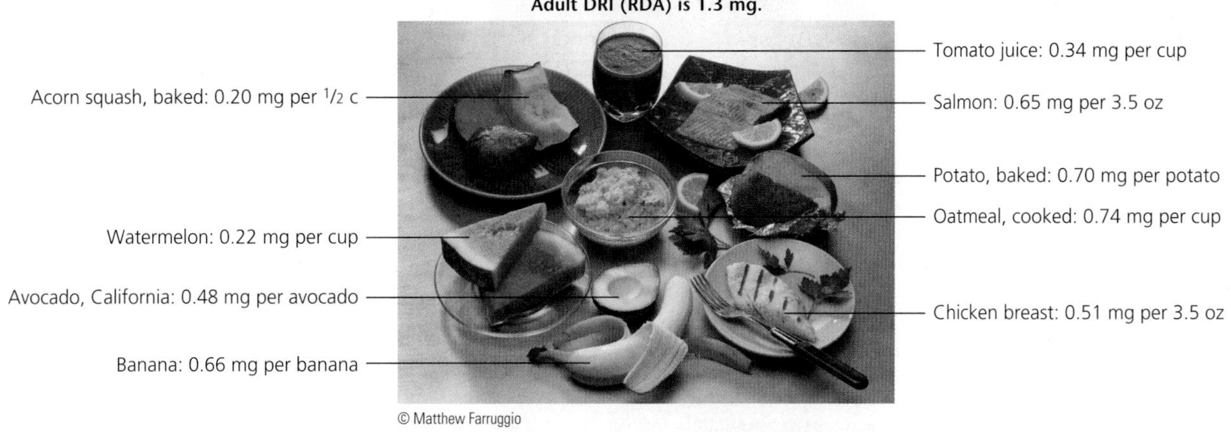

© Matthew Farruggio

Vitamin B_6 is also widely reputed as a cure for **premenstrual syndrome (PMS).** Some people have claimed that a deficiency of the vitamin goes hand in hand with imbalances of hormones (particularly estrogen), which cause the depression, mood swings, and other symptoms characteristic of PMS. Although this theory has never been proven to be scientifically sound, women have taken **megadoses** of B_6—as much as 2,000 times the RDA in some cases—in an effort to treat PMS. However, a number of these women began to experience symptoms associated with damage to the nervous system, such as numb feet and loss of sensation in the hands and mouth.[4] Granted, not everyone is likely to suffer toxicity symptoms as a result of swallowing megadoses of vitamin B_6, because excess amounts are excreted in the urine. But the problems seen in these women underscore the potential hazards of taking megadoses of any vitamin or nutritional supplement.

TABLE 6-6
Vitamin B_6 in Foods

(mg)	Sources
0.70	Baked potato (1)
0.66	Banana (1)
0.65	Salmon (3.5 oz)
0.51	Chicken breast (3.5 oz)
0.42	Figs, dried (10)
0.35	Pork chop (3 oz)
0.34	Sirloin steak (3 oz)
0.31	Cantaloupe (½)
0.30	Tuna (3 oz)
0.26	Ground beef (3 oz)
0.22	Spinach, cooked (½ c)
0.22	Watermelon (1 c)
0.20	Flounder/sole (3 oz)
0.20	Soybeans (½ c)
0.15	Navy beans (½ c)
0.14	Brown rice, cooked (½ c)
0.14	Sunflower seeds (2 tbsp)
0.11	Asparagus, cooked (½ c)
0.11	Broccoli, cooked (½ c)
0.10	Fat-free milk (1 c)
0.07	Zucchini, cooked (½ c)

Folate

Folate (also called *folic acid* or *folacin*) is a coenzyme with many functions in the body. It is particularly important in the synthesis of DNA and the formation of red blood cells. A folate deficiency creates misshapen red blood cells that are unable to carry sufficient oxygen to the body's other cells, thereby causing a certain kind of **anemia.** Thus, folate deficiency results in a generalized malaise with many symptoms, including fatigue, diarrhea, irritability, forgetfulness, lack of appetite, and headache. Folate deficiency can easily be confused with general ill health, depressed mood, and senility in the elderly. Folate deficiency may also elevate a person's risk for certain cancers—notably cervical cancer in women and colon cancer.[5]

Derived from the word *foliage*, folate occurs naturally in fresh green, leafy vegetables, but it is easily lost when foods are overcooked, canned, dehydrated, or otherwise processed. People who are growing rapidly run a high risk of folate deficiency because folate is needed to promote the rapid multiplication of cells that occurs during growth. That's why, for example, the need for folate increases

premenstrual syndrome (PMS) a cluster of physical, emotional, and psychological symptoms that some women experience seven to ten days before menstruation. Symptoms can include acne, anxiety, food cravings (especially for sweets), back pain, breast tenderness, cramps, depression, fatigue, headaches, irritability, moodiness, water retention, and weight gain. Because a clear-cut treatment for the symptoms of PMS has not been identified, women who suffer from the problem rank as prime targets for unproved nutritional remedies for the condition.

megadose a dose of ten or more times the amount normally recommended. An overdose is an amount high enough to cause toxicity symptoms. Megadoses taken over a long period often result in an overdose.

anemia any condition in which the blood is unable to deliver oxygen to the cells of the body. Examples include a shortage or abnormality of the red blood cells. Many nutrient deficiencies and diseases can cause anemia.

neural tube defects malformations of the brain and/or spinal cord during embryonic development.

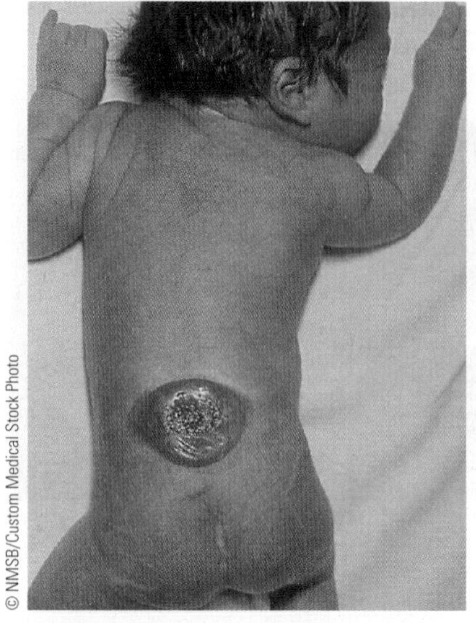

▲ *This child has spina bifida—a birth defect characterized by the incomplete closing of the casing around the spinal cord. Afflicting some 3,000 U.S. infants born each year, the problem causes partial paralysis.*

during pregnancy, when large amounts of the vitamin are needed to support the growth of the fetus.

Folate plays a crucial role in a healthy pregnancy. A growing body of evidence indicates that consuming a generous amount of folate reduces the risk of bearing a baby with a type of birth defect called **neural tube defect.** Afflicting some 3,000 infants born in the United States each year, neural tube defects include *spina bifida*—the incomplete closing of the bony casing around the spinal cord that causes partial paralysis—and *anencephaly*—a condition in which major parts of the brain are missing. All women of childbearing age are advised to consume the recommended amount of folate because adequate levels of the nutrient must be ingested before and during the first few weeks of pregnancy—the period during which the neural tube of the embryo is closing but when most women are not aware of being pregnant.

Although folate is abundant in vegetables, legumes, and seeds, as shown in Table 6-7, most women typically consume foods that provide only about half the 400-microgram amount recommended. For this reason, the Food and Drug Administration (FDA) has mandated that all enriched grain products be fortified with folic acid, an easily absorbed synthetic form of folate, to improve intakes in the United States.[6]* As noted in Figure 6-3, the United States has recorded significant declines in neural tube defects since folic acid fortification of all enriched cereal grain products began. Labels on fortified products may make the health claim "adequate intake of folate has been shown to reduce the risk of neural tube defects." A person can obtain the recommended 400 micrograms of folate by increasing consumption of foods naturally rich in folate (orange juice and green vegetables) and foods fortified with folic

TABLE 6-7
Folate in Foods

(μg)	Sources
179	Black-eyed peas (½ c)
179	Lentils, cooked (½ c)
146	Asparagus, cooked (½ c)
113	Spinach, cooked (½ c)
104	Broccoli, cooked (1 c)
100	Oatmeal, instant (½ c)
≥97	Fortified cereal (1 c)
85	Turnip greens, cooked (½ c)
76	Romaine lettuce (1 c)
71	Peanuts (⅓ c)
65	Kidney beans (½ c)
58	Spinach, fresh (1 c)
55	Lima beans (½ c)
47	Cantaloupe (½)
41	Sunflower seeds (2 tbsp)
40	Orange (1)
27	Cauliflower, cooked (½ c)
19	Tofu (soybean curd) (½ c)
14	Whole-wheat bread (1 slice)
13	Fat-free milk (1 c)
13	Strawberries (½ c)
8	Sirloin steak (3 oz)

SOURCES OF FOLATE
Adult DRI (RDA) is 400 μg.

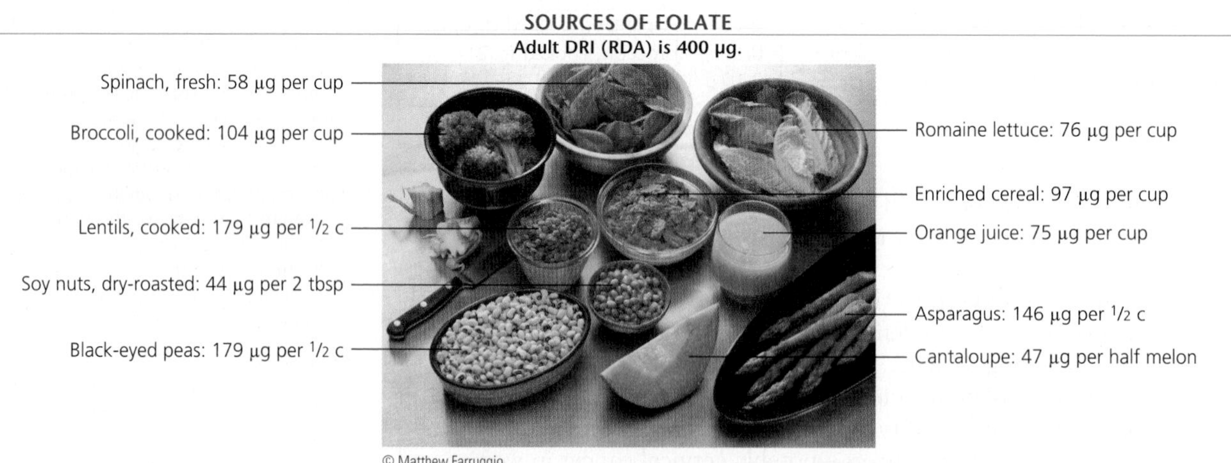

*All enriched cereal grain products are fortified with 1.4 milligrams of folic acid per 100 grams of food. Recommended folate intakes are stated in micrograms (μg) DFE. A microgram is a thousandth of a milligram. DFE stands for Dietary Folate Equivalent, a unit of measure expressing the amount of folate available to the body from naturally occurring food sources. The measure accounts for the differences in absorption between food folate and the more absorbable synthetic folic acid added to foods and supplements. Appendix B offers a conversion factor for calculating DFE.

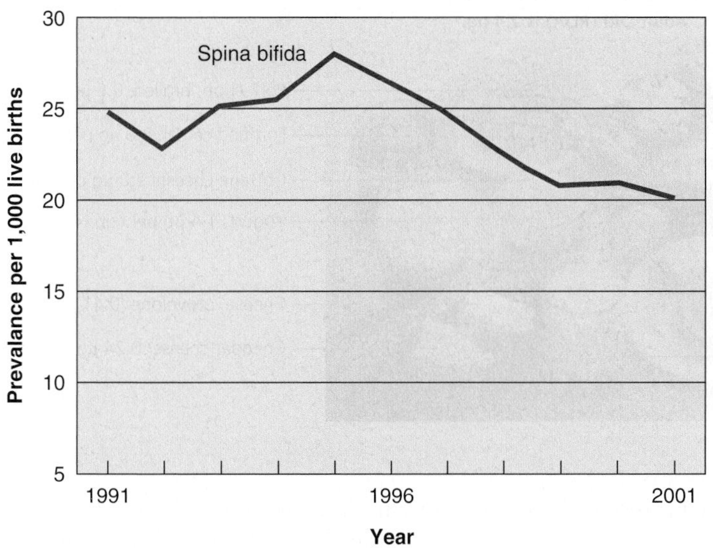

FIGURE 6-3
INCIDENCE OF NEURAL TUBE DEFECTS (SPINA BIFIDA), 1991–2001

A decline in neural tube defects has occurred in the United States since fortification of grain products with folic acid began in 1996.

SOURCE: Centers for Disease Control and Prevention, Folic acid and prevention of spina bifida and anencephaly, *Morbidity & Mortality Weekly Report* 51 (2002): 11–13.

acid (breakfast cereals, breads, rice, or pasta), or by taking a folic acid supplement daily.[7] Because high levels of blood folate can mask a true vitamin B_{12} deficiency, total folate intake should not exceed 1 milligram daily.[8]

B Vitamins and Heart Disease

Low intakes of three B vitamins—folate, vitamin B_{12}, and vitamin B_6—are linked with increased risk of fatal heart disease in both men and women.[9] People with low blood levels of these B vitamins tend to have high blood levels of the protein related compound **homocysteine.**[10] High levels of homocysteine seem to enhance blood-clot formation and damage to arterial walls as well as raising the risk of suffering a heart attack or stroke as much as fourfold.[11] In addition, scientists suspect that high homocysteine levels may not only be damaging to the cardiovascular system but may also be toxic for brain tissue and impair cognitive ability.[12] The B vitamins clear homocysteine from the blood and prevent its toxic buildup. The research linking a vitamin B-poor diet with increased homocysteine levels highlights the importance of consuming generous amounts of these nutrients.

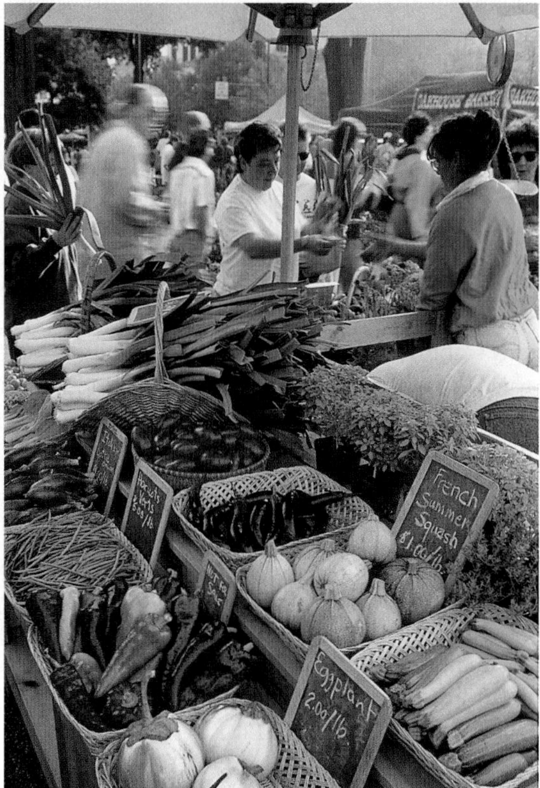

© Zane Williams/Getty Images

◀ *When you think of the B vitamin folate, think of fresh produce. Folate derives its name from the word* foliage, *and it is naturally abundant in fresh dark green vegetables (such as spinach and broccoli), some fresh fruits (oranges and melons), and legumes (lentils and black-eyed peas). Other sources include foods made from grains enriched with folic acid such as breakfast cereals, breads, rice, or pasta.*

Vitamin B_{12}

Vitamin B_{12} maintains the sheaths that surround and protect nerve fibers. The nutrient also works closely with folate, enabling it to manufacture red blood cells. When a B_{12}

homocysteine (ho-mo-SIS-teen) a chemical that appears to be toxic to the blood vessels of the heart. High blood levels of homocysteine have been associated with low blood levels of vitamin B_{12}, vitamin B_6, and folate.

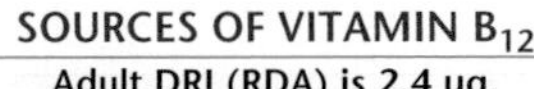

SOURCES OF VITAMIN B_{12}

Adult DRI (RDA) is 2.4 µg.

TABLE 6-8
Vitamin B_{12} in Foods

(µg)	Sources
14.00	Chicken liver (3 oz)
7.50	Sardines (3 oz)
3.00	Tuna (3 oz)
2.01	Ground beef (3 oz)
2.00	Cottage cheese (1 c)
1.39	Plain nonfat yogurt (1 c)
1.27	Shrimp (3 oz)
1.18	Haddock (3 oz)
0.93	Fat-free milk (1 c)
0.50	Egg (1)
0.35	Cheddar cheese (1.5 oz)
0.29	Chicken breast (½)

deficiency is present, folate is unable to do its work building red blood cells. As a result, a person suffering a lack of vitamin B_{12} ends up with the same sort of anemia seen in people with a folate deficiency and characterized by large, immature red blood cells. Although extra folate will clear up the anemia, it will not take care of the other problems resulting from a B_{12} deficiency, namely a creeping paralysis of the nerves and muscles that can cause permanent nerve damage if left untreated. Thus, because excess folate can clear up the blood problems that signal an otherwise hard-to-diagnose vitamin B_{12} deficiency, the amount of folate that can be added to enriched foods is limited by law.

To be sure, dietary deficiencies of vitamin B_{12} are unlikely for people who eat animal foods such as meat, milk, cheese, and eggs, all of which supply generous amounts of the nutrient (see Table 6-8). Strict vegetarians who eschew meat, eggs, and dairy products, however, need to find alternative sources of the nutrient, such as vitamin B_{12}-fortified soy beverages, fortified cereals, or B_{12} supplements.

Several other groups of people are also at high risk for vitamin B_{12} deficiency, not because of a lack of the vitamin in their diets, but because of physical conditions that hamper the body's ability to make use of the nutrient. One such group is people who inherit a genetic defect that leaves the body unable to make a compound known as **intrinsic factor.** Produced in the stomach, intrinsic factor enables the body to absorb and make use of vitamin B_{12}; without the compound, vitamin B_{12} deficiency develops. In this instance or in the case of stomach damage that interferes with the production of intrinsic factor, people must get vitamin B_{12} injections.

Another group likely to experience vitamin B_{12} deficiencies is the elderly. An estimated 20 percent of seniors in their sixties and 40 percent of those in their eighties develop **atrophic gastritis,** an age-related condition characterized by the stomach's inability to produce enough acid, which in turn hampers the body's ability to use vitamin B_{12}. In severe cases, the condition also limits the stomach's ability to make intrinsic factor. Vitamin B_{12} deficiencies resulting from atrophic gastritis appear to be easily treated with vitamin B_{12} supplements or injections.[13]

Pantothenic Acid and Biotin

Two other B vitamins—pantothenic acid and biotin—are needed to synthesize coenzymes that are active in a multitude of body systems. Biotin is also required for cell growth, synthesis of DNA (the genetic "blueprint" present in every cell), and maintenance of blood glucose levels. Because both pantothenic acid and biotin are widespread in foods, people who eat a varied diet are not at risk for deficiencies.

intrinsic factor a compound made in the stomach that is necessary for the body's absorption of vitamin B_{12}.

atrophic gastritis an age-related condition characterized by the stomach's inability to produce enough acid, which in turn leads to vitamin B_{12} deficiencies.

SOURCES OF BIOTIN AND PANTOTHENIC ACID*

Adult DRI (AI): **Biotin: 30 µg**
Pantothenic acid: 5 mg

*Information concerning biotin and pantothenic acid in foods is incomplete: Deficiencies are rare.

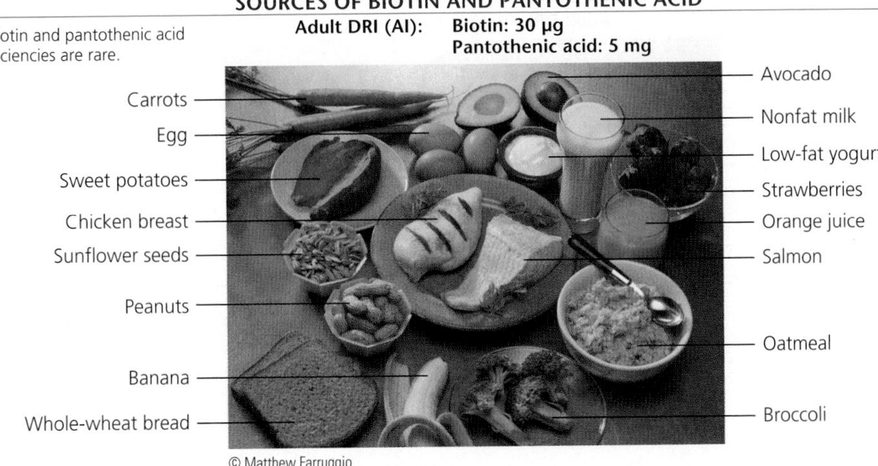

Vitamin C

Vitamin C is required for the production and maintenance of **collagen,** the protein foundation material for the body's connective tissue, including bones, teeth, skin, and tendons. Vitamin C has also been touted as a nutrient that can help fight stress. In times of stress, the body does use more vitamin C than usual because the vitamin is involved in the release of stress hormones. However, the amount of extra vitamin C used as a result of, say, work-related stress or the stress of ending a significant relationship is minuscule and is more than accounted for by a diet that regularly includes vitamin C-rich foods (see Table 6-9).

Vitamin C also boosts the body's ability to fight infections, and a growing body of research suggests that it may protect against heart disease and certain types of cancer. Vitamin C's potential role as a chronic disease fighter stems from its function as an **antioxidant.**[14]

As their name suggests, antioxidants are "antioxygen"—they fight oxygen, in a manner of speaking. Some chemical reactions that occur in the body involve the use of oxygen. Whereas these reactions are essential to the body's ability to function, they also lead to the creation of highly toxic compounds called **free radicals.** Environmental pollutants such as cigarette smoke and ozone also prompt the formation of free radicals. Left unchecked, these compounds can cause severe cell injury and ultimately may contribute to the development of chronic diseases such as cancer and heart disease.

Fortunately, the body has a built-in defense system to protect against potential damage from free radicals. That defense system makes use of the antioxidant nutrients—vitamin C, vitamin E, and the carotenoids (discussed later with vitamin A). In addition, the body manufactures certain enzymes, one of which contains the mineral selenium, that help to fight free radicals.

The antioxidants all work in one way or another to squelch free radicals before they injure the body (see Figure 6-4 on page 185). Vitamin C helps stop free radicals in their tracks, working with vitamin E to block damaging chain reactions that appear to promote heart disease and cancer. In addition, vitamin C is a powerful scavenger of environmental air pollutants. In fact, the National Academy of Sciences advises smokers to consume an additional 35 milligrams of vitamin C a day as compared to nonsmokers. The more smoke a person inhales, the more free radicals that are produced, and the more vitamin C that is needed to fight them.

collagen (COLL-a-jen) the characteristic protein of connective tissue.

kolla = glue
gennan = to produce

antioxidant a substance, such as a vitamin, that is "anti-oxygen"—that is, it helps to prevent damage done to the body as a result of chemical reactions that involve the use of oxygen.

free radicals highly toxic compounds created in the body as a result of chemical reactions that involve oxygen. Environmental pollutants such as cigarette smoke and ozone also prompt the formation of free radicals.

"A little vitamin C ought to clear that up in no time."

SOURCES OF VITAMIN C
Adult DRI (RDA) is 75 to 90 mg.

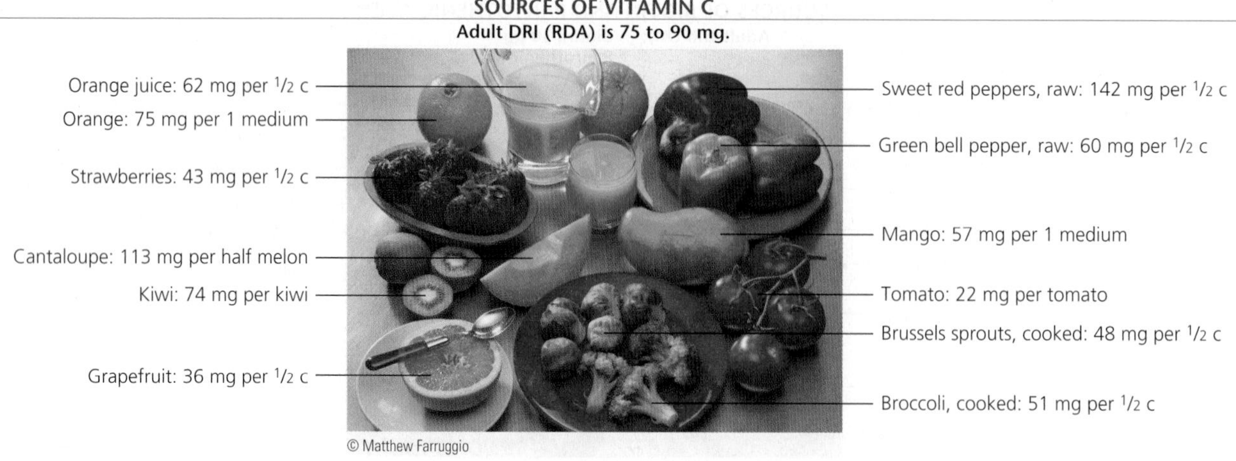

TABLE 6-9
Vitamin C in Foods

(mg)	Sources
187	Papaya (1)
113	Cantaloupe (½)
75	Orange (1)
74	Kiwi (1)
71	Grapefruit juice (¾ c)
62	Orange juice, fresh (½ c)
60	Green pepper (½ c)
57	Mango (1)
51	Broccoli, cooked (½ c)
48	Brussels sprouts, cooked (½ c)
43	Strawberries (½ c)
36	Pink/red grapefruit (½)
27	Cauliflower, cooked (½ c)
26	Baked potato (1)
22	Bok choy, cooked (½ c)
22	Cabbage, raw (1 c)
22	Tomato, fresh (1)
16	Raspberries (½ c)
10	Asparagus, cooked (½ c)
9	Spinach, cooked (½ c)

Of course, vitamin C is most famous for its long-standing notoriety as a cure for the common cold. Ever since the publication of the controversial book *Vitamin C and the Common Cold* by the award-winning scientist Linus Pauling, millions of Americans have followed Dr. Pauling's advice and swallowed megadoses of vitamin C. Despite the popularity of vitamin C as a cold remedy, however, many carefully controlled studies have shown that it plays an insignificant, if indeed any, role in preventing colds. At best, the nutrient in amounts up to 1 gram per day may shorten the duration of a cold by one day and slightly reduce the severity of cold symptoms in some people.[15] To be sure, many people swear by vitamin C, and it may be that their belief in the nutrient is so strong that they experience a placebo effect as a result of their faith in its curative powers.

Vitamin C-rich foods are widely available in the United States and include not only oranges and other citrus fruits but also broccoli, Brussels sprouts, cantaloupe, and strawberries. A single serving of any of those foods provides more than half the DRI for the vitamin. Potatoes also contribute significant amounts of vitamin C to the American diet because they are eaten so often.

Vitamin C is widespread in the food supply. Still, deficiencies arise in infants who are not given a source of vitamin C as well as in children and the elderly who have inadequate consumption of fruits and vegetables.

Fat-Soluble Vitamins

The four fat-soluble vitamins—A, D, E, and K—are absorbed from the digestive tract with the aid of fats in the diet and bile produced by the liver. Any disorder that interferes with fat digestion or absorption can precipitate a deficiency of the fat-soluble vitamins. Once in the bloodstream, these vitamins are escorted by protein carriers because they are *insoluble* in water. Because they are stored in the liver and in body fat, you do not need to consume them daily unless your intakes are usually marginal.

Vitamin A

Vitamin A has the distinction of being the first fat-soluble vitamin to be identified. It is one of the most versatile vitamins, playing roles in several important body processes.

The best known function of vitamin A is in vision. For a person to see, light reaching the eye must be transformed into nerve impulses that the brain interprets to produce visual images. The *transformers* are molecules of **pigment** in the cells of the **retina,** a paper-thin tissue lining the back of the eye. A portion of each pigment molecule is **retinal,** a compound the body can synthesize only if vitamin A is supplied by the diet in some form. Thus, when vitamin A is defi-

pigment a molecule capable of absorbing certain wavelengths of light. Pigments in the eye permit us to perceive different colors.

retina (RET-in-uh) the paper-thin layer of light-sensitive cells lining the back of the inside of the eye.

retinal (RET-in-al) one of the active forms of vitamin A that functions in the pigments of the eye. Other active forms of vitamin A include retinol and retinoic acid.

FIGURE 6-4
THE ANTIOXIDANTS VERSUS FREE RADICALS IN THE BODY

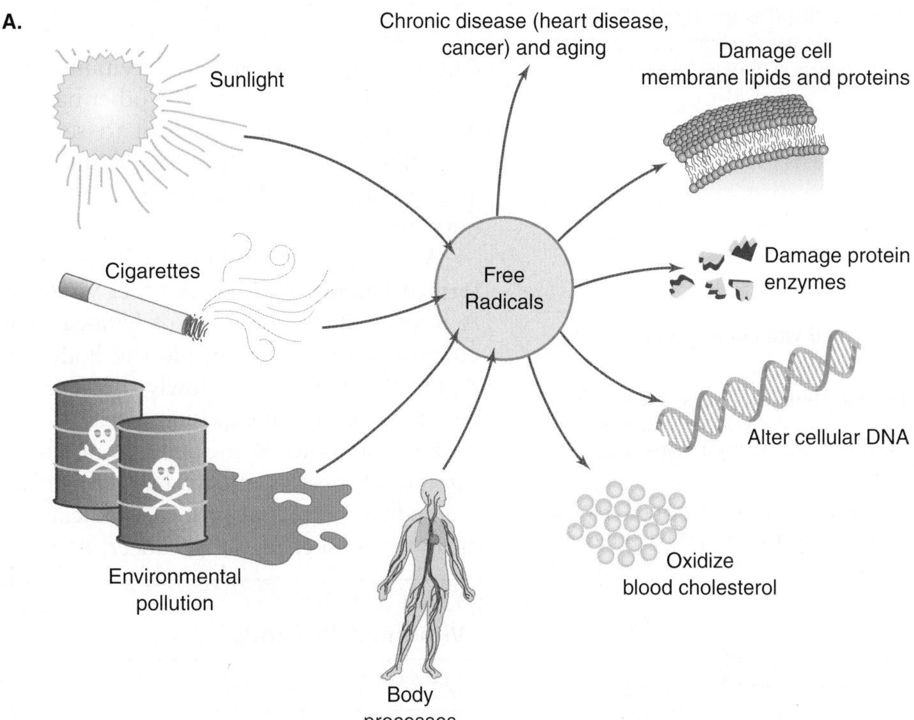

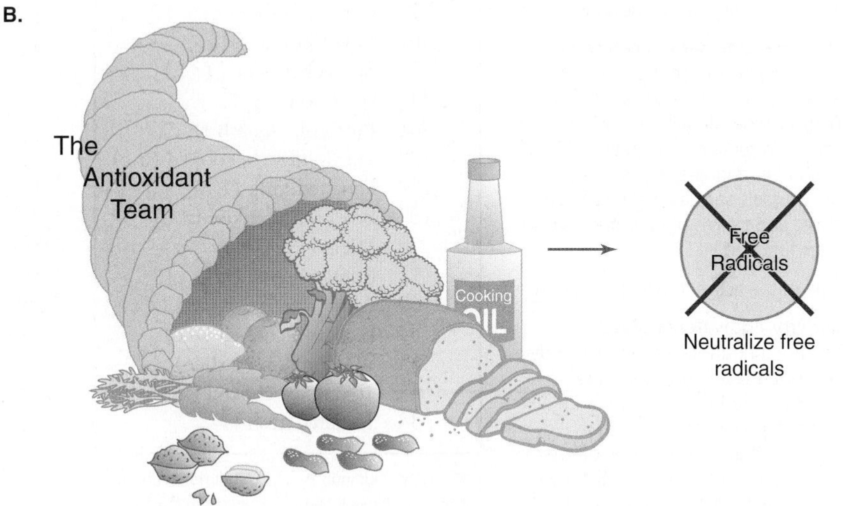

A. Free radicals—unstable oxygen molecules—can be formed from sunlight, in cigarette smoke and environmental pollution, and as a result of many normal chemical reactions involving oxygen in the body. These free radicals attack healthy molecules in the body in hopes of stealing an electron to help stabilize themselves, which in turn can cause cell and tissue damage to the body. Free radical activity can cause damage to the body's enzymes, cell membranes, and nuclear DNA, or they can result in the formation of oxidized LDL-cholesterol in the arteries.

B. The antioxidant team includes vitamin C, vitamin E, the carotenoids (for example, beta-carotene, lutein, zeaxanthin, lycopene), selenium (a trace mineral), and many naturally occurring nonnutrients—called phytochemicals—found in fruits, vegetables, legumes, and whole grains. The antioxidants prevent free radicals from attacking cells and causing damage by neutralizing the free radicals and converting them back into stable oxygen molecules.

SOURCE: Adapted from *Antioxidant Nutrients* (Mount Olive, NJ: BASF Corporation, 1997).

cient, vision is impaired. Specifically, the eye has difficulty adapting to changing light levels. For a person deficient in vitamin A, a flash of bright light at night (after the eye has adapted to darkness) is followed by a prolonged spell of **night blindness.** Because night blindness is easy to diagnose, it aids in the diagnosis of vitamin A deficiency. (Night blindness is only a symptom, however, and it may indicate a condition other than vitamin A deficiency.)

Vitamin A serves other roles in the body. It helps to maintain healthy **epithelial tissue**—skin and the cells (called *epithelial cells*) lining such body cavities as the small intestine and lungs. Vitamin A is also involved in the production of sperm, the normal development of fetuses, the immune response, hearing, taste, and growth.

As much as a year's supply of vitamin A can be stored in the body—90 percent of it in the liver. If you stop eating good food sources of vitamin A, deficiency symptoms will not begin to appear until your stores are depleted. Then, however, the consequences are profound and include blindness and reduced resistance to infection. Although vitamin A deficiency is rarely seen in developed countries

night blindness slow recovery of vision following flashes of bright light at night; an early symptom of vitamin A deficiency.

epithelial (ep-ih-THEE-lee-ul) **tissue** the cells that form the outer surface of the body and line the body cavities and the principal passageways leading to the exterior. Examples include the cornea, digestive tract lining, respiratory tract lining, and skin. The epithelial cells produce mucus to protect these tissues from bacteria and other potentially harmful substances. Without this mucus, infections become more likely.

See this chapter's Spotlight feature for more information about the role of the carotenoids in disease prevention.

such as the United States and Canada, it is a serious public health problem in developing countries, where millions of children suffer from blindness, infections, and the other consequences of vitamin A deficiencies.

Vitamin A toxicity, on the other hand, is not nearly as widespread as deficiency. Nevertheless, it can also lead to severe health consequences, including joint pain, dryness of skin, hair loss, irritability, fatigue, headaches, weakness, nausea, and liver damage. Thus, it's especially important not to take megadoses of this nutrient.

Although toxicity poses a hazard to people who take supplements of **preformed vitamin A,** toxicity poses virtually no risk to people who obtain vitamin A from foods in the form of **beta-carotene,** an orange plant pigment that is a vitamin A **precursor.** Inside the body, beta-carotene is converted into vitamin A, but this happens so slowly that excess amounts are not stored as vitamin A, but are stored in fat deposits instead.

Beta-carotene is a member of the **carotenoid** family of pigments. The carotenoids possess antioxidant properties and work with vitamins C and E in the body to protect against free radical damage that leads to diseases of the respiratory tract, such as lung cancer, as well as other chronic conditions. Certain carotenoids with the antioxidant properties found in dark green leafy vegetables, such as spinach, kale, collard greens, and Swiss chard may help prevent **age-related macular degeneration** as well as lower the risk of cataracts.[16] These carotenoids work by filtering out harmful light rays that could cause free-radical damage to the eye.

Because the body uses both the preformed vitamin A and the beta-carotene in foods to make **retinol,** the amount of vitamin A that comes from foods is usually expressed in **retinol activity equivalents (RAE)**—a measure of the amount of retinol the body will derive from the food. Table 6-10 shows the amount of vitamin A, expressed in RAE, that comes from various foods.

The major sources of vitamin A (in the form of beta-carotene) are almost all brightly colored in hues of green, yellow, orange, and red. Any plant food with significant vitamin A activity must have some color,

preformed vitamin A vitamin A in its active form.

beta-carotene an orange pigment found in plants that is converted into vitamin A inside the body. Beta-carotene is also an antioxidant.

precursor a compound that can be converted into another compound. For example, beta-carotene is a precursor of vitamin A.

pre = before
cursor = runner, forerunner

carotenoids (kah-ROT-eh-noyds) a group of pigments (yellow, orange, and red) found in foods. (See the discussion on phytochemicals later in the chapter for more about this family of compounds.)

age-related macular degeneration oxidative damage to the central portion of the eye—called the macula—that allows you to focus and see details clearly (peripheral vision remains unimpaired). The carotenoids lutein and zeaxanthin—found in broccoli, Brussels sprouts, kale, spinach, corn, lettuce, and peas—may protect normal macular function.

retinol one of the active forms of vitamin A.

retinol activity equivalents (RAE) a measure of the amount of retinol the body will derive from a food containing preformed vitamin A or beta-carotene and other vitamin A precursors. Note that some tables list vitamin A in terms of *International Units (IU).* (See Appendix B for methods of converting from one measure to another.)

TABLE 6-10
Vitamin A in Foods

(µg RAE)	Sources
6,582	Beef liver (3 oz)*
971	Carrot, fresh (½ c)
961	Sweet potato (½ c)
472	Spinach, cooked (½ c)
444	Cantaloupe (½)
361	Butternut squash (½ c)
198	Turnip greens, cooked (½ c)
150	Fortified milk (1 c)*
109	Bok choy, cooked (½ c)
78	Mango (1)
72	Romaine lettuce (1 c)
60	Apricot halves, dried (¼ c)
54	Broccoli, cooked (½ c)
53	Watermelon (1 slice)
45	Tomatoes, cooked (½ c)
38	Tomato, fresh (1)

*Preformed vitamin A. The rest of the items on the chart derive vitamin A from beta-carotene.

SOURCES OF VITAMIN A AND BETA CAROTENE
Adult DRI (RDA) is 700 to 900 µg RAE.

*Preformed vitamin A
†Beta-carotene

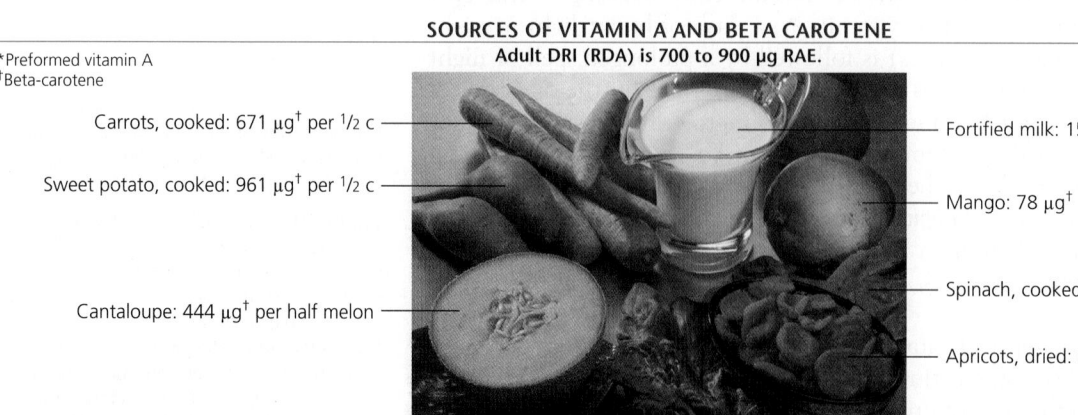

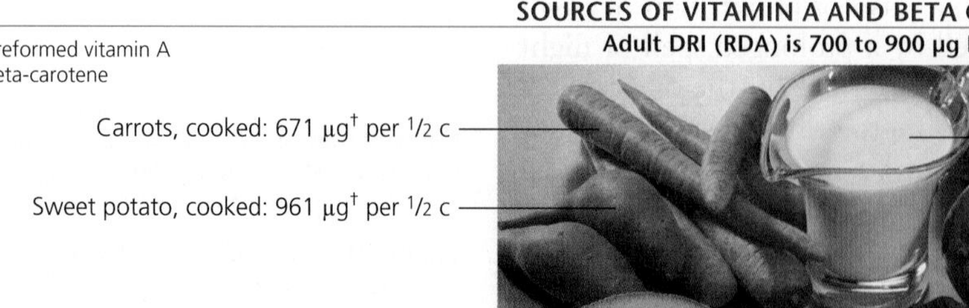

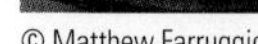

because beta-carotene is a rich, deep yellow, almost orange color. (Preformed vitamin A is pale yellow.) The dark green leafy vegetables contain large amounts of the green pigment **chlorophyll,** which masks the carotene in them.

chlorophyll the green pigment of plants that traps energy from sunlight and uses this energy in photosynthesis (the synthesis of carbohydrate by green plants).

In the United States, about half of the vitamin A consumed in foods comes from fruits and vegetables, and about half of that comes from *dark* leafy greens, such as broccoli and spinach and *rich* yellow or *deep* orange vegetables, such as winter squash, carrots, and sweet potatoes. The other half of the vitamin A comes from milk, cheese, butter, and other dairy products, eggs, and a few meats, such as liver. When whole milk is processed to produce fat-free milk, the vitamin is lost along with the fat that is skimmed. (Remember that vitamin A is found in the fat.) In the United States and Canada, fat-free milk is fortified with vitamin A to compensate for its loss. Likewise, margarine is fortified to provide the amount of vitamin A typically found in butter. (Milk and margarine are also fortified with vitamin D.)

Thus, the best and easiest way to ensure that you meet your vitamin A needs is to consume generous amounts of a variety of dark green and deep orange vegetables and fruits. Because these foods provide such an abundance of beta-carotene, along with other nutrients, most dietary guidelines advise eating at least five servings of fruits and vegetables daily, including at least one dark green or deep orange item every other day.

Vitamin D

Vitamin D is a member of a large bone-making and bone maintenance team composed of several nutrients and other compounds, including vitamin C and vitamin K; hormones; the protein collagen; and the minerals calcium, phosphorus, magnesium, and fluoride. Vitamin D's special role involves assisting in the absorption of dietary calcium and helping to make calcium and phosphorus available in the blood that bathes the bones so that these minerals can be deposited as the bones harden. In addition, vitamin D acts very much like a hormone—a compound manufactured by one organ of the body that affects another. Indeed, vitamin D exerts an influence on a number of organs, including the kidneys and the intestines.

Another particularly unique feature of vitamin D is that the body can synthesize it with the help of sunlight, regardless of dietary consumption. Vitamin D is commonly called the sunshine vitamin because the liver uses cholesterol to make a vitamin D precursor, which is converted to vitamin D with the help of the sun's ultraviolet rays. The liver alters the molecule, and the kidney alters it further to produce the active form of the vitamin. This is why diseases affecting either the liver or the kidneys, which in turn upset vitamin production, may ultimately lead to bone deterioration.

TABLE 6-11
Vitamin D in Foods

(µg)	Sources
4.3	Salmon (3 oz)
3.0	Shrimp (3 oz)
2.5	Fat-free milk (1 c)
0.9	Cod liver oil (1 tbsp)
0.6	Egg (1)
0.5	Margarine (1 tsp)
≥0.5	Fortified cereals

SOURCES OF VITAMIN D
Adult DRI (AI) is 5 to 10 µg.

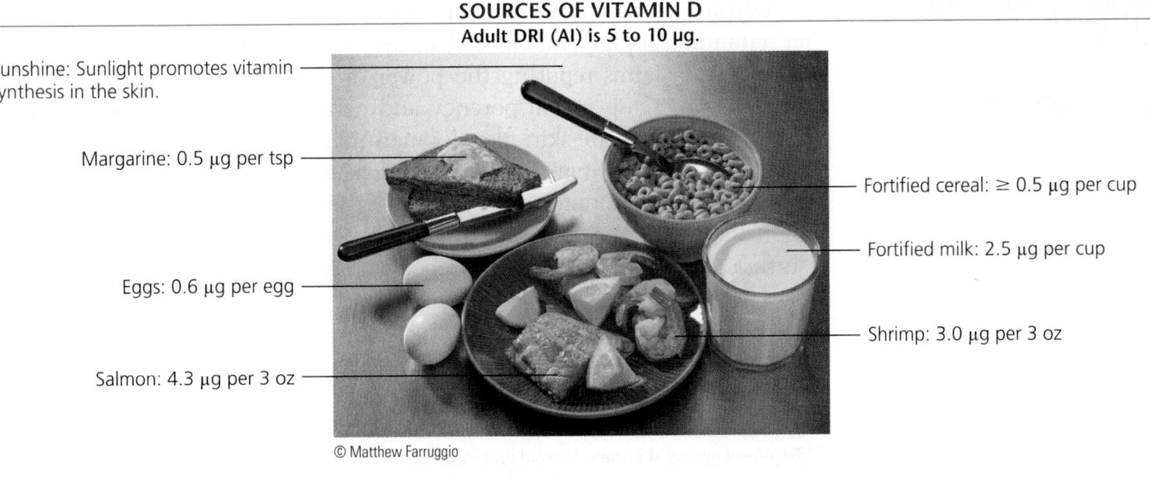

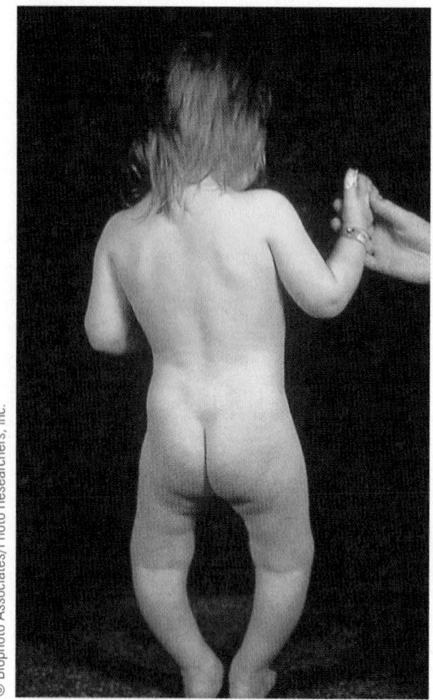

© Biophoto Associates/Photo Researchers, Inc.

▲ *This child has the bowed legs characteristic of rickets. Worldwide, rickets affects many children who live in poverty and do not have access to sunlight or adequate foods containing vitamin D.*

© David Young-Wolff/PhotoEdit

▲ *Vitamin D is the sunshine vitamin. A 15-minute walk, on a clear summer day, two to three times per week, can help supply much of your needed vitamin D.*

osteomalacia (os-tee-o-mal-AY-shuh) the disease resulting from vitamin D deficiency in adults and characterized by softening of the bones. (Its counterpart in children is called rickets.) Symptoms include bowed legs and a curved spine.

Because the body can make vitamin D with the help of sunlight, you can meet your needs for the nutrient either via sun exposure or through diet.* However, significant amounts of the nutrient come from only a few animal foods—notably eggs, liver, and some fish (see Table 6-11 on page 187). And, even in these foods, the vitamin D content varies greatly.

Although vitamin D is not prevalent in the diet, most adults, especially those living in sunny, southern regions, need not worry about the vitamin D content of the foods they eat because their bodies are getting plenty of the nutrient as a result of sun exposure. Sun exposure of the face, hands, and arms for just 5 to 15 minutes several times a week is usually all it takes to meet vitamin D needs. However, people who live in northern parts of the country—above an imaginary line drawn between Boston and the Oregon–California border—are not exposed from November through February to enough ultraviolet rays from the sun to synthesize vitamin D. The same holds true for housebound or institutionalized elderly people, who not only get outside less often than younger people but also tend to be much less efficient at producing vitamin D via the skin/sun.[17] For these people, eating vitamin D-rich foods, such as fortified milk, fatty fish (including sardines, herring, mackerel, and salmon), eggs, and some fortified cereals, is particularly important.

Children who fail to get enough vitamin D characteristically develop bowed legs, which are often the most obvious sign of the deficiency disease rickets. In adults, vitamin D deficiency causes **osteomalacia,** most often in women whose diets lack calcium, who get little exposure to the sun, and who go through several closely spaced pregnancies and prolonged periods of breastfeeding. Osteomalacia causes the bones, particularly the leg bones and spine, to become soft, porous, and weak.

Although vitamin D deficiency depresses calcium absorption, resulting in low blood calcium levels and abnormal bone development, an excess of vitamin D does just the opposite. It increases calcium absorption, causing abnormally high concentration of the mineral in the blood, which then tends to be deposited in the soft tissues. This is especially likely to happen in the kidneys, resulting in the formation of calcium-containing stones called kidney stones. The Tolerable Upper Intake Level for vitamin D is set at 50 micrograms per day.

Vitamin E

Vitamin E is known as a vitamin in search of a disease.[18] That's because vitamin E is widespread in the food supply, and deficiencies of the nutrient are rare. The great majority of vitamin E in the diet comes from vegetable oils and products such as margarine, salad dressings, and shortenings (animal fats such as butter and lard contain negligible amounts of the nutrient). Soybean, cottonseed, corn, and safflower oils contain generous amounts of vitamin E, as do nuts and seeds. Smaller amounts come from fruits, vegetables, grains, and other foods. Wheat germ, for example, is an excellent source of vitamin E. Table 6-12 lists sources of vitamin E.

Despite the rarity of deficiency, however, vitamin E is one of the most popular vitamin supplements. For decades, people have swallowed all manner of extravagant claims reputing the power of the nutrient to improve athletic performance; increase sexual potency and performance; and prevent graying of the hair, wrinkling of the skin, development of age spots, and other signs of aging, to name just a few. Vitamin E has never been proven to be a panacea for any of those problems. Vitamin E supplements are also often touted as a remedy for nighttime leg cramps. Again, the evidence to support that claim is limited, so the nutrient should not be self-prescribed as a treatment.[19]

The link between vitamin E and heart disease and other chronic diseases is an area of active scientific research. Because vitamin E performs a key role as an antioxidant in the body, scientists suspect that it is involved in protecting the membranes of the lungs, heart, brain, and other organs against damage from pol-

*To protect against skin cancer, avoid prolonged exposure to the sun.

SOURCES OF VITAMIN E
Adult DRI (RDA) is 15 mg.

Mango, fresh: 2.3 mg per mango
Olive oil: 1.7 mg per tbsp
Safflower oil: 6.0 mg per tbsp
Wheat germ: 2.6 mg per 2 tbsp
Almonds: 2.2 mg per 1/2 c
Canola oil: 2.9 mg per tbsp
Mayonnaise: 1.7 mg per tbsp
Sunflower seeds: 17.0 mg per 1/4 c
Shrimp: 1.0 mg per 3 oz
Avocado: 2.8 mg per avocado

lutants and other environmental hazards. Scientists believe that the vitamin E residing in the fatty cell membranes that surround cells acts as a scavenger of free radicals that enter the area.[20] When vitamin E is absent, the free radicals can attack the cell and start a chemical chain reaction that damages the cell membrane, making it leaky and ultimately causing it to break down completely. For more than a decade, research suggested that vitamin E may protect against heart disease because it could thwart the free radicals that might otherwise damage the walls of blood vessels and contribute to coronary artery disease.[21] However, except for persons with low levels of vitamin E in their blood, recent evidence is inconclusive and does not support the notion of taking routine vitamin E supplements with the hope of preventing chronic heart disease.[22]

The cell membranes harbor vitamin E, and a deficiency of the nutrient causes those membranes, particularly those in red blood cells, to rupture and cause a type of anemia. Although extremely rare in healthy adults, this scenario sometimes occurs in premature infants who are born before vitamin E is transferred to them from their mothers. Other groups of people who run the risk of deficiency include those who cannot absorb fats as a result of diseases and those with certain blood disorders.

Vitamin E toxicity appears to be rare, occurring only in people who take extremely high doses. Suspected symptoms include alteration of the body's blood-clotting mechanisms and interference with the function of vitamin K.

TABLE 6-12
Vitamin E in Foods

(mg)	Sources
8.5	Sunflower seeds (2 tbsp)
6.0	Safflower oil (1 tbsp)
3.0	Corn oil (1 tbsp)
3.0	Peanut butter (2 tbsp)
3.0	Peanuts (1 oz)
2.9	Canola oil (1 tbsp)
2.8	Avocado (1)
2.6	Wheat germ (2 tbsp)
2.4	Mango (1)
1.7	Mayonnaise (1 tbsp)
1.7	Olive oil (1 tbsp)
1.6	Peanut oil (1 tbsp)
1.0	Shrimp (3 oz)

For additional perspectives on the use of vitamin and mineral supplements, see the Eat Well Be Well feature in Chapter 7 on page 239.

Vitamin K

The "K" in vitamin K stands for the Danish word *koagulation* (coagulation or clotting). The key function of vitamin K is its role in the blood-clotting system of the body, where its presence can mean the difference between life and death. It is essential for the synthesis of at least 4 of the 13 proteins involved, along with calcium, in making the blood clot. When *any* of these blood-clotting factors is absent, blood cannot clot, leaving a person vulnerable to excessive bleeding upon injury.

Accumulating evidence also supports an active role for vitamin K in the maintenance of bone health. Vitamin K works in conjunction with vitamin D to synthesize a bone protein that helps to regulate the calcium levels in the blood. Low levels of vitamin K in the blood have been associated with low bone-mineral density, and researchers have noted a lower risk of hip fracture in older women who have high intakes of vitamin K than in those who have low intakes of the vitamin.[23]

Vitamin K can be synthesized by the **intestinal flora**—the bacteria that reside in the digestive tract. In addition, as Table 6-13 shows, many foods supply ample amounts of the vitamin—green leafy vegetables and members of the cabbage family in particular.

intestinal flora the normal bacterial inhabitants of the digestive tract.
flora = plant inhabitants

SOURCES OF VITAMIN K
Adult DRI is 90 to 120 µg.

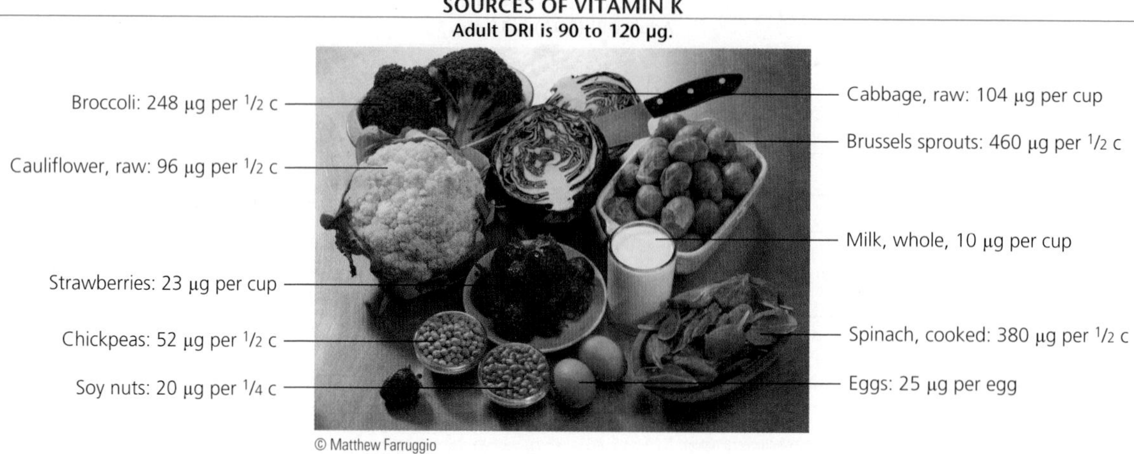

© Matthew Farruggio

TABLE 6-13
Vitamin K in Foods

(µg)	Sources
380	Spinach, cooked (½ c)
364	Turnip greens, cooked (½ c)
104	Cabbage, raw (1 c)
96	Cauliflower, raw (½ c)
60	Lettuce (1 c)
58	Broccoli, raw (½ c)
52	Chickpeas (½ c)
25	Egg (1)
20	Soybeans, dry roasted (¼ c)
19	Canola oil (1 tbsp)
10	Strawberries (½ c)

Because vitamin K is obtained both in the diet and via the intestinal bacteria, deficiencies are rare and occur only under unusual circumstances. Taking antibiotics for an extended period of time, for instance, could kill some of the intestinal bacteria and thereby prompt a deficiency.

Newborn babies are the one group that is commonly susceptible to a vitamin K deficiency because a baby's digestive tract contains no bacteria before it is born. After birth, the infant's intestinal tract gradually becomes populated with bacteria, but this happens over time. What's more, baby formula or breast milk generally doesn't contain adequate amounts of vitamin K. Thus, newborns are given a dose of vitamin K to prevent the possibility of a life-threatening hemorrhage in the case of injury.

Vitamin K toxicity is rare, but it can occur when supplemental doses are taken. In particular, adults who must pay attention to the amount of vitamin K in their diet are those who take anticoagulant drugs designed to prevent the blood from clotting and possibly causing a stroke or heart attack. People taking such medications are advised to keep their consumption of vitamin K fairly constant from day to day because large fluctuations can limit the effectiveness of the anticlotting drugs.[24]

Nonvitamins

In addition to vitamins, **choline**—a substance needed by the body to make lecithin and other molecules—is considered a "conditionally" essential nutrient because the body becomes unable to make sufficient amounts of choline when fed a choline-free diet.[25] However, because choline is found in milk, eggs, peanuts, and many other foods, deficiencies are rare. The recommended intake for choline is listed in the DRI table on the inside front cover of this book.

Various other substances have been mistaken as essential nutrients for humans because bacteria or animals need them to live. Sometimes called nonvitamins, some of the substances are important to cell membranes (inositol) and cellular activities (carnitine) but are not essential in human diets because our bodies can make these compounds as needed. Moreover, these substances are abundant in common foods. Research about the role of these nonvitamins for humans is ongoing.

Other substances that are sometimes added to vitamin supplements because they are essential for growth in certain nonhuman species are truly nonessential in human diets. These include PABA (para-aminobenzoic acid), the bioflavonoids ("vitamin P" or hesperidin) and ubiquinone (coenzyme Q_{10}), vitamin B_{15} (a hoax), and vitamin B_{17} (laetrile, a falsely touted cancer cure).

choline a nonessential nutrient used by the body to synthesize various compounds, including the phospholipid lecithin; the body can make choline from the amino acid methionine.

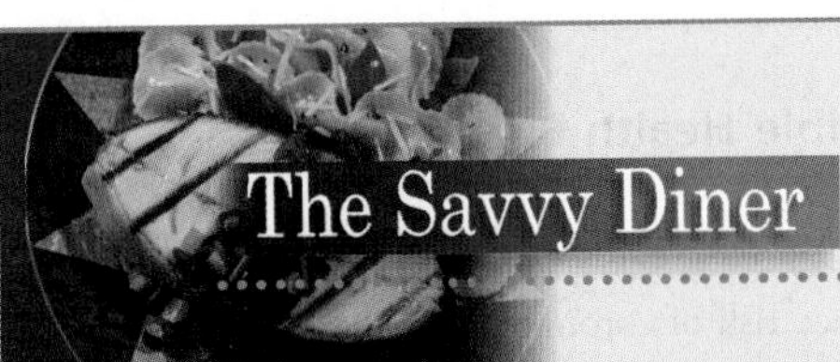

Color Your Plate with Vitamin-Rich Foods—and Handle Them with Care

▲ *Once you've taken the time to select fresh, vitamin-rich fruits and vegetables, be sure to handle them with care when you get them home.*

Buying a variety of vitamin-rich fruits and vegetables at the market is one of the first steps to obtaining plenty of those all-important nutrients in your diet. The next step is storing and cooking those foods in ways that minimize the loss of vitamins that can occur as a result of improper storage and preparation. To help get the most from the produce you buy, use the following tips on fruit and vegetable storage and preparation.

- Shop for produce at least once a week. The longer fruits and vegetables stay in your refrigerator before eating, the more nutrients that are likely to be lost.
- Store fruits and vegetables (other than bananas, tomatoes, and potatoes) in your refrigerator rather than in a fruit bowl or on the kitchen counter. Chilling slows the metabolic rate of the cells of a fruit or vegetable, which in turn causes the cells to use less of their own nutrients. Thus, chilling prevents nutrient depletion.[26]
- Store fruits and vegetables whole, peeling and cutting only what you need immediately before cooking or eating. Once you cut into the skin of an item and expose it to air, vitamin loss begins. After slicing, the vitamin C content of oranges, grapefruits, tomatoes, and strawberries begins to decline. If you do have leftover cut produce, wrap it tightly in airtight plastic or store it in an airtight container inside the refrigerator.
- Try to eat frozen vegetables within a month or two of purchase, because nutrient losses occur over time in stored vegetables.
- Cook vegetables in the least amount of water and for the shortest period of time possible. Water-soluble vitamins readily dissolve into cooking water, and heat destroys some as well. To minimize such losses, steam vegetables over water, or cook vegetables in a microwave oven.

Phytonutrients in Foods: The Phytochemical Superstars

One of the newest and most exciting areas of nutrition research today focuses on a class of substances in plant foods called **phytochemicals,** nonnutritive substances in plants that possess health-protective benefits. Phytochemicals are the compounds that give plants their brilliant colors—for example, lycopene, a pigment that makes tomatoes red and watermelon pink—and distinctive aromas—for example, the allium compounds that give us garlic breath. These natural compounds also protect plants from the ravages of overexposure to sunlight and other environmental threats and insects.

Most recently, the term phytochemicals has been popularized to refer in particular to plant chemicals that may affect health and prevent disease. Of particular interest are phytochemicals in edible plants, including fruits and vegetables, whole grains, legumes, herbs, and seeds. See Table 6-14 for a sampling of the phytochemicals that have been identified, as well as common food sources and beneficial effects and attributes of each phytochemical group. In the classic sense, the naturally occurring phytochemicals are not nutrients; they do not provide energy or building materials. However, ongoing research shows that phytochemicals might perform important functions by acting as powerful antioxidants, decreasing blood pressure and cholesterol, preventing cataracts, reducing menopause symptoms and risk for osteoporosis, and slowing, or even reversing certain cancers.

phytochemicals (FIGH-toe-CHEM-icals) physiologically active compounds found in plants that are not essential nutrients but appear to help promote health and reduce risk for cancer, heart disease, and other conditions. Also called phytonutrients.

phyto = plant

TABLE 6-14
A Sampling of Phytochemicals: Classifications, Food Sources, and Possible Health Benefits

Phytochemical	Food Sources	Possible Health Benefits
Beta glucan	Oat bran, rolled oats, oat flour	May reduce risk of coronary heart disease.
Capsaicin	Hot peppers	May reduce risk of fatal clotting in heart disease.
Carotenoids	Deeply colored fruits and vegetables	Act as antioxidants.
• Beta-carotene	Orange fruits and vegetables (apricots, cantaloupe, carrots, mango, pumpkin, sweet potato, winter squash) and dark green vegetables (spinach, kale, turnip greens)	May help reduce risk of many cancers and strengthen the immune system.
• Lutein and zeaxanthin	Pumpkin, summer squash, corn, eggs, broccoli, and dark green leafy vegetables, such as kale, collard greens, and spinach	May reduce risk of age-related eye disorders by protecting the retina from harmful ultraviolet radiation and by neutralizing free radicals.[a]
• Lycopene	Tomatoes, tomato products, watermelon, red grapefruit, and red peppers	May help reduce risk of prostate and other cancers.
Curcumin	Turmeric, a spice	May inhibit enzymes that activate carcinogens.
Flavonoids and other phenols[b]		
• Anthocyanin • Ellagic acid • Resveratrol	Fruits (apples, berries, cherries, citrus, grapes, pears, pomegranates, prunes), whole grains, nuts, chocolate, black or green tea, eggplant, potato peel, red cabbage, beets, celery, peppers, onions, soy, and red wine	May act as antioxidants, decrease inflammation, reduce plaque buildup in arteries, increase HDL-cholesterol levels, deactivate carcinogens, and inhibit cancer development.
Indoles	Cruciferous vegetables such as broccoli, kale, cauliflower, cabbage, turnip, and Brussels sprouts	May stimulate enzymes that make the hormone estrogen less effective, possibly reducing breast cancer risk.
Isothiocyanates (most notably, sulforaphane)	Broccoli, kale, and other cruciferous vegetables, horseradish	May help stimulate protective enzymes that detoxify carcinogens, bolstering the body's natural ability to ward off cancer.
Monoterpenes (such as limonene)	Citrus (fruits, juices, peels, oils)	May act as antioxidants and increase production of enzymes that may help the body dispose of carcinogens.
Organosulfur compounds (such as allicin)	Onions, garlic, chives, and leeks	May block the action of cancer-causing chemicals; may offer heart protection by decreasing production of cholesterol by the liver.
Phytic acid	Whole grains	May prevent free radical formation by binding to minerals and thereby reduce cancer risk.
Phytoestrogens (part of flavonoid family)		
• Isoflavones: (genistein and daidzein)	Soy foods and other legumes	May protect against heart disease by lowering blood cholesterol; may lower risk of breast, ovarian, and other cancers by blocking the action of the hormone estrogen.
• Lignans	Flaxseed, whole grains	Exhibits estrogen-blocking activity and may lower risk of breast, ovarian, colon, and prostate cancer.
Saponins	Sprouts, potatoes, green vegetables, tomatoes, nuts, whole grains, soy foods, and legumes	May strengthen the immune system and interfere with DNA replication, preventing cancer cells from multiplying.
Tannins	Grapes, red and white wine, tea	Act as antioxidants; may inhibit enzymes that activate carcinogens.

[a]Disorders causing vision loss in older adults, including age-related macular degeneration (AMD) and cataracts.
[b]Flavonoids are a subset of phenols. Flavonoids of interest include anthocyanins, catechins, ferulic acid, flavones, glycosides, quercetin, resveratrol, rutin, tangeretin, and nobiletin. The larger class of phenolic phytochemicals includes phenols, ellagic acid, capsaicin (in chili peppers), coumarin, curcumin, and others.

Scorecard — Are You Reaping the Power of Produce?

Overwhelming evidence points to the health benefits derived from diets rich in fruits and vegetables, including an enhanced immune system and reduced risk for many chronic conditions, such as heart disease, high blood pressure, certain types of cancer, and age-related vision loss. To achieve the maximum benefits from the vitamins, minerals, antioxidants, and phytochemicals available in produce, aim for a *minimum* of five servings every day. Be adventurous! Select from as wide a variety of fruits and vegetables as possible. Be sure to frequently include members of the "superstar" categories of fruits and vegetables listed below. To evaluate your intake of fruits and vegetables, take the 5 A Day challenge by answering the following questions and checking the appropriate boxes.

Evaluate Your Intake of the Superstar Fruits and Vegetables

	Daily	Occasionally	Seldom	Never
I eat citrus fruits (oranges, grapefruits, etc.).	☐	☐	☐	☐
I eat berries (strawberries, blueberries, etc.), kiwi, or grapes.	☐	☐	☐	☐
I eat cruciferous vegetables such as cabbage, broccoli, cauliflower, Brussels sprouts, and kale.	☐	☐	☐	☐
I eat dark green leafy vegetables such as spinach, turnip greens, collard greens, mustard greens, or beet greens.	☐	☐	☐	☐
I eat deep yellow, orange, and red fruits and vegetables such as sweet potatoes, carrots, winter squash, pumpkin, cantaloupe, nectarines, peaches, apricots, mangos, papayas, tomatoes, red peppers, and watermelon.	☐	☐	☐	☐

Take the 5 A Day Challenge

	NUMBER OF SERVINGS (EACH BOX REPRESENTS ONE SERVING)*				
I start the day with a serving of fruit at breakfast.	☐	☐	☐	☐	☐
I snack on fruits or vegetables during the day.	☐	☐	☐	☐	☐
I include a serving of fruit or a vegetable at lunch.	☐	☐	☐	☐	☐
I eat one or more vegetable servings at dinner.	☐	☐	☐	☐	☐
I choose berries, melon, pears, apples, or other fruit for a sweet dessert.	☐	☐	☐	☐	☐

Total Boxes Checked: ______________

SCORING FOR THE 5 A DAY CHALLENGE

Fewer than five boxes checked:

Ouch! You're missing out on the health benefits from produce. Take the 5 A Day challenge.

At least five boxes checked:

Good! You're getting the *minimum* number of servings of fruits and vegetables for a healthy diet. Be sure to select a variety of fruits and vegetables and to include servings from the superstar groups as well.

Five to nine boxes (or more) checked:

Excellent! You're getting the recommended number of servings for fruits and vegetables in a healthy diet. Be sure to select a variety of fruits and vegetables and to include servings from the superstar groups as well.

*A serving of fruit equals one medium piece of fruit, ½ cup cut or cooked fruit, ¼ cup dried fruit, or ½ cup fruit juice. A serving of vegetables equals 1 cup leafy vegetables, ½ cup cooked or cut vegetables, or ½ cup vegetable juice.

An individual fruit or vegetable may contain many (50 or more) of the thousands of known phytochemicals, but one, or only a select few, are usually present in a large amount. For example, garlic contains more than 160 identified compounds. When a clove of garlic is cut or crushed it produces sulfur compounds,

Visit your local farmers' market or grocery store produce aisle for a cornucopia of health-promoting compounds: carotenoids in carrots, spinach, and tomatoes; sulphoraphane in broccoli and bok choy, capsaicin in chili peppers. To find the farmers' market nearest you, log on to the USDA website at www.ams.usda.gov/farmersmarkets.

such as *allicin*, which scientists believe is partly responsible for the health benefits attributed to garlic. Finding the specific chemical in a food that offers the disease-protecting potential is not easy, however. Clinical studies in which people eat foods rich in certain phytochemicals are currently under way. If health benefits, such as protection against cancer, are observed, then the active ingredient in the food must be determined. Is the phytochemical, a vitamin, fiber, or the low-fat content of the increased fruit or vegetable diet responsible for the benefit? Based on current research, any one or a combination of these factors might be responsible.

Mechanisms of Actions of Phytochemicals

Many foods contain numerous phytochemicals, each one acting on one or several mechanisms. Some have antioxidant properties (protecting against harmful cell damage), others have anticancer properties (preventing initiation and promotion of cancer), and some have anti-estrogen properties (blocking the action of estrogen and lowering the risk of some cancers).

Different phytochemicals have different modes of action, and one individual phytochemical may exhibit more than one mechanism of action. A phytochemical may influence one or more stages of cancer, from initiation to promotion and progression to a malignant tumor. Phytochemicals act by both *direct* and *indirect* mechanisms. They may act directly to *inhibit* enzymes that activate carcinogens or to *induce* enzymes that detoxify carcinogens. They may act indirectly by stimulating the immune response or scavenging free radicals to prevent DNA damage. See this chapter's Spotlight feature for a discussion of nutrition's role in cancer.

Although many phytochemicals act on cells to suppress cancer development, they may also help protect against other diseases, notably heart disease. Some phytochemicals may influence blood pressure and blood clotting, whereas others reduce the synthesis and absorption of cholesterol. Certain pigments (especially carotenoids) in plant foods may protect the eye against free radical damage and thus prevent or postpone macular degeneration, which can lead to blindness in older adults. One of the most widely studied groups of phytochemicals, the isoflavones found in soy products, is currently being researched for its potential role in preventing diseases such as osteoporosis, heart disease, and various cancers (mainly prostate, breast, and ovarian), and for alleviating symptoms associated with menopause. See the Spotlight feature in Chapter 5 for more about the benefits of soy.

How to Optimize Phytochemicals in a Daily Eating Plan

Research indicates that pure extracts of phytochemicals in supplements are less effective than phytochemicals in whole foods. Some phytochemicals might not be metabolized in pure form, and some might not function by themselves. Thus, phytochemicals have less protective power when ingested as concentrated extracts, such as in pills.

We know that the absorption, metabolism, and distribution of some nutrients are dependent upon the presence of other nutrients. Likewise, it appears that the absorption, metabolism, distribution, and function of phytochemicals also depend on the combination of one phytochemical with other phytochemicals or other substances that occur naturally in food. An individual phytochemical does not necessarily provide protection against disease, but rather the combination of the phytochemical with other phytochemicals or food components provides the protection.

Until more is known about phytochemicals and how they function, it is best to follow the recommendations of the MyPyramid food guide and consume the recommended amounts of vegetables and fruits per day along with a variety of whole grains, soy foods, and other legumes, nuts, and seeds. Use the Scorecard on page 193 to evaluate your current intake of fruits and vegetables. In addition, consider the tips offered in the Eat Well Be Well feature that follows.

Eat Well Be Well Let Food Be Your Medicine

Lycopene in your tomato sauce? Beta-carotene in your soup? Today, one of the hottest areas in food science and nutrition policy is **functional foods.** Nutritionists, food scientists, food marketers, and others are exploring how today's traditional foods, and perhaps new formulations, may open doors to a healthier tomorrow.[27] The fact that food is intimately linked to optimal health is not a novel concept. "Let food be your medicine and medicine be your food" was a tenet espoused by Hippocrates in approximately 400 B.C.[28] Almost 2,500 years later, this philosophy is once more of utmost importance, as it is the "food as medicine" philosophy that underpins the paradigm of functional foods. In this feature, we look at the promising benefits of functional foods in disease prevention, as well as how we can incorporate functional foods into our daily eating plan so we can reap the benefits.

▲ *Visit your nearest grocery store for a wide assortment of functional foods.*

© Scott Hirko/Hespenheide Design

1 Visit the Local Grocery Store.

Functional foods are not really a new concept; grocery stores are already filled with numerous items that meet the definition of functional foods (see Figure 6-5).[29] One of the first to appear was calcium-fortified orange juice, which gave people who don't like milk as much bone-healthy calcium per glass as milk. We now have grain products fortified with folate to prevent birth defects and possibly heart disease. Many cereal grains, fruits, vegetables, nuts, and dairy products are touted for their potential for cancer prevention benefits as well as cardiovascular protection.

The Quaker Oats Man continues to smile as the cholesterol-lowering effects of the soluble fiber beta-glucan in oats are well documented, and omega-3 fatty acids in flaxseed oil and fatty fish such as salmon or halibut appear to reduce the risk of heart disease as well as having anti-inflammatory properties. Research indicates that the organosulfur compounds in garlic may have it all—cancer-fighting, cholesterol-lowering, antibiotic, and antihypertensive properties—which far outweigh the herb's ability to cause bad breath. Probiotics (friendly bacteria) in fermented milk products such as yogurt and prebiotics (fermentable dietary fiber) in oatmeal, flax, and legumes also appear to have multiple effects, especially in promoting a healthy gastrointestinal tract. Probiotics are also associated with reducing the risk of colon cancer, lowering cholesterol, and out-competing potentially disease-causing bacteria in the gastrointestinal tract. Soy protein has been the focus of intense research efforts because of its ability to lower cholesterol and possibly reduce the risk of cancer, osteoporosis, and symptoms associated with menopause.[30] Current research is investigating a certain type of fat—called conjugated linoleic acid (CLA)—found primarily in dairy products for numerous potential health benefits including improvements in immune function, body composition, and bone mineralization. Identification of CLA underscores the importance of meeting nutrient needs through foods from all the major food groups.[31]

2 Follow the Research.

So much is still unknown when it comes to functional foods that additional research is needed to clarify the role of phytochemicals in health promotion. Such research will also provide plant physiologists and human nutritionists with the vital information they need to develop an enhanced food supply and better target dietary recommendations for the general public.[32] For years, the food industry focused on "taking out the bad stuff" (fat, sodium, cholesterol, calories) and restoring or enhancing favorable nutrients (calcium, folate, fiber). We may soon see the day where we have tomatoes with extra lycopene, sulforaphane-enriched broccoli, and sweet potatoes with extra carotene.

---

functional food a general term for foods that provide an *additional* physiological or psychological benefit beyond that of meeting basic nutritional needs. Also called *medical foods* or *designer foods*—foods "fortified" with phytochemicals or plants bred to contain high levels of phytochemicals. (Genetic engineering of foods, also called *biotechnology,* is discussed in Chapter 12.)

FIGURE 6-5
FUNCTIONAL FOOD PYRAMID

SOURCE: Adapted from University of Illinois, Functional Foods for Health.

GRAINS	VEGETABLES	FRUITS	MILK	MEAT & BEANS
Psyllium-containing bread and cereal, Corn products, Flaxseed, Oat products, Rye products, Wheat bran products	Broccoli, Brussel sprouts, Cabbage, Carrots, Cauliflower, Celery, Garlic, Horseradish, Jerusalem artichokes, Leeks, Onions, Scallions/Shallots, Soybeans, Tomatoes, Watercress	Apples, Bananas, Blueberries, Cranberries, Grapefruit, Grapes/Juice, Lemons, Limes, Oranges, Raspberries	Cheese, Milk, Milk products, Soy milk products, Yogurt	Beans, Beef, Eggs, Mackerel, Salmon, Soy nuts, Soy protein, Sardines, Tuna, Walnuts

Here are examples of some foods that are potentially associated with maintaining health.

Antioxidant
Broccoli family
Carrots
Citrus fruit
Cocoa/chocolate
Flaxseed
Grapes/juice
Honey
Horseradish
Raspberries
Tomatoes

Maintains Vision
Blueberries
Broccoli
Carrots
Corn products
Eggs
Leafy greens

Improves Gastrointestinal Health
Bananas
Honey
Horseradish
Jerusalem artichokes
Milk
Onion family
Yogurt

Reduces Cancer Risk
Apples
Beans
Berries
Broccoli family
Citrus fruit
Corn products
Flaxseed
Garlic
Grapes/juice
Green or black tea
Milk products
Onion family
Rye products
Salmon
Soy products
Tomatoes
Wheat bran products

Improves Heart Health
Apples
Beans
Berries
Fish (salmon, tuna, mackerel, sardines)
Flaxseed
Garlic
Grapes/juice
Green or black tea
Milk products
Oat products
Onion family
Psyllium-containing bread and cereals
Soy products
Walnuts

Maintains Urinary Tract Health
Blueberries
Cranberries

Reduces Blood Pressure
Bananas
Celery
Cheese
Garlic
Milk
Soy products

Improves Bone Health
Cheese
Milk
Milk products
Soy milk products
Soy nuts

3 Incorporate Functional Foods into Your Daily Diet.

You can start by incorporating the MyPyramid food guide into your daily food planning if you are not already using it (see the box that follows). Begin with the Grains group, and consume the recommended amount of whole-grain breads and cereals, especially oatmeal. Grind flaxseed in a coffee grinder and add it to your cereal.

Consume at least nine servings of vegetables and fruits daily. Be sure to include a variety of vegetables and fruits, including dark orange vegetables and fruits such as sweet potatoes, carrots, winter squash, apricots, cantaloupe, and peaches. Add dark green leafy vegetables such as kale, mustard greens, collards, and spinach, along with broccoli and Brussels sprouts. Include tomatoes as often as possible and also add garlic and onions to your diet. Eat plenty of purple vegetables such as eggplant and red cabbage. Try experimenting with soy milk, tofu, miso, and tempeh, and include them regularly in your diet.

Salmon and other fatty fish (sardines, mackerel, herring, tuna) contain omega-3 fatty acids, which are shown to offer heart protection. Try selecting two fish meals per week.

Yogurt, cheese, and milk products go a long way in offering the calcium needed for healthy bones; be sure to have at least three servings daily. Many dairy products are functional foods. The live bacteria used to ferment milk products such as yogurts and yogurt-based drinks enhance intestinal flora.

Last but not least, top your meal off with a cup of green or black tea, a source of phenolic compounds, which many researchers are convinced protect against cancer, heart disease, and stroke. As you can see, the best advice is to consume a variety of foods that contain both known beneficial compounds and those still awaiting discovery.

Tips for adding functional foods to your daily diet:[33]

At Breakfast

- Mix **oatmeal** with **blueberries.**
- Top **whole-grain dry** or **hot cereal** with **yogurt.**
- Spread **soy nut butter** on **whole-grain toast.**
- Drink a glass of sparkling **purple grape juice** with breakfast.
- Blend **soy milk** with fresh **pineapple** or **frozen berries.**

Healthy Meal Ideas

- Mix **tuna** salad with grated **carrots, red peppers, onions,** and **garlic.**
- Serve **whole-grain pasta** with **tomato sauce** and **fresh herbs.**
- Cook **leeks** and **onions** with **tomatoes** as a side dish.
- Grill **salmon** and serve with **fresh greens** and **yogurt** salad dressing.
- Try low-fat cream of **carrot, spinach,** and **broccoli** soups.
- Enjoy a cup of **green tea** with a marinated **tofu** sandwich.
- Stir-fry fresh **vegetables** with extra **garlic.**

Snacks on the Go

- Grab a piece of fresh **fruit.**
- Mix **soy nuts, whole-grain cereal,** and **dried fruit** together and hit the trail.
- Grab a glass of **tomato, cranberry,** or **orange juice.**
- Try fresh **broccoli, cauliflower,** and **carrots** with **tofu** dip.
- Mix **bananas** with fresh **raspberries.**

Easy Phytochemical-Rich Minestrone Soup

Four antioxidant-rich veggies make this soup a real winner.

1 16-oz package frozen broccoli, cauliflower, and carrot blend
2 15-oz cans stewed tomatoes
2 14½-oz cans broth (beef, vegetable, or poultry)
1 15-oz can great northern beans
2 oz uncooked whole-wheat vermicelli (break into 2-inch pieces)
Grated Parmesan cheese

In a large saucepan, combine vegetables, tomatoes, broth, beans, and pasta and bring to a boil. Reduce heat, cover, and simmer 6 to 8 minutes or until vegetables and pasta are tender.
Sprinkle with Parmesan cheese.

Makes four to six (1½ c) servings.
Nutrition information per serving: 210 calories and 2 grams fat

Nutrition Action Medicinal Herbs

© Steve Terrill/Corbis

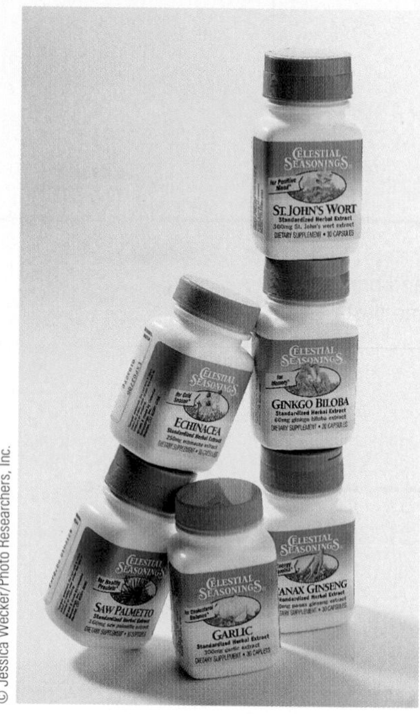

© Jessica Wecker/Photo Researchers, Inc.

▲ *Research is currently under way in the United States to test the safety and efficacy of a few of the most popular herbs on the market today.*

Throughout human history, people have relied on herbal medicines, and the use of herbs and medicinal plants for any number of ailments is a universal phenomenon. The World Health Organization estimates that about 80 percent of the world's population depends on traditional herbal medicine for primary health care.[34] Only with the development of 20th-century Western medicine have synthetic chemicals found their place in the medical system. Yet, even in a modern pharmacy in the United States, more than 25 percent of medicines are extracted from plants or are synthetic copies or derivatives of plant chemicals.[35]

In the United States, plant medicines composed of whole plants (crude drugs) or complex extracts are sold as dietary supplements because natural (or herbal) medicines are not economically viable candidates for drug research and development. Most botanicals contain one or several relatively dilute compounds, and thus they tend to have milder actions than the more concentrated chemicals found in most drugs. Therefore, herbal medicines usually take longer to act than regular medicinal products and few herbs have the potency of a prescription drug.

Pharmaceutical companies are less willing to spend the millions of dollars to fund research on plants that grow in the wild (and therefore cannot be patented), and most herb manufacturers don't have the funds to support large research studies. Despite this, herbal products are becoming increasingly popular in the United States, and natural food stores are reporting annual growth rates as high as 60 to 80 percent for medicinal herbs in bulk, capsules, extracts, tinctures, tablets, and teas.[36] Since 1990, the use of herbal medicines in the United States has increased by 380 percent, and by the year 2010, sales of herbal products are projected to be in the range of $25 billion annually.[37] See Table 6-15 on page 200 for a list of herbs thought to be effective and those that should be avoided.

What is driving this trend toward increased use of herbal products? Primarily the growth is consumer driven.[38] Most consumers learn of herbal products through the media, either in magazines, television or radio commercials, or by word of mouth from others. Other factors responsible for this trend are an interest in returning to a more natural lifestyle; dissatisfaction with the current state of Western health care; the unwanted side effects of prescription drugs; the spiraling cost and disarray of managed health care; aging baby boomers who want a better quality of health; a strong interest in alternative and complementary therapies; and, finally, a large arena of sales and marketing campaigns, often making use of famous personalities to market herbal products to consumers.

Many herbal product users believe that herbal medicines are the "natural" way to good health. However, natural is not always synonymous with safe because there are no regulations to oversee the manufacture and marketing of herbal supplements. Also, when it comes to botanicals, several plant species may look identical, but one may, in fact, be toxic. If the person collecting the herbs is not entirely knowledgeable, there is a danger that the toxic herb may be mixed with the medicinal herb, and such mistakes have been made. A further concern is that consumers have no way of knowing if the product they purchase has an "effective" amount of the active compound.

In 1994, Congress passed the Dietary Supplement Health Education Act (DSHEA), which severely restricted the FDA's authority over virtually any product labeled "supplement" as long as the product made no claim to affect a disease. DSHEA allowed herbal medicines to be marketed without prior approval from FDA. What DSHEA does allow manufacturers to state on a label is how it affects a structure or function of the body, such as the claim that the herbal product can "support," "promote," or "maintain" health. DSHEA states that a product cannot claim that it affects disease, and a manufacturer cannot state on the label that the herbal product will "prevent," "treat," "diagnose," "mitigate," or "cure" disease. A disclaimer must always be included on the label, stating: "This product has not been evaluated by the Food and Drug Administration. This product is not intended to diagnose, treat, cure, or prevent any disease." Therefore, herbal products are not obliged to meet any standards of effectiveness or safety that have been established for other medicines, which require extensive laboratory and clinical trials before FDA approval. Today, a supplement is presumed safe until the FDA receives well-documented reports of adverse reactions.

In October 1998, Congress established the National Center for Complementary and Alternative Medicine (NCCAM), a division of the National Institutes of Health. The center is devoted to conducting and supporting basic and applied research and training, and it disseminates information on complementary and alternative medicine to practitioners and the public. Some of the herbs that are currently undergoing research here in the United States are garlic, St. John's wort, Ginkgo biloba, saw palmetto, echinacea, hawthorn, and cranberry.

Until we have further research studies from which to evaluate the safety and efficacy of herbal medicines, physicians, health professionals, and consumers must continue to seek valid information and further education from reliable sources. Use the following guidelines for choosing and using herbal medicines:

- Be informed; seek out unbiased, scientific sources. Inform your physician, especially if taking prescribed medications.
- Do not exceed recommended doses or use herbal medicines for prolonged periods. Call your physician or the FDA Med Alert hotline at 800-332-1088 if you experience adverse effects.

To find out more about herbal medicines, some recommended publications and websites are as follows:

- American Botanical Council, *The Complete German Commission E Monographs: Therapeutic Guide to Herbal Medicines*
- S. Foster and V. E. Tyler, *Tyler's Honest Herbal: A Sensible Guide to the Use of Herbs and Related Remedies*, 2000
- American Botanical Council, *HerbalGram*, a peer-reviewed journal
- American Botanical Council: www.herbalgram.org
- Herb Research Foundation: www.herbs.org
- U.S. Food and Drug Administration: www.fda.gov
- National Institutes of Health/NCCAM: nccam.nih.gov
- U.S. Pharmacopeia: www.usp.org

TABLE 6-15
The Garden of Herbal Remedies

These herbs are thought to be effective:*

Herb	Why It Is Used	How It Works	Cautions
 Black Cohosh	Reduces symptoms of premenstrual syndrome, painful menstruation, and the hot flashes associated with menopause.	Appears to function as an estrogen substitute and a suppressor of luteinizing hormone.	Occasional stomach pain or intestinal discomfort. Because no long-term studies have been done, use of black cohosh should be limited to six months.
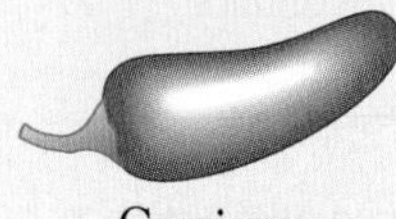 Capsicum	Best known as the hot red peppers cayenne and chili. Applied topically for chronic pain from conditions such as shingles and trigeminal neuralgia.	The active ingredient, capsaicin, works as a counterirritant and decreases sensitivity to pain by depleting substance P, a neurotransmitter that facilitates the transmission of pain impulses to the spinal cord.	Overuse can result in a prolonged insensitivity to pain. More concentrated products can cause a burning sensation. Users must avoid contact with eyes, genitals, and other mucous membranes.
 Echinacea	These species of purple cornflower appear to shorten the intensity and duration of colds and flus, may help to control urinary tract infections, and, when applied topically, speed the healing of wounds.	Though echinacea lacks direct antibiotic activity, it helps the body muster up its own defenses against invading micro-organisms.	Experts warn against using echinacea for more than eight weeks at a time and against its use by people with autoimmune diseases like multiple sclerosis and rheumatoid arthritis or by those who are infected with HIV.
 Feverfew	The dried leaves of this plant have been shown to reduce the frequency and severity of migraine as well as its frequently associated symptoms of nausea and vomiting.	The active ingredient, parthenolide, appears to act on the blood vessels of the brain, making them less reactive to certain compounds.	Most commercial preparations recommend doses that are much too high. 250 micrograms a day of parthenolide—or 125 milligrams of the herb—is an adequate dose.
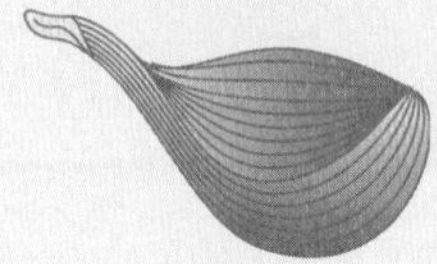 Garlic	Active against viruses, fungi, and parasites. It may also lower cholesterol and inhibit the formation of blood clots, actions that might help to prevent heart attacks.	When fresh garlic is crushed, enzymes convert alliin to allicin, a potent antibiotic. Garlic tablets and capsules containing alliin and the enzyme can be absorbed when dissolved in the intestines and not the stomach.	More than five cloves of garlic a day can result in heartburn, flatulence, and other gastrointestinal problems. People taking anticoagulants should be cautious about taking garlic.
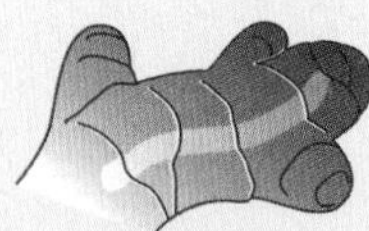 Ginger	A time-honored remedy for settling an upset stomach, ginger has been shown in clinical studies to prevent motion sickness and nausea following surgery.	Components in the aromatic oil and resin of ginger have been found to strengthen the heart and to promote secretion of saliva and gastric juices.	May prolong postoperative bleeding, aggravate gallstones, and cause heartburn. There is also debate about its safety when used to treat morning sickness.
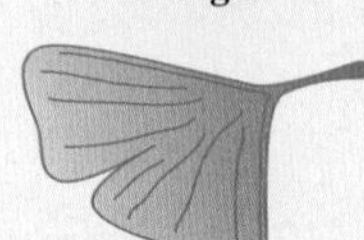 Ginkgo	Used medicinally in China for hundreds of years, Ginkgo biloba was recently reported to improve short-term memory and concentration in people with early Alzheimer's disease.	Appears to work by increasing the brain's tolerance for low levels of oxygen and by enhancing blood flow to the brain and extremities.	Possible side effects include indigestion, headache, and allergic skin reactions.
 Milk Thistle	One of the few herbs that has been demonstrated to protect the liver against toxins. Also encourages regeneration of new liver cells.	The seeds contain a compound called silymarin, which helps liver cells keep out toxins and may promote formation of new liver cells.	When used as capsules containing 200 milligrams of concentrated extract (140 milligrams of silymarin), no harmful effects have been reported.
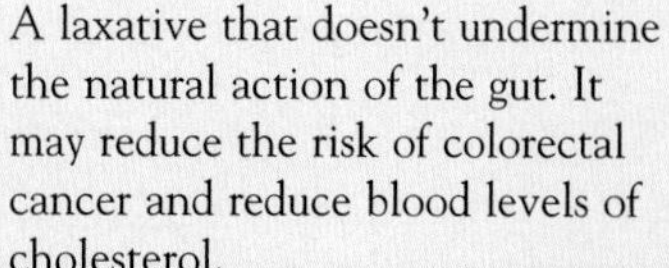 Psyllium	A laxative that doesn't undermine the natural action of the gut. It may reduce the risk of colorectal cancer and reduce blood levels of cholesterol.	The seeds of this herb have husks that are filled with mucilage, a fiber that swells with water in the intestines to add bulk and lubrication to the stool.	Increase in flatulence in some people, especially if a great deal is consumed.

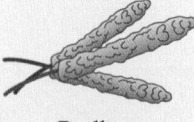

TABLE 6-15

The Garden of Herbal Remedies (continued)

These herbs are thought to be effective:*

Herb	Why It Is Used	How It Works	Cautions
Saw Palmetto	Studies on patients with an enlarged prostate have shown that extracts of this palm tree can reduce urinary symptoms even though the gland may not shrink.	Nonhormonal chemicals in saw palmetto appear to work through their antiandrogen and anti-inflammatory activity.	Some experts are concerned that those taking the herb may have inaccurate PSA readings, used as an early warning sign of prostate cancer.
St. John's Wort	Some reports attest to this herb's ability to relieve mild depression. It may also have sedating and antianxiety activity.	Inhibits uptake of serotonin by nerve cells. Though products are standardized for hypericin, hyperforin now appears to be a more potent antidepressant than hypericin.	Based on the sun-induced toxicity of hypericin in animals, people are advised to avoid exposure to bright sunlight.
Valerian	Perhaps best characterized as a mild tranquilizer.	Parts of this plant have antianxiety effects, making it potentially useful in treating nervousness and insomnia.	Long-term use can cause headache, restlessness, sleeplessness, and heart function disorders.

These herbs should be avoided:

Herb	Why It Is Used	Reasons for Caution	Herb	Why It Is Used	Reasons for Caution
Aconite	Pain, rheumatism, headaches	Numerous poisonings in China	**Kava**	Sedative, anxiety, insomnia, restlessness	Possible liver damage; should not be used with alcohol or other depressants, or by pregnant women
Belladonna	Spasms, gastrointestinal pain	Contains three toxic alkaloids, including atropine	**Kombucha Tea**	AIDS, insomnia, acne	Can cause liver damage, intestinal problems, and death
Blue Cohosh	Menstrual ailments, worms	Can induce labor	**Lobelia**	Mood booster	Can cause rapid heartbeat, coma, and death
Borage	Coughs, diuretic, mood booster	May contain liver toxins and carcinogens	**Pennyroyal**	Stimulant, gastric distress	Liver damage, convulsions, and deaths
Broom	Intoxicant, diuretic, heart problems	May slow heart rhythm; contains toxic alkaloids	**Poke Root**	Emetic, rheumatism	Extremely toxic; can cause low blood pressure, respiratory depression
Chaparral	Arthritis, cancer, pain, colds	Can cause severe hepatitis and liver failure	**Sassafras**	Stimulant, sweat producers, syphilis	Contains the carcinogen safrole; banned from use in food
Comfrey	Cuts, bruises, ulcers	Contains toxins linked to liver disease and death	**Scullcap**	Tranquilizer	Can cause liver damage
Ephedra	Stimulant, decongestant	Contains cardiac toxins resulting in dozens of deaths	**Wormwood**	Tonic, digestion	Can cause convulsions, loss of consciousness, and hallucinations
Germander	Digestion, fever	Stimulant can cause heart problems			

*In general, much more controlled research is needed to examine the potential active components, efficacy, and long-term safety of the herbs listed here.

SOURCES: *The Physician's Desk Reference for Herbal Medicines*; V. E. Tyler with S. Foster, *Tyler's Honest Herbals*; *Tyler's Herbs of Choice*; *Journal of the American Dietetic Association* (October 1997); The Mayo Clinic Food and Nutrition Center @ www.mayoclinic.com.

Spotlight Nutrition and Cancer Prevention

You probably know someone who has cancer, who has recovered from it, or who has died of it. After heart disease, cancer is our most prevalent disease and can be expected, in one form or another, to affect one out of every three Americans living today. Given what is known now about the link between diet and cancer, you are well advised to learn about this connection. Unlike so many factors in our environment, the food you choose to eat is a factor you can control to a great extent. This discussion attempts to answer questions about the connection between diet and cancer.

How is diet associated with cancer?

Numerous studies conducted in both laboratory animals and humans over the past two decades have shown that many connections exist between diet and cancer. Constituents in foods may be responsible for starting the cancer (a process called *initiation*) or for speeding its development (a two-step process that includes tumor *promotion* and *progression*), or they may protect against cancer. Also, for the person who has cancer, diet can make a crucial difference in recovery by helping to restore body weight and improve nutritional status.

Not all studies have shown a firm relationship between cancer and food and nutrient intake. In particular, some epidemiologic studies—that is, studies of disease rates and food patterns of groups of people—have failed to demonstrate such a link. Even when a positive association seems likely, interpreting data that may link dietary components with cancer or other chronic diseases requires a great deal of caution. Remember that an increase in one component of the diet can cause increases or decreases in others. For example, if a close correlation is shown between cancer and the consumption of animal protein by a human population, can we be sure that the critical factor is the animal protein? The real factor could be increased fat consumption, because fat is consumed with animal protein. Or the cancer might be caused by what is crowded out of the diet: vitamins, minerals, fiber, or nonnutrients that would otherwise have been supplied by the missing fruits, vegetables, legumes, and whole grains.

These issues must be considered when examining the results of studies describing a connection between diet and cancer. Our diets are complex and diverse, making it difficult to separate the effect of a single dietary component from the hundreds of other constituents in foods. In addition, the difficulty of evaluating the diet–cancer link is compounded by the fact that many cancers take up to 20 years to develop. Thus, assessing a cancer patient's diet today is not as helpful as knowing how that patient ate 10, 20, or even 30 years before the cancer was diagnosed. Finally, our eating pattern is only one of many factors that contribute to the development of cancer (see Figure 6-6).

Is nutrition related to causing cancer the way other environmental factors are—such as smoking or air pollution?

Yes. The National Cancer Institute estimates that more than 85 percent of all cancers are associated with lifestyle and environmental factors, including nutrition (see Figure 6-7). Lifestyle factors can usually be controlled by the individual, and they include tobacco use, diet, exercise, alcohol consumption, exposure to sunlight, patterns of sexual behavior, and personal hygiene. Individuals typically have little control over environmental factors, such as the exposure to carcinogens in the workplace, radiation during medical and

FIGURE 6-6
THE IMPACT OF GENETIC, ENVIRONMENTAL, AND LIFESTYLE RISK FACTORS ON CANCER

SOURCE: American Institute for Cancer Research, *Stopping Cancer Before It Starts: The American Institute for Cancer Research's Program for Cancer Prevention* (New York: Golden Books, 1999), 10.

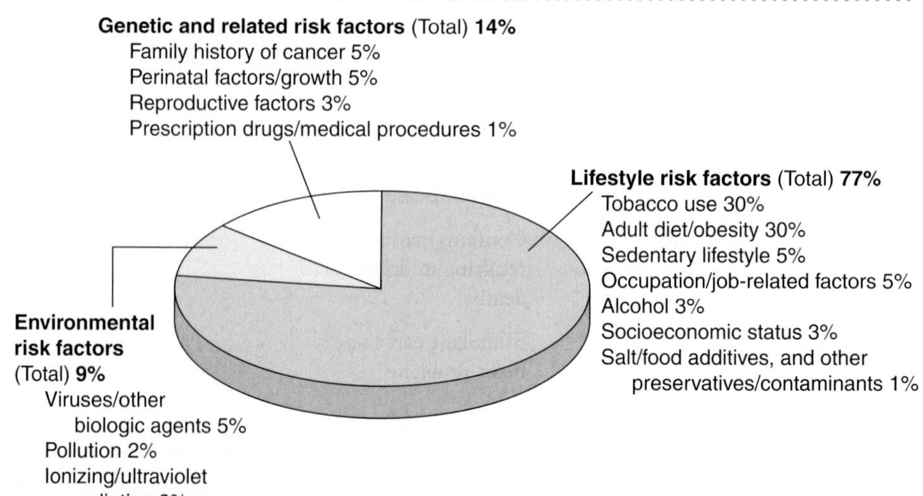

FIGURE 6-7
HOW CANCER DEVELOPS

Examples of Carcinogens
A. Cancer initiation. Initiation by a carcinogen causes cancerous alterations in previously healthy body cells.
Ultraviolet light/ radiation
Cigarettes
Toxic environmental pollution/contaminants
Healthy body cells
Cancer initiation
Damaged cellular DNA and altered genetic material
Development of cancerous cells
B. Cancer promotion. Cancer promoters enhance the growth of abnormal cancerous cells.
Cancerous cells
Cancer promotion
Cancer progression
Examples of Cancer Promoters:
Ultraviolet light/radiation
Toxic environmental pollution/contaminants
Lack of physical activity and obesity
Cigarettes
Excess alcohol
Excess dietary fat
C. Cellular repair. Cancer antipromoters squelch free radical damage and enhance the body's ability to repair damaged DNA strands.
Damaged cellular DNA and altered genetic material
Cellular repair
Healthy body cells
Examples of Cancer Antipromoters:
The antioxidant, fiber, and phytochemical team
Strong immune system and healthy body weight

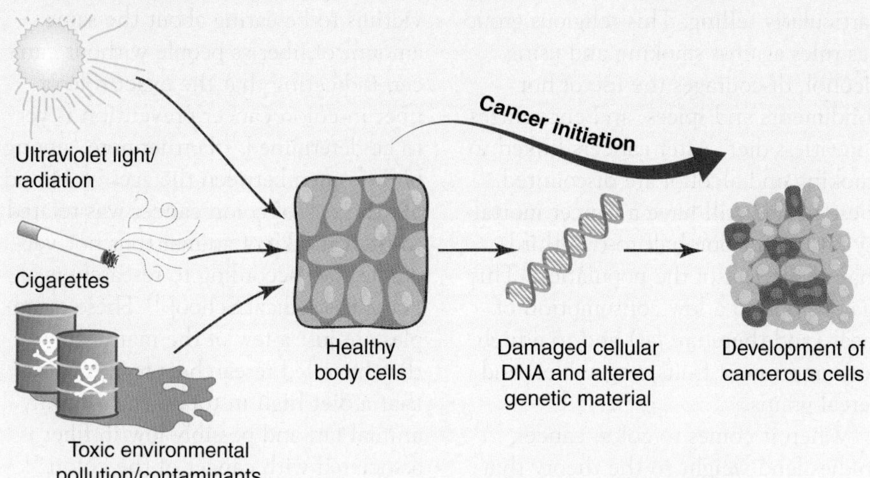

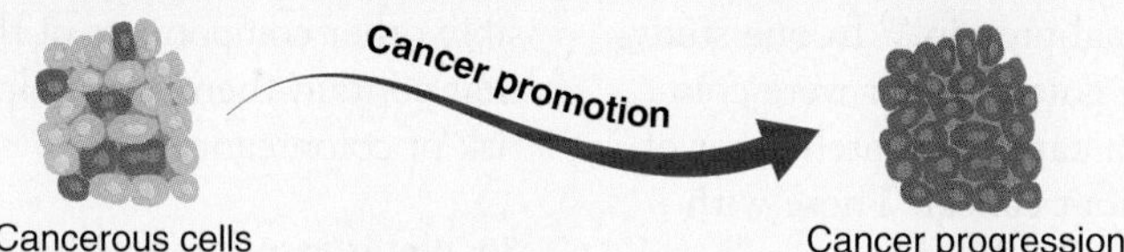

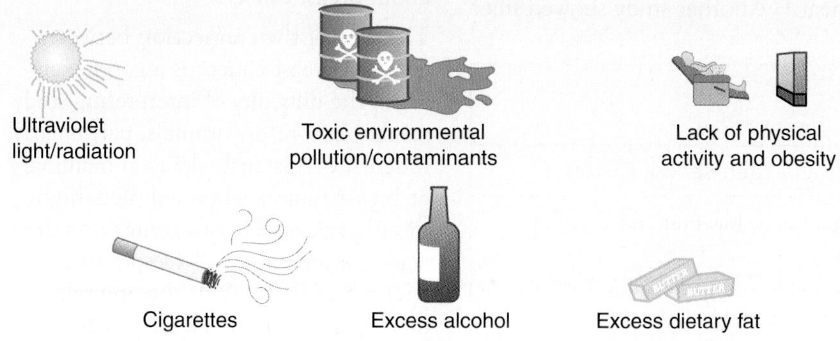

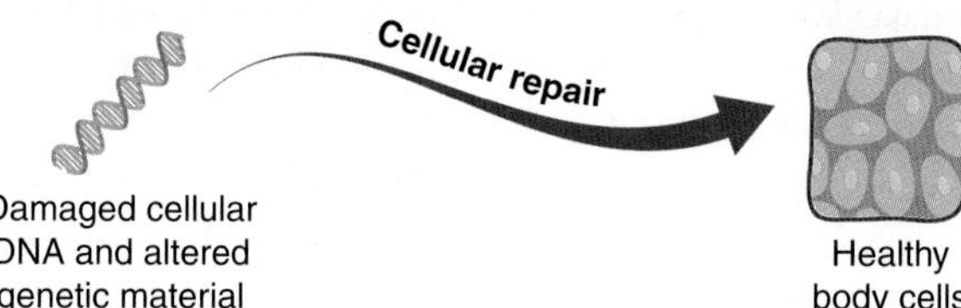

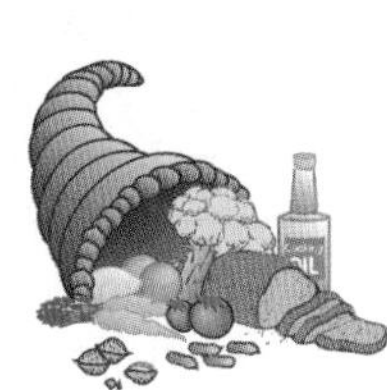

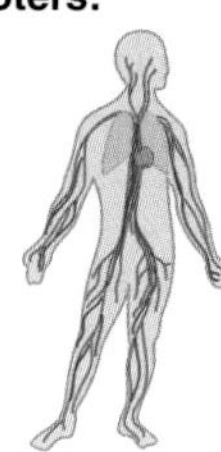

dental procedures, or contaminants (either naturally occurring or artificially created) present in the soil, air, and water. Of course, we also have no control over the genetic factors that contribute to the development of cancer.

Nutrition is a lifestyle factor that may account for as much as 35 percent of all cancer deaths.[39] Thus, researchers take the study of nutrition and cancer seriously. They are attempting to discover what dietary differences exist between people who do and do not get cancer. In the process, they are trying to identify the various dietary factors that may contribute to or protect against many different types of cancer, including cancer of the esophagus, stomach, liver, pancreas, colon, rectum, breast, ovary, prostate, and lung (see Table 6-16).

What is some of the evidence linking diet and cancer?

Studies of the eating habits of different population groups provide some of the knowledge we already have about the diet–cancer connection. Research involving Seventh-Day Adventists, a group of people with a remarkably low death rate from cancers of all kinds, is particularly telling. This religious group has rules against smoking and using alcohol, discourages the use of hot condiments and spices, and encourages a meatless diet. After cancers linked to smoking and alcohol are discounted, these people still have a cancer mortality rate about one-half to two-thirds that of the rest of the population. This may be due to a low consumption of meat (and therefore *fat*) and to a high consumption of fruits, vegetables, and cereal grains.

When it comes to colon cancer, studies lend weight to the theory that colon cancer is associated with indicators of affluence, such as a high-fat diet rich in animal protein.[40] In one study, people with colon cancer were compared with a carefully matched set of people without cancer. Those with cancer were found to have a strikingly higher consumption of meat, especially red meat.[41] Another study showed fiber consumption to be lower in colon cancer victims than in comparable people who did not have cancer. However, another study showed colon cancer victims to be eating about the same amount of fiber as people without cancer, indicating that the exact role of fiber in colon cancer prevention is yet to be determined.[42] Furthermore, among U.S. women between the ages of 34 and 59, the risk of colon cancer was related to their intake of animal (but not vegetable) fat, according to researchers at Harvard Medical School.[43] These examples are just a few of the many studies that have led researchers to the view that a diet high in total fat, especially animal fat, and possibly low in fiber is associated with cancer of the colon.[44] Other results suggest that beta-carotene, vitamin C, calcium, and folate or possibly other components of the foods that contain them, can help reduce the risk of colon cancer.[45]

So, diet is associated with the development of colon cancer. What about breast cancer?

The case of the connection between diet and breast cancer is a good example of the difficulty of interpreting study results. Laboratory animals, particularly rodents, consistently develop mammary or breast tumors when fed diets high in either vegetable oils (omega-6 fatty acids) or animal fats.[46] Despite the strength of this link in the animal model, similar results have not been seen consistently in human studies.[47]

A group of researchers in Athens, Greece, who studied 120 women with breast cancer and 120 women without cancer, have reported no association between breast cancer and the consumption of fats and oils.[48] By comparison, a research group in China reported that women with a high intake of fat and calories and a low intake of vegetables and dietary fiber had an increased risk of breast cancer.[49] As you can see from these apparently conflicting results, the role of dietary fat in breast cancer development remains unclear. Most studies of human populations fail to confirm an association between dietary fat and breast cancer. Labora-

TABLE 6-16
Diet and Cancer

Cancer Site	Associated with These Diet and Exercise Risk Factors
Breast	High intakes of calories and alcohol; obesity, low fruit and vegetable intake; low level of physical activity
Colon or Rectum	High red meat intake; excessive alcohol; low-fiber diets; low-calcium and vitamin D intake; obesity; low level of physical activity
Esophagus	Excessive alcohol intakes; low intakes of vitamins and minerals; obesity
Lung	Low fruit and vegetable intake; low level of physical activity
Ovary and Endometrium	High-fat diets; obesity; low fruit and vegetable intake; low level of physical activity
Prostate	High saturated fat diets may promote tumor growth; obesity; low level of physical activity
Pancreas	Low fruit and vegetable intake
Stomach	Regular consumption of smoked foods and foods cured with salt or nitrite compounds; low fruit and vegetable intake
Liver	Ingestion of *aflatoxin*-contaminated grains; regular consumption of smoked foods and foods cured with salt or nitrite compounds; alcohol abuse
Mouth and Throat	Excessive alcohol intake; low fruit and vegetable intake

SOURCES: *Food, Nutrition, and Prevention of Cancer: A Global Perspective* (Washington, D.C.: American Institute for Cancer Research, 1997); and Proceedings of the International Research Conference on Food, Nutrition, and Cancer, *The Journal of Nutrition* 132 (2002): 3449S–3534S.

tory studies, however, show that eating a low-fat diet can reduce estrogen levels and, theoretically, may reduce the risk of breast cancer. Many ongoing studies continue to examine the possible role of dietary fat in breast cancer.

Of course, excess fat in the diet can contribute to overweight, and recent studies have indicated that women who have gained excess weight as adults may have an increased risk of breast cancer.[50] Health experts advise us to limit weight gain during adulthood to no more than ten pounds and to get regular exercise. In fact, regular exercise throughout the reproductive years may significantly lower breast cancer risk. Body fat is involved in the production of estrogen. Therefore, by discouraging the accumulation of excess body fat (especially around the abdomen), exercise may decrease the total amount of estrogen a woman is exposed to over her lifetime.

What should consumers do?

The American Cancer Society and the National Cancer Institute have reviewed the evidence independently and have pointed out the following specific concerns.[51]

- *Total calorie intake*. Studies on animals show that reduced food intake reduces cancer incidence at any age, but the evidence is less clear for human beings. Obesity, however, does increase risks for some cancers in both animals and humans (refer to Chapter 9).[52]
- *Fat*. Both animal studies and population studies support the view that high fat intake increases the incidence of cancers of the ovaries, colon, and prostate.
- *Protein*. High protein intake may be associated with increased risks of certain kinds of cancer, but the evidence is not yet firm enough to permit a definitive statement.
- *Carbohydrate*. There is little evidence that carbohydrates as such play a role in cancer development.
- *Beta-carotene*. Inadequate intakes of beta-carotene correlate with a high incidence of cancers of the lung, bladder, and larynx; by inference, adequate intakes may help protect against these cancers.[53]
- *Vitamin C*. Vitamin C may help prevent the formation of cancer-causing agents and thereby protect against cancers of the esophagus and stomach.[54]
- *Cruciferous vegetables*. The consumption of cauliflower, cabbage, Brussels sprouts, broccoli, kohlrabi, and rutabagas is associated with a reduced incidence of cancer at several sites.[55] Cruciferous vegetables, a group of vegetables named for their cross-shaped blossoms, have been shown to protect against cancer in laboratory animals.
- *Other vitamins and minerals*. Bits of evidence suggest that other nutrients may protect against certain types of cancer, but no firm conclusions can yet be made. The effect of the antioxidant nutrients—vitamin C, beta-carotene, vitamin E, and selenium—on cancer risk is an area of active research.[56] The trace mineral selenium, used in the body's production of its own antioxidants, is believed to protect against cancer of the esophagus, stomach, colon, and rectum.[57] Likewise, vitamin E may protect against cancer, particularly cancer of the gastrointestinal tract.[58]
- *Calcium*. Low calcium intake has been associated with increased colon cancer. Conversely, people who consume more calcium tend to develop less colon cancer.[59]
- *Fiber*. Fiber may help protect against some cancers by, for example, speeding up the passage of all materials through the colon so that its walls are not exposed for long to cancer-causing substances.

Fat is linked to certain cancers, and fiber is associated with cancer prevention. Do vegetarians have a lower incidence of those cancers?

Yes, they do. The Seventh-Day Adventists have already been mentioned.

▲ *Cruciferous vegetables, such as cauliflower, broccoli, and Brussels sprouts, contain nutrients and nonnutrients that protect against cancer.*

Vegetarian women also have less breast cancer than do women who eat meat.

A number of studies have examined the relationship between cancer and vegetable consumption. Many of them have shown that people with colon cancer eat vegetables less frequently than do others. One study revealed that colon cancer victims specifically consumed less cabbage, broccoli, and Brussels sprouts than did people free of cancer. Similarly, comparisons of stomach cancer victims' diets with those of carefully matched people without cancer showed lower consumption of vegetables in the cancer group—in one case, vegetables in general; in another case, fresh vegetables; in still others, lettuce and other fresh greens or vegetables containing vitamin C. Some of the suspects for causing stomach cancer are chemicals known as nitrosamines, which are produced in the stomach and intestines from nitrites found in foods. Vegetables may help in cancer prevention by contributing vitamin C, which inhibits the conversion of nitrites to nitrosamines.[60]

Another healthy aspect of many plant-based diets is the use of soy foods. A number of studies conducted in China and Japan have shown that consumers of tofu, soy milk, and other soy foods have lower cancer rates than those who rarely consume these foods (see the Spotlight feature in Chapter 5).

The data suggest that consuming one or two servings of soy a day (for example, 1 cup of soy milk or 4 ounces of tofu) may reduce risk of breast, prostate, lung, and colon cancers.[61]

What about alcohol? Is it connected with cancer?

Yes. Environmental causes of head, neck, and esophageal cancer have been studied, and the major factor appears to be the combination of alcohol and tobacco use. However, dietary factors have turned up, pointing to a low intake of fruits and raw vegetables, specifically of the fruits and vegetables that contribute the orange pigment beta-carotene (which converts to vitamin A in the body) and the vitamin riboflavin. Beta-carotene—noted for giving carrots, winter squash, sweet potatoes, apricots, cantaloupe, and other fruits and vegetables their familiar colors—and its relatives may also be important in reducing the risk of skin cancer.

Researchers are currently probing the possible association of alcohol consumption with increased risk of breast cancer.[62] Although some studies have shown a modest but significant increase in risk of breast cancer for women who were classified as heavy drinkers (more than 40 grams of alcohol or more than four drinks per day), other studies have found no relationship.[63] More research is needed to clarify this issue. However, if a causal relationship between relatively heavy drinking and increased risk of breast-cancer is confirmed, such evidence would lend additional support to the societal benefits derived from not consuming excessive amounts of alcohol.

What is the association between beta-carotene and other carotenoids and cancer?

Among the known actions of vitamin A (beta-carotene) are its important roles in maintaining immune function. A strong immune system may be able to prevent cancers from gaining control even after they have been initiated in the body.

The carotenoids are a family of powerful antioxidants found in dark yellow, orange, and red fruits and vegetables and in dark green vegetables. They are potent free-radical fighters that may play important roles in preventing cancer. The family includes alpha- and beta-carotene, lycopene, lutein, zeaxanthin, and others. Alpha- and beta-carotene appear to protect against the progression of cancer, whereas other carotenoids may offer more protection at earlier stages of cancer development. Carotenoids may work against cancer by boosting the immune system and supporting the enzymes that detoxify carcinogens. Beta-carotene and lycopene may be effective in preventing damaged cells from proliferating and becoming malignant. Some scientists now believe lycopene, which is found in red fruits and vegetables such as tomatoes, tomato products, red peppers, and watermelon, is the most powerful of the antioxidant carotenoids. Research is currently under way to examine a link between lycopene and lowered risk of prostate and other cancers.[64] Meanwhile, we are advised to consume five to nine servings of a variety of fruits and vegetables each day. Be sure to include deep green and brightly colored fruits and vegetables every day. By doing so, your ability to repair free-radical damage in the body is likely to be enhanced by the carotenoids or possibly by something else in these plant foods.

Would that "something" in the vegetables be a vitamin?

Not necessarily. Both fruits and vegetables appear to have a protective effect beyond those already discussed for beta-carotene, vitamin C, and fiber. Researchers have identified substances known as *phytochemicals*—naturally occurring plant compounds—in fruits and vegetables that may play a role in decreasing cancer risk and strengthening the immune system.[65] (Refer to Table 6-14 for a sampling of these compounds.) These vegetables also contain folate, a vitamin, which is involved in cell multiplication, and may prove to play an important role in colon and cervical cancer prevention.[66] The effects of the members of the cabbage family may be due to substances known as indoles, a type of phytochemical, which may act by inducing an enzyme in the host that destroys cancer-causing agents.

Do these findings have any implications for the way a person should eat right now, today?

Although clearly there is still much to learn, many experts believe that enough is known to take the first preventive steps. An exhaustive review of some 4,500 previous studies on cancer has led scientists at The American Institute for Cancer Research to release a comprehensive set of recommendations for cancer prevention.[67] To put these recommendations into action, consider these simple steps:

1. Eat a variety of healthful foods, with an emphasis on plant sources.
   - Eat five or more servings of a variety of vegetables and fruits each day.
   - Choose whole grains in preference to processed (refined) grains and sugars.
   - Limit consumption of red meats, especially high-fat meats and processed meats.
   - Choose foods that help maintain a healthful weight.
2. Adopt a physically active lifestyle.
   - Engage in at least moderate activity for 30 minutes or more on 5 or more days of the week; 45 minutes or more of moderate to vigorous activity on 5 or more days per week may further enhance reductions in the risk of breast and colon cancer.
3. Maintain a healthful weight throughout life.
   - Balance caloric intake with physical activity.
   - Lose weight if currently overweight or obese.

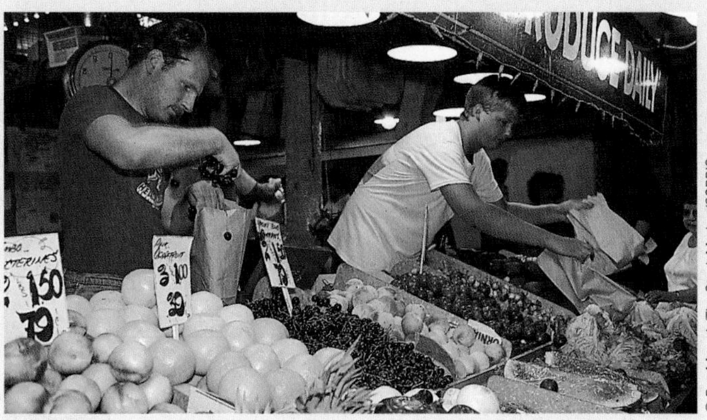

© Roy Morsch/The Stock Market/CORBIS

▲ *The National Cancer Institute's National 5 A Day for Better Health program urges consumers to eat five or more servings of fruits and vegetables rich in fiber, vitamins, minerals, antioxidants, and phytochemicals every day to help protect against cancer.*

4. If you drink alcoholic beverages, limit consumption.
5. Prepare and store food safely.
6. Do not use tobacco in any form.

Your lifestyle choices regarding diet, exercise, and smoking can be powerful tools in reducing your risk for cancer (see Figure 6-8 on page 209). Taken together, this advice agrees with most recommendations made to the public for helping prevent heart disease, diabetes, and many other ills, as well as cancer.

Public efforts are now under way to reduce the major risk factors for cancer.[68] The National Cancer Institute (NCI) designed its National 5 A Day for Better Health program to increase per capita fruit and vegetable consumption. The NCI's 5 A Day program promotes a simple nutrition message: *Eat five or more servings of fruits and vegetables every day for better health.*[69]

Does it make a difference if I choose to take supplements instead of eating vegetables?

Yes. Fiber, vitamin C, beta-carotene, riboflavin, indoles, and other phytochemicals found in *foods* appear to have preventive effects on cancer development. And because it is obvious that researchers do not have the answer—just many partial answers—it is best to stick with foods. Supplements may not contain some as yet unidentified components found in foods that may help to protect against cancer. Also, with vitamin/mineral supplementation, there is always the risk of an excessively high intake. (Recall that vitamin A, selenium, and other vitamins and minerals can be toxic at high doses.) *It is best to rely on foods.* Fruits and vegetables add vitamins, minerals, and fiber to your diet—as well as color and flavor.

Are any anticancer benefits derived from drinking tea?

Scientists note that people in Asia, where large quantities of tea—both green and black—are consumed, have a lower incidence of esophageal, colon, stomach, and other cancers. Laboratory studies suggest that polyphenols—phytochemicals having potent antioxidant properties—found in tea may help protect against these cancers by blocking the formation of carcinogenic compounds in the body. However, the implications for humans are not yet known.

Does cooking food at high temperatures increase the risk of cancer?

Some research suggests that frying or charcoal-broiling meats—especially fatty meats—at very high temperatures creates chemicals on the surface of the food that may increase cancer risk. Preserving meats by methods involving smoke also increases their content of potentially carcinogenic chemicals. We are advised to eat grilled, smoked, or cured foods only occasionally. When you do, consume them with fruits and vegetables that contain protective antioxidant and phytochemical factors. Choose techniques such as braising, poaching, stewing, baking, or roasting instead.

Why should I vary my diet?

The recommendation to eat a varied diet is based on an important cancer-prevention strategy—dilution. The standard advice to eat a variety of foods takes on new meaning in this quote: "The wider the variety of food intake, the greater the number of different chemical substances consumed, and the less is the chance that any one chemical will reach a hazardous level in the diet."[70] In other words, when you add new foods to the diet, you dilute what is in one food with what is in another.

The variety principle traditionally means to eat foods from each of the various food groups. This principle must also be applied within each of the food groups. Don't alternate just between corn and potatoes. Select different vegetables each time you go to the store—broccoli, peas, green beans, squash, and many others.

Although you cannot control many cancer-causing factors, you can decide which food habits you will keep and which ones you will change. By using these guidelines in making your choices, you will have every reason to feel confident that you are providing your body with the best nutrition at the lowest possible risk. Remember that in the final analysis, your risk of developing cancer can be reduced significantly by not smoking, consuming alcohol in moderation if at all, and by adopting healthful eating and exercise habits.

FIGURE 6-8
FACTORS THAT DECREASE OR INCREASE CANCER RISK

*From: Cancer Facts and Figures, 2005; available at www.cancer.org.

SOURCE: Adapted from World Cancer Research Fund and American Institute for Cancer Research, *Food, Nutrition, and the Prevention of Cancer: A Global Perspective* (Washington, D.C.: American Institute for Cancer Research, 1997), 506–507; Proceedings of the International Research Conference on Food, Nutrition, and Cancer, *The Journal of Nutrition* Vol. 132 (S), November, 2002.

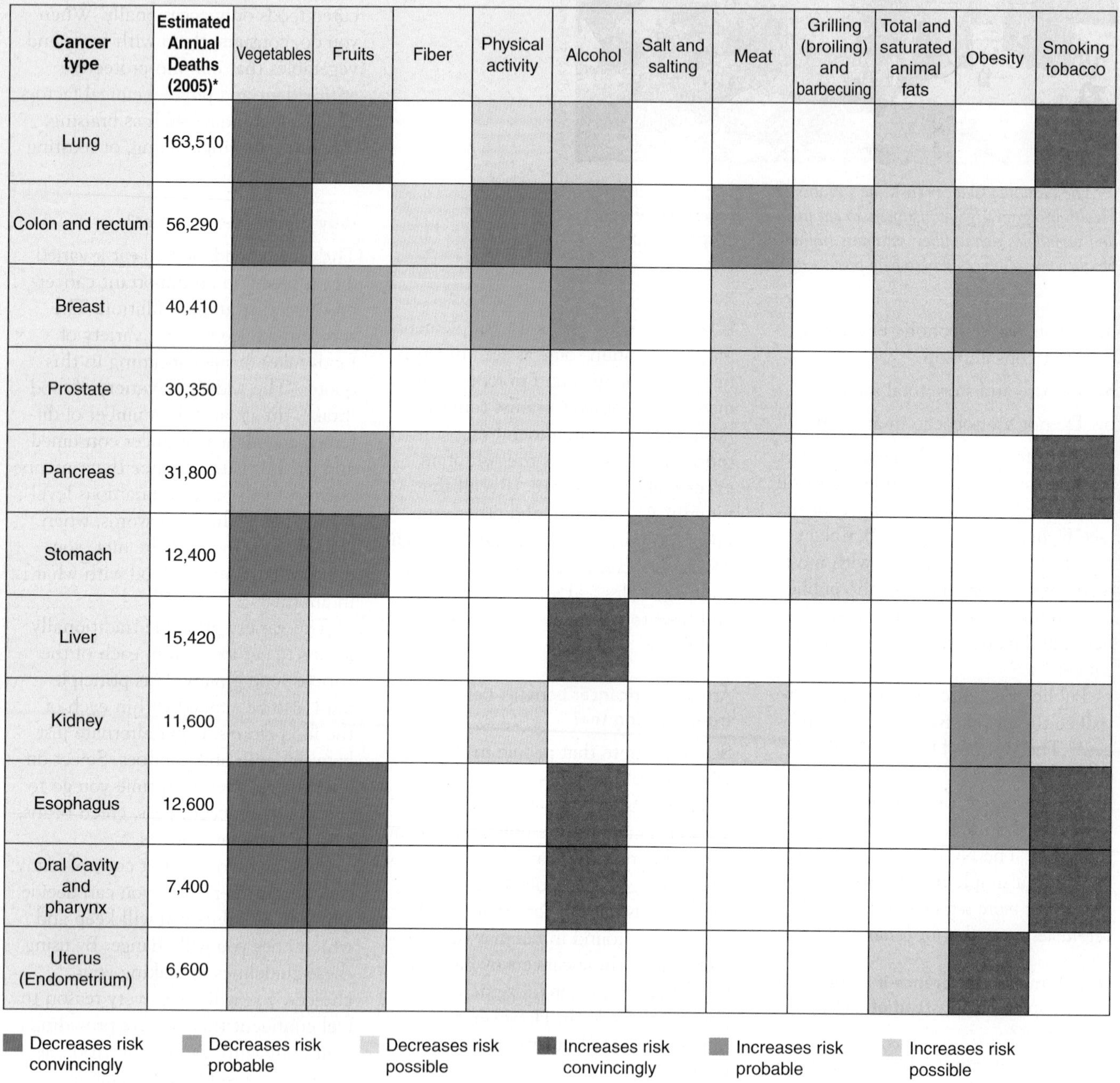

Cancer type	Estimated Annual Deaths (2005)*	Vegetables	Fruits	Fiber	Physical activity	Alcohol	Salt and salting	Meat	Grilling (broiling) and barbecuing	Total and saturated animal fats	Obesity	Smoking tobacco
Lung	163,510											
Colon and rectum	56,290											
Breast	40,410											
Prostate	30,350											
Pancreas	31,800											
Stomach	12,400											
Liver	15,420											
Kidney	11,600											
Esophagus	12,600											
Oral Cavity and pharynx	7,400											
Uterus (Endometrium)	6,600											

■ Decreases risk convincingly ■ Decreases risk probable ■ Decreases risk possible ■ Increases risk convincingly ■ Increases risk probable ■ Increases risk possible

In Review

1. Describe some general differences between water-soluble and fat-soluble vitamins.

2. Vitamin deficiencies:
   a. are easily corrected because the body creates the vitamins it needs.
   b. can be fatal.
   c. have physical and metabolic consequences.
   d. cannot be reversed.
   e. Both b and c are correct.

3. Which vitamin can be made by the body?
   a. Vitamin A
   b. Vitamin E
   c. Vitamin C
   d. Vitamin D
   e. The body can make none of the vitamins.

4. Which of the following factors can destroy vitamins in the foods we prepare and store at home?
   a. oxygen
   b. water
   c. heat
   d. ultraviolet light
   e. all of the above

5. Which of the following substances may have anti-carcinogenic properties (protect against cancer)?
   a. beta-carotene
   b. vitamin C
   c. vitamin E
   d. all of the above

6. Vitamin D helps your body use calcium more effectively. You can get vitamin D from:
   a. being outdoors in the sunlight.
   b. drinking vitamin D fortified milk.
   c. eating broccoli.
   d. a and b
   e. a, b, and c

7. Which of the following is *not* a B vitamin?
   a. folate
   b. biotin
   c. pellagra
   d. riboflavin

8. Briefly explain what a phytochemical is and how it is used in the body.

9. Which of the following is *not* an effect of any vitamin?
   a. provides calories to the body
   b. helps release energy
   c. acts as an antioxidant
   d. regulates blood clotting

10. Name the two classes of vitamins and list the vitamins in each group.

Menu of Online Study Tools

A variety of study tools for this chapter are available at our website to deepen your understanding of chapter concepts. Go to
www.thomsonedu.com/nutrition/boyle
to find

- Practice tests
- Flashcards
- Glossary
- Web links
- Animations
- Chapter summaries, learning objectives, and crossword puzzles

Nutrition on the Web

www.thomsonedu.com/nutrition/boyle
Go to the *Personal Nutrition* site to check for the latest updates to chapter topics or to access links to related websites.

www.healthfinder.gov
Search for reliable consumer information on individual vitamins.

www.iom.edu/fnb
Search for updates regarding new Dietary Reference Intakes.

www.nal.usda.gov/fnic
Search USDA's Food and Nutrition Information Center for individual vitamins, food composition, and vitamin-related topics.

www.nlm.nih.gov
Free access to national Library of Medicine's Medline for information searches on a variety of health-related topics.

www.eatright.org
The American Dietetic Association's site with position papers on vitamin supplements, functional foods, and many resources.

www.5aday.org
Information about the National 5 A Day program.

http://vm.cfsan.fda.gov/~dms/supplmnt.html
FDA Center for Food Safety and Applied Nutrition; search for supplements.

http://ods.od.nih.gov
Access information from the National Institutes of Health—Office of Dietary Supplements.

www.herbs.org
Information on herbs from the Herb Research Foundation.

http://nccam.nih.gov
National Center for Complementary and Alternative Medicine.

www.consumerlab.com
Evaluates consumer products relating to health, including vitamins, minerals, herbal products, ergogenic aids, other supplements, and functional foods.

www.cancer.org
Information about cancer from the American Cancer Society.

www.aicr.org
Information and research updates about cancer.

Water and the Minerals

7

ASK YOURSELF . . .

Which of the following statements about nutrition are true, and which are false? For each false statement, what is true?

1. Calcium is the most important mineral in human nutrition.
2. Milk is nature's most nearly perfect food because it is rich in every nutrient.
3. It is generally harder for women than for men to obtain diets that are adequate in calcium.
4. Milk is necessary for children, but adults can find replacements for it.
5. Sodium is bad for the body and should be avoided.
6. When a person becomes deficient in iron, the very first symptom to appear is anemia.
7. Zinc is toxic in excess.
8. Both too little and too much iodine in the diet can cause swelling of the thyroid gland, known as goiter.
9. A diet high in salt is associated with high blood pressure in some individuals.
10. Osteoporosis is a disease that can affect men and women at any age.

Answers found on the following page.

The pleasure of eating . . . is of all times, all ages, all conditions. . . . Because it may be enjoyed with other enjoyments, and even console us for their absence. . . . Because its impressions are more durable and more dependent on our will. . . . Because in eating we experience a certain indescribably keen sensation of pleasure, by what we eat we repair the losses we have sustained, and prolong life.

Anthelme Brillat-Savarin (1755–1826, French politician and gourmet; author of *Physiology of Taste*)

CONTENTS

Throughout history humans have known that water is indispensable to life itself as well as essential for health. Likewise, for hundreds, if not thousands, of years, the physical and chemical properties of minerals such as gold, silver, lead, and copper were known to metallurgists and alchemists. Even so, the role of some minerals in biological processes was recognized only within the past few hundred years. The discovery of iron in blood, for example, occurred in 1713. The identification of calcium in bone was made in 1771.[1] The role of minerals in human nutrition was not fully appreciated until the late 19th century.

The previous chapter described the water-soluble and fat-soluble vitamins, their biological roles, food sources, and human requirements. This chapter discusses the minerals known to be important in human nutrition. In some respects, minerals are similar to vitamins. Like the vitamins, minerals themselves do not contribute energy (calories) to the diet. Of the minerals important in human nutrition, most have diverse functions within the body and work with enzymes to facilitate chemical reactions. Also like the vitamins, most minerals are required in the diet in very small amounts.

In other respects, minerals are different from vitamins. Whereas vitamins are organic compounds, **minerals** are **inorganic** compounds that occur naturally in the earth's crust. And unlike the vitamins, some minerals (such as calcium) contribute to the building of body structures (such as bone).

minerals small, naturally occurring, inorganic, chemical elements; the minerals serve as structural components of the body and in many vital body processes.

inorganic being or composed of matter other than plant or animal.

As with other areas of research in human nutrition, many questions related to mineral metabolism remain unanswered. Scientists continue to study the biochemical functions of minerals, the mechanisms by which they activate enzymes, the factors that control their blood and tissue levels, and the ways in which the composition of the diet affects the body's ability to make use of these dietary components. The many complex metabolic interactions of minerals make this a challenging area of research.

Although we can survive for months or even years without some vitamins and minerals, we can last only a few days without water. Thus, this chapter begins with a discussion of water—the most essential nutrient of all.

Water—The Most Essential Nutrient

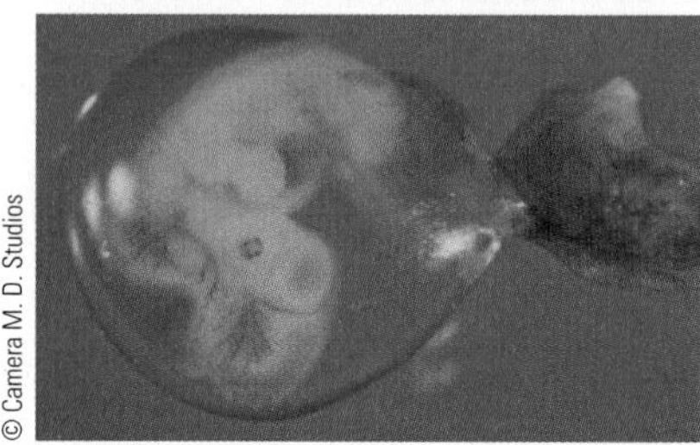

▲ *Life begins in water.*

Although we often take it for granted, water is by far the nutrient that is most needed by the body. Consisting entirely of a combination of hydrogen and oxygen atoms, water makes up part of every cell, tissue, and organ in the body and accounts for about 60 percent of body weight (see Figure 7-1), even contributing to body parts thought of as "dry." For example, bone is more than 20 percent water, muscle is 75 percent water, and teeth are about 10 percent water.

FIGURE 7-1
WATER—THE NUMBER ONE NUTRIENT

SOURCE: C. Lecos, "Water: The Number One Nutrient," *FDA Consumer* (November 1983).

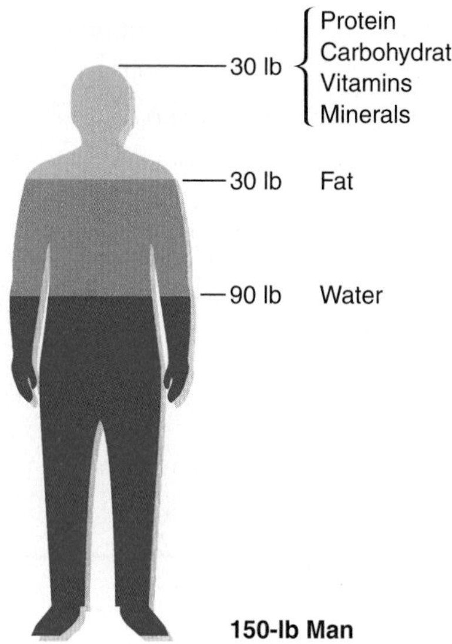

Inside the body, water performs many tasks vital to life (see Table 7-1), such as helping to transport the nutrients needed to nourish the cells. The blood is a river of water that flows through the arteries, capillaries, and veins, bringing each cell the exact substances and particles it requires. This same river carries away waste products formed during the reactions that take place in the cells. In addition, water acts as a shock absorber in joints and around the spinal cord, lubricates the digestive tract as well as all the tissues moistened with mucus, surrounds and cushions an unborn child, and plays a key role in maintaining body temperature.

TABLE 7-1
The Functions of Water in the Body

- Transports nutrients
- Carries away waste
- Moistens eyes, mouth, and nose
- Hydrates skin
- Ensures adequate blood volume
- Forms main component of body fluids
- Participates in many chemical reactions
- Helps maintain normal body temperature
- Acts as a lubricant around joints
- Serves as a shock absorber inside the spinal cord and in the amniotic sac surrounding a fetus

Water and Exercise

The water in your blood—known as *plasma volume*, or just *plasma*—serves a function similar to the function of the water in your car's radiator. This particular function of water is critical to humans during exercise because as the blood continually circulates throughout your body, it picks up the tremendous amount of heat generated by your working muscles. The plasma then transports this heat to your skin, through which it is expelled from the body primarily by evaporation of sweat. You can think of sweating as your body's air-conditioning system. As sweat evaporates from your skin, it expels large amounts of heat, helping to keep your body cool. However, sweating only cools your body when the sweat evaporates from your skin. If the sweat simply rolls down your face or down your back, body heat is not released. On a humid day, for example, the air is already saturated with water, impairing the evaporation of sweat. Hot, humid days, then, are doubly dangerous because you continue to sweat and lose precious body water, but your body temperature doesn't fall. When exercising on humid days, it is extremely important to pay particular attention to your fluid needs.

Ask Yourself Answers: **1.** False. No one essential mineral is more important than any other. **2.** False. Milk is an excellent food, but it is poor in several nutrients, including iron. **3.** True. **4.** False. Strictly speaking, milk is not absolutely necessary in anyone's diet, but its nutrients are hard to obtain from other foods, and it is recommended for both children and adults. **5.** False. Sodium is an essential nutrient, but excess sodium should be avoided. **6.** False. When a person becomes deficient in iron, one of the last symptoms to appear is anemia; fatigue and weakness appear first. **7.** True. **8.** True. **9.** True. **10.** True.

Thus, as your body heats up due to the energy released during exercise, it loses water by sweating. How much sweat you lose depends on the intensity and duration of the activity. The more intense the exercise, the more heat you generate and the more you sweat. If you don't replace this lost water, your plasma volume will decrease. In its attempt to maintain plasma volume, your body will pull water from your muscles and organs. As water is pulled from muscles, cramps may occur, along with premature fatigue and a noticeable decline in performance.

When plasma levels are low (meaning that blood volume is lower), your heart is forced to beat faster to supply sufficient oxygen to your muscles. Finally, because less plasma is circulating to transport the heat to your skin, the heat builds up, and your body's internal temperature continues to rise. All these changes force your body to work at a higher intensity, leading to early exhaustion. A water loss equal to 2 percent of body weight can reduce muscular work capacity by 20 to 30 percent.

The recommended amount of fluid sufficient to prevent dehydration and **heat stroke** can be surprising. Athletes can lose two or more quarts of fluid during every hour of heavy exercise, and they must rehydrate before, during, and after exercise to replace the lost fluid. Even casual exercisers must drink some fluids while exercising. Unfortunately, thirst is not a reliable indicator of how much to drink because it signals too late, after fluid stores are already depleted.[2] Chapter 10 presents one schedule of hydration before, during, and after exercise (see page 331). To know how much water is needed to replenish fluid losses after a workout, weigh yourself before and after—the difference is all water. One pound equals about 2 cups of fluid.

Water in the Diet

Adults consume and excrete some 1½ to 3 quarts of water a day (see Figure 7-2). Although most of the water we take in comes from juice, milk, soft drinks, and other beverages, including tap water, foods also add considerable amounts of water to the diet. For example, as much as 85 to 95 percent of fruits and vegetables is water.[3]

Just as the sources of water in our diet vary, so too does the water we drink "straight." The makeup of water differs depending on where it comes from and how it is processed—variations that can have significant health implications. One of the most basic distinctions, hard versus soft water, is based on the concentrations of three minerals: calcium, magnesium, and sodium. **Hard water** usually comes from shallow ground and contains relatively high levels of minerals,

heat stroke an acute and dangerous reaction to heat buildup in the body, requiring emergency medical attention; also called *sun stroke*.

hard water water with a high concentration of minerals such as calcium and magnesium.

Signs of heat stroke include:

- Very high body temperature (104° Fahrenheit or higher)
- Hot, dry, red skin
- Sudden cessation of sweating
- Deep breathing and fast pulse
- Blurred vision
- Confusion, delirium, hallucinations
- Convulsions
- Loss of consciousness

To prevent heat stroke, drink plenty of fluid before, during, and after exercise; avoid over-exercising in hot weather; and stop exercising at any sign of heat exhaustion.

Signs of heat exhaustion include:

- Cool, clammy, pale skin
- Dizziness
- Dry mouth
- Fatigue/weakness
- Headache
- Muscle cramps
- Nausea
- Sweating
- Weak and rapid pulse

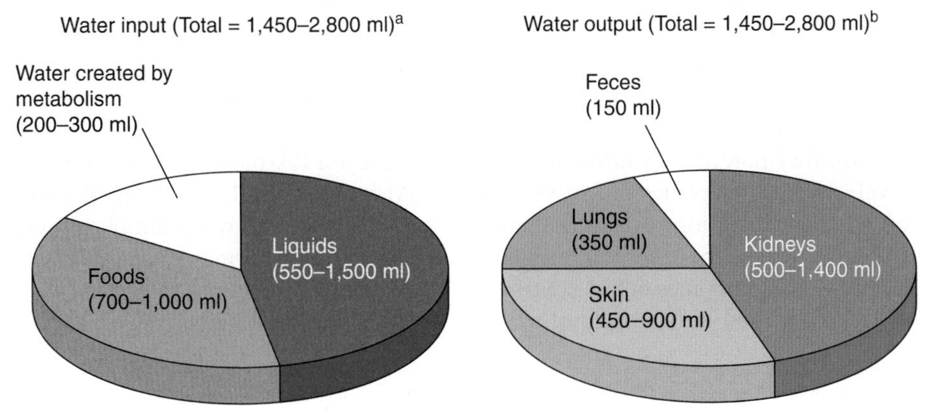

FIGURE 7-2
WATER BALANCE IN THE BODY

Water enters the body in liquids and foods, and some water is created in the body as a by-product of metabolic processes. Water leaves the body through the evaporation of sweat, in the moisture of exhaled breath, in the urine, and in the feces.

[a] This amount equals 1½ to 3 quarts (1 oz equals approximately 30 ml).

[b] Adults are advised to consume 1.0 to 1.5 ml of water from all sources for each calorie expended. For instance, if you require 2,000 calories a day, you need approximately 2 quarts of fluid.
2,000 cal × 1 ml/cal = 2,000 ml
2,000 ml ÷ 30 ml/1 oz = 67 oz
67 oz ÷ 32 oz/qt = 2 qt

SOURCE: Adapted from F. Sizer and E. Whitney, *Nutrition: Concepts and Controversies*, 10th ed. (Belmont, CA: Thomson Wadsworth, 2006): 267. Information on fluid requirements from S. Kleiner, "Water: An Essential but Overlooked Nutrient," *Journal of the American Dietetic Association* 99 (1999): 200.

soft water water containing a high sodium concentration.

primarily calcium and magnesium. **Soft water,** on the other hand, generally flows from deep in the earth and has a higher concentration of sodium.

Although your water utility company can tell you whether your water is soft or hard, you can probably distinguish between the two based on your own experience. Soft water helps soap lather better than hard water and leaves less of a ring on the bathtub. Hard water, to the contrary, doesn't clean clothes as thoroughly as soft water and leaves a residue of rocklike crystals on the inside of the teakettle over time. That's why many consumers prefer soft water.

From a health standpoint, however, hard water seems to be the better alternative.[4] One reason is that the excess sodium carried in soft water, even in small amounts, adds more of the mineral to our already sodium-laden diets. More importantly, soft water dissolves potentially toxic substances such as lead from pipes. Therefore, people who install water softeners in their homes for the purpose of getting cleaner laundry and better mileage from soap would do well to connect them only to their hot water lines. That way, they can use hot, soft water for washing and bathing and use cold, hard water for drinking and cooking.

Keeping Water Safe

In addition to having varying concentrations of minerals, water taken from the earth contains different levels of bacteria, microorganisms, and heavy metals such as lead. To ensure that the water that flows from the tap is safe to drink, the Environmental Protection Agency (EPA), the arm of the U.S. government responsible for monitoring municipal water supplies, sets limits for potential contaminants such as mercury, nitrate, and silver in drinking water. By law, the public must be notified by a water utility company within 24 hours of discovering any potentially dangerous contaminants in drinking water. The EPA also mandates that tap water be disinfected if bacteria levels run high to prevent the spread of waterborne diseases such as typhoid and dysentery. Such precautionary measures go a long way in keeping the U.S. water supply one of the safest in the world.

One potential contaminant is a parasite called *Cryptosporidium*. Found in lakes and rivers that have come into contact with sewage or animal waste, *Cryptosporidium* has emerged as a health threat to vulnerable people because it is highly resistant to chlorine and other disinfectants used in municipal water supplies. In healthy people, the parasite can cause diarrhea and other flu-like symptoms that typically subside within a week to 10 days. In people with weak immune systems, however, this parasite can lead to severe, long-lasting gastrointestinal problems that can even cause death.

Because of the *Cryptosporidium* hazard, the Environmental Protection Agency and the Centers for Disease Control and Prevention recommend special precautions for people with severely weakened immune systems—those with HIV infection, cancer and transplant patients taking immunosuppressive drugs, and people born with a weakened immune system. Public health groups advise such individuals to talk to a health care provider about boiling tap water for 1 minute (if it is meant to be used for drinking water) to destroy any *Cryptosporidium* present, or about buying a special water filter system.[5]*

Another potential health threat over which the EPA has little control is the level of lead that comes out of your faucet. Although the EPA has put a ceiling on the concentration of lead that may be present in public water supplies, once the water leaves the reservoir, unhealthy levels of the metal may dissolve into it. Before the water comes out of your faucet, it may flow through pipes that are made of lead or joined by lead solder, which can leach the metal into the water as it

*For more information, visit the EPA Drinking Water Homepage at www.epa.gov. To obtain a list of suitable filters, contact NSF International at info@nsf.org or visit their website at www.nsf.org.

passes through. Granted, the U.S. government banned the use of lead-containing plumbing systems back in 1986, but dwellings built before that time may not have lead-free pipes.

The issue of possible lead-contaminated water has generated a good deal of recent publicity and concern. Certainly, the attention is warranted, given that when lead accumulates in the body, it begins to damage the nerves, kidneys, and liver along with the cardiovascular, reproductive, immunologic, and gastrointestinal systems. Lead is especially toxic to children and fetuses, in whom it can cause neurologic problems as severe as brain damage.[6]

Luckily, lead levels can usually be kept to a minimum simply by "flushing" the tap, that is, letting the water run until it becomes as cold as possible, thereby ridding it of water that has been sitting in pipes and dissolving lead for any length of time. Pipes should be flushed with cold water only, because cold water is less likely than warm or hot water to dissolve lead as it flows through them. To be sure, in some instances, flushing is not enough to reduce harmful lead levels. Some homes or apartments may need special water treatment systems, the necessity of which can only be determined by subjecting tap water to a lead test. Names of certified laboratories that analyze water can be obtained from a local branch of the EPA, or by calling the EPA's Safe Drinking Water Hotline (800–426–4791).

Bottled Water

Bottled water has become one of America's most popular beverages today, at a total cost of more than $6 billion a year.[7] Although the reasons for this trend are many, bottled water's perceived health benefits fall near the top of the list. Surveys have found that about 25 percent of bottled water drinkers choose the beverage for health and safety reasons, and another 25 percent believe that bottled water is pure and free of contaminants.[8]

Regardless of its pristine image, bottled water is not necessarily any more pure or healthful than water that flows out of the tap. Consider that the Food and Drug Administration (FDA), the bottled water industry watchdog, does not require that bottled water meet higher standards for quality, such as the maximum level of contaminants, than public water supplies regulated by the EPA. For the most part, the FDA simply follows EPA's regulatory lead. Typically, however, bottled water is filtered to remove chemicals such as chlorine that may impart a certain taste. But that doesn't make it any safer. In fact, about 25 to 40 percent of bottled water comes from the same municipal water supplies as tap water.[9] Furthermore, some bottled waters do not contain any or enough of the fluoride needed to fight cavities.[10] The only way to determine whether a certain water contains the mineral is to check with the company that bottles it. If you use bottled water, look for the trademark of the International Bottled Water Association (IBWA)—a trade organization that supports FDA regulations.

This is not to say that bottled water is necessarily any better or worse, from a health standpoint, than tap water. It's certainly preferable to tap water for those who like its taste. The problem is that the demand for bottled water increases the use of non-renewable resources to make the plastic bottles, which in turn creates an energy demand for recycling the used bottles. Additionally, many consumers pay 250 to 10,000 times more per gallon for bottled water than for tap water because they think bottled water is more healthful. Bottlers add to the confusion by sprinkling terms such as "pure," "crystal pure," and "premium" on labels illustrated with pictures of glaciers, mountain streams, and waterfalls, even when the water inside comes from a public reservoir. However, the FDA has set forth regulations mandating clear labeling of bottled waters. The miniglossary for bottled water explains what some of the terms used on bottles actually mean.

MINIGLOSSARY FOR BOTTLED WATER

© Michael A. Keller 2000/The Stock Market/CORBIS

artesian water or **artesian well water** water drawn from a well that taps a confined water-bearing rock or rock formation.

ground water water that comes from an underground body of water that does not come into contact with any surface water.

mineral water water that is drawn from an underground source and contains at least 250 parts per million of dissolved solids. If the water contains between 250 and 500 parts per million total dissolved solids, the statement "low mineral content" must appear. If it contains more than 1,500 parts per million, the statement "high mineral content" must appear. If a cup of the water contains at least 20 milligrams of calcium, 0.36 milligram of iron, or 5 milligrams of sodium, the product must carry nutrition labeling.

purified water (also known as **demineralized water, distilled water, deionized water,** or **reverse osmosis water**) water from which all the minerals have been removed, thereby eliminating the possibility that the minerals might corrode, say, a steam iron.

sparkling bottled water water whose carbon dioxide (the ingredient that makes soda pop bubbly) is naturally present. That is, carbonation is not added from an outside source.

spring water water derived from an underground formation from which water flows naturally to the surface of the earth and to which minerals have not been added or taken away. It may be collected either at the spring itself or through a hole tapping the underground formation feeding the spring.

well water ground water derived from a rock formation by way of a hole bored, drilled, or otherwise constructed in the ground.

from a community water system or from a municipal source statement that must appear on bottles containing water derived from a municipal water supply. The phrase must conspicuously precede or follow the name of the brand.

seltzer* tap water injected with carbon dioxide and containing no added salts.

club soda* artificially carbonated water containing added salts and minerals.

tonic water* artificially carbonated water with added sugar and/or high-fructose corn syrup, sodium, and quinine.

*These items are not considered bottled water in government parlance. The FDA defines bottled water as water that is sealed in bottles or other containers and is intended for human consumption, excluding soda, seltzer, flavored, and vended water products.

The Major Minerals

major mineral an essential mineral nutrient found in the human body in amounts greater than 5 grams.

trace mineral an essential mineral nutrient found in the human body in amounts less than 5 grams.

Nutritionists traditionally divide minerals into two large classes: the **major minerals** and the **trace minerals.** The distinction between them is that the major minerals occur in relatively large quantities in the body and are needed in the daily diet in relatively large amounts—on the order of a gram or so each. The trace minerals occur in the body in minute quantities and are needed in smaller amounts in the daily diet. Figure 7-3 offers tips for locating the minerals in the

FIGURE 7-3
GOOD SOURCES OF MINERALS IN THE USDA MYPYRAMID FOOD GUIDE[a]

[a] Serving sizes shown here are based on a 2,000 calorie diet. Go to www.mypyramid.gov for a quick estimate of how much food you should eat from the different food groups based on your age, gender, and activity level.
[b] In whole-grain choices.

GRAINS	VEGETABLES	FRUITS	MILK	MEAT & BEANS
6 oz	2 1/2 cups	2 cups	3 cups	5 1/2 oz
chromium[b], iron, zinc, magnesium[b], sodium, sulfur, chloride, iodine, copper[b], potassium[b], phosphorus	iron, magnesium, selenium, calcium, potassium, sodium	potassium, magnesium, iron	calcium, phosphorus, magnesium, sodium, zinc, potassium, iodine	chromium, iron, copper, sulfur, iodine, zinc, phosphorus, magnesium, sodium, potassium, fluoride

MyPyramid food guide, and Table 7-2 lists the minerals known to be essential in human nutrition. The discussions that follow focus on the minerals of particular interest in human nutrition, primarily because deficiencies in these minerals can cause human disease and suffering.

Calcium

Calcium is the most abundant mineral in the body. Ninety-nine percent of the body's calcium is found in the bones, which play two important roles. First, the bones support and protect the body's soft tissues. Second, they serve as a calcium bank, providing calcium to the body fluids whenever the supply is running low.

Although only a small part (about 1 percent) of the body's calcium is in its fluids, circulating calcium is vital to life. Calcium is required for the transmission of nerve impulses. It is essential for muscle contraction and thus helps maintain the heartbeat. It appears to be essential for the integrity of cell membranes and for the maintenance of normal blood pressure. Calcium must also be present if blood clotting is to occur, and it is a **cofactor** for several enzymes.

Everyone knows that children need calcium daily to support the growth of their bones and teeth, but not everyone is aware of adults' needs for daily intakes of calcium. Abundant evidence now supports the importance of calcium for adults, especially women, who need about as much calcium in their later years as they did when they were adolescents.[11] A calcium deficit during the growing years and in adulthood contributes to gradual bone loss, **osteoporosis,** which can totally cripple a person in later life.

Other nutrients are also important to bone growth and maintenance. Fluoride and vitamin D deficiencies, like calcium deficiencies, can cause loss of bone density. So can heredity, abnormal hormone levels, alcohol, prescription medications, other drugs, and lack of exercise (especially weight-bearing exercises), but dietary calcium is one of the most important factors. This chapter's Spotlight feature discusses osteoporosis and its possible causes and prevention.

Table 7-3 shows that calcium appears almost exclusively in three classes of foods: milk and milk products, green vegetables such as broccoli, kale, bok choy, collards, and turnip greens, and a few fish and shellfish. Milk and milk products typically contain the most calcium per serving. Many greens are also good choices, but a complication enters in—absorption. It is not clear to what extent calcium is absorbed from certain green vegetables (notably, spinach and Swiss chard) whereas calcium is known to be very well absorbed from milk.[12] Milk contains both vitamin D and lactose, both of which enhance calcium absorption and promote bone health. Milk and milk products also normally supply about 40 percent of people's intake of riboflavin. Figure 7-4 on page 221 shows the DRI Committee's current recommendations for calcium intakes.

cofactor a mineral element that, like a coenzyme, works with an enzyme to facilitate a chemical reaction.

osteoporosis (OSS-tee-oh-pore-OH-sis) also known as *adult bone loss;* a disease in which the bones become porous and fragile.

osteo = bones
poros = porous

TABLE 7-2
A Guide to the Minerals

Mineral	Best Sources	Chief Roles	Deficiency Symptoms	Toxicity Symptoms
Major Minerals				
Calcium	Milk and milk products, small fish (with bones), tofu, certain green vegetables, legumes, fortified juices	Principal mineral of bones and teeth; involved in muscle contraction and relaxation, nerve function, blood clotting, blood pressure	Stunted growth in children; bone loss (osteoporosis) in adults	Excess calcium is usually excreted except in hormonal imbalance states
Phosphorus	Meat, poultry, fish, dairy products, soft drinks, processed foods	Part of every cell; involved in acid–base balance and energy transfer	Muscle weakness and bone pain (rarely seen)	May cause calcium excretion
Magnesium	Nuts, legumes, whole grains, dark green vegetables, seafoods, chocolate, cocoa	Involved in bone mineralization, protein synthesis, enzyme action, normal muscular contraction, nerve transmission	Weakness, confusion, depressed pancreatic hormone secretion, growth failure, hallucinations, muscle spasms	Excess intakes (from overuse of laxatives) has caused low blood pressure, lack of coordination, coma, and death
Sodium	Salt, soy sauce; processed foods such as cured, canned, pickled, and many boxed foods	Helps maintain normal fluid and acid–base balance; nerve impulse transmission	Muscle cramps, mental apathy, loss of appetite	High blood pressure
Chloride	Salt, soy sauce; processed foods	Part of hydrochloric acid found in the stomach, necessary for proper digestion, fluid balance	Growth failure in children, muscle cramps, mental apathy, loss of appetite	Normally harmless (the gas chlorine is a poison but evaporates from water); vomiting
Potassium	All whole foods: meats, milk, fruits, vegetables, grains, legumes	Facilitates many reactions, including protein synthesis, fluid balance, nerve transmission, and contraction of muscles	Muscle weakness, paralysis, confusion; can cause death; accompanies dehydration	Causes muscular weakness; triggers vomiting; if given into a vein, can stop the heart
Sulfur	All protein-containing foods	Component of certain amino acids; part of biotin, thiamin, and insulin	None known; protein deficiency would occur first	Would occur only if sulfur amino acids were eaten in excess; this (in animals) depresses growth

© Terry Heffernan

▸ *Got milk? If not, be sure to consume the recommended amount of calcium—the amount found in three to four glasses of milk—from the many alternate sources of calcium available on the market today (see Table 7-3).*

Some foods contain **binders** that combine chemically with calcium and other minerals such as iron and zinc to prevent their absorption, carrying them out of the body with other wastes. For example, **phytic acid** renders the calcium, iron, zinc, and magnesium in certain foods less available than they might be otherwise; **oxalic acid** binds calcium and iron. Phytic acid is found in oatmeal and other whole-grain cereals; oxalic acid is found in beet greens, rhubarb, and spinach, among other foods. These binders seem to depress absorption of the calcium present in the *same* food as the binder but do not depress the absorption of the calcium from other foods consumed at the same time. Because fiber in general seems to hinder calcium absorption, the higher your diet is in fiber, the higher it should also be in calcium. This fact doesn't diminish the overall value of high-fiber foods, which are important and nutritious for many reasons.

binders in foods, chemical compounds that can combine with nutrients (especially minerals) to form complexes the body cannot absorb. Examples of such binders are **phytic** (FIGHT-ic) **acid** and **oxalic** (ox-AL-ic) **acid.**

TABLE 7-2
A Guide to the Minerals (continued)

Mineral	Best Sources	Chief Roles	Deficiency Symptoms	Toxicity Symptoms
Trace Minerals				
Iodine	Iodized salt, seafood, bread	Part of thyroxine, which regulates metabolism	Goiter, cretinism	Depressed thyroid activity
Iron	Red meats, fish, poultry, shellfish, eggs, legumes, dried fruits, fortified cereals	Hemoglobin formation; part of myoglobin; energy utilization	Anemia: weakness, pallor, headaches, reduced immunity, inability to concentrate, cold intolerance	Iron overload: infections, liver injury, acidosis, shock
Zinc	Protein-containing foods: meats, fish, shellfish, poultry, grains, vegetables	Part of insulin and many enzymes; involved in making genetic material and proteins, immunity, vitamin A transport, taste, wound healing, making sperm, fetal development	Growth failure in children, delayed development of sexual organs, loss of taste, poor wound healing	Fever, nausea, vomiting, diarrhea, kidney failure
Copper	Meats, seafood, nuts, drinking water	Helps make hemoglobin; part of several enzymes	Anemia, bone changes (rare in human beings)	Nausea, vomiting, diarrhea
Fluoride	Drinking water (if fluoride containing or fluoridated), tea, seafood	Formation of bones and teeth; helps make teeth resistant to decay	Susceptibility to tooth decay	Fluorosis (discoloration of teeth); nausea, vomiting, diarrhea
Selenium	Seafood, meats, grains, vegetables (depending on soil conditions)	Helps protect body compounds from oxidation; works with vitamin E	Fragile red blood cells, cataracts, growth failure, heart damage	Nausea, abdominal pain; nail and hair changes; liver and nerve damage
Chromium	Meats, unrefined foods, vegetable oils	Associated with insulin needed for release of energy from glucose	Abnormal glucose metabolism	Occupational exposures damage skin and kidneys
Molybdenum	Legumes, cereals, organ meats	Facilitates, with enzymes, many cell processes	Unknown	Enzyme inhibition
Manganese	Widely distributed in foods	Facilitates, with enzymes, many cell processes	In animals: poor growth, nervous system disorders, abnormal reproduction	Poisoning, nervous system disorders

Protein also affects calcium status by affecting excretion. The higher the diet is in protein, the greater the amount of calcium excreted. This is why people in the United States and Canada are told to ingest more calcium than people in countries whose protein intakes are lower.

Age (years)	Calcium needed (mg) per day	Number of milk or milk product servings per day (or the equivalent)
1–3	500	2
4–8	800	3
9–18	1,300	4
19–50	1,000	3
51+	1,200	4

FIGURE 7-4
CALCIUM RECOMMENDATIONS*

*In general, 1 cup of milk or yogurt, 1½ ounces of natural cheese, or 2 ounces of processed cheese can be considered as 1 cup from the milk group. Equivalent choices include nondairy servings of beverages or food such as calcium-fortified juices that provide a similar amount of calcium as found in a 1 cup serving of milk. The USDA MyPyramid Food Guide sets a minimum of 3 cups of milk each day for a 2,000 calorie diet and 2 cups per day for children under 8 years of age.

TABLE 7-3
Calcium in Foods

(mg)	Sources
448	Yogurt, plain, low-fat (1 c)
408	Swiss cheese (½ oz)
350	Orange juice, calcium fortified (1 c)
348	American cheese (2 oz)
345	Yogurt with fruit (1 c)
325	Sardines (with bones) (3 oz)
316	Fat-free milk (1 c)
306	Cheddar cheese (1½ oz)
300	Parmesan cheese (1 oz)
300	Rice drink, calcium-fortified (1 c)
275	Shrimp (3 oz)
250	Pizza (1 slice)
248	Frozen yogurt (1 c)
≥200	Soy milk, calcium-fortified (1 c)
≥200	Cereal and snack bars, calcium-fortified (1 bar)
197	Turnip greens, cooked (1 c)
186	Cream soup (1 c)
182	Salmon (with bones) (3 oz)
180	Kale, cooked (1 c)
179	Collard greens, cooked (½ c)
165	Instant oatmeal (1 packet)
160	Bok choy (1 c)
154	Cottage cheese (1 c)
144	Pudding, chocolate (½ c)
130*	Tofu (½ c)
126	Almonds (⅓ c)
122	Waffle (1)
94	Broccoli (1 c)
87	Ice cream (½ c)
52	Tortilla, corn (1)
50	Dried beans, cooked (½ c)

*The calcium content of tofu varies depending on processing methods. Look for tofu processed with calcium salts.

Alternative Sources of Calcium Milk need not be your only source of calcium. Yogurt is an acceptable substitute for regular milk, and puddings, custards, and baked goods can be prepared in such a way that they also contain appreciable amounts of milk. Powdered nonfat milk, which is an excellent and inexpensive source of protein, calcium, and other nutrients, can be added to many foods (such as cookies, soups, casseroles, and meatloaf) during preparation. Nonfat yogurt fortified with extra milk solids is another excellent calcium source. Also, small fish such as Atlantic sardines, or canned, pink salmon with soft edible bones are similar to milk and milk products in calcium richness.

A number of calcium-fortified foods are available, including calcium-fortified juices, fruit drinks, soy drinks, breads, cereals, waffles, snack bars, and hot cocoa mixes. Many times, these foods are fortified to match or exceed the amount of calcium in a cup of milk. For example, an 8-ounce serving of fortified orange juice provides about 350 milligrams of calcium. Remember to check the Nutrition Facts panel on food labels for the amount of calcium contained in the foods you eat.

The word *daily* must be stressed with respect to calcium intake. Because of the body's limited ability to absorb calcium, it cannot handle massive doses periodically but needs frequent opportunities to take in small amounts. To evaluate your own diet for sources of calcium, see the Scorecard on page 223.

Milk Substitutes Some people have **milk allergy** or **lactose intolerance** and can't drink milk. For them, calcium-rich substitutes must be found. Among the possible substitutes for persons with milk allergy are boiled milk, goat's milk, calcium-fortified soy milk or nondairy foods, and calcium supplements. People with lactose intolerance can choose enzyme-treated milk, calcium-fortified soy milk, small amounts of milk products such as plain yogurt and aged cheese, as well as nondairy foods containing calcium, or calcium supplements.

Phosphorus

Phosphorus is second to calcium in mineral abundance in the body. About 85 percent of it is found combined with calcium in the crystals of the bones and teeth as calcium phosphate, the chief compound that gives them strength and rigidity. Phosphorus is also a part of DNA and RNA, the genetic code material present in every cell. Phosphorus is thus necessary for all growth because DNA and RNA provide the instructions for new cells to be formed.

Phosphorus plays many key roles in the cells' use of energy nutrients. Many enzymes and the B vitamins become active only when a phosphate group is

SOURCES OF CALCIUM
Adult DRI (AI):* **19 to 50 years: 1,000 mg**
51 years and older: 1,200 mg

*Dietary Reference Intakes (DRI) for all age groups are listed on the inside front cover.

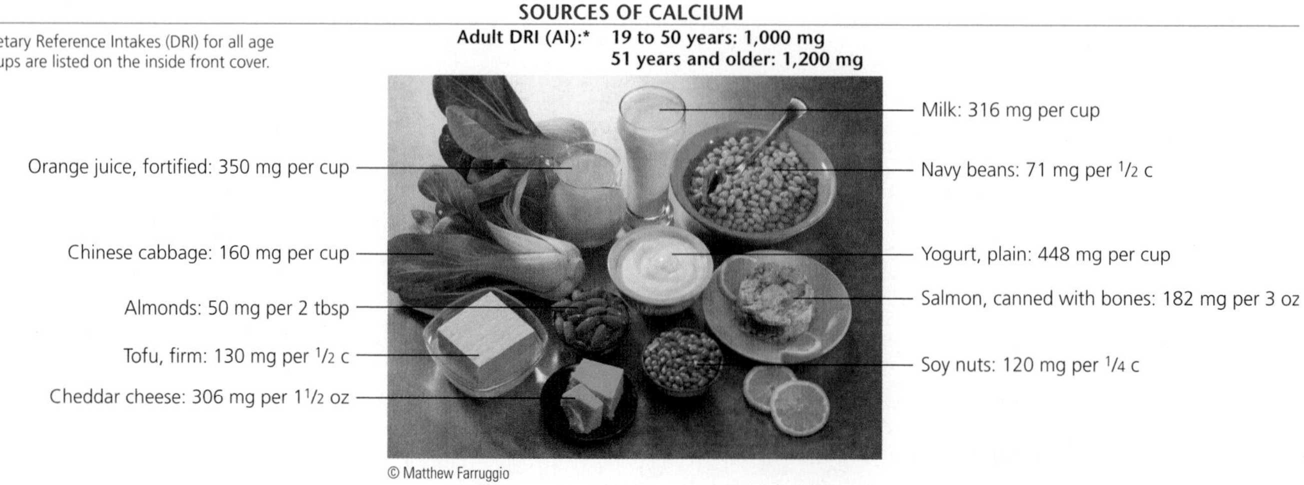

Scorecard

Calcium Sources

We need at least 1,000 milligrams of calcium daily. Answer these questions by considering what you ate yesterday.

1. Did you drink milk (nonfat, low-fat, or whole) yesterday? If so, give yourself 3 points for every 8-ounce glass (1 cup). ________
2. Did you eat yogurt? Give yourself 4 points for each 8-ounce serving. ________
3. Did you eat (1 cup) calcium-fortified cereal with ½ cup of milk? Give yourself 4 points for every serving. ________
4. Did you eat 1 cup other type of cereal with ½ cup of milk? Give yourself 2 points for every serving. ________
5. Did you drink juice that is fortified with calcium? For every 6-ounce serving give yourself 2 points. ________
6. Did you eat canned salmon with bones or tofu (that's been processed with calcium) yesterday? Give yourself 3 points for each 3-ounce portion eaten (or ½ cup tofu). ________
7. Did you eat cheese yesterday? For every 1 ounce eaten, give yourself 2 points. ________
8. Did you eat cottage cheese? For each ½-cup serving, give yourself 1 point. ________
9. Did you eat broccoli, kale, collards, or bok choy? For every 1 cup, raw or cooked, give yourself 1 point. ________
10. Did you have ice cream, pudding, or frozen yogurt yesterday? For a 1-cup serving give yourself 1 point. ________

Now add up all your points. *Total Points:* ________

Multiply your total points × 100: ________

This gives you an idea of how many milligrams of calcium you are getting each day.

SOURCE: Adapted from *The Calcium Connection,* © 1994 Continental Baking Company.

attached. The B vitamins, you will recall, play a major role in energy metabolism. Phosphorus is critical in energy exchange.

Some lipids (phospholipids) contain phosphorus as part of their structure. They help to transport other lipids in the blood; they also form part of the structure of cell membranes, where they affect the transport of nutrients and wastes into and out of the cells.

Animal protein is the best source of phosphorus because phosphorus is so abundant in the energetic cells of animals. People who eat large amounts of animal protein have high phosphorus intakes. People who regularly consume carbonated beverages also have high phosphorus intakes because they contain phosphoric acid. Soft drink consumption is now estimated to be about 40 gallons per person each year. However, when soft drinks replace milk in the diet, fracture risks increase, especially for girls and women.[13]

The recommended intake of phosphorus is lower than that for calcium. It is believed to provide sufficient phosphorus to ensure adequate absorption and retention of calcium. Higher intakes of phosphorus can interfere with calcium absorption. The good news is that people do not need to make a special effort to eat foods containing phosphorus because it is present in virtually all foods (see Table 7-4).

milk allergy the most common food allergy; caused by the protein in raw milk.

lactose intolerance as described in Chapter 3, an inherited or acquired inability to digest lactose as a result of a failure to produce the enzyme lactase.

SOURCES OF PHOSPHORUS

Adult DRI (RDA) is 700 mg.

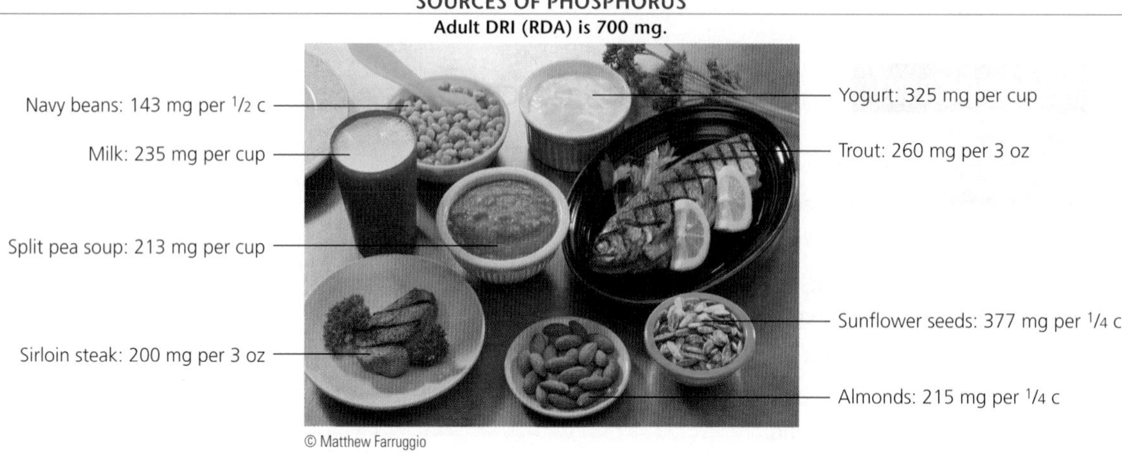

TABLE 7-4
Phosphorus in Foods

(mg)	Sources
422	American cheese (2 oz)
341	Cottage cheese (1 c)
325	Yogurt (1 c)
280	Salmon (canned) (3 oz)
242	Pork (3 oz)
235	Fat-free milk (1 c)
200	Sirloin steak (3 oz)
189	Sunflower seeds (2 tbsp)
186	Turkey (3 oz)
174	Peanuts (⅓ c)
161	Hamburger (3 oz)
149	Shredded wheat (1 c)
143	Navy beans (½ c)
139	Tuna (3 oz)
115	Potato (1)
104	Peanut butter (2 tbsp)
67	Corn (½ c)
51	Cola (12 oz)
46	Broccoli (½ c)
46	Wheat bread (1 slice)
45	Diet cola (12 oz)

Sulfur and Magnesium

Sulfur is present in some amino acids and in all proteins. Its most important role is in helping strands of protein assume and hold a particular shape, thus enabling them to do their specific jobs, such as enzyme work. Skin, hair, and nails contain some of the body's more rigid proteins, and they have a high sulfur content. There is no recommended intake for sulfur, and no deficiencies are known. Only a person who lacks dietary protein to the point of severe protein deficiency will lack the sulfur-containing amino acids.

Magnesium acts in all the cells of the muscles, heart, liver, and other soft tissues, where it forms part of the protein-making machinery and is necessary for the release of energy. Magnesium also helps to relax muscles after contraction and promotes resistance to tooth decay by helping to hold calcium in tooth enamel. Bone magnesium seems to be a reservoir to ensure that some will be on hand for vital reactions regardless of recent dietary intake. Areas of the country that have hard water—higher in magnesium and calcium—have lower rates of death from cardiovascular disease. A deficiency of magnesium may be related to sudden death from heart failure and to high blood pressure.[14] A dietary deficiency of magnesium is not likely but may occur as a result of vomiting, diarrhea, alcohol abuse, or protein malnutrition. Also, magnesium deficiency may be found in people who have been fed magnesium-poor fluids into a vein for too long or in people using diuretics. Good food sources of magnesium include nuts, legumes, cereal grains, dark green vegetables, seafoods, chocolate, and cocoa (see Table 7-5).

Sodium, Potassium, and Chloride

About 40 percent of the body's water weight is inside the cells, and about 15 percent bathes the outsides of the cells. The remainder is in the blood vessels. Special conditions are needed to regulate the amount of water inside and outside the cells so that the cells do not collapse from water leaving them or swell up under the stress of too much water entering them. The cells cannot manage this by pumping water across their membranes, because water slips back and forth freely. However, they can pump minerals across their membranes, and these minerals attract the water to come along with them. This is how the cells maintain water balance. Minerals are used for this purpose in a special form—as **ions** or **electrolytes.** In this form, as single, electrically charged particles, the minerals play many roles, including helping to maintain water balance and acid–base balance.

Sodium, potassium, and chloride are examples of electrolytes—dissolved substances in blood and body fluids that carry electric charges. Sodium is the chief positively charged ion needed to maintain the volume of fluid outside cells;

ions (EYE-ons) electrically charged particles, such as sodium (positively charged) and chloride (negatively charged).

electrolytes compounds that partially dissociate in water to form ions; examples are sodium, potassium, and chloride.

SOURCES OF MAGNESIUM

Adult DRI (RDA) is 310 to 420 mg.

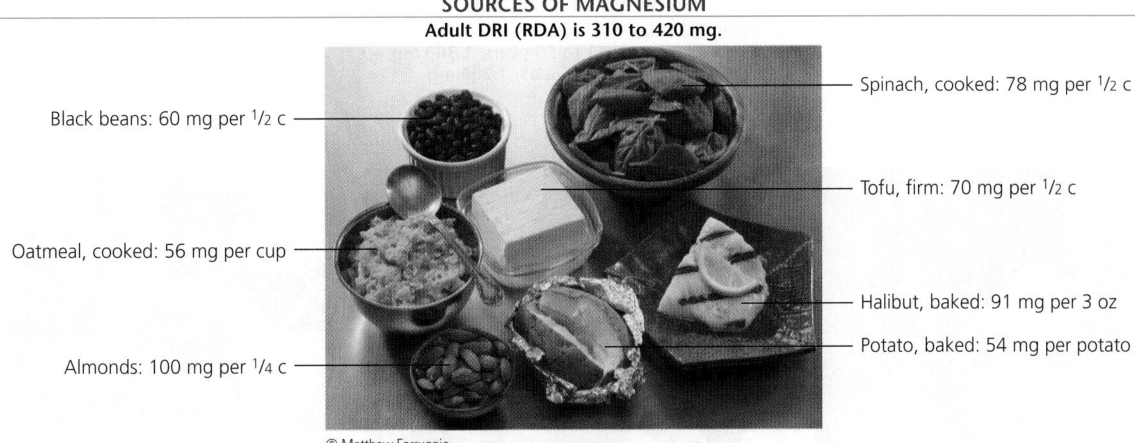

potassium is the chief positively charged ion inside the cells. Chloride is the major negatively charged ion in the fluids outside the cells, where it is found mostly in association with sodium. Electrolytes influence the distribution of fluids among the various body compartments. Specifically, when present in healthy amounts, electrolytes serve to keep the fluids inside and outside the cells in balance so that the cells can function properly and allow body fluids to bring cells the nutrients they need and remove waste products from them. With a normal diet, many factors in addition to the intake of sodium and chloride found in salt work together to keep the fluid volume fairly constant inside and outside of cells.

Electrolytes also help to create the environment in which the cells' work takes place—work such as nerve-to-nerve communication, heartbeats, and contraction of muscles. When a person's body loses fluid, whether it is sweat, blood, or urine, the person also loses electrolytes. Maintaining healthy concentrations of electrolytes is crucial to the life-sustaining activities of the vital organs. When large amounts of body fluid are lost, as in heat stroke, infant diarrhea, or injury, replacing them is a critical medical task. As already mentioned, people who exercise also lose fluids, and they must replace them to avoid dehydration. Chapter 10 discusses fluid needs during and after exercise as well as popular sports drinks that can help to replace electrolytes that also may have been lost.

Fluid inside cells is rich in potassium (K^+).

Fluid outside cells is rich in sodium (Na^+) and chloride Cl^-).

▲ ***Electrolytes***
Sodium, potassium, and chloride are examples of body electrolytes. Potassium, which is usually found in the fluids inside the cells, carries a positive charge. Sodium and chloride are usually found in the fluids outside the cells. Sodium carries a positive charge, whereas chloride carries a negative charge.

TABLE 7-5
Magnesium in Foods

(mg)	Sources
140	Almonds (⅓ c)
119	Cashews (⅓ c)
95	Raisin bran (1 c)
85	Peanuts (⅓ c)
78	Spinach, cooked (½ c)
70	Tofu (½ c)
60	Black beans (½ c)
55	Oysters (steamed) (3 oz)
54	Baked potato (1)
46	Soy milk (1 c)
45	Avocado (½ c)
45	Black-eyed peas (½ c)
43	Yogurt, plain (1 c)
42	Brown rice, cooked (½ c)
40	Lima beans (½ c)
33	Dried figs (¼ c)
28	Fat-free milk (1 c)
27	Chicken (3 oz)
21	Sunflower seeds (2 tbsp)
21	Hamburger (3 oz)
20	Pork (3 oz)
17	Milk chocolate (1 oz)

Sodium Sodium is part of sodium chloride, ordinary table **salt,** a food seasoning and preservative. The recommended sodium intake is set at 1,500 milligrams for young adults, 1,300 milligrams for adults ages 51 through 70, and 1,200 milligrams for older adults. Because average sodium intakes are about 3,300 milligrams per day, substantially higher than recommended, the *Dietary Guidelines for Americans* recommend consuming little sodium and salt and staying below the upper limit of 2,300 milligrams of sodium per day (approximately 1 teaspoon of

salt a pair of charged mineral particles, such as sodium (Na^+) and chloride (Cl^-), that associate together. In water, they dissociate and help to carry electric current—that is, they become electrolytes.

SOURCES OF SODIUM

*Tolerable Upper Intake Level is no more that 2,300 mg/day.

Adult DRI (AI):* **19 to 50 years: 1,500 mg**
51 to 70 years: 1,300 mg
71+ years: 1,200 mg

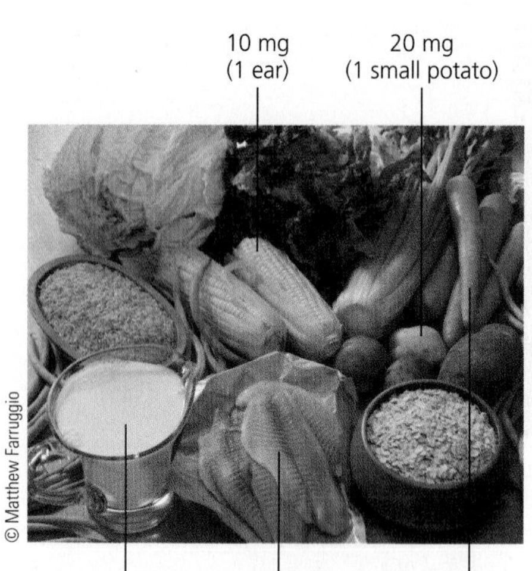

▲ *Many whole, unprocessed foods are low in sodium. These foods contribute less than 10 percent of the sodium in the U.S. diet.*

68 mg (1 tsp) · 167 mg (1 tbsp) · 300 mg (1 tsp) · 2,000 mg (1 tsp) · 56 mg (1 tsp)

▲ *The salt added during cooking or at the table contributes about 15 percent of the sodium in the U.S. diet.*

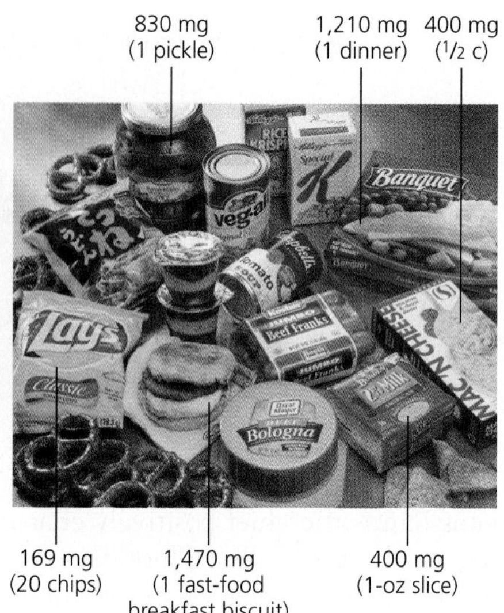

▲ *Most of the sodium (75 percent) in our diet is added by food manufacturers to processed foods such as these.*

hypertension sustained high blood pressure.

hyper = too much
tension = pressure

diuretics (dye-you-RET-ics) medications and other substances causing increased water excretion.

dia = through
ouron = urine

salt). The use of highly salted foods can contribute to high blood pressure **(hypertension)** in those who are genetically susceptible. (A discussion of the causes and prevention of hypertension appears in the Nutrition Action feature starting on page 229.)

If any members of your family have high blood pressure, you are advised to curtail your sodium consumption and make sure that your potassium and calcium intakes are ample. Table 7-6 shows a sampling of the sodium content of common foods and reveals that, in general, the more processed a food is, the more sodium it contains. As shown in the table below, whole, unprocessed foods, on the other hand, tend to be high in potassium and low in sodium.

People who wish to choose and prepare foods with less salt need to know that what they pour from the salt shaker may be only a sixth of the total salt they consume. Up to 75 percent of the salt in the diet has been added to foods by food processors. Because processed foods don't always taste salty, eating can sometimes be a guessing game. Remember to check food labels for sodium content. See the Savvy Diner feature on page 233 for suggested ways to reduce salt intake.

Whole Foods versus Processed Foods

Food	Potassium (mg)	Sodium (mg)	Potassium-to-Sodium Ratio
Corn (cooked), 1 c	242	8	30:1
Corn flakes, 1 c	25	300	1:12
Peaches (fresh), 1	193	0	171:1
Peach pie, 1 piece	131	253	1:2

Potassium Potassium is critical to maintaining the heartbeat. The sudden deaths that occur during fasting, severe diarrhea, or severe vomiting are thought to be due to heart failure caused by potassium loss. As the principal positively charged ion inside body cells, potassium plays a major role in maintaining water balance and cell integrity.

In dehydration, sodium is lost along with water that is lost from the body, but more critical damage comes because potassium is lost from inside cells at the same time. Potassium deficiency is especially dangerous because it affects the brain cells, making the victim unaware of the need for water. Adults are warned not to take **diuretics,** except under the direction of a physician, because some of them cause potassium excretion. Physicians prescribing such diuretics tell their

TABLE 7-6
Where's the Sodium?

Food Groups	Sodium (mg)
Grains and Grain Products	
Cooked cereal, rice, pasta, unsalted, ½ c	0–5
Ready-to-eat cereal, 1 c	100–360
Bread, 1 slice	110–175
Vegetables	
Fresh or frozen, cooked without salt, ½ c	1–70
Canned or frozen with sauce, ½ c	140–460
Tomato juice, canned, ¾ c	820
Fruit	
Fresh, frozen, canned, ½ c	0–5
Low-fat or Fat-free Dairy Foods	
Milk, 1 c	120
Yogurt, 8 oz	160
Natural cheeses, 1½ oz	110–450
Processed cheeses, 1½ oz	600
Nuts, Seeds, and Dry Beans	
Peanuts, salted, ⅓ c	120
Peanuts, unsalted, ⅓ c	0–5
Beans, cooked from dried or frozen, without salt, ½ c	0–5
Beans, canned, ½ c	400
Meats, Fish, and Poultry	
Fresh meat, fish, poultry, 3 oz	30–90
Tuna, canned, water pack, no salt added, 3 oz	35–45
Tuna, canned, water pack, 3 oz	250–350
Ham, lean, roasted, 3 oz	1,020

SOURCE: National Heart, Lung, and Blood Institute, "The DASH Eating Plan," *NIH Publication No. 03-4082,* Bethesda, MD: National Institutes of Health (May 2003): 7.

patients to eat potassium-rich foods to compensate for the losses and, depending on the diuretic, may also advise a lowered sodium intake.

The relationship of potassium and sodium in maintaining the blood pressure is not entirely clear. Abundant evidence supports the simple view that the two minerals have opposite effects. In any case, it is clear that increasing the potassium in the diet can promote sodium excretion under most circumstances and thereby lower the blood pressure.[15] A lifelong intake of foods that are low in

SOURCES OF POTASSIUM
Adult DRI (AI) is 4,700 mg.

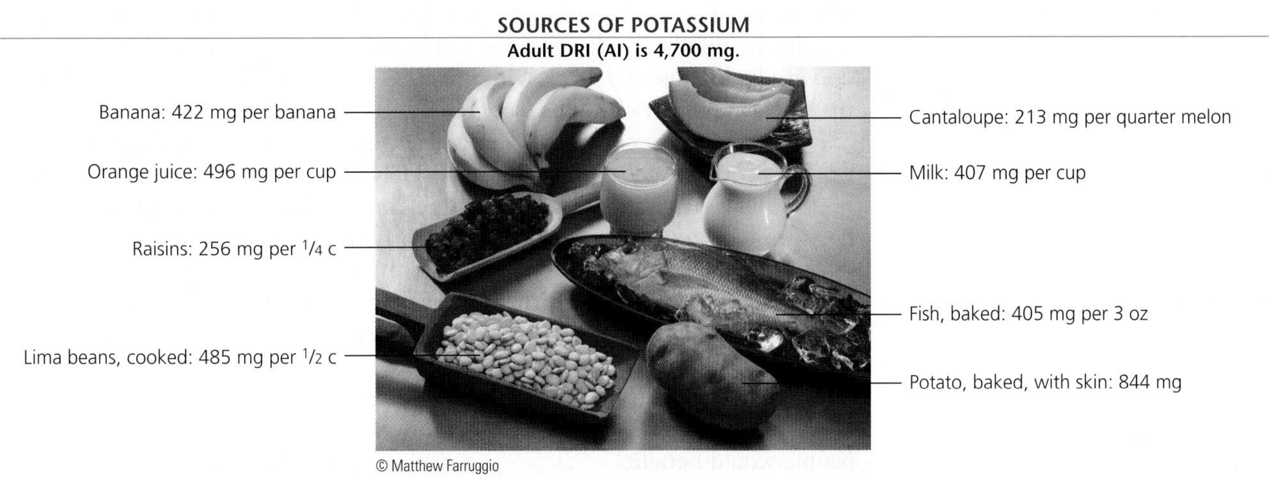

TABLE 7-7
Potassium in Foods

(mg)	Sources
844	Baked potato (1)
529	Yogurt (1 c)
496	Orange juice (1 c)
485	Lima beans (½ c)
422	Banana (1 medium)
419	Spinach, cooked (½ c)
407	Fat-free milk (1 c)
400	Pinto beans (½ c)
329	Kidney beans (½ c)
319	Salmon (3 oz)
315	Bok choy, cooked (½ c)
299	Pork (3 oz)
297	Hamburger (3 oz)
273	Tomato (1 medium)
256	Raisins (¼ c)
254	Raisin bran (1 c)
237	Chicken (3 oz)
232	Carrots (½ c)
228	Broccoli (½ c)
213	Cantaloupe (¼ melon)

sodium and high in potassium protects against hypertension and is thought to play a role in the low blood pressure seen in vegetarians.[16]

A dietary deficiency of potassium is unlikely, but high-sodium diets that are also high in processed foods and low in fresh fruits and vegetables can make it a possibility. Whole foods of all kinds, including fruits, vegetables, grains, meats, fish, and poultry, are among the richest sources of potassium. Potassium is also abundant in milk. Table 7-7 shows the potassium content of foods.

Some people have medical reasons for needing potassium supplementation, but these people need to be medically supervised. For example, people on medically supervised, very low-calorie weight-loss diets may be advised to take a potassium supplement. A physician must monitor the potassium status of these people and order supplements commensurate with the degree of depletion.

Potassium supplements should never be self-prescribed. Potassium toxicity from potassium in supplement form is a greater concern than potassium deficiency. The body protects itself from this eventuality as best it can. If you consume more than you need, the kidneys accelerate their excretion and so maintain control. Should their limit be exceeded (if you ingest too much potassium too fast), a vomiting reflex is triggered. However, if the digestive tract is bypassed and potassium is injected into a vein, the heart can stop.

Chloride The negative ion, chloride, accompanies sodium in the fluids outside the cells. Because chloride can move freely across membranes, it is also found inside the cells in association with potassium. In the blood, chloride helps in maintaining the acid–base balance. In the stomach, the chloride ion is part of hydrochloric acid, which maintains the strong acidity of the stomach—needed for protein digestion. Nearly all dietary chloride comes from sodium chloride or salt.

The Trace Minerals

If you could extract all of the trace minerals from the body, you would obtain only a bit of dust, hardly enough to fill a teaspoon. As tiny as their quantities are, though, each of the trace minerals performs several vital roles for which no substitute will do. A deficiency of any of them may be fatal, and an excess of many can be equally deadly.

Iron

Iron is the body's oxygen carrier. Bound into the protein **hemoglobin** in the red blood cells, it helps transport oxygen from lungs to tissues and thus permits the release of energy from fuels to do the cells' work. When the iron supply is too low, **iron-deficiency anemia** occurs, characterized by weakness, tiredness, apathy, headaches, increased sensitivity to cold, and a paleness that reflects the reduction in the number and size of the red blood cells. A person with this anemia can do very little muscular work without disabling fatigue but can replenish iron status by eating iron-rich foods (see Table 7-8 on page 234).

It is difficult to convey the extent and severity of iron deficiency among the world's people. According to the World Health Organization, dietary intake of iron is inadequate in more than 75 percent of the world's population.[17] People begin to feel iron deficiency's impact, without knowing it, long before anemia is diagnosed. They don't appear to have an obvious deficiency disease; they just appear unmotivated and apathetic. Because they work and play less, they are less physically fit. Prevalence rates for iron-deficiency anemia in developed countries range from 5 to 20 percent. In the United States, iron deficiency remains relatively prevalent among toddlers, adolescent girls, and women of childbearing age.[18] If this one worldwide malnutrition problem could be alleviated, millions of people would benefit.

(Text continues on page 233.)

hemoglobin (HEEM-oh-globe-in) the oxygen-carrying protein of the blood; found in the red blood cells.

iron-deficiency anemia a reduction of the number and size of red blood cells and a loss of their color because of iron deficiency.

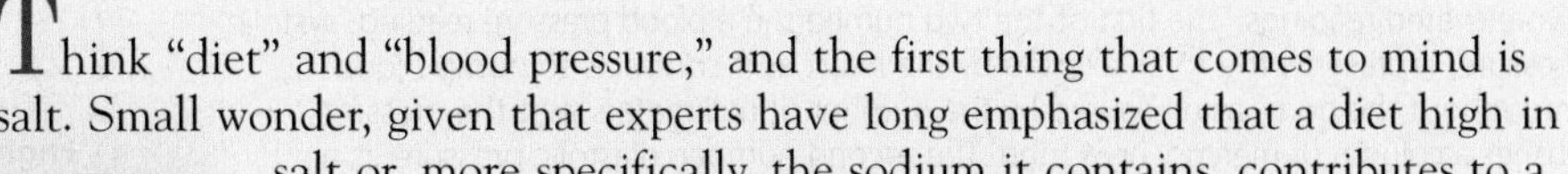

Nutrition Action Diet and Blood Pressure—The Salt Shaker and Beyond

Think "diet" and "blood pressure," and the first thing that comes to mind is salt. Small wonder, given that experts have long emphasized that a diet high in salt or, more specifically, the sodium it contains, contributes to a public health problem known as hypertension (or high blood pressure as seen in Figure 7-5).[19] Nevertheless, a number of other, albeit less notorious, dietary factors play a role in the unhealthful rise in blood pressure that afflicts some 50 million Americans.

© PhotoDisc/GettyImages

Risk factors for high blood pressure:

- Obesity
- Family history of high blood pressure
- Race (African Americans are more likely than whites to develop hypertension.)
- Age (In the United States, blood pressure tends to rise with age.)
- Excess alcohol consumption
- Sedentary lifestyle

Obesity, for instance, inarguably ranks as the number one dietary culprit linked to hypertension.[20] That's because excess weight forces the heart to work harder to supply blood to the extra pounds of fat tissue. High blood pressure occurs about twice as often among obese people as among thinner people. Fortunately, losing as little as 5 percent of the excess weight has been shown, in some cases, to lower blood pressure to the point that antihypertensive medications may become unnecessary.[21] In other words, a little weight loss, regardless of sodium intake, can be enough to get high blood pressure back under control. With increasing obesity, age, and blood pressure levels comes an increasing probability that a person will become *salt sensitive*, meaning that their blood pressure rises in proportion to the amount of sodium they consume.[22] For them, of course, keeping dietary sodium to a minimum can help lower blood pressure.

Drinking excessive amounts of alcohol is another dietary factor that may lead to high blood pressure (not to mention poor compliance with drug regimens and other treatments for the problem). In fact, consuming more than a couple of ounces of alcohol daily has been blamed for leading to 5 to 11 percent of cases of hypertension among men.[23] The National High Blood Pressure Education Program thus recommends that to control high blood pressure, those who drink should do so only in moderation.[24]

In addition to recommending that hypertensive people lose weight and drink alcohol in moderation, it is clear that modest changes in lifestyle can add up to significant gains in controlling high blood pressure. It appears that attention to a healthful lifestyle can also help to prevent the problem from arising in the first place. Consider that researchers at Northwestern University Medical School in Chicago put one group of adults prone to high blood pressure on a preventive program that included losing weight, lowering sodium intake, cutting down on alcohol, and exercising more. They then compared them to a similar group left to their own devices. The researchers found that over the course of five years, only one in eleven persons who changed their lifestyle ended up with hypertension. But of those who didn't alter their habits, one in five became hypertensive.[25]

Along with the factors just explained, scientists are exploring other nutrients that may influence blood pressure.[26] Some researchers believe, for instance, that it is not dietary sodium per se that alters blood pressure, but rather the ratio between it and potassium, sodium's partner in regulating the body's water balance. In other words, the more potassium and less sodium in the diet, the greater the likelihood that the body will maintain a normal blood pressure. Keep in

FIGURE 7-5
ARE YOUR NUMBERS UP?

One of the characteristics of hypertension is that it has been called a "silent killer" that cannot be felt and may go undetected for years. That's why it is crucial to have your blood pressure checked on a regular basis. Diagnosis of hypertension requires at least two elevated readings. The first of the two numbers in a blood pressure reading, systolic pressure, represents the force exerted by the heart as it contracts to pump blood throughout the body, as measured by the number of millimeters that the pressure pushes a column of mercury up a tube. The second number, diastolic pressure, is a measure of the pressure the blood exerts on artery walls between heart beats. A reading of below 120 over 80 millimeters of mercury is considered "optimal" for adults 18 years and older. Measurements between 120 over 80 and 139 over 89 are classified as prehypertension. Individuals with prehypertension are at increased risk for progression to hypertension and are encouraged to make lifestyle changes to prevent hypertension. Individuals in the 130 over 80 to 139 over 89 range are at twice the risk to develop hypertension as those with lower values. Beyond those levels, the risks of heart attacks and strokes rise in direct proportion to increasing blood pressure.*

*The categories are for adults not taking high blood pressure medications. If your systolic and diastolic numbers fall into different categories, your overall status depends on the higher category.

SOURCE: National Institutes of Health, National Heart, Lung, and Blood Institute, *The Seventh Report of the Joint National Committee on Prevention, Detection, Evaluation, and Treatment of High Blood Pressure* (Washington, D.C.: U.S. Department of Health and Human Services, May 2003).

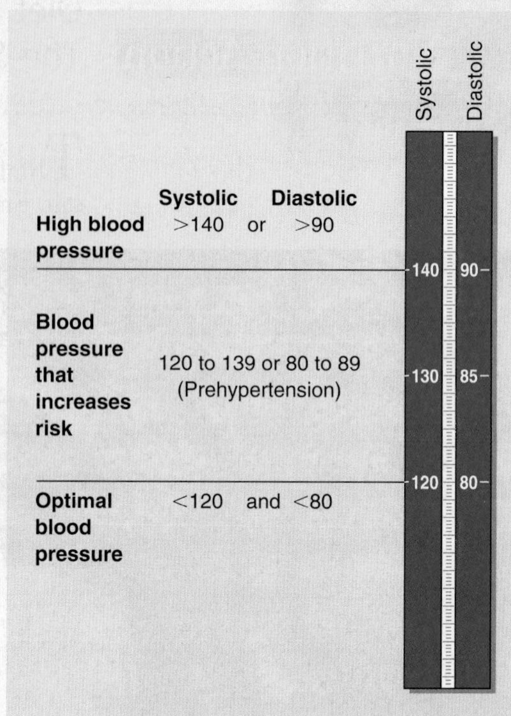

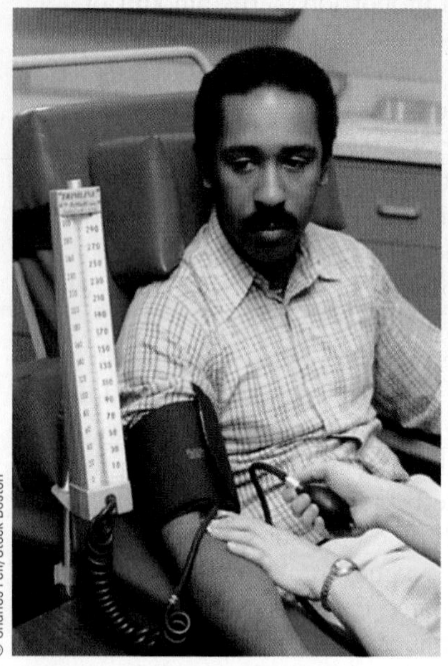

▲ *The most effective single measure you can take to protect yourself against high blood pressure is to know what your blood pressure is.*

© Charles Feil/Stock Boston

mind that diets rich in high-potassium foods such as fresh fruits and vegetables tend to go hand in hand with low sodium consumption.[27]

In addition, some studies indicate that people who eat high-calcium diets are less likely to have high blood pressure; conversely, people with inadequate calcium intakes are more likely to have hypertension.[28] It is recommended, therefore, that people eat a diet that contains two to three servings daily of calcium-rich foods such as milk or calcium-fortified orange juice.

Finally, some evidence suggests that eating a diet that includes monounsaturated fat and is low in saturated fat helps to lower blood pressure. It certainly makes sense for those with high blood pressure to eat such a diet if for no other reason than to help prevent heart disease, a condition that hypertensives run a high risk of developing (refer to the Spotlight feature in Chapter 4).[29]

An Eating Plan to Reduce High Blood Pressure

Two recent studies have shown that following a particular eating plan—called the *DASH diet*—and reducing sodium intake lowers blood pressure. The first study, Dietary Approaches to Stop Hypertension (DASH), examined the effects of overall diet on 459 adult participants with normal to high blood pressures. After just 8 weeks, those people with mild hypertension who were on the DASH diet—a diet high in fruits, vegetables, whole grains, and low-fat dairy products and low in total fat, saturated fat, and cholesterol—saw their blood pressures decrease.[30] A second landmark study, DASH-Sodium, has shown that a combination of the DASH diet and sodium reduction can lower blood pressure even more. The DASH-Sodium study looked at the effect on blood pressure of a reduced dietary sodium intake as 412 participants followed *either* the DASH diet or an eating pattern resembling the typical American diet. Participants were followed for a month at each of three sodium levels: 3,300 milligrams per day (a level consumed by many Americans), 2,400 milligrams per day (maximum amount recommended on food labels), and the lower intake DASH diet of

1,500 milligrams per day. Results showed that reducing dietary sodium reduced blood pressure for both the DASH diet and the typical American diet, with the biggest reductions in blood pressure seen for the DASH diet. Although those with hypertension saw the biggest reductions, those without the disease also saw declines in blood pressure. Figure 7-6 illustrates the DASH diet in pyramid form along with a sample reduced-sodium DASH diet meal pattern.

Overall, the results of the DASH studies echo the recommendations for an overall healthful diet: Choose low-fat dairy products, smaller portions of meat, plenty of fruits and vegetables, and ample servings of high-fiber, whole-grain products. This eating pattern may not only reduce the risk for hypertension but also for heart disease (low in saturated fat and cholesterol), diabetes (high in complex carbohydrates and fiber), osteoporosis (adequate in calcium), and cancer (lots of fruits and vegetables). The take-home message seems to be that if

FIGURE 7-6
THE DASH DIET PYRAMID

Sample DASH Menu: The DASH eating pattern is lower in fat, saturated fat, cholesterol, and sodium, and higher in complex carbohydrates, potassium, magnesium, and calcium than the typical American diet. The sample menu provides: 2,010 calories, 5 servings fruits, 7 servings vegetables, 3 servings dairy foods, 59 grams fat, 121 milligrams cholesterol, and 1,356 milligrams sodium.

*Sweets should be low in fat (for example, jelly, jam, sugar, maple syrup, fruit-flavored gelatin, jelly beans, hard candy, sorbet, fruit ices).

SOURCE: NHLBI, *Facts About the DASH Eating Plan,* May 2003; available at www.nhlbi.nih.gov.

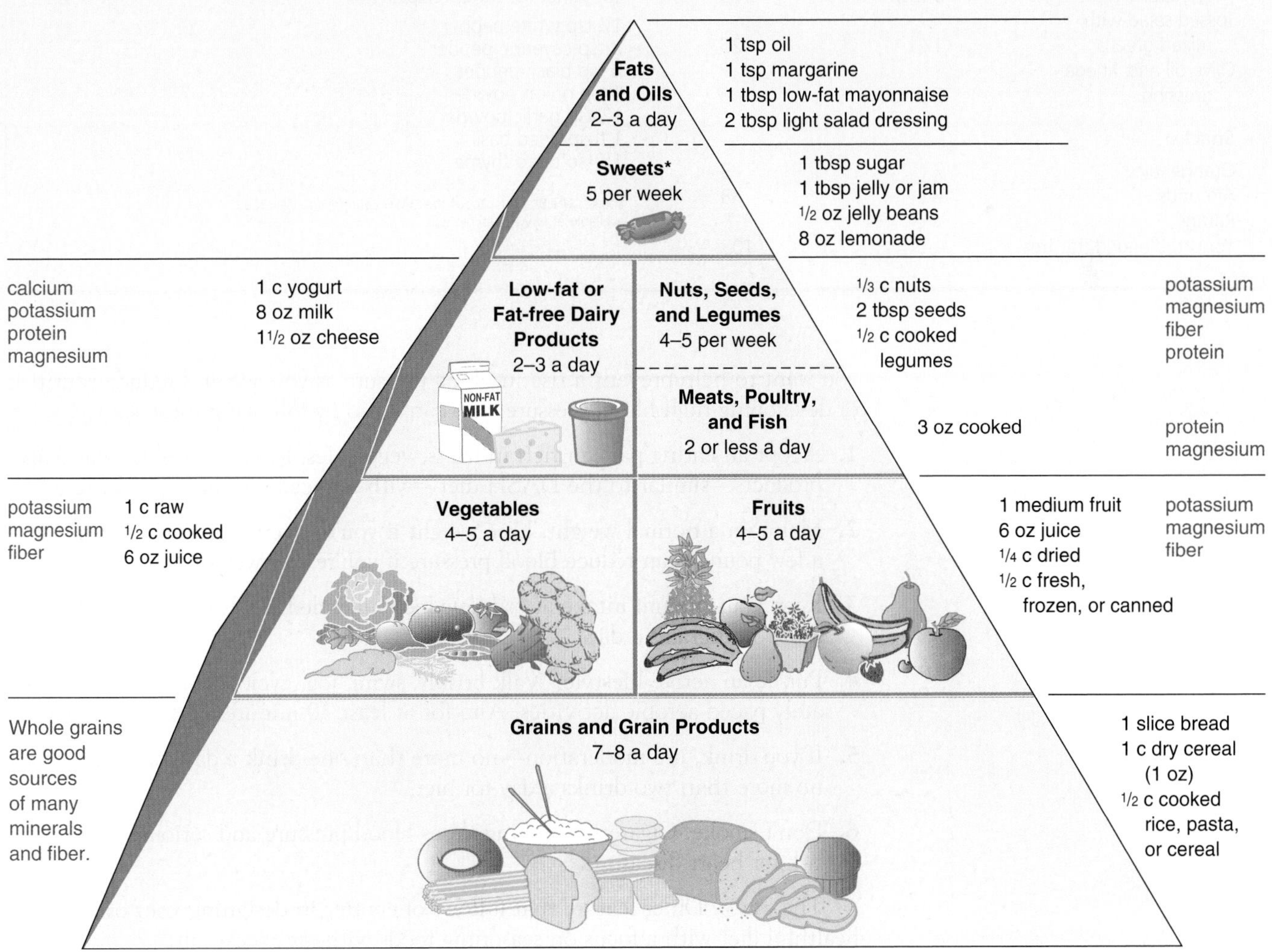

Sample Menu	Amount	Amount of Sodium (mg)
Breakfast:		
Cereal, shredded wheat	½ c	2
Skim milk	1 c	126
Orange juice	1 c	5
Banana, raw	1 medium	1
Bread, 100% whole wheat	1 slice	149
Lunch:		
Chicken salad*	¾ c	151
Whole-wheat bread	2 slices	298
Dijon mustard	1 tsp	125
Tomatoes, raw, fresh	2 large slices	6
Mixed cooked vegetables	1 c	25
Fruit cocktail, juice pack	½ c	5
Dinner:		
Spicy Baked Cod*	3 oz	93
Beans, snap, green, cooked from frozen, without salt	1 c	11
Potato, baked, with skin	1 large	15
Sour cream, low fat	2 tbsp	30
Chives (or scallions)	1 tbsp	0
Fat-free natural cheddar cheese	3 tbsp	169
Tossed salad with mixed greens	1½ c	25
Olive oil and vinegar dressing	2 tbsp	0
Snacks:		
Orange juice	½ c	3
Almonds	⅓ c	3
Raisins	¼ c	5
Yogurt, blended, fat free	1 c	103

***Recipes**

Chicken Salad
(makes 5 servings)

3¼ c chicken breast, cooked, cubed
3 tbsp light mayonnaise
¼ c celery, chopped
1 tbsp lemon juice
½ tsp onion powder

Steps: Mix ingredients in a large bowl and serve.
Serving size: ¾ cup

Spicy Baked Cod
(makes 4 servings)

1 lb cod or other fish fillet, fresh or thawed from frozen
1 tbsp olive oil
1 tsp spicy seasoning mix (see below)

Steps:

1. Preheat oven to 350°F. Spray small baking dish with cooking oil spray.
2. Wash and dry cod. Place in dish and drizzle with oil/seasoning mix.
3. Bake uncovered for 12 minutes or until fish flakes with fork.
4. Cut into 4 pieces and serve.

Spicy seasoning mix
Mix together the following ingredients and store in airtight container for other recipes:

1½ tsp white pepper
½ tsp cayenne pepper
½ tsp black pepper
1 tsp onion powder
1¼ tsp garlic powder
1 tbsp dried basil
1½ tsp dried thyme

SOURCE: NHLBI, *Facts About the DASH Eating Plan,* May 2003; available at www.nhlbi.nih.gov.

you want to help prevent a rise in blood pressure as you age and reduce your risk of developing high blood pressure, you can do so by following these six tips:

1. Adopt an eating pattern rich in fruits, vegetables, legumes, and low-fat dairy products—similar to the DASH diet—with reduced saturated fat content.
2. Maintain a normal weight. Lose weight if you're overweight; even losing just a few pounds can reduce blood pressure if you're overweight.
3. Keep your sodium intake at *or below* recommended levels—not more than 2,300 milligrams a day.
4. Pursue an active lifestyle: Walk briskly, swim, jog, cycle, or do other moderately paced aerobic activities. Aim for at least 30 minutes of activities daily.
5. If you drink, use moderation—no more than one drink a day for women, and no more than two drinks a day for men.
6. Don't smoke. Cigarette smoking raises blood pressure and seriously increases risk for heart disease.

The Savvy Diner feature that follows offers tips in designing your own healthful diet with a focus on seasoning foods without excess salt.

The Savvy Diner Choose and Prepare Foods with Less Salt

You've decided to cut your salt intake. How do you know which foods to buy? How can you cook foods with less salt and good flavor? Fortunately, cutting some of the salt out of your diet isn't difficult. Here are a few basic principles.

At the Supermarket

- Read the Nutrition Facts label to compare the amount of sodium in processed foods—such as frozen dinners, cereals, soups, salad dressings, and sauces. The amount in different types and brands often varies widely. Look for labels that say "low sodium." They contain 140 mg (about 5% of the Daily Value) or less sodium per serving. Look for the key words salt and sodium in the ingredients list.
- Limit cured foods (such as bacon and ham), foods packed in brine (such as pickles, pickled vegetables, olives, and sauerkraut), and condiments (such as MSG, mustard, horseradish, catsup, soy sauce, and barbecue sauce).
- Cut back on frozen dinners, mixed dishes such as pizza, packaged mixes, canned soups or broths, and salad dressings—these products often have a lot of sodium. Choose reduced-sodium or no-salt-added products when possible. Many commercially prepared products are available in sodium-reduced versions. For example, most supermarkets carry reduced-sodium soy sauce, canned tuna, soups, tomato sauce, nuts, canned vegetables, crackers, and pretzels.
- Buy fresh, natural foods more frequently than processed foods, which tend to be high in salt.

At Home

- Explore what herbs and spices complement specific foods. Add flavor with herbs, spices, lemon, lime, vinegar, or salt-free seasoning blends.
- Cook rice, pasta, and hot cereals without salt. Cut back on instant or flavored rice, pasta, and cereal mixes, which usually have added salt, and use only half of the seasoning packet provided.
- Rinse canned foods, such as black beans or tuna, to remove some sodium.
- Consult the *DASH Eating Plan* for these and more sodium-reducing tips at www.nhlbi.nih.gov.

When Eating Out

- Ask how foods are prepared. Ask that they be prepared without added salt or salt-containing ingredients.
- Move the salt shaker away.
- Limit condiments, such as mustard, catsup, and pickles.
- Choose fruit or vegetables, instead of salty snack foods.

The cause of iron deficiency is usually malnutrition—that is, inadequate intake, either from limited access to food or from high consumption of foods low in iron. Among causes other than malnutrition, blood loss is the primary one, caused in many countries by parasitic infections of the digestive tract.

The typical Western mixed diet provides only about 5 to 6 milligrams of iron in every 1,000 calories. The recommended daily intake for an adult man is 8 milligrams. Because most men easily eat more than 2,000 calories, they can meet their iron needs without special effort.

The situation for women is different. Women may have normal blood cell counts or hemoglobin levels and yet may need more iron because their body stores may be depleted, a factor that doesn't show up in standard tests. Because most women typically eat less food than men, their iron intakes are lower. And because women menstruate, their iron losses are greater. These two factors may put women much closer to the borderline of deficiency.

The recommended intake for a woman before menopause is 18 milligrams per day; pregnant women need 30 milligrams daily. Because women typically consume fewer than 2,000 calories per day, they understandably have trouble achieving adequate iron intakes. A woman who wants to meet her iron needs from foods must increase the iron-to-calorie ratio of her diet so that she will receive about double the average amount of iron—about 10 milligrams per 1,000 calories. This means she must emphasize iron-rich foods in her daily diet.

SOURCES OF IRON

Adult DRI (RDA) is 8 to 18 mg.

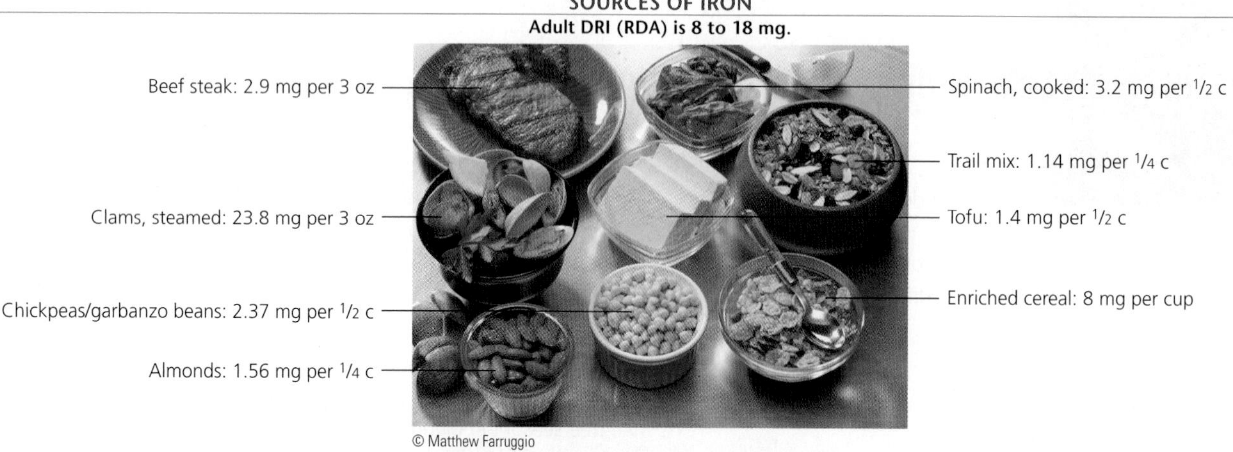

© Matthew Farruggio

TABLE 7-8

Iron in Foods

(mg)	Sources
23.80	Clams, steamed (3 oz)
8.20	Instant Cream of Wheat (¾ c)
8.00	Enriched cereal (¾ c)
3.20	Spinach, cooked (½ c)
3.13	Beef pot roast (3 oz)
2.90	Sirloin steak (3 oz)
2.75	Baked potato (1)
2.65	Shrimp (3 oz)
2.48	Sardines (3 oz)
2.35	Hamburger (3 oz)
2.30	Navy beans (½ c)
2.25	Lima beans (½ c)
2.25	Prune juice (¾ c)
2.15	Black-eyed peas (½ c)
2.00	Swiss chard (½ c)
1.61	Kidney beans (½ c)
1.59	Oatmeal, cooked (1 c)
1.40	Tofu (½ c)
1.30	Tuna (3 oz)
1.30	Dried figs (¼ c)
1.26	Green peas (½ c)
0.94	Wheat bread (1 slice)
0.82	Apricot halves, dried (5)
0.76	Raisins (¼ c)
0.65	Broccoli, cooked (½ c)

© Polara Studios

▲ *Iron cookware adds supplemental iron to foods.*

© Michael Newman/PhotoEdit

▲ *You can combine foods to achieve maximum iron absorption—the heme iron in the meat and the vitamin C in the tomatoes in this chili help you absorb the nonheme iron from the beans.*

Table 7-8 shows the amounts of iron in foods. Meats, fish, and poultry are superior sources on a per-serving basis. People must select foods carefully to obtain enough iron because it is present in such small quantities in most foods. The best meat sources are liver, red meats, poultry, fish, oysters, and clams. Among the grains, whole grains and enriched and fortified breads and cereals are best, and dried beans are a good source. Some fruits and vegetables contain appreciable amounts of iron, as shown in the table. Foods in the milk group are notoriously poor iron sources.

Ways to Enhance Iron Absorption Iron occurs in two forms in foods—as **heme iron,** bound into the iron-carrying proteins such as hemoglobin in meats, poultry, and fish, and as **nonheme iron** in both plant and animal foods. Heme iron is much more reliably absorbed than is nonheme iron. Nonheme iron's absorption is affected by many factors, including the amount of vitamin C consumed with meals.[31] Vitamin C promotes iron absorption and can triple the amount of nonheme iron absorbed from foods eaten at the same meal.

Contamination iron—that is, iron obtained from cookware or soil—can also increase iron intake significantly.[32] Consumers who cook their foods in iron cookware can contribute to their iron intake. For example, the iron content of a half-cup of spaghetti sauce simmered in a glass dish is 3 milligrams, but it is 87 milligrams when the sauce is cooked in an iron skillet. Similarly, dried apricots and raisins contain more iron than the fresh fruit because they are dried in iron pans. Admittedly, this form of iron is not as well absorbed as the iron from meat, but every little bit helps.

heme (HEEM) **iron** the iron-holding part of the hemoglobin protein, found in meat, fish, and poultry. About 40 percent of the iron in meat, fish, and poultry is bound into heme. Meat, fish, and poultry also contain a factor (MFP factor) other than heme that promotes the absorption of iron, even of the iron from other foods eaten at the same time as the meat.

nonheme iron the iron found in plant foods.

Some food components interfere with iron absorption. Phytic acid is one example; it occurs in some fruits, vegetables, and whole grains, as mentioned earlier, as well as in nonherbal tea. Other examples are the tannins, which occur in black teas, coffee, cola drinks, chocolate, and red wines. Fiber also can reduce iron absorption because it speeds up the transit of materials through the intestines.

contamination iron iron found in foods as the result of contamination by inorganic iron salts from iron cookware, iron-containing soils, and the like.

iron overload a condition in which the body contains more iron than it needs or can handle; excess iron is toxic and can damage the liver. The most common cause of iron overload is the genetic disorder hemochromatosis.

Iron Toxicity Large amounts of iron can be toxic to the body. **Iron overload** is a condition in which the body absorbs excessive amounts of iron, and it is more common in men than in women. As a result, tissue damage occurs, especially in organs such as the liver that store iron. Infections are more likely to occur when this condition exists because bacteria thrive on iron-rich blood.

Researchers are currently investigating a possible link between excess iron stores in the body and increased risk of chronic conditions such as heart disease. Researchers in Finland found that in a three-year study of 2,000 healthy men, the risk of heart attack was twice as great for men with the highest levels of stored iron in their bodies.[33] Although iron is an important essential nutrient in the body, it can also act as a powerful oxidizing agent in reactions that produce free radicals in the body. As noted in Chapter 4, free radicals can initiate the changes in LDL-cholesterol that eventually damage artery walls and lead to heart disease. Much more research is needed in this area before the relationship between iron and heart disease can be clarified. Until all the answers are in, avoid taking extra supplemental iron unless you have been diagnosed as deficient. Be sure to keep iron supplements out of the reach of children, too, who can fatally overdose on such tablets.

Zinc

Zinc is found in every cell of the body and plays a major role with more than 50 enzymes that regulate cell multiplication and growth, normal metabolism of protein, carbohydrate, fat, and alcohol, and the disposal of damaging free radicals. Zinc is associated with the hormone insulin, which regulates the body's fuel supply. It is involved in the utilization of vitamin A, taste perception, thyroid function, wound healing, the synthesis of sperm, and the development of sexual organs and bone.

More recently, zinc has become known for its role in promoting a healthy immune system.[34] A flurry of zinc lozenges have appeared on the market following the results of a study seeking an antidote for the common cold.[35] People who used zinc gluconate lozenges every 2 hours starting within 24 hours of a cold's onset reduced the length of their colds by 3 days. However, other studies have not reported similar results, and whether zinc lozenges are any more effective than placebos remains to be determined. Furthermore, the long-term effects of such supplemental use of zinc are unknown.

Zinc deficiencies were first reported in the 1960s in the Middle East, where studies on adolescent boys revealed severe growth retardation and delayed sexual maturation—symptoms responsive to zinc supplementation. The native diets were typically low in animal protein and zinc but were high in fiber and other compounds that bind minerals. The researchers learned that the binders were carrying zinc out of the boys' bodies, thus causing the deficiency.

Since then, cases of zinc deficiency have been discovered closer to home.[36] Zinc deficiency can cause night blindness, hair loss, poor appetite, susceptibility to infection, delayed healings of cuts or abrasions, decreased taste and smell sensitivity, and poor growth in children. Because zinc is lost from the body daily in much the same way as protein is, it must be replenished daily.

People who are building new tissue have the highest zinc needs—infants, children, teenagers, and pregnant women. The pregnant teenager is at particular risk because she needs zinc for her own growth as well as for her fetus's growth. Pregnant vegetarians are at risk, too, because their diets are high in fiber and zinc binding factors. Dieters also need to be reminded that very low-calorie diets

© Harold Sanstead, M.D.

▲ *The Egyptian boy in this picture is 17 years old but is only 4 feet tall, the height of an average 7-year-old in the United States. His genitalia are like those of a 6-year-old. The retardation, known as dwarfism, is rightly ascribed to zinc deficiency because it is partially reversible when zinc is restored to the diet.*

SOURCES OF ZINC

Adult DRI (RDA) is 8 to 11 mg.

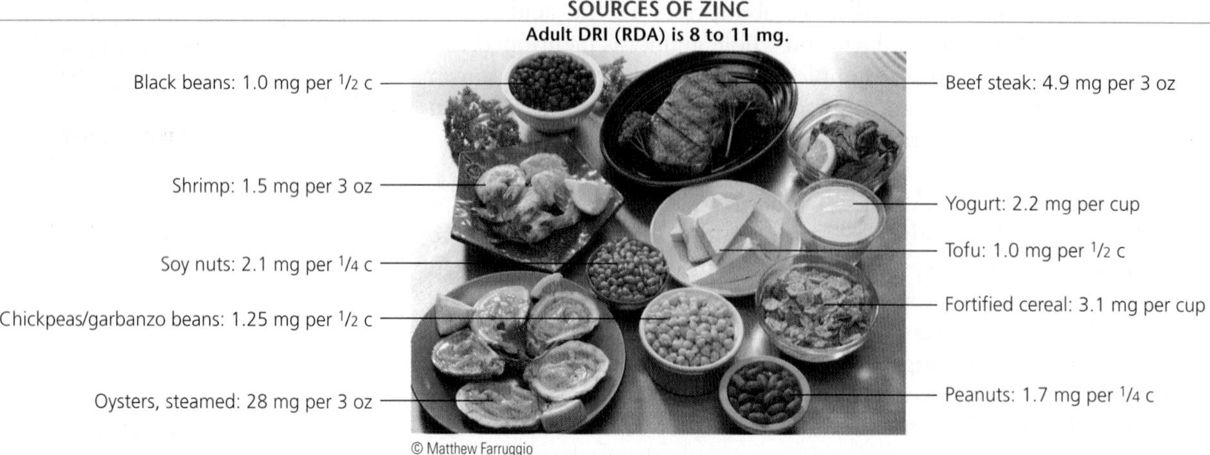

© Matthew Farruggio

TABLE 7-9
Zinc in Foods

(mg)	Sources
28.00	Oysters, cooked (3 oz)
6.50	Crabmeat (3 oz)
5.27	Ground beef (3 oz)
4.90	Sirloin steak (3 oz)
4.65	Beef pot roast (3 oz)
4.00	Soybeans, dry-roasted (½ c)
3.10	Enriched cereal (¾ c)
2.20	Yogurt (1 c)
1.73	Turkey (3 oz)
1.70	Peanuts (¼ c)
1.66	Swiss cheese (½ oz)
1.50	Shrimp (3 oz)
1.32	Cheddar cheese (½ oz)
1.10	Black-eyed peas (½ c)
1.00	Black beans (½ c)
1.00	Tofu (½ c)
0.98	Fat-free milk (1 c)
0.75	Green peas (½ c)
0.70	Kidney beans (½ c)
0.68	Spinach, cooked (½ c)
0.55	Whole-wheat bread (1 slice)

cause not only a low zinc intake but also a loss of zinc from body tissues as they break down to release fuel.

Zinc is a relatively nontoxic element. However, it can be toxic if consumed in large enough quantities. Consumption of high levels of zinc can cause a host of symptoms, including vomiting, diarrhea, fever, and exhaustion.[37] The hazards of overconsumption are greatest when consumers dose themselves with supplements. Excess supplemental zinc can cause imbalances of both copper and iron in the body. Chronic consumption of zinc exceeding 15 milligrams per day is not recommended without close medical supervision.

Table 7-9 shows the zinc amounts in foods. An average 1,500-calorie diet provides about 6.3 milligrams of zinc per day, or about 60 to 80 percent of the recommended intake. Zinc is highest in foods of high protein content, such as shellfish (especially oysters), meats, and liver. As a rule of thumb, two servings a day of animal protein will provide most of the zinc a healthy person needs. Whole-grain products are good sources of zinc if large quantities are eaten (the phytate in grains does not inhibit the absorption of zinc in people consuming ordinary diets). Cow's milk protein (casein) binds zinc avidly and seems to prevent its absorption somewhat; infants absorb zinc better from human breast milk. Fresh and canned vegetables vary in zinc content, depending on the soil in which they are grown. The zinc content of cooking water varies from region to region as well.

Iodine

Iodine occurs in the body in an infinitesimal quantity, but its principal role in human nutrition is well known. It is part of the thyroid hormones, which regulate body temperature, metabolic rate, reproduction, and growth. The hormones enter every cell of the body to control the rate at which the cells use oxygen and release energy.

When the iodine level of the blood is low, the thyroid gland may enlarge until it causes swelling in the throat area—a condition called **goiter.** Goiter is estimated to affect 200 million people the world over.

In addition to causing sluggishness and weight gain, an iodine deficiency can have serious effects on fetal development. Severe thyroid undersecretion by a woman during pregnancy causes the extreme and irreversible mental and physical retardation of the child known as **cretinism.** A person with this condition has mental retardation and a face and body with many abnormalities. Much of the mental retardation associated with cretinism can be averted by early diagnosis and treatment of the mother's iodine deficiency.

The amount of iodine in foods reflects the amount present in the soil in which plants are grown or on which animals graze. Soil iodine is greatest along the coastal regions. In the United States, in areas where the soil is low in iodine

goiter (GOY-ter) enlargement of the thyroid gland caused by iodine deficiency.

cretinism (CREE-tin-ism) severe mental and physical retardation of an infant caused by iodine deficiency during pregnancy.

(most notably in the plains states and around the Great Lakes and St. Lawrence River areas and in the Willamette Valley of Oregon), widespread goiter and cretinism appeared in the local people during the 1930s. Iodized salt was introduced as a preventive measure, and these scourges disappeared.

By the 1970s, a dramatic increase in iodine intakes had occurred—well above the recommended 150 micrograms a day.[38] The toxic level at which detectable harm results is thought to be only a few times higher than the average consumption levels at that time. Excessive intakes of iodine can cause an enlargement of the thyroid gland resembling a goiter; in infants it can be so severe as to block the airways and cause suffocation. Most of the excess iodine came from iodates (dough conditioners used in the baking industry), from milk produced by cows exposed to iodine-containing medications, and from disinfectants used during the milk treatment process.

Recent data have shown a surprising decline in iodine intake during the last two decades of the twentieth century, especially in women of reproductive age.[39] Although the average American dietary iodine intake is above the recommended intake, some 15 percent of women may be iodine deficient. The recent decrease in iodine intake may be attributed to several factors, including the replacement of iodine with bromine salts as the dough conditioner in commercial bread production, reduced intake of iodine-rich egg yolks for cholesterol concerns, reduced use of iodized table salt for high blood pressure concerns, the dairy industry's effort to reduce iodine in milk, and the increasing use of noniodized salt in manufactured foods.[40] The sudden emergence of this problem points to a need for continued surveillance of the food supply to prevent the effects associated with an iodine deficiency, including miscarriages, goiter, and mental retardation in babies of iodine-deficient mothers.

People sometimes ask if they should be sure to buy *iodized* salt in the grocery store to ensure adequate iodine intake. Although most consumers now have access to fruits and vegetables grown in coastal areas rich in iodine, health experts state the importance of using iodized salt to maintain an adequate iodine intake.

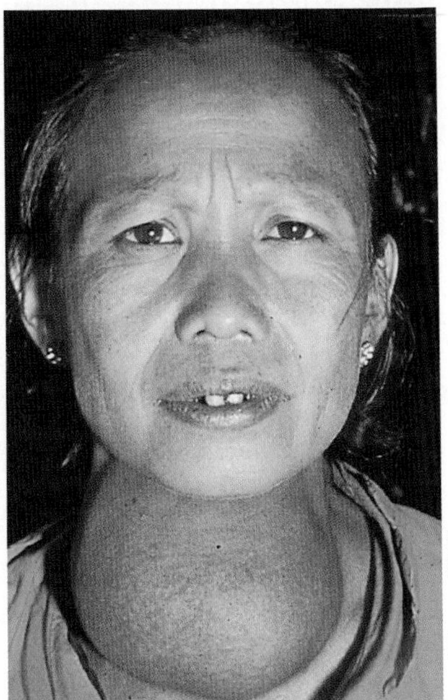

Lester V. Bergman/CORBIS

▲ *In iodine deficiency, the thyroid gland enlarges—a condition known as* simple goiter.

Fluoride

Only a trace of fluoride occurs in the human body, but studies have demonstrated that for people who live where diets are high in fluoride, the crystalline deposits in their teeth and bones are larger and more perfectly formed than in people who live where diets are low in fluoride. Fluoride not only protects children's teeth from decay, but it also makes the bones of older people resistant to adult bone loss (osteoporosis). Thus, its continuous presence in body fluids is desirable.

Drinking water is the usual source of fluoride. Where fluoride is lacking in the water supply, the incidence of dental decay is very high. Fluoridation of community water where needed to raise its fluoride concentration to one part per million (ppm) is thus an important public health measure (see Figure 7-7).

In some communities, the natural fluoride concentration in water is high (2 to 8 ppm), and children's teeth develop with mottled enamel—a condition called **fluorosis.** True toxicity from fluoride overdoses can occur, but usually only after years of chronic daily intakes of 20 to 80 times the amounts normally consumed from fluoridated water.

Although drinking fluoridated water ranks as one of the most effective ways to prevent dental cavities, many Americans continue to drink water not treated with the mineral, putting themselves at a high risk of developing tooth decay. Part of the problem is that the practice of adding fluoride to public water systems has been the subject of bitter controversy ever since it was introduced in 1945. Antifluoridation groups claim that drinking fluoride-containing water can cause everything from cancer to birth defects, despite reports to the contrary issued by major health organizations, including the National Institute of Dental Health and the American Dietetic Association.[41] In fact, a report from the National Research Council—based on hundreds of studies and the most comprehensive

Alternatives to water fluoridation:

- Fluoride toothpastes
- Fluoride treatments for teeth
- Fluoride tablets and drops

fluorosis (floor-OH-sis) discoloration of the teeth from ingestion of too much fluoride during tooth development.

FIGURE 7-7
FLUORIDATION IN THE UNITED STATES

SOURCE: Centers for Disease Control and Prevention, Recommendations for using fluoride to prevent and control dental caries in the United States, *Morbidity and Mortality Weekly Report* (August 17, 2002): 10.

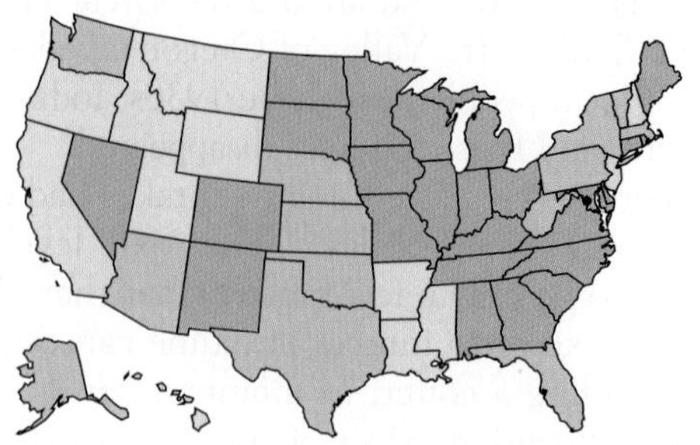
75% or more of the population using public water in these states is using fluoridated water.
50%–74% of the population is using fluoridated water.
< 49% of the population is using fluoridated water.

investigation of the health effects of fluoride to date—concluded that current levels of fluoride in drinking water are not associated with cancer, kidney disease, stomach and intestinal problems, infertility, birth defects, genetic mutations, or any other health problems for which fluoride has been blamed. However, fluoride in water, when combined with excess fluoride from toothpaste, mouth rinses, supplements, and other sources, can lead to fluorosis, the irreversible condition mentioned earlier which only occurs during tooth development. For this reason, children living in areas with fluoridated water should not be given fluoride supplements unless prescribed by a physician.[42]

Copper, Manganese, Chromium, Selenium, and Molybdenum

Several trace minerals have been found to have important roles in a variety of metabolic and physiological processes. Copper, for example, is involved in making red blood cells, manufacturing collagen, healing wounds, and maintaining the sheaths around nerve fibers. Chromium works closely with the hormone insulin to help the cells take up glucose and break it down for energy.[43] Selenium functions as part of an antioxidant enzyme and can substitute for vitamin E in some of that vitamin's antioxidant activities. Research is currently under way to investigate a possible role for selenium in protecting against the development of prostate and other forms of cancer.[44] Manganese and molybdenum both function as working parts of several enzymes. (Refer to Table 7-2 on page 220 for a brief description and common food sources of each of these minerals.)

Trace Minerals of Uncertain Status

None of the trace minerals has been known for very long, and some are extremely recent newcomers. Some researchers consider the trace mineral boron to be one factor that contributes to the risk of osteoporosis because of its effects on calcium metabolism.[45] Nickel is now recognized as important for the health of many body tissues, and nickel deficiency can harm the liver and other organs. Silicon is known to be involved in bone calcification, at least in animals. Tin is necessary for growth in animals and probably in humans. Vanadium, too, is necessary for growth and bone development as well as for normal reproduction. Cobalt is recognized as the mineral in the large vitamin B_{12} molecule; the alternative name for this vitamin—cobalamin—reflects the presence of cobalt. In the future, we may discover that many other trace minerals, for example, silver, mercury, lead, barium, and cadmium, also play key roles in human nutrition. Even arsenic, famous as the poisonous instrument of death in many murder mysteries and known to be a carcinogen, may turn out to be an essential nutrient in tiny quantities.

Eat Well Be Well Choosing a Vitamin/Mineral Supplement

For years, health experts have been saying that most healthy people can meet their vitamin and mineral needs with a balanced diet. Nevertheless, almost half of the U.S. population, including about half of young adults and college students, pop vitamin and mineral pills.[46] Americans spend more than $20 billion annually on pills, powders, liquids, and other vitamin and mineral supplements, often in the mistaken belief that such preparations will ensure proper nutrition, help reduce stress, decrease fatigue, and increase pep and energy.

But before you buy a supplement, remember that most major health organizations—from the American Dietetic Association to the American Medical Association to the American Academy of Pediatrics—essentially agree that healthy children and adults should be able to get all the nutrients they need by eating a variety of foods. However, those organizations and other experts say that taking a multivitamin/mineral supplement, under the guidance of a physician or dietitian, may be in order for these particular groups of people:

- People following very-low-calorie diets
- People with certain diseases or those taking medications that interfere with appetite, absorption, or excretion of nutrients
- Strict vegetarians, whose diets may fall short in vitamin B_{12}, vitamin D, calcium, iron, and zinc
- Women who are pregnant or breastfeeding, phases that bolster the need for nutrients, including iron and folate
- Women with excessive menstrual bleeding, who may need iron supplements
- Women during their childbearing years who do not consume folate-rich or folic acid–fortified foods may need more folate in their diets to prevent neural tube defects in infants
- Anyone with lactose intolerance or who does not consume milk or other dairy products needs a source of calcium; those with inadequate exposure to sunlight may also need vitamin D
- Elderly people who may have difficulty choosing an adequate diet, chewing problems, or a reduced ability to absorb and metabolize certain nutrients (see Chapter 11)
- People who are recovering from surgery, burn injuries, or other illnesses that increase nutrient needs
- People with heart disease or who are at risk for heart disease and consume diets inadequate in the B vitamins (folate, vitamin B_6, and vitamin B_{12})
- People with chronic diseases of the digestive tract or other conditions that lead to poor intake or deplete nutrient stores
- People with alcohol or other drug addictions are likely to have a shortage of vitamins and minerals in their diets

◀ This symbol on a dietary supplement label helps you know that the supplement actually contains the listed ingredients in the declared amount, does not contain harmful levels of contaminants, will break down and release its ingredients in the body, and has been made under good manufacturing practices. See also: www.uspverified.org.

If you decide to start taking a supplement, keep the following points in mind when choosing one:

1. Remember that price is not an indication of quality. Many products sold at major retail chains and drugstores are just as high in quality as pricier versions sold in health food stores. Look for a product that meets high standards for manufacturing. One way to do this is to check the label to see whether the product meets USP standards—manufacturing practices set forth by the U.S. Pharmacopeia, the organization that establishes drug standards. The organization's standards require that a supplement be able to disintegrate and dissolve thoroughly in the stomach within a certain period of time, thereby increasing the chances that the nutrients inside are absorbed and used by the body. Also look for a bottle or package that carries an expiration date. If it doesn't, you run the risk of buying a product that has been sitting on a shelf for an indefinite period of time. After a while, the product may lose its potency.
2. Look for a supplement that contains both vitamins and minerals, with no more than 100 percent to 150 percent of the recommended Daily Values for each. For the most part, nutrients work in concert with one another, promoting the body's ability to make use of them. Products that include a balanced mix of vitamins and minerals are the best bet for most people. Steer clear of products containing extraneous substances such as PABA, hesperidin, inositol, and bee pollen. These nonvitamin substances have never been proved essential to humans and only add to the price of the supplement. Be wary of taking multivitamins that also contain herbs. Although herbal products are considered to be dietary supplements, the unregulated herbal industry of today is a buyer-beware market (see page 198).
3. Buy products sold in childproof bottles or packages if you have children around. Vitamins and minerals, especially iron, can be highly toxic to children. Every year, tens of thousands of children swallow excess vitamin/mineral supplements, and iron-tablet overdoses alone are one of the top causes of accidental death in youngsters.

Spotlight Osteoporosis—The Silent Stalker of the Bones

© Larry Mulvehill/Photo Researchers, Inc.

▸ *Osteoporosis strikes the bones. This 75-year-old woman has lost height—not because her legs have grown shorter but because vertebrae in her back have collapsed. These fractures occur on the front side of the spinal bones, resulting in a hunched-over posture.*

- As she waited in a grocery checkout line, Nancy Miller Friesen felt a stab of pain in her left hip. The 40-year-old Fort Worth woman had always been healthy and active—suddenly she could barely stand. When she went to her family doctor, she learned that several of her vertebrae had fractured. The radiologist studying her X rays thought he was looking at the bones of a 70-year-old woman.
- Lila Rubin had just teed off at a New Jersey golf course. Putting down her club, the robust 63-year-old felt a searing back pain. When Rubin met with her orthopedic surgeon, she was told she'd suffered a serious compression fracture.
- Debra Epstein of New Haven, Connecticut, wasn't worried when her physician diagnosed a mild curvature of her upper spine 7 years ago. She was only 31 and in otherwise excellent health. Then last year, her right wrist began aching and her back tired easily. After tests, she learned that at age 37, she had the brittle bones of a woman twice her age.

The three women just described share the potentially crippling and sometimes deadly condition: osteoporosis.[47] Osteoporosis threatens the integrity of the skeleton in some 44 million adults in the United States. One out of two women and one out of four men over 50 will suffer an osteoporosis-related fracture in their lifetime. Osteoporosis is more prevalent in women than in men after age 50 for several reasons:[48]

- Women generally have less bone mass than men.
- Women typically have lower calcium intakes than men.
- Women more often use weight-loss diets, which tend to be low in calcium and lead to bone loss.[49]
- Bone loss begins earlier in women because of women's different hormonal makeup, and the loss is accelerated at **menopause,** when their protective **estrogen** secretion declines.
- Pregnancy and lactation decrease the calcium reserves in bones whenever calcium intake is inadequate.
- Women live longer than men, and bone loss continues with aging.

What is bone loss?

Adult bone loss affects the entire skeleton, but it occurs first in the pelvis and the spine. Unfortunately, osteoporosis establishes itself silently in its victims' lives. Symptoms do not usually occur until late in life. The disease often results in fractures of the hip, spine, and wrists—over 1.5 million such fractures a year—that occur with very little or no stress or pressure (see Figure 7-8).[50] Often, osteoporosis first becomes apparent when someone's hip suddenly gives way. Fewer than 50 percent of all women with hip fractures return to previous activity levels, and many are confined to wheelchairs for the remainder of their lives. Approximately 20 percent of these women die of complications within the first year after hip fracture.

Bones are made up of a complex matrix based on the protein collagen, into which the crystals of the bone minerals—principally calcium and phosphorus—are deposited. Bone tissue has two forms: **trabecular bone** and

FIGURE 7-8
BONE LOSS AND MOST COMMON TYPES OF BONE FRACTURES IN WOMEN
A total of 1.5 million fractures occur each year including 250,000 fractures at sites not shown here.

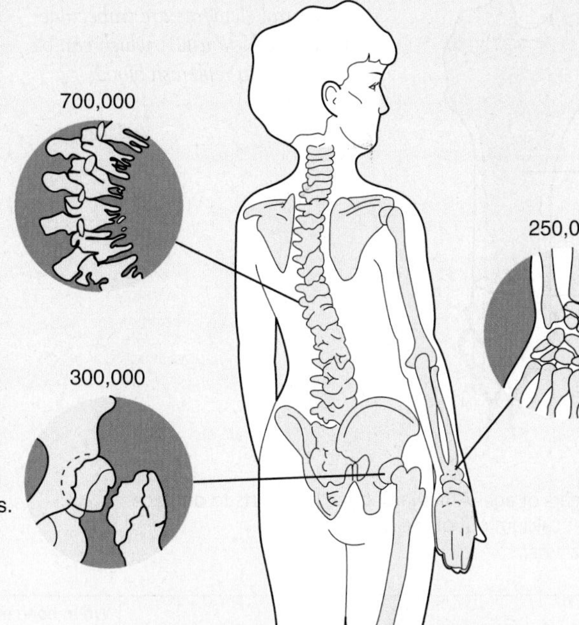
Spinal Vertebrae Fracture
• More likely at ages 55 to 75 years.
• Fracture can occur from bending, lifting, or spontaneously.
• Bone weakens and collapses, leading to loss of height and chronic back pain.
• Fractures occur on front side of vertebrae, and a hunched-over posture results.
700,000
250,000
300,000
Hip Fracture
• Most serious type of fracture.
• Most occur at 70 years or older.
• May lead to serious complications.
• 50 percent of victims become institutionalized.
Wrist or Forearm Fracture
• Most occur at 50 years or older.
• Fracture often occurs when hands are used to break a fall.
• Can be early warning sign for osteoporosis.

cortical bone. The thick ivory-like outer portion of a bone is the cortical bone. Cortical bone provides a covering for the inner trabecular bone—a lacy network of calcium-containing crystals that are almost spongelike in appearance (see Figure 7-9). When calcium intakes are low, hormones call first upon the trabecular bone to release calcium into the blood for use by the rest of the body whenever levels fall too low.* Over time, the lacy network of bones becomes less dense and more fragile as calcium deposits are withdrawn.

Although bones undergo remodeling throughout life—adding and losing bone minerals—the total amount of bone mass in the human body reaches a peak by about age 30.[51] Afterward, bones lose strength and density as bone minerals are lost. The principal determinant of bone health is peak bone mass. Think of bone mass as money in the bank: the larger the savings account, the easier it will be to withstand some loss of bone as one ages. To attain a healthful peak bone mass, it is necessary to have an optimal intake of calcium during the years of

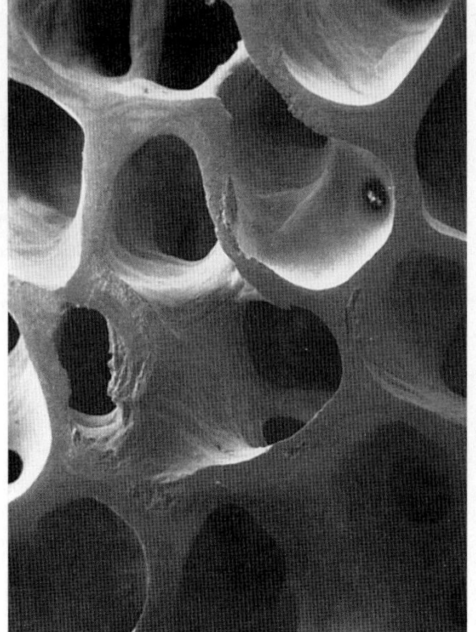

▲ *Electron micrograph of healthy trabecular bone.*

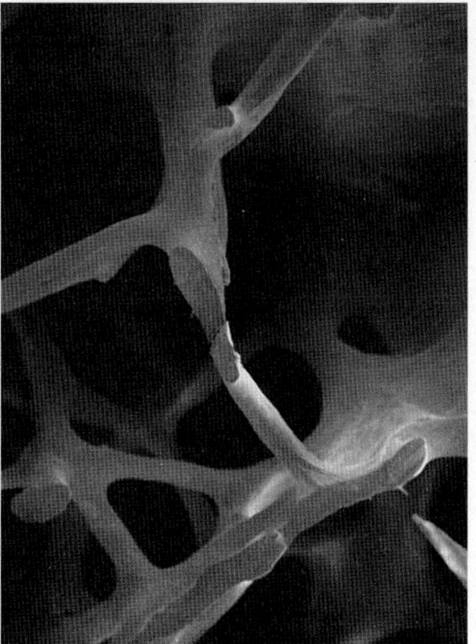

▲ *Electron micrograph of trabecular bone affected by osteoporosis.*

*Vitamin D and parathyroid hormone circulate in the blood whenever the calcium concentration in the blood falls too low. These hormones cause the release of calcium from the bones, a decreased excretion of calcium from the kidneys, and an increased absorption of dietary calcium from the intestines. The hormone calcitonin is released from the thyroid gland when levels of blood calcium get too high; it acts to stop the release of calcium from bone and to slow intestinal absorption of the mineral.

FIGURE 7-9
A BONE'S LIFE

A. Bone is living tissue that continuously remodels itself. Bone contains cells embedded in a hard mineralized matrix. The osteoblast cells make new bone; the osteoclast cells break down old or damaged bone.

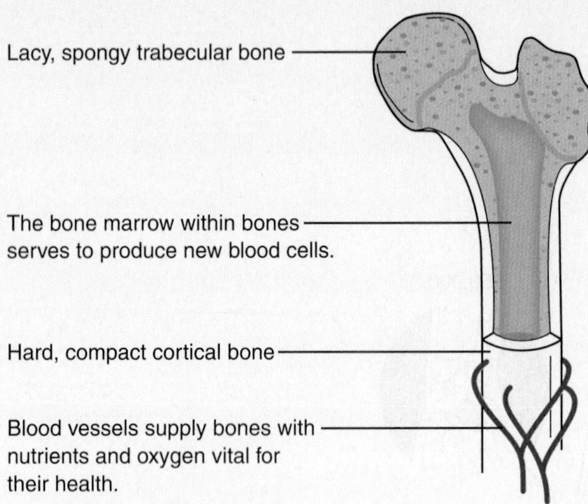
Lacy, spongy trabecular bone
The bone marrow within bones serves to produce new blood cells.
Hard, compact cortical bone
Blood vessels supply bones with nutrients and oxygen vital for their health.

▸ *Cross section of bone. The lacy structural elements are trabeculae (tra-BECK-you-le), which can be drawn on to replenish blood calcium.*

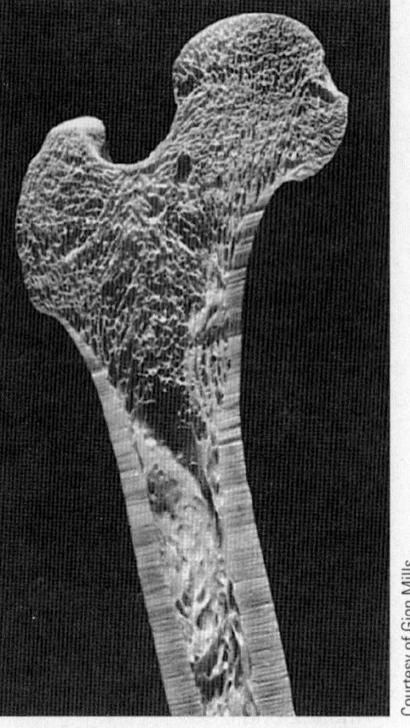
Courtesy of Gjon Mills

B. Peak bone mass occurs at approximately 30 years of age. Afterward, bone loss starts to outpace bone deposition; at menopause there is a surge of calcium out of the bones.

SOURCE: Adapted from The Stay Well Company, San Bruno, CA. © 1998.

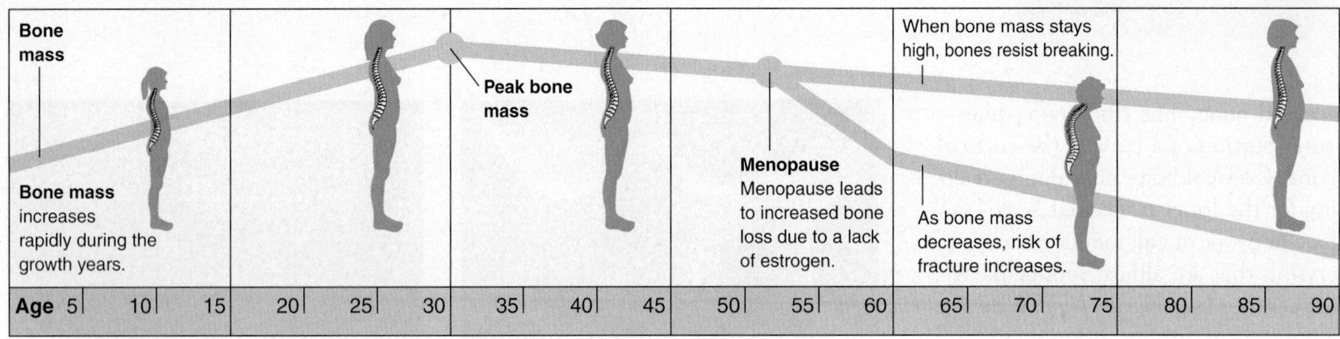
Bone mass
Bone mass increases rapidly during the growth years.
Peak bone mass
Menopause
Menopause leads to increased bone loss due to a lack of estrogen.
When bone mass stays high, bones resist breaking.
As bone mass decreases, risk of fracture increases.
Age 5 10 15 20 25 30 35 40 45 50 55 60 65 70 75 80 85 90

bone growth—all through the growing years and on into the years of young adulthood.[52]

Can osteoporosis be prevented?

Although there is no cure for osteoporosis, the good news is that you can take steps to minimize your risk. First, determine your risk by considering the following list of factors that play a role in osteoporosis:

- *Age*. During childhood, adolescence, and the young adult years, the cells that build bone **(osteoblasts)** form more bone than the bone-dismantling cells **(osteoclasts)** take away. With aging, the bone-building cells become less active, but the bone dismantlers continue to work—causing bone tissue and strength to decline.
- *Gender*. Osteoporosis is four times more common in women. Men achieve a higher peak bone mass than women do, and the rate of age-related bone loss is lower in men than in women (see Figure 7-10).
- *Age-related decline in hormones*. Hormones are important in bone health for both men and women. Menopause deprives women of the protective effects of estrogen. Estrogen improves calcium absorption from the intestines and reduces excretion of the mineral by the kidneys.[53] Bone loss accelerates in women for the five to ten years following menopause. The earlier that menopause occurs, the greater a woman's risk of osteoporosis. Women who have ceased menstruating (a reliable indicator of reduced estrogen levels) comprise the largest segment of people with osteoporosis. Menopause occurs naturally in women typically between the ages of 45 and 55, but it can occur earlier in life with the removal of diseased ovaries. Risk of osteoporosis increases in men with an age-related decline in testosterone; low testosterone levels increase risk of fractures in men.[54]
- *Abnormal absence of menstrual periods (estrogen deficiency)*. Menstruation

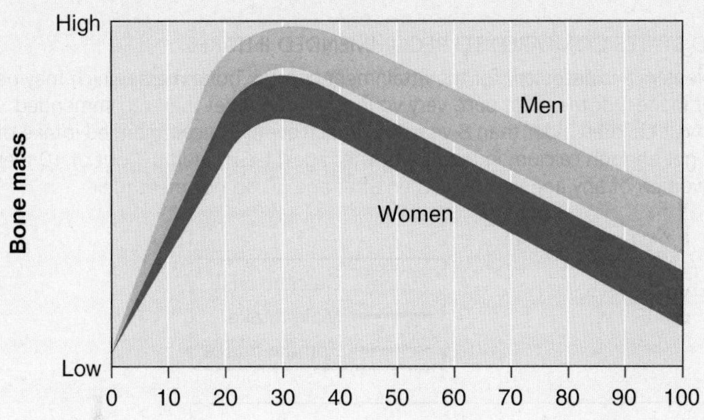

FIGURE 7-10
BONE LOSS THAT OCCURS WITH AGE

SOURCE: National Osteoporosis Foundation, 1997.

can also cease in women who overexercise and are underweight, a condition called athletic amenorrhea (discussed in Chapter 10). For the same reason, eating disorders (anorexia nervosa and bulimia) increase the risk of osteoporosis later in life.[55]

- *Family history*. The greater the history of skeletal fractures of the hips and vertebrae among elderly relatives, the greater your risk for osteoporosis.
- *Race and ethnic background*. Those of British, northern European, Chinese, Japanese, or Mexican American background, or Hispanic people from Central and South America are at the highest risk. African American women tend to have denser bones and are at a lower risk.[56] Undoubtedly, environmental factors—such as calcium intakes, physical activity, smoking, body weight, and alcohol intake—can influence the ultimate outcome of one's genetic heritage.
- *Body build*. The smaller the frame and the thinner the person (under 127 pounds), the greater the risk. Petite women have less bone to lose than larger-boned women.
- *Sedentary lifestyle*. Inactivity leads to bone loss.[57]
- *Smoking and alcohol*. Smoking and excessive alcohol intake increase the risk. People who smoke tend to have lower body weights than nonsmokers. Some women smokers have lower levels of estrogen in their blood and experience an earlier onset of menopause compared to nonsmokers.[58] Alcohol decreases the activity of the bone-building cells, interferes with the absorption of calcium, and may increase excretion of calcium from the kidneys. People who abuse alcohol also tend to have poor nutrition status, including low intakes of calcium.
- *Medical conditions*. Certain medical conditions including type 1 diabetes, thyroid disorders, rheumatoid arthritis, asthma, seizures, and organ transplants are associated with increased risk, primarily because of the drugs used to treat them. Prolonged use of medications such as excessive thyroid hormone, antiseizure drugs, and anti-inflammatory drugs—such as prednisone—used to treat asthma, arthritis, and some cancers may reduce calcium absorption, impair bone formation, and accelerate bone loss.
- *A bone-healthy diet that includes calcium, vitamin D, and other nutrients*. Calcium intake early in life affects the attainment of peak bone mass, achieved by about age 30, which may be the most important determinant of the risk of a fracture in later life.[59] Hence, ensuring an adequate calcium intake throughout life appears to be a sensible strategy. Adequate intake of Vitamin D is required for the absorption of calcium. Many older adults, who typically have lower intakes of vitamin D and reduced ability to synthesize the vitamin from sunshine, absorb less calcium as they age.

The calcium intake of a sizeable portion of the U.S. population, especially female adolescents and female adults, is estimated to be below the recommended intake (see Figure 7-11).[60] Between the ages of 18 and 30 years, the time of the attainment of peak bone mass, more than two-thirds of all American women consume less calcium than they need. One explanation is the obsession (especially among adolescent girls) with dieting. Consumption of diet sodas has increased dramatically over the past three decades, often displacing milk as an accompanying beverage to meals.[61]

Other nutrients also play a role in bone health.[62] For example, high intakes of phosphorus, protein, vitamin A (as retinol), and sodium may adversely affect calcium in the body.[63] Adequate intakes of vitamins C and K are needed to maintain healthy bone. Low vitamin K intakes have recently been associated with increased risk of hip fractures in the elderly.[64] Potassium, magnesium, and fruit and vegetable intakes are also associated with greater bone density in older adults.[65] The many phytochemicals found in fruits, vegetables, and legumes also offer protective benefits to bone (as discussed in Chapter 6).

Should I consider taking calcium supplements if I am unable to meet the recommended intake for calcium using foods? If so, what should I consider when choosing a supplement?

Although it is preferable to meet calcium needs by consuming calcium-rich foods, calcium supplements are an

FIGURE 7-11

THE CALCIUM GAP: DAILY CALCIUM INTAKES IN THE UNITED STATES COMPARED TO RECOMMENDED INTAKES

Maintaining adequate calcium intake during childhood and adolescence is necessary for the attainment of peak bone mass, which may be important in reducing the risk of fractures and osteoporosis later in life. For the most part, very young children meet their recommended calcium intakes. In contrast, according to national survey data, most children older than 8 years fail to achieve the recommended intake of calcium. Some 81 percent of girls and 48 percent of boys fail to get enough calcium in their diets after age 11—including 7 out of 10 teen boys and 9 out of 10 teen girls. Likewise, the majority of adult women of any age and adult men after age 50 do not meet their recommended calcium intakes.

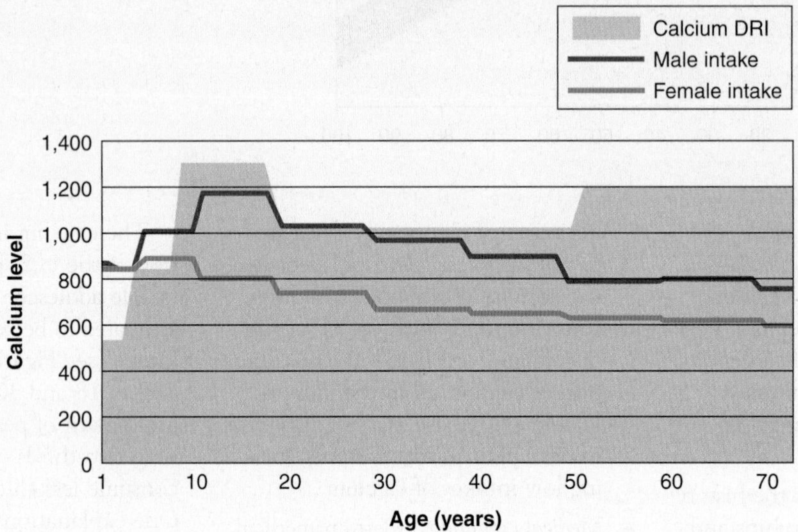

alternate way to obtain calcium.[66] Supplements may be needed by those who either cannot or will not adjust their diets to get enough calcium from food. Supplements have the advantage of being easy to take, but they have the disadvantage of not offering other nutrients, such as the thiamin, riboflavin, niacin, potassium, phosphorus, magnesium, zinc, vitamin A, and vitamin D found naturally in milk.

Choosing a calcium supplement can be confusing. Calcium comes in a number of different salts, and the cost of a one-month supply of 1,000 milligrams per day can vary widely. The organic salts include calcium citrate, calcium citrate-malate, and calcium lactate; the inorganic salts include calcium carbonate and calcium phosphate. Calcium carbonate is the least expensive and contains the most calcium per pill, but an organic salt, such as calcium citrate, is better absorbed,

Soft drinks often replace milk as a beverage in the diet of teenagers, accounting for about 10 percent of their caloric intake.

especially by older people whose secretion of stomach acid—a factor known to enhance calcium absorption—tends to be reduced.[67] A strategy to overcome this problem is to take calcium carbonate two or three times a day in divided doses, with meals, to improve absorption. Or, try a chewable form, or take the calcium carbonate pill with a glass of orange juice to help it dissolve. Because regular vitamin-mineral pills contain only small amounts of calcium, read the label carefully. Note, too, that many calcium supplements come with added nutrients, such as vitamin D and magnesium. You may not need these additional nutrients if you are consuming a nutritious diet, taking a multivitamin-mineral pill, or eating highly fortified cereals.

Keep in mind that a cup of milk contains about 300 milligrams of calcium. If you are using supplements to replace the calcium found in milk, choose a supplement that contains at least 300 milligrams of calcium. Also, because calcium is best absorbed in doses of 500 milligrams or less, consuming supplements with more than 500 milligrams of calcium per pill is not advantageous. For example, if you need to take 1,000 milligrams of calcium in supplement form, choose a 500-milligram tablet and take one in the morning and one in the evening. A generic brand is fine to use—just be sure to check the label to see how much elemental calcium and which form of calcium it contains. Avoid calcium supplements derived from dolomite, oyster shell, or bone meal because some have been found to be contaminated with lead.[68] Table 7-10 offers tips for selecting from among the many popular calcium supplements available today. See also the Eat Well Be Well feature earlier in this chapter.

Is there a way to detect bone loss before the onset of a fracture?

A **bone density** test can detect low bone density before a fracture occurs, and it can predict your chances of fracturing a bone in the future. Since 1998, bone-density screening has been covered by Medicare for all women over age 65. The National Osteoporosis Foundation encourages women to discuss with their physician the possibility of having their bone density measured at the onset of menopause, when bone loss sharply accelerates, or if they have other medical conditions that increase their risk for osteoporosis. The more risk factors you have for osteoporosis, the more you need to evaluate your bone density as you age.

TABLE 7-10
Tips for Selecting and Using a Calcium Supplement

Supplement[a]	Calcium (milligrams/pill)	Advantages	How to Take[b]
Calcium Carbonate		Least expensive and most available	Take with meals
Caltrate, One A Day,	600, 500		
Os-Cal,[c] Tums,[d]	500, 500		
VIACTIV,[d]	500		
GNC, Solgar	400, 600		
Calcium Citrate		Easily absorbed, regardless of the amount of stomach acid	Take on an empty stomach
Citracal	315		
Solgar	250		
Calcium Phosphate		May be less likely to cause constipation	Take with meals
Posture-D	600		
Your Life[c]	250		

[a] Calcium with other supplements:
- If you also take an iron supplement, take it at a different time of day. Each mineral is better absorbed alone.
- Very high calcium intakes may limit zinc absorption—a concern for the many Americans who already get too little zinc. High intakes may also increase the risk of kidney stones, especially when taken on an empty stomach. The safe, upper limit for total calcium (from foods and supplements) is 2,500 milligrams per day.
- Although vitamin D is necessary for calcium absorption, excessive amounts may be harmful. Supplemental vitamin D in the amount of 400 to 600 IU per day should be sufficient for most individuals.

[b] Calcium with medications:
- Calcium should be taken separately from some medications such as Fosamax or the drug's effectiveness can be diminished. Ask your doctor or pharmacist if you have any questions about food–drug interactions.

[c] Calcium source is oyster shells.
[d] Chewable.

SOURCE: Adapted from "How to Choose and Use a Calcium Supplement," *Nutrition in Clinical Care* 1(2) (1998): 100.

Are any drug treatments available to reverse bone loss?

Yes. Among the bone-conserving drugs currently approved by the Food and Drug Administration are **estrogen replacement therapy, designer estrogens,** and **bisphosphonates** for osteoporosis prevention (see the accompanying Miniglossary). These options and the hormones **calcitonin** and **parathyroid hormone** are also available for the treatment of osteoporosis.[69] These drugs work by slowing bone loss, allowing bones to slowly rebuild and increase bone density, and by reducing the risk of fractures by as much as 50 percent. To be effective, these treatments should be accompanied by the recommended intakes of calcium (1,200 mg for women over 50 years of age) and vitamin D (10 micrograms for adults 51 to 70 years), as well as regular weight-bearing exercise.[70]

What can be done to prevent the debilitating effects of osteoporosis?

Not surprisingly, the same recommendations have been made before for lowering risks for other chronic diseases. Eat a nutritious diet and exercise *throughout life*. Consider the following steps for building bone mass and lowering your risk for osteoporosis:[71]

- Maximize peak bone mass and be vigilant about keeping the bones well supplied with calcium. Consume the recommended amount of calcium (and vitamin D) for your age group (refer back to Figure 7-4 on page 221). According to researcher Robert Heaney, only a few simple changes can add an extra

© David Madison/GettyImages

▲ *Incorporate regular sessions of weight-bearing exercise, such as brisk walking, into your weekly schedule to slow and possibly even prevent bone loss (for more about exercise, see Chapter 10).*

MINIGLOSSARY

bisphosphonates drugs that decrease the risk of fractures by acting on the bone-dismantling cells (osteoclasts) and inhibiting their resorption of bone tissue; examples are Fosamax, Actonel, and Boniva. The side effects of long-term use of these drugs have yet to be determined.

bone density a measure of bone strength that reflects the degree of bone mineralization. The DEXA (dual-energy X-ray absorptiometry) bone density test compares your bone density to that of a healthy young adult. Bone density tests—using a dual beam of low-level X rays—take a snapshot of bone density in the spine, wrist, and hip. Simpler, less precise tests use ultrasound to measure bone density in the wrist or heel, and they can be done in a physician's office to possibly identify individuals in need of more precise testing.

calcitonin a hormone used as a drug to decrease the rate of bone loss in osteoporosis. Administered as a nasal spray (Miacalcin) or by injection, calcitonin works by inhibiting the bone resorption activity of osteoclasts.

cortical bone the dense outer ivory-like layer of bone that provides an exterior shell over trabecular bone.

designer estrogens (Selective Estrogen Receptor Modulators—SERMs) drugs that act on *estrogen receptors* in osteoblasts to promote an increase in bone mass; an example is raloxifene (Evista). Unlike estrogen, SERMs have little effect on reproductive tissues of the breast or uterus. *Estrogen receptors* are cellular molecules that bind to estrogen, selective estrogen receptor modulators, or phytoestrogens and deliver these compounds to the nucleus of the cell. Phytoestrogens, estrogen-like compounds found in soy foods, may act as SERMs (see the Spotlight in Chapter 5).

estrogen a major female hormone; important in connection with nutrition because it maintains calcium balance and because its secretion abruptly declines at menopause.

estrogen replacement therapy (ERT) administration of estrogen to replace the natural hormone that declines with menopause. Because ERT may increase the risk of uterine and breast cancer, *hormone replacement therapy,* the administration of a combination of estrogen with the hormone progesterone, is often used. However, the FDA recommends that HRT and ERT be considered only for women with significant risk of osteoporosis that outweighs the risk of the drugs (increased risk for breast cancer, stroke, and heart disease).

menopause the time of life at which a woman's menstrual cycle ceases, usually at about 45 to 50 years of age.

osteoblast a bone-building cell; responsible for formation of bone.

osteoclast a bone-destroying cell; responsible for resorption and removal of bone.

parathyroid hormone a hormone used as a drug (Teriparatide) to stimulate new bone formation; administered by injection once a day in the thigh or abdomen. Daily injections of Teriparatide (brand name Forteo) lead to increased bone mineral density.

peak bone mass the highest bone density achieved for an individual; accumulated over the first three decades of life; typically occurs by 30 years of age. After age 30, bone resorption slowly begins to exceed bone formation.

trabecular (tra-BECK-you-lar) **bone** the lacy inner network of calcium-containing crystals; spongelike in appearance, it supports the bone's structure.

Do You Have a Bone-Healthy Lifestyle?

No matter what your age, your food and lifestyle choices make a difference to your bone health. How are you "banking" on bones? Give yourself 10 points for each **always**, 5 points for each **sometimes**, and zero for each **never**.

Do You . . . ?	ALWAYS	SOMETIMES	NEVER
1. Eat three or more servings of milk, yogurt, cheese, or other calcium-rich foods?	☐	☐	☐
2. Drink vitamin D-fortified milk or get moderate exposure to sunlight?	☐	☐	☐
3. Keep your body moving—with at least 30 minutes of weight-bearing activity, at least three times weekly?	☐	☐	☐
4. Consume enough calories to maintain a healthy weight (not too thin)?	☐	☐	☐
5. Avoid cigarettes?	☐	☐	☐
6. Go easy on alcoholic beverages?	☐	☐	☐
Score	___ +	___ +	___ = ___

SCORING

A perfect 60: You are banking on beautiful bones as a lifelong investment. Toast to a perfect score with an ice-cold glass of milk!
35–55 points: For healthier bones, you'd be wise to improve your investment strategies. Otherwise, there may be future penalties.
30 or less points: Your calcium withdrawals may exceed your deposits. Make changes now before your window of opportunity closes.

SOURCE: *Banking on Beautiful Bones: A Lifelong Commitment to Calcium* (Rosemont, Il: National Dairy Council, 1998). Reprinted by permission.

1,000 milligrams of calcium to an ordinary diet. Put a couple of heaping teaspoons of nonfat dried milk powder in each cup of coffee or tea, drink calcium-fortified orange juice, and eat 1 cup of low-fat yogurt, a dark green leafy vegetable, and a 1-inch cube of hard cheese.[72]

- Consume alcohol only in moderation, if at all, and avoid cigarettes altogether.
- Exercise regularly, because exercise can reduce the risk of developing osteoporosis by making bones stronger and increasing their ability to absorb calcium.[73] Incorporate regular sessions of weight-bearing exercise, such as brisk walking, weight training, stair climbing, rope jumping, dancing, hiking, tennis, or aerobic dance classes, into your weekly schedule to slow and possibly even prevent bone loss.[74] Swimming and cycling are less effective at building bone mass, because they put less weight on the bones. Exercises designed to strengthen the back muscles and improve posture are also recommended, because they help the skeleton bear its burden of weight. Increased muscle strength and flexibility improve balance and helps prevent debilitating falls.[75]
- For women at or nearing menopause, talk to your health care provider about the need for bone density testing, medications, and non-drug treatments to slow bone loss after menopause.

Taking these steps to lower your risk of osteoporosis or to treat it if you already have it improves the chances of enjoying a quality life in your later years. Determine your own risk for osteoporosis using the box above, which examines your current lifestyle for osteoporosis risk factors.

In Review

1. Which statement does *not* describe the role of major minerals in the body?
   a. They give teeth and bones their rigidity and strength.
   b. They regulate body processes.
   c. They provide energy.
   d. They help maintain fluid balance.
2. Trace minerals are less essential than the major minerals.
   a. true
   b. false
3. Almost all (99%) of the calcium in the body is used to:
   a. provide energy for cells.
   b. provide rigidity for the bones and teeth.
   c. regulate the transmission of nerve impulses.
   d. maintain the blood level of calcium within very narrow limits.
4. Minerals are divided into two categories. Name the two categories and list the minerals in each one.
5. Briefly explain how cells maintain water balance and name some minerals involved in this process.
6. __________ is the most abundant mineral in the body.
   a. Potassium
   b. Phosphorus
   c. Sodium
   d. Calcium
7. Which of the following is *not* a good way to enhance iron absorption?
   a. Take large doses of iron supplements.
   b. Cook with iron cookware.
   c. Eat vitamin C-rich foods when eating iron-rich foods.
   d. Eat foods containing heme iron.
8. From which source might people unintentionally consume excessive sodium?
   a. fresh vegetables
   b. processed foods
   c. fast-food restaurants
   d. fresh fruits
   e. b and c
9. Briefly explain why fluoride is added to drinking water.
10. Why is water "the most essential nutrient of all"?
    a. It accounts for 60% of body weight.
    b. It carries nutrients throughout the body.
    c. It aids in the regulation of body temperature.
    d. all of the above

Menu of Online Study Tools

A variety of study tools for this chapter are available at our website to deepen your understanding of chapter concepts. Go to

www.thomsonedu.com/nutrition/boyle

to find

- Practice tests
- Flashcards
- Glossary
- Web links
- Animations
- Chapter summaries, learning objectives, and crossword puzzles

Nutrition on the Web

www.thomsonedu.com/nutrition/boyle
Go to the *Personal Nutrition* site to check for the latest updates to chapter topics or to access links to related websites.

www.healthfinder.gov/library
Search for information on water and the minerals.

www.iom.edu/fnb
Search for updates regarding new Dietary Reference Intakes for Electrolytes and Water.

www.nal.usda.gov/fnic
Search here for mineral-related topics.

www.nlm.nih.gov
Free access to National Library of Medicine's Medline for information searches on a variety of health-related topics.

www.eatright.org
American Dietetic Association's resources on mineral topics.

www.nationaldairycouncil.org
Search here for the latest in calcium and nutrition research.

www.whymilk.com
Check out the milk tips, recipes, and contests for consumers.

www.osteo.org
Fact sheets on many osteoporosis-related topics.

www.nof.org
Information from the National Osteoporosis Foundation.

www.4women.gov
Information on health issues for women in question-and-answer format; includes updates surrounding hormone replacement therapy and menopause.

www.epa.gov
Information from the Environmental Protection Agency regarding water quality and water safety issues.

www.nhlbi.nih.gov/health/public/heart/hbp/dash/
Information on the DASH diet, recipes, menus, and other resources for people with high blood pressure.

www.cdc.gov
Information about water safety from the Centers for Disease Control and Prevention.

Alcohol and Nutrition

8

ASK YOURSELF . . .

Which of the following statements about nutrition are true, and which are false? For each false statement, what is true?

1. A 12-ounce beer, a 5-ounce glass of wine, and a 1-ounce shot of tequila all contain the same amount of alcohol.
2. The impact of alcohol on health depends partly on whether you're a man or a woman.
3. Burnt toast is a good hangover remedy.
4. Alcohol is calorie free.
5. Regular drinkers become more tolerant of the effects of alcohol, so they must drink more to feel the effects of alcohol.
6. Reflexes are not impaired if your blood alcohol concentration is below the legal limits of intoxication.
7. It is safe for a pregnant woman to have one alcoholic beverage a day.
8. Drinking alcohol may be associated with an increased risk of breast cancer.
9. Moderate drinking can reduce the risk of heart disease.
10. Heavy drinking is defined as more than two drinks a day for women and more than four drinks a day for men.

Answers found on the following page.

Drink moderately, for drunkenness neither keeps a secret, nor observes a promise.

Miguel de Cervantes

CONTENTS

What Is Alcohol?

Is alcohol a nutrient? Or is it a drug? **Alcohol** is *not* a nutrient and is actually a general term used to describe a group of organic chemicals with common properties. Members of this group include ethanol, methanol, and isopropanol, among others. This chapter discusses the most commonly ingested of this group—ethyl alcohol or ethanol (EtOH). Therefore, for the remainder of this chapter, *alcohol* refers to ethanol.

Alcohol is a clear, volatile liquid that burns easily. In fact, it is so flammable that it can easily be used for fuel. It has a slight, characteristic odor and is very soluble in water. Alcohol can be made by four different methods (Table 8-1).

Physiologically, alcohol is a sedative and central nervous system depressant. Just like other drugs, the extent to which central nervous system function is impaired by alcohol consumption is directly related to the amount of alcohol in the blood. Similarly, like other drugs, any health benefits resulting from alcohol consumption depend on the amount of alcohol consumed.

It's difficult to classify alcohol when it comes to nutrition. Like the energy nutrients (carbohydrates, proteins, and fats), alcohol supplies energy (7 calories/gram). But unlike the energy nutrients, alcohol is not an essential nutrient, nor is it stored in the body.

alcohol clear, colorless volatile liquid; the most commonly ingested form is ethyl alcohol or ethanol (EtOH).

alcohol dehydrogenase a liver enzyme that mediates the metabolism of alcohol.

acetaldehyde (ass-et-AL-duh-hide) a substance into which drinking alcohol (ethanol) is metabolized.

TABLE 8-1
Methods of Making Alcohol

Method	Ingredients	Used to Make:
Fermentation	Fruit or grain mixtures	Beer, wine
Distillation	Fermented fruit or grain mixtures	Spirits (such as whiskey, rum, vodka, and gin)
Chemical modification of fossil fuels	Oil, natural gas, or coal	Industrial alcohol
Chemical combination of hydrogen with carbon monoxide	Methanol (wood alcohol)	Paints, paint strippers, duplicator fluid, model airplane fuel, dry gas, and windshield washer fluids

Absorption and Metabolism of Alcohol

When alcohol is consumed, it passes down the esophagus, through the stomach, and into the small intestine. About 20 percent of the alcohol is absorbed in the stomach, and about 80 percent is absorbed in the small intestine (see Figure 8-1 on page 255).[1] Alcohol absorbed from the small intestine passes into the portal vein, where it is transported directly to the liver, where a portion of the alcohol is then metabolized by enzymes. One enzyme in particular, **alcohol dehydrogenase,** facilitates conversion of alcohol into **acetaldehyde** and water. Acetaldehyde is quickly transformed to acetate by other enzymes and then ultimately metabolized to form carbon dioxide and water. Until all the alcohol consumed has been metabolized, it is disseminated throughout the body via the bloodstream, affecting the brain and other tissues.[2]

Nearly all alcohol ingested is metabolized in the liver, but a small amount remains unmetabolized and can be measured in breath (lungs exhale about 5 percent of alcohol, which can be detected by Breathalyzer devices) and urine (kidneys eliminate approximately 5 percent of alcohol in the urine).[3] Alcohol is also eliminated through sweat, feces, breast milk, and saliva.

The liver can metabolize only a limited amount of alcohol per hour, no matter how much is consumed. A healthy average person can eliminate ½ ounce (15 milliliters) of alcohol per hour. Rate of alcohol metabolism depends, in part, on the amount of alcohol dehydrogenase formed in the liver. The amount of alcohol dehydrogenase produced in the liver varies among individuals and appears to depend on a person's genetic makeup.[4] Nonetheless, alcohol is metabolized more slowly than it is absorbed,[5] and it takes approximately one hour to metabolize one standard drink[6] (see the Savvy Diner feature on page 265). Controlled consumption can prevent accumulation of alcohol in the body and, therefore, it can prevent intoxication as well.[7]

Factors Influencing Absorption and Metabolism

Food Presence of food in the stomach slows absorption of alcohol.[8] The rate at which alcohol is absorbed depends on how quickly or slowly the stomach empties its contents into the small intestine. Dietary fat delays emptying time of the stomach and consequently slows absorption of alcohol.

Gender Men and women absorb and metabolize alcohol differently. Women have higher blood alcohol concentrations (BAC) after consuming the same amount of alcohol as men. A woman will absorb 30 percent more alcohol into

Ask Yourself Answers: **1.** True. **2.** True. **3.** True. **4.** False. Alcohol supplies 7 cal/gram. **5.** True. **6.** False. Having a measurable amount of alcohol in the blood can mean judgment and reflexes are impaired. **7.** False. No amount of alcohol use during pregnancy has been proven safe. **8.** True. **9.** True. **10.** True.

Nutrition Action How It All Adds Up

© PhotoDisc/Getty Images

Did you know the following facts?[9]

1. Almost 4 percent of college students drink alcohol on a daily basis.
2. College students spend over $5.5 billion a year (averaging over $450 per student per year) on alcohol, mostly beer. This is more than they spend on books, soft drinks, coffee, juice, and milk combined.
3. The total volume of alcohol consumed by college students each year (about 430 million gallons) would fill an Olympic-size swimming pool on every college and university campus in the United States.
4. A daily glass of wine can increase your weight by as much as 10 pounds a year.
5. Of the 12 million undergraduate students enrolled in U.S. colleges and universities, as many as 360,000 will die from alcohol-related causes while in school. Not everyone who dies from "alcohol-related causes" actually consumes alcohol. A majority of those who die are victims (in fatal motor vehicle accidents or involved in other crimes) of others who do consume alcohol.

Most college students either know someone who has experienced an alcohol-related problem or have had one themselves—problems such as vomiting or hangovers due to excessive drinking, driving under the influence citations (DUIs), fender benders, date rape, STDs (sexually transmitted diseases), and maybe even long-term difficulties such as pancreatitis. As a college student, you can make the choice to drink responsibly, also known as "knowing when to say when" or "low-risk drinking." If you choose to drink, the following recommendations come on high authority:

- Restrict yourself to two drinks a day if you're a man and one a day if you're a woman (the effects of alcohol are different for men and women).

- Drink alcohol no more than four days a week.
- Drink slowly. Restrict yourself to one drink per hour—the liver can't metabolize alcohol faster than this.

Imbibing more alcohol than described in these guidelines will increase the risk of finding yourself on your knees in front of your (or a friend's, or a perfect stranger's) toilet, with long-term alcohol-related problems like pancreatitis, liver damage, or both. And the risks increase the more the guidelines are exceeded.

You can decrease your risk of alcohol-related problems with a little common sense. Ask yourself these questions:

1. What is my reaction to alcohol? When I drink, do I often lose control or do something stupid? Do I pass out or get into arguments or fights?
2. Does anyone in my family have a history of alcohol-related problems?
3. What's my situation? Do I have an exam or final in the morning? Am I meeting the parents of that special someone for the first time? Will I be driving within the next 24 hours?

If you lose control when drinking, take a closer look at your drinking habits. Problems with alcohol tend to run in families, so this is another good reason to think about whether to drink or not to drink. You don't have to be a rocket scientist to realize you should take it easy if you have important obligations to tend to (and if you don't believe that getting behind the wheel of a motorcycle, car, truck, or SUV is a responsibility, then maybe you should drop out of college—you're not mature enough to be here). Sometimes, circumstances call for abstinence as the only alternative.

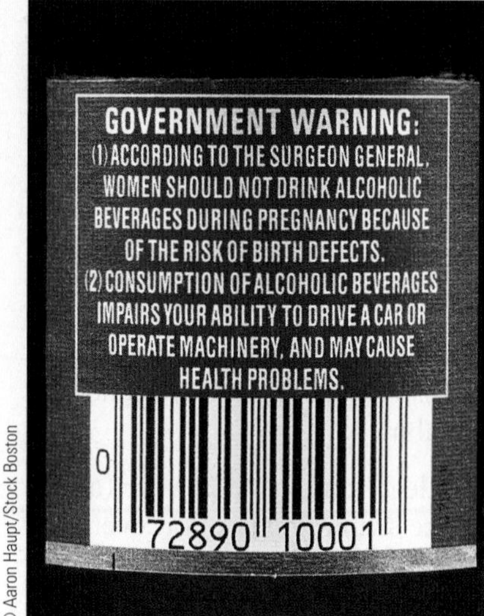

GOVERNMENT WARNING:
(1) ACCORDING TO THE SURGEON GENERAL, WOMEN SHOULD NOT DRINK ALCOHOLIC BEVERAGES DURING PREGNANCY BECAUSE OF THE RISK OF BIRTH DEFECTS.
(2) CONSUMPTION OF ALCOHOLIC BEVERAGES IMPAIRS YOUR ABILITY TO DRIVE A CAR OR OPERATE MACHINERY, AND MAY CAUSE HEALTH PROBLEMS.
0
72890 10001

O.B.S.E.R.V.E.

Alcohol can be dangerous in particular situations. Recalling the letters in the word *OBSERVE* will facilitate your memory to significantly lower your probability of experiencing an alcohol-related problem by simply not drinking.

It is best not to drink if you are:

- On certain medications or have certain illnesses (check with your doctor). Medicines mixed with alcohol may give you a buzz you didn't bargain for.
- Behind the wheel or engaged in tasks requiring full mental or physical functioning. Drinking and driving is one of *the* most dangerous things anyone can do. But you already knew that.
- Stressed out or tired. Drinking doesn't relieve stress; it only complicates it further. It could also lead to depression. Talk to a friend or counselor instead.
- Either the son, daughter, or sibling of someone with alcoholism. Problems with alcohol are often a family affair.
- Recovering from alcohol abuse or drug dependency. If you've succeeded in getting yourself off an addictive substance, the last thing you want is to fall off the wagon. Stay clean.
- Violating laws, policies, or personal values. If drinking may mean legal trouble or expulsion from school, don't do it. It's not worth it.
- Expecting, nursing, or considering pregnancy. You're abstaining for two.

If you fall into any of these categories, it's best *not* to drink. By OBSERVEing the times not to drink, you will significantly reduce your risk of having alcohol-related problems.

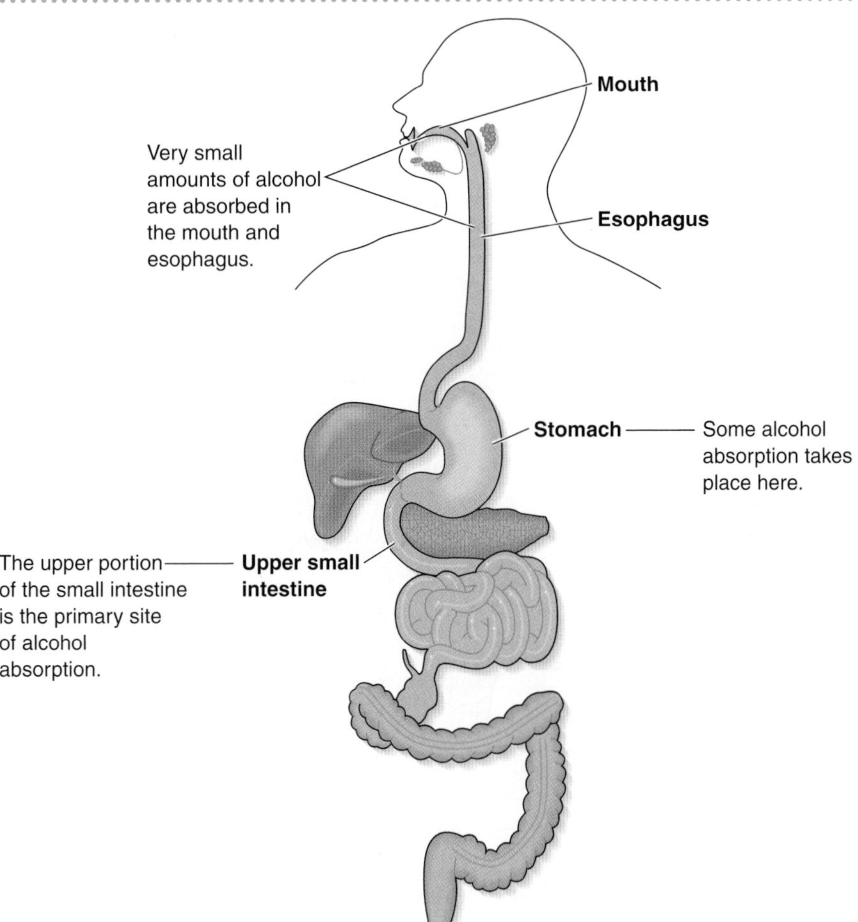
Mouth
Very small amounts of alcohol are absorbed in the mouth and esophagus.
Esophagus
Stomach
Some alcohol absorption takes place here.
The upper portion of the small intestine is the primary site of alcohol absorption.
Upper small intestine

FIGURE 8-1
ALCOHOL ABSORPTION

her bloodstream than a man of the same weight from the same amount of alcohol. One drink for a woman could potentially have the same effect as two drinks for a man. Additionally, women are more susceptible to alcoholic liver disease, heart muscle damage,[10] and brain damage.[11] These metabolic differences can be attributed to smaller amounts of body water[12] in women's bodies and a lower activity of alcohol dehydrogenase in the stomach, causing a larger proportion of ingested alcohol to reach the blood.[13]

Ethnicity Native Americans have higher rates of liver damage due to alcohol consumption than other ethnic groups.

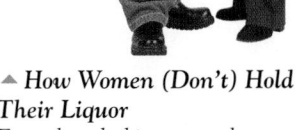

© Amos Morgan/GettyImages

▲ ***How Women (Don't) Hold Their Liquor***

Even though this man and woman are the same height and weight, the woman has a lower capacity for metabolizing alcohol.

- Body composition: *Because women have less water in their bodies, the concentration of alcohol in their bodies is higher (drink for drink, more alcohol gets into women's bodies).*
- Enzymes: *Alcohol dehydrogenase is about 40 percent less active in a woman's stomach, allowing more undiluted alcohol to enter the bloodstream.*
- Hormones: *Alcohol can change estrogen levels during a woman's menstrual cycle, increasing the risk for breast cancer.*

Alcohol and Its Effects

Because alcohol is distributed so quickly and thoroughly in the body, it can affect the central nervous system even in small concentrations. Even small amounts of alcohol in the blood can slow reactions (see Figure 8-2). The body responds to alcohol in stages; therefore, higher BAC leads to increased loss of mental and physical control (see Table 8-2 on page 257). If a large amount of alcohol is consumed over a short period of time, a person may lose consciousness or even die.[14]

Alcohol and Medications Use of prescription or over-the-counter medications can increase the effects of alcohol. Chronic, heavy drinking (more than two drinks a day for women or more than four drinks a day for men for a couple of years) appears to activate an enzyme that may be responsible for changing the over-the-counter pain reliever acetaminophen (for example, Tylenol) and many others into chemicals that can produce liver damage, even when acetaminophen

FIGURE 8-2
ALCOHOL'S EFFECTS ON THE BRAIN

Alcohol is rightly termed an anesthetic because it puts brain centers to sleep in this order: first the cortex, then the emotion governing centers, then the centers that govern muscular control, and finally the deep centers that control respiration and heartbeat.

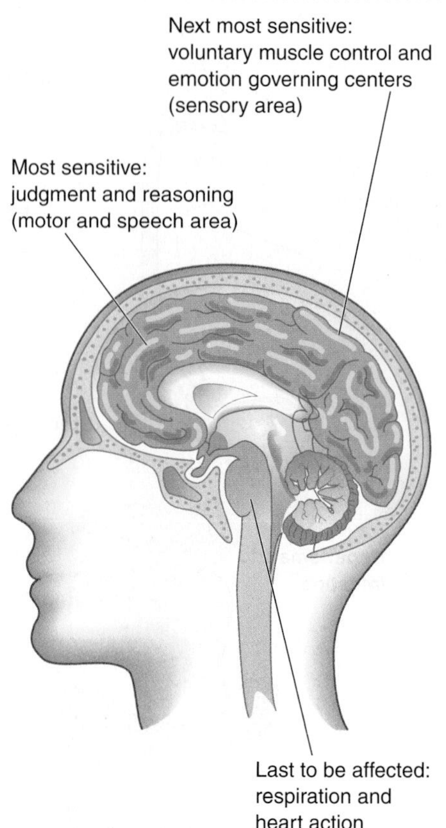

is taken in recommended doses[15] (four to five "extra-strength" pills taken over the course of the day).[16] This interaction is more likely to occur when acetaminophen is taken after, rather than before, the alcohol has been metabolized.

Alcohol and Sex Hormones Alcohol metabolism alters the balance of sex hormones in men and women.[17] In men, alcohol metabolism contributes to testicular injury and impairs testosterone synthesis and sperm production.[18] Prolonged testosterone deficiency may contribute to feminization in males such as breast enlargement.[19]

Alcohol and Urine Output Although some people believe it is simply from the amount of extra liquid they are ingesting, excessive urination when drinking alcohol really has more to do with the alcohol itself. In fact, an ounce of hard liquor will hypothetically cause the same amount of urination as a 12-ounce beer. Alcohol blocks the antidiuretic hormone (ADH), which is secreted from the pituitary gland and is responsible for regulating urine production. ADH works by causing the kidneys to conserve fluids. Thus, when this hormone is blocked, urine output increases considerably because the kidneys are not properly absorbing fluids. This leads to significant loss of water and eventually to dehydration.

Alcohol and Hangovers A hangover is usually thought of as a group of ailments usually including but not limited to: headache, nausea, vomiting, sensitivity to light and sound, dry mouth, and irritability. As explained above, dehydration results from excessive consumption of alcohol, and it is thought to be the main factor contributing to what is referred to as a hangover. In addition, alcohol affects the stomach lining, causing considerable distention. As its depressive effects wear off, sensitivity to noise and sound are amplified. Although theories (and myths) abound about how and why hangovers occur and how to subdue their effects, the best way to avoid these unwanted side effects is to drink in moderation.

TABLE 8-2
Effects of Drinking Alcohol

Approximate BAC[a]	Stage	Recognizable Consequence[b]
0.01–0.02	Subclinical	Behavior nearly normal by ordinary observation
0.03–0.12 BAC 0.08–0.10 or greater means legal intoxication in most states	Euphoria	Increased self-confidence or daring Attention span shortens Skin flushes Judgment diminishes; may say first thought that comes to mind rather than appropriate comment for a given situation Trouble with fine movements such as writing or signing name
0.09–0.25	Excitement	Becomes sleepy Trouble understanding or remembering things (even recent events); impaired perception and comprehension Decreased coordination, increased reaction times Blurred vision, reduced peripheral vision and glare recovery Decreased senses (hearing, tasting, feeling, etc.)
0.18–0.30	Confusion	Disoriented, mental confusion (might not know where they are or what they are doing) Dizziness and staggering Exaggerated emotional states (aggressive, withdrawn, or overly affectionate) Vision disturbances Sleepiness Slurred speech Uncoordinated movements (trouble catching a thrown object) Increased pain threshold (may not feel pain as readily as a sober person)
0.25–0.4	Stupor	Can barely move at all Cannot respond to stimuli Cannot stand or walk Vomiting, incontinence Lapse in and out of consciousness
0.35–0.50	Coma	Unconscious Reflexes depressed (pupils do not respond appropriately to light) Body temperature lowers Breathing slows and becomes more shallow Heart rate slows Possible death
>0.50	Death	Death from respiratory arrest

[a] Grams per 100 ml of blood or grams/210 liters of breath.
[b] BAC and effects of drinking alcohol vary from person to person and depend on body weight, amount of food eaten while drinking, and each person's ability to tolerate alcohol.

SOURCE: C. C. Freudenrich, *How Alcohol Works*, available online at www.howstuffworks.com/alcohol2.htm.

Alcohol and Blood Alcohol Level Blood alcohol level (BAL) indicates the amount of alcohol in the bloodstream and is referenced in milligrams of alcohol per 100 milliliters of blood. This system does not directly relate to how many drinks you consume but, instead, to the amount of alcohol actually circulating in the blood at a given time. BAL is affected by amount and speed at which alcohol is consumed. Because the liver can metabolize only one drink per hour, any amount beyond that will hypothetically enter into circulation at full strength. As mentioned already, many variables determine the effects alcohol will have on a person. Height, weight, and gender all have strong associations with what BAL will be after a given amount of alcohol is consumed. Although a Breathalyzer or urinalysis provide a far more accurate reading, Table 8-3 provides an estimate of BAL for different people at different levels of alcohol consumption. (See Table 8-2 for a reminder of what these levels represent.)

tolerance decrease of effectiveness of a drug after a period of prolonged or heavy use.

metabolic tolerance increased efficiency of removing high levels of alcohol from the blood due to long-term exposure leading to more drinking and possible addiction.

TABLE 8-3
DRINK/WEIGHT INDEX CHART

Drinks Consumed/Sex		Weight 100	120	140	160	180	200	220	240
1	Male	.04	.04	.03	.03	.02	.02	.02	.02
	Female	.05	.04	.04	.03	.03	.03	.02	.02
2	Male	.09	.07	.06	.05	.05	.04	.04	.04
	Female	.10	.08	.07	.06	.06	.05	.05	.04
3	Male	.13	.11	.09	.08	.07	.07	.06	.05
	Female	.15	.13	.11	.10	.08	.08	.07	.06
4	Male	.17	.15	.13	.11	.10	.09	.08	.07
	Female	.20	.17	.15	.13	.11	.10	.09	.09
5	Male	.22	.18	.16	.14	.12	.11	.10	.09
	Female	.25	.21	.18	.16	.14	.13	.12	.11
6	Male	.26	.22	.19	.16	.15	.13	.12	.11
	Female	.30	.26	.22	.19	.17	.15	.14	.13
7	Male	.30	.25	.22	.19	.17	.15	.14	.13
	Female	.36	.30	.26	.22	.20	.18	.16	.15
8	Male	.35	.29	.25	.22	.19	.17	.16	.15
	Female	.41	.33	.29	.26	.23	.20	.18	.16
9	Male	.39	.35	.28	.25	.22	.20	.18	.16
	Female	.46	.38	.33	.29	.26	.23	.21	.19
10	Male	.39	.35	.28	.25	.22	.20	.18	.16
	Female	.51	.42	.36	.32	.28	.25	.23	.21
11	Male	.48	.40	.34	.30	.26	.24	.22	.20
	Female	.56	.46	.40	.35	.31	.27	.25	.23
12	Male	.53	.43	.37	.32	.29	.26	.24	.21
	Female	.61	.50	.43	.37	.33	.30	.28	.25
13	Male	.57	.47	.40	.35	.31	.29	.26	.23
	Female	.66	.55	.47	.40	.36	.32	.30	.27
14	Male	.62	.50	.43	.37	.34	.31	.28	.25
	Female	.71	.59	.51	.43	.39	.35	.32	.29
15	Male	.66	.54	.47	.40	.36	.34	.30	.27
	Female	.76	.63	.55	.46	.42	.37	.35	.32

SOURCE: From Beer Booze and Books, http://www.beerboozebooks.com/bal.htm.

Alcohol and Driving The best advice is to *never* drink and drive. Studies have shown that even one drink can considerably impair a driver's response time. Whereas driving is a very normal task, every driver must remember that behind the wheel of a car, he or she can do substantial damage to property and can even take people's lives, including their own. In the year 2004, 248,000 people were injured (one person injured every two minutes) and 16,700 people died as a result of drunk driving (that is one alcohol-related fatality every 31 minutes). Drinking and driving is simply not worth the risk. However, in most states, the legal limit is 0.08 and for those under 21 there is zero tolerance for any trace of alcohol. If this limit is exceeded, the driver could receive a DUI (driving under the influence) or DWI (driving while intoxicated), which in some states is a felony.

Alcohol and Tolerance **Tolerance** is the decrease of effectiveness of a drug after a period of prolonged or heavy use. Continued exposure to alcohol causes physiological adaptations that increase tolerance. These adaptations change a person's behavior. Two types of tolerance are at work with alcohol adaptation. **Metabolic tolerance** occurs because, with continued exposure, alcohol is metabolized at a higher rate (up to 72 percent more quickly). Long-term exposure to alcohol increases levels of alcohol dehydrogenase in the liver, resulting in lower peak blood alcohol concentrations. This means that the body becomes more effective in removing high levels of alcohol in the blood. Even so, it also means the person must drink more alcohol to experience comparable effects as before, which leads to more drinking and can contribute to addiction.[20]

Functional tolerance results from an actual change in sensitivity to a drug. Compared to the average person, chronic alcohol users can have twice the tolerance for alcohol as the normal chemical and electrical functions of their nerve cells increase to counteract the inhibitory effects of alcohol exposure. The increased nerve activity helps chronic alcohol users function normally when they have a higher BAC, but it also tends to make them irritable when they are not drinking. Furthermore, the increased nerve activity may make them crave alcohol. Undoubtedly, the increased nerve activity contributes to hallucinations and convulsions (such as delirium tremens) when alcohol is removed, making it challenging to overcome **alcohol abuse** and **alcohol dependency.**[21]

functional tolerance actual change in sensitivity to a drug resulting in hallucinations and convulsions when alcohol is removed.

alcohol abuse continued use of alcohol in spite of negative psychological, social, family, employment, or school problems because of alcohol.

alcohol dependency (alcoholism) a dependency on alcohol marked by compulsive uncontrollable drinking with negative effects on physical health, family relationships, and social health.

Impact of Alcohol on Nutrition

If you are a light drinker, in good health, and otherwise well nourished, the occasional consumption of alcohol will probably have little effect on your nutritional status. The biggest risk to you will come most likely from the additional calories provided by alcohol, which may contribute to unwanted weight gain. See Table 8-4 for the calorie content of commonly consumed alcoholic beverages. This is not to say that alcohol has *no effect* on your nutritional status, because it does. Alcohol causes fundamental changes in metabolism that occur whenever you consume alcohol. The extent to which alcohol affects your nutritional status depends on how much alcohol you consume and your current nutritional and health status.

If you drink excessively on a regular basis, your nutritional status will become compromised. Protein deficiency can develop, both from the depression of protein synthesis in the cells and, in the drinker who substitutes alcohol for food, from poor diet. Alcohol affects every tissue's nutrient metabolism in different ways. Stomach cells become inflamed and vulnerable to ulcer formation. Intestinal cells fail to absorb thiamin, folate, and vitamin B_{12}. Liver cells lose efficiency in activating vitamin D, and they alter their production and excretion of bile. Rod cells in the retina, which normally process vitamin A alcohol (retinol) into the form needed for vision (retinal), find themselves processing drinking alcohol instead. The kidneys excrete increased quantities of magnesium, calcium, potassium, zinc, and folate. Acetaldehyde interferes with metabolism, too. It dislodges vitamin B_6 from its protective binding protein so that it is destroyed, causing a vitamin B_6 deficiency and thereby lower production of red blood cells.

TABLE 8-4
Calories in Alcoholic Beverages and Mixers

Beverage	Amount (oz)	Calories
Beer	12	150
Light beer	12	100
Gin, rum, vodka, whiskey (80 proof)	1½	100
Dessert wine	3½	140
Table wine (red, rosé, white)	5	100
Wine cooler	12	170
Liqueurs	1½	155–185
Tonic, ginger ale	8	80
Cola	8	100
Fruit juice	8	110
Club soda, plain seltzer, diet soda	8	0

Health Benefits of Alcohol

Drinking moderate amounts (see The Savvy Diner on page 265) of alcohol appears to be safe for healthy people who do not have alcohol abuse or dependency problems.[22] People who consume one to two drinks daily have lower mortality rates than nondrinkers.[23] Alcohol, like any other drug, has a beneficial dose and a level (dose) that will cause harm. Whether alcohol consumption will produce beneficial or adverse effects depends on a person's alcohol consumption, age, and background for cardiovascular risk. Rates of death from all causes are lowest among those who report consuming one drink per day.[24] And, although most research indicates wine consumption to be most beneficial, it appears that the benefits are from the alcohol itself, not the other components of each type of alcoholic beverage.[25]

Alcohol and Cardiovascular Health

Cardiovascular disease is the leading cause of death worldwide.[26] More than a dozen research studies have demonstrated a consistent, positive correlation between moderate alcohol consumption and decreasing incidence of heart disease.[27] The protective effect of alcohol is the result of increased levels of high-density lipoprotein (HDL) cholesterol.[28] HDL-cholesterol removes cholesterol from the lining of artery walls and carries it back to the liver for excretion. Alcohol also inhibits blood from forming clots, reducing risk of death from heart attack.[29] This anticlotting effect of moderate drinking reduces risk of thrombotic or ischemic stroke (blockage of blood vessel in the brain) and increases risk of hemorrhagic stroke (rupture of a blood vessel within the brain).

Health Risks of Alcohol

Moderate drinking is not risk free. Deaths reduced by moderate alcohol consumption are generally found in age groups with high rates of coronary heart disease—in other words, in people 45 years and older. However, most deaths due to alcohol consumption occur in people younger than 45 years.[30] Among young adults, the risks (alcohol abuse and dependence, alcohol-related violent behavior and injuries) of alcohol consumption far outweigh any benefits that may accrue later in life.[31]

Accidents

Alcohol affects judgment and slows reflexes, which can lead to falls or accidents with vehicles or other heavy machinery. It can also increase the likelihood of homicide and suicide. Alcohol makes men less able to control impulses toward violence and makes women less capable of recognizing cues to potential violent behavior.

Drug Interactions

Like alcohol, many other **drugs** are metabolized in the liver. The liver has limited processing capacity, and alcohol and drugs may compete with each other for metabolism. Alcohol may hinder another drug's metabolism, keeping the medication in the system longer than intended and possibly increasing risk of side effects. See Table 8-5 for specific alcohol–drug interactions.

Other Risks

Night Blindness Alcohol blocks formation of retinal, a compound in the eye responsible for vision in low light.

drugs substances that can modify one or more of the body's functions.

TABLE 8-5
Specific Alcohol–Drug Interactions

Drug	Interaction with Alcohol
Antibiotics	Nausea, vomiting, headache, and possible convulsions
Antidepressants	Increases sedative effect, impairing mental skills used for driving; other interactions can produce a dangerous increase in blood pressure
Oral hypoglycemic drugs (used to treat type 2 diabetes)	Can prolong effects causing nausea, headache, or a dangerous decrease in blood glucose levels; chronic alcohol consumption can decrease the drug's effect
Over-the-counter antihistamines (such as Benadryl) used to treat symptoms of allergy and insomnia	May intensify the sedation causing dizziness or drowsiness
Barbiturates	Prolongs sedative effect; chronic alcohol use decreases sedative effect
Cardiovascular medications	Can cause dizziness or fainting upon standing up; chronic alcohol use may decrease therapeutic effect of some cardiovascular drugs
Cocaine/crack	Increases heart rate three to five times as much as when either drug is used alone (this increases metabolism, which in turn increases blood alcohol levels)
Ecstasy/MDMA	Increases level of dehydration to dangerous levels
GHB	Slows rate of breathing to dangerous levels
Heroin	Can induce heroin overdose, leading to cessation of breathing
Marijuana	Increases suppression of gag reflex (inhibiting vomiting), which can increase risk of alcohol poisoning
Narcotic pain relievers (opiates, morphine, codeine, Darvon, Demerol)	Enhances sedative effect of both substances, increasing risk of death by overdose
Nonnarcotic pain relievers (aspirin and similar nonprescription pain relievers)	Increases stomach bleeding and inhibition of blood clotting; increases risk of liver damage (acetaminophen [Tylenol])
Oral contraceptives	May hinder absorption of alcohol into bloodstream, delaying onset of intoxication
PCP/Special K	Compounds central nervous system (CNS) depression
Sedatives and hypnotics (sleeping pills, Valium)	Severe drowsiness, increasing risk of household and automotive accidents
Tobacco	Doubles risk of heart attack; increases risk for lung, throat, mouth, and bladder cancers, stroke, high blood pressure, miscarriage, and low birth-weight infants

SOURCE: *Drinking: A Student's Guide*, "Alcohol and Legal Drugs," available online at www.mcneese.edu/community.alcohol/legal.html.

Breast Cancer Even consumed in moderate amounts (one drink per day for women), alcohol may increase the risk of breast cancer in women.

Other Cancers Heavy alcohol use, especially when combined with smoking, appears to increase risk of cancer of the throat and esophagus. Risk of cancer increases for people who have hepatitis and cirrhosis. Risk of colon cancer also increases with alcohol use.

Liver Damage **Alcoholic hepatitis** (inflammation of the liver) and **cirrhosis** (scarring of liver tissue that interferes with blood flow and liver function) are two common consequences of heavy drinking. Hepatitis can lead to permanent liver damage if alcohol consumption continues.

High Blood Pressure and Stroke Heavy drinking causes high blood pressure (hypertension) as the heart compensates for the initially reduced blood pressure caused by alcohol consumption. Hypertension is a risk factor for stroke and heart disease.

alcoholic hepatitis inflammation and injury to the liver due to excess alcohol consumption.

cirrhosis a chronic, degenerative disease of the liver in which the liver cells become infiltrated with fibrous tissues; blood flow through the liver is obstructed, causing back pressure and eventually leading to coma and death unless the cause of the disease is removed; the most common cause of cirrhosis is chronic alcohol abuse.

Pancreatitis Long-term alcohol consumption can result in recurrent attacks of severe pain caused by inflammation of the pancreas. The pancreas is responsible for production of many digestive enzymes. Continued use of alcohol can ultimately cause permanent damage to the pancreas.

Gastrointestinal Symptoms Alcohol irritates the lining of the stomach and impairs intestinal enzymes and transport systems. As such, alcohol consumption can cause a wide range of common, uncomfortable but reversible problems including gastritis (inflammation of the lining of the stomach), stomach and intestinal ulcers, diarrhea, and weight loss.

Brain Damage Brain cells in various parts of the brain die, reducing total brain mass.

Decreased Sex Hormone Production Alcohol causes decreased testosterone secretion from the hypothalamus/pituitary and testes, resulting in decreased sperm production. It can cause infertility in women by disrupting or changing the menstrual cycle.

Anemia Poor nutrition as a result of excessive alcohol intake decreases iron and vitamin B levels, leading to anemia.

Emotional and Social Problems Alcohol affects emotional centers in the limbic system, causing anxiety and depression. Emotional and physical effects of alcohol can contribute to marital and family problems including domestic violence, as well as work-related problems such as excessive absences and poor work performance.

Weighing the Pros and Cons of Alcohol Consumption

So how do you decide if low to moderate alcohol consumption will provide health benefits or health risks for you? One way is to compare your age and gender to the leading causes of death for those of similar ages and gender. The leading causes of death for men under the age of 40 and women under the age of 50 (premenopausal) are accidents and breast cancer, respectively. In these cases, the risks of low to moderate alcohol consumption outweigh the benefits. The leading cause of death for men over the age of 40 and women over the age of 50 is heart disease. Thus, for these people, the benefits of low to moderate alcohol consumption outweigh the risks.

What Is Alcohol Abuse or Alcoholism?

Alcoholism is a dependency on alcohol characterized by craving (a strong need to drink), loss of control (being unable to stop drinking despite a desire to do so), physical dependence and withdrawal symptoms, and tolerance (increasing difficulty of becoming drunk).

Although many people stereotype persons with alcoholism as those who spend their lives at a bar or as skid row drunks, they can, in fact, be anyone. Previous notions of alcoholism as a moral flaw have basically been replaced with the theory that alcoholism is a disease with genetic, psychosocial, and environmental factors and influences. Successful people such as lawyers, doctors, professors, and even college students can suffer from what is commonly known as alcoholism.

The following is a scale of alcohol addiction from Narcotics Anonymous (NA) and Alcoholics Anonymous (AA) (http://nickscape.net/recoveryzone/).

Use

Alcohol "use" is considered to be the ingestion of alcohol or other drugs without experiencing any negative consequences. If a college student drinks a beer at a fraternity party, she has used alcohol. This can also apply to any drug.

Misuse

When a person experiences negative consequences from his or her use of alcohol or other drugs, it is "misuse." A large percentage of the population misuses drugs or alcohol at some point in their lives. However, this does not imply that the negative consequences are minor. For example, the same college student continues to use alcohol on an infrequent basis. Her friends throw her a surprise party and she drinks more than usual. On the way home, she is arrested for DUI. Although she does not really have a problem with alcohol, in this instance, the consequences are not minor.

Abuse

Abuse constitutes continued use of alcohol or other drugs in spite of negative consequences. Let's continue to observe the college student who was arrested for DUI. If she does *not* have a substance abuse problem, she will probably cut way back on her alcohol consumption if not giving it up altogether. Getting a DUI would be enough of a deterrent. However, if shortly thereafter she goes to another party, drinks in excess, and then drives her car, it is alcohol abuse.

Dependency/Addiction

Compulsive use of alcohol or other drugs, regardless of adverse or negative consequences, presents the problem of dependency or addiction. For example, suppose that our college student has received three DUIs in one year. She is now on probation and will be sentenced to one year in prison if caught using alcohol again. However, she continues to drink at parties and at local establishments. This young woman is clearly addicted to alcohol because the negative consequences do not affect her decisions or deter her use.

When an individual is clearly in the first or second stage of alcoholism (use or misuse), there are no indications about whether or not he or she will naturally progress to the final stages. However, once an individual reaches the abuse stage, a high probability exists that he or she will progress into dependency on the drugs or alcohol. Such individuals should seek professional help to stop the progression.

What Is AA?

Alcoholics Anonymous (AA) is perhaps the most well-known method to deal with alcohol addiction. Alcoholics Anonymous defines itself as "a fellowship of men and women who share their experience, strength, and hope with each other that they may solve their common problem and help others to recover from alcoholism." This not-for-profit organization, established in 1935, has helped millions of people deal with this life-altering disease. Through group meetings, self-proclaimed alcoholics come together for counseling, guidance, and support with the primary purpose of staying sober and helping other alcoholics achieve sobriety. Their well-known "12-Step Program" focuses not only on the physical act of abstaining from alcohol, but the psychosocial aspects as well. For the purpose of anonymity, membership records do not exist, and people are welcome from all walks of life and at all levels of alcoholism.

Alcohol Assessment Questionnaire

This is a screening test designed to help you determine if you have a problem with alcohol. Read through the following questions about your use of alcoholic beverages during the past year. In the questions, a "drink" is defined as 12 ounces of beer, 5 ounces of wine, or 1 to 1½ ounces of 80-proof liquor. Select the most appropriate answer for each question by checking yes or no.

EARLY WARNING SIGNS OF ALCOHOL ABUSE

	YES	NO
1. Have you missed classes more than once due to a hangover?	☐	☐
2. Have you thought that you should cut down on your drinking?	☐	☐
3. Have you decided to cut down on your drinking and found out you could *not?*	☐	☐
4. Have you been angered by the criticism of others about your drinking?	☐	☐
5. Have you been in fights while drinking?	☐	☐
6. Is excessive alcohol use a significant part of your weekly social and recreational activities?	☐	☐
7. Have you had problems with resident assistants or campus police because of your drinking?	☐	☐
8. Do you routinely "binge" drink? (Binge drinking for women is defined as drinking four or more drinks during an episode of drinking. For men, five or more drinks is considered binge drinking.)	☐	☐
9. Have you ever had periods of time you cannot account for while you were drinking or after drinking occurrences?	☐	☐
10. Have you had sexual experiences after drinking that you felt bad about later?	☐	☐

GIVE YOURSELF 1 POINT FOR EACH YES RESPONSE.

0 Congratulations! Your response on this test suggests that your use of alcohol is not causing you any ongoing negative experiences that indicate early warning signs of alcohol abuse or dependence.

1 Now is the time to evaluate how much you are drinking, how often, and the impact your alcohol consumption is having on you. A score of 1 also indicates that you should probably reduce the quantity of alcohol you consume.

2 or greater More than one yes response indicates the definite need for you to limit your alcohol use by either abstaining or limiting your use to responsible levels of consumption. If you are unable to control your use, then it's time to abstain completely.

What Is Binge Drinking?

What would you think if you were at a party or restaurant and saw someone nearby drink an entire six-pack of a soft drink in one sitting? That's 72 ounces of pop. Strange, right? What will all that sugar do to the person's teeth? What will all of those calories do to his or her waistline? What about all of that caffeine at one time?

Now let's say that same person is drinking a six-pack of beer instead of soft drinks. Sound weird, too? Probably, but many college students do it all the time. It's called **"binge drinking"**—drinking at least five drinks at one time if you are a man or four drinks at one time if you are a woman, in one sitting. The short-term reactions caused by consuming large amounts of alcohol in a brief period of time are serious: vomiting, dizziness, impaired mental capabilities, and hangover. Other effects can include risky sexual behavior, alcohol-related injuries, or death.

Think it doesn't happen all that often? About 50 percent of college men and 37 percent of college women report drinking more than five drinks at one time in one sitting.

SOURCE: *Drinking: A Student's Guide*, "Binge Drinking," available online at www.mcneese.edu/community.alcohol/binge.html.

SOURCE: Reprinted from L. Hickman, *Self Tests*, "Early Signs of Alcohol Abuse," available online at www.nd.edu/~ucc/ucc_alcohol2.html.

What Is a Drink?

Ancient Persians ate five almonds to prevent hangovers. For Romans and Greeks, celery was thought to cure a hangover. By the late 1800s, celery still had a medicinal use, advertised by Sears, Roebuck & Company as a celery tonic to calm the nerves.[32]

According to the *Dietary Guidelines for Americans* (2005), "If you drink alcoholic beverages, do so in moderation." But what is drinking in moderation? If you decide to drink alcohol, you should have as much information as possible so you can drink responsibly.

Moderation is defined as the following:

- Men: no more than two drinks per day
- Women: no more than one drink per day

These limits are based on the disparity between men and women in both weight and alcohol metabolism. Whereas drinking in moderation provides little, if any, health benefits for young adults, moderate drinking for men over 45 and women over 55 may lower their risk for coronary heart disease.

More than one drink a day for women or two drinks a day for men can increase risk for motor vehicle accidents, other injuries, high blood pressure, stroke, violence, suicide, and certain types of cancer. Even one drink per day can slightly increase risk of breast cancer. Alcohol consumption during pregnancy (see the Spotlight on page 266) increases risk of birth defects.

What Is a Drink?

Alcohol is alcohol is alcohol. It does not matter if the beverage of choice is beer, wine, a wine cooler, a cocktail, or a mixed drink. A standard serving of beer, distilled spirits, and wine each contain the same amounts of alcohol (see the photo). A standard serving is as follows:

- 12 ounces of regular beer (150 calories)
- 5 ounces of wine (100 calories)
- 1½ ounces of 80-proof distilled spirits (100 calories)
- 12 ounces of wine/malt or spirit-based cooler (210–230 calories)
- 3 ounces of sherry or port (130 calories)
- 9¾ ounces of malt liquor (135 calories)

When it comes to drinking alcohol, the old adage is true: It doesn't matter what you drink—it's really *how much* you drink that counts.

Sometimes responsible drinking means not drinking at all. Some people who should not drink alcoholic beverages include:

- Children and adolescents
- Individuals of any age who cannot restrict their drinking to moderate levels
- Women who *may* become pregnant or who are pregnant. A safe alcohol intake has not been established for women at any time during pregnancy, including the first few weeks (see the Spotlight in this chapter).
- Individuals who plan to drive, operate machinery, or take part in any activities that require attention, skill, or coordination. Most people retain some alcohol in their blood up to two to three hours after a single drink.
- Individuals taking prescription or over-the-counter medications that can interact with alcohol

Sources: American Dietetic Association, Alcohol Beverages: Making Responsible Drinking Choices, available online at www.eatright.org; Dietary Guidelines for Americans, available online at www.health.gov/dietaryguidelines; National Consumers League, Alcohol: How It All Adds Up, available online at www.nclnet.org.

Spotlight

Fetal Alcohol Syndrome

What is fetal alcohol syndrome?

Scientists first created the term *fetal alcohol syndrome* (FAS) more than 30 years ago to describe a pattern of birth defects found in children of mothers who drank alcohol during pregnancy.[33] In the United States, FAS is one of the primary causes of birth defects and is considered to be the most common cause of preventable mental retardation. It is defined by four criteria:[34]

- Maternal drinking during pregnancy
- Characteristic pattern of facial abnormalities (see Table 8-6)
- Growth retardation
- Brain damage, often manifested by intellectual difficulties or behavioral problems (see Table 8-6). Studies have demonstrated abnormalities of certain brain regions in babies exposed to alcohol during their mothers' pregnancy (see Figure 8-3).

TABLE 8-6
Signs of FAS

Characteristic Facial Features of FAS
Small eyes with drooping upper lids
Short, upturned nose
Flattened cheeks
Small jaw
Thin upper lip
Flattened groove in middle of upper lip
Central Nervous System Problems
Mental retardation
Hyperactivity
Delayed development of gross motor skills (rolling over, crawling, walking)
Delayed development of fine motor skills (grasping objects with the thumb and index finger, and transferring objects from one hand to another)
Impaired language development
Memory problems, poor judgment, distractibility, impulsiveness
Learning problems
Seizures

SOURCE: National Institutes of Health, "Highlights from the 10th Special Report to Congress," *Alcohol Research and Health* 24, no. 1 (2000).

How does alcohol get into the baby's body?

As you can see, the effects of fetal exposure to alcohol last throughout the child's life. Alcohol consumed by a pregnant woman travels through her bloodstream and across the placenta to her baby. The unborn baby's body (fetus) can metabolize the alcohol but does so at a much slower rate than the adult body. As a result, the alcohol level in the baby's blood is higher than in the mother's. And, the alcohol remains in the baby's blood longer.

How much alcohol does a woman have to drink to cause FAS?

Women who drink frequently (more than four alcoholic beverages a day) seriously increase the likelihood that their babies will have FAS. Moreover, binge drinking (four or more drinks per occasion) is an especially hazardous drinking pattern in terms of FAS risk.[35] On the other hand, no quantity of alcohol use during pregnancy has been established to be safe. Effects of FAS have been seen in children whose mothers drank moderately or lightly during pregnancy. An average of one drink a day increases a baby's risk of FAS.

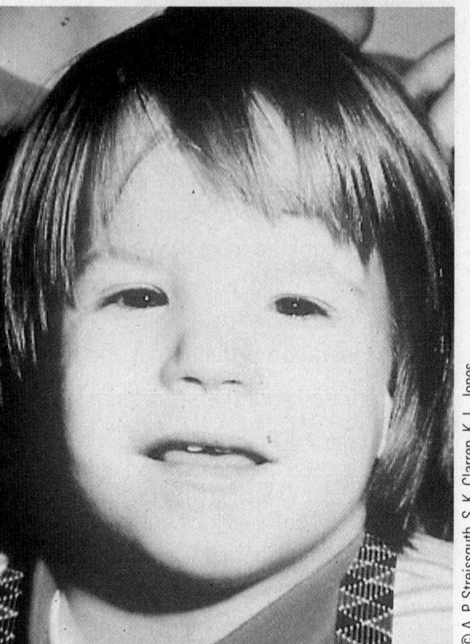

▲ *These facial traits (low nasal bridge, short eyelid opening, small head circumference, undeveloped groove in center of upper lip) are typical of fetal alcohol syndrome, caused by drinking alcohol during pregnancy. Irreversible abnormalities of the brain and other organs accompany these facial features.*

How can FAS be prevented?

Unlike most birth defects, FAS is completely preventable, because its direct cause—maternal drinking—is a controllable behavior.[36] Children with FAS do not outgrow the signs listed on Table 8-6. These problems will last for a child's whole life. Pregnant women can prevent FAS by not consuming alcohol. No amount of alcohol use during pregnancy has been proven to be safe. For that reason, any woman who suspects she might be pregnant should *stop drinking immediately*. Women who are attempting to get pregnant *should not drink* alcohol. Because many women of childbearing age drink regularly, it's likely that their babies are exposed to alcohol before pregnancy is detected. It is common for a woman to be pregnant for four to six weeks before she knows she is pregnant. Alcohol can hurt a baby even during the first one to two months of pregnancy, and no type of alcoholic beverage—beer, wine, wine coolers, and liquor (whiskey, vodka, tequila, gin, and rum)—is exempt.

The bottom line? Everything a woman eats or drinks affects her baby. Fetal alcohol syndrome could be completely eliminated if pregnant women did not consume alcohol.[37]

FIGURE 8-3
AREAS OF THE BRAIN THAT CAN BE DAMAGED IN THE WOMB BY THE MOTHER'S CONSUMPTION OF ALCOHOL

SOURCE: *Alcohol Health & Research World* 18, no. 1 (1994).

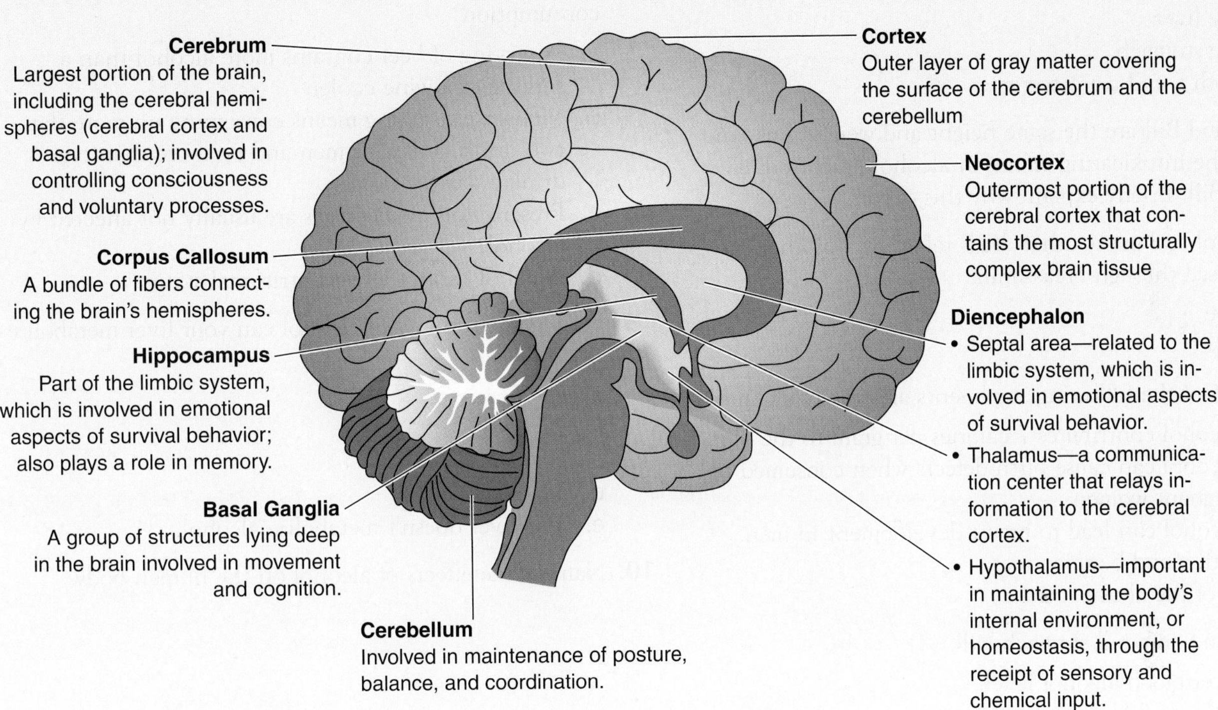
Cerebrum
Largest portion of the brain, including the cerebral hemispheres (cerebral cortex and basal ganglia); involved in controlling consciousness and voluntary processes.
Corpus Callosum
A bundle of fibers connecting the brain's hemispheres.
Hippocampus
Part of the limbic system, which is involved in emotional aspects of survival behavior; also plays a role in memory.
Basal Ganglia
A group of structures lying deep in the brain involved in movement and cognition.
Cerebellum
Involved in maintenance of posture, balance, and coordination.
Cortex
Outer layer of gray matter covering the surface of the cerebrum and the cerebellum
Neocortex
Outermost portion of the cerebral cortex that contains the most structurally complex brain tissue
Diencephalon
• Septal area—related to the limbic system, which is involved in emotional aspects of survival behavior.
• Thalamus—a communication center that relays information to the cerebral cortex.
• Hypothalamus—important in maintaining the body's internal environment, or homeostasis, through the receipt of sensory and chemical input.

You Can't Drink to Your Baby's Health!
Alcohol can have a bad influence. Especially on an unborn baby. The Arc, a national organization on mental retardation, warns that drinking alcohol during pregnancy may cause birth defects.
You wouldn't put alcohol in your baby's bottle. Why put it in your unborn baby?
For more information on how you can help prevent mental retardation, call your local chapter of The Arc.
The Arc
Formerly, Association for Retarded Citizens

Courtesy of the ARC

◂ *Targeted media campaigns can help increase public awareness of the adverse effects of alcohol use during pregnancy. Because levels of binge and frequent drinking among nonpregnant women have not declined, all women of childbearing age should be warned about the adverse effects of alcohol use to avert early prenatal exposure before women become aware of pregnancy. Additional information about CDC's activities to prevent alcohol-exposed pregnancies is available at www.cdc.gov/ncbddd/fas.*

In Review

1. Symptoms of alcohol poisoning include:
   a. becoming unconscious.
   b. rapid pulse.
   c. low blood pressure.
   d. dilated pupils.
   e. all of the above

2. Alcohol is absorbed in:
   a. the small intestine.
   b. the liver.
   c. the stomach.
   d. Both a and c are correct.

3. Jane and Bill are the same height and weight, but Jane feels the intoxicating effects of alcohol much sooner than Bill. Briefly explain why this occurs.

4. Alcohol and many other drugs taken by a mother can be passed through breast milk.
   a. true
   b. false

5. Which of the following statements are true of alcohol?
   a. Alcohol contributes 7 calories per gram to the diet.
   b. Alcohol can cause birth defects when consumed by pregnant women.
   c. Alcohol can lead to breast development in men.
   d. Both a and b are correct.
   e. all of the above

6. Drinking on a full stomach will:
   a. lower blood alcohol level.
   b. prevent all hangovers.
   c. slow absorption of alcohol into the bloodstream.
   d. decrease the amount of alcohol metabolized.

7. Fetal alcohol syndrome is:
   a. a condition a child will grow out of.
   b. completely preventable.
   c. associated with facial abnormalities and growth retardation.
   d. Both b and c are correct.
   e. all of the above

8. Which of the following are true about alcohol consumption?
   a. A serving of beer contains more alcohol than a serving of a wine cooler.
   b. Moderate drinking means consuming no more than two drinks a day for men and no more than one drink a day for women.
   c. Prescription medications are usually not affected by alcohol.
   d. Alcohol cannot affect nutritional status.

9. How many servings of alcohol can your liver metabolize in one hour?
   a. ½
   b. 1
   c. 2
   d. 3
   e. The liver doesn't metabolize alcohol.

10. Name some effects of alcohol on the human body.

Menu of Online Study Tools

A variety of study tools for this chapter are available at our website to deepen your understanding of chapter concepts. Go to
www.thomsonedu.com/nutrition/boyle
to find

- Practice tests
- Flashcards
- Glossary
- Web links
- Animations
- Chapter summaries, learning objectives, and crossword puzzles

Nutrition on the Web

www.thomsonedu.com/nutrition/boyle
Go to the *Personal Nutrition* site to check for the latest updates to chapter topics or to access links to related websites.

www.edc.org/hec/thisweek/quiz1.htm
Take the quiz about alcohol and other drug myths and misconceptions.

www.alcoholics-anonymous.org
Information from Alcoholics Anonymous.

http://alcoholstudies.rutgers.edu
Learn more about alcohol use and alcohol-related problems from the Center of Alcohol Studies, an interdisciplinary research center at Rutgers University.

www.niaaa.nih.gov
Search this site to learn more about alcohol abuse and alcoholism research.

www.intox.com/about_alcohol.asp
Don't believe alcohol is a drug? Don't believe alcohol affects your motor skills? Use this website to actually view how just a couple of drinks of alcohol can affect your handwriting. And, if alcohol can affect your handwriting, how do you think it affects your ability to drive a two-ton automobile?

www.pbs.org/newshour/bb/health/jan-june99/bingedrinking_6-2.html
Search this website for information regarding binge drinking.

www.cdc.gov/ncbddd/fas
Information about fetal alcohol syndrome from the Centers for Disease Control and Prevention with links to related websites.

www.modimes.org
March of Dimes (MOD) provides a health library with fact sheets on a variety of subjects, including fetal alcohol syndrome.

www.thearc.org
The ARC of the United States (formerly known as Association of Retarded Citizens) has an alcohol policies site that offers fact sheets, action alerts, and press releases pertaining to alcohol-related issues including fetal alcohol syndrome and drinking during pregnancy.

Weight Management

9

ASK YOURSELF . . .

Which of the following statements about nutrition are true, and which are false? For each false statement, what is true?

1. The less you weigh, the better it is for your health.
2. Obese people pay higher insurance premiums than thin people.
3. If you weigh too much according to the scales and the so-called ideal weight tables, you are too fat.
4. If you are too fat, it is because you eat too much.
5. Basal metabolism contributes only a small percentage of a person's daily energy output.
6. Probably the most important contributor to the obesity problem in our country is underactivity.
7. Any food can make you fat, even carrot sticks, if you eat enough of them.
8. You can lose weight faster on a properly designed diet and exercise program than on a total fast.
9. Fad diets are popular because their followers achieve quick and permanent weight loss.
10. Anorexia nervosa is a disease in which a person has no appetite.

Answers found on the following page.

Before you begin a thing remind yourself that difficulties and delays quite impossible to foresee are ahead. . . . You can see only one thing clearly, and that is your goal. Form a mental vision of that and cling to it through thick and thin.

K. Norris

CONTENTS

Say the word *diet*, and most people think "starvation," "deprivation," "hunger," and something to go "on" and "off." But, as this chapter shows, this thinking is unrealistic and self-defeating. Strict, temporary diets—and the attitudes that go with them—don't work for the great majority of overweight people. In fact, although Americans spend more than $37 billion a year on programs and products aimed at weight loss, obesity rates are higher than ever.[1]

Currently, 67 percent of adults and approximately 17 percent of children and adolescents in the United States are either overweight or obese—exceeding their healthy weight range (see Figure 9-1). The annual costs of overweight and obesity are staggering: more than $117 billion, including $61 billion in direct costs (treatment of related disease) and $56 billion in indirect costs (lost productivity due to disability, morbidity, and mortality).[2]

A Closer Look at Obesity

The World Health Organization describes obesity as "an escalating epidemic" and one of the greatest neglected public health problems of our time (see Figure 9-2). Because obesity is a disease with multiple health risks such as type 2 diabetes, heart disease, and some forms of cancer, the increasing rates of obesity are

FIGURE 9-1
TRENDS IN PREVALENCE OF OVERWEIGHT AND OBESITY AMONG CHILDREN AND ADULTS, UNITED STATES, 1988–2004

*See Figure 9-3 on page 276 for how to calculate and interpret Body Mass Index (BMI).

SOURCE: Adapted from A. A. Hedley and coauthors, Prevalence of overweight and obesity among U.S. children, adolescents, and adults, 1999–2002, *Journal of the American Medical Association* 201 (2004): 2847–50; and C. L. Ogden and coauthors, Prevalence of overweight and obesity in the United States, 1999–2004, *Journal of the American Medical Association* 295 (2006): 1549–55.

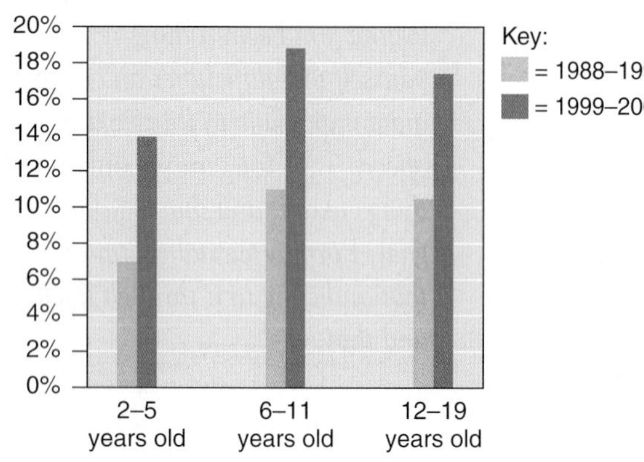

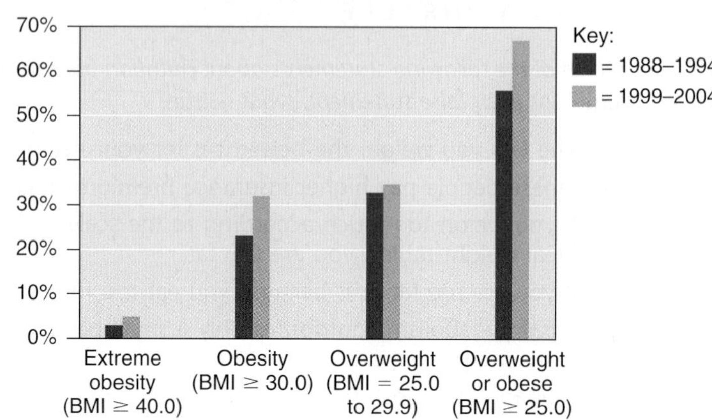

© Chabruken/Getty Images

▲ *Bodies come in many shapes and sizes. Which are healthy?*

projected to result in increased rates of disability and preventable deaths.[3] One of the national health objectives for *Healthy People 2010* is to reduce the prevalence of obesity among adults to less than 15%.

Although many factors (including genetics) influence body weight, excess energy intake and physical inactivity are the leading causes of **overweight** and **obesity,** and they represent the best opportunities for prevention and treatment.[4] Consider how the following trends in our society have increased opportunities for poor nutrition (particularly excess calories) and decreased opportunities for physical activity:

- Food portion sizes and obesity rates have grown at the same rate. In the 1960s, an average fast-food meal of a hamburger, fries, and a 12-ounce cola provided 590 calories; today, supersized, fast-food meals deliver 1,500 calories or more.[5] Even an extra 150 calories a day—the calorie cost of supersizing a soft drink—can convert to about 16 extra pounds of body fat every year.
- Vending machines selling soft drinks, high-fat snacks, and sweet snacks are common in schools and workplaces. Milk, juices, water, and healthy snacks are far less accessible.
- Adults spend more time in sedentary activities, such as watching television, working on the computer, or commuting to and from work and school.
- Children watch 12 to 14 hours of television a week and spend 7 hours playing video games.[6]
- Schools offer fewer physical education classes for children.

overweight conventionally defined as weight between 10 and 20 percent above the desirable weight for height, or a body mass index (BMI) of 25.0 through 29.9 (see page 276).

obesity conventionally defined as weight 20 percent or more above the desirable weight for height, or a BMI of 30 or greater.

underweight weight 10 percent or more below the desirable weight for height, or a BMI of less than 18.5.

Ask Yourself Answers: **1.** False. Being thin is good for health only to a point; being too thin is as risky as being too fat. **2.** True. **3.** False. A high weight according to the scales and the so-called ideal weight tables may reflect heavy bones and muscles rather than excess fatness. **4.** False. If you are too fat, it could be because you exercise too little. **5.** False. Basal metabolism contributes about 60 percent or more of the average person's daily energy output. **6.** True. **7.** True. **8.** True. **9.** False. Although you may lose weight quickly on a fad diet, much of the weight you lose may be muscle or water and the weight may soon be regained when the diet ends. **10.** False. People with anorexia nervosa are constantly hungry but control their eating.

- More families live in communities that are designed for car use but are unsuitable (lack of green space for recreation) and often unsafe (lack of sidewalks, inadequate street lighting) for activities such as walking, biking, and running.

Problems Associated with Weight

Clearly, each individual can weigh too much or too little to be healthy. The **underweight** person has minimal body fat stores and is at a disadvantage in situations where energy reserves are needed, such as a prolonged period of physiological stress or injury. Other problems for the underweight person may include menstrual irregularity, infertility, and osteoporosis.

The physical risks of overweight and obesity are greater for some people than for others, depending on inherited susceptibilities to conditions such as high blood pressure, high blood cholesterol, and diabetes. High blood pressure is made worse by weight gain and can often be normalized merely by weight loss. Diabetes can be precipitated in genetically susceptible people who become overweight.

Obesity also increases the risk of heart disease partly because excess fat pads crowd the heart muscle and the lungs within the body cavity. These fat pads encumber the heart as it beats, requiring it to work harder to deliver oxygen and nutrients to the rest of the body. Also, the lungs cannot expand fully, thus limiting the oxygen intake of each breath and causing the heart to work even harder. Furthermore, because each extra pound of fat tissue demands to be fed by way of miles of capillaries, the heart must work extra hard in obese people to pump blood through a network of blood vessels that is vastly larger than that of a thin person. Even a healthy heart is strained by excess fat. When a diseased heart finds itself in this bind, a sudden increase in workload may be more than it can handle.

Gallbladder disease, too, can be brought on in susceptible people merely by excess weight.[7] Similarly, obesity increases a woman's risk of developing breast cancer.[8] Table 9-1 shows additional conditions that can be caused, or made worse, by obesity.

In addition to being at risk for these health hazards, millions of obese people incur risks from ill-advised, misguided diet programs. Some fad diets are actually more hazardous to health than obesity! Many claims, treatments, devices, and gadgets for losing weight are, at best, simply ineffective, whereas others are truly dangerous. Over the centuries, "magic" weight-loss plans have been offered time and again, and their success is in their popularity, not in their results.

Although some obese people suffer none of these physical health hazards, no

TABLE 9-1
Problems Associated with Obesity

- Abdominal hernias
- Accidents
- Certain cancers: colon, rectal, prostate, breast, uterus, cervical, ovarian
- Complications during pregnancy
- Complications with surgical procedures
- Decreased longevity
- Decreased quality of life
- Depression
- Diabetes (type 2)
- Fertility problems
- Gallbladder and liver disease
- Gout
- Heart disease
- High blood cholesterol levels
- Hormonal imbalances
- Hypertension
- Injury to weight-bearing joints
- Kidney abnormalities
- Metabolic syndrome*
- Osteoarthritis (knees, hips, lower spine)
- Poor self-esteem
- Respiratory problems
- Sleep disturbances
- Varicose veins

*Metabolic syndrome is a clustering of risk factors associated with the development of type 2 diabetes and heart disease. These risk factors include excess body weight (especially around the abdomen), insulin resistance, high blood pressure, high levels of triglycerides, low levels of HDL-cholesterol, and increased blood glucose levels.

FIGURE 9-2
THE EPIDEMIC OF OBESITY AMONG U.S. ADULTS

Note the increasingly upward trend of obesity among the 50 states. In 1991, no states had obesity rates of greater than 20 percent, and only 4 states had obesity rates greater than 15 percent. By 2000, only 1 state (Colorado) had an obesity rate of less than 15 percent, and 22 states had obesity rates greater than 20 percent. In 2004, 7 states had obesity rates of 15–19 percent; 33 states had rates of 20–24 percent; and 9 states had rates more than 25 percent (data unavailable for one state).

Source: U.S. Obesity Trends 1985–2004, Centers for Disease Control and Prevention, available online at www.cdc.gov.

1991

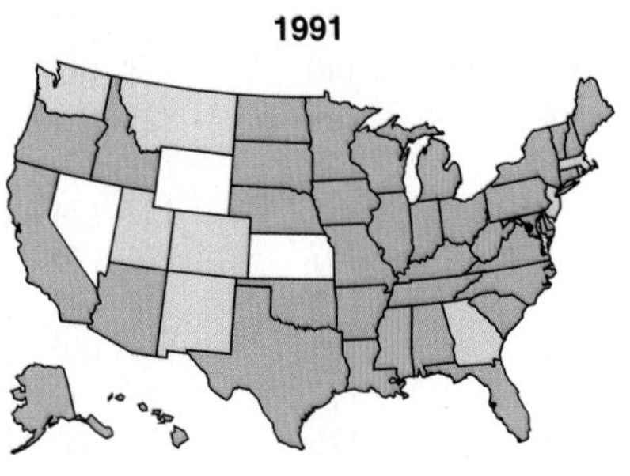

1995

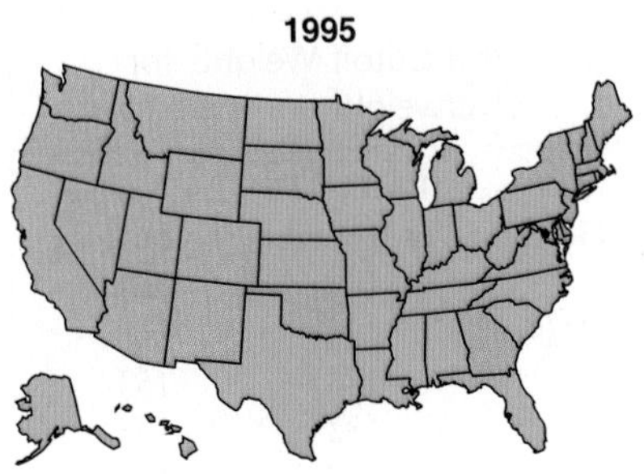

2000

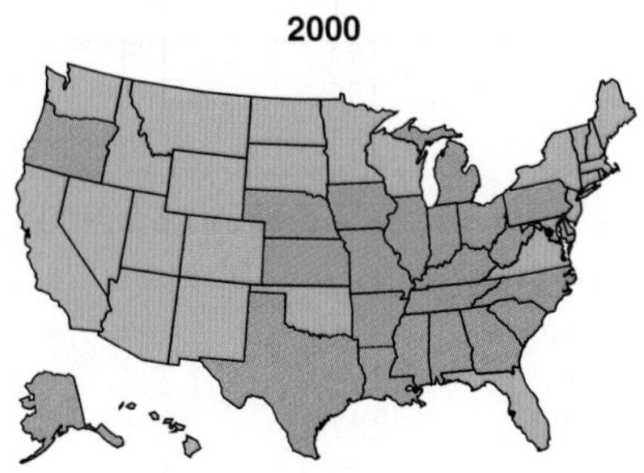

2004

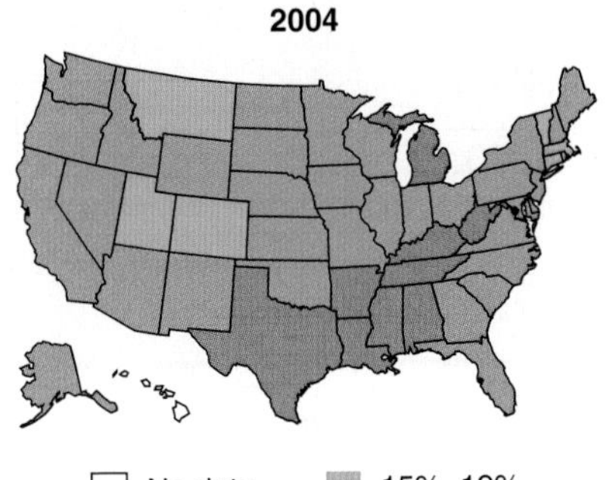

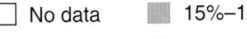

TABLE 9-2
Weight for Height Standards

Height	Healthy Weight Ranges[a] Weight (lb)[b] Midpoint	Range
4′10″	105	91–119
4′11″	109	94–124
5′0″	112	97–128
5′1″	116	101–132
5′2″	120	104–137
5′3″	124	107–141
5′4″	128	111–146
5′5″	132	114–150
5′6″	136	118–155
5′7″	140	121–160
5′8″	144	125–164
5′9″	149	129–169
5′10″	153	132–174
5′11″	157	136–179
6′0″	162	140–184
6′1″	166	144–189
6′2″	171	148–195
6′3″	176	152–200
6′4″	180	156–205

BMI Cutoff Weights for Overweight and Obesity

Height	BMI of 25 overweight (lb)	BMI of 30 obese (lb)
4′11″	124	148
5′0″	128	153
5′1″	132	158
5′2″	136	164
5′3″	141	169
5′4″	145	174
5′5″	150	180
5′6″	155	186
5′7″	159	191
5′8″	164	197
5′9″	169	203
5′10″	174	207
5′11″	179	215
6′0″	184	221
6′1″	189	227
6′2″	194	233
6′3″	200	240
6′4″	205	246

[a] The higher weights in the ranges generally apply to men, who tend to have more muscle and bone; the lower weights more often apply to women, who have less muscle and bone.
[b] Without shoes or clothes.

underwater weighing (hydrostatic weighing) a measure of density and volume; the less a person weighs underwater compared to the person's out-of-water weight, the greater the proportion of body fat (fat is less dense or more buoyant than lean tissue).

one in our society quite escapes one disadvantage of obesity. In many parts of North America, obesity is a social and economic handicap.[9] Obese individuals suffer from discrimination in many areas, including social relationships and the job market. Psychologically, a body size that embarrasses or shames a person can be a private anguish. For people who perceive themselves as fat in a society that prizes thinness, real or imagined obesity can thrust them into withdrawal, shame, humiliation, and isolation.

How thin, then, is too thin—and how fat is too fat? The following section discusses the concept of a healthful weight.

What Is a Healthful Weight?

Defining a "healthful weight" has many problems. Although Table 9-2 provides healthy weight ranges, the real question is: Healthy for what? For long-distance runners, every unneeded pound is a disadvantage; the lowest amount of body fat that doesn't compromise hormonal balance and fuel availability is desirable. Weight matters less for swimmers, and fat contributes to their buoyancy and insulates them against the cold. In the case of swimmers, to a point, more is desirable. Dancers and models may value thinness so highly that they compromise their health to attain it.

Some societies value fatness, equating it with prosperity; others value thinness to the point of obsession (our own being an example). This chapter first asks what range of weights is compatible with wellness and long life and then suggests that personal preferences dictate the choice of a weight within that range.

Body Weight versus Body Fat

The question of what weight is healthful is harder to answer than you might at first think. To think merely in terms of weight oversimplifies the issue of body fatness and health. Two people of the same sex, age, and height may both weigh the same, yet one may be too fat and the other too thin. The difference lies in their body composition. One may have small, light bones and minimally developed muscles, while the other has big, heavy bones and well-developed muscles. The first person could have too much body fat and the second person too little. For example, football players, body builders, and other athletes may weigh in as overweight for their height according to the weight tables. However, they typically have less body fat than would pose a health risk. Likewise, many sedentary persons who weigh in as "normal" may actually be too fat. Thus, obesity must be defined by amount of body fat rather than by weight.

The health risks of obesity refer to people who have too much body fat. On the average, men having over 25 percent body fat and women having over 33 percent body fat are considered obese.[10] Desirable measures are 12–20 percent body fat for most men and 20–30 percent body fat for most women.

Measuring Body Fat

Body fatness is hard to measure. One very accurate way is to obtain a measure of the body's density—that is, weight divided by volume. Lean tissue is more dense than fat tissue. Weight is easy enough; just step on an accurate scale. But to calculate body density, you have to immerse the whole body in a tank of water and measure the amount of water displaced. However, not many health professionals have space in their offices for the equipment needed for this procedure, known as *hydrostatic weighing* or **underwater weighing.** Body density can more easily be determined by air displacement methods such as the *BodPod,* which measures the volume of air displaced by a person when seated in a sealed device of known vol-

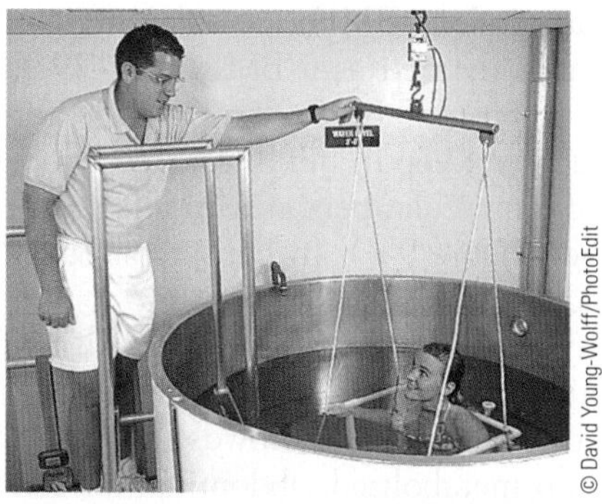
© David Young-Wolff/PhotoEdit

▲ *Hydrostatic (underwater) weighing.*

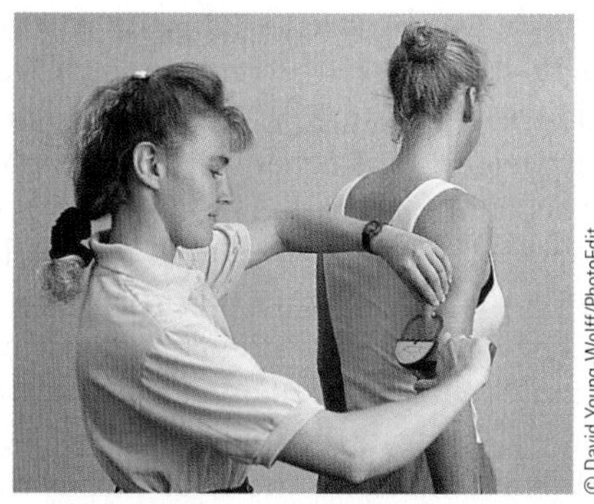
© David Young-Wolff/PhotoEdit

▲ *The fatfold test gives a fair approximation of total body fat.*

ume.[11]* Most professionals employ the **skinfold test,** using a caliper—a pinching device that measures the thickness of a fold of fat in such areas as the back of the arm, below the shoulder blade, and the side of the waist. About 50 percent of the body's fat lies beneath the skin, and its thickness at these locations can be compared with standard tables to give a fair approximation of total body fat, at least for most people.

Although costly, the same technology used to measure bone density—*dual energy X-ray absorptiometry* (the DEXA test) can yield an accurate image of the body's fat-free tissue and total fat content. An alternate method for assessing body fatness is **bioelectrical impedance,** in which electrodes are attached to a person's hand and foot. This method measures a person's fat by measuring how fast a slight electrical current is conducted through the body (from the ankle to the wrist). Because fat is a poor conductor of electricity, the more fat one has, the more resistance this current encounters in the body.[12]

Distribution of Body Fat

© Alan Oddie/PhotoEdit

▲ *Central obesity, characterized by an "apple-shaped" body with large abdominal-type fat stores, is a strong risk factor for type 2 diabetes, heart disease, and other problems.*

Not everyone carries his or her body fat in the same places. To complicate matters further, how body fat is distributed has health implications. Excess fat around the middle—**central obesity**—is associated with increased health hazards.[13] Body types are described as being either apple shaped or pear shaped. Apple-shaped people, those who store most of their excess fat around the abdomen (typically men), are at a greater risk for developing diabetes, hypertension, elevated levels of blood cholesterol, and heart disease than pear-shaped people, who store excess fat on the hips, thighs, and buttocks (typically women). A simple calculation of a person's waist circumference can be used to assess abdominal fat.

Weighing in for Health

Obesity presents one of the most serious health risks that people face—and one that people should, theoretically, be able to control. In response to the rising epidemic of excessive weights, new guidelines for the assessment and treatment of overweight and obesity have been developed.[14] A person's health risk due to weight depends on three factors: body weight, amount and location of body fat, and current health status.

*The air displacement technique used by the *BodPod* relies on the physics of Boyle's Law, which states that pressure and volume vary inversely with one another (for example, as pressure goes up, volume goes down and vice versa). Measuring air displacement (pressure changes) caused by a person seated in the sealed *BodPod* chamber of known volume provides the data needed to calculate a person's body density.

© Courtesy of Life Measurement, Inc.

▲ *The BodPod measures body density.*

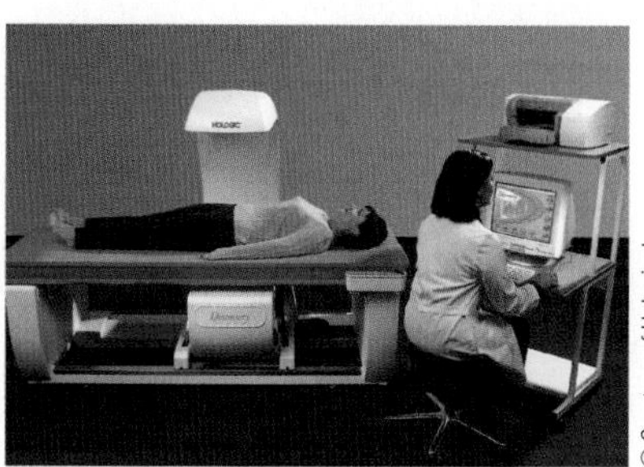
© Courtesy of Hologic, Inc.

▲ *Dual energy X-ray absorptiometry (DEXA).*

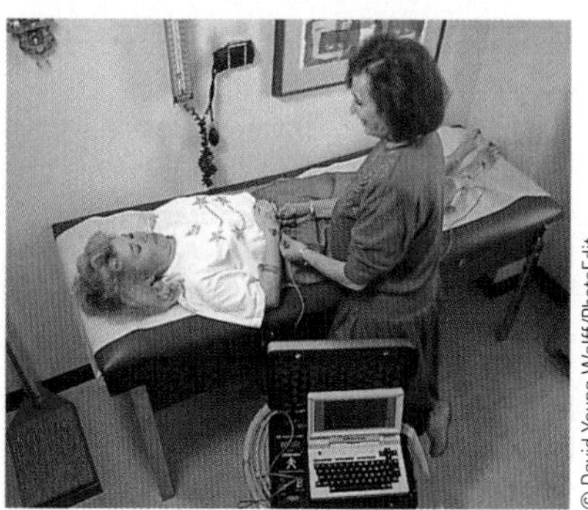
© David Young-Wolff/PhotoEdit

▲ *Bioelectrical impedance measures body fat.*

skinfold test a method in which the thickness of a fold of skin on the back of the arm (the triceps), below the shoulder blade (subscapular), or in other areas is measured with an instrument called a caliper. Obesity is defined by triceps skinfold thickness equal to or greater than 18–19 mm in adult men or 25–26 mm in women.

bioelectrical impedance estimation of body fat content made by measuring how quickly electrical current is conducted through the body.

central obesity excess fat on the abdomen and around the trunk. *Peripheral obesity* is excess fat on the arms, thighs, hips, and buttocks.

body mass index an index of a person's weight in relation to height that correlates with total body fat content.

waist circumference a measure used to assess abdominal (visceral) fat; excess fat in the abdomen increases a person's risk for health problems.

Waist circumference denoting risk of obesity-related health problems:

Substantially Increased Risk
Men > 40"
Women > 35"

Body weight is assessed using the **body mass index (BMI),** an index of your weight in relation to your height. Overweight is defined as a BMI of 25–29.9, and obesity is defined as a BMI of 30 or above (refer to Table 9-2 on page 274 for body weights that equate with these BMI values).[15] Keep in mind that BMI does not account for location of fat in the body, and a muscular person with a low percentage of body fat may have a high BMI (see Figure 9-3). In consideration of this shortcoming of the BMI, the other two factors are taken into account.

The **waist circumference** measurement provides information about the distribution of fat in the abdomen. Excess abdominal fat poses a greater health risk than excess fat in the hips and thighs. The extra abdominal fat crowds the organs and its proximity to the liver means that, when metabolized, abdominal fat can raise blood cholesterol levels and lower the body's sensitivity to insulin. Disease risk rises significantly with a waist circumference of over 35 inches in women and over 40 inches in men.[16]

The third factor to consider is current health status—the presence of weight-related health problems and risk factors for diseases. These may include family health history, heart disease, type 2 diabetes, high blood cholesterol, high blood pressure, cigarette smoking, osteoarthritis, gallstones, or sleep apnea (irregular breathing during sleep). According to the new treatment guidelines, an initial goal for treatment of overweight and obese people with risk factors is to reduce body weight by about 10 percent at a rate of about 1 to 2 pounds per week.[17] For overweight individuals, losing as little as 5 to 10 percent of their body weight may improve many of the problems linked to being overweight, such as diabetes and high blood pressure. The accompanying Healthy Weight Scorecard helps you make sense of these guidelines and evaluate your own weight status.

Energy Balance

Suppose you decide that you are too fat or too thin and that you got that way by having an unbalanced energy budget—that is, by eating either more or less food energy than you spent. Fatness and thinness are reflections of excessive or deficient energy stores. You store extra energy as fat only if you eat *more* food energy in a

FIGURE 9-3
UNDERSTANDING BODY MASS INDEX (BMI)

SOURCE: Adapted from G. Bray, *Contemporary Diagnosis and Management of Obesity,* © 1998 Handbooks in Healthcare, Co., a division of AAM Co., Inc.

Benefits of Using BMI
BMI correlates strongly with body fatness and risk of disease and death.

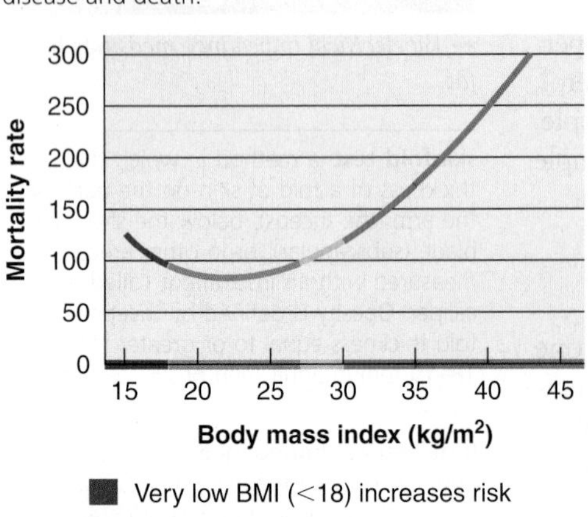

Cautions in Using BMI
May overestimate body fat in athletes and underestimate body fat in adults over 65.

Body Mass Index (BMI) can be calculated as follows:

BMI = [weight (pounds) ÷ height (inches)²] × 703

For example, the BMI for a woman who is 5′5″ tall and weighs 145 lbs would be:

[145 ÷ (65 × 65)] × 703 = 24

- BMI < 18.5 = underweight
- BMI 18.5 to 24.9 = normal weight
- BMI 25 to 29.9 = overweight
- BMI ≥ 30 = obesity
- BMI ≥ 40 = severe obesity

◀ *A muscular person, such as this bodybuilder, often has a low percentage of body fat but a high BMI.*

Scorecard

Healthy Weight

A wide range of weights is compatible with good health. Within this range, the definition of desirable or healthful weight is up to the individual, depending on such factors as family history, occupation, physical and recreational activities, and personal preferences. To determine if your weight is a healthful weight for you:

1. *Calculate your body mass index (BMI).* Find your height and weight in the following chart.* Your BMI is at the top of the column that contains your weight. Record your BMI here: ________

BMI:	19	20	21	22	23	24	25	26	27	28	29	30	35	40
HEIGHT	WEIGHT (POUNDS)													
5'0"	97	102	107	112	118	123	128	133	138	143	148	153	179	204
5'1"	100	106	111	116	122	127	132	137	143	148	153	158	185	211
5'2"	104	109	115	120	126	131	136	142	147	153	158	164	191	218
5'3"	107	113	118	124	130	135	141	146	152	158	163	169	197	225
5'4"	110	116	122	128	134	140	145	151	157	163	169	174	204	232
5'5"	114	120	126	132	138	144	150	156	162	168	174	180	210	240
5'6"	118	124	130	136	142	148	155	161	167	173	179	186	216	247
5'7"	121	127	134	140	146	153	159	166	172	178	185	191	223	255
5'8"	125	131	138	144	151	158	164	171	177	184	190	197	230	262
5'9"	128	135	142	149	155	162	169	176	182	189	196	203	236	270
5'10"	132	139	146	153	160	167	174	181	188	195	202	209	243	278
5'11"	136	143	150	157	165	172	179	186	193	200	208	215	250	286
6'0"	140	147	154	162	169	177	184	191	199	206	213	221	258	294
6'1"	144	151	159	166	174	182	189	197	204	212	219	227	265	302
6'2"	148	155	163	171	179	186	194	202	210	218	225	233	272	311
	Healthy Weight						Overweight					Obese		

SOURCE: National Heart, Lung and Blood Institute.

2. Determine if your fat distribution is associated with health risks. Measure your waist circumference by placing a tape measure around your waist just above your belly button. Record your waist in inches here. ________

3. Is your weight affecting your health? Do you have any of these weight-related health problems or risk factors?
   - Heart disease
   - Type 2 diabetes
   - High blood pressure
   - High LDL-cholesterol
   - Low HDL-cholesterol
   - High triglycerides
   - Osteoarthritis
   - Recurrent gallstones
   - Sleep disturbances
   - Cigarette smoking
   - Sedentary lifestyle
   - Male ≥ 45 years or post-menopausal female

4. How does your current weight measure up to these considerations?
   - If your BMI is acceptable for good health and if your waist measurement is not high (see page 276), you will want to maintain this weight. If you need to lose weight or gain weight, consider the tips offered throughout this chapter for healthfully changing your weight.
   - You should consider losing weight if:
     Your BMI is 30 or greater
     Your BMI is 25 to 29 *and* you have two or more of the weight-related health problems or risk factors listed above.
     Your waist circumference exceeds 40 inches (for men) or 35 inches (for women) *and* you have two or more weight-related health problems or risk factors.
   - Weight loss is optional for you if your BMI is 25 to 29 and you do *not* have two or more weight-related health problems (particularly if your BMI is under 27 or you have large muscles and bones).

*Additional heights and weights are listed on the inside back cover, or you can use this formula to calculate your BMI:

$$\frac{\text{weight (in pounds)}}{\text{height}^2 \text{ (in inches)}} \times 703$$

Example: A 5'8" person weighing 145 pounds has a BMI of 22

$$\text{BMI} = \frac{145}{68^2} \times 703$$

$$\text{BMI} = \frac{145}{4624} \times 703$$

BMI = 22 (rounded)

FIGURE 9-4
HOW THE BODY EXPENDS ENERGY

The body expends most of its energy on *basal metabolism*—maintaining basic physiological processes such as breathing, heartbeat, and other involuntary activities. The second largest amount of energy is expended for voluntary *physical activities*—an amount that will vary by activity level. A minor amount of energy is also used for the *thermic effect of food*—the energy needed to digest, absorb, and process the food you eat.

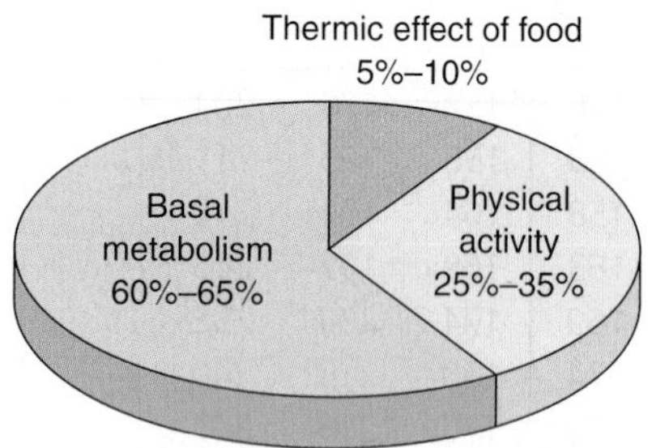

day than you use to fuel your metabolic and other activities. Similarly, you lose stored fat only if you eat *less* food energy in a day than you use as fuel. A day's energy balance can be stated like this:

Change in energy stores = Energy in − Energy out.

You already know about the "energy in" side of this equation. An apple brings you 80 calories; a candy bar, 290 calories. You may also know that for each 3,500 *excess* calories consumed ("excess" meaning more than your body needs), you store 1 pound of body fat.

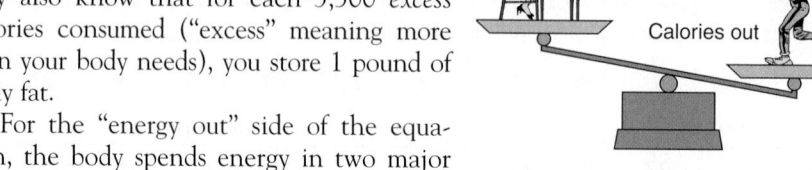

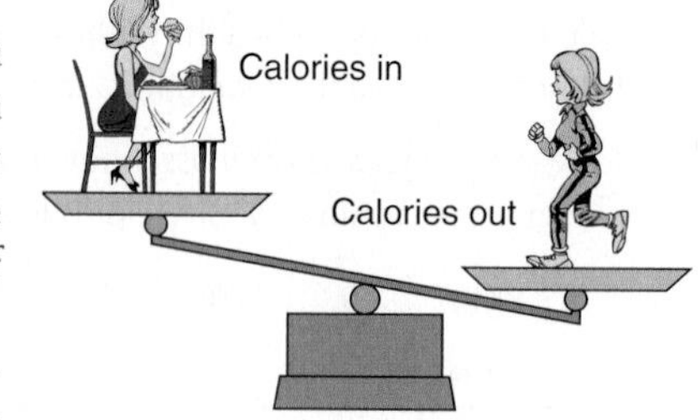

For the "energy out" side of the equation, the body spends energy in two major ways: to fuel its metabolic activities and to fuel its muscle activities. As shown in Figure 9-4, you also spend a small amount of energy to digest, absorb, transport, process, and store the food you eat, which is known as the *thermic effect of food*. The following three sections discuss the body's needs for energy.

Basal Metabolism

About 60 percent or more of the energy the average person spends goes to support the ongoing metabolic work of the body's cells, the **basal metabolism.**[18] This is the work that goes on all the time, without conscious awareness. The beating of the heart, the inhaling and exhaling of air, the maintenance of body temperature, and the sending of nerve and hormonal messages to direct these activities are the basal processes that maintain life. Basal metabolic needs are surprisingly large. A person whose total energy expenditure amounts to 2,000 calories per day spends as many as 1,200 to 1,400 of them to support basal metabolism.

The **basal metabolic rate** (often abbreviated **BMR**) is influenced by a number of factors (see Table 9-3). In general, the younger a person is, the higher the basal metabolic rate, partly because of the increased activity of cells undergoing division. The BMR is most pronounced during the growth spurts that take place during infancy, puberty, and pregnancy. Body composition also influences metabolic rate. Muscle tissue is highly active even when it is resting, whereas fat tissue is comparatively inactive. Therefore, when comparing two people of the same body weight, the person with more lean muscle tissue has a higher BMR, and the person with more fat tissue would have a lower BMR. Lean body mass decreases with age, but consistent physical activity—especially endurance and strength-building activities—may prevent some of the decline.

Gender correlates roughly with body composition. Men generally have a faster metabolic rate than women, and researchers believe that this is because of men's greater percentage of lean tissue. (A woman athlete has a greater percentage of lean tissue than a sedentary man of the same weight and so would have a higher metabolic rate.) Fever increases the energy needs of cells as their increased activities to fight off infection and generate heat speed up the metabolic rate.

Fasting and constant malnutrition lower the metabolic rate because of the loss of lean tissue and the slowdown of activities the body can't afford to support fully. This slowing of metabolism seems to be a protective mechanism to conserve energy when there is a shortage, and it hampers weight loss in a person who fasts or undertakes a very low-calorie diet.

Some hormones—the stress hormones, for example—influence metabolism. They increase the energy demands of every cell and thus raise the metabolic rate. This raised metabolism partly accounts for the weight loss sometimes seen in people experiencing extreme stress in their life, although other factors, such as upset digestion and loss of appetite, also enter in. The activity of the thyroid gland also influences the basal metabolic rate. The less thyroid hormone secreted, the lower the energy requirement for maintenance of basal functions.

TABLE 9-3
Factors That Influence the Basal Metabolic Rate

Factors That Increase BMR:
- Caffeine
- Fever
- Growth (higher in children and pregnant women)
- Height (higher in tall, thin people)
- High thyroid hormone
- Male gender (more lean tissue)
- Muscle mass (the more lean tissue, the higher the BMR)
- Smoking (nicotine)
- Stress

Factors That Decrease BMR:
- Age (slows down with age)
- Low thyroid hormone
- Reduced energy intake (fasting, starvation, low-calorie diets)
- Sleep (BMR is lowest when sleeping)

basal metabolism the sum total of all the chemical activities of the cells necessary to sustain life, exclusive of voluntary activities—that is, the ongoing activities of the cells when the body is at rest.

basal metabolic rate (BMR) the rate at which the body spends energy to support its basal metabolism. The BMR accounts for the largest component of a person's daily energy (calorie) needs.

Voluntary Activities

Muscular activity does not contribute as much to most people's energy outputs as basal metabolism does. On the average, it amounts to only about 30 percent of the total. But unlike basal metabolism, which cannot be changed immediately, physical activity can be changed at will. If you want to tinker with your energy balance, this is the component—on the output side—that you can alter significantly in the short term. If you increase physical activity consistently, you will also ultimately increase the energy your body spends on metabolic activity because you will have an increase in lean body mass.

The energy spent on physical activity is the energy spent moving the body's skeletal muscles—the muscles of the arms, back, abdomen, legs, and so forth—and the extra energy needed to speed up the heartbeat and respiration rate. The number of calories spent depends on three factors: (1) the amount of muscle mass required, (2) the amount of weight being moved, and (3) the amount of time the activity takes. Thus, an activity involving both the arms and the legs requires more calories than an activity of the same intensity involving only the legs; an activity performed by a heavier person requires more energy than the same activity performed by a lighter person; and an activity performed for 40 minutes requires twice as much energy as that same activity performed for 20 minutes.

It may be disheartening for a college student to discover that mental activity requires little energy, even though it may be tiring. Studying for an exam may be hard work, but it doesn't burn body fat. People who are very, very busy—surfing the Internet, making phone calls, riding in their cars—may wonder why they gain weight. Even though they are socially or intellectually active, such activities involve few muscles and therefore little energy expenditure.

▲ *Body composition influences metabolic rate. Weight training can help shift your body composition toward more lean tissue, thereby speeding up your metabolism. See Chapter 10 for guidelines on how to get started.*

Total Energy Needs

A typical breakdown of the total energy spent by a lightly active person (for example, a student who walks back and forth to classes) might look like this:

Energy for basal metabolism:	1,400 calories
Energy for physical activity:	560 calories
Total:	1,960 calories

The first component (BMI) is larger, and you cannot change it much. You can, however, change the second component—physical activity—and use more calories. To increase your basal metabolic output, make exercise a daily habit. Your body composition will gradually change, and your basal energy output will pick up the pace as well. You can figure out roughly how much energy you need in a day by using the method illustrated in the box on page 280.

In summary, the amount of fat stored in a person's body depends on the balance between the total food energy the person has taken in and the total energy the person has expended. But why do so many people have excessive fat stores?

Causes of Obesity

Some people eat more than they need or exercise less than they should to maintain their body weight, and they get fat. Some eat less or exercise more, and they get thin. Amazingly, many people eat exactly what they need and stay at the same weight year after year. A single extra pat of butter each day would make them gain 5 pounds in a year, but if these people overeat by that much on one day, they apparently undereat by the same amount the next day. How do they do this, and (in contrast) why do some people fail to maintain their weight? In general, two schools of thought address the causes of obesity.[19] One school of thought attributes it to inside-the-body causes and the other to environmental factors.

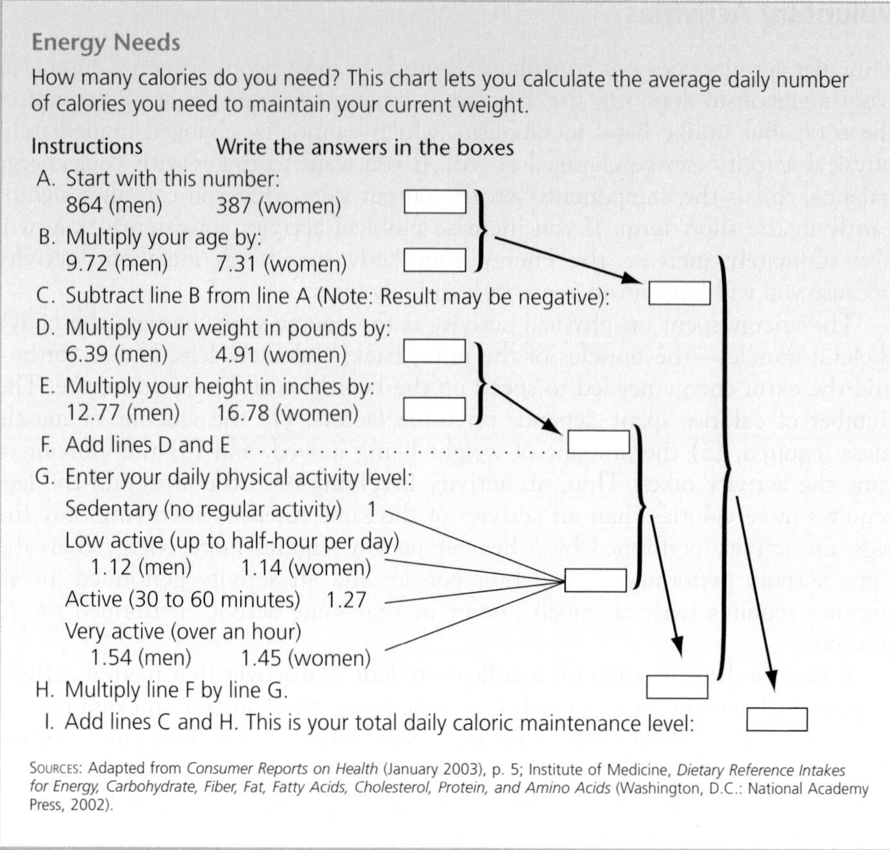

Energy Needs

How many calories do you need? This chart lets you calculate the average daily number of calories you need to maintain your current weight.

Instructions — Write the answers in the boxes

A. Start with this number:
864 (men) 387 (women)

B. Multiply your age by:
9.72 (men) 7.31 (women)

C. Subtract line B from line A (Note: Result may be negative):

D. Multiply your weight in pounds by:
6.39 (men) 4.91 (women)

E. Multiply your height in inches by:
12.77 (men) 16.78 (women)

F. Add lines D and E . . . .

G. Enter your daily physical activity level:
Sedentary (no regular activity) 1
Low active (up to half-hour per day)
1.12 (men) 1.14 (women)
Active (30 to 60 minutes) 1.27
Very active (over an hour)
1.54 (men) 1.45 (women)

H. Multiply line F by line G.

I. Add lines C and H. This is your total daily caloric maintenance level:

SOURCES: Adapted from *Consumer Reports on Health* (January 2003), p. 5; Institute of Medicine, *Dietary Reference Intakes for Energy, Carbohydrate, Fiber, Fat, Fatty Acids, Cholesterol, Protein, and Amino Acids* (Washington, D.C.: National Academy Press, 2002).

Genetics

The theory that a hereditary, inside-the-body factor for obesity may exist is supported by the existence of animal strains that are genetically fat. Such animals tend to be fat in any environment—that is, they are fat regardless of the kind or variety of food that is available. In humans, studies have shown that identical twins—whether raised together or apart—tend to have similar weight-gain patterns. Also, twins raised by adoptive parents tend to have body shapes similar to their biological parents. Not all studies confirm this, however.[20] Moreover, pairs of twins purposefully overfed in clinical experiments tend to respond similarly to the extra calories. Some sets of twins gain considerable weight when fed a certain number of calories, whereas other pairs put on relatively few pounds even on the same diet.[21]

Set-Point Theory One popular inside-the-body theory is the so-called **set-point theory.** Noting that many people who lose weight on reducing diets subsequently return to their original weight, some researchers have suggested that the body "wants" to maintain a certain amount of fat and regulates eating behaviors and hormonal actions to defend its "set point." The theory implies that science should search inside obese people to find the causes of their problems—perhaps in their hunger-regulating mechanisms.

set-point theory the theory that the body tends to maintain a certain weight by adjusting hunger, appetite, and food energy intake on the one hand and metabolism (energy output) on the other so that a person's conscious efforts to alter weight may be foiled.

Leptin Researchers have identified a gene—named *ob* (for *obese*)—that appears to produce a hormone called *leptin*, after the Greek word for slender. When it is released from fat cells, leptin seems to tell the body to stop eating.[22] Researchers report that as body fat stores increase, blood leptin also increases. The brain responds by decreasing appetite and increasing energy expenditure. Likewise, when body fat stores decrease, blood leptin decreases, and the brain responds by stimulating appetite and decreasing energy expenditure. Mice who have a defective form of the gene fail to produce leptin and can weigh as much as three times more than normal mice. Overweight people, too, may have a defective form of this gene (or may be unresponsive to leptin).[23] More research is needed to clarify this mechanism.

Fat Cell Theory Some overweight infants become overweight adults, but most grow out of their obesity in childhood. An overweight child, however, is more likely to remain overweight into adulthood.[24] Some researchers propose the **fat cell theory**—that childhood obesity is persistent because early overfeeding (during the growing years) may cause fat cells to increase abnormally in *number*. According to this theory, a person's *number* of fat cells is relatively fixed by adulthood; afterwards, a gain or loss of weight either increases or diminishes the size of the fat cells. Unfortunately, persons with greater numbers of fat cells are less likely to lose weight successfully. Some researchers suggest that the body triggers hunger signals when the fat stored in these cells begins to decrease. Because fat cells increase in number during childhood, prevention of obesity is critical during the growing years.

Additionally, fat cells of obese people contain higher levels of the enzyme **lipoprotein lipase (LPL),** which determines the rate at which adipose cells store fat. The larger the fat cell (and the greater the number of fat cells), the more LPL and the more easily the body can pull triglycerides into fat cells for storage. Unfortunately, LPL activity rises further with weight loss, enhancing the body's ability to regain the lost weight.[25] A question still to be answered is whether some people develop obesity because their fat cells contain an abnormal amount of LPL from birth.

fat cell theory states that during the growing years, fat cells respond to overfeeding by producing additional fat cells; the number of fat cells eventually becomes relatively fixed, and overfeeding from this point on causes the body to enlarge existing fat cells.

lipoprotein lipase (LPL) an enzyme located on the surfaces of fat cells that enables the cell to convert blood triglycerides into fatty acids and glycerol to be pulled into the cell for reassembly and storage as body fat.

external cue theory the theory that some people eat in response to such external factors as the presence of food or the time of day rather than to such internal factors as hunger.

hunger the physiological drive to find and eat food, experienced as an unpleasant sensation.

appetite the psychological desire to find and eat food, experienced as a pleasant sensation, often in the absence of hunger.

Environment

The other point of view is that obesity is environmentally determined. Proponents of this view hold that people overeat or underexercise because they are pushed to do so by factors in their surroundings—foremost among them, the availability of a multitude of delectable foods and a lack of opportunity for vigorous physical activity. The two views—inside-the-body versus environment—are not mutually exclusive, and they may both be at work, even within the same person.

Some people seem to have inherited or learned a way of resisting external stimuli to eat, but others have not. In a classic experiment with "cafeteria rats," ordinary rats who were fed regular rat chow maintained normal weight (for rats), but when those very same rats were offered free access to a wide variety of tempting, rich, highly palatable foods, they greatly overate and became obese. Similarly, one study found a positive correlation between overfatness and a diet offering a wide variety of snacks and sweets.[26] This is the basis of the **external cue theory**—the theory that, at least in some people, the internal regulatory systems are easily overridden by environmental influences. Does this mean obesity is hereditary, environmental, or both?

It seems likely that both hereditary and environmental factors influence obesity in human beings. The tendency to obesity is probably inherited, but the environment is probably influential in the sense that it can prevent or permit the development of obesity when the potential is there.

A Closer Look at Eating Behavior

In human beings, learning plays an important role. Although we have genetically inborn instincts, things learned during early childhood experiences combined with our current environments may conflict with these instincts. Thus, the **hunger** drive is programmed by heredity, but **appetite** is influenced by learned responses that may cause us to ignore or over-respond to our hunger. Another way to say this is to say that hunger is physiological, whereas appetite is psychological, and

satiety the feeling of fullness or satisfaction that people feel following a meal.

hypothalamus (high-poh-THALL-ah-mus) a part of the brain that senses a variety of conditions in the blood, such as temperature, salt content, and glucose content, and then signals other parts of the brain or body to change those conditions when necessary.

arousal as used in this context, heightened activity of certain brain centers associated with excitement and anxiety.

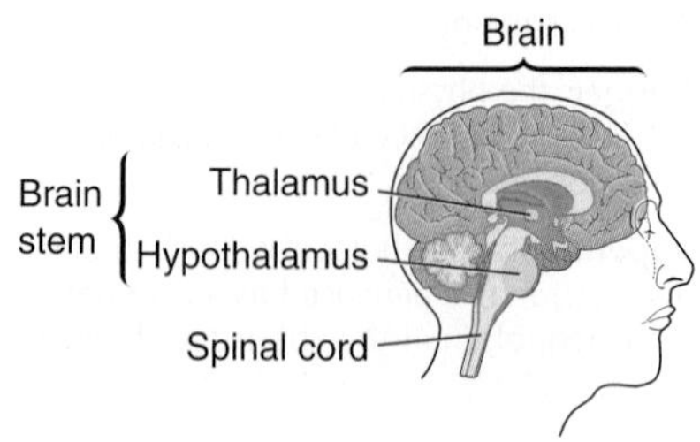

▲ *Researchers believe that the hypothalamus controls the sensations of hunger and satiety.*

the two don't always coincide. We have all experienced appetite without hunger: "I'm not hungry, but I'd love to have some." We also often experience the reverse: "I know I'm hungry, but I don't feel like eating."

The ways people respond to hunger and appetite determine whether they eat too much, too little, or just enough to maintain their weight. A third factor, **satiety,** which signals that it is time to stop eating, also affects our eating patterns and weight issues.

In human physiology, research is beginning to find possible answers to what regulates eating behavior. The stomach's nerves perceive stretching, and you stop eating when your stomach feels stretched full. Blood glucose level is also thought to be involved. You get hungry when your blood glucose level falls—or perhaps when your liver glycogen is beginning to be exhausted. Blood lipids, and possibly amino acids and other molecules, also play a role. When you eat, you secrete hormones to regulate digestive activity; these hormones may also convey the message to the brain that it is time to start or stop eating.*

Where in the brain are these messages received (whatever they are)? One brain area stands out as a regulator for food behavior—the **hypothalamus.** The hypothalamus communicates with the hormone system and the nervous system. It integrates many signals received from the rest of the body, including information about the blood's temperature, sodium content, and glucose content. We know it is important in regulating eating because damage to the hypothalamus produces derangements in eating behavior and body weight. In some cases it can cause severe weight loss, and in others it can cause extreme overeating. In the person with a normal hypothalamus, however, appropriate eating behavior seems to be a response to a whole host of signals rather than a response to a single signal arriving at some one location in the hypothalamus. Somehow these many inputs become integrated into a final common path—the act of eating.

A person who eats inappropriately may have established a habitual behavior pattern that inappropriately links certain stimuli to the act of eating. The study of behavior offers insight into the causes of overeating by viewing it as a conditioned response. Sometimes eating behavior is initiated by the wrong triggers. For example, a crying child with a skinned knee who is offered a lollipop may learn to associate food with comfort and inappropriately feel the need for food when experiencing emotional pain later in life.

Eating behavior, then, may be a response not only to hunger or appetite but also to complex human sensations such as yearning, craving, addiction, or compulsion. Often, eating is used to relieve boredom or to ward off depression. Some people respond to anxiety—or, in fact, to any kind of **arousal**—by eating. Significantly, however, when these people are able to name their aroused condition, they often gain a feeling that they have some control over it and are not as likely to overeat.

Stress may also directly promote the accumulation of body fat. The stress hormones favor the breakdown of energy stores (glycogen and fat) into glucose and fatty acids, which can be used to fuel the human "fight or flight" response. If a person doesn't use the fuel in physical exertion, however, the body cannot turn these fragments back into glycogen. Thus, its only alternative is to convert them to fat. Each time glucose is pulled out of storage in response to stress and then transformed into fat, the lowered glucose level or exhausted glycogen will signal hunger, and the person will eat again soon after.

Stress eating may appear in different patterns. Some people eat excessively at night, whereas others binge during emotional crises. Some people react oppositely and reject food when stressed. We do not know why these behaviors occur, but research continues.

*Researchers have recently discovered ghrelin (GRELL-in), a protein produced by the stomach cells that acts as a hormone affecting the hypothalamus and stimulating appetite. Ghrelin levels are high before a meal (reflecting hunger) and fall rapidly after a meal (reflecting satiety). Researchers continue to investigate ghrelin's exact role in food intake regulation.

The many possible causes of obesity mentioned so far all relate to the input side of the energy equation. What about output? Probably the most important contributor to the obesity problem in our country is underactivity.[27] Some obese people eat less than lean people, but due to extraordinary inactivity, they still manage to store surplus calories. Some people move more efficiently than others, too. Two people of the same age, height, and weight might use different amounts of calories walking five miles because of the different ways in which they move their muscles.

No two people are alike, either physically or psychologically, and the causes of obesity are no doubt as varied as the obese people themselves. Many causes may contribute to the problem in a single person. Given this complexity, it is obvious that there is no panacea. The top priority should be prevention, but where prevention has failed, the treatment of obesity must involve a simultaneous attack on many fronts.[28]

Weight Gain and Loss

When you step on the scale and see that you weigh a pound more or less than you did the last time, you haven't necessarily gained or lost body fat. Changes in body weight reflect shifts in many different materials—not only fat but also fluid, bone minerals, and lean tissues such as muscles. The gain of a pound, for example, may reflect gained muscle and bone and an overall shift toward a leaner body type. Because it is so important for people concerned with weight control to realize this, this section discusses the various changes that take place with gains and losses of weight.

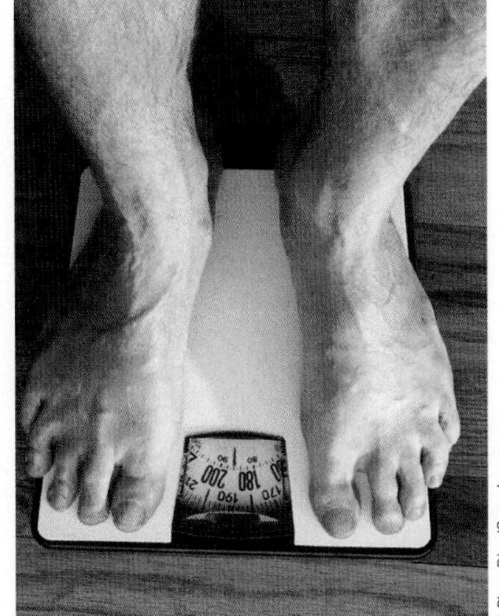

© PhotoDisc/Getty Images

▲ *The loss (or gain) of a pound does not always reflect the loss (or gain) of body fat.*

A healthy man or woman about 5 feet, 10 inches tall who weighs 150 pounds carries about 90 of those pounds as water and 30 as fat. The other 30 pounds are the so-called lean tissues: muscles; organs such as the heart, brain, and liver; and the bones of the skeleton.* Stripped of water and fat, then, the person weighs only 30 pounds. This lean tissue is the body's vital machinery that maintains health and life. When a person who is too fat seeks to lose weight, it should be fat—not this precious lean tissue—that is lost. For someone who wants to gain weight, it is desirable to gain lean *and* fat, not just fat.

Weight is gained or lost in different body tissues, depending on how a person goes about it. Most quick-weight-loss diet schemes promote large losses of fluid that create large, temporary changes in the scale's weight with little or no real loss of body fat. The rest of this chapter underscores this distinction, and a later section on weight-loss strategy stresses exercise as a means of supporting lean tissue during weight loss.

Weight Gain

When you eat more calories than you need, where does the excess go? The energy nutrients—carbohydrate, fat, and protein—contribute to body stores as follows:

- Carbohydrate is broken down to glucose for absorption. Inside the body, glucose may be built up to glycogen or converted to and stored as fat.
- Fat is broken down to its component parts (including fatty acids) for absorption. Inside the body, these components are easily converted to storage fat.
- Protein, too, is broken down to its basic units (amino acids) for absorption. Inside the body, these units may be used to replace body proteins. The amino acids that are not used lose their nitrogen and are converted to fat.

*For a healthy woman or man 5 feet tall who weighs 100 pounds, the comparable figures would be 60 pounds of water, 20 pounds of fat, and 20 pounds of lean tissue.

Notice in Figure 9-5 that although three kinds of materials enter the body, they are stored for later use in only two forms, glycogen and fat. Also notice that when protein is stored in the form of fat, it cannot be converted back into protein later. The amino acids lose their nitrogen, which is then excreted in the urine. It does not matter whether you are eating hamburgers, brownies, or carrot sticks; if you eat enough of any food, the excess will be turned to fat within hours.

Weight Loss and Fasting

When the tables are turned and you stop eating altogether, your body must draw on its stored supplies of nutrients to keep going. Nothing is wrong with this; in fact, it is a great advantage that you can eat periodically, store fuel, and then use up that fuel between meals. The between-meal interval is ideally about 4 to 6 hours—about the length of time it takes to use up most of the available liver glycogen—or 12 to 14 hours at night, when body systems are slowed down and the need for energy is lower. If a person doesn't eat for, say, three days or even a week, the body makes one adjustment after another.

The body's first adjustment to fasting is to use the liver's glycogen for needed fuel. (The glycogen in muscles is reserved for the muscles' own use—and they are using it.) The liver's glycogen, remember, is the body's source of blood glucose to fuel brain and nerve activities. Ordinarily, the brain and nerves can use no other fuel, but after about a day without food, the primary supply is gone. Where, then, does the body turn to keep its nervous system going? Whatever it has to do, it will do, for the nervous system runs the body, and when it stops, the body dies.

An obvious alternative source of energy would be the abundant fat stores most people carry. At first, this fat is of no use to the nervous system. The muscles and other organs can use fat as fuel, but the nervous system ordinarily cannot. Nor can the body convert this fat to glucose, because it possesses no enzymes to do so. It does, however, possess enzymes to convert protein to glucose.

When the fast continues, the body turns to its own lean tissues to provide the necessary supply of glucose (see Figure 9-5). One reason people lose weight so dramatically within the first three days of a fast is that they are devouring their own protein tissues as fuel. Because protein contains only half as many calories per pound as fat, it disappears twice as fast. Also, with each pound of body protein, three or four pounds of associated water are lost. This same process accounts for the rapid weight loss seen in the early stages of a low-carbohydrate diet.

If the body were to continue to consume itself at this rate, death would ensue within about ten days. After all, the liver, the heart and skeletal muscles, the lung tissue, and the blood—all vital tissues—are being burned as fuel. (In fact, fasting or starving people remain alive only until their body fat is gone or until half their lean tissue is gone, whichever comes first.) But now the body plays its last ace. It begins converting fat stores into a form it can use to help feed the nervous system and so forestall the end. This is known as **ketosis.**

Ketosis is an adaptation to fasting or carbohydrate deprivation. Instead of breaking fat molecules all the way down to carbon dioxide and water as it normally does, the body takes partially broken-down fat fragments, combines them into ketone bodies (compounds that are normally rare in the blood), and lets them circulate in the bloodstream. The advantage for the body is that about half of the brain cells can use these compounds for energy. Thus, indirectly, the nervous system begins to feed on the body's fat stores. This reduces the nervous system's need for glucose, spares the muscle and other lean tissue from being devoured so quickly, and prolongs the starving person's life. Because of ketosis, an initially healthy person who is totally deprived of food can live for as long as six to eight weeks.

ketosis (kee-TOE-sis) an adaptation of the body to prolonged (several days) fasting or carbohydrate restriction: body fat is converted to ketones, which can be used as fuel for some brain cells.

FIGURE 9-5
FEASTING AND FASTING
In A, the person is storing energy. In B, the person is drawing on stored energy. In C, the person is in ketosis.

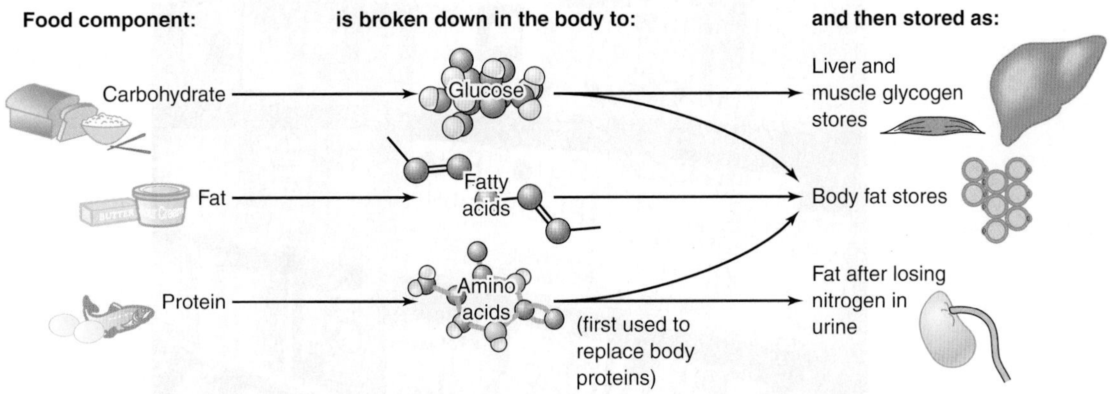

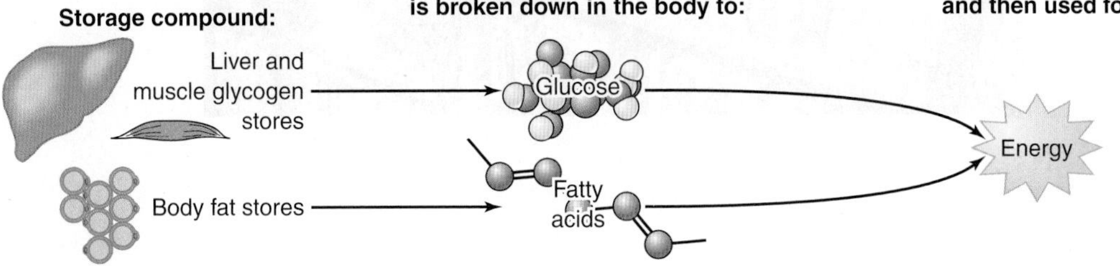

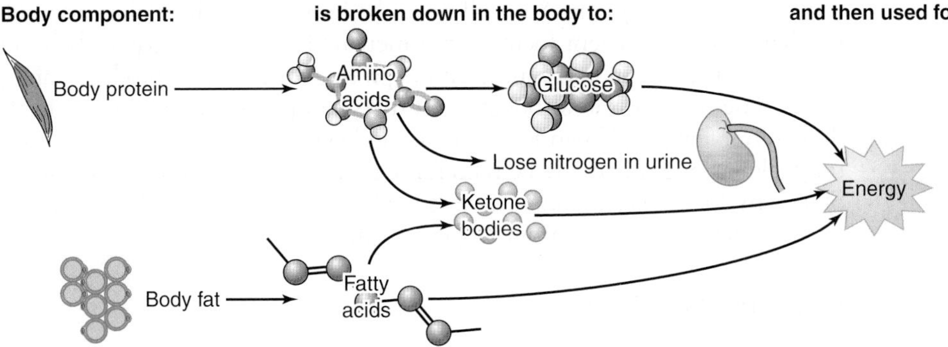

Fasting has been practiced as a periodic discipline by respected, wise people in many cultures. However, ketosis may be harmful to the body by upsetting the acid–base balance of the blood. For the person who merely wants to lose weight, then, fasting is not the best way. For one thing, even in ketosis, the body's lean tissue continues to be lost at a rapid rate to supply glucose to those nervous system cells that cannot use ketones as fuel. For another, the body becomes conservative during a fast and slows its metabolism so as to lose as little energy as it possibly can. A well-designed, low-calorie diet, accompanied by the appropriate exercise program, has actually been observed to promote the same rate of *weight* loss as, and a faster rate of *fat* loss, than a total fast. The subject of the Nutrition Action feature that follows is what to look for in a weight-loss diet—including tips for how *not* to design a diet.

Nutrition Action Diet Confusion—Weighing the Evidence

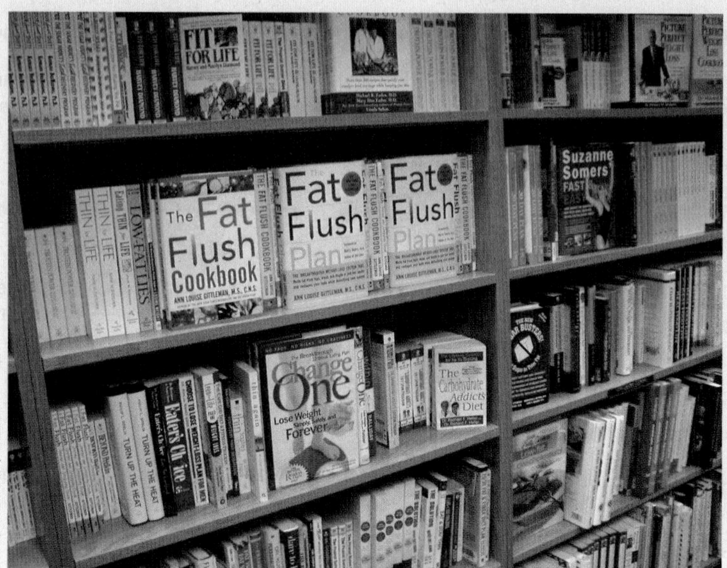

© Susan Van Etten/PhotoEdit

Lose weight while you sleep! Lose 30 pounds in just 20 days! Eat the foods you love and lose weight! You will never be hungry! Do these claims sound familiar? With the recent focus on the increased rates of obesity in the United States and the world, burgeoning efforts promote diet books, products, and programs. The truth is that although most diets can provide a weight loss in the short term, few people can lose weight and keep it off permanently, and some of these claims might actually be harmful. Dieting is big business in the United States. In 2004, one national survey found that 33 percent of American adults are on a diet, an increase from 24 percent in 2000.[29]

How Do Diets Work?

Diets work because people limit their food consumption. Excess weight is the consequence of an energy imbalance, caused by overconsumption of food or decreased physical activity relative to individual requirements. Limiting dietary intake can take several forms:

- Elimination or restriction of certain food groups, such as carbohydrates
- Portion control through prepackaged meals, snacks, or drinks
- Alteration of meal patterns or content
- Control of food intake through point systems or monitoring

A comparison of the approximate caloric content and macronutrient distribution of several types of diets is provided in Table 9-4.

TABLE 9-4
Comparison of Diet Programs/Eating Plans to Typical American Diet

Type of Diet	Example	General Dietary Characteristics	Comments
Typical American diet		Carb.: 50% Protein: 15% Fat: 35% Average of 2,200 cal/day	• Low in fruits and vegetables, dairy and whole grains • High in saturated fat and unrefined carbohydrates
Balanced nutrient, moderate-calorie approach	DASH Diet or Diet based on MyPyramid food guide; commercial plans such as Diet Center, Jenny Craig, Nutri/System, Physician's Weight Loss, Shapedown Pediatric Program, Weight Watchers, Setpoint Diet, Sonoma Diet, Volumetrics	Carb.: 55–60% Protein: 15–20% Fat: 20–30% Usually 1,200 to 1,700 cal/day	• Based on set pattern of selections from food lists using regular grocery store foods or prepackaged foods supplemented by fresh food items • Low in saturated fat and ample in fruits, vegetables, and fiber • Recommend reasonable weight-loss goal of 0.5 to 2.0 lb/week • Prepackaged plans may limit food choices • Most recommend exercise plan • Many encourage dietary record-keeping • Some offer weight-maintenance plans/support
Very low-fat, high-carbohydrate approach	Ornish Diet (Eat More, Weigh Less), Pritikin Diet, T-Factor Diet, Choose to Lose, Fit or Fat	Carb.: 65% Protein: 10–20% Fat: ≤ 10–19% Limited intake of animal protein, nuts, seeds, other fats	• Long-term compliance with some plans may be difficult because of low level of fat • Can be low in calcium • Some plans restrict healthful foods (seafood low-fat dairy, poultry) • Some encourage exercise and stress-management techniques
Low-carbohydrate, high-protein, high-fat approach	Atkins New Diet Revolution, Protein Power, Stillman Diet (The Doctor's Quick Weight Loss Diet), the Carbohydrate Addict's Diet, Scarsdale Diet	Carb.: ≤ 20% Protein: 25–40% Fat: ≥ 55–65% Strictly limits Carb. to less than 100–125 g/d	• Promote quick weight loss (much is water loss rather than fat loss) • Ketosis causes loss of appetite • Can be too high in saturated fat • Low in carbohydrates, vitamins, minerals, and fiber • Not practical for long-term because of rigid diet or restricted food choices
Moderate-carbohydrate, high-protein, moderate-fat approach	The Zone Diet, Sugar Busters, South Beach Diet	Carb.: 40–50% Protein: 25–40% Fat: 30–40%	• Diet rigid and difficult to maintain • Enough carbohydrates to avoid ketosis • Low in carbohydrates; can be low in vitamins and minerals
Novelty diets	Immune Power Diet, Rotation Diet, Cabbage Soup Diet, Beverly Hills Diet, Dr. Phil	Most promote certain foods, or combinations of foods, or nutrients as having unique (magical) qualities	• No scientific basis for recommendations
Very low-calorie diets	Health Management Resources (HMR), Medifast, Optifast	Less than 800 cal/day	• Requires medical supervision • For clients with BMI ≥ 30 or BMI ≥ 27 with other risk factors; may be difficult to transition to regular meals
Weight-loss online diets	Cyberdiet, DietWatch, eDiets, Nutrio.com	Meal plans and other tools available online	• Recommend reasonable weight-loss goal of 0.5 to 2.0 lb/week • Most encourage exercise • Some offer weight-maintenance plans/support

SOURCE: Adapted from Weighing the Diet Books, *Nutrition Action Newsletter*, January/February 2004: 3–8; M. Freedman and coauthors, Popular diets: A scientific review, *Obesity Research* 9 (2001): 1S–39S; and A guide to rating the weight-loss websites, *Tufts University Health and Nutrition Letter*, May 2001, pp. 1–4.

What Are Common Diets?

Although diet fads change quickly, certain types of diets have appeared during the past few years. Here are some of the most common:[30]

- ***Dr. Atkins New Diet Revolution*** In this diet, consumption of high-fat meats, cheeses, and fats is encouraged and consumption of carbohydrates (such as fruit, breads, and cereals) is severely limited. The underlying premise of the diet is that elimination of these foods will produce a "benign dietary ketoacidosis," which leads to decreased hunger and slows excessive food consumption. Ketosis can be accompanied by bad breath, nausea, headaches, and fatigue. High protein intake may exacerbate gout and kidney disease, and high saturated fat intake can increase blood cholesterol levels.
- ***The Zone Diet*** This rigid eating plan separates foods into "macronutrient blocks."
- ***The South Beach Diet*** This regimen is a more healthful version of the Atkins high-protein, low-carbohydrate diet, incorporating lower fat protein sources such as chicken and fish, whole grains, and vegetables and fruits. The plan does limit some foods, such as carrots, bananas, pineapple, and watermelon, and the first phase of the diet is more restrictive than later phases.
- ***Weight Watchers*** Dieters may use a list of core foods or a point system to select and eat foods to reduce caloric intake and lose weight. In Weight Watchers, no food is forbidden, but all must be balanced with other choices.
- ***Dr. Ornish Eat More, Weigh Less*** Weight loss is based on consuming a very low-fat diet (10 percent of calories from fat), with little meat, oils, nuts, butter, dairy (except non-fat), sweets, or alcohol. The original Ornish plan included diet together with exercise and stress reduction.
- ***Eat Right for Your Blood Type*** This diet is based on the claim that your blood type determines the types of foods you should eat and how your body absorbs nutrients. For example, people with type O blood should consume meat, seafood, fruits, and vegetables, but less wheat and beans. There is no scientific basis for this claim.
- ***Dr. Phil's Ultimate Weight Solution*** The book describing this diet focuses on "Keys to Weight-loss Freedom," but these concepts do not include defined meal plans or recipes. The diet promotes seafood, poultry, meat, low-fat dairy, whole grains, fruits, vegetables, and some oils. Supplements, weight-loss bars, and shakes are also promoted.
- ***The New Glucose Revolution*** This eating plan encourages consumption of low-glycemic foods, such as beans, pasta, most fruits, vegetables, low-fat dairy, and meats. Unfortunately, the glycemic index is not always a reliable measure of increases in blood glucose, which can vary with the food itself and with other foods consumed at the same time. In addition, changes in blood glucose level depend on the amount of food consumed.

A recent study published in the *Journal of the American Medical Association* evaluated four of these diets (the Atkins, Ornish, Weight Watchers, and Zone diets) and found that, after one year on the diet, all four modestly reduced body weight and some cardiac risk factors.[31] Adherence to each diet for the 12-month period varied, ranging from 50 percent for the Ornish diet and 53 percent for the Atkins diet to 65 percent for both the Weight Watchers and Zone diets. The subjects who had the best adherence to the diets had the best results, and cardiac risk factors were more closely associated with weight loss than with diet type. In general, the subjects had more difficulty following the more restrictive diets (the Ornish and Atkins diets). Although this is just one study with small sample sizes, it does suggest that there are many ways to lose weight, that people find it difficult to adhere to very restrictive diets for a long time, and that we need to find methods of keeping people motivated to stay on any new eating plan.

RELAX! ANOTHER ONE WILL BE ALONG ANY SECOND!
DIET FAD

SOURCE: The St. Petersburg Times. Reprinted with permission.

How Can You Determine If a Diet Is Healthful?

What can you do to determine whether a particular diet plan is useful *and* healthful? Use the checklist that follows.[32]

1. *Does the weight-loss program systematically eliminate one group of foods from a person's eating pattern?* For example, are all carbohydrates systematically eliminated from a person's diet? Are dairy products eliminated? In general, a diet that eliminates a certain food group is probably lacking in important nutrients and dietary variety, and it will be difficult for a person to adhere to that eating plan.
2. *Does the weight-loss program encourage specific supplements or foods that can be purchased only from selected distributors?* These supplements or foods often contain ingredients that may be harmful or unproven.
3. *Does the weight-loss program tout magic or miracle foods or products that burn fat?* The only way to burn fat is to increase your physical activity levels or decrease the amount of total food that you consume. You cannot "burn" fat with sauna belts, body wraps, thigh-reducing creams, or similar products. If you consume more than you expend or if you lower your physical activity level and keep your food intake the same, your body will store the extra calories as fat.
4. *Does the weight-loss program promote bizarre quantities of only one food or one type of food?* Some diets include eating only one food each day or unlimited amounts of certain foods, such as grapefruit or cabbage soup. Such advice runs counter to everything we know about the broad spectrum of human nutritional needs.
5. *Does the weight-loss program have rigid menus?* If a diet has specific meal plans and times to eat, it will be difficult to incorporate individual preferences. People are unique, so no one diet plan will work for everyone. A person who loves Thai food will not succeed on a diet if there is no way to incorporate Thai food into his or her eating plan.
6. *Does the weight-loss program promote specific food combinations?* Some diets include combinations of foods that should or should not be eaten at the same time. These food combinations have no basis in fact and needlessly restrict the dieter's options for reasonable food choices.
7. *Does the weight-loss program promise a weight loss of more than 2 pounds per week for an extended period of time?* If so, the initial weight loss will probably be due to water loss. A more realistic diet plan will aim for a weight loss of 0.5 to 2.0 pounds per week.
8. *Does the weight-loss program provide a warning to people with diabetes, high blood pressure, or other health conditions?* People with preexisting health conditions should consult a physician or other health care provider before beginning any diet. Elimination of certain food groups or eating excessive amounts of certain foods can exacerbate these problems and may interfere with the effectiveness of certain medications.
9. *Does the weight-loss program encourage or promote increased physical activity?* Although people can lose weight by limiting food intake alone, research

TABLE 9-5
Risks Associated with the Very Low-Calorie Diets

- Blood sugar imbalance
- Cold intolerance
- Constipation
- Decreased basal metabolic rate
- Dehydration
- Diarrhea
- Emotional problems
- Fatigue/weakness
- Gallstones and kidney stones
- Headaches
- Heart irregularity
- Ketosis
- Loss of lean body tissue
- Kidney infection
- Menstrual irregularity
- Mineral and electrolyte imbalances
- Sleeplessness
- Sudden death

SOURCE: Adapted from Position of the American Dietetic Association: Very low-calorie weight loss diets, *Journal of the American Dietetic Association* (May 1990), p. 722.

has shown that the most successful weight-loss plans include lifestyle changes, such as increasing exercise.

10. *Does the weight-loss program encourage an intake that is very low in calories (below 800 calories/day)* without supervision of medical experts? Very low-calorie diets should only be used by persons with severe obesity or obesity with other health-related problems. Because the calorie intake is so low, the diet must be supplemented with vitamins and minerals. In addition, the dieter must be strictly observed by a physician for any of the adverse health effects listed in Table 9-5. Finally, the person needs dietary counseling to handle "real" food choices before the end of the diet, or weight gain can quickly ensue.

Don't Give Up!

For some folks, weight loss may seem like a lost cause, but don't give up! Several strategies and diets have been proven successful. The strategies supported by the most evidence are detailed in a recent analysis by the USDA and backed up by data from the National Weight Control Registry, a study that examines people who have lost at least 30 pounds and have maintained that loss for at least a year.[33]

You can take several steps to help you make wise decisions about dieting:

1. Be familiar with the current fad diets, and study any diet that interests you. These food plans often include scientifically based statements intermingled with inaccuracies, so you have to know the literature to refute any incorrect claims.
2. Seek out appropriate weight-loss strategies and programs. A recent evidence-based review indicates that most weight loss is associated with diets that include about 1,400 to 1,500 calories per day. It is essential to control energy intake for any weight-loss plan.[34] Weight Watchers has been cited as a good option in many recent studies because of the variety of foods offered and because its principles are based on scientific evidence. Internet-based programs are available for people who like to keep records and need support but cannot attend group sessions. The DASH diet has been found to significantly reduce high blood pressure and improve other chronic disease outcomes, and it is free on the NIH website. About half of the people in the National Weight Control Registry (NWCR) lost weight without any formal program, indicating that the more individualized a program, the more likely it is for people to adhere to it for longer periods of time. Finally, most successful attempts to lose weight include some type of regular exercise.
3. The following websites contain good information or handouts that you can use to determine whether following a particular diet will be harmful or not: (1) the Federal Trade Commission (FTC) website, www.ftc.gov, including Weighing the Evidence in Diet Ads, and (2) the American Heart Association *Fad Diets at* www.americanheart.org.
4. Report fraudulent or deceptive weight-loss claims. Any weight-loss claims that are distributed via the Internet, television, or print media can be reported at www.ftc.gov or by calling 1-877-FTC-HELP (1-877-382-4357).

SOURCE: This feature is adapted from: D. M. Hoelscher and C. McCullum-Gomez, Addressing the obesity epidemic: An issue for public health policy, in M. Boyle and D. Holben, *Community Nutrition in Action: An Entrepreneurial Approach*, 4th ed. (Wadsworth/Thomson Learning, 2006), pp. 270–275.

Websites

DASH (Dietary Approaches to Stop Hypertension) diet: www.nhlbi.nih.gov/health/public/heart/hbp/dash
eDiets: www.ediets.com
Weight Watchers: www.weightwatchers.com
Jenny Craig: www.jennycraig.com
Diet Center: www.dietcenter.com
DietWatch: www.dietwatch.com
Cyberdiet: www.cyberdiet.com/
Nutrisystem: www.nutrisystem.com
Nutrio.com: http://nutrio.com/
The Diet Detective: http://thedietdetective.com/
Sonoma Diet: www.sonomadiet.com
South Beach Diet online: www.southbeachdiet.com/public/
Health Management Resources: www.yourbetterhealth.com
Learning tool for fad diets: Go to http://wemarket4u.net/fatfoe/ to see an ad for FatFoe™ Eggplant Extract, and click on the "order now" button.

Drugs and Weight Loss

The search is on to find a safe and effective drug solution to the problem of obesity. The ideal drug needs to be safe, free of undesirable side effects and abuse potential, and effective at reducing body fat. Additionally, as with the drug treatment for other chronic disorders (for example, high blood pressure), the ideal drug should be safe and effective for long-term use. To foster long-term success, any such drug treatment should be combined with lifestyle changes, including exercise and a healthy diet.[35] Until recently, most of the available drugs were appetite suppressants, which were available as prescriptions or over-the-counter (see Table 9-6).

Numerous other diet aids on the market include products with mysterious sounding ingredients, such as spirulina, chitosan, chromium picolinate, bitter orange extract, "carb blockers," "fat blockers," ginseng, and many others. Manufacturers typically suggest such products can aid in weight loss, but their ingredients often serve as little more than fillers. To date, none have proven effective in aiding weight loss. For example, manufacturers of products that include chromium picolinate praise its drug-like abilities to reduce fat, build lean tissue, suppress appetite, and increase metabolism, and they imply that our diets lack chromium. However, chromium picolinate has not been approved for weight loss by the FDA, and its claims are not backed by scientific data.[36] The FDA and the Federal Trade Commission have begun taking action against diet aids and products touting false claims for weight loss.[37]

The popularity of herbal supplements for weight loss is on the rise, partly because consumers have the mistaken notion that herbs are a "natural" alternative to prescription drugs. Thus far, herbal preparations in the United States are produced and marketed without regulations to ensure their safety or effectiveness, and some have been linked to side effects ranging from nausea and headaches to heart attacks and death. For example, the Chinese herb ma huang—commonly called ephedra—contains ephedrine, a stimulant that mimics

TABLE 9-6
Drugs for Weight Loss

	Action in the Body	Potential Side Effects and/or Comments
Prescription Drugs (Trade Name)		
Sibutramine (Meridia)	Appetite-suppressing drug; enhances the effects of serotonin by slowing the body's breakdown of the serotonin it naturally produces.*	Dry mouth, insomnia, nausea, abdominal pain, and increased heart rate and blood pressure. People who are not obese or anyone with high blood pressure are cautioned not to use the drug.
Orlistat (Xenical)	A lipase inhibitor; acts by inhibiting the enzymes—gastric and pancreatic lipase—needed for fat digestion; binds to the enzymes, making them unavailable to digest dietary fat. Because fat absorption is reduced by as much as 30 percent, people often lose weight.	Bloating, gas, frequent bowel movements, and anal leakage of undigested fats in people who do not adhere to a low-fat diet. Because Xenical may block the absorption of fat-soluble vitamins, users are advised to take a multivitamin supplement. The drug is intended only for the obese person—with a BMI greater than 30.
Phentermine (Adipex-P or Fastin)	Raises the level of norepinephrine, signals satiety and acts as a stimulant to increase the rate at which you burn calories.	Increased heart rate and blood pressure, dry mouth, insomnia, nausea, agitation, diarrhea, and constipation.
Over-the-Counter Drugs		
Benzocaine	An anesthetic found in gum or candy form that numbs the taste buds and reduces the desire for food.	Trade names: Diet Ayds candy or Slim Mint gum; the only over-the-counter appetite-suppressant ingredient currently approved by FDA without prescription.

*Serotonin, a brain chemical (neurotransmitter), acts to curb the appetite and thus reduces food intake.

the action of the drug phentermine. Ephedrine can cause tremors, insomnia, severe headaches, high blood pressure, heart attacks, and stroke. Some herbal preparations combine ephedra with other substances (for example, caffeine-containing guarana), to enhance ephedrine's effects. More than 800 reports of illness and dozens of deaths are linked to ephedra, and both the FDA and Canada prohibit its use.[38]

Other herbal preparations sold as dieter's tea contain senna, rhubarb root, aloe, cascara, or buckthorn. The "tea" has a laxative effect that can cause dehydration, diarrhea, nausea, fainting, and in some cases, even death.[39] For more about herbs thought to be effective and those that should be avoided, see the Nutrition Action feature in Chapter 6.

Surgery and Weight Loss

Sheer desperation prompts some obese people to request weight-loss surgery. Surgery may be an option for people who cannot lose weight by traditional means and are severely obese (BMI > 40) or have a BMI of at least 35 with one or more severe health problems.[40] Three common types of surgery incude gastroplasty (stomach stapling), gastric bypass, and gastric banding (see Figure 9-6). Common side effects include nausea, vomiting, diarrhea, heartburn, abdominal pain, and band slippage or pouch enlargement. Other more severe effects, including complications during surgery, infections, and death, may occur as well.

Gastroplasty involves stapling the stomach to reduce its volume, thus forcing the person to eat less. *Gastric bypass* is the most common surgical procedure for obesity in the United States. First, the surgeon creates a small pouch in the stomach with staples or a plastic band. Next, a Y-shaped section of the small intestine is attached to the pouch to allow food to bypass the lower stomach and the upper portion of the small intestine. This procedure restricts both food intake and the amount of calories and nutrients the body absorbs. The second most common operation in the United States for weight loss is called *gastric banding* and involves placement of an adjustable silicone band around the upper part of the stomach to create a small pouch which limits food consumption and creates an earlier feeling of fullness.

FIGURE 9-6
SURGICAL PROCEDURES FOR OBESITY

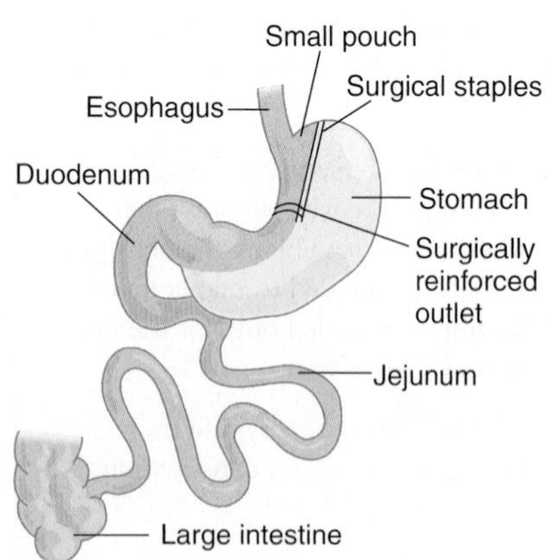

In vertical-banded gastroplasty, a small pouch is constructed at the top of the stomach that both limits the amount of food that the stomach can hold and restricts the passage of food to the small intestine.

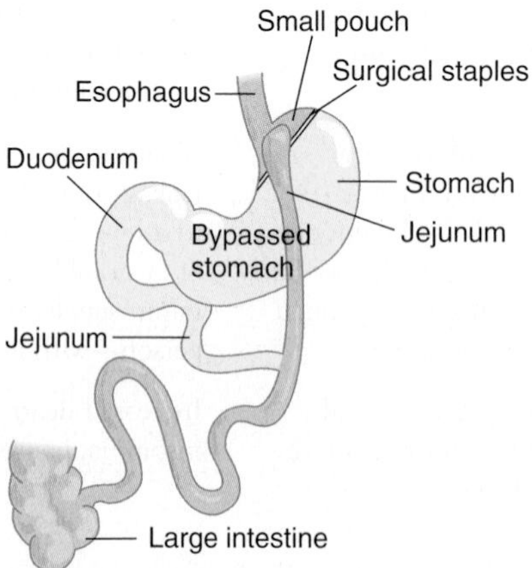

In gastric bypass (Roux-en-Y) operations, a small pouch near the top of the stomach is created with staples or a plastic band and connected directly to the middle portion of the small intestine (jejunum).

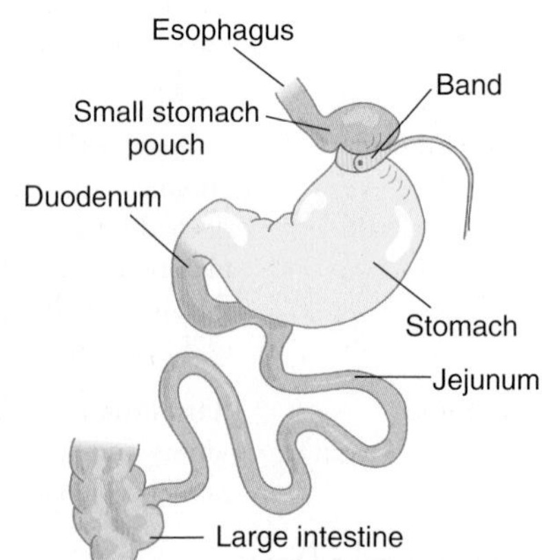

Gastric banding reduces the size of the stomach by using an adjustable silicone band or cuff near the top of the stomach to create a pouch that fills quickly and empties slowly. The inflatable band is connected to an access port placed close to the skin, allowing surgeons to tighten or loosen it to meet an individual's needs.

At first, the small stomach pouch that results from weight-loss surgery holds about 1 ounce of food and later may stretch to 2–3 ounces. Nausea and vomiting occur if the person continues to overeat following surgery. Most people can eat about ½ to 1 cup of food without discomfort or nausea, but the food must be soft, moist, and well chewed. Most people lose more than 50 percent of their excess weight and keep that weight off long term. As in other treatments for obesity, the best results are achieved with healthy eating behaviors and regular physical activity.

Another approach involves cosmetic surgery. One such procedure is **liposuction,** in which the surgeon uses a small hollow tube to suction out fatty tissue from beneath the skin. People who wish to remove the fat from a particular area can elect this procedure, which sometimes brings pleasing results but sometimes produces a figure in which one part of the body is disproportionately thin relative to the others.

liposuction a type of surgery (also called *lipectomy*) that vacuums out fat cells that have accumulated, typically in the buttocks and thighs. If the person continues to eat more calories than are expended through physical activity, fat will return to the fat cells that remain in those regions.

Surgery is appropriate in some instances. Cosmetic surgery can minimize disfigurements, improve self-confidence, and ease the way toward concentration on life issues that are more important than external appearance. After surgery, however, the same person resides within the skin as before. A changed appearance does not guarantee changed eating habits, a better personality, reduced interpersonal conflicts, or any other improvements in the quality of one's life.

Successful Weight-Loss Strategies

Given that so many approaches are likely to fail, what weight-loss strategies work? How can a person lose weight safely and permanently? The secret is a sensible (not to say *easy*) three-pronged approach involving healthful eating habits, exercise, and behavior change. Such an approach takes tremendous dedication, especially at first, for a person whose habits have promoted obesity to make new habits out of the hundred or so behaviors necessary to promote a healthful weight. Even the most effective weight-loss programs reveal a grim pattern: Many people complete the program, lose about 10 percent of their body weight, regain two-thirds of it within 1 year, and regain almost all of it within 5 years.[41] Still, many others who resolve to lose weight do so, and they manage to keep it off. The University of Pittsburgh School of Medicine maintains a National Weight Control Registry of some 3,000 persons who have successfully achieved and maintained weight loss for a number of years. A recent survey of dieters found that 20 percent of respondents had lost up to 42 pounds and kept it off for an average of 7 years.[42] Those who typically succeed do so because they have employed many of the techniques described in this chapter as well those in the Eat Well Be Well feature that follows.

Eating Plan Strategies

1. Get personally involved.
2. Adopt a realistic plan, and then keep track of calories.
3. Make the eating plan adequate.
4. Emphasize high nutrient density.
5. Individualize. Eat foods you like.
6. Stress "dos," not "don'ts."
7. Eat regular meals.
8. Take a positive view of yourself.
9. Visualize a changed future self.
10. Take well-spaced weighings to avoid discouragement.

Profile of Successful Dieters

They know their weight. (They check their scale weight weekly.)
They are motivated from within.
They know what they eat. (They keep records.)
They engage in regular exercise (60 to 90 minutes of moderate exercise on most days).
They have social support (relationships/groups).
They control their intake of alcohol, fat, and sugar.
They follow a personalized diet plan—one that they can enjoy permanently.
They lose no more than a pound or two a week.
They set reasonable goals—such as losing just 10 percent of their weight as a start.
They accept an occasional lapse rather than aim for perfection.
They view weight management as a long-term commitment.

Personalize Your Weight-Loss Plan

The way a particular person loses weight is a highly individual matter. Two different weight-loss plans may be equally successful and yet have little or nothing in common. You have to find a plan that's right for you. To emphasize the personal nature of weight-loss plans, the following sections are written as advice to "you" even though you may not need to lose weight at all. To help you understand the issues of weight loss, you can pretend that you have 10 to 200 pounds to lose and are being competently counseled by someone familiar with the techniques known to be effective. Notes in the margin highlight the principles involved.

No particular diet is magical, and no particular food must be either included or avoided. Because you are the one who will have to live with the eating plan, you had better be involved in creating it. Don't think of it as going "on" a diet—because then you may be tempted to go "off" it. Lifestyle changes are only successful if the pounds do not return. Think of it as an eating plan that you will

Eat Well Be Well Never Say "Diet"

The problem with going on a "crash" diet with a goal of, say, a 15-pound weight loss in 3 weeks is that it's a quick fix—a temporary solution to what is typically a chronic problem. Often, the dieter tries to rigidly restrict eating by, for example, skipping meals or eating salads all day. But trying not to eat is like trying not to breathe, as one group of psychologists put it.[43] The body and mind soon rebel and, like a person gasping for air, the dieter loses control and binges. In fact, experts believe that rigid diets play a strong role in the development of binge-eating problems.[44] After the binge, the person feels that he has failed, gives up the diet altogether, eats more to make himself feel better, puts on some weight, feels even worse, gains even more weight, decides to try another restrictive diet, and begins the whole cycle all over again. The solution, say many experts, is *not* to focus on losing a certain amount of weight within a set period of time. This feature shows you how.

1 Adopt a Nondiet Approach to Weight Loss.

A more healthful, nondiet approach to weight loss is to gradually develop habits that you can live with permanently and that will help you shed pounds and keep them off over the long run. Instead of measuring your success by the needle on the scale, gauge your progress by the strides you make in adopting good eating and exercise habits as well as healthful attitudes about yourself and your body.

2 Set Achievable Goals.

Most people recognize the importance of the eating and exercise habits discussed in this chapter, but few dieters recognize that the attitude problems, which often accompany restrictive dieting, can stand in the way of long-term weight loss. One of the most common problems is "all-or-nothing" thinking in which people view the world as black or white, right or wrong, good or bad, and so forth. When it comes to diets, this attitude translates into "good or bad" foods and "on or off" limits to diet food or junk food.[45] Such thinking sets the stage for failure as the dieter tries to live up to extremely rigid, unrealistic goals: "Ice cream is bad, so I must never eat it," "I will jog three miles every day," or "I will never order anything but a salad when I go out to eat." Typically, a person who has an all-or-nothing attitude might go several days without, say, ice cream, which she views as "bad junk food," and then splurge on a double-dip cone. Instead of recognizing that ice cream isn't "bad" and that enjoying some every now and then won't undo all her previous efforts, she tells herself, "I've blown my diet completely—I have no willpower." Consequently, feeling guilty and depressed because she's "blown it" anyway, she eats another pint of ice cream and some cookies to make herself feel better. If she had a different attitude—"I've been eating lots of fruits and vegetables and go bicycling several times a week, so one ice cream cone won't hurt me"—she may have avoided this bout of despair and binging.

3 Focus on Health Rather than Appearance.

Another attitude that can thwart efforts to achieve a healthy weight is the "lookist" attitude—that is, the notion that weight and appearance determine a person's worth and happiness.[46] Nearly everyone holds lookist attitudes in one form or another and wishes that he could change something about his appearance in the hopes that it would make life more enjoyable. But in our "thin-is-in" society, overweight people suffer

adopt for life. The diet must consist of foods that you like or can learn to like, that are available to you, and that are within your means.

If you want to lose weight, a deficit of 500 calories a day for 7 days (3,500 calories a week) is enough to lose a pound of body fat each week. If you choose to spend an extra 250 calories a day in some form of exercise, you could increase this energy deficit.

Choose a calorie level you can live with. The "10-calorie rule" will enable you to lose a pound or two a week while still supporting your basal metabolism: Allow 10 calories a day for each pound of your present body weight. As you lose weight, you can gradually adjust calories downward to keep losing at this rate. Thus, a person who weighs 220 pounds should begin with 2,200 calories a day; one who weighs 150 pounds should begin with 1,500 calories a day.

Put nutritional adequacy high on your list of priorities. This is a way of putting yourself first—"I like me, and I'm going to take good care of me." This means including foods that are rich in valuable nutrients such as vegetables and tremendous prejudice and tend to be painfully self-conscious about their bodies. As a result, they may mistakenly believe that all their dreams will come true once they lose weight and that all their problems will be lost along with the pounds. By pinning all their hopes for happiness on losing weight, they put themselves under enormous pressure to shed pounds. In addition, people with this attitude often do not live in the present because they are so caught up in fantasizing about what they think life will be like once they lose weight.

Using weight and appearance as a measure of self-worth and happiness can be extremely destructive. The person who does so may desperately try to lose weight with restrictive, unrealistic diets or unhealthful, strenuous exercise regimens. Each "slip-up" whittles away at the person's self-esteem, which in turn may lead to feelings of rejection, depression, and social isolation, which in turn may prompt a binge, and so forth. Even the person who loses the desired number of pounds may find that he or she still has many of the same life problems as before, and this realization can lead to depression and loss of self-esteem.

The way around lookist thinking is to focus on health—both mental and physical—rather than appearance. Table 9-7 lists some common characteristics of the obese self (which holds lookist attitudes) and the healthy self. Although it can be difficult to overcome society's prejudices about weight, striving to adopt the ideals of the healthy self, regardless of your weight, can be a major step in helping you take care of your mental and physical health.

TABLE 9-7

The Obese Self versus the Healthy Self

The Obese Self	The Healthy Self
Uses appearance as measure of self-worth	Uses caring for others/self as measure of self-worth
Is socially isolated	Is socially involved
Rejects self	Accepts self
Goes on and off extreme diets	Eats healthfully
Rarely exercises (or overdoes it and quits)	Exercises regularly and sensibly
Focuses on mistakes; feels like a failure	Views mistakes as learning experiences and recognizes that nobody is perfect
Uses obesity as excuse for failures in life	Doesn't allow obesity to interfere with life
Focuses on past	Focuses on present
Allows negative emotions to reduce self-control	Uses positive attitude to enhance self-control
Focuses on appearance	Focuses on health
Wants to improve appearance to get love/affection	Wants to be healthy to participate fully in loving relationships

SOURCE: Adapted from J. P. Foreyt and G. K. Goodrick, *Living without Dieting* (New York: Warner Books, 1992), 55.

fruits, whole-grain breads and cereals, and a reasonable amount of protein-rich foods such as lean meats, skinless poultry, fish, eggs, legumes, low-fat cheeses, and fat-free milk. Within these categories, learn what foods you like, and eat them often. If you resolve to include a certain number of servings of food from each of these groups each day, you may be so busy making sure you get what you need that you will have little time or appetite left for high-calorie or empty-calorie foods. Researchers have shown that reducing the intake of fat alone can promote significant weight loss, especially when coupled with a high complex-carbohydrate diet.[47]

A small amount of fat should be included in each meal to make it satisfying and keep you from getting hungry again too soon. You don't have to use pure fat such as butter, margarine, or oil. Rather, most of the fat should come from protein-rich foods, such as lean meats, eggs, poultry, fish, and low-fat cheeses. Add any pure fat with extra caution. A slip of the butter knife adds more calories than a slip of the sugar spoon. Keep concentrated sweets to a minimum, and let

Weight-Loss Readiness Quiz

Are you ready to lose weight? Your attitude affects your ability to succeed. Take this readiness quiz to see if you're mentally ready before you begin.

Mark each statement as "true" or "false." Be honest with yourself! The answers should reflect the way you really think—not how you'd like to be!

1. I have thought a lot about my eating habits and physical activities, and I know what I might change.
2. I know that I need to make permanent, not temporary, changes in my eating and activity patterns.
3. I will feel successful only if I lose a lot of weight.
4. I know that it's best if I lose weight slowly.
5. I'm thinking about losing weight now because I really want to, not because someone else thinks I should.
6. I think losing weight would solve other problems in my life.
7. I am willing and able to increase my regular physical activity.
8. I can lose weight successfully if I have no slipups.
9. I am willing to commit time and effort each week to organize and plan my food and activity choices.
10. Once I lose a few pounds but reach a plateau (and can't seem to lose more), I usually lose the motivation to keep going toward my weight goal.
11. I want to start a weight-loss program, even though my life is unusually stressful right now.

Now Score Yourself

Look at your answers for items 1, 2, 4, 5, 7, and 9. Score "1" if you answered "true" and "0" if you answered "false." For items 3, 6, 8, 10, and 11, score "0" for each "true" answer and "1" for each false answer.

No single item indicates if you're ready to start losing weight. But the higher your total score, the more likely you are to be successful.

If you scored 8 or higher, you probably have good reasons to lose weight now. And you know some of the steps that can help you succeed.

If you scored 5 to 7 points, you may need to reevaluate your reasons for losing weight and the strategies you'd follow.

If you scored 4 or less, now may not be the right time for you to lose weight. You may be successful initially, but you may not be able to sustain the effort to reach or maintain your weight goal. Reconsider your reasons and approach.

Interpret Your Score

Your answers can be clues to some stumbling blocks in your weight-management success. Any item you scored as "0" suggests a misconception about weight loss, or a problem area for you. So let's look at each item a bit more closely.

1. You can't change what you don't understand, and that includes your eating habits and activity pattern. Keep records for a week to pinpoint when, what, why, and how much you eat—as well as patterns and obstacles to regular physical activity.
2. You may be able to lose weight in the short run with drastic or highly restrictive changes in your eating habits or activity pattern. But they may be hard to live with permanently. Your food and activity plans should be healthful ones that you can enjoy and sustain.
3. Many people fantasize about reaching a weight goal that's unrealistically low. If that sounds like you, rethink your meaning of success. A reasonable goal takes body type into consideration—and sets smaller, achievable "mile markers" along the way.
4. If you equate success with fast weight loss, you'll have problems keeping weight off. This "quick fix" attitude can backfire when you face the challenges of weight maintenance. The best and healthiest approach is to lose weight slowly while learning strategies to keep weight off permanently.
5. To be successful, the desire for and commitment to weight loss must come from you—not your best friend or a family member. People who lose weight, then keep it off, take responsibility for their weight goals and choose their own approach.
6. Being overweight may contribute to some social problems, but it's rarely the single cause. Whereas body image and self-esteem are strongly linked, thinking you can solve all your problems by losing weight isn't realistic. And it may set you up for disappointment.
7. A habit of regular, moderate physical activity is a key factor to successfully losing weight—and keeping it off. For weight control, physical activity doesn't need to be strenuous to be effective. Any moderate physical activity that you enjoy and will do regularly counts.
8. Most people don't expect perfection in their daily lives, yet they often feel they must stick to a weight-loss program perfectly. Perfection at weight loss isn't realistic. Rather than viewing lapses as catastrophes, see them as opportunities to discover what triggers your problems and to develop strategies for the future.
9. To successfully lose weight, you must take time to assess your problem areas, then develop the approach that's best for you. Success requires planning, commitment, and time.
10. First, a plateau in an ongoing weight loss program is perfectly normal, so don't give up too soon! Before you lose your motivation, think about any past efforts that have failed, then identify strategies that can help you overcome those hurdles.
11. Weight loss itself can be a source of stress, so if you're already under stress, you may find a weight loss program somewhat difficult to implement right now. Try to resolve other stressors in your life before starting your weight-loss effort.

Source: From ADA/*The American Dietetic Association's Complete Food and Nutrition Guide*, pp. 32–33, © John Wiley & Sons, Inc. Reprinted with permission.

your carbohydrate come from fruits, vegetables, and whole-grain foods. Figure 9-7 shows three suitable patterns for weight-loss eating plans. Table 9-8 offers more tips for cutting back on fat and calories, and the Savvy Diner feature on page 301 gives suggestions for defensive dining.

If you include alcohol or other empty-calorie items in your eating plan, limit them to no more than 150 calories a day. Budget this amount into your chosen calorie level, and reconcile yourself to a slower rate of weight loss.

Three meals a day is standard for our society, but no law says you shouldn't have four or five meals—only be sure that they are smaller, of course. What is

FIGURE 9-7
SAMPLE BALANCED WEIGHT-LOSS DIETS USING THE MyPYRAMID FOOD GUIDE[a]

[a]Assumes no alcohol intake.
[b]Not recommended for pregnant or lactating women, children (depending on age), or those who have special dietary needs. At or below this low level of calorie intake, it may not be possible to obtain recommended amounts of all nutrients from foods; therefore, it is important to make careful food choices, and the need for dietary supplements should be evaluated.
[c]The 1,200 and 1,400 plans include a 170 discretionary calorie allowance; the 1,800 plan includes 195 discretionary calories.
[d]For maximum nutritional value, make whole-grain, high-fiber choices.
[e]Choose fat-free or low-fat (1%) milk products.
[f]Select lean meat and use cooking methods that do not require added fat.

Daily recommendations for a 1,200-calorie diet[b,c]

GRAINS	VEGETABLES	FRUITS
4 oz[d]	1.5 cups	1 cup

OILS	MILK	MEAT & BEANS
4 tsp	2 cups[e]	3 oz[f]

Daily recommendations for a 1,400-calorie diet[c]

GRAINS	VEGETABLES	FRUITS
5 oz[d]	1.5 cups	1.5 cups

OILS	MILK	MEAT & BEANS
4 tsp	2 cups[e]	4 oz[f]

Daily recommendations for a 1,800-calorie diet[c]

GRAINS	VEGETABLES	FRUITS
6 oz[d]	2.5 cups	1.5 cups

OILS	MILK	MEAT & BEANS
5 tsp	3 cups[e]	5 oz[f]

most important is to eat regularly and, if at all possible, to eat before you become very hungry.

Keep a record of what you have eaten each day for at least a week or two until your habits are automatic. Resume record keeping whenever you need to.

At first it may seem as if you are spending all your waking hours thinking about and planning your meals. Such a massive effort is always required when you are learning a new skill. After about 3 weeks, however, it will be much easier as your new eating pattern becomes a habit. Some characteristics of successful dieters are listed in the margin on page 293.[48]

Aim for Gradual Weight Loss

Do not weigh yourself more than once a week. Although 3,500 calories roughly equals a pound of body fat, there is no simple relationship between calorie balance and weight loss over short intervals. A gain or loss of a pound or more in a few days can reverse quickly; the smoothed-out average is what is real. Don't expect to lose continuously as fast as you did at first. A sizable water loss is common during the first week, but it will not happen again.

Expect to Reach a Plateau Many dieters experience a temporary plateau after about 3 weeks—not because they are slipping but because they have gained water weight temporarily while they are still losing body fat. The fat they are hoping to lose must be used for energy. To use it, the body must combine it with oxygen (oxidize it) to make carbon dioxide and water. These compounds are heavier than the fat they are made from because oxygen has been added to them.* The carbon dioxide will be exhaled quickly, but the water takes longer to leave the cells. First, it makes its way into the spaces between the cells and finally enters the bloodstream. Only after the water has arrived in the blood do

*Water weight accumulates during fat oxidation because one fatty acid weighing 284 units leaves behind water weighing 324 units—14 percent more.

TABLE 9-8
Healthful Tips for Weight Loss

- Eat breakfast.
- Watch out for second helpings of higher-calorie foods.
- Choose low-fat and low-calorie versions of foods you like.
- Go easy on foods that are high in fat or sugar.
- Limit alcoholic beverages.
- Roast, broil, boil, steam, or poach foods rather than fry them.
- Select lean cuts of meat and trim visible fat.
- Plan ahead for snacks of fruit, vegetables, or low-fat yogurt.
- Use spices and herbs instead of sauces, butter, or other fats.
- Consume low-fat or fat-free dairy products.
- Try fresh fruit for dessert or baked products made with less fat and sugar (e.g., angel food cake).
- Use alternatives to foods as rewards (such as long walks, relaxing baths, a visit with a friend, a hobby, gardening, a good book).
- Keep a food diary.
- Aim for 60 minutes of exercise on most days of the week; try wearing a pedometer and walking 10,000 to 12,000 steps a day.

the kidneys remove it and send it to the bladder for excretion. Meanwhile, this excess water creates a temporary weight gain, but one day the plateau will break. The signal that this is happening is frequent urination.

Aim for a Positive Gain in Lean Body Mass If you have been working out lately, successive weighings may show a slight occasional gain when you expect a loss. This may reflect a welcome development: the gain of lean body mass—just what you want to be healthy. In fact, weight loss without exercise can have a negative effect on body composition. No doubt you've heard someone say as a joke, "I've lost 200 pounds, but I've never been more than 20 pounds overweight." Typically, this is the person who diets, loses weight, regains the weight, and diets again throughout life—*without* exercising. When this pattern becomes a lifetime of weight fluctuations, it can have a lasting impact on the makeup of the body and the way it burns calories. Dubbed the "yo-yo effect," this pattern of losing, gaining, and losing weight again may actually cause the body to accumulate a greater percentage of fat and less lean muscle with each round of dieting.[49]

If you cut back drastically the number of calories you consume without exercising for, say, a month or two, you'll probably lose a great deal of weight each week in the form of not only fat but also lean muscle. Then, if you later put the weight back on, the regained pounds may be primarily fat—not muscle. Thus, you may weigh the same as you did when you began the diet, but your body is now composed of more fat and less muscle. Because muscle burns more calories just to sustain itself than fat does, your body needs fewer calories to maintain its weight than it did before. If you go back to eating the way you used to, chances are you will gain even more weight.

Maintain Bone Health by Increasing Calcium Intake Weight loss—even modest weight loss—can also result in a depletion of bone mineral density.[50] Fat loss may lower estrogen levels, a hormone necessary for healthy bone maintenance. Recent research emphasizes the importance of including both adequate calcium intakes and exercise in any weight-loss regimen, particularly such weight-bearing exercises as walking, jogging, stair climbing, and weight lifting. Researchers at the University of Tennessee provide additional evidence for the benefits of dietary calcium in weight-loss regimens. New research suggests that low calcium intakes favor increased deposition of body fat, whereas higher calcium intakes (at levels generally recommended for health) may not only inhibit storage of body fat but also stimulate the use of body fat for energy.[51]

In addition to the risk of altering the body's overall composition during repeated bouts of weight loss, consider that each unsuccessful attempt to keep weight off is often viewed by the dieter as a personal failure, which can erode the person's self-esteem and trigger painful feelings of guilt and depression.[52] A more healthful alternative to crash dieting is to gradually develop healthful, permanent lifestyle habits that will help you to lose weight and keep it off.

Adopt a Physically Active Lifestyle

Exercise makes three contributions to a weight-management program. It increases calorie expenditure, alters body composition in a desirable direction, and alters metabolism.[53] Exercise also offers the psychological benefits of looking and feeling healthy, and the increased self-esteem that accompanies these benefits can enhance the motivation to maintain a healthful lifestyle for the long run.

Compared with lean tissue, fat tissue is relatively inactive metabolically. Metabolic activity burns calories—lots of them. Thus, the more lean tissue you develop, the faster your metabolism becomes, the more calories you spend, and the more you can afford to eat. This brings you both pleasure and nutrients. Exercise, by shifting body composition toward more lean tissue, speeds up the metabolism *permanently*—that is, for as long as you keep your body conditioned.

Strategies for Using Exercise for Weight Control

1. Make it active exercise; move your muscles.
2. Think in terms of quantity, not speed.
3. Exercise informally, in daily routines.

Furthermore, the more muscle and lean tissue you have, the more fat you will burn—all day long, even when you are resting.

The next chapter offers many pointers about becoming fit, but a few notes on strategy are in order here. For one thing, keep in mind that if exercise is to help with weight loss, it must be *active* exercise of moderate intensity—voluntary moving of muscles. Being moved passively, such as by a machine at a health spa or by a massage, does not increase calorie expenditure. The more muscles you move, the more calories you spend (see Table 9-9). You can tell that the exercise is moderate if you are breathing a little faster than normal but can still easily carry on a conversation.

People sometimes ask about spot reducing. Can you lose fat in particular locations? Unfortunately, muscles don't "own" the fat that surrounds them. Since all body fat is shared by all the muscles and organs, spot-reducing exercises that work only the flabby parts won't help reduce the fat located there. The good news is, however, that tightening muscles in trouble spots by way of a balanced, all-over exercise program may improve the appearance of the fatty areas.

Another thing to keep in mind is that the number of calories spent in an activity depends more on how much a person weighs than on how fast the person can do the exercise (see Table 9-9). For example, a person who weighs 125 pounds burns off 78 calories by running a 9-minute mile. That same person, walking a mile in 15 minutes, burns almost the same amount—82 calories. Similarly, a 200-pound person spends 125 calories on the 9-minute mile, and a similar amount—131 calories—on the 15-minute walk. The rule seems to be that you don't have to work fast to use up calories effectively. If you choose to walk rather than run the distance, you will use up about the same energy; it will just take you longer.

A good suggestion that you may have heard already is to incorporate more exercise into your daily schedule in many simple, small-scale ways. Park the car at the far end of the parking lot; use the stairs instead of the elevator; do a round of sit-ups before you get up in the morning. If you do these types of things and also incorporate regular aerobic exercise and strength training into your schedule, your heart and lungs as well as your skeletal muscles will be fit (see Chapter 10).

Weight-Gain Strategies

It is as hard for a person who tends to be thin to gain a pound as it is for a person who tends to be fat to lose one. The person who wants to gain weight is faced with some of the same challenges as the one who wants to lose weight—learning new habits, learning to like new foods, and establishing discipline related to meals and mealtimes. But there are major differences.

Knowing that vigorous physical activity costs calories, an active person may wonder whether it is advisable to curtail activity. The answer is no, not unless the underweight condition is so extreme as to be endangering your health. The healthful way to gain weight is to build yourself up with patient and consistent training while eating nutritious foods containing enough extra calories to support the weight gain. If you add extra snacks of high-calorie nutritious foods, such as a peanut butter sandwich with a glass of milk before bedtime, and a healthful mid-afternoon fruit smoothie or milkshake, you can add 700 to 800 extra calories a day and achieve a healthful weight gain of 1 to 1½ pounds per week. Choose calorie-dense snacks, such as fruit yogurt, granola, dried fruits, apple or grape juice, nuts, sunflower seeds, fig bars, and baked potatoes topped with low-fat cheese.

A person wanting to gain weight often has to learn to eat different foods. No matter how many helpings of carrots you consume, you won't gain weight very fast because carrots simply don't offer enough calories. Thus, the person who can't consume much food volume is encouraged to eat calorie-dense foods at meals (the very ones the dieter tries to avoid). For example, if you are trying to

▲ *Exercise is essential for weight control.*

TABLE 9-9
Calories Spent during Various Activities

Activity	Calories Expended per Hour[a] Man[b]	Woman[b]
Sitting quietly	100	80
Standing quietly	120	95
Light activity: Cleaning house, Office work, Playing baseball, Playing golf	300	240
Moderate activity: Walking briskly (4 mph), Gardening, Cycling (5.5 mph), Dancing, Playing basketball	460	370
Strenuous activity: Jogging (9 min/mile), Playing football, Swimming	730	580
Very strenuous activity: Running (7 min/mile), Racquetball, Skiing	920	740

[a] May vary depending on environmental conditions.
[b] Healthy man, 175 lbs; healthy woman, 140 lbs. (To determine caloric expenditure at a different weight, both men and women should divide their body weight by 175, then multiply by the man's caloric expenditure for the activity.)

SOURCE: U.S. Department of Agriculture, U.S. Department of Health and Human Services, *Dietary Guidelines for Americans*.

© Polara Studios

▲ *Add extra calories to the milk or juice you drink by tossing in a frozen banana or other fruit and whipping up a smoothie in the blender.*

© Polara Studios

▲ *Try eating more food at each meal. If you normally eat one sandwich for lunch, try eating an extra sandwich between meals.*

gain weight, you should choose a milkshake instead of milk, peanut butter instead of lean meat, an avocado instead of a cucumber, and a blueberry bran muffin instead of whole-wheat bread. When you do eat carrots, dip them in hummus and use creamy dressings on salads, yogurt on fruit, cottage cheese on potatoes, and so forth. Because fat contains twice as many calories per teaspoon as sugar, it adds calories without adding bulk. For heart health, be sure to choose the "good" fats, found in foods such as avocados, olives, peanuts, pistachios, and other nuts and seeds, and limit your intake of saturated fats.

Also, eat more frequently. Make three sandwiches in the morning, and eat them between classes in addition to the day's three regular meals. Spend more time eating each meal. If you fill up fast, eat the highest-calorie items first. Don't start with soup or salad; start with the main course. Finish with dessert. Many underweight people have simply been too busy (for months) to eat enough to gain or maintain weight. These strategies will help you change this behavior pattern.

Whether you need to gain, lose, or maintain weight, paying attention to what you eat will pay off in long-term wellness benefits. This chapter emphasizes the relationship of food and eating to body weight and has the same message as other parts of the book: To support wellness, eat regular, balanced meals composed of a wide variety of foods you enjoy.

Breaking Old Habits

"Just do it," urge the makers of Nike sportswear on billboards, in magazines, and wherever else they promote their products, implying that adhering to a fitness program is just a matter of donning sneakers and heading for the track or gym. But as most of us know from experience, sticking to a regular exercise program for the first time or changing eating habits isn't always easy. Most people who successfully quit smoking make three or four attempts before they finally kick the habit. And most people make the same New Year's resolution at least five years in a row before they stick to it permanently.[55] Breaking old patterns of behavior and developing new ones involves a number of stages before reaching the point at which you're able to change for good.

First, you must accept the belief that you *can* change: "I can lose some weight and try to become more physically fit," you might tell yourself. This might inspire you to *want* to change, which in turn will motivate you to find ways to actually make the change—for example, talking to a registered dietitian about diet strategies and reading about various forms of exercise. After you've gone through those steps, you will be ready to take action—that is, to "just do it." Finally, to maintain your behavior change, you can create a game plan to help you handle the inevitable lapses that will occur over time. When you follow these steps and then stick with your plan for a good deal of time, you're home free.

Elements of Behavior Change[54]

1. *Precontemplation:* You need to change, but you're not yet ready to accept that fact.
2. *Contemplation:* You want to change, but you're not sure how.
3. *Preparation:* You gain knowledge to set up a plan of action for change.
4. *Action:* You jump in and "just do it."
5. *Maintenance:* You work on sticking to your plan of action.
6. *Termination:* You have achieved lasting change and experience few, if any, temptations or relapses.

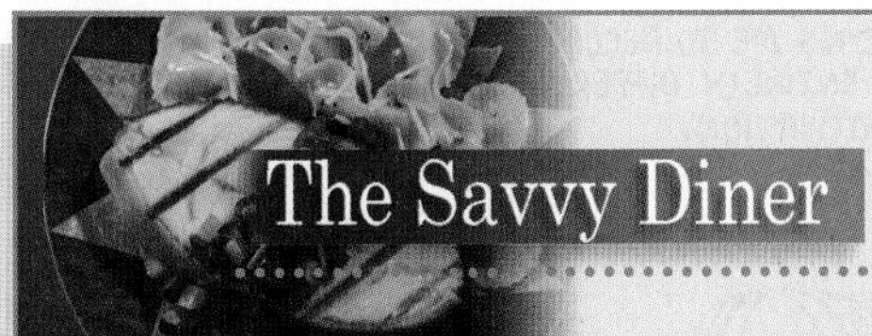

Aiming for a Healthy Weight While Dining Out

Pizza Yankee . . . The story goes that a pizza parlor waitress asked Yogi Berra whether he wanted his pizza cut into six or eight slices. "Six, please," he said. "No way am I hungry enough to eat eight."[56]

Dining out can be a pleasant time to relax, socialize, taste the foods of different cultures, and provide a break from your usual schedule. And it does *not* necessarily mean that you overindulge in high-calorie foods. With practice, you can maximize the benefits of eating out by adopting some of the following strategies:

- Examine your options before selecting foods at buffets, cafeterias, or food courts. Otherwise, you may find yourself halfway through the line with a full plate and end up with a mountain of food on your plate when you reach the end of the line.
- Take the edge off hunger by starting with a broth-based soup, small salad, fruit, or light appetizer.
- Make specific requests. Ask for salad dressings, sauces, or gravies *on the side.* This puts you in control of the portions. Ask if reduced-fat or fat-free salad dressings are available.
- Ask for entrées to be broiled, baked, grilled, steamed, or roasted. See "Eating Well on the Run" for tips on ordering at quick-service restaurants.
- Request fresh fruit, sorbet, or low-fat frozen yogurt for dessert. It will give you the sweetness you want without excess calories from fat. If you choose to splurge on your favorite chocolate dessert, enjoy it by sharing it with a friend. After the dinner out, go back to your healthful eating habits and exercise program.
- Be a menu sleuth: The following terms typically describe high-fat items: au gratin, Alfredo, batter-dipped, breaded, bearnaise, carbonara, creamy, crispy, croquette, deep-fried, flaky, fritters, parmigiana, tempura, with gravy, or with hollandaise sauce. Order red pasta sauces rather than white.
- Downsize your order. Many restaurants serve portions that are two or more times the recommended serving size. Request a doggie bag so you can take half the amount home. Or, consider sharing an entrée with a friend and ordering a salad to go with it. You can also order à la carte—one or two nutritious, low-fat appetizers such as steamed shrimp or raw vegetables and a low-fat dip along with a salad instead of a huge main meal.
- Finally, eat slowly and enjoy your meal, along with the company and conversation. You'll be giving your body time to digest the food, you'll feel satisfied, and you'll have enjoyed yourself at the same time.

Eating Well on the Run*

Smart Choices

Sandwich Shop: Fresh sliced veggies in a pita with low-fat dressing, cup of minestrone soup, turkey breast sandwich with mustard, lettuce, tomato, fresh fruit.

Rotisserie Chicken: Chicken breast (remove skin), steamed vegetables, mashed sweet potatoes, tossed salad, fruit salad. Select plain rolls instead of cornbread or biscuits.

Fast Food: Grilled chicken breast sandwich (no sauce), single hamburger without cheese, grilled chicken salad, garden salad, low-fat or nonfat yogurt, fat-free muffin, cereal, low-fat milk.

Salad Bars: Broth-based soups, fresh bread or bread sticks, fresh greens, chopped veggies, beans, low-fat dressing, fresh fruit salad. Avoid marinated beans and oily pasta salads.

Asian Take-Out: Wonton soup, pho (Vietnamese noodle soup), hot and sour soup, steamed vegetable dumplings, steamed vegetable mixtures over rice or noodles.

Pizza Night: Choose flavorful, low-fat toppings such as peppers, onions, sliced tomatoes, spinach, broccoli or mushrooms. Ask for your pizza with less cheese.

*SOURCE: Adapted from *Healthy Eating Away from Home* (Washington, D.C.: American Institute for Cancer Research, 1996).

Behavior Modification

Making lasting behavioral changes poses ongoing challenges. A process known as **behavior modification** can help. Developed by psychologists to help people change their habits, behavior modification techniques are similar to mapping out a game plan by following the specific steps listed in the margin on page 302.

Set a Goal and Get to Know Yourself For *Step 1,* identify your goals. You might decide, for example, that your goal is to lower the fat and calorie content of your diet. Next, in *Step 2,* you record your present behavior pattern and think

behavior modification a process developed by psychologists for helping people make lasting behavior changes.

about the reasons for those behaviors. In this case, for example, you might keep a food diary—writing down everything you eat for five days and describing how you feel each time you eat (see Table 9-10 for a sample food diary). This technique will help you understand why you behave the way you do in certain situations. You might find, for instance, that every time you are angry at your significant other or get a bad grade you eat a candy bar or two to comfort yourself. Or perhaps without even thinking about it, you nibble on whatever happens to be handy while watching television.

Behavior Modification Steps

Step 1: Identify the goal.
Step 2: Record your present behavior pattern. Identify the reasons you practice these behaviors.
Step 3: Identify the behaviors that will lead to the goal and the rewards of those behaviors.
Step 4: Commit yourself to changing. Face what you'll have to give up or change to make the desired behavior a reality. Envision your changed future self.
Step 5: Plan. Divide the behavior into manageable portions. Set small, achievable goals and plan periodic rewards.
Step 6: Try out the plan. Modify the plan, if necessary, in ways that will help you succeed.
Step 7: Evaluate your progress on a regular basis.

Change Behaviors and Reward Yourself Having identified your goal and current patterns of behavior, you can now move to *Step 3*, which is to determine strategies to help you meet your goal and to choose some rewards for when you follow these strategies. An example of a possible behavior change strategy might be that you decide to talk to a friend instead of heading to the kitchen cupboard or vending machine when you're angry. You could save hundreds of calories and probably experience emotional relief by getting your feelings off your chest.

This is also the time to remind yourself of the progress you're making and the benefits you're gaining—weight loss, increased energy, better health. And it's also the time to give yourself a tangible reward, such as a night out at the movies or a new item of clothing. This step is key because behavior research indicates that once you have instituted a change in your life, you will maintain it only by positive reinforcement. Rewards don't need to be expensive. For example, you can take time for a long, candlelit bath, or celebrate with a leisurely Sunday morning—curled up with a book you've been wanting to read.

In *Step 4*, after you've pinpointed behaviors you want to alter, you must *commit* yourself to make the changes and take the time to envision the healthier person you'll become as a result.

TABLE 9-10
Sample Entries in Food Diary

Food Eaten	Time and Place	With Whom	Mood	Other Activity
1 large blueberry muffin 1 cup coffee with 1 oz cream and 1 tsp sugar	7:30 a.m., donut shop	Alone	Stressed—late for class	Skimming notes for upcoming class
1 hot dog with 2 tbsp ketchup 1 small bag potato chips 12-oz can diet soda	12:45 p.m., cafeteria	2 friends	Relaxed	Talking with friends
1 candy bar	2:35 p.m., bedroom	Alone	Angry about argument with significant other	Watching TV

Plan to Take Small Steps *Step* 5 is planning, and the way to do this is to take small steps. Divide your primary goal into smaller goals that can be achieved one at a time. Although you might have identified 10 behaviors that you want to change to meet your overall goal, trying to accomplish all 10 in the first week or two is probably very unrealistic, and it may simply set the stage for disappointment and failure (and cause you to give up completely). To give yourself a more realistic chance for success, choose only one or two of the smaller goals, do your best to stick to them for, say, two weeks, and then reward yourself for doing so before going on to work on your next goals.

Give It a Try and Adapt Your Plan! In *Step* 6 of behavior modification, you put your plan into action. As you meet your goals and reward yourself for doing so, also be aware of when you are not meeting them. If something is not working, don't give up. Instead, go on to *Step* 7 and feel free to modify your goals—perhaps those that you initially set were too difficult or too simple to achieve. You may, for example, decide to choose the smaller goal of cutting some of the fat out of your diet for the first two weeks and wait for a later time to work on reducing the calorie count. As you can see, when you evaluate your progress on a regular basis, you offer yourself the flexibility to adapt your plan so that you can continue to move forward to meeting your goal.

Be Prepared for a Relapse While you're working on modifying your behavior, it is critical to keep in mind that you're only human and there will be days when you slip and revert back to an old pattern of behavior. Instead of dwelling on it or berating yourself, realize that relapses are par for the course, forgive yourself, and simply move on. Keep in mind that even though you may have a setback, you're still moving forward if you learn from your mistakes and get right back on track again. That's why some experts prefer the term *recycle* to the term *relapse;* people who make a mistake and then learn from it can reuse, or recycle, their new knowledge.

Set Priorities Setting priorities helps to ensure that you adopt one or two behaviors that you stick to all the time. Also, pay attention to how other areas of your life may be affecting your readiness to make changes and meet your goals. Many people waste a great deal of time and energy trying to make changes at difficult times. Someone who is going through the breakup of a long-term relationship or having difficulties at work or school, for example, may be under too much stress to deal with making a behavior change on top of everything else.

Consider the "Rule of Three" Keep in mind that you might need to try new behaviors more than once before you begin to like them. Dietitians sometimes use a "rule of three" in counseling. Try fat-free milk once, and you may not like it. Try it a second time, and you may be able to tolerate drinking it, even though it doesn't taste as good as whole milk. Try it a third time, and you may conclude that although it's not as good as whole milk, you don't really mind drinking it. Then you'll be able to maintain your new habit more easily.

Strategies for Changing the Way You Eat

The steps in this section are designed specifically to help with weight control. (Chapter 10 includes strategies for modifying exercise behavior.) First, begin by keeping a record of your present eating behavior that you can compare with your future progress. Your eating records will eventually help you see how far you've come in changing your eating habits.

This record will also help you with the second step, which is to identify cues and situations that trigger your behavior of reaching for food. For example, some cues may prompt you to eat when you're not hungry, such as watching television

Continued Motivation:

- Persist long enough to experience the rewards, such as improved self-image and enhanced self-esteem.
- Remember the price of the old behavior.
- Keep in mind where you started.
- Tune in to the benefits of the new behavior.

A Recipe for Success

Be realistic.
Make small changes over time in what you eat and your activity level. Aim for losing no more than 1 to 2 pounds per week. Even a small amount of weight loss can bring major health benefits.

Be adventurous.
Expand your tastes to enjoy a wide variety of foods and activities. Get fewer calories by going "large" on flavor and small on portions.

Be flexible.
You don't have to give up your favorite foods to manage your weight. Just balance what you eat and what you do over several days.

Be sensible.
Enjoy all foods, just don't overdo any of them. Slow down when you're eating and listen to your body's signals to know when you're hungry or full.

Be active.
It's not *what* activity you choose, it's whether you enjoy doing it. If you enjoy it, you are likely to do it more often!

Source: Adapted from Weighing In on Health, *Food and Nutrition News* 70 (1998): 6.

or walking by a vending machine. Once the cues are identified, you can resolve to stop responding to them and try to eat only in one place in a certain room, such as at the table in the dining room. In addition, try the following tips to eliminate the temptation to eat when you really don't want to:

- Don't buy hard-to-resist foods such as cakes, cookies, and ice cream.
- Don't shop when you're hungry and thereby more likely to buy tempting foods.
- Don't leave large amounts of food within easy reach. Keep serving dishes off the table, for instance, and put the cookie jar out of sight.
- Make small portions of food look large by spreading food out and putting it on smaller plates. Garnish empty space with low-calorie vegetables.
- Try to eat regular meals and snacks instead of skipping them, thereby reducing the likelihood of becoming uncomfortably hungry and then overeating later.
- Ask your family and friends to encourage you to eat a healthful diet and not to criticize you if you splurge occasionally.

Third, make it easy for yourself to eat the way you want to eat.

- Keep a variety of nutritious foods such as fruits and vegetables readily available.
- If you like to eat between meals, plan snacks that fit easily into your diet, such as low-fat yogurt and fruit.

Fourth, take a look at and then, if necessary, alter the manner in which you eat. If you eat quickly, slow down by chewing food thoroughly, pausing between bites, putting down your utensils, and swallowing before reloading your fork or spoon. Eating slowly will give your body a chance to feel full and satisfied.

Finally, reward yourself for positive behaviors.

- Plan to buy something new to wear or do something you enjoy, such as attending a sporting event or a concert, each time you meet your goals.

Spotlight The Eating Disorders

The relentless pursuit of thinness and fear of being fat are a haunting nightmare that drives millions of American teens and adults to starve, vomit, or purge.[57] The illness of **anorexia nervosa**—self-starvation—has been recognized as a psychiatric syndrome since the 1870s. Its companion disease **bulimia nervosa**—gorging on food and then purging—was not recognized as a separate eating disorder until the 1960s and 1970s. Some researchers suspect that a complex interplay between environmental, social, and perhaps genetic factors triggers the disorder in victims, mostly women. Others question whether or not the fear of fatness isn't a mask for underlying emotional problems. Experts speculate that focusing on the body diverts these people's attention away from and suppresses the painful emotion of anger, feelings of low self-esteem, the inability to express their feelings, or poor family relationships. The acute focus on the body develops into an intense fear of fatness, a characteristic intrinsically linked with food.[58] As a result, for many people, food is seen only as a source of body fat and so becomes carefully controlled. But why food? Why not use some other method of coping with stress?

Enter the societal link. In a society where thinness is equated with material success and even self-worth, especially for women, becoming thin appears to be the yellow brick road that leads to happiness. Unfortunately, as victims of eating disorders come to learn, practicing self-starvation or gorging and purging leads instead to physical and emotional pain.

What is considered an eating disorder?

The term **eating disorder** comprises a wide spectrum of conditions including **anorexia nervosa, bulimia nervosa,** and **binge-eating disorder.**[59] Although the various conditions differ in their origin and consequences, they appear to have similarities—all of the conditions exhibit excessive preoccupation with weight, a fear of fatness, and a distorted body image. Many times the person with an eating disorder falls short of the diagnostic criteria shown in Table 9-11 for anorexia nervosa or bulimia nervosa. Some of these people are described as having an **unspecified eating disorder** and can include people who:[60]

- Meet all the criteria for anorexia nervosa except irregular menses.
- Meet all the criteria for anorexia nervosa except that their weight remains within a normal range.
- Meet all the criteria for bulimia nervosa except that their binges are less frequent.
- Have recurrent episodes of binge eating but do not compensate using the methods of those with bulimia nervosa, a condition known as binge-eating disorder (see Table 9-12 on page 307).[61]

What is the difference between an eating disorder and disordered eating?

Disordered eating occurs when a person eats (or doesn't eat) because of an external stimulus rather than an internal one. For example, an external stimulus might be, "I'm lonely," "I'm angry," or "I'm anxious," and "these chocolate cookies will distract me and help me feel better." An internal stimulus is a physical sensation—a message from your brain about your current state of hunger or satiety, such as "my stomach is growling," or "I feel full." People with eating disorders spend more time in "external eating" than in responding to internal hunger cues.[62] Many people eat for reasons other than hunger from time to time, but if this type of behavior controls someone's thoughts and interferes with their everyday routines, it becomes an eating disorder.

© Tony Freeman/PhotoEdit

▲ *Looking in the mirror, a woman with anorexia nervosa distorts her body image and overestimates her body fatness.*

What are the symptoms of anorexia nervosa?

Anorexics deprive themselves of food except for controlled amounts of very low-calorie foods such as unbuttered toast or popcorn, apples, and green beans. And even intake of these foods is painstakingly limited. After three to four days of eating very small amounts of food, hunger pangs subside. Once their appetite is suppressed, anorexics report feeling quite energetic, as if on a high, making the strict fast easier to stick to. But the body needs fuel to run. To compensate for the lack of fuel from food, the body turns inward for its fuel and begins to slowly destroy muscle and fat tissue for energy. The following are some of the physical symptoms associated with the starvation seen in anorexics:[63]

TABLE 9-11
Diagnosis of Eating Disorders

Anorexia Nervosa	Bulimia Nervosa
A person with anorexia nervosa demonstrates the following: 1. Refusal to maintain body weight at or above a minimal normal weight for age and height 2. Intense fear of weight gain or becoming fat, even though underweight 3. Distorted body image, undue influence of body weight or shape on self-evaluation, or denial of the seriousness of the current low body weight 4. In females past puberty, amenorrhea, that is, the absence of at least three consecutive menstrual cycles **Two types:** *Restricting type:* During the episode of anorexia nervosa, the person does not regularly engage in binge-eating or purging behavior (that is, self-induced vomiting or the misuse of laxatives, diuretics, or enemas). *Binge eating/purging type:* During the episode of anorexia nervosa, the person regularly engages in binge-eating or purging behavior (that is, self-induced vomiting or the misuse of laxatives, diuretics, or enemas).	A person with bulimia nervosa demonstrates the following: 1. Recurrent episodes of binge eating characterized by: a. eating in a discrete period of time an amount of food that is definitely larger than most people would eat during a similar period of time and under similar circumstances b. a sense of lack of control over eating during the episode 2. Recurrent compensatory behavior to prevent weight gain, such as self-induced vomiting; misuse of laxatives, diuretics, enemas, or other medications; fasting; or excessive exercise 3. Binge eating and compensatory behaviors that occur, on an average, at least twice a week for three months 4. Self-evaluation unduly influenced by body shape and weight 5. Does not occur exclusively during episodes of anorexia nervosa. **Two types:** *Purging type:* The person regularly engages in self-induced vomiting or the misuse of laxatives, diuretics, or enemas. *Nonpurging type:* The person uses other behaviors, such as fasting or excessive exercise, but does not regularly engage in self-induced vomiting or the misuse of laxatives, diuretics, or enemas.

SOURCE: Adapted from American Psychiatric Association, *Diagnostic and Statistical Manual of Mental Disorders*, 4th ed. (Washington, D.C.: American Psychiatric Association, 2000).

- Wasting of the whole body, including muscle tissue and bones
- Arrested sexual development and stopping of menstruation due to loss of body fat
- Drying and yellowing of the skin from an accumulation of a stored vitamin A compound released from body fat
- Intolerance to cold weather due to loss of subcutaneous fat
- Growth of hair on the body, perhaps in response to a decrease in body temperature
- Loss of health and texture of hair
- Pain on touch
- Lowered blood pressure and metabolic rate
- Anemia
- Severe sleep disturbance[64]
- Depression, possibly related to changes in neurotransmitter function in the brain[65]

At the same time the physical symptoms appear, distorted psychological symptoms develop. When looking in the mirror, anorexics do not see the emaciated body others see but continue to see themselves as being too fat.[66] A preoccupation with death develops, accompanied by a frantic pursuit of physical fitness by means of stringent exercise routines. The anorexic deals with parents and family in a manipulative way so as to become the center of attention. Diet becomes so totally engrossing that the anorexic may be quite socially isolated except for friends who stick close by and worry without knowing how to help.

By this time, the anorexic has reached absolute minimum body weight—for example, 65 to 70 pounds for a woman of average height. The person is on the verge of incurring permanent brain damage and chronic debilitation or death. The National Association of Anorexia Nervosa and Associated Disorders (ANAD) estimates that of those with severe eating disorders, 6 percent die, usually because major organs—heart and kidneys—fail.

What are the symptoms of bulimia?

Unlike anorexics, bulimics don't shrink to skeleton-like proportions. They usually have a healthy body weight or may even be slightly overweight. Bulimics also follow rigid rules of dietary restraint, but their routine is not as rigorous as the anorexic's starvation routine, and they occasionally break their own rigid rules.

TABLE 9-12
Criteria for Diagnosis of Binge-Eating Disorder

A person with a binge-eating disorder demonstrates the following:

1. Recurrent episodes of binge eating. An episode of binge eating is characterized by both of the following:
   a. Eating, in a discrete period of time (for example, within any 2-hour period), an amount of food that is definitely larger than most people would eat in a similar period of time under similar circumstances.
   b. A sense of lack of control over eating during the episode (such as, a feeling that one cannot stop eating or control what or how much one is eating).
2. Binge-eating episodes are associated with at least three of the following:
   a. Eating much more rapidly than normal.
   b. Eating until feeling uncomfortably full.
   c. Eating large amounts of food when not feeling physically hungry.
   d. Eating alone because of being embarrassed by how much one is eating.
   e. Feeling disgusted with oneself, depressed, or very guilty after overeating.
3. The binge eating causes marked distress.
4. The binge eating occurs, on average, at least twice a week for six months.
5. The binge eating is not associated with the regular use of inappropriate compensatory behaviors (purging, fasting, excessive exercise) and does not occur exclusively during the course of anorexia nervosa or bulimia nervosa.

SOURCE: Adapted from American Psychiatric Association, *Diagnostic and Statistical Manual of Mental Disorders*, 4th ed. (Washington, D.C.: American Psychiatric Association, 2000).

Quickly, and usually privately, bulimics gorge on foods that are often sweet, starchy, and high in fat or calories and require little chewing. The binge ends when it would hurt to eat any more, when they are interrupted, or when they go to sleep or induce vomiting to expel the food just eaten.

Bulimics often feel controlled by the vicious circle that develops: anxiety about being "fat" leads to rigid dietary restraint. Mounting hunger from not eating and an increased preoccupation with food cause them to break their rules, and binging results. After gorging, an intense fear of fatness overtakes the person, who then vomits to get rid of the food and release the fear of becoming "fatter." Feelings of guilt and shame follow the purge, building a new level of anxiety over the body, and the cycle begins again. Each binge reinforces the idea that additional rigidity of dietary restraint is required to prevent weight gain. Excessive use of laxatives or diuretics or bouts of vigorous exercise replace vomiting in some cases.

Binge eating is seldom life threatening, but at the extreme, it can be physically damaging, causing lacerations of the stomach, tearing or irritation of the esophagus (in those who vomit frequently), dental caries (from acidic vomit attacking the teeth), electrolyte imbalances and malnutrition (in those who vomit and those who take laxatives).

Bulimics also suffer from a distorted body image.[67] They see themselves as fat and needing to restrict food, even though they usually have a healthy body weight. Bulimics prefer a body size somewhat smaller than normal.

How are anorexia nervosa and bulimia treated?

There are several philosophies regarding the treatment of eating disorders. The four major approaches include individual psychotherapy, hospitalization, family therapy, and behavior modification therapy.[68] Some therapists use more than one approach. Most treatment methods focus on identifying the societal and environmental pressures that triggered the eating disorder and on exposing the emotions masked by it. An interdisciplinary team made up of a psychologist, a social worker, a family therapy counselor, and a dietitian work with the family and the patient to reestablish emotional and nutritional health.[69] Length of treatment varies from two months to two to three years, depending on the patient's readiness for change and the type of treatment. American researcher Hilde Bruche, M.D., focuses on the problems of low self-esteem, guilt, anxiety, depression, and a sense of helplessness. In addition, family therapy focuses on changing patterns of family interaction.

Normal nutrition must be restored in the anorexic. After some progress is made in counseling, the dietitian can help the patient gain a new understanding of a healthful eating pattern and clear up misconceptions about food and nutrition.[70]

© Michael Newman/PhotoEdit

▲ *For many people with anorexia nervosa, a full day's diet may consist of no more than three or four items.*

© Felicia Martinez/PhotoEdit

▲ *For many people with bulimia, guilt, depression, and self-condemnation follow a binge-eating episode.*

What are the early warning signs for eating disorders?

Table 9-13 shows the similarities and differences among the eating disorders. Families and friends can be alerted to several possible warning signs of eating disorders (see Table 9-14). Severe dieting often precedes the illness. Anorexics develop an exaggerated interest in food but at the same time deny their hunger and stop eating. Their distorted body image makes them feel fat even as weight loss continues. The anorexic begins to have sleep problems, shows unusual devotion to schoolwork, and often undertakes a program of unrelenting exercise. Bulimics may binge and self-induce vomiting or use excessive amounts of laxatives. Reduced food intake usually causes sufficient weight loss to stop menstrual cycles in women. People with eating disorders were usually good children who did not indulge in rebellion. Not all anorexics and bulimics exhibit all symptoms. Early detection is vital. Use the Eating Attitudes Quiz on page 310 to see if your own eating attitudes and behaviors are within a normal range.

What can be done to prevent eating disorders?

Consider the following suggestions for helping people to remain healthy and for preventing the occurrence of eating disorders altogether.[71]

- Discourage restrictive dieting and meal skipping. Severe dieting may lead to eating disorders, especially in adolescent girls. Model a lifestyle of healthy eating and exercise, and never encourage unhealthy "quick-fix" weight loss.
- Promote fitness and a healthy body rather than thinness and numbers on a scale.
- Help teens understand and accept the normal physiological changes in body composition and weight that occur with puberty—changes that may be interpreted as "getting fat."
- Don't plant unhealthy seeds. Avoid assigning "good" or "bad" labels to food. Don't offer food as a reward. Don't imply that the shape or size of your body, or someone else's, has anything to do with self-worth (for example, "I'd be happy if I were thin").
- Be sensitive and careful about making weight-related comments or recommendations. Even playful teasing may trigger long-term struggles with disordered eating or eating disorders.
- Encourage normal expression of the basic emotions: joy, anger, fear, sadness, loneliness, guilt, and shame.

TABLE 9-13
How the Eating Disorders Compare

	Anorexia Nervosa	Bulimia Nervosa	Binge-Eating Disorder
Estimated Prevalence*	Up to 0.5%–1.0%	Up to 1.0%	Up to 2.0%
Male versus Female Incidence	5%–10% male versus 90%–95% female	5%–10% male versus 90%–95% female	Unknown
Typical Age of Onset	Early to middle adolescence	Late teens to early twenties	Any age, but not usually recognized until adulthood
Weight	Extremely thin and emaciated; <85% of desirable body weight	Near-desirable body weight, but often has weight fluctuations	Usually overweight or obese
Self-Esteem	Low	Low	Low
Depression	Common	Common	Common
Substance Abuse	Rare	Common	Rare
Rate of Weight Loss	Rapid	Repeatedly loses and gains weight or chronically diets without losing weight	Repeatedly loses and gains weight or chronically diets without losing weight
Past Dieting	Yes	Yes	Yes

*Determining accurate statistics is difficult because physicians are not required to report eating disorders to a health agency and because people who have eating disorders tend to deny that they have a problem and are very secretive about their behaviors.

SOURCE: J. Kirby, for the American Dietetic Association, *Dieting for Dummies* (Foster City, CA: IDG Books Worldwide, 1998), 67–68.

What can I do if I am worried that someone I know may have an eating disorder?[72]

If you or others observe behaviors in a friend, roommate, or family member that suggest the possibility of an eating disorder, consider the following:

- Make a plan to approach the person in a private place when there is no immediate stress and time to talk.
- Present in a caring but straightforward way what you have observed and what your concerns are. Tell her or him that you are worried and want to help.
- Give the person time to talk and encourage them to verbalize their feelings. Listen carefully and accept what is said in a nonjudgmental manner.
- Do not argue about whether there is or is not a problem—power struggles

TABLE 9-14
Warning Signs of an Eating Disorder

- Preoccupation with weight, food, calories, and dieting, to the extent that it consistently intrudes on conversations and interferes with other activities
- Excessive exercise—despite weather, fatigue, illness, and injury
- Dramatic weight loss; arrested growth in children
- Anxiety about being fat which does not diminish with weight loss
- Reluctance to be weighed
- Inability to gain weight
- Difficulty eating in social situations
- Deceptive or secretive behavior
- Alternating periods of severely restrictive dieting and overeating
- Evidence of binge eating or consumption of large quantities of food inconsistent with the person's weight
- Evidence of use of laxatives, diuretics, or agents that cause vomiting (emetics)
- Extreme concern about appearance as a defining feature of self-esteem (for example, I am "thin and good," or "gross and bad")
- Social withdrawal or avoidance of activities because of weight and shape concerns
- Absence from school or work
- Depression
- Disrupted menstruation
- Constipation or diarrhea
- Paleness or lightheadedness not accounted for by other medical conditions
- Susceptibility to fractures

SOURCES: Adapted from Eating Disorders Awareness and Prevention, Inc. materials and D. Rosen, Dieting Disorder, *New England Journal of Medicine,* April 8, 1999.

are not helpful. Perhaps you can say, "I hear what you are saying, and I hope you are right that this is not a problem. But I am still very worried about you and your health, and that is not going to go away."

- Provide information about resources for treatment. Offer to go with the person and wait while they have their first appointment with a counselor, physician, or dietitian.
- If you are concerned that the eating disorder is severe or life threatening, enlist help from the counseling center, a relative, a friend, or the person's roommate *before* you intervene. Present a united and supportive front together.
- If the person denies the problem, becomes angry, or refuses treatment, understand that this is often part of the illness. Besides, they have a right to refuse treatment (unless their life is in danger). You may feel helpless, angry, and frustrated with them. You might say, "I know you can refuse to go for help, but I am still your friend, and I am worried about you. I may bring this up again later, and maybe we can talk more about it then." Follow through on that—and on any other promise you make.
- Do not try to be a hero or a rescuer; you will probably be resented. Eating disorders are stubborn problems, and treatment is most effective when the person is truly ready for it. You may have planted a seed that helps them get ready.

For more information about anorexia nervosa, bulimia, and associated disorders, log on to the following websites:

- *Anorexia Nervosa and Related Eating Disorders:* www.anred.com
- *National Eating Disorders Association:* www.nationaleatingdisorders.org
- *National Institute of Mental Health:* www.nimh.nih.gov
- *Gurze Books' Eating Disorders Resource Catalog:* www.gurze.com
- *Eating Disorder Referral and Information Center:* www.edreferral.com
- *Canadian National Eating Disorder Information Centre:* www.nedic.ca

MINIGLOSSARY

anorexia nervosa literally, "nervous lack of appetite," a disorder (usually seen in teenage girls) involving self-starvation to the extreme.

an = without
orexis = appetite

binge-eating disorder an eating disorder characterized by uncontrolled chronic episodes of overeating (compulsive overeating) without other symptoms of eating disorders. Typically, the episodes of binge eating occur at least twice a week for a period of six months or more.

bulimia nervosa, bulimarexia (byoo-LEE-me-uh, byoo-lee-ma-REX-ee-uh) binge eating (literally, "eating like an ox"), combined with an intense fear of becoming fat and usually followed by self-induced vomiting or taking laxatives.

buli = ox

disordered eating eating food as an outlet for emotional stress rather than in response to internal physiological cues.

eating disorder general term for several conditions (anorexia nervosa, bulimia nervosa, binge-eating disorder) that exhibit an excessive preoccupation with body weight, a fear of body fatness, and a distorted body image.

unspecified eating disorders some people suffer from unspecified eating disorders; that is, they exhibit some but not all of the criteria for specific eating disorders.

Eating Attitudes Quiz

Answer the following questions to evaluate your own eating attitudes and behaviors. Do they fall within a normal range? Use these responses:

A = always
U = usually
O = often
S = sometimes
R = rarely
N = never

__________ **1.** I am terrified about being overweight.
__________ **2.** I avoid eating when I am hungry.
__________ **3.** I find myself preoccupied with food.
__________ **4.** I have gone on eating binges where I feel that I may not be able to stop.
__________ **5.** I cut my food into very small pieces.
__________ **6.** I am aware of the calorie content of the foods I eat.
__________ **7.** I particularly avoid foods with high carbohydrate content.
__________ **8.** I feel that others would prefer that I eat more.
__________ **9.** I vomit after I eat.
__________ **10.** I feel extremely guilty after eating.
__________ **11.** I am preoccupied with a desire to be thinner.
__________ **12.** I think about burning up calories when I exercise.
__________ **13.** Other people think I am too thin.
__________ **14.** I am preoccupied with the thought of having fat on my body.
__________ **15.** I take longer than other people to eat my meals.
__________ **16.** I avoid foods with sugar in them.
__________ **17.** I eat diet foods.
__________ **18.** I feel that food controls my life.
__________ **19.** I display self-control around food.
__________ **20.** I feel that others pressure me to eat.
__________ **21.** I give too much time and thought to food.
__________ **22.** I feel uncomfortable after eating sweets.
__________ **23.** I engage in dieting behavior.
__________ **24.** I like my stomach to be empty.
__________ **25.** I enjoy trying new rich foods.
__________ **26.** I have the impulse to vomit after meals.

SCORING
Never = 3
Rarely = 2
Sometimes = 1
Always, usually, and often = 0

A total score under 20 points may indicate abnormal eating behavior and the risk of having or developing an eating disorder. Share your results with a health professional for further evaluation.

SOURCE: J. A. McSherry, Progress in the diagnosis of anorexia nervosa, *Journal of the Royal Society of Health* 106 (1986): 8–9; *Eating Attitudes Test* and scoring developed by Dr. P. Garfinkel.

In Review

1. What is a safe rate of weight loss on a long-term basis for most overweight people?
   a. 0.5 to 2 pounds/week
   b. 3–4 pounds/week
   c. 5% body weight per month
   d. 10% body weight per month

2. Herbal supplements used for weight loss:
   a. are strictly regulated by the FDA.
   b. are hard to obtain.
   c. do not cause side effects like prescription medications.
   d. are safe because they are natural.
   e. can cause serious side effects, including death.

3. Which energy-yielding nutrient is most easily stored as fat tissue by the body?
   a. fat
   b. protein
   c. carbohydrate
   d. There is no difference in how they are stored.

4. Briefly explain basal metabolism as it compares to physical or voluntary activity.

5. Obesity is:
   a. a growing epidemic.
   b. influenced by activity level.
   c. a disease with multiple health risks.
   d. all of the above

6. Briefly explain the "energy in, energy out" theory related to weight gain and weight loss.

7. Healthfully gaining weight involves:
   a. loading the diet with fat.
   b. adding calories to the diet.
   c. ceasing all physical activity.
   d. eating healthy snacks throughout the day.
   e. Both b and d are correct.

8. List three problems associated with fad diets.

9. Which of the following is *not* true of a healthy weight?
   a. It can be gauged by BMI.
   b. It is a weight that is compatible with wellness and long life.
   c. It is thought of as "the thinner the better."
   d. It is difficult to gauge.

10. What is the first source of fuel the body uses when fasting?
    a. water
    b. fat
    c. protein
    d. glycogen

Menu of Online Study Tools

A variety of study tools for this chapter are available at our website to deepen your understanding of chapter concepts. Go to
www.thomsonedu.com/nutrition/boyle
to find

- Practice tests
- Flashcards
- Glossary
- Web links
- Animations
- Chapter summaries, learning objectives, and crossword puzzles

Nutrition on the Web

www.thomsonedu.com/nutrition/boyle
Go to the *Personal Nutrition* site to check for the latest updates to chapter topics or to access links to related websites.

www.eatright.org
Find useful links to weight management resources.

www.ars.usda.gov/ba/bhnrc/ndl
Look up calories in foods and beverages.

www.healthfinder.gov
Search for information on obesity on this U.S. government site.

http://win.niddk.nih.gov/publications/choosing.htm
Weight-control Information Network on choosing a safe and successful diet.

www.nal.usda.gov/fnic/etext/000060.html
USDA Food and Nutrition Information Center on weight control and obesity.

www.wheatfoods.org
Wheat Council "Setting the Record Straight."

www.healthyweight.net
Weight management resources by scientific experts.

www.quackwatch.org
Evaluate weight-loss gimmicks, products, and promotions.

www.fda.gov/cder
Information on drugs for weight loss from the Center for Drug Evaluation and Research.

www.ncbi.nlm.nih.gov/PubMed/
Use this search engine to locate information on obesity and related topics.

www.hugs.com
HUGS International, Inc., offers books, tapes, and training guides for teen and adult eating disorder programs and nondieting approach to weight management.

www.nwcr.ws/
The National Weight Control Registry.

www.intelihealth.com
Information about weight management, obesity,"the freshman 15," and eating disorders, plus numerous interactive tools.

www.mayoclinic.com
Visit the Mayo Clinic's Food and Nutrition Healthy Living Center; check out the virtual cookbook for healthy recipes and find tools to help with weight management.

www.caloriecontrol.org
The Calorie Control Council provides healthy recipes, an online calorie counter, and virtual calculators for figuring body mass index and calories expended during various exercises.

Nutrition and Fitness

10

ASK YOURSELF . . .

Which of the following statements about nutrition are true, and which are false? For each false statement, what is true?

1. Regular exercise can help people increase their lean body mass and reduce their fat tissue.
2. Less than 25 percent of U.S. adults exercise adequately.
3. People who fail to exercise regularly are more likely to fall prey to degenerative diseases such as heart disease, osteoporosis, and diabetes.
4. Essentially, to be fit means to be at desirable weight and to have strong muscles.
5. People should never push themselves to exercise longer or harder than they can easily manage to do.
6. Of all the components of fitness, cardiovascular endurance has the most impact on health and longevity.
7. If you run out of breath, it is a sign that your heart and lungs are not strong enough to perform the desired tasks.
8. When a muscular athlete stops exercising, much of his or her muscle tissue turns to fat.
9. The use of steroid hormones can cause a disfiguring disease.
10. Athletes can lose two or more quarts of fluid during every hour of heavy exercise and must rehydrate before, during, and after exercising.

Answers found on the following page.

Get health. No labor, effort, nor exercise that can gain it must be grudged.

R. W. Emerson (1803–1882, American essayist and poet)

CONTENTS

Get moving! The benefits of regular exercise make up an impressive list, and the list keeps growing longer as new discoveries are made. People who begin a fitness program to trim fat or add muscle are soon pleasantly surprised to find that they have more energy, feel less tense, sleep better, and feel healthier.[1] And so they keep it up! Most people do not realize the many benefits of regular exercise. Read the list in the margin on page 316 and imagine how getting more exercise might change your life for the better.

In addition, mounting evidence suggests that our bodies need regular, moderate exercise that gets our hearts beating and forces our muscles to work harder than they usually do to stay healthy. Physiologically speaking, overall fitness is a balance between different body systems. With respect to the joints, flexibility is important. With respect to muscles, strength and endurance are important.

This chapter illustrates the effects of nutrition on fitness—and vice versa, for a two-way relationship exists. Optimal nutrition contributes to athletic performance, and, conversely, regular exercise contributes to a person's ability to use and store nutrients optimally. Together, the two are indispensable to a high quality of life. **Fitness,** like good nutrition, is an essential component of good health.

No one can promise that you will receive all of these benefits if you exercise, but almost everyone who exercises reaps at least some of them. Physical activity

© David Young-Wolff/PhotoEdit

▲ *Being fit is more than being free of disease; it is feeling full of vitality and enthusiasm for life.*

need not be strenuous to achieve health benefits. Moderate amounts of physical activity are recommended for people of all ages. However, at present, only about 33 percent of adults engage regularly in sustained physical activity of any intensity during leisure time. About 38 percent of adults report no physical activity at all.

Getting Started on Lifetime Fitness

The notion that a certain minimum daily average amount of activity is indispensable to health is just now reaching public consciousness. The Institute of Medicine recommends that we spend a total of at *least* 60 minutes on most days of the week engaged in any one of numerous forms of physical activities (see Table 10-1).[2] This 60 minutes can be accumulated in relatively brief sessions of activity—just mix and match your preferred activities in periods as short as 8 to 10 minutes that total 60 minutes by the end of the day. For example, walk your dog for 20 minutes in the morning, enjoy a 20-minute bike ride through the neighborhood after classes, and take a 20-minute walk after dinner. The new guidelines stress the value of moderate activity and suggest that the total *amount* of activity is more important than the manner in which it is carried out.

For total fitness, an exercise program that incorporates aerobic activity, strength training, and stretching is best. The more active you are, the more fit you are likely to be.

In proceeding with a fitness program, keep in mind that fitness builds slowly, and so activity should increase gradually. For beginners, consistency is very important. Establish a regular pattern of physical activity first (for example, 30 minutes cumulative to start) and plan to increase that amount over time. View your exercise time as a lifelong commitment.

If you are just starting on a fitness program, a few precautions are important. If you are an apparently healthy male older than 40 years of age or an apparently healthy female older than 50 years of age, the American College of Sports Medicine recommends that you have a medical examination and diagnostic **exercise stress test** before you start a *vigorous* exercise program. Beginning a *moderate* program, such as walking, however, does not require a physician's exam.[3]

For most people, physical activity should not pose any problem or hazard. However, medical advice concerning suitable type of activity is necessary for anyone with two or more of the risk factors shown in the margin or for anyone diagnosed with cardiac or other known diseases.

The term *fitness* is not restricted to the seasoned athlete. With a basic understand-

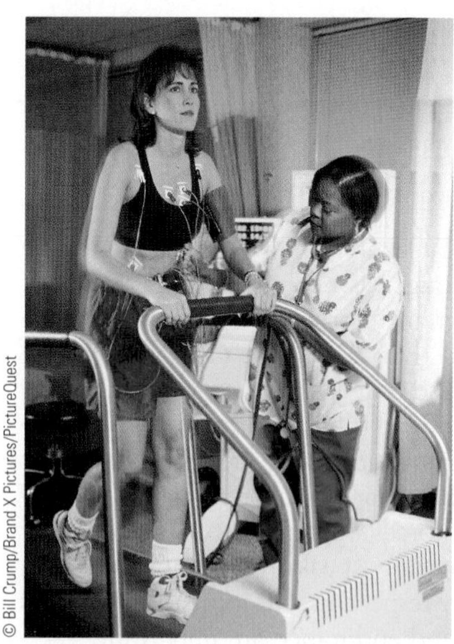

© Bill Crump/Brand X Pictures/PictureQuest

▲ *The exercise stress test measures heart function during exercise.*

Benefits of physical activity include:

- Increased self-confidence
- Easier weight control
- More energy
- Less stress and anxiety
- Improved sleep
- Enhanced immunity
- Lowered risk of heart disease
- Lowered risk of certain cancers
- Stronger bones
- Lowered risk of diabetes
- Lowered risk of high blood pressure
- Increased quality of life
- Increased independence in life's later years

An "apparently healthy" individual has *no more than one* of the following risk factors:

- Sedentary lifestyle
- Age (men > 40 years of age; women > 50 years of age)
- Family history of heart disease
- Cigarette smoking
- High blood pressure
- High blood cholesterol (>200 mg/dL)
- Diabetes

fitness the body's ability to meet physical demands, composed of four components: flexibility, strength, muscle endurance, and cardiovascular endurance.

exercise stress test a test that monitors heart function during exercise to detect abnormalities that may not show up under ordinary conditions; exercise physiologists and trained physicians or health care professionals can administer the test.

Ask Yourself Answers: **1.** True. **2.** True. **3.** True. **4.** False. To be fit means not only to be at desirable weight and to have strong muscles but also to be flexible and, most importantly, to have muscular and cardiovascular endurance. **5.** False. The overload principle states that people should push themselves to exercise longer or harder than they can easily manage to do—although not, of course, to the point of strain. **6.** True. **7.** False. If you run out of breath, it is not a sign that your heart and lungs are weak but a sign that you are going into oxygen debt. **8.** False. Muscle tissue does not turn to fat, but for a muscular athlete who stops exercising, muscle tissue is lost and fat is gained. **9.** True. **10.** True.

TABLE 10-1
Exercise Guidelines for Fitness

Mode/Activity	Examples	Duration/Frequency	Benefits
General physical (or leisure) activity	• Walking (3–4 mph) • Gardening • Golfing (no cart) • Active play (volleyball, basketball, softball, Ping-Pong)	Engage in at least 30 minutes of moderate-intensity activity, above usual activity, at work or home on most days of the week. For most people, greater health benefits can be obtained by engaging in physical activity of more vigorous intensity or longer duration.*	Reduces the risk of chronic disease in adulthood and improves overall health.
		Engage in approximately 60 minutes of moderate- to vigorous-intensity activity on most days of the week while not exceeding caloric intake requirements.	Helps manage body weight and prevent gradual, unhealthy body weight gain in adulthood.
		Participate in at least 60–90 minutes of daily moderate-intensity physical activity while not exceeding caloric intake requirements.	Sustains weight loss in adulthood
Achieve physical fitness by including cardiovascular conditioning, resistance exercises or calisthenics for muscle strength and endurance, and stretching exercises for flexibility.			
Cardiovascular (or aerobic) exercise	• Jogging • Tennis • Biking • Walking (4–5 mph) • Swimming • Rowing • Dancing • Stair climbing	20+ minutes of sustained aerobic activity 3 to 5 times per week.	Helps lower blood pressure and cholesterol levels and reduce risk for heart disease, diabetes, obesity, and certain types of cancer
Resistance (strength) training	• Lifting weights • Push-ups • Sit-ups • Pull-ups • Chair stands	20–45 minutes per session, 2 to 3 nonconsecutive days per week (eight or more exercises, 1–3 sets, 10–15 reps)	Maintains muscle mass and strength; promotes strong bones; reduces risk and symptoms of arthritis; improves glycemic control
Stretching	• Standing or seated toe touch • Overhead reach • Yoga	Hold each positioned stretch for 20–30 seconds; do on most, preferably all, days of the week	Reduces risk for injuries and falls; maintains and increases muscle and joint flexibility

*A *moderate* physical activity is any activity that requires about as much energy as walking 2 miles in 30 minutes. Active everyday chores and aerobic activities count toward your 60 minutes, too.

SOURCE: Adapted from Physical activity for every age and stage, *Nutrition Update*, Fall/Winter 2002 (East Hanover, NJ: The Nutrition Update Group); *Dietary Guidelines for Americans*, 2005.

ing of the concept of total fitness and a personal commitment to a physically active lifestyle, anyone can become fit (see Figure 10-1). To be fit, you don't have to be able to finish the local marathon, nor do you have to develop the muscles of a Mr. Universe or Miss Olympia. Rather, what you need is a reasonable weight (refer to Chapter 9) and enough flexibility, muscle strength, muscle endurance, and cardiovascular endurance to meet the everyday demands that life places on you, plus some to spare. The Scorecard on page 324 helps you evaluate your own level of physical activity.

FIGURE 10-1
THE ACTIVITY PYRAMID

Physical activity helps you feel and perform at your best. The key to sticking with a regular exercise program is to choose activities that you enjoy. Be flexible. Remember that you don't have to spend 60 consecutive minutes engaged in physical activity—a few minutes here and a few minutes there add to the benefits of moderate exercise. Keep in mind that *any* activity is good (for example, "just do something"), but *more* is better. The goal is to have a *physically active* lifestyle.

SOURCE: Adapted from J. Norstrom, *Ten Tips to Healthy Eating and Physical Activity for You,* © 1995 The American Dietetic Association National Center for Nutrition and Dietetics, International Food Information Council, and President's Council on Physical Fitness and Sports.

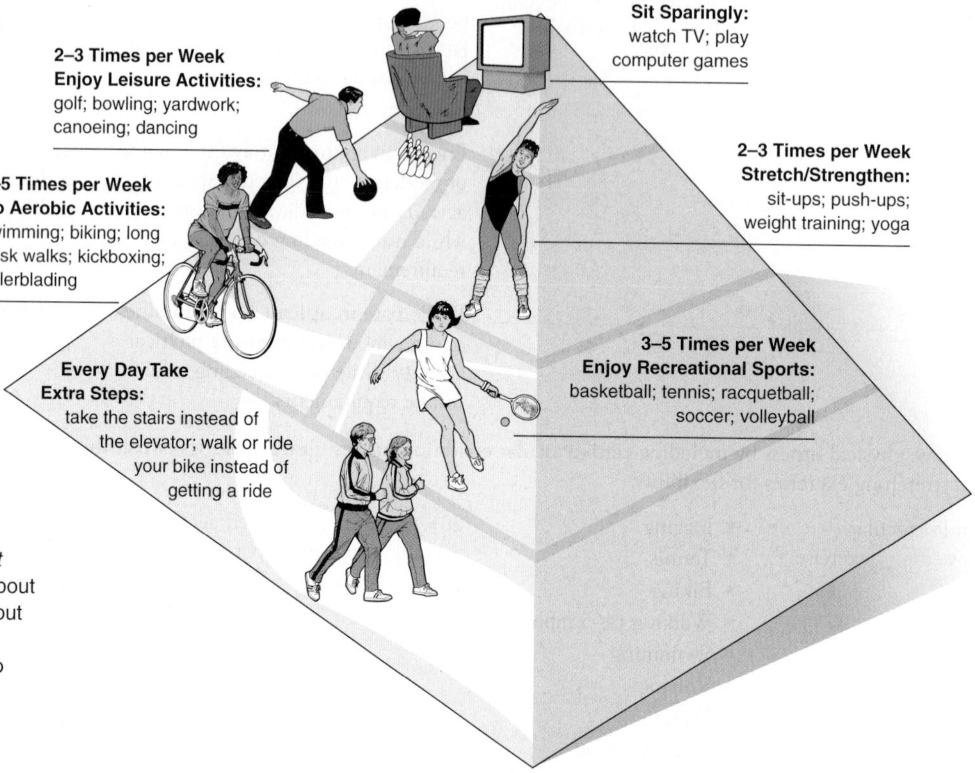

If you are generally *inactive*, increase daily activities at the base of the Activity Pyramid.
- Take the stairs.
- Hide the TV remote control.
- Make extra trips around the house or yard.
- Stretch while standing in line.
- Walk whenever you can.

If your activity routines are *sporadic* (active in summer but not in the winter), become consistent with activity by increasing activity in the middle of the pyramid.
- Find activities you enjoy.
- Plan activities in your day.
- Set realistic goals.

If your physical activity is *consistent* (on most days of the week), think about the long term as you move throughout the pyramid.
- Change your routine if you start to get bored.
- Explore new activities.

Sit Sparingly: watch TV; play computer games

2–3 Times per Week Enjoy Leisure Activities: golf; bowling; yardwork; canoeing; dancing

2–3 Times per Week Stretch/Strengthen: sit-ups; push-ups; weight training; yoga

3–5 Times per Week Do Aerobic Activities: swimming; biking; long brisk walks; kickboxing; rollerblading

3–5 Times per Week Enjoy Recreational Sports: basketball; tennis; racquetball; soccer; volleyball

Every Day Take Extra Steps: take the stairs instead of the elevator; walk or ride your bike instead of getting a ride

For fitness, remember the FIT principle:

F—frequency: number of exercise sessions per week; at least three to five sessions per week are recommended.
I—intensity: how hard you exercise (for example, the degree of exertion while exercising); it is recommended that you exercise at 55 to 90 percent of your maximum heart rate per minute—known as your target heart rate.
T—time: duration or length of time that you exercise with your heart rate elevated into your target heart rate zone (the minimum amount is thought to be 20 to 30 minutes per session).

overload an extra physical demand placed on the body. A principle of training is that for a body system to improve, its workload must be increased by increments over time.

hypertrophy an increase in size in response to use.

atrophy a decrease in size in response to disuse.

The Components of Fitness

Physical Conditioning

Physical conditioning refers to a planned program of exercise directed toward improving the function of a particular body system. Placing regular, physical demand on your body and forcing it to do more than it usually does will cause it to adapt and function more efficiently. In terms of exercise, this is called **overload.** Muscles respond to the overload (more than normal amount) of exercise by gaining strength and ability to endure. The overload principle applies equally to all aspects of fitness: flexibility, muscle strength, muscle endurance, and cardiovascular endurance.

You can apply overload to your exercise program in several ways:

- You can do the activity more often—that is, increase its **frequency.**
- You can do the activity more strenuously—that is, increase its **intensity.**
- Or you can do it for longer periods of **time**—that is, increase its duration.

All three strategies work well, and you can pick one or a combination of the three, depending on your fitness goals. Gaining strength may not be visible in all cases. But for some, such as male bodybuilders, muscles increase in strength and size. This response is called **hypertrophy.** The converse is also true. If muscles are not called on to perform, they decrease in size. This response is called **atrophy.**

Strength

Strength is the ability of the muscles to work against resistance, such as pulling yourself out of a swimming pool, carrying a backpack full of books, or opening a jar of pickles. The purpose of strength training is to build well-toned muscles that let you accomplish daily activities at work and during recreation as well as to prevent injury. As muscles get stronger, individual fibers thicken and enlarge. Our ability to respond to strength training continues to a very old age.[4] The connective tissues making up muscles, tendons, and ligaments also strengthen and become more efficient at using energy. The benefit is that strong muscles, tendons, and ligaments play a key role in preventing injury. For example, strong quadriceps—the muscles on the front of the thigh—stabilize your knee as you bike. Strong calf muscles and ankle ligaments decrease the risk of an ankle sprain when you walk briskly or jog. Strength training also helps with weight loss by increasing lean muscle mass and thus increasing your basal metabolic rate.

Many of today's mechanical aids that have been invented to make life easier actually rob us of the opportunity to develop strength. For example, in the past we would have gained strength by chopping firewood instead of simply turning up the thermostat. Today, we must put forth conscious effort to develop strength. No matter what exercises you choose, safety in strength training is essential. Get proper instruction before you set up a strength-training program.

strength the ability of muscles to work against resistance.

flexibility the ability to bend or extend without injury; depends on the elasticity of the muscles, tendons, and ligaments and on the condition of the joints.

static stretches stretches that lengthen tissues without injury; characterized by long-lasting, painless, pleasurable stretches.

endurance the ability to sustain an effort for a long time. One type, **muscle endurance,** is the ability of a muscle to contract repeatedly within a given time without becoming exhausted. Another type, **cardiovascular endurance,** is the ability of the cardiovascular system to sustain effort over a period of time.

Flexibility

Keeping your muscles and joints pliable is critical for developing a fit body. A flexible body can move as it was designed to move and will bend rather than tear or break in response to sudden stress. **Flexibility** (range of motion) depends on the condition and interrelationships of bones, ligaments, muscles, and tendons.

Flexibility tends to decrease with age but improves in response to stretching, and it can be maintained in most people by frequent stretching. Stretching exercises improve flexibility by increasing muscle and tendon elasticity and length. Stretching should be done slowly—called **static stretches.** When you feel a slight strain in the muscle, hold the position for 20 to 30 seconds. Bouncy, rapid stretches can cause minute tears in the muscle and also set up a reaction in the muscle that makes it resist the stretch. Avoid painful stretches. They are clearly excessive and often do more damage than good.

▲ *People's bodies are shaped by what they do.*

Stretching routines are commonly part of a warm-up routine before exercise. Low-intensity preliminary exercise allows your heart—also a muscle—to slowly accelerate and make necessary adjustments in blood flow and oxygen supply, preparing the heart for the work it is about to perform. Using calisthenics as a warm-up, such as walking, marching in place, or doing some other moderate rhythmic activity, prepares your heart muscle for action. After your light warm-up, stretch the muscles that you will be using in your main exercise activity. Waiting to stretch until after your warm-up allows blood to move into the muscles, making them easier to stretch. Doing stretches after your exercise session gives your heart a chance to gradually slow its pace. It also allows you to lengthen those muscles that have become tight and tense from the exercise. You can make greater gains in flexibility by stretching after your workout because your muscles are warm and easier to stretch.

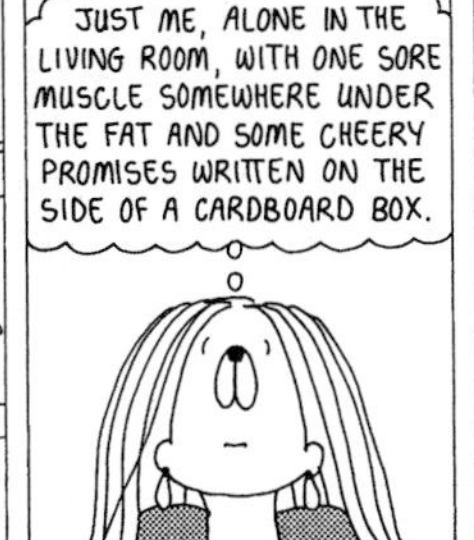

Muscle Endurance

Muscle endurance, the third component of fitness, is the power of a muscle to keep working for long periods. Your muscle endurance influences your performance in the last set of a tennis match, your swing on the 18th hole of a golf game, or your ability to pedal during the last 10 miles of a 100-mile bike tour. Endurance of certain muscles can be tested by the number of sit-ups or push-ups you can accomplish in a certain period of time. But remember, these tests evaluate only the abdominal and upper arm muscles.

Cardiovascular Endurance

To develop cardiovascular fitness, choose an aerobic activity such as:

Aerobic dancing	Roller skating
Bench stepping	Rope jumping
Bicycling	Rowing
Cross-country skiing	Running
Fast walking	Speed skating
Jogging	Stair climbing
Roller blading	Swimming
	Treadmill walking or running

Another important aspect of endurance is the length of time that you can keep going with an elevated heart rate—that is, how long your heart can endure a given demand. This kind of endurance is called **cardiovascular endurance.** The heart is a muscle, and, like your other muscles, it can respond to repeated demands by becoming larger and stronger.

Exercises that promote cardiovascular endurance are the best for making short-term fitness gains and for long-term health—as well as for weight control. The best exercises to develop cardiovascular endurance are those that repetitively use large muscle groups—arms and legs—and that last for a continuous 20 to 60 minutes. Examples include brisk walking, aerobic dance, running, cycling, cross-country skiing, and rowing. The American College of Sports Medicine recommends that people participate in cardiovascular conditioning activities at least three times a week for a continuous 20 to 60 minutes.[5]

Energy for Exercise

Your body runs on water, oxygen, and food—primarily carbohydrate and fat. The chemical process that converts these substances to energy is called metabolism. Two interrelated energy-producing systems are at work in your body. One system, which depends on oxygen, is called **aerobic** metabolism. The other system functions without oxygen and is called **anaerobic** metabolism. Understanding how the two systems work is important because it explains why you would choose certain exercises over others to strengthen your heart, why you eat what you do, and what factors influence your performance during sporting events.

Aerobic and Anaerobic Metabolism

At rest, your muscles burn mostly fat and some carbohydrate for energy. During exercise, though, the amount of energy the muscles use depends on the interplay between fuel availability and oxygen availability. To an exercising muscle, oxygen is everything. With ample oxygen, muscles can extract all available energy from carbohydrate and fat by means of aerobic metabolism. During moderate exercise, your lungs and circulatory system have no trouble keeping up with the muscles' need for oxygen. You breathe deeply and easily, and your heart beats steadily—the exercise is aerobic. But the heart and lungs can supply only so much oxygen so fast.

When the muscles' exertion becomes great enough that their energy demand outstrips the oxygen supply, they must also rely on anaerobic metabolism for energy. Because the anaerobic metabolic pathway can only burn carbohydrate for fuel, it draws heavily on your limited body stores of carbohydrate. Nevertheless, this system provides an immediate energy source without requiring oxygen. Thanks to anaerobic metabolism, you can dash out of the way of an oncoming car or sprint ahead of your competitor at the finish line. Unfortunately, however, this energy-yielding system is extremely inefficient. Only 5 percent of carbohydrate's energy-producing potential is harnessed by this pathway.[6]

Because the anaerobic metabolic pathway only partially burns your carbohydrate, it also litters your muscle with lactic acid—partly broken down portions of glucose. When lactic acid builds up in the muscles, it causes burning pain and can lead to muscle exhaustion within seconds if it is not drained away. An effective

(Text continues on page 325.)

aerobic requiring oxygen.

anaerobic not requiring oxygen.

Nutrition Action

Nutrition and Fitness—Forever Young—or You're Not Going to Take Aging Lying Down, Are You?

© David Young-Wolff/PhotoEdit

No matter how old, the body of modern humans (and how it works) was designed over 100,000 years ago. One could even conclude that we humans have inherited genes that have been fine-tuned to support a physically active lifestyle. In fact, scientists from the University of Missouri, University of Pennsylvania, and East Carolina University not only hypothesize this to be true but go on to say "that physical inactivity (defined as the activity equivalent of less than 30 minutes of brisk walking each day) in sedentary societies directly contributes to multiple chronic health disorders."[7] They conclude that physical inactivity is an abnormal state because our bodies have been "programmed" to expect physical activity, thus causing the metabolic dysfunctions that lead to a host of chronic health conditions such as heart disease, cancer, obesity, diabetes, and high blood pressure.

12 Easy Ways to Be Sedentary

Cellular phones
Computer games
Dishwashers
Drive-thru windows
E-mail/Internet
Escalators and elevators
Food delivery services
Garage door openers
Housekeeping and lawn services
Moving sidewalks
Remote controls
Shopping by phone

Yesterday's Genes, Today's Lifestyle

Nearly all of your biochemistry and physiology (in other words, how you process foods and metabolize energy) was fine-tuned to conditions of life that existed more than 10,000 years ago. But what we eat has changed more in the last 40 years than in the previous 40,000 years. So what's the big problem? Our genes don't know it! We still "process" food the same way our ancestors did—very efficiently (sometimes called the "thrifty gene" hypothesis). To understand this concept you only need to compare "yesterday's" diet and physical activity patterns to "today's" diet and exercise habits (see the photos at the top of page 322).

"Now all we need is fake exercise."

Choose Your Weapon

You have your choice of weapons against the life-threatening diseases associated with sedentary aging, as listed in Table 10-2. Recommendations for exercise include 60 minutes of moderate-intensity exercise (brisk walking between

Preagricultural Hunter-Gatherers

Burned ~3,000 calories per day
Moderate physical activity >30 minutes per day to provide basic necessities (food, water, shelter, gathering materials for warmth, etc.)
Feast or famine
Lean wild game or fish
Uncultivated fruits and vegetables

Industrialized Modern Humans

Burn ~1,800 calories per day
Sedentary (desk jobs, computer games, TV, dishwashers, drive-thru windows, etc.)
Abundance of food (continuous feeding)
Grain-fattened meats
Refined sugar

classes, bicycling from your dorm to classes, climbing the stairs instead of taking the elevators) on all or most days of the week and resistance exercise (using free weights, exercise bands, weight machines, or push-ups and squats). Again, it doesn't matter if the 60 minutes is accumulated in small increments or at one time.[8] Resistance exercise should average three sessions (20 to 45 minutes per session) per week, working one to two muscle groups during each session. The different muscle groups are the chest, shoulders, arms, back, abdominals, and legs. Each session should include one to three sets (10 to 15 repetitions per set), resting 2 to 3 minutes between sets.[9] The photos below clearly demonstrate the need for physical activity throughout life.

So you don't have free weights? Don't have any of that fancy exercise equipment advertised on TV? Not to worry. For no-frills weight equipment, simply

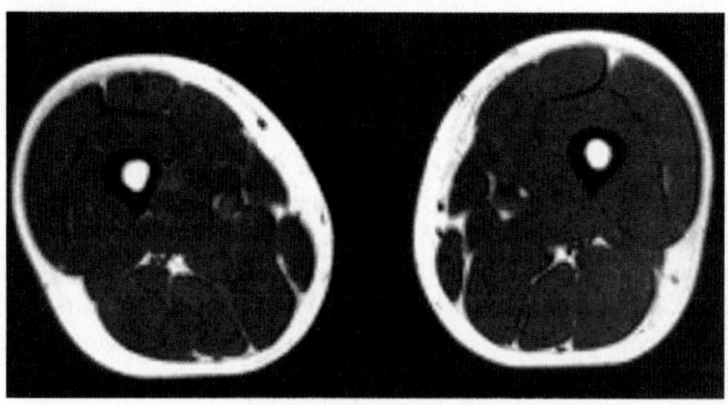
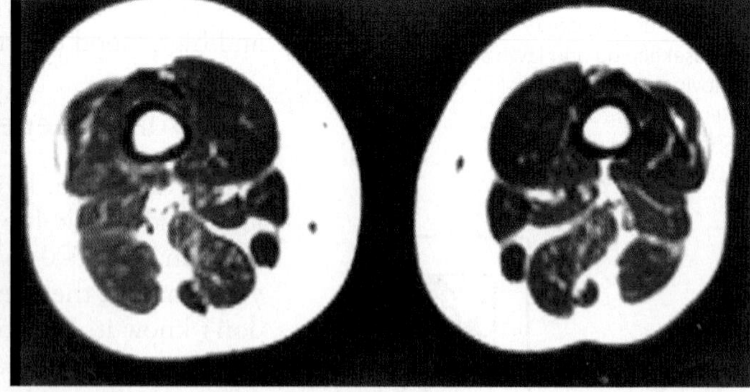

▲ *The photo at the left shows cross-sections of the thighs of an active 20-year-old woman. The small white circle is the femur or thighbone. The photo at the right shows the thigh muscles of an older, sedentary woman in her late 60s. The older woman's thighs have lost muscle and gained fat. Strength training activities such as walking, stair climbing, knee extensions, and leg curls can help prevent as well as reverse at least some of the muscle loss that accompanies aging and a sedentary lifestyle.[10] In other words, such a loss of muscle mass and the accompanying decrease in basal metabolism are not inevitable changes that occur with healthy aging. However, as the photos illustrate, these changes will accompany sedentary aging.*

take two empty one-gallon milk containers and add 1 pint (2 cups) of water (1 pint = 1 pound). Increase the amount of water in the milk containers to increase the amount of weight you lift. For a makeshift barbell, hang the milk containers on each end of a broomstick. Yes, it's a little hokey. But it's cheap!

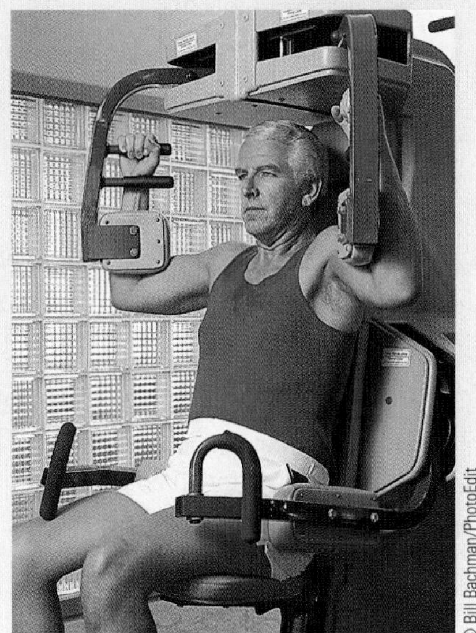

▲ *Strength training promotes healthy bones and muscles at any age.*

Get Off Your "Buts"

Perhaps you are thinking, "*But* I can't afford membership to a gym." Or "*But* I don't have any exercise equipment." Or maybe your first thought is, "*But* I go to school full-time and work part-time, and I have to study when I'm not working or going to class, and I have friends and a family. . . . I don't have time to exercise." Don't let those thoughts in! Remember, instead, that you don't need special equipment, special clothes, or membership to a fitness club. All you need to do is what we are (physiologically) supposed to be doing—short bouts of activity throughout the day. For example, walk to your classes instead of driving. When you do have to drive to campus or the mall, park at the far end of the parking lot. Take stairs instead of elevators. Is the weather bad? Go out to the mall and walk. You will be active and will get to window-shop at the same time. Remember, our ancient ancestors did not have exercise equipment or fitness clubs, either. But the required activities of their lives—hunting food, building shelters, chopping wood—kept them alive, strong, and healthy. The need for physical activity to stay alive hasn't changed today.

TABLE 10-2
Health Consequences of Physical Inactivity

Cardiovascular diseases	Coronary artery disease Congestive heart failure Hypertension Stroke Peripheral artery disease
Metabolic diseases	Type 2 diabetes Obesity Low levels of HDL-cholesterol High levels of triglycerides, LDL-cholesterol, and total cholesterol Gallbladder disease
Cancers	Breast cancer Colon cancer Prostate cancer Pancreatic cancer Melanoma
Pulmonary diseases	Asthma
Immune dysfunction	Susceptibility to viral infections
Musculoskeletal disorders	Osteoporosis Physical frailty
Neurological disorders	Cognitive dysfunction

Physical Activity

How physically active are you? For each question answered yes, give yourself the number of points indicated. Then total your points to determine your score.

A. Vigorous Exercise Routines

1. I participate in active recreational sports such as tennis or racquetball for an hour or more:
a. about once a week (2 points)
b. about twice a week (4 points)
c. three times a week (6 points)
d. four times a week (8 points)
e. not at all (0 points)

2. I participate in vigorous fitness activities like aerobic dancing, roller blading, jogging, or swimming (at least 20 minutes each session):
a. about once a week (3 points)
b. about twice a week (6 points)
c. three times a week (9 points)
d. four times a week (12 points)
e. not at all (0 points)

B. Other Exercise Routines

3. At least two times a week, I work out with weights for at least 10 minutes:
a. two sessions a week (2 points)
b. three sessions a week (3 points)
c. four or more sessions a week (4 points)
d. not at all (0 points)

4. At least two times a week, I perform floor workouts (sit-ups, push-ups) for at least 10 minutes:
a. two sessions a week (2 points)
b. three sessions a week (3 points)
c. four or more sessions a week (4 points)
d. not at all (0 points)

5. At least two times a week, I participate in yoga or perform stretching exercises for at least 10 minutes:
a. two sessions a week (2 points)
b. three sessions a week (3 points)
c. four or more sessions a week (4 points)
d. not at all (0 points)

C. Occupation and Daily Activities

6. I walk to and from school, work, and shopping (½ mile or more each way), two or three times a week or more. (1 point)

7. I climb stairs rather than using elevators or escalators, every other day or more. (1 point)

8. My school, job, or household routine involves physical activity that fits the following description:
a. Most of my day is spent in desk work or light physical activity. (0 points)
b. Most of my day is spent in farm activities, moderate physical activity, brisk walking, or comparable activities. (4 points)
c. My typical day includes several hours of heavy physical activity (shoveling, lifting, etc.). (2 points per day)

D. Leisure Activities

9. I do several hours of gardening, lawn work, or similar hobby work each week. (1 point)

10. At least once a week I dance vigorously (folk or line dancing) for an hour or more. (1 point)

11. In season, I play 9 to 18 holes of golf at least once a week, and I do not use a power cart. (2 points)

12. I walk for exercise or recreation:
a. 1–2 hours a week (1 point)
b. 3–4 hours a week (2 points)
c. 5 hours or more a week (3 points)
d. not at all (0 points)

13. In *addition* to the above, I engage in other forms of physical activity:
a. 1–2 hours a week (1 point)
b. 3–4 hours a week (2 points)
c. 5 hours or more a week (3 points)

SCORING

Record your point scores here.

CATEGORY	SCORE
A. Vigorous exercise routines	________
B. Other exercise routines	________
C. Occupation and daily activities	________
D. Leisure activities	________
Total:	________

Evaluation of total score (circle one).

- Inactive (0–5 points).
- Moderately active (6–11 points).
- Active (12–20 points).
- Very active (21 points or over).

If your score categorizes you as inactive or only moderately active, think of activities that you could realistically engage in on a regular basis to raise your score to "active" (12 points).

SOURCE: Adapted with permission of Russell Pate (University of South Carolina, Human Performance Laboratory).

Eat Well Be Well Use It AND Lose It!

The word "routine" almost always accompanies "exercise plan" when trying to enhance health through physical activity. It should be no surprise that positive health effects cannot be realized with inconsistent exercise. Sporadic binges of activity are of little or no benefit and often do more harm than good. But how do you find time and motivation to get fit and stay fit in today's hectic lifestyle? Many people are dying to tell you how, and they claim to have just the plan for you! Look at any late-night infomercial or shopping channel to see the many exercise gurus with videos, equipment, and supplements to allegedly keep you motivated and provide great results in a short amount of time. But just as fad diets come and go, the promised success with these gizmos and gadgets is usually unsustainable, and the equipment that works for one person may not work for another.

What's usually missing from these infomercials is the whole story. Although many products and programs promise to burn loads of fat in minimal time, they typically fail to mention that success with their product *must be* combined with a total fitness routine—which would, then, defeat the purpose of buying their product in the first place! Don't be swayed by commercials that liberally use the phrase "scientific research." If their "scientific research" is not peer reviewed and published in a respected scientific journal, the results are most likely invalid or, at the very least, biased. Those who market these programs and products are skilled with their presentations, and their real goal is to separate you from your hard-earned money. If their product sounds too good to be true, it probably is!

© Esbin-Anderson/The Image Works

© Tim Davis/Photo Researchers, Inc.

▲ *Make exercise a habit. Choose an activity you enjoy.*

Instead, pay attention to the guidelines in this text, do additional reading, consult experts, talk to your physician, and let education and common sense guide your fitness routine. Even though exercise is not always easy, it is worthwhile and can change your life! In your exercise program, consistency is the key, and no amount of gizmos or gadgets can replace that. Here are some tips for sustaining an exercise program:

1. Check with your physician before starting any exercise program or weight loss plan. Your doctor can assess your ability for a given level of activity and refer you to qualified professionals for assistance, such as physical therapists, exercise physiologists, dietitians, certified personal trainers.
2. Find an exercise buddy. You can motivate each other. Use a gradual approach and set realistic goals. Don't jump out of bed tomorrow morning and expect to be able to run 10 miles. Set a realistic smaller first goal and use it as motivation to set your next goal. Experiment a little to find the time that is right for you to exercise. Many people like to exercise in the morning before their day begins. If you try that and it doesn't work, don't give up. You might have more luck with a lunchtime routine or an after-work exercise schedule.
3. Don't overdo it, especially in the beginning! Listen to your body and let it set the pace while still challenging yourself. "No pain, no gain" is a myth that can be dangerous.
4. Don't focus on weight loss. Focus on your new energy level and how much better you feel as your clothes start to fit again!

strategy for dealing with lactic acid buildup is to relax the muscles at every opportunity, allowing the circulating blood to carry it away and bring in oxygen to support aerobic metabolism. Fortunately, lactic acid is not a waste product. The blood delivers it to your liver, where it is converted back into glucose.

Neither the aerobic nor the anaerobic metabolic pathway functions exclusively to supply energy to your body. The two work together, complementing and supporting each other. Keep in mind, however, that carbohydrate is absolutely essential for exercise. Without it, your muscles can't perform. When you exercise aerobically, muscles burn fat and extract energy from carbohydrate more efficiently

cardiovascular conditioning or **training effect** the effect of regular exercise on the cardiovascular system—including improvements in heart, lung, and muscle function and increased blood volume.

in the presence of oxygen, thereby conserving your body's limited store of carbohydrate. Thus, you want to exercise at an intensity that allows your heart and lungs to keep pace with the oxygen needs of your working muscles.

Aerobic Exercise—Exercise for the Heart

To meet your body's increased oxygen needs during aerobic exercise, your heart must pump oxygen-rich blood to muscles at a faster pace than normal. This increased demand on the heart makes the heart stronger and increases its endurance. In addition, aerobic exercise improves the endurance of the lungs and the muscles along the arteries and in the walls of the digestive tract and, of course, the muscles directly involved in the activity. These all-over improvements are called **cardiovascular conditioning** or the **training effect.** In cardiovascular conditioning, the total blood volume increases so that the blood can carry more oxygen. The heart muscle becomes stronger and larger. Therefore, because each beat of the heart pumps more blood, it needs to pump less often. The muscles that work the lungs gain strength and endurance, and breathing becomes more efficient. Circulation through the body's arteries and veins improves. Blood moves easily, and the blood pressure falls. Muscles throughout the body become firmer. Figure 10-2 shows the major relationships among the heart, circulatory system, and lungs.

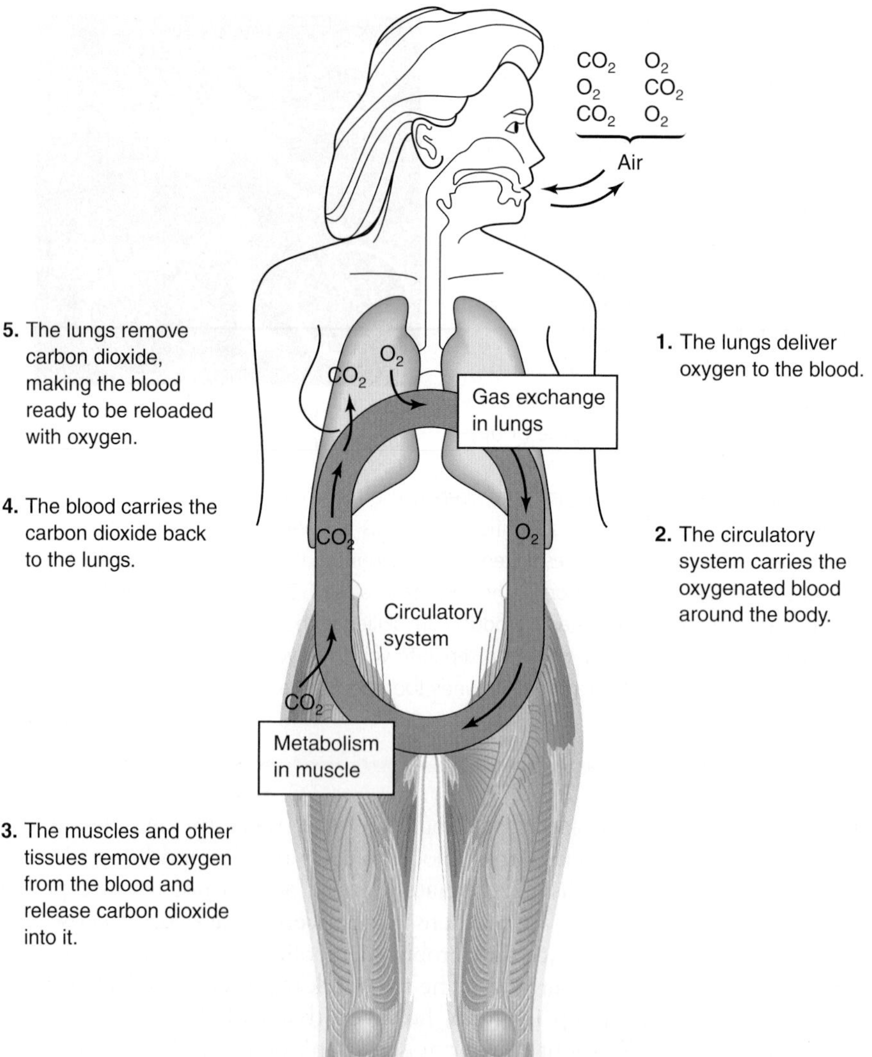

FIGURE 10-2
DELIVERY OF OXYGEN BY THE HEART AND LUNGS TO THE MUSCLES

The more fit a muscle is, the more oxygen it draws from the blood. That oxygen is drawn from the lungs, so the person with more fit muscles extracts more oxygen from the inhaled air than a person with less fit muscles. The cardiovascular system responds to the demand for oxygen by building up its capacity to deliver oxygen. Researchers can measure cardiovascular fitness by measuring the amount of oxygen a person consumes per minute while working out, a measure called *VO_2 max.*

To make these gains in cardiovascular conditioning, you must elevate your heart rate (pulse). This elevated heart rate—called your **target heart rate**—must be considerably faster than the resting rate to push (overload) the heart but not so fast as to strain it. To achieve this goal you must work up to the point at which you can exercise aerobically for at least 20 minutes or longer.

target heart rate the heartbeat rate that will achieve a cardiovascular conditioning effect for a given person—fast enough to push the heart but not so fast as to strain it.

An informal pulse check can give you some indication of how conditioned your heart is to start with. As a rule of thumb, the average resting pulse rate for adults is around 70 beats per minute, but the rate can be higher or lower. Active people may have resting pulse rates of 50 or even lower.

For cardiovascular conditioning, you can calculate your target heart rate using your age as the starting point. The older you are, the lower your maximum target heart rate. As your heart becomes stronger, more intense exercise will be required to reach the same target rate. For example, at first, walking at a pace of 3 miles per hour may cause you to reach your target heart rate. After 6 to 8 weeks of walking at this pace, you may notice you no longer reach your target heart rate. That's because your heart is stronger. It now needs more of a challenge to beat faster. Increasing the intensity of your workout by walking faster can provide this challenge.

To calculate your target heart rate range, take the following steps:

1. *Estimate your maximum heart rate (MHR).* Subtract your age from 220. This provides an estimate of the absolute maximum heart rate possible for a person your age. You should never exercise at this rate, of course.
2. *Determine your target heart rate range.* Multiply your MHR by 55 percent and 90 percent to find your upper and lower limits (see margin example).

When you can work out at your target heart rate for 20 to 60 minutes, you know that you have arrived at your cardiovascular fitness goal.

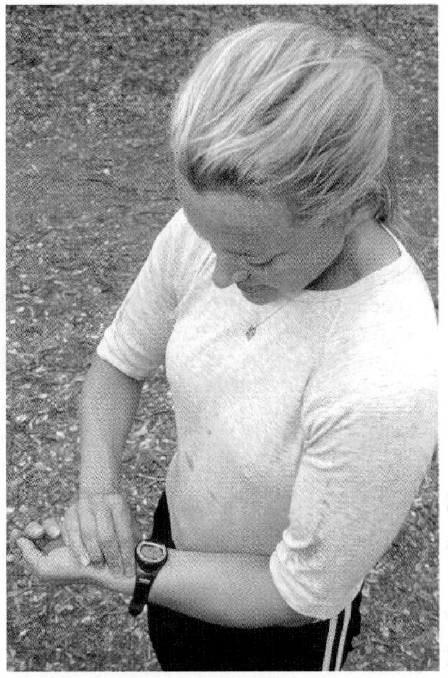

▲ *To take your pulse and monitor your heart rate during exercise, lightly press your middle and index fingers on the radial artery (on the thumb side of the wrist), as shown here. Count your pulse for 10 seconds, and then multiply by 6 to give beats per minute.*

Example: *Jennifer, age 25*
Maximum heart rate: 220 − 25 = 195
Lower limit (55%) of target heart rate range: 0.55 × 195 = 107
Upper limit (90%) of target heart rate range: 0.90 × 195 = 176

Target heart rate range: 107 to 176 beats per minute. Therefore, when Jennifer exercises aerobically, her heart should beat at least as fast as 107 beats per minute but no faster than 176 beats per minute.

Fuels for Exercise

Your energy-producing pathways require oxygen and the two muscle fuels, glucose and fatty acids. As Figure 10-2 shows, the oxygen comes from the lungs, which pass it to the blood, which carries it to the muscles. Your muscles, and to some extent your liver, supply carbohydrate to your muscles from their carbohydrate supply (see Figure 10-3). The fatty acids come mostly from fat inside the muscles but partly from fat that is released from the body's fat stores, and the blood delivers these fatty acids to the muscles.

Glucose Use during Exercise

Glucose comes from carbohydrate-rich foods—breads, pasta, rice, legumes, fruits, vegetables, milk, and yogurt. Your body stores glucose in your liver and muscles in the form of glycogen, a long chain of glucose molecules linked together.

During exercise, the body supplies glucose to the muscles from the stores of glycogen in the liver and in the muscles themselves. The longer the exercise lasts or the more intense it is, the more glucose a person uses. Recall that exercise done at an intensity that outstrips the ability of the heart and lungs to supply oxygen to working muscles relies primarily on glucose for fuel. Thus, activities such as sprinting quickly deplete the body's stores of glycogen. Other activities, such as jogging or brisk walking, in which the body can meet the muscles' oxygen demands, use glycogen more conservatively. Nonetheless, joggers and walkers still use glycogen, and eventually they can run out of it.

When a person begins exercising, for the first 20 minutes or so, about one-fifth of the body's total glycogen store is rapidly used.[11] If exercise continues beyond 20 minutes, glycogen use slows down (see Figure 10-4 on page 329). To conserve the remaining glycogen supply, the body begins to rely more on fat for fuel. At some point, if exercise continues long enough, glycogen will run out

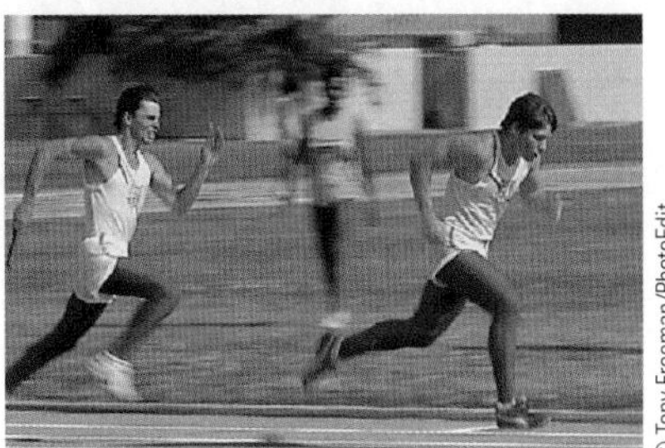

▲ *Anaerobic exercise. Glucose is the principal source of energy for activities of high intensity.*

FIGURE 10-3
THE USE OF GLYCOGEN AND BODY FAT FOR ENERGY DURING EXERCISE
Training can increase the amount of glycogen a muscle can conserve during exercise. The more fit a muscle is, the more fat it can burn for energy when oxygen is present—sparing the valuable glycogen.

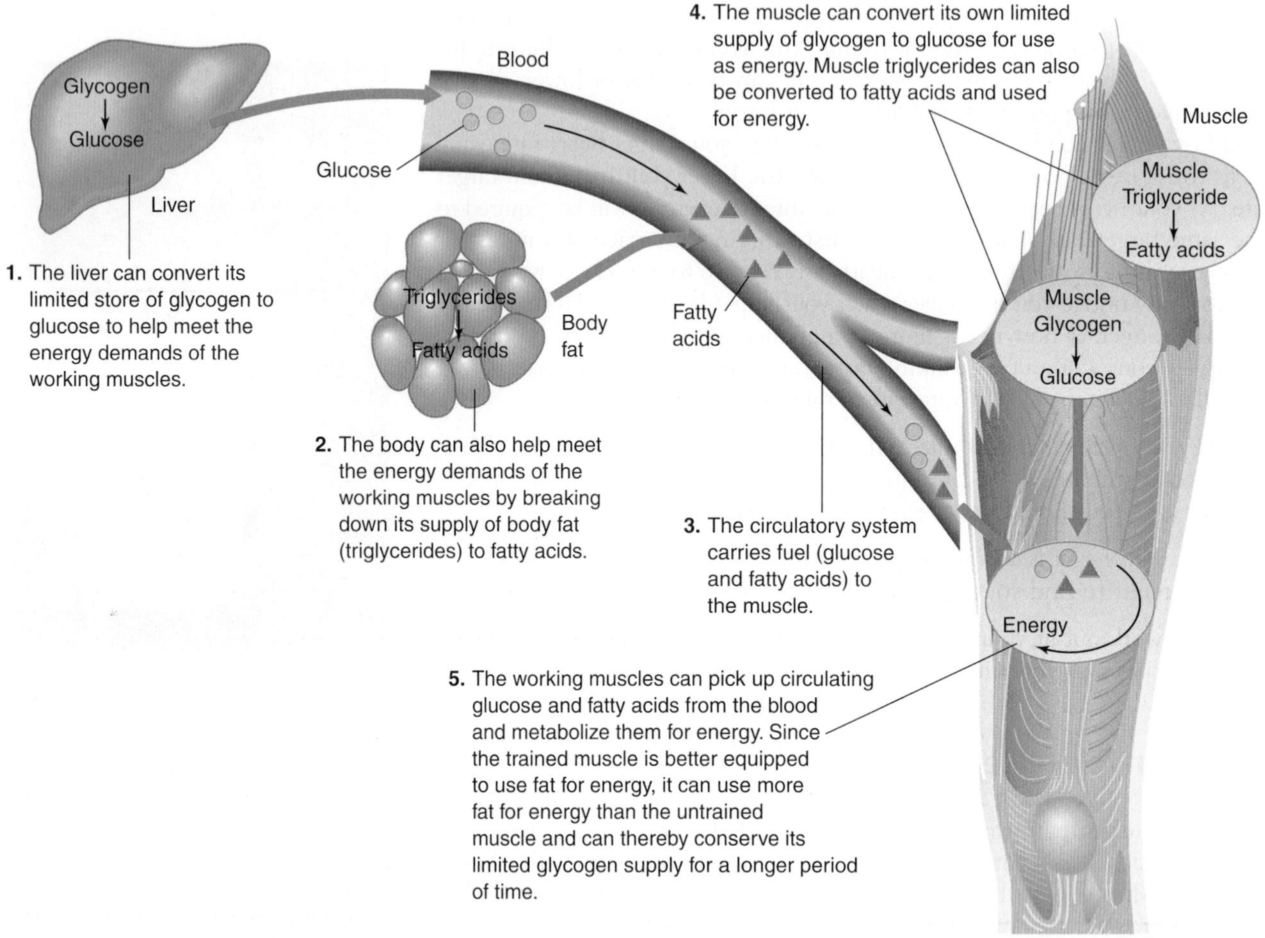

Glycogen → Glucose
Liver

1. The liver can convert its limited store of glycogen to glucose to help meet the energy demands of the working muscles.

Blood
Glucose

Triglycerides → Fatty acids
Body fat

2. The body can also help meet the energy demands of the working muscles by breaking down its supply of body fat (triglycerides) to fatty acids.

Fatty acids

3. The circulatory system carries fuel (glucose and fatty acids) to the muscle.

4. The muscle can convert its own limited supply of glycogen to glucose for use as energy. Muscle triglycerides can also be converted to fatty acids and used for energy.

Muscle
Muscle Triglyceride → Fatty acids
Muscle Glycogen → Glucose
Energy

5. The working muscles can pick up circulating glucose and fatty acids from the blood and metabolize them for energy. Since the trained muscle is better equipped to use fat for energy, it can use more fat for energy than the untrained muscle and can thereby conserve its limited glycogen supply for a longer period of time.

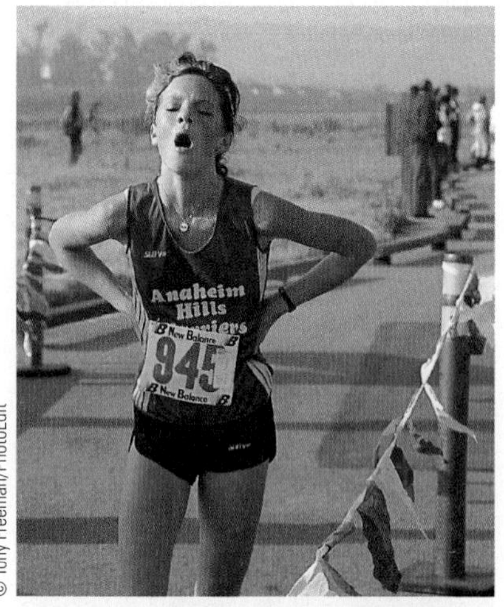

▲ *People who participate in endurance events know to build up their reserves of muscle glycogen before an event, so that they do not run out or "hit the wall" before the finish line.*

almost completely. People who run out of muscle glycogen during an event (for example, before the finish line in a marathon) "hit the wall." They must slow down their pace because muscle glycogen is no longer available as fuel. Exercise can continue for a short time after that, only because the liver scrambles to produce the minimum amount of glucose needed to briefly forestall body shutdown. When blood sugars dip too low, the nervous system function comes almost to a halt, making exercise difficult, if not impossible, even though there is still plenty of fat left to burn.

Another factor that influences glycogen use during exercise is how well trained the person is. In the beginning stages of building an exercise program, a person uses more glucose than a trained athlete. This is because their "untrained" muscles can quickly and easily extract energy from glucose. However, for muscles to easily extract energy from fat requires that the muscle cells contain abundant fat-burning enzymes. With training, the muscles adapt and pack their cells with more fat-burning enzymes. Thus, trained muscles are able to use more fat and conserve their glucose.

The amount of glycogen that is present in the muscles before exercise also influences glycogen use. By following the diet prescription in Figure 10-5 athletes can provide their muscles with enough glycogen to support exercise. In low- to moderate-intensity activities (walking, bicycling, dancing), the casual exerciser rarely depletes the glycogen in their stores. Carbohydrate loading—a practice

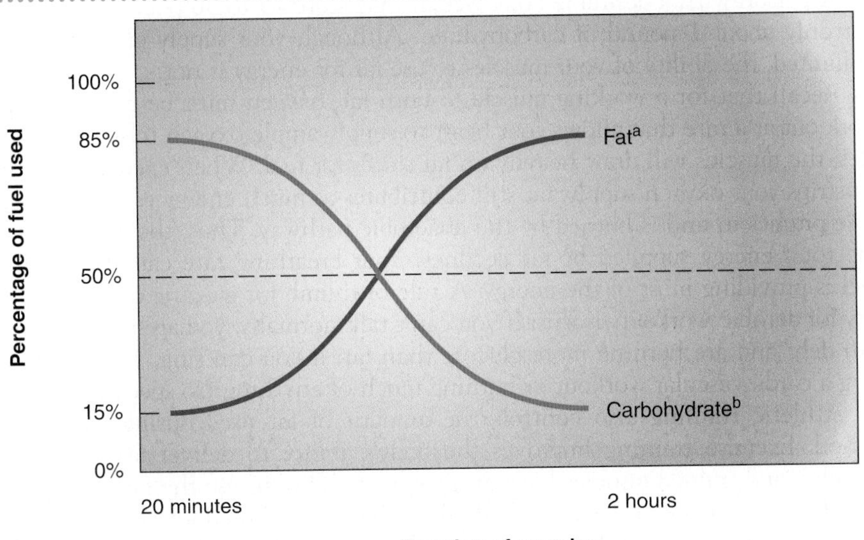

FIGURE 10-4
FUEL USE AND DURATION/INTENSITY OF EXERCISE

For most people, fat isn't used much as a fuel for exercise until you've been working out aerobically for at least 20 minutes, and it is not used as a primary fuel until after 2 hours.

[a] The more moderate the intensity of activity (brisk walking, jogging, aerobic dancing), and the longer the duration, the greater the use of fat for fuel.
[b] The higher the intensity of activity (sprinting, hurdles, rowing), the greater the use of carbohydrate for fuel.

endurance athletes follow to trick their muscles into storing extra glycogen—may not be beneficial for people who exercise less than 90 minutes per workout at a low intensity, although competitive athletes who exercise at a high intensity for more than 90 minutes at a time may benefit from carbohydrate loading. Muscles typically have enough glycogen to fuel 1½- to 2-hour bouts of activity.

An athlete who follows the glycogen-loading technique in preparation for an upcoming event will first exercise intensely without restricting carbohydrates, then gradually cut back on exercise the week before the competition, rest completely the day before, and eat a very high-carbohydrate diet. Endurance athletes who follow this plan can keep going longer than their competitors without ill effects.[12] In a hot climate, extra glycogen offers an additional advantage. As glycogen breaks down, it releases water, which helps to meet the athlete's fluid needs.

Fat Use during Exercise

When you exercise, the fat your muscles burn comes from the fatty deposits all over the body, especially from those with the greatest amounts of fat to spare. That is why physically fit people look trim all over—they reduce their fat stores all over the body, not just those overlaying the working muscles.

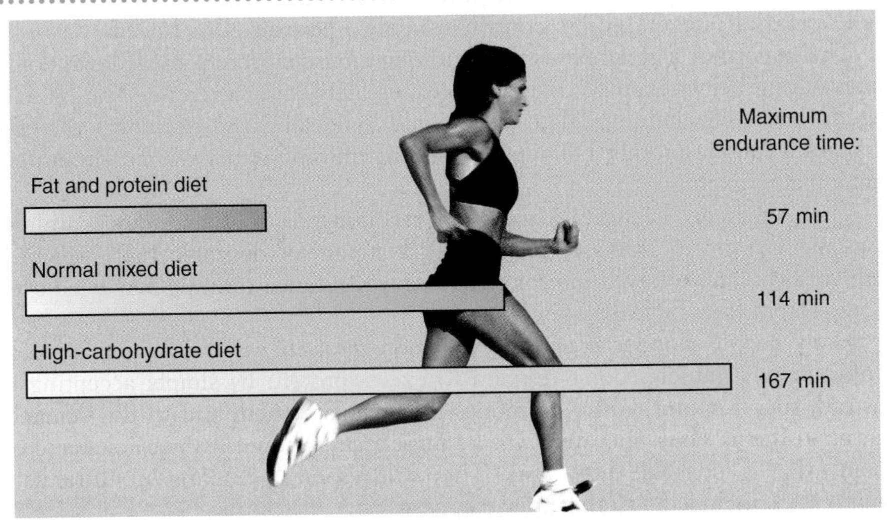

FIGURE 10-5
THE EFFECT OF DIET ON PHYSICAL ENDURANCE

A high-carbohydrate diet can increase an athlete's endurance. In this study, the fat and protein diet provided 94 percent of calories from fat and 6 percent from protein; the normal mixed diet provided 55 percent of calories from carbohydrate; and the high-carbohydrate diet provided 83 percent of calories from carbohydrate.

Photo: © Cory Sorensen/CORBIS

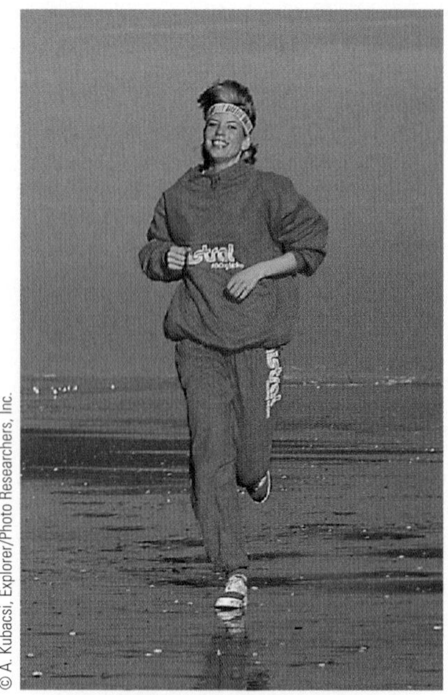

▲ *Aerobic exercise. Fats are the main source of energy for activities of low to moderate intensity.*

A person with a desirable body weight may store 25 to 30 pounds of body fat but only about 1 pound of carbohydrate. Although your supply of fat is almost unlimited, the ability of your muscles to use fat for energy is not.

Recall that for a working muscle to burn fat, oxygen must be present. If you work out at a rate that allows your heart to supply ample oxygen to working muscles, the muscles will draw heavily on fat stores for fuel. When exercise intensity outstrips your oxygen supply, fat still contributes as much energy as ever, but glucose pitches in and is burned by the anaerobic pathway. Thus, the percentage of the total energy supplied by fat declines. Your breathing rate can signal which fuel is providing most of the energy. A rule of thumb for gauging exercise intensity for aerobic workouts is this: If you can't talk normally, you are incurring oxygen debt and are burning more glucose than fat; if you can sing, you aren't getting a cardiovascular workout or burning much of anything (so speed up).

Athletic training also controls the amount of fat used during an exercise period. Exercise training improves the body's ability to deliver fat to working muscles, and trained muscles have an increased ability to use the fat.

Much attention has been focused on the type of fuel used for varying exercise intensities and duration. Research shows that when athletes exercise at a moderate intensity, they initially use more carbohydrate than fat for fuel.[13] Gradually, as exercise continues for more than 20 minutes, the fuel ratio shifts, and the athletes use more fat. For athletes participating in endurance sports, such as marathon runners and long-distance cyclists, who want to conserve their limited supply of carbohydrate, switching to a fat-burning energy system is crucial.

Protein Needs for Fitness

Fit people have more muscle than fat; exercise involves muscles; muscles are made largely of protein. It seems logical, then, that to become or stay fit, an athlete might need more protein. Although it's true that fat and glucose are the primary fuels for working muscles, 5 to 10 percent of energy needs for weightlifters and endurance-sport athletes comes from muscle protein. So, do athletes need more protein?

The athlete's body may use slightly more protein, especially during the early stages of training. Initial increases in muscle mass, numbers of red blood cells to carry oxygen, and amounts of aerobic enzymes in muscles to use fuel efficiently may elevate an athlete's protein needs. In addition, hormonal changes during exercise can temporarily slow the amount of protein the muscle makes and can encourage the muscle to break down its protein stores.[14] How much protein an athlete uses for fuel during hard exercise (endurance exercise and heavy weightlifting) depends on exercise intensity and duration, the athlete's fitness level, and the glycogen stores in the athlete's muscles. When glycogen stores are well stocked, however, protein contributes only 5 percent of fuel needs.

The important factor here is that, although muscle protein breakdown dominates during heavy exercise, muscle *growth* escalates after exercise. The muscles use the available amino acids to repair and build, and the net effect of these changes is muscle protein buildup. Consistent training enhances muscle protein buildup after exercise.

The American Dietetic Association recommends that endurance athletes consume 1.2 to 1.6 grams of protein per kilogram of desirable body weight.[15] Athletes who are involved in prolonged heavy resistance training may need even more protein (see Table 10-3).

Many people wonder if eating even more protein will help build muscles. Unfortunately, muscles don't respond to excess protein by simply accepting it. Instead, they respond to the hormones that regulate them and to the demands put upon them. Thus, the way to make muscle cells grow is to put a demand on them—that is, to make them work. They will respond by taking up nutrients—amino acids included—so that they can grow.

TABLE 10-3
Protein Recommendations for Athletes

	Recommendations (g/kg/day)
RDA for adults	0.8
Endurance athletes	1.2–1.6
Resistance training (bodybuilders, strength athletes)	1.6–1.7

SOURCE: Position of the American Dietetic Association, Dietitians of Canada, and the American College of Sports Medicine: Nutrition and athletic performance, *Journal of the American Dietetic Association* 100 (2002): 1543–1556.

Fluid Needs and Exercise

Replenishing fluid lost during exercise is easily accomplished. Yet many athletes and fitness enthusiasts either don't drink enough or don't drink at all.[16] Ignoring body fluid needs can hinder performance and increase risk of heat-related injury.[17]

The amount of fluid needed to prevent dehydration and **heat stroke** can be surprising. Athletes, who can lose 2 or more quarts of fluid during every hour of heavy exercise, must rehydrate *before*, *during*, and *after* exercise to replace it. Even casual exercisers must drink some fluids while exercising. Thirst is unreliable as an indicator of how much to drink—it signals too late, after fluid stores are depleted.[18] Table 10-4 presents one schedule of hydration. To know how much water is needed to replenish fluid losses after a workout, weigh yourself before and after—the difference is all water. One pound equals roughly 2 cups of fluid.

Water and Fluid Replacement Drinks

For fitness enthusiasts, the choice between water and a sports drink is primarily a matter of personal taste and desired performance abilities. But for endurance events (continuous exercise for longer than 60 minutes), mounting evidence indicates that consuming a properly balanced sports drink during exercise will enhance energy status and endurance and help maintain plasma volume levels better than drinking water does.[19]

How the body manages water and carbohydrate use during exercise determines how well it performs. Sports drinks are designed to enhance the body's use of carbohydrate and water (see Table 10-5). The carbohydrate in a sports beverage serves three purposes during exercise: (1) It becomes an energy source for working muscles, (2) it helps maintain blood glucose at an optimum level, and (3) it helps increase the rate of water absorption from the small intestine, helping better maintain plasma volume. In addition, the drink can supply water and minerals lost from sweating.[20]

There are many factors to consider when choosing a sports drink. The ideal beverage should leave the digestive tract rapidly and enter circulation, where it is needed. Carbohydrate solutions don't all empty from the stomach at the same rate. The drink should contain at least 4 percent but no more than 8 percent carbohydrate by volume. Drinks containing more than 10 percent carbohydrate, such as sodas, fruit juice, Kool-Aid types of drinks, and some sports drinks, take longer to absorb. Some can cause cramps, nausea, bloating, and diarrhea. Drinks

© Michael Newman/PhotoEdit

▲ *Sports drinks can enhance fluid and energy status during endurance events.*

TABLE 10-4
Schedule of Hydration Before, During, and After Exercise*

When to Drink	Amount of Fluid
2–3 hours before exercise	About 2–3 cups
Every 15 minutes during exercise beginning at the start of exercise	6–12 oz
After exercise	2 cups fluid for each pound of body weight lost

*These guidelines are for exercise lasting less than 1 hour. During intense exercise lasting more than 1 hour, the consumption of approximately 1 liter of sports drink per hour (containing 4–8 percent carbohydrate per liter) is recommended to maintain oxidation of carbohydrates and delay fatigue.

SOURCE: Adapted from American College of Sports Medicine, American Dietetic Association, and Dietitians of Canada. Position stand on nutrition and athletic performance, *Medicine and Science in Sports and Exercise* 32 (2000): 2130–2145.

heat stroke an acute and dangerous reaction to heat buildup in the body, requiring emergency medical attention; also called *sun stroke.* (See page 215 for ways to recognize and prevent heat stroke.)

© David Young-Wolff/PhotoEdit

▲ *Plan to drink fluids before, during, and after exercise.*

TABLE 10-5
Fluid Replacement Drinks

Sports Drink	Calories/Cup	Carbohydrate Percentage	Sodium (mg)
All Sport	70	9.0	55
Gatorade Thirst Quencher	50	6.0	110
Isostar	70	8.0	150
Met-Rx	75	8.0	125
Accelerade	60	7.0	127
PowerAde	72	8.0	55

with less than 4 percent carbohydrate may not offer an endurance-enhancing effect. Drinks using a blend of glucose polymers—short chains of carbohydrate—and fructose leave the stomach at the same rate as water, speeding the availability of the carbohydrate and water to working muscles.[21]

Sodium is another important factor to consider when exercising.[22] Because most people consume enough salt in their regular diet to replace the sodium they lose during exercise, it's not essential for a fluid replacement drink to provide large amounts of sodium. In fact, too much sodium can delay muscles' receipt of water.

Research shows that about 50 milligrams of sodium per cup will help stimulate water absorption from your gut. Other studies have found that people who drink a beverage with some sodium tend to drink more of it. If the drink tastes good, athletes and exercisers will want to drink it and thereby meet their fluid needs.

For people who are exercising to lose weight, however, drinking a full quart of a sports drink may in fact simply resupply the amount of calories they expended during a 40-minute aerobic class or 30 minutes of swimming or biking. For these people, plain water is a better choice.

Vitamins and Minerals for Exercise

© David Young-Wolff/PhotoEdit

▲ *Muscles grow in response to work, not to eating protein. Several vitamins regulate the muscle-building pathways.*

Your muscles burn food and oxygen to make energy. How well they burn these fuels depends, however, on your supply of vitamins and minerals. Without small amounts of these potent substances, your muscles' ability to work is compromised.

The Vitamins

Vitamins are the links and regulators of energy-producing and muscle-building pathways. Without them, your muscles' ability to convert food energy to body energy is hindered, and muscle protein formation is slowed. Table 10-6 lists a few vitamins and minerals and their exercise-supporting functions.

The B vitamins are of special interest to athletes and exercisers because they govern the energy-producing reactions of metabolism. Needs for these vitamins increase proportionally with energy expenditure. A person who expends 4,000 calories per day needs twice as much of the B vitamins as someone who expends 2,000 calories. A well-balanced diet that meets athletes' energy needs and that features complex carbohydrate-rich foods will ensure B vitamin intakes proportional to energy intake.

Researchers are presently studying the protective effects of antioxidants on recovery from exercise and performance. Because the body uses oxygen at a

TABLE 10-6
Exercise-Related Functions of Vitamins and Minerals

Vitamin or Mineral	Function
Thiamin, riboflavin, pantothenic acid, niacin, magnesium	Energy-releasing reactions
Vitamin B_6, zinc	Building of muscle protein
Folate, vitamin B_{12}, copper	Building of red blood cells to carry oxygen
Biotin	Fat and glycogen synthesis
Vitamin C	Collagen formation for joint and other tissue integrity; antioxidant ability may reduce oxidative tissue damage
Vitamin E	Protect cell membranes from oxidative damage
Iron	Transport of oxygen in blood and in muscle tissue
Calcium, vitamin D, vitamin A, phosphorus	Building of bone structure; muscle contractions; nerve transmissions
Sodium, potassium, chloride	Maintenance of fluid balance; transmission of nerve impulses for muscle contraction
Chromium	Assistance in insulin's glucose-storage function
Magnesium	Cardiac and other muscle contraction

NOTE: This is just a sampling. All vitamins and minerals play indispensable roles in exercise.

higher rate during exercise, the generation of free radicals and the potential for exercise-induced tissue damage increase in the body. Although more research is needed, preliminary studies support a role for the antioxidant nutrients, which may enhance recovery from exercise by reducing exercise-induced oxidative injury.[23] Meeting the recommendation of eating five or more fruits and vegetables per day will help athletes meet recommended intakes for the antioxidant nutrients.

The Minerals

Iron is a core component of the body's oxygen taxi service: hemoglobin and myoglobin. A lack of oxygen compromises the muscles' ability to perform. Iron deficiency has not been reported as a problem for fitness enthusiasts who exercise moderately.[24] Male and female endurance athletes, though, may be prone to developing mild iron deficiency, diagnosed by low blood ferritin levels, a measure of the body's store of iron. Menstruating female athletes are at particular risk—growth and menstruation combined with strenuous training can take a toll on a woman's iron stores.[25]

A combination of factors increases an athlete's chances of depleting his or her iron stores. Inadequate dietary intakes of iron-rich foods combined with iron losses due to physical activity can compromise iron status. Physical activity may cause increased iron losses in sweat, feces, and urine, plus increased destruction of red blood cells that occurs during exercise. (Chapter 7 contains numerous suggestions for obtaining sufficient iron from foods.)

Sometimes iron deficiencies can be corrected only with iron supplements. If you are concerned about your iron level, see a physician. Iron supplements should not be taken without medical supervision. High iron intakes can induce

sports anemia a temporary condition of low blood hemoglobin level, associated with the early stage of athletic training.

stress fracture bone damage or breakage caused by stress on bone surfaces during exercise.

amenorrhea cessation of menstruation associated with strenuous athletic training

deficiencies of trace minerals, such as copper and zinc, and can produce an iron overload in some people.

An apparent anemia—sometimes called "**sports anemia**"—also can occur in athletes that does not reflect a reduction in the blood's iron supply, but rather indicates an increase in the blood plasma volume.[26] This occurs because athletic training causes the kidneys to conserve sodium and water. In other words, the extra blood volume dilutes the concentration of iron, thereby making it seem as if the blood does not contain enough of the mineral. Sports anemia is considered a temporary state and probably reflects a normal adaptation to physical training.

The Bones and Exercise

Bones absorb great stresses during exercise, and like the muscles, they respond by growing thicker and stronger. Weight-bearing exercises—running, walking, dancing, rope skipping, or activities such as strength training in which significant muscular force can be generated against the long bones of the body—encourage bone development. A bone that is not strong enough to withstand the strain of an athletic exertion can break, causing what has become known as a **stress fracture.** When a person suffers such a break, there are three probable causes. The first is unbalanced muscle development, which allows strong muscles to pull against the bone opposed only by weaker, undeveloped muscles, thereby leaving the bone susceptible to fractures. The second cause of stress fractures is bone weakness caused by inadequate calcium intake. A possible third cause, which occurs in women who have ceased menstruating, is when reduced estrogen concentration leads to bone mineral loss and therefore to fragile bones.

Balanced muscle development can protect the bones from undue stresses. Each set of muscles pulling against bone should be kept in check by an equally strong set of opposing muscles. Thus, when you work one set of muscles in training, you should also work the opposing muscles. For example, if you work your back and leg muscles (by jogging or walking, for example), also work your abdominal muscles, too (do sit-ups). Bones, like muscles, take time to develop strength. To avoid the likelihood of stress fractures, give your bones and muscles plenty of time to build up to one level of performance before moving up to the next level. Maintaining adequate calcium intake throughout life may be one of the primary defenses against developing weak bones.

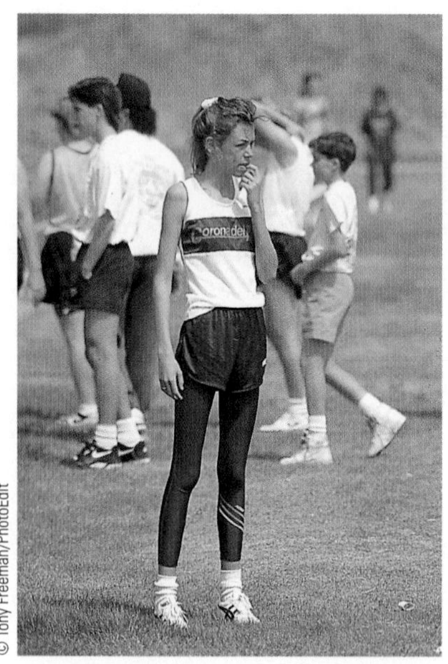

© Tony Freeman/PhotoEdit

▲ *The condition characterized by the potentially fatal combination of disordered eating, amenorrhea, and low bone density is referred to as the* female athlete triad.

Some women who exercise strenuously cease to menstruate, a condition called **amenorrhea.** Such women have lower than normal amounts of estrogen, a hormone essential for maintaining the integrity of the bones. With low estrogen levels, the mineral structures of the bones are rapidly dismantled, weakening the skeleton. Women who have athletic amenorrhea are at risk for stress fractures now and adult bone loss later in life.[27] To reverse the condition, they should not stop exercising altogether because reasonable amounts of exercise may be a key defense against bone depletion. They should, however, seek evaluation from a health care provider who specializes in sports medicine to find the cause of and receive treatment for their amenorrhea.

Eating disorders are sometimes related to athletic amenorrhea, and a logical part of diagnosis is to look carefully at the woman's diet. It could be that a diet too low in calories, coupled with low body fat stores and strenuous exercise, sets the stage for amenorrhea to develop. In such cases, calcium intakes between 1,000 and 1,300 milligrams per day may help protect the bones.

The Savvy Diner Food for Fitness

The best nutrition prescription for peak performance is a well-balanced diet. Although no eating plan meets every athlete's needs, certain components are common to all well-balanced diets. The athlete's diet should account for increased energy needs, vitamin and mineral needs, the relative efficiency of various foods as fuels, and current knowledge about long-term health. An eating plan that supplies 60 percent of calories from complex carbohydrate, 15 percent of calories from protein, and 25 percent of calories from fat supplies a proper fuel mix to the muscles of athletes and fitness enthusiasts and will also maintain health.[28] Two critical nutrition periods for the athlete are the training diet and the pre-competition diet.

Planning the Diet

- A diet rich in complex carbohydrate and low in fat provides the best balance of nutrients for health and also the best support for physical activity. The table shows some sample balanced eating plans for athletes who wish to increase their carbohydrate intake along with their calories.
- Choose foods to provide nutrients as well as calories—extra milk for calcium and riboflavin, many vegetables for B vitamins, meat or alternates for iron and other vitamins and minerals, and whole grains for magnesium and chromium. The photos on page 336 show examples of high-carbohydrate meals for the athlete.
- An athlete may be able to eat more food by consuming it in six or eight meals each day rather than in three or four meals. Large snacks of milkshakes, dried fruits, peanut butter sandwiches, or cheese and crackers can add substantial calories and nutrients.

The Pre-Game Meal

- The best choices for the meal before a competitive event are foods that are high in carbohydrate and low in fat, protein, and fiber. Fat and protein slow the stomach's emptying, and the waste products generated during protein metabolism require that too much water be excreted with them.
- Fiber intake is not desirable immediately before physical exertion because it stays in the digestive tract too long and draws water out of the blood.
- A high-carbohydrate meal supports blood glucose levels during competition. Olympic training tables are laden with foods such as breads, whole-grain cereals, pasta, rice, potatoes, and fruit juices.
- For pre-game meals and snacks, choose foods such as grape juice, apricot nectar, pineapple juice, Jell-O, sherbet, popsicles, raisins, apricots, figs, dates, jams and toast, pancakes with syrup, honey, pasta, baked white or sweet potatoes, steamed vegetables, low-fat frozen yogurt, angel food or sponge cake with fresh fruit, and graham crackers.
- Stay away from higher-fat foods such as meats, cheese, nuts, gravies, cream, French fries, muffins, croissants, biscuits, butter, potato chips, pies, and ice cream.

High-Carbohydrate Eating Patterns for Various Energy Levels

Use the number of servings indicated to arrive at the specified energy levels.[a]

Food Group	Calorie Level 1,500	2,000	2,500	3,000	3,500	4,000
Milk	3	3	4	4	4	4
Fruits	5	6	7	9	10	12
Vegetables	3	3	3	5	6	7
Grains	7	11	16	18	20	24
Oils[b]	2	3	5	6	8	10
Meat[c] & Beans	5	5	5	5	6	6
Percentage of carbohydrate	58	58	63	64	60	62

[a] Refer to Chapter 2, Table 2-6 on page 48 for serving sizes.
[b] A serving of fat is equivalent to 1 tsp butter, margarine, or oil.
[c] Meat servings are given as total ounces of meat; a typical serving includes 2 to 3 ounces.

- Include plenty of fluids—two or more 8-ounce glasses of water or juice per meal—to ensure adequate hydration.
- Any meal should be finished at least 2 to 4 hours before the event because digestion requires routing the blood supply to the digestive tract to pick up nutrients. By the time the contest begins, the circulating blood should be freed from that task and, instead, should be available for carrying oxygen and fuel to the muscles.

Sample Meals for High-Carbohydrate Intakes for the Exercise Enthusiast at Two Calorie Levels

Breakfast
1 c coffee
8 oz low-fat milk
2 pieces whole-wheat toast
4 tsp jelly
½ c strawberries
½ c orange juice
1 c oatmeal and raisins with 2 tsp brown sugar

Morning Snack
4 tsp trail mix

Lunch
12 oz iced tea with sugar
1 orange
1 banana
2 beef and bean burritos

Afternoon Snack
A smoothie made from 12 oz nonfat milk 1 frozen banana
1 apple
4 rye wafers with 1 oz low-fat cheese

Dinner
8 oz low-fat milk
1 c sherbet
1 c spinach salad with 1 tbsp dressing
1 dinner roll with 2 tsp butter
¼ tomato
1 c broccoli
4 oz salmon
¾ c noodles with parsley and 2 tsp butter

Total Calories: 3,119
61% cal from carbohydrates, 24% cal from fat, 15% cal from protein

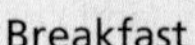

Breakfast
1 c coffee
½ c strawberries
8 oz nonfat milk
1 c oatmeal and raisins

Morning Snack
4 tsp trail mix

Lunch
12 oz iced tea with sugar
1 orange
1 beef and bean burrito

Dinner
½ c sherbet
1 c spinach salad with 1 tbsp dressing
8 oz nonfat milk
¼ tomato
1 c broccoli
½ c noodles with parsley and 2 tsp butter
4 oz salmon

Total Calories: 1,759
57% cal from carbohydrates, 24% cal from fat, 19% cal from protein

Spotlight Athletes and Supplements—Help or Hype?

Competitors in the ancient Greek Olympiad reportedly used mushrooms and herbs.[29] Since then, virtually every food has at one time or another been touted as the "magic bullet" that will enhance performance. Athletes have been known to swallow everything from bee pollen to brewer's yeast to kelp to wheat germ in their quest to gain the competitive edge. Although the idea of using pills and potions to achieve peak performance (to run faster or jump higher) may be seductive, scientific evidence supporting such claims is sorely lacking.[30]

Can nutritional supplements enhance the benefits I achieve from my everyday workouts?

Most so-called **ergogenic aids**—that is, substances that increase the ability to exercise harder—are costly versions of vitamins, minerals, sugar, and other substances that can be provided easily by a balanced diet.[31] Table 10-7 describes some of the many substances currently promoted as ergogenic aids.

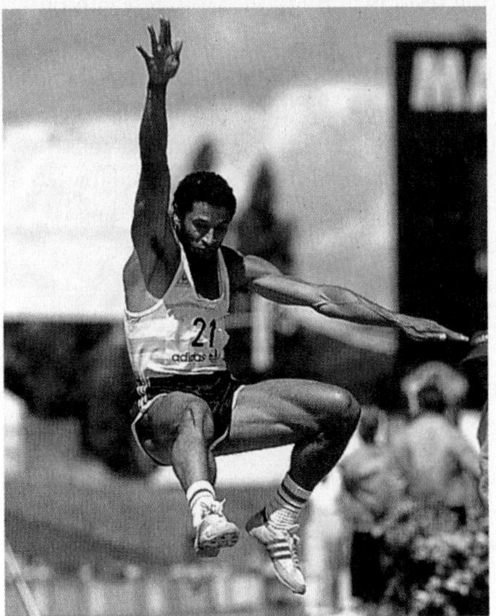

▲ *Athletes rank as prime targets of amino acid supplement manufacturers, whose products promote false hopes.*

© Felicia Martinez/PhotoEdit

For example, bee pollen is simply a mixture of protein, carbohydrate, a bit of fat, and a few vitamins and minerals.[32] Though touted by one manufacturer as a "natural and balanced source of extra energy" appreciated by "athletes worldwide," it has been tested at Louisiana State University among both runners and swimmers and found to confer no benefit whatsoever on an athlete's training or performance abilities.[33] The same goes for chromium picolinate, promoted by some to increase lean body mass and delay fatigue due to its role in glucose utilization. Although chromium is necessary for muscle function by transferring glucose from the blood to the muscle cells, true chromium deficiencies are rare to nonexistent. To date, there is no evidence that chromium supplements improve athletic performance in healthy individuals who do not have a chromium deficiency.[34]

Surveys have found that 53–80 percent of athletes use a vitamin or mineral supplement, although no evidence exists that doing so improves performance.[35] More than 40 years of research has provided no strong evidence that popping vitamins and minerals increases energy or athletic prowess of adequately nourished people. Except for iron, vitamin and mineral deficiencies are rare among athletes. Because most athletes eat more food than non-athletes eat to meet their increased energy demands, the extra food intake usually provides the additional vitamins, minerals, and other beneficial substances they need, if they choose a well-balanced diet.

Of course, when athletes firmly hold that one or another ergogenic aid does indeed improve performance, convincing them otherwise can be extremely difficult. One reason for this is their profound belief that a particular substance can actually produce a psychological benefit. This effect is known as the **placebo effect.**

> **MINIGLOSSARY**
>
> **ergogenic aids** anything that helps to increase the capacity to work or exercise.
>
> *ergo* = work
> *genic* = give rise to
>
> **placebo effect** an improvement in a person's sense of well-being or physical health in response to the use of a placebo (a substance having no medicinal properties or medicinal effects).
>
> **anabolic steroids** synthetic male hormones with a chemical structure similar to that of cholesterol; such hormones have wide-ranging effects on body functioning.

I see many claims about amino acids and athletic ability. Can I improve my performance by using amino acid supplements?

Athletes rank as prime targets for amino acid supplement manufacturers. Pick up any copy of one of the bodybuilding magazines, and you'll probably see ads for supplements packed with "free-form," "predigested," and "peptide-bond" amino acids touted as optimum sources of protein for athletes. Scientific-sounding names notwithstanding, such products have never been proven to increase muscle size or enhance athletic prowess. Consider that one comparison of U.S. Marine officer candidates given protein supplements with another set of trainees who received a placebo indicated that the groups

TABLE 10-7
A Sampling of Popular Ergogenic Aids [a,b]

Ergogenic Aids with Unproven Claims

Amino acids (for example, arginine, ornithine, glycine) nonessential amino acids falsely promoted to increase muscle mass and strength by stimulating growth hormone and insulin. Individual amino acids do not significantly increase muscle mass or growth hormone. Weight lifting and endurance training do.

Anabolic steroids synthetic male hormones (related to testosterone) that stimulate growth of body tissues, with many adverse effects as listed in the footnote on page 339.

Bee pollen mixture of bee saliva, plant nectar, and pollen touted falsely to enhance athletic performance. May cause allergic reactions in people with a sensitivity to bee stings and honey allergies.

Carnitine a compound synthesized in the body from two amino acids (lysine and methionine) and required in fat metabolism. Falsely claimed to increase the use of fatty acids and spare glycogen during exercise, delay fatigue, and decrease body fat. The body produces sufficient amounts on its own. No evidence that supplementation in healthy people improves energy or enhances fat loss.

Chromium picolinate Chromium is an essential component of the glucose tolerance factor, which facilitates the action of insulin in the body. Picolinate is a natural derivative of the amino acid tryptophan. Falsely promoted to increase muscle mass, decrease body fat, enhance energy, and promote weight loss. Choose instead, a diet rich in whole, unprocessed foods.

Coenzyme Q10 a lipid made by the body and used by cells in energy metabolism; falsely touted to increase exercise performance and stamina in athletes; potential antioxidant role. May increase oxygen use and stamina in heart disease patients, but no significant effect is seen in healthy athletes.

DHEA (Dehydroepiandrosterone) a precursor of the hormones testosterone and estrogen; falsely promoted to increase production of testosterone, build muscle, burn fat, and delay the effects of aging. Long-term effects unknown; self-supplementation not recommended.

Ginseng a collective term used to describe several species of plants, belonging to the genus *Panax,* containing bioactive compounds in their roots. Falsely touted to enhance exercise endurance and boost energy. A lack of well-controlled research has yielded inconclusive evidence for the benefits of ginseng. The potential for adverse drug–herb or herb–herb interactions with ginseng exists.[a]

Pyruvate a three-carbon compound derived from the breakdown of glucose for energy in the body. Falsely promoted to increase fat burning and endurance. Side effects include intestinal gas and diarrhea, which could interfere with performance.

Ergogenic Aids with Some (Not All) Scientific Support for Claims

Creatine A nitrogen-containing substance made by the body that combines with phosphate to form the high-energy compound creatine phosphate (CP). CP is stored and used by muscle for ATP production. Some (not all) studies show that creatine may increase CP content in muscles and improve short-term (<30 seconds) strenuous exercise performance (for instance, sprinting, weight lifting). Long-term effects are unknown.

Caffeine a stimulant that increases blood levels of epinephrine. Caffeine is promoted for improved endurance and utilization of fatty acids during exercise. Consuming 2 to 3 cups of coffee (equal to 3 to 6 milligrams of caffeine per kilogram body weight) 1 hour before exercise may improve endurance performance. High caffeine consumption may cause dehydration, headache, nausea, muscle tremors, and fast heart rate.

HMB (beta-hydroxy-beta-methylbutyrate) a metabolite of the branched-chain amino acid leucine that is promoted to increase muscle mass and strength by preventing muscle damage or speeding up muscle repair during resistance training. More research on long-term safety and effectiveness is needed.

Phosphate salt a salt with claims for improved endurance. Found to increase a substance in red blood cells (diphosphoglycerate) and enhance the cell's ability to deliver oxygen to muscle cells and reduce levels of disabling lactic acid in elite athletes. However, more research on safety and efficacy of phosphate loading is needed. Excess can cause loss of bone calcium.

Sodium bicarbonate baking soda is touted to buffer lactic acid in the body and thereby reduce pain and improve maximal-level anaerobic performance. May cause diarrhea in users due to the high sodium load. Effects of repeated ingestions are unknown.

[a] See also Table 6-15 on page 200 for information on herbal supplements.
[b] For more information, see E. Coleman, *Eating for Endurance* (Palo Alto, CA: Bull, 1997); S. A. Sarubin, *The Health Professional's Guide to Popular Dietary Supplements*, 2nd ed. (Chicago: The American Dietetic Association, 2002).

© Richard Anderson

▲ *An endless array of ergogenic aids are marketed to athletes and other sports enthusiasts. Although big on claims, few are based on scientific evidence.*

performed equally well before, during, and after the program.[36] (Refer to the Spotlight feature in Chapter 1 for some tips on how to spot fraudulent nutritional products.)

An additional important fact is that your body can't store extra amino acids, whether they come from food you eat or from supplements. Your body converts the excess into fat. This conversion generates urea, which increases your body's need for water. Increased urination of urea can lead to dehydration, which impedes training and performance.

What risks are associated with amino acid supplements?

The FDA recently asked a panel of experts to review the safety of amino acids currently available in the marketplace. The panel's conclusions underscore the need for better regulation of these supplements. For instance, the panel found that the labels of most amino acid supplements failed to carry vital information including suggested doses, shelf life, and contraindications for use of the product. In addition, the panel identified certain groups of people who may be at particularly high risk for suffering health problems as a result of swallowing amino acid supplements. Children and teenagers, for example, may not grow properly if they take amino acid pills or powders. That's because young, underdeveloped bodies may metabolize amino acids differently than adult bodies, possibly leaving young people more vulnerable to harmful effects of excess amino acids.

The panel also found that much of the advertising and product label information about the effectiveness of amino acid supplements is based on anecdotes rather than grounded in careful, scientific research. In fact, the lack of data to support the usefulness of many amino acid supplements has been a concern for years.

Health risks aside, consumers should also note that special protein supplements are typically very expensive. Foods, on the other hand, supply ample amounts of protein at a fraction of the cost. One glass of milk, a serving of rice and beans, or a 3-ounce portion of chicken, for example, provides a generous helping of all nine essential amino acids for less than half the price of a dose of most amino acid tablets, liquids, or powders.

What about arginine and other amino acid products advertised for weight control? Do these products work?

Arginine is an amino acid that has been promoted as "causing weight loss overnight" by stimulating secretion of a substance called *human growth hormone,* which in turn supposedly promotes weight loss. Although it's true that arginine can prompt the release of the hormone, it will do so only when people take whopping doses that are unlikely to be found in supplements. And, even if a person were to take enough arginine to prompt a surge of the hormone into the body, he or she wouldn't automatically shed pounds. Human growth hormone has not been found to cause weight loss. Thus, claims that arginine "burns fat" are spurious, at best. An FDA advisory panel on over-the-counter weight loss products investigated arginine along with 11 other amino acids touted as diet aids—namely, cystine, histidine, isoleucine, leucine, L-lysine, methionine, phenylalanine, threonine, tryptophan, tyrosine, and valine—and found no basis for the claims about the effectiveness of any of these products in controlling weight.

Can anabolic steroids help me increase the size and strength of my muscles?

Psychological impact aside, pill popping may be harmful in some cases. Swallowing supplements known as **anabolic steroids**—synthetic hormones that appear to help build muscle—can be dangerous.

A particularly popular practice among weight lifters and bodybuilders, steroid abuse often begins around the age of 18 years.[37] Although steroids may help to increase muscular size and strength in some people, they can also bring about numerous side effects, including acne, liver abnormalities, temporary infertility, and offensive outbursts, often referred to as "roid rages."*

*Side effects of steroids include acne, anxiety, blood clots, blood poisoning, cancer, diarrhea, dizziness, fatigue, heart disease, hypertension, irreversible baldness in women, jaundice, kidney damage, liver damage, male pattern baldness, mood swings, nausea, oily skin, prostate enlargement, psychotic depression, shrunken testicles, sterility (reversible), stroke, stunted growth in adolescents, swelling of feet or lower legs, and yellowing of the eyes or skin.

Among adults, many effects of steroid use are reversible. Unfortunately, adolescents aren't so lucky. Several studies show that adolescent steroid users may suffer the serious consequences of premature skeletal maturation, decreased spermatogenesis, and elevated risk of injury.[38]

Along with being unhealthful, steroid use is considered unethical by domestic and international sports organizations such as the American College of Sports Medicine and the International Olympic Committee. As a case in point, track star Ben Johnson lost his Olympic gold medal for the 100-meter sprint in 1988 after officials discovered he had been using steroids.[39]

What about energy drinks? They seem perfect for athletic performance!

Carbohydrates, vitamins, minerals, and hydration all in one bottle . . . sounds too good to be true! Well, for the athlete using "energy drinks" for an extra edge, it probably is.[40] For the past ten years energy drinks have increased in variety and market share of beverage sales across the United States. Their formulation of high amounts of caffeine, simple carbohydrates, and mixtures of vitamins and minerals has caused many people to use them as a "pick me up" to get through their day. In particular, manufacturers target athletes because of their desire to gain an advantage over the competition. Unfortunately, however, energy drinks are not properly formulated to increase athletic performance.

Caffeine. Thought of as the central ingredient in almost all "energy drinks," caffeine is a central nervous system stimulant that has been shown to increase athletic ability slightly when consumed in moderate amounts. However, higher doses are considered "doping" in most athletic organizations, and they cause unwanted side effects such as light-headedness and twitchiness along with laxative and diuretic effects. Even though energy drinks do not all have exactly the same ingredients, nearly all contain high concentrations of caffeine.[41]

Carbohydrates. Although carbohydrates are the preferred source of fuel for athletes, high levels of carbohydrates in an energy drink pose undesirable side effects. One problem is that absorption of water is slowed considerably, putting the athlete's hydration status at risk. Furthermore, like caffeine, simple carbohydrate ingestion immediately before physical exertion might cause gastrointestinal distress and create a laxative effect.

Vitamins and minerals. There is little chance an athlete who consumes a normal diet would be at risk for vitamin or mineral deficiency. Accordingly, adding nutrients to beverages has never been proven to markedly increase athletic performance. A daily multivitamin and a balanced diet are far less costly than relying on fortified energy drinks as a source of vitamins and minerals.

Other ingredients. It is important to realize that these drinks sometimes include additional ingredients such as amino acids and herbal supplements. Even though they may be catchy items to list on the label, few, if any, have been proven to increase athletic performance. In fact, herbal ingredients may increase the risk of serious drug–nutrient interactions if the athlete is also taking medication.[42]

What questions should I ask if I'm considering consuming energy drinks?[43]

- Does the drink contain herbal ingredients that will affect a medication I am currently taking?
- Does any research back up proposed benefits of this beverage?
- Does the beverage label include a nutrition facts panel? If not, avoid this product!
- Can I do anything else to positively affect my energy level when training? (For example, am I eating a balanced diet, getting enough rest and hydration, and training for the proper length of time?)
- Is the cost of the product really worth what I am getting?
- How high is the caffeine level? Will it cause me to fail a doping test?

Some athletes use caffeine or alcohol instead of popping pills to promote athletic prowess. Can these substances improve my athletic performance?

Many exercisers drink caffeine-containing beverages such as coffee, tea, or cola to enhance performance and endurance. Caffeine apparently stimulates the release of fats into the blood that the body can then use instead of glycogen as a source of energy. Thus, the glycogen is "spared," or saved, for later use, and the amount of time an exerciser can endure physical activity before running out of fuel is prolonged.

The glycogen-sparing effect of caffeine, however, is beneficial only for athletes who exercise for more than 1½ to 2 hours at a time. As we said earlier, the muscles generally store enough glycogen to fuel as many as 90 minutes of activity. Moreover, even endurance athletes can experience certain downsides to consuming caffeine. Because caffeine is a diuretic, it promotes frequent urination and fluid loss that can lead to dehydration. In addition, caffeine can induce rapid heart rate and jitters, which can interfere with performance. Athletes would also do well to remember that caffeine is a drug that neither the American College of Sports Medicine nor the International Olympic Committee condones for use among athletes.

Along with caffeinated beverages, alcoholic drinks are often touted as choice fluids for athletes. Beer, for example, is sometimes portrayed as the perfect carbohydrate-containing complement to both before- and after-competition meals. Despite such images, alcoholic drinks rank as poor sources of fluid and energy for several reasons. For one, alcohol is a diuretic that can bring about fluid loss and dehydration. More importantly, the amount of alcohol in just one beer or glass of wine depresses the nervous system, thereby slowing an athlete's reaction time and interfering with reflexes and coordination. Also, one can of beer provides only 50 carbohydrate calories. The rest of the calories come from alcohol, which must be metabolized by your liver, not your muscles. The American College of Sports Medicine and the American Dietetic Association both conclude that use of alcohol hinders performance. (Chapter 8 presents a detailed explanation of how alcohol affects the body.)

The special supplements discussed here are just a sampling of the many "magic" pills and potions promoted to athletes. A nutritious diet and regular physical activity enhance performance far better than these products for gaining the competitive edge. If you're in doubt about a particular product you see boasted as an ergogenic aid, ask yourself some of the following questions:

1. Is the promised action of the product based on magical thinking? ("Develop a trim body with no exercise.")
2. Does the promotion claim that "doctors agree" or "research has determined," without clarification? (Which doctors? What research?)
3. Does the promoter use scare tactics to pressure you into buying the product? ("It's the only one available without poisons.")
4. Is the product advertised as having a multitude of different beneficial effects? ("Makes bigger muscles; gives that pumped-up feeling; improves digestion, coordination, and breathing.")
5. Is the product available only from the sponsor by mail order and with payment in advance?
6. Does the promoter use many case histories or testimonials from grateful users?

Every yes answer is a point against the claimant—a warning signal that you are dealing with misinformation. Three or more points is a sure sign of quackery.

In Review

1. The components of fitness include:
   a. flexibility
   b. strength
   c. muscle endurance
   d. cardiovascular endurance
   e. all of the above

2. __________ is the ability of the muscles to work against resistance.
   a. Flexibility
   b. Physical condition
   c. Strength
   d. Endurance

3. Which of the following is true of the heart?
   a. It cannot be affected by exercise.
   b. It is a muscle that can become larger and stronger.
   c. Lifting weights and other strength building exercises are the best activities for heart health.
   d. none of the above

4. Differentiate between anaerobic and aerobic exercise, giving examples of each.

5. Which nutrients are used for physical activity?
   a. proteins
   b. carbohydrates
   c. fats
   d. All of the above have specific roles.

6. Athletes—bodybuilders in particular—often take protein supplements. Briefly explain the role of protein in fitness and how much protein intake is recommended.

7. When involved in physical activity, fluids like sports drinks and water should be consumed:
   a. never, soda and juices are better.
   b. before, during, and after exercise.
   c. in low amounts during activity.
   d. only if high amounts of sodium are added to them.

8. Jen is the star of her high school cross-country team and competes in 2–3 meets weekly. Although she tries to eat a balanced diet, she often grabs whatever she can on the way to and from events. What deficiency is Jen at risk for and why?

9. The more fit the muscle:
   a. the more oxygen it uses.
   b. the less energy it uses.
   c. the stronger it gets.
   d. Both a and c are correct.
   e. none of the above

10. Who would be most likely to deplete their glycogen levels quickly?
   a. an experienced endurance runner
   b. a person who is sprinting for the first time
   c. an experienced sprinter
   d. a bodybuilder

Menu of Online Study Tools

A variety of study tools for this chapter are available at our website to deepen your understanding of chapter concepts. Go to www.thomsonedu.com/nutrition/boyle to find

- Practice tests
- Flashcards
- Glossary
- Web links
- Animations
- Chapter summaries, learning objectives, and crossword puzzles

Nutrition on the Web

www.thomsonedu.com/nutrition/boyle
Go to the *Personal Nutrition* site to check for the latest updates to chapter topics or to access links to related websites.

www.cdc.gov/nccdphp/sgr/sgr.htm
Look here for the Surgeon General's Report on Physical Activity and Health.

www.fitness.gov
Information from the President's Council on Fitness and Sports.

www.sportsci.org
Web page of *Sportscience News.*

www.acsm.org
The American College of Sports Medicine website.

www.shapeup.org/fitness.html
Provides practical advice on fitness and nutrition.

www.cdc.gov/nccdphp/dnpa/
Provides links to many nutrition and physical activity sites.

www.scandpg.org
SCAN: Sports, Cardiovascular, and Wellness Nutritionists—a dietetic practice group of the American Dietetic Association.

www.gssiweb.com
The Gatorade Sports Science Institute provides updates on exercise science, dietary supplements, sports drinks, and eating disorders in athletes.

www.nal.usda.gov/fnic/etext/fnic.html
See the "Fitness, Sports, and Sports Nutrition" link for nutrition and exercise information.

www.ncahf.org
The National Council Against Health Fraud website offers current information on ergogenic aids and nutrition fads.

www.quackwatch.org
Quackwatch:Your Guide to Health Fraud and Quackery.

www.ncbi.nlm.nih.gov/PubMed/
A search engine to help you locate information about exercise.

www.acefitness.org
The American Council on Exercise provides fact sheets, an online newsletter, and an information resource center.

www.runnersworld.com
Nutrition and fitness information and online calculators for beginners to elite athletes.

www.kidshealth.org
Information about nutrition and fitness for children and teens.

The Life Cycle: Conception through the Later Years

11

ASK YOURSELF . . .

Which of the following statements about nutrition are true, and which are false? For each false statement, what is true?

1. The poor nutrition of a pregnant woman can impair the health of her grandchild, even after that child has grown up.
2. A woman needs twice as many calories per day in late pregnancy as she did before she was pregnant.
3. Even one alcoholic beverage, if taken at the wrong time during pregnancy, can damage the development of the nervous system in the unborn infant.
4. A woman who craves a food during pregnancy instinctively knows that she needs the nutrients in that food.
5. Substances in a mother's milk can protect the infant against certain diseases to which the mother has been exposed.
6. If a child loses his or her appetite, the caretaker must insist that the child eat his or her meals anyway.
7. School lunches provide all of the nutrients children need in a day.
8. For older adults, age-related weight gain is inevitable.
9. The number of older people is declining.
10. As you grow older, you need more calories, but fewer vitamins and minerals, to stay healthy.

Answers found on the following page.

How far you go in life depends on your being tender with the young, compassionate with the aged, sympathetic with the striving, and tolerant of the weak and the strong. Because someday in life you will have been all of these.

George Washington Carver
(1864–1943, American botanist)

CONTENTS

Nutrition shares with other lifestyle factors the responsibility for maintaining good health. The complete prescription for good health presented in Chapter 1 reads as follows: avoid excess alcohol, don't smoke, maintain a desirable weight, exercise regularly, get regular sleep, and eat nutritious, regular meals. A person who practices good health habits can expect to delay the onset of even minimal disability by several years, compared with a person who practices few or none of them.[1] If you believe in accepting the things you can't control and controlling the things you can, you are in luck. Your nutritional health can be controlled, and it deserves your conscientious attention. This chapter follows people through the life cycle, considering their special nutritional needs at each stage.

Pregnancy: Nutrition for the Future

The only way nutrients can reach the developing fetus in the uterus is through the **placenta,** the special organ that grows inside the uterus to support the new life. If the mother's nutrient stores are inadequate early in pregnancy when the placenta is developing, the fetus will develop poorly, no matter how well the

Nutritional risk factors in pregnancy

- Age 15 or under
- Unwanted pregnancy
- Many pregnancies close together (depletes nutrient stores)
- History of poor pregnancy outcome
- Poverty
- Lack of access to health care
- Low education level
- Inadequate diet (such as that due to food faddism or dieting)
- Iron deficiency anemia early in pregnancy
- Cigarette smoking
- Alcohol or drug abuse
- Chronic disease requiring special diet (for example, diabetes)
- Underweight or overweight
- Insufficient or excessive weight gain in pregnancy
- Carrying twins or triplets

mother eats later. After getting such a poor start on life, a female child may grow up poorly equipped to support a normal pregnancy, and she, too, may bear a poorly developed infant. Thus, a woman who has poor nutrition habits during her early pregnancy can even impair the health of her *grandchild*.

Infants born to malnourished mothers are more likely to become ill, to have birth defects, and to suffer retarded mental or physical development than infants who are born to healthy women. Malnutrition in the **prenatal** and early **postnatal** periods also affects learning ability and behavior. According to the fetal origins hypothesis, if a woman's nutrient intake is under- or oversupplied—particularly at critical phases of fetal development—long-term alterations in tissue function may occur. For example, if a woman's energy intake is low during the third trimester of pregnancy, pancreatic cell development may be hindered, resulting in impaired glucose tolerance and increased risk of developing diabetes later in life.[2] Clearly, it is critical to provide the best nutrition at the early stages of life. Ideally, a woman will start pregnancy at a healthful weight, with filled nutrient stores, and with the firmly established habit of eating a balanced and varied diet. The Pregnancy Readiness Scorecard on page 349 helps women evaluate their nutritional readiness for pregnancy and identify which of their eating habits might need improvement.

Nutritional Needs of Pregnant Women

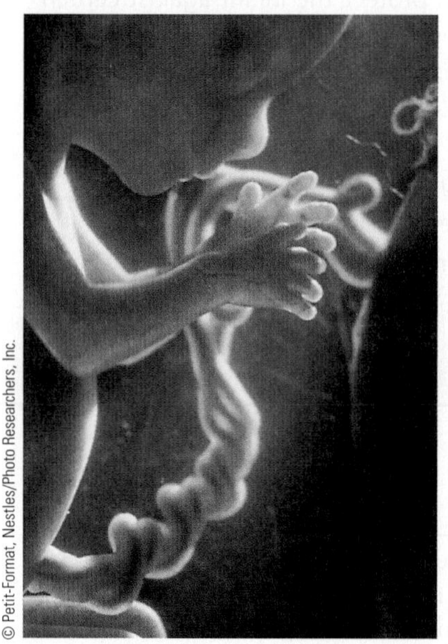
© Petit-Format, Nestles/Photo Researchers, Inc.

▸ *Notice the umbilical cord connecting this 16-week-old fetus with the placenta. The placenta is the organ inside the uterus in which maternal blood vessels lie side by side with fetal blood vessels entering it through the umbilical cord. This close association between the two circulatory systems permits the mother's bloodstream to deliver nutrients and oxygen to the fetus and to carry away fetal waste products.*

For pregnant and lactating women, their nutrient needs are higher than at any other time in their adult life, and they have greater need for certain nutrients than for others (see Figure 11-1). Notice that although their nutrient needs are much higher than usual, their energy needs are not. To support the metabolic demands of pregnancy and fetal development, the recommended average increase is only about 17 percent higher than recommended maintenance calories. An additional 350 calories during the second **trimester** and an additional 450 calories per day during the third trimester are recommended.[3]

Nearly all nutrients are recommended in increased amounts during pregnancy and lactation. The nutrient needs of pregnancy are best met by the routine intake of a variety of foods (see Table 11-1 on page 348). The nutrients deserving special attention in the diets of pregnant women include protein, folate, iron, zinc, and calcium, as well as vitamins known to be toxic in excess amounts.[4]

The recommended protein intake is about an additional 20 grams per day more than nonpregnant requirements. Many women already eat enough protein to cover the increased demand of pregnancy.

The pregnant woman's recommended folate intake is 50 percent greater than

placenta (pla-SEN-tuh) the organ inside the uterus in which the mother's and fetus's circulatory systems intertwine and in which exchange of materials between maternal and fetal blood takes place. The fetus receives nutrients and oxygen across the placenta; the mother's blood picks up carbon dioxide and other waste materials to be excreted via her lungs and kidneys.

prenatal prior to birth.

postnatal after birth.

trimester one-third of the normal duration of pregnancy; the first trimester is 0 to 13 weeks, the second is 13 to 26 weeks, and the third trimester is 26 to 40 weeks.

Ask Yourself Answers: **1.** True. **2.** False. A woman needs only 15 percent more calories per day during pregnancy than she did before. **3.** True. **4.** False. A woman's cravings during pregnancy do not seem to reflect real physiological needs. **5.** True. **6.** False. A child's appetite regulates food intake to meet need; caretakers should not force food on children because this will only create conflict. **7.** False. School lunch is designed to provide one-third of the nutrients schoolchildren need in a day. **8.** False. Although age-related weight gain is a fact of life for many people, it is not inevitable. Consuming low-fat meals within your calorie allowance and making physical activity part of your daily routine can help. **9.** False. Today, 12.4 percent of the U.S. population is 65 or older. By the year 2030, 20 percent of the population will be more than 65 years of age. The fastest growing group is those over age 85. **10.** False. Older people need fewer calories (due to lower basal metabolic rates) but the same amounts of most vitamins and minerals as younger people. Some authorities recommend higher intakes of certain nutrients (calcium, vitamin D, antioxidants, B vitamins) for older adults.

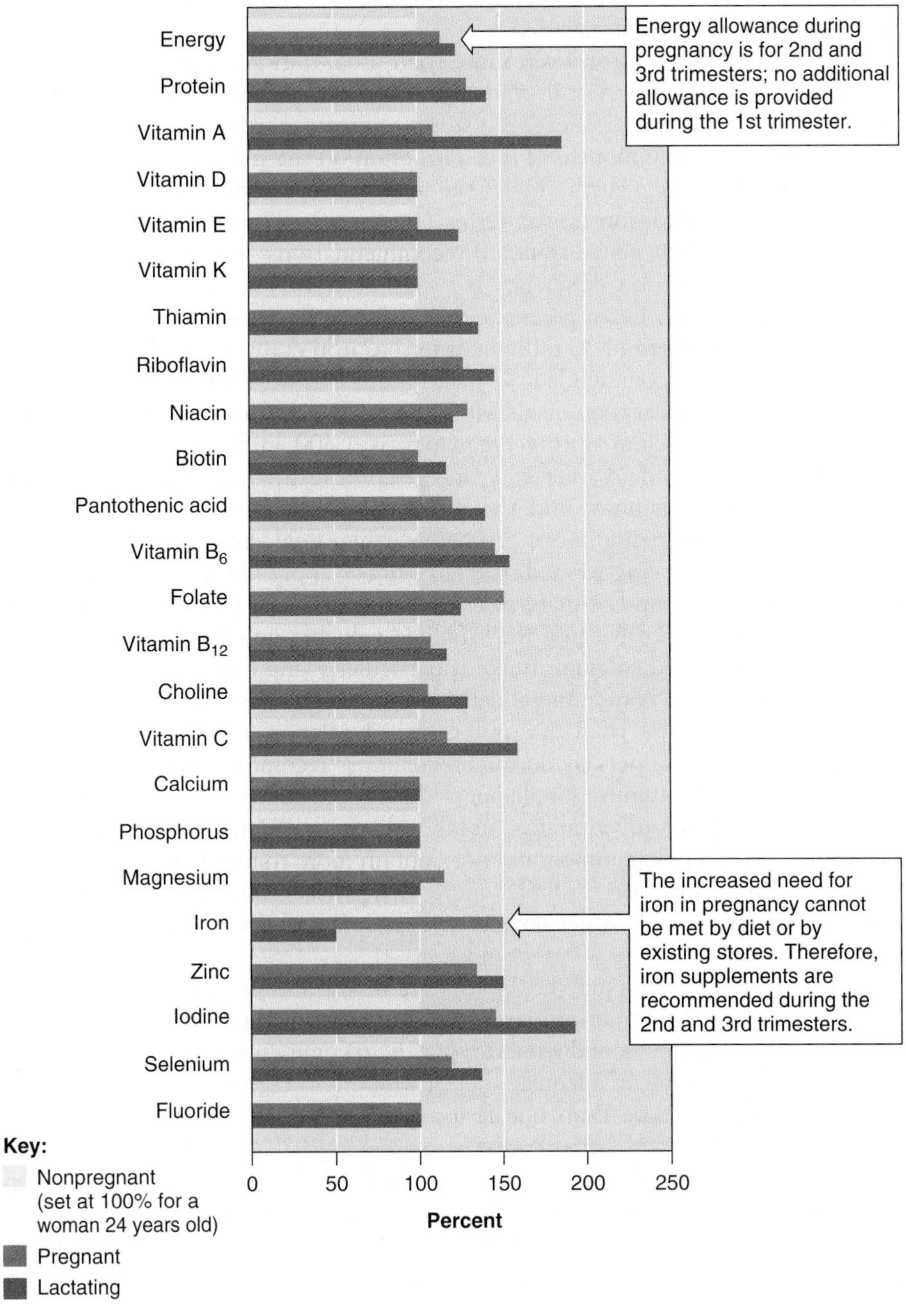

FIGURE 11-1
COMPARISON OF NUTRIENT NEEDS OF NONPREGNANT, PREGNANT, AND LACTATING WOMEN*

*For actual values, turn to the table on the inside front cover.

Energy
Protein
Vitamin A
Vitamin D
Vitamin E
Vitamin K
Thiamin
Riboflavin
Niacin
Biotin
Pantothenic acid
Vitamin B_6
Folate
Vitamin B_{12}
Choline
Vitamin C
Calcium
Phosphorus
Magnesium
Iron
Zinc
Iodine
Selenium
Fluoride

Energy allowance during pregnancy is for 2nd and 3rd trimesters; no additional allowance is provided during the 1st trimester.

The increased need for iron in pregnancy cannot be met by diet or by existing stores. Therefore, iron supplements are recommended during the 2nd and 3rd trimesters.

0 50 100 150 200 250
Percent

Key:
Nonpregnant (set at 100% for a woman 24 years old)
Pregnant
Lactating

the normal requirement due to her large increase in blood volume and the rapid growth of the fetus. Certain studies have shown that folate supplements given around the time of conception reduce the recurrence of **neural tube defects,** such as spina bifida, in the infants of women who previously have had such births.[5] To lower the risk of neural tube birth defects, women are advised to get the recommended amounts of folate—especially *before* becoming pregnant (400 micrograms) and during the first trimester of pregnancy (600 micrograms).[6]

As of 1999, all refined grain products (bread, cereal, cornmeal, farina, flour, grits, pasta, and rice) are fortified with folate. The folate contained in fortified foods and supplements is almost twice as well absorbed as the folate that is naturally available in such foods as green leafy vegetables, citrus fruits, whole-grain breads, or legumes. Women are advised to choose a variety of foods naturally high in folate as well as foods that have been fortified with folate. Because high folate intake can mask a vitamin B_{12} deficiency, folate intake should not exceed

neural tube defects include any of a number of birth defects in the orderly formation of the neural tube during early gestation. Both the brain and the spinal cord develop from the neural tube and defects result in various central nervous system disorders. The two main types are *spina bifida* (incomplete closure of the bony casing around the spinal cord) and *anencephaly* (a partially or completely missing brain). Because the neural tube closes before the sixth week of pregnancy, women are advised to consume adequate folate from as early as 3 months prior to conception.

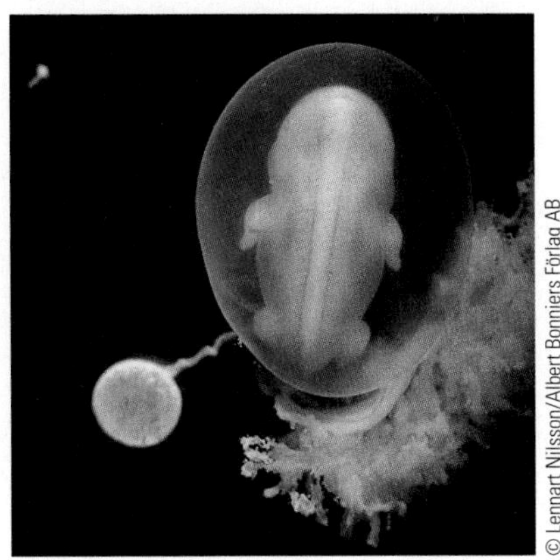

▲ *The neural tube (outlined by the delicate red arteries) has successfully closed after only six weeks of pregnancy.*

1 milligram per day. If the woman's dietary intake of folate is low, a 300-microgram folate supplement is recommended.[7]

The body conserves iron even more than usual during pregnancy. Menstruation ceases, and absorption of iron increases up to threefold. However, the developing fetus draws on its mother's iron stores to create stores of its own to carry it through the first 3 to 6 months of life. This drain on the mother's iron supply can precipitate a deficiency. Furthermore, she will lose blood when she gives birth.

Foods containing iron:

- Red meat, fish, and other meat
- Dried fruits
- Legumes
- Whole-grain and fortified breads and cereals
- Dark green vegetables

The recommended iron intake during pregnancy is 27 milligrams per day—an increase of 50 percent above standard recommendations—to meet maternal and fetal needs. Because iron deficiency is a common problem among nonpregnant women, many women begin pregnancy with diminished iron stores. For this reason, an iron supplement of 30 milligrams ferrous iron daily during the second and third trimesters is recommended. To facilitate absorption from the supplement, iron should be taken between meals with vitamin C-rich fruit juices or at bedtime.

The DRI for calcium during pregnancy is 1,300 milligrams for teens and 1,000 milligrams for adults over 18 years of age. Intestinal absorption of calcium doubles early in pregnancy, and the mineral is stored in the mother's bones. Later, during the last trimester of pregnancy, when fetal skeletal growth is maximum and teeth are being formed, the fetus draws approximately 300 milligrams of calcium per day from the maternal blood supply.[8] Dairy products are a recommended calcium source because they include valuable vitamin D and riboflavin as well. This increased calcium intake is particularly important for women under 25 years of age whose bone mineral density is still increasing. For the woman who normally consumes less than 600 milligrams of calcium a day, a 600-milligram supplement of calcium per day during pregnancy is recommended.

Foods containing calcium:

- Milk
- Other dairy products (yogurt, cheese)
- Green leafy vegetables
- Legumes
- Fortified juice or soy milk
- Certain brands of tofu

Routine use of vitamin supplements during pregnancy is not advised, and excess intakes of certain vitamins, notably vitamins A and D, can cause fetal malformations.[9] Supplements should not contain more than one to two times the recommended levels. For example, more than three times the recommended intake of vitamin A taken per day early in pregnancy may cause malformations of the newborn.[10]

Nutrient supplements may be appropriate in certain circumstances, however. For example, a multivitamin-mineral supplement (providing 100 percent of the DRI) beginning in the second trimester may be recommended for women who do not ordinarily consume an adequate diet or who are in high-risk categories—such as women carrying more than one fetus, heavy cigarette smokers, and alcohol and other drug abusers.[11]

Because energy needs increase less than nutrient needs, the pregnant woman must select foods of high nutrient density. A woman who already eats well can simply increase her servings of nutritious foods to meet her increasing nutrient needs (refer to Table 11-1).

TABLE 11-1
Food Guide for Pregnant and Lactating Women

	Number of Servings[a]	
Food Group	Nonpregnant Woman	Pregnant or Lactating Woman[b]
Grains	6 to 11	7 to 11 (7+)
Vegetables	3 to 5	4 to 5 (5+)
Fruits	2 to 4	3 to 4 (4+)
Meat & Beans	2 to 3	3 (3+)
Milk	3	3 to 4 (4+)

[a] Refer to Table 2-6 and Figure 2-5 in Chapter 2 for a summary of foods in each group and serving sizes.
[b] Numbers in parentheses indicate numbers of servings recommended for the pregnant teenager.

Scorecard: Pregnancy Readiness

Are you nutritionally ready for pregnancy? Score each question. A score of 21 is perfect; scores below 3 per question identify areas for improvement.

1. My body weight is desirable for my height:
   - **a.** Right on target (3 points)
   - **b.** Within 10 percent (2 points)
   - **c.** 10 to 20 percent above (1 point)
   - **d.** More than 20 percent above or 20 percent below (0 points)
2. I drink milk or use milk alternatives every day:
   - **a.** Equivalent of 3 cups or more a day (3 points)
   - **b.** About 2 cups a day (2 points)
   - **c.** About 1 cup a day (1 point)
   - **d.** No milk or milk substitutes (0 points)
3. I eat vegetables daily:
   - **a.** Five servings a day (3 points)
   - **b.** Four servings a day (2 points)
   - **c.** Three servings a day (1 point)
   - **d.** Two or fewer servings a day (0 points)
4. I eat fruits daily:
   - **a.** Four servings a day (3 points)
   - **b.** Three servings a day (2 points)
   - **c.** Two servings a day (1 point)
   - **d.** One or fewer servings a day (0 points)
5. I eat folate-rich foods, such as green leafy vegetables, orange juice, cantaloupe, legumes, and fortified grain products, daily:
   - **a.** Three to four servings, or enough to provide 400 μg daily (3 points)
   - **b.** Two to three servings, or enough to meet half the recommended intake (1 point)
   - **c.** One or fewer servings, or less than half the current recommended intake (0 points)
6. I eat iron-rich foods, such as meats, legumes, or fortified cereals, daily:
   - **a.** Two servings, or enough to meet the recommended intake (3 points)
   - **b.** One serving, or enough to meet about half the recommended intake (1 point)
   - **c.** Less than one serving, or less than half the recommended intake (0 points)
7. I am physically fit because I have a well-established habit of exercising daily, and I will be able to continue exercising during pregnancy:
   - **a.** I am as fit as I can be. (3 points)
   - **b.** I am fairly fit. (2 points)
   - **c.** I am not fit. (0 points)

Maternal Weight Gain

Normal weight gain and adequate nutrition support the health of the mother and the development of the fetus. The recommendations for weight gain take into account a mother's prepregnancy weight for height or body mass index (BMI), as shown in Table 11-2. A woman who begins pregnancy at a healthful weight should gain between 25 and 35 pounds. A woman pregnant with twins needs to gain 35 to 45 pounds. An underweight woman needs to gain between 28 and 40 pounds; an obese woman, between 16 and 25 pounds. Weight gains at the upper end of the range are recommended for pregnant teenagers because of their increased risk of low weight gains and delivery of low-birthweight infants.

Low weight gain in pregnancy is associated with increased risk of delivering a **low-birthweight (LBW)** infant. Not all small babies are unhealthy, but birthweight and length of gestation are the primary indicators of an infant's future health status. A low-birthweight baby is more likely than a normal-weight baby to experience complications during delivery and has a statistically greater chance of having physical and mental birth defects, developing diseases, and dying early in life.

Excessive weight gain during pregnancy increases the risk of complications during labor and delivery as well as postpartum obesity.[12] Obese women also have an increased risk for complications during pregnancy, including hypertension and gestational diabetes.

TABLE 11-2
Recommended Weight Gain for Pregnant Women

Weight Category*	Recommended Gain (lbs)
Underweight	28–40
Normal weight	25–35
Overweight	15–25
Obesity	≥ 15

*Underweight is defined as BMI < 18.5; normal weight as BMI 18.5 to 24.9; overweight as BMI 25.0 to 29.9; and obesity as BMI ≥ 30.0.

SOURCE: Adapted from Food and Nutrition Board, *Nutrition During Pregnancy* (Washington, D.C.: National Academy Press, 1990).

TABLE 11-3
An Example of the Pregnant Woman's Weight Gain

Development	Weight Gain (lb)
Infant at birth	7–8
Placenta	1
Increase in mother's blood volume to supply placenta	4
Increase in mother's fluid volume	4
Increase in size of mother's uterus and the muscles to support it	2
Increase in size of mother's breasts	2
Fluid to surround infant in amniotic sac	2
Mother's fat stores (varies)	3–12
Total	25–35

NOTE: The pattern of gain should be about 1 pound a month for the first 3 months and 1 pound a week thereafter. Different patterns of weight gain are suggested for underweight, normal-weight, and overweight women.

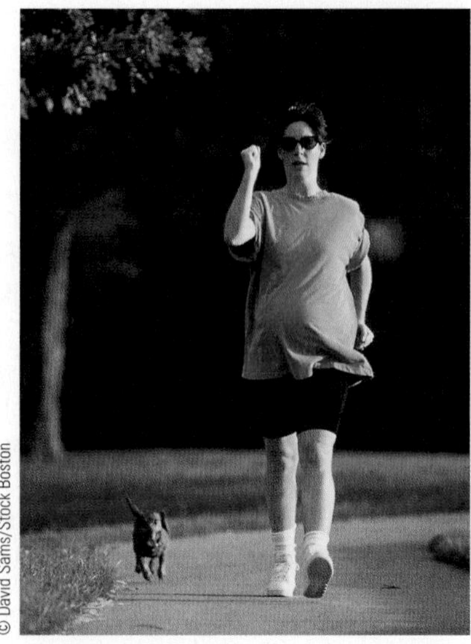

© David Sams/Stock Boston

▲ *For most women, the surest way to have a healthy baby is to follow a healthy lifestyle: Get early prenatal care, eat a well-balanced diet, exercise regularly with your doctor's permission, and avoid cigarettes, alcohol, and other drugs.*

An infant at birth weighs only about 6½ to 8 pounds, but the mother's built-up body tissues (blood, blood vessels, muscle, fat stores, and others) that provided a healthful environment for the developing fetus weighs more than 20 pounds (see Table 11-3). Weight gain should be lowest during the first trimester—3 to 4 pounds for the entire trimester—followed by a steady gain of about 1 pound per week thereafter. If a woman gains more than the recommended amount of weight early in pregnancy, however, she should not try to diet in the last weeks. Weight-loss dieting during pregnancy is not recommended. A *sudden* large weight gain may indicate the onset of pregnancy-induced hypertension (discussed later). A woman who experiences this type of weight gain should see her health care provider.

Some of the weight a woman gains in pregnancy is lost at delivery. For the woman who has gained only the recommended 25 to 35 pounds, the remainder is generally lost within a few months as her blood volume returns to normal and she loses the fluids she has accumulated.

Practices to Avoid

Optimal pregnancy outcome is influenced by maternal nutrient intake, but it can also be affected by maternal use of nonfood substances, excess caffeine, low-calorie diets, megadoses of certain vitamins, tobacco, alcohol, and illicit drugs. See this chapter's Nutrition Action feature for recommendations regarding caffeine and pregnancy.

The term **pica** refers to the craving of nonfood items that have little or no nutritional value. When pica occurs during pregnancy, it typically involves the consumption of dirt, clay, or laundry starch, but episodes of pica have included compulsive ingestion of such things as ice, paper, and coffee grounds.[13] The medical consequences of pica can include malnutrition (as nonfood items replace nutritious foods in the diet), obesity (from overconsumption of items such as starch), poisoning (from ingestion of toxic compounds), or intestinal obstruction (due to consuming large amounts of clay or starch).

Some other practices are also truly harmful, and their potential impact on pregnancy outcome is too great to risk. For example, low-carbohydrate or low-calorie diets that cause ketosis can deprive the fetus's brain of needed glucose and cause congenital deformity. Protein deprivation can cause children's height and head circumference to diminish markedly and irreversibly.

Another harmful maternal practice is smoking, which restricts the blood supply to the growing fetus, thereby limiting the delivery of nutrients and removal of wastes. Smoking stunts fetal growth, thus increasing the risk of premature delivery, low birthweight, retarded development, and spontaneous abortions. Smoking is responsible for 20 to 30 percent of all low-birthweight deliveries in the United States.[14] Sudden infant death syndrome (SIDS) has been linked to smoking during pregnancy and to postnatal exposure to second-hand smoke.[15]

As discussed in Chapter 8, research confirms that alcohol consumption adversely affects fetal development. Even as few as one or two drinks daily can

low birthweight (LBW) a birthweight of 5½ lb (2,500 g) or less, used as a predictor of poor health in the newborn and as a probable indicator of poor nutrition status of the mother during and/or before pregnancy. Normal birthweight for a full-term baby is 6½ to 8¾ lb (about 3,000 to 4,000 g). LBW infants are either born premature or have suffered growth failure in the uterus—they may or may not be born early, but they are small.

pica the craving of nonfood items such as clay, ice, and laundry starch; does not appear to be limited to any particular geographic area, race, sex, culture, or social status.

cause **fetal alcohol syndrome (FAS)**—irreversible brain damage and mental and physical retardation in the fetus. The most severe impact of maternal drinking is likely to occur in the first month, before the woman even is sure she is pregnant. This preventable condition (FAS) is estimated to occur in approximately 1 to 2 infants per 1,000 live births and is the leading known cause of mental retardation in the United States.[16] Birth defects, low birthweight, and spontaneous abortions occur more often in pregnancies of women who drink even as little as 2 ounces of alcohol daily during pregnancy. Accumulating evidence that even one drink may be too much has led the American Academy of Pediatrics to take the position that women should stop drinking as soon as they *plan* to become pregnant.[17]

fetal alcohol syndrome (FAS) the cluster of symptoms seen in an infant or child whose mother consumed excess alcohol during pregnancy, including retarded growth, impaired development of the central nervous system, and facial malformations. A lesser condition—called *fetal alcohol effect (FAE)*—causes learning impairment and other more subtle abnormalities in infants exposed to alcohol during pregnancy.

Drugs other than alcohol taken during pregnancy can also cause birth defects. A particularly dramatic example is the acne medication Accutane (isotretinoin), which causes major deformities during fetal development. Pregnant women should avoid taking all drugs except on the advice of their physician.

Although many pregnant women view herbal products as safe and natural alternatives to over-the-counter medications, most herbal products have not been formally evaluated in terms of how they might affect pregnancy. Therefore, pregnant women are advised to avoid using herbal supplements during pregnancy.[18] See Chapter 6 for more information on the use of herbal remedies, including safety and regulatory issues.

Pregnant women, as well as young children, are particularly vulnerable to the effects of environmental contaminants, notably lead and mercury. These substances can severely impair an unborn child's developing nervous system. To reduce exposure to lead during pregnancy, the Food and Drug Administration (FDA) advises pregnant women to avoid frequent use of ceramic mugs for hot beverages such as coffee or tea and not to use lead crystal glassware daily.

To reduce exposure to mercury, the FDA advises that children under 6 years of age, pregnant or lactating women, and women planning pregnancy within a year should not eat large ocean fish such as shark, swordfish, king mackerel, and tilefish. These long-lived predatory fish accumulate the highest levels of mercury and pose the greatest risk to people who eat them regularly. FDA advises pregnant women to select a variety of other kinds of fish—including shellfish, canned fish, and smaller ocean fish or farm-raised fish—and that these women can safely eat 12 ounces per week of cooked fish.[19] Consumption of freshwater fish should be limited to no more than one fish meal per week. For information about the risks of mercury in seafood, call 1 (888) SAFEFOOD or visit www.cfsan.fda.gov.

Common Nutrition-Related Problems of Pregnancy

Common physical problems in pregnancy include morning sickness and, later, constipation. The nausea of morning sickness seems unavoidable because it arises from the hormonal changes taking place early in pregnancy, but it can sometimes be alleviated. Suggested strategies to alleviate the nausea and vomiting of morning sickness include eating soda crackers, hard candies, or other dry starchy foods before getting up in the morning, eating small frequent meals as soon as you feel hungry, and avoiding any specific food (especially highly seasoned foods or foods with strong odors) causing nausea or vomiting.[20]

Later, as the hormones of pregnancy alter her muscle tone and the growing fetus crowds her intestinal organs, an expectant mother may complain of constipation. A high-fiber diet, plentiful fluid intake, and regular exercise will help relieve this condition.

Women's cravings during pregnancy reflect alterations in taste and smell sensitivities, rather than real physiological needs. If a woman craves pickles and chocolate sauce at two o'clock in the morning, for example, it is probably not because she lacks a combination of nutrients uniquely supplied by these foods.

pregnancy-induced hypertension (PIH) high blood pressure that develops during the second half of pregnancy.

preeclampsia a condition characterized by hypertension, fluid retention, and protein in the urine.

eclampsia a severe extension of preeclampsia characterized by convulsions; may lead to coma.

gestational diabetes the appearance of abnormal glucose tolerance during pregnancy.

The woman is, however, expressing a need as real and as important as her need for nutrients—the need for support, understanding, and love. More serious problems needing control during pregnancy include hypertension and diabetes.

Hypertension in Pregnancy Ideally, a woman with preexisting hypertension has her blood pressure under control before becoming pregnant. Otherwise, she may have an increased risk of delivering a low-birthweight baby. Some women develop a *transient hypertension of pregnancy* during the second half of their pregnancy. Usually, this is a mild form of hypertension with no adverse effects on pregnancy outcome, and her blood pressure returns to normal shortly after the baby is born. Sometimes, however, high blood pressure in a pregnant woman signals the onset of **pregnancy-induced hypertension (PIH).** Preeclampsia and eclampsia are hypertensive conditions induced by pregnancy. **Preeclampsia** is characterized by high blood pressure, protein in the urine, and generalized edema that may cause sudden, large weight gain from retained water. Fluid retention alone, which is quite common in pregnant women, is not sufficient to diagnose preeclampsia.[21] Warning signs of preeclampsia include severe and constant headaches; sudden weight gain (1 lb/day); swelling of face, hands, and feet; dizziness; and blurred vision. **Eclampsia,** the most severe form of pregnancy-induced hypertension, is characterized by convulsions that may lead to coma. PIH can retard fetal growth and cause the placenta to separate from the uterus, resulting in stillbirth. Both conditions present serious health risks to mother and fetus and demand careful medical treatment.

Women meeting one or more of the following criteria are screened for gestational diabetes:

- ≥ 25 years of age
- Family history of diabetes
- Member of an ethnic or racial group with a high prevalence of diabetes (Hispanic, Native American, Asian, African American, or Pacific Islander)
- Obesity (especially central obesity)
- History of glucose intolerance
- History of gestational diabetes
- History of delivery of newborn >10 lbs (>4500 g)

Diabetes Infants born to women with diabetes are at greater risk for premature birth, congenital defects, excessively high birthweight, and respiratory distress syndrome.[22] Metabolic control of diabetes before and throughout pregnancy is critical. In some women, pregnancy can alter carbohydrate metabolism and precipitate a condition known as **gestational diabetes.** The abnormal blood glucose levels usually occur during the second half of pregnancy and return to normal after pregnancy for most women with the condition. However, some women with gestational diabetes are at increased risk for developing type 2 diabetes after pregnancy, especially if too much weight is gained. Risk factors include being age 25 or older, a previous history of gestational diabetes, obesity, and a family history of diabetes. Some women with gestational diabetes have the classic symptoms of diabetes—increased thirst, hunger, urination, weakness—but other women have no warning signs. For this reason, pregnant women who are at risk for developing gestational diabetes are screened for the condition between the twenty-fourth and twenty-eighth weeks of gestation.

Adolescent Pregnancy

More than 700,000 teenagers become pregnant in the United States each year—one out of every eight babies is born to a teenager—and more than a tenth of these mothers are under age 15.[23] The complexity of social, emotional, and physical factors makes teen pregnancy one of the most challenging situations for meeting nutritional needs. According to a position paper from the American Dietetic Association, pregnant adolescents are nutritionally at risk and require early intervention and special care throughout pregnancy.[24] Medical and nutritional risks are particularly high when the teenager is within 2 years of menarche (usually 15 years of age or younger). Risks for pregnant teens include higher rates of pregnancy-induced hypertension, iron-deficiency anemia, premature birth, stillbirths, low-birthweight infants, and prolonged labor.

Pregnancy places adolescent girls, who are already at risk for nutrition problems, at even greater risk because of the increased energy and nutrient demands of pregnancy. To support the needs of both mother and infant, adolescents are encouraged to strive for pregnancy weight gains at the upper end of the ranges

(Text continues on page 355.)

Nutrition Action Filtering the Evidence about Caffeine

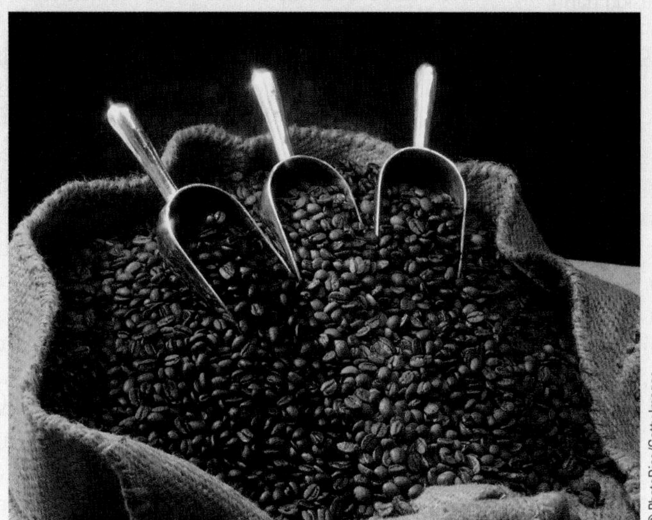
© PhotoDisc/Getty Images

In 1657, when merchants first introduced Londoners to a Middle Eastern brew known as coffee, they boasted it as a "wholesome and physical drink," an elixir of health suitable for treating colds, coughs, gout, and many other ills.[25] Today at least eight out of ten Americans consume caffeine, the most widely used behaviorally active drug in the world. Modern-day coffee drinkers have been subject to a barrage of reports linking coffee and other **caffeine** containing products with more than 100 diseases. Fingers have repeatedly pointed at caffeine as the culprit behind breast disease, cancer, heart disease, birth defects, and high blood pressure, to name just a few.

Despite the brouhaha, the jury is still out as to whether caffeine is truly to blame. Scientists have yet to confirm long-standing suspicions that caffeine contributes to any health problems other than jitteriness. One reason for all the controversy is that much of the evidence linking the substance with different diseases has been clouded by a number of issues. Some studies do not measure caffeine sources except for coffee and tea, such as soft drinks, chocolate, and certain medications. In addition, the amount of caffeine and other substances in coffee or tea can vary considerably depending on how the beverage is brewed, a fact most studies fail to take into account.

For example, consider the widely debated question of whether drinking coffee raises blood cholesterol. Granted, a great deal of strong evidence suggests that the beverage does contribute to high blood cholesterol and therefore to heart disease. The hitch is that most of the evidence comes from Scandinavia, where coffee is boiled rather than brewed in automatic drip coffeemakers or electric percolators. Subsequent research has found that whereas boiled coffee appears to boost blood cholesterol levels, filtered coffee does not, most likely because substances in boiled coffee other than caffeine may be the cholesterol-raising culprits.[26]

Another issue that most research has not filtered out is that coffee drinking itself may not contribute to ill health; however, it seems to be part of a lifestyle that does. After questioning some 2,600 men and women about their health habits, a group of Boston-based researchers found that women who opted for

caffeine a type of compound, called a *methylxanthine,* found in coffee beans, cola nuts, cocoa beans, and tea leaves. A central nervous system stimulant, caffeine's effects include increasing the heart rate, boosting urine production, and raising the metabolic rate.

caffeine dependence syndrome dependence on caffeine characterized by at least three of the four following criteria: withdrawal symptoms such as headache and fatigue; caffeine consumption despite knowledge that it may be causing harm; repeated, unsuccessful attempts to cut back on caffeine; and tolerance to caffeine.

decaffeinated coffee exclusively were more likely than regular coffee drinkers to, among other things, eat vegetables frequently and exercise regularly. Male decaf drinkers also tended to have adopted more healthful habits, such as eating low-fat diets, than did their regular coffee-drinking counterparts. Thus, it may not be the coffee but rather the poor health habits that often go along with the lifestyle that contribute to health problems.[27]

None of this is meant to say that caffeine is necessarily good for people. As anyone who can't get going in the morning without a cup of coffee or can of caffeinated soda is well aware, consuming caffeine day in and day out can be habit forming. In fact, some researchers have identified a condition called **caffeine dependence syndrome,** characterized by at least three of the four following criteria: withdrawal symptoms such as headache and fatigue; caffeine consumption despite knowledge that it may be causing harm; repeated, unsuccessful attempts to cut back on caffeine; and tolerance to caffeine.[28]

Along with people who are dependent on caffeine, people with certain medical conditions would do well to consume caffeine in moderation or avoid it completely. Pregnant women, for example, should limit their caffeine intake. Although it has never been proven to cause birth defects, caffeine does cross the placenta and enter the fetus, where large amounts can affect the unborn baby's heart rate and breathing.[29] Some research also suggests that the amount of caffeine in three or four cups of coffee could raise the risk of suffering a miscarriage, perhaps by decreasing blood flow through the placenta.[30] Women who drink caffeine during pregnancy are generally advised to limit consumption to less than 150 milligrams per day—the equivalent of about 16 ounces of coffee.[31] In addition, anyone with ulcers should steer clear of caffeinated *and* decaffeinated coffee, both of which stimulate the secretion of acid, which can irritate the stomach's lining. Chapter 10 discusses caffeine's effect on physical performance.

For a healthy person, drinking one or two cups of coffee, tea, or cola a day does not seem to pose any hazard. Only those who are particularly sensitive to caffeine and suffer symptoms such as headaches, nervousness, and insomnia after consuming it really need to consider avoiding it—or at least cutting back. (See Table 11-4 to figure out how much caffeine you're taking in each day.)[32]

If you drink coffee or a can or two of cola every day and decide to quit, do it gradually. Even moderate caffeine users who try to stop cold turkey often suffer from withdrawal symptoms, such as splitting headaches, fatigue, moodiness, and nausea. Try instead to cut back gradually by, say, drinking no more than one cup every two or three days. Another way to cut back on caffeine is to mix some decaffeinated grounds in with your regular coffee and gradually use more and more decaf and less regular. Likewise, you can drink a can of decaffeinated soda now and then until you've weaned yourself off the caffeine.

TABLE 11-4
Caffeine Countdown

Drinks and Foods	Average Caffeine Content (mg)
Coffee (8-oz cup)	
Brewed, drip method	85
Instant	75
Decaffeinated, brewed or instant	3
Espresso (2-oz cup)	80
Tea (8-oz cup)	
Brewed, black, steeped for 3 minutes	47
Iced tea (8-oz glass)	25
Soft drinks (12-oz can)	
Colas, regular or diet	36–50
Mountain Dew, Mello Yello, Kick	52–56
Jolt	104
Cocoa beverage (8-oz cup)	6
Chocolate milk beverage (8 oz)	5–8
Milk chocolate candy (1 oz)	6–15
Drugs*	**Average Caffeine Content (mg)**
Pain relievers (standard extra-strength dose)	
Excedrin, Bayer, Midol	120–130
Stimulants	
NoDoz, Vivarin	200

*Because products change, contact the manufacturer for an update on products you use regularly.

recommended for pregnant women (refer to Table 11-2). Those who gain between 30 and 35 pounds during pregnancy have lower risks of delivering low-birthweight infants.[33] Adequate nutrition can substantially improve the course and outcome of adolescent pregnancy.[34] A model program for providing nutritional help to teenage mothers is, among others, the WIC (Women's, Infants' and Children's) program, a federally funded program that provides nutrition education and low-cost nutritious foods to low-income pregnant women, mothers, and their children.

Nutrition of the Breastfeeding Mother

Adequate nutrition for the mother makes a highly significant contribution to successful lactation. A nursing mother produces 30 ounces of milk a day, on the average, with wide variations possible. Current recommendations suggest that 400 calories to support this milk production come from added food and that the rest come from the stores of fat the mother's body has accumulated during pregnancy for this purpose. (Table 11-1 on page 348 shows a food pattern that will meet the lactating woman's nutrient needs.)

The period of lactation is the natural time for a woman to lose the extra body fat she has accumulated during pregnancy. Once lactation is established, if her food choices are judicious, a nursing mother can tolerate a calorie deficit and a gradual loss of weight (1 pound per week) without any effect on her milk output. Fat can only be mobilized slowly, however, and an energy deficit that is too large will inhibit lactation.

Healthy Infants

The growth of infants directly reflects their nutritional well-being and is the major indicator of their nutrition status. A baby grows faster during the first year of life than ever again, doubling its birthweight during the first 4 to 6 months, and tripling its birthweight by the end of the first year. Adequate nutrition during infancy is critical to support this rapid rate of growth and development. Clearly, from the point of view of nutrition, the first year is the most important year of a person's life. This section provides an overview of nutrient requirements, current recommendations for feeding healthy infants, and the relationship between infant feeding and selected pediatric nutrition issues.

Milk for the Infant: Breastfeeding

Breastfeeding has both emotional and physical health advantages. Emotional bonding is facilitated by many events and behaviors of mother and infant during the early months and years; one of the first can be breastfeeding.

During the first 2 or 3 days of lactation, the breasts produce **colostrum,** a premilk substance containing antibodies and white cells from the mother's blood. Because it contains immunity factors, colostrum helps to protect the newborn infant from those infections against which the mother has developed an immunity—precisely those in the environment against which the infant needs protection. Entering the infant's body with the milk, these antibodies inactivate bacteria within the digestive tract, where they could otherwise cause intestinal infections. Breast milk also contains antibodies, although not as much as colostrum contains.

Colostrum and breast milk both contain the **bifidus factor** that favors the growth of the "friendly" probiotic bacteria such as *Lactobacillus bifidus* in the infant's digestive tract so that other, harmful bacteria cannot grow there. (Probiotics are discussed in Chapter 6.) Breast milk also contains the powerful antibacterial agent **lactoferrin,** as well as other factors, including several enzymes, several

colostrum (co-LAHS-trum) a milk-like secretion from the breast, rich in protective factors, present during the first day or so after delivery and before milk appears.

bifidus factor (BIFF-id-us) a factor in colostrum and breast milk that favors the growth in the infant's intestinal tract of the "friendly" bacteria *Lactobacillus bifidus* so that other, less desirable intestinal inhabitants will not flourish.

lactoferrin (lak-toe-FERR-in) a factor in breast milk that binds and helps absorb iron and keeps it from supporting the growth of the infant's intestinal bacteria.

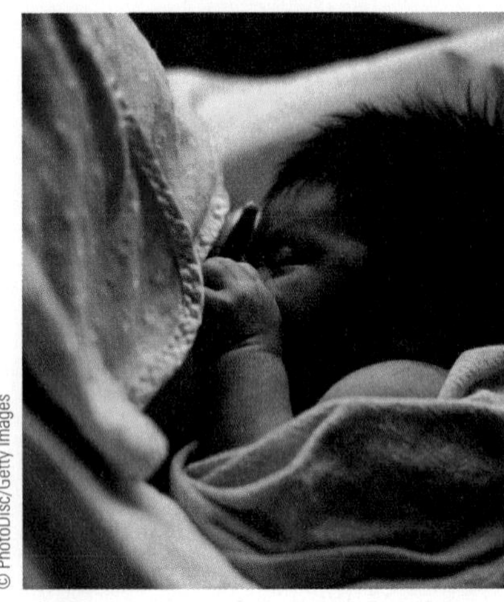

▲ *Breast milk is a very special substance.*

hormones, and lipids that help to protect the infant against infection. Breastfed infants have lower rates of hospital admissions, ear infections, diarrhea, rashes, allergies, asthmatic disease, and other health problems than bottle-fed infants.[35]

Breast milk is tailor-made to meet the nutrient needs of the young infant. It offers its carbohydrate in the easy-to-assimilate form of lactose; its fat contains a generous proportion of the essential omega-6 fatty acid linoleic acid; and its protein, alpha-lactalbumin, is one that the infant can easily digest. Depending on the mother's diet, breast milk can also deliver the beneficial omega-3 fatty acids to the infant. With the exception of vitamin D, its vitamin contents are ample. As for minerals, calcium, phosphorus, and magnesium are present in amounts appropriate for the rate of growth expected in a human infant, and breast milk is low in sodium. Its iron is highly absorbable, and the presence of a zinc-binding protein favors the absorption of the zinc it contains.

The American Academy of Pediatrics recommends that infants receive breast milk for the first 12 months of life.[36] Despite the health benefits, however, the incidence of breastfeeding in the United States declined from the mid-1980s until the early 1990s.[37] Today, approximately 69 percent of mothers currently initiate breastfeeding—lower than the *Healthy People 2010* goal, which is "to increase to at least 75 percent the proportion of mothers who breastfeed their babies." Only 29 percent of mothers, however, are still breastfeeding after 6 months. Analysis of data from a survey of mothers indicates that breastfeeding rates continue to be the highest among women who are older, well educated, relatively affluent, and/or live in the western United States. Among those least likely to breastfeed are women who are in low-income groups, black, under age 20, and/or live in the southeastern United States.[38] Table 11-5 lists ten steps hospitals can take to promote breastfeeding.

A number of barriers to achieving the nation's health objective for increasing the incidence of breastfeeding have been identified. They include lack of knowledge, an absence of work policies and facilities that support lactating women (for example, extended maternity leave, part-time employment, facilities for pumping

TABLE 11-5
Baby-Friendly Hospitals: Ten Steps to Successful Breastfeeding

To promote breastfeeding, every maternity facility should:

- Develop a written breastfeeding policy that is routinely communicated to all health care staff.
- Train all health care staff in the skills necessary to implement the breastfeeding policy.
- Inform all pregnant women about the benefits and management of breastfeeding.
- Help mothers initiate breastfeeding within ½ hour of birth.
- Show mothers how to breastfeed and how to maintain lactation, even if they need to be separated from their infants.
- Give newborn infants no food or drink other than breast milk, unless medically indicated.
- Practice rooming-in, allowing mothers and infants to remain together 24 hours a day.
- Encourage breastfeeding on demand.
- Give no artificial nipples or pacifiers to breastfeeding infants.*
- Foster the establishment of breastfeeding support groups and refer mothers to them at discharge from the facility.

*Compared with nonusers, infants who use pacifiers breastfeed less frequently and stop breastfeeding at a younger age.

SOURCE: United Nations Children's Fund and World Health Organization, *Barriers and Solutions to the Global Ten Steps to Successful Breastfeeding*, 1994.

breast milk or breastfeeding, and on-site child care), and the portrayal of bottle feeding rather than breastfeeding as the norm in the American society.

Contraindications to Breastfeeding

Breastfeeding is not recommended in certain circumstances. A woman must not breastfeed if she has a communicable disease such as tuberculosis or hepatitis or if she takes a medication that is secreted in breast milk and is known to affect the infant. Drug addicts—including alcohol abusers—are capable of ingesting such high doses of their drug that their infants can become addicts by way of breast milk. In such cases, breastfeeding is also contraindicated.

Because the human immunodeficiency virus (HIV), which is responsible for AIDS, can be passed to an infant through breast milk, HIV-infected mothers should not breastfeed. Unfortunately, sometimes the nutritional and immunologic benefits of breast milk outweigh the risks of HIV transmission through breastfeeding. Such is the case in many developing countries, where lack of safe drinking water to make formula increases the risk of diarrhea and disease for the infant. However, the World Health Organization encourages HIV-positive mothers in developing countries to feed formula whenever their circumstances assure a regular supply of infant formula and safe drinking water.[39]

Most prescription drugs taken by mothers do not reach nursing infants in sufficiently large quantities to affect them adversely. As a precaution, however, a nursing mother should consult with her physician before taking any drug. *Minimal* use of alcohol between feedings is compatible with breastfeeding, but both alcohol and nicotine can enter breast milk. These substances also adversely affect maternal production and composition of breast milk. It is important not to expose the infant to secondhand smoke due to the risk of health problems such as impaired growth, respiratory problems, and sudden infant death. Coffee drinking is fine in moderation (two to three cups a day), as is eating foods such as garlic and spices. Sometimes, a particular food may affect the baby's liking for the mother's milk, but this matter requires individual detective work. (Examples are chocolate for some babies, excess caffeine for others, and foods that cause gas in the mother for still others.) If a woman has an ordinary cold, she can continue nursing without worry. The infant may catch it from her anyway but may actually be less susceptible than a bottle-fed baby would be, thanks to the immunologic protection offered by breast milk.

Feeding Formula

Like the breastfeeding mother, the mother who offers formula to her baby has reasons for making her choice, and her feelings should be honored. Infant formulas are manufactured to approximate the nutrient composition of breast milk. National and international standards have been set for the nutrient content of infant formulas. The immunologic protection of breast milk, however, cannot be duplicated. One major advantage of formula feeding is the relief it provides to the mother whose attempts at breastfeeding have met with frustration. Formula provides adequate nourishment for the infant, and a mother can choose this alternative with confidence.

Many mothers breastfeed at first and then wean the baby within the first 1 to 6 months. When a woman chooses to wean her infant during the first 12 months of life, it is imperative that she shift to *infant formula*. Cow's milk, both whole and reduced-fat, is not recommended during the first year of life, according to the American Academy of Pediatrics (AAP).[40] Cow's milk is an inappropriate replacement for breast milk or infant formula because it provides insufficient vitamin C and iron and excessive sodium and protein. Feeding cow's milk to infants may increase the risk of iron-deficiency anemia and cow's milk protein allergy.

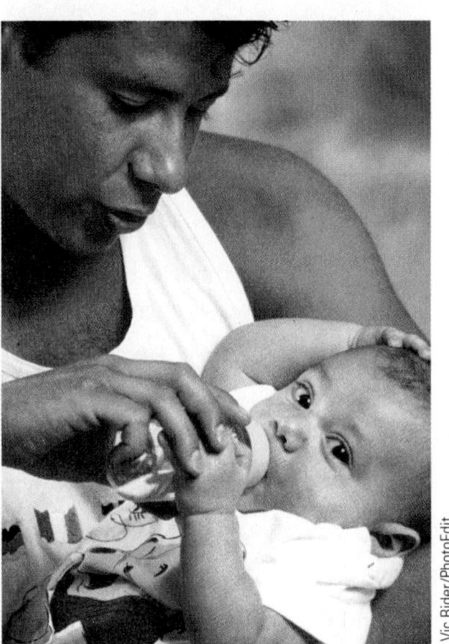

▲ *Using formula feeding or pumping breast milk into a bottle allows other family members to enjoy feeding the infant.*

For infants with special problems, many variations of infant formulas are available. Special formulas based on soy protein are available for infants allergic to milk protein, and formulas with the lactose replaced can be used for infants with lactose intolerance.

Supplements for the Infant

Breast milk or formula and the infant's own internal stores will meet most nutrient needs for the first 4 to 6 months. Thereafter, the introduction of properly chosen juices and foods will normally keep up with the infant's changing requirements. At 4 to 6 months, infants require additional iron, preferably in the form of iron-fortified cereal.

Breast milk does not provide enough vitamin D for the infant, and vitamin D deficiency causes impaired bone mineralization in children. Manufacturers fortify infant formulas with vitamin D, but due to the low concentration of vitamin D in breast milk, pediatricians routinely prescribe a vitamin D supplement for breast-fed infants whose mothers are vitamin D deficient or those who do not receive adequate exposure to sunlight.

If the water supply is deficient in fluoride, the breast-fed infant may also need supplemental fluoride after 6 months of age. The pediatrician should prescribe it, if appropriate.

Food for the Infant

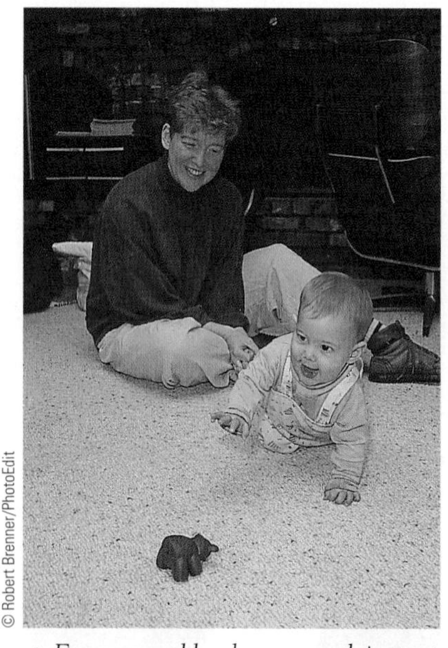

© Robert Brenner/PhotoEdit

▲ *Energy saved by slower growth is spent in increased activity.*

The infant's rapid growth and metabolism demand an adequate supply of all essential nutrients.[41] Because of their small size, infants need smaller total amounts of the nutrients than adults do, but when comparisons are made based on body weight, infants need over twice as much of many of the nutrients. Figure 11-2 compares a 5-month-old baby's needs with those of an adult man. As you can see, some of the differences are extraordinary. After 6 months, calorie needs increase less rapidly as the growth rate begins to slow down, but some of the energy saved by slower growth is spent on increased activity.

The most important nutrient of all—for infants as for everyone—is the one easiest to forget: water. The younger a child, the greater the percentage of the body weight is water and the more inefficient their kidneys are at concentrating waste, making water easy to lose. Conditions that cause fluid loss, such as vomiting, diarrhea, sweating, or normal urination, can rapidly propel an infant into life-threatening dehydration unless it is replaced. Fluid and electrolyte imbalances caused by diarrhea and infection kill more of the world's children than any disease or disaster. Because infants can only cry and cannot tell you what they are crying for, it is important to remember that they may need fluid and to let them drink plain water until their thirst is quenched.

Solid foods may normally be added to a baby's diet when the baby is about 6 months old, depending on readiness. The infant is developmentally ready when he or she can sit upright with support and can control head movements.

Solids should not be introduced too early because infants are more likely to develop allergies to them in the early months. But all babies are different, and the program of food additions should depend on the individual baby, not on any rigid schedule. Table 11-6 presents a suggested sequence for feeding infants.

The addition of foods to a baby's diet should be governed by three considerations: the baby's nutrient needs, the baby's physical readiness to handle different forms of foods, and the need to detect and control allergic reactions. Nutrients needed early are iron and vitamin C. Because a baby's stored iron supply from before birth runs out after the birthweight doubles, breast milk or iron-fortified formula, followed by iron-fortified cereals, and, later, meat or meat alternates such as legumes are recommended. Fruits that contain vitamin C can be introduced to enhance the absorption of iron. Fruit juices can be introduced at 6 months of age and should be served in a cup rather than a bottle.

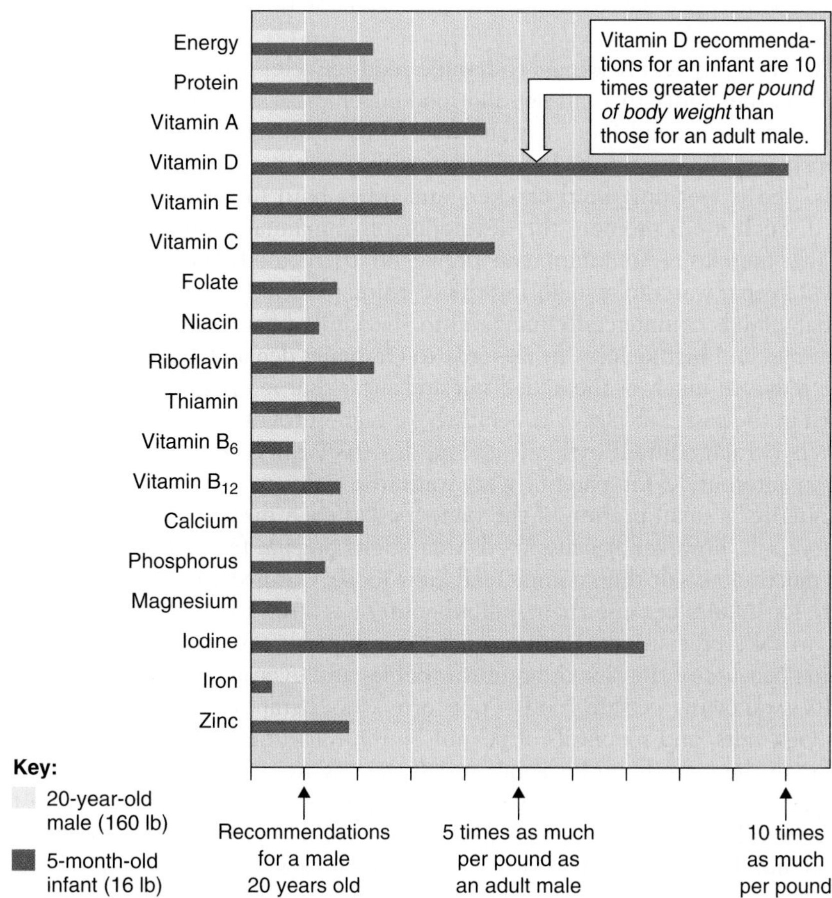

FIGURE 11-2
RECOMMENDED INTAKES OF NUTRIENTS FOR AN INFANT AND AN ADULT COMPARED ON THE BASIS OF BODY WEIGHT

SOURCE: Adapted with permission from E. Whitney and S. Rolfes, *Understanding Nutrition,* 10th ed. Copyright © 2005 by Wadsworth Publishing Co. All rights reserved.

As for sweets (soda pop, baby food desserts, candy, rich pies, and cakes), there is no room in the baby's diet for these empty-calorie foods. In contrast, naturally sweetened fruits and juices supply not only calories, but they also supply needed nutrients to support normal growth and development. However, the AAP recommends limiting juice consumption by infants and young children to

TABLE 11-6
First Foods for the Infant

Age (Months)	Food Additions
4 to 6	Iron-fortified rice cereal, followed by other cereals (baby can swallow nonliquid foods now)[a]
6 to 8	Mashed vegetables and fruits, infant breads and crackers; unsweetened fruit juices[b]
8 to 10	Protein foods (soft cheeses, yogurt, tofu, mashed cooked beans, finely chopped meat, fish, chicken, egg yolk), toast, teething crackers (for emerging teeth), soft-cooked vegetables, and fruit
10 to 12	Whole egg (allergies are less likely now), whole milk (at 1 year)

[a] Mix with breast milk, formula, or water. Later, other cereals can be introduced, but they should still be iron-fortified varieties.
[b] All baby juices are fortified with vitamin C. Orange juice may cause allergies; apple juice may be a better juice to feed first. Offer juices in a cup to prevent nursing bottle syndrome.

SOURCE: Adapted from Committee on Nutrition, American Academy of Pediatrics, *Pediatric Nutrition Handbook,* 5th ed., ed. R. E. Kleinman (Elk Grove Village, IL: American Academy of Pediatrics, 2004).

4 to 6 ounces per day to avoid the displacement of other important nutrients from their diets.[42]

A baby's physical readiness to handle foods develops in many small steps. For example, the ability to swallow solid food develops at around 4 to 6 months, and experience with solid food at that time helps to develop swallowing ability by desensitizing the gag reflex. Later still, when a baby can sit up, can handle finger foods, and is teething, hard crackers and other hard finger foods may be introduced. Such foods promote the development of manual dexterity and control of the jaw muscles. (An infant can choke on these foods, however, so an adult should keep a watchful eye during the learning process.)

Baby foods commercially prepared in the United States and Canada are safe, nutritious, and high quality. In response to consumer demand, baby food companies have removed much of the added salt and sugar that many of their products contained in the past. Baby foods generally have high nutrient density, except for mixed dinners (which contain little meat) and desserts (which are heavily sweetened).

An alternative for parents who want the baby to have family foods is to "blenderize" a small portion of the table food at each meal. This requires cooking without salt, however, because foods that adults prepare for themselves often contain much more salt than commercial baby foods. Canned vegetables are inappropriate for infants because their sodium content is often too high. It is also important to take precautions against food poisoning. Honey should *never* be fed to infants because of the risk of botulism. Babies and even young children have difficulty swallowing certain foods—popcorn, whole grapes, hard candies, bite-size hot dogs, nuts, and spoonfuls of peanut butter, for instance. An infant can easily choke on these foods, and it is not worth the risk to give such foods to infants.

▲ *Let infants handle food as they become ready.*

At 1 year of age, the obvious food to supply most of the nutrients the baby needs is still milk; 2 to 3½ cups a day are now sufficient. Infants less than 2 years old should drink whole milk rather than low-fat or fat-free milk. They need the fat and vitamins A and D of fortified whole milk until 2 years of age. The other foods—meat, iron-fortified cereal, enriched or whole-grain bread, fruits, and vegetables—should be supplied in variety and in sufficient amounts to round out total calorie needs. Ideally, the 1-year-old is sitting at the table and eating many of the same foods everyone else eats. A meal plan that meets the requirements for the 1-year-old is shown in Table 11-7.

The wise parent of a 1-year-old offers nutrition and love together. Both promote growth. It is literally true that "feeding with love" produces better growth in both weight and height of children than feeding the same food in an emotionally negative climate. It also promotes better brain development. The formation of nerve-to-nerve connections in the brain depends both on nutrients and on environmental stimulation.

The person feeding a 1-year-old should keep in mind that the baby is also developing eating habits that will persist throughout life.[43] Mealtimes should be relaxed and leisurely. Children should learn to eat slowly, pause and enjoy their table companions, and stop eating when they are full. The "clean your plate" dictum should be stamped out for all time, and in its place, parents who wish to avoid waste should learn to serve smaller portions or teach their children to serve themselves as much as they truly want to eat. Physical activity should be encouraged on a daily basis to promote strong skeletal and muscular development and to establish habits that will promote good health throughout life.

Nutrition-Related Problems of Infancy

Iron deficiency and food allergies are two of the most significant nutrition-related problems of infants.

Iron Deficiency Iron deficiency remains a prevalent nutritional problem for infants, although it has declined in recent years in large part due to the increasing use of iron-fortified formulas. Feeding cow's milk earlier than recommended in infancy can cause iron deficiency because of its poor iron content and the potential to cause gastrointestinal blood loss in susceptible infants.[44] Other factors contributing to iron deficiency in infancy include breastfeeding for more than 6 months without providing supplemental iron, feeding infant formula that is not fortified with iron, the infant's rapid rate of growth, low birthweight, and low socioeconomic status. To prevent iron deficiency, the American Academy of Pediatrics recommends that infants be fed breast milk or iron-fortified formula for the first year of life, with appropriate foods added as shown in Table 11–6.

Food Allergies Genetics is probably the most significant factor affecting an infant's susceptibility to food allergies. At-risk infants can be identified by means of careful skin testing and by a family history. Breast milk is recommended for those infants allergic to cow's milk protein and is preferable to soy or goat's milk formulas because infants are sometimes allergic to these proteins as well. To reduce the risk of food sensitivity or allergic reactions to other foods, new foods should be introduced one at a time to facilitate prompt detection of allergies. For example, when cereals are introduced, try rice cereal first for 5 to 7 days; it causes allergy least often. Try wheat cereal last; it is the most common offender. If a cereal causes irritability from skin rash, digestive upset, or respiratory discomfort, discontinue its use before going on to the next food. Allergies aren't easy to detect. About nine times out of ten, the allergy won't be evident immediately but will manifest itself in vague symptoms occurring up to 5 days after the offending food is eaten. Allowing several days to elapse between the introduction of each new food will provide time for the clinical symptoms to appear so that the offending food may be identified. The Nutrition Action feature in Chapter 5 provides more information about food allergies and intolerances in children and adults.

TABLE 11-7
Meal Plan for a 1-Year-Old

Breakfast
½ c whole milk
½ c iron-fortified cereal
1–2 tbsp fruit

Snack
½ c yogurt
Teething crackers
1–2 tbsp fruit

Lunch
1 c whole milk
2 to 3 tbsp vegetables
1 egg or 1 oz chopped meat or well-cooked mashed legumes
½ c noodles

Snack
½ c whole milk
½ slice toast
1 tbsp peanut butter

Supper
1 c whole milk
2 oz chopped meat or well-cooked, mashed legumes
½ c potato or rice
2 to 3 tbsp vegetables
2 to 3 tbsp fruit

Early and Middle Childhood

Childhood is a critical time in human development. Children typically grow taller by two to three inches and heavier by five or more pounds each year between the age of one and adolescence. They master fine motor skills (including those related to eating and drinking), become increasingly independent, and learn to express themselves appropriately. This section describes the nutrient requirements of children and the primary nutritional problems of this population.

Growth and Nutrient Needs of Children

After age one, a child's growth rate slows, but the body continues to change dramatically. At age 1, most babies have just learned to stand and toddle; by age 2, they can take long strides with solid confidence and are learning to run, jump, and climb. The internal change that makes these new accomplishments possible is the accumulation of a larger body mass and greater density of bone and muscle tissue. These changes are obvious in Figure 11-3.

Children generally become leaner between the ages of 6 months and 6 years, after which time there is a gradual increase in fat thickness in both males and females until puberty. Females have a greater body fat content than males at all stages of development. The energy requirements of children are determined by their individual basal metabolic rates, activity patterns, and rates of growth. Toddlers (ages 1 to 3 years) need about 1,000 calories per day. By the age of 10, children need about 2,000 calories per day. Appetite decreases markedly around the age of 1 year, in line with the great reduction in growth rate. Thereafter, the appetite fluctuates; a child will need and demand much more food during periods of rapid growth than during slow periods.

To provide the gradually increasing needs for all nutrients during the growing years, the MyPyramid Food Guide recommends a balance among milk and milk

FIGURE 11-3
1-YEAR-OLD AND 2-YEAR-OLD
The 2-year-old has lost much of the baby fat; the muscles (especially in the back, buttocks, and legs) have firmed and strengthened, and the leg bones have lengthened and increased in density.

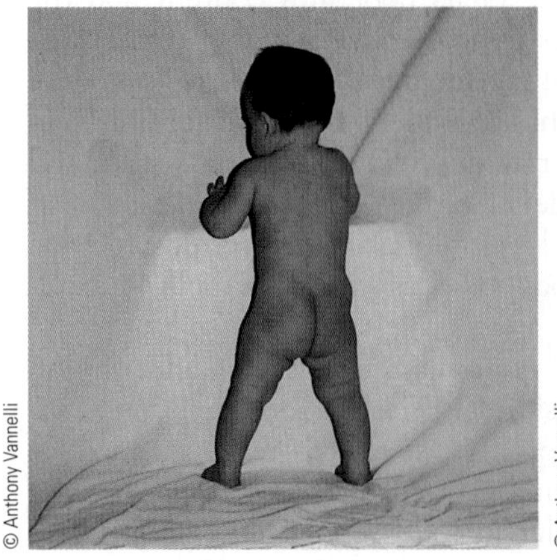

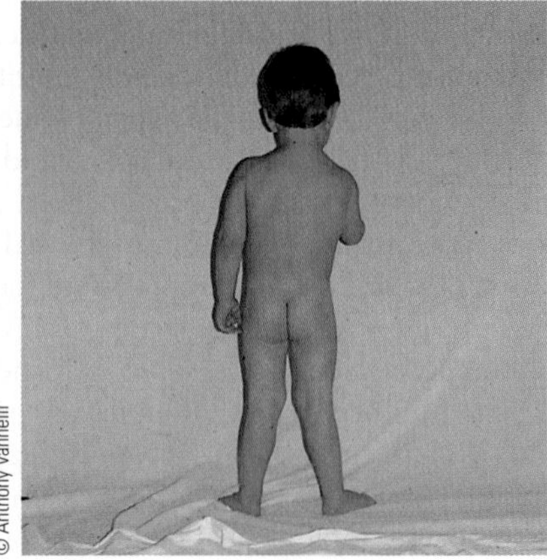

products, meats and meat alternates, fruits, vegetables, and grains. The interactive *MyPyramid for Kids* food guide presents the recommended proportion of food from each food group and focuses on the importance of making healthy food choices every day (see Figure 11-4). It also provides a general guide regarding which foods should be offered on a daily basis. In addition, the www.mypyramid.gov website offers interactive games and guides to help children and parents along the way.

After the crucial first year, a parent can still do a great deal to foster the development of healthful eating habits.[45] Table 11-8 offers tips to make feeding times enjoyable for both parent and child. The goal is to teach children to like nutritious foods in all food categories. (Nutrient intake recommendations for children are given on the inside front cover.)

Candy, cola, and other concentrated sweets must be limited in a child's diet to help ensure that their needed nutrients will be supplied by other foods. A child can't be expected to choose nutritious foods on the basis of taste alone because our human preference for sweets is innate. On the other hand, an active

TABLE 11-8
Strategies to Foster Healthful Eating Habits and Happy Mealtimes

These tips may make feeding time easier and more relaxing for both parent and child:

- Schedule regular meals and snacks for toddlers because they require frequent feeding to ensure adequate intake of calories and nutrients.
- Offer a variety of foods that includes at least one that the child likes.
- Remain calm if the child leaves an entire meal untouched.
- Do not be concerned about short food jags, stretches of time when the child wants the same food over and over.
- Allow the child to eat slowly.
- Offer healthy food in a relaxed manner, and children will eat what they need. Try these suggestions for healthful snacking:

  Keep plenty of washed and cut raw vegetables in the refrigerator. Team up with a yogurt or bean dip.

  Top frozen waffles and pancakes with fresh fruit for a tasty and refreshing snack.

  Make your own frozen juice pops in an ice cube tray.

  Put together a large batch of cereal, pretzel, and nut mix. Divide into individual plastic bags.

SOURCE: Adapted from M. G. Hermann, *The ABCs of Children's Nutrition* (Chicago: American Dietetic Association, 1991), 3–7.

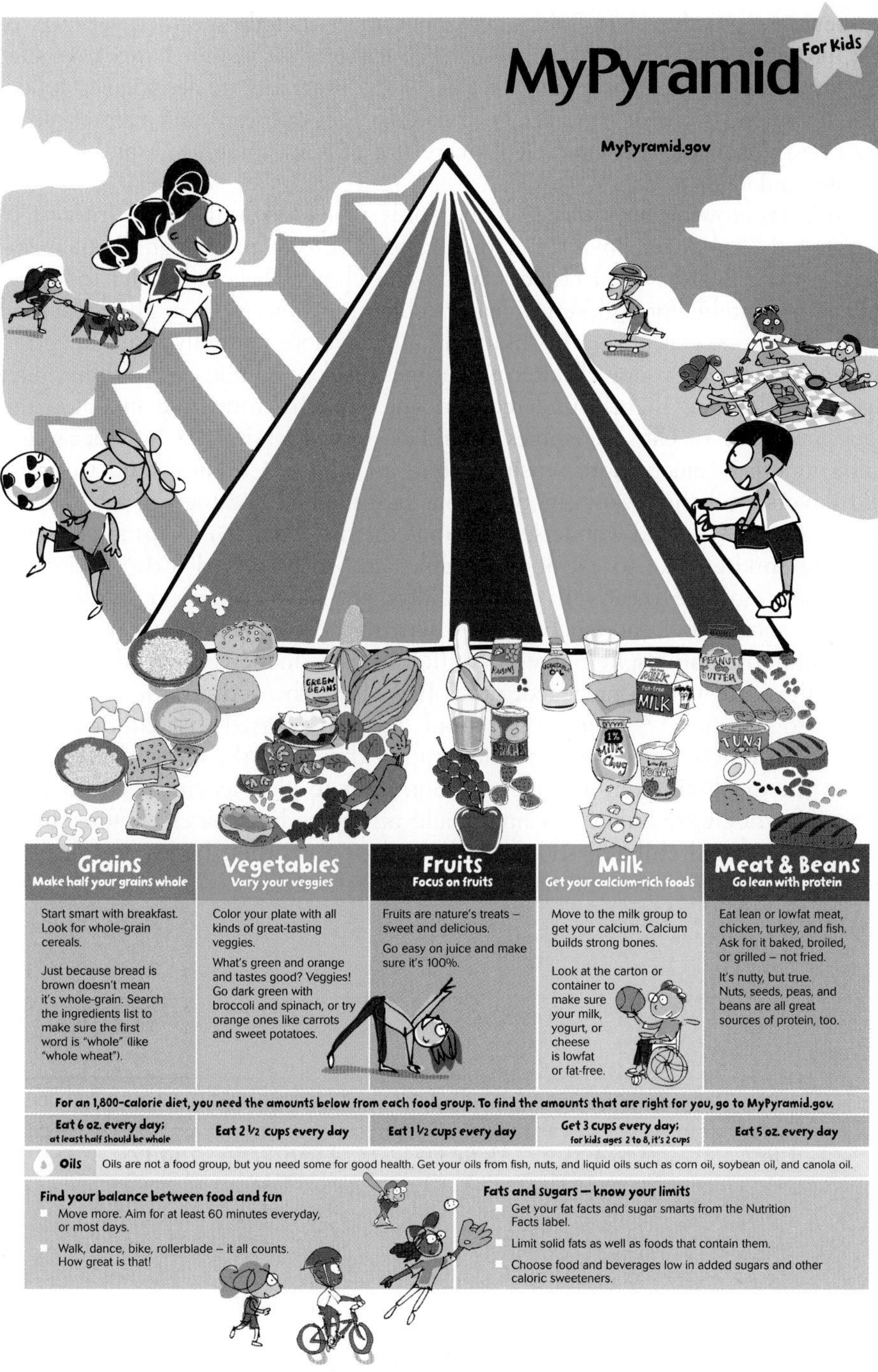

FIGURE 11-4
MYPYRAMID FOR KIDS

SOURCE: United States Department of Agriculture, Food and Nutrition Service September 2005; available at www.mypyramid.gov/kids/index.html

child can enjoy the higher-calorie nutritious foods in each category: ice cream or pudding in the milk group and whole-grain crackers in the grains group. These foods, made from milk and grain, carry valuable nutrients and encourage a child to learn, appropriately, that eating is fun.

Other Factors That Influence Childhood Nutrition

Although parents do what they can to establish favorable eating behaviors during the transition from infancy to childhood, when their children enter preschool or grade school they encounter foods prepared and served by outsiders. The U.S. government funds several programs to provide nutritious, high-quality meals for children at school. School lunches are designed to meet certain requirements. As shown in Table 11-9, they must include specified servings of milk, protein-rich foods (meat, cheese, eggs, legumes, or peanut butter), vegetables, fruits, and bread or other grain foods. These programs are designed to follow the recommendations of the *Dietary Guidelines for Americans* and to provide at least one-third of the recommended intakes for protein, vitamins A and C, iron, calcium, and calories.[46]

Children growing up today need not only to be fed well in the interest of their growth and development, but they also need to learn enough about nutrition to make healthy food choices when the choices become theirs to make. Thus, it is desirable for children to learn to like nutritious foods in all of the food groups. With one exception, this liking usually develops naturally. The exception is vegetables, which young children sometimes dislike and refuse. Children prefer vegetables that are slightly undercooked and crunchy, attractive in color and shape, and easy to eat. Vegetables should be served warm, not hot, because a child's mouth is much more sensitive than an adult's. Children tend to prefer mild flavors, and they want smooth foods like mashed potatoes or pea soup to have *no lumps*. (A child wonders, with some disgust, what the lumps might be.)

Little children like to eat at little tables and to be served little portions of food. They also love to eat with other children and have been observed to stay at the table longer and eat much more when in the company of their peers. A bright, unhurried atmosphere free of conflict is also conducive to good appetite.

Ideally, each meal is preceded, not followed, by the activity the child looks forward to the most. A number of schools have discovered that children eat a much better lunch if recess occurs before rather than after the meal. When recess follows the meal, children are likely to hurry out to play, leaving food on their plates that they were hungry for and would otherwise have eaten. Before sitting down to eat, small children should be helped to wash their hands and faces to decrease likelihood of contaminating the food with bacteria.

TABLE 11-9
School Lunch Patterns for Different Ages

	Preschool (Age)		Grade School Through High School (Grade)		
Food Group	**1 to 2**	**3 to 4**	**k to 3**	**4 to 6**	**7 to 12**
Meat or meat alternate					
1 serving:					
Lean meat, poultry, or fish	1 oz	1½ oz	1½ oz	2 oz	3 oz
Cheese	1 oz	1½ oz	1½ oz	2 oz	3 oz
Large egg(s)	½	¾	¾	1	1½
Cooked dry beans or peas	¼ c	⅜ c	⅜ c	½ c	¾ c
Peanut butter or other nut or seed butters	2 tbsp	3 tbsp	3 tbsp	4 tbsp	6 tbsp
Peanuts, soy nuts, tree nuts, seeds[a]	½ oz	¾ oz	¾ oz	1 oz	1½ oz
Yogurt, plain or flavored	4 oz	6 oz	6 oz	8 oz	12 oz
Vegetable and/or fruit					
2 or more servings, both to total	½ c	½ c	½ c	¾ c	¾ c
Bread or bread alternate					
Servings[b]	5 per week	8 per week	8 per week	8 per week	10 per week
Milk					
1 serving of fluid milk	¾ c	¾ c	1 c	1 c	1 c

[a] Can be used to meet up to ½ serving of meat, but must be accompanied by other meat/meat alternate in the meal.
[b] A serving is 1 slice bread; 1 biscuit, roll, or muffin; ½ cup cooked rice, pasta, or cereal grain, minimum of ½ serving per day for ages 1–2 and 1 serving per day for all others.

SOURCE: U.S. Department of Agriculture, *Food Program Facts—National School Lunch Program*, 2003.

Many little children, both boys and girls, enjoy helping in the kitchen. Their participation provides many opportunities to encourage good food habits. Vegetables are pretty, especially when fresh, and provide opportunities to learn about color, about growing things and their seeds, about shapes and textures—all of which are fascinating to young children. Measuring, stirring, decorating, cutting, and arranging vegetables are skills even a very small child can practice with enjoyment and pride.

When introducing new foods at the table, parents are advised to offer them one at a time—and only a small amount at first. Whenever possible, the new food should be presented at the beginning of the meal, when the child is hungry. If the child is irritable, or feeling sick, don't insist, but withdraw the new food and try it again a few days later. Remember, parents have inclinations and dislikes to which they feel entitled; children should be accorded the same privilege.

Nutrition-Related Problems of Childhood

Nutrition plays a critical role in the development and growth of children. However, as shown in Figure 11-5, the diet quality of most children ages 2 to 9 is less than optimal.[47] As indicated by the **Healthy Eating Index (HEI),** children ages 7 to 9 have a lower diet quality than younger children, and the lower quality is associated with low fruit and poor sodium scores—perhaps because as children get older, they consume more fast food and salty snacks. Dietary quality continues to decline from childhood to adolescence, especially with the decreased consumption of vegetables, fruits, and milk, and the increased intake of soft drinks. More than two-thirds of children consume well above the recommended levels of total fat and saturated fat.[48] For all children, the percent of total fat from milk and eggs has decreased, whereas the percentage of fat from cheese and snacks has increased.[49]

FIGURE 11-5
A HEALTHY EATING REPORT CARD FOR CHILDREN, AGES 2 TO 9

SOURCE: Center for Nutrition Policy and Promotion, USDA, 2001.

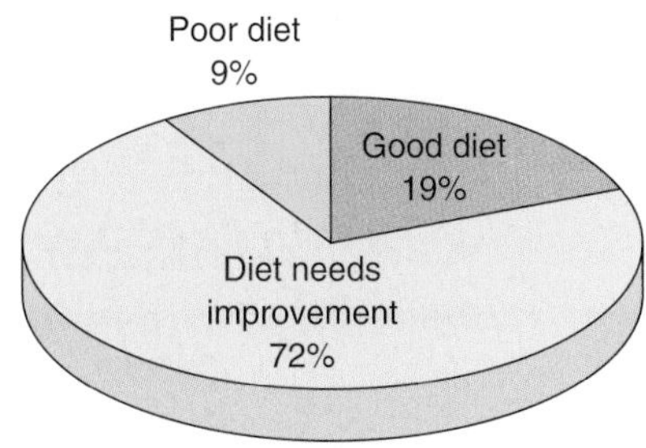

National surveys show that beverage choices for children of all ages has changed from whole milk to lower-fat milk, soft drinks, and fruit and fruit-flavored drinks.[50] These changes are especially pronounced for adolescents. On any given day, a majority of all children consume soft drinks, which are the second leading energy source for children ages 2 to 18. The primary energy source for children and youth comes from a high consumption of grain products, found in dough-based dishes such as pizza, pastas, and Mexican food.[51] These foods are not only high in calories but are often high in total fat and sodium and low in fiber.

The most common nutrition-related problems among U.S. children include overweight and obesity, iron-deficiency anemia, and high blood cholesterol levels. The epidemic of overweight among U.S. children requires a multifaceted strategy to address the health, social, physical, and environmental issues that may prevent and treat this escalating public health problem.[52] This chapter's Spotlight feature addresses the problem of overweight in children.

Iron-Deficiency Anemia Of all nutritional disorders, other than obesity, in U.S. children, the most common is iron-deficiency anemia. It is most prevalent in low-birthweight infants, babies from 6 months to 2 years of age, and in children and adolescents from low-income families.[53] To ensure adequate iron nutrition, parents should offer an abundance of iron-rich foods such as lean meats, fish, poultry, eggs, and legumes. Grain products should be whole-grain or enriched only. Milk, beneficial as it is, is a poor iron source, and dairy products should be consumed only in the amounts needed to ensure optimal calcium intakes.

High Blood Cholesterol Considerable evidence exists that atherosclerosis begins in childhood and that this process is related to high blood cholesterol levels.[54] Children and adolescents in the United States have higher blood cholesterol levels and higher dietary intakes of saturated fat and cholesterol than children in other countries.[55] A panel of experts from major U.S. health and

Healthy Eating Index (HEI) a summary measure of the quality of one's diet. The HEI provides an overall picture of how well one's diet conforms to the nutrition recommendations contained in the Dietary Guidelines for Americans and the USDA Pyramid. The Index factors in such dietary practices as consumption of total fat, saturated fat, cholesterol, and sodium, and the variety of foods in the diet. Check out the Interactive HEI at www.cnpp.usda.gov.

professional organizations has recommended that children age 2 and older eat diets containing no more than 30 percent of total calories from fat, less than 10 percent of total calories from saturated fat, and less than 300 milligrams of cholesterol daily. However, from birth to 2 years of age, a child's fat consumption should *not* be restricted because fat is a concentrated source of the calories needed to ensure proper physical development.

The panel also advised that youngsters and teens should have their blood cholesterol measured if they have one parent with a high blood cholesterol level.[56] For children and adolescents, a total of 200 milligrams per deciliter of blood or more is considered high; 170 to 199 is borderline high; and less than 170 is acceptable.

The Importance of Teen Nutrition

Adolescence is a time of change. Between the ages of about 10 and 18 years in girls and 12 and 20 years in boys, marked changes take place in physical, intellectual, and emotional growth and development. The maturation process is initiated and controlled by a variety of hormones, including, among others, growth hormone, prolactin, estrogen, testosterone, and the thyroid hormones. Many aspects of the maturation process are influenced by dietary intake and nutritional status. This section reviews the nutrient requirements and special nutrition-related problems of teenagers.

Nutrient Needs of Adolescents

The dramatic changes in body composition and the rate of growth that occur during early adolescence give rise to the familiar phrase "the adolescent growth spurt." The magnitude of these changes is such that the linear growth increments during adolescence can contribute about 15 to 25 percent of adult stature. The rate of weight gain can contribute anywhere from 40 to 50 percent of the adult body mass. This remarkable growth rate requires adequate intakes of energy and nutrients.[57] The individual energy needs for teenagers is influenced by body size, activity levels, and biological factors affecting growth. Consult the DRI table shown on the inside front cover for the energy and nutrient intake recommendations for adolescents.

The rates and patterns of growth for individual teenagers varies widely.[58] The only way to be sure teenagers are growing normally is to compare their heights and weights with previous measures taken at intervals and note whether reasonable progress is being made. Physical growth charts for children and adolescents and information on assessing children's growth status can be accessed at www.cdc.gov/growthcharts.

Nutrition-Related Problems of Adolescents

Most adolescents in the United States are perceived as "healthy." However, many U.S. adolescents experience a variety of health and nutrition problems, some related to their risk-seeking behaviors and an inability to deal with abstract notions such as "good health" and the link between current behaviors and long-term health. Nutrition-related problems among U.S. adolescents include overweight and obesity (discussed in this chapter's Spotlight feature), undernutrition, iron-deficiency anemia, low dietary calcium intakes, high blood cholesterol levels, dental caries, and eating disorders.

Undernutrition Some groups of adolescents are at risk for reduced energy and food intakes, such as adolescents from low-income families and those who have run away from home or abuse alcohol or other drugs. Approximately 17 percent of youth ages 18 or younger are estimated to be living below the poverty line. African American and Hispanic teenagers are nearly twice as likely to live in poverty as are white youth. Chronic dieters are also at risk. Thirteen percent of the 17,354 females in grades 7 through 12 who were interviewed in the Minnesota Adolescent Survey reported being chronic dieters, defined as always on a diet or having been on a diet for 10 of the previous 12 months.[59]

▲ *Nutritious snacks can supply added nutrients to the active teenager's diet.*

Iron-Deficiency Anemia Iron needs increase during adolescence, especially for females as they start to menstruate. In boys, the requirements for absorbed iron increase because of an expanding blood volume and rise in hemoglobin concentration that accompanies the development of larger muscles and sexual maturation. (After the male adolescent growth spurt, their need for iron falls off.) Whereas most males have an adequate iron intake during adolescence, many females between 12 and 22 years of age have iron intakes below the current recommended intake. Girls typically consume fewer total calories and less meat than boys do.

Low Calcium Intakes Another problem nutrient during the teen years is calcium. Low intakes of calcium-rich foods during adolescence compromise peak bone mass development and increase the risk of osteoporosis later in life. Adolescents need a minimum of 1,300 milligrams of calcium each day to achieve an optimal peak bone mass and healthy bones. However, the calcium intake for many teenagers, especially girls, is below the recommended amount.[60] Adolescent girls on low-calorie diets who drink diet soft drinks instead of milk are at particular risk for age-related bone loss and subsequent fractures. See the Spotlight feature in Chapter 7 for more about calcium and osteoporosis.

High Blood Cholesterol Teenagers have many of the same risk factors for high blood cholesterol as adults: family history of coronary heart disease; diets high in total fat, saturated fat, and cholesterol; hypertension; low activity levels; and smoking. Although the process of atherosclerosis is not completely understood, it is believed that the fatty streaks in blood vessels that develop in young people progress to the fibrous plaques of adulthood.[61] For more information about diet and heart disease, see the Spotlight feature in Chapter 4.

Dental Caries Although dental caries are largely preventable, this remains the most common chronic disease of children aged 5 to 17 years.[62] By age 5, 60 percent of all children have had tooth decay, and more than 90 percent of 18-year-olds have experienced decay.[63] Children in low-income households, and especially those who are American Indian, African American, or Hispanic, have three times the risk of tooth decay because they lack access to or encounter barriers to accessing dental services. Fortunately, the incidence of dental caries has decreased by as much as 30 to 50 percent over the past two decades, partly because of fluoridation of public drinking water, improved dental hygiene, and the use of fluoride in toothpastes and mouthwashes.

Eating Disorders Eating disorders have become a serious health problem in recent years. The most common eating disorders are anorexia nervosa and bulimia nervosa. A constellation of individual, familial, sociocultural, and biological factors contribute to these disorders, which threaten physical health and psychological well-being. See the Spotlight feature in Chapter 9 for more information about eating disorders.

Some individuals are more predisposed to developing an eating disorder than others. For example, about 90 percent of people with eating disorders are female. Most are Caucasian, with few cases seen among blacks and other minority groups. Most individuals who develop eating disorders are adolescents or young adults who typically began experiencing food-related and self-image problems between the ages of 14 and 30 years. Because these syndromes are surrounded by secrecy, their prevalence is not known with certainty, although it has increased dramatically within the past three decades.[64] Estimates in the general population range from 1 percent for anorexia nervosa to 1 to 5 percent for bulimia nervosa. These two eating disorders are also sometimes seen in adolescent athletes, many of whom compete in sports such as gymnastics, wrestling, distance running, diving, horse racing, and swimming, which demand a rigid control of body weight.

Nutrition in Later Life

In the United States, the life expectancy is now 73 years for men and 79 years for women.[65] To what extent is aging inevitable? Apparently, aging is an inevitable, natural process programmed into our genes at conception. Nevertheless, we can adopt lifestyle habits, such as consuming a healthful diet, exercising, and paying attention to our work and recreational environments, that will slow the aging process—within the natural limits set by heredity.[66] Clearly, good nutrition can slow down and ease the aging process in many significant ways. However, no potions, foods, or pills will prolong youth.

One approach to the prevention of aging has been to study other cultures in the hope of finding an extremely long-lived race of people and then learning their secrets. (See the *Eat Well Be Well* feature on page 371 for more insights regarding nutrition and longevity.) The views of the experts are summed up best by simply saying that disease can *shorten* people's lives and poor nutrition practices make diseases more likely to occur. Thus, by postponing and slowing disease processes, optimal nutrition can help to prolong life and can promote health to make later years more enjoyable.[67] This section focuses on the nutrient requirements of older adults, the diseases that seem to come with age, their risk factors, and the relevance of nutrition to these diseases.

FIGURE 11-6
THE AGING OF THE POPULATION

In 1900, 4 percent of the U.S. population (3.1 million people) were more than 65 years of age; in 2000, 12.4 percent (35 million people) were more than 65; by 2030, 20 percent (about 70 million people) will have reached age 65.

SOURCE: Data from *A Profile of Older Americans, 2004;* available at www.aoa.gov.

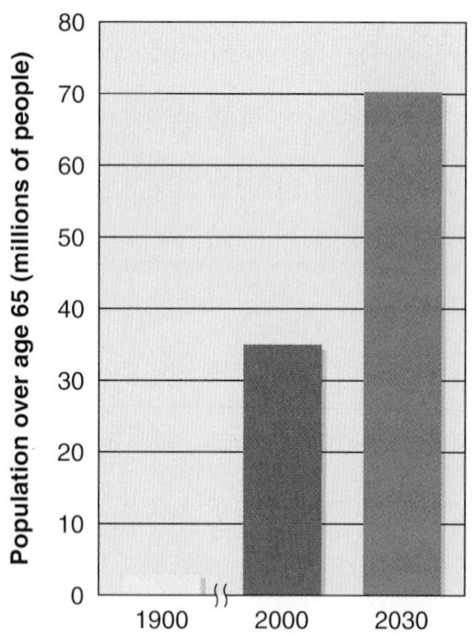

Demographic Trends and Aging

The number of elderly (aged 65 years and older) in the United States will double by 2030 to more than 70 million people. In 2000, people aged 65 and over accounted for 12.4 percent of the population, and this proportion is expected to rise to approximately 14 percent in 2010 and increase to nearly 20 percent by 2030. Nearly 12 percent of the population will be more than age 74 by 2030. The increased growth in the elderly population in the United States is illustrated in Figure 11-6.[68]

The baby boom that took place between 1946 and 1964 and improved life expectancy are important contributors to the growing elderly population in the United States. Baby boomers will increase the numbers of the older middle-aged (ages 46 through 63) until 2010, when they will begin to swell the ranks of the

retired population. This improved life expectancy has resulted from better prenatal and postnatal care and improved means of combating disease in older adults. For example, the death rate from heart disease began to decline in the 1960s and continues to fall today. More than half of the drop is attributed to a decline in smoking and fewer people with high blood pressure or high blood cholesterol.

▲ *Any form of physical activity—even a 10-minute walk around the block—is better than nothing.*

Healthy Adults

An individual's current health profile is substantially determined by behavioral risk factors. Among the leading causes of death for adults ages 25 and older are heart disease, cancer, stroke, injuries, chronic lung disease, diabetes, and liver disease, all of which have been associated with behavioral risk factors. To what extent can nutrition prevent or retard the development of these diseases?

Many of the health problems associated with the later years are preventable or can be controlled.[69] For example, eliminating certain risk behaviors and adopting healthy ones can improve the quality of life for older persons and reduce their risk of disability. For example, incorporating exercise and a healthful diet into one's lifestyle can contribute to weight loss and to controlling three important risk factors for heart disease: high saturated fat intake, overweight, and a sedentary lifestyle. Figure 11-7 illustrates how the same set of dietary recommendations can promote general health and help to prevent a broad spectrum of chronic diseases.[70]

Figure 11-8 puts nutrition (a factor you *can* control) in perspective with respect to heredity (a factor you can't control). It illustrates the point that some diseases are much more responsive to nutrition than others and that some are not responsive at all. At one extreme are diseases that can be completely cured by supplying missing nutrients, and at the other extreme are certain genetic, or inherited, diseases that are unaltered by nutrition. Most diseases fall in between, being influenced by inherited susceptibility but responsive to dietary manipulations that help to counteract the disease process. Thus, diabetes may be controlled by means of a diet low in fat and high in complex carbohydrate, arthritis may be somewhat relieved by weight reduction, and cardiovascular disease may respond favorably to a diet low in saturated fat and cholesterol.

FIGURE 11-7
CONSISTENCY OF RECOMMENDATIONS FOR REDUCING THE RISK OF CHRONIC DISEASES OR THEIR COMPLICATIONS

Change diet →	Reduce fats	Control calories	Increase starch[a] and fiber	Reduce sodium	Control alcohol	Increase antioxidants[b]
Reduce risk ↓						
Heart disease	●	●	●	●		●
Cancer	●	●	●	●	●	●
Stroke	●	●		●	●	●
Diabetes	●	●	●			●
Gastrointestinal diseases[c]	●	●	●		●	

[a] Starch refers to complex carbohydrates provided by fruits, vegetables, and whole-grain products.

[b] Aim for a minimum of 5 to 9 servings of fruits and vegetables daily.

[c] Gastrointestinal diseases affected by dietary factors are primarily gallbladder disease (fat and calories), diverticular disease (fiber), and cirrhosis of the liver (alcohol).

SOURCE: J. M. McGinnis and M. Nestle, The Surgeon General's Report on Nutrition and Health: Policy implications and implementation strategies, *American Journal of Clinical Nutrition* 49 (1989): 26. © American Society for Clinical Nutrition; AHA Conference Proceedings, Summary of a Scientific Conference on Preventive Nutrition: Pediatrics to Geriatrics, *Circulation* 100 (1999): 450–456.

FIGURE 11-8
NUTRITION AND DISEASE

Not all diseases are equally influenced by diet. Some are purely hereditary, such as sickle-cell anemia. Some may be inherited (or the tendency to develop them may be inherited) but may also be influenced by diet, such as some forms of diabetes. Some are purely dietary, such as the vitamin and mineral deficiency diseases. Nutrition alone is certainly not enough to prevent many diseases, but it helps.

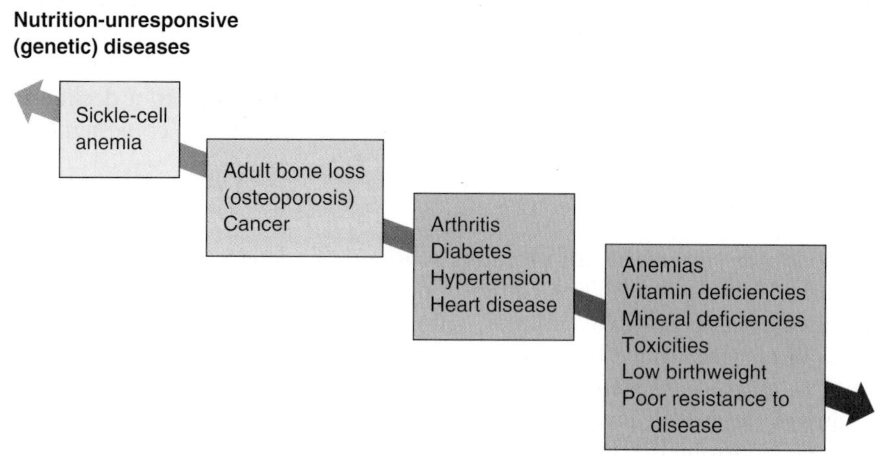

Aging and Nutrition Status

Growing old is often associated with frailty, sickness, and a loss of vitality. Although the aging members of our society do experience chronic illness and the associated disabilities, there is great heterogeneity among this population. That is, older people vary greatly in their social, economic, and lifestyle situations as well as their capacity to function and their physical condition.[71] Each person ages at a different rate, sometimes making chronological age different from biological age. Most older persons live at home, are fully independent, and have lives of good quality.[72] Older persons who have problems with the **activities of daily living (ADL)** are known as the frail elderly. Because they depend on others to perform these essential activities, they are likely to be at risk for malnutrition.

To meet nutritional needs, diets of older adults should include:

- At least three servings of low-fat dairy products or suitable alternatives
- Two or more servings from the Meat & Beans group that feature dried beans (rich in fiber), fish, eggs, and lean cuts of meat and poultry
- At least five servings of fruits and vegetables featuring foods that are richly colored (dark green, orange, red, or yellow)
- At least six servings of nutrient-dense, fiber-rich whole grains and fortified cereals
- At least eight glasses of water or noncaffeinated fluids per day to prevent dehydration

Nutritional Needs and Intakes

Many of the nutrient needs of the elderly are the same as for younger persons, but some special considerations deserve emphasis. Calorie needs decline with age because of a decrease in basal metabolism related to loss of lean tissue and a decrease in physical activity. The recommended energy intake decreases by 10 calories per day for males and 7 calories per day for females for each year of age above 19.[73] Given their lower energy allowances, older adults are advised to select mostly nutrient-dense foods. General nutrition guidelines for older adults are listed in the margin.[74] The recommended eating pattern reflects the lower caloric needs of most healthy older adults and emphasizes the need for adequate fluid intake. The recommended energy and nutrient intakes for older adults are shown inside the front cover.

On one side of the energy budget, energy is taken in, and on the other side, it is spent. If you are motivated to maintain your good health into the later years, you should plan regular exercise into your days. Ideally, the exercise should be intense enough to increase the heartbeat and respiration rate and to prevent the atrophy of all muscles (not only the heart) that otherwise takes place. Many older people believe that they can't participate in strenuous exercise, but studies have shown that they can do more than they think they can. Any exercise—even a ten-minute walk each day—is better than none. With persistence, great improvement is possible. Modest endurance training can improve cardiovascular and respiratory function and promote good muscle tone as well as controlling the accumulation of body fat.

activities of daily living (ADL) include bathing, dressing, grooming, transferring from bed to chair, going to the bathroom, and feeding one's self.

Although caloric needs may decrease with age, the need for certain nutrients such as calcium, vitamin D, vitamin C, vitamin B_{12}, and vitamin B_6 may actually

Eat Well Be Well: Eating Pattern for Longevity

Need proof that good dietary choices can have a profound effect on your health and longevity? Look no further than the Okinawans! These people dwell on a group of islands—known collectively as Okinawa—that lie southwest of mainland Japan. Okinawans enjoy one of the longest life spans of anyone on earth, and they do so while maintaining a very high quality of life. This fact has sparked much interest in the Okinawan culture and resulted in the 25-year Okinawa Centenarian Study that began in 1976 to investigate what makes these people so healthy.[75] Researchers studied more than six hundred Okinawan centenarians and numerous others in their seventies, eighties, and nineties.

When the study began, researchers were surprised to find that many centenarians were still very active, devoid of health problems, and looked years younger than their chronological age. Upon further investigation, they also discovered that these people have low levels of cancer-causing free radicals, low cardiovascular risks, extremely healthy bone densities, and a lower prevalence of dementia than those of the same age in other countries. Indeed, Okinawan elders have among the lowest mortality rates in the world from many chronic diseases, including cancer, stroke, osteoporosis, and heart disease. Although there is no magic bullet that results in the Okinawans' longevity and health, Okinawan centenarians have a number of variables in common.[76]

1 Enough Is Enough.

Okinawan elders have an average body mass index (BMI) that ranges from 18 to 22. The Okinawans stay lean by eating a low-calorie diet and practicing calorie control in a cultural habit known as *hara hachi bu* (only eating until they are 80 percent full) and keeping physically active every day. In contrast, middle-aged Okinawans, who have a less traditional lifestyle, have a BMI of 26, the highest in Japan and similar to Americans. As discussed in Chapter 9, a BMI of 25 and greater is considered overweight and places you at greater risk for chronic diseases, especially heart disease and stroke.

2 Moderation and a Healthy Lifestyle Are Key Cultural Values.

Okinawan elders never smoke. They consume a diet that is 80 percent plant-based and naturally high in unrefined whole grains, soy, vegetables, and fruits—all of which are rich in antioxidants and phytochemicals. They consume higher intakes of good fats from omega-3 rich fish and monounsaturated fats, and they have rather low amounts of saturated and *trans* fat in their diet. They keep active every day throughout their lives in a variety of activities such as gardening, traditional dance, and martial arts. In fact, many of those studied still participated in competitive games and karate past the age of 100!

◂ Okinawan twin sisters—Gin Kanie and Kin Narita at age 106—celebrate with family and friends. Okinawan centenarians maintain optimistic attitudes and strong social bonds throughout their lives.

Even though these concepts may seem rather simple, the average American is not following suit. With high stress, sedentary lifestyles, and diets high in saturated fats and low in protective nutrients such as antioxidants, it seems that the American way of life is the opposite of the Okinawans. It might be hard to drastically change your entire lifestyle to follow that of the Okinawans; however, making some dietary changes would be a good place to start. Who knows, maybe you too can live to be a centenarian! Here is a general overview of the Okinawan elders' way of eating:

- They eat about 500 calories less per day than Americans.
- They eat plenty of whole grains, seven servings of vegetables, and about four servings of fruits a day.
- They eat seaweed, coldwater fish, and other seafood choices at least three times a week.
- They include soy foods in their daily diet, and consume poultry, eggs, pork, beef, and other meats in moderation.
- They drink plenty of water and tea, and drink alcohol in moderation.
- They consume very few sweets.

3 Psychological and Spiritual Health Matters.

Okinawans put family first. They keep socially engaged, maintaining strong bonds with friends and family. They have an easy-going approach to life. Centenarians score high on optimistic attitudes and adaptability. They possess a strong sense of purpose, which translates roughly to "that which makes one's life worth living."

TABLE 11-10
The Impact of Aging on Nutrient Needs

Nutritional Concern	Age-Related Finding	Implications for Healthy Aging
Energy	Energy needs decrease due to decline in lean tissue.	Get regular physical activity, including strength training exercises (see Chapter 10).
Water	Reduced sense of thirst; increased urine output.	To prevent dehydration, drink 6 to 8 glasses of fluid a day.
Fiber	Constipation is a common complaint due to low fiber and fluid intakes and lack of exercise.	Consume 21 to 38 grams of fiber from a variety of sources such as fresh fruits, vegetables, legumes, and whole-grain products; get adequate amounts of fluid and exercise.
Vitamin D	Intakes may be low; skin becomes less able to synthesize vitamin from sunlight; less exposure to sunlight.	Choose fortified milk and cereals, salmon and other fatty fish; get daily exposure to sunlight if possible; or use a supplement (400 to 600 IU per day).
Calcium	Intakes typically low; reduced gastric acidity impairs absorption, reduced bone mass.	To offset bone loss, choose calcium-rich foods (low-fat or fat-free milk, yogurt, cheese), fortified sources of calcium (orange and other juices), or take a calcium supplement (see the Spotlight feature in Chapter 7 for guidelines); get regular weight-bearing exercise.
Iron	Anemia less common after menopause; reduced gastric acidity may impair absorption; chronic blood loss or low energy intakes may increase risk of deficiency.	Consume adequate calories from a variety of nutrient-dense foods with a source of vitamin C to enhance absorption of dietary iron.
Protein, vitamin E, vitamin B_6, zinc	Intakes may be low; decline in immune function.	Meet nutrient needs from low-fat servings of meat, fish, poultry, and legumes; use nuts in moderation; choose at least six servings of whole-grain products.
Antioxidants: vitamin C, vitamin E, carotenoids (beta-carotene, lutein)	Decline in vision (cataracts, macular degeneration); increased oxidative stress to body tissues.	Eat at least 5 servings of fruits and vegetables a day; choose regular servings of green leafy vegetables, brightly colored fruits and vegetables, and citrus fruits; choose whole-grain products; use nuts in moderation.
Vitamin A	Decline in liver's uptake of vitamin A; increased absorption may lower the aging body's requirement.	Avoid supplements of vitamin A.
Vitamin B_{12}	Intakes may be low; reduced gastric acidity impairs absorption of food-bound vitamin B_{12}.	Choose foods fortified with vitamin B_{12} (e.g., breakfast cereals), or take a supplement that contains vitamin B_{12}.
Folate, vitamin B_6, vitamin B_{12}	Intakes may be low—causing elevated blood levels of homocysteine and increased risk of heart disease; use of certain medications may impair B-vitamin status.	Consume citrus and other fresh fruits and green leafy vegetables, legumes, whole grains, and foods fortified with folate and vitamin B_{12} (such as breakfast cereals).

increase with the effects of aging (see Table 11-10).[77] For example, as many as 30 percent of persons older than 50 years may experience reduced stomach acidity, which can interfere with their ability to absorb vitamin B_{12}, calcium, and iron effectively from foods. In addition, older adults who have low intakes of fortified dairy products may have increased needs for vitamin D because their skin has become less efficient at making vitamin D when exposed to sunlight.[78]

Assessing the diet quality of older adults is important for identifying issues relevant to their health and nutrition status. According to the Healthy Eating Index, a composite measure of overall dietary quality based on the *Dietary Guidelines for Americans* and the USDA Pyramid, only 13 percent of the population aged 45 to 64 years consumes a "good" diet, with the dairy and fruit groups need-

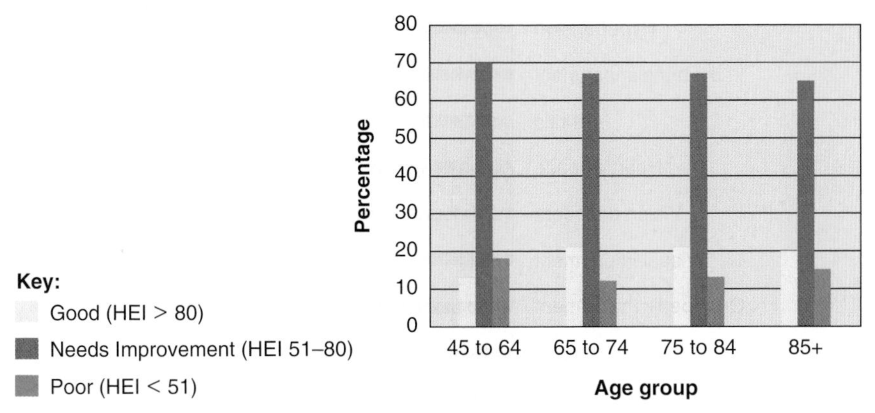

FIGURE 11-9
A HEALTHY EATING REPORT CARD FOR ADULTS AGE 45 OR OLDER

SOURCE: Center for Nutrition Policy and Promotion, United States Department of Agriculture, *Nutrition Insights* (July 1999).

*Dietary quality was measured by the Healthy Eating Index (HEI). The HEI is a summary measure of people's overall diet quality. The HEI is expressed as one score on a scale of 1 to 100, but it is comprised of the sum of 10 components. Each component score can range from 0 to 10. Components 1–5 measure the degree to which a person's diet conforms to the serving recommendations from the USDA MyPyramid's five major food groups: Grains, Vegetables, Fruits, Milk, and Meat & Beans. A high score for these components is reached by maximizing consumption of recommended amounts. Components 6–9 measure compliance of total fat and saturated fat intake according to the *Dietary Guidelines for Americans* and of cholesterol and sodium from the *Daily Values* listed on the Nutrition Facts label. A high score is reached by consuming at or below recommended amounts. The last component evaluates variety in the diet. A person consuming eight or more different foods each day will score 10 points. The HEI is available at www.usda.gov/cnpp.

ing the most improvement. Figure 11-9 summarizes the overall dietary quality of independent, free-living middle-aged and elderly adults.

Some health practitioners recommend nutritional supplements for older adults, particularly for chronic conditions such as osteoporosis, arthritis, or anemia. However, advertisers often target older adults by recommending supplements to reduce the effects of aging and increase longevity. Such advertising makes the elderly vulnerable to exploitation by quacks, when their money is better spent on nutritious foods. Surveys designed to find out what kinds of nutrient supplements older people are using tend to support the view that in many cases, older people are wasting their money.[79]

Nutrition-Related Problems of Older Adults

© Tom Ives

▲ *The evidence for using dietary supplements to slow the aging process is not as strong as the evidence supporting a diet rich in fruits, vegetables, and other nutrient-dense foods along with a physically active lifestyle.*

Although aging is not completely understood, we know that it involves progressive changes in every body tissue and organ: the brain, heart, lungs, digestive tract, and bones. After age 35, functional capacity declines in almost every organ system. Some changes, including oral problems, interfere with nutrient intake; others affect absorption, storage, and utilization of nutrients; and still others increase the excretion of and need for specific nutrients. Examples of various changes associated with aging that can affect nutritional status are listed in the margin and include sensory impairments, altered endocrine, gastrointestinal, and cardiovascular functions, as well as changes in the renal and musculoskeletal systems. Both genetic and environmental factors contribute to these declines. Many of the changes are inevitable, but a healthful lifestyle that combines regular physical activity with adequate intakes of all essential nutrients can forestall degeneration and improve the quality of life into the later years.

As a person gets older, the chances of suffering a chronic illness or functional impairment become greater. Among the diseases that befall some people in later life are heart disease, hypertension, cancer, diverticulosis, osteoporosis, dementia, diabetes, and gum disease. Chronic conditions contributing to disability include arthritis, heart disease, strokes, disorders of vision and hearing, nutritional deficiencies, and oral-dental problems (see Figure 11-10). Dementia (especially Alzheimer's disease) is a major contributor to disability and nursing home placement for those over age 75.[80] As noted in Table 11-11 malnutrition can occur secondary to these conditions, many of which require special diets that can further compromise nutrition status in the older adult.

Changes in biological function between the ages of 30 and 70

- Cardiac output: ↓30%
- Maximum heart rate: ↓25%
- Vital capacity: ↓40%
- Maximum O_2 uptake: ↓60%
- Muscle mass: ↓30%
- Hand grip, flexibility: ↓30%
- Bone mineralization: ↓20–30%
- Renal function: ↓40%
- Taste and smell: ↓90%
- Basal metabolic rate: ↓15%

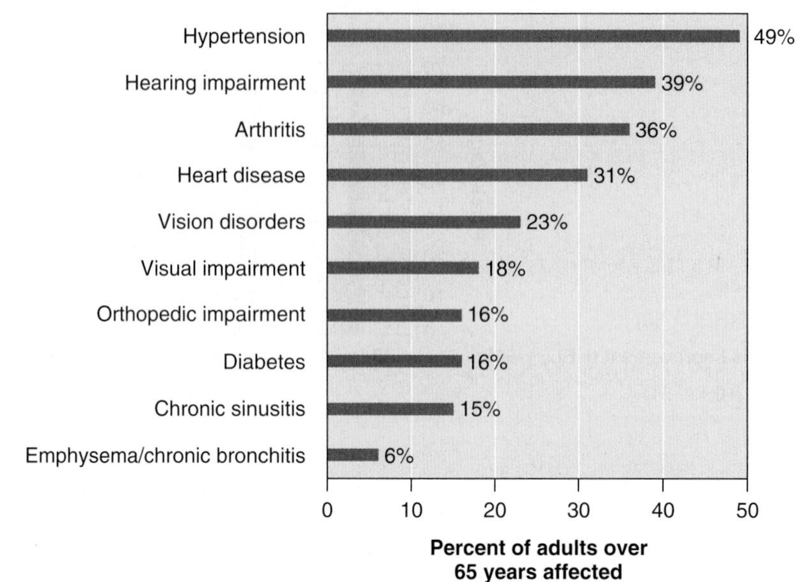

FIGURE 11-10
COMMON CAUSES OF DISABILITY IN OLDER PERSONS

SOURCE: National Center for Health Statistics, 2003.

Factors Affecting Nutrition Status of Older Adults

Individually or in combination, the social, economic, psychological, cultural, and environmental factors associated with aging may interact with the physiological changes and further affect nutrition status in older adults. These factors are listed in the margin on page 375.

Up to one-quarter of all elderly adults and one-half of all hospitalized elderly may be suffering from malnutrition.[81] About 3.6 million elderly adults (10.4 percent) have incomes at or below the poverty level; another 2.2 million are classi-

TABLE 11-11
Malnutrition That Is Secondary to Disease or Physiological State

Disease or Condition	Effects on Nutrition Status
Atherosclerosis	May increase difficulties in regulating fluid balances if caused by congestive heart failure.
Cancer	Weight loss, lack of appetite, and secondary malnutrition are common.
Dental and oral disease	May alter the ability to chew and thus reduce dietary intake. Increased likelihood of choking and aspiration.
Depression and dementia	Increased or decreased food intakes are common. A person with dementia may have decreased ability to get food, or the appetite may be very small or very great. Judgment and balance in meal planning are generally absent.
Diabetes mellitus	Increased risk of other diet-related diseases. Weight loss is needed if obesity is present.
Emphysema	May be difficult to eat because of breathing problems.
End-stage kidney disease	Alters fluid and electrolyte needs. Infections and low-grade fever may increase energy output and weight loss.
Gastrointestinal disorders	Increased risk of malabsorption of nutrients and consequent undernutrition.
High blood pressure	Weight gain may exacerbate high blood pressure.
Osteoarthritis	Makes motion difficult, including those activities related to purchasing, serving, eating, and cleaning up after meals. Predisposes people to a sedentary lifestyle and may give rise to obesity.
Osteoporosis	Limits the ability to purchase and prepare food if mobility is affected. If severe scoliosis is present, the appetite may be altered.
Stroke	May alter abilities in the cognitive and motor realms related to food and eating.

SOURCE: Adapted from Institute of Medicine, *The Second Fifty Years: Promoting Health and Preventing Disability* (Washington, D.C.: National Academy Press, 1992), pp. 168–69.

fied as "near-poor." In addition, a national survey found that about 30 percent of all noninstitutionalized people over the age of 65 live alone, 45 percent take multiple prescription drugs that can interfere with appetite and nutrient absorption, and 22 percent have difficulties with basic activities (for example, preparing meals and taking medications). All of these factors place older persons at risk for malnutrition.[82] Identifying older adults who are at nutritional risk is the first step in maintaining their quality of life and functional status.

The American Dietetic Association, the American Academy of Family Physicians, and the National Council on Aging have collaborated since 1990 on an effort—called the Nutrition Screening Initiative—to promote nutrition screening and early intervention as part of routine health care. A key premise to the Nutrition Screening Initiative is that nutrition status is a "vital sign"—as vital to health assessment as blood pressure and pulse rate.[83] The Nutrition Screening Initiative has identified 10 signs of malnutrition in the elderly. The first letter of each item in the list spells out the word *DETERMINE*, as shown below. Using these signs, they developed a 10-question, self-assessment checklist (see Figure 11-11) to help individuals identify and score factors that place them at nutritional risk. This checklist addresses disease, eating status, tooth loss or mouth pain, economic hardship, reduced social contact, multiple medications, involuntary weight loss or gain, and need for assistance with self-care.[84]

Financial resources, living arrangements, and a social support network, including availability of caregivers, can also directly affect a person's nutrition status. Poverty and social isolation particularly impair the nutrition status of many older adults, as noted perceptively by a professor of psychiatry:[85]

> It is not what the older person eats but with whom that will be the deciding factor in proper care for him. The oft-repeated complaint of the older patient that he has little incentive to prepare food for only himself is not merely a statement of fact but also a rebuke to the questioner for failing to perceive his isolation and aloneness and to realize that food . . . for one's self lacks the condiment of another's presence which can transform the simplest fare to the ceremonial act with all its shared meaning.

Factors influencing nutritional status of older adults

Physiological
- Dietary intake
- Lack of appetite
- Inactivity/immobility
- Poor taste and smell acuity
- Alcohol or drug abuse
- Presence of chronic disease
- Polypharmacy
- Disability
- Oral health problems

Socioeconomic
- Cultural beliefs
- Poverty
- Limited education
- Limited access to health care
- Institutionalization

Psychological
- Loneliness
- Cognitive impairment
- Dementia
- Depression
- Loss of spouse
- Social isolation

Environmental
- Inadequate housing
- Inadequate cooking facilities
- Lack of transportation
- Lack of access to health services

The DETERMINE Checklist

	Yes
I have an illness or condition that makes me eat different kinds and/or amounts of food.	2
I eat fewer than 2 meals per day.	3
I eat few fruits or vegetables and use few milk products.	2
I have 3 or more drinks of beer, liquor, or wine almost every day.	2
I have tooth or mouth problems that make it hard for me to eat.	2
I don't always have enough money to buy the food I need.	4
I eat alone most of the time.	1
I take 3 or more different prescribed or over-the-counter drugs a day.	1
Without wanting to, I have lost or gained 10 pounds in the past 6 months.	2
I am not always physically able to shop, cook, and/or feed myself.	2
Total:	

Score:

0–2: Good! Recheck your score in 6 months.

3–5: Moderate nutritional risk. Visit your local office on aging, senior nutrition program, senior citizens center, or health department for tips on improving eating habits.

6 or more: High nutritional risk. See your doctor, dietitian, or other health care professional for help in improving your nutrition status.

FIGURE 11-11
CHECKLIST TO DETERMINE YOUR NUTRITIONAL HEALTH

The warning signs of poor nutritional health are often overlooked. To see whether you (or people you know) are at nutritional risk, take this simple quiz.

SOURCE: Nutrition Screening Initiative.

The DETERMINE signs of malnutrition in the elderly

- **D**isease
- **E**ating poorly
- **T**ooth loss or oral pain
- **E**conomic hardship
- **R**educed social contact
- **M**ultiple medications
- **I**nvoluntary weight loss or gain
- **N**eed of assistance with self-care
- **E**lderly person older than 80 years

Sources of Nutritional Assistance

In response to the socioeconomic problems—low income, inadequate facilities for preparing food, lack of transportation, and inability to afford dental care, among others—that trouble many older adults and may lead to malnutrition, federal, state, and local agencies have mandated nutrition programs for the elderly.

The Food Stamp program enables people who qualify to obtain electronic benefits in the form of a "credit" card with which to buy food. In many areas, food banks allow older people on limited incomes to buy good food for less money. A food bank buys industry "irregulars"—products that have been mislabeled, underweighted, redesigned, or mispackaged that would otherwise be thrown away. Nothing is wrong with this food, the industry can credit it as a donation, and the buyer (often a food-preparing site) can obtain the food for a small handling fee and make it available at a greatly reduced price.

The federal Elderly Nutrition Program (ENP) is intended to improve older people's nutrition status and help them avoid medical problems, continue living in communities of their own choice, and stay out of institutions. Its specific goals are to provide:

- Low-cost, nutritious meals
- Opportunities for social interaction
- Nutrition education and shopping assistance
- Counseling and referral to other social and rehabilitation services
- Transportation services

The program makes one hot meal available at noon five days a week, supplying a third of the recommended nutrient intakes. There is no cost for these meals, but participants sometimes make voluntary contributions.

A part of ENP is the Congregate Meal Program. Administrators select sites for congregate meals to feed as many of the eligible elderly as possible. The congregate meal sites are often community centers, senior citizen centers, religious facilities, schools, extended care facilities, or elderly housing complexes. Volunteers may also deliver meals to those who are homebound either permanently or temporarily; these efforts are known as the Home-Delivered Meals Program. The home delivery program ensures nutrition, but its recipients miss out on the social benefit of the congregate meal sites. Every effort is made to persuade recipients to attend the shared meals if they can.

Evaluations of the congregate and home-delivered meals programs generally show that they improve the dietary intake and nutrition status of their clients.[86] Participants generally have greater diversity in their diets and higher intakes of essential nutrients and are less likely to report food insecurity than nonparticipants. Additional benefits come as a result of screening and the referrals generated by such programs as well as from the activities associated with the congregate meals—diet counseling, exercise, adult education, and activities. Participants also benefit from the opportunity for improved socialization.

Looking Ahead and Growing Old

As a nation, we tend to value the future more than the present, putting off enjoying today so that we will have money, prestige, or time to have fun tomorrow. The elderly feel this loss of future. The present is their time for leisure and enjoyment, but they often have no experience using leisure time.

The solution is to begin to prepare for old age early in life, both psychologically and nutritionally. Preparation for this period should, of course, include financial planning, but other lifelong habits should be developed as well (see Figure 11-12). Every adult needs to learn to reach out to others to forestall the loneliness that may otherwise ensue. Adults need to develop some skills or activities,

The Savvy Diner Meals for One

Eating a food you haven't tasted before prolongs your life by 75 days, according to an old Japanese saying.[87] Perhaps finding someone to share that food with might extend your life even more.

Planning nourishing meals for one person that offer variety and optimal nutrition can be challenging. The following tips can help singles of any age buy and prepare foods.

- Keep cupboards and refrigerator stocked with milk, eggs, bread, tortillas, pita bread, canned beans, jars of spaghetti sauce, rice, pasta or noodles, potatoes, onions, canned soups and broth, margarine, cooking oil, and frozen vegetables.
- Keep fresh fruits or vegetables, low-fat yogurt or cheese, and popcorn on hand for easy-to-grab snacking.
- Buy large bags of frozen vegetables if you have sufficient freezer space. Take out the exact amount you need at mealtime.
- Keep an assortment of whole-grain breads, bagels, and muffins in the freezer. Take out individual servings as you need them.
- Buy large packages of meat and poultry when they are on sale. Divide the package into individual servings and freeze them separately.
- Buy fruits and vegetables in season; they will be cheaper and most flavorful at these times.[88]

  Winter: oranges, grapefruits, sweet potatoes, rutabagas, greens

  Spring: asparagus, green beans, sweet peas, rhubarb

  Summer: berries, peaches, zucchini, melons

  Fall: apples, pears, acorn and butternut squash
- Keep it simple and full of vegetables. A heaping bowl of hot soup or stew can make a delightfully healthy and satisfying meal on a cold night. Search the Internet for healthy recipes such as Corn and Shrimp Chowder, Sweet Potato and Peanut Soup, or Autumn Harvest Stew.

© Japack Company/CORBIS

- Cook for several meals at a time. Roast a turkey breast, or skinless, boneless chicken breasts, and use half for dinner and the rest for lunches—in sandwiches, tortillas, or salads.
- Double a favorite recipe. Label and store the extra servings in the freezer. Date these so that you will use the oldest first.
- Add your own steamed vegetables to a frozen entrée. Add a tossed salad, whole-grain roll, and a naturally sweet fruit.
- Choose frozen entrees that contain no more than 10 grams of fat per 300 calories and fewer than 800 milligrams of sodium per serving.
- When you cook for yourself, imagine that you are cooking for special guests.

such as volunteer work, reading, games, hobbies, or intellectual pursuits, that they can continue into their later years to give meaning to their lives. Every adult also needs to develop the habit of adjusting to change, especially when it comes without consent, so that it is not seen as a loss of control over his or her life. The goal is to arrive at maturity with as healthy a mind and body as possible; this means cultivating good nutrition status and maintaining a program of daily exercise.

In general, the ability of the elderly to function well varies from person to person and depends on several factors. The following "life advantages" seem to contribute to good physical and mental health in later years:[89]

- Genetic potential for extended longevity. Some persons seem to have inherited a reduced susceptibility to degenerative diseases.
- A continued desire for new knowledge and new experiences. Some studies suggest that active minds, ever involved in learning new things, may be more resistant to decline.
- Socialization, intimacy, and family integrity. Older persons thrive in situations that nurture love, understanding, shared responsibility, and mutual respect.

FIGURE 11-12
THE AGING WELL PYRAMID

The time to prepare for old age is early in life. Practice the items found at the base of the pyramid to achieve an optimal sense of well-being. Use the inner four compartments of the pyramid to create a balance among all aspects of your life: nutrition, physical activity, social health, and emotional well-being. Use the top of the pyramid to manage everyday stresses such as traffic gridlock, exams, and work deadlines.

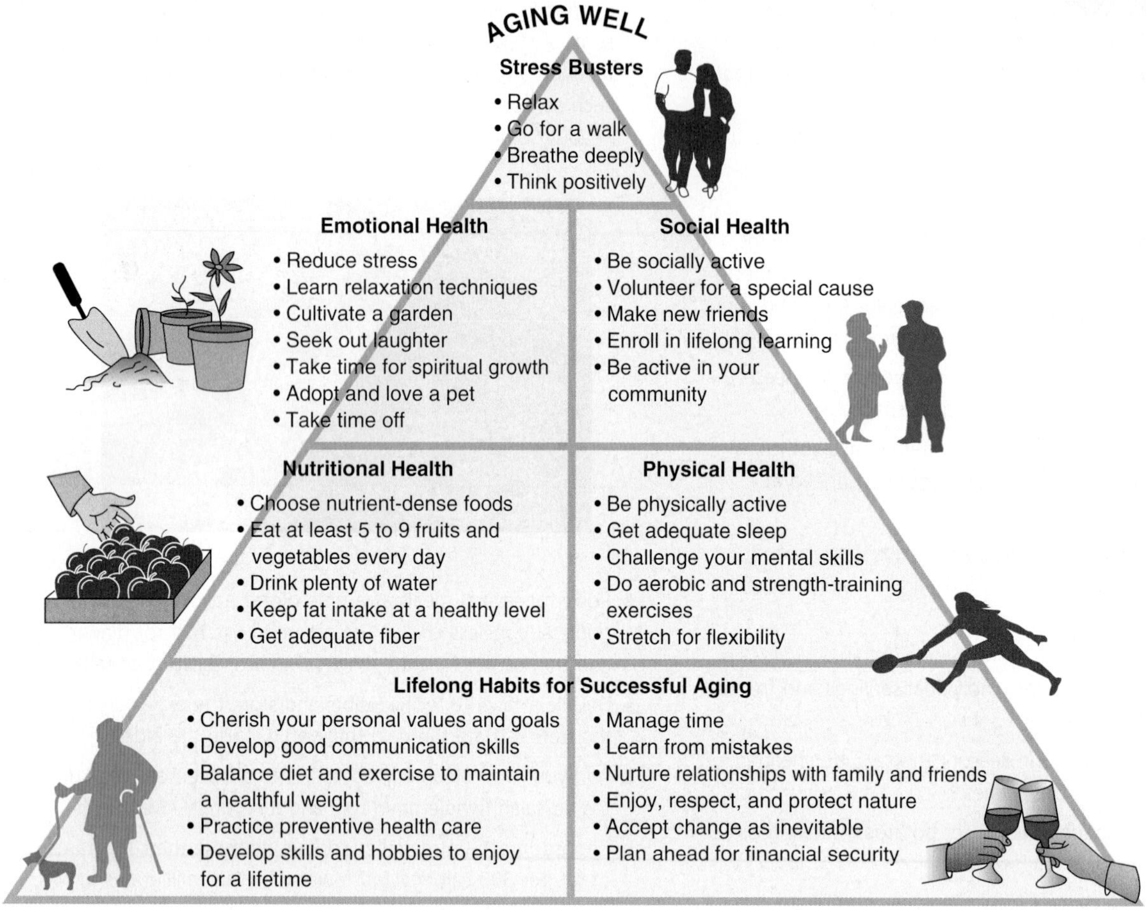

AGING WELL

Stress Busters
- Relax
- Go for a walk
- Breathe deeply
- Think positively

Emotional Health
- Reduce stress
- Learn relaxation techniques
- Cultivate a garden
- Seek out laughter
- Take time for spiritual growth
- Adopt and love a pet
- Take time off

Social Health
- Be socially active
- Volunteer for a special cause
- Make new friends
- Enroll in lifelong learning
- Be active in your community

Nutritional Health
- Choose nutrient-dense foods
- Eat at least 5 to 9 fruits and vegetables every day
- Drink plenty of water
- Keep fat intake at a healthy level
- Get adequate fiber

Physical Health
- Be physically active
- Get adequate sleep
- Challenge your mental skills
- Do aerobic and strength-training exercises
- Stretch for flexibility

Lifelong Habits for Successful Aging
- Cherish your personal values and goals
- Develop good communication skills
- Balance diet and exercise to maintain a healthful weight
- Practice preventive health care
- Develop skills and hobbies to enjoy for a lifetime
- Manage time
- Learn from mistakes
- Nurture relationships with family and friends
- Enjoy, respect, and protect nature
- Accept change as inevitable
- Plan ahead for financial security

- Adhering to a prudent diet while avoiding excess food energy, fat, cholesterol, and sodium. A diet with adequate intakes of all essential nutrients has a positive impact on health and weight management.
- Avoiding substance abuse.
- Acceptable living arrangements.
- Financial independence.
- Access to health care, including: a family physician, health clinic, public health nursing service providing home health care, dentist, podiatrist, physical therapist, pharmacist, and registered dietitian.

Everyone knows older people who have maintained many contacts—through relatives, friends, church, synagogue, or fraternal orders—and have not allowed themselves to drift into isolation. Upon analysis, you will find that their favorable environment came from a lifetime of effort. These people spent their entire lives reaching out to others and practicing the art of weaving others into their own lives. Likewise, a lifetime of effort is required for good nutrition status in the later years. Those who have eaten a wide variety of foods, stayed trim, and remained physically active are best able to withstand the assaults of change.

Spotlight Addressing Overweight in Childhood

For the past three decades, the health status of U.S. children and adolescents has generally improved as seen in reduced infant mortality rates and a decline in the nutrient deficiency diseases of the past. However, approximately 9 million children over 6 years of age are overweight.[90] The increasing prevalence of overweight* is now among the most important public health challenges in the United States.

▲ *The popularity of fast-food meals poses a challenge to the nutritional quality of children's diets, because these meals are typically high in fat, sodium, sugar, and calories and lacking in fruits and vegetables.*

© Tony Freeman/PhotoEdit

What is the prevalence of overweight in children in the United States?

There are two terms used to refer to children and adolescents whose excess body weight could pose medical risks: *overweight* (BMI-for-age = 95th percentile) and *at risk of overweight* (BMI-for-age = 85th to 95th percentile). The number of children who are overweight has more than tripled in the United States, from about 5 to 6 percent in the 1970s and 1980s to 17 percent today among children aged 6 to 19, and 13.9 percent among children aged 2 to 5 years.[91] These increases are greater in certain ethnic and racial groups, particularly among Hispanics and African Americans, as well as in certain regions of the country—notably the South. Figure 9-1 on page 272 illustrates the increasing trend toward overweight in U.S. children.

What problems are associated with overweight in children?

Overweight in childhood involves significant risks to physical and emotional health. In 2000, it was estimated that 30 percent of boys and 40 percent of girls born in the United States are at risk for being diagnosed with type 2 diabetes at some point in their lives.[92] The same conditions associated with overweight adults, such as type 2 diabetes, high blood lipids, high blood pressure, and gallbladder disease are now appearing in young children and teens with greater frequency. Overweight children are also at risk for injury to weight-bearing joints, decreased quality of life, depression, poor self-esteem, respiratory problems, and sleep disturbances, and other additional serious health problems. Since overweight children and adolescents are more likely to become overweight and obese adults, experts project that the rate of chronic diseases will skyrocket even higher in the United States if something isn't done about this epidemic.[93]

Why is overweight so prevalent in children and adolescents today?[94]

The rapid increases in overweight over the past three decades are due primarily to societal and environmental factors.[95] Factors in the environment that may contribute to overweight include increased calorie intake due to increased portion sizes; increased consumption of foods and meals away from home; excessive consumption of soft drinks; lack of positive role models; and physical inactivity. According to one recent study, the strongest predictors of overweight and overfatness in fifth graders was their prior status in third grade.[96] Second to prior overweight, the strongest predictor of subsequent overweight in children is television viewing.[97]

What are children and adolescents eating, and how is it affecting their health?

Eating practices influence a child's physical growth, cognitive development, and overall health.[98] Reports from a number of organizations have shown that children are failing to meet the recommended nutrition guidelines by not consuming enough fruits and

*The term overweight is used in preference to the term *obesity* and refers to children and youth between the ages of 2 and 18 years who have body mass indexes (BMIs) equal to or greater than the 95th percentile of the age- and gender-specific BMI charts developed by the Centers for Disease Control and Prevention (CDC), available at www.cdc.gov.

vegetables and by eating too many foods high in fat and added sugars. Children's eating habits have changed over the past three decades. Dietary data, collected in large nationwide surveys to determine trends in nutrient intakes, have shown that most children and adolescents have either poor diets or diets that need improvement.[99]

Over the past 25 years, the portion sizes of commonly consumed foods, such as soft drinks and hamburgers, have increased. Large food portions that provide more calories than smaller portions may be contributing to the increasing prevalence of overweight in children and teens. For example, children 3 to 5 years of age consumed 25 percent more of an entrée when they were presented with portions that were double an age-appropriate standard size.[100]

Today, with the demise of traditional family meals, many busy parents rely on fast food meals to nourish their children. Research shows that children who ate fast food consumed more total energy and had poorer diet quality (e.g., more added sugars, more sugar-sweetened beverages, less milk, and fewer fruits and non-starchy vegetables) on days with fast food, compared to days without fast food.[101]

What impact do the media have on a child's nutrition status?

Children today spend a large amount of their leisure time using various media, including television, DVDs, video games, computers, and cell phones. Children hear a great deal about foods via television. Many authorities are concerned that television commercials may have an undesirable impact on children. It is estimated that the average child sees more than 30,000 commercials a year. A recent Institute of Medicine report, "*Preventing Childhood Obesity: Health in the Balance*," noted that more than half of television advertisements directed at

© Michael Newman/PhotoEdit

▲ *Television can have adverse effects on children by influencing their selection of foods, snacking habits, and activity levels.*

children promote foods and beverages such as candy, fast food, snack foods, soft drinks, and sweetened breakfast cereals that are high in calories and fat and low in fiber and other essential nutrients.[102] Billions of advertising dollars are spent each year in the effort to sell these foods to children.

The more time children spend watching television, the less time they have for physical activity. Research results show that children who watch television more often had lower activity levels and were less likely to participate in organized sports or community activities. Parents can encourage physical activity in children to help prevent overweight by participating in activities with their children, such as cycling, swimming, rollerblading, skating, hiking, and helping them find physical activities that they enjoy![103]

What are some simple tips for helping children learn good eating habits?

- Parents can set a good example with their own eating habits.
- Discourage children from eating while watching TV or doing homework.
- Whenever possible, eat meals together as a family, and keep fast-food meals to a minimum.
- Encourage children to "listen to their stomachs" and eat only when they are hungry.
- Limit consumption of high-fat or high-sugar foods, including soft drinks.
- Include children in food preparation before meals as a way of helping them become more interested in and aware of what they are eating.

What impact do schools have on children's diets?[104]

Even though school meal programs appear to have a positive effect on children's consumption of milk, fruit, vegetables, and some vitamins and minerals, there is evidence that school meals contribute too much fat in menu items such as pizza, chicken sandwiches, French fries, and baked goods. In light of these findings, the USDA launched a reform of the School Lunch Program in 1994 that directed schools to upgrade the nutritional value of their meals to adhere to the *Dietary Guidelines for Americans*. To meet the nutrition standards, menus now include more fruits, vegetables, whole grains, and low-fat and reduced-fat foods. More vegetarian options are available, as well as salad bars and a variety of prepared salads.

If the School Lunch Program promotes foods that meet the *Dietary Guidelines for Americans,* why don't all children participate in the program?

The environment in some schools discourages students from eating meals provided by the School Lunch Program and encourages food choices and eating habits that are not consistent with the *Dietary Guidelines for Americans*. Efforts to promote healthful eating habits and provide nutrition instruction may be contradicted in school settings where the sale of food and beverages, many with low nutritional appeal, is promoted in snack bars, school stores, and vending machines. Even less healthful à la carte items sold by the school cafeteria may "compete" with school meals for a student's meal money.

Is anything being done to help build more healthful school environments?

Many states, and many school districts, are developing policies that limit the sale of competitive foods—foods sold in competition with USDA school meals that are categorized as "foods of minimal nutritional value" such as soft drinks, candies, and candy-coated popcorn. Some states have already implemented sweeping changes to nutrition policies in schools such as eliminating deep-fat frying; restricting portion sizes on chips and French fries, fruit drinks, and certain snacks and sweets; limiting fats and sugar; and offering fruits and vegetables daily at every point of service.[105]

The Surgeon General's call to action to prevent and decrease overweight and obesity outlines the following actions for creating healthful school environments:[106]

- Provide age-appropriate nutrition and health education to help students develop lifelong, healthful lifestyle habits.
- Adopt policies that require all foods and beverages available on school campuses and school events to contribute toward eating patterns that are consistent with the *Dietary Guidelines for Americans*.
- Ensure that healthful snacks and foods are provided in vending machines, school stores, and other venues.

What types of prevention efforts are recommended for addressing overweight in children?

Prevention of overweight involves a focus on *energy balance*—calories consumed versus calories expended—so taking action against overweight in childhood must address the factors that influence both eating and physical activity.[107] In general, most interventions for prevention of overweight in children have been conducted through the schools.[108] Effective interventions among children tend to include a component on decreasing television viewing. School-based programs that increase physical activity also appear effective in reducing body size, especially when physical activity is for one hour or more.[109] Most interventions involve a family component. For small children, the parents are often "gatekeepers" for diet and physical activity because they control the types of foods in the house, the access and availability of those foods, opportunities for activity or inactivity, and meal patterns. In addition, parents often serve as role models for their children. Although family is important for preschool and elementary-aged children, adolescents are often more interested in peer relationships and influences, so interventions for adolescents should include a peer component as well.[110]

See the box below for additional recommendations from the Institute on Medicine regarding the Prevention of Childhood Obesity. For additional information, visit the following websites:

- *Dietary Guidelines 2005:* www.healthierus.gov/dietaryguidelines
- *University of California Berkeley Center for Weight and Health:* www.cnr.berkeley.edu/cwh/
- *USDA Food and Nutrition Information Center, Weight Control and Obesity:* www.nal.usda.gov/fnic/etext/000060.html
- *Coordinated Approach to Child Health:* www.sph.uth.tmc.edu/chppr/catch/
- *Planet Health:* www.hsph.harvard.edu/prc/proj_planet.html
- *Stanford's Student Media Awareness to Reduce Television (SMART):* hprc.stanford.edu/pages/store/itemDetail.asp?169
- *HUGS International, Inc. for teens and adults:* www.hugs.com
- *Team Nutrition:* www.fns.usda.gov/tn

Key Recommendations from the Institute of Medicine for the Prevention of Childhood Obesity[111]

- The food industry should make obesity prevention in children and youth a priority by developing and promoting food products that are low in dietary fat and energy.
- Nutrition labeling should be clear and useful so that parents and youth can make informed product comparisons and decisions to achieve and maintain energy balance at a healthy weight.
- The media should incorporate positive behavior change messages into television programs and popular magazines, and they should regulate television advertising aimed at children to minimize the risk of obesity in children and youth.
- Local governments, public health agencies, schools, and community organizations should collaboratively develop and promote programs that encourage healthful eating behaviors and regular physical activity.
- Local governments, private developers, and community groups should expand opportunities for physical activity including recreational facilities, parks, playgrounds, sidewalks, bike paths, and safe routes for walking or bicycling to school.
- Schools should provide a consistent environment that is conducive to healthful eating behaviors and regular physical activity.
- Parents should promote healthful eating behaviors and regular physical activity for their children.

In Review

1. What is the recommended range of weight gain during pregnancy for a normal-weight woman?
   a. 10–18 lbs
   b. 19–24 lbs
   c. 25–35 lbs
   d. 38–44 lbs
2. Which nutrients are needed in greatest amounts during pregnancy? Describe wise food choices for each.
3. List the benefits of breastfeeding for the infant.
4. Peak bone density occurs at what age?
   a. around 15
   b. around 50
   c. around 30
   d. at birth
5. Malnutrition has the greatest impact on mental development when it occurs:
   a. in the elderly.
   b. in the physically impaired elementary school child.
   c. during the preschool years.
   d. during pregnancy and infancy.
6. Which of the following nutrition problems are common with teenagers?
   a. undernutrition
   b. low calcium intakes
   c. eating disorders
   d. all of the above
7. Children should:
   a. be forced to eat everything on their plates.
   b. be encouraged to try a variety of foods.
   c. be given supplements to meet vitamin and mineral requirements.
   d. not be allowed to help prepare food of any type; it's too dangerous.
8. Which of the following is true of infants?
   a. Hydration is vital because a high percentage of their body weight is water.
   b. Calorie needs increase less rapidly after the first 6 months.
   c. Solid food should not be introduced too soon.
   d. all of the above
9. Name some factors that affect the health status of older adults.
10. What factors contribute to childhood weight problems?

Menu of Online Study Tools

A variety of study tools for this chapter are available at our website to deepen your understanding of chapter concepts. Go to
www.thomsonedu.com/nutrition/boyle
to find

- Practice tests
- Flashcards
- Glossary
- Web links
- Animations
- Chapter summaries, learning objectives, and crossword puzzles

Nutrition on the Web

www.thomsonedu.com/nutrition/boyle
Go to the *Personal Nutrition* site to check for the latest updates to chapter topics or to access links to related websites.

www.mayoclinic.com/health/pregnancy/PR99999
Information on pregnancy from the Mayo Clinic Pregnancy Center.

www.marchofdimes.org
Information on birth defects from the March of Dimes.

www.nofas.org
A site for information about fetal alcohol syndrome.

www.lalecheleague.org
Information about breastfeeding available from La Leche League.

www.aap.org
Website of the American Academy of Pediatrics.

www.ific.org
Information on promoting healthy lifestyles for children.

www.kidshealth.org
Information on health promotion for children.

www.kidnetic.com
An interactive website combining food, fitness, and fun for children.

www.eatright.org
The American Dietetic Association offers information on nutrition and pregnancy and the nutrient needs of children and older adults.

www.healthfinder.gov/justforyou/
Search for nutrition topics related to stages of the life cycle.

www.aarp.org
Resources from the American Association for Retired Persons.

www.nia.nih.gov
Information on aging from the National Institute on Aging.

www.fns.usda.gov/fns
Facts about the U.S. food assistance programs.

Food Safety and the Global Food Supply

12

ASK YOURSELF . . .

Which of the following statements about nutrition are true, and which are false? For each false statement, what is true?

1. Pesticides rank as the number one hazard in the U.S. food supply.
2. The most frequent cause of foodborne illness in homes and restaurants is inadequate cooling of foods.
3. If a food contains a toxic substance, a person who eats it will become ill.
4. Tainted mayonnaise frequently causes food poisoning.
5. Imported foods may contain residues of pesticides that are illegal in the United States.
6. A USDA rule on organic crops allows the use of genetically engineered ingredients and irradiation in organic foods.
7. Hunger in the United States almost exclusively afflicts unemployed homeless people.
8. Legal pesticides are poisonous only to pests, not to people.
9. Most foods that cause food poisoning are contaminated by the manufacturer or processor.
10. Food additives are a major cause of cancer in the United States.

Answers found on the following page.

The role of the infinitely small is infinitely large.

Louis Pasteur (1822–1895, French chemist and microbiologist who developed the pasteurization process)

So far this book has dealt primarily with the nutrients and how your body handles them. This chapter takes the study of nutrition one step further and examines some nonnutrient components of food—bacteria, additives, and pesticides, to name a few—and how they affect the food supply. In addition, the chapter takes a global view of the science of nutrition, looking at how foodways in different countries influence each other as well as how some of our food habits affect the environment.

What additives do foods contain, and what are the effects of those additives? Are foods ever contaminated? How can you reduce your risk of food poisoning? What potential do new food technologies, such as irradiation and genetic engineering, hold for our lives and for the health of the environment? This chapter addresses these and other questions.

CONTENTS

Foodborne Illnesses and the Agents That Cause Them

North America has the safest, most plentiful food supply in the world, thanks to the concerted efforts of food suppliers, food processors, and federal, state, and local governments, all of which are concerned with food safety. North Americans also enjoy the most diverse food supply in the world, consisting of an incredible

foodborne illness illness occurring as a result of ingesting food or water contaminated with a poisonous substance, such as a toxin or chemical *(food intoxication)* or an infectious agent, such as bacteria, viruses, or parasites *(foodborne infection);* commonly called *food poisoning.*

pasteurization the process of sterilizing food via heat treatment.

array of fresh and processed foods. Most American supermarkets stock a variety of foods from other countries—cookies and crackers from Denmark and the United Kingdom, Belgian chocolates, Italian cheeses, beef and veal products from New Zealand and Australia, goose liver paté from France, and specialty foods from Japan, Mexico, and even China. Our international food interests can be seen in the produce section as well, where exotic star fruit, papaya, and mango from overseas markets are widely available. Still, the diverse mix of fresh and processed foods coming from local, national, and international markets underscores the need to understand where the culprits behind foodborne illnesses originate.

Within hours of the September 11, 2001, attacks on the Pentagon and New York City, the nation's food and water supplies were identified as likely targets of terrorists. Food safety today refers to a food supply that is free of foodborne pathogens as well as one that is safe from bioterrorism.[1] As part of a heightened awareness of bioterrorism threats, the Food and Drug Administration (FDA) has created a special website that links to information about bioterrorism as well as to other information sources. See www.fda.gov/oc/opacom/hottopics/bioterrorism.html.

Foodborne illness is one of the greatest concerns of public health experts and the food industry. Each year, as many as 76 million Americans experience foodborne illness, and an estimated 5,000 deaths are linked to tainted foods.[2] Incredible as these figures are, they probably represent only a fraction of the whole picture. For a number of reasons, many mild cases of foodborne illness are never reported: The victims pass off the symptoms as flu and do not seek medical attention, the illness is misdiagnosed as another problem with similar symptoms, the victim fails to recognize food as the source of the illness, or the physician doesn't report the illness to local health agencies. Diarrhea, nausea, abdominal pain, or vomiting without fever or upper respiratory distress is often taken to be flu, but people who experience such symptoms are highly likely to be suffering from foodborne illness.

The fact that more people aren't afflicted with foodborne illness is surprising because disease-producing microorganisms proliferate in our environment. Consider that all raw foods contain microbes, and foods often pick up more microbes during production, processing, packaging, transport, storage, and preparation. For this reason, the food industry uses various control measures to limit the risk of foodborne illness. Bacteria or their spores are destroyed or inactivated by heat treatments such as **pasteurization** and canning. Freezing, dehydrating, and refrigerating food also halts or slows down bacterial growth. In addition, special packaging techniques and antimicrobial preservatives help to control food-related pathogens.

What causes most cases of foodborne illness? Many people believe that chemical additives and pesticide residues added during the growing and processing of food pose the greatest risk. However, most foodborne disease is actually caused by mishandling of food, either in food service establishments such as restaurants or in homes. In fact, food from restaurants, cafeterias, and other food service establishments accounts for about two-thirds of all reported outbreaks of foodborne illness, whereas food eaten at home accounts for about one-fourth.[3] Most cases of foodborne disease are caused by faulty handling, cooking, and storing of food

Ask Yourself Answers: **1.** False. The greatest hazard present in the U.S. food supply today is not pesticides but bacteria, viruses, mold, and other microorganisms that cause food poisoning. **2.** True. **3.** False. If a food contains a toxic substance, a person who eats it *may* become ill, depending on whether enough of the toxin to cause illness is present. **4.** False. Mayonnaise rarely carries high levels of the bacteria that cause illness. **5.** True. **6.** False. A March 2000 USDA rule on organic crops rejects the use of genetically engineered ingredients and irradiation in organic foods. **7.** False. Hunger in the United States affects not only the unemployed homeless but also homeowners and the working poor. **8.** False. Legal pesticides can be poisonous to people, animals, and plants, depending on the amount of the pesticide. **9.** False. Most cases of food poisoning are the result of improper handling of food *after* it leaves the processor or manufacturer. **10.** False. Food additives, which pose minimal health risks, are not a major cause of cancer.

long after it has left the manufacturer or processor. Table 12-1 ranks the most common food hazards.

Experts classify foodborne diseases into two types: intoxications and infections. **Food intoxications** occur when a chemical or toxin transmitted by way of food causes the body to malfunction. An example of food intoxication is the vomiting, nausea, abdominal cramping, sweating, and chills that result from eating food contaminated with a strain of bacteria called *Staphylococcus aureus*. This bacterium produces what is known as an **enterotoxin,** a toxin that causes severe gastrointestinal distress. Other types of bacteria can produce **neurotoxins,** or toxins that afflict the nervous system. Food intoxication can also be caused by eating food that has been contaminated with a chemical, such as lead or some other heavy metal.

Foodborne infections, on the other hand, occur as a result of eating a food that contains living microorganisms, such as bacteria, viruses, or parasites that are capable of multiplying and thriving in the body. When ingested in large amounts, such microorganisms can wreak havoc in the digestive tract or other areas of the body. An example of this type of foodborne illness is infection with *Vibrio* bacteria, which often reside in raw seafood such as oysters and clams. Inside the body, the bacteria settle in quickly and cause abrupt onset of chills, fever, or vomiting. The following section provides an overview of the various agents that can cause a foodborne illness—be it an intoxication or infection.

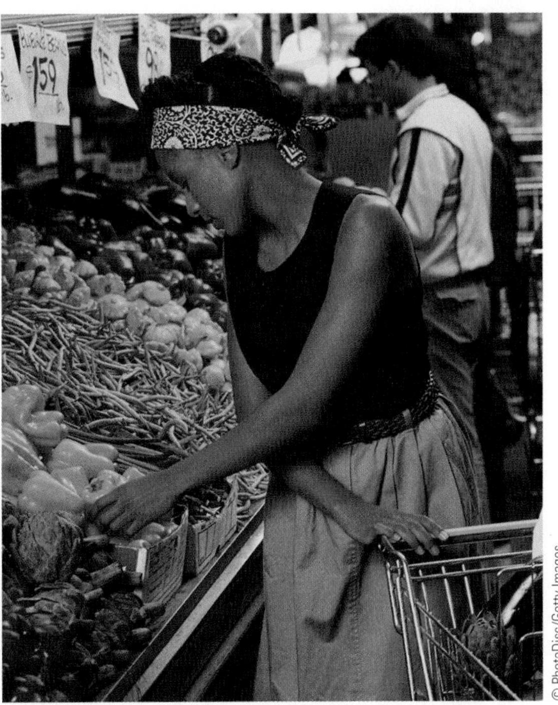

▲ *Americans enjoy the safest, most diverse food supply in the world.*

Microbial Agents

When we order a meal at a restaurant or reach for an egg or glass of milk from the refrigerator at home, we don't usually think about the microorganisms that might be lurking in the foods or their potential to cause illness. We tend to assume that the foods and beverages we consume are safe to eat or drink. Granted, most of the time we don't need to worry. Typically, a food must harbor thousands of microorganisms before it causes nausea, diarrhea, cramps, or other symptoms of foodborne illness. What's more, a healthy body can usually defend itself against small amounts of the "bad bugs."

However, when proper food-handling procedures are not followed carefully, the risk of food poisoning from bacteria or other microbial agents soars. When food items are not cooked or stored properly, they provide the perfect medium in

TABLE 12-1

FDA's Rank of Areas of Concern in the Food Supply

Although most people think that chemicals such as pesticides and additives rank as the most dangerous, illness-producing substances in our food supply, the real hazard comes from naturally occurring bacteria and other microbes.

Most dangerous:

1. Microbial foodborne illness
2. Naturally occurring toxins in foods
3. Residues in foods, including:
   - Environmental contaminants, such as industrial chemicals
   - Pesticides
   - Animal drugs, such as hormones or antibiotics
4. Food processing and nutrients in foods

Least dangerous:

5. Intentional food additives
6. Genetic modification of foods

food intoxication illness caused by eating food that contains a harmful toxin or chemical.

enterotoxin a toxic compound, produced by microorganisms, that harms mucous membranes, as in the gastrointestinal tract.

entero = intestine

neurotoxin a poisonous compound that disrupts the nervous system.

neuro = nerve

foodborne infection illness caused by eating a food containing bacteria or other microorganisms capable of growing and thriving in a person's tissues.

which microorganisms can flourish. For children, the elderly, people with compromised immune systems, and other vulnerable people, it might only take small amounts of the offending microorganisms to cause trouble. Table 12-2 summarizes the common food sources of the microbial agents responsible for most outbreaks of foodborne illness.

TABLE 12-2
Microbial Food Agents: Organisms That Can Bug You

Disease and Organism That Causes It	Annual Incidence (Deaths)	Source of Illness	Usual Onset after Eating	Symptoms
Bacteria				
Botulism (*Botulinum* toxin produced by *Clostridium botulinum* bacteria)	60 (4 deaths)	Spores of these bacteria are widespread. But these bacteria produce toxin only in an anaerobic environment of little acidity. Found in low-acid canned foods such as corn, green beans, soups, beets, asparagus, mushrooms, tuna, and liver paté. Also in luncheon meats, ham, sausage, stuffed eggplant, lobster, smoked and salted fish, and herb-flavored oils.	4–36 hours	Neurotoxic symptoms, including double vision, inability to swallow, speech difficulty, and progressive paralysis of the respiratory system. Get medical help immediately; can be fatal.
Campylobacteriosis (*Campylobacter jejuni*)	2 million (100 deaths)	Raw (or undercooked) poultry, meat, eggs, and unpasteurized milk.	2–5 days	Diarrhea, abdominal cramping, fever, and sometimes bloody stools. Lasts 7–10 days.
Listeriosis (*Listeria monocytogenes*)	2,500 (500 deaths)	Found in soft cheese, raw meat and seafood, unpasteurized milk, imported seafood products, frozen cooked crab meat, cooked shrimp, and cooked surimi (imitation shellfish).	48–72 hours (though symptoms can strike 7–30 days after eating)	Fever, headache, nausea, and vomiting
Perfringens food poisoning (*Clostridium perfringens*)	250,000 (7 deaths)	In most instances, caused by failure to keep food hot. A few organisms are often present after cooking and multiply to toxic levels during cooldown and storage of prepared foods. Meats and meat products are most frequently implicated.	8–12 hours	Abdominal pain and diarrhea and sometimes nausea and vomiting. Symptoms last a day or less and are usually mild.
Salmonellosis (*Salmonella* bacteria)	1,400,000 (600 deaths)	Raw or undercooked eggs, meats, poultry, milk and other dairy products, shrimp, frog legs, yeast, coconut, pasta, chocolate, and unpasteurized juices are most frequently involved.	6–48 hours	Nausea, abdominal cramps, diarrhea, fever, and headache
Hemolytic-uremic syndrome Hemorrhagic colitis (Shiga Toxin-producing *E. coli* [STEC] released by *Escherichia coli* O157:H7 and others)	110,000 (78 deaths)	Undercooked hamburger and roast beef, raw milk, raw apple cider, contaminated water, mayonnaise, and vegetables (especially alfalfa sprouts)	12–72 hours	Abdominal cramps, vomiting, nausea, watery diarrhea that often turns bloody, low-grade fever, and, in severe cases, kidney failure, strokes, and seizures
Shigellosis (*Shigella* bacteria)	90,000 (14 deaths)	Found in milk and dairy products, poultry, and potato salad. Food becomes contaminated when a human carrier does not wash hands and then handles liquid or moist food.	1–2 days	Abdominal cramps, diarrhea, fever, sometimes vomiting, and blood, pus, or mucus in stools

Of the microbial pathogens, *Campylobacter jejuni* ranks as one of the most prevalent pathogens and one of the leading causes of foodborne illness in the United States. In fact, two pathogens—*Campylobacter jejuni* and *Salmonella*—account for about 70 percent of diagnosed cases of bacterial foodborne illness in

TABLE 12-2

Microbial Food Agents: Organisms That Can Bug You (continued)

Disease and Organism That Causes It	Annual Incidence (Deaths)	Source of Illness	Usual Onset after Eating	Symptoms
Staphylococcal food poisoning (Staphylococcal toxin produced by *Staphylococcus aureus* bacteria)	185,000 (2 deaths)	Toxin produced when food contaminated with the bacteria is left too long at room temperature. Meats, poultry, egg products, tuna, potato and macaroni salads, and cream-filled pastries are good environments for these bacteria to produce toxin.	30 minutes to 8 hours	Diarrhea, vomiting, nausea, abdominal pain, cramps, and prostration. Lasts 24–48 hours. Rarely fatal.
Vibrio infection (*Vibrio vulnificus*)	47 (18 deaths)	The bacteria live in coastal waters and can infect humans either through open wounds or through consumption of contaminated seafood.	24 hours	Chills, fever, abdominal cramps, vomiting, diarrhea. At high risk are people with liver conditions, low gastric (stomach) acid, and weakened immune systems.
Parasite				
Cryptosporidiosis (*Cryptosporidium parvum*)	30,000 (7 deaths)	Swimming in or drinking contaminated water; contaminated raw fruits and vegetables, unpasteurized milk, juices and ciders.	2–10 days	Diarrhea, nausea, fever; can be symptomless
Cyclosporiasis (*Cyclospora cayetanensis*)	16,000 (no deaths)	Contaminated water or produce.	7 days	Anorexia, diarrhea, weight loss, fatigue, vomiting, muscle aches; can be symptomless
Trichinosis (*Trichinella spiralis*)	50 (no deaths)	Raw or undercooked pork or wild game. Worms burrow into muscle tissue.	24 hours	Nausea, vomiting, abdominal pain, diarrhea, fever; after 2 weeks, muscle pain, fluid retention, weight loss, and fever
Giardiasis "Traveler's Diarrhea" (*Giardia duodenalis*) (formerly *G. lamblia*)	200,000 (1 death)	Most frequently associated with consumption of contaminated water. May be transmitted by uncooked foods that become contaminated while growing or after cooking by infected food handlers.	12 hours to several days	Sudden onset of explosive watery stools, abdominal cramps, anorexia, nausea, and vomiting
Virus				
Hepatitis (Hepatitis A virus)	4,200 (4 deaths)	Mollusks (oysters, clams, mussels, scallops, and cockles) become carriers when their beds are polluted by untreated sewage. Raw shellfish are especially potent carriers, although cooking does not always kill the virus.	15–50 days	Begins with malaise, appetite loss, nausea, vomiting, and fever. After 3–10 days, patient develops jaundice with darkened urine. Severe cases can cause liver damage and death.
"Stomach flu"* Norwalk-type viruses	9,200,000 (124 deaths)	Salads, sandwiches, or other foods contaminated by infected food handlers; raw oysters	18–72 hours	Headache, fever, vomiting, diarrhea, abdominal pain

*A misnomer; unrelated to influenza.

SOURCES: Adapted from A. Hecht, *The Unwelcome Dinner Guest: Preventing Foodborne Illness* (Washington, D.C.: U.S. Government Printing Office, DHHS Publication No. (FDA) 91-2244); and Economic Research Service, *Economics of Foodborne Disease: Food and Pathogens*, available at www.ers.usda.gov.

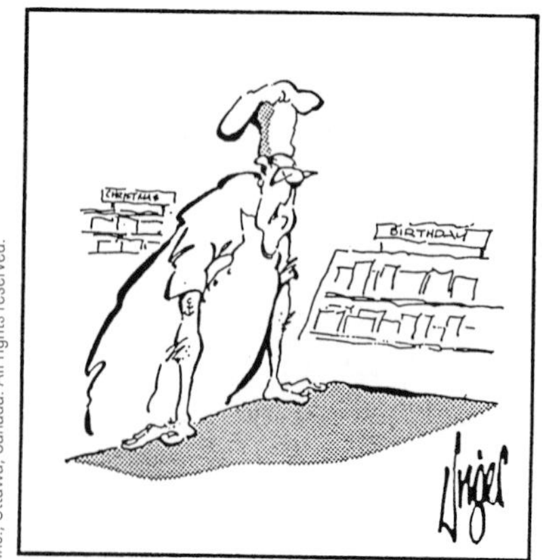

"I need 148 get-well cards."

the United States.[4] Symptoms of the illness caused by *Campylobacter jejuni*, called campylobacteriosis, usually begin 2 to 5 days after eating the contaminated food and last for 7 to 10 days. Many cases of campylobacteriosis probably go unreported because the illness's characteristic diarrhea, abdominal pain, and fever mimic flulike gastrointestinal ills.

In the past, egg products were a major source of salmonellosis—an illness caused by *Salmonella* bacteria. However, this is no longer the case due to the mandatory pasteurization of eggs used in the commercial preparation of ice cream, baked goods, and other egg-containing food products. Raw eggs, however, have been implicated in outbreaks of salmonellosis, as has raw, unpasteurized milk. (Because of this concern, the sale of raw milk is banned in most states.) Today, most outbreaks of salmonellosis are caused by faulty handling of raw meat and poultry, unpasteurized juice, raw sprouts, and mangos.

Staphylococcus aureus is another strain of bacteria responsible for many of the reported cases of foodborne illnesses that occur in the United States. The bacterium, found in the nose and throat and on the skin of most people, can be transmitted to food when an individual with an infected wound or boil or a respiratory infection handles food improperly. The bacterium itself isn't directly responsible for the illness, however. Rather, it produces staphylococcal enterotoxin, which causes food poisoning within ½ to 8 hours of eating a contaminated food. The foods typically implicated in *S. aureus* intoxication include meat and poultry products, egg products, tuna, potato and macaroni salads, and cream-filled pastries.

Another particularly deadly foodborne pathogen is *Clostridium botulinum*, the agent that causes the nausea, vomiting, dizziness, and muscle paralysis known as botulism. This severe illness results from eating food in which the bacterium has flourished and produced a neurotoxin. The toxin binds irreversibly to nerve endings and causes progressive paralysis, which makes swallowing and breathing difficult. Botulism typically develops within 4 to 36 hours of eating the contaminated food. Because it can be fatal, botulism requires immediate medical attention.

Spores of *C. botulinum* are ubiquitous, having been detected in everything from shellfish to fruits and vegetables to honey to corn syrup.* But the *botulinum* bacteria only produce the deadly neurotoxin if they are in a warm, oxygen-free, low-acid environment. That's why improperly sterilized low-acid canned goods are the most common culprits in cases of botulism. Note also that trendy oils flavored with garlic or herbs can harbor *C. botulinum*. Manufacturers of such mixtures must add antibacterial agents to their preparations, but people who make their own flavored oils should be sure to store them in the refrigerator.

Another especially virulent pathogen that is emerging as a major public health concern is *Escherichia coli* O157:H7.† First recognized as a cause of foodborne illness in 1982, *E. coli* garnered national attention about a decade later when it caused a major outbreak of foodborne illness in the northwest part of the United States. Undercooked, contaminated hamburgers sold at a major fast-food chain prompted the 1993 scare, which ultimately led to more than 500 reported cases of illness, some 150 hospitalizations, and 3 deaths. The outbreak sparked a national debate about the safety of the U.S. meat and poultry supply.[5]

Found in the intestinal tracts of mammals, *E. coli* is usually transmitted via animal feces. It poses special concern because it is so dangerous. Unlike most other illness-producing microorganisms, *E. coli* need not be present in large numbers to make a person sick; just a few bacteria seem to do the trick. Once ingested, the bacteria clings to the intestinal wall, where it releases an entero-

▲ *Covered, bottled garlic-in-oil left at room temperature provides the perfect warm, oxygen-free, low-acid environment for the toxin-releasing spores of* Clostridium botulinum.

*Honey has been found to contain dormant bacterial spores, which can awaken in the human body to produce botulism. In adults, this is not a hazard, but infants under one year of age should never be fed honey. Honey has been implicated in several cases of sudden infant death.

†*Escherichia coli* O157:H7 is a particular strain of *E. coli* bacteria. When *E. coli* is mentioned throughout the rest of the chapter, it refers to the O157:H7 strain.

toxin that causes abdominal pain, watery diarrhea that often turns bloody, and, in vulnerable people, such as children and the elderly, *E. coli* may cause serious complications, including kidney failure and death.

Most *E. coli* outbreaks have been linked to undercooked beef, particularly hamburger. Alfalfa sprouts, raw milk, and fresh apple cider, presumably made from apples exposed to tainted animal manure, have also been implicated in some outbreaks.[6] As with all illness-causing microorganisms, the best defense against *E. coli* is careful food handling.

Mold is another potential microbial food contaminant. Certain molds produce poisonous compounds called **aflatoxins.** These compounds are powerful liver toxins in animals, are known to be carcinogenic in some species, and can be lethal if consumed in large doses. Aflatoxins have been found on peanuts, wheat, corn, meat pies, dry beans, and even refrigerated and frozen pastries. Molds that produce the aflatoxins typically flourish when foods such as corn and peanuts get wet and are then stored in a warm place, such as a grain silo or railroad boxcar. Controlling aflatoxins in food is difficult, but it is given high priority by the food industry and regulatory agencies.

To be sure, not all microorganisms are bad. A mold called *Penicillium roquefortii* imparts the special, pungent flavor of Roquefort cheese. Another mold strain, *P. camembertii,* lends flavor to Camembert cheese. Likewise, a strain of mold called *Bacteria aceti* causes the alcohol in wine and hard cider to turn to vinegar.[7] What's more, yogurt owes its existence to active cultures of bacteria added during processing, and a new area of research—called probiotics—suggests that the "good bugs" in the yogurt may help fight "bad bugs" that cause yeast infections and other ills.[8]

aflatoxin a poisonous toxin produced by molds.

toxicants poisons, that is, agents that cause physical harm or death when present in large amounts.

Natural Toxins

Most people assume that products derived from plants are safe because they are "natural." But natural food **toxicants** in plants—especially herbs—are sometimes to blame for poisonings. Even familiar foods that are generally safe sometimes harbor potential toxicants. For example, potatoes contain a substance called *solanine,* a powerful inhibitor of nerve impulses. The green substance accumulates just beneath the vegetable's skin, usually in harmless amounts. When potatoes are exposed to light, however, they sometimes develop excess solanine. When this happens, the green area and about ½ inch of potato around it should be removed before cooking.

When plants are transformed into powders and potions, their components become more concentrated, as do their potentially harmful effects. Nevertheless, people often assume that because they come from plants, these substances must be safe. Unlike the chemical composition of standard pharmaceutical drugs, however, the composition of herbal products is not regulated by the government. This lack of safety standards for herbal pills, powders, teas, and other potions, which have become increasingly popular in recent years, ranks as a major concern among public health officials. The potential risks are amplified when an herb is mislabeled, misidentified, or mixed with another potentially toxic substance.

For example, several children and adults suffered life-threatening respiratory problems and liver malfunctions as a result of taking a Chinese herb called Jin Bu Huan. Investigators found, among other problems, that the plant from which the product had been derived was misidentified and that the product carried false and misleading medical claims.[9] Along the same lines, the FDA has banned the use of a plant derivative called ma huang—which contains ephedrine—an amphetamine-like substance often used in weight-loss products—because it has been linked to rapid heart rate, stroke, nerve damage, and several deaths.[10]

These are just a few of the herb-related problems that continue to surface. They highlight the need to be cautious about using herbs of any sort. Adults should not take large amounts of a particular herbal product or take more than

Numerous herbs have been implicated in liver failure and other health problems. Here are just a few:[11]

Chaparral
Kava
Jin Bu Huan
Ma huang
Germander
Comfrey
Mistletoe
Skullcap
Margosa oil
Maté tea
Gordolobo yerba tea
Pennyroyal (squawmint) oil
(See Chapter 6 for more information about herbal remedies.)

one at a time without consulting a competent professional regarding the product's various effects. Pregnant and breastfeeding women should avoid using herbs because they can expose a fetus or an infant to a toxic dose.[12] In addition, people with any type of liver disease or condition should be wary of herbal products; liver failure is one of the hallmarks of an herbal overdose, because toxins accumulate in that organ.

In addition to plant foods, other types of food, notably seafood, often harbor natural toxicants. For example, a type of fish called puffers, long considered a delicacy in Japan, contains a potent poison—tetrodotoxin—which doesn't harm the fish but can be lethal to people and other animals. Tetrodotoxin, which is 275 times deadlier than cyanide, works by blocking nerve impulses. Over a period of several hours, it will eventually close down a person's entire nervous system. The toxin is so deadly that sale of puffers is illegal in the United States.[13]

Puffer poisoning is just one example of a foodborne disease traced to eating a toxic sea creature. Others also exist. For instance, paralytic shellfish poisoning can occur after eating mollusks (clams, mussels, oysters, and scallops) contaminated with marine algae that produce a neurotoxin. The mollusks themselves don't become ill, but people who eat them do, experiencing such symptoms as nausea, vomiting, cramps, and muscle weakness. In each of these cases, the toxin is not destroyed by the heat of cooking. Fish can also harbor viruses, including hepatitis viruses, worms, and other parasites. For these reasons, it is especially important to buy seafood from reputable vendors and to handle it with care.

Safe Food Storage and Preparation

Common food safety mistakes:

- Excessive store-to-refrigerator lag time
- Not washing hands before food handling
- Unclean equipment or utensils
- Countertop thawing
- Room-temperature marinating
- Using same spoon to stir and taste
- Cross-contamination:
  Using the same board or knife to cut raw meat and vegetables or fruits
  Using the same plate for raw and cooked foods
- Inadequate cooking or reheating
- Failure to keep hot foods at high temperature
- Improper cooling: leftovers and "doggie bags" left out

Commercially prepared, canned, or packaged food is usually free of harmful microbial agents when it leaves the manufacturer. When a batch of food is contaminated, batch numbering ensures that it can be recalled quickly, and the public can be forewarned. Although tampering with food products is unlikely to occur in grocery store foods, your best protection is to carefully inspect the seals and wrappers of every purchase. Jars should be firmly sealed (many have safety buttons—areas of the lid designed to pop up once opened). Packages should be free of holes or tears. A broken seal or mangled package does not provide protection against microorganisms or other contaminants. Likewise, do not purchase any canned goods that have dents and cracks or bulges, which can indicate possible contamination with *Clostridium botulinum*.

Most cases of food poisoning occur in the home or the restaurant and are caused by improper food storage or handling. Food poisoning caused by kitchen mistakes can be avoided by doing three simple things: keep cold food cold; keep your hands, the utensils, and the kitchen clean; and keep hot food hot. Most bacteria flourish in warm environments, but heat kills them and cold temperatures halt their growth.

Keep Cold Foods Cold The first step, keeping cold food cold, starts when you leave the grocery store. If you are running errands in a car, do the grocery shopping last so that the food does not stay in the car too long, especially in hot weather. As soon as you get home, put foods in a refrigerator set at 40 degrees Fahrenheit or a freezer kept at 0 degrees Fahrenheit, and be sure not to leave food in the refrigerator too long before eating it (see Figure 12-1). Place packages of raw meat, poultry, or fish on a plate before refrigerating or store them in plastic storage bags to prevent bacteria-containing juices from dripping onto other foods. Thaw frozen food in the refrigerator or microwave oven—not on the kitchen counter. Because bacteria can multiply at room temperature, they can thrive in the relatively warm exterior of a food before the interior has thawed.

FIGURE 12-1
TAKE CONTROL OF HOME FOOD SAFETY

24-hour bug? Or something you ate? Very often what seems like the flu may be foodborne illness. The following four simple actions will help you take control of food safety in your kitchen.

SOURCE: Adapted from the American Dietetic Association and the ConAgra Foundation's *Home Food Safety . . . It's In Your Hands* program. For more information, visit *www.homefoodsafety.org* and www.fightbac.org.

1. CLEAN: Wash Hands and Surfaces Often Using Hot Soapy Water.

Hand washing may eliminate half of all cases of foodborne illness and reduce the spread of the common cold and flu.

- Wash hands in warm, soapy water for a minimum of 20 seconds before preparing foods and after handling raw meat, poultry, and seafood.

2. SEPARATE: Keep Raw Meats and Ready-to-Eat Foods Separate.

- Never work with raw meat, poultry, or seafood on the same surface that you use for other foods without cleaning the surface after you've finished.
- Use different cutting boards: one to cut raw meat, poultry, and seafood and another for ready-to-eat foods, such as breads and vegetables. Discard old cutting boards that have cracks, crevices, and excessive knife scars.

3. COOK: Cook to Proper Temperatures.

Harmful bacteria are destroyed when food is cooked to proper temperatures. Buy a meat thermometer and use it!

4. CHILL: Refrigerate Promptly Below 40°F.

- Defrost foods in the refrigerator or in the microwave. Marinate foods in the refrigerator. Refrigerate foods quickly to slow the growth of bacteria and prevent foodborne illness. Set your refrigerator below 40°F to keep perishable foods out of the "danger zone"—40° to 140°F.

General Guidelines for Leftovers

Perishable food	Keeps up to
Cooked fresh vegetables	3–4 days
Cooked pasta	3–5 days
Cooked rice	1 week
Deli counter meats	5 days
Greens	1–2 days
Meat	
Ham, cooked and sliced	3–4 days
Hot dogs, opened	1 week
Lunch meats, prepackaged, opened	3–5 days
Cooked beef, pork, poultry, fish, meat casseroles	3–4 days
Cooked patties and nuggets, gravy and broth	1–2 days
Seafood, cooked	2 days
Soups and stews	3–4 days
Stuffing	1–2 days

When in doubt, throw it out!

Safe Cooking Temperatures

°F: 0, 20, 40, 60, 80, 100, 120, 140, 160, 180, 200, 220

- Medium Rare Beef and Lamb; Seafood
- Ground Beef, (patties, meatballs, meatloaf), Veal, Lamb, Pork; Medium Beef and Pork Roasts, Steaks, Chops, Egg dishes, Casseroles
- Ground Poultry, Leftovers, Stuffing (cooked separately)
- Well-done Beef, Lamb, Steaks, Chops, Poultry Breasts, Pork Roasts
- Whole Chicken, Turkey

Wash Hands and Surfaces Often Along with keeping foods properly chilled, keeping the kitchen clean prevents contamination of otherwise wholesome foods. Before you handle food, wash your hands in warm, soapy water. In addition, be sure to wash your hands after touching meat, poultry, or fish to prevent the spread of any bacteria that your hands have picked up. Likewise, keep countertops and all kitchen equipment clean with soap and hot water. Because bacteria love to nestle down in the fibers of kitchen cloths, sponges, and wooden cutting boards, take particular care to keep such items clean.

Keep Hot Foods Hot Keeping hot food hot means first cooking food thoroughly to ensure that the heat destroys any bacteria that might have been present in the uncooked food. See Figure 12-1 for proper cooking temperatures. After cooking, foods must be kept hot until serving to prevent bacterial growth. Never leave perishable food at room temperature for more than 2 hours. Before refrigerating a large quantity of hot food, such as a pot of chili or a large casserole, divide it up and place it in shallow containers to allow easy cooling. Otherwise the food inside the pot may stay warm for a dangerously long time, even in the refrigerator.

© Felicia Martinez/PhotoEdit

▲ *Never thaw food on a kitchen counter. Bacteria can flourish at room temperature.*

cross-contamination the inadvertent transfer of bacteria from one food to another that occurs, for instance, by chopping vegetables on the same cutting board that was used to skin poultry.

mad cow disease (bovine spongiform encephalopathy or BSE) a rare and fatal degenerative disease first diagnosed in 1986 in cattle in the United Kingdom. The bovine disease may be passed to humans who eat the meat of infected animals and may lead to death due to brain and nerve damage.

Prevent Cross-Contamination Meat, poultry, and fish require special handling because they often harbor high levels of bacteria. In addition, they provide a moist, nutritious environment—just right for microbial growth. Wash anything that has come into contact with such foods to prevent **cross-contamination.** For instance, after marinating raw meat in a dish, don't put the meat back in the same dish after cooking it, and don't use the marinade unless it has been cooked thoroughly. Wash the dish in hot, soapy water before reusing it, or the bacteria inevitably left in the dish from the raw meat can contaminate and grow in the cooked product or other food—a classic example of cross-contamination. Similarly, wash a cutting board (and your hands) after, say, skinning chicken on it. If you don't, and you use the contaminated board to chop raw vegetables for a salad, the vegetables can pick up the bacteria from the poultry. Because the salad won't be heated, the bacteria won't be killed.

Food Safety for Meats Ground meat is especially susceptible to bacterial contamination. Steaks and roasts are not as risky because bacteria usually settle on the outside of the cuts and are easily destroyed when the outside is heated. But, because the process of grinding meat spreads the bacteria throughout, foods made with ground meat, such as hamburger patties or meat loaf, must be cooked all the way through to kill the bacteria in the middle. To decrease your risk of eating contaminated ground beef, cook burgers until the juices run clear and not a trace of pink is left on the inside, and always order burgers well done at restaurants. When baking a meat loaf, use a thermometer to test the internal temperature. Be especially careful when cooking ground meat in a microwave oven, which sometimes cooks foods unevenly.

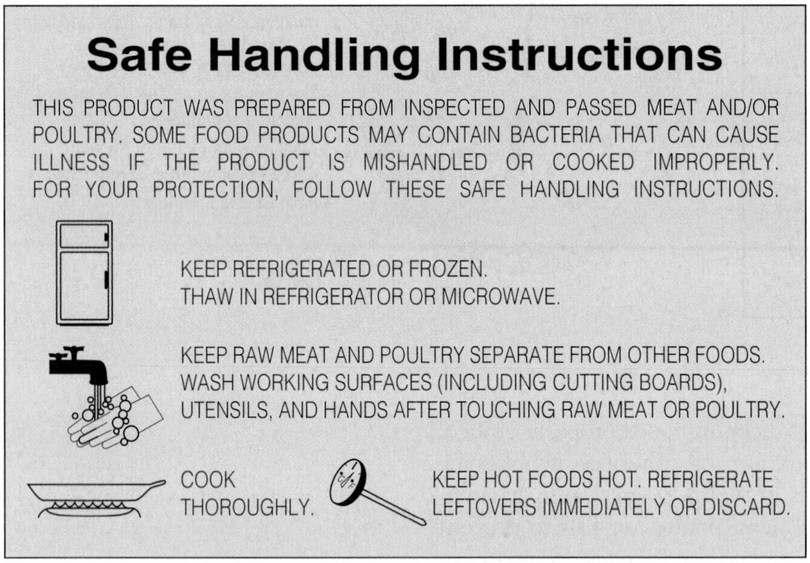

In the 1990s, outbreaks of **mad cow disease** in the United Kingdom received exaggerated media coverage, raising concerns among American consumers regarding the safety of consuming meat from infected animals. Agricultural officials in the United Kingdom now prohibit the addition of mammalian meat and bonemeal in the feed of any food-producing animals, because these tissues are suspected to be the source of the disease. As a result, the rate of newly reported cases of the disease is decreasing. Mad cow disease poses little or no concern to consumers in the United States, however, because the USDA banned cattle imports from Great Britain and other countries affected by the disease in 1989. USDA also bans the use of most mammalian protein tissues in the manufacture of animal feed given to ruminant animals (for example, cows, sheep, and goats).[14]

Safe Handling Instructions

THIS PRODUCT WAS PREPARED FROM INSPECTED AND PASSED MEAT AND/OR POULTRY. SOME FOOD PRODUCTS MAY CONTAIN BACTERIA THAT CAN CAUSE ILLNESS IF THE PRODUCT IS MISHANDLED OR COOKED IMPROPERLY. FOR YOUR PROTECTION, FOLLOW THESE SAFE HANDLING INSTRUCTIONS.

KEEP REFRIGERATED OR FROZEN.
THAW IN REFRIGERATOR OR MICROWAVE.

KEEP RAW MEAT AND POULTRY SEPARATE FROM OTHER FOODS.
WASH WORKING SURFACES (INCLUDING CUTTING BOARDS),
UTENSILS, AND HANDS AFTER TOUCHING RAW MEAT OR POULTRY.

COOK THOROUGHLY.

KEEP HOT FOODS HOT. REFRIGERATE LEFTOVERS IMMEDIATELY OR DISCARD.

▲ *The U.S. Department of Agriculture requires that all fresh meat and poultry products carry this label as a reminder to handle the products carefully.*

Food Safety for Seafood Seafood also should be handled with care, especially fish intended to be eaten raw or only lightly steamed. The dangerous foodborne diseases that may lurk in normal-appearing seafood—worms, parasites, severe viral intestinal disorders, and hepatitis—can be much worse than those of normal bacterial spoilage. Raw fish dishes such as sushi and sashimi are safe for most healthy people to eat if they are prepared with fresh fish that has been commercially frozen at temperatures lower than most home freezers can attain. This type of freezing kills any parasites that might be present. However, eating raw or undercooked oysters, clams, mussels, and whole scallops is especially risky because these types of seafood sometimes carry a strain of bacteria known as *Vibrios*, which can multiply even during refrigeration. Although *Vibrios* bacteria are

TABLE 12-3
Individuals at High Risk for Foodborne Illness and Foods They Should Avoid

High-Risk Patient Categories	Foods to Avoid
Young children Pregnant women Elderly individuals Immunocompromised individuals	Raw fish or shellfish Raw or unpasteurized milk or cheeses Soft, French-style cheeses and patés Raw or undercooked eggs or foods containing raw or lightly cooked eggs (e.g., certain salad dressings, cookie and cake batters, sauces, and beverages such as unpasteurized eggnog) Raw or undercooked meat or poultry Precooked processed meats that have not been reheated (e.g., deli meats, hot dogs) Raw sprouts (alfalfa, clover, and radish) Unpasteurized fruit or vegetable juices

SOURCE: Data from *Diagnosis and Management of Foodborne Illnesses: A Primer for Physicians*; www.ama-assn.org ama/pub/article/ 3707-3938.html; *Nutrition and the MD* (Hagerstown, MD: Lippincott and Wilkins, March 2003), p. 8.

destroyed by thorough cooking, they can thrive in raw shellfish and cause serious illness, sometimes even causing a deadly blood poisoning that can kill a person within a day or two.[15] Thus, the hazards of eating raw or undercooked seafood are especially serious for vulnerable people, such as those with liver disease, diabetes, gastrointestinal disorders, and HIV infection. People with these and other diseases that may compromise the body's immune system and, hence, its ability to defend itself against food poisoning (see Table 12-3) need to be especially cautious when preparing and consuming seafood.

© Martin Chaffer/GettyImages

▲ *Handle raw meat and poultry with care and cook it thoroughly to destroy any bacteria present. Place it on a clean plate when it is cooked.*

Food Safety for Picnics When planning picnics, choose foods that last without refrigeration, such as fresh fruits and vegetables, breads and crackers, and canned spreads and cheeses that can be opened and used on the spot. Aged cheeses, such as cheddar and Swiss, do well for an hour or so, but they should be kept in an ice chest for longer periods. Chill dishes containing mayonnaise or other types of dressing before, during, and after the picnic. Burgers, chicken breasts, and other foods intended for grilling, should not be partially cooked ahead of time. Partial cooking may not kill all the bacteria, and if these half-cooked items are not heated thoroughly later, chances of bacterial contamination run high. Partial cooking is a safe method only when, for example, you take a burger or other food *directly* from the microwave oven to the grill.[16]

In general, remember that fresh food smells fresh. Do not eat any food that carries an "off" odor because the smell is probably the result of bacterial wastes and indicates a dangerous number of bacteria in the food. To be sure, not all types of food poisoning are detectable by odor, but if a food smells bad, chances are high that it is spoiled. Refer to the Savvy Diner feature for recommendations about preventing foodborne illness. See also the Food Safety Scorecard to help you rate your food safety knowledge.

Pesticides and Other Chemical Contaminants

Food producers and food processors exert major efforts to maintain a safe food supply. Some risk of consuming undesirable substances, or **contaminants,** however, is unavoidable. Our industrial society's reliance on chemical processes means that foods may become contaminated by a variety of chemicals introduced into the environment. In addition, pesticides used in agricultural techniques affect the food supply. The following section examines some of the major chemical players in the food supply and looks at some ways that scientists hope to reduce them.

contaminants potentially dangerous substances, such as lead, that can accidentally get into foods.

Keep Food Safe to Eat

The tradition of tea drinking has been popular for more than 2,000 years in China. Contaminated water may have been one of the reasons. A hot drink made with boiled water was less likely to cause digestive problems than plain water.[17]

To Prevent Foodborne Illness

- When running errands, make the grocery store your last stop. When you get home, refrigerate perishables immediately.
- Refrigerate leftovers promptly and heat them thoroughly to at least 165°F before serving.
- Use clean eggs with intact shells and cook eggs before eating them (soft-boiled for 7 minutes, poached for 5, or fried for 3 minutes on each side).
- Keep hot foods hot (140°F or above). Keep cold foods cold (40°F or below).
- Do not prepare food if you have a skin infection or infectious disease. Anyone, though, may be a carrier of bacteria and should avoid coughing or sneezing over food.
- When in doubt, throw it out. Do not even taste food that is suspect. (An off odor, however, is not necessarily detectable in a food containing toxins.)
- Discard food from cans that leak or bulge. Dispose of the food in a manner that will protect other people and animals from its accidental use.
- Cook all meat and poultry to 160°F or higher—until the flesh is no longer pink and all the juices run clear. Order hamburgers and other beef items well done when eating out.
- Wash fruits and vegetables thoroughly before eating.

To Ensure a Safe Catch: Seafood Safety Tips

- Choose fresh fish steaks and fillets that are moist, with no drying or browning around the edges.
- Buy seafood only from reputable dealers and cook fish within 2 days of purchase. Use packaged, frozen seafood before the expiration date.
- Make sure the fish odor is fresh and mild. Fresh fish and shellfish should not have an overly "fishy" odor. Refrigerate seafood immediately below 40°F after buying.
- Cook fish to 145°F or until the flesh is opaque and separates easily with a fork.
- Refrigerate smoked, pickled, vacuum-packed, and modified-atmosphere-packed fish products.

▲ *Many of the chemicals that contaminate foods are the waste products of industry.*

organic halogens compounds that contain one or more of a class of atoms called halogens, including fluorine, chlorine, iodine, or bromine.

heavy metals any of a number of mineral ions, such as mercury and lead, so named because of their relatively high atomic weight. Many heavy metals are poisonous.

toxicity the ability of a substance to harm living organisms. All substances are toxic if present in high enough concentrations.

hazard state of danger; used to refer to any circumstance in which harm is possible.

Chemical Agents

Some of the problem industrial chemicals prevalent in the environment and food supply are **organic halogens,** such as polychlorinated biphenyl (PCB) and polybrominated biphenyl (PBB), and **heavy metals,** such as lead and mercury. Fortunately, episodes of direct, excessive chemical contamination such as chemical spills are rare. However, when contamination *does* occur, the effects can be far-reaching. A list of several chemical contaminants of particular concern in foods is presented in Table 12-4 on page 399.

Under normal circumstances, even though a chemical substance is toxic, it tends to pose only a small hazard because it is usually carefully controlled. However, when a chemical spill or other accident occurs, the risk can suddenly soar. That's why scientists differentiate between **toxicity** (a property of all substances) and **hazard** (the likelihood of a substance's actually causing harm). All substances are potentially toxic, but they are hazardous only if consumed in large enough quantities. In other words, the dose makes the poison. Thus, a chemical that is present in foods in minuscule amounts does not pose a significant hazard until for some reason it becomes concentrated in the food, that is, in toxic amounts.

Most experts agree that although chemicals in foods pose a small hazard, the chief concern is accidental gross contamination. In some instances, however, chemical contamination can be subtle and insidious. The problem of chronic low-level lead poisoning is a prime example. In the United States, lead poisoning ranks as one of the most common childhood environmental health problems. More than 400,000 children under the age of 6 have blood lead levels high enough to cause mental and other health problems.[18]

Food Safety

Improper storage of food not only increases the risk of food poisoning, but it also almost always results in a loss of nutrients and good taste. The following quiz is designed to measure your food safety savvy.

1. If you're packing a picnic, it's okay for the cold foods to be at room temperature prior to packing as long as they are placed in a cooler with ice or ice packs. *True or False?*
2. Foods prepared with mayonnaise—macaroni salad, potato salad, and cole slaw—are common sources of food poisoning. *True or False?*
3. Raw ground meat or poultry can be stored in the refrigerator, but they should be used within:
   - **a.** 1–2 days.
   - **b.** 2–3 days.
   - **c.** 3–4 days.
   - **d.** 1 week.
4. What signs indicate that canned foods may be contaminated?
   - **a.** bulging
   - **b.** leaking
   - **c.** spurting of liquid when opened
   - **d.** All of the above.
5. Which of the following foods have been linked to food poisoning?
   - **a.** cooked rice
   - **b.** apple cider
   - **c.** shellfish
   - **d.** All of the above.
6. The best place in the refrigerator to store milk is in the door. *True or False?*
7. When bringing home groceries from the market, it's a good idea to rewrap meat and poultry before placing them in the refrigerator or freezer. *True or False?*
8. Fresh fish should smell "fishy." *True or False?*
9. The best way to handle green-skinned potatoes is to:
   - **a.** throw them away.
   - **b.** soak them in cold water.
   - **c.** peel the skin and remove some of the flesh prior to cooking.
   - **d.** remove the green section after cooking.
10. Canned foods can be stored in the pantry indefinitely. *True or False?*
11. You can tell if a food is contaminated by the way it looks, smells, or tastes. *True or False?*
12. What is the maximum time perishable foods can be kept at room temperature?
    - **a.** ½ hour
    - **b.** 1 hour
    - **c.** 2 hours
    - **d.** 24 hours
13. It's okay to thaw frozen ham on the kitchen counter because salted and smoked meats are free of bacteria. *True or False?*
14. You can reduce your exposure to chemical contaminants in whole fish by proper cleaning. *True or False?*
15. Cloudy liquid around packaged hot dogs indicates bacteria have started growing. *True or False?*

Lead contamination usually does not poison a person all at once; rather, low levels build up gradually in the soft tissues of the kidneys, bone marrow, liver, and brain. Over time, the accumulated lead can cause such health problems as diminished intelligence and impaired development. Pregnant women and young children are particularly vulnerable to the effects of lead because their bodies

Answers

1. False. Be sure food is cold or frozen before placing it in a cooler. This minimizes the chance of microbial growth. Use ice packs between food items. Frozen juice boxes can also be packed and enjoyed later in the day after they have thawed.
2. False. Adding mayonnaise to food does not increase the chance of food poisoning. Most store-bought mayonnaise contains ingredients (vinegar, lemon juice, and salt) that actually slow bacterial growth.
3. (a) Ground meat and ground poultry are more perishable than other meats. They should be refrigerated and cooked (or frozen) within 1–2 days of purchase.
4. (d) Never buy or use products with bulging lids, leaking cans, or cracked jars. All are warning signs that a product may be contaminated with the bacteria that cause deadly botulism.
5. (d) Although the most common offenders are poultry, meat, and shellfish, any food can cause food poisoning, including cooked rice, if mishandled.
6. False. The refrigerator door does not stay as cold as the rest of the refrigerator. Highly perishable foods like milk and eggs should be stored on an inside shelf. Use the door for condiments.
7. False. Leave meat and poultry in store wrapping. The less you handle it, the better. This is especially true for ground meat and ground poultry.
8. False. Fresh fish should smell like a fresh sea breeze. If it smells "fishy," don't buy it.
9. (c) Green-skinned potatoes contain a natural toxin called solanine, which develops when potatoes are exposed to sunlight. It imparts a bitter taste and may cause stomach upset if eaten in large quantities. Before cooking, simply remove the green area and about ½ inch of flesh around it.
10. Canned foods have an extended but not infinite shelf life. Remember to place newly purchased cans behind older ones. To be safe, throw out beans and high-acid foods (for example, pineapple, tomatoes) after 1 year and other canned foods after 2 years. Use canned fish within 6 months.
11. False. Food spoilage may leave tell-tale signs such as changes in looks, smell, or taste, but contaminated food does not. Remember this rule of thumb: "When in doubt, throw it out."
12. (c) The longer a perishable food is kept at room temperature, the greater the likelihood that bacteria will multiply to dangerous levels. Food kept unrefrigerated for more than two hours is a prime target for bacterial growth.
13. False. Many people think that salted or smoked meats are immune to bacterial contamination. That's not so, especially since many manufacturers have gradually lowered the salt content of cured meats. With any meat or poultry, the safest way to thaw it is in the refrigerator overnight.
14. True. By removing the skin and trimming the fatty tissue along the back and belly, where chemicals tend to concentrate, you can reduce your exposure to contaminants.
15. True. Although hot dogs are processed to last longer than other meat products, *Listeria monocytogenes* can grow even under refrigeration. If you notice cloudy liquid, discard the franks. Freeze hot dogs if you don't plan to use them within 2 weeks. If opened, use within a week.

HOW DID YOU SCORE?

Count up the number of questions you answered correctly.

12–15	Congratulations! You're quite the food safety scholar.
8–11	You're fairly savvy when it comes to food safety, but don't push your luck. Brush up on the questions you had trouble with.
7 or below	You're a likely target for food poisoning. Mend your ways; it's never too late.

SOURCE: Adapted with permission from A. Schepers, What's Your Food Safety I.Q.?, *Environmental Nutrition,* 22 (1999): 1, 6. © Copyright 1999 by Environmental Nutrition, Inc., 52 Riverside Drive, New York, NY 10024.

Getting the Lead Out

To protect against overexposure to lead:
- Do not store fruit juices or other acidic foods in ceramic containers.
- Do not store beverages in lead crystal containers.
- Do not eat or drink from items that show a dusty or chalky gray residue on the glaze after they are washed.
- Do not feed babies from crystal bottles.
- Run cold water for a minute before using it for drinking or cooking.

For more information call the National Lead Information Center at 1-800-424-LEAD.

absorb high levels of calcium to meet their growth needs but the body cannot distinguish between lead and calcium. As a result, they absorb lead more readily than other people do.

Historically, lead has entered the food and water supply largely through leaded gasoline exhaust, which contaminates rainfall, which in turn pervades crop soil and water supplies. Lead contamination of the water supply continues to be a source of concern (refer to the discussion on water in Chapter 7). In addition, lead solder used to seal the seams of cans was a source of the contaminant for many years. Fortunately, a 25-year process to phase out lead in gasoline reached its goal in 1995, and lead-soldered cans have been eliminated. Today the levels of lead in foods and beverages are the lowest in history.

Nevertheless, lead paint abounds in older housing, and new sources of lead contamination have surfaced during recent years. When it comes to lead exposure via food and beverages, scientists have identified ceramic hollowware, lead crystal,

TABLE 12-4
Examples of Contaminants in Foods

Name and Description	Sources	Toxic Effects	Typical Route to Food Chain
Cadmium (heavy metal)	Used in industrial processes, including electroplating, plastics, batteries, alloys, pigments, smelters, and burning fuels; present in cigarette smoke	No immediately detectable symptoms; slowly and irreversibly damages kidneys and liver	Enters air in smokestack emissions, settles on ground and is absorbed into plants, consumed by farm animals, and eaten in meat and produce by people. Sewage sludge and fertilizers leave large amounts in the soil; runoff contaminates shellfish.
Lead (heavy metal)	Lead crystal, improperly manufactured and old ceramic ware, paint, old plumbing, and leaded gasoline	Displaces calcium, iron, zinc, and other minerals from their sites of action in the nervous system, bone marrow, kidneys, and liver, causing failure to function; causes breakage of red blood cells (anemia) and interferes with the immune response	Air pollution, leaded gasoline, water pipes, improperly manufactured and old ceramic ware, and lead crystal
Mercury (heavy metal)	Widely dispersed in gases from the Earth's crust; local high concentrations from industry, electrical equipment, paints, and agriculture	Poisons the nervous system, especially in fetuses	Inorganic mercury released into waterways by industry and acid rain is converted to methylmercury by bacteria and ingested by fish*
Polychlorinated biphenyl (PCB) (organic compound)	No natural source; produced for use in electrical equipment	Causes long-lasting skin eruptions, eye irritation, growth retardation in children of exposed mothers, anorexia, fatigue, and other effects	Discarded electrical equipment, accidental industrial leakage, or reuse of PCB containers for food

*The FDA recommends that pregnant women and lactating women eliminate from their diets certain species of large, predatory fish that can contain high levels of mercury which can cause neurological damage during sensitive stages of fetal brain development. These include shark, swordfish, king mackerel, tilefish. and tuna steaks. As an extra precaution, women who plan to become pregnant should avoid those species for 6 months before conception. Visit www.cfsan.fda.gov for more information and updates.

and foil capsules on wine bottles as potential sources. Lead can leach from the glaze of ceramic bowls, mugs, and pitchers that have not been properly formulated or fired, especially when the food or beverage inside is hot and acidic, such as coffee, tea, or tomato soup. Wine and other alcoholic beverages also promote the leaching of lead, so experts advise against storing alcohol in pitchers and other containers made with lead crystal. In addition, it's a good idea to use a damp cloth to wipe the rims of wine bottles capped with foil, which has been shown to leach lead. For help on lowering your own risk, see the tips in the margin on page 398.

Although chemical contamination is not the greatest hazard posed by the food supply, there are still many unknowns. For example, what are the effects of prolonged exposure? Despite the uncertainties, keep in mind that more often than not, healthful eating habits and overall good health protect against the toxicity of food contaminants and other environmental pollutants. Eating a wide variety of food ensures an adequate supply of essential nutrients and minimizes exposure to potential environmental contaminants.

▲ *Some antique ceramic and crystal items are best left on the shelf. The older the piece, the higher the chances that it leaches lead.*

Pesticide Residues

Pesticides are substances used to prevent, destroy, or repel harmful pests, including insects, spiders, bacteria, weeds, molds and mildews, rodents, and other living things. Unlike many other chemicals present in the food supply, pesticides are intended specifically to poison living things. The trouble with pesticides, of course, is that they can inadvertently harm wildlife, people, and other species.

pesticides chemicals applied intentionally to plants, including foods, to prevent or eliminate pest damage. Pests include all living organisms that destroy or spoil foods: bacteria, molds and fungi, insects, and rats and other rodents, to name a few.

regulation a legal mandate that must be obeyed. Failure to follow a regulation brings about legal consequences.

risk for pesticide residues, the harm a substance may confer. Scientists estimate risk by assessing the amount of a chemical that each person in a population might consume over time (also called *exposure*) and by considering how toxic the substance might be (*toxicity*).

risk = exposure × toxicity

exposure = amount of substance in food × amount of food eaten

When farmers first began using pesticides, they chose potent ones—chemicals designed to keep on killing for as long as they remained in the soil. Unfortunately, after years of widespread pesticide use, some unsettling side effects of the chemicals surfaced. They washed from the soil into lakes, rivers, and oceans, contaminating water supplies; they poisoned farm workers, who breathed the chemicals day in and day out; they endangered many species of wildlife, disrupting the animals' abilities to reproduce and wiping out entire populations; and, most disturbing, they would not go away.

For example, in the 1960s, scientists observed that a popular pesticide called DDT had begun accumulating in the body fat of animals. DDT threatened the survival of the American eagle by weakening eggshells to the point of collapse, killing the developing chicks inside. DDT also appeared in big fish, carnivorous animals, people, and human breast milk. Finally, after years of widespread agricultural use, the United States banned DDT. However, the pesticide still lingers in the environment. Many foreign countries continue to use DDT, including countries from which the United States buys produce. Regrettably, despite the U.S. ban on DDT, U.S. companies are still allowed to sell it (as well as other banned pesticides) to countries where DDT use remains legal. This practice may come back to haunt us, however, when we buy imported produce that has been exposed to DDT. In addition, the DDT situation illustrates the need to consider environmental issues from a global perspective; residues of chemicals banned in one country can travel the entire world not only through the imported and exported foods but also through wind, rain, and waterways.

DDT taught us another lesson about pesticides: A contaminant that builds up in the body carries the greatest potential risk. In the body, a contaminant that is quickly broken down to some harmless compound poses the least risk to health. Likewise, if the body can easily and rapidly excrete a pesticide residue, the body may not be harmed by it. But if the residue enters the body and stays there, all the while interacting with the body's cells and systems, it may wreak havoc. Additional doses piled on top of the first ones compound the damaging effects. Moreover, when a substance resists breakdown either inside the body (by the body's own enzymes) or outside (by microorganisms) and furthermore accumulates from one species to the next, it builds up in the food chain (see Figure 12-2). DDT causes all these problems, which is why it is so deadly.

Even so, pesticides do have a place in the farming community. Farmers would face daunting obstacles if they tried to grow crops without them. Careful use of pesticides often boosts crop yields, which in turn helps to keep the price of fruits and vegetables down and the availability of a wide variety of produce high. Still, as the DDT experience illustrates, pesticides must be evaluated scrupulously and used judiciously.

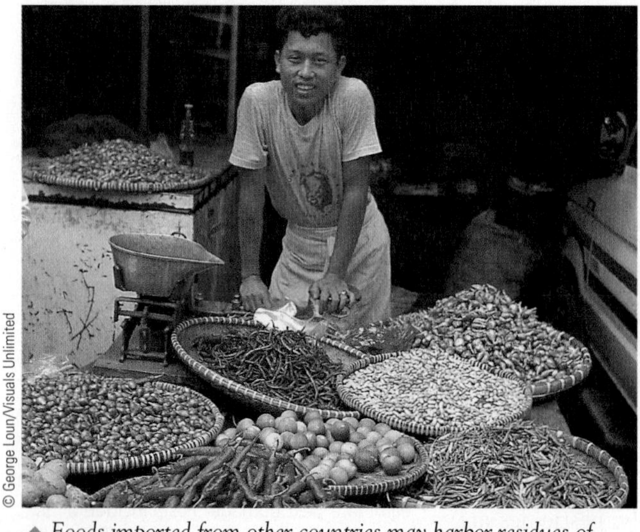

▲ *Foods imported from other countries may harbor residues of pesticides that have been banned for use in the United States.*

The ideal pesticide destroys the target pest and then breaks down quickly into other products that pose minimal hazard to people and other animals. Scientists have made many strides in developing relatively innocuous pesticides over the years, and the search for better, safer pesticides continues. What's more, since the introduction of pesticides decades ago, many national and international agencies have adopted strict **regulations** for pesticide use.

In the United States, the Environmental Protection Agency (EPA) determines whether a particular chemical may be used on U.S. crops. In deciding whether to approve a pesticide, the EPA scrutinizes dozens of studies that assess the substance's possible effects on people, wildlife, fish, or plants as well as its potential to cause such problems as cancer and birth defects. To do so, the EPA examines the estimated amount of a pesticide residue that a person might be exposed to during the course of a 70-year lifetime and the **risk** of

FIGURE 12-2
HOW A FOOD CHAIN WORKS

A person who eats fish regularly may consume about 100 pounds of it in a year. These fish will, in turn, have eaten a few tons of small plant-eating fish during their lifetime. The little plant eaters will have ingested several tons of plants. If the plants have been contaminated with toxic chemicals, the bodies of the small fish that eat them will contain high concentrations of the chemicals; the larger fish that eat the little fish will harbor even higher amounts of the chemicals; and so on through the food chain. If none of the chemicals are lost along the way, the person at the top of the food chain ultimately eats the same amount of chemical contaminants that was contained in the original *several tons* of plants at the bottom of the food chain.

harm that amount of exposure might pose. If the risk is unacceptably high, the pesticide is ruled out. This extensive research process often requires years of research and costs millions of dollars.[19]

Once the EPA approves a pesticide for use, it sets forth safety standards. For those pesticides it allows, the EPA determines which crops may be treated with it and how much may be applied. It also establishes a **tolerance**—that is, the maximum amount of a pesticide residue allowed in or on a food (see Table 12-5). The EPA identifies what is called a **reference dose** for the pesticide—the amount that could be consumed daily without posing any health risk. First, scientists use animal studies to estimate the maximum amount of the chemical that a person could take in daily without suffering harm. Next, to ensure an extra **margin of safety,** they calculate a small fraction of this amount, usually 1⁄100, which becomes the reference dose. In other words, the reference dose is 1⁄100 of the maximum amount of the substance that appears to be safe. Scientists factor in this large margin of safety as a precaution to help ensure that even highly vulnerable people won't be harmed by the substance in question (see Figure 12-3).[20]

TABLE 12-5
Pesticides in Perspective

When you hear about pesticide residues in food, the amounts measured are either parts per million (ppm), parts per billion (ppb), or parts per trillion (ppt). The following comparisons show just how tiny these amounts are.

1 ppm:	1 gram of residue in 1 million grams of food
	1 inch in 16 miles
	1 minute in 2 years
	1 cent in $10,000
1 ppb:	1 gram of residue in 1 billion grams of food
	1 inch in 16,000 miles
	1 second in 32 years
	1 cent in $10 million
1 ppt:	1 gram of residue in 1 trillion grams of food
	1 inch in 16 million miles
	1 second in 32,000 years
	1 square foot on floor tile the size of Indiana

SOURCE: Adapted from International Food Information Council, *Pesticides and Food Safety* (Washington, D.C.: International Food Information Council, 1995), 2.

tolerance the maximum amount of a particular substance allowed in food.

reference dose the estimated amount of a chemical that could be consumed daily without causing harmful effects.

margin of safety from a food safety standpoint, the margin is a zone between the maximum amount of a substance that appears to be safe and the amount allowed in the food supply.

FIGURE 12-3
MARGIN OF SAFETY

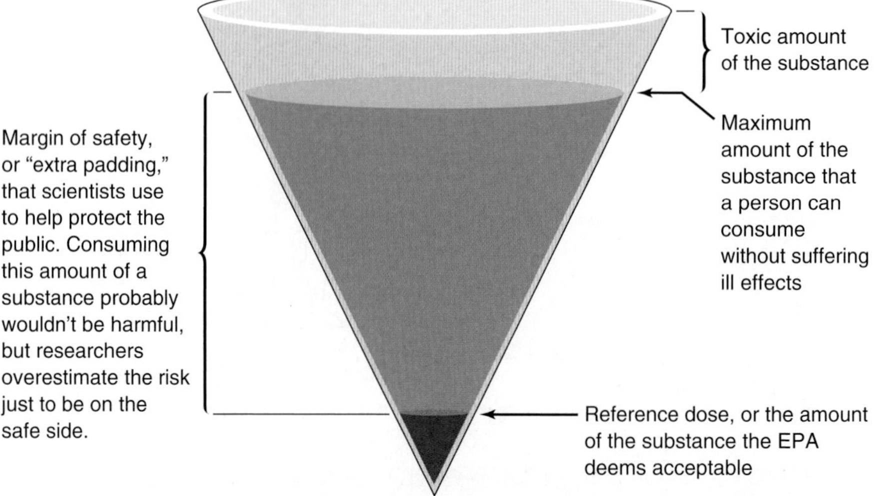
Toxic amount
of the substance
Maximum
amount of the
substance that
a person can
consume
without suffering
ill effects
Margin of safety,
or "extra padding,"
that scientists use
to help protect the
public. Consuming
this amount of a
substance probably
wouldn't be harmful,
but researchers
overestimate the risk
just to be on the
safe side.
Reference dose, or the amount
of the substance the EPA
deems acceptable

After the EPA approves a pesticide, the FDA, in its ongoing monitoring program, begins to check for residues. Inspectors collect samples from packers, shippers, and other food handlers and test them for pesticide residues. If a food contains residues that exceed the EPA's tolerance limits, the FDA can seize the entire shipment and press criminal charges. Meanwhile, how can you protect yourself against unacceptably high levels of pesticides? The Nutrition Action feature on page 405 offers a perspective.

Although EPA and FDA regulations go a long way in protecting the public from harmful pesticide residues, problems with the monitoring system still exist. As explained earlier, other countries may use pesticides banned in the United States, and imported foods might not be tested for the presence of those pesticides. In addition, consider that in 1993 the National Academy of Sciences issued the results of a large-scale, 5-year study on the risk of pesticide residues in the diets of infants and children. Although it concluded that the U.S. food supply is safe for children, it called for changes in the methods used to assess health risks from pesticides to better account for differences between adults and children.[21]

Despite the uncertainties, however, major health organizations, including the American Academy of Pediatrics, the American Cancer Society, and the American Medical Association, agree that the health risks posed by pesticide residues are minimal and that the health risks of *not* eating fruits and vegetables for fear of consuming pesticide residues far outweigh the slight risk linked with those substances. To keep pesticide residues from all types of produce to a minimum, use the produce-handling tips in Table 12-6.

Food Additives

food additive any substance added to food, including substances used in production, processing, treatment, packaging, transportation, or storage.

intentional food additives substances intentionally added to food; examples include nutrients, colors, spices, and herbs.

incidental food additives (or indirect additives) substances that accidentally get into food as a result of contact with it during growing, processing, packaging, storing, or some other stage before the food is consumed.

From a safety standpoint, **food additives** rank among the least hazardous substances in food, although consumers tend to rank them high on their list of food-related concerns. The great majority of food additives enhance the color, flavor, texture, or stability of foods or even improve the nutritional value of certain items, as shown in Table 12-7. Many additives are common substances such as vitamins, herbs, and spices, deliberately added to foods and called direct or **intentional food additives.**

On the other hand, **incidental food additives,** such as packaging materials or processing chemicals, often accidentally get into foods during processing. The federal government regulates all food additives and requires that food processors perform tests to determine whether additives are present in safe levels.

Manufacturers must go through a lengthy, costly process to get FDA approval for a new food additive. The manufacturer must conduct extensive research to

TABLE 12-6
Handling Produce Properly*

To keep pesticide residues to a minimum, use the following tips:

- Rinse produce thoroughly with water and scrub with a vegetable brush. When present, most residues reside on the surface of a product. Peel produce to which wax has been applied.
- Discard the outer leaves of lettuce, cabbage, and other leafy vegetables. Rinse the interior leaves thoroughly to remove dirt and debris.
- Use a knife to peel an orange or grapefruit; do not bite into the peel.
- Eat a variety of fruits and vegetables. Farmers use different chemicals for different crops, so eating a wide variety of produce helps to cut down on exposure to any particular pesticide residue.

*To reduce pesticide residue intake from meats, trim fat from meat and remove the skin from poultry and fish (pesticide residues concentrate in the animals' fat tissues). Discard fats and oils in broths and pan drippings.

SOURCE: Adapted from C. F. Chaisson and coauthors, *Pesticides in Food: A Guide for Professionals* (Chicago: American Dietetic Association, 1991), 18.

show that the additive in question does what it is supposed to do, that it can be detected and measured in foods to which it has been added, and that it is safe in the amounts in which it will be used.

As with pesticide residues, many food additives are allowed only in amounts that ensure a wide margin of safety. Most additives that pose any potential risk are allowed in foods only at levels 1⁄100 of those at which the risk is still known to be zero. Even nutrient food additives are subject to the margin of safety concept. For example, even though iodine has long been added to salt to prevent iodine deficiency, the amounts added have been controlled because excess iodine can be deadly.

© Polara Studios

▲ *Food additives extend the shelf life of many commonly eaten foods.*

The GRAS List and the Delaney Clause

Attention to two aspects of food law will help you to better understand the issues surrounding the use of additives in foods.

TABLE 12-7
Major Uses of Food Additives

Function of Additive	Examples	Foods in Which Often Used
Impart or maintain consistency	Stabilizers, thickeners, anticaking agents, and emulsifiers including alginates, lecithin, mono- and di-glycerides, glycerine, pectin, guar gum	Baked goods, cake mixes, salad dressings, ice cream, processed cheese, table salt, chocolate
Improve or maintain nutritional value	Vitamins A and D, thiamin, niacin, folic acid, ascorbic acid (vitamin C), calcium citrate, zinc oxide, iron, iodine	Flour, bread, biscuits, breakfast cereals, pasta, margarine, milk, iodized salt, juices
Maintain palatability and wholesomeness	Ascorbic acid, butylated hydroxyanisole (BHA), butylated hydroxytoluene (BHT), benzoates, sulfites	Bread, crackers, frozen and dried fruit, margarine, lard, potato chips, cake mixes, meat
Produce light texture; control acidity or alkalinity	Yeast, sodium bicarbonate, citric acid, lactic acid, phosphoric acid	Cakes, cookies, quick breads, crackers, butter, soft drinks
Enhance flavor or impart desired color	Cloves, ginger, fructose, aspartame, MSG, FD&C Red No. 40, caramel, turmeric	Spice cake, gingerbread, soft drinks, yogurt, soup, candy, cheese, jams, gum

SOURCE: Adapted from *Food Additives* (Rockville, Md.; Washington, D.C.: Food and Drug Administration in cooperation with International Food Information Council, 1992), 7–8.

GRAS (Generally Recognized As Safe) list a list of ingredients, established by the FDA, that had long been in use and were believed safe. The list is subject to revision as new facts become known.

Delaney clause a provision in the 1958 Food Additives Amendment that prohibited manufacturers from using any substance that was known to cause cancer in animals or humans at any dose level.

Substances Generally Recognized As Safe (GRAS) For the first half of the 20th century, scientists evaluated the safety of food additives using a simple approach: An added substance was either "safe" and therefore permitted for use in foods or "poisonous and deleterious" and therefore banned. As the study of toxic agents advanced, scientists realized that eventually they would be able to show that virtually every substance poses a health hazard if the dose is large enough. Consequently, they recognized that simply classifying an additive as "safe" or "poisonous" was not an effective means of evaluation.

To get around the problem, Congress proposed the bill that later became the Food Additives Amendment of 1958. As the members of Congress debated the bill, a question arose as to how to deal with the additives already in use. Congress decided that a "safe" substance in use prior to 1958 would be deemed a "Substance Generally Recognized as Safe" (a GRAS substance) and be put on the **GRAS list.** Substances not in use before that time would be classified as food additives and subject to regulation under the Food Additives Amendment.

With the establishment of the amendment, the FDA put hundreds of substances on the GRAS list. Everything from vegetable oils, salt, pepper, sugar, caffeine, vinegar, and baking powder to meat, poultry, eggs, milk, seafood, cereals, fruit, and vegetables were—and still are—classified as GRAS substances.

The GRAS list came under scrutiny in 1969, however, when safety questions arose about the GRAS substance cyclamate, an artificial sweetener. After reviewing hundreds of studies, the FDA decided to ban cyclamate from use in foods and beverages. As a result of this incident, President Richard Nixon ordered a reevaluation of the safety of all substances on the GRAS list. The FDA conducted a sweeping review and removed about 300 substances from the list. Today, the more than 400 substances classified as GRAS are continually subject to reexamination as new facts and concerns arise.

Delaney Clause Another piece of legislation came about during the debate about the Food Additives Amendment of 1958. At that time, James J. Delaney (U.S. Representative from New York) sponsored a provision to the bill that forbid the approval of any additive found to cause cancer in animals no matter how small the dose. The rationale for the provision stemmed from the widely held belief that it was possible to completely eliminate cancer-causing agents from the food supply. Congress voted to add the **Delaney clause** into the Food Additives Amendment of 1958, despite considerable debate.

In recent years, the Delaney clause often put regulatory agencies in a legal bind because it is virtually impossible to eliminate potential cancer-causing agents from the food supply. The problem is that many substances, even those found naturally in a number of foods, can cause cancer when given to animals in large enough amounts. In fact, critics charged that the Delaney clause encouraged the use of studies in which animals were fed substances in doses hundreds of thousands of times greater than the dose a person could consume via food. The results of such studies can be meaningless. In addition, as scientists continued to develop better techniques for detecting chemical residues in foods, the task of completely eliminating minuscule amounts became even more complex and unrealistic. As a result, the Food Quality Protection Act of 1996 eliminated the Delaney clause from law.

New Technologies on the Horizon

Back in the 1860s, a French scientist named Louis Pasteur came up with a radical, "newfangled" process by which the disease-producing microorganisms in a food could be destroyed by exposing it to heat. Dubbed pasteurization, after Dr. Pasteur, the process marked a major breakthrough in the science of food safety.

(Text continues on page 408.)

Nutrition Action

Should You Buy Organically Grown Produce or Meats?

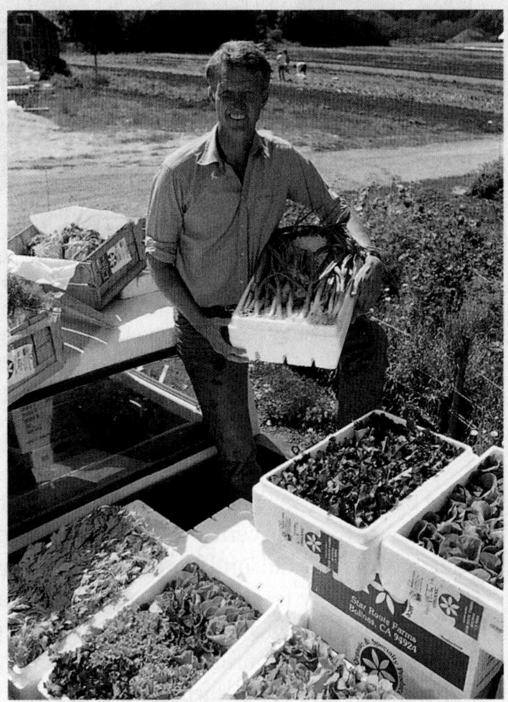

© Nancy Richmond/The Image Works

Once sold only in health food stores, organic fruits, vegetables, meats, and other **certified organic foods** can now be found in most mainstream markets nationwide. Sales of organic food products have grown from $78 million in 1980 to about $15 billion today, and sales of organically grown crops are predicted to increase fourfold during the current decade. Fresh produce is the best-selling organically grown food worldwide, but organic dairy products, soy foods, eggs, meat, and poultry are also popular.

Organically grown foods have been available for more than 40 years, but they have only recently been regulated. The United States Department of Agriculture (USDA) has created a national reference standard to define what is and what is not organic. Organic crops must be raised without using pesticides, petroleum-based fertilizers, or sewage sludge-based fertilizers. Animals raised on organic farms must be fed organic feed and given access to the outdoors. Additionally, they may not be given antibiotics or growth hormones.[22]

These regulations were put into place in October 2002 to regulate foods labeled "organic" whether grown in the United States or imported from other countries. The regulations mandate certification of organic products by USDA-approved state or private regulating authorities and set strict labeling standards for processed organic food. To use the USDA organic seal on their labels, raw products must be 100 percent organic, and processed foods must contain 95 percent organic ingredients. If a food contains between 70 and 95 percent organic contents, the label can read "product made with organic ingredients." Additionally, irradiated or genetically engineered products cannot be labeled as organic.[23] See Figure 12-4 for more about consumer information on food labels.

certified organic foods crops or livestock grown and processed according to USDA regulations concerning use of pesticides, herbicides, fungicides, fertilizers, preservatives, other synthetic chemicals, growth hormones, antibiotics, or other drugs.

FIGURE 12-4
CONSUMER INFORMATION ABOUT ORGANIC FOODS AND PRODUCT LABELS

These sample cereal boxes show the four labeling categories. Look for the name and address of the government-approved certifier on all packaged products that contain at least 70 percent organic ingredients. For more detailed information on the USDA organic standards, visit www.ams.usda.gov/nop.

SOURCE: Adapted from *The National Organic Program: Labeling Packed Products,* January 9, 2003; available at www.ams.usda.gov/nop/ProdHandlers/LabelTable.htm.

Foods made with 100 percent organic ingredients may claim "100% organic" and use the USDA seal.

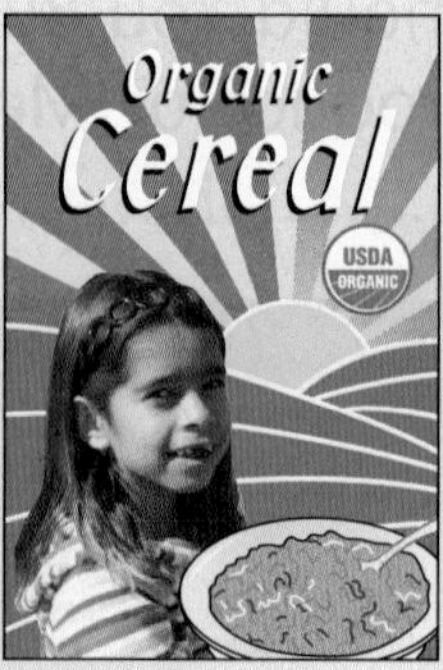

Foods made with at least 95 percent organic ingredients may claim "organic" and use the USDA seal.

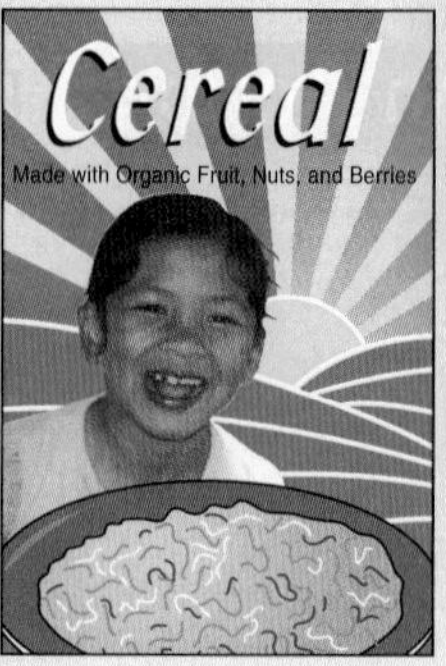

Foods made with at least 70 percent organic ingredients may list up to three of those ingredients on the front panel.

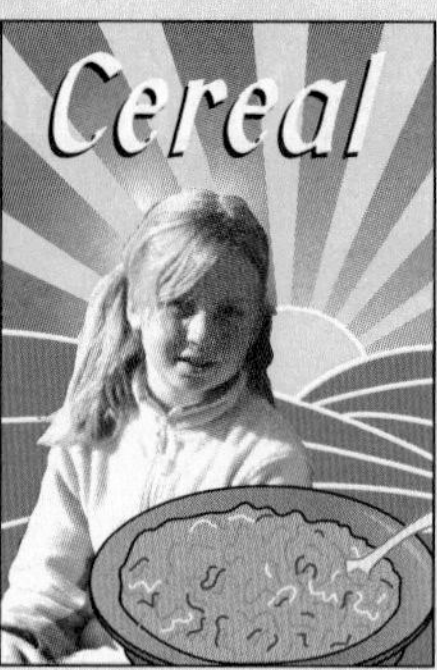

Foods made with less than 70 percent organic ingredients may list them on the side panel but may not make any claims on the front.

▲ *The USDA organic seal is intended to provide information to consumers. It will help organic farmers and ranchers further expand their already growing markets.*

The USDA makes no claims that organically grown food is safer or more nutritious than conventionally produced foods. It is important to understand that organic agricultural practices cannot guarantee that organically grown products are entirely free of residues. Additionally, organic produce is more expensive because it is grown without synthetic pesticides or chemicals and is more labor intensive. Organic crop harvests are often not as high as those grown conventionally, and fewer farmers use organic methods and sustainable agriculture practices. As a result, the cost of organically grown foods reflects the greater demands placed on the farmer.[24]

However, organically grown food casts a vote in favor of agricultural techniques that promote the well-being of the environment and farming communities over the long run. Reputable organic farmers typically use more "eco-friendly" techniques such as crop rotation to help keep pests under control than conventional farmers. These practices help protect the soil, the groundwater, and the farm workers themselves against chemical contamination. The box below lists some of the pros and cons of buying organic foods.

Still, many farmers who don't meet "organic" standards per se use other techniques to keep pesticide use to a minimum. For example, more and more farmers have adopted a system called **integrated pest management**—a technique by which farmers cut back on chemicals by combining strategies such as crop rotation, genetic engineering (discussed at the end of the chapter), and biological controls such as the release of a predator insect on a crop to get rid of another pest.

Advantages of Certified Organic Foods

- Less synthetic pesticide, hormone, and other chemical fertilizer residues in foods and waterways
- Beneficial to the environment: Improved soil fertility and crop diversity; increased water and energy conservation
- Farmers may follow more humane animal welfare standards

Disadvantages of Certified Organic Foods

- More expensive and tend to spoil faster since they contain no preservatives
- May be fertilized with improperly composted animal manure containing potentially harmful organisms, such as *E. coli* O157:H7
- May be cross-contaminated with pesticides from nearby conventionally sprayed fields

integrated pest management the use of biological controls, crop rotation, genetic engineering, and other tactics to reduce chemical use in the growing of crops.

Eat Well Be Well Eat Fresh Eat Local

It's Saturday morning in Anytown, USA, and in the center of the old "downtown" a large gathering of community members bustle around, chat, and enjoy their morning out. No, it's not a flashback to the 1950s. These people are buying locally grown produce and goods from farmers in their community.

Farmers' markets have existed in different forms for many years and have enjoyed a 111% growth over the last decade. Through these markets, farmers are able to build a local clientele and reap profits without a "middle man." Buyers also benefit from consuming produce that is ultra fresh and in season. In addition, programs like the USDA's Seniors Farmers' Market and WIC Farmers' Market provide vouchers for low-income participants to buy fresh, nutritious produce from farmers' markets. In addition, communities as a whole benefit from farmers' markets due to the small investment and potential boost to their economy.

Farmers' markets are much more than a place to buy produce, however. In an age when mega supermarkets have overtaken the landscape, these markets provide a great locale for a community gathering place. Farmers' markets give customers the unique experience of being able to converse with the actual grower of the produce. In doing so, these consumers can make suggestions about produce they would like to buy in the future and learn how different food items are grown. Try doing that at a supermarket!

Why Choose a Farmers' Market?

1. They help develop a food culture—a variety of foods that are commonly grown in that area can be displayed and purchased seasonally.
2. Foods are less processed and less packaging is used. This creates less waste, which is good for the environment. Moreover, energy resources used to transport more distantly produced foods can be saved and begin to diminish the size of the region's ecological footprint (see Figure 12-5).
3. Foods are usually fresher. Most of the foods have been picked within 24 hours of purchase and have not been frozen or treated with preservatives.
4. Farmers' markets encourage healthier food choices by promoting fresh-picked seasonal fruit and vegetables.
5. Farmers' markets boost local economy because the food distributor ("middle man") is eliminated. Thus, the farmer's income is more likely to be reinvested within the community.

To locate your local farmers market, visit www.ams.usda.gov/farmersmarkets for information, dates, and times of a market near you!

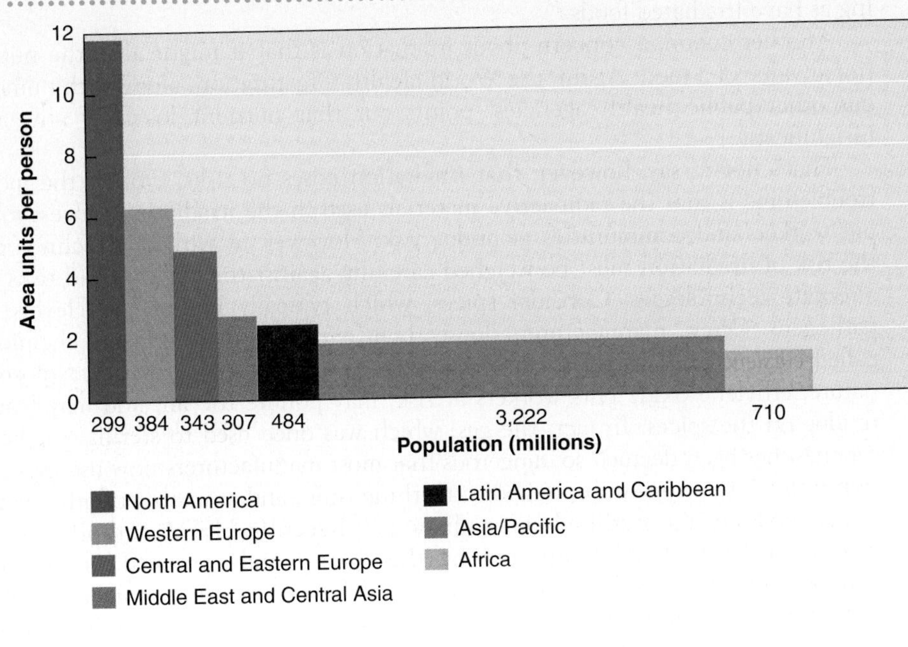

SOURCE: United Nations Population Fund, www.unfpa.org/swp/2001/english/ch03.html.

FIGURE 12-5
ECOLOGICAL FOOTPRINTS
According to the United Nations, the ecological footprint is a measure of the resources needed to support a region's consumption of food, materials, and energy—in other words, its impact on the environment. In this figure, each region is represented by a rectangle in which the width is proportional to the population and the height represents per capita resource consumption in terms of the area of productive land or sea required to produce those natural resources. The ecological footprint takes into account both the per capita resource consumption and the population growth. Thus, Asia, which has a population more than ten times that of North America but a per capita resource consumption level only one-sixth as large, has a footprint only slightly larger than North America. Take the *Ecological Footprint Quiz* at *www.myfootprint.org* to estimate how much productive land and water you need to support what you use and what you discard.

irradiation the process of exposing a substance to low doses of radiation, using gamma rays, X-rays, or electricity (electron beams) to kill insects, bacteria, and other potentially harmful microorganisms.

unique radiolytic products substances unique to irradiated food and apparently created during the process of irradiation.

At the time, however, Dr. Pasteur's discovery met with widespread fear and opposition. In fact, it wasn't accepted as a vital public health measure until 1909, when the city of Chicago set forth the first U.S. law requiring pasteurization of milk.

Today, of course, few consumers even question the wisdom of drinking pasteurized milk. Still, according to many experts, new technologies such as irradiation have prompted a public outcry similar to the turn-of-the-century resistance to pasteurization.[25] As public health organizations strive to feed a fast-growing world population while ensuring a safe food supply, debate about new ways of doing so is sure to heat up. The following sections explore some of the controversial food technologies under consideration as alternatives to help improve the safety of our food supply and to maintain the nutritional value of the foods available in the marketplace.

Irradiation

FDA approves irradiation of:

- Citrus fruits
- Flour
- Fresh or frozen red meats
- Mushrooms
- Onions
- Potatoes
- Poultry
- Spices
- Strawberries
- Tropical fruits
- Wheat

Irradiation is one of the foremost technologies earmarked by food safety experts for increased future use in the United States. The process involves exposing food to low doses of radiation, which destroys insects and several types of bacteria, including *Salmonella*. Contrary to popular belief, irradiation does not make a food radioactive. Rather, the radiation rays pass through food, leaving behind no radioactive residues.

Aside from the misunderstanding that irradiation makes food radioactive, many of the criticisms about the process center around substances called **unique radiolytic products**—compounds that are not present in food naturally or after conventional processing. Critics charge that these products may pose health hazards to consumers who eat irradiated food. According to a comprehensive report from the World Health Organization, however, concern about unique radiolytic products is "probably unfounded" because many of the substances found in irradiated foods are similar, if not identical, to compounds found in other foods. Even those that appear to be unique may not have been detected in conventionally processed foods because those items haven't undergone the same extensive testing as have irradiated foods.[26]

Another common concern about irradiation is that it might alter the nutritional value of a food. Again, the World Health Organization, along with numerous other public health agencies, points out that nutrient losses, if any, are insignificant.

▲ *The FDA requires that the labels of irradiated foods carry this internationally known radura symbol for radiation. The circle in the middle represents an energy source; the five breaks in the outer circle symbolize rays generated by the energy source; and the two petals signify food. For unpackaged irradiated meat, the statement and logo must be displayed at the point of sale for consumers. The labeling requirements do not pertain to foodservice establishments such as restaurants.*

That's not to say, however, that irradiation poses no risks. One of the most troublesome is that the radioactive materials used in the irradiation process may put workers and communities at undue risk. However, as with any technology, the risk of irradiation must be weighed carefully against the benefits and risks of alternate technologies. Consider spices, which typically harbor high levels of pathogens. Often, manufacturers douse them with a toxic, explosive chemical called *ethylene oxide* to rid them of bacteria. Yet because of its toxic, explosive nature, ethylene oxide puts workers at risk, may pollute the air, and may leave residue on the spices. In fact, the gas, which was once used to sterilize medical supplies, has been deemed so dangerous that most manufacturers now use irradiation instead. (Cotton swabs, tampons, teething rings, and a number of other consumer goods are also sterilized via irradiation.)[27] Recently, USDA authorized irradiation of meat already inspected by the agency and approved as safe for consumption. As a result, food companies now have the option of using irradiation on raw meat and meat products (for example, ground beef, frozen hamburger patties, or frozen poultry).

The World Health Organization, the Food and Agriculture Organization of the United Nations, the Food and Drug Administration, and numerous other agencies encourage the use of irradiation in the fight against foodborne disease and food loss, and some 35 countries have approved it for use. According to

WHO, each year spoilage, insect infestation, and the like lead to losses of as much as 50 percent of the world's food supply, losses that could be eliminated with irradiation. What's more, the process may help prevent deaths resulting from foodborne illness. Clearly, as the pressures to feed the world safely continue to mount, the public and scientific community will continue to examine the pros and cons of irradiation.

Biotechnology and Genetic Engineering

In May 1994, a new breed of tomato made headlines. Called the Flavr Savr, the product garnered national repute as the first food created via **genetic engineering** to hit the market. It wasn't so much the product itself, which is simply a slow-ripening tomato, that caused such a stir, but rather the opening of the regulatory door that can pave the way to a whole new crop of genetically engineered products. What are these high-tech foods, and will we see more of them in the future?

Just as a person's genes determine hair color, eye color, and other characteristics, a plant's genes dictate the plant's structure, resistance to spoilage, and other qualities. With genetic engineering, scientists can alter a plant's genes to make a particular trait more desirable. To create the Flavr Savr tomato, scientists first identified the gene that causes ripe tomatoes to soften and rot. Then they isolated the gene, reversed it (or turned it backward, so to speak), and inserted the reversed gene back into tomato plants. The backward gene in the new tomatoes suppresses, or "turns off," the rotting gene. As a result, the Flavr Savr tomato stays ripe longer than regular tomatoes (see Figure 12-6). This allows growers more time to ship it to the consumer without refrigeration.[28]

Although the idea of altering a plant's genes to create a different version sounds new, scientists have used a crude version of the concept for centuries. In the 1500s, farmers crossed, say, a good food crop with another plant, which was resistant to disease, to form a hybrid that contained traits of both plants. Many of the fruits and vegetables we enjoy today were produced by this sort of cross breeding. For example, corn as we know it today came from cross-breeding it

genetic engineering the use of biotechnology to alter the genes of a plant in an effort to create a new plant with different traits; also called genetic modification. In some cases, a plant's gene(s) may be deleted or altered, or a gene(s) may be introduced from different organisms or species. The foods or crops produced are called genetically engineered (GE) or genetically modified (GM) foods or crops.

FIGURE 12-6
CREATING THE FLAVR SAVR TOMATO

SOURCE: Adapted from Calgene's recipe for genetically engineered tomatoes, *FDA Consumer* (April 1995): 9.

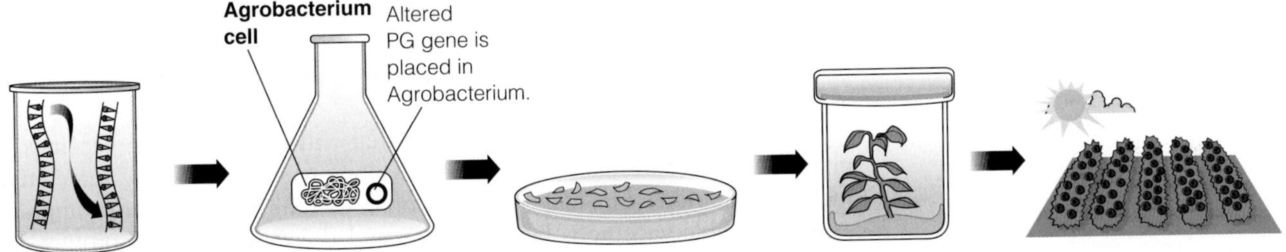

1. The PG (polygalacturonase) gene, which causes ripe tomatoes to soften and rot, is isolated and cloned. The sequence of PG gene is reversed so that the gene is backwards (in what scientists call the "antisense" orientation).

2. The reversed PG gene is put into Agrobacterium, which is a bacterium that infects plants and is commonly used by genetic engineers to insert modified or foreign genes into other target cells.

3. The Agrobacteria are then placed in a petri dish with pieces of leaf from a tomato plant. The cut edges of the leaves absorb the Agrobacteria and antisense PG gene. Antisense gene thus becomes part of the genetic material of the tomato plant cells.

4. The leaf cuttings regenerate into tomato plants that contain the reversed PG gene. The new plants sprout roots, are transplanted to soil, and grow to mature tomato plants. The seeds collected from these greenhouse tomatoes are planted outdoors for field trials and more seed production.

5. In the genetically engineered tomato, the natural PG gene's production of the fruit-rotting PG enzyme is repressed by the reversed gene. This gives the commercial tomato extended shelf life, allowing it to ripen more fully on the vine and still have time to get to market before it spoils.

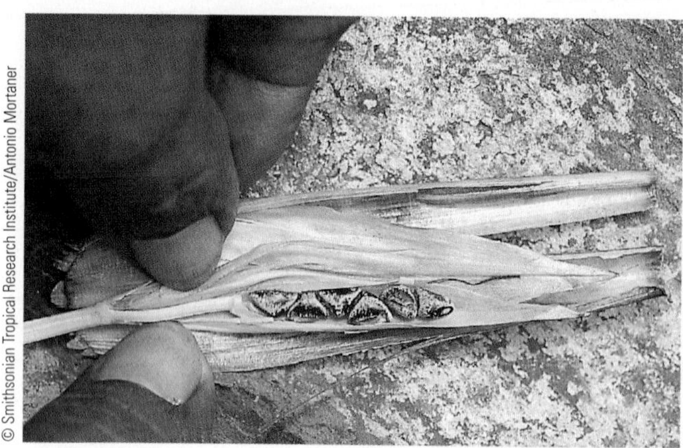

▲ *Farmers bred corn as we know it today from this wild corn.*

with a type of corn that has a much harder outer shell on the kernel; kiwi was originally a small, hard berry; and nectarines are modified peaches.[29] Genetic engineering is different, however, because of its speed and the precision it affords scientists. Whereas crossbreeding typically takes more than a decade, genetic engineering can yield new products in about five years.

In addition, genetic engineering allows scientists to "mix and match" genes from different species—say, mixing a gene from a fish with the genes of a vegetable. This type of research, called **transgenetics,** raises many ethical questions, especially among vegetarians and other people who fear that inserting an animal gene into a fruit or vegetable makes a food that is on some level both vegetable and animal. The possibilities created by this technology are seemingly endless. It has been suggested that ultimately, the world will obtain most of its food, fuel, fiber, and some of its pharmaceuticals from genetically altered vegetation and trees.[30] Major companies spend billions of dollars annually on **biotechnology** as researchers continue to develop new applications for genetically engineered products.

The U.S. Department of Agriculture reports that, since 1987, more than 40 new genetically engineered agricultural products have completed all the federal regulatory requirements and are now sold commercially. Recently released government data on acreage of biotechnology-derived crops indicates that genetically engineered soybean, cotton, and corn account for up to 44 percent of total acreage planted.[31] The majority of these plants have been engineered with genes from bacteria that destroy pests or protect plants against herbicides. The resistant plants created through genetic engineering lower the farmers' dependence on the use of pesticides and weed killers while increasing overall production.

Genetic engineering is used to boost the nutritional value of foods. Advances in biotechnology can increase levels of natural health-enhancing substances in plants known as *phytochemicals*. These compounds, many of which are found in soybeans, tomatoes, garlic and other plant foods, have been shown to protect against cardiovascular disease, cancer, and free-radical damage to tissues. One such compound, beta-carotene, belongs to a class of compounds known as carotenoids. In 1998, the USDA scientists released three new tomato-breeding lines that contain about 10 to 25 times more beta-carotene than typical tomatoes.[32] Also, researchers recently succeeded at manipulating a gene that improves the quality of the proteins made by sweet potato, an important food crop in poorer countries where high-quality protein foods are at a shortage. Early study results show the protein content of the genetically engineered potatoes increased as much as fivefold.[33]

Scientists are now developing plants that will produce edible vaccines. Such plants could be extremely valuable in developing countries where some environmental conditions can make conventional vaccine administration impractical. Also, an edible vaccine is being investigated that will hopefully protect against diarrhea, a major cause of infant mortality in developing countries.[34]

The introduction of genetically engineered food has encountered several problems. Some people are concerned about the potential impact of an organism that would not have evolved under normal conditions. We cannot know the long-term environmental consequences as these genetically engineered plants multiply and mutate. Some people fear that naturally occurring cross-pollination between genetically engineered plants with nearby weeds may spread traits from plants to weeds. Potentially, this so-called "superweed" would be resistant to insects and herbicides.

People with food allergies express concern that new varieties of food produced by transgenetics may introduce allergens not found in the food before it

transgenetics the process of transferring genes from one species to another unrelated species.

biotechnology the use of biological systems or living organisms to make or modify products. Includes traditional methods used in making products such as wine, beer, yogurt, and cheese; cross breeding to enhance crop production; and modification of living plants, animals, and fish through the manipulation of genes (genetic engineering).

was altered. Indeed, researchers have shown that allergens can be transferred through bioengineering, as in one case where an allergic protein showed up in soybeans that had been genetically altered with proteins from Brazil nuts.[35]

Another widely debated issue surrounding genetic engineering involves labeling. Many consumer groups have called for across-the-board labels on genetically engineered foods. But according to the FDA, with some exceptions, the food must carry distinct consequences to consumers who eat it before it requires special labeling. When genes from peanuts and other foods known to be common causes of allergies are put into a food, the label must indicate that the food contains an allergen unless the manufacturer can prove that the item's potential to cause allergies has not been transferred via the gene. In addition, a food that has been genetically engineered to significantly change, say, its fiber content or nutrient composition must bear a label that states the nature of the change.[36] A number of petitions currently calling for labeling of foods with genetically engineered ingredients may lead to FDA approval of required labeling for these foods in the future.[37] The accompanying box lists some examples of currently available genetically engineered crops and highlights some of the potential benefits and risks associated with them.

Despite the concerns, many scientists are hopeful that careful use of genetic engineering will confer long-term benefits. For instance, the development of insect- and disease-resistant plants may allow farmers to grow crops with fewer chemicals. The possibilities are enormous, and each new product considered for entry into the marketplace will require careful scrutiny.

Risks vs. Benefits: The Debate over Genetically Engineered (GE) Foods and Crops

Some Currently Available Genetically Engineered (GE) Crops*

- **Maize:** Insect resistance (reduced insect damage)
- **Soybean:** Herbicide tolerance (greater weed control)
- **Cotton:** Insect resistance (reduced insect damage)
- **Rice:** Pro-Vitamin A (increase vitamin A supply)

Potential Benefits of Genetically Engineered (GE) Foods and Crops

- **Increased nutritional value of staple foods:** Genes are being inserted into rice to make it produce beta-carotene, which the body converts into vitamin A. This experimental transgenic "golden rice" has the potential to reduce vitamin A deficiency, a leading cause of blindness and a significant factor in many child deaths.
- **Reduced environmental impact:** Scientists are developing trees with modified cell lignin content. When used to make pulp and paper, the modified wood requires less processing with harsh chemicals.
- **Increased fish yield:** Researchers have modified the gene that governs growth hormones in tilapia, a farmed fish, offering the prospect of increased yield and greater availability of fish protein in local diets.
- **Increased nutrient absorption by livestock:** Animal feed under development will improve animals' absorption of phosphorus. This reduces the phosphorus in animal waste, which pollutes groundwater.
- **Tolerance of poor environmental conditions:** Scientists are working to produce drought-resistant or salt-tolerant transgenic crops, which can then be grown on marginal land.

Potential Risks of Genetically Engineered (GE) Foods and Crops

- **Inadequate controls:** Although safety regimes are being improved, control over GE crop releases is not completely effective. In 2000, for example, a maize variety cleared only for animal consumption was found in food products.
- **Transfer of allergens:** Allergens can be transferred inadvertently from an existing to a target organism, and new allergens can be created. For example, when a Brazil nut gene was transferred to soybean, tests found that a known allergen had also been transferred. However, the danger was detected in testing, and the soybean was not released.
- **Unpredictability:** GE crops may have unforeseen effects on farming systems—for example, by taking more resources from the soil or using more water than normal crops use.
- **Undesired gene movement:** Genes brought into a species artificially may cross accidentally to an unintended species. For example, resistance to herbicide could spread from a GE crop into weeds, which could then become herbicide-resistant themselves.
- **Environmental hazards:** GE fish might alter the composition of natural fish populations if they escape into the wild. For example, fish that have been genetically modified to eat more in order to grow faster might invade new territories and displace native fish populations.

*A full biotechnology glossary is available at www.fao.org/biotech/gloss.htm.

SOURCE: The Food and Agriculture Organization of the United Nations; available at www.fao.org.

Spotlight

Domestic and World Hunger

This book emphasizes the problems of *overnutrition*—obesity, heart disease, cancer, and others—diseases of economically developed nations. However, many people in developing nations as well as people in the less privileged parts of developed nations suffer from problems of **undernutrition,** which is characterized by chronic debilitating hunger and malnutrition. These conditions are most visible in times of **famine,** but they are widespread and persistent even when famine does not occur.

▸ *Children can be the first to show the signs of undernutrition due to their high nutrient needs. No famine, no flood, no earthquake, no war has ever claimed the lives of 200,000 children in a single week. Yet UNICEF estimates that malnutrition and disease claim more than that number of child victims under the age of five years every week at the rate of one every three seconds.*

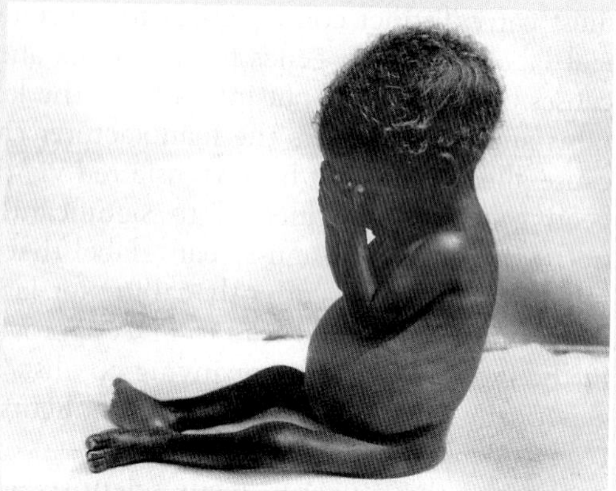

© Louise Gubb/The Image Works

All people need food. Regardless of our race, religion, gender, or nationality, our bodies experience similarly the effects of hunger and its companion **malnutrition**—listlessness, weakness, failure to thrive, stunted growth, mental retardation, muscle wastage, scurvy, pellagra, beriberi, anemia, rickets, osteoporosis, goiter, tooth decay, blindness, and a host of other effects, including death. Apathy and shortened attention span are only two of a number of behavioral symptoms often mistaken for laziness, lack of intelligence, or mental illness in undernourished people.

How are hunger and poverty related?

The phenomenon of *hunger* today is being discussed in terms of **food security** or **food insecurity.** Food security is defined as access by all people at all times to enough food for an active, healthy life and at a minimum includes the following: (1) the ready availability of nutritionally adequate and safe foods and (2) the ability to acquire personally acceptable foods in a socially acceptable way.[38] Food insecurity was once viewed as a problem of overpopulation and inadequate food production, but now many people recognize it as a problem of **poverty.** Food is *available* but is not *accessible* to the poor, who have neither land nor money. Poverty is much more than an economic condition and exists for many reasons, including overpopulation, greed, unemployment, and the lack of productive resources such as land, tools, and credit. Consequently, if we are to provide adequate nutrition for all the Earth's hungry people, we must transform the economic, political, and social structures that limit food production, distribution, and consumption and create a gap between rich and poor.

Approximately how many people worldwide are affected by food insecurity?

The Food and Agriculture Organization (FAO) estimates that of the more than 6 billion people in the world, at least 852 million—18 percent of the developing world's population—suffer from chronic, severe undernutrition, consuming too little food each day to meet even minimum energy requirements.[39] Three micronutrient deficiencies are of particular concern worldwide: vitamin A deficiency, the world's most common cause of preventable child blindness and vision impairment; iron-deficiency anemia; and iodine deficiency, which causes high levels of goiter and child retardation.[40]

Worldwide, about 40,000 to 50,000 people die each day as a result of undernutrition. Millions of children die each year from the diseases of poverty: parasitic and infectious diseases such as dysentery, whooping cough, measles, tuberculosis, cholera, and malaria. These diseases interact with poor nutrition to form a vicious cycle in which the outcome for many is death.[41]

When is the risk for undernutrition highest?

Undernutrition runs high when the body's nutrient needs are high, as in times of rapid growth. If family food is limited, pregnant and lactating women, infants, and children are the first to show the signs of undernutrition. Low birthweight contributes to more than half of the deaths worldwide of children less than 5 years of age. **UNICEF** refers to the **under-5 mortality rate (U5MR)** as the best indicator of children's overall health and well-being.[42] UNICEF argues that the U5MR reflects a country's overall resources directed at children:

. . . the U5MR reflects the nutritional health and the health knowledge of mothers; the level of immunization and use of ***oral rehydration therapy;*** *the availability of maternal and health services (including prenatal care); income and food availability in the family; the availability of clean water and safe sanitation; and the overall safety of the child's environment.*[43]

Until the middle of the 20th century in most of the developing countries, babies were breastfed for their first year of life, with supplements of other milk and cereal gruel added to their diets after the first several months. Today, only about half of all infants are exclusively breastfed to the age of 4 months.[44] A number of factors contributed to this unfortunate decline, including the aggressive promotion and sale of infant formula to new mothers; the encouragement by health care practitioners for mothers to bottle feed (with free samples sent home from the hospital after delivery of the newborn); and the global pattern of urbanization and the accompanying loss of cultural ties supporting breastfeeding combined with more women working outside the home. Overall, WHO estimates that more than 1.5 million children's lives could be saved each year if all mothers gave their babies nothing but breast milk for the first 6 months of life.[45]

Breastfeeding permits infants in many developing countries to achieve weight and height gains equal to those of children in developed countries until about 6 months of age. Even if infants are protected by breastfeeding at first, they must eventually be weaned. The weaning period is one of the most dangerous periods for children in developing countries. Newly weaned infants often receive nutrient-poor diluted cereals or starchy root crops, and the infants' foods are often prepared with contaminated water, making infection almost inevitable.

▸ Feeding the hungry—in the United States

© Bettman/CORBIS

What is the status of food insecurity in the United States?

In 2004, more than one in ten households experienced food insecurity. This represents almost 37 million people. Food insecurity rates are higher than average in female-headed households, households with children, especially black and Hispanic households, and households in inner-city areas.[46] A recent analysis of state hunger surveys found consistent results across all surveys and evidence for several broad conclusions:[47]

- Food insecurity is a chronic problem in the United States.
- Food insecurity is not due to food shortages. Hunger results from unequal distribution of economic resources—poverty.
- People who lack access to a variety of resources—not just food—are most at risk of hunger. When income is inadequate to meet the costs of housing, utilities, health care, and other fixed expenses, these items compete with and may take precedence over food.
- Private charity cannot solve the food insecurity problem. Voluntary activities are limited in expertise, time, and resources and are likely to require government support to continue.

What are the causes of food insecurity in the United States?

The most compelling single reason is poverty. Poverty and food insecurity are interdependent. Nutrition surveys investigating people's nutritional health in the United States have demonstrated consistently that the lower a family's income, the less adequate the family's nutrition status.

Although poverty is the major cause of food insecurity in the United States, other problems contribute as well, including alcoholism and chronic substance abuse, mental illness, homelessness, the reluctance of people to accept what they perceive as "welfare" or "charity"; delays in receiving public assistance benefits; an increase in the number of single mothers without the means to care for their children; health problems of old age; lack of access to assistance programs; insufficient community food resources for the hungry; and insufficient community transportation systems to deliver food to hungry people who have no transportation.

What is being done to help reduce problems of food insecurity?

The U.S. Department of Agriculture's (USDA) Food and Nutrition Service (FNS) implements an array of programs as a "food safety net," to provide children and low-income people with food or the means to purchase food.[48] The programs served an estimated 1 out of every 5 Americans during 2004. Five programs—Food Stamp Program; National School Lunch Program; Special Supplemental Nutrition Program for Women, Infants, and Children (WIC); Child and Adult Care Food Program; and School Breakfast Program—account for 94 percent of all federal expenditures for food assistance. Other food assistance programs for seniors (Elderly Nutrition Program) are administered by the Department of Health and Human Services (DHHS).

The terms **food recovery** and **gleaning** refer to programs that collect excess wholesome food for delivery to hungry people.* Approximately 96 billion pounds—or more than one-quarter of the 356 billion pounds of food produced in this country for human consumption—are lost at the retail and food service levels. In an effort to reduce food waste, the 1997 National Summit on Food Recovery and Gleaning set a goal of providing an additional 500 million pounds of food a year to feeding organizations.

Additionally, concerned citizens work through community programs and churches to provide meals to the hungry. **Second Harvest,** the nation's largest supplier of surplus food, distributed over 2 billion pounds of food to nearly 200 **food banks** and some 50,0000 agencies for direct distribution around the nation in 2003.[49]

How does world hunger differ from hunger in the United States?

World hunger is more extreme than domestic hunger. In fact, most people would find it hard to imagine the severity of poverty in the developing world:

> *Many hundreds of millions of people in the poorest countries are preoccupied solely with survival and elementary needs. For them, work is frequently not available, or pay is low, and conditions barely tolerable. Homes are constructed of impermanent materials and have neither piped water nor sanitation. Electricity is a luxury. Health services are thinly spread, and in rural areas only rarely within walking distance. Permanent insecurity is the condition of the poor . . . in the wealthy countries, ordinary men and women face genuine economic problems. . . . But they rarely face anything resembling the total deprivation found in the poor countries.*[50]

World hunger is a problem of supply and demand, of inappropriate technology, of environmental abuse, of demographic distribution, of unequal access to resources, of extremes in dietary patterns, and of unjust economic systems. People who are poor are often powerless to change their situation because they have less access to vital resources such as education, training, food, and health services.

How are international trade and debt connected to hunger?

Over the years, developing countries have seen the prices of imported fuels and manufactured items rise much faster than the prices they receive for their export goods (such as bananas, coffee, and various raw materials) on the international market. The combination of high import costs with low export profits often pushes a developing country into accelerating international debt that sometimes leads to bankruptcy.

Debt and trade are closely related to the progress a country can make toward achieving an adequate diet for its people. As import prices increase relative to export prices, more of a country's total money base moves abroad to pay for the imports. With more and more of its money abroad, the country is forced to borrow money, usually at high interest rates, to continue functioning at home. As more and more of its financial resources are being used to pay off interest on the country's trade debts, less and less money is available to deal with food insecurity at home. Each year, the debt crisis worsens and leads to further problems with hunger.

What about the role of multinational corporations in this issue?

Typically, large landowners and **multinational corporations** hire indigenous people for below-subsistence wages to work in the fertile farmlands growing crops to be exported for profit—leaving little fertile land for the local farmers to use to grow their own food. The local people work hard cultivating cash crops for others, not food crops for themselves. The money they earn is not even enough to buy the products they help to produce. They do not adequately share in the profits realized from the marketing of products grown with their labor. The results: imported foods—bananas, beef, cocoa, coconuts, coffee, pineapples, sugar, tea, winter tomatoes, and others—fill the grocery stores of developed countries, while the poor who labored to grow these foods have less food and fewer resources than when they farmed the land for their own use. Additional cropland is diverted for nonfood, cash crops—tobacco, rubber, cotton, and other agricultural products. These practices have also had an adverse effect on the financial status of many U.S. farmers. The foreign cash crops often undersell the same U.S.-grown produce. The U.S. farmer cannot compete against these lower-priced imported foods and may be forced out of business.

How does overpopulation fit into the picture?

The current world population is approximately 6 billion, and the United Nations projects this figure to be approximately 9 billion by the year

*The Good Samaritan Food Donation Act of 1996 encourages the donation of food and grocery products to nonprofit organizations such as homeless shelters, soup kitchens, and churches for distribution to needy individuals. The law provides uniform national protection to citizens, businesses, and nonprofit groups that in good faith donate, recover, and distribute excess food. The law limits the liability of donors to instances of gross negligence or intentional misconduct.

2050. The Earth may not be able to adequately support this many people. The world's present population is certainly a concern, as is the projected population increase. As serious as the population problem is, it is only one cause of the world food problem. Poverty seems to be at the root of both hunger and overpopulation.

Three major factors affect population growth: birthrates, death rates, and standards of living. Low-income countries have high birthrates, high death rates, and low standards of living.

When people's standard of living rises, giving them better access to health care, family planning, and education, the death rate falls. In time, the birth rate also falls. As the standard of living continues to improve, the family earns sufficient income to risk having smaller numbers of children. A family depends on its children to cultivate the land, secure food and water, and provide for the adults in their old age. Under conditions of ongoing poverty, parents choose to have many children to ensure that some will survive to adulthood. Children represent the "social security" of the poor. Improvements in economic status help relieve the need for this "insurance" and thus help reduce the birth rate.

Is there a better way to distribute world resources?

Land reform—for example, giving people a meaningful opportunity to produce food for local consumption—can combine with population control to increase everyone's assets. Poor nations must be allowed to increase their agricultural productivity. Much is involved, but to put it simply, poor nations must gain greater access to five things simultaneously: land, capital, water, technology, and knowledge.[51] Equally important, each nation must adopt the political priority of improving the conditions of all its people. International food aid may be required temporarily during the development period, but eventually this aid will become less and less necessary.

Governments can learn from recent history the importance of developing local agricultural technology. A major effort made in the 1960s and 1970s—the green revolution—demonstrated

MINIGLOSSARY

appropriate technology a technology that utilizes locally abundant resources in preference to locally scarce resources. For developing countries, which usually have a large labor force and little capital, the appropriate technology would therefore be labor intensive.

famine widespread lack of access to food caused by natural disasters, political factors, or war; characterized by a large number of deaths due to starvation and malnutrition.

food banks nonprofit community organizations that collect surplus commodities from the government and edible but often unmarketable foods from private industry for use by nonprofit charities, institutions, and feeding programs at nominal cost.

food insecurity the inability to acquire or consume an adequate quality or sufficient quantity of food in socially acceptable ways, or the uncertainty that one will be able to do so.

food recovery such activities as salvaging perishable produce from grocery stores and wholesale food markets; rescuing surplus prepared food from restaurants, corporate cafeterias, and caterers; and collecting non-perishable, canned or boxed processed food from manufacturers, supermarkets, or people's homes. The items recovered are donated to hungry people.

food security access by all people at all times to enough food for an active and healthy life. Food security has two aspects: ensuring that adequate food supplies are available and ensuring that households whose members suffer from undernutrition have the ability to acquire food, either by producing it themselves or by being able to purchase it.

gleaning the harvesting of excess food from farms, orchards, and packing houses to feed the hungry.

GOBI an acronym formed from the elements of UNICEF's Child Survival campaign—**G**rowth charts, **O**ral rehydration therapy, **B**reast milk, and **I**mmunization.

malnutrition the impairment of health resulting from a relative deficiency or excess of food energy and specific nutrients necessary for health.

multinational corporations international companies with direct investments and/or operative facilities in more than one country. U.S. oil and food companies are examples.

oral rehydration therapy (ORT) the treatment of dehydration (usually due to diarrhea caused by infectious disease) with an oral solution; ORT as developed by UNICEF is intended to enable a mother to mix a simple solution for her child from substances that she has at home.

poverty the state of having too little money to meet minimum needs for food, clothing, and shelter. The U.S. Department of Agriculture defines the poverty level in the United States as an annual income of $20,000 for a family of four in 2006.

Second Harvest a national food-banking network to which the majority of food banks belong.

under-5-mortality rate (U5MR) the number of children who die before the age of five for every 1,000 live births.

undernutrition (also called **hunger**) as used in this discussion, a term that describes the domestic and world food problem of a continuous lack of the food energy and nutrients necessary to achieve and maintain health and protection from disease.

UNICEF the United Nations International Children's Emergency Fund, now referred to as the United Nations Children's Fund.

the potential for increased grain production in Asia. It was an effort to bring the agricultural technology of the industrial world to the developing countries, but the high-yielding strains of wheat and rice that were selected required irrigation, chemical fertilizers, and pesticides—all costly and beyond the economic means of too many of the farmers in the developing world.

Instead of transplanting industrial technology into the developing countries, small, efficient farms and local structures for marketing, credit, transportation, food storage, and agricultural education should be developed. International research centers need to examine the conditions of tropical countries and orient their research toward **appropriate technology**—labor-intensive rather than energy-intensive agricultural methods. For example, labor-intensive technology, such as the use of manual grinders for grains, is appropriate in some places because it makes the best use of human, financial, and natural resources. A manual grinder can process 20 pounds of grain per hour, replacing the mortar and pestle, which in the same time can pound a maximum of only 3 pounds.[52] The specific technology that is appropriate varies from situation to situation.

Environmental concerns must be taken more seriously as well. As important as the amount of land available for crop production is the condition of the soil and the availability of water. Soil erosion is now accelerating on every continent at a rate that threatens the world's ability to continue feeding itself.[53] Erosion of soil has always occurred; it is a natural process. But in the past, processes that build the soil up, such as the growth of trees, have compensated for erosion.

Is there hope for a world without hunger for women and children?

Women make up 50 percent of the world's population. Any solution to the problems of poverty and hunger is incomplete and even hopeless if it fails to address the role of women in developing countries, for women and their children represent the majority of those living in poverty.

Women play a vital role in the nutrition of their nation's people. Their nutrition during pregnancy and lactation determines the future health of their children. If women are weakened by malnutrition or are ignorant about how to feed their families, the consequences ripple outward to affect many other individuals. The importance of the role women play in these countries is increasingly appreciated, and many countries now offer development programs with women in mind.

Seven basic strategies are at the heart of women's programs:

- Removing barriers to financial credit
- Providing access to time-saving technologies
- Providing appropriate training to promote self-reliance
- Teaching management and marketing skills
- Making health and day care services available
- Forming women's support groups[54]
- Providing information and technology to promote planned pregnancies

The recognition of women's needs by some development organizations is an encouraging trend in the efforts to fight the world hunger crisis.

There is hopeful news for children in developing countries—the group that is most strongly affected by poverty, malnutrition, and food insecurity and its relationship to the environment.[55] **GOBI,** a child survival plan set forth by UNICEF, has made outstanding progress in cutting the number of hunger-related child deaths. GOBI is an acronym formed from four simple, but profoundly important, elements of UNICEF's Child Survival campaign: growth charts, oral rehydration therapy (ORT), breast milk, and immunization.

In this program, a mother learns to weigh her child every month and chart the child's growth on a specially designed paper growth chart. She can learn to detect the early stages of hidden malnutrition.

The importance of oral rehydration therapy (ORT) is that most children who die of malnutrition do not starve to death—they die because their health has been compromised by dehydration due to infections causing diarrhea. The spread of ORT is preventing an estimated one million dehydration deaths each year.[56] Oral rehydration therapy is the administration of a simple solution that mothers can make up themselves, using locally available ingredients, which increases a body's ability to absorb fluids 25-fold.[57] International development groups also provide mothers with packets of premeasured salt and sugar to be mixed with water in rural and urban areas. A safe and sanitary supply of drinking water is a prerequisite for the success of the ORT program.

The promotion of breastfeeding among mothers in developing countries has many benefits. Breast milk is hygienic, is readily available, is nutritionally sound, and provides infants with immunologic protection specific for their environment. In the developing world, the advantages of breastfeeding over formula feeding can mean the difference between life and death.

Immunizations (the *I* of GOBI) could prevent most of the five million deaths each year from measles, diphtheria, tetanus, whooping cough, poliomyelitis, and tuberculosis. The immunization achievements of the last two decades are credited with the prevention of approximately three million deaths a year as well as the protection of many millions more from disease, malnutrition, blindness, deafness, and polio.[58]

The first World Summit for Children in history was convened by UNICEF in September of 1990 for the purpose of making a renewed commitment to ending child deaths and child malnutrition. Significantly, *nutrition* was mentioned for the first time in world history as an internationally recognized human right.[59] An immediate result of this summit has been an increase in the number of governments actively adopting the child survival strategies of UNICEF: universal immunization; oral rehydration therapy; a massive effort to promote breastfeeding as the ideal food for at least the first six months of an infant's life; an attack on malnutrition involving nutrition surveillance focusing on growth monitoring and weighing of infants at least once every month for the first 18 months of the child's life; and nutrition and literacy education that will empower women in developing countries and lead to a reduction in nutrition-related diseases among vulnerable children.[60]

What can I do to help alleviate hunger problems?

The problems of hunger can appear so great that they sometimes seem approachable only by way of worldwide political decisions. To this end, many individuals and groups are working to improve the chances of the future well-being of the world and its people through a number of national and international organizations.

Regardless of the type and level of involvement a person chooses, each person can make a difference. Individual people can do any of the following:

- Assist in government and community programs as volunteers.
- Help to increase the visibility and accessibility of existing programs and services to those who need them.
- Document the needs that exist in their own communities.
- Join with others in their community who have similar interests.
- Follow current hunger legislation; call and write legislators about hunger issues.

Individuals can also help change the world through the personal choices they make each day. Our choices have an impact on the way the rest of the world's people live and die. People in affluent nations have the freedom and means to choose their lifestyles. We can find ways to reduce our consumption of the world's nonrenewable resources and use only what is absolutely required. As one person put it, "the widespread simplification of life is vital to the well-being of the entire human family."[61] Personal lifestyles do matter, for a society is nothing more than the sum of its individuals. As we go, so goes our world.

George McGovern, former U.S. senator and current U.S. ambassador to the UN Agencies on Food and Agriculture in Rome, calls us all to action with the following words—excerpted from his book, *The Third Freedom: Ending Hunger in Our Time*.[62]

> *Hunger is a political condition. The Earth has enough knowledge and resources to eradicate this ancient scourge. Hunger has plagued the world for thousands of years. But ending it is a greater moral imperative now than ever before, because for the first time humanity has the instruments in hand to defeat this cruel enemy at a very reasonable cost.*
>
> *What will it cost if we don't end the hunger that now afflicts so many of our fellow humans? The World Bank has concluded that each year malnutrition causes the loss of 46 million years of productive life, at a cost of $16 billion annually, several times the cost of ending hunger and turning this loss into productive gain.*
>
> *Of course it is impossible to evaluate with dollars the real cost of hunger. What is the value of a human life? The twentieth century was the most violent in human history. With nearly 150 million people killed by war. But in just the last half of that century nearly three times as many died of malnutrition or related causes. How does one put a dollar figure on this terrible toll silently collected by the Grim Reaper? What is the cost of 850 million hungry people dragging through shortened and miserable lives, unable to study, work, play, or otherwise function normally because of the ever-present drain of hunger and malnutrition on body, mind, and spirit? What is the cost of millions of young mothers breaking under the despair of watching their children waste away and die from malnutrition? This is a problem we can resolve at a fraction of the cost of ignoring it. We need to be about that task now. I give you my word that anyone who looks honestly at world hunger and measures the cost of ending it for all time will conclude that this is a bargain well worth seizing. More often than not, those who look at the problem and the cost of its solution will wonder why humanity didn't resolve it long ago.*

In Review

1. Your children have been playing outside. When they come in for dinner, you remind them to wash their hands with soap. How long must your children wash to be effective?
   a. 10 seconds
   b. 20 seconds
   c. 3 minutes
   d. 1 hour
2. Foods are on the stove from the last meal. Leftover foods should not stay out of the refrigerator for more than __________ hours.
   a. 1 to 2
   b. 1 to 3
   c. 3 to 4
   d. 4 to 6
3. You and your roommates are getting ready to make chicken and vegetable stir-fry using the same cutting board for both. You suggest using two cutting boards, one for the chicken and one for the ready to eat foods.
   a. This is necessary.
   b. This is unnecessary.
   c. As long as you wipe the board with a paper towel it's safe.
   d. You should never argue with your roommates; do whatever they want.
4. Which of the following is true of food additives?
   a. They rank as one of the most hazardous substances found within food.
   b. They must be approved by the FDA before use.
   c. They are safe at any amount.
   d. none of the above
5. Which foods cause the most cases of foodborne illness?
   a. fruits and vegetables
   b. milk and dairy products
   c. meat, seafood, and eggs
   d. grains
6. Which item should be purchased first when you shop at the super store?
   a. milk and potato salad
   b. shoes
   c. cheese
   d. hamburger
7. It is best to keep hot food over __________ degrees F and cold foods under __________ degrees F.
   a. 140/40
   b. 100/0
   c. 98.6/32
   d. 65/20
8. Which of the following is the FDA's highest food supply concern?
   a. animal drugs such as hormones or antibiotics
   b. naturally occurring toxins in foods
   c. microbial foodborne illness
   d. pesticides
9. The primary cause of hunger and food insecurity in the world is:
   a. lack of education about purchasing nutritious foods on a budget.
   b. abuse of alcohol and other drugs.
   c. mental illness.
   d. poverty.
10. Briefly explain the difference between food intoxication and foodborne infection.

Menu of Online Study Tools

A variety of study tools for this chapter are available at our website to deepen your understanding of chapter concepts. Go to www.thomsonedu.com/nutrition/boyle to find

- Practice tests
- Flashcards
- Glossary
- Web links
- Animations
- Chapter summaries, learning objectives, and crossword puzzles

Nutrition on the Web

www.thomsonedu.com/nutrition/boyle
Go to the Personal Nutrition site to check for the latest updates to chapter topics or to access links to related websites.

www.homefoodsafety.org
Home food safety tips, an interactive quiz to test your food safety knowledge, and useful links to related sites.

www.foodsafety.gov
The National Food Safety Database site provides food safety information, daily news stories, and many useful links.

www.fda.gov/search.html
Search for information on food safety topics.

www.nal.usda.gov/foodborne
Information from the USDA Foodborne Illness Education Center.

www.pueblo.gsa.gov
Resources from the Consumer Information Center.

www.fightbac.org
Information from the Partnership for Food Safety Education.

www.cdc.gov/foodsafety
Food safety information and tips for international travelers.

www.cfsan.fda.gov
Information on food safety from the Center for Food Safety and Applied Nutrition. The toll-free information hotline is 1-888-SAFEFOOD.

www.healthfinder.gov
Use this search engine to locate home food safety information for handling meat, poultry, and seafood.

www.epa.gov
Information from the Environmental Protection Agency.

www.ams.usda.gov/nop
Information on organic foods.

www.organic-center.org
Learn more about the benefits of organic farming and products to society.

www.myfootprint.org
Take the Ecological Footprint Quiz to estimate how much productive land and water you need to support what you use and what you discard.

www.fsis.usda.gov
Information on food safety in the marketplace.

www.extension.iastate.edu
Click on search and look for information on food safety.

www.hendpg.com
Information from the Hunger & Environmental Dietetic Practice group of the American Dietetic Association to raise awareness and action on hunger and environmental issues.

www.ift.org
Search the Institute of Food Technologists site for information about biotechnology and other chapter topics.

www.ucsusa.org
Information regarding biotechnology and sustainable agriculture from the Union of Concerned Scientists.

www.who.org
Resources on world hunger, poverty, and overpopulation.

www.secondharvest.org
Information about food distribution, community food banks, and food security topics.

www.frac.org
Information on U.S. food security issues and food assistance programs from the Food Research and Action Center.

www.thehungersite.com
You can click a button that says "Donate free food," and a food donation is made to the United Nations World Food Program.

www.wfp.org
A resource for information about world hunger from the United Nations World Food Programme.

www.worldhungeryear.org
Information about food insecurity in the United States.

www.fns.usda.gov/fns
Information on the U.S. food assistance and food recovery programs.

www.fao.org/sd
Information about sustainable development provided by FAO.

www.oxfamamerica.org
Information regarding global hunger and poverty.

www.bread.org
Advocacy information on domestic and world hunger issues.

www.unicef.org
Information on child survival programs worldwide.

Appendixes

An Introduction to the Human Body

The brief anatomy lesson that follows is a lesson in "anatomy for nutrition's sake" to review the body systems and terminology referred to in this book. To make the body's design understandable, the first few paragraphs are devoted to the essential needs of the cells and the body mechanisms that ensure these needs are met.

The Cells

The body is composed of millions of cells, and not one of them knows anything about food. Although you get hungry for meat, milk, or bread, each cell of your body sits in its place waiting until the nutrients it needs pass by. Each of the body's **cells** is a self-contained, living entity (Figure A-1), although each depends on the rest of the body to supply its needs. Each cell keeps itself alive just as its single-celled ancestors did, living alone in the ocean three billion years ago, by taking up the substances it needs from the surrounding fluid and releasing the wastes it produces into that fluid.

The body cells' most basic need, always, is for energy fuel and the oxygen with which to burn it. Next, cells need water, the environment in which they live. Then they need building blocks to maintain themselves—especially the materials they can't make for themselves. These building blocks—the **essential nutrients**—must be supplied preformed from food. The need for these nutrients is a limitation

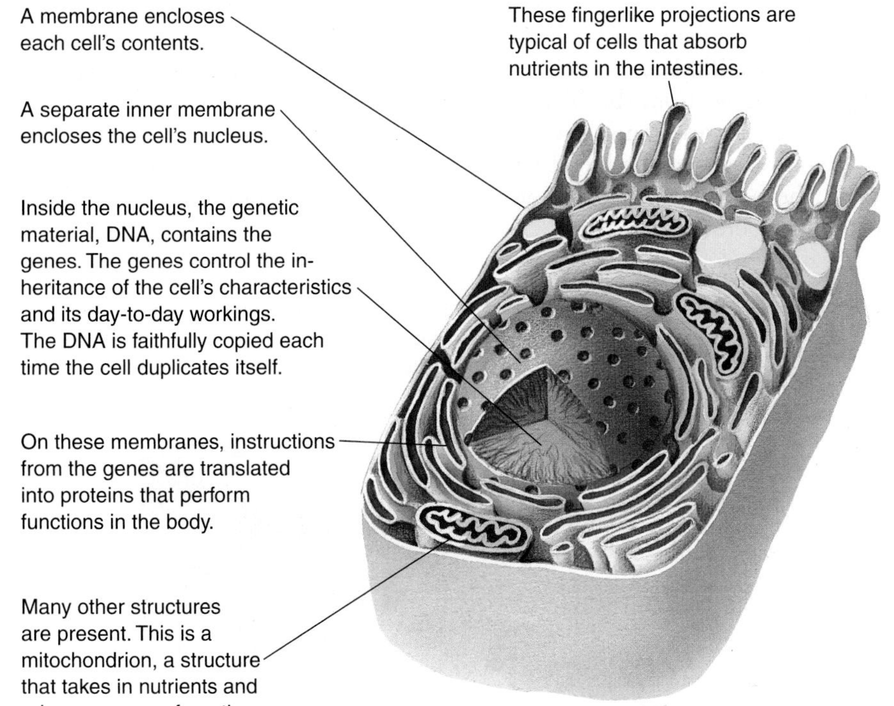
A membrane encloses each cell's contents.
These fingerlike projections are typical of cells that absorb nutrients in the intestines.
A separate inner membrane encloses the cell's nucleus.
Inside the nucleus, the genetic material, DNA, contains the genes. The genes control the inheritance of the cell's characteristics and its day-to-day workings. The DNA is faithfully copied each time the cell duplicates itself.
On these membranes, instructions from the genes are translated into proteins that perform functions in the body.
Many other structures are present. This is a mitochondrion, a structure that takes in nutrients and releases energy from them.

FIGURE A-1
A TYPICAL CELL (SIMPLIFIED DIAGRAM)

of our heredity from which there is no appeal, and this need underlies the first principle of diet planning. Whatever foods we choose, they must provide energy, water, and the essential nutrients. In a sense, the body is merely a system that is organized to provide for these needs of its cells.

In the human body every cell works in cooperation with every other cell to support the whole. The cell's **genes** determine the nature of that work. Each gene is a blueprint that directs the making of a piece of protein machinery—most often an **enzyme**—that helps to do the cell's work. Each cell contains a complete set of genes, but different genes are active in different types of cells. For example, in some intestinal cells, the genes for making digestive enzymes are active; in some of the body's fat cells, the genes for making enzymes that make and break down fat are active.

Cells are organized into *tissues* that perform specialized tasks governed by the genes that are active in them. For example, some cells are joined together to form muscle tissue, which can contract. Tissues also are organized in sets to form whole *organs*. In the heart organ, for example, muscle tissues, nerve tissues, connective tissues, and other types all work together to pump blood. Some jobs around the body require that several related organs cooperate. The organs that join together to work on a function are parts of a *body system*. For example, the heart, lungs, and blood vessels all work to deliver oxygen and nutrients to the body tissues as parts of the cardiovascular system. The next few sections present some body systems with special significance to nutrition.

The Body Fluids

Every cell in the body needs a continuous supply of water, oxygen, energy, and building materials. The body fluids supply these necessities, bathing the outside of all the cells (see Figure A-2). Every cell continuously uses up oxygen (producing carbon dioxide) and nutrients (producing waste products). The body fluids are the transport canals for these materials, carrying oxygen and nutrients to the cells and carbon dioxide and waste away from them. These fluids must circulate to pick up fresh supplies and deliver the wastes to points of disposal.

The fluids that bathe the cells and circulate around the body are the extracellular fluids, the **blood** and **lymph** (Figure A-3). Blood travels within the **arteries, veins,** and **capillaries,** as well as within the heart's chambers (Figure A-4). Lymph is derived from the blood in the capillaries; it squeezes out across their walls and circulates around the cells, permitting exchange of materials. Some of the lymph returns to the blood farther along the capillaries, and the rest travels around the body by way of its own vessels, eventually returning to the bloodstream elsewhere.

FIGURE A-2
ONE CELL AND ITS ASSOCIATED FLUIDS

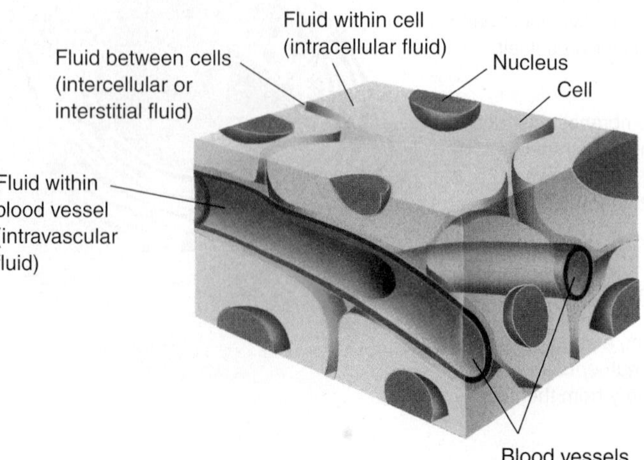

FIGURE A-3
HOW THE BODY FLUIDS CIRCULATE AROUND CELLS
The upper left-hand box shows a tiny portion of tissue with blood flowing through its network of capillaries (greatly enlarged). The boxes at the right illustrate the movement of the extracellular fluid.

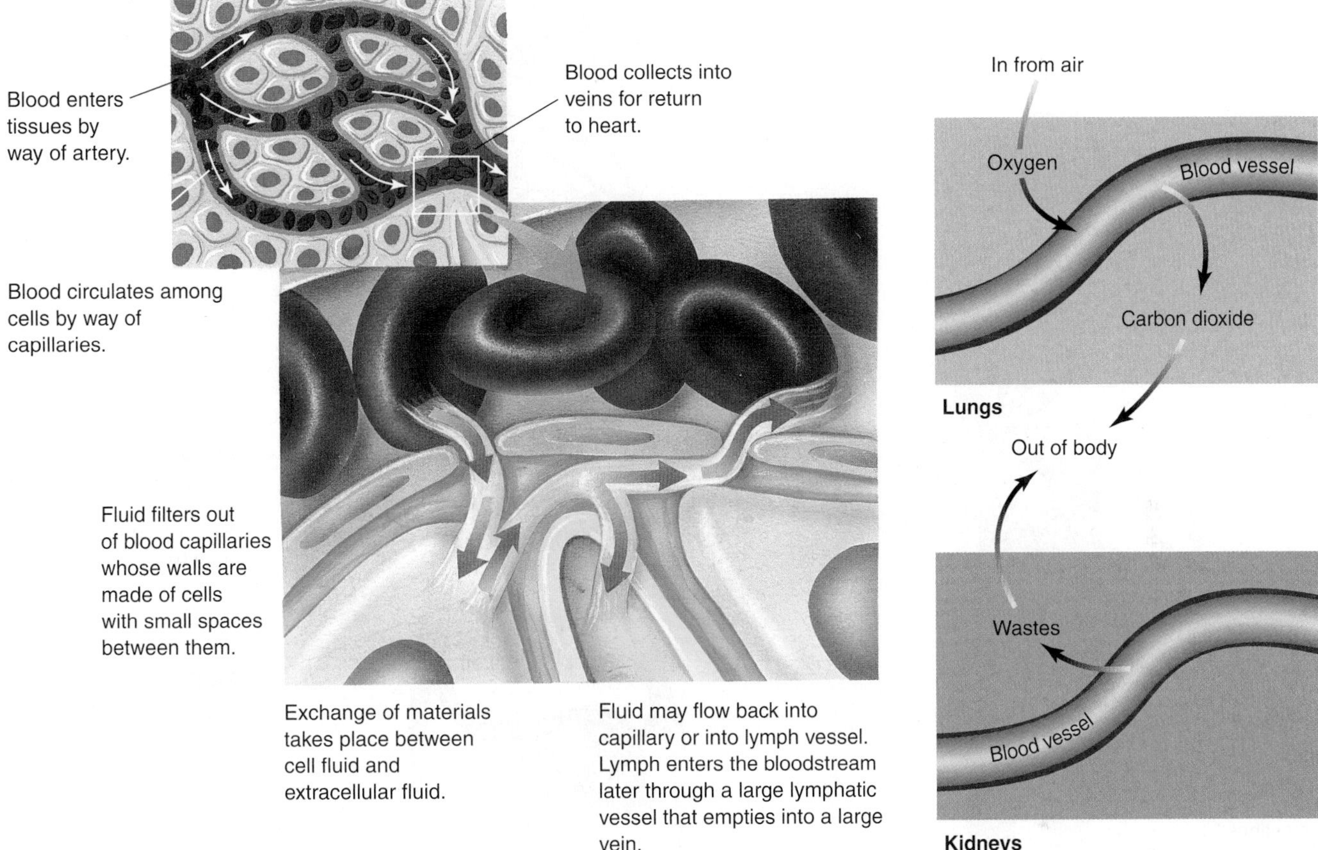
Blood enters tissues by way of artery.
Blood collects into veins for return to heart.
Blood circulates among cells by way of capillaries.
Fluid filters out of blood capillaries whose walls are made of cells with small spaces between them.
Exchange of materials takes place between cell fluid and extracellular fluid.
Fluid may flow back into capillary or into lymph vessel. Lymph enters the bloodstream later through a large lymphatic vessel that empties into a large vein.
In from air
Oxygen
Blood vessel
Carbon dioxide
Lungs
Out of body
Wastes
Blood vessel
Kidneys

The Circulatory System

As the blood, pumped by the heart, travels through the circulatory system, it picks up and delivers materials as needed. Its routing ensures that all cells will be served. Oxygen is picked up and carbon dioxide is released in the **lungs,** and all blood that circulates to the lungs is returned to the heart. From there, it must go to the other body tissues. Thus, all tissues receive freshly oxygenated blood.

As it passes the digestive system, the blood delivers oxygen to the cells there and picks up nutrients from the **intestine** for distribution elsewhere. All blood leaving the digestive system must go next to the **liver,** which has the special task of chemically altering the absorbed materials to make them better suited for use by other tissues. Then, when passing through the **kidneys,** the blood is cleansed of its wastes.

As it flows through the skin, the blood is cooled by radiating heat to the surroundings, helping to maintain the temperature of the body's internal organs. Fluid leaving the blood as lymph may ultimately evaporate from the lungs and skin or be used to make body secretions, such as digestive juices, which will be used within the body for various purposes. On its return to the heart, the blood has delivered most of its oxygen and picked up carbon dioxide from the body cells. Its next stop is the lungs once again, to release its carbon dioxide and replenish its oxygen.

In summary, the routing of the blood is as shown in Figure A-4:

- Heart to body to heart to lungs to heart (repeat).

The portion of the blood that flows by the digestive tract travels from:

- Heart to digestive tract to liver to heart.

FIGURE A-4
THE CARDIOVASCULAR SYSTEM

Blood leaves right side of heart, picks up oxygen in lungs, and returns to left side of heart. Blood leaves left side of heart, goes to the head, or to the digestive tract and then to the liver, or to the lower body, and then returns to right side of heart.

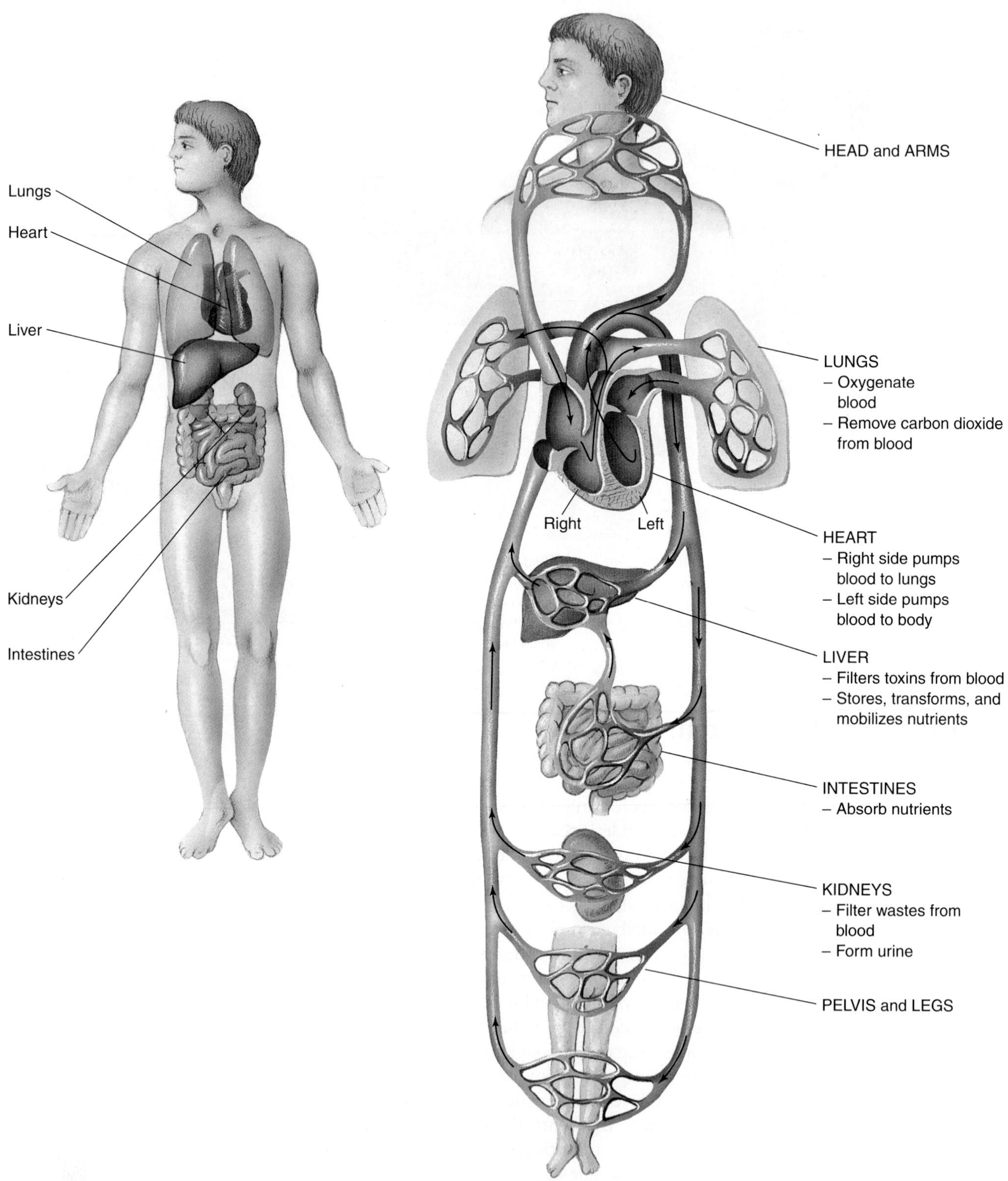

The Immune System

Many of the body's cells cooperate to maintain its defenses against infection. The skin presents a physical barrier, and the body's cavities (lungs, digestive tract, and others) are lined with membranes that resist penetration by invading **microbes** or unwanted substances. The body's linings are easily damaged by nutrient deficiencies, and clinicians inspect both the skin and the inside of the mouth to detect signs of malnutrition. (The chapters on protein, vitamins, and minerals present detailed information about the signs of nutrient deficiencies.)

When a wound or infection penetrates these first lines of defense (the skin and linings), the lymph and blood present internal defenses: cells and proteins that can inactivate, remove, or destroy microbes and foreign substances. Specialized cells recognize the chemical structures of some foreign materials and remember them for a time so that they can quickly mobilize their defenses when they encounter them again. This ability confers **immunity** against many diseases that you have previously fought and conquered. Some immune cells produce proteins that act as ammunition (**antibodies**) designed to destroy specific targets (**antigens**), and still other cells can gobble up and digest the invaders.

Immune system components reside in tissues throughout the body—in the linings of the bones, in the digestive tract, in the blood vessels, in the lymph glands, and in glands of their own. They are in constant flux, being made and dismantled rapidly, and their maintenance requires a continuous supply of nutrients. A deficiency or an overdose of any nutrient is likely to affect the immune system adversely, and a deficiency of nutrients early in an infant's development can weaken that individual's defenses against infection for years.

The Hormonal and Nervous Systems

The blood also carries messengers, chemical signals from one system of cells to another, that communicate the changing needs of the living system. These chemical messengers, or **hormones,** are secreted and released into the blood by the **endocrine** glands. For example, when the **pancreas** (a gland) experiences a too-high concentration of glucose in the blood, it releases **insulin** (a hormone). Insulin stimulates the liver, muscles, and fat cells to remove glucose from the blood and put it away. When the blood glucose level falls too low, the pancreas secretes another hormone, glucagon. The liver responds by releasing glucose into the blood once again. For more about the blood glucose level, see Chapter 3.

Glands and hormones abound in the body, each gland a detector system to monitor a condition in the body that needs regulation and each hormone a messenger to stimulate certain tissues to take appropriate action. Examples of the working of these hormones appear throughout this book.

The body's other major communication system is, of course, the nervous system. With the brain and spinal cord as central controllers, the system receives and integrates messages from sensory receptors throughout the body—sight, hearing, touch, smell, taste, and others—which all communicate to the brain about the state of the outer and inner worlds, including the availability of food and the need to eat. The system then returns instructions to the muscles and glands, telling them what to do.

The nervous system's part in hunger regulation is coordinated by the brain. The sensations of hunger and appetite are experienced in the **cortex** of the brain, the thinking, outer layer. However, much of the brain's regulatory work goes on in the deep brain centers, without the person's (or the cortex's) awareness. An organ there, the **hypothalamus** (Figure A-5), monitors many body conditions, including the availability of nutrients and water.

FIGURE A-5
THE BRAIN'S HYPOTHALAMUS AND CORTEX

The hypothalamus monitors the body's conditions and sends signals to the brain's thinking portion, the cortex, which decides on actions.

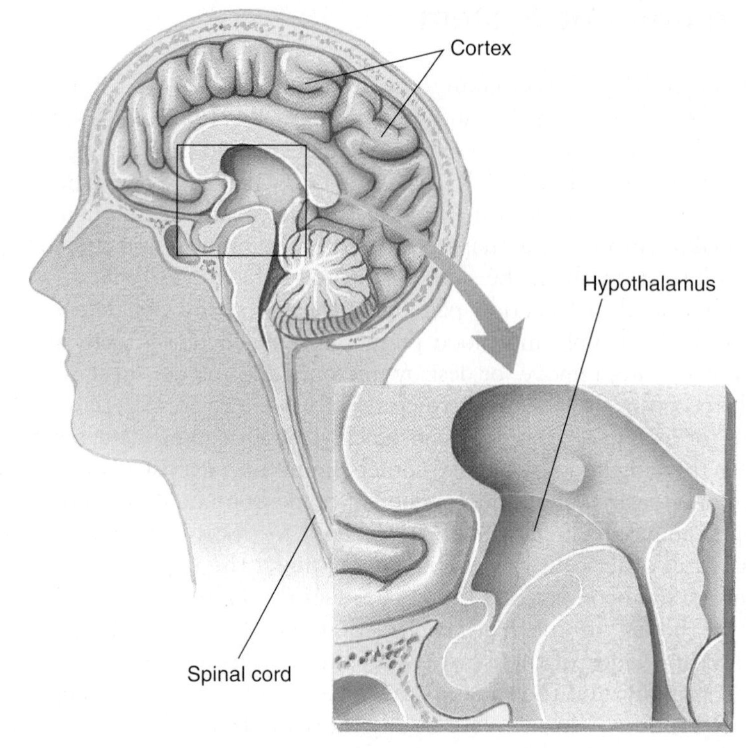

FIGURE A-6
THE EXCRETORY SYSTEM

1. Blood enters the kidneys by way of arteries and disperses into capillaries.
2. The kidneys filter waste from blood and send it as urine to the bladder.
3. The bladder periodically eliminates urine.

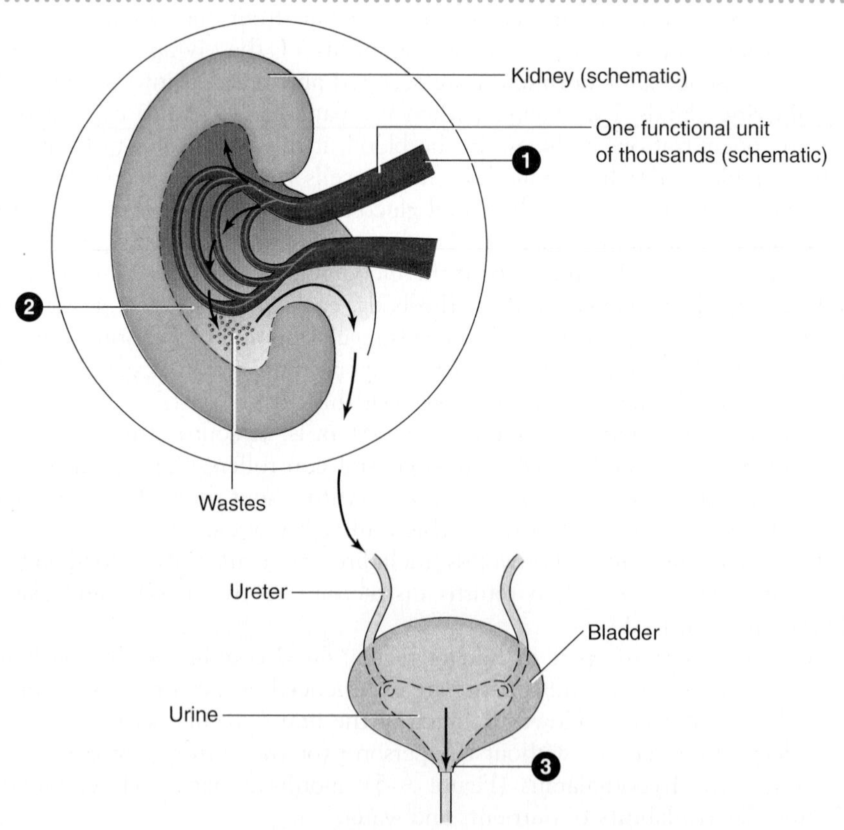

The Excretory System

To dispose of waste, the kidneys straddle the circulatory system and filter each pass of the blood (see Figure A-6). Waste materials removed with water are collected as urine in tubes that deliver them to the urinary bladder, which is periodically emptied. Thus, the blood is purified continuously throughout the day, and dissolved minerals are excreted as necessary (including sodium, to keep blood pressure from rising too high). As you might expect, the kidneys' work is regulated by hormones secreted by glands that respond to conditions in the blood (such as the sodium concentration).

Temperature Regulation

All the body's cells obtain energy by breaking down the nutrients—carbohydrate, fat, and to some extent protein—and one of the ways this energy is released is as heat. The heat is lost to the air through the skin surface. Temperature regulation involves speeding up or slowing down cellular heat production (**metabolism**) and increasing or decreasing heat loss through the skin. Specialized nerve cells in an area of the brain called the hypothalamus serve as a thermostat, measuring the temperature of the blood. These cells signal other cells, near the body surface, to respond appropriately. When the body is too hot, blood vessels just below the surface of the skin dilate, allowing warm blood to flow near the surface, where its heat can radiate away. The sweat glands also are activated to secrete warm fluid onto the skin surface, where its heat can be lost by evaporation. When the body is cold, these mechanisms shut down, which triggers shivering and generates heat.

By means of these systems of transportation, communication, waste disposal, and heat regulation, the cells of the multicellular human body cooperate to provide one another with a circulating bath of warm, clean, nutritive fluid whose composition is finely regulated to meet their needs.

The Digestive System

You may eat meals only two or three times a day, but your body's cells need their nutrients 24 hours a day. Providing the needed nutrients requires the cooperation of millions of specialized cells. When the body's cells are deprived of fuel, certain nerve cells in the brain (the hypothalamus) detect this condition and generate nerve impulses that signal hunger to the conscious part of the brain, the cortex. They also stimulate the stomach to intensify its contractions, creating hunger pangs. Becoming conscious of hunger, then, you eat, thus delivering a complex mixture of chewed and swallowed food to the intestinal tract.

Many of the cells lining the intestinal tract secrete powerful juices and enzymes to disintegrate nutrients (especially carbohydrate and protein) into their component parts. Two organs outside the digestive tract—the liver with its associated gallbladder and the pancreas—also contribute digestive juices through a common duct into the small intestine. The presence of these digestive juices and enzymes requires that still other cells specialize in protecting the digestive system. They secrete a thick, viscous substance known as **mucus,** or the **mucous membrane,** which coats the intestinal tract lining and ensures that it will not itself be digested.

The process of digestion is diagrammed in Figure A-7. The first part, the mouth, is designed for physically breaking down foods. The teeth cut off a bite-size portion and then, aided by the tongue, grind it finely enough to be mixed with saliva and swallowed. The esophagus carries the mixture to the stomach. The stomach is supplied with several sets of muscles to mix and grind it further and secretes acid and enzymes that begin to break it down chemically.

Digestive tract secretions:

Salivary glands:
- Saliva
- Salivary amylase (enzyme that breaks down starch)

Stomach (gastric) glands:
- Gastric juice
- Hydrochloric acid (uncoils protein)
- Gastric protease (enzyme that breaks down protein)
- Mucus (thick coating that protects the stomach wall from these secretions)

Intestinal cells:
- Enzymes (break down carbohydrate, fat, and protein)
- Mucus (thin coating that protects the intestinal wall)

Liver and gallbladder:
- Bile (emulsifier that separates fat into small particles enzymes can attack)

Pancreas:
- Bicarbonate (neutralizes acid fluid from stomach so intestinal and pancreatic enzymes can work on its contents)
- Enzymes (break down carbohydrate, fat, and protein)

FIGURE A-7
THE DIGESTIVE SYSTEM

Fiber	Carbohydrate
Mouth The mechanical action of the mouth and teeth crushes and tears fiber in food and mixes it with saliva to moisten it for swallowing.	The salivary glands secrete a watery fluid into the mouth to moisten the food. The salivary enzyme amylase begins digestion: Starch —amylase→ small polysaccharides, maltose. . .
Esophagus Fiber is unchanged.	Digestion of starch continues as swallowed food moves down the esophagus.
Stomach Fiber is unchanged.	Stomach acid and enzymes start to digest salivary enzymes, halting starch digestion. To a small extent, stomach acid hydrolyzes maltose and sucrose.
Small intestine Fiber is unchanged.	The pancreas produces enzymes and releases them through the pancreatic duct into the small intestine: Polysaccharides —pancreatic amylase→ disaccharides. Then enzymes on the surfaces of the small intestinal cells break disaccharides into monosaccharides, and the cells absorb them: Maltose —maltase→ glucose + glucose. Sucrose —sucrase→ fructose + glucose. Lactose —lactase→ galactose + glucose.
Colon (large intestine) Most fiber passes intact through the digestive tract to the colon. Here, bacterial enzymes digest some fiber: Some fiber —bacterial enzymes→ fatty acids, gas. Fiber holds water; regulates bowel activity; and binds cholesterol and some minerals, carrying them out of the body as it is excreted with feces.	

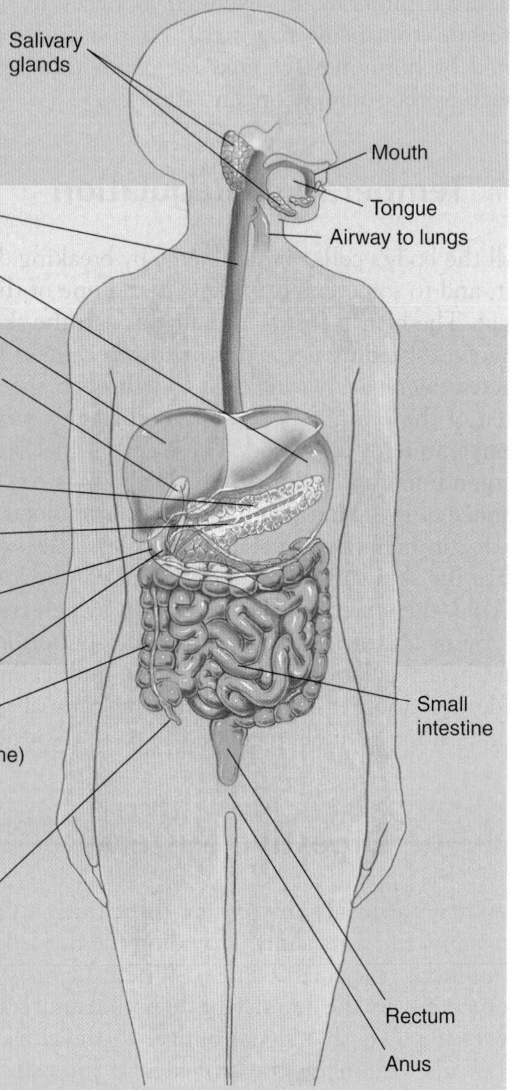

FIGURE A-7
THE DIGESTIVE SYSTEM—Continued

Fat	Protein	Vitamins	Minerals and Water
Mouth Glands in the base of the tongue secrete a fat-digesting enzyme known as lingual lipase. Some hard fats begin to melt as they reach body temperature.	In the mouth, chewing crushes and softens protein-rich foods and mixes them with saliva to be swallowed.	No action.	The salivary glands add water to disperse and carry food.
Esophagus Fat is unchanged.	No action.	No action.	No action.
Stomach The degree of hydrolysis is slight for most fats but may be appreciable for milk fats. The stomach's churning action mixes fat with water and acid. A gastric enzyme accesses and hydrolyzes a small percentage of fat.	Stomach acid works to uncoil protein strands and activate stomach enzymes. Then the enzymes break the strands into smaller fragments: $\text{Protein} \xrightarrow[\text{HCl}]{\text{pepsin}} \text{smaller polypeptides}$	Water-soluble vitamins need little action by the digestive organs except absorption in the small intestine. However, vitamin B_{12} requires "intrinsic factor" produced by the stomach in order to be absorbed.	The stomach secretes enough watery fluid to turn a moist, chewed mass of swallowed food into a liquid. Stomach acid acts on iron to make it more absorbable. Vitamin C and a factor in meat also increase iron absorption.
Small intestine The liver secretes bile; the gallbladder stores it and releases it through the common bile duct into the small intestine when fat arrives there. The bile emulsifies the fat, making it ready for enzyme action. The pancreas produces fat-digesting enzymes and releases them through the common bile duct into the small intestine. These enzymes split triglycerides into monoglycerides, free fatty acids, and glycerol, which are absorbed.	In the small intestine, the fragments of protein are split into free amino acids, dipeptides, and tripeptides with the help of enzymes from the pancreas and small intestine. Enzymes on the surface of the small intestinal cells break these peptides into amino acids, and they are absorbed through the cells into the blood. The large intestine carries any undigested protein residue out of the body. Normally, practically all the protein is digested and absorbed.	Bile emulsifies fat-soluble vitamins and aids in their absorption with other fats. Water-soluble vitamins are absorbed.	The small intestine, pancreas, and liver add enough fluid so that approximately 2 gallons are secreted into the intestine in a day. Many minerals are absorbed. Vitamin D aids in the absorption of calcium.
Colon Some fat and cholesterol, trapped in fiber, exit in feces.		Bacteria produce vitamin K, which is absorbed.	More minerals and most of the water are absorbed.

During the preparatory stage, as the complex carbohydrate known as starch is released from a food (such as bread), an enzyme present in the saliva starts to break it down chemically to smaller units. But this action is stopped when the carbohydrate units reach the stomach, because glands in the stomach wall exude hydrochloric acid. The salivary enzyme that breaks up starch is digested in the stomach, together with other proteins. Further dismantling of carbohydrate occurs after it leaves the stomach.

Fats and oils, taken as part of such complex foods as meats or nuts or in relatively pure form as butter or oil, are not much affected until after they leave the stomach.

Proteins are eaten as part of foods such as meat, milk, and legumes. Although no chemical action on them takes place in the mouth, chewing and mixing protein with saliva is an important part of preparing it for the chemical action that begins in the stomach. There, enzymes and hydrochloric acid break apart the large, complex protein molecules into smaller pieces known as *peptides* and finally into dipeptides, tripeptides, and amino acids.

For further study of digestion, visit

- National Digestive Diseases Information Clearinghouse (NDDIC) at http://digestive.niddk.nih.gov
- National Institute of Diabetes, Digestive, and Kidney Diseases at www.niddk.nih.gov/health/health.htm
- American College of Gastroenterology at www.acg.gi.org

The complicated chemical dismantling that takes place beyond the stomach requires that only small amounts be processed at one time. To accomplish this, the **pylorus,** a circular muscle surrounding the lower end of the stomach, controls the exit of the contents, allowing only a little at a time to be squirted forcefully into the small intestine. Gradually the stomach empties itself by means of these powerful squirts.

The small intestine is *the* organ of digestion and absorption; it finishes the job the mouth and stomach have started. It is actually about 20 feet long, but it is called small because its diameter is small compared with that of the large intestine. Its contents must touch its walls to make contact with the secretions and to be absorbed at the proper places. At the end of the small intestine, a circular muscle (similar in function to the pylorus at the end of the stomach) controls the flow of the contents going into the large intestine (colon).

The small intestine works with the precision of a laboratory chemist. As the thoroughly liquefied and partially digested nutrient mixture arrives there, hormonal messages tell the gallbladder to send its **emulsifier, bile,** in amounts matched to the amount of fat present. Other hormones notify the pancreas to release **bicarbonate** in amounts precisely adjusted to neutralize the stomach acid as well as enzymes of the appropriate kinds and quantities to continue dismantling whatever large molecules remain. Such messages also keep the strong muscles embedded in the walls of the intestine contracting, in a squeezing activity called **peristalsis,** so that the contents will be pressed along to the next region. Peristalsis is stimulated by the presence of roughage or fiber and is quieted by the presence of fat, which requires a longer time for digestion.

Meanwhile, as the pancreatic and intestinal enzymes act on the bonds that hold the large nutrients together, smaller and smaller units make their appearance in the intestinal fluids. Finally, units that cells can use—glucose, glycerol, fatty acids, and amino acids, among others—are released.

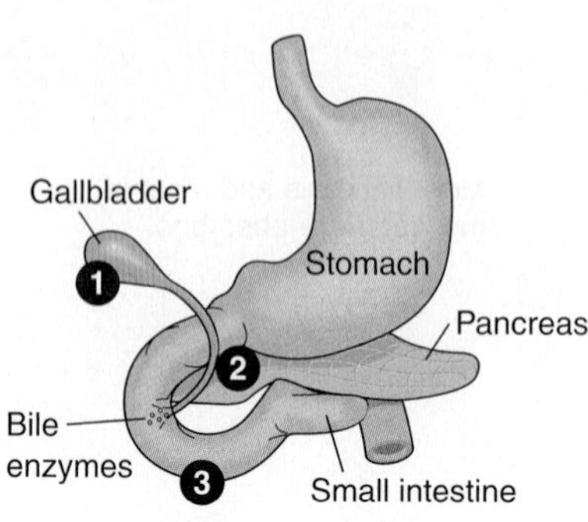

Small Intestine—details

1. The gallbladder sends bile into the small intestine by way of a duct.
2. The pancreas sends enzymes (and bicarbonate).
3. The small intestine also secretes enzymes.

Once the digestive system has broken food down to its nutrient components, it must deliver them to the rest of the body. The cells of the intestinal lining absorb nutrients from the mixture within the intestine and deposit them in the blood and lymph. Every molecule of nutrient must traverse one of these cells if it is to enter the body fluids. The cells are selective: they can recognize the nutrients needed by the body. The cells are also extraordinarily efficient: they absorb enough nutrients to nourish all the body's other cells.

The intestinal tract lining is composed of a single sheet of cells, and the sheet pokes out into millions of finger-shaped projections called **villi.** Each villus has its own capillary network and a lymph vessel so that as nutrients move across the cells, they can immediately mingle into the body fluids. On every villus every cell has a brushlike covering of tiny hairs, called **microvilli,** that can trap the nutrient particles. Figure A-8 provides a close look at these details.

The small intestine's lining, villi and all, is wrinkled into thousands of folds, so that its absorbing surface is enormous. If the folds and the villi that cover

FIGURE A-8
DETAILS OF THE LINING OF THE SMALL INTESTINE

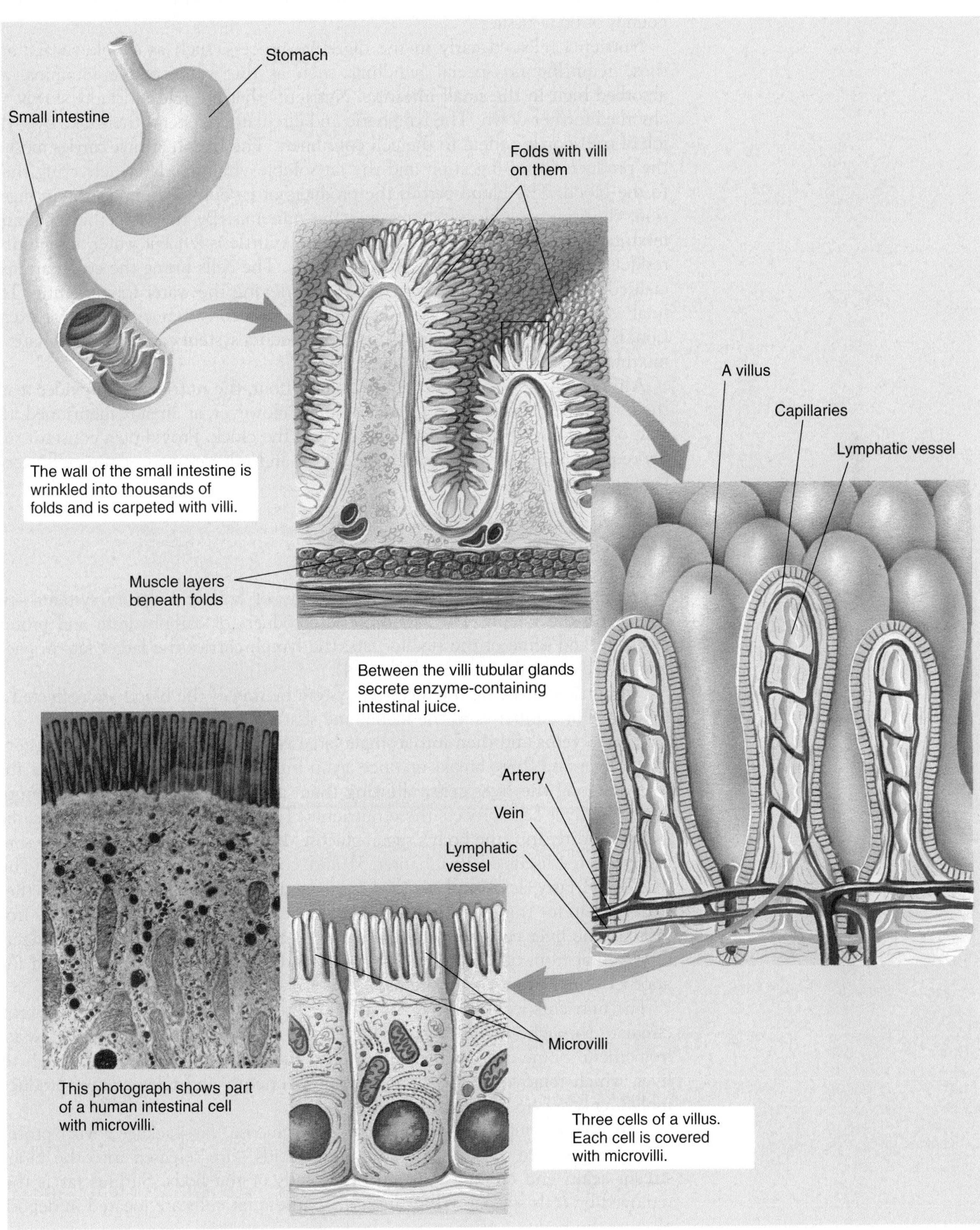

them were spread out flat, the total area would equal a third of a football field in size. The billions of cells of that surface, although they weigh only 4 to 5 pounds, absorb enough nutrients in a few hours a day to nourish the other 150 or so pounds of body tissues.

Nutrients released early in the digestive process, such as simple sugars, and those requiring no special handling, such as the water-soluble vitamins, are absorbed high in the small intestine. Nutrients that are released more slowly are absorbed further down. The lymphatic and circulatory systems then take over the job of transporting them to the cell consumers. The lymph at first carries most of the products of fat digestion and the fat-soluble vitamins, later delivering them to the blood. The blood carries the products of carbohydrate and protein digestion, the water-soluble vitamins, and the minerals. By the time the remaining mixture reaches the end of the small intestine, little is left but water, indigestible residue (mostly fiber), and dissolved minerals. The cells lining the colon are specialized for absorbing these minerals and retrieving the water for recycling. The final waste product, the feces, a smooth paste of a consistency suitable for excretion, is stored in the colon until excretion. Such a system can adjust to whatever mixture of foods is presented.

Although a meal may be eaten in half an hour, the nutrients it provides reach the body fluids over a span of about 4 hours. However, as already mentioned, the cells of the body need their nutrients around the clock. Providing a constant supply requires that there be systems of storage and release to meet the cells' needs between meals.

Storage Systems

Nutrients leave the digestive system by way of both circulatory systems—the blood and the lymph. The blood carries products of carbohydrate and protein digestion and some of the smaller fats; the lymph carries the larger fats in packages called *chylomicrons* (see Chapter 4).

All nutrients leaving the digestive system by way of the blood are collected in thousands of capillaries in the membrane that supports the intestine. These converge into veins and then into a single large vein. This vein conveys its contents to the liver and there breaks up once again into a vast network of capillaries that weave among the liver cells, allowing them access to the newly arriving nutrients. The liver cells process these nutrients. They convert the sugars from carbohydrate mostly into the body's sugar, glucose. If there is a surplus, they store some as glycogen (discussed in Chapter 3) and convert the remainder to fat. They reassemble fatty acids and glycerol from fat into larger fats and package them with protein for transport to other parts of the body. As for the amino acids from protein, the liver cells alter these as needed, making glucose from some if necessary and fat from others if there is an excess, or converting one amino acid into another to use in making proteins.

The nutrients leaving the digestive tract by way of the lymph as chylomicrons circulate throughout the body, giving all cells the opportunity to withdraw fats from them. Some also find their way into the blood and circulate through the liver, which removes them, alters their components, and releases new products, including other lipoproteins.

The new products of liver metabolism—glucose, fat packaged with protein (lipoproteins—see Chapter 4), and amino acids—are released into the bloodstream again and circulated to all other cells of the body. Surplus fat is then removed by cells specialized for its storage; these fat cells are located in deposits all over the body.

The liver's glycogen provides a reserve supply of the body's sugar, glucose, and thus can sustain cell activities if the intervals between meals become so long that glucose absorbed from ingested foods is used up. When the body is depending solely on liver glycogen, however, the supply is used up within 3 to 6 hours. Similarly, the fat cells store reserves of fat, the body's other principal energy nutrient. Unlike the liver, however, the fat cells have virtually infinite storage capacity and can continue to supply fat for days, weeks, or even months when no food is eaten. (Chapter 9 contains more about fasting.)

These storage systems for glucose and fat ensure that the cells will not go without energy nutrients even if the body is hungry for food, except under extreme conditions. Body stores also exist for many other nutrients, each with a characteristic capacity. For example, the third energy nutrient, protein, is held in an available pool (the amino acids in the liver and blood) that is rather rapidly depleted during protein deficiency (see Chapter 5). The liver and fat cells store many vitamins, and the bones provide reserves of calcium, sodium, and other minerals that can be drawn on to keep the blood levels constant and to meet cellular demands.

Metabolism: Breaking Down Nutrients for Energy

The breaking down of body compounds is known as **catabolism.** These reactions usually release energy and are represented, wherever possible, by "down" arrows in chemical diagrams (see Figure A-9). Glycogen can be broken down to glucose, triglycerides to fatty acids and glycerol, and protein to amino acids. When the body needs energy, it breaks any or all of the four basic units—glucose, fatty acids, glycerol, and amino acids—into even smaller units. When the body does not require energy, the end-products of digestion (glucose, amino acids, glycerol, and fatty acids) are used to build body compounds in a process called **anabolism** (see Figure A-9). Anabolic reactions involve the conversion of glucose to glycogen or fat, the conversion of amino acids to body proteins or fat, and the synthesis of body fat from glycerol and fatty acids. Catabolism and anabolism are examples of **energy metabolism.**

FIGURE A-9
REACTIONS OF ENERGY METABOLISM COMPARED

SOURCE: Reprinted with permission from E. N. Whitney and coauthors, *Nutrition for Health and Health Care* (St. Paul, MN: West Publishing, 1995), p. 120.

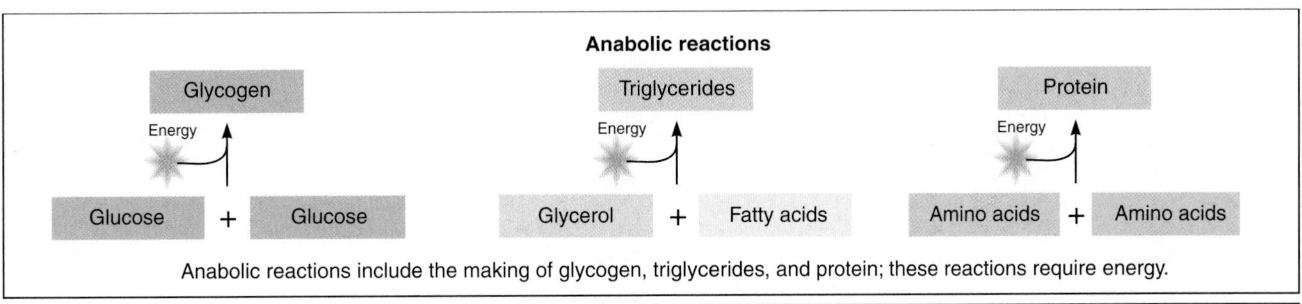

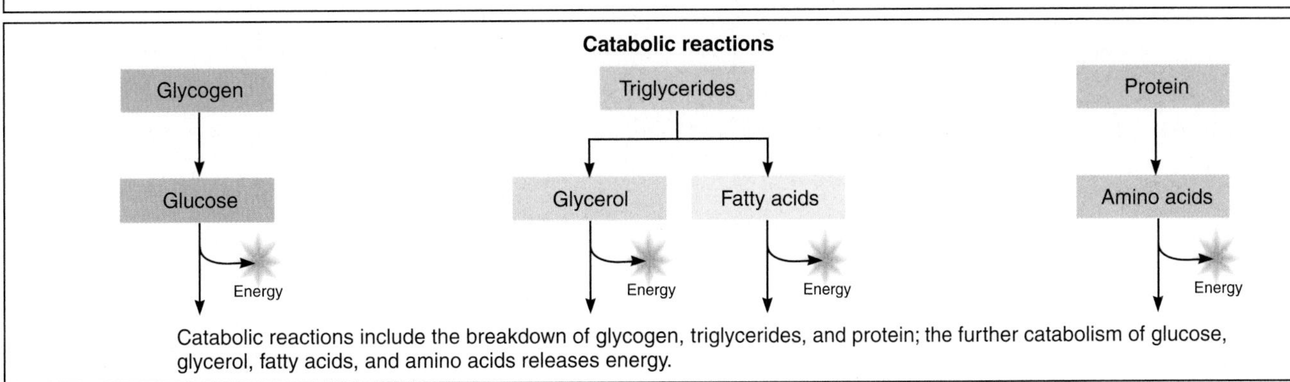

Other Systems

In addition to the systems already described, the body has many more: the bones, the muscles, the nerves, the lungs, the reproductive organs, and others. All of these systems cooperate so that each cell can carry on its own life. Each assures, through hormonal or nerve-mediated messages, that its needs will be met by the others, and each contributes to the welfare of the whole by doing its specialized work.

Of the millions of cells in the body, only a small percentage comprise the cortex of the brain, in which the conscious mind resides. These cells receive messages from other cells when they require you to "become conscious" of a need for decision and action. In modern life, the need may be as complex as, for example, to notice that you feel anxious and to decide to consult an advisor, or it may be a "simple" need such as "I'm tired, I think I'll go to bed," or "I'm hungry, I guess I'd better eat."

Most of the body's work is done automatically and is finely regulated to achieve a state of well-being. But when your cortex does become involved, you would do well to "listen" to your body and to cultivate an understanding and appreciation of its needs. Then, when you make decisions, you will act to promote your health.

MINIGLOSSARY

anabolism (ann-ABB-o-lism) reactions in which small molecules are put together to build larger ones. Anabolic reactions consume energy.
ana = up

antibodies proteins made by the immune system, expressly designed to combine with and to inactivate specific antigens.

antigens microbes or substances that are foreign to the body.

arteries blood vessels that carry blood containing fresh oxygen supplies from the heart to the tissues.

bicarbonate a chemical that neutralizes acid; a secretion of the pancreas.

bile a compound made from cholesterol by the liver, stored in the gallbladder, and secreted into the small intestine. It emulsifies lipids to ready them for enzymatic digestion.

blood the fluid of the circulatory system—water, red and white blood cells and other formed particles, proteins, nutrients, oxygen, and other constituents.

capillaries minute, weblike blood vessels that connect arteries to veins and permit transfer of materials between blood and tissues.

catabolism (ca-TAB-o-lism) reactions in which large molecules are broken down to smaller ones. Catabolic reactions usually release energy.
kata = down

cells the smallest units in which independent life can exist. All living things are single cells or organisms made of cells.

cortex an outer covering; in the brain, that part in which conscious thought takes place.

emulsifier (ee-MULL-sih-fire) a compound with both water-soluble and fat-soluble portions that can attract lipids into water to form an emulsion.

endocrine (EN-doh-crin) a term to describe a gland secreting or a hormone being secreted into the blood.
endo = into

energy metabolism all the reactions by which the body obtains and spends the energy from food or body stores.

enzyme a protein catalyst. A catalyst is a compound that facilitates (speeds up the rate of) a chemical reaction without itself being altered in the process.

essential nutrients compounds that can't be synthesized by the body in amounts sufficient to meet physiological needs.

gene a unit of a cell's inheritance, made of a chemical, DNA, that is copied faithfully so that every time the cell divides, both its offspring get identical copies. Genes direct the cells' machinery to make the proteins that form each cell's structures and to do its work.

hormone a chemical messenger, secreted by one organ (a gland) in response to a condition in the body, that acts on another organ or organs to change that condition.

hypothalamus (high-poh-THALL-uh-mus) a part of the brain that senses a variety of conditions in the blood, such as temperature, salt content, glucose content, and others, and signals other parts of the brain or body to change those conditions when necessary.

immunity the ability to successfully resist a disease, conferred on the body by way of the immune system's memory of previous exposure to that disease and its ability to mount a specific defense promptly and swiftly.

insulin a hormone from the pancreas that helps glucose get into cells.

intestine a long, tubular organ of digestion and the site of nutrient absorption.

kidneys the organs that filter the blood to remove waste material and forward it to the bladder for excretion.

liver the large, lobed organ that lies under the ribs and filters the blood, removing, processing, and readying for redistribution many of its materials.

lungs the organs of gas exchange. Blood circulating through the lungs releases its carbon dioxide and picks up fresh oxygen to carry to the tissues.

lymph (LIMF) the fluid outside the circulatory system that bathes the cells, derived from the blood by being pressed through the capillary walls; similar to the blood in composition but without red blood cells.

metabolism (meh-TAB-o-lism) total of all chemical reactions that go on in living cells.

microbes bacteria, viruses, or other organisms invisible to the naked eye; some cause disease.

microvilli (MY-croh-VILL-ee, MY-croh-VILL-eye) tiny hairlike projections on each cell of the intestinal tract lining that can trap nutrient particles and translocate them into the cells (singular: **microvillus**).

mucus (MYOO-cus) a thick, slippery coating of the intestinal tract lining (and other body linings) that protects the cells from exposure to digestive juices. The adjective form is *mucous,* and the coating is often called the mucous membrane.

pancreas a gland that secretes the endocrine hormone insulin and also produces the exocrine secretions that aid digestion in the small intestine. (An *exocrine* secretion is one that is expelled through a duct into a body cavity or onto the surface of the skin; *exo* means "out." See also *endocrine.*)

peristalsis (perri-STALL-sis) the wavelike squeezing motions of the stomach and intestines that push their contents along the digestive tract.

pylorus (pye-LORE-us) muscle that regulates the opening of the bottom of the stomach.

veins blood vessels that carry used blood from the tissues back to the heart.

villi (VILL-ee, VILL-eye) fingerlike projections of the sheet of cells that line the GI tract; the villi make the surface area much greater than it would otherwise be (singular: **villus**).

Aids to Calculations and the Food Exchange System

Many mathematical problems have been worked out for you as examples at appropriate places in the text. This appendix aims to help with the use of the metric system and with those problems not fully explained elsewhere. Chapter 2 introduced dietary guidelines, food group plans, and the exchange system. This appendix also provides more detail about the U.S. food exchange system.

Conversion Factors*

Conversion factors are useful mathematical tools in everyday calculations, like the ones encountered in the study of nutrition. A conversion factor is a fraction in which the numerator (top) and the denominator (bottom) express the same quantity in different units. For example, 2.2 pounds and 1 kilogram are *equivalent;* they express the same weight. The conversion factor used to change pounds to kilograms or vice versa is:

$$\frac{2.2 \text{ lb}}{1 \text{ kg}} \quad \text{or} \quad \frac{1 \text{ kg}}{2.2 \text{ lb}}$$

Because either of these factors equals 1, a measurement can be multiplied by the factor without changing the value of the measurement. Thus its units can be changed.

The correct factor to use in a problem is the one with the unit you are seeking in the numerator (top) of the fraction. Following are three examples of problems commonly encountered in nutrition study; they illustrate the usefulness of conversion factors.

Example 1

Convert ¼ cup to an approximate number of milliliters for use in a recipe.

1. The conversion factor is:

$$\frac{1 \text{ c}}{250 \text{ ml}} \quad \text{or} \quad \frac{250 \text{ ml}}{1 \text{ c}}$$

2. Multiply ¼ cup by the factor:

$$\tfrac{1}{4}\not{c} \times \frac{250 \text{ ml}}{1\not{c}} = 62.5 \text{ ml, or about 60 ml}$$

Example 2

Convert the weight of 130 pounds to kilograms.

1. Choose the conversion factor in which the unit you are seeking is on top:

$$\frac{1 \text{ kg}}{2.2 \text{ lb}}$$

*For a listing of specific conversion factors, see Table B-1 on page A-22.

2. Multiply 130 pounds by the factor:

$$130 \text{ lb} \times \frac{1 \text{ kg}}{2.2 \text{ lb}} = \frac{130 \text{ kg}}{2.2} = 59 \text{ kg (rounded off to nearest whole number)}$$

Example 3

How many grams of saturated fat are contained in a 3-ounce hamburger? A 4-ounce hamburger contains 7 grams of saturated fat.

1. You are seeking grams of saturated fat; therefore, the conversion factor is:

$$\frac{7 \text{ g saturated fat}}{4 \text{ oz hamburger}}$$

2. Multiply 3 ounces of hamburger by the conversion factor:

$$3 \text{ oz hamburger} \times \frac{7 \text{ g saturated fat}}{4 \text{ oz hamburger}} = \frac{3 \times 7\text{g}}{4} = \frac{21}{4}$$
$$= 5 \text{ g saturated fat (rounded off to nearest whole number)}$$

Percentages

A *percentage* is a comparison between a number of items (perhaps your intake of calories) and a standard number (perhaps the number of calories recommended for your age and gender). The standard number is the number you divide by. The answer you get after the division must be multiplied by 100 to be stated as a percentage (*percent* means "per 100").

Example 4

What percentage of the 2002 DRI for energy (EER) is your calorie intake?

1. We'll use 2,400 calories to demonstrate. (The EER are lised on the inside front cover.)
2. Total your calorie intake for a day—for example, 1,200 calories.
3. Divide your calorie intake by the EER calories:

$$1{,}200 \text{ cal (your intake)} \div 2{,}400 \text{ cal (EER)} = 0.50$$

4. Multiply your answer by 100 to state it as a percentage:

$$0.50 \times 100 = 50.0 = 50\%$$

In some problems in nutrition, the percentage may be more than 100. For example, suppose your daily intake of vitamin C is 500 mg and your RDA (male) is 90 mg. Your intake as a percentage of the RDA is more than 100 percent (that is, you consume more than 100 percent of your vitamin C RDA). The following calculations show your vitamin C intake as a percentage of the RDA:

$$500 \div 90 = 5.55$$
$$5.55 \times 100 = 555\% \text{ of RDA}$$

Sometimes the comparison is between a part of a whole (for example, your calories from protein) and the total amount (your total calories). In this case, the total number is the one you divide by as shown in Example 5.

Example 5

What percentages of your total calories for the day come from protein, fat, and carbohydrate?

1. Using Appendix F and your diet record, find the total grams of protein, fat, and carbohydrate you consumed—for example, 60 grams protein, 80 grams fat, and 285 grams carbohydrate.

2. Multiply the number of grams by the number of calories from 1 gram of each energy nutrient (conversion factors):

$$60 \text{ g protein} \times \frac{4 \text{ cal}}{1 \text{ g protein}} = 240 \text{ cal}$$

$$80 \text{ g fat} \times \frac{9 \text{ cal}}{1 \text{ g fat}} = 720 \text{ cal}$$

$$285 \text{ g carbohydrate} \times \frac{4 \text{ cal}}{1 \text{ g carbohydrate}} = 1{,}140 \text{ cal}$$

$$240 + 720 + 1{,}140 = 2{,}100 \text{ cal}$$

3. Find the percentage of total calories from each energy nutrient (see Example 4):

$$\text{Protein: } 240 \div 2{,}100 = 0.114$$
$$0.114 \times 100 = 11.4 = 11\% \text{ of calories}$$

$$\text{Fat: } 720 \div 2{,}100 = 0.343$$
$$0.343 \times 100 = 34.3 = 34\% \text{ of calories}$$

$$\text{Carbohydrate: } 1{,}140 \div 2{,}100 = 0.543$$
$$0.543 \times 100 = 54.3 = 54\% \text{ of calories}$$

$$11\% + 34\% + 54\% = 99\% \text{ of calories (total)}$$

The percentages total 99 percent rather than 100 percent because a little was lost from each number in rounding off. Either 99 or 101 is a reasonable total.

Nutrient Units and Unit Conversions

International Units (IU)

To convert IU to:

µg vitamin D: divide by 40 or multiply by 0.025.

mg αTE:[a] divide by 1.5

Folate

To convert micrograms of synthetic folate in supplements and enriched foods to Dietary Folate Equivalents (µg DFE):

µg synthetic folate × 1.7 = µg DFE

For naturally occurring folate, assign each microgram folate a value of 1 µg DFE:

µg folate = µg DFE

Sodium

To convert milligrams of sodium to grams of salt:

mg sodium ÷ 400 = g of salt

The reverse is also true:

g salt × 400 = mg sodium

Niacin

1 mg NE (niacin equivalent) = 1 mg niacin = 60 mg dietary tryptophan

Vitamin A

1 µg RAE[b] = 1 µg retinol
= 12 µg beta-carotene
= 24 µg other vitamin A carotenoids

1 IU = 0.3 µg retinol
= 3.6 µg beta-carotene
= 7.2 µg other vitamin A carotenoids

To convert older RE[c] values to micrograms RAE:

1 µg RE retinol = 1 µg RAE retinol
6 µg RE beta-carotene = 12 µg RAE beta-carotene
12 µg RE other vitamin A carotenoids = 24 µg RAE other vitamin A carotenoids

[a]Alpha-tocopherol equivalents (vitamin E).
[b]Retinol activity equivalents (vitamin A).
[c]Retinol equivalents (vitamin A).

TABLE B-1
Conversion Factors

Length	Volume	Temperature
1 inch = 2.54 cm 1 ft = 30.48 cm 1 m = 39.37 inches	1 tbsp = 3 tsp or 15 ml 1 oz = 2 tbsp or 30 ml ½ c = about 125 ml 1 c = 16 tbsp or about 250 ml 1 qt = 4 c or .95 liter 1 qt = 32 oz (fluid) 1 l = 1.06 qt or 1000 ml 1 gal = 16 c or 4 qt or 3.79 liter	Steam — 100°C 212°F — Steam Body temperature — 37°C 98.6°F — Body temperature Ice — 0°C 32°F — Ice Centigrade Fahrenheit $t_F = 9/5\ t_c + 32$ $t_c = 5/9\ (t_F - 32)$

Weight		Energy
1 oz = about 30 g (28.35 g) 1 lb = about 454 g 1 kg = 1000 g 1 kg = 2.2 lb 1 g = 1000 mg 1 mg = 1000 µg 1 µg = 1/1000 mg	1 tsp any powder or liquid = about 5 g or 5 ml 1 tbsp any powder or liquid = about 15 g or 15 ml ½ c any vegetable, fruit, or fluid weighs about 100 g	1 cal = 4.2 kJ = 0.004 mJ 1 mJ = 240 cal 1 kJ = 0.24 cal 1 g carbohydrate = 4 cal = 17 kJ 1 g fat = 9 cal = 37 kJ 1 g protein = 4 cal = 17 kJ 1 g alcohol = 7 cal = 29 kJ

Body Weights: Quick Estimation of Desirable Body Weight

Men

For 5 ft, consider 106 lb a reasonable weight.

For each inch over 5 ft, add 6 lb.

Subtract 6 lb for each inch under 5 ft. Add 10% for a large-framed individual; subtract 10% for a small-framed individual.

Example: A man 5 ft 8 inches tall (medium frame) would start at 106 lb, add 48, and arrive at a reasonable weight of 154 lb.

Women

For 5 ft, consider 100 lb a reasonable weight.

For each inch over 5 ft, add 5 lb.

Subtract 5 lb for each inch under 5 ft. Add 10% for a large-framed individual; subtract 10% for a small-framed individual.

For each year under 25 (down to 18), subtract 1 lb.

Example: A woman 21 years old, 5 ft 4 inches tall (medium-frame), would start at 100 lb, add 20, and subtract 4, arriving at a reasonable weight of 116 lb.

Table B-2 shows how to determine frame size.

TABLE B-2
Frame Size

To make a simple approximation of your frame size:

Extend your arm and bend the forearm upward at a 90-degree angle. Keep the fingers straight and turn the inside of your wrist away from your body. Place the thumb and index finger of your other hand on the two prominent bones on either side of your elbow. Measure the space between your fingers against a ruler or a tape measure.[a] Compare the measurements with the following standards. (These standards represent the elbow measurements for medium-framed men and women of various heights. Measurements smaller than those listed indicate you have a small frame, and larger measurements indicate a large frame.[b])

Men		Women	
Height in 1-in. Heels	Elbow Breadth	Height in 1-in. Heels	Elbow Breadth
5 ft 2 in. to 5 ft 3 in.	2½ to 2⅞ in.	4 ft 10 in. to 4 ft 11 in.	2¼ to 2½ in.
5 ft 4 in. to 5 ft 7 in.	2⅝ to 2⅞ in.	5 ft 0 in. to 5 ft 3 in.	2¼ to 2½ in.
5 ft 8 in. to 5 ft 11 in.	2¾ to 3 in.	5 ft 4 in. to 5 ft 7 in.	2⅜ to 2⅝ in.
6 ft 0 in. to 6 ft 3 in.	2¾ to 3⅛ in.	5 ft 8 in. to 5 ft 11 in.	2⅜ to 3⅝ in.
6 ft 4 in. and over	2⅞ to 3¼ in.	6 ft 0 in. and over	2½ to 3¾ in.

[a]For the most accurate measurement, have your health care provider measure your elbow breadth with a caliper.

[b]A simple estimate of frame size can be derived by circling your wrist with the thumb and middle finger of your other hand. If the thumb and finger do not meet, you most likely have a large frame. If the thumb and finger just meet, you most likely have a medium frame, and if the fingers overlap greatly, you may have a small frame.

The U.S. Exchange System

The U.S. exchange system divides the foods suitable for use in planning a healthy diet into seven lists—the starch, fruit, milk, other carbohydrates, vegetable, meat and meat substitutes, and fat lists.* The exchange lists are shown in Table B-3. Figure B-1 provides examples of foods, portion sizes, and energy-nutrient contributions for each of the exchange lists.

TABLE B-3

The Food Exchange System

List	Portion Size	Carbohydrate (g)	Protein (g)	Fat (g)	Energy (cal)
Carbohydrate Group					
Starch	1 slice; ½ c	15	3	1 or less	80
Fruit	varies	15	—	—	60
Milk	1 c				
Nonfat*		12	8	0–3	90
Low-fat		12	8	5	120
Whole		12	8	8	150
Other carbohydrates	varies	15	varies	varies	varies
Vegetable	½ c	5	2	—	25
Meat and Meat Substitute Group	1 oz				
Very lean		—	7	0–1	35
Lean		—	7	3	55
Medium-fat		—	7	5	75
High-fat		—	7	8	100
Fat Group	1 tsp pure fat	—	—	5	45

*Nonfat is the same as fat-free or skim.

*The U.S. Exchange System presented here is based on material in *Exchange Lists for Meal Planning*, 1995, prepared by committees of the American Diabetes Association and the American Dietetic Association.

FIGURE B-1
THE EXCHANGE SYSTEM: EXAMPLE FOODS, PORTION SIZES, AND ENERGY-NUTRIENT CONTRIBUTIONS

Starch
1 starch exchange is like:
1 slice bread
¾ c ready-to-eat cereal
½ c cooked pasta, rice noodles, or bulgur
⅓ c cooked rice
½ c cooked beans[a]
½ c corn, peas, or yams
1 small (3 oz) potato
½ bagel, English muffin, or bun
1 tortilla, waffle, roll, taco, or matzoh
(1 starch = 15 g carbohydrate, 3 g protein, 0–1 g fat, and 80 cal.)

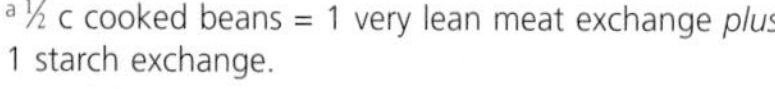
[a] ½ c cooked beans = 1 very lean meat exchange *plus* 1 starch exchange.

Vegetables
1 vegetable exchange is like:
½ c cooked carrots, greens, green beans, brussels sprouts, beets, broccoli, cauliflower, or spinach
1 c raw carrots, radishes, or salad greens
1 lg tomato
(1 vegetable = 5 g carbohydrate, 2 g protein, and 25 cal.)

Fruits
1 fruit exchange is like:
1 small banana, nectarine, apple, or orange
½ large grapefruit, pear, or papaya
½ c orange, apple, or grapefruit juice
17 small grapes
⅓ cantaloupe (or 1 c cubes)
2 tbs raisins
1½ dried figs
3 dates
1½ carambola (star fruit)
(1 fruit = 15 g carbohydrate and 60 cal.)

Meat and substitutes (very lean)
1 very lean meat exchange is like:
1 oz chicken (white meat, no skin)
1 oz cod, flounder, or trout
1 oz tuna (canned in water)
1 oz clams, crab, lobster, scallops, shrimp, or imitation seafood
1 oz fat-free cheese
½ c cooked beans, peas, or lentils
¼ c nonfat or low-fat cottage cheese
2 egg whites (or ¼ c egg substitute)
(1 very lean meat = 7 g protein, 0–1 g fat, and 35 cal.)

Meats and substitutes (lean)
1 lean meat exchange is like:
1 oz beef or pork tenderloin
1 oz chicken (dark meat, no skin)
1 oz herring or salmon
1 oz tuna (canned in oil, drained)
1 oz low-fat cheese or luncheon meats
(1 lean meat = 7 g protein, 3 g fat, and 55 cal.)

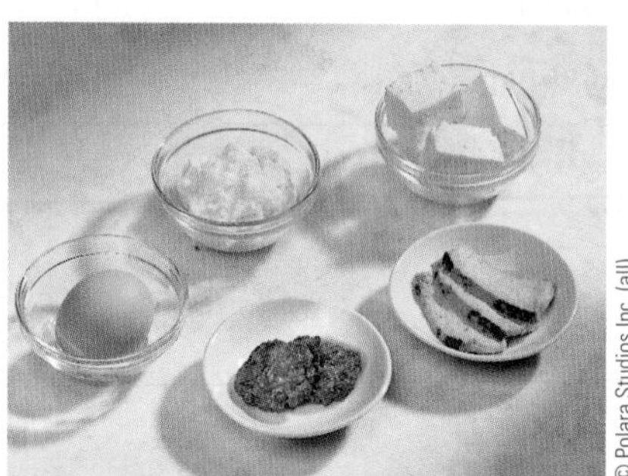

Meats and substitutes (medium-fat)
1 medium-fat meat exchange is like:
1 oz ground beef
1 oz pork chop
1 egg
¼ c ricotta
4 oz tofu
(1 medium-fat meat = 7 g protein, 5 g fat, and 75 cal.)

FIGURE B-1

THE EXCHANGE SYSTEM: EXAMPLE FOODS, PORTION SIZES, AND ENERGY-NUTRIENT CONTRIBUTIONS—Continued

Other carbohydrates

1 other carbohydrates exchange is like:
2 small cookies
1 small brownie or cake
5 vanilla wafers
1 granola bar
½ c ice cream
(1 other carbohydrate = 15 g carbohydrate and may be exchanged for 1 starch, 1 fruit, or 1 milk. Because many items on this list contain added sugar and fat, their fat and calorie values vary, and their portion sizes are small.)

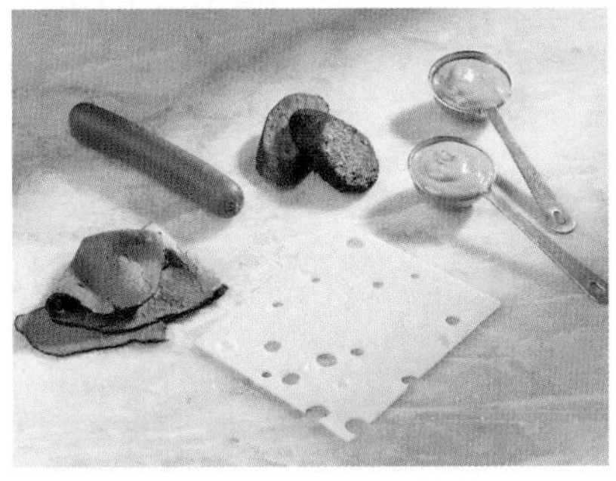

Meats and substitutes (high-fat)

1 high-fat meat exchange is like:
1 oz pork sausage
1 oz luncheon meat (such as bologna)
1 oz regular cheese (such as cheddar or swiss)
1 small hot dog (turkey or chicken)[b]
2 tbs peanut butter[c]
(1 high-fat meat = 7 g protein, 8 g fat, and 100 cal.)

[b]A beef or pork hot dog counts as 1 high-fat meat exchange *plus* 1 fat exchange.
[c]Peanut butter counts as 1 high-fat meat exchange *plus* 1 fat exchange.

Milks (nonfat and very low-fat)

1 nonfat milk exchange is like:
1 c nonfat or 1% milk
¾ c nonfat yogurt, plain
1 c nonfat or low-fat buttermilk
½ c evaporated nonfat milk
⅓ c dry nonfat milk
(1 nonfat milk = 12 g carbohydrate, 8 g protein, 0–3 g fat, and 90 cal.)

Milks (low-fat)

1 low-fat milk exchange is like:
1 c 2% milk
¾ c low-fat yogurt, plain
(1 low-fat milk = 12 g carbohydrate, 8 g protein, 5 g fat, and 120 cal.)

Milks (whole)

1 whole milk exchange is like:
1 c whole milk
½ c evaporated whole milk
(1 whole milk = 12 g carbohydrate, 8 g protein, 8 g fat, and 150 cal.)

THE FAT GROUP

Fats

1 fat exchange is like:
1 tsp butter
1 tsp margarine or mayonnaise (1 tbs reduced fat)
1 tsp any oil
1 tbs salad dressing (2 tbs reduced fat)
8 large black olives
10 large peanuts
⅛ medium avocado
1 slice bacon
2 tbs shredded coconut
1 tbs cream cheese (2 tbs reduced fat)
(1 fat = 5 g fat and 45 cal.)

Canadian Nutrition and Physical Activity Guidelines

Canada's Guidelines for Healthy Eating

Canada's Guidelines for Healthy Eating encourage healthy Canadians over 2 years of age to

- Enjoy a variety of foods.
- Emphasize cereals, breads, other grain products, vegetables, and fruits.
- Choose lower-fat dairy products, leaner meats, and foods prepared with little or no fat.
- Achieve and maintain a healthy body weight by enjoying regular physical activity and healthy eating.
- Limit salt, alcohol, and caffeine.

Canada's Recommended Nutrient Intakes (RNI)

A major revision of the nutrient recommendations is under way in the United States and Canada. The Dietary Reference Intakes (DRI) reports have replaced both the 1989 RDA in the United States and the 1990 RNI in Canada. The new DRI are presented on the inside front cover.

Canada's Food Guide to Healthy Eating

Canada's Food Guide to Healthy Eating gives detailed information for selecting foods to meet *Canada's Guidelines for Healthy Eating* (see Figure C-1). The *Food Guide* was designed to meet the nutritional needs of all Canadians 4 years of age and older and takes a total diet approach, rather than emphasizing a single food, meal, or day's meals and snacks. The rainbow side of the *Food Guide* shows the four food groups with pictorial examples of foods in each group. The bar side shows the number of servings recommended for each group. Figure C-2 presents Canada's Physical Activity Guide.

Canada's Guidelines for Healthy Eating and *Food Guide* are being revised to complement the new Dietary Reference Intakes. Updates are available from Health Canada's Office of Nutrition Policy and Promotion at www.hc-sc.gc.ca/hpfb-dgpsa/onpp-bppn/.

FIGURE C-1
CANADA'S FOOD GUIDE TO HEALTHY EATING

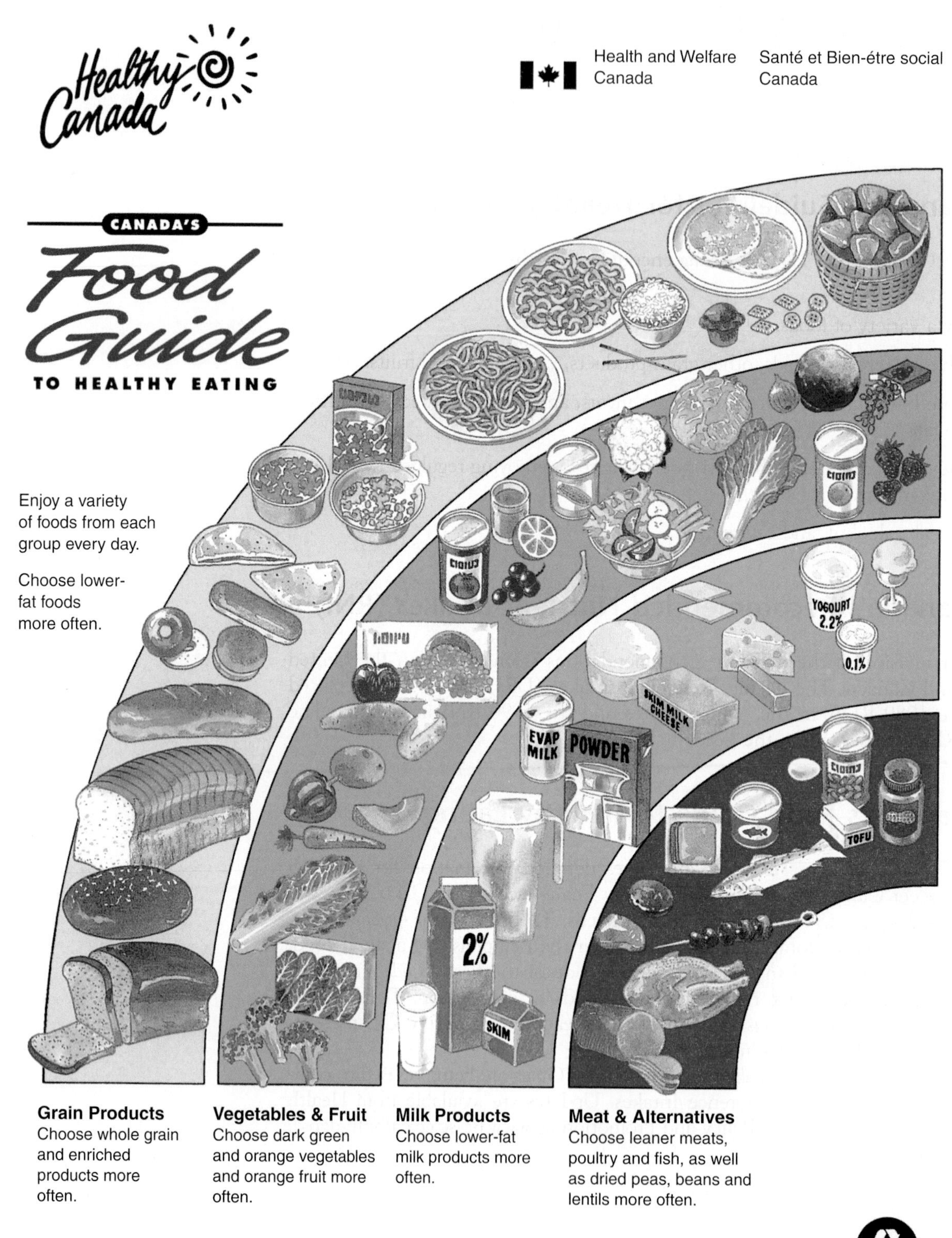

Healthy Canada
Health and Welfare Canada
Santé et Bien-étre social Canada
CANADA'S
Food Guide
TO HEALTHY EATING
Enjoy a variety of foods from each group every day.
Choose lower-fat foods more often.
YOGOURT 2.2%
0.1%
SKIM MILK CHEESE
EVAP MILK
POWDER
2%
SKIM
TOFU
Grain Products
Choose whole grain and enriched products more often.
Vegetables & Fruit
Choose dark green and orange vegetables and orange fruit more often.
Milk Products
Choose lower-fat milk products more often.
Meat & Alternatives
Choose leaner meats, poultry and fish, as well as dried peas, beans and lentils more often.

FIGURE C-1
CANADA'S FOOD GUIDE TO HEALTHY EATING—Continued

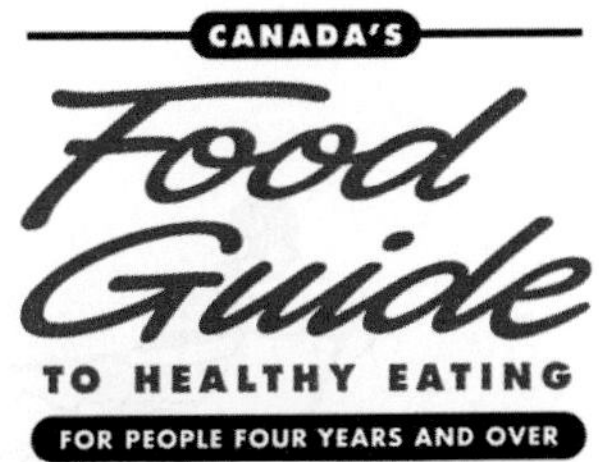

CANADA'S
Food Guide
TO HEALTHY EATING
FOR PEOPLE FOUR YEARS AND OVER

Different People Need Different Amounts of Food

The amount of food you need every day from the 4 food groups and other foods depends on your age, body size, activity level, whether you are male or female and if you are pregnant or breast-feeding. That's why the Food Guide gives a lower and higher number of servings for each food group. For example, young children can choose the lower number of servings, while male teenagers can go to the higher number. Most other people can choose servings somewhere in between.

Grain Products
5-12
SERVINGS PER DAY

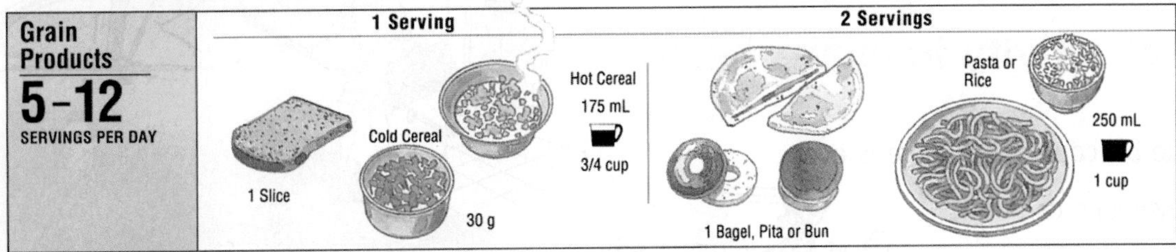

Vegetables & Fruit
5-10
SERVINGS PER DAY

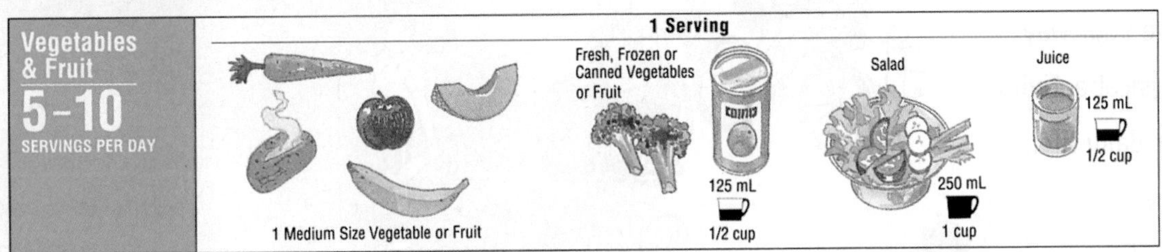

Milk Products
SERVINGS PER DAY
Children 4–9 years: 2–3
Youth 10–16 years: 3–4
Adults: 2–4
Pregnant & Breast-feeding Women: 3–4

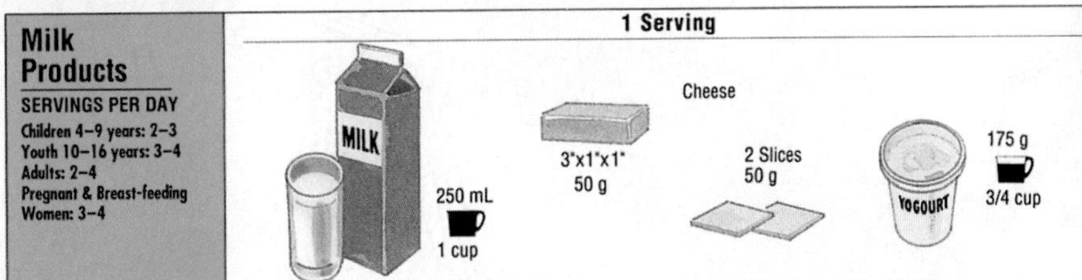

Meat & Alternatives
2-3
SERVINGS PER DAY

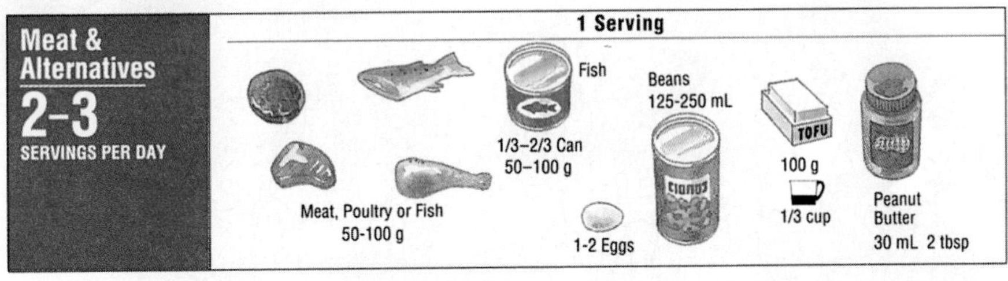

Other Foods

Taste and enjoyment can also come from other foods and beverages that are not part of the 4 food groups. Some of these foods are higher in fat or Calories, so use these foods in moderation.

Enjoy eating well, being active and feeling good about yourself. That's VITALIT

ISBN 0-662-19648-1

FIGURE C-2
CANADA'S PHYSICAL ACTIVITY GUIDE

CANADA'S
Physical Activity Guide
to Healthy Active Living
Physical activity improves health.
Every little bit counts, but more is even better – everyone can do it!
Get active your way – build physical activity into your daily life...
• at home
• at school
• at work
• at play
• on the way
...that's active living!
TAKE THE STAIRS
Increase
Endurance Activities
Increase
Flexibility Activities
Increase
Strength Activities
Reduce
Sitting for long periods
Health Canada
Santé Canada
Canadian Society for Exercise Physiology

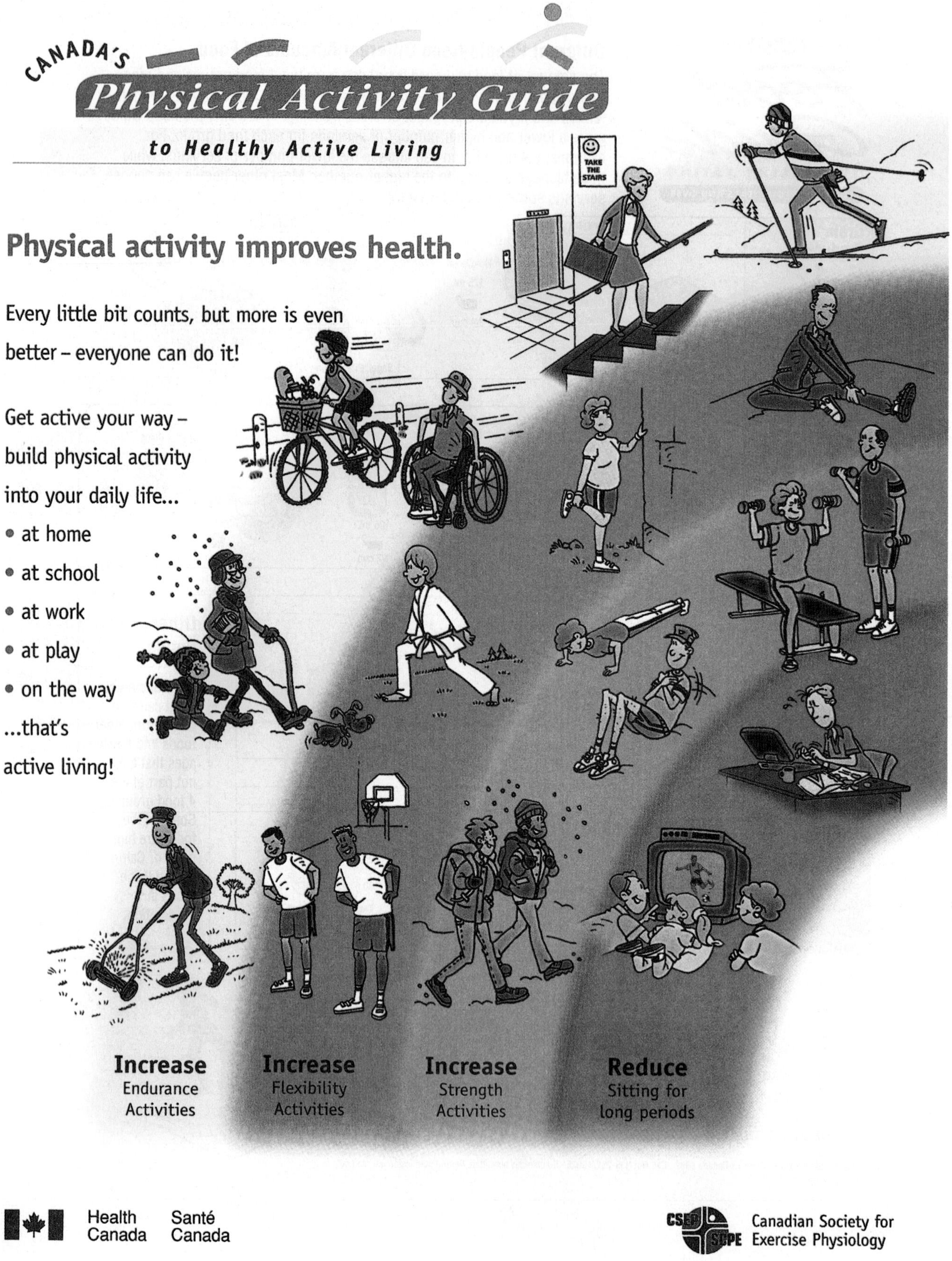

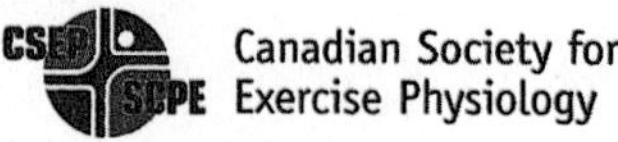

FIGURE C-2
CANADA'S PHYSICAL ACTIVITY GUIDE—Continued

Choose a variety of activities from these three groups:

Endurance

4-7 days a week
Continuous activities for your heart, lungs and circulatory system.

Flexibility

4-7 days a week
Gentle reaching, bending and stretching activities to keep your muscles relaxed and joints mobile.

Strength

2-4 days a week
Activities against resistance to strengthen muscles and bones and improve posture.

Starting slowly is very safe for most people. Not sure? Consult your health professional.

For a copy of the *Guide Handbook* and more information: **1-888-334-9769**, or **www.paguide.com**

Eating well is also important. Follow *Canada's Food Guide to Healthy Eating* to make wise food choices.

Get Active Your Way, Every Day—For Life!

Scientists say accumulate 60 minutes of physical activity every day to stay healthy or improve your health. As you progress to moderate activities you can cut down to 30 minutes, 4 days a week. Add-up your activities in periods of at least 10 minutes each. Start slowly... and build up.

Time needed depends on effort

Very Light Effort	Light Effort *60 minutes*	Moderate Effort *30-60 minutes*	Vigorous Effort *20-30 minutes*	Maximum Effort
• Strolling • Dusting	• Light walking • Volleyball • Easy gardening • Stretching	• Brisk walking • Biking • Raking leaves • Swimming • Dancing • Water aerobics	• Aerobics • Jogging • Hockey • Basketball • Fast swimming • Fast dancing	• Sprinting • Racing

Range needed to stay healthy

You Can Do It – Getting started is easier than you think

Physical activity doesn't have to be very hard. Build physical activities into your daily routine.

- Walk whenever you can–get off the bus early, use the stairs instead of the elevator.
- Reduce inactivity for long periods, like watching TV.
- Get up from the couch and stretch and bend for a few minutes every hour.
- Play actively with your kids.
- Choose to walk, wheel or cycle for short trips.
- Start with a 10 minute walk–gradually increase the time.
- Find out about walking and cycling paths nearby and use them.
- Observe a physical activity class to see if you want to try it.
- Try one class to start, you don't have to make a long-term commitment.
- Do the activities you are doing now, more often.

Benefits of regular activity:	Health risks of inactivity:
• better health • improved fitness • better posture and balance • better self-esteem • weight control • stronger muscles and bones • feeling more energetic • relaxation and reduced stress • continued independent living in later life	• premature death • heart disease • obesity • high blood pressure • adult-onset diabetes • osteoporosis • stroke • depression • colon cancer

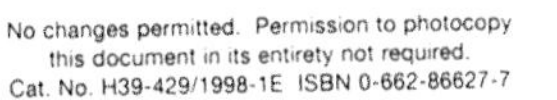

Cat. No. H39-429/1998-1E ISBN 0-662-86627-7

Chapter Notes

CHAPTER 1

1. American Society for Nutritional Sciences, *Nutrition Notes* 38 (2002): 2.
2. A. E. Sloan, Top ten global food trends, *Food Technology* 59 (2005): 31.
3. Position of the American Dietetic Association: Food and Nutrition Misinformation, *Journal of the American Dietetic Association* 106 (2006): 601–607.
4. Federal Trade Commission, Bureau of Consumer Protection. *Deception in Weight-Loss Advertising Workshop: Seizing Opportunities and Building Partnerships to Stop Weight-Loss Fraud.* Washington, DC: 2003.
5. A. E. Sloan, Top ten global food trends, *Food Technology* 59 (2005): 20–32.
6. Centers for Disease Control and Prevention, *National Vital Statistics Reports* 53, no. 15 (2005).
7. *The Surgeon General's Call to Action to Prevent and Decrease Overweight and Obesity, 2001* (Washington, DC: U.S. Department of Health and Human Services, 2001); E. E. Calle and coauthors, Body mass index and mortality in a prospective cohort of U.S. adults, *New England Journal of Medicine* 341 (1999): 1097–1105; and T. L. S. Visscher and coauthors, Obesity and unhealthy life-years in adult Finns, *Archives of Internal Medicine* 164 (2004): 1413–1420; as cited in M. Boyle and D. Holben, *Community Nutrition in Action: An Entrepreneurial Approach* (Belmont, CA: Thomson Wadsworth, 2006), 237.
8. J. M. McGinnis and W. H. Foege, Actual causes of death in the United States, *Journal of the American Medical Association* 270 (1993): 2207–2212.
9. M. Minkler, Health education, health promotion, and the open society: An historical perspective, *Health Education Quarterly* 16 (1989): 17–30; Position of the American Dietetic Association: The role of dietetics professionals in health promotion and disease prevention, *Journal of the American Dietetic Association* 102 (2002): 1680–1687.
10. U.S. Department of Health and Human Services, Public Health Service, *Healthy People 2010: Understanding and Improving Health* (Washington, DC: U.S. Government Printing Office, November 2000).
11. L. Breslow and N. Breslow, Health practices and disability: Some evidence from Alameda County, *Preventive Medicine* 22 (1993): 86–95.
12. National Cancer Institute, *5 A Day for Better Health Program Evaluation Report* (Bethesda, MD: National Institutes of Health), 2006, www.cancercontrol.cancer.gov/5ad_exec.html.
13. U.S. Department of Health and Human Services and U.S. Department of Agriculture, *Dietary Guidelines for Americans, 2005*, 6th ed. (Washington, DC: U.S. Government Printing Office, January 2005).
14. L. Breslow and N. Breslow, Health practices and disability: Some evidence from Alameda County, *Preventive Medicine* 22 (1993): 86–95; A. J. Vita and coauthors, Aging, health risks, and cumulative disability, *New England Journal of Medicine* 338 (1998): 1035–1041.
15. Public Health Service, U.S. Department of Health and Human Services, *HP 2010 Progress Review: Nutrition and Overweight*, January 21, 2004; C. L. Ogden and coauthors, Prevalence of overweight and obesity in the United States, 1999–2004, *Journal of the American Medical Association* 295 (2006): 1549–1555.
16. Fact of the Day Archive (December 13, 2002): National Restaurant Association, Meal Consumption Behavior, www.restaurant.org/research/fact_archives.cfm.
17. Ibid.
18. The food composition data in the photos on page 16 are from McDonalds USA Nutrition Facts, www.mcdonalds.com/countries/usa/food/ nutrition_facts.
19. L. R. Young and M. Nestle, The contribution of expanding portion sizes to the U.S. obesity epidemic, *American Journal of Public Health* 92 (2002): 246–248; S. J. Nielsen and B. M. Popkin, Patterns and trends in food portion sizes, 1977–1998, *Journal of the American Medical Association* 289 (2003): 450–453.
20. L. R. Young and M. Nestle, Portion sizes in dietary assessment: Issues and policy implications, *Nutrition Reviews* 53 (1995): 149–158.
21. L. R. Young and M. Nestle, 2002; B. J. Rolls, D. Engell, and L. L. Birch, Serving portion size influences 5-year-old but not 3-year-old children's food intake, *Journal of the American Dietetic Association* 100 (2000): 232–234; B. J. Rolls, The supersizing of America: Portion size and the obesity epidemic, *Nutrition Today* 38 (2003): 42–53.
22. National Center for Chronic Disease Prevention and Health Promotion, Centers for Disease Control and Prevention (CDC), *Obesity and Overweight—A Public Health Epidemic*, www.cdc.gov/nccdphp/dnpa/obesity.index.htm.
23. The National Alliance for Nutrition and Activity (NANA), *From Wallet to Waistline: The Hidden Cost of Super Sizing* (Washington, DC: NANA, June 2002).
24. Wendy's, www.wendys.com.
25. American Dietetic Association, *Fast food. Can fast foods fit into a healthy eating plan?* www.eatright.org/healthy/fastfood.html.
26. The list of barriers in the margin is from Princeton Survey Research Associates, *Shopping for Health 2002: Self Care Perspectives* (Washington, DC: Food Marketing Institute and Emmaus, PA: Rodale, 2002), 14.
27. H. A. Raynor and coauthors, A cost-analysis of adopting a healthful diet in a family-based obesity treatment program, *Journal of the American Dietetic Association* 102 (2002): 645–656.
28. Ibid.
29. K. Collins, Eating healthy on the cheap. MSNBC Health, Nutrition Notes, www.msnbc.com/news/806791.asp.
30. B. W. Hickman and coauthors, Nutrition claims in advertising: A study of four women's magazines, *Journal of Nutrition Education* 25 (1993): 227–235.
31. K. Kotz and M. Story, Food advertisements during children's Saturday morning television programming: Are they consistent with dietary recommendations? *Journal of the American Dietetic Association* 94 (1994): 1296–1300; C. J. Crespo and coauthors, Television watching, energy intake, and obesity in U.S. children: Results from the third National Health and Nutrition Examination Survey, 1988–1994, *Archives of Pediatrics and Adolescent Medicine* 155 (2001): 360–365.
32. Consumers Union, *Out of Balance*, 2005, www.consumersunion.org/pub/core_health_care/002657html.
33. American Dietetic Association, *Nutrition and You: Trends 2002* (Chicago: American Dietetic Association, 2002).
34. J. MacClancy, *Consuming Culture: Why You Eat What You Eat* (New York: Henry Holt and Company, 1992), 38.
35. The discussion of sustainability in the food system is adapted from C. McCullum and coauthors, Evidence-based strategies to build community food security, *Journal of the American Dietetic Association* 105 (2005): 278–283.

36. M. Boyle and L. Aomari, Position of the American Dietetic Association: Addressing world hunger, malnutrition, and food insecurity, *Journal of the American Dietetic Association* 103 (2003): 1046–1057; J. Wilkins, Seasonal and local diets: Consumers' role in achieving a sustainable food system. *Research in Rural Sociology and Development* 6 (1995): 150–152; G. Feenstra, Local food systems and sustainable communities, *American Journal of Alternative Agriculture* 12 (1997): 26–28; and J. Peters, Community food systems: Working toward a sustainable future, *Journal of the American Dietetic Association* 97 (1997): 955–956.
37. B. Storper, Moving toward healthful sustainable diets, *Nutrition Today* 38 (2003): 57–59.
38. J. Walsh and N. de Beaufort, Dietitians and local farmers: A unique alliance to change the way people think about food, *Hunger and Environmental Nutrition Dietetic Practice Group Newsletter* 1 (Spring 2003): 3; Position of the American Dietetic Association: Dietetics professionals can implement practices to conserve natural resources and protect the environment, *Journal of the American Dietetic Association* 101 (2001): 1221–1227.
39. M. Eure, Healthy eating starts with healthy food shopping. About.com Senior health, seniorhealth.about.com/library/nutrition/ bl_food_shop .htm.
40. United States Department of Agriculture, Center for Nutrition Policy and Promotion, *Recipes and Tips for Healthy, Thrifty Meals* (May 2000).
41. W. H. Glinsmann and G. K. Beauchamp, Babies need sugars in moderation, *Pediatric Basics* 69 (1994): 19–21.
42. World Cancer Research Fund and American Institute for Cancer Research, *Food, Nutrition, and the Prevention of Cancer: A Global Perspective* (Washington, DC: American Institute for Cancer Research, 1997).
43. C. S. Fuchs and coauthors, Dietary fiber and the risk of colorectal cancer and adenoma in women, *New England Journal of Medicine* 340 (1999): 169–176.
44. World Cancer Research Fund, 1997.
45. D. R. Jacobs and coauthors, Whole grain intake may reduce the risk of ischemic heart disease in post-menopausal women: The Iowa Women's Health Study, *American Journal of Clinical Nutrition* 68 (1998): 248–257.
46. The discussion about finding credible sources of information on the Internet is adapted from M. Boyle and D. Holben, *Community Nutrition in Action: An Entrepreneurial Approach*, 4th ed. (Belmont, CA: Thomson Wadsworth, 2006), 198.
47. R. Harris, Evaluating Internet research sources online, www.virtualsalt .com/evalu8it.htm; see also *Bibliography on Evaluating Web Information*, www.lib.vt.edu/help/instruct/evaluate/evalbiblio.html.
48. FDA *Consumer*, October 1989.

CHAPTER 2

1. K. Kant and coauthors, Dietary diversity and subsequent mortality in the First National Health and Nutrition Examination Survey Epidemiologic Follow-up Study, *American Journal of Clinical Nutrition* 57 (1993): 434–440.
2. S. L. Anderson, A look at the Japanese dietary guidelines, *Journal of the American Dietetic Association* 90 (1990): 1527.
3. The discussion of Nutrient Recommendations is adapted from Food and Nutrition Board, Institute of Medicine, *Dietary Reference Intakes for Calcium, Phosphorus, Magnesium, Vitamin D, and Fluoride* (Washington, DC: National Academy Press, 1997), S-1–S-13; Institute of Medicine, *Dietary Reference Intakes for Energy, Carbohydrate, Fiber, Fat, Fatty Acids, Cholesterol, Protein, and Amino Acids* (Washington, DC: National Academy Press, 2002); B. L. Devaney and S. I. Barr, DRI, EAR, RDA, AI, UL: Making sense of this alphabet soup, *Nutrition Today* 37 (2002): 226–232.
4. Food and Nutrition Board, Institute of Medicine, 2002.
5. Report of a Joint WHO/FAO Expert Consultation, *Diet, Nutrition, and the Prevention of Chronic Diseases*, WHO Technical Report Series 916 (Geneva: World Health Organization, 2003).
6. American Dietetic Association, Snacking, www.eatright.org/erm/ erm032002.html; American Dietetic Association, Incorporate healthy snacking in your eating plan, www.eatright.org/erm/ erm072202.html.
7. L. Jahns, A. M. Siega-Riz, and S. M. Popkins, The increasing prevalence of snacking among U.S. children from 1977 to 1996, *The Journal of Pediatrics* 138 (2001): 493–498.
8. D. Voyatzis, Sensible snacking, *Tufts Daily*, September 30, 2002; Tufts Nutrition, nutrition.tufts.edu/publications/matters/2002-09-30.html.
9. American Heart Association, *Nutritious Nibbles* (Dallas: American Heart Association, 1984), 5; U.S. Department of Agriculture, *Making Bag Lunches, Snacks, and Desserts Using the Dietary Guidelines*, USDA HNS Home and Garden Bulletin No. 232–9 (Washington, DC: U.S. Government Printing Office), 2–25.
10. L. S. Sims, A special issue (food labeling reform) deserves a special issue (of *Nutrition Today*)! *Nutrition Today* (September/October 1993): 4.
11. K. Baghurst, Fruits and vegetables: Why is it so hard to increase intakes? *Nutrition Today* 38 (2003): 11–20.
12. Food and Drug Administration, *FDA Backgrounder: The New Food Label*, April 1994.
13. C. J. Geiger, Health claims: History, current regulatory status, and consumer research, *Journal of the American Dietetic Association* 98 (1998): 1312–1322.
14. Food and Drug Administration, Center for Food Safety and Applied Nutrition, *A Food Labeling Guide*, Appendix C: Health Claims, www .cfsan.fda.gov/~dms/flg-6c.html; K. C. Ellwood, What are qualified health claims? *Nutrition Today* 41 (2006): 56–61; M. Ternus and coauthors, Qualified health claim for nuts and heart disease prevention: Development of consumer-friendly language, *Nutrition Today* 41 (2006): 62–66.
15. J. F. Mariani, *Dictionary of American Food and Drink* (New York: Hearst Books, 1994).
16. S. J. Algert, E. Brzenzinski, and T. H. Ellison, *Mexican American Food Practices, Customs, and Holidays*, Ethnic and Regional Food Practices. A Series (Chicago: American Dietetic Association and American Diabetes Association, 1998), 1–9, 23–26.
17. K. M. Ma, *Chinese American Food Practices, Customs, and Holidays*, Ethnic and Regional Food Practices. A Series (Chicago: American Dietetic Association and American Diabetes Association, 1998), 1–10, 27–31.
18. T. C. Campbell and J. Chen, Diet and chronic degenerative diseases: A summary of results from an ecologic study in rural China, in N. Totowa, *Western Diseases* (Humana Press, 1994), 67–118.
19. Should Americans be eating more Chinese food? *The Tufts University Diet & Nutrition Letter* 8 (1990): 5.
20. M. Nestle, Mediterranean diets: Historical and research overview, *The American Journal of Clinical Nutrition* 61 (1995): 1313S–1329S.
21. P. G. Kittler and K. Sucher, *Food and Culture in America*, 3rd ed. (Belmont, CA: Wadsworth Publishing, 2000).
22. C. Higgins and coauthors, *Jewish Food Practices, Customs, and Holidays*, Ethnic and Regional Food Practices: A Series (Chicago: American Dietetic Association and American Diabetes Association, 1998), 1–7, 18–20.

CHAPTER 3

1. M. Lee and S. D. Krasinski, Human adult-onset lactase decline: An update, *Nutrition Reviews* 56 (1998): 1–8; F. L. Suarez and S. D. Savaiano, Diet, genetics and lactose intolerance, *Food Technology* 51 (1997): 74–76; L. G. Tolstoi, Adult-type lactase deficiency, *Nutrition Today* 35 (2000): 134–141.
2. L. Suarez and coauthors, Tolerance to the daily ingestion of two cups of milk by individuals claiming lactose intolerance, *American Journal of Clinical Nutrition* 65 (1997): 1502–1506; B. A. Pribila and coauthors, Improved lactose digestion and intolerance among African-American adolescent girls fed a dairy-rich diet, *Journal of the American Dietetic Association* 100 (2000): 524–528.
3. Position of the American Dietetic Association: Health implications of dietary fiber, *Journal of the American Dietetic Association* 102 (2002): 993; T. T. Fung and coauthors, Whole-grain intake and the risk of type-2 diabetes: A prospective study in men, *American Journal of Clinical Nutrition* 76 (2002): 535–540.
4. E. Giovannucci and coauthors, Relationship of diet to risk of colorectal adenoma in men, *Journal of the National Cancer Institute* 84 (1992):

91–98; J. L. Slavin and coauthors, The role of whole grains in disease prevention, *Journal of the American Dietetic Association* 101 (2001): 780–785.

5. L. Brown and coauthors, Cholesterol-lowering effects of dietary fiber: A meta-analysis, *American Journal of Clinical Nutrition* 69 (1999): 30–42; M. L. Fernandez, Soluble fiber and non-digestible carbohydrate effects on plasma lipids and cardiovascular risk, *Current Opinion in Lipidology* 12 (2001): 35–40.
6. J. L. Slavin and coauthors, Plausible mechanisms for the protectiveness of whole grains, *American Journal of Clinical Nutrition* 70 (1999): 459S–463S; F. B. Hu and W. C. Willett, Optimal diets for prevention of coronary heart disease, *Journal of the American Medical Association* 288 (2002): 2569–2578.
7. J. Salmeron and coauthors, Dietary fiber, glycemic load, and risk of noninsulin-dependent diabetes mellitus in women, *Journal of the American Medical Association* 277 (1997): 472–477; T. T. Fung and coauthors, Whole-grain intake and the risk of type 2 diabetes: A prospective study in men, *American Journal of Clinical Nutrition* 76 (2002): 535–540.
8. U.S. Department of Health and Human Services and U.S. Department of Agriculture, *Dietary Guidelines for Americans, 2005*, 6th ed. (Washington, DC: U.S. Government Printing Office, January 2005).
9. *Healthy People 2010 Progress Review: Nutrition and Overweight*, January 21, 2004.
10. *Dietary Guidelines for Americans, 2005*, January 2005.
11. Position of the American Dietetic Association: Health implications of dietary fiber, *Journal of the American Dietetic Association* 102 (2002): 993.
12. Report of a Joint WHO/FAO Expert Consultation, *Diet, Nutrition, and the Prevention of Chronic Diseases*, WHO Technical Report Series 916 (Geneva: World Health Organization, 2003).
13. Food and Nutrition Board, Institute of Medicine, *Dietary Reference Intakes for Energy, Carbohydrate, Fiber, Fat, Fatty Acids, Cholesterol, Protein, and Amino Acids* (Washington, DC: National Academy Press, 2002).
14. A. D. Liese and coauthors, Whole-grain intake and insulin sensitivity: The Insulin Resistance Atherosclerosis Study, *American Journal of Clinical Nutrition* 78 (2003): 965–971.
15. E. S. Ford, W. H. Giles, and W. H. Dietz, Prevalence of the metabolic syndrome among U.S. adults: Findings from the third National Health and Nutrition Examination Survey, *Journal of the American Medical Association* 287 (2002): 356–359; S. Cook and coauthors, Prevalence of a metabolic syndrome phenotype in adolescents: Findings from the third National Health and Nutrition Examination Survey, 1988–1994, *Archives of Pediatrics and Adolescent Medicine*, 157 (2003): 821–227, as cited in M. Boyle and D. Holben, *Community Nutrition in Action: An Entrepreneurial Approach* (Belmont, CA: Thomson Wadsworth, 2006), 242.
16. Food and Nutrition Board, Institute of Medicine, 2002.
17. U.S. Department of Agriculture and U.S. Department of Health and Human Services, *MyPyramid: Steps to a Healthier You* (2005), www.mypyramid.gov.
18. U.S. Department of Health and Human Services and U.S. Department of Agriculture, *Dietary Guidelines for Americans, 2005*, 6th ed. (Washington, DC: U.S. Government Printing Office, January 2005).
19. D. J. Jenkins and coauthors, Glycemic index: Overview of implications in health and disease, *American Journal of Clinical Nutrition* 76 (2002): 266S–273S.
20. F. Q. Nuttall, Dietary fiber in the management of diabetes, *Diabetes* 42 (1993): 503–508.
21. J. P. Kirwan and coauthors, Effects of a moderate glycemic meal on exercise duration and substrate utilization, *Medicine and Science in Sports and Exercise* 33 (2001): 1517–1523.
22. The Expert Committee on the Diagnosis and Classification of Diabetes Mellitus, Report on the Expert Committee on the Diagnosis and Classification of Diabetes Mellitus, *Diabetes Care* 20 (1997): 1183–1197; American Diabetes Association, Nutrition recommendations and principles for people with diabetes mellitus, *Diabetes Care* 23 (2000): 543S–546S.
23. R. Butler and coauthors, Type 2 diabetes: Causes, complications, and new screening recommendations, *Geriatrics* 53 (1998): 47–54; American Diabetes Association Position Statement: Evidence-based nutrition principles and recommendations for the treatment and prevention of diabetes and related complications, *Journal of the American Dietetic Association* 102 (2002): 109–118.
24. Centers for Disease Control and Prevention, Age-Adjusted Prevalence of Diagnosed Diabetes per 100 Adult Population, by State, United States, 2004, www.cdc.gov/diabetes; A. H. Mokdad and coauthors, Prevalence of obesity, diabetes, and obesity-related health risk factors, *Journal of the American Medical Association* 289 (2003): 76–79.
25. G. Wang and W. H. Dietz, Economic burden of obesity in youths aged 6 to 17 years, *Pediatrics* 109 (2002): E81; American Diabetes Association, Type 2 diabetes in children and adolescents, *Pediatrics* 105 (2000): 671–680.
26. American Diabetes Association, Nutrition recommendations and principles for people with diabetes mellitus, *Diabetes Care* 23 (2000): 543S–546S.
27. A. M. Coulston and R. K. Johnson, Sugar and sugar myths: Myths and realities, *Journal of the American Dietetic Association* 102 (2002): 351–353.
28. U.S. Department of Health and Human Services and U.S. Department of Agriculture, *Dietary Guidelines for Americans, 2005*, 6th ed. (Washington, DC: U.S. Government Printing Office, January 2005).
29. J. H. Shaw, Causes and control of dental caries, *New England Journal of Medicine*, 317 (1987): 996–1004.
30. S. Kashket and coauthors, Lack of correlation between food retention on the human dentition and consumer perception of food stickiness, *Journal of Dental Research* 70 (1991): 1314–1319.
31. R. Harris and coauthors, Risk factors for dental caries in young children: A systematic review of the literature, *Community Dental Health* 21 (2004): 71–85; C. C. Mobley, Nutrition needs and oral health in children, *Topics in Clinical Nutrition* 20 (2005): 200–210.
32. American Dental Association, *Diet & Dental Health* (Chicago: American Dental Association, 1993), 2.
33. M. E. Jensen, Responses of interproximal plaque pH to snack foods and effect of chewing sorbitol-containing gum, *Journal of the American Dental Association* 113 (1986): 262–266.
34. C. A. Palmer, Important relationships between diet, nutrition, and oral health, *Nutrition in Clinical Care* 4 (2001): 4–14; R. Touger-Decker, Nutrition and oral health in older adults, *Topics in Clinical Nutrition* 20 (2005): 211–218.
35. C. O. Enwonwu, Interface of malnutrition and periodontal diseases, *American Journal of Clinical Nutrition* 61 (1995): 430S–436S; Position of the American Dietetic Association: Oral health and nutrition, *Journal of the American Dietetic Association* 103 (2003): 615–625.
36. J. O. Alvarez, Nutrition, tooth development, and dental caries, *American Journal of Clinical Nutrition* 61 (1995): 410S–416S.
37. Personal communication with Marketing Data Enterprises, Inc., Valley Stream, New York.
38. S. Cohen and coauthors, Saccharin and urothelial proliferation: A threshold phenomenon, *Journal of the American Societies for Experimental Biology* 6 (1992): A-1594.
39. Position of the American Dietetic Association: Use of Nutritive and Nonnutritive Sweeteners, *Journal of the American Dietetic Association* 104 (2004): 255–275.
40. Council on Scientific Affairs, Saccharin: Review of safety issues, *Journal of the American Medical Association* 254 (1985): 2622–2644.
41. U.S. Department of Health and Human Services, National Toxicology Program, *The Tenth Report on Carcinogens*, 9th ed., 2001, ehis.niehs.nih.gov/roc/toc9.html.
42. S. S. Schiffman and coauthors, Aspartame and susceptibility to headache, *New England Journal of Medicine* 317 (1987): 1181–1185.
43. P. A. Spiers and coauthors, Aspartame: Neuropsychologic and neurophysiologic evaluation of acute and chronic effects, *American Journal of Clinical Nutrition* 68 (1998): 531–537; American Council on Science and Health, *Low-Calorie Sweeteners* (New York: American Council on Science and Health, 1993), 7–13; Position of the American Dietetic Association: Use of nutritive and nonnutritive sweeteners, *Journal of the American Dietetic Association* 104 (2004): 255–275.

44. Position of the American Dietetic Association: Use of nutritive and nonnutritive sweeteners, *Journal of the American Dietetic Association* 104 (2004): 255–275.
45. J. E. Blundell and A. J. Hill, Paradoxical effects of an intense sweetener (aspartame) on appetite, *The Lancet* 1 (1986): 1092–1093.
46. B. J. Rolls, Effects of intense sweeteners on hunger, food intake, and body weight: A review, *American Journal of Clinical Nutrition* 53 (1991): 872–878; A. Drewnowski, Comparing the effects of aspartame and sucrose on motivational ratings, taste preferences, and energy intakes in humans, *American Journal of Clinical Nutrition* 59 (1994): 338–345.
47. Position of the American Dietetic Association: Use of nutritive and nonnutritive sweeteners, *Journal of the American Dietetic Association* 104 (2004): 255–275.

CHAPTER 4

1. P. M. Kris-Etherton, W. S. Harris, and L. J. Appel, Fish consumption, fish oil, omega-3 fatty acids, and cardiovascular disease, *Circulation* 106 (2002): 2747–2757; P. Nestel and coauthors, The n-3 fatty acids eicosapentaenoic acid and docosahexaenoic acid increase systemic arterial compliance in humans, *American Journal of Clinical Nutrition* 76 (2002): 326–330.
2. A. P. Simopoulos, Omega-3 fatty acids in health and disease and in growth and development, *American Journal of Clinical Nutrition* 54 (1991): 438–463; S. B. Eaton and coauthors, Dietary intake of long-chain polyunsaturated fatty acids during the paleolithic, *World Reviews Nutrition & Dietetics* 83 (1998): 12–23; H. Gerster, N–3 fish oil polyunsaturated fatty acids and bleeding, *Journal of Nutrition and Environmental Medicine* 5 (1995): 281–296.
3. P. M. Kris-Etherton, W. S. Harris, and L. J. Appel, Fish consumption, fish oil, omega-3 fatty acids, and cardiovascular disease, *Circulation* 106 (2002): 2747–2757; Expert Panel on Detection, Evaluation, and Treatment of High Blood Cholesterol in Adults, Summary of the Third Report of the National Cholesterol Education Program (NCEP) Expert Panel on Detection, Evaluation, and Treatment of High Blood Cholesterol in Adults, *Journal of the American Medical Association* 285 (2001): 2486–2498.
4. L. Van Horn and coauthors, Dietary management of cardiovascular disease: A year 2002 perspective, *Nutrition in Clinical Care* 4 (2001): 314–331.
5. G. Wolf, The role of oxidized low-density lipoprotein in the activation of peroxisome proliferator-activated receptor (gamma): Implications for atherosclerosis, *Nutrition Reviews* 57 (1999): 88–91; R. Ross, Atherosclerosis: An inflammatory disease, *New England Journal of Medicine* 340 (1999): 115–126.
6. P. Kris-Etherton and coauthors, Summary of the scientific conference on dietary fatty acids and cardiovascular health, *Circulation* 10 (2001): 1034–1039.
7. R. M. Krauss and coauthors, AHA Dietary Guidelines, Revision 2000: A statement for healthcare professionals from the nutrition committee of the American Heart Association, *Circulation* 102 (2000): 2284–2299.
8. Adult Treatment Panel III, *Journal of the American Medical Association* 285 (2001): 2486–2498.
9. J. W. Anderson and coauthors, Long-term cholesterol-lowering effects of psyllium as an adjunct to diet therapy in the treatment of hypercholesterolemia, *American Journal of Clinical Nutrition* 71 (2000): 1433–1438.
10. E. B. Rimm and coauthors, Vitamin E consumption and the risk of coronary heart disease in men, *New England Journal of Medicine* 328 (1993): 1450–1456; M. J. Stampfer and coauthors, Vitamin E consumption and the risk of coronary heart disease in women, *New England Journal of Medicine* 328 (1993): 1444–1449; P. Knekt, Antioxidant vitamin intake and coronary mortality in a longitudinal population study, *American Journal of Epidemiology* 139 (1994): 1180–1189.
11. D. A. Tribble, AHA Science Advisory: Antioxidant consumption and risk of coronary heart disease: Emphasis on vitamin C, vitamin E, and β-carotene: A statement for health care professionals from the American Heart Association, *Circulation* 99 (1999): 591–595.
12. Food and Nutrition Board, Institute of Medicine, *Dietary Reference Intakes for Energy, Carbohydrate, Fiber, Fat, Fatty Acids, Cholesterol, Protein, and Amino Acids* (Washington, DC: National Academy Press, 2002).
13. H. Farah and J. Buzby, U.S. food consumption up 16 percent since 1970, *Amber Waves* (November 2005).
14. S. Yusuf and coauthors, Effect of potentially modifiable risk factors associated with myocardial infarction in 52 countries (the INTERHEART study): Case control study, *Lancet* 364 (2004): 937–952; Food and Nutrition Board, Institute of Medicine, *Dietary Reference Intakes for Energy, Carbohydrate, Fiber, Fat, Fatty Acids, Cholesterol, Protein, and Amino Acids* (Washington, DC: National Academy Press, 2002).
15. A. Ascherio and coauthors, *Trans* fatty acids and coronary heart disease, *New England Journal of Medicine* 340 (1999): 1994–1998.
16. A. H. Lichtenstein and coauthors, Effects of different forms of dietary hydrogenated fats on serum lipoprotein cholesterol levels, *New England Journal of Medicine* 340 (1999): 1933–1940.
17. W. C. Willett and A. Ascherio, *Trans* fatty acids: Are the effects only marginal? *American Journal of Public Health* 84 (1994): 722–724.
18. Revealing *Trans* fats, *FDA Consumer* (September/October 2003): 20–26.
19. Food and Nutrition Board, Institute of Medicine, *Dietary Reference Intakes for Energy, Carbohydrate, Fiber, Fat, Fatty Acids, Cholesterol, Protein, and Amino Acids* (Washington, DC: National Academy Press, 2002); P. A. Cotton and coauthors, Dietary sources of nutrients among U.S. adults, 1994–1996, *Journal of the American Dietetic Association* 104 (2004): 921–930.
20. American Heart Association, Fat: American Heart Association Scientific Position, www.americanheart.org.
21. M. Clarke, The Mediterranean food guide, *Timely Topics from the Department of Human Nutrition*. Kansas State Research and Extension, www.oznet.ksu.edu/ext_F&N/_Timely/medrdiet.htm.
22. W. S. Harris and L. J. Appel, New guidelines focus on fish, fish oils, omega-3 fatty acids, *American Heart Association Statement*. November 18, 2002, www.americanheart.org.
23. F. Fuentes and coauthors, Mediterranean and low-fat diets improve endothelial function in hypercholesterolemic men, *Annals of Internal Medicine* 134 (2001): 1115–1119; F. B. Hu and W. C. Willett, Optimal diets for prevention of coronary heart disease, *Journal of the American Medical Association* 288 (2002).
24. American Heart Association, Fish oil and omega-3 fatty acids: American Heart Association Recommendation, www.americanheart.org.
25. P. M. Kris-Etherton, W. S. Harris, and L. J. Appel, Fish consumption, fish oil, omega-3 fatty acids, and cardiovascular disease, *Circulation* 106 (2002): 2737–2757.
26. E. M. Ward, Balancing essential dietary fats: When more fat might be better, *Environmental Nutrition* 24 (2001): 1, 6.
27. H. Blackburn, Co-investigator, Seven Countries Study, as quoted in D. Schardt, B. Liebman, and S. Schmidt, Going Mediterranean, *Nutrition Action Health Letter* 21 (1994): 1–5.
28. Personal communication with Consumer Affairs, Entenmann's, Bay Shore, NY.
29. Personal communication with Consumer Affairs, Kraft, Glenview, IL.
30. G. E. Ruoff, Reducing fat intake with fat substitutes, *American Family Physician* 43 (1991): 1235–1242.
31. *Simplesse All Natural Fat Substitute: A Scientific Overview* (Deerfield, IL: The Simplesse Company, 1991), 3–10.
32. U.S. Department of Health and Human Services, HHS News, FDA Approves Fat Substitute, Olestra, January 24, 1996; FDA changes labeling requirement for olestra, *FDA Talk Paper* #T03-59, August 2003.
33. Position paper of the American Dietetic Association: Fat replacers, *Journal of the American Dietetic Association* 105 (2005): 266–275.
34. Summary of the Third Report of the National Cholesterol Education Program (NCEP) Expert Panel on Detection, Evaluation, and Treatment of High Blood Cholesterol in Adults, *Journal of the American Medical Association* 285 (2001): 2486–2498.
35. Adult Treatment Panel III, 2001.

36. L. Van Horn and coauthors, Dietary management of cardiovascular disease: A year 2002 perspective, *Nutrition in Clinical Care* 4 (2001): 314–331.
37. R. M. Krauss and coauthors, AHA Dietary Guidelines, Revision 2000: A statement for healthcare professionals from the nutrition committee of the American Heart Association, *Circulation* 102 (2000): 2284–2299.
38. B. S. Levine and C. Cooper, Plant stanol esters: A new tool in the dietary management of cholesterol, *Nutrition Today* 35 (2000): 61–66.
39. Adult Treatment Panel III, 2001; F. B. Hu and coauthors, Dietary fat intake and the risk of coronary heart disease in women, *New England Journal of Medicine* 337 (1997): 1491–1499.
40. R. H. Eckel and R. M. Krauss, American Heart Association Call to Action: Obesity as a major risk factor for coronary heart disease, *Circulation* 97 (1998): 2099–2100; R. R. Wing and coauthors, Change in waist-hip ratio with weight loss and its association with change in cardiovascular risk factors, *American Journal of Clinical Nutrition* 55 (1992): 1086–1092.
41. P. M. Kris-Etherton, W. S. Harris, and L. J. Appel, Fish consumption, fish oil, omega-3 fatty acids, and cardiovascular disease, *Circulation* 106 (2002): 2747–2757.
42. K. M. Fairfield and R. H. Fletcher, Vitamins for chronic disease prevention in adults, *Journal of the American Medical Association* 287 (2002): 3116–3126; G. S. Omenn and coauthors, Preventing coronary heart disease: B vitamins and homocysteine, *Circulation* 97 (1998): 421–424; O. Nygard and coauthors, Major lifestyle determinants of plasma total homocysteine distribution: The Hordaland Homocysteine Study, *American Journal of Clinical Nutrition* 67 (1998): 263–270.
43. K. Reynolds and coauthors, Alcohol consumption and risk of stroke, a meta-analysis, *Journal of the American Medical Association* 289 (2003): 579–588.
44. J. M. Gaziano and coauthors, Moderate alcohol intake, increased levels of high-density lipoprotein and its subfractions, and decreased risk of myocardial infarction, *New England Journal of Medicine* 329 (1993): 1829–1834; K. J. Mukamal and coauthors, Roles of drinking pattern and type of alcohol consumed in coronary heart disease in men, *New England Journal of Medicine* 348 (2003): 109–118.
45. S. L. Englebardt, Eat, drink, go back to work, *American Health* (June 1994): 90.
46. T. A. Pearson and P. Terry, What to advise patients about drinking alcohol: The clinician's conundrum, *Journal of the American Medical Association* 272 (1994): 967–968.
47. K. Breithaupt and coauthors, Protective effect of chronic garlic intake on elastic properties of the aorta in the elderly, *Circulation* 96 (1997): 2649–2655; A. Orekhov and J. Grunwald, Effects of garlic on atherosclerosis, *Nutrition* 13 (1997): 656–663; J. Isaacsohm and coauthors, Garlic powder and plasma lipids and lipoproteins, *Archives of Internal Medicine* 158 (1998): 1189–1194.
48. R. M. Krauss and coauthors, AHA Dietary Guidelines, Revision 2000: A statement for healthcare professionals from the nutrition committee of the American Heart Association, *Circulation* 102 (2000): 2284–2299.
49. J. P. Stong and coauthors, Prevalence and extent of atherosclerosis in adolescents and young adults, *Journal of the American Medical Association* 281 (1999): 727–735; American Heart Association Scientific Statement: Cardiovascular health in childhood, *Circulation* 106 (2002): 143–160; American Heart Association, Guidelines for cardiovascular risk reduction for high-risk children, *Circulation* 107 (2003): 1562–1566.

CHAPTER 5

1. R. K. Chandra, Nutrition and the immune system: An introduction, *American Journal of Clinical Nutrition* 66 (1997): S460–S463.
2. V. R. Young, Soy protein in relation to human protein and amino acid nutrition, *Journal of the American Dietetic Association* 91 (1991): 828–835.
3. D. G. Schroeder and R. Martorell, Enhancing child survival by preventing malnutrition, *American Journal of Clinical Nutrition* 65 (1997): 1080–1081.
4. Food and Nutrition Board, Institute of Medicine, *Dietary Reference Intakes for Energy, Carbohydrate, Fiber, Fat, Fatty Acids, Cholesterol, Protein, and Amino Acids* (Washington, DC: National Academy Press, 2002).
5. B. Torun and F. Chew, Protein-energy malnutrition, in M. E. Shils, J. A. Olson, and M. Shike, eds., *Modern Nutrition in Health and Disease* (Baltimore, MD: Williams and Wilkins, 1999), 963–988.
6. Position of the American Dietetic Association: Food insecurity and hunger in the United States, *Journal of the American Dietetic Association* 106 (2006): 446–458; M. Nord, M. Andrews, and S. Carlson, *Household Food Security in the United States, 2004* (ERR-11) (Alexandria, VA: US Department of Agriculture, Economic Research Services, 2005).
7. L. K. Massey, Does excess dietary calcium adversely affect bone? Symposium overview, *Journal of Nutrition* 128 (1998): 1048–1050.
8. The discussion on adding legumes to the diet is adapted from Eat Right Montana's Celebrating Healthy Families 2005, August 2005.
9. Ibid.
10. I. Chalmers, *The Great Food Almanac* (San Francisco: Collins Publishers, 1994).
11. American Institute for Cancer Research, *Stopping Cancer Before It Starts* (New York: Golden Books, 1999), 214–217.
12. C. L. Vecchia, Vegetable consumption and risk of chronic disease, *Epidemiology* 9 (1998): 208–210; American Institute for Cancer Research (AICR), *The New American Plate* (Washington, DC: AICR, 2002).
13. J. Carol and J. Sobal, Model of the process of adopting vegetarian diets: Health vegetarians and ethical vegetarians, *Journal of Nutrition Education* 30 (1998): 196–202; E. Lea and A. Worsley, The cognitive contexts of beliefs about the healthiness of meat, *Public Health Nutrition* 5 (2002): 37–45.
14. Position of the American Dietetic Association and Dietitians of Canada: Vegetarian diets, *Journal of the American Dietetic Association* 103 (2003): 748–765.
15. V. R. Young and P. L. Pellett, Plant proteins in relation to human protein and amino acid nutrition, *American Journal of Clinical Nutrition* 59 (1994): 1203S–1212S; Position of the American Dietetic Association and Dietitians of Canada, 2003.
16. C. Lamberg-Allardt and coauthors, Low serum 25-hydroxyvitamin D concentrations and secondary hyperparathyroidism in middle-aged white strict vegetarians, *American Journal of Clinical Nutrition* 58 (1993): 684–689.
17. J. R. Hunt and coauthors, Zinc absorption, mineral balance, and blood lipids in women consuming controlled lacto-ovo-vegetarian and omnivorous diets for 8 weeks, *American Journal of Clinical Nutrition* 67 (1998): 421–430; Position of the American Dietetic Association and Dietitians of Canada, 2003.
18. S. I. Barr and coauthors, Spinal bone mineral density in premenopausal vegetarian and nonvegetarian women: Cross-sectional and prospective comparisons, *Journal of the American Dietetic Association* 98 (1998): 760–765.
19. V. Messina, V. Melina, and R. Mangels, A new food guide for North American vegetarians, *Journal of the American Dietetic Association* 103 (2003): 771–775.
20. K. Burke, The use of soy foods in a vegetarian diet, *Topics in Clinical Nutrition* 10 (1995): 37–43.
21. T. J. Key and coauthors, Dietary habits and mortality in 11,000 vegetarians and health-conscious people: Results of a 17-year follow up, *British Medical Journal* 313 (1996): 775–779; T. J. Key and coauthors, Mortality in vegetarians and nonvegetarians: Detailed findings from a collaborative analysis of five prospective studies, *American Journal of Clinical Nutrition* 70 (1999): 516S–524S.
22. P. Walter, Effects of vegetarian diets on aging and longevity, *Nutrition Reviews* 55 (1997): S61–S65; P. N. Appleby and coauthors, The Oxford vegetarian study: An overview, *American Journal of Clinical Nutrition* 70 (1999): 525S–531S.
23. B. Hunter, Food allergies: No trivial matter, *Consumers' Research Magazine* 82 (1999): 2.

24. Food allergies: How worried should you be? *Tufts University Health & Nutrition Letter* 17 (1999): 2.
25. Food Allergy Network, *Information About Food Allergies*, 2003.
26. H. Sampson, Food Allergy, *Journal of the American Medical Association* 278 (1997): 22–26; S. Sicherer, Manifestations of food allergy: Evaluation and management, *American Family Physician* 59 (1999): 2–8.
27. FDA to require food manufacturers to list food allergens, *FDA News*, December 20, 2005.
28. S. Gottlieb, Scientists develop vaccine strategy for peanut allergy, *British Medical Journal* 318 (1999): 71–88.
29. Soy and health: Discovering the health benefits of isoflavones fact sheet, United Soybean Board, 1999; S. M. Potter, Soy—New health benefits associated with an ancient food, *Nutrition Today* 35 (2000): 53–59; M. Friedman and D. L. Brandon, Nutritional and health benefits of soy proteins, *Journal of Agriculture and Food Chemistry* 49 (2001): 1069–1086; M. Messina and V. Messina, Provisional recommended soy protein and isoflavone intakes for healthy adults, *Nutrition Today* 38 (2003): 100–109.
30. *U.S. Soyfoods Directory: Alphabetical List of Soyfoods* (Indiana: Indiana Soybean Board, 1998).
31. J. R. Crouse and coauthors, A randomized trial comparing the effect of casein with that of soy protein containing varying amounts of isoflavones on plasma concentrations of lipids and lipoproteins, *Archives of Internal Medicine* 159 (1999): 2070–2076.
32. J. W. Anderson and coauthors, Meta-analysis of the effects of soy protein intake on serum lipids, *New England Journal of Medicine* 333 (1995): 276–282.
33. R. M. Bakhit and coauthors, Intake of 25 g of soybean protein with or without fiber alters plasma lipids in men with elevated cholesterol concentrations, *Journal of Nutrition* 124 (1994): 213–222; S. M. Potter and coauthors, Depression of plasma cholesterol in men by consumption of baked products containing soy protein, *American Journal of Clinical Nutrition* 58 (1993): 501–506.
34. S. M. Potter and coauthors, Soy protein and isoflavones: Their effects on blood lipids and bone density in postmenopausal women, *American Journal of Clinical Nutrition* 68 (1998): 1375S–1379S; D. L. Alekel and coauthors, Isoflavone-rich soy protein isolate exerts significant bone sparing in the lumbar spine in perimenopausal women, Third International Symposium on the Role of Soy in Preventing and Treating Chronic Disease, Washington, DC, October 31–November 4, 1999; B. H. Arjmandi and B. J. Smith, Soy isoflavones' osteoprotective role in postmenopausal women: Mechanism of action, *Journal of Nutritional Biochemistry* 13 (2002): 130–137.
35. National Osteoporosis Foundation, *Fast Facts on Osteoporosis* (Washington, DC: National Osteoporosis Foundation, 1997).
36. M. Lock, Menopause in cultural context, *Experimental Gerontology* 29 (1994): 307–317; American Heart Association, 1997.
37. L. Bren, The estrogen and progestin dilemma: New advice, labeling guidelines, *FDA Consumer* 37 (2003): 10–11; M. Glazier, M. B. Gina, and M. A. Bowman, Review of the evidence for the use of phytoestrogens as a replacement for traditional estrogen replacement therapy, *Archives of Internal Medicine* 161 (2001): 1161–1172; North American Menopause Society (NAMS), Report from the NAMS Advisory Panel on Postmenopausal Hormone Therapy, October 6, 2002, www.menopause.org.
38. A. L. Murkies and coauthors, Dietary flour supplementation decreases postmenopausal hot flushes: Effect of soy and wheat, *Maturitas* 221 (1995): 189–195.
39. D. C. Knight and J. A. Eden, A review of the clinical effects of phytoestrogens, *Obstetrics and Gynecology* 87 (1996): 897–904; M. Messina, Soy foods and soybean isoflavones and menopausal health, *Nutrition in Clinical Care* 5 (2002): 272–282; K. K. Han and coauthors, Benefits of soy isoflavone therapeutic regimen on menopausal symptoms, *Obstetrics and Gynecology* 99 (2002): 389–394.
40. American Cancer Society, *Cancer Facts & Figures 2005* (New York: ACS, Inc., 2005).
41. American Cancer Society, 2005.
42. L. N. Kolonel, Variability in diet and its relation to risk in ethnic and migrant groups, *Basic Life Sciences* 43 (1988): 129–135.
43. A. Nomura and coauthors, Breast cancer and diet among the Japanese in Hawaii, *American Journal of Clinical Nutrition* 31 (1978): 2020–2025.
44. R. K. Severson and coauthors, A prospective study of demographics, diet, and prostate cancer among men of Japanese ancestry in Hawaii, *Cancer Research* 49 (1989): 1857–1860.
45. M. Messina and S. Barnes, The role of soy products in reducing risk of cancer, *Journal of the National Cancer Institute* 83 (1991): 541–546; D. Li and coauthors, Soybean isoflavones reduce experimental metastasis in mice, *Journal of Nutrition* 129 (1999): 1075–1078; J. R. Zhou and coauthors, Soybean components inhibit orthotopic growth of human prostate cancer cells in SCID mice, Third International Symposium on the Role of Soy in Preventing and Treating Chronic Disease, Washington, DC, October 31–November 4, 1999.
46. K. B. Bouker and L. Hilakivi-Clarke, Genistein: Does it prevent or promote breast cancer? *Environmental Health Perspective* 108 (2000): 701–708; A. Wu and coauthors, Adolescent and adult soy intake and risk of breast cancer in Asian-Americans, *Carcinogenesis* 23 (2003): 1491–1496.
47. M. Messina, C. Gardener, and S. Barnes, Gaining insight into the health effects of soy but a long way still to go: Commentary on the Fourth International Symposium on the Role of Soy in Preventing and Treating Chronic Disease, *Journal of Nutrition* 132 (2002): 547S–551S.
48. M. C. Librenti and coauthors, Effects of soya and cellulose fibers on postprandial glycemic response in type 2 diabetic patients, *Diabetes Care* 15 (1992): 111–113; T. J. Stephonson and J. W. Anderson, Isoflavone-rich protein healthy addition to diabetic diet, *The Soy Connection* (Summer 2002).

CHAPTER 6

1. J. Jaramillo-Arango, The conquest of nutritional diseases (vitamins), in *The British Contribution to Medicine* (Edinburgh: E. & S. Livingstone, Ltd., 1953), 140–162.
2. R. Hill and coauthors, The discovery of vitamins, in *The Chemistry of Life* (Cambridge: Cambridge University Press, 1970), 156–170.
3. M. B. Elam and coauthors, Effect of niacin on lipid and lipoprotein levels and glycemic control in patients with diabetes and peripheral artery disease: The Arterial Disease Multiple Intervention Trial (ADMIT), a randomized trial, *Journal of the American Medical Association* 284 (2000): 1263–1270; L. Lasagna, Over-the-counter niacin, *Journal of the American Medical Association* 271 (1994): 709–710.
4. M. K. Berman and coauthors, Vitamin B_6 in premenstrual syndrome, *Journal of the American Dietetic Association* 90 (1990): 859–860.
5. L. Bailey, G. Rampersaud, and G. Kauwell, Folic acid supplements and fortification affect the risk for neural tube defects, vascular disease, and cancer: Evolving science, *Journal of Nutrition* 133 (2003): 1961S–1968S; G. Rampersaud, L. Bailey, and G. Kauwell, Relationship of folate to colorectal and cervical cancer: Review and recommendations for practitioners, *Journal of the American Dietetic Association* 102 (2002): 1273–1282.
6. E. P. Quinlivian and J. F. Gregory, III, Effect of food fortification on folic acid intake in the United States, *American Journal of Clinical Nutrition* 77 (2003): 221–225; Centers for Disease Control and Prevention, Spina bifida and anencephaly before and after folic acid mandate—United States, 1995–1996 and 1999–2000, *Morbidity & Mortality Weekly Report* 53 (2004): 362–365.
7. C. W. Suitor and L. B. Bailey, Dietary folate equivalents: Interpretation and application, *Journal of the American Dietetic Association* 100 (2000): 88–94.
8. Food and Nutrition Board, Institute of Medicine, Dietary Reference Intakes for Thiamin, Riboflavin, Niacin, Vitamin $B_{6,}$ Folate, Vitamin $B_{12,}$ Pantothenic Acid, Biotin, and Choline (Washington, DC: National Academy Press, 1998).
9. K. M. Riggs and coauthors, Relations of vitamin B_{12}, vitamin B_6, folate, and homocysteine to cognitive performance in the Normative Aging Study, *American Journal of Clinical Nutrition* 63 (1996): 306–314; K. M. Fairfield and R. H. Fletcher, Vitamins for chronic disease prevention in adults, *Journal of the American Medical Association* 287 (2002): 3116–3126.

10. G. S. Omenn and coauthors, Preventing coronary heart disease: B vitamins and homocysteine, *Circulation* 97 (1998): 421–424; O. Nygard and coauthors, Major lifestyle determinants of plasma total homocysteine distribution: The Hordaland Homocysteine Study, *American Journal of Clinical Nutrition* 67 (1998): 263–270.
11. J. Selhub and coauthors, Association between plasma homocysteine concentrations and extracranial carotid-artery stenosis, *New England Journal of Medicine* 332 (1995): 286–291; M. J. Stampfer and M. R. Malinow, Can lowering homocysteine levels reduce cardiovascular risk? *New England Journal of Medicine* 332 (1995): 328–329; H. I. Morrison and coauthors, Serum folate and risk of fatal coronary heart disease, *Journal of the American Medical Association* 275 (1996): 1893–1896.
12. S. Seshadri and coauthors, Plasma homocysteine as a risk factor for dementia and Alzheimer's disease, *New England Journal of Medicine* 346 (2002): 466; R. LeBoeuf, Homocysteine and Alzheimer's disease, *Journal of the American Dietetic Association* 103 (2003): 304–307; I. H. Rosenberg, B vitamins, homocysteine, and neurocognitive functions, *Nutrition Reviews* 59 (2001): S69–S74; M. S. Morris, Folate, homocysteine, and neurological function, *Nutrition in Clinical Care* 5 (2002): 124–132.
13. S. P. Stabler and R. H. Allen, Vitamin B_{12} deficiency as a worldwide problem, *Annual Review of Nutrition* 24 (2004): 299–326.
14. R. A. Jacob and G. Sotoudeh, Vitamin C function and status in chronic disease, *Nutrition in Clinical Care* 5 (2002): 66–74; S. J. Padayatty and coauthors, Vitamin C as an antioxidant: Evaluation of its role in disease prevention, *Journal of the American College of Nutrition* 22 (2003): 18–35.
15. L. C. Pauling, *Vitamin C and the Common Cold* (San Francisco: W. H. Freeman, 1970); H. Hemilia, Vitamin C supplementation and common cold symptoms: Factors affecting the magnitude of the benefit, *Medical Hypotheses* 52 (1999): 171–178; C. Audera and coauthors, Mega-dose vitamin C in treatment of the common cold: A randomized controlled trial, *Oncology* 175 (2001): 359–362.
16. E. J. Johnson, The role of the carotenoids in human health, *Nutrition in Clinical Care* 5 (2002): 56–65; Age Related Eye Disease Research Group, A randomized, placebo-controlled clinical trial of high dose supplementation with vitamins C and E, beta carotene, and zinc for age-related macular degeneration and vision loss, AREDS Report No. 8, *Archives Ophthalmology* 119 (2001): 1417–1436; Age Related Eye Disease Research Group, A randomized, placebo-controlled clinical trial of high dose supplementation with vitamins C and E and beta carotene for age-related cataract and vision loss, AREDS Report No. 9, *Archives Ophthalmology* 119 (2001): 1439–1452.
17. M. F. Holic, Sunlight and vitamin D for bone health and prevention of autoimmune diseases, cancers, and cardiovascular disease, *American Journal of Clinical Nutrition* 80 (2004): 1678S–1688S; M. S. Calvo and S. J. Whiting, Prevalence of vitamin D insufficiency in Canada and the United States: Importance to health status and efficacy of current food fortification and dietary supplement use, *Nutrition Reviews* 61 (2003): 107–113; C. Moore and coauthors, Vitamin D intake in the United States, *Journal of the American Dietetic Association* 104 (2004): 980–983.
18. W. Mertz, A balanced approach to nutrition for health: The need for biologically essential minerals and vitamins, *Journal of the American Dietetic Association* 94 (1994): 1259–1262.
19. P. S. Connolly, Treatment of nocturnal leg cramps, *Archives of Internal Medicine* 152 (1992): 1877–1880.
20. L. Mosca and coauthors, Antioxidant nutrient supplementation reduces the susceptibility of low density lipoprotein to oxidation in patients with coronary artery disease, *Journal of the American College of Cardiology* 30 (1997): 392–399; P. M. Kris-Etherton and coauthors, Antioxidant vitamin supplements and cardiovascular disease, *Circulation* 110 (2004): 637– 641.
21. J. Blumberg, An update: Vitamin E supplementation and heart disease, *Nutrition in Clinical Care* 5 (2002): 50–55;
22. E. R. Miller and coauthors, Meta-analysis: High dosage vitamin E supplementation may increase all-cause mortality, *Annals of Internal Medicine* 142 (2005).
23. S. L. Booth and coauthors, Vitamin K intake and bone mineral density in women and men, *American Journal of Clinical Nutrition* 77 (2003): 512–516; P. Weber, The role of vitamins in the prevention of osteoporosis: A brief status report, *International Journal for Vitamin and Nutrition Research* 69 (1999): 194–197.
24. J. Hirsh and V. Fuster, Guide to anticoagulant therapy, part 2: Oral anticoagulants, *Circulation* 89 (1994): 1473.
25. Food and Nutrition Board, Institute of Medicine, *Dietary Reference Intakes for Thiamin, Riboflavin, Niacin, Vitamin B_6, Folate, Vitamin B_{12}, Pantothenic Acid, Biotin, and Choline* (Washington, DC: National Academy Press, 1998).
26. S. J. VanGarde and M. Woodburn, *Food Preservation and Safety* (Ames, Iowa: Iowa State University Press, 1994), 109.
27. C. M. Hasler, Functional foods: Benefits, concerns, and challenges—A position paper from the American Council on Science and Health, *Journal of Nutrition* 132 (2002): 3772–3781; Parts of this discussion are adapted from Functional foods: Opening the door to better health, *Food Insight* (November/December 1995).
28. W. H. S. Jones, ed., *Hippocrates* (1932): 351.
29. Position of the American Dietetic Association: Functional foods, *Journal of the American Dietetic Association* 104 (2004): 814–826; K. Klotzbach-Shimomura, Functional foods: The role of physiologically active compounds in relation to disease, *Topics in Clinical Nutrition* 16 (2001): 68–78.
30. J. W. Anderson and coauthors, Meta-analysis of the effects of soy protein intake on serum lipids, *New England Journal of Medicine* 333 (1995): 276–282; M. Messina and S. Barnes, The role of soy products in reducing risk of cancer, *Journal of the National Cancer Institute* 83 (1991): 541–546; B. H. Arjmandi and coauthors, Dietary soybean protein presents bone loss in an ovariectomized rat model of osteoporosis, *Journal of Nutrition* 126 (1996): 162–167; P. Albertazzi and coauthors, The effect of dietary soy supplementation on hot flushes, *Obstetrics and Gynecology* 91 (1997): 6–11.
31. J. Raloff, The good *trans* fat: Will one family of animal fats become a medicine? *Science News* 159 (2001): 136.
32. C. L. Taylor, Regulatory frameworks for functional foods and dietary supplements, *Nutrition Reviews* 62 (2004): 55–59.
33. University of Illinois Functional Foods for Health Program, www.ag.uiuc.edu.
34. N. R. Farnsworth and coauthors, Medicinal plants in therapy, *Bulletin of the World Health Organization* 63 (1985): 965–1170.
35. N. R. Farnsworth, The role of medicinal plants in drug development, chapter in *Natural Products and Drug Development*, P. Krogsgaard-Larsen, S. Brogger Christensen, and H. Kofod, eds., Proceedings of Alfred Benzon Symposium 20 (Munksgaard: Copenhagen, 1984), 17–30.
36. P. Goldman, Herbal medicines today and the roots of modern pharmacology, *Annals of Internal Medicine* 135 (2001): 594–600; D. M. Marcus and A. P. Grollman, Botanical medicines: The need for new regulations, *New England Journal of Medicine* 347 (2002): 2073–2076.
37. P. Brevort, The economics of botanicals: The U.S. experience. Presentation at the NIH/OAM Conference on Botanicals: A Role in U.S. Health Care? Washington, DC, December 16, 1994; L. G. Tolstoi, Herbal remedies: Buyer beware! *Nutrition Today* 36 (2001): 223–230.
38. G. B. Mahady, Herbal remedies: The promise scrutinized by science, Program for Collaborative Research in the Pharmaceutical Sciences, College of Pharmacy, University of Illinois at Chicago, April 1999.
39. World Cancer Research Fund and American Institute for Cancer Research, *Food, Nutrition, and the Prevention of Cancer: A Global Perspective* (Washington, DC: American Institute for Cancer Research, 1997), 506–507; P. Lichtenstein and coauthors, Environmental and heritable factors in the causation of cancer—Analyses of cohorts of twins from Sweden, Denmark, and Finland, *New England Journal of Medicine* 343 (2000): 78–81; National Cancer Policy Board, Institute of Medicine, *Fulfilling the Potential of Cancer Prevention and Early Detection* (Washington, DC: National Academies Press, 2003); G. N. Wogan and coauthors, Environmental and chemical carcinogenesis, *Seminars in Cancer Biology* 14 (2004): 473–486.

40. E. Giovannucci and B. Goldin, The role of fat, fatty acids, and total energy intake in the etiology of human colon cancer, *American Journal of Clinical Nutrition* 66 (1997): S1564–S1571; E. Giovannucci, Diet, body weight, and colorectal cancer: A summary of the epidemiologic evidence, *Journal of Women's Health* 12 (2003): 173–182.

41. E. Giovannucci and coauthors, Intake of fat, meat, and fiber in relation to risk of colon cancer in men, *Cancer Research* 54 (1994): 2390–2397.

42. C. S. Fuchs and coauthors, Dietary fiber and the risk of colorectal cancer and adenoma in women, *New England Journal of Medicine* 340 (1999): 169–176.

43. W. C. Willett and coauthors, Relation of meat, fat, and fiber intake to the risk of colon cancer in a prospective study among women, *New England Journal of Medicine* 323 (1990): 1664–1672.

44. American Gastroenterological Association, Medical position statement: Impact of dietary fiber on colon cancer occurrence, *Gastroenterology* 118 (2000): 1233–1234; B. S. Reddy, Role of dietary fiber in colon cancer: An overview, *American Journal of Medicine* 106 (1999): S16–S19; M. C. Jansen and coauthors, Dietary fiber and plant foods in relation to colorectal cancer mortality: The Seven Countries Study, *International Journal of Cancer* 81 (1999): 174–179.

45. E. Giovannucci, Epidemiologic studies of folate and colorectal neoplasia: A review, *Journal of Nutrition* 132 (2002): 2350S–2354S; M. Ferraroni and coauthors, Selected micronutrient intake and the risk of colorectal cancer, *British Journal of Cancer* 70 (1994): 1150–1155; P. R. Holt and coauthors, Modulation of abnormal colonic epithelial cell proliferation and differentiation by low-fat dairy foods, *Journal of the American Medical Association* 280 (1998): 1074–1079.

46. K. K. Carroll, Dietary fats and cancer, *American Journal of Clinical Nutrition* 53 (1991): 1064S–1067S; C. W. Welsch, Dietary fat, calories, and mammary gland tumorigenesis, in *Advances in Experimental Medicine and Biology* (New York: Plenum Press, 1992), 203–221.

47. P. L. Zock, Dietary fats and cancer, *Current Opinions in Lipidology* 12 (2001): 5–10; P. Toniolo and coauthors, Consumption of meat, animal products, protein, and fat and risk of breast cancer: A prospective cohort study in New York, *Epidemiology* 5 (1994): 391–397.

48. K. Katsouyanni and coauthors, The association of fat and other macronutrients with breast cancer: A case-controlled study from Greece, *British Journal of Cancer* 70 (1994): 537–541.

49. Xiu-Ying Qi and coauthors, The association between breast cancer and diet and other factors, *Asia Pacific Journal of Public Health* 7 (1994): 98–104.

50. M. D. Holmes and coauthors, Association of dietary intake of fat and fatty acids with risk of breast cancer, *Journal of the American Medical Association* 281 (1999): 914–920; J. B. Magen, The relationship between obesity and breast cancer risk and mortality, *Nutrition Reviews* 61 (2003): 73–76; E. E. Calle and M. J. Thun, Obesity and cancer, *Oncogene* 23 (2004): 6365–6378.

51. American Institute for Cancer Research, *Diet and Health Recommendations for Cancer Prevention* (Washington, DC: American Institute for Cancer Research, 1998); American Cancer Society guidelines on nutrition and physical activity for cancer prevention: Reducing the risk of cancer with healthy food choices and physical activity, *CA—A Cancer Journal for Clinicians* 52 (2002): 92–118.

52. G. A. Bray, The underlying basis for obesity: Relationship to cancer, *Journal of Nutrition* 132 (2002): 3451S–3455S; C. M. Friedenreich and M. R. Orenstein, Physical activity and cancer prevention: Etiologic evidence and biological mechanisms, *Journal of Nutrition* 132 (2002): 3456S–3466S; J. Verloop and coauthors, Physical activity and breast cancer risk in women aged 20 to 54 years, *Journal of the National Cancer Institute* 92 (2000): 128–135; B. J. Caan and coauthors, Body size and the risk of colon cancer in a large case-control study, *International Journal of Obesity and Related Metabolic Disorders* 22 (1998): 178–184; L. C. Yong and coauthors, Prospective study of relative weight and risk of breast cancer: The breast cancer detection demonstration project follow-up study, 1979 to 1987–1989, *American Journal of Epidemiology* 143 (1996): 985–995; Z. Huang and coauthors, Dual effects of weight and weight gain on breast cancer risk, *Journal of the American Medical Association* 278 (1997): 1407–1411.

53. K. A. Steinmetz and J. D. Potter, Vegetables, fruit, and cancer prevention: A review, *Journal of the American Dietetic Association* 96 (1996): 1027–1039.

54. G. Block, Vitamin C and cancer prevention: The epidemiologic evidence, *American Journal of Clinical Nutrition* 53 (1991): 270S–282S.

55. P. Talalay and J. W. Fahey, Phytochemicals from cruciferous plants protect against cancer by modulating carcinogen metabolism, *Journal of Nutrition* 131 (2001): 3027S; W. J. Craig, Phytochemicals: Guardians of our health, *Journal of the American Dietetic Association* 97 (1997): S199–S204; Z. Djuric and coauthors, Oxidative DNA damage levels in blood from women at high risk for breast cancer are associated with dietary intakes of meats, vegetables, and fruits, *Journal of the American Dietetic Association* 98 (1998): 524–528.

56. K. A. Steinmetz and J. D. Potter, Vegetables, fruit, and cancer prevention: A review, *Journal of the American Dietetic Association* 96 (1996): 1027–1039.

57. L. C. Clark and coauthors, Reducing cancer risk with selenium, *Journal of the American Medical Association* 276 (1995): 1957–1963.

58. P. Knekt and coauthors, Vitamin E and cancer prevention, *American Journal of Clinical Nutrition* 53 (1991): 283S–286S.

59. J. A. Baron and coauthors, Calcium supplements for the prevention of colorectal adenomas, *New England Journal of Medicine* 340 (1999): 101–107.

60. M. A. Rogers and coauthors, Consumption of nitrate, nitrite, and nitrosodimethylamine and the risk of upper aerodigestive tract cancer, *Cancer Epidemiology, Biomarkers, and Prevention* 4 (1995): 29–36; H. Chen and coauthors, Dietary patterns and adenocarcinoma of the esophagus and distal stomach, *American Journal of Clinical Nutrition* 75 (2002): 137–141.

61. M. Messina and S. Barnes, The role of soy products in reducing risk of cancer, *Journal of the National Cancer Institute* 83 (1991): 541–546.

62. S. A. Smith-Warner and coauthors, Alcohol and breast cancer in women: A pooled analysis of cohort studies, *Journal of the American Medical Association* 279 (1998): 535–540.

63. G. Howe and coauthors, The association between alcohol and breast cancer risk: Evidence from the combined analysis of six dietary case-control studies, *International Journal of Cancer* 47 (1991): 707–710; L. Holmberg and coauthors, Diet and breast cancer risk: Results from a population-based, case-control study in Sweden, *Archives of Internal Medicine* 154 (1994): 1805–1811; J. L. Freudenheim and coauthors, Lifetime alcohol consumption and risk of breast cancer, *Nutrition and Cancer* 23 (1995): 1–11.

64. E. Giovannucci, Tomatoes, tomato-based products, lycopene, and cancer: Review of the epidemiologic literature, *Journal of the National Cancer Institute* 91 (1998): 317–331.

65. A. Bloch and C. A. Thompson, Position of the American Dietetic Association: Phytochemicals and functional foods, *Journal of the American Dietetic Association* 95 (1995): 493–496; Position of the American Dietetic Association: Functional foods, *Journal of the American Dietetic Association* 104 (2004): 814–826; A. S. Bloch, Phytochemicals and functional foods for cancer risk reduction, *Topics in Clinical Nutrition* 15 (2000): 24–28.

66. J. Childers and coauthors, Chemoprevention of cervical cancer with folic acid: A Phase III Southwest Oncology Group Intergroup Study, *Cancer Epidemiology, Biomarkers, and Prevention* 4 (1995): 155–159.

67. American Cancer Society 2001 Nutrition and Physical Activity Guidelines Advisory Committee, *American Cancer Society Guidelines on Nutrition and Physical Activity for Cancer*, www.cancer.org.

68. National Cancer Policy Board, Institute of Medicine, *Fulfilling the Potential of Cancer Prevention and Early Detection* (Washington, DC: National Academies Press, 2003).

69. J. Dwyer, Diet and nutritional strategies for cancer risk reduction: Focus on the 21st century, *Cancer* 72 (1993): 1024–1031.

70. Committee on Comparative Toxicity of Naturally Occurring Carcinogens, *Carcinogens and Anticarcinogens in the Human Diet* (Washington, DC: National Academy Press, 1996).

CHAPTER 7

1. C. M. McCay, Anorganic substances, in *Notes on the History of Nutrition Research,* ed. F. Verzar (Vienna: Hans Huber Publishers, 1973), 156–184.
2. Position of the American Dietetic Association, Dietitians of Canada, and the American College of Sports Medicine: Nutrition and athletic performance, *Journal of the American Dietetic Association* 100 (2000): 1543–1556.
3. S. M. Kleiner, Water: An essential but overlooked nutrient, *Journal of the American Dietetic Association* 99 (1999): 200–206; Standing Committee on the Scientific Evaluation of Dietary Reference Intakes, Food and Nutrition Board, Institute of Medicine, *Dietary Reference Intakes: Water, Potassium, Sodium, Chloride, and Sulfate* (Washington, DC: National Academy Press, 2004), 4–26.
4. M. P. Sauvant and D. Pepin, Geographic variation of the mortality from cardiovascular disease and drinking water in a French small area (Puy de Dome), *Environmental Research* 84 (2000): 219–227; R. Maheswaran and coauthors, Magnesium in drinking water supplies and mortality from acute myocardial infarction in northwest England, *Heart* 82 (1999): 455–460; C. Y. Yang and H. F. Chiu, Calcium and magnesium in drinking water and the risk of death from hypertension, *American Journal of Hypertension* 12 (1999): 894–899.
5. U.S. Environmental Protection Agency and Centers for Disease Control and Prevention, *Guidance for People with Severely Weakened Immune Systems* (Washington, DC: U.S. Environmental Protection Agency, June 1999).
6. J. F. Rosen and P. Mushak, Primary prevention of childhood lead poisoning—The only solution, *New England Journal of Medicine* 344 (2001): 1470–1471; National Center for Environmental Health, CDC Childhood Lead Poisoning Prevention Program, www.cdc.gov/nceh/lead/lead.htm.
7. Sales of bottled water show no signs of drying up, *The Food Institute Report* (October 29, 2001): 4.
8. L. M. Posnick and H. Kim, Bottled water regulation and the FDA, *Food Safety Magazine* (August/September 2002).
9. A. C. Bullers, Bottled water: Better than the tap? *FDA Consumer Magazine* 36, no. 4 (2002).
10. J. Stannard and coauthors, Fluoride content of some bottled waters and recommendation for fluoride supplementation, *Journal of Pedodontics* 14 (1990): 103–107; Position of the American Dietetic Association: The impact of fluoride on health, *Journal of the American Dietetic Association* 105 (2005): 1620–1628.
11. Committee on Dietary Reference Intakes, *Dietary Reference Intakes for Calcium, Phosphorus, Magnesium, Vitamin D, and Fluoride* (Washington, DC: National Academy Press, 1997).
12. R. P. Heaney and coauthors, Food factors influencing calcium availability, in *Nutritional Aspects of Osteoporosis,* eds., P. Burckharat and R. P. Heaney, Proceedings of the 2nd International Symposium on Osteoporosis, Lausanne, Switzerland, May 1994 (New York: Raven Press, 1995).
13. G. Wyshak and R. E. Frisch, Carbonated beverages, dietary calcium, the dietary calcium-phosphorus ratio, and bone fractures in girls and boys, *Journal of Adolescent Health* 15 (1994): 210–215; K. L. Tucker, Does milk intake in childhood protect against later osteoporosis? *American Journal of Clinical Nutrition* 77 (2003): 10–11; C. D. Frary, R. K. Johnson, and M. Q. Wang, Children and adolescents' choice of foods and beverages high in added sugars are associated with intakes of key nutrients and food groups, *Journal of Adolescent Health* 34 (2004): 56–63.
14. E. Rubenowitz and coauthors, Magnesium in drinking water and death from myocardial infarction, *American Journal of Epidemiology* 143 (1996): 456–462; L. A. Martini and R. J. Wood, Assessing magnesium status: A persisting problem, *Nutrition in Clinical Care* 4 (2001): 332–337.
15. Standing Committee on the Scientific Evaluation of Dietary Reference Intakes, Food and Nutrition Board, Institute of Medicine, *Dietary Reference Intakes: Water, Potassium, Sodium, Chloride, and Sulfate* (Washington, DC: National Academy Press, 2004), 4–26.
16. Position of the American Dietetic Association and Dietitians of Canada: Vegetarian diets, *Journal of the American Dietetic Association* 103 (2003): 748–765.
17. J. B. Mason and coauthors, *The Micronutrient Report: Current Progress and Trends in the Control of Vitamin A, Iron, and Iodine Deficiencies* (Ottawa, Ontario: The Micronutrient Initiative, 2001).
18. Centers for Disease Control and Prevention, Iron deficiency—United States, 1999–2000, *Morbidity and Mortality Weekly Report* 51 (2002): 897–899; U.S. Department of Health and Human Services, Public Health Service, *Healthy People 2010: Understanding and Improving Health* (Washington, DC: U.S. Government Printing Office, 2000); E. M. Ross, Evaluation and treatment of iron deficiency in adults, *Nutrition in Clinical Care* 5 (2002): 220–224.
19. P. M. Suter, C. Sierro, and W. Vetter, Nutritional factors in the control of blood pressure and hypertension, *Nutrition in Clinical Care* 5 (2002): 9–19; A. Avic, Salt and hypertension, *Archives of Internal Medicine* 161 (2001): 507–509.
20. S. A. Corrigan and coauthors, Weight reduction in the prevention and treatment of hypertension: A review of representative clinical trials, *American Journal of Health Promotion* 5 (1991): 208–214; P. K. Whelton and coauthors, Sodium reduction and weight loss in the treatment of hypertension in older persons, *Journal of the American Medical Association* 279 (1998): 839–846.
21. The Joint National Committee on Prevention, Detection, Evaluation, and Treatment of High Blood Pressure, The Seventh Report of the Joint National Committee on Prevention, Detection, Evaluation, and Treatment of High Blood Pressure, *Journal of the American Medical Association* 289 (2003): 2560–2572.
22. P. M. Suter, C. Sierro, and W. Vetter, Nutritional factors in the control of blood pressure and hypertension, *Nutrition in Clinical Care* 5 (2002): 9–19.
23. Joint National Committee, 2003; P. M. Suter and W. Vetter, The effect of alcohol on blood pressure, *Nutrition in Clinical Care* 3 (2000): 24–34.
24. Joint National Committee, 2003.
25. R. Stamler and coauthors, Primary prevention of hypertension by nutrional-hygienic means, *Journal of the American Medical Association* 262 (1989): 1801–1807.
26. T. Kotchen and D. McCarron, Dietary electrolytes and blood pressure: A statement for healthcare professionals from the American Heart Association, *Circulation* 98 (1998): 613–617; T. A. Kotchen and J. M. Kotchen, Dietary sodium and blood pressure: Interactions with other nutrients, *American Journal of Clinical Nutrition* 65 (1997): 708S–711S.
27. P. Suter, Potassium and hypertension, *Nutrition Reviews* 56 (1998): 151–153.
28. C. G. Osborne and coauthors, Evidence for the relationship of calcium to blood pressure, *Nutrition Reviews* 54 (1996): 365–381; L. M. Resnick, The role of dietary calcium in hypertension: A hierarchical overview, *American Journal of Hypertension* 12, part 1 (1999): 99–112; D. A. McCarron and M. E. Reusser, Finding consensus in the dietary calcium blood pressure debate, *Journal American College of Nutrition* 18 (1999): 398S–405S.
29. Joint National Committee, 2003.
30. The DASH clinical trial, *New England Journal of Medicine* 336 (1997): 1117; F. M. Sacks and coauthors, Effects on blood pressure of reduced dietary sodium and the dietary approaches to stop hypertension (DASH) diet, *New England Journal of Medicine* 334 (2001): 3–10.
31. J. D. Cook and M. B. Reddy, Effect of ascorbic acid intake on nonheme iron absorption from a complete diet, *American Journal of Clinical Nutrition* 73 (2001): 93–98; M. B. Reddy, R. F. Hurrell, and C. D. Cook, Estimation of nonheme iron bioavailability from meal composition, *American Journal of Clinical Nutrition* 71 (2000): 937–943.
32. E. V. M. Borigato and F. E. Martinez, Iron nutritional status is improved in Brazilian preterm infants fed food cooked in iron pots, *The Journal of Nutrition* 128 (1998): 855–859.
33. J. T. Salonen and coauthors, High stored iron levels are associated with excess risk of myocardial infarction in Eastern Finnish men, *Circulation* 86 (1992): 803–811; K. Klipstein-Grobusch and coauthors,

Dietary iron and risk of myocardial infarction in the Rotterdam Study, *American Journal of Epidemiology* 149 (1999): 421–428; Food and Nutrition Board, Institute of Medicine, *Dietary Reference Intakes for Vitamin A, Vitamin K, Arsenic, Boron, Chromium, Copper, Iodine, Iron, Manganese, Molybdenum, Nickel, Silicon, Vanadium, and Zinc* (Washington, DC: National Academy Press, 2001), 357–378.

34. A. H. Shankar and A. S. Prasad, Zinc and immune function: The biological basis to altered resistance to infection, *American Journal of Clinical Nutrition* 69 (1998): S447–S463; N. W. Solomons, Mild human zinc deficiency produces an imbalance between cell-mediated and humoral immunity, *Nutrition Reviews* 56 (1998): 27–32; R. Bahl and coauthors, Plasma zinc as a predictor of diarrheal and respiratory morbidity in children in an urban slum setting, *American Journal of Clinical Nutrition* 68 (1998): 414S–417S.

35. Zinc lozenges reduce the duration of common cold symptoms, *Nutrition Reviews* 55 (1997): 82–88.

36. Food and Nutrition Board, Institute of Medicine, *Dietary Reference Intakes for Vitamin A, Vitamin K, Arsenic, Boron, Chromium, Copper, Iodine, Iron, Manganese, Molybdenum, Nickel, Silicon, Vanadium, and Zinc* (Washington, DC: National Academy Press, 2001).

37. J. C. King and C. L. Keen, Zinc, in M. E. Shils and J. A. Olson, eds., *Modern Nutrition in Health and Disease*, 10th ed. (Baltimore, MD: Williams and Wilkins, 2005).

38. J. A. Pennington, A review of iodine toxicity reports, *Journal of the American Dietetic Association* 90 (1990): 1571–1581.

39. K. Lee and coauthors, Too much versus too little: The implications of current iodine intake in the United States, *Nutrition Reviews* 57 (1999): 177–181.

40. J. G. Hollowell and coauthors, Iodine nutrition in the United States. Trends and public health implications: Iodine excretion data from National Health and Nutrition Examination Surveys I and III, *Journal of Clinical Endocrinology and Metabolism* 83 (1998): 3401–3408.

41. Position of the American Dietetic Association: The impact of fluoride on health, *Journal of the American Dietetic Association* 105 (2005): 1620–1628.

42. Centers for Disease Control and Prevention, Recommendations for using fluoride to prevent and control dental caries in the United States, *Morbidity and Mortality Weekly Report* 50 (2001): 1–42.

43. B. J. Stoecker, Chromium, in M. E. Shils and coauthors, *Modern Nutrition in Health and Disease*, 10th ed. (Baltimore, MD: Williams and Wilkins, 2005); Food and Nutrition Board, Institute of Medicine, 2001.

44. D. H. Holben and A. M. Smith, The diverse role of selenium within selenoproteins: A review, *Journal of the American Dietetic Association* 99 (1999): 836–843; K. Yoshizawa and coauthors, Study of prediagnostic selenium level in toenails and the risk of advanced prostate cancer, *Journal of the National Cancer Institute* 90 (1998): 1219–1224; Food and Nutrition Board, Institute of Medicine, *Dietary Reference Intakes for Vitamin C, Vitamin E, Selenium, and Carotenoids* (Washington, DC: National Academy Press, 2000); R. F. Burk, Selenium, an antioxidant nutrient, *Nutrition in Clinical Care* 5 (2002): 75–79.

45. Food and Nutrition Board, Institute of Medicine, 2001; S. Meacham and coauthors, Effect of boron supplementation on blood and urinary calcium, magnesium, and phosphorus, and urinary boron in athletic and sedentary women, *American Journal of Clinical Nutrition* 61 (1995): 341–345.

46. Position of the American Dietetic Association: Food fortification and dietary supplementation, *Journal of the American Dietetic Association* 101 (2001): 115–125.

47. The opening vignettes are from P. Ola and E. D'Aulaire, The health risk women can no longer ignore, *Reader's Digest* (August 1994): 91–95.

48. NIH Consensus Development Panel, Osteoporosis prevention, diagnosis, and therapy, *Journal of the American Medical Association* 285 (2001): 785–795; L. W. Turner and coauthors, Osteoporotic fracture among older U.S. women, *Journal of Aging and Health* 10 (1998): 372–391; E. Siris and coauthors, Design of NORA, the National Osteoporosis Risk Assessment Program, A longitudinal U.S. Registry of postmenopausal women, *Osteoporosis International* 8 (1998): S62–S69.

49. T. A. Ricci and coauthors, Calcium supplementation suppresses bone turnover during weight reduction in postmenopausal women, *Journal of Bone and Mineral Research* 13 (1998): 1045–1050.

50. National Center for Injury Prevention and Control, Falls and hip fractures among the elderly, *Unintentional Injury Fact Sheet*, 2003, www.cdc.gov.

51. Food and Nutrition Board, Institute of Medicine, 1997; V. Matkovic, Nutrition, genetics, and skeletal development, *Journal of the American College of Nutrition* 15 (1996): 556–569; G. M. Chan, K. Hoffman, and M. McMurray, Effects of dairy products on bone and body composition in pubertal girls, *Journal of Pediatrics* 128 (1995): 551–556.

52. J. B. Anderson and P. A. Rondano, Peak bone mass development of females: Can young adult women improve their peak bone mass? *Journal of the American College of Nutrition* 15 (1996): 570–574.

53. R. L. Smith and coauthors, Prevention of postmenopausal osteoporosis: A comparative study of exercise, calcium supplementation, and hormone replacement therapy, *New England Journal of Medicine* 325 (1991): 1189–1195.

54. F. H. Anderson, Osteoporosis in men, *International Journal of Clinical Practice* 52 (1998): 176–180.

55. M. L. Rencken and coauthors, Bone density at multiple skeletal sites in amenorrheic athletes, *Journal of the American Medical Association* 276 (1996): 238–240.

56. C. J. Strange, Boning up on osteoporosis, *FDA Consumer*, reprint (August 1997).

57. E. L. Smith and C. Gilligan, Effects of inactivity and exercise on bone, *Physician and Sportsmedicine* 15 (1997): 91–102.

58. M. N. Hadley and S. V. Reddy, Smoking and the human vertebral column: A review of the impact of cigarette use on vertebral bone metabolism and spinal fusion, *Neurosurgery* 41 (1997): 116–124; C. W. Slemenda, Cigarettes and the skeleton, *New England Journal of Medicine* 330 (1994): 430–431.

59. F. R. Greer and N. F. Krebs, Optimizing bone health and calcium intakes of infants, children, and adolescents, *Pediatrics* 117 (2006): 578–585; H. J. Kalkwarf, J. C. Khoury, and B. P. Lamphear, Milk intake during childhood and adolescence, adult bone density, and osteoporotic fractures in U.S. women, *American Journal of Clinical Nutrition* 77 (2003): 257–265; R. P. Heaney and coauthors, Dietary changes favorably affect bone remodeling in older adults, *Journal of the American Dietetic Association* 99 (1999): 1228–1233.

60. F. R. Greer and N. F. Krebs, Optimizing bone health and calcium intakes of infants, children, and adolescents, *Pediatrics* 117 (2006): 578–585.

61. D. Neumark-Sztainer and coauthors, Correlates of inadequate consumption of dairy products among adolescents, *Journal of Nutrition Education* 29 (1997): 12–20; L. Harnack, J. Stang, and M. Story, Soft drink consumption among U.S. children and adolescents: Nutritional consequences, *Journal of the American Dietetic Association* 99 (1999): 436–441; S. A. Bowman, Beverage choices of young females: Changes and impact on nutrient intakes, *Journal of the American Dietetic Association* 102 (2002): 1234–1239.

62. P. Weber, The role of vitamins in the prevention of osteoporosis: A brief status report, *International Journal for Vitamin and Nutrition Research* 69 (1999): 194–197; G. Ferland, Vitamin K-dependent proteins: An update, *Nutrition Reviews* 56 (1998): 223–230.

63. D. E. Sellmeyer and coauthors, A high ratio of dietary animal to vegetable protein increases the rate of bone loss and the risk of fracture in postmenopausal women, *American Journal of Clinical Nutrition* 73 (2001): 118–122; U. S. Barzel and L. K. Massey, Excess dietary protein can adversely affect bone, *Journal of Nutrition* 128 (1998): 1051–1053; S. J. Whiting and B. Lemke, Excess retinol intake may explain the high incidence of osteoporosis in northern Europe, *Nutrition Reviews* 57 (1999): 192–195; F. Ginty, A. Flynn, and K. D. Cashman, The effect of dietary sodium intake on biochemical markers of bone metabolism in young, *British Journal of Nutrition* 79 (1998): 343–350.

64. D. Feskanich and coauthors, Vitamin K intake and hip fractures in women: A prospective study, *American Journal of Clinical Nutrition* 69 (1999): 74–79.
65. K. L. Tucker and coauthors, Bone mineral density and dietary patterns in older adults: The Framingham Osteoporosis Study, *American Journal of Clinical Nutrition* 76 (2002): 245–252.
66. B. Dawson-Hughes and coauthors, Effect of calcium and vitamin D supplements on bone density in men and women 65 years of age or older, *New England Journal of Medicine* 337 (1997): 670–675; B. Dawson-Hughes, Calcium, vitamin D, and risk of osteoporosis in adults: Essential information for the clinician, *Nutrition in Clinical Care* 1 (1998): 63–70; K. M. Chiu, Efficacy of calcium supplements on bone mass in postmenopausal women, *The Journal of Gerontology* 54 (1999): 275–280.
67. J. R. Saltzman, Nutritionally significant changes in gastrointestinal functioning with aging, *Nutrition in Clinical Care* 1 (1998): 20–29.
68. C. C. Collins and M. A. Summa, Clinical significance of lead content in dietary calcium supplements, *Nutrition in Clinical Care* 1 (1998): 156–159.
69. Food and Drug Administration, FDA approves new labels for estrogen and estrogen with progestin therapies for postmenopausal women following review of Women's Health Initiative data, *FDA News* (January 8, 2003); L. Pachucki-Hyde, Cutting your risk of osteoporosis, *Diabetes Self-Management*, (May/June, 1998): 36–43; E. Barrett-Connor, Hormone replacement therapy, *British Medical Journal* 317 (1998): 457–461; R. Lindsay, The role of estrogen in the prevention of osteoporosis, *Endocrinology and Metabolism Clinics of North America* 27 (1998): 399–405; M. McClung and coauthors, Alendronate prevents postmenopausal bone loss in women without osteoporosis, *Annals of Internal Medicine* 128 (1998): 253–261.
70. J. W. Nieves and coauthors, Calcium potentiates the effect of estrogen and calcitonin on bone mass: Review and analysis, *American Journal of Clinical Nutrition* 67 (1998): 18–24.
71. P. Taxel, Osteoporosis: Detection, prevention, and treatment in primary care, *Geriatrics* 53 (1998): 22–40; NIH Consensus Development Panel, Osteoporosis prevention, diagnosis, and therapy, *Journal of the American Medical Association* 285 (2001): 785–795.
72. Top ten advances of 1993, *Harvard Health Letter* 19, no. 5 (1994): 7.
73. R. D. Lewis and C. M. Modlesky, Nutrition, physical activity, and bone health in women, *International Journal of Sports Nutrition* 8 (1998): 250–284; T. V. Nguyen and coauthors, Bone loss, physical activity, and weight change in elderly women: The Dubbo Osteoporosis Epidemiology Study, *Journal of Bone Mineral Research* 13 (1998): 1458–1467.
74. R. L. Prince and coauthors, Prevention of postmenopausal osteoporosis: A comparative study of exercise, calcium supplementation, and hormone replacement therapy, *New England Journal of Medicine* 325 (1991): 1189–1195; E. Ernst, Exercise for female osteoporosis: A systematic review of randomized clinical trials, *Sports Medicine* 25 (1998): 359–368.
75. N. K. Henderson and coauthors, The roles of exercise and fall risk reduction in the prevention of osteoporosis, *Endocrinology and Metabolism Clinics of North America* 27 (1998): 369–387.

CHAPTER 8

1. C. C. Freudenrich, *How Alcohol Works*, www.howstuffworks.com/alcohol2.htm.
2. W. F. Bosron, T. Ehrig, and T. K. Li, Genetic factors in alcohol metabolism and alcoholism, *Seminars in Liver Disease* 13, no. 2 (1993): 126–135; H. Wallgren, Absorption, diffusion, distribution and elimination of ethanol: Effect on biological membranes, in *International Encyclopedia of Pharmacology and Therapeutics*. Vol. 1 (Oxford: Pergamon, 1970), 161–188.
3. C. C. Freudenrich, *How Alcohol Works*, National Institute on Alcohol Abuse and Alcoholism, *Alcohol Alert* No. 35, PH 371 (Bethesda, MD: National Institute on Alcohol Abuse and Alcoholism, 1997), www.niaaa.nih.gov/publications/aa35-text.htm.
4. Bosron and coauthors, Genetic factors in alcohol metabolism; L. Z. Benet, D. L. Kroetz, and L. B. Sheiner, Pharmacokinetics: The dynamics of drug absorption, distribution, and elimination, in P. B. Molinoff and R. W. Ruddon, eds., *Goodman and Gillman's The Pharmacological Basis of Therapeutics*, 9th ed. (New York: McGraw-Hill, 1996), 3–27.
5. C. C. Freudenrich, *How Alcohol Works;* National Institute on Alcohol Abuse and Alcoholism, *Alcohol Alert* No. 35, PH 371 (Bethesda, MD: National Institute on Alcohol Abuse and Alcoholism, 1997).
6. C. C. Freudenrich, *How Alcohol Works*.
7. C. C. Freudenrich, *How Alcohol Works*.
8. W. F. Bosron, T. Ehrig, and T. K. Li, Genetic factors in alcohol metabolism and alcoholism. *Seminars in Liver Disease* 13, no. 2 (1993): 126–135; H. Wallgren, Absorption, diffusion, distribution and elimination of ethanol: Effect on biological membranes, in *International Encyclopedia of Pharmacology and Therapeutics*. Vol. 1 (Oxford: Pergamon, 1970), 161–188; A. G. Fraser and coauthors, Individual and intra-individual variability of ethanol concentration-time profiles: Comparison of ethanol ingestion before and after an evening meal, *British Journal of Clinical Pharmacology* 40 (1996): 387–392.
9. C. C. Freudenrich, *How Alcohol Works; Drinking: A Student's Guide:* "Alcohol Facts and Stats," and *Drinking: A Student's Guide:* "Alcohol Risk Reduction," www.mcneese.edu/community/alcohol.stats.html; U.S. Department of Transportation, National Highway Traffic Safety Administration, National Center for Statistics and Analysis, Traffic Safety Facts, 2004 data, www.nrd.nhtsa.dot.gov.
10. A. Urbano-Márquez and coauthors, The greater risk of alcoholic cardiomyopathy and myopathy in women compared with men, *Journal of the American Medical Association* 274 (1995):149–154.
11. S. J. Nixon, Cognitive deficits in alcoholic women, *Alcohol Health and Research World* 18, no. 3 (1994): 228–232.
12. National Institute on Alcohol Abuse and Alcoholism, *Alcohol Alert: Alcohol and Women*, No. 10, PH 290 (Bethesda, MD: National Institute on Alcohol Abuse and Alcoholism, 1990), www.niaaa.nih.gov/publications/aa10.htm.
13. C. C. Freudenrich, *How Alcohol Works;* National Institute on Alcohol Abuse and Alcoholism, *Alcohol Alert* No. 35, PH 371 (Bethesda, MD: National Institute on Alcohol Abuse and Alcoholism, 1997).
14. R. Spendler, *WebMD Health Guide: Blood Alcohol*, www.webmd.com/content/healthwise/143/35577.htm?printing=true.
15. C. S. Lieber, Metabolic consequence of ethanol, *Endocrinologist* 4, no. 2 (1994): 127–139; National Institute on Alcohol Abuse and Alcoholism, *Alcohol Alert: Alcohol Medication Interactions* No. 27, PH 290 (Bethesda, MD: National Institute on Alcohol Abuse and Alcoholism, 1995), www.niaaa.nih.gov/publications/aa27.htm; M. Black, Acetaminophen hepatotoxicity, *Annual Review of Medicine* 35 (1984): 577–593.
16. National Institute on Alcohol Abuse and Alcoholism. *Alcohol Alert: Alcohol Medication Interactions* No. 27, PH 290 (Bethesda, MD: National Institute on Alcohol Abuse and Alcoholism, 1995); L. B. Seeff and coauthors, Acetaminophen hepatotoxicity in alcoholics: A therapeutic misadventure, *Annals of Internal Medicine* 104 (1986): 399–404.
17. S. Andersson and coauthors, Effects of alcohol metabolism, *Alcoholism: Clinical and Experimental Research* 10, no. 6 (1986): 55S–63S; H. I. Wright, J. S. Gavaler, and D.Van Thiel, Effects of alcohol on the male reproductive system, *Alcohol Health & Research World* 15, no. 2 (1991): 110–114; T. J. Cicero and R. D. Bell, Effects of ethanol and acetaldehyde on the biosynthesis of testosterone in the rodent testes, *Biochemical and Biophysical Research Communications* 94, no. 3 (1980): 814–819; D. E. Johnston, Inhibition of testosterone synthesis by ethanol and acetaldehyde, *Biochemical Pharmacology* 30, no. 13 (1981): 1827–1830; Y. B. Chiao and D. H. Van Thiel, Biochemical mechanisms that contribute to alcohol-induced hypogonadism in the male, *Alcoholism: Clinical and Experimental Research* 7, no. 2 (1983): 131–134; N. K. Mello, J. H. Mendelson, and S. K. Tech, An overview of the effects of alcohol on neuroendrocrine function in women, in S. Zakhari, ed., *Alcohol and the Endocrine System*, National Institute on Alcohol Abuse and Alcoholism Research Monograph No. 23, NIH Publication No. 93–3533 (Bethesda, MD: National Institute on Alcohol Abuse and Alcoholism, 1993), 139–169.
18. H. I. Wright, J. S. Gavaler, and D. Van Thiel, Effects of alcohol on the male reproductive system, *Alcohol Health & Research World* 15, no. 2

(1991): 110–114; D. H. Van Thiel, J. Gavaler, and R. Leser, Ethanol inhibition of vitamin A metabolism in the testes: Possible mechanism for sterility in alcoholics, *Science* 186 (1974): 941–942.

19. P. Bannister and M. S. Lowosky, Ethanol and hypogonadism, *Alcohol and Alcoholism* 22, no. 3 (1987): 213–217.
20. C. C. Freudenrich, *How Alcohol Works;* National Institute on Alcohol Abuse and Alcoholism, *Alcohol Alert* No. 35, PH 371 (Bethesda, MD: National Institute on Alcohol Abuse and Alcoholism, 1997).
21. C. C. Freudenrich, *How Alcohol Works.*
22. D. M. Goldberg, Health effects of moderate alcohol consumption, *Patient Care* 7, no. 5 (1996): 56–73.
23. M. J. Thun and coauthors, Alcohol consumption and mortality among middle-aged and elderly U.S. adults, *New England Journal of Medicine* 337 (1997): 1705–1714.
24. Ibid.
25. E. B. Rimm and coauthors, Review of moderate alcohol consumption and reduced risk of coronary heart disease: Is the effect due to beer, wine, or spirits? *British Medical Journal* 312 (1996): 731–736.
26. D. M. Goldberg, Health effects of moderate alcohol consumption, *Patient Care* 7, no. 5 (1996): 56–73.
27. T. A. Pearson, American Heart Association Advisory: Alcohol and heart disease, *Circulation* 94 (1996): 3023–3025.
28. T. Gordon and coauthors, Alcohol and high-density lipoprotein cholesterol, *Circulation* 64, suppl. III (1981): III-63–III-67.
29. P. M. Ridker and coauthors, Association of moderate alcohol consumption and plasma concentration of endogenous tissue-type plasminogen activator, *Journal of the American Medical Association* 272 (1994): 929–933; S. Renaud and M. de Lorgeril, Wine, alcohol, platelets, and the French paradox for coronary heart disease, *Lancet* 339 (1992): 1523–1526.
30. National Institute on Alcohol Abuse and Alcoholism, *Alcohol Alert: Alcohol and Women*, No. 45 (Bethesda, MD: National Institute on Alcohol Abuse and Alcoholism, 1999).
31. T. A. Pearson, American Heart Association Advisory: Alcohol and heart disease, *Circulation* 94 (1996): 3023–3025.
32. E. Lehner and J. Lehner, *Folklore and Odysseys of Food & Medicinal Plants* (New York: Tudor Publishing Co., 1962), and M. Toussant-Samat, *A History of Food* (Cambridge, MA: Blackwell Publishers, 1994), as cited in The American Dietetic Association, *Food Folklore: Tales and Truths about What We Eat* (Minneapolis, MN: Chronimed Publishing, 1999), 16.
33. National Institute on Alcohol Abuse and Alcoholism, *Alcohol Alert: Fetal Alcohol Exposure and the Brain*, No. 50 (Bethesda, MD: National Institute on Alcohol Abuse and Alcoholism, 2000).
34. National Institute on Alcohol Abuse and Alcoholism, *Alcohol Alert: Fetal Alcohol Exposure and the Brain*, No. 50 (Bethesda, MD: National Institute on Alcohol Abuse and Alcoholism, 2000); K. L. Jones and D. W. Smith, Recognition of the fetal alcohol syndrome in early infancy, *Lancet* 2 (1973): 999–1001.
35. National Institutes of Health, Highlights from the 10th Special Report to Congress, *Alcohol Research and Health* 24, no. 1 (2000); K. Stratton, C. Howe, and F. Battaglia, eds., *Fetal Alcohol Syndrome: Diagnosis, Epidemiology, Prevention, and Treatment* (Washington, DC: National Academy Press, 1996).
36. National Institutes of Health, Highlights from the 10th Special Report to Congress, *Alcohol Research and Health* 24, no. 1 (2000).
37. National Institute on Alcohol Abuse and Alcoholism, *Alcohol Alert: Fetal Alcohol Exposure and the Brain*, No. 50 (Bethesda, MD: National Institute on Alcohol Abuse and Alcoholism, 2000).

CHAPTER 9

1. Position of the American Dietetic Association: Food and nutrition misinformation, *Journal of the American Dietetic Association* 106 (2006): 601–607; A. A. Hedley and coauthors, Prevalence of overweight and obesity among U.S. children, adolescents, and adults, 1999–2002, *Journal of the American Medical Association* 201 (2004): 2847–2850; and C. L. Ogden and coauthors, Prevalence of overweight and obesity in the United States, 1999–2004, *Journal of the American Medical Association* 295 (2006): 1549–1555.
2. *The Surgeon General's Call to Action to Prevent and Decrease Overweight and Obesity, 2001* (Washington, DC: U.S. Department of Health and Human Services, 2001).
3. *The Surgeon General's Call to Action to Prevent and Decrease Overweight and Obesity, 2001*; E. E. Calle and coauthors, Body-mass index and mortality in a prospective cohort of U.S. adults, *New England Journal of Medicine* 341 (1999): 1097–1105; T. S. Visscher and coauthors, Obesity and unhealthy life years in adult Finns, *Archives of Internal Medicine* 164 (2004): 1413–1420.
4. The discussion of trends is adapted from M. Fierro, The Obesity Epidemic—How States Can Trim the "Fat," *Issue Brief* (Washington, DC: National Governors Association Center for Best Practices, June 13, 2002).
5. B. J. Rolls, The supersizing of America: Portion size and the obesity epidemic, *Nutrition Today* 38 (2003): 42–53; The National Alliance for Nutrition and Activity (NANA), *From Wallet to Waistline: The Hidden Cost of Super Sizing* (Washington, DC: NANA, June 2002); L. R. Young and M. Nestle, The contribution of expanding portion sizes to the U.S. obesity epidemic, *American Journal of Public Health* 92 (2002): 246–248; S. J. Nielsen and B. M. Popkin, Patterns and trends in food portion sizes, 1977–1998, *Journal of the American Medical Association* 289 (2003): 450–453; M. Nestle, Increasing portion sizes in American diets: More calories, more obesity, *Journal of the American Dietetic Association* 103 (2003): 39–40; H. Smiciklas Wright and coauthors, Foods commonly eaten in the United States, 1989–1991 and 1994–1996: Are portion sizes changing? *Journal of the American Dietetic Association* 103 (2003): 41–47.
6. A. Hedley and coauthors, Prevalence of overweight and obesity among U.S. children, adolescents, and adults, 1999–2002, *Journal of the American Medical Association* 291 (2004): 2847–2850; C. J. Crespo and coauthors, Television watching, energy intake, and obesity in U.S. children: Results from the third National Health and Nutrition Examination Survey, 1988–1994, *Archives of Pediatrics and Adolescent Medicine* 155 (2001): 360–365.
7. D. Festi and coauthors, Gallbladder motility and gallstone formation in obese patients following very low-calorie diets: Use it (fat) to lose it (well), *International Journal of Obesity and Related Metabolic Disorders* 22 (1998): 592–600.
8. G. A. Bray, The underlying basis for obesity: Relationship to cancer, *The Journal of Nutrition* 132 (2002): 3451S–3455S.
9. A. Must and S. E. Anderson, Effects of obesity on morbidity in children and adolescents, *Nutrition in Clinical Care* 6 (2003): 4–12; R. S. Strauss and H. A. Pollak, Social marginalization of overweight children, *Archives of Pediatrics and Adolescent Medicine* 157 (2003): 746–752; M. E. Eisenberg, D. Neumark-Sztainer, and M. Story, Associations of weight-based teasing and emotional well-being among adolescents, *Archives of Pediatrics and Adolescent Medicine* 157 (2003): 733–738.
10. G. A. Bray, *Contemporary Diagnosis and Management of Obesity* (Newtown, PA: Handbooks in Healthcare Co., 1998).
11. Baylor College of Medicine, Undergoing a BodPod test: Densitometry theory, www.bcm.tmc.edu/bodycomplab/bpodprocesspage.htm; see also G. A. Bray, 1998; C. J. Hoffman and L. A. Hildebrandt, Use of air displacement plethysmography to monitor body composition: A beneficial tool for dietitians, *Journal of the American Dietetic Association* 101 (2001): 986–987.
12. R. Roubenoff and coauthors, Predicting body fatness: The body mass index versus estimation by bioelectrical impedance, *American Journal of Public Health* 85 (1995): 726–728.
13. P. Bjorntorp, Obesity, *Lancet* 350 (1997): 423–426.
14. The National Heart, Lung, and Blood Institute Expert Panel on the Identification, Evaluation, and Treatment of Overweight and Obesity in Adults, Executive summary of the clinical guidelines on the identification, evaluation, and treatment of overweight and obesity in adults, *Journal of the American Dietetic Association* 98 (1998): 1178–1181.
15. R. J. Kuczmarski and coauthors, Varying body mass index cutoff points to describe overweight prevalence among U.S. adults: NHANES III (1988–1994), *Obesity Research* 5 (1997): 542–545.

16. National Heart, Lung, and Blood Institute expert panel, 1998; S. K. Zhu, Waist circumference and obesity-associated risk factors among whites in the third National Health and Nutrition Examination Survey: Clinical action thresholds, *American Journal of Clinical Nutrition* 7 (2002): 743.

17. L. Van Horn and coauthors, The dietitian's role in developing and implementing the first federal obesity guidelines, *Journal of the American Dietetic Association* 98 (1998): 1115–1117.

18. E. T. Poehlman and E. S. Horton, Energy needs: Assessment and requirements in humans, in M. E. Shils, J. A. Olson, M. Shike, and A. C. Ross, eds., *Modern Nutrition in Health and Disease*, 10th ed. (Baltimore, MD: Williams and Wilkins, 2005).

19. G. A. Bray, *Contemporary Diagnosis and Management of Obesity* (Newtown, PA: Handbooks in Healthcare Co., 1998); J. O. Hill and coauthors, Obesity and the environment: Where do we go from here? *Science* 299 (2003): 853; P. A. Lachance, Human obesity, *Food Technology* 48 (1994): 127–138; J. T. Travis, The hunger hormone? *Science News* 161 (2002): 107; E. M. Blass, Biological and environmental determinants of childhood obesity, *Nutrition in Clinical Care* 6 (2003): 13–19; D. M. Cultler, E. K. Glaeser, and J. M. Shapiro, Why have Americans become more obese? In *The Economics of Obesity/E-FAN-04-004* (Washington, DC: Economic Research Service, US Department of Agriculture, May 2004), 2–5.

20. A. J. Stunkard and coauthors, A twin study of human obesity, *Journal of the American Medical Association* 256 (1986): 51–54; A. J. Stunkard and coauthors, An adoption study of human obesity, *New England Journal of Medicine* 314 (1986): 193–198; A. J. Stunkard and coauthors, The body mass index of twins who have been reared apart, *New England Journal of Medicine* 322 (1990): 1483–1487; J. O. Hill, H. R. Wyatt, E. L. Melanson, Genetic and environmental contributions to obesity. *Medical Clinics of North America* 84 (2000): 333–346; T. Rankinen and coauthors, The human obesity gene map: The 2001 update, *Obesity Research* 10 (2002): 196–243.

21. C. Bouchard and coauthors, The response to long-term overfeeding in identical twins, *New England Journal of Medicine* 322 (1990): 1477–1482.

22. J. M. Friedman, The function of leptin in nutrition, weight, and physiology, *Nutrition Reviews* 60 (2002): S1.

23. C. S. Mantzoros, The role of leptin in human obesity and disease: A review of the current evidence, *Annals of Internal Medicine* 130 (1999): 671–680; R.V. Considine, Serum immunoreactive-leptin concentrations in normal weight and obese humans, *New England Journal of Medicine* 334 (1996): 292–295; J. M. Friedman, Leptin, leptin receptors, and the control of body weight, *Nutrition Reviews* 56 (1998): S38–S41.

24. R. C. Whitaker and coauthors, Predicting obesity in young adulthood from childhood and parental obesity, *New England Journal of Medicine* 337 (1997): 869–873; M. C. Bellizzi and W. H. Dietz, Workshop on childhood obesity: Summary of the discussion, *American Journal of Clinical Nutrition* 70 (1999): S173–S175; R. J. Hancox, B. J. Milne, and R. Poulton. Association between child and adolescent television viewing and adult health: A longitudinal birth cohort study, *Lancet* 364 (2004): 257–62.

25. F. Xavier Pi-Sunyer, Obesity, in M. E. Shils, J. A. Olson, M. Shike, and A. C. Ross, eds., *Modern Nutrition in Health and Disease*, 10th ed. (Baltimore, MD: Williams and Wilkins, 2005).

26. M. A. McCrory and coauthors, Dietary variety within food groups: Association with energy intake and body fatness in men and women, *American Journal of Clinical Nutrition* 69 (1999): 440–447.

27. K. Patrick and coauthors, Diet, physical activity, and sedentary behaviors as risk factors for overweight in adolescence, *Archives in Pediatric and Adolescent Medicine* 158 (2004): 385–390; Position of the American Dietetic Association: Weight management, *Journal of the American Dietetic Association* 102 (2002): 1145–1155.

28. K. D. Brownell and T. A. Wadden, Etiology and treatment of obesity: Understanding a serious, prevalent, and refractory disorder, *Journal of Consulting and Clinical Psychology* 60 (1992): 505–517; Position of the American Dietetic Association: Weight management, *Journal of the American Dietetic Association* 102 (2002): 1145–1155; D. Riehe and coauthors, Evaluation of a healthy lifestyle approach to weight management, *Preventive Medicine* 36 (2003): 45–54; W. S. Poston and J. P. Foreyt, Successful management of the obese patient, *American Family Physician* 61 (2000): 3615; J. M. Lyznicki and coauthors, Obesity: Assessment and management in primary care, *American Family Physician* 63 (2001): 2185; S. Z. Yanovski and J. A. Yanovski, Obesity, *The New England Journal of Medicine* 346 (2002): 591.

29. Calorie Control Council National Consumer Survey, 2004, www.caloriecontrol.org/trndstat.html.

30. Adapted from Weighing the diet books, *Nutrition Action Newsletter* (January/February 2004): 3–8; M. Freedman and coauthors, Popular diets: A scientific review, *Obesity Research* 9 (2001): 1S–39S; M. L. Dansinger and coauthors, Comparison of the Atkins, Ornish, Weight Watchers, and Zone diets for weight loss and heart disease risk reduction, *Journal of the American Medical Association* 293 (2005); 43–53; V. S. Retelny, Fad diet review, www.foodandhealth.com, 2005; Wheat Council, "Setting the Record Straight," www.wheatfoods.org/.

31. M. L. Dansinger and coauthors, 2005.

32. Adapted from American Heart Association, "Quick Weight Loss or Fad Diets," www.americanheart.org/presenter.jhtml?identifier_4584; Weighing the diet books, *Nutrition Action Newsletter* (January/February 2004): 3–8; Weight-control Information Network on choosing a safe and successful diet, win.niddk.nih.gov/publications/choosing.htm.

33. M. Freedman and coauthors, Popular diets: A scientific review, *Obesity Research* 9 (2001): 1S–39S; National Weight Control Registry, www.uchsc.edu/nutrition/WyattJortberg/nwcr.htm.

34. M. Freedman and coauthors, 2001.

35. J. M. Rippe, The obesity epidemic: Challenges and opportunities, *Journal of the American Dietetic Association* 98 (1998): 85S; National Center for Chronic Disease Prevention and Health Promotion, Centers for Disease Control and Prevention (CDC), *Obesity and Overweight—A Public Health Epidemic*, www.cdc.gov/nccdphp/dnpa/ obesity.index.htm; J. Dausch, Determining when obesity is a disease, *Journal of the American Dietetic Association* 101 (2001): 293.

36. S. Schurgin and R. D. Siegel, Pharmacotherapy of obesity: An update, *Nutrition in Clinical Care* 6 (2003): 27–37; F. M. Berg, Chromium picolinate: Scam of the hour, *Obesity and Health* 7 (1993): 54–55.

37. Federal Trade Commission, Weighing the Evidence in Diet Ads, www.ftc.gov.

38. Food and Drug Administration, Consumer alert: FDA plans regulation prohibiting sale of ephedra-containing dietary supplements and advises consumers to stop using these products, December 2003, www.fda.gov; M. H. Pittler and E. Ernst, Dietary supplements for body weight reduction: A systematic review, *American Journal of Clinical Nutrition* 79 (2004): 529–536.

39. P. Kurtzweil, Dieter's brews make tea time a dangerous affair, *FDA Consumer* (July/August 1997): 6–11.

40. National Institute of Diabetes and Digestive and Kidney Diseases, *Gastric Surgery for Severe Obesity* (Bethesda, MD: National Institutes of Health, 1996), 1–6; E. H. Livingston, Obesity and its surgical management, *American Journal of Surgery* 184 (2002): 103–113; E. Lin, Gastric bypass, *Archives of Surgery* 139 (2004): 780–784; H. Buchwald and coauthors, Bariatric surgery: A systematic review and meta-analysis, *Journal of the American Medical Association* 292 (2004): 1724–1737.

41. A. Groin and coauthors, Promoting long-term weight control: Does dieting consistency matter? *International Journal of Obesity and Related Metabolic Disorders* 28 (2004): 278–281; J. W. Anderson and coauthors, Long-term weight loss maintenance: A meta-analysis of U.S. studies, *American Journal of Clinical Nutrition* 74 (2001): 579–584; L. Bren, Losing weight: More than counting calories, *FDA Consumer* 36 (2002): 18–25.

42. M. T. McGuire, R. R. Wing, and J. O. Hill, The prevalence of weight loss maintenance among American adults, *International Journal of Obesity* 23 (1999): 1314–1319; 1,800 calories, 4 miles keep the weight off, *Diabetes Today* (November 17, 2004).

43. J. P. Foreyt and G. K. Goodrick, *Living Without Dieting* (New York: Warner Books, 1992), 28.

44. Ibid., 30.
45. K. D. Brownell, *The LEARN Program for Weight Control* (Dallas, TX: American Health Publishing Company, 1994), 102–103.
46. Foreyt and Goodrick, 1992, 43–58.
47. M. Shah and coauthors, Comparison of a low-fat, ad libitum complex carbohydrate diet with a low energy diet in moderately obese women, *American Journal of Clinical Nutrition* 59 (1994): 980–984; S. M. Shick and coauthors, Persons successful at long-term weight loss and maintenance continue to consume a low-energy, low-fat diet, *Journal of the American Dietetic Association* 98 (1998): 408–413; J. P. Foreyt and G. K. Goodrick, Dieting and weight loss: The energy perspective, *Nutrition Reviews* 59 (2001): S25–S28.
48. M. T. McGuire and coauthors, Long-term maintenance of weight loss: Do people who lose weight through various weight loss methods use different methods to maintain their weight? *International Journal of Obesity and Related Metabolic Disorders* 22 (1998): 572–577; R. R. Wing and J. O. Hill, Successful weight loss maintenance, *Annual Review of Nutrition* 21 (2001): 323.
49. K. Brownell, Yo-yo dieting, in *Nutrition 91/92*, ed., C. C. Cook-Fuller with S. Barrett (Guilford, CT: The Dushkin Publishing Group, 1991), 132–134; National Task Force on the Prevention and Treatment of Obesity, Weight cycling, *Journal of the American Medical Association* 275 (1994): 1196–1202.
50. R. E. Anderson and coauthors, Changes in bone mineral content in obese dieting women, *Metabolism: Clinical and Experimental* 46 (1997): 857–861; K. M. Rourke and coauthors, Effect of weight change on bone mass in female adolescents, *Journal of the American Dietetic Association* 103 (2003): 369–372.
51. M. B. Zemel and coauthors, Regulation of adiposity by dietary calcium, *FASEB Journal* 14 (2000): 1132–1138.
52. C. E. Ross, Overweight and depression, *Journal of Health and Social Behavior* 35 (1994): 63–78; M. E. Lean and coauthors, Impairment of health and quality of life in people with large waist circumference, *The Lancet* 351 (1998): 853–856.
53. H. R. Wyatt and J. O. Hill, Let's get serious about promoting physical activity, *American Journal of Clinical Nutrition* 75 (2002): 449; J. M. Rippe and S. Hess, The role of physical activity in the prevention and management of obesity, *Journal of the American Dietetic Association* 98 (1998): 31S–38S; P. D. Wood, Clinical applications of diet and physical activity in weight loss, *Nutrition Reviews* 54 (1998): S131–S135; R. L. Weinsier and coauthors, Free-living activity energy expenditure in women successful and unsuccessful at maintaining a normal body weight, *American Journal of Clinical Nutrition* 75 (2002): 499; M. Gilliat-Wimberly and coauthors, Effects of habitual physical activity on the resting metabolic rates and body compositions of women aged 35 to 50 years, *Journal of the American Dietetic Association* 101 (2001): 1181–1188.
54. J. O. Prochaska, *Changing for Good* (New York: William Morrow and Company, 1994), 47.
55. Ibid., 38–50.
56. I. Chalmers, *The Great Food Almanac* (San Francisco: Collins Publishers, 1994).
57. H. Steiner and J. Lock, Anorexia nervosa and bulimia nervosa in children and adolescents: A review of the past ten years, *Journal of the American Academy of Child & Adolescent Psychiatry* 37 (1998): 352–357; G. C. Patton and coauthors, Onset of adolescent eating disorders: Population-based cohort study over 3 years, *British Journal of Medicine* 318 (1999): 765–768.
58. D. Neumark-Sztainer and P. J. Hannan, Weight-related behaviors among adolescent girls and boys: Results from a national survey, *Archives of Pediatric and Adolescent Medicine* 154 (2000): 569–577; D. Neumark-Sztainer and coauthors, Obesity, disordered eating, and eating disorders in a longitudinal study of adolescents: How do dieters fare 5 years later? *Journal of the American Dietetic Association* 106 (2006): 559–568.
59. Position of the American Dietetic Association: Nutrition intervention in the treatment of anorexia nervosa bulimia nervosa, and eating disorders not otherwise specified (EDNOS), *Journal of the American Dietetic Association* 101 (2001): 810; A. E. Becker and coauthors, Eating disorders, *New England Journal of Medicine* 340 (1999): 1092–1098.
60. Task force on DSM-IV, 307.50 Eating Disorders Not Otherwise Specified, *DSM-IV Draft Criteria* (Washington, DC: American Psychiatric Association, 1993), 2.
61. C. W. Baker and K. D. Brownell, Binge eating disorder: Identification and management, *Nutrition in Clinical Care* 2 (1999): 344–353.
62. N. I. Hahn and M. M. Woolsey, When food becomes a cry for help, *Journal of the American Dietetic Association* 98 (1998): 395–398.
63. C. L. Rock, Nutritional and medical assessment and management of eating disorders, *Nutrition in Clinical Care* 2 (1999): 332–343.
64. T. Pryor and W. Wiederman, Personality features and expressed concerns of adolescents with eating disorders, *Adolescence* 33 (1998): 291–298.
65. W. Kaye and coauthors, Serotonin neuronal function and selective serotonin reuptake inhibitor treatment in anorexia and bulimia nervosa, *Biological Psychiatry* 44 (1998): 825–838.
66. A. Gila and coauthors, Subjective body-image dimensions in normal and anorexic adolescents, *British Journal of Medical Psychology* 71 (1998): 175–184.
67. A. Gila, 1998.
68. F. Klapper and coauthors, Psychiatric management of eating disorders, *Nutrition in Clinical Care* 2 (1999): 354–360.
69. D. Williamson, *Assessment of Eating Disorders: Obesity, Anorexia, and Bulimia Nervosa* (New York: Pergamon Press, 1990); Practice guidelines for eating disorders, *American Journal of Psychiatry* 150 (1993): 212–218.
70. Position of the American Dietetic Association: Nutrition intervention in the treatment of anorexia nervosa, bulimia nervosa, and eating disorders not otherwise specified (EDNOS), *Journal of the American Dietetic Association* 101 (2001): 810.
71. Klapper, 1999; B. Abramovitz and L. L. Birch, Five-year-old girls' ideas about dieting are predicted by their mothers' dieting, *Journal of the American Dietetic Association* 100 (2000): 1157–1163.
72. The guidelines listed are from M. Herrin, Dartmouth College Nutrition Education Program; for more information, contact Eating Disorders Awareness and Prevention, Inc., www.edap.org.

CHAPTER 10

1. C. Rosenbloom and M. Bahns, What can we learn about diet and physical activity from the master athletes? *Nutrition Today* 40 (2005): 267–272; Y. A. Kesaniemi and coauthors, Dose response issues concerning physical activity and health: An evidence-based symposium, *Medicine & Science in Sports & Exercise* 33 (2001): S351; C. M. Friedenreich and M. R. Orenstein, Physical activity and cancer prevention: Etiologic evidence and biological mechanisms, *Journal of Nutrition* 132 (2002): 3456S–3466S; J. Verloop and coauthors, Physical activity and breast cancer risk in women aged 20 to 54 years, *Journal of the National Cancer Institute* 92 (2000): 128–135; W. D. Schmidt and coauthors, Effects of long versus short bouts of exercise on fitness and weight loss in overweight females, *Journal of the American College of Nutrition* 20 (2001): 494.
2. Institute of Medicine, *Dietary Reference Intakes for Energy, Carbohydrate, Fiber, Fat, Fatty Acids, Cholesterol, Protein, and Amino Acids* (Washington, DC: National Academy Press), 2002; J. McCaffree, Physical activity: How much is enough? *Journal of the American Dietetic Association* 103 (2003): 153–154.
3. American College of Sports Medicine, *ACSM's Guidelines for Exercise Testing and Prescription*, 6th ed. (Philadelphia: Lippincott, Williams & Wilkins, 2000).
4. M. A. Fiatarone and coauthors, Exercise training and nutritional supplementation for physical frailty in very elderly people, *New England Journal of Medicine* 330 (1994): 1769–1775; M. A. F. Singh, Exercise for disease prevention in the geriatric population, *Nutrition in Clinical Care* 4 (2001): 296–305.
5. American College of Sports Medicine, *ACSM's Guidelines for Exercise Testing and Prescription*, 6th ed. (Philadelphia: Lippincott, Williams & Wilkins, 2000).

6. M. H. Williams, *Nutrition for Health, Fitness, and Sport,* 7th ed. (Boston: McGraw Hill, 2005).
7. F. W. Booth and coauthors, Waging war on physical inactivity: Using modern molecular ammunition against an ancient enemy, *Journal of Applied Physiology* 93 (2002): 3–30.
8. To work up a sweat, or not? *Tufts University Health and Nutrition Letter* 20, no. 10 (December 2002): 6.
9. New strength-training guidelines for older people: Too cautious? *Tufts University Health and Nutrition Letter* 20, no. 1 (March 2002): 8.
10. Are you doing all you can to fight sarcopenia? *Tufts University Health and Nutrition Letter* 21, no. 1 (March 2003): 1, 4–5; C. E. Broeder, The effects of either high-intensity resistance or endurance training on resting metabolic rate, *American Journal of Clinical Nutrition* 55 (1992): 802–810.
11. E. Hultman and coauthors, Work and exercise, in *Modern Nutrition in Health and Disease,* 10th ed., M. E. Shils, J. A. Olson, and M. Shike, eds. (Baltimore, MD: Williams and Wilkins, 2005).
12. E. Coleman, Carbohydrate and exercise, in C. A. Rosenbloom, ed., *Sports Nutrition: A Guide for the Professional Working with Active People,* 3rd ed. (Chicago: American Dietetic Association, 2000), 13–31; A. N. Bosch, S. C. Dennis, and T. D. Noakes, Influence of carbohydrate loading on fuel substrate turnover and oxidation during prolonged exercise, *Journal of Applied Physiology* 74 (1993): 1921–1927; L. M. Burke and coauthors, Muscle glycogen storage after prolonged exercise: Effect of frequency of carbohydrate feedings, *American Journal of Clinical Nutrition* 64 (1996): 115–119.
13. J. A. Romijn and coauthors, Regulation of endogenous fat and carbohydrate metabolism in relation to exercise intensity and duration, *American Journal of Physiology* 265 (1993): E380–E391; G. A. Brooks and J. Mercier, Balance of carbohydrate and lipid utilization during exercise: The "crossover" concept, *Journal of Applied Physiology* 76 (1994): 2253–2261; J. H. Wilmore, Physical energy: Fuel metabolism, *Nutrition Reviews* 59 (2001): S13.
14. R. A. Fielding and J. A. Parkington, What are the dietary protein requirements of physically active individuals? New evidence on the effects of exercise on protein utilization during post-exercise recovery, *Nutrition in Clinical Care* 5 (2002): 191–196.
15. Position of the American Dietetic Association, Dietitians of Canada, and the American College of Sports Medicine: Nutrition and athletic performance, *Journal of the American Dietetic Association* 100 (2000): 1543–1556.
16. C. M. Cumming and coauthors, Recreational runners' beliefs and practices concerning water intake, *Journal of Nutrition Education* 26 (1994): 195–197.
17. American College of Sports Medicine, American Dietetic Association, and Dietitians of Canada, Position stand on nutrition and athletic performance, *Medicine and Science in Sports and Exercise* 32 (2000): 2130–2145.
18. M. Millard-Stafford, Fluid replacement during exercise in the heat, *Sports Medicine* 13 (1992): 223–233; R. Wexler, Evaluation and treatment of heat-related illnesses, *American Family Physician* 65 (2002): 2037–2039.
19. C. V. Gisolfi, Fluid balance for optimal performance, *Nutrition Reviews* 54 (1996): S159–S168; L. Bonci, "Energy" drinks: Help, harm, or hype? *Sports Science Exchange* 15 (2002): 1–4, www.gssiweb.com.
20. J. S. Coombes and K. L. Hamilton, The effectiveness of commercially available sports drinks, *Sports Medicine* 29 (2000): 181–209.
21. K. B. Wheeler and A. M. Cameron, Plasma volume: The hidden key to performance, *American Fitness Quarterly* (April 1990): 24–26.
22. M. F. Bergeron, Sodium: The forgotten nutrient, *Sports Science Exchange* 13 (2000): 1–4, www.gssiweb.com.
23. M. Meydani and coauthors, Protective effect of vitamin E on exercise-induced oxidative damage in young and older adults, *American Journal of Physiology* 264 (1993); A. K. Adams and T. M. Best, The role of antioxidants in exercise and disease prevention, *The Physician and Sportsmedicine* 30 (2002): 37.
24. J. Beard and B. Tobin, Iron status and exercise, *American Journal of Clinical Nutrition* 72 (2000): 594S.
25. L. M. Weight, P. Jacobs, and T. D. Noakes, Dietary iron deficiency and sports anemia, *British Journal of Nutrition* 68 (1992): 253–260.
26. Y. I. Shu and J. D. Haas, Iron depletion without anemia and physical performance in young women, *American Journal of Clinical Nutrition* 66 (1997): 334–341; S. P. Bourque and coauthors, Twelve weeks of endurance exercise training does not affect iron status in women, *Journal of the American Dietetic Association* 97 (1997): 1116–1121.
27. K. Beals and M. M. Manore, The prevalence and consequence of subclinical eating disorders in female athletes, *International Journal of Sports Nutrition* 4 (1994): 175–195; L. E. Thrash and J. B. Anderson, The female athlete triad: Nutrition, menstrual disturbances, and low bone mass, *Nutrition Today* 35 (2000): 168–174.
28. W. S. Holt, Jr., Nutrition and athletes, *American Family Physician* 47 (1993): 1757–1764; D. M. Ahrendt, Ergogenic aids: Counseling the athlete, *American Family Physician* 63 (2001): 913.
29. T. H. Murray, The ethics of drugs and sports, in *Drugs and Performance in Sports,* R. H. Strauss, ed. (Philadelphia: W. B. Saunders Company, 1987), 11–21.
30. L. E. Armstong and C. M. Maresh, Vitamin and mineral supplements as nutritional aids to exercise performance and health, *Nutrition Reviews* 54 (1996): S149–S158; E. A. Applegate and L. E. Grivetti, Search for the competitive edge: A history of dietary fads and supplements, *Journal of Nutrition* 127 (1997): 869S–873S; D. M. Ahrendt, Ergogenic aids: Counseling the athlete, *American Family Physician* 63 (2001): 913; A. Sarubin, *The Health Professional's Guide to Popular Dietary Supplements,* 2nd ed. (Chicago: The American Dietetic Association, 2002); S. Nelson Steen and E. Coleman, Selected ergogenic aids used by athletes, *Nutrition in Clinical Practice* 14 (1999): 287–295.
31. Position of the American Dietetic Association, Dietitians of Canada, and the American College of Sports Medicine: Nutrition and athletic performance, *Journal of the American Dietetic Association* 100 (2000): 1543–1556.
32. S. Barrett, Don't buy phony ergogenic aids, *Nutrition Forum* (May/June 1997): 19–21, 24.
33. G. Mirkin, Can bee pollen benefit health? *Journal of the American Medical Association* 262 (1989): 1854.
34. L. S. Walker and coauthors, Chromium picolinate and body composition and muscular performance in wrestlers, *Medicine and Science in Sports and Exercise* 30 (1998): 1730–1737; K. E. Grant and coauthors, Chromium and exercise training: Effect on obese women, *Medicine and Science in Sports and Exercise* 29 (1997): 992–998.
35. E. A. Applegate, 1997.
36. P. J. Rasch and coauthors, Protein dietary supplementation and physical performance, *Medicine and Science in Sports* 1 (1969): 195–199.
37. C. E. Yesalis and coauthors, Anabolic steroid use in the United States, *Journal of the American Medical Association* 270 (1993): 1217–1221.
38. Committee on Sports Medicine and Fitness, Adolescents and anabolic steroids: A subject review, *Pediatrics* 99 (1997): 443–447.
39. D. C. Nieman, *Fitness and Sports Medicine: An Introduction* (Palo Alto, CA: Bull Publishing Company, 1990), 246–258.
40. L. Boci, "Energy" drinks: Help, harm or hype? *Gatorade Sports Science Institute,* 15 (2002):1.
41. M. Wahlcrist, FAQ on caffeine and energy drinks, *Nutrition Australia* (2001), www.nutritionaustralia.org.
42. A. Izzo and E. Ernst, Interactions between herbal medicines and prescribed drugs: A systematic review, *Drugs* 61 (2005): 2163–2175.
43. D. Edell, Are energy drinks safe? *The Athletic Advisor* (2005), www.athleticadvisor.com.

CHAPTER 11

1. L. Breslow and N. Breslow, Health practices and disability: Some evidence from Alameda County, *Preventive Medicine* 22 (1993): 86–95; A. J. Vita and coauthors, Aging, health risks, and cumulative disability, *New England Journal of Medicine* 338 (1998): 1035–1041.
2. K. M. Godfrey and D. J. Barker, Fetal nutrition and adult disease, *American Journal of Clinical Nutrition* 71 (2000): 1344S–1352S; M. W. Gillman, Developmental origins of health and disease, *The New England Journal of Medicine* 353 (2005): 1848–1850.

3. Institute of Medicine, *Dietary Reference Intakes for Energy, Carbohydrate, Fiber, Fat, Fatty Acids, Cholesterol, Protein, and Amino Acids* (Washington, DC: National Academy Press, 2002).
4. Position of the American Dietetic Association: Nutrition and lifestyle for a healthy pregnancy outcome, *Journal of the American Dietetic Association* 102 (2002): 1479–1490; S. Brundage, Preconception healthcare, *American Family Physician* 65 (2002): 2507.
5. G. Kauwell, Folic acid supplements and fortification affect the risk for neural tube defects, vascular disease, and cancer: Evolving science, *Journal of Nutrition* 133 (2003): 1961S–1968S.
6. G. Vozenilek, What they don't know could hurt them: Increasing public awareness of folic acid and neural tube defects, *Journal of the American Dietetic Association* 99 (1999): 20–22.
7. J. Erikson, Folic acid and prevention of spina bifida and anencephaly, *Morbidity and Mortality Weekly Report* 51 (2002): 1–3; A. A. Yates, Dietary Reference Intakes: The new basis for recommendations for calcium and related nutrients, B vitamins, and choline, *Journal of the American Dietetic Association* 98 (1998): 699–706.
8. A. Prentice, Maternal calcium metabolism and bone mineral status, *American Journal of Clinical Nutrition* 71 (2000): 312S; M. F. Picciano, Pregnancy and lactation: Physiological adjustments, nutritional requirements, and the role of dietary supplements, *Journal of Nutrition* 133 (2003): 1997S–2002S.
9. Position of the American Dietetic Association: Nutrition and lifestyle for a healthy pregnancy outcome, *Journal of the American Dietetic Association* 102 (2002): 1479–1490.
10. V. Azais-Braesco and G. Pascal, Vitamin A in pregnancy: Requirements and safety limits, *American Journal of Clinical Nutrition* 71 (2000): 1325S.
11. Position of the American Dietetic Association: Nutrition and lifestyle for a healthy pregnancy outcome, 2002.
12. Ibid.
13. A. J. Rainville, Pica practices of pregnant women are associated with lower maternal hemoglobin level at delivery, *Journal of the American Dietetic Association* 98 (1998): 293–296.
14. U.S. Department of Health and Human Services, *Healthy People 2010* (Washington, DC: U.S. Government Printing Office, January 2000); S. J. Ventura and coauthors, Trends and variations in smoking during pregnancy and low birthweight: Evidence from the birth certificate, 1990–2000, *Pediatrics* 111 (2003): 1176–1180.
15. J. R. Di Franza, C. A. Aligne, and M. Weitzman, Prenatal and postnatal environmental tobacco smoke exposure and children's health, *Pediatrics* 113 (2004): 1007–1015.
16. M. L. Plant and coauthors, Alcohol and pregnancy, in I. MacDonald, ed., *Health Issues Related to Alcohol Consumption* (Washington, DC: ILSI Press, 1999), 182–213.
17. American Academy of Pediatrics, Fetal alcohol syndrome and alcohol-related neurodevelopmental disorders, *Pediatrics* 106 (2000): 358–360.
18. Position of the American Dietetic Association: Nutrition and lifestyle for a healthy pregnancy outcome, 2002; Center for Evaluation of Risks to Human Reproduction, 2003, cerhr.niehs.nih.gov.
19. FDA and EPA announce the revised consumer advisory on methylmercury in fish, *News* (2004), www.cfsan.fda.gov.
20. N. I. Hahn and M. Erick, Battling morning (noon and night) sickness: New approaches for treating an age-old problem, *Journal of the American Dietetic Association* 94 (1994): 147–148.
21. G. Dekker and B. Sibai, Primary, secondary, and tertiary prevention of preeclampsia, *Lancet* 357 (2001): 209–215; F. B. Pipkin, Risk factors for preeclampsia, *New England Journal of Medicine* 344 (2001): 926; B. M. Sibai, Diagnosis and management of gestational hypertension and preeclampsia, *Obstetrics and Gynecology* 102 (2003): 181–192.
22. American Diabetes Association, Position statement: Gestational diabetes mellitus, *Diabetes Care* 26 (2003): S103–S105; Report of the Expert Committee on the Diagnosis and Classification of Diabetes Mellitus, *Diabetes Care* 26 (2003): S5–S20.
23. M. Story and J. Stang, eds., *Nutrition and the Pregnant Adolescent: A Practical Reference Guide* (Washington, DC: Maternal and Child Health Bureau, Health Resources and Services Administration, 2000).
24. R. J. Trissler, The child within: A guide to nutrition counseling for pregnant teens, *Journal of the American Dietetic Association* 99 (1999): 916–918; see also V. A. Long, T. Martin, and C. Janson-Sand, The Great Beginnings Program: Impact of a nutrition curriculum on nutrition knowledge, diet quality, and birth outcomes in pregnant and parenting teens, *Journal of the American Dietetic Association* 102 (2002): S86–S89.
25. R. Tannahil, *Food in History* (New York: Crown Publishers, 1988), 275.
26. R. Urgert and coauthors, Effects of cafestol and kahweol from coffee grounds on serum lipids and serum liver enzymes in humans, *American Journal of Clinical Nutrition* 61 (1995): 149–154.
27. A. Leviton and E. N. Allred, Correlates of decaffeinated coffee choice, *Epidemiology* 5 (1994): 537–540.
28. E. C. Strain and coauthors, Caffeine dependence syndrome, *Journal of the American Medical Association* 272 (1994): 1043–1048; E. H. Hogan, B. A. Hornick, and A. Bouchoux, Communicating the message: Clarifying the controversies about caffeine, *Nutrition Today* 37 (2002): 28–35.
29. S. G. Oei and coauthors, Fetal arrhythmia caused by excessive intake of caffeine by pregnant women, *British Medical Journal* 298 (1989): 1075–1076.
30. S. Cnattingius and coauthors, Caffeine intake and the risk of first trimester spontaneous abortion, *New England Journal of Medicine* 343 (2000): 1839–1845.
31. Position of the American Dietetic Association: Nutrition and lifestyle for a healthy pregnancy outcome, 2002.
32. E. H. Hogan, B. A. Hornick, and A. Bouchoux, Communicating the message: Clarifying the controversies about caffeine, *Nutrition Today* 37 (2001): 28–35.
33. Position of the American Dietetic Association: Nutrition and lifestyle for a healthy pregnancy outcome, 2002.
34. M. Story and J. Stang, eds., *Nutrition and the Pregnant Adolescent: A Practical Reference Guide* (Washington, DC: Maternal and Child Health Bureau, Health Resources and Services Administration, 2000).
35. Position of the American Dietetic Association: Promoting and supporting breastfeeding, *Journal of the American Dietetic Association* 105 (2005): 810–818.
36. Committee on Nutrition, American Academy of Pediatrics, *Pediatric Nutrition Handbook*, 5th ed., R. E. Kleinman, ed. (Elk Grove, IL: American Academy of Pediatrics, 2004), 110–111.
37. Office of Women's Health, *Breastfeeding: HHS Blueprint for Action on Breastfeeding* (Washington, DC: Department of Health and Human Services, 2000), 1–21.
38. U.S. Department of Health and Human Services, *Healthy People 2010* (Washington, DC: U.S. Government Printing Office, January 2000).
39. World Health Organization, *HIV and Infant Feeding* (Geneva, Switzerland: WHO, 1998).
40. Committee on Nutrition, American Academy of Pediatrics, *Pediatric Nutrition Handbook*, 5th ed., R. E. Kleinman, ed. (Elk Grove, IL: American Academy of Pediatrics, 2004).
41. S. F. Fomon, Feeding normal infants: Rationale for recommendations, *Journal of the American Dietetic Association* 101 (2001): 1002–1005; W. C. Heird, Nutritional requirements during infancy, in B. A. Bowman and R. M. Russell, *Present Knowledge in Nutrition*, 9th ed. (Washington, DC: International Life Sciences Institute, 2006); J. Stang, Improving the eating patterns of infants and toddlers, *Journal of the American Dietetic Association* 106 (2006): 87–89.
42. American Academy of Pediatrics Committee on Nutrition, The use and misuse of fruit juice in pediatrics, *Pediatrics* 107 (2001): 1210–1213.
43. S. B. Roberts and M. B. Heyman, How to feed babies and toddlers in the 21st century, *Zero to Three* (August/September 2000): 24–28.
44. L. A. Kazal, Prevention of iron deficiency in infants and toddlers, *American Family Physician* 66 (2002): 1217–1221.
45. Position of the American Dietetic Association: Dietary guidance for healthy children aged 2 to 11 years, *Journal of the American Dietetic Association* 104 (2004): 660–677.
46. U.S. Department of Agriculture, Menu planning in the national school lunch program, *Food Program Facts* (Washington, DC: U.S. Department of Agriculture, 2003).
47. Position of the American Dietetic Association, Society for Nutrition Education, and American School Food Service Association: Nutrition

services: An essential component of comprehensive school health programs, *Journal of the American Dietetic Association* 103 (2003): 505–514; M. Lino and coauthors, The quality of young children's diets, *Family Economics and Nutrition Review* 14 (2002): 52–60; Center for Nutrition Policy and Promotion, Report card on the diet quality of children ages 2 to 9, *Nutrition Insights* (September 2001).

48. HP 2010 Progress Review: Nutrition and Overweight, January 21, 2004; P. M. Gleason and C. W. Suitor, Changes in children's diets: 1989–1991 to 1994–1996 (Washington, DC: U.S. Department of Agriculture, 2001), Report No. CN-01-CD1; E. C. Wilkinson, S. J. Mickle, and J. D. Goldman, Trends in food and nutrient intakes by children in the United States, *Family Economics and Nutrition Review* 14, no. 9 (2002): 56–62.

49. Ibid; L. M. Fiorito, Dairy and dairy-related nutrient intake during middle childhood, *Journal of the American Dietetic Association* 106 (2006): 534–542.

50. R. Rajeshwari and coauthors, Secular trends in children's sweetened-beverage consumption (1974–1994): The Bogalusa Heart Study, *Journal of the American Dietetic Association* 105 (2005): 208–214; S. J. Nielsen and B. Popkin, Changes in beverage intake between 1977 and 2001, *American Journal of Preventive Medicine* 27 (2004): 205–210.

51. Position of the American Dietetic Association: Dietary guidance for healthy children ages 2 to 11 years, *Journal of the American Dietetic Association* 104 (2004): 660–677.

52. C. L. Ogden and coauthors, Prevalence of overweight and obesity in the United States, 1999–2004, *Journal of the American Medical Association* 295 (2006): 1549–1555; D. B. Johnson and coauthors, Preventing obesity: A life cycle perspective, *Journal of the American Dietetic Association* 106 (2006): 97–102.

53. Centers for Disease Control and Prevention, Iron deficiency—United States, 1999–2000, *Morbidity and Mortality Weekly Report* 51 (2002): 897–899.

54. S. Morey, American Academy of Pediatrics releases report on cholesterol levels in children and adolescents, *American Family Physician* 57 (1998): 2266–2268.

55. American Heart Association Scientific Statement: Cardiovascular health in childhood, *Circulation* 106 (2002): 143–160; American Heart Association, Guidelines for cardiovascular risk reduction for high-risk children, *Circulation* 107 (2003): 1562–1566.

56. Ibid.

57. B. A. Spear, Adolescent growth and development, *Journal of the American Dietetic Association* 102 (2002): S23–S29.

58. M. Mascarenhas and coauthors, *Adolescence,* in B. A. Bowman and R. M. Russell, *Present Knowledge in Nutrition,* 9th ed. (Washington, DC: International Life Sciences Institute, 2006).

59. D. Neumark-Sztainer and P. J. Hannan, Weight-related behaviors among adolescent girls and boys: Results from a national survey, *Archives of Pediatric and Adolescent Medicine* 154 (2000): 569–577; D. Neumark-Sztainer and coauthors, Obesity, disordered eating, and eating disorders in a longitudinal study of adolescents: How do dieters fare 5 years later? *Journal of the American Dietetic Association* 106 (2006): 559–568.

60. L. A. Lytle, Nutritional issues for adolescents, *Journal of the American Dietetic Association* 102 (2002): S8–S12; F. R. Greer and N. F. Krebs, Optimizing bone health and calcium intakes of infants, children, and adolescents, *Pediatrics* 117 (2006): 578–585.

61. Expert Panel on Detection, Evaluation, and Treatment of High Blood Cholesterol in Adults, Summary of the Third Report of the National Cholesterol Education Program (NCEP) Expert Panel on Detection, Evaluation, and Treatment of High Blood Cholesterol in Adults (Adult Treatment Panel III), *Journal of the American Medical Association* 285 (2001): 2486–2498.

62. Centers for Disease Control and Prevention, *Recommendations for Using Fluoride to Prevent and Control Dental Caries in the United States,* August 2001; Center for Disease Control, Fact Sheet, *Preventing Dental Caries,* October 2002; www.cdc.gov/OralHealth/factsheets/dental_caries.htm.

63. Ibid.

64. Position of the American Dietetic Association: Nutrition intervention in the treatment of anorexia nervosa, bulimia nervosa, and eating disorders not otherwise specified (EDNOS), *Journal of the American Dietetic Association* 101 (2001): 810.

65. Administration on Aging, *Profile of Older Americans, 2005* (Hyattsville, MD: U.S. Department of Health and Human Services, 2005).

66. D. Christensen, Making sense of centenarians, *Science News* 159 (2001): 156; Aging: Living to 100: What's the secret? *Harvard Health Letter* 27, no. 3 (2002): 1.

67. Position Paper of the American Dietetic Association: Nutrition across the spectrum of aging, *Journal of the American Dietetic Association* 105 (2005): 616–633.

68. Administration on Aging, *Profile of Older Americans, 2005* (Hyattsville, MD: U.S. Department of Health and Human Services, 2005).

69. H. Kerschener, Productive aging: A quality of life agenda, *Journal of the American Dietetic Association* 98 (1998): 1445–1448; C. M. Wellington and B. A. Piet, Living younger: A lifestyle and nutrition program for seniors 54 and better, *Journal of the American Dietetic Association* 101 (2001): A-79; N. Sahyoun, Nutrition education for the healthy elderly population: Isn't it time? *Journal of Nutrition Education and Behavior* 34 (2002): S42–S47; Report of a Joint WHO/FAO Expert Consultation, *Diet, Nutrition, and the Prevention of Chronic Diseases,* WHO Technical Report Series 916 (Geneva: World Health Organization, 2003); J. G. Dausch, Aging issues moving mainstream, *Journal of the American Dietetic Association* 103 (2003): 683–684.

70. AMA Conference Proceedings, Unified dietary recommendations, summary of a scientific conference on preventive nutrition: Pediatrics to geriatrics, *Circulation* 100 (1999): 450–455.

71. K. E. Miller and coauthors, The geriatric patient: A systematic approach to maintaining health, *American Family Physician* 61 (2000): 1089–1092.

72. Position Paper of the American Dietetic Association: Nutrition across the spectrum of aging, *Journal of the American Dietetic Association* 105 (2005): 616–633.

73. Institute of Medicine, *Dietary Reference Intakes for Energy, Carbohydrate, Fiber, Fat, Fatty Acids, Cholesterol, Protein, and Amino Acids* (Washington, DC: National Academy Press, 2002).

74. American Society for Nutritional Science, Modified food guide pyramid for people over 70 years of age, *Journal of Nutrition* 129 (1999): 751–753.

75. Parts of the following discussion were adapted from B. J. Willcox, D. C. Willcox, and M. Suzuki, *The Okinawa Program: Learn the Secrets to Healthy Longevity* (New York: Three Rivers Press, 2001), and D. Buettner, The secrets of long life, *National Geographic* 208 (2005): 2–27.

76. Ibid.

77. Position Paper of the American Dietetic Association: Nutrition across the spectrum of aging, *Journal of the American Dietetic Association* 105 (2005): 616–633; L. McBean and coauthors, Healthy eating in later years, *Nutrition Today* 36 (2001): 192–201.

78. K. M. Fairfield and R. H. Fletcher, Vitamins for chronic disease prevention in adults, *Journal of the American Medical Association* 287 (2002): 3116–3126; R. M. Russell and H. W. Baik, Clinical implications of vitamin B_{12} deficiency in the elderly, *Nutrition in Clinical Care* 4 (2001): 214–220; R. Semba and coauthors, Vitamin D deficiency among older women with and without disability, *American Journal of Clinical Nutrition* 72 (2000): 1529.

79. A. Sarubin, *The Health Professional's Guide to Popular Dietary Supplements,* 2nd ed. (Chicago: The American Dietetic Association, 2002); F. Tripp, The use of dietary supplements in the elderly: Current issues and recommendations, *Journal of the American Dietetic Association* 97 (1997): S181–S183; M. Freeman and coauthors, Cognitive, behavioral, and environmental correlates of nutrient supplement use among independently living older adults, *Journal of Nutrition for the Elderly* 17 (1998): 19–37; J. Howard and coauthors, Investigating relationships between nutrition knowledge, attitudes, and beliefs and dietary adequacy of the elderly, *Journal of Nutrition for the Elderly* 17 (1998): 38–51.

80. American Dietetic Association, *Nutrition Care of the Older Adult: A Handbook for Dietetics Professionals Working Throughout the Continuum of Care,* 2nd ed. (Chicago: American Dietetic Association, 2004);

J. L. Cummings and G. Cole, Alzheimer disease, *Journal of the American Medical Association* 287 (2002): 2335–2338.

81. Administration on Aging, *Profile of Older Americans, 2005* (Hyattsville, MD: U.S. Department of Health and Human Services, 2005).
82. Ibid.
83. J. V. White and coauthors, Nutrition Screening Initiative: Development and implementation of the public awareness checklist and screening tools, *Journal of the American Dietetic Association* 92 (1992): 163–167.
84. Ibid.
85. J. Weinberg, Psychologic implications of the nutritional needs of the elderly, *Journal of the American Dietetic Association* 60 (1972): 293–296.
86. N. Wellman, L. Y. Rosenzweig, and J. L. Lloyd, Thirty years of the Older American Nutrition Program, *Journal of the American Dietetic Association* 102 (2002): 348–350; Mathematica Policy Research, Inc., *Serving Elders at Risk, The Older Americans Act Nutrition Programs: National Evaluation of the Elderly Nutrition Program 1993–1995, Volume 1: Title III, Evaluation Findings* (Washington, DC: U.S. Department of Health and Human Services, 1996); B. E. Millen and coauthors, The Elderly Nutrition Program: An effective national framework for preventive nutrition interventions, *Journal of the American Dietetic Association* 102 (2002): 234–240.
87. P. G. Kittler and K. Sucher, *Food and Culture in America: A Nutrition Handbook*, 2nd ed. (Belmont, CA: Wadsworth Publishing Co., 1998).
88. American Institute for Cancer Research, *Cooking Solo: Cooking Ideas for One or Two that Keep Cancer Prevention in Mind* (Washington, DC: American Institute for Cancer Research, 2001), 9.
89. The list of advantages is adapted from D. A. Roe, *Geriatric Nutrition*, 3rd ed. (Englewood Cliffs, NJ: Prentice Hall, 1992), 1–9.
90. Institute of Medicine, *Preventing Childhood Obesity: Health in the Balance* (Washington, DC: National Academy of Medicine, 2005); Childhood obesity in the United States: Facts and Figures, September 2004, www.iom.edu.
91. C. L. Ogden and coauthors, Prevalence of overweight and obesity in the United States, 1999–2004, *Journal of the American Medical Association* 295 (2006): 1549–1555.
92. Childhood obesity in the United States: Facts and Figures, September 2004; American Diabetes Association, Obesity speeds onset of both types of diabetes in kids, Statistics about Youth & Diabetes, www.diabetes.org/diabetes-statistics/children.jsp.
93. Institute of Medicine, *Preventing Childhood Obesity: Health in the Balance* (Washington, DC: The National Academies Press, 2005).
94. Parts of the following discussion are from D. M. Hoelscher and C. McCullum-Gomez, Addressing the obesity epidemic, in M. Struble and D. Holben, *Community Nutrition in Action: An Entrepreneurial Approach* (Belmont, CA: Thomson Wadsworth, 2006).
95. Ibid; M. P. Galvez, T. R. Frieden, and P. J. Landrigan, Obesity in the 21st century, *Environmental Health Perspective* 111 (2003): 684–685; C. B. Ebbeling, D. B. Pawlak, and D. S. Ludwig, Childhood obesity: Public-health crisis, common sense cure, *Lancet* 360 (2002): 473–482.
96. J. T. Dwyer and coauthors, Predictors of overweight and overfatness in a multiethnic pediatric population, *American Journal of Clinical Nutrition* 67 (1998): 602–610.
97. C. J. Crespo and coauthors, Television watching, energy intake, and obesity in U.S. children: Results from the third National Health and Nutrition Examination Survey, 1988–1994, *Archives of Pediatrics and Adolescent Medicine* 155 (2001): 360–365.
98. C. Ebbeling and coauthors, Childhood obesity: Public health crisis, common sense cure, *Lancet* 360 (2002): 473–475; G. Wang and W. H. Dietz, Economic burden of obesity in youths aged 6 to 17 years: 1979–1999, *Pediatrics* 109 (2002): E81.
99. P. Basiotis and coauthors, The healthy eating index: 1999–2000, U.S. Department of Agriculture, Center for Nutrition Policy and Promotion, CNPP-12; Center for Nutrition Policy and Promotion, Report card on the diet quality of children ages 2 to 9, *Nutrition Insights* (September 2001), www.cnpp.usda.gov.
100. Position of the American Dietetic Association: Dietary guidance for healthy children ages 2 to 11 years, *Journal of the American Dietetic Association* 104 (2004): 660–677.
101. S. A. Bowman and coauthors, Effects of fast-food consumption on energy intake and diet quality among children in a national household survey, *Pediatrics* 113 (2004): 112–118, as cited in D. M. Hoelscher and C. McCullum-Gomez, Addressing the obesity epidemic, in M. Struble and D. Holben, *Community Nutrition in Action: An Entrepreneurial Approach* (Belmont, CA: Thomson Wadsworth, 2006).
102. Institute of Medicine, *Preventing Childhood Obesity: Health in the Balance* (Washington, DC: The National Academies Press, 2005); K. A. Coon and coauthors, Watching television at meals is related to food consumption patterns in children, *Pediatrics* 107 (2001): E7; C. Byrd-Bredbenner and coauthors, Nutrition messages on prime-time television programs, *Topics in Clinical Nutrition* 16 (2001): 61–72.
103. M. Golan, Parents as the exclusive agents of change in the treatment of childhood obesity, *The American Journal of Clinical Nutrition* 67 (1998): 1130–1135; S. St. Jeor and coauthors, Family-based interventions for the treatment of childhood obesity, *Journal of the American Dietetic Association* 102 (2002): 640–644.
104. The following discussion is adapted in part from M. Struble and D. Holben, *Community Nutrition in Action: An Entrepreneurial Approach* (Belmont, CA: Thomson Wadsworth, 2006), 367–369.
105. K. Anderson and coauthors, Eat smart: North Carolina's recommended standards for all foods available in school. North Carolina Department of Health and Human Services, North Carolina Division of Public Health, 2004; Texas Department of Agriculture. Texas public school nutrition policy, revised, August 1, 2004, Healthykids@agr.state.tx.us.
106. U.S. Department of Health and Human Services, *The Surgeon General's Call to Action to Prevent and Decrease Overweight and Obesity* (U.S. Department of Health and Human Services, Public Health Service, Office of the Surgeon General, 2001).
107. The following discussion is adapted from D. M. Hoelscher and C. McCullum-Gomez, Addressing the obesity epidemic, in M. Struble and D. Holben, *Community Nutrition in Action: An Entrepreneurial Approach* (Belmont, CA: Thomson Wadsworth, 2006), 249.
108. T. Baranowski and coauthors, School-based obesity prevention: A blueprint for taming the epidemic, *American Journal of Health Behavior* 26 (2002): 486–493.
109. T. Dwyer and coauthors, An investigation of the effects of daily physical activity on the health of primary school students in South Australia, *International Journal of Epidemiology* 12 (1983): 308–313.
110. D. M. Hoelscher and coauthors, Designing effective interventions for adolescents, *Journal of the American Dietetic Association* 102 (2002): 552–563.
111. Institute of Medicine, *Preventing Childhood Obesity: Health in the Balance* (Washington, DC: The National Academies Press, 2005).

CHAPTER 12

1. B. Bruemmer, Food biosecurity, *Journal of the American Dietetic Association* 103 (2003): 687–691.
2. Preliminary FoodNet data on the incidence of foodborne illnesses—Selected sites, United States, 2001, *Morbidity and Mortality Weekly Report* 51 (2002): 325.
3. Economic Research Service, *Economics of Foodborne Disease: Food and Pathogens*, www.ers.usda.gov/Briefing/FoodborneDisease; J. C. Buzby, Children and microbial foodborne illness, Food Review 24 (2001): 32–37; Position of the American Dietetic Association: Food and water safety, *Journal of the American Dietetic Association* 97 (1997): 1048–1053.
4. D. W. K. Acheson, Emerging foodborne pathogens, *Nutrition & the M.D.* 29, no. 3 (2003): 1–4; A. Hingley, Campylobacter: Low-profile bug is food poisoning leader, *FDA Consumer* (September/October 1999): 14–17.
5. B. P. Bell and coauthors, A multi-state outbreak of *Escherichia coli* O157:H7-associated bloody diarrhea and hemolytic uremic syndrome from hamburgers: The Washington experience, *Journal of the Ameri-*

can Medical Association 272 (1994): 1349–1353; U.S. Department of Agriculture Food Safety and Inspection Service, *E. coli* O157:H7 at a glance, *Food News for Consumers* 10 (1993): 5.

6. R. E. Besser and coauthors, An outbreak of diarrhea and hemolytic uremic syndrome from *Escherichia coli* O157:H7 in fresh-pressed apple cider, *Journal of the American Medical Association* 269 (1993): 2217–2219; M. Neill, Foodborne illness and food safety, in B. A. Bowman and R. M. Russell, eds., *Present Knowledge in Nutrition,* 8th ed. (Washington, DC: International Life Sciences Institute, 2001), 717–724.
7. S. C. Witt, *Biotechnology, Microbes, and the Environment* (San Francisco: Center for Science Information, 1990), 182–183.
8. L. Kopp-Hoolihan, Prophylactic and therapeutic uses of probiotics: A review, *Journal of the American Dietetic Association* 101 (2001): 229–238, 241; G. Reid, Probiotics for urogenital health, *Nutrition in Clinical Care* 5 (2002): 3–8.
9. R. S. Horowitz and coauthors, Jin Bu Huan toxicity in children—Colorado 1993, *Morbidity and Mortality Weekly Report* 42 (1993): 633–635.
10. U.S. Department of Health and Human Services, Public Health Service, FDA Warns Consumers Against Nature's Nutrition Formula One, statement issued February 28, 1995; W. K. Jones, Safety of Dietary Supplements Containing Ephredrine Alkaloids, FDA Public Meeting Summary, August 2000, www.cfsan.fda.gov/list.html; Food and Drug Administration, Evidence on the Safety and Effectiveness of Ephedra: Implications for Regulation, February 28, 2003; Food and Drug Administration, HHS acts to reduce potential risks of dietary supplements containing ephedra, *FDA News* (February 28, 2003).
11. R. S. Koff, Herbal hepatotoxicity: Revisiting a dangerous alternative, *Journal of the American Medical Association* 273 (1995): 502; Food and Drug Administration, Center for Food Safety and Applied Nutrition, Kava-containing dietary supplements may be associated with severe liver injury, *Consumer Advisory* (March 25, 2002).
12. Position of the American Dietetic Association: Nutrition and lifestyle for a healthy pregnancy outcome, *Journal of the American Dietetic Association* 102 (2002): 1479–1490.
13. N. D. Vietmeyer, The preposterous puffer, *National Geographic* 166 (1984): 260–270.
14. L. Bren, Trying to keep mad cow disease out of U.S. herds, *FDA Consumer* (March/April 2001): 12; R. McCarty, Managing bovine spongiform encephalopathy risk in the United States, *Nutrition Today* 37 (2002): 17–18; D. W. K. Acheson, Bovine spongiform encephalopathy (mad cow disease), *Nutrition Today* 37 (2002): 19–25; Food and Drug Administration, Bovine spongiform encephalopathy (BSE), May 2003 update, www.cfsan.fda.gov/list.html.
15. P. Kurtzweil, Critical steps toward safer seafood, *FDA Consumer,* Publication No. (FDA) 99-2317 (November/December 1997, revised February 1999).
16. Food Safety and Inspection Service, *Food Safety Facts: Barbecue Food Safety* (Washington, DC: United States Department of Agriculture, April 2003).
17. J. Trager, *The Food Chronology* (New York: Henry Holt and Company, 1995), as cited in *Food Folklore: Tales and Truths about What We Eat* (Minneapolis, MN: Chronimed Publishing, 1999).
18. J. F. Rosen and P. Mushak, Primary prevention of childhood lead poisoning—the only solution, *New England Journal of Medicine,* 344 (2001): 1470–1472.
19. International Food Information Council, *Pesticides and Food Safety* (Washington, DC: International Food Information Council Foundation, 1995), 2.
20. Ibid., 2–4; C. F. Chaisson and coauthors, *Pesticides in Food: A Guide for Professionals* (Chicago: The American Dietetic Association, 1991), 4–8.
21. National Academy of Sciences, National Research Council, *Pesticides in the Diets of Infants and Children* (Washington, DC: National Academy Press, 1993); U.S. Environmental Protection Agency, Why children may be especially sensitive to pesticides (January 2003).
22. United States Department of Agriculture, *The National Organic Program: Background Information,* www.ams.usda.gov/nop.
23. United States Department of Agriculture, *The National Organic Program: Background Information,* www.ams.usda.gov/nop.
24. CNN: In-Depth-Food—Organic Foods. The organic explainer, www.cnn.com/HEALTH/indepth.food/organic/explainer.html.
25. World Health Organization, *Safety and Nutritional Adequacy of Irradiated Food* (Geneva, Switzerland: World Health Organization, 1994), 4–5; WHO Technical Report Series 890, *High-Dose Irradiation: Wholesomeness of Food Irradiated with Doses above 10kGy* (Geneva, Switzerland: World Health Organization, 1999).
26. Position of the American Dietetic Association: Food irradiation, *Journal of the American Dietetic Association* 100 (2000): 246–253.
27. Position of the American Dietetic Association: Food irradiation, 2000; Community Nutrition Institute, Irradiation of meat begins amidst failure of safe food safety policy, *Nutrition Week* 8 (2000): 1–7.
28. J. Henkel, Genetic engineering: Fast forwarding to the future, *FDA Consumer* (April 1995): 6–11.
29. H. I. Miller, Foods of the future: The new biotechnology and FDA regulation, *Journal of the American Medical Association* 269 (1993): 910–912.
30. P. Abelson, A third technological revolution, *Science* 279 (1998): 5359; D. D. Stadler, A. F. Reeder, and C. H. Strohbehn, Application of genetic engineering to foods, *Topics in Clinical Nutrition* 14 (1999): 39–50; C. McCullum, Food biotechnology in the new millennium: Promises, realities, and challenges, *Journal of the American Dietetic Association* 100 (2000): 1311–1315; B. C. Babcock and C. A. Francis, Solving nutrition challenges requires more than new biotechnologies, *Journal of the American Dietetic Association* 100 (2000): 1308–1311; Position of the American Dietetic Association: Biotechnology and the future of food, *Journal of the American Dietetic Association* 106 (2006): 285–293.
31. United States Department of Agriculture, Agricultural Research Service, *USDA and Biotechnology,* www.usda.gov.
32. T. Weaver, *USDA Releases New Tomatoes with Increased Beta-Carotene* (Washington, DC: USDA Agricultural Research Service, 1998).
33. A. Moffaat, Toting up the harvest of transgenic plants, *Science* 282 (1998): 5397; U.S. Department of Energy Office of Science, Human Genome Program, Genetically Modified Foods and Organisms, updated March 12, 2003.
34. A. Moffaat, Toting up the harvest of transgenic plants, *Science* 282 (1998): 5397; M. A. Mackey and C. R. Santerre, Biotechnology and our food supply, *Nutrition Today* 35 (2000): 120–128; M. Mackey, The application of biotechnology to nutrition: An overview, *Journal of the American College of Nutrition* 21 (2002): 157S; M. Falk and coauthors, Food biotechnology: Benefits and concerns, *Journal of Nutrition* 132 (2002): 1384–1386.
35. J. A. Nordlee and coauthors, Identification of a Brazil-nut allergen in transgenetic soybeans, *New England Journal of Medicine* 334 (1996): 688–692.
36. L. Thompson, Are bioengineered foods safe? *FDA Consumer* 34 (2000): 18–23.
37. Community Nutrition Institute, Consumers Union calls for GE testing, labeling, *Nutrition Week* 34 (1999): 1–2; C. Silva and R. Leonard, U.S. prepares to OK labels on GE foods, *Nutrition Week* 45 (1999): 1–2; Community Nutrition Institute, Biotech debate heats up in Boston: Strong grassroots movement forming, *Nutrition Week* 30 (2000): 1–3.
38. G. Bickel, M. Andrews, and S. Carlson, The magnitude of hunger: In a new national measure of food security, *Topics in Clinical Nutrition* 13 (1998): 15–30; Position of the American Dietetic Association: Domestic food and nutrition security, *Journal of the American Dietetic Association* 106 (2006): 446–458.
39. Food and Agriculture Organization, *The State of Food Security in the World 2002* (Rome, Italy, 2002).
40. U. Ramakrishnan, Prevalence of micronutrient malnutrition worldwide, *Nutrition Reviews* 60 (2002): S46; J. B. Mason and coauthors, *The Micronutrient Report: Current Progress and Trends in the Control of Vitamin A, Iron, and Iodine Deficiencies* (Ottawa, Ontario: The

Micronutrient Initiative, 2001); United Nations Administrative Committee on Coordination Sub-Committee on Nutrition, *Fourth Report on the World Nutrition Situation* (Geneva, Switzerland: ACC/SCN in collaboration with the International Food Policy Research Institute, 2000); U. Kapil and A. Bhavna, Adverse effects of poor micronutrient status during childhood and adolescence, *Nutrition Reviews* 60 (2002): S84.

41. Community Nutrition Institute, Malnutrition accounts for over half of child deaths worldwide, says UNICEF, *Nutrition Week* 28 (1998): 1–2; G. H. Brundtland, Nutrition and infection: Malnutrition and mortality in public health, *Nutrition Reviews* 58, suppl. (2000): S1–S6.

42. United Nations Children's Fund, *The State of the World's Children 2001* (New York: UNICEF, 2001).

43. Ibid.

44. United Nations Children's Fund, *We the Children: Meeting the Promises of the World Summit for Children* (New York: UNICEF, 2001), www.unicef.org.

45. World Health Organization, *The Optimal Duration of Exclusive Breastfeeding: A Systematic Review* (Geneva, Switzerland: WHO, 2002).

46. M. Nord, M. Andrews, and J. Winicki, Frequency and duration of food insecurity and hunger in U.S. households, *Journal of Nutrition Education and Behavior* 34 (2002): 194–201; J. F. Guthrie and M. Nord, Federal activities to monitor food security, *Journal of the American Dietetic Association* 102 (2002): 906; K. Alaimo and coauthors, Food insufficiency, family income, and health in U.S. preschool and school-aged children, *American Journal of Public Health* (May 2001); J. T. Cook, Clinical implications of household food security: Definitions, monitoring, and policy, *Nutrition in Clinical Care* 5 (2002): 152–167.

47. The United States Conference of Mayors, *A Status Report on Hunger and Homelessness in America's Cities* (December 2005), http://usmayors .org.

48. Economic Research Service, Food program costs, 1970–2001, *Food Review* 25 (2002): 45; E. Kennedy and E. Cooney, Development of child nutrition programs in the United States, *Journal of Nutrition* 131 (2001): 431S; Position of the American Dietetic Association: Domestic food and nutrition security, *Journal of the American Dietetic Association* 106 (2006): 446–458.

49. America's Second Harvest, *Hunger in America, 2005*, November 2005, www.secondharvest.org.

50. P. Uvin, The state of world hunger, *Nutrition Reviews* 52 (1994): 151–161.

51. P. Foster and H. D. Leathers, *The World Food Problem: Tackling the Causes of Undernutrition in the Third World*, 2nd ed. (Boulder, CO: Lynne Rienner Publishers, 1999).

52. Bread for the World Institute, *Hunger 1998: Hunger in a Global Economy* (Silver Spring, MD: Bread for the World Institute, 1997).

53. L. R. Brown, *Who Will Feed China? Wake Up Call for a Small Planet* (New York: W. W. Norton, 1995); G. Gardner, *Shrinking Fields: Cropland Loss in a World of Eight Billion* (Washington, DC: Worldwatch Institute, 1996); M. W. Rosegrant, *Water Resources in the Twenty-First Century*, Discussion Paper 20 (Washington, DC: International Food Policy Research Institute, 1997).

54. Oxfam America, *Facts for Action: Women Creating a New World*, No. 3 (Boston: Oxfam America, 1991), 2–3; A. R. Quisumbing and coauthors, *Women: The Key to Food Security* (Washington, DC: The International Food Policy Research Institute, Food Policy Report, 1995).

55. Position of the American Dietetic Association: Addressing world hunger, malnutrition, and food insecurity, *Journal of the American Dietetic Association* 103 (2003).

56. United Nations Children's Fund, *We the Children: Meeting the Promises of the World Summit for Children* (New York: UNICEF, 2001); United Nations Children's Fund, *The State of the World's Children 2002* (New York: Oxford University Press, 2002).

57. C. D. Williams, N. Baumslag, and D. B. Jelliffe, *Mother and Child Health: Delivering the Services* (New York: Oxford University Press, 1994).

58. United Nations Children's Fund, *We the Children: Meeting the Promises of the World Summit for Children* (New York: UNICEF, 2001).

59. S. Lewis, Food security, environment, poverty, and the world's children, *Journal of Nutrition Education* 24 (1992): 35–55.

60. United Nations Children's Fund, *We the Children: Meeting the Promises of the World Summit for Children* (New York: UNICEF, 2001); Millennium Development Goals, 2002, in Bread for the World Institute, *Hunger 2002* (Washington, DC: Bread for the World Institute, 2002).

61. D. Elgin, *Voluntary Simplicity: Toward a Way of Life That Is Outwardly Simple, Inwardly Rich* (New York: William Morrow, 1981), 25.

62. G. McGovern, *The Third Freedom: Ending Hunger in our Time* (New York: Simon and Schuster, 2001), 11, 14–15.

Answers to In Review Questions

Chapter 1

1. d
2. a
3. b
4. d
5. d
6. b
7. a
8. b
9. b
10. d

Chapter 2

1. Calorie dense: c, d, f, h, j; nutrient dense: a, b, e, g, i
2. a
3. a
4. a
5. **A** stands for adequacy, which refers to your diet providing the nutrients, fiber, and energy that you require to live a healthy life. **B** stands for balance in the diet, meaning that no one food or food type is emphasized at the expense of another. **C** stands for calorie control, which means watching your energy intake and balancing it with the amount of energy you exert.
6. c
7. b
8. d
9. d
10. d

Chapter 3

1. b
2. Monosaccharides: glucose, fructose, galactose. Disaccharides: sucrose = glucose + fructose; lactose = glucose + galactose; maltose = glucose + glucose.
3. d
4. c
5. d
6. If the blood delivers more glucose than cells need, the liver and muscles take up the surplus to build the polysaccharide glycogen. Excess glucose can also be stored as body fat.
7. 7a = 1
   7b = 3
   7c = 2
8. b
9. c
10. Refer to Table 3-8.

Chapter 4

1. a
2. a
3. d
4. d

5. Saturation refers to the chemical structure—specifically to the number of hydrogens the fatty acid chain is holding. If every available bond from the carbons is holding hydrogen, the chain is referred to as saturated. However, if there is one point of unsaturation, it is referred to as monounsaturated. If there are two or more points of unsaturation, it is referred to as polyunsaturated.
6. e
7. Hydrogenation is a process that forces hydrogen into an unsaturated oil to create a firmer product (like spreadable margarine). This process enables some points of unsaturation to accept hydrogen causing them to become more saturated.
8. This means that the human body cannot synthesize these fatty acids from sources within the body. These two fatty acids are required by the body so they must be supplied by the diet.
9. c
10. b

Chapter 5

1. b
2. a
3. c
4. b
5. Help with acid–base balance; act as enzymes to help chemical reactions take place; help regulate fluid balance; work as antibodies within the immune system, can act as hormones; build tissue; provide energy; and act as transporters in various capacities in the body.
6. a
7. c
8. a
9. d
10. The body cannot store protein, therefore it must have the correct building blocks (amino acids) at hand to synthesize complete proteins for various uses. A complete protein is one containing all the essential amino acids in the right proportion relative to need. The quality of a food protein is judged by the proportions of essential amino acids it contains relative to our needs. Animal and soy proteins are the highest in quality.

Chapter 6

1. Water-soluble vitamins are carried in the blood, excreted in the urine, needed in frequent, small doses, and are less likely to reach toxic levels in the body. Fat-soluble vitamins are absorbed into the lymph and carried in the blood by protein carriers, stored in body fat, needed in periodic doses, and are more likely to be toxic when consumed in excess of needs.
2. e
3. d
4. e
5. d
6. d
7. c
8. A phytochemical is a nonnutritive substance found mostly in plant-based foods that may possess health-protective benefits. Some phytochemicals are believed to work as powerful antioxidants and are thought to reduce risk for cancer, heart disease, and other conditions.
9. a
10. The two classes of vitamins are: water-soluble (B vitamins and C) and fat-soluble (A, D, E, and K).

Chapter 7

1. c
2. b
3. b

4. *Major Minerals:* Calcium, Phosphorus, Magnesium, Sodium, Chloride, Potassium, Sulfur. *Trace Minerals:* Iodine, Iron, Zinc, Copper, Fluoride, Selenium, Chromium, Molybdenum, Manganese.
5. Cells regulate water balance by pumping minerals across their membranes. Minerals attract water to come with them wherever they go, so when the cells pump minerals in or out of the cell, water follows them. Sodium, potassium, and chloride are minerals used in this process and they are known as electrolytes.
6. d
7. a
8. e
9. Fluoride is indigenous to many water supplies, but water is further treated with the mineral because studies have shown that diets containing adequate amounts of fluoride promote healthy teeth and bones.
10. d

Chapter 8

1. e
2. d
3. The female body contains less water than the male body and therefore requires less alcohol to achieve a given blood alcohol level compared to a man. Alcohol dehydrogenase, the enzyme used to metabolize alcohol, is less abundant in the female body causing more undiluted alcohol to enter the bloodstream.
4. a
5. e
6. c
7. d
8. b
9. b
10. Alcohol can affect the way prescription medications or over-the-counter medications react in the body, alter the balance of sex hormones, and negatively impact nutrition status—to name a few. On the other hand, a reduced risk for heart disease in those who consume alcohol in moderation may occur.

Chapter 9

1. a
2. e
3. a
4. About 60% or more of the energy used by the body is used for basal metabolism. This involves beating of the heart, breathing, and other processes that go on without conscious awareness. Voluntary activities, such as running, walking, and general exercise, do not make as much of a contribution to the total energy used by the body, but they are more easily changed at will.
5. d
6. Energy balance is the relationship between amounts of calories consumed compared to the amount of calories expended. The balance between energy in and energy out determines whether a person stores or uses body fat. Weight loss means that calories eaten are less than calories expended. Weight gain means that calories eaten are greater than calories expended.
7. e
8. Usually are not sustainable; usually label "bad" and "good" foods; usually facilitate loss of water, minerals, vitamins, and lean body mass; and can set the dieter up for failure.
9. c
10. d

Chapter 10

1. e
2. c
3. b

4. Anaerobic exercise uses carbohydrate for energy and does not require oxygen for metabolism. An example is sprinting. Aerobic exercise requires oxygen for fat metabolism. Examples include: jogging, brisk walking, and swimming.
5. d
6. The body of an athlete may use slightly more protein because of increased muscle mass. Increased protein intake may be needed to maintain and gain muscle mass. The RDAs per kilogram of body weight are: Healthy adults, 0.8 g/kg/day; endurance athletes, 1.2–1.6 g/kg/day; and resistance training, 1.6–1.7 g/kg/day.
7. b
8. Jen is at risk for iron deficiency (anemia, sports anemia). Her risk factors include being a menstruating and growing female, being an athlete, and the fact that she probably doesn't eat a diet rich in iron.
9. b
10. b

Chapter 11

1. c
2. A pregnant woman needs additional calories and protein to support growth of new tissues, B vitamins to help with the metabolism of the increased caloric intake, folate and vitamin B_{12} for new cell growth, vitamin D and the bone minerals for skeletal growth, iron for fetal needs, and zinc for new cell growth and protein synthesis. A wise selection of foods meets the recommendations of the MyPyramid Food Guide with additional servings needed from all five food groups.
3. Breast milk is tailor-made to meet the nutrient needs of the young infant. The infant receives colostrum during the first 2 or 3 days of lactation, which contains antibodies and other immunity factors that protect the infant from infections. Other factors in breast milk, including several enzymes, hormones, and lipids, also protect against infection. Bifidus factor in breast milk favors the growth of the "friendly" probiotic bacteria in the infant's digestive tract.
4. c
5. d
6. d
7. b
8. d
9. Multiple prescriptions, lower income, living alone, GI disorders, and others listed in Table 11-11.
10. Family eating patterns, genetic susceptibility, television viewing, lack of activity, increased fast food consumption.

Chapter 12

1. b
2. a
3. a
4. b
5. c
6. b
7. a
8. c
9. d
10. Food intoxication occurs when a harmful chemical or toxin is transmitted by food and causes illness. Foodborne infections, on the other hand, occur as a result of eating a food that contains living microorganisms such as bacteria, viruses, or parasites capable of multiplying, thriving, and causing adverse effects on the body.

Table of Food Composition

This edition of the table of food composition has been updated to reflect current nutrient data for foods, to remove outdated foods, and to add foods that are new to the marketplace.* The nutrient database for this appendix is compiled from a variety of sources, including the USDA Standard Release database (Release 17) and manufacturers' data. The USDA database provides data for a wider variety of foods and nutrients than other sources. Because laboratory analysis for each nutrient can be quite costly, manufacturers tend to provide data only for those nutrients mandated on food labels. Consequently, data for their foods are often incomplete; any missing information on this table is designated as a dash. Keep in mind that a dash means only that the information is unknown and should not be interpreted as a zero. A zero means that the nutrient is not present in the food.

Whenever using nutrient data, remember that many factors influence the nutrient contents of foods. These factors include the mineral content of the soil, the diet fed to the animal or the fertilizer used on the plant, the season of harvest, the method of processing, the length and method of storage, the method of cooking, the method of analysis, and the moisture content of the sample analyzed. With so many influencing factors, users should view nutrient data as a close approximation of the actual amount.

For updates, corrections, and a list of 8,000 additional foods and codes found in the diet analysis software that accompanies this text, visit **www.thomsonedu.com/nutrition** and click on *Diet Analysis Plus*.

- ***Fats*** Total fats, as well as the breakdown of total fats to saturated, monounsaturated, and polyunsaturated fats, are listed in the table. The fatty acids seldom add up to the total in part due to rounding but also because values are derived from a variety of laboratories and other fatty acid components.
- **Trans *Fats*** *Trans* fat data has been listed in the table. Because food manufacturers have been required to report *trans* fats on food labels only since January 1, 2006, much of the data is incomplete. Missing *trans* fat data is designated with a dash. As additional *trans* fat data becomes available, the table will be updated.
- ***Vitamin A and Vitamin E*** In keeping with the 2001 RDA for vitamin A, this appendix presents data for vitamin A in micrograms (μg) RAE. Similarly, because the 2000 RDA for vitamin E is based only on the alpha-tocopherol form of vitamin E, this appendix reports vitamin E data in milligrams (mg) alpha-tocopherol, listed on the table as Vit E (mg α).
- ***Bioavailability*** Keep in mind that the availability of nutrients from foods depends not only on the quantity provided by a food, but also on the amount absorbed and used by the body—the bioavailability. The bioavailability of folate from fortified foods, for example, is greater than from naturally occurring sources. Similarly, the body can make niacin from the amino acid tryptophan, but niacin values in this table (and most databases) report preformed niacin only.
- ***Using the Table*** The foods and beverages in this table have been organized into several categories, which are listed at the head of each right-hand page. Page numbers have been provided, and each group has been color-coded to make it easier to find individual foods.

*This food composition table has been prepared by Wadsworth Publishing Company. The nutritional data are supplied by Axxya Systems.

TABLE F-1

Food Composition (Computer code number is for Wadsworth Diet Analysis program) (For purposes of calculations, use "0" for t, <1, <.1, <.01, etc.)

DA + Code	Food Description	Quantity	Measure	Wt (g)	H_2O (g)	Ener (kcal)	Prot (g)	Carb (g)	Dietary Fiber (g)	Fat (g)	Fat Breakdown (g) Sat	Mono	Poly	Trans
	BREADS, BAKED GOODS, CAKES, COOKIES, CRACKERS, CHIPS, PIES													
	Bagels													
8534	Cinnamon & raisin	1	item(s)	71	23	195	7	39	2	1	0.19	0.12	0.48	—
4910	Enriched, all varieties	1	item(s)	71	23	195	7	38	2	1	0.16	0.09	0.49	0
4911	Plain, enriched, toasted	1	item(s)	66	18	195	7	38	2	1	0.16	0.09	0.49	0
8538	Oat bran	1	item(s)	71	23	181	8	38	3	1	0.14	0.18	0.35	—
12079	Whole grain	1	item(s)	85	—	170	9	35	6	2.5	0	—	—	0
	Biscuits													
25008	Biscuits	1	item(s)	41	16	121	3	16	1	5	1.40	1.41	1.82	0
16729	Scone	1	item(s)	42	11	149	4	19	1	6	2.01	2.55	1.26	—
25166	Wheat biscuits	1	item(s)	55	21	162	4	22	1	7	1.90	1.92	2.51	0
	Bread													
325	Boston brown, canned	1	slice(s)	45	21	88	2	19	2	1	0.13	0.09	0.25	—
8716	Bread sticks, plain	4	item(s)	24	1	99	3	16	1	2	0.34	0.86	0.87	—
25176	Cornbread	1	piece(s)	55	26	141	5	18	1	5	2.09	1.44	1.50	0
327	Cracked wheat	1	slice(s)	25	9	65	2	12	1	1	0.23	0.48	0.17	—
9079	Croutons, plain	¼	cup(s)	8	<1	31	1	6	<1	<1	0.11	0.23	0.10	—
8582	Egg	1	slice(s)	40	14	115	4	19	1	2	0.64	0.92	0.44	—
8585	Egg, toasted	1	slice(s)	37	10	117	4	19	1	2	0.60	1.11	0.43	—
329	French	1	slice(s)	25	9	69	2	13	1	1	0.16	0.30	0.17	—
8591	French, toasted	1	slice(s)	23	7	69	2	13	1	1	0.16	0.30	0.17	—
8597	Indian fry	1	item(s)	90	24	296	6	48	2	9	2.08	3.59	2.33	—
332	Italian	1	slice(s)	30	11	81	3	15	1	1	0.26	0.24	0.42	—
1393	Mixed grain	1	slice(s)	26	10	65	3	12	2	1	0.21	0.40	0.24	—
8604	Mixed grain, toasted	1	slice(s)	24	8	65	3	12	2	1	0.21	0.40	0.24	—
8605	Oat bran	1	slice(s)	30	13	71	3	12	1	1	0.21	0.48	0.51	—
8608	Oat bran, toasted	1	slice(s)	27	10	70	3	12	1	1	0.21	0.47	0.50	—
8609	Oatmeal	1	slice(s)	27	10	73	2	13	1	1	0.19	0.43	0.46	—
8613	Oatmeal, toasted	1	slice(s)	25	8	73	2	13	1	1	0.19	0.43	0.46	—
1409	Pita	1	item(s)	60	19	165	5	33	1	1	0.10	0.06	0.32	—
7905	Pita, whole wheat	1	item(s)	64	20	170	6	35	5	2	0.26	0.22	0.68	—
338	Pumpernickel	1	slice(s)	32	12	80	3	15	2	1	0.14	0.30	0.40	—
334	Raisin, enriched	1	slice(s)	26	9	71	2	14	1	1	0.28	0.60	0.18	—
8625	Raisin, toasted	1	slice(s)	24	7	71	2	14	1	1	0.28	0.60	0.18	—
10168	Rice, white	1	slice(s)	42	—	140	1	21	1	6	0.50	—	—	0
8653	Rye	1	slice(s)	32	12	83	3	15	2	1	0.20	0.42	0.26	—
8654	Rye, toasted	1	slice(s)	29	9	82	3	15	2	1	0.20	0.42	0.25	—
336	Rye, light	1	slice(s)	25	9	65	2	12	2	1	0.20	0.30	0.30	—
8588	Sourdough	1	slice(s)	25	9	69	2	13	1	1	0.16	0.30	0.17	—
8592	Sourdough, toasted	1	slice(s)	23	7	69	2	13	1	1	0.16	0.30	0.17	—
491	Submarine or hoagie roll	1	item(s)	135	41	400	11	72	4	8	1.80	3.00	2.20	—
8596	Vienna, toasted	1	slice(s)	23	7	69	2	13	1	1	0.16	0.30	0.17	—
8670	Wheat	1	slice(s)	25	9	65	2	12	1	1	0.22	0.43	0.23	—
8671	Wheat, toasted	1	slice(s)	23	7	65	2	12	1	1	0.22	0.43	0.23	—
340	White	1	slice(s)	25	9	67	2	13	1	1	0.18	0.17	0.34	—
1395	Whole wheat	1	slice(s)	46	15	128	4	24	3	2	0.37	0.53	1.35	—
	Cakes													
386	Angel food, from mix	1	slice(s)	50	16	129	3	29	<1	<1	0.02	0.01	0.06	—
8772	Butter pound, ready to eat, commercially prepared	1	slice(s)	75	18	291	4	37	<1	15	8.67	4.43	0.80	—
8737	Carrot, cream cheese frosting, from mix	1	slice(s)	111	23	484	5	52	1	29	5.43	7.24	15.10	—
4931	Chocolate, chocolate icing, commercially prepared	1	slice(s)	64	15	235	3	35	2	10	3.05	5.61	1.18	—
8756	Chocolate, from mix	1	slice(s)	95	23	340	5	51	2	14	5.16	5.74	2.62	—
393	Devil's food cupcake, chocolate frosting	1	item(s)	35	8	120	2	20	1	4	1.80	1.60	0.60	—
8757	Fruitcake, ready to eat, commercially prepared	1	piece(s)	43	11	139	1	26	2	4	0.45	1.81	1.43	—
1397	Pineapple upside down, from mix	1	slice(s)	115	37	367	4	58	1	14	3.35	5.97	3.77	—
411	Sponge, from mix	1	slice(s)	63	19	187	5	36	<1	3	0.82	0.99	0.41	—
8817	White, coconut frosting, from mix	1	slice(s)	112	23	399	5	71	1	12	4.36	4.14	2.42	—
8819	Yellow, chocolate frosting, ready to eat, commercially prepared	1	slice(s)	64	14	243	2	35	1	11	2.98	6.14	1.35	—
8822	Yellow, vanilla frosting, ready to eat, commercially prepared	1	slice(s)	64	14	239	2	38	<1	9	1.52	3.91	3.30	—
	Snack cakes													
8791	Chocolate snack cake, creme filled, w/frosting	1	item(s)	50	10	188	2	30	<1	7	1.43	2.85	2.62	—
25010	Cinnamon coffee cake	1	piece(s)	72	23	231	4	36	1	8	2.19	2.65	2.99	0

PAGE KEY: A–58 = Breads/Baked Goods A–62 = Cereal/Rice/Pasta A–66 = Fruit A–70 = Vegetables/Legumes A–80 = Nuts/Seeds A–82 = Vegetarian
A–84 = Dairy A–90 = Eggs A–90 = Seafood A–92 = Meats A–96 = Poultry A–96 = Processed Meats A–98 = Beverages A–102 = Fats/Oils
A–104 = Sweets A–106 = Sauces/Condiments/Spices A–108 = Mixed Foods/Soups/Sandwiches A–114 = Fast Food A–130 = Convenience A–132 = Baby Foods

Chol (mg)	Calc (mg)	Iron (mg)	Magn (mg)	Pota (mg)	Sodi (mg)	Zinc (mg)	Vit A (RAE) (µg)	Thia (mg)	Vit E (mg α)	Ribo (mg)	Niac (mg)	Vit B_6 (mg)	Fola (µg)	Vit C (mg)	Vit B_{12} (µg)	Sele (µg)
0	13	2.70	20	105	229	0.80	15	0.27	0.22	0.20	2.19	0.04	79	<1	0	22
0	53	2.53	21	72	379	0.62	0	0.38	0.07	0.22	3.24	0.04	75	0	0	23
0	53	2.52	20	72	379	0.62	0	0.31	0.08	0.20	2.91	0.03	64	0	0	23
0	9	2.19	22	82	360	0.64	1	0.24	0.23	0.24	2.10	0.03	70	<1	0	24
0	200	1.08	120	0	200	4.5	0	0.44	—	0.5	8	0.6	—	0	1.79	0
<1	33	1.01	6	37	205	0.27	9	0.13	0.01	0.12	1.08	0.01	26	0	<.1	7
49	80	1.31	7	48	288	0.29	—	0.15	0.43	0.16	1.20	0.03	8	<.1	<1	—
<1	57	1.22	16	81	321	0.42	12	0.16	0.01	0.13	1.49	0.03	29	<.1	<.1	12
<1	32	0.95	28	143	284	0.23	11	0.01	0.14	0.05	0.50	0.04	5	0	<.1	10
0	5	1.03	8	30	158	0.21	0	0.14	0.24	0.13	1.27	0.02	39	0	0	9
21	88	1.01	10	59	209	0.57	38	0.13	0.33	0.16	0.98	0.04	36	2	<1	6
0	11	0.70	13	44	135	0.31	0	0.09	—	0.06	0.92	0.08	15	0	<.1	6
0	6	0.31	2	9	52	0.07	0	0.05	—	0.02	0.41	0.00	10	0	0	3
20	37	1.22	8	46	197	0.32	25	0.18	0.10	0.17	1.94	0.03	42	0	<.1	12
21	38	1.24	8	47	200	0.32	26	0.14	0.11	0.16	1.77	0.02	36	0	<.1	12
0	19	0.63	7	28	152	0.22	0	0.13	0.08	0.08	1.19	0.01	37	0	0	8
0	19	0.63	7	28	152	0.22	0	0.10	0.07	0.07	1.07	0.01	22	0	0	8
0	210	3.24	14	67	626	0.45	0	0.39	—	0.27	3.27	0.02	67	0	0	21
0	23	0.88	8	33	175	0.26	0	0.14	0.09	0.09	1.31	0.01	57	0	0	8
0	24	0.90	14	53	127	0.33	0	0.11	0.09	0.09	1.13	0.09	31	<.1	<.1	8
0	24	0.90	14	53	127	0.33	0	0.08	0.08	0.08	1.02	0.08	28	<.1	<.1	8
0	20	0.94	11	44	122	0.27	1	0.15	0.13	0.10	1.45	0.02	24	0	0	9
0	19	0.93	9	33	121	0.28	1	0.12	0.13	0.09	1.29	0.01	19	0	0	9
0	18	0.73	10	38	162	0.28	1	0.11	0.13	0.06	0.85	0.02	17	0	<.1	7
0	18	0.74	10	39	163	0.28	1	0.09	0.13	0.06	0.77	0.02	13	<.1	<.1	7
0	52	1.57	16	72	322	0.50	0	0.36	0.18	0.20	2.78	0.02	64	0	0	16
0	10	1.96	44	109	340	0.97	0	0.22	0.39	0.05	1.82	0.17	22	0	0	28
0	22	0.92	17	67	215	0.47	0	0.10	0.13	0.10	0.99	0.04	30	0	0	8
0	17	0.75	7	59	101	0.19	0	0.09	0.07	0.10	0.90	0.02	28	<.1	0	5
0	17	0.76	7	59	102	0.19	0	0.07	0.07	0.09	0.81	0.02	24	<.1	0	5
0	40	1.08	—	45	160	—	0	0.23	—	0.14	1.20	—	40	0	—	—
0	23	0.91	13	53	211	0.36	0	0.14	0.11	0.11	1.22	0.02	35	<1	0	10
0	23	0.90	12	53	210	0.36	0	0.11	0.11	0.10	1.09	0.02	30	<.1	0	10
0	20	0.70	4	51	175	0.18	0	0.10	—	0.08	0.80	0.01	5	0	<.1	8
0	19	0.63	7	28	152	0.22	0	0.13	0.08	0.08	1.19	0.01	37	0	0	8
0	19	0.63	7	28	152	0.22	0	0.10	0.07	0.07	1.07	0.01	22	0	0	8
0	100	3.80	—	128	683	—	0	0.54	—	0.33	4.50	0.05	—	0	—	42
0	19	0.63	7	28	152	0.22	0	0.10	0.07	0.07	1.07	0.01	22	0	0	8
0	26	0.83	12	50	133	0.26	0	0.10	0.07	0.07	1.03	0.02	23	0	0	8
0	26	0.83	12	50	132	0.26	0	0.08	0.07	0.06	0.93	0.02	19	0	0	8
0	38	0.94	6	25	170	0.19	0	0.11	0.05	0.08	1.10	0.02	28	0	0	4
0	15	1.43	37	144	159	0.69	0	0.14	0.35	0.10	1.83	0.09	30	0	0	18
0	42	0.12	4	68	255	0.07	0	0.05	0.00	0.10	0.09	0.00	10	0	<.1	8
166	26	1.04	8	89	299	0.35	112	0.10	—	0.98	0.03	31	0	<1	7	
60	28	1.39	20	124	273	0.54	—	0.15	—	0.17	1.13	0.08	13	1	<1	—
27	28	1.41	22	128	214	0.44	—	0.02	—	0.09	0.37	0.03	11	<.1	<.1	2
55	57	1.53	30	133	299	0.66	38	0.13	—	0.20	1.08	0.04	26	<1	<1	11
19	21	0.70	—	46	92	—	—	0.04	—	0.05	0.30	—	2	0	—	2
2	14	0.89	7	66	116	0.12	3	0.02	0.39	0.04	0.34	0.02	9	<1	<.1	1
25	138	1.70	15	129	367	0.36	71	0.18	—	0.18	1.37	0.04	30	1	<.1	11
107	26	1.00	6	89	144	0.37	49	0.10	—	0.19	0.76	0.04	25	0	<1	12
1	101	1.30	13	111	318	0.37	13	0.14	0.13	0.21	1.19	0.03	35	<1	<.1	12
35	24	1.33	19	114	216	0.40	21	0.08	—	0.10	0.80	0.02	14	0	<1	2
35	40	0.68	4	34	220	0.16	12	0.06	—	0.04	0.32	0.02	17	0	<.1	4
9	37	1.68	21	61	213	0.26	3	0.11	1.09	0.15	1.21	0.01	20	0	<.1	1
26	50	1.46	10	81	277	0.38	35	0.14	0.23	0.16	1.17	0.02	30	<.1	<1	10

TABLE F-1

Food Composition (Computer code number is for Wadsworth Diet Analysis program) (For purposes of calculations, use "0" for t, <1, <.1, <.01, etc.)

DA + Code	Food Description	Quantity	Measure	Wt (g)	H_2O (g)	Ener (kcal)	Prot (g)	Carb (g)	Dietary Fiber (g)	Fat (g)	Fat Breakdown (g) Sat	Mono	Poly	Trans
	BREADS, BAKED GOODS, CAKES, COOKIES, CRACKERS, CHIPS, PIES—Continued													
16777	Funnel cake	1	item(s)	90	37	278	7	29	1	14	2.77	4.46	6.33	—
8794	Sponge snack cake, creme filled	1	item(s)	43	9	155	1	27	<1	5	1.09	1.73	1.40	—
	Snacks, chips, pretzels													
29428	Bagel chips, plain	3	item(s)	29	—	130	3	19	1	5	0.50	—	—	—
29429	Bagel chips, toasted onion	3	item(s)	29	—	130	4	20	1	5	0.50	—	—	—
38192	Chex traditional snack mix	1	cup(s)	46	—	198	3	33	2	6	0.76	—	—	—
654	Potato chips, salted	20	item(s)	28	1	152	2	15	1	10	3.11	2.79	3.46	—
8816	Potato chips, unsalted	20	item(s)	28	1	152	2	15	1	10	3.11	2.79	3.46	—
4641	Tortilla chips, plain	6	item(s)	28	1	142	2	18	2	7	1.43	4.39	1.03	—
5096	Pretzels, plain, hard, twists	5	item(s)	30	1	114	3	24	1	1	0.23	0.41	0.37	—
4632	Pretzels, whole wheat	1	ounce(s)	28	1	103	3	23	2	1	0.16	0.29	0.24	—
	Cookies													
8859	Animal crackers	12	piece(s)	30	0	134	2	22	<1	4	1.03	2.29	0.56	—
8876	Brownie, prepared from mix	1	item(s)	24	3	112	1	12	1	7	1.76	2.60	2.26	—
25207	Chocolate chip cookies	1	item(s)	30	4	140	2	16	1	8	2.09	3.26	2.09	0
8915	Chocolate sandwich cookie, extra creme filling	1	item(s)	13	<1	65	<1	9	<1	3	0.50	1.39	1.22	1.10
14145	Fig Newtons	1	item(s)	16	—	55	1	10	1	1	0.50	0.50	0.00	0.50
8920	Fortune cookie	1	item(s)	8	1	30	<1	7	<1	<1	0.05	0.11	0.04	—
25208	Oatmeal cookies	1	item(s)	69	12	234	6	45	3	4	0.70	1.28	1.85	0
25213	Peanut butter cookies	1	item(s)	35	4	163	4	17	1	9	1.65	4.72	2.43	0
33095	Sugar cookies	1	item(s)	16	4	61	1	7	<1	3	0.63	1.27	0.87	0
9002	Vanilla sandwich cookie, creme filling	1	item(s)	10	<1	48	<1	7	<1	2	0.30	0.84	0.76	—
	Crackers													
9008	Cheese crackers (mini)	30	item(s)	30	1	151	3	17	1	8	2.81	3.63	0.74	—
9010	Cheese crackers (mini), low salt	30	item(s)	30	1	151	3	17	1	8	2.82	2.70	1.44	—
9012	Cheese cracker sandwich w/peanut butter	4	item(s)	28	1	139	3	16	1	7	1.23	3.64	1.43	—
8928	Honey graham crackers	4	item(s)	28	1	118	2	22	1	3	0.43	1.14	1.07	—
9016	Matzo crackers, plain	1	item(s)	28	1	112	3	24	1	<1	0.06	0.04	0.17	—
9024	Melba toast	3	item(s)	15	1	59	2	11	1	<1	0.07	0.12	0.19	—
14189	Ritz crackers	5	item(s)	16	<1	80	1	10	1	4	0.50	1.50	0.00	—
9014	Rye crispbread crackers	1	item(s)	10	1	37	1	8	2	<1	0.01	0.02	0.06	—
9028	Rye melba toast	3	item(s)	15	1	58	2	12	1	1	0.07	0.14	0.20	—
9040	Rye wafer	1	item(s)	11	1	37	1	9	3	<.1	0.01	0.02	0.04	—
432	Saltine crackers	5	item(s)	15	1	65	1	11	<1	2	0.44	0.96	0.25	0.54
9046	Saltine crackers, low salt	5	item(s)	15	1	65	1	11	<1	2	0.44	0.96	0.25	—
9048	Snack crackers, round	10	item(s)	30	1	151	2	18	<1	8	1.13	3.19	2.86	—
9050	Snack crackers, round, low salt	10	item(s)	30	1	151	2	18	<1	8	1.13	3.19	2.86	—
9052	Snack cracker sandwich, cheese filling	4	item(s)	28	1	134	3	17	1	6	1.72	3.15	0.72	—
9054	Snack cracker sandwich, peanut butter filling	4	item(s)	28	1	138	3	16	1	7	1.38	3.86	1.30	—
9044	Soda crackers	5	item(s)	15	1	65	1	11	<1	2	0.44	0.96	0.25	0.54
9055	Wheat crackers	10	item(s)	30	1	142	3	19	1	6	1.55	3.43	0.84	—
9057	Wheat crackers, low salt	10	item(s)	30	1	142	3	19	1	6	1.55	3.43	0.84	—
9059	Wheat cracker sandwich, cheese filling	4	item(s)	28	1	139	3	16	1	7	1.16	2.90	2.57	—
9061	Wheat cracker sandwich, peanut butter filling	4	item(s)	28	1	139	4	15	1	7	1.29	3.29	2.48	—
9022	Whole wheat crackers	7	item(s)	28	1	124	2	19	3	5	0.95	1.65	1.85	—
	Pastry													
16754	Apple fritter	1	item(s)	17	6	62	1	6	<1	4	0.87	1.69	1.13	—
5118	Cinnamon sweet roll w/icing, from refrigerator dough	1	item(s)	30	7	109	2	17	1	4	1.00	2.23	0.52	—
4945	Croissant, butter	1	item(s)	57	13	231	5	26	1	12	6.59	3.15	0.62	—
9096	Danish pastry, nut	1	item(s)	65	13	280	5	30	1	16	3.78	8.90	2.78	—
4947	Doughnut, cake	1	item(s)	47	10	198	2	23	1	11	1.70	4.37	3.70	—
9105	Doughnut, cake, chocolate glazed	1	item(s)	42	7	175	2	24	1	8	2.16	4.74	1.04	—
9115	Doughnut, creme filling	1	item(s)	85	32	307	5	26	1	21	4.62	10.27	2.62	—
437	Doughnut, glazed	1	item(s)	60	15	242	4	27	1	14	3.49	7.72	1.74	—
9117	Doughnut, jelly filling	1	item(s)	85	30	289	5	33	1	16	4.12	8.69	2.02	—
10617	Toaster pastry, brown sugar cinnamon	1	item(s)	50	5	210	3	35	1	6	1.00	4.00	1.00	—
30928	Toaster pastry, cream cheese	1	item(s)	54	—	200	3	23	1	11	3.50	—	—	—
	Muffins													
25015	Blueberry	1	item(s)	63	30	160	3	23	1	6	0.87	1.48	3.25	0
4997	Bran, from mix	1	item(s)	50	18	138	3	23	2	5	1.18	2.34	0.72	—
9189	Corn, ready to eat	1	item(s)	57	19	174	3	29	2	5	0.77	1.20	1.83	—

PAGE KEY: A–58 = Breads/Baked Goods A–62 = Cereal/Rice/Pasta A–66 = Fruit A–70 = Vegetables/Legumes A–80 = Nuts/Seeds A–82 = Vegetarian A–84 = Dairy A–90 = Eggs A–90 = Seafood A–92 = Meats A–96 = Poultry A–96 = Processed Meats A–98 = Beverages A–102 = Fats/Oils A–104 = Sweets A–106 = Sauces/Condiments/Spices A–108 = Mixed Foods/Soups/Sandwiches A–114 = Fast Food A–130 = Convenience A–132 = Baby Foods

Chol (mg)	Calc (mg)	Iron (mg)	Magn (mg)	Pota (mg)	Sodi (mg)	Zinc (mg)	Vit A (RAE) (µg)	Thia (mg)	Vit E (mg α)	Ribo (mg)	Niac (mg)	Vit B_6 (mg)	Fola (µg)	Vit C (mg)	Vit B_{12} (µg)	Sele (µg)
63	128	1.86	18	154	273	0.64	—	0.24	1.55	0.32	1.86	0.05	14	<1	<1	—
7	19	0.55	3	37	155	0.12	2	0.07	0.50	0.06	0.52	0.01	17	<.1	<.1	1
0	0	0.72	—	45	70	—	0	—	—	—	—	—	—	0	0	—
0	0	0.72	—	50	300	—	0	—	—	—	—	—	—	0	0	—
0	0	0.55	0	76	623	0.00	0	0.09	—	0.05	1.22	0.00	12	0	—	—
0	7	0.46	19	362	169	0.31	0	0.05	1.91	0.06	1.09	0.19	13	9	0	2
0	7	0.46	19	362	2	0.31	0	0.05	2.59	0.06	1.09	0.19	13	9	0	2
0	44	0.43	25	56	150	0.43	1	0.02	1	0.05	0.36	0.08	3	0	0	2
0	11	1.30	11	44	515	0.26	0	0.14	—	0.19	1.58	0.03	51	0	0	2
0	8	0.76	9	122	58	0.18	0	0.12	—	0.08	1.86	0.08	15	<1	0	—
0	13	0.82	5	30	1118	0.19	—	0.10	0.04	0.09	1.04	0.00	50	0	<.1	—
18	14	0.44	13	42	82	0.23	42	0.03	—	0.05	0.24	0.02	7	<.1	<.1	3
13	11	0.70	12	62	109	0.24	27	0.07	0.54	0.06	0.82	0.02	16	<.1	<.1	4
0	3	0.37	4	16	64	0.08	0	0.01	0.25	0.02	0.20	0.00	6	0	<.1	<1
0	5	0.36	—	40	60	—	4	0.03	—	0.04	0.22	—	—	<1	—	—
<1	1	0.12	1	3	22	0.01	<.1	0.01	0.00	0.01	0.15	0.00	5	0	<.1	<1
<.1	26	1.94	49	177	311	1.43	48	0.23	0.23	0.12	1.24	0.09	30	<1	<.1	17
13	28	0.67	22	104	157	0.46	51	0.08	0.74	0.09	1.81	0.05	21	<.1	<.1	5
18	5	0.32	2	13	50	0.08	31	0.04	0.28	0.04	0.28	0.01	8	<.1	<.1	3
0	3	0.22	1	9	35	0.04	0	0.03	0.16	0.02	0.27	0.00	5	0	0	<1
4	45	1.43	11	44	299	0.34	9	0.17	0.66	0.13	1.40	0.17	46	0	<1	3
4	45	1.44	11	32	137	0.33	—	0.18	—	0.12	1.41	0.18	8	0	<1	—
0	14	0.76	16	61	199	0.29	0	0.15	0.16	0.08	1.63	0.04	26	0	<.1	2
0	7	1.04	8	38	169	0.23	0	0.06	0.09	0.09	1.15	0.02	13	0	0	3
0	4	0.90	7	32	1	0.19	0	0.11	0.02	0.08	1.11	0.03	5	0	0	10
0	14	0.56	9	30	124	0.30	0	0.06	0.06	0.04	0.62	0.01	19	0	0	5
0	20	0.72	3	10	135	0.23	—	0.07	—	0.04	0.45	0.01	10	1	0	—
0	3	0.24	8	32	26	0.24	0	0.02	0.08	0.01	0.10	0.02	5	0	0	4
0	12	0.55	6	29	135	0.20	0	0.07	—	0.04	0.71	0.01	13	0	0	6
0	4	0.65	13	54	87	0.31	0	0.05	0.09	0.03	0.17	0.03	5	<.1	0	3
0	18	0.81	4	19	195	0.12	0	0.08	0.15	0.07	0.79	0.01	19	0	0	2
0	18	0.81	4	109	95	0.12	0	0.08	0.02	0.07	0.79	0.01	19	0	0	3
0	36	1.08	8	40	254	0.20	0	0.12	0.61	0.10	1.21	0.02	27	0	0	2
0	36	1.08	8	107	112	0.20	0	0.12	0.61	0.10	1.21	0.02	27	0	0	2
1	72	0.67	10	120	392	0.17	5	0.12	0.06	0.19	1.05	0.01	28	<.1	<.1	6
0	23	0.78	15	60	201	0.32	0	0.14	0.58	0.08	1.71	0.04	24	0	<.1	3
0	18	0.81	4	19	195	0.12	0	0.08	0.15	0.07	0.79	0.01	19	0	0	2
0	15	1.32	19	55	239	0.48	0	0.15	0.15	0.10	1.49	0.04	35	0	0	2
0	15	1.32	19	61	85	0.48	0	0.15	0.15	0.10	1.49	0.04	15	0	0	10
2	57	0.73	15	86	256	0.24	5	0.10	—	0.12	0.89	0.07	18	<1	<.1	7
0	48	0.75	11	83	226	0.23	0	0.11	—	0.08	1.65	0.04	20	0	0	6
0	14	0.86	28	83	185	0.60	0	0.06	0.24	0.03	1.27	0.05	8	0	0	4
14	9	0.25	2	24	7	0.09	—	0.03	0.07	0.04	0.23	0.01	2	<1	<.1	—
0	10	0.80	4	19	250	0.10	—	0.12	—	0.07	1.09	0.01	14	<.1	<.1	—
38	21	1.16	9	67	424	0.43	101	0.22	—	0.14	1.25	0.03	35	<1	<.1	13
30	61	1.17	21	62	236	0.57	6	0.14	0.53	0.16	1.50	0.07	54	1	<1	9
17	21	0.92	9	60	257	0.26	—	0.10	—	0.11	0.87	0.03	22	<.1	<1	0
24	89	0.95	14	45	143	0.24	5	0.02	0.09	0.03	0.20	0.01	19	<.1	<.1	2
20	21	1.56	17	68	263	0.68	9	0.29	0.25	0.13	1.91	0.06	60	0	<1	9
4	26	0.36	13	65	205	0.46	2	0.53	—	0.04	0.39	0.03	13	<.1	<.1	5
22	21	1.50	17	67	249	0.64	14	0.27	0.37	0.12	1.82	0.09	58	0	<1	11
0	0	1.80	—	70	190	—	—	0.15	—	0.17	2.00	0.20	40	0	0	—
15	0	1.08	—	—	230	—	—	—	—	—	—	—	—	0	—	—
20	50	1.15	7	56	288	0.39	20	0.14	0.76	0.15	1.14	0.03	29	<1	<1	9
34	16	1.27	29	74	234	0.57	—	0.10	—	0.12	1.44	0.09	33	0	<.1	—
15	42	1.60	18	39	297	0.31	30	0.16	0.46	0.19	1.16	0.05	46	0	<.1	9

TABLE F-1

Food Composition (Computer code number is for Wadsworth Diet Analysis program) (For purposes of calculations, use "0" for t, <1, <.1, <.01, etc.)

DA + Code	Food Description	Quantity	Measure	Wt (g)	H_2O (g)	Ener (kcal)	Prot (g)	Carb (g)	Dietary Fiber (g)	Fat (g)	Fat Breakdown (g) Sat	Mono	Poly	Trans
	BREADS, BAKED GOODS, CAKES, COOKIES, CRACKERS, CHIPS, PIES—Continued													
9121	English muffin, plain, enriched	1	item(s)	57	24	134	4	26	2	1	0.15	0.17	0.51	—
29582	English, toasted	1	item(s)	50	19	128	4	25	1	1	0.14	0.16	0.48	—
9145	English, wheat	1	item(s)	57	24	127	5	26	3	1	0.16	0.16	0.48	—
	Granola bars													
38161	Kudos milk chocolate w/fruit & nuts	1	item(s)	28	—	90	2	15	1	3	1.00	—	—	—
38196	Nature Valley banana nut crunchy	1	item(s)	21	—	95	2	14	1	4	0.50	—	—	—
38187	Nature Valley fruit n nut trail mix	1	item(s)	35	—	140	3	25	2	4	0.50	—	—	—
1383	Plain, hard	1	item(s)	25	1	115	2	16	1	5	0.58	1.07	2.95	—
4606	Plain, soft	1	item(s)	28	2	126	2	19	1	5	2.06	1.08	1.51	—
	Pies													
454	Apple pie, from home recipe	1	slice(s)	155	73	411	4	58	2	19	4.73	8.36	5.17	—
470	Pecan pie, from home recipe	1	slice(s)	122	24	503	6	64	0	27	4.87	13.64	6.97	—
472	Pumpkin pie, from home recipe	1	slice(s)	155	91	316	7	41	0	14	4.92	5.73	2.81	—
9007	Pie crust, frozen, ready to bake, enriched, baked	1	slice(s)	16	2	82	1	8	<1	5	1.69	2.51	0.65	—
5052	Pie crust, prepared w/water, baked	1	slice(s)	20	2	100	1	10	<1	6	1.54	3.46	0.77	—
	Rolls													
8555	Crescent dinner roll	1	item(s)	28	10	80	2	14	1	1	0.34	0.70	0.25	—
489	Hamburger roll or bun, plain	1	item(s)	43	15	120	4	21	1	2	0.47	0.48	0.85	—
490	Hard roll	1	item(s)	57	18	167	6	30	1	2	0.35	0.65	0.98	—
5127	Kaiser roll	1	item(s)	57	18	167	6	30	1	2	0.35	0.65	0.98	—
5130	Whole wheat roll or bun	1	item(s)	28	9	76	2	15	2	1	0.24	0.34	0.62	—
	Sport bars													
37026	Balance original chocolate	1	item(s)	50	—	200	14	22	1	6	3.50	—	—	—
37024	Balance original peanut butter	1	item(s)	50	—	200	14	22	1	6	2.50	—	—	—
36580	Clif Bar chocolate brownie energy bar	1	item(s)	68	—	240	10	41	6	4	1.00	—	—	—
36583	Clif Bar crunchy peanut butter energy bar	1	item(s)	68	—	240	12	39	5	5	0.50	—	—	—
36584	Clif Luna tropical crisp energy bar	1	item(s)	48	—	180	10	24	2	5	3.50	0.00	0.00	—
12005	Powerbar apple cinnamon	1	item(s)	65	—	230	10	45	3	3	0.50	1.50	0.50	—
16078	Powerbar banana	1	item(s)	65	—	230	9	45	3	2	0.50	1.00	0.50	—
16080	Powerbar chocolate	1	item(s)	65	—	230	10	45	3	2	0.50	0.50	1.00	—
16079	Powerbar mocha	1	item(s)	65	—	230	10	45	3	3	1.00	1.00	0.50	—
	Tortillas													
1391	Corn tortillas, soft	1	item(s)	26	11	58	1	12	1	1	0.09	0.17	0.29	—
1669	Flour tortilla	1	item(s)	32	9	104	3	18	1	2	0.56	1.21	0.34	—
1390	Taco shells, hard	1	item(s)	13	1	62	1	8	1	3	0.43	1.19	1.13	—
	Pancakes, waffles													
8926	Pancakes, blueberry, from recipe	3	item(s)	114	61	253	7	33	1	10	2.26	2.64	4.74	—
5037	Pancakes, from mix w/egg & milk	3	item(s)	114	60	249	9	33	2	9	2.33	2.36	3.33	—
9219	Waffle, plain, frozen, toasted	2	item(s)	66	28	174	4	27	2	5	0.95	2.12	1.84	—
500	Waffle, plain, from recipe	1	item(s)	75	<.1	218	6	25	2	11	2.14	2.64	5.08	—
30311	Waffle, 100% whole grain	1	item(s)	75	32	201	7	25	2	8	2.35	3.38	2.06	—
	CEREAL, FLOUR, GRAIN, PASTA, NOODLES, POPCORN													
	Grain													
2861	Amaranth, dry	½	cup(s)	98	10	365	14	65	15	6	1.62	1.40	2.82	—
1953	Barley, pearled, cooked	½	cup(s)	79	54	97	2	22	3	<1	0.07	0.04	0.17	—
1956	Buckwheat groats, cooked, roasted	½	cup(s)	84	64	77	3	17	2	1	0.11	0.16	0.16	—
1957	Bulgur, cooked	½	cup(s)	91	71	76	3	17	4	<1	0.04	0.03	0.09	—
1963	Couscous, cooked	½	cup(s)	79	57	88	3	18	1	<1	0.02	0.02	0.05	—
1967	Millet, cooked	½	cup(s)	120	86	143	4	28	2	1	0.21	0.22	0.61	—
1969	Oat bran, dry	½	cup(s)	47	3	116	8	31	7	3	0.62	1.12	1.30	—
1972	Quinoa, dry	½	cup(s)	85	8	318	11	59	5	5	0.50	1.30	1.99	—
	Rice													
129	Brown, long grain, cooked	½	cup(s)	98	71	108	3	22	2	1	0.18	0.32	0.31	—
2863	Brown, medium grain, cooked	½	cup(s)	97.5	0.07	109.19	2.26	22.92	1.75	0.8	0.16	0.29	0.28	—
37488	Jasmine, saffroned, cooked	½	cup(s)	280	—	340	8	78	0	0	0.00	—	—	0
30280	Pilaf, cooked	½	cup(s)	103	74	129	2	22	1	3	0.67	1.61	0.95	—
28066	Spanish, cooked	½	cup(s)	120	3	25	2	1	<1	<1	0.33	0.07	18.31	0
2867	White glutinous, cooked	½	cup(s)	87	67	84	2	18	1	<1	0.03	0.06	0.06	—
482	White, instant long grain, enriched, boiled	½	cup(s)	83	63	81	2	18	<1	<1	0.04	0.04	0.04	—
484	White, long grain, boiled	½	cup(s)	79	54	103	2	22	<1	<1	0.06	0.07	0.06	—
486	White, long grain, enriched, parboiled, cooked	½	cup(s)	88	63	100	2	22	<1	<1	0.06	0.07	0.06	—
1194	Wild brown, cooked	½	cup(s)	82	0.06	82.81	3.27	17.49	1.47	0.27	0.04	0.04	0.17	—

PAGE KEY: A–58 = Breads/Baked Goods A–62 = Cereal/Rice/Pasta A–66 = Fruit A–70 = Vegetables/Legumes A–80 = Nuts/Seeds A–82 = Vegetarian A–84 = Dairy A–90 = Eggs A–90 = Seafood A–92 = Meats A–96 = Poultry A–96 = Processed Meats A–98 = Beverages A–102 = Fats/Oils A–104 = Sweets A–106 = Sauces/Condiments/Spices A–108 = Mixed Foods/Soups/Sandwiches A–114 = Fast Food A–130 = Convenience A–132 = Baby Foods

Chol (mg)	Calc (mg)	Iron (mg)	Magn (mg)	Pota (mg)	Sodi (mg)	Zinc (mg)	Vit A (RAE) (µg)	Thia (mg)	Vit E (mg α)	Ribo (mg)	Niac (mg)	Vit B_6 (mg)	Fola (µg)	Vit C (mg)	Vit B_{12} (µg)	Sele (µg)
0	30	1.43	12	75	264	0.40	0	0.25	—	0.16	2.21	0.02	42	0	<.1	—
0	95	1.36	11	72	252	0.38	0	0.19	0.17	0.14	1.90	0.02	15	<.1	<.1	—
0	101	1.64	21	106	218	0.61	0	0.25	0.26	0.17	1.91	0.05	36	0	0	17
0	200	0.36	—	—	60	—	0	—	—	—	—	—	—	0	0	—
0	10	0.54	—	60	80	—	0	—	—	—	—	—	—	0	—	—
0	0	0.00	—	—	95	—	0	—	—	—	—	—	—	0	—	—
0	15	0.72	24	82	72	0.50	2	0.06	—	0.03	0.39	0.02	6	<1	0	4
<1	30	0.73	21	92	79	0.43	0	0.08	—	0.05	0.15	0.03	7	0	<1	5
0	11	1.74	11	122	327	0.29	17	0.23	—	0.17	1.91	0.05	37	3	0	12
106	39	1.81	32	162	320	1.24	100	0.23	—	0.22	1.03	0.07	32	<1	<1	15
65	146	1.97	29	288	349	0.71	660	0.14	—	0.31	1.21	0.07	33	3	<1	11
0	3	0.36	3	18	104	0.05	0	0.04	0.42	0.06	0.39	0.01	9	0	<.1	<1
0	12	0.43	3	12	146	0.08	0	0.06	—	0.04	0.47	0.01	20	0	0	—
0	39	0.89	6	39	157	0.17	0	0.14	0.02	0.09	1.10	0.01	—	0	<.1	—
0	59	1.43	9	40	206	0.28	0	0.17	0.03	0.14	1.79	0.03	48	0	<.1	8
0	54	1.87	15	62	310	0.54	0	0.27	0.24	0.19	2.42	0.02	54	0	0	22
0	54	1.87	15	62	310	0.54	0	0.27	—	0.19	2.42	0.02	54	0	0	22
0	30	0.69	24	78	136	0.57	0	0.07	—	0.04	1.05	0.06	9	0	0	14
3	100	4.50	40	160	180	3.75	—	0.38	—	0.43	5.00	0.50	100	60	2	18
3	100	4.50	40	130	230	3.75	—	0.38	—	0.43	5.00	0.50	100	60	2	18
0	250	5.40	120	260	150	3.75	—	0.38	—	0.26	4.00	0.40	80	60	1	18
0	250	5.40	120	300	290	3.75	—	0.38	—	0.34	6.00	0.40	100	60	1	14
0	350	6.30	140	120	135	5.25	—	1.50	—	1.70	20.00	2.00	400	60	6	25
0	300	6.30	140	110	90	5.25	0	1.50	—	1.70	20.00	2.00	400	60	6	—
0	300	6.30	140	200	90	5.25	0	1.50	—	1.70	20.00	2.00	400	60	6	—
0	300	6.30	140	150	90	5.25	0	1.50	—	1.70	20.00	2.00	400	60	6	—
0	300	6.30	140	150	90	5.25	0	1.50	—	1.70	20.00	2.00	400	60	6	—
0	46	0.36	17	40	42	0.24	0	0.03	0.07	0.02	0.39	0.06	26	0	0	1
0	40	1.06	8	42	153	0.23	0	0.17	0.06	0.09	1.14	0.02	33	0	0	7
0	21	0.33	14	24	49	0.19	0	0.03	0.22	0.01	0.18	0.04	17	0	0	2
64	235	1.96	18	157	470	0.62	57	0.22	—	0.31	1.74	0.06	41	3	<1	16
81	245	1.48	25	227	576	0.86	82	0.23	—	0.36	1.40	0.12	105	1	<1	—
16	153	2.95	15	84	519	0.38	253	0.25	0.65	0.31	2.93	0.59	36	0	2	11
52	191	1.73	14	119	383	0.50	49	0.19	—	0.26	1.55	0.04	51	<1	<1	35
71	196	1.56	30	173	374	0.85	—	0.15	0.32	0.25	1.47	0.09	14	<1	<1	—
0	149	7.40	259	357	20	3.10	0	0.08	—	0.20	1.25	0.22	48	4	0	—
0	9	1.04	17	73	2	0.64	0	0.07	0.01	0.05	1.62	0.09	13	0	0	7
0	6	0.67	43	74	3	0.51	0	0.03	0.08	0.03	0.79	0.06	12	0	0	2
0	9	0.87	29	62	5	0.52	0	0.05	0.01	0.03	0.91	0.08	16	0	0	1
0	6	0.30	6	46	4	0.20	0	0.05	0.10	0.02	0.77	0.04	12	0	0	22
0	4	0.76	53	74	2	1.09	0	0.13	0.02	0.10	1.60	0.13	23	0	0	1
0	27	2.54	110	266	2	1.46	0	0.55	0.47	0.10	0.44	0.08	24	0	0	21
0	51	7.86	179	629	18	2.81	0	0.17	—	0.34	2.49	0.19	42	0	0	—
0	10	0.41	42	42	5	0.61	0	0.09	0.03	0.02	1.49	0.14	4	0	0	10
0	9.75	0.51	42.9	77.02	0.97	0.6	0	0.09	—	0.01	1.29	0.14	3.9	0	0	38
0	—	2.16	—	—	780	—	—	—	—	—	—	—	—	—	—	—
0	13	1.16	9	55	403	0.38	—	0.13	0.28	0.02	1.24	0.06	4	<1	<.1	—
1	47	0.78	48	1	13	0.13	<1	0.03	0.06	0.19	8.71	0.14	<.1	7	<.1	9
0	2	0.12	4	9	4	0.36	0	0.02	0.03	0.01	0.25	0.02	1	0	0	5
0	7	0.52	4	3	2	0.20	0	0.06	0.01	0.04	0.73	0.01	58	0	0	3
0	8	0.95	9	28	1	0.39	0	0.13	0.03	0.01	1.17	0.07	46	0	0	6
0	17	0.99	11	32	3	0.27	0	0.22	0.01	0.02	1.23	0.02	67	0	0	7
0	2.46	0.49	26.23	82.81	2.46	1.09	0	0.04	—	0.07	1.05	0.11	21.31	0	0	0.65

TABLE F–1
Food Composition (Computer code number is for Wadsworth Diet Analysis program) (For purposes of calculations, use "0" for t, <1, <.1, <.01, etc.)

DA + Code	Food Description	Quantity	Measure	Wt (g)	H_2O (g)	Ener (kcal)	Prot (g)	Carb (g)	Dietary Fiber (g)	Fat (g)	Fat Breakdown (g) Sat	Mono	Poly	Trans
	CEREAL, FLOUR, GRAIN, PASTA, NOODLES, POPCORN—Continued													
	Flour & grain fractions													
505	All purpose flour, self rising, enriched	½	cup(s)	63	7	221	6	46	2	1	0.10	0.05	0.26	—
503	All purpose flour, white, bleached, enriched	½	cup(s)	63	7	228	6	48	2	1	0.10	0.05	0.26	—
1643	Barley flour	½	cup(s)	56	6	198	4	45	2	1	0.16	0.10	0.38	—
383	Buckwheat flour, whole groat	½	cup(s)	60	7	201	8	42	6	2	0.41	0.57	0.57	—
504	Cake wheat flour, enriched	½	cup(s)	55	7	197	4	43	1	<1	0.07	0.04	0.21	—
426	Cornmeal, degermed, enriched	½	cup(s)	69	8	253	6	54	5	1	0.16	0.28	0.49	—
424	Cornmeal, yellow whole grain	½	cup(s)	61	6	221	5	47	4	2	0.31	0.58	1.00	—
1644	Masa corn flour, enriched	½	cup(s)	57	5	208	5	43	5	2	0.30	0.57	0.98	—
1976	Rice flour, brown	½	cup(s)	79	9	287	6	60	4	2	0.44	0.80	0.79	—
1645	Rice flour, white	½	cup(s)	79	9	289	5	63	2	1	0.30	0.35	0.30	—
1978	Rye flour, dark	½	cup(s)	64	7	207	9	44	14	2	0.20	0.21	0.77	—
1980	Semolina, enriched	½	cup(s)	84	11	301	11	61	3	1	0.13	0.10	0.36	—
2827	Soy flour, raw	½	cup(s)	43	2	186	15	15	4	9	1.27	1.94	4.96	—
1990	Wheat germ, crude	2	tablespoon(s)	14	2	52	3	7	2	1	0.24	0.20	0.86	—
506	Whole wheat flour	½	cup(s)	60	6	203	8	44	7	1	0.19	0.14	0.47	—
	Breakfast bars													
39230	Atkins Morning Start apple crisp	1	item(s)	37	—	170	11	12	6	9	4.00	—	—	—
10574	Health Valley fat free apple	1	item(s)	38	—	110	2	26	3	0	0.00	0.00	0.00	0
10647	Nutri-Grain blueberry cereal bar	1	item(s)	37	5	140	2	27	1	3	0.50	2.00	0.50	—
10648	Nutri-Grain raspberry cereal bar	1	item(s)	37	5	140	2	27	1	3	0.50	2.00	0.50	—
10649	Nutri-Grain strawberry cereal bar	1	item(s)	37	5	140	2	27	1	3	0.50	2.00	0.50	—
	Breakfast cereals, hot													
363	Corn grits, white, regular & quick, enriched, cooked w/water & salt	½	cup(s)	121	103	71	2	16	<1	<1	0.03	0.06	0.10	—
8636	Corn grits, yellow, regular & quick, enriched, cooked w/salt	½	cup(s)	121	103	71	2	16	<1	<1	0.03	0.06	0.10	—
1260	Cream of Wheat, instant, prepared	½	cup(s)	121	106	61	2	13	<1	<.1	0.01	0.01	0.04	0
365	Farina, enriched, cooked w/water & salt	½	cup(s)	117	102	56	2	12	<1	<.1	0.01	0.01	0.03	—
8657	Oatmeal, cooked w/water	½	cup(s)	117	100	74	3	13	2	1	0.19	0.37	0.44	—
5500	Oatmeal, maple & brown sugar, instant, prepared	1	item(s)	198	150	200	5	40	2	2	0.42	0.74	0.85	—
5510	Oatmeal, ready to serve, packet	1	item(s)	186	158	112	4	20	3	2	0.38	0.66	0.76	—
	Breakfast cereals, ready to eat													
1197	All-Bran	1	cup(s)	62	2	160	8	46	20	2	0.00	0.00	1.00	0
1200	All-Bran Buds	1	cup(s)	91	3	212	6	73	42	3	—	—	—	0
1199	Apple Jacks	1	cup(s)	33	1	130	1	30	1	1	—	—	—	0
13633	Bran Flakes, Post	1	cup(s)	40	1	133	4	32	7	1	0.00	0.00	0.71	—
1204	Cap'n Crunch	1	cup(s)	36	1	144	2	30	1	2	0.53	0.39	0.27	—
1205	Cap'n Crunch Crunchberries w/wildberry colors	1	cup(s)	35	1	139	2	29	1	2	0.49	0.39	0.28	—
1206	Cheerios	1	cup(s)	30	1	110	3	22	3	2	0.00	0.50	0.50	—
3415	Cocoa Puffs	1	cup(s)	30	1	120	1	26	0	1	—	—	—	—
1207	Cocoa Rice Krispies	1	cup(s)	41	1	160	1	36	1	1	0.67	0.00	0.00	—
5522	Complete wheat bran flakes	1	cup(s)	39	1	120	4	31	7	1	—	—	—	0
1211	Corn Flakes	1	cup(s)	28	1	100	2	24	1	0	0.00	0.00	0.00	0
1247	Corn Pops	1	cup(s)	31	1	120	1	28	0	0	0.00	0.00	0.00	0
1937	Cracklin' Oat Bran	1	cup(s)	65	0	266	5	47	7	9	2.70	4.70	1.33	0
1220	Froot Loops	1	cup(s)	32	1	120	1	28	1	1	0.50	0.00	0.00	—
38214	Frosted Cheerios	1	cup(s)	30	—	120	2	25	1	1	0.00	0.00	0.00	—
372	Frosted Flakes	1	cup(s)	41	1	160	1	37	1	0	0.00	0.00	0.00	0
38215	Frosted Mini Chex	1	cup(s)	40	—	146	1	36	0	0	0.00	0.00	0.00	0
10268	Frosted Mini-Wheats	5	item(s)	51	3	180	5	41	5	1	0.00	0.00	0.50	0
38216	Frosted Wheaties	1	cup(s)	40	—	146	1	36	<1	0	0.00	0.00	0.00	0
1223	Granola, prepared	½	cup(s)	61	0	299	9	32	5	15	2.76	4.7	6.53	—
13334	Granola, Quaker 100% natural, oats & honey	½	cup(s)	48	0	219	5	31	3	9	3.83	4.0	1.19	—
13335	Granola, Quaker 100% natural, oats, honey & raisins	½	cup(s)	51	0	225	5	34	3	9	3.57	3.80	1.10	—
2415	Honey Bunches of Oats honey roasted	1	cup(s)	40	1	160	3	33	1	2	0.67	1.20	0.13	—
1227	Honey Nut Cheerios	1	cup(s)	30	1	120	3	24	2	2	0.00	0.50	0.00	—
2424	Honeycomb	1	cup(s)	22	<1	83	2	20	<1	<1	0.00	—	—	—
10286	Kashi puffed	1	cup(s)	25	—	70	3	13	2	1	0.00	—	—	—
1231	Kix	1	cup(s)	23	<1	90	2	20	1	<1	0.00	0.00	0.00	—
30569	Life	1	cup(s)	43	2	160	4	33	3	2	0.35	0.64	0.61	—
1233	Lucky Charms	1	cup(s)	30	1	120	2	25	1	1	0.00	0.00	0.00	—

PAGE KEY: A–58 = Breads/Baked Goods A–62 = Cereal/Rice/Pasta A–66 = Fruit A–70 = Vegetables/Legumes A–80 = Nuts/Seeds A–82 = Vegetarian A–84 = Dairy A–90 = Eggs A–90 = Seafood A–92 = Meats A–96 = Poultry A–96 = Processed Meats A–98 = Beverages A–102 = Fats/Oils A–104 = Sweets A–106 = Sauces/Condiments/Spices A–108 = Mixed Foods/Soups/Sandwiches A–114 = Fast Food A–130 = Convenience A–132 = Baby Foods

Chol (mg)	Calc (mg)	Iron (mg)	Magn (mg)	Pota (mg)	Sodi (mg)	Zinc (mg)	Vit A (RAE) (µg)	Thia (mg)	Vit E (mg α)	Ribo (mg)	Niac (mg)	Vit B_6 (mg)	Fola (µg)	Vit C (mg)	Vit B_{12} (µg)	Sele (µg)
0	211	2.92	12	78	794	0.39	0	0.42	0.03	0.26	3.65	0.03	123	0	0	22
0	9	2.90	14	67	1	0.44	0	0.49	0.04	0.31	3.69	0.03	114	0	0	21
0	16	0.71	45	186	4	1.05	0	0.07	—	0.03	2.57	0.16	13	0	0	2
0	25	2.44	151	346	7	1.87	0	0.25	0.19	0.11	3.69	0.35	32	0	0	3
0	8	3.99	9	57	1	0.34	0	0.49	0.01	0.23	3.70	0.02	101	0	0	3
0	3	2.85	28	112	2	0.50	8	0.49	0.10	0.28	3.47	0.18	161	0	0	5
0	4	2.10	77	175	21	1.11	7	0.23	0.26	0.12	2.22	0.19	15	0	0	9
0	80	4.11	63	170	3	1.01	0	0.81	0.09	0.43	5.61	0.21	133	0	0	9
0	9	1.56	88	228	6	1.94	0	0.35	0.95	0.06	5.01	0.58	13	0	0	—
0	8	0.28	28	60	0	0.63	0	0.11	0.09	0.02	2.05	0.34	3	0	0	12
0	36	4.13	159	467	1	3.60	1	0.20	0.90	0.16	2.73	0.28	38	0	0	23
0	14	3.64	39	155	1	0.88	0	0.68	0.22	0.48	5.00	0.09	153	0	0	75
0	88	2.71	183	1070	6	1.67	3	0.25	0.93	0.49	1.84	0.20	147	0	0	3
0	6	0.90	34	128	2	1.77	0	0.27	—	0.07	0.98	0.19	40	0	0	11
0	20	2.33	83	243	3	1.76	0	0.27	0.49	0.13	3.82	0.20	26	0	0	42
0	200	—	—	90	70	—	—	0.23	—	0.26	3.00	—	—	9	—	—
0	0	0.72	—	160	25	—	—	0.09	—	0.03	0.40	—	—	1	—	—
0	200	1.80	8	75	110	1.50	—	0.38	—	0.43	5.00	0.50	40	0	0	—
0	200	1.80	8	70	110	1.50	—	0.38	—	0.43	5.00	0.50	40	0	0	—
0	200	1.80	8	55	110	1.50	—	0.38	—	0.43	5.00	0.50	40	0	0	—
0	4	0.73	6	25	270	0.08	0	0.10	0.02	0.07	0.87	0.03	40	0	0	4
0	4	0.73	6	25	270	0.08	2	0.10	0.02	0.07	0.87	0.03	40	0	0	3
0	27	8.60	2	17	1	0.10	0	0.07	—	0.04	0.60	0.01	357	0	0	—
0	5	0.58	2	15	383	0.09	0	0.07	0.01	0.05	0.57	0.01	40	0	0	11
0	9	0.80	28	66	1	0.57	0	0.13	0.12	0.02	0.15	0.02	5	0	0	9
0	26	6.84	50	126	404	1.04	0	1.02	—	0.05	1.57	0.31	30	0	0	11
0	21	3.96	45	112	241	0.93	0	0.60	—	0.05	0.78	0.19	19	0	0	4
0	300	9.00	200	700	160	3.00	300	0.75	—	0.85	10.00	4.00	800	12	12	6
0	0	13.64	182	909	606	4.55	455	1.14	—	1.29	15.15	6.06	1212	18	18	26
0	0	4.50	8	35	150	1.50	150	0.38	—	0.43	5.00	0.50	100	15	2	2
0	0	10.77	80	253	293	2.00	—	0.50	—	0.57	6.65	0.67	133	0	2	—
0	5	6.00	20	72	269	4.99	3	0.51	—	0.57	6.66	0.67	133	0	0	7
<.1	7	6.14	19	71	242	5.12	2	0.51	—	0.57	6.66	0.67	133	<.1	0	7
0	100	8.10	40	95	280	3.75	150	0.38	—	0.43	5.00	0.50	200	6	2	11
0	100	4.50	8	50	170	3.75	0	0.38	—	0.43	5.00	0.50	100	6	2	2
0	53	5.99	11	67	253	2.00	200	0.50	—	0.57	6.65	0.67	133	20	2	6
0	0	23.94	53	226	279	19.95	299	2.00	—	2.26	26.60	2.66	532	80	8	4
0	0	8.10	3	25	200	0.17	150	0.38	—	0.43	5.00	0.50	100	6	2	1
0	0	1.80	2	25	120	1.50	150	0.38	—	0.43	5.00	0.50	100	6	2	2
0	27	2.38	80	293	186	2.00	299	0.49	—	0.56	6.65	0.67	218	20	2	14
0	0	4.50	8	35	150	1.50	150	0.38	—	0.43	5.00	0.50	100	15	2	2
0	100	4.50	16	55	210	3.75	—	0.38	—	0.43	5.00	0.50	100	6	2	—
0	0	5.99	4	27	200	0.21	200	0.50	—	0.57	6.65	0.67	133	8	2	2
0	133	11.97	—	33	266	3.99	—	0.50	—	0.57	6.65	0.67	266	8	2	—
0	0	15.30	60	170	5	1.50	0	0.38	—	0.43	5.00	0.50	100	0	2	2
0	133	10.77	0	47	266	9.98	—	1.00	—	1.13	13.30	1.33	532	8	4	—
0	48	2.59	107	328	13	2.5	2	0.44	3.59	0.17	1.29	0.18	51	1	0	16.95
1	61	1.21	51	225	20	1.04	1	0.12	—	0.11	0.81	0.07	17	<1	0.1	8.3
1	59	1.24	49	250	19	0.99	<1	0.12	—	0.11	0.8	0.07	16	<1	0.1	8.82
0	0	3.59	21	67	253	0.40	—	0.50	—	0.57	6.65	0.67	133	0	2	—
0	100	4.50	24	95	270	3.75	—	0.38	—	0.43	5.00	0.50	200	6	2	7
0	0	2.03	6	26	165	1.13	—	0.28	—	0.32	3.74	0.37	75	0	1	—
0	0	0.72	—	35	0	—	0	0.03	—	0.03	0.80	0.00	—	0	—	—
0	113	6.08	6	26	203	2.81	113	0.28	—	0.32	3.75	0.38	150	5	1	5
0	124	11.92	41	121	218	5.32	1	0.53	—	0.60	7.10	0.70	142	0	0	11
0	100	4.50	16	60	210	3.75	—	0.38	—	0.43	5.00	0.50	200	6	2	6

TABLE F–1

Food Composition (Computer code number is for Wadsworth Diet Analysis program) (For purposes of calculations, use "0" for t, <1, <.1, <.01, etc.)

DA + Code	Food Description	Quantity	Measure	Wt (g)	H_2O (g)	Ener (kcal)	Prot (g)	Carb (g)	Dietary Fiber (g)	Fat (g)	Fat Breakdown (g) Sat	Mono	Poly	Trans
	CEREAL, FLOUR, GRAIN, PASTA, NOODLES, POPCORN—Continued													
1201	Multi-Bran Chex	1	cup(s)	58	1	200	4	49	7	2	0.00	0.00	0.00	0
38220	Multi Grain Cheerios	1	cup(s)	30	—	110	3	24	3	1	0.00	0.00	0.00	—
1238	Nutri-Grain golden wheat	1	cup(s)	40	—	133	4	31	5	1	0.00	0.00	0.67	—
1241	Product 19	1	cup(s)	30	1	100	2	25	1	0	0.00	0.00	0.00	0
32432	Puffed rice, fortified	1	cup(s)	14	<1	56	1	13	<1	<.1	0.02	—	—	—
32433	Puffed wheat, fortified	1	cup(s)	12	0	43.68	1.76	9.55	0.52	0.14	0.02	—	—	—
2420	Raisin Bran	1	cup(s)	59	5	190	4	47	8	1	0.00	0.10	0.36	—
1244	Rice Chex	1	cup(s)	25	1	96	2	22	<1	0	0.00	0.00	0.00	0
1245	Rice Krispies	1	cup(s)	26	1	96	2	23	0	0	0.00	0.00	0.00	0
5593	Shredded Wheat	1	cup(s)	25	1	88	3	20	3	1	0.04	0.01	0.10	—
1248	Smacks	1	cup(s)	36	1	133	3	32	1	1	0.00	0.00	0.00	—
1246	Special K	1	cup(s)	31	1	110	7	22	1	0	0.00	0.00	0.00	0
3428	Total, corn flakes	1	cup(s)	23	1	83	2	18	1	0	0.00	0.00	0.00	0
1253	Total whole grain	1	cup(s)	40	1	146	3	31	4	1	0.00	0.00	0.00	—
1254	Trix	1	cup(s)	30	1	120	1	27	1	1	0.00	0.00	0.00	—
382	Wheat germ, toasted	2	tablespoon(s)	14	0	53.95	4.11	7	2.13	1.51	0.25	0.21	0.93	—
1257	Wheaties	1	cup(s)	30	1	110	3	24	3	1	0.00	0.00	0.00	—
	Pasta, noodles													
449	Chinese chow mein noodles, cooked	½	cup(s)	23	<1	119	2	13	1	7	0.99	1.73	3.90	—
1995	Corn pasta, cooked	½	cup(s)	70	48	88	2	20	3	1	0.07	0.13	0.23	—
448	Egg noodles, enriched, cooked	½	cup(s)	80	55	106	4	20	1	1	0.25	0.34	0.33	0.02
440	Macaroni, enriched, cooked	½	cup(s)	70	46	99	3	20	1	<1	0.07	0.06	0.19	—
1996	Pasta, plain, fresh-refrigerated, cooked	½	cup(s)	64	44	84	3	16	0	1	0.10	0.08	0.27	—
1725	Ramen noodles, cooked	½	cup(s)	114	95	104	3	15	1	4	0.19	0.22	0.21	—
2878	Soba noodles, cooked	½	cup(s)	95	69	94	5	20	0	<.1	0.02	0.02	0.03	—
2879	Somen noodles, cooked	½	cup(s)	88	60	115	4	24	0	<1	0.02	0.02	0.06	—
493	Spaghetti, al dente, cooked	½	cup(s)	65	42	95	4	20	1	1	0.05	0.05	0.15	—
2884	Spaghetti, whole wheat, cooked	½	cup(s)	70	47	87	4	19	3	<1	0.07	0.05	0.15	—
1563	Spinach egg noodles, enriched, cooked	½	cup(s)	80	55	105	4	19	2	1	0.29	0.39	0.28	—
2000	Tricolor vegetable macaroni, enriched, cooked	½	cup(s)	67	46	86	3	18	3	<.1	0.01	0.01	0.03	—
	Popcorn													
476	Air popped	1	cup(s)	8	<1	31	1	6	1	<1	0.05	0.09	0.15	—
4619	Caramel	1	cup(s)	35	1	152	1	28	2	5	1.27	1.01	1.58	—
4620	Cheese flavored	1	cup(s)	37	1	196	3	19	4	12	2.38	3.61	5.72	—
477	Popped in oil	1	cup(s)	33	1	165	3	19	3	9	1.61	2.70	4.43	—
	FRUIT AND FRUIT JUICES													
	Apples													
223	Raw medium, w/peel	1	item(s)	138	118	72	<1	19	3	<1	0.04	0.01	0.07	—
224	Slices	½	cup(s)	55	47	29	<1	8	1	<.1	0.02	0.00	0.03	—
946	Slices w/o skin, boiled	½	cup(s)	85	73	45	<1	12	2	<1	0.05	0.01	0.09	—
948	Dried, sulfured	½	cup(s)	22	7	52	<1	14	2	<.1	0.01	0.00	0.02	—
952	Juice, from frozen concentrate	½	cup(s)	120	105	56	<1	14	<1	<1	0.02	0.00	0.04	—
225	Juice, unsweetened, canned	½	cup(s)	124	109	58	<.1	14	<1	<1	0.02	0.01	0.04	—
226	Applesauce, sweetened, canned	½	cup(s)	128	101	97	<1	25	2	<1	0.04	0.01	0.07	—
227	Applesauce, unsweetened, canned	½	cup(s)	122	108	52	<1	14	1	<.1	0.01	0.00	0.02	—
38492	Crabapples	1	item(s)	35	28	27	<1	7	1	<1	0.02	0.00	0.03	—
	Apricot													
228	Fresh w/o pits	4	item(s)	140	121	67	2	16	3	1	0.04	0.24	0.11	—
230	Halves, dried, sulfured	¼	cup(s)	33	10	79	1	21	2	<1	0.01	0.02	0.02	—
229	Halves w/skin, canned in heavy syrup	½	cup(s)	129	100	107	1	28	2	<1	0.01	0.04	0.02	—
	Avocado													
233	California, whole, w/o skin or pit	1	item(s)	170	<1	284	3	15	12	26	3.59	16.61	3.42	—
234	Florida, whole, w/o skin or pit	1	item(s)	304	<1	365	7	24	17	31	5.90	16.70	5.00	—
2998	Pureed	⅛	cup(s)	29	21	46	1	2	2	4	0.61	2.82	0.52	—
	Banana													
235	Fresh whole, w/o peel	1	item(s)	118	88	105	1	27	3	<1	0.13	0.04	0.09	—
4580	Dried chips	¼	cup(s)	55	2	287	1	32	4	19	16.00	1.08	0.35	—
	Blackberries													
237	Raw	½	cup(s)	72	63	31	1	7	4	<1	0.01	0.03	0.20	—
958	Unsweetened, frozen	½	cup(s)	76	62	48	1	12	4	<1	0.01	0.03	0.18	—
	Blueberries													
238	Raw	½	cup(s)	72	61	41	1	10	2	<1	0.02	0.03	0.11	—
959	Canned in heavy syrup	½	cup(s)	128	98	113	1	28	2	<1	0.03	0.06	0.18	—
960	Unsweetened, frozen	½	cup(s)	78	67	40	1	10	2	1	0.04	0.07	0.22	—

PAGE KEY: A–58 = Breads/Baked Goods A–62 = Cereal/Rice/Pasta A–66 = Fruit A–70 = Vegetables/Legumes A–80 = Nuts/Seeds A–82 = Vegetarian A–84 = Dairy A–90 = Eggs A–90 = Seafood A–92 = Meats A–96 = Poultry A–96 = Processed Meats A–98 = Beverages A–102 = Fats/Oils A–104 = Sweets A–106 = Sauces/Condiments/Spices A–108 = Mixed Foods/Soups/Sandwiches A–114 = Fast Food A–130 = Convenience A–132 = Baby Foods

Chol (mg)	Calc (mg)	Iron (mg)	Magn (mg)	Pota (mg)	Sodi (mg)	Zinc (mg)	Vit A (RAE) (μg)	Thia (mg)	Vit E (mg α)	Ribo (mg)	Niac (mg)	Vit B_6 (mg)	Fola (μg)	Vit C (mg)	Vit B_{12} (μg)	Sele (μg)
0	100	16.20	60	220	390	3.75	158	0.38	—	0.03	5.00	0.50	100	6	2	5
0	100	18.00	24	85	200	15.00	—	1.50	—	1.70	20.00	2.00	400	15	6	—
0	0	1.46	32	146	279	4.99	0	0.50	—	0.57	6.65	0.67	133	20	2	9
0	0	18.00	16	50	210	15.00	225	1.50	—	1.70	20.00	2.00	400	60	6	4
0	1	4.44	4	16	<1	0.14	0	0.36	—	0.25	4.94	0.01	3	0	0	1
0	3.35	3.8	17.39	41.75	0.47	0.28	0	0.31	—	0.21	4.23	0.02	3.83	0	0	14.77
0	20	10.80	80	340	300	2.25	—	0.53	—	0.60	7.00	0.70	140	0	2	—
0	80	7.20	7	28	232	3.00	—	0.30	—	0.34	4.00	0.40	160	5	1	1
0	0	1.44	13	32	256	0.48	120	0.30	—	0.34	4.80	0.40	80	5	1	4
0	10	1.08	31	92	2	0.70	0	0.07	—	0.06	1.77	0.10	12	0	0	1
0	0	0.48	11	53	67	0.40	200	0.50	—	0.57	6.65	0.67	133	8	2	17
0	0	8.70	16	60	220	0.90	225	0.53	—	0.60	7.00	2.00	400	15	6	7
0	750	13.50	0	23	158	11.25	113	1.13	22.50	1.28	15.00	1.50	300	45	5	1
0	1330	23.94	32	120	253	19.95	200	2.00	31.24	2.26	26.60	2.66	532	80	8	2
0	100	4.50	0	15	190	3.75	150	0.38	—	0.43	5.00	0.50	100	6	2	6
0	6.35	1.28	45.2	133.76	0.56	2.35	0	0.23	—	0.11	0.78	0.13	49.72	0.84	0	9.18
0	0	8.10	32	110	220	7.50	150	0.75	2.26	0.85	10.00	1.00	200	6	3	1
0	5	1.06	12	27	99	0.32	0	0.13	—	0.09	1.34	0.02	20	0	0	10
0	1	0.18	25	22	0	0.44	2	0.04	0.78	0.02	0.39	0.04	4	0	0	2
26	10	1.27	15	22	6	0.49	5	0.15	0.14	0.07	1.19	0.03	51	0	<.1	17
0	5	0.98	13	22	1	0.37	0	0.14	0.04	0.07	1.17	0.02	54	0	0	15
21	4	0.73	12	15	4	0.36	4	0.13	—	0.10	0.63	0.02	41	0	<.1	—
18	9	0.89	9	34	415	0.31	—	0.08	—	0.05	0.71	0.03	4	<.1	<.1	—
0	4	0.45	9	33	57	0.11	0	0.09	—	0.02	0.48	0.04	7	0	0	—
0	7	0.46	2	25	141	0.19	0	0.02	—	0.03	0.09	0.01	2	0	0	—
0	7	1.00	12	52	1	0.35	0	0.12	0.04	0.07	0.90	0.04	8	0	0	40
0	11	0.74	21	31	2	0.57	0	0.08	0.21	0.03	0.49	0.06	4	0	0	18
26	15	0.87	19	30	10	0.50	4	0.20	0.46	0.10	1.18	0.09	51	0	<1	17
0	7	0.33	13	21	4	0.29	3	0.08	0.06	0.04	0.72	0.02	44	0	0	13
0	1	0.22	11	24	<1	0.28	1	0.02	0.02	0.02	0.16	0.02	2	0	0	1
2	15	0.61	12	38	73	0.20	1	0.02	0.42	0.02	0.77	0.01	2	0	<.1	1
4	42	0.83	34	97	331	0.75	14	0.05	—	0.09	0.54	0.09	4	<1	<1	4
0	3	0.92	36	74	292	0.87	3	0.04	—	0.04	0.51	0.07	6	<.1	0	2
0	8	0.17	7	148	1	0.06	4	0.02	—	0.04	0.13	0.06	4	6	0	0
0	3	0.07	3	59	1	0.02	2	0.01	—	0.01	0.05	0.02	2	3	0	0
0	4	0.16	3	75	1	0.03	2	0.01	0.04	0.01	0.08	0.04	1	<1	0	<1
0	3	0.30	3	97	19	0.04	0	0.00	0.11	0.03	0.20	0.03	0	1	0	<1
0	7	0.31	6	151	8	0.05	0	0.00	0.01	0.02	0.05	0.04	0	1	0	<1
0	9	0.46	4	148	4	0.04	0	0.03	0.01	0.02	0.12	0.04	0	1	0	<1
0	5	0.45	4	78	4	0.05	1	0.02	0.27	0.04	0.24	0.03	1	2	0	<1
0	4	0.15	4	92	2	0.04	1	0.02	0.26	0.03	0.23	0.03	1	1	0	<1
0	6	0.13	2	68	<1	—	0	0.01	—	0.01	0.04	—	2	3	0	—
0	18	0.55	14	363	1	0.28	134	0.04	1.25	0.06	0.84	0.08	13	14	0	<1
0	18	0.88	11	383	3	0.13	59	0.00	1.43	0.02	0.85	0.05	3	<1	0	1
0	12	0.39	9	181	5	0.14	80	0.03	0.77	0.03	0.49	0.07	3	4	0	<1
0	22	1.00	49	861	14	1.12	104	0.12	3.35	0.24	3.24	0.47	105	15	0	1
0	30	0.50	73	1067	6	1.20	185	0.00	0.09	0.10	2.00	0.20	106	53	0	0
0	3	0.16	8	139	2	0.18	2	0.02	0.60	0.04	0.50	0.07	17	3	0	<1
0	6	0.31	32	422	1	0.18	4	0.04	0.12	0.09	0.78	0.43	24	10	0	1
0	10	0.69	42	296	3	0.41	2	0.05	0.13	0.01	0.39	0.14	8	3	0	1
0	21	0.45	14	117	1	0.38	8	0.01	0.84	0.02	0.47	0.02	18	15	0	<1
0	22	0.60	17	106	1	0.19	5	0.02	0.88	0.03	0.91	0.05	26	2	0	<1
0	4	0.20	4	55	1	0.12	2	0.03	0.41	0.03	0.30	0.04	4	7	0	<.1
0	6	0.42	5	51	4	0.09	3	0.04	0.49	0.07	0.14	0.05	3	1	0	<1
0	6	0.14	4	42	1	0.06	2	0.03	0.37	0.03	0.41	0.05	6	2	0	0

TABLE F–1

Food Composition (Computer code number is for Wadsworth Diet Analysis program) (For purposes of calculations, use "0" for t, <1, <.1, <.01, etc.)

DA + Code	Food Description	Quantity	Measure	Wt (g)	H_2O (g)	Ener (kcal)	Prot (g)	Carb (g)	Dietary Fiber (g)	Fat (g)	Fat Breakdown (g) Sat	Mono	Poly	Trans
	FRUIT AND FRUIT JUICES—Continued													
	Boysenberries													
961	Canned in heavy syrup	½	cup(s)	128	98	113	1	29	3	<1	0.01	0.02	0.09	—
962	Unsweetened, frozen	½	cup(s)	66	57	33	1	8	3	<1	0.01	0.02	0.10	—
35576	**Breadfruit**	1	item(s)	384	271	396	4	104	17	1	0.00	0.00	0.00	—
	Cherries													
3000	Sour red, raw	½	cup(s)	78	67	39	1	9	1	<1	0.05	0.06	0.07	—
967	Sour red, canned in water	½	cup(s)	122	110	44	1	11	1	<1	0.03	0.03	0.04	—
240	Sweet, raw	½	cup(s)	73	60	46	1	12	2	<1	0.03	0.03	0.04	—
3004	Sweet, canned in heavy syrup	½	cup(s)	127	98	105	1	27	2	<1	0.04	0.05	0.06	—
969	Sweet, canned in water	½	cup(s)	124	108	57	1	15	2	<1	0.03	0.04	0.05	—
	Cranberries													
3007	Chopped, raw	½	cup(s)	55	48	25	<1	7	3	<.1	0.01	0.01	0.03	—
1638	Cranberry juice cocktail	½	cup(s)	127	108	72	0	18	<1	<1	0.01	0.02	0.06	—
241	Cranberry juice cocktail, low calorie, w/saccharin	½	cup(s)	127	120	24	<.1	6	0	<.1	0.00	0.00	0.00	—
1717	Cranberry apple juice drink	½	cup(s)	123	100	87	<.1	22	<1	<.1	0.00	0.00	0.00	—
242	Cranberry sauce, sweetened, canned	¼	cup(s)	69	42	105	<1	27	1	<1	0.01	0.01	0.05	—
	Dates													
244	Domestic, chopped	¼	cup(s)	44.5	0	126	1	33	4	<1	0.01	0.01	0	—
243	Domestic, whole	¼	cup(s)	44.5	0	126	1	33	4	<1	0.01	0.01	0	—
	Figs													
973	Raw, medium	2	item(s)	101	80	74	1	19	3	<1	0.06	0.07	0.14	—
975	Canned in heavy syrup	½	cup(s)	130	99	114	<1	30	3	<1	0.03	0.03	0.06	—
974	Canned in water	½	cup(s)	124	106	66	<1	17	3	<1	0.02	0.03	0.06	—
	Fruit cocktail & salad													
245	Fruit cocktail, canned in heavy syrup	½	cup(s)	124	100	91	<1	23	1	<.1	0.01	0.02	0.04	—
978	Fruit cocktail, canned in juice	½	cup(s)	119	104	55	1	14	1	<.1	0.00	0.00	0.00	—
977	Fruit cocktail, canned in water	½	cup(s)	119	108	38	<1	10	1	<.1	0.01	0.01	0.02	—
979	Fruit salad, canned in water	½	cup(s)	123	112	37	<1	10	1	<.1	0.01	0.02	0.03	—
	Gooseberries													
981	Raw	½	cup(s)	75	66	33	1	8	3	<1	0.03	0.04	0.24	—
982	Canned in light syrup	½	cup(s)	126	101	92	1	24	3	<1	0.02	0.02	0.14	—
	Grapefruit													
3022	Raw, pink or red	½	cup(s)	115	<.1	48	1	12	2	<1	0.02	0.02	0.04	—
247	Raw, white	½	item(s)	118	107	39	1	10	1	<1	0.02	0.02	0.03	—
251	Juice, pink, sweetened, canned	½	cup(s)	125	109	58	1	14	<1	<1	0.02	0.02	0.03	—
249	Juice, white	½	cup(s)	124	111	48	1	11	<1	<1	0.02	0.02	0.03	—
248	Sections, canned in light syrup	½	cup(s)	127	106	76	1	20	1	<1	0.02	0.02	0.03	—
983	Sections, canned in water	½	cup(s)	122	<.1	44	1	11	<1	<1	0.02	0.02	0.03	—
	Grapes													
255	American, slip skin	½	cup(s)	46	37	31	<1	8	<1	<1	0.05	0.01	0.05	—
256	European, red or green, adherent skin	½	cup(s)	80	<.1	55	1	14	1	<1	0.04	0.01	0.04	—
259	Juice, sweetened, added vitamin C, from frozen concentrate	½	cup(s)	125	109	64	<1	16	<1	<1	0.04	0.01	0.03	—
3159	Juice drink, canned	½	cup(s)	125	109	63	<1	16	0	0	0.00	0.00	0.00	—
3060	Raisins, seeded, packed	¼	cup(s)	41	7	122	1	32	3	<1	0.07	0.01	0.07	—
987	**Guava, raw**	1	item(s)	90	77	46	1	11	5	1	0.15	0.05	0.23	—
35593	**Guava, strawberry**	1	item(s)	6	5	4	<.1	1	<1	<.1	0.01	0.00	0.02	—
3027	**Jackfruit**	½	cup(s)	83	61	78	1	20	1	<1	0.05	0.04	0.07	—
8458	**Kiwi fruit**	1	item(s)	77	63	53	1	11	3	1	0.02	0.03	0.19	—
	Lemon													
992	Raw	1	item(s)	108	94	22	1	12	5	<1	0.04	0.01	0.10	—
262	Juice	1	tablespoon(s)	15	14	4	<.1	1	<.1	0	0.00	0.00	0.00	—
993	Peel	1	teaspoon(s)	2	2	1	<.1	<1	<1	<.1	0.00	0.00	0.00	—
	Lime													
994	Raw	1	item(s)	67	61	15	<1	6	2	<.1	0.01	0.01	0.02	—
269	Juice	1	tablespoon(s)	15	14	4	<.1	1	<.1	<.1	0.00	0.00	0.00	—
995	**Loganberries, frozen**	½	cup(s)	74	62	40	1	10	4	<1	0.01	0.02	0.13	—
	Mandarin orange													
1038	Canned in juice	½	cup(s)	125	111	46	1	12	1	<.1	0.00	0.01	0.01	—
1039	Canned in light syrup	½	cup(s)	126	105	77	1	20	1	<1	0.02	0.02	0.03	—
999	**Mango**	½	item(s)	104	85	67	1	18	2	<1	0.07	0.10	0.05	—
1005	**Nectarine, raw, sliced**	½	cup(s)	69	60	30	1	7	1	<1	0.02	0.06	0.08	—
	Melons													
271	Cantaloupe	½	cup(s)	80	72	27	1	7	1	<1	0.04	0.00	0.07	—
1000	Casaba melon	½	cup(s)	85	78	24	1	6	1	<.1	0.02	0.00	0.03	—

PAGE KEY: A–58 = Breads/Baked Goods A–62 = Cereal/Rice/Pasta A–66 = Fruit A–70 = Vegetables/Legumes A–80 = Nuts/Seeds A–82 = Vegetarian A–84 = Dairy A–90 = Eggs A–90 = Seafood A–92 = Meats A–96 = Poultry A–96 = Processed Meats A–98 = Beverages A–102 = Fats/Oils A–104 = Sweets A–106 = Sauces/Condiments/Spices A–108 = Mixed Foods/Soups/Sandwiches A–114 = Fast Food A–130 = Convenience A–132 = Baby Foods

Chol (mg)	Calc (mg)	Iron (mg)	Magn (mg)	Pota (mg)	Sodi (mg)	Zinc (mg)	Vit A (RAE) (μg)	Thia (mg)	Vit E (mg α)	Ribo (mg)	Niac (mg)	Vit B_6 (mg)	Fola (μg)	Vit C (mg)	Vit B_{12} (μg)	Sele (μg)
0	23	0.55	14	115	4	0.24	3	0.03	—	0.04	0.29	0.05	44	8	0	1
0	18	0.56	11	92	1	0.15	2	0.03	0.57	0.02	0.51	0.04	42	2	0	<1
0	65	2.07	96	1882	8	0.46	8	0.42	—	0.12	3.46	0.00	54	111	0	2
0	12	0.25	7	134	2	0.08	50	0.02	0.05	0.03	0.31	0.03	6	8	0	0
0	13	1.67	7	120	9	0.09	46	0.02	0.28	0.05	0.22	0.05	10	3	0	0
0	9	0.26	8	161	0	0.05	2	0.02	0.05	0.02	0.11	0.04	3	5	0	0
0	11	0.44	11	183	4	0.13	10	0.03	0.29	0.05	0.50	0.04	5	5	0	0
0	14	0.45	11	162	1	0.10	10	0.03	0.29	0.05	0.51	0.04	5	3	0	0
0	4	0.14	3	47	1	0.06	2	0.01	0.66	0.01	0.06	0.03	1	7	0	<.1
0	4	0.19	3	23	3	0.09	0	0.01	0.28	0.01	0.04	0.02	0	45	0	0
0	11	0.05	3	32	4	0.03	0	0.00	0.06	0.00	0.01	0.00	0	41	0	0
0	6	0.15	2	34	9	0.22	0	0.01	0.15	0.02	0.07	0.03	0	39	0	0
0	3	0.15	2	18	20	0.03	1	0.01	0.57	0.01	0.07	0.01	1	1	0	<1
0	17	0.45	19	292	1	0.12	1	0.02	0.02	0.02	0.56	0.07	9	<1	0	1
0	17	0.45	19	292	1	0.12	1	0.02	0.02	0.02	0.56	0.07	9	<1	0	1
0	35	0.37	17	233	1	0.15	7	0.06	0.11	0.05	0.40	0.11	6	2	0	<1
0	35	0.36	13	128	1	0.14	3	0.03	0.16	0.05	0.55	0.09	3	1	0	<1
0	35	0.36	12	128	1	0.15	2	0.03	0.10	0.05	0.55	0.09	2	1	0	<1
0	7	0.36	6	109	7	0.10	12	0.02	0.50	0.02	0.46	0.06	4	2	0	1
0	9	0.25	8	113	5	0.11	18	0.01	0.47	0.02	0.48	0.06	4	3	0	1
0	6	0.30	8	111	5	0.11	15	0.02	0.47	0.01	0.43	0.06	4	2	0	1
0	9	0.37	6	96	4	0.10	27	0.02	—	0.03	0.46	0.04	4	2	0	1
0	19	0.23	8	149	1	0.09	11	0.03	0.28	0.02	0.23	0.06	5	21	0	<1
0	20	0.42	8	97	3	0.14	9	0.03	—	0.07	0.19	0.02	4	13	0	1
0	25	0.09	10	155	0	0.08	30	0.05	0.15	0.03	0.23	0.06	15	36	0	<1
0	14	0.07	11	175	0	0.08	2	0.04	0.15	0.02	0.32	0.05	12	39	0	2
0	10	0.45	13	203	3	0.08	0	0.05	0.05	0.03	0.40	0.03	13	34	0	<1
0	11	0.25	15	200	1	0.06	2	0.05	0.27	0.02	0.25	0.05	12	47	0	<1
0	18	0.51	13	164	3	0.10	0	0.05	0.11	0.03	0.31	0.03	11	27	0	1
0	18	0.50	12	161	2	0.11	0	0.05	0.11	0.03	0.30	0.02	11	27	0	1
0	6	0.13	2	88	1	0.02	2	0.04	0.09	0.03	0.14	0.05	2	2	0	<.1
0	8	0.29	6	153	2	0.06	6	0.06	0.15	0.06	0.15	0.07	2	9	0	<.1
0	5	0.13	5	26	3	0.05	0	0.02	0.00	0.03	0.16	0.05	1	30	0	<1
0	4	0.13	4	41	1	0.03	0	0.01	0.00	0.02	0.09	0.02	1	20	0	<1
0	12	1.07	12	340	12	0.07	0	0.05	—	0.08	0.46	0.08	1	2	0	<1
0	18	0.28	9	256	3	0.21	28	0.05	0.66	0.05	1.08	0.13	13	165	0	1
0	1	0.01	1	18	2	—	—	0.00	—	0.00	0.04	0.00	—	2	0	—
0	28	0.50	31	251	2	0.35	12	0.02	—	0.09	0.33	0.09	12	6	0	<1
0	30	0.38	14	251	2	0.10	4	—	—	0.02	0.25	0.05	<.1	74	0	—
0	66	0.76	13	157	3	0.11	2	0.05	—	0.04	0.22	0.12	—	83	0	1
0	1	0.00	1	19	<1	0.01	<1	0.00	0.02	0.00	0.02	0.01	2	7	0	<.1
0	3	0.02	<1	3	<1	0.01	<.1	0.00	0.00	0.00	0.01	0.00	<1	3	0	<.1
0	9	0.06	5	78	1	0.05	1	0.02	0.15	0.01	0.10	0.03	7	20	0	<.1
0	1	0.00	1	17	<1	0.01	<1	0.00	0.03	0.00	0.02	0.01	1	5	0	<.1
0	19	0.47	15	107	1	0.25	1	0.04	0.64	0.02	0.62	0.05	19	11	0	<1
0	14	0.34	14	166	6	0.63	54	0.10	0.12	0.04	0.55	0.05	6	43	0	<1
0	9	0.47	10	98	8	0.30	53	0.07	0.13	0.06	0.56	0.05	6	25	0	1
0	10	0.13	9	161	2	0.04	39	0.06	1.16	0.06	0.60	0.14	14	29	0	1
0	4	0.19	6	139	0	0.12	12	0.02	0.53	0.02	0.78	0.02	3	4	0	0
0	7	0.17	10	215	13	0.14	136	0.03	0.04	0.02	0.59	0.06	17	30	0	<1
0	9	0.29	9	155	8	0.06	0	0.01	0.04	0.03	0.20	0.14	7	19	0	<1

TABLE F-1
Food Composition (Computer code number is for Wadsworth Diet Analysis program) (For purposes of calculations, use "0" for t, <1, <.1, <.01, etc.)

DA + Code	Food Description	Quantity	Measure	Wt (g)	H_2O (g)	Ener (kcal)	Prot (g)	Carb (g)	Dietary Fiber (g)	Fat (g)	Fat Breakdown (g) Sat	Mono	Poly	Trans
	FRUIT AND FRUIT JUICES—Continued													
272	Honeydew	½	cup(s)	89	80	32	<1	8	1	<1	0.03	0.00	0.05	—
318	Watermelon	½	cup(s)	77	71	23	<1	6	<1	<1	0.01	0.03	0.04	—
	Orange													
273	Raw	1	item(s)	131	114	62	1	15	3	<1	0.02	0.03	0.03	—
3040	Peel	1	teaspoon(s)	2	1	2	<.1	1	<1	<.1	0.00	0.00	0.00	—
274	Sections	½	cup(s)	90	78	43	1	11	2	<1	0.01	0.02	0.02	—
275	Juice	½	cup(s)	124	109	56	1	13	<1	<1	0.03	0.04	0.05	—
29630	Juice, fresh squeezed	½	cup(s)	124	109	56	1	13	<1	<1	0.03	0.04	0.05	—
14414	Juice w/calcium & extra vitamin C	½	cup(s)	125	55	1	13	<1	0	0.00	0.00	0.00	0	—
278	Juice, unsweetened, from frozen concentrate	½	cup(s)	125	110	56	1	13	<1	<.1	0.01	0.01	0.01	—
	Papaya													
282	Raw	½	cup(s)	70	62	27	<1	7	1	<.1	0.03	0.03	0.02	—
16830	Dried, strips	2	item(s)	46	12	119	2	30	5	<1	0.13	0.12	0.09	—
35640	**Passion fruit, purple**	1	item(s)	18	13	17	<1	4	3	<1	0.00	0.00	0.00	—
	Peach													
283	Raw, medium	1	item(s)	98	87	38	1	9	1	<1	0.02	0.07	0.08	—
285	Halves, canned in heavy syrup	½	cup(s)	131	104	97	1	26	2	<1	0.01	0.05	0.06	—
286	Halves, canned in water	½	cup(s)	122	114	29	1	7	2	<.1	0.01	0.03	0.03	—
290	Slices, sweetened, frozen	½	cup(s)	125	93	118	1	30	2	<1	0.02	0.06	0.08	—
	Pear													
291	Raw	1	item(s)	166	139	96	1	26	5	<1	0.01	0.04	0.05	—
8672	Asian	1	item(s)	122	108	51	1	13	4	<1	0.01	0.06	0.07	—
293	Danjou	1	item(s)	200	168	120	1	30	5	1	0.00	0.20	0.20	—
294	Halves, canned in heavy syrup	½	cup(s)	133	107	98	<1	25	2	<1	0.01	0.04	0.04	—
1012	Halves, canned in juice	½	cup(s)	124	107	62	<1	16	2	<.1	0.00	0.02	0.02	—
1017	**Persimmon**	1	item(s)	25	16	32	<1	8	0	<1	0.01	0.02	0.02	—
	Pineapple													
295	Raw, diced	½	cup(s)	78	67	37	<1	10	1	<.1	0.01	0.01	0.03	—
3053	Canned in extra heavy syrup	½	cup(s)	130	101	108	<1	28	1	<1	0.01	0.02	0.05	—
1019	Canned in juice	½	cup(s)	125	104	75	1	20	1	<.1	0.01	0.01	0.04	—
296	Canned in light syrup	½	cup(s)	126	108	66	<1	17	1	<1	0.01	0.02	0.05	—
1018	Canned in water	½	cup(s)	123	112	39	1	10	1	<1	0.01	0.01	0.04	—
299	Juice, unsweetened, canned	½	cup(s)	125	107	70	<1	17	<1	<1	0.01	0.01	0.04	—
1024	**Plantain, cooked**	½	cup(s)	77	52	89	1	24	2	<1	0.05	0.01	0.03	—
300	**Plum, raw, large**	1	item(s)	83	72	38	1	9	1	<1	0.01	0.11	0.04	—
1027	**Pomegranate**	1	item(s)	154	125	105	1	26	1	<1	0.06	0.07	0.10	—
	Prunes													
5644	Dried	2	item(s)	17	5	40	<1	11	1	<.1	0.01	0.06	0.02	—
305	Dried, stewed	½	cup(s)	119	<.1	128	1	33	4	<1	0.00	0.15	0.04	—
306	Juice, canned	1	cup(s)	256	208	182	2	45	3	<.1	0.01	0.05	0.02	—
	Raisins, *see* grapes													
	Raspberries													
309	Raw	½	cup(s)	62	53	32	1	7	4	<1	0.01	0.04	0.23	—
310	Red, sweetened, frozen	½	cup(s)	125	91	129	1	33	6	<1	0.01	0.02	0.11	—
311	**Rhubarb, cooked with sugar**	½	cup(s)	120	82	140	1	38	3	<.1	0.00	0.00	0.05	—
	Strawberries													
313	Raw	½	cup(s)	72	65	23	<1	6	1	<1	0.01	0.03	0.11	—
315	Sweetened, frozen, thawed	½	cup(s)	128	100	99	1	27	2	<1	0.01	0.02	0.09	—
16828	**Tangelo**	1	item(s)	95	82	45	1	11	2	<1	0.01	0.02	0.02	—
	Tangerine													
316	Raw	1	item(s)	84	74	37	1	9	2	<1	0.02	0.03	0.03	—
1040	Juice	½	cup(s)	124	110	53	1	12	<1	<1	0.03	0.04	0.05	—
	VEGETABLES, LEGUMES													
	Amaranth													
1042	Leaves, raw	1	cup(s)	28	26	6	1	1	0	<.1	0.03	0.02	0.04	—
1043	Leaves, boiled, drained	½	cup(s)	66	60	14	1	3	0	<1	0.03	0.03	0.05	—
8683	**Arugula leaves, raw**	1	cup(s)	20	18	5	1	1	<1	<1	0.02	0.01	0.06	—
	Artichoke													
1044	Boiled, drained	1	item(s)	120	101	60	4	13	6	<1	0.04	0.01	0.08	—
2885	Hearts, boiled, drained	½	cup(s)	84	71	42	3	9	5	<1	0.03	0.00	0.06	—
	Asparagus													
566	Boiled, drained	½	cup(s)	90	0.08	20	2	4	2	0.19	0.06	0	0.12	—
568	Canned, drained	½	cup(s)	121	114	23	3	3	2	1	0.18	0.03	0.34	—
565	Tips, frozen, boiled, drained	½	cup(s)	90	82	25	3	4	1	<1	0.09	0.01	0.17	—

PAGE KEY: A–58 = Breads/Baked Goods A–62 = Cereal/Rice/Pasta A–66 = Fruit A–70 = Vegetables/Legumes A–80 = Nuts/Seeds A–82 = Vegetarian A–84 = Dairy A–90 = Eggs A–90 = Seafood A–92 = Meats A–96 = Poultry A–96 = Processed Meats A–98 = Beverages A–102 = Fats/Oils A–104 = Sweets A–106 = Sauces/Condiments/Spices A–108 = Mixed Foods/Soups/Sandwiches A–114 = Fast Food A–130 = Convenience A–132 = Baby Foods

Chol (mg)	Calc (mg)	Iron (mg)	Magn (mg)	Pota (mg)	Sodi (mg)	Zinc (mg)	Vit A (RAE) (μg)	Thia (mg)	Vit E (mg α)	Ribo (mg)	Niac (mg)	Vit B_6 (mg)	Fola (μg)	Vit C (mg)	Vit B_{12} (μg)	Sele (μg)
0	5	0.15	9	203	16	0.08	3	0.03	0.02	0.01	0.37	0.08	17	16	0	1
0	5	0.19	8	86	1	0.08	22	0.03	0.04	0.02	0.14	0.03	2	6	0	<1
0	52	0.13	13	237	0	0.09	14	0.11	0.24	0.05	0.37	0.08	39	70	0	1
0	3	0.02	<1	4	<.1	0.01	<1	0.00	0.00	0.00	0.02	0.00	1	3	0	<.1
0	36	0.09	9	164	0	0.06	10	0.08	0.16	0.04	0.26	0.05	27	48	0	<1
0	14	0.25	14	248	1	0.06	12	0.11	0.05	0.04	0.50	0.05	37	62	0	<1
0	14	0.25	14	248	1	0.06	—	0.11	0.05	0.04	0.50	0.05	38	62	0	—
—	175	—	—	225	0	—	5	0.08	—	—	0.40	0.06	30	54	0	—
0	11	0.12	12	237	1	0.06	6	0.10	0.25	0.02	0.25	0.05	55	48	0	<1
0	17	0.07	7	180	2	0.05	39	0.02	0.51	0.02	0.24	0.01	27	43	0	<1
0	73	0.30	30	783	9	0.21	—	0.06	2.22	0.09	0.93	0.05	58	38	0	—
0	2	0.29	5	63	5	—	—	0.00	—	0.02	0.27	—	3	5	0	<1
0	6	0.25	9	186	0	0.17	16	0.02	0.72	0.03	0.79	0.02	4	6	0	<.1
0	4	0.35	7	121	8	0.12	22	0.01	0.64	0.03	0.80	0.02	4	4	0	<1
0	2	0.39	6	121	4	0.11	33	0.01	0.60	0.02	0.64	0.02	4	4	0	<1
0	4	0.46	6	163	8	0.06	18	0.02	0.77	0.04	0.82	0.02	4	118	0	1
0	15	0.28	12	198	2	0.17	2	0.02	0.20	0.04	0.26	0.05	12	7	0	<1
0	5	0.00	10	148	0	0.02	0	0.01	0.15	0.01	0.27	0.03	10	5	0	<1
0	22	0.50	12	250	0	0.24	—	0.04	1.00	0.08	0.20	0.04	15	8	0	1
0	7	0.29	5	86	7	0.11	0	0.01	0.11	0.03	0.32	0.02	1	1	0	0
0	11	0.36	9	119	5	0.11	0	0.01	0.10	0.01	0.25	0.02	1	2	0	0
0	7	0.63	—	78	<1	—	—	—	—	—	—	—	—	17	0	0
0	10	0.22	9	89	1	0.08	2	0.06	0.02	0.02	0.38	0.09	12	28	0	<.1
0	18	0.49	20	133	1	0.14	1	0.12	—	0.03	0.37	0.10	7	9	0	—
0	17	0.35	17	152	1	0.12	2	0.12	0.01	0.02	0.35	0.09	6	12	0	<1
0	18	0.49	20	132	1	0.15	3	0.11	0.01	0.03	0.37	0.09	6	9	0	1
0	18	0.49	22	156	1	0.15	2	0.11	0.01	0.03	0.37	0.09	6	9	0	<1
0	21	0.33	16	168	1	0.14	0	0.07	0.03	0.03	0.32	0.12	29	13	0	<1
0	2	0.45	25	358	4	0.10	35	0.04	0.10	0.04	0.58	0.18	20	8	0	1
0	5	0.14	6	130	0	0.08	14	0.02	0.21	0.02	0.34	0.02	4	8	0	0
0	5	0.46	5	399	5	0.18	8	0.05	0.92	0.05	0.46	0.16	9	9	0	1
0	9	0.42	8	125	1	0.09	17	0.01	0.00	0.03	0.33	0.04	1	1	0	<1
0	23	0.46	21	383	1	0.19	37	0.00	0.23	0.12	0.85	0.23	0	3	0	<1
0	31	3.02	36	707	10	0.54	0	0.04	0.31	0.18	2.01	0.56	0	10	0	2
0	15	0.42	14	93	1	0.26	1	0.02	0.54	0.02	0.37	0.03	13	16	0	<1
0	19	0.81	16	143	1	0.23	4	0.02	0.90	0.06	0.29	0.04	33	21	0	<1
0	174	0.25	16	115	1	—	—	0.02	—	0.03	0.25	—	—	4	0	—
0	12	0.30	9	110	1	0.10	1	0.02	0.21	0.02	0.28	0.03	17	42	0	<1
0	14	0.60	8	125	1	0.06	1	0.02	0.31	0.10	0.37	0.04	5	50	0	1
0	38	0.10	10	172	0	0.07	—	0.08	0.17	0.04	0.27	0.06	29	51	0	—
0	12	0.08	10	132	1	0.20	29	0.09	0.17	0.02	0.13	0.06	17	26	0	<1
0	22	0.25	10	220	1	0.04	16	0.07	0.16	0.02	0.12	0.05	6	38	0	<1
0	60	0.65	15	171	6	0.25	0	0.01	—	0.04	0.18	0.05	24	12	0	<1
0	138	1.49	36	423	14	0.58	92	0.01	—	0.09	0.37	0.12	38	27	0	1
0	32	0.29	9	74	5	0.09	24	0.01	0.09	0.02	0.06	0.01	19	3	0	<.1
0	54	1.55	72	425	114	0.59	11	0.08	0.23	0.08	1.20	0.13	61	12	0	<1
0	38	1.08	50	297	80	0.41	8	0.05	0.16	0.06	0.84	0.09	43	8	0	<1
0	20.7	0.81	12.6	201.6	12.6	0.54	48.59	0.14	1.35	0.12	0.97	0.07	134.1	6.92	0	5.48
0	19	0.73	12	208	347	0.48	50	0.07	0.38	0.12	1.15	0.13	116	22	0	2
0	21	0.58	12	196	4	0.50	—	0.06	1.08	0.09	0.93	0.02	121	22	0	4

TABLE F–1

Food Composition (Computer code number is for Wadsworth Diet Analysis program) (For purposes of calculations, use "0" for t, <1, <.1, <.01, etc.)

DA + Code	Food Description	Quantity	Measure	Wt (g)	H_2O (g)	Ener (kcal)	Prot (g)	Carb (g)	Dietary Fiber (g)	Fat (g)	Fat Breakdown (g) Sat	Mono	Poly	Trans
	VEGETABLES, LEGUMES—Continued													
	Bamboo shoots													
1048	Boiled, drained	½	cup(s)	60	58	7	1	1	1	<1	0.03	0.00	0.06	—
1049	Canned, drained	½	cup(s)	65	62	12	1	2	1	<1	0.06	0.01	0.12	—
	Beans													
1801	Adzuki beans, boiled	½	cup(s)	115	76	147	9	28	8	<1	0.04	—	—	—
511	Baked beans w/franks, canned	½	cup(s)	129	89	182	9	20	9	8	3.02	3.64	1.07	—
512	Baked beans w/pork in tomato sauce, canned	½	cup(s)	127	92	124	7	25	6	1	0.50	0.56	0.17	—
513	Baked beans w/pork in sweet sauce, canned	½	cup(s)	127	89	140	7	27	7	2	0.71	0.80	0.24	0
1805	Black beans, boiled	½	cup(s)	86	57	114	8	20	7	<1	0.12	0.04	0.20	—
14597	Chickpeas, garbanzo beans, or bengal gram, boiled	½	cup(s)	82	49	134	7	22	6	2	0.22	0.48	0.95	—
569	Fordhook lima beans, frozen, boiled, drained	½	cup(s)	85	62	88	5	16	5	<1	0.07	0.02	0.14	—
1806	French beans, boiled	½	cup(s)	89	59	114	6	21	8	1	0.07	0.05	0.40	—
2773	Great northern beans, boiled	½	cup(s)	89	61	104	7	19	6	<1	0.12	0.02	0.17	—
2736	Hyacinth beans, boiled, drained	½	cup(s)	44	38	22	1	4	0	<1	0.05	0.06	0.00	—
515	Lima beans, boiled, drained	½	cup(s)	85	57	105	6	20	5	<1	0.06	0.02	0.13	—
570	Lima beans, baby, frozen, boiled, drained	½	cup(s)	90	65	95	6	18	5	<1	0.06	0.02	0.13	—
579	Mung beans, sprouted, boiled, drained	½	cup(s)	62	<.1	13	1	3	<1	<.1	0.02	0.00	0.02	—
510	Navy beans, boiled	½	cup(s)	91	57	129	8	24	6	1	0.13	0.05	0.22	0
32816	Pinto beans, boiled, drained, no salt added	½	cup(s)	114	106	25	2	5	0	<1	0.04	0.03	0.21	—
1052	Pinto beans, frozen, boiled, drained	½	cup(s)	47	27	76	4	15	4	<1	0.03	0.02	0.13	—
514	Red kidney beans, canned	½	cup(s)	128	99	109	7	20	8	<1	0.06	0.03	0.24	—
1810	Refried beans, canned	½	cup(s)	127	96	119	7	20	7	2	0.60	0.71	0.19	—
1053	Shell beans, canned	½	cup(s)	123	111	37	2	8	4	<1	0.03	0.02	0.13	—
1670	Soybeans, boiled	½	cup(s)	86	54	149	14	9	5	8	1.12	1.70	4.36	—
1108	Soybeans, green, boiled, drained	½	cup(s)	90	62	127	11	10	4	6	0.67	1.09	2.71	—
1807	White beans, small, boiled	½	cup(s)	90	57	127	8	23	9	1	0.15	0.05	0.25	—
574	Green string beans, canned, fat added in cooking	½	cup(s)	93	<.1	41	1	4	2	3	0.51	1.23	0.75	—
575	Yellow snap, string or wax beans, boiled, drained	½	cup(s)	62	<.1	22	1	5	2	<1	0.04	0.00	0.09	—
576	Yellow snap, string or wax beans, frozen, boiled, drained	½	cup(s)	68	<.1	19	1	4	2	<1	0.02	0.00	0.05	—
	Beets													
580	Whole, boiled, drained	2	item(s)	100	87	44	2	10	2	<1	0.03	0.04	0.06	—
581	Sliced, boiled, drained	½	cup(s)	85	74	37	1	8	2	<1	0.02	0.03	0.05	—
583	Sliced, canned, drained	½	cup(s)	85	77	26	1	6	1	<1	0.02	0.02	0.04	—
2730	Pickled, canned with liquid	½	cup(s)	114	93	74	1	18	3	<.1	0.01	0.02	0.03	—
584	Beet greens, boiled, drained	½	cup(s)	72	64	19	2	4	2	<1	0.02	0.03	0.05	—
585	**Cowpeas or black-eyed peas, boiled, drained**	½	cup(s)	83	0.06	80	2.61	16.76	4.12	0.31	0.07	0.02	0.13	—
	Broccoli													
587	Raw, chopped	½	cup(s)	44	39	15	1	3	1	<1	0.02	0.00	0.02	—
588	Chopped, boiled, drained	½	cup(s)	78	70	27	2	6	3	<1	0.06	0.03	0.13	—
590	Frozen, chopped, boiled, drained	½	cup(s)	92	83	26	3	5	3	<1	0.02	0.01	0.05	—
16848	**Broccoflower, raw, chopped**	½	cup(s)	32	29	10	1	2	1	<.1	0.01	0.01	0.04	—
	Brussels sprouts													
591	Boiled, drained	½	cup(s)	78	69	28	2	6	2	<1	0.08	0.03	0.20	—
592	Frozen, boiled, drained	½	cup(s)	78	67	33	3	6	3	<1	0.06	0.02	0.16	—
	Cabbage													
594	Raw, shredded	1	cup(s)	70	65	17	1	4	2	<.1	0.01	0.01	0.04	—
595	Boiled, drained, no salt added	1	cup(s)	150	140	33	2	7	3	1	0.08	0.05	0.29	—
35611	Chinese (pak choi or bok choy), boiled w/salt, drained	1	cup(s)	170	162	20	3	3	2	<1	0.04	0.02	0.13	—
16869	Kim chee	1	cup(s)	150	138	31	2	6	2	<1	0.04	0.02	0.15	—
596	Red, shredded, raw	1	cup(s)	70	63	22	1	5	1	<1	0.02	0.01	0.09	—
597	Savoy, shredded, raw	1	cup(s)	70	64	19	1	4	2	<.1	0.01	0.00	0.03	—
11710	**Capers**	1	teaspoon(s)	5	—	0	0	0	0	0	0.00	0.00	0.00	0
	Carrots													
600	Raw	½	cup(s)	61	54	25	1	6	2	<1	0.02	0.01	0.06	0
8691	Raw, baby	8	item(s)	80	72	28	1	7	1	<1	0.02	0.01	0.05	0
601	Grated	½	cup(s)	55	49	23	1	5	2	<1	0.02	0.01	0.06	0
602	Sliced, boiled, drained	½	cup(s)	78	0.07	27.29	0.59	6.41	2.33	0.14	0.02	0	0.08	—

PAGE KEY: A–58 = Breads/Baked Goods A–62 = Cereal/Rice/Pasta A–66 = Fruit A–70 = Vegetables/Legumes A–80 = Nuts/Seeds A–82 = Vegetarian A–84 = Dairy A–90 = Eggs A–90 = Seafood A–92 = Meats A–96 = Poultry A–96 = Processed Meats A–98 = Beverages A–102 = Fats/Oils A–104 = Sweets A–106 = Sauces/Condiments/Spices A–108 = Mixed Foods/Soups/Sandwiches A–114 = Fast Food A–130 = Convenience A–132 = Baby Foods

Chol (mg)	Calc (mg)	Iron (mg)	Magn (mg)	Pota (mg)	Sodi (mg)	Zinc (mg)	Vit A (RAE) (µg)	Thia (mg)	Vit E (mg α)	Ribo (mg)	Niac (mg)	Vit B_6 (mg)	Fola (µg)	Vit C (mg)	Vit B_{12} (µg)	Sele (µg)
0	7	0.14	2	320	2	0.28	0	0.01	—	0.03	0.18	0.06	1	0	0	<1
0	5	0.21	3	52	5	0.43	1	0.02	0.41	0.02	0.09	0.09	2	1	0	<1
0	32	2.30	60	612	9	2.04	0	0.13	—	0.07	0.82	0.11	139	0	0	1
8	62	2.22	36	302	553	2.40	5	0.07	0.59	0.07	1.16	0.06	39	3	0	8
9	71	4.15	44	380	557	7.41	5	0.07	0.13	0.06	0.63	0.09	29	4	0	6
9	77	2.10	43	336	425	1.90	1	0.06	0.04	0.08	0.44	0.11	47	4	0	6
0	23	1.81	60	305	1	0.96	0	0.21	—	0.05	0.43	0.06	128	0	0	1
0	40	2.37	39	239	6	1.25	1	0.10	0.29	0.05	0.43	0.11	141	1	0	3
0	26	1.55	36	258	59	0.63	9	0.06	0.25	0.05	0.91	0.10	18	11	0	1
0	56	0.96	50	327	5	0.57	0	0.12	—	0.05	0.48	0.09	66	1	0	1
0	60	1.89	44	346	2	0.78	0	0.14	—	0.05	0.60	0.10	90	1	0	4
0	18	0.33	18	114	1	0.17	3	0.02	—	0.04	0.21	0.01	20	2	0	1
0	27	2.08	63	485	14	0.67	16	0.12	0.12	0.08	0.88	0.16	22	9	0	2
0	25	1.76	50	370	26	0.50	7	0.06	0.58	0.05	0.69	0.10	14	5	0	2
0	7	0.40	9	63	6	0.29	1	0.03	0.04	0.06	0.50	0.03	18	7	0	<1
0	64	2.26	54	335	1	0.96	0	0.18	0.01	0.06	0.48	0.15	127	1	0	5
0	17	0.75	20	111	58	0.19	0	0.08	—	0.07	0.82	0.06	146	7	0	1
0	24	1.27	25	304	39	0.32	0	0.13	—	0.05	0.30	0.09	16	<1	0	1
0	31	1.61	36	329	436	0.70	0	0.13	0.77	0.11	0.58	0.03	65	1	0	2
10	44	2.10	42	338	378	1.48	0	0.03	0.00	0.02	0.40	0.18	14	8	0	2
0	36	1.21	18	134	409	0.33	13	0.04	0.04	0.07	0.25	0.06	22	4	0	1
0	88	4.42	74	443	1	0.99	0	0.13	0.30	0.25	0.34	0.20	46	1	0	6
0	131	2.25	54	485	13	0.82	7	0.23	—	0.14	1.13	0.05	100	15	0	1
0	65	2.54	61	414	2	0.98	0	0.21	—	0.05	0.24	0.11	123	0	0	1
0	24	0.81	12	100	266	0.26	129	0.01	0.40	0.05	0.18	0.03	—	4	0.00	—
0	29	0.80	16	187	2	0.22	5	0.05	0.28	0.06	0.38	0.03	21	6	0	<1
0	33	0.59	16	85	6	0.32	7	0.02	0.24	0.06	0.26	0.04	16	3	0	<1
0	16	0.79	23	305	77	0.35	2	0.03	0.04	0.04	0.33	0.07	80	4	0	1
0	14	0.67	20	259	65	0.30	2	0.02	0.03	0.03	0.28	0.06	68	3	0	1
0	13	1.55	14	126	165	0.18	1	0.01	0.03	0.03	0.13	0.05	26	3	0	<1
0	12	0.47	17	168	300	0.30	1	0.01	—	0.05	0.28	0.06	31	3	0	1
0	82	1.37	49	654	174	0.36	276	0.08	1.30	0.21	0.36	0.10	10	18	0	1
0	105.59	0.92	42.9	344.85	3.29	0.84	65.17	0.08	0.18	0.12	1.15	0.05	104.77	1.81	0	2.06
0	21	0.32	9	139	15	0.18	15	0.03	0.34	0.05	0.28	0.08	28	39	0	1
0	31	0.52	16	229	32	0.35	76	0.05	1.13	0.10	0.43	0.16	84	51	0	1
0	30	0.56	12	131	10	0.26	52	0.05	1.21	0.07	0.42	0.12	52	37	0	1
0	11	0.23	6	96	7	0.20	0	0.03	0.01	0.03	0.23	0.07	18	28	0	—
0	28	0.94	16	247	16	0.26	30	0.08	0.34	0.06	0.47	0.14	47	48	0	1
0	20	0.37	14	225	12	0.19	36	0.08	0.40	0.09	0.42	0.22	78	35	0	<1
0	33	0.41	11	172	13	0.13	6	0.04	0.10	0.03	0.21	0.07	30	23	0	1
0	47	0.26	12	146	12	0.14	11	0.09	0.18	0.08	0.42	0.17	30	30	0	1
0	158	1.77	19	631	459	0.29	360	0.05	0.15	0.11	0.73	0.28	70	44	0	1
0	145	1.28	27	375	995	0.36	—	0.07	0.08	0.10	0.75	0.34	88	80	0	—
0	32	0.56	11	170	19	0.15	39	0.04	0.12	0.05	0.29	0.15	13	40	0	<1
0	25	0.28	20	161	20	0.19	35	0.05	—	0.02	0.21	0.13	56	22	0	1
0	—	—	—	—	105	—	—	—	—	—	—	—	—	—	0	—
0	20	0.18	7	195	42	0.15	367	0.04	0.40	0.04	0.60	0.08	12	4	0	<.1
0	26	0.71	8	190	62	0.14	552	0.02	—	0.03	0.44	0.08	26	7	0	1
0	18	0.17	7	177	38	0.13	333	0.04	0.36	0.03	0.54	0.08	11	3	0	<.1
0	23.39	0.26	7.8	183.3	45.24	0.15	1914.9	0.05	0.80	0.03	0.5	0.11	10.92	2.8	0	0.54

TABLE F–1

Food Composition (Computer code number is for Wadsworth Diet Analysis program) (For purposes of calculations, use "0" for t, <1, <.1, <.01, etc.)

DA + Code	Food Description	Quantity	Measure	Wt (g)	H_2O (g)	Ener (kcal)	Prot (g)	Carb (g)	Dietary Fiber (g)	Fat (g)	Fat Breakdown (g) Sat	Mono	Poly	Trans
	VEGETABLES, LEGUMES—Continued													
1055	Juice, canned	½	cup(s)	123	109	49	1	11	1	<1	0.03	0.01	0.09	—
32725	**Cassava or manioc**	½	cup(s)	103	61	165	1	39	2	<1	0.08	0.08	0.05	—
	Cauliflower													
605	Raw, chopped,	½	cup(s)	50	46	13	1	3	1	<1	0.02	0.01	0.05	—
606	Boiled, drained	½	cup(s)	62	58	14	1	3	2	<1	0.04	0.02	0.13	—
607	Frozen, boiled, drained	½	cup(s)	90	85	17	1	3	2	<1	0.03	0.01	0.09	—
	Celery													
609	Diced	½	cup(s)	60	58	8	<1	2	1	<1	0.03	0.02	0.05	—
608	Stalk	2	item(s)	80	76	11	1	2	1	<1	0.03	0.03	0.06	—
	Chard													
1056	Swiss chard, raw	1	cup(s)	36	33	7	1	1	1	<.1	0.01	0.01	0.03	—
1057	Swiss chard, boiled, drained	½	cup(s)	88	81	18	2	4	2	<.1	0.01	0.01	0.02	—
	Collard greens													
610	Boiled, drained	½	cup(s)	95	87	25	2	5	3	<1	0.04	0.02	0.16	—
611	Frozen, chopped, boiled, drained	½	cup(s)	85	75	31	3	6	2	<1	0.05	0.02	0.18	—
	Corn													
29614	Yellow corn, fresh, cooked	1	item(s)	100	0.06	107.37	3.3	24.96	2.78	1.27	0.19	0.37	0.59	—
612	Yellow sweet corn, boiled, drained	½	cup(s)	82	57	89	3	21	2	1	0.16	0.31	0.49	—
614	Yellow sweet corn, frozen, boiled, drained	½	cup(s)	82	63	66	2	16	2	1	0.08	0.16	0.26	—
615	Yellow creamed sweet corn, canned	½	cup(s)	128	101	92	2	23	2	1	0.08	0.16	0.25	—
618	**Cucumber**	¼	item(s)	75	72	11	<1	3	<1	<.1	0.03	0.00	0.04	—
16870	**Cucumber, kim chee**	½	cup(s)	75	68	16	1	4	1	<.1	0.02	0.00	0.03	—
	Dandelion greens													
2734	Raw	1	cup(s)	55	47	25	1	5	2	<1	0.09	0.01	0.17	—
620	Chopped, boiled, drained	½	cup(s)	53	47	17	1	3	2	<1	0.08	0.01	0.14	—
1066	**Eggplant, boiled, drained**	½	cup(s)	48	43	17	<1	4	1	<1	0.02	0.01	0.04	—
621	**Endive or escarole, chopped, raw**	1	cup(s)	53	49	9	1	2	2	<1	0.03	0.00	0.05	—
8784	**Jicama or yambean**	½	cup(s)	65	59	25	<1	6	3	<.1	0.01	0.00	0.03	—
	Kale													
29313	Raw	1	cup(s)	67	57	34	2	7	1	<1	0.06	0.03	0.23	—
623	Frozen, chopped, boiled, drained	½	cup(s)	65	59	20	2	3	1	<1	0.04	0.02	0.15	—
	Kohlrabi													
1071	Raw	1	cup(s)	135	123	36	2	8	5	<1	0.02	0.01	0.06	—
1072	Boiled, drained	½	cup(s)	83	74	24	1	6	1	<.1	0.01	0.01	0.04	—
	Leeks													
1073	Raw	1	cup(s)	89	74	54	1	13	2	<1	0.04	0.00	0.15	—
1074	Boiled, drained	½	cup(s)	52	47	16	<1	4	1	<1	0.01	0.00	0.06	—
	Lentils													
522	Boiled	½	cup(s)	99	69	115	9	20	8	<1	0.05	0.06	0.17	—
1075	Sprouted	1	cup(s)	77	52	82	7	17	0	<1	0.04	0.08	0.17	—
	Lettuce													
624	Butterhead, boston, or bibb	1	cup(s)	55	53	7	1	1	1	<1	0.02	0.00	0.06	—
625	Butterhead leaves	11	piece(s)	83	79	11	1	2	1	<1	0.02	0.01	0.10	—
626	Iceberg	1	cup(s)	55	53	6	<1	1	1	<.1	0.01	0.00	0.03	—
628	Iceberg, chopped	1	cup(s)	55	53	6	<1	1	1	<.1	0.01	0.00	0.03	—
629	Looseleaf	1	cup(s)	56	54	8	1	2	1	<.1	0.01	0.00	0.05	—
1665	Romaine, shredded	1	cup(s)	56	53	10	1	2	1	<1	0.02	0.01	0.09	—
	Mushrooms													
15585	Crimini (about 6)	3	ounce(s)	85	28	4	3	2	0	0.00	0.00	0.00	0	0
8700	Enoki	30	item(s)	90	80	31	2	6	2	<1	0.04	0.01	0.14	—
630	Mushrooms, raw	½	cup(s)	35	32	8	1	1	<1	<1	0.02	0.00	0.05	—
1079	Mushrooms, boiled, drained	½	cup(s)	78	71	22	2	4	2	<1	0.05	0.01	0.14	—
1080	Mushrooms, canned, drained	½	cup(s)	78	71	20	1	4	2	<1	0.03	0.00	0.09	—
15587	Portobello, raw	1	item(s)	85	30	3	4	3	0	0.00	0.00	0.00	0	0
2743	Shiitake, cooked	½	cup(s)	73	61	40	1	10	2	<1	0.04	0.05	0.02	—
	Mustard greens													
29319	Raw	1	cup(s)	56	51	15	2	3	2	<1	0.01	0.05	0.02	—
2744	Frozen, boiled, drained	½	cup(s)	75	70	14	2	2	2	<1	0.01	0.08	0.04	—
	Okra													
632	Sliced, boiled, drained	½	cup(s)	80	74	18	1	4	2	<1	0.04	0.02	0.04	—
32742	Frozen, boiled, drained, no salt added	½	cup(s)	92	84	26	2	5	3	<1	0.07	0.05	0.07	—
16866	Batter coated, fried	11	piece(s)	83	55	160	2	13	2	11	1.50	2.80	6.37	—
	Onions													
633	Raw, chopped	½	cup(s)	80	71	34	1	8	1	<.1	0.02	0.02	0.05	—
635	Chopped, boiled, drained	½	cup(s)	106	93	47	1	11	1	<1	0.03	0.03	0.08	—

PAGE KEY: A–58 = Breads/Baked Goods A–62 = Cereal/Rice/Pasta A–66 = Fruit A–70 = Vegetables/Legumes A–80 = Nuts/Seeds A–82 = Vegetarian A–84 = Dairy A–90 = Eggs A–90 = Seafood A–92 = Meats A–96 = Poultry A–96 = Processed Meats A–98 = Beverages A–102 = Fats/Oils A–104 = Sweets A–106 = Sauces/Condiments/Spices A–108 = Mixed Foods/Soups/Sandwiches A–114 = Fast Food A–130 = Convenience A–132 = Baby Foods

Chol (mg)	Calc (mg)	Iron (mg)	Magn (mg)	Pota (mg)	Sodi (mg)	Zinc (mg)	Vit A (RAE) (µg)	Thia (mg)	Vit E (mg α)	Ribo (mg)	Niac (mg)	Vit B_6 (mg)	Fola (µg)	Vit C (mg)	Vit B_{12} (µg)	Sele (µg)
0	30	0.57	17	359	36	0.22	1176	0.11	1.43	0.07	0.47	0.27	5	10	0	1
0	16	0.28	22	279	14	0.35	1	0.09	0.20	0.05	0.88	0.09	28	21	0	1
0	11	0.22	8	152	15	0.14	1	0.03	0.04	0.03	0.26	0.11	29	23	0	<1
0	10	0.20	6	88	9	0.11	1	0.03	0.04	0.03	0.25	0.11	27	27	0	<1
0	15	0.37	8	125	16	0.12	0	0.03	0.05	0.05	0.28	0.08	37	28	0	1
0	24	0.12	7	157	48	0.08	13	0.01	0.16	0.03	0.19	0.04	22	2	0	<1
0	32	0.16	9	208	64	0.10	18	0.02	0.22	0.05	0.26	0.06	29	2	0	<1
0	18	0.65	29	136	77	0.13	110	0.01	0.68	0.03	0.14	0.04	5	11	0	<1
0	51	1.98	75	480	157	0.29	268	0.03	1.65	0.08	0.32	0.07	8	16	0	1
0	133	1.10	19	110	15	0.22	386	0.04	0.84	0.10	0.55	0.12	88	17	0	<1
0	179	0.95	26	213	43	0.23	489	0.04	1.06	0.10	0.54	0.10	65	22	0	1
0	2.12	0.6	31.81	247.59	242.45	0.47	21.87	0.21	0.09	0.07	1.6	0.05	—	6.16	0	—
0	2	0.50	26	204	14	0.39	11	0.18	0.07	0.06	1.32	0.05	38	5	0	<1
0	2	0.39	23	191	1	0.52	8	0.02	0.06	0.05	1.08	0.08	29	3	0	1
0	4	0.49	22	172	365	0.68	5	0.03	0.09	0.07	1.23	0.08	58	6	0	1
0	12	0.21	10	111	2	0.15	4	0.02	0.02	0.02	0.07	0.03	5	2	0	<1
0	7	3.62	6	88	766	0.38	—	0.02	0.36	0.02	0.35	0.08	17	3	0	—
0	103	1.71	20	219	42	0.23	137	0.11	2.65	0.14	0.45	0.14	15	19	0	<1
0	74	0.95	13	122	23	0.15	260	0.07	1.79	0.09	0.27	0.08	7	9	0	<1
0	3	0.12	5	59	<1	0.06	1	0.04	0.20	0.01	0.29	0.04	7	1	0	<.1
0	27	0.44	8	165	12	0.41	57	0.04	0.23	0.04	0.21	0.01	75	3	0	<1
0	8	0.39	8	98	3	0.10	1	0.01	0.30	0.02	0.13	0.03	8	13	0	<1
0	90	1.14	23	299	29	0.29	515	0.07	—	0.09	0.67	0.18	19	80	0	1
0	90	0.61	12	209	10	0.12	478	0.03	0.60	0.07	0.44	0.06	9	16	0	1
0	32	0.54	26	473	27	0.04	3	0.07	0.65	0.03	0.54	0.20	22	84	0	1
0	21	0.33	16	281	17	0.26	2	0.03	0.43	0.02	0.32	0.13	10	45	0	1
0	53	1.87	25	160	18	0.11	74	0.05	0.82	0.03	0.36	0.21	57	11	0	1
0	16	0.57	7	45	5	0.03	1	0.01	—	0.01	0.10	0.06	13	2	0	<1
0	19	3.30	36	365	2	1.26	0	0.17	0.11	0.07	1.05	0.18	179	1	0	3
0	19	2.47	28	248	8	1.16	2	0.18	—	0.10	0.87	0.15	77	13	0	<1
0	19	0.69	7	132	3	0.11	92	0.03	0.10	0.03	0.20	0.05	40	2	0	<1
0	29	1.02	11	196	4	0.17	137	0.05	0.15	0.05	0.29	0.07	60	3	0	<1
0	11	0.19	4	84	5	0.09	9	0.02	0.10	0.01	0.07	0.03	31	2	0	<1
0	11	0.19	4	84	5	0.09	9	0.02	0.10	0.01	0.07	0.03	31	2	0	<1
0	20	0.48	7	109	16	0.10	208	0.04	0.16	0.05	0.21	0.05	21	10	0	<1
0	19	0.55	8	139	5	0.13	163	0.04	0.07	0.04	0.18	0.04	77	14	0	<1
0	0.67	—	—	33	—	0	—	—	—	—	—	—	—	0	0	—
0	1	0.80	14	343	3	0.51	0	0.08	0.01	0.09	3.28	0.04	27	11	0	14
0	1	0.18	3	110	1	0.18	0	0.03	0.00	0.15	1.35	0.04	6	1	<.1	3
0	5	1.36	9	278	2	0.68	0	0.06	0.01	0.23	3.48	0.07	14	3	0	9
0	9	0.62	12	101	332	0.56	0	0.07	0.01	0.02	1.24	0.05	9	0	0	3
40	0.36	—	—	10	—	0	—	—	—	—	—	—	—	0	0	—
0	2	0.32	10	85	3	0.96	0	0.03	0.01	0.12	1.09	0.12	15	<1	0	18
0	58	0.82	18	199	14	0.11	295	0.05	1.13	0.06	0.45	0.10	105	39	0	1
0	76	0.84	10	104	19	0.15	266	0.03	1.01	0.04	0.19	0.08	53	10	0	<1
0	62	0.22	29	108	5	0.34	11	0.11	0.22	0.04	0.70	0.15	37	13	0	<1
0	88	0.62	47	215	3	0.57	16	0.09	0.29	0.11	0.72	0.04	134	11	0	1
2	54	1.13	32	170	110	0.44	—	0.16	1.51	0.13	1.29	0.11	34	9	<.1	—
0	18	0.15	8	115	2	0.13	0	0.04	0.02	0.02	0.07	0.12	15	5	0	<1
0	23	0.26	12	177	3	0.22	0	0.04	0.02	0.02	0.18	0.14	16	6	0	1

TABLE F-1

Food Composition (Computer code number is for Wadsworth Diet Analysis program) (For purposes of calculations, use "0" for t, <1, <.1, <.01, etc.)

DA + Code	Food Description	Quantity	Measure	Wt (g)	H_2O (g)	Ener (kcal)	Prot (g)	Carb (g)	Dietary Fiber (g)	Fat (g)	Fat Breakdown (g) Sat	Mono	Poly	Trans
	VEGETABLES, LEGUMES—Continued													
2748	Frozen, boiled, drained	½	cup(s)	106	98	30	1	7	2	<1	0.02	0.01	0.04	—
16850	Red onions, sliced, raw	½	cup(s)	58	52	22	1	5	1	<.1	0.02	0.01	0.04	—
636	Scallions, green or spring onions	2	item(s)	30	27	10	1	2	1	<.1	0.01	0.01	0.02	—
1081	Onion rings, breaded & pan fried, frozen, heated	11	item(s)	78	22	318	4	30	1	21	6.70	8.49	3.99	—
16860	**Palm hearts, cooked**	½	cup(s)	73	51	75	2	19	1	<1	0.03	0.00	0.07	—
637	**Parsley, chopped**	1	tablespoon(s)	4	3	1	<1	<1	<1	<.1	0.01	0.01	0.00	—
638	**Parsnips, sliced, boiled, drained**	½	cup(s)	78	63	55	1	13	3	<1	0.04	0.09	0.04	—
	Peas													
639	Green peas, canned, drained	½	cup(s)	85	69	59	4	11	3	<1	0.05	0.03	0.14	—
641	Green peas, frozen, boiled, drained	½	cup(s)	80	64	62	4	11	4	<1	0.04	0.02	0.10	—
35694	Pea pods, boiled w/salt, drained	½	cup(s)	80	71	34	3	6	2	<1	0.04	0.02	0.08	—
1082	Peas & carrots, canned w/liquid	½	cup(s)	128	112	48	3	11	3	<1	0.06	0.03	0.16	—
1083	Peas & carrots, frozen, boiled, drained	½	cup(s)	80	69	38	2	8	2	<1	0.06	0.03	0.16	—
640	Snow or sugar peas, raw	½	cup(s)	32	28	13	1	2	1	<.1	0.01	0.01	0.03	—
2750	Snow or sugar peas, frozen, boiled, drained	½	cup(s)	80	69	42	3	7	2	<1	0.06	0.03	0.13	—
29324	Split peas, sprouted	½	cup(s)	60	37	77	5	17	0	<1	0.07	0.04	0.20	—
	Peppers													
643	Green bell or sweet, raw	½	cup(s)	75	70	15	1	3	1	<1	0.04	0.01	0.05	—
644	Green bell or sweet, boiled, drained	½	cup(s)	68	62	19	1	5	1	<1	0.02	0.01	0.07	—
1664	Green hot chili	1	item(s)	45	39	18	1	4	1	<.1	0.01	0.00	0.05	—
1663	Green hot chili, canned w/liquid	½	cup(s)	68	63	14	1	3	1	<.1	0.01	0.00	0.04	—
1086	Jalapeno, canned w/liquid	½	cup(s)	68	60	18	1	3	2	1	0.07	0.04	0.35	—
8703	Yellow bell or sweet	1	item(s)	186	171	50	2	12	2	<1	0.06	0.03	0.21	—
1087	**Poi**	½	cup(s)	122	87	136	<1	33	<1	<1	0.04	0.01	0.07	—
	Potatoes													
5791	Baked, flesh & skin	1	item(s)	202	144	220	5	51	4	<1	0.05	0.00	0.09	—
645	Baked, flesh only	½	cup(s)	61	46	57	1	13	1	<.1	0.02	0.00	0.03	—
1088	Baked, skin only	1	item(s)	58	27	115	2	27	5	<.1	0.02	0.00	0.02	—
5794	Boiled, drained, skin & flesh	1	item(s)	150	116	129	3	30	2	<1	0.04	0.00	0.06	—
647	Boiled, flesh only	½	cup(s)	78	60	67	1	16	1	<.1	0.02	0.00	0.03	—
5795	Boiled in skin, drained, flesh only	1	item(s)	136	105	118	3	27	2	<1	0.04	0.00	0.06	—
2759	Microwaved	1	item(s)	202	146	212	5	49	5	<1	0.05	0.00	0.09	—
5804	Microwaved, skin only	1	item(s)	58	37	77	3	17	4	<.1	0.02	0.00	0.02	—
2760	Microwaved in skin, flesh only	½	cup(s)	78	57	78	2	18	1	<.1	0.02	0.00	0.03	—
1089	Au gratin, prepared w/butter	½	cup(s)	123	91	162	6	14	2	9	5.80	2.63	0.34	—
1090	Au gratin mix, prepared w/water, whole milk, & butter	½	cup(s)	114	90	106	3	15	1	5	2.94	1.34	0.15	—
648	French fried, deep fried, prepared from raw	14	item(s)	70	32	190	3	24	2	10	1.93	4.21	2.97	—
649	French fried, frozen, heated	14	item(s)	70	40	140	2	22	2	5	0.88	3.33	0.55	—
1091	Hashed brown	½	cup(s)	78	37	207	2	27	2	10	1.11	3.13	2.78	—
653	Mashed, from dehydrated granules w/milk, water, & margarine	½	cup(s)	105	80	122	2	17	1	5	1.27	2.05	1.41	—
652	Mashed, w/margarine & whole milk	½	cup(s)	105	79	119	2	18	2	4	1.05	1.83	1.27	—
1097	Potato puffs, frozen, heated	½	cup(s)	64	34	142	2	20	2	7	3.26	2.79	0.51	—
1093	Scalloped, prepared w/butter	½	cup(s)	123	99	105	4	13	2	5	2.76	1.27	0.20	—
1094	Scalloped mix, prepared w/water, whole milk, & butter	½	cup(s)	114	90	106	2	15	1	5	2.99	1.38	0.22	—
	Pumpkin													
1773	Boiled, drained	½	cup(s)	123	115	25	1	6	1	<.1	0.05	0.01	0.00	—
656	Canned	½	cup(s)	123	110	42	1	10	4	<1	0.18	0.05	0.02	—
	Radicchio									<.1				
2498	Raw	1	cup(s)	40	37	9	1	2	<1	<1	0.02	0.00	0.04	—
8731	Raw, leaves	10	item(s)	80	75	18	1	4	1	<1	0.05	0.01	0.09	—
657	**Radishes**	6	item(s)	27	26	4	<1	1	<1	<.1	0.01	0.00	0.01	—
1099	**Rutabaga, boiled, drained**	½	cup(s)	85	76	33	1	7	2	<1	0.02	0.02	0.08	—
658	**Sauerkraut, canned**	½	cup(s)	114	105	22	1	5	3	<1	0.04	0.01	0.07	—
	Seaweed													
1102	Kelp	½	cup(s)	41	33	17	1	4	1	<1	0.10	0.04	0.02	—
1104	Spirulina, dried	½	cup(s)	8	<1	22	4	2	<1	1	0.20	0.05	0.16	—
1106	**Shallots**	3	tablespoon(s)	30	24	22	1	5	0	<.1	0.01	0.00	0.01	—
	Soybeans													
1670	Boiled	½	cup(s)	86	0.05	148.77	14.31	8.53	5.15	7.71	1.11	1.7	4.35	—
2825	Dry roasted	½	cup(s)	86	1	388	34	28	7	19	2.69	4.11	10.50	—
2824	Roasted, salted	½	cup(s)	86	2	405	30	29	15	22	3.16	4.82	12.33	—

PAGE KEY: A–58 = Breads/Baked Goods A–62 = Cereal/Rice/Pasta A–66 = Fruit A–70 = Vegetables/Legumes A–80 = Nuts/Seeds A–82 = Vegetarian A–84 = Dairy A–90 = Eggs A–90 = Seafood A–92 = Meats A–96 = Poultry A–96 = Processed Meats A–98 = Beverages A–102 = Fats/Oils A–104 = Sweets A–106 = Sauces/Condiments/Spices A–108 = Mixed Foods/Soups/Sandwiches A–114 = Fast Food A–130 = Convenience A–132 = Baby Foods

Chol (mg)	Calc (mg)	Iron (mg)	Magn (mg)	Pota (mg)	Sodi (mg)	Zinc (mg)	Vit A (RAE) (µg)	Thia (mg)	Vit E (mg α)	Ribo (mg)	Niac (mg)	Vit B_6 (mg)	Fola (µg)	Vit C (mg)	Vit B_{12} (µg)	Sele (µg)
0	17	0.32	6	115	13	0.07	0	0.02	0.01	0.03	0.15	0.07	14	3	0	<1
0	11	0.13	6	90	2	0.11	0	0.02	0.01	0.01	0.09	0.07	11	4	0	—
0	22	0.44	6	83	5	0.12	15	0.02	0.17	0.02	0.16	0.02	19	6	0	<1
0	24	1.32	15	101	293	0.33	9	0.22	—	0.11	2.82	0.06	52	1	0	3
0	13	1.23	7	1318	10	2.72	—	0.03	0.37	0.13	0.62	0.53	15	5	0	—
0	5	0.24	2	21	2	0.04	16	0.00	0.03	0.00	0.05	0.00	6	5	0	<.1
0	29	0.45	23	286	8	0.20	0	0.06	0.78	0.04	0.56	0.07	45	10	0	1
0	17	0.81	14	147	214	0.60	23	0.10	0.03	0.07	0.62	0.05	37	8	0	1
0	19	1.22	18	88	58	0.54	84	0.23	0.02	0.08	1.18	0.09	47	8	0	1
0	34	1.58	21	192	192	0.30	43	0.10	0.31	0.06	0.43	0.12	23	38	0	1
0	29	0.96	18	128	332	0.74	368	0.09	—	0.07	0.74	0.11	23	8	0	1
0	18	0.75	13	126	54	0.36	374	0.18	0.42	0.05	0.92	0.07	21	6	0	1
0	14	0.66	8	63	1	0.09	17	0.05	0.12	0.03	0.19	0.05	13	19	0	<1
0	47	1.92	22	174	4	0.39	53	0.05	0.38	0.10	0.45	0.14	28	18	0	1
0	22	1.36	34	229	12	0.63	5	0.14	—	0.09	1.85	0.16	86	6	0	<1
0	7	0.25	7	130	2	0.10	13	0.04	0.28	0.02	0.36	0.17	8	60	0	0
0	6	0.31	7	113	1	0.08	10	0.04	0.36	0.02	0.32	0.16	11	51	0	<1
0	8	0.54	11	153	3	0.14	27	0.04	0.31	0.04	0.43	0.13	10	109	0	<1
0	5	0.34	10	127	798	0.12	24	0.01	0.47	0.03	0.54	0.10	7	46	0	<1
0	16	1.28	10	131	1136	0.23	58	0.03	0.47	0.03	0.27	0.13	10	7	0	<1
0	20	0.86	22	394	4	0.32	19	0.05	—	0.05	1.66	0.31	48	341	0	1
0	19	1.07	29	223	15	0.27	4	0.16	2.80	0.05	1.34	0.33	26	5	0	1
0	20	2.75	55	844	16	0.65	0	0.22	—	0.07	3.32	0.70	22	26	0	2
0	3	0.21	15	239	3	0.18	0	0.06	0.02	0.01	0.85	0.18	5	8	0	<1
0	20	4.08	25	332	12	0.28	1	0.07	0.02	0.06	1.78	0.36	13	8	0	<1
0	13	1.27	34	572	7	0.47	0	0.15	—	0.03	2.13	0.44	15	18	0	—
0	6	0.24	16	256	4	0.21	0	0.08	0.01	0.01	1.02	0.21	7	6	0	<1
0	7	0.42	30	515	5	0.41	0	0.14	—	0.03	1.96	0.41	14	18	0	<1
0	22	2.50	55	903	16	0.73	0	0.24	—	0.06	3.46	0.69	24	31	0	1
0	27	3.45	21	377	9	0.30	0	0.04	—	0.04	1.29	0.29	10	9	0	<1
0	4	0.32	20	321	5	0.26	0	0.10	—	0.02	1.27	0.25	9	12	0	<1
28	146	0.78	25	485	530	0.85	78	0.08	—	0.14	1.22	0.21	13	12	0	3
17	94	0.36	17	249	499	0.27	59	0.02	—	0.09	1.07	0.05	8	4	0	3
0	9	1.02	28	731	8	0.53	0	0.10	0.09	0.05	1.90	0.33	13	21	0	—
0	6	0.87	15	293	21	0.28	0	0.08	0.08	0.02	1.46	0.22	8	7	0	<1
0	11	0.43	27	449	267	0.37	0	0.13	0.01	0.03	1.80	0.37	12	10	0	<1
2	34	0.22	21	163	181	0.25	49	0.09	0.54	0.09	0.91	0.17	8	7	<1	6
1	21	0.27	20	342	350	0.32	43	0.10	0.44	0.05	1.23	0.26	9	11	<.1	1
0	19	1.00	12	243	477	0.19	0	0.13	0.15	0.05	1.38	0.15	11	4	0	<1
15	70	0.70	23	463	410	0.49	39	0.08	—	0.11	1.29	0.22	13	13	0	2
13	41	0.43	16	231	388	0.28	40	0.02	—	0.06	1.17	0.05	11	4	0	2
0	18	0.70	11	282	1	0.28	306	0.04	0.98	0.10	0.51	0.05	11	6	0	<1
0	32	1.70	28	252	6	0.21	953	0.03	1.30	0.07	0.45	0.07	15	5	0	<1
0	8	0.23	5	121	9	0.25	<1	0.01	0.90	0.01	0.10	0.02	24	3	0	<1
0	15	0.46	10	242	18	0.50	1	0.01	1.81	0.02	0.20	0.05	48	6	0	1
0	7	0.09	3	63	11	0.08	0	0.00	0.00	0.01	0.07	0.02	7	4	0	<1
0	41	0.45	20	277	17	0.30	0	0.07	0.27	0.03	0.61	0.09	13	16	0	1
0	34	1.67	15	193	751	0.22	1	0.02	0.11	0.02	0.16	0.15	27	17	0	1
0	68	1.16	49	36	94	0.50	2	0.02	0.35	0.06	0.19	0.00	73	1	0	<1
0	9	2.14	15	102	79	0.15	2	0.18	0.38	0.28	0.96	0.03	7	1	0	1
0	11	0.36	6	100	4	0.12	18	0.02	—	0.01	0.06	0.10	10	2	0	<1
0	87.72	4.42	73.95	442.89	0.86	0.98	0.86	0.13	0.30	0.24	0.34	0.2	46.43	1.46	0	6.27
0	120	3.40	196	1173	2	4.10	0	0.37	—	0.65	0.91	0.19	176	4	0	17
0	119	3.35	125	1264	140	2.70	9	0.09	0.78	0.12	1.21	0.18	181	2	0	16

TABLE F–1
Food Composition (Computer code number is for Wadsworth Diet Analysis program) (For purposes of calculations, use "0" for t, <1, <.1, <.01, etc.)

DA + Code	Food Description	Quantity	Measure	Wt (g)	H_2O (g)	Ener (kcal)	Prot (g)	Carb (g)	Dietary Fiber (g)	Fat (g)	Fat Breakdown (g) Sat	Mono	Poly	Trans
	VEGETABLES, LEGUMES—Continued													
30282	Soup (miso)	1	cup(s)	240	218	85	6	8	2	3	0.59	1.05	1.47	—
8739	Sprouted, stir fried	3	ounce(s)	85	57	106	11	8	1	6	0.84	1.37	3.41	—
	Soy products													
1813	Soy milk	1	cup(s)	240	214	118	9	11	3	5	0.51	0.78	2.00	—
2838	Tofu, dried, frozen (koyadofu)	3	ounce(s)	85	5	408	41	12	6	26	3.73	5.70	14.57	—
13844	Tofu, extra firm	3	ounce(s)	79	—	80	8	2	1	4	0.50	0.87	2.60	—
13843	Tofu, firm	3	ounce(s)	79	—	80	8	2	1	4	0.50	0.87	2.17	—
1816	Tofu, firm, w/calcium sulfate & magnesium chloride (nigari)	3	ounce(s)	85	0.07	65.48	6.83	2.52	0.34	3.79	0.54	0.83	2.14	—
1817	Tofu, fried	3	ounce(s)	85	43	230	15	9	3	17	2.48	3.79	9.69	—
13841	Tofu, silken	3	ounce(s)	91	—	30	6	0	1	1	0.50	0.51	1.52	—
13842	Tofu, soft	3	ounce(s)	91	—	30	6	1	1	1	0.50	1.00	2.00	—
1671	Tofu, soft, w/calcium sulfate & magnesium chloride (nigari)	3	ounce(s)	85	0.07	51.88	5.57	1.53	0.17	3.13	0.45	0.69	1.76	—
	Spinach													
659	Raw, chopped	1	cup(s)	30	27	7	1	1	1	<1	0.02	0.00	0.05	—
663	Canned, drained	½	cup(s)	108	100	25	3	4	3	1	0.09	0.02	0.23	—
660	Chopped, boiled, drained	½	cup(s)	90	82	21	3	3	2	<1	0.04	0.01	0.10	—
661	Chopped, frozen, boiled, drained	½	cup(s)	95	84	30	4	5	4	<1	0.09	0.00	0.20	—
662	Leaf, frozen, boiled, drained	½	cup(s)	95	84	30	4	5	4	<1	0.09	0.00	0.20	—
8470	Trimmed leaves	1	cup(s)	32	27	3	1	<.1	3	<.1	—	—	—	—
	Squash													
1662	Acorn, baked	½	cup(s)	103	85	57	1	15	5	<1	0.03	0.01	0.06	—
29702	Acorn, boiled, mashed	½	cup(s)	123	110	42	1	11	3	<.1	0.02	0.01	0.04	—
1661	Butternut, baked	½	cup(s)	103	90	41	1	11	3	<.1	0.02	0.01	0.04	—
29451	Butternut, frozen, boiled	½	cup(s)	132	116	51	2	13	2	<.1	0.02	0.01	0.04	—
32773	Butternut, frozen, boiled, mashed, no salt added	½	cup(s)	122	<1	47	1	12	0	<.1	0.02	0.00	0.03	—
29700	Crookneck & straightneck, boiled, drained	½	cup(s)	90	0.08	18	0.81	3.87	1.25	0.27	0.05	0.02	0.11	—
29703	Hubbard, baked	v	cup(s)	103	87	51	3	11	0	1	0.13	0.05	0.27	—
1660	Hubbard, boiled, mashed	½	cup(s)	118	107	35	2	8	3	<1	0.09	0.03	0.18	—
29704	Spaghetti, boiled, drained, or baked	½	cup(s)	78	72	21	1	5	1	<1	0.05	0.02	0.10	—
664	Summer, all varieties, sliced, boiled, drained	½	cup(s)	90	84	18	1	4	1	<1	0.06	0.02	0.12	—
665	Winter, all varieties, baked, mashed	½	cup(s)	103	91	38	1	9	3	<1	0.13	0.05	0.27	—
1112	Zucchini, boiled, drained	½	cup(s)	90	85	14	1	4	1	<.1	0.01	0.00	0.02	—
1113	Zucchini, frozen, boiled, drained	½	cup(s)	113	107	19	1	4	1	<1	0.03	0.01	0.06	—
	Sweet potatoes													
666	Baked, peeled	½	cup(s)	100	76	90	2	21	3	<1	0.03	0.00	0.06	—
667	Boiled, mashed	½	cup(s)	166	133	126	2	29	4	<1	0.05	0.00	0.10	—
668	Candied, home recipe	½	cup(s)	84	56	115	1	23	2	3	1.13	0.53	0.12	—
670	Canned, vacuum pack	½	cup(s)	100	76	91	2	21	2	<1	0.04	0.01	0.09	—
2765	Frozen, baked	½	cup(s)	88	65	88	2	21	2	<1	0.02	0.00	0.05	—
1136	Yams, baked or boiled, drained	½	cup(s)	68	48	79	1	19	3	<.1	0.02	0.00	0.04	—
32785	**Taro shoots, cooked, no salt added**	½	cup(s)	70	67	10	1	2	0	<.1	0.01	0.00	0.02	—
	Tomatillo													
8774	Raw	2	item(s)	68	62	22	1	4	1	1	0.09	0.11	0.28	—
8777	Raw, chopped	½	cup(s)	66	60	21	1	4	1	1	0.09	0.10	0.28	—
	Tomato													
671	Fresh, ripe, red	1	item(s)	123	0.11	22.13	1.08	4.82	1.47	0.24	0.05	0.06	0.16	—
16846	Fresh, cherry	5	item(s)	85	0.07	17.85	0.72	3.94	0.93	0.28	0.03	0.04	0.11	—
3952	Diced, red	½	cup(s)	90	85	16	1	4	1	<1	0.04	0.05	0.12	—
1118	Boiled, red	½	cup(s)	120	113	22	1	5	1	<1	0.02	0.02	0.05	—
675	Juice, canned	½	cup(s)	122	115	21	1	5	<1	<.1	0.01	0.01	0.03	—
75	Juice, no salt added	½	cup(s)	122	115	21	1	5	<1	<.1	0.01	0.01	0.03	—
1699	Paste, canned	2	tablespoon(s)	33	24	27	1	6	1	<1	0.04	0.03	0.07	—
1700	Puree, canned	¼	cup(s)	63	55	24	1	6	1	<1	0.02	0.02	0.05	—
1125	Sauce, canned	¼	cup(s)	61	55	20	1	5	1	<1	0.02	0.02	0.06	—
1120	Stewed, canned, red	½	cup(s)	128	117	33	1	8	1	<1	0.03	0.04	0.10	—
8778	Sun dried	½	cup(s)	27	4	70	4	15	3	1	0.12	0.13	0.30	—
8783	Sun dried in oil, drained	¼	cup(s)	28	15	59	1	6	2	4	0.52	2.38	0.57	—
	Turnips													
677	Turnips, cubed, boiled, drained	½	cup(s)	78	73	17	1	4	2	<.1	0.01	0.00	0.03	—
678	Turnip greens, chopped, boiled, drained	½	cup(s)	72	67	14	1	3	3	<1	0.04	0.01	0.07	—
679	Turnip greens, frozen, chopped, boiled, drained	½	cup(s)	82	74	24	3	4	3	<1	0.08	0.02	0.14	—

PAGE KEY: A–58 = Breads/Baked Goods A–62 = Cereal/Rice/Pasta A–66 = Fruit A–70 = Vegetables/Legumes A–80 = Nuts/Seeds A–82 = Vegetarian A–84 = Dairy A–90 = Eggs A–90 = Seafood A–92 = Meats A–96 = Poultry A–96 = Processed Meats A–98 = Beverages A–102 = Fats/Oils A–104 = Sweets A–106 = Sauces/Condiments/Spices A–108 = Mixed Foods/Soups/Sandwiches A–114 = Fast Food A–130 = Convenience A–132 = Baby Foods

Chol (mg)	Calc (mg)	Iron (mg)	Magn (mg)	Pota (mg)	Sodi (mg)	Zinc (mg)	Vit A (RAE) (µg)	Thia (mg)	Vit E (mg α)	Ribo (mg)	Niac (mg)	Vit B_6 (mg)	Fola (µg)	Vit C (mg)	Vit B_{12} (µg)	Sele (µg)
0	64	1.89	37	361	988	0.87	—	0.06	0.96	0.16	2.61	0.17	57	4	<1	—
0	70	0.34	82	482	12	1.79	1	0.36	—	0.16	0.94	0.14	108	10	0	1
0	10	1.39	46	338	29	0.55	5	0.39	3.24	0.17	0.35	0.10	5	0	0	3
0	310	8.28	50	17	5	4.17	22	0.42	—	0.27	1.01	0.24	78	1	0	46
0	60	1.08	78	—	0	—	0	—	0.03	—	—	—	—	0	0	—
0	60	1.08	52	—	0	—	0	—	—	—	—	—	—	0	0	—
0	137.78	1.23	39.12	149.68	6.8	0.85	0.85	0.07	—	0.08	0	0.05	28.06	0.17	0	7.99
0	316	4.14	51	124	14	1.69	1	0.14	0.03	0.04	0.09	0.08	23	0	0	24
0	300	0.73	35	—	65	—	0	—	—	—	—	—	—	0	2	—
0	300	0.72	33	—	65	—	0	—	—	—	—	—	—	0	2	—
0	94.4	0.94	22.96	102.05	6.8	0.54	0.85	0.03	0.01	0.03	0.45	0.04	37.42	0.17	0	7.56
0	30	0.81	24	167	24	0.16	141	0.02	0.61	0.06	0.22	0.06	58	8	0	<1
0	138	2.49	82	375	29	0.50	531	0.02	2.10	0.15	0.42	0.11	106	16	0	2
0	122	3.21	78	419	63	0.68	472	0.09	1.87	0.21	0.44	0.22	131	9	0	1
0	145	1.86	78	287	92	0.47	573	0.07	3.36	0.17	0.42	0.13	115	2	0	5
0	145	1.86	78	287	92	0.47	573	0.07	3.36	0.17	0.42	0.13	115	2	0	5
0	25	2.13	25	134	38	0.18	—	0.03	—	0.06	0.18	0.07	<.1	8	0	—
0	45	0.95	44	448	4	0.17	22	0.17	—	0.01	0.90	0.20	19	11	0	1
0	32	0.69	32	322	4	0.13	50	0.12	—	0.01	0.65	0.14	13	8	0	<1
0	42	0.62	30	291	4	0.13	572	0.07	1.32	0.02	0.99	0.13	19	15	0	1
0	25	0.77	12	176	3	0.16	—	0.07	—	0.05	0.61	0.09	22	5	0	1
0	23	0.70	11	162	2	0.14	406	0.05	—	0.05	0.56	0.08	19	4	0	1
0	18.2	0.41	18.2	183.73	1.73	0.25	29.46	0.04	—	0.03	0.39	0.09	19.93	7.28	0	0.17
0	17	0.48	23	367	8	0.15	310	0.08	—	0.05	0.57	0.18	16	10	0	1
0	12	0.33	15	253	6	0.12	236	0.05	0.14	0.03	0.39	0.12	12	8	0	<1
0	16	0.26	9	91	14	0.16	5	0.03	0.09	0.02	0.63	0.08	6	3	0	<1
0	24	0.32	22	173	1	0.35	10	0.04	0.13	0.04	0.46	0.06	18	5	0	<1
0	23	0.45	13	448	1	0.23	268	0.02	0.12	0.07	0.51	0.17	21	10	0	<1
0	12	0.32	20	228	3	0.16	50	0.04	0.11	0.04	0.39	0.07	15	4	0	<1
0	19	0.54	15	219	2	0.23	11	0.05	0.14	0.05	0.44	0.05	9	4	0	<1
0	38	0.69	27	475	36	0.32	961	1.45	0.71	0.11	1.49	0.29	6	20	0	<1
0	45	1.20	30	382	45	0.33	1310	0.09	1.56	0.08	0.89	0.27	10	21	0	<1
7	22	0.95	9	159	59	0.13	176	0.02	—	0.04	0.33	0.03	9	6	0	1
0	22	0.89	22	312	53	0.18	399	0.04	1.00	0.06	0.74	0.19	17	26	0	1
0	31	0.48	18	332	7	0.26	722	0.06	0.68	0.05	0.49	0.16	19	8	0	1
0	10	0.36	12	458	5	0.14	4	0.06	0.26	0.02	0.38	0.16	11	8	0	<1
0	10	0.29	6	241	1	0.38	2	0.03	—	0.04	0.57	0.08	2	13	0	1
0	5	0.42	14	182	1	0.15	4	0.03	0.26	0.02	1.26	0.04	5	8	0	<1
0	5	0.41	13	177	1	0.15	4	0.03	0.25	0.02	1.22	0.04	5	8	0	<1
0	12.3	0.33	13.52	291.51	6.15	0.2	76.26	0.04	0.66	0.02	0.73	0.09	18.45	15.62	0	0
0	4.25	0.37	9.35	188.69	7.65	0.07	52.7	0.05	0.46	0.03	0.53	0.07	—	16.23	0	—
0	9	0.24	10	213	5	0.15	38	0.03	0.49	0.02	0.53	0.07	14	11	0	0
0	13	0.82	11	262	13	0.17	29	0.04	0.67	0.03	0.64	0.09	16	27	0	1
0	12	0.52	13	279	328	0.18	28	0.06	0.39	0.04	0.82	0.14	24	22	0	<1
0	12	0.52	13	279	12	0.18	28	0.06	0.39	0.04	0.82	0.14	24	22	0	<1
0	12	0.98	14	333	259	0.21	25	0.02	1.41	0.05	1.01	0.07	4	7	0	2
0	11	1.11	14	274	249	0.23	16	0.02	1.23	0.05	0.92	0.08	7	7	0	3
0	8	0.62	10	203	321	0.12	10	0.01	1.27	0.04	0.60	0.06	6	4	0	<1
0	43	1.70	15	264	282	0.22	11	0.06	1.06	0.04	0.91	0.02	6	10	0	1
0	30	2.45	52	925	566	0.54	12	0.14	0.00	0.13	2.44	0.09	18	11	0	1
0	13	0.74	22	430	73	0.21	18	0.05	—	0.11	1.00	0.09	6	28	0	1
0	26	0.14	7	138	12	0.09	0	0.02	0.02	0.02	0.23	0.05	7	9	0	<1
0	99	0.58	16	146	21	0.10	274	0.03	1.35	0.05	0.30	0.13	85	20	0	1
0	125	1.59	21	184	12	0.34	441	0.04	2.18	0.06	0.38	0.05	32	18	0	1

TABLE F–1
Food Composition (Computer code number is for Wadsworth Diet Analysis program) (For purposes of calculations, use "0" for t, <1, <.1, <.01, etc.)

DA + Code	Food Description	Quantity	Measure	Wt (g)	H_2O (g)	Ener (kcal)	Prot (g)	Carb (g)	Dietary Fiber (g)	Fat (g)	Fat Breakdown (g) Sat	Mono	Poly	Trans
	VEGETABLES, LEGUMES—Continued													
	Vegetables, mixed													
1132	Canned, drained	½	cup(s)	82	71	40	2	8	2	<1	0.04	0.01	0.10	—
680	Frozen, boiled, drained	½	cup(s)	91	76	59	3	12	4	<1	0.03	0.01	0.07	—
7489	Vegetable juice, V8 100%	½	cup(s)	120	113	25	1	5	1	0	0.00	0.00	0.00	0
7490	Vegetable juice, V8 low sodium	½	cup(s)	120	113	25	0	7	1	0	0.00	0.00	0.00	0
7491	Vegetable juice, V8 spicy hot	½	cup(s)	120	113	25	1	5	1	0	0.00	0.00	0.00	0
	Water chestnuts													
31073	Sliced, drained	½	cup(s)	75	70	20	<1	5	1	0	0.00	0.00	0.00	0
31087	Whole	½	cup(s)	75	70	20	<1	5	1	0	0.00	0.00	0.00	0
1135	**Watercress**	1	cup(s)	34	32	4	1	<1	<1	<.1	0.01	0.00	0.01	—
	NUTS, SEEDS, AND PRODUCTS													
	Almonds													
32886	Blanched	¼	cup(s)	36	2	211	8	7	4	18	1.41	11.70	4.37	—
32887	Dry roasted, no salt added	¼	cup(s)	35	1	206	8	7	4	18	1.40	11.61	4.36	—
29724	Dry roasted, salted	¼	cup(s)	35	1	206	8	7	4	18	1.40	11.61	4.36	—
29725	Oil roasted, salted	¼	cup(s)	39	1	238	8	7	4	22	1.65	13.66	5.31	—
508	Slivered	¼	cup(s)	34	2	195	7	7	4	17	1.31	10.85	4.12	—
1137	Almond butter, no salt added	1	tablespoon(s)	16	<1	101	2	3	1	9	0.90	6.14	1.98	—
32940	Almond butter, salt added	1	tablespoon(s)	16	<1	101	2	3	1	9	0.90	6.14	1.98	—
1138	**Beechnuts, dried**	¼	cup(s)	57	4	327	4	19	5	28	3.25	12.43	11.41	—
517	**Brazil nuts, unblanched, dried**	¼	cup(s)	35	1	230	5	4	3	23	5.30	8.59	7.20	—
1166	**Breadfruit seeds, roasted**	¼	cup(s)	57	28	118	4	23	3	2	0.41	0.20	0.82	—
1139	**Butternuts, dried**	¼	cup(s)	30	1	184	7	4	1	17	0.39	3.13	12.82	—
	Cashews													
1140	Dry roasted	¼	cup(s)	34	1	197	5	11	1	16	3.14	9.36	2.68	—
518	Oil roasted	¼	cup(s)	33	1	189	5	10	1	16	2.76	8.42	2.78	—
32889	Cashew butter, no salt added	1	tablespoon(s)	16	<1	94	3	4	<1	8	1.56	4.66	1.34	—
32931	Cashew butter, salt added	1	tablespoon(s)	16	<1	94	3	4	<1	8	1.56	4.66	1.34	—
	Coconut													
32896	Dried, not sweetened	¼	cup(s)	60	2	393	4	14	10	38	34.06	1.63	0.42	—
1153	Dried, shredded, sweetened	¼	cup(s)	24	3	122	1	12	1	9	7.68	0.37	0.09	—
520	Shredded	¼	cup(s)	21	10	75	1	3	2	7	6.27	0.30	0.08	—
	Chestnuts													
1152	Chinese, roasted	¼	cup(s)	57	23	136	3	30	0	1	0.10	0.35	0.17	—
32895	European, boiled & steamed	¼	cup(s)	57	39	74	1	16	0	1	0.15	0.27	0.31	—
32911	European, roasted	¼	cup(s)	57	23	139	2	30	3	1	0.23	0.43	0.49	—
32922	Japanese, boiled & steamed	¼	cup(s)	57	49	32	<1	7	0	<1	0.02	0.06	0.03	—
32923	Japanese, roasted	¼	cup(s)	57	28	114	2	26	0	<1	0.07	0.24	0.12	—
4958	**Flaxseeds or linseeds**	¼	cup(s)	57	5	276	11	19	16	19	1.79	3.85	12.54	—
32904	**Ginkgo nuts, dried**	¼	cup(s)	57	7	197	6	41	0	1	0.22	0.42	0.42	—
	Hazelnuts or filberts													
32901	Blanched	¼	cup(s)	57	3	357	8	10	6	35	2.65	27.32	3.15	—
32902	Dry roasted, no salt added	¼	cup(s)	57	1	366	9	10	5	35	2.56	26.43	4.80	—
1156	**Hickorynuts, dried**	¼	cup(s)	30	1	197	4	5	2	19	2.11	9.78	6.57	—
	Macadamias													
1157	Raw	¼	cup(s)	34	<1	241	3	5	3	25	4.04	19.72	0.50	—
32905	Dry roasted, no salt added	¼	cup(s)	34	1	241	3	4	3	25	4.00	19.86	0.50	—
32932	Dry roasted, salt added	¼	cup(s)	34	1	240	3	4	3	25	4.00	19.86	0.50	—
	Mixed nuts													
1159	With peanuts, dry roasted	¼	cup(s)	34	1	203	6	9	3	18	2.36	10.75	3.69	—
32933	With peanuts, dry roasted, salt added	¼	cup(s)	34	1	203	6	9	3	18	2.36	10.75	3.69	—
32906	Without peanuts, oil roasted, no salt added	¼	cup(s)	36	1	221	6	8	2	20	3.27	11.93	4.12	—
	Peanuts													
2807	Dry roasted	¼	cup(s)	37	0	214	9	8	3	18	2.51	8.99	5.72	—
2806	Dry roasted, salted	¼	cup(s)	37	0	214	9	8	3	18	2.51	8.99	5.72	—
1763	Oil roasted, salted	¼	cup(s)	36	0	216	10	5	3	19	3.12	9.33	5.49	—
2804	Raw	¼	cup(s)	37	2	207	9	6	3	18	2.49	8.92	5.68	—
1884	Peanut butter, chunky	1	tablespoon(s)	16	<1	94	4	3	1	8	1.53	3.77	2.27	—
30303	Peanut butter, low sodium	1	tablespoon(s)	16	<1	95	4	3	1	8	1.66	3.88	2.21	—
30305	Peanut butter, reduced fat	1	tablespoon(s)	18	<1	94	5	6	1	6	1.33	2.91	1.85	—
524	Peanut butter, smooth	1	tablespoon(s)	16	<1	96	4	3	1	8	1.60	3.96	2.38	—
	Pecans													
32907	Dry roasted, no salt added	¼	cup(s)	57	1	403	5	8	5	42	3.56	24.92	11.66	—

PAGE KEY: A–58 = Breads/Baked Goods A–62 = Cereal/Rice/Pasta A–66 = Fruit A–70 = Vegetables/Legumes A–80 = Nuts/Seeds A–82 = Vegetarian A–84 = Dairy A–90 = Eggs A–90 = Seafood A–92 = Meats A–96 = Poultry A–96 = Processed Meats A–98 = Beverages A–102 = Fats/Oils A–104 = Sweets A–106 = Sauces/Condiments/Spices A–108 = Mixed Foods/Soups/Sandwiches A–114 = Fast Food A–130 = Convenience A–132 = Baby Foods

Chol (mg)	Calc (mg)	Iron (mg)	Magn (mg)	Pota (mg)	Sodi (mg)	Zinc (mg)	Vit A (RAE) (µg)	Thia (mg)	Vit E (mg α)	Ribo (mg)	Niac (mg)	Vit B_6 (mg)	Fola (µg)	Vit C (mg)	Vit B_{12} (µg)	Sele (µg)
0	22	0.86	13	237	121	0.33	474	0.04	0.28	0.04	0.47	0.06	20	4	0	<1
0	23	0.75	20	154	32	0.45	195	0.06	0.40	0.11	0.77	0.07	17	3	0	<1
0	20	0.54	13	270	310	0.24	50	0.05	—	0.03	0.87	0.17	—	30	0	—
0	20	0.36	—	420	70	—	63	0.02	—	0.02	0.75	—	—	30	0	—
0	20	0.36	13	255	370	0.24	50	0.05	—	0.03	0.88	0.17	—	18	0	—
0	7	0.23	—	—	6	—	0	—	—	—	—	—	—	2	—	—
0	7	0.23	—	—	6	—	0	—	—	—	—	—	—	2	—	—
0	41	0.07	7	112	14	0.04	80	0.03	0.34	0.04	0.07	0.04	3	15	0	<1
0	78	1.35	100	249	10	1.13	0	0.07	8.96	0.20	1.33	0.04	11	0	0	1
0	92	1.56	99	257	<1	1.22	0	0.03	8.97	0.30	1.33	0.04	11	0	0	1
0	92	1.56	99	257	117	1.22	0	0.03	8.97	0.30	1.33	0.04	11	0	0	1
0	114	1.44	108	274	133	1.20	0	0.04	10.19	0.31	1.44	0.05	11	0	0	1
0	84	1.45	93	246	<1	1.13	0	0.08	8.73	0.27	1.32	0.04	10	0	0	1
0	43	0.59	48	121	2	0.49	0	0.02	—	0.10	0.46	0.01	10	<1	0	—
0	43	0.59	48	121	72	0.49	0	0.02	—	0.10	0.46	0.01	10	<1	0	1
0	1	1.40	0	578	22	0.20	0	0.17	—	0.21	0.50	0.39	64	9	0	4
0	56	0.85	132	231	1	1.42	0	0.22	2.01	0.01	0.10	0.04	8	<1	0	671
0	49	0.51	35	615	16	0.59	9	0.23	—	0.14	4.20	0.24	34	4	0	8
0	16	1.21	71	126	<1	0.94	2	0.11	—	0.04	0.31	0.17	20	1	0	5
0	15	2.06	89	194	5	1.92	0	0.07	0.32	0.07	0.48	0.09	24	0	0	4
0	14	1.97	89	205	4	1.74	0	0.12	0.30	0.07	0.56	0.10	8	<.1	0	7
0	7	0.80	41	87	2	0.83	0	0.05	—	0.03	0.26	0.04	11	0	0	2
0	7	0.80	41	87	98	0.83	0	0.05	0.15	0.03	0.26	0.04	11	0	0	2
0	15	1.98	54	323	22	1.20	0	0.04	0.26	0.06	0.36	0.18	5	1	0	11
0	4	0.47	12	82	64	0.44	0	0.01	0.10	0.00	0.12	0.07	2	<1	0	4
0	3	0.51	7	75	4	0.23	0	0.01	0.05	0.00	0.11	0.01	5	1	0	2
0	11	0.85	51	271	2	0.53	0	0.09	—	0.05	0.85	0.25	41	22	0	4
0	26	0.98	31	405	15	0.14	1	0.08	—	0.06	0.41	0.13	22	15	0	—
0	16	0.52	19	336	1	0.32	1	0.14	0.28	0.10	0.76	0.28	40	15	0	1
0	6	0.30	10	67	3	0.23	1	0.07	—	0.03	0.31	0.06	10	5	0	—
0	20	1.19	36	242	11	0.81	2	0.26	—	—	0.40	0.24	33	16	0	—
0	111	3.48	203	381	19	2.34	0	0.10	—	0.09	0.78	0.52	156	1	0	3
0	11	0.91	30	566	7	0.38	31	0.24	—	0.10	6.65	0.36	60	17	0	—
0	84	1.87	91	373	0	1.25	1	0.27	9.92	0.06	0.88	0.33	44	1	0	2
0	70	2.48	98	428	0	1.42	2	0.19	8.66	0.07	1.16	0.35	50	2	0	2
0	18	0.64	52	131	<1	1.29	2	0.26	—	0.04	0.27	0.06	12	1	0	2
0	28	1.24	44	123	2	0.44	0	0.40	0.18	0.05	0.83	0.09	4	<1	0	1
0	23	0.89	40	122	1	0.43	0	0.24	0.19	0.03	0.76	0.12	3	<1	0	1
0	23	0.89	40	122	89	0.43	0	0.24	0.19	0.03	0.76	0.12	3	<1	0	4
0	24	1.27	77	204	4	1.30	<1	0.07	—	0.07	1.61	0.10	17	<1	0	1
0	24	1.27	77	204	229	1.30	0	0.07	3.75	0.07	1.61	0.10	17	<1	0	3
0	38	0.93	90	196	4	1.68	<1	0.18	—	0.17	0.71	0.06	20	<1	0	—
0	20	0.82	64	240	2	1.20	0	0.15	2.56	0.03	4.93	0.09	53	0	0	3
0	20	0.82	64	240	297	1.20	0	0.15	2.89	0.03	4.93	0.09	53	0	0	3
0	22	0.54	63	261	115	1.18	0	0.03	2.50	0.03	4.97	0.16	43	<1	0	1
0	34	1.67	61	257	7	1.19	0	0.23	3.04	0.05	4.40	0.13	88	0	0	3
0	8	0.33	31	101	75	0.52	0	0.02	1.01	0.02	2.19	0.07	15	0	0	1
0	6	0.29	25	107	3	0.47	0	0.01	1.23	0.02	2.14	0.07	12	0	0	—
0	6	0.34	31	120	97	0.50	0	0.05	1.20	0.01	2.63	0.06	11	0	0	—
0	8	0.30	28	88	80	0.47	0	0.01	1.44	0.02	2.14	0.07	12	0	0	1
0	41	1.59	75	240	1	2.87	4	0.26	0.74	0.06	0.66	0.11	9	<1	0	2

TABLE F-1

Food Composition (Computer code number is for Wadsworth Diet Analysis program) (For purposes of calculations, use "0" for t, <1, <.1, <.01, etc.)

DA + Code	Food Description	Quantity	Measure	Wt (g)	H_2O (g)	Ener (kcal)	Prot (g)	Carb (g)	Dietary Fiber (g)	Fat (g)	Fat Breakdown (g) Sat	Mono	Poly	Trans
	NUTS, SEEDS, AND PRODUCTS—Continued													
32936	Dry roasted, salt added	¼	cup(s)	57	1	403	5	8	5	42	3.56	24.92	11.66	—
1162	Halves, oil roasted	¼	cup(s)	28	<1	197	3	4	3	21	1.99	11.27	6.49	—
526	Raw	¼	cup(s)	27	1	187	2	4	3	19	1.67	11.02	5.84	—
12973	**Pine nuts or pignolia, dried**	1	tablespoon(s)	9	<1	58	1	1	<1	6	0.42	1.61	2.93	—
	Pistachios													
1164	Dry roasted	¼	cup(s)	32	1	183	7	9	3	15	1.78	7.75	4.45	—
32938	Dry roasted, salt added	¼	cup(s)	32	1	182	7	9	3	15	1.78	7.75	4.45	—
1167	**Pumpkin or squash seeds, roasted**	¼	cup(s)	57	4	296	19	8	2	24	4.52	7.43	10.90	—
	Sesame													
1169	Sesame seeds, whole, roasted, toasted	3	teaspoon(s)	9	<1	51	2	2	1	4	0.60	1.63	1.89	—
32912	Sesame butter paste	1	tablespoon(s)	16	<1	95	3	4	1	8	1.14	3.07	3.57	—
32941	Tahini or sesame butter	1	tablespoon(s)	15	<1	89	3	3	1	8	1.11	3.00	3.48	—
	Soy nuts													
34173	Deep sea salted	¼	cup(s)	56	—	240	24	18	10	8	2.00	—	—	—
34174	Unsalted	¼	cup(s)	56	—	240	24	18	10	8	2.00	—	—	—
	Sunflower seeds													
528	Kernels, dried	¼	cup(s)	36	2	205	8	7	4	18	1.87	3.41	11.78	—
29721	Kernels, dry roasted, salted	¼	cup(s)	32	<1	186	6	8	3	16	1.67	3.04	10.52	—
29723	Kernels, toasted, salted	¼	cup(s)	34	<1	207	6	7	4	19	1.99	3.63	12.56	—
32928	Sunflower seed butter, salt added	1	tablespoon(s)	16	<1	93	3	4	0	8	0.80	1.46	5.04	—
	Trail mix													
4646	Trail mix	¼	cup(s)	38	3	173	5	17	2	11	2.08	4.70	3.62	—
4647	Trail mix with chocolate chips	¼	cup(s)	38	2	182	5	17	0	12	2.29	5.08	4.23	—
4648	Tropical trail mix	¼	cup(s)	35	3	142	2	23	0	6	2.97	0.87	1.81	—
	Walnuts													
529	Dried black, chopped	¼	cup(s)	31	1	193	8	3	2	18	1.05	4.69	10.96	—
531	English or persian	¼	cup(s)	30	1	196	5	4	2	20	1.84	2.68	14.15	—
	VEGETARIAN FOODS													
	Prepared													
34222	Brown rice & tofu stir-fry (vegan)	8	ounce(s)	227	183	228	12	13	3	16	1.25	4.03	9.54	0
34368	Cheese enchilada casserole (lacto)	8	ounce(s)	227	86	410	18	41	4	19	10.06	6.54	1.24	0
34247	Five bean casserole (vegan)	8	ounce(s)	228	178	178	6	26	5	6	1.11	2.49	1.96	0
34261	Lentil stew (vegan)	8	ounce(s)	228	152	125	8	24	7	<1	0.08	0.07	0.21	0
34397	Macaroni & cheese (lacto)	8	ounce(s)	226	163	181	8	17	<1	9	4.37	2.88	0.89	0
34238	Steamed rice & vegetables (vegan)	8	ounce(s)	228	100	265	5	40	3	10	1.84	3.91	4.07	0
34308	Tofu rice burgers (ovo-lacto)	1	piece(s)	218	78	435	22	68	6	8	1.69	2.39	3.52	—
34276	Vegan spinach enchiladas (vegan)	1	piece(s)	82	59	93	5	15	2	2	0.34	0.55	1.27	—
34243	Vegetable chow mein (vegan)	8	ounce(s)	227	163	166	6	22	2	6	0.65	2.66	2.47	0
34454	Vegetable lasagna (lacto)	8	ounce(s)	225	154	177	12	25	2	4	1.92	0.93	0.34	0
34339	Vegetable marinara (vegan)	8	ounce(s)	229	182	94	3	15	1	3	0.36	1.32	0.92	0
34356	Vegetable rice casserole (lacto)	8	ounce(s)	227	172	230	9	24	4	12	4.67	3.48	2.96	—
34311	Vegetable strudel (ovo-lacto)	8	ounce(s)	227	100	756	19	51	4	54	18.24	26.38	6.17	0
34371	Vegetable taco (lacto)	1	item(s)	227	147	365	13	43	9	17	6.45	5.81	4.02	—
34282	Vegetarian chili (vegan)	8	ounce(s)	227	196	116	6	21	7	2	0.24	0.29	0.74	0
34367	Vegetarian vegetable soup (vegan)	8	ounce(s)	226	204	92	3	14	2	4	0.77	1.67	1.30	0
	Boca burger													
32067	All American flamed grilled patty	1	item(s)	71	—	110	14	6	4	4	1.00	—	—	0
32070	Bigger chef max's favorite	1	item(s)	99	—	130	18	11	5	4	1.00	1.00	1.50	—
32069	Bigger vegan	1	item(s)	99	—	120	18	11	6	0	0.00	0.00	0.00	0
32074	Boca chik'n nuggets	4	item(s)	87	—	190	16	16	2	7	2.00	—	—	0
32075	Boca meatless ground burger	½	cup(s)	57	—	70	11	7	4	1	0.00	—	—	—
32073	Boca tenders	1	item(s)	85	—	140	20	9	3	3	0.00	2.00	1.00	—
32072	Breakfast links	2	item(s)	45	—	100	10	6	5	4	0.00	—	—	0
32071	Breakfast patties	1	item(s)	38	—	80	8	5	3	4	0.00	—	—	0
32068	Roasted garlic patty	1	item(s)	71	—	100	14	7	5	2	0.50	—	—	0
32066	Vegan original patty	1	item(s)	71	—	90	13	4	0	1	0.00	—	—	0
	Gardenburger													
37810	Bbq chik'n with sauce	1	item(s)	142	—	250	14	30	5	8	1.00	—	—	0
39661	Black bean burger	1	item(s)	71	—	80	8	11	4	2	0.00	—	—	0
39666	Buffalo chick'n wing	3	item(s)	95	—	180	9	8	5	12	1.50	—	—	0
37808	Chik'n grill	1	item(s)	71	—	100	13	5	3	3	0.00	—	—	0
39665	Country fried chicken w/creamy pepper gravy	1	item(s)	142	—	190	9	16	2	9	1.00	—	—	0
37805	Crispy nuggets	6	item(s)	82	—	180	4	22	3	9	1.50	—	—	—
39663	Homestyle classic burger	1	item(s)	71	—	110	12	6	4	5	0.50	—	—	0

PAGE KEY: A–58 = Breads/Baked Goods A–62 = Cereal/Rice/Pasta A–66 = Fruit A–70 = Vegetables/Legumes A–80 = Nuts/Seeds A–82 = Vegetarian A–84 = Dairy A–90 = Eggs A–90 = Seafood A–92 = Meats A–96 = Poultry A–96 = Processed Meats A–98 = Beverages A–102 = Fats/Oils A–104 = Sweets A–106 = Sauces/Condiments/Spices A–108 = Mixed Foods/Soups/Sandwiches A–114 = Fast Food A–130 = Convenience A–132 = Baby Foods

Chol (mg)	Calc (mg)	Iron (mg)	Magn (mg)	Pota (mg)	Sodi (mg)	Zinc (mg)	Vit A (RAE) (μg)	Thia (mg)	Vit E (mg α)	Ribo (mg)	Niac (mg)	Vit B_6 (mg)	Fola (μg)	Vit C (mg)	Vit B_{12} (μg)	Sele (μg)
0	41	1.59	75	240	217	2.87	4	0.26	0.74	0.06	0.66	0.11	9	<1	0	2
0	18	0.68	33	108	<1	1.23	1	0.13	0.70	0.03	0.33	0.05	4	<1	0	2
0	19	0.68	33	111	0	1.22	1	0.18	0.38	0.04	0.32	0.06	6	<1	0	1
0	1	0.48	22	51	<1	0.55	<.1	0.03	0.80	0.02	0.38	0.01	6	<.1	0	<.1
				<1												
0	35	1.34	38	333	<1	0.74	4	0.27	0.62	0.05	0.46	0.41	16	1	0	3
0	35	1.34	38	333	<1	0.74	4	0.27	0.62	0.05	0.46	0.41	16	1	0	3
0	24	8.48	303	457	10	4.22	11	0.12	0.00	0.18	0.99	0.05	32	1	0	3
0	89	1.33	32	43	1	0.64	0	0.07	—	0.02	0.41	0.07	9	0	0	1
0	154	3.07	58	93	2	1.17	<1	0.04	—	0.03	1.07	0.13	16	0	0	1
0	21	0.66	14	69	5	0.69	<1	0.24	—	0.02	0.85	0.02	15	1	0	<1
0	120	2.16	—	—	300	—	0	—	—	—	—	—	—	0	—	—
0	120	2.16	—	—	20	—	0	—	—	—	—	—	—	0	—	—
0	42	2.44	127	248	1	1.82	1	0.82	12.42	0.09	1.62	0.28	82	1	0	21
0	22	1.22	41	272	250	1.69	<1	0.03	8.35	0.08	2.25	0.26	76	<1	0	25
0	19	2.28	43	164	205	1.78	0	0.11	—	0.10	1.41	0.27	80	<1	0	21
0	20	0.76	59	12	83	0.85	<1	0.05	—	0.05	0.85	0.13	38	<1	0	—
0	29	1.14	59	257	86	1.21	<1	0.17	—	0.07	1.77	0.11	27	1	0	—
2	41	1.27	60	243	45	1.18	1	0.15	—	0.08	1.65	0.10	24	<1	0	—
0	20	0.92	34	248	4	0.41	1	0.16	—	0.04	0.52	0.11	15	3	0	—
0	19	0.98	63	163	1	1.05	1	0.02	0.56	0.04	0.15	0.18	10	1	0	5
0	29	0.87	47	132	1	0.93	<1	0.10	0.21	0.05	0.34	0.16	29	<1	0	1
0	266	4.73	88	375	112	1.51	121	0.14	0.07	0.12	1.08	0.28	32	18	0	11
42	468	2.58	37	204	1219	1.96	107	0.33	0.06	0.38	2.38	0.11	77	22	<1	22
0	48	1.71	42	367	618	0.60	54	0.09	0.53	0.08	0.93	0.11	33	8	<.1	4
0	23	2.35	31	380	289	0.87	18	0.14	0.14	0.10	1.50	0.16	61	13	0	9
22	187	0.77	20	120	768	1.11	82	0.15	0.29	0.24	1.02	0.04	39	<.1	<1	16
0	41	1.43	68	358	1403	0.91	86	0.16	3.05	0.12	2.76	0.30	28	13	<.1	8
51	468	4.78	90	455	2454	2.07	82	0.27	0.12	0.27	3.43	0.30	99	2	<1	43
0	117	1.13	40	168	134	0.68	26	0.07	—	0.07	0.54	0.11	46	1	0	5
0	190	3.65	28	302	371	0.74	8	0.13	0.06	0.12	1.43	0.15	47	7	0	6
10	144	1.91	33	393	637	1.06	31	0.20	0.05	0.27	2.07	0.21	64	15	<1	19
0	15	0.85	17	180	378	0.35	18	0.13	0.50	0.08	1.25	0.11	41	20	0	10
16	176	1.72	28	395	609	1.19	121	0.16	0.35	0.29	1.93	0.18	92	54	<1	6
46	318	3.36	39	299	813	1.98	288	0.45	0.21	0.50	4.52	0.16	123	27	<1	31
21	231	2.58	83	550	893	1.80	81	0.23	0.10	0.18	1.48	0.25	132	12	<1	10
<1	68	2.42	41	532	383	0.78	46	0.13	0.15	0.13	1.26	0.18	58	16	0	5
0	37	1.32	28	443	503	0.44	109	0.11	0.55	0.08	1.54	0.22	38	24	<.1	1
3	150	1.80	—	—	370	—	0	—	—	—	—	—	—	0	—	—
5	150	2.70	—	—	400	—	—	—	—	—	—	—	—	0	—	—
0	60	1.80	—	—	380	—	0	—	—	—	—	—	—	2	—	—
0	80	1.80	—	220	570	—	0	—	—	—	—	—	—	0	—	—
0	80	1.44	—	—	220	—	0	—	—	—	—	—	—	0	—	—
0	80	1.08	—	—	440	—	0	—	—	—	—	—	—	0	—	—
0	60	1.44	—	—	330	—	0	—	—	—	—	—	—	0	—	—
0	60	1.44	—	—	260	—	0	—	—	—	—	—	—	0	—	—
3	100	1.80	—	—	400	—	0	—	—	—	—	—	—	1	—	—
0	80	1.80	—	—	350	—	0	—	—	—	—	—	—	1	—	—
0	150	1.08	—	—	890	—	—	—	—	—	—	—	—	0	—	—
0	40	1.44	—	—	330	—	—	—	—	—	—	—	—	0	—	—
0	40	0.72	—	—	1000	—	—	—	—	—	—	—	—	0	—	—
0	60	3.60	—	—	360	—	—	—	—	—	—	—	—	0	—	—
5	40	1.44	—	—	550	—	—	—	—	—	—	—	—	0	—	—
5	60	0.72	—	—	570	—	—	—	—	—	—	—	—	5	—	—
0	80	1.44	—	—	380	—	—	—	—	—	—	—	—	0	—	—

TABLE F–1

Food Composition (Computer code number is for Wadsworth Diet Analysis program) (For purposes of calculations, use "0" for t, <1, <.1, <.01, etc.)

DA + Code	Food Description	Quantity	Measure	Wt (g)	H_2O (g)	Ener (kcal)	Prot (g)	Carb (g)	Dietary Fiber (g)	Fat (g)	Fat Breakdown (g) Sat	Mono	Poly	Trans
	VEGETARIAN FOODS—Continued													
37807	Meatless breakfast sausage	1	item(s)	43	—	50	5	2	2	4	0.00	—	—	0
37809	Meatless meatballs	6	item(s)	85	—	110	12	8	4	5	1.00	—	—	0
37806	Meatless riblets w/sauce	1	item(s)	142	—	210	17	11	4	5	0.00	—	—	0
29913	Original	3	ounce(s)	85	—	132	7	19	4	4	1.80	1.80	0.60	0
31707	Santa Fe	3	ounce(s)	85	—	156	—	24	5	3	1.20	—	—	—
29915	Veggie medley	3	ounce(s)	85	—	108	6	22	4	0	0.00	0.00	0.00	0
	Loma Linda													
9311	Big franks	1	item(s)	51	30	110	10	2	2	7	1.00	2.00	4.00	0
9315	Chik'n nuggets	5	item(s)	85	40	240	14	13	4	15	2.00	4.50	8.00	0
9317	Corn dogs	1	item(s)	71	31	150	7	22	3	4	0.50	1.00	2.50	0
9323	Fried chik'n with gravy	2	piece(s)	80	46	150	12	5	2	10	1.50	2.50	5.00	0
9326	Linketts, canned	1	item(s)	35	21	70	7	1	1	5	0.50	1.00	2.50	0
9336	Redi-Burger patties, canned	1	slice(s)	85	50	120	18	7	4	3	0.50	0.50	1.50	0
9354	Tender Rounds meatball substitute, canned in gravy	6	piece(s)	80	54	120	13	6	1	5	0.50	1.00	2.50	0
	Morningstar Farms													
33707	America's Original Veggie Dog links	1	item(s)	57	—	80	11	6	1	1	0.00	0.00	0.00	0
9362	Better n Eggs egg substitute	¼	cup(s)	57	50	20	5	0	0	0	0.00	0.00	0.00	0
9368	Breakfast links	2	item(s)	45	27	80	9	3	2	3	0.50	0.50	2.00	0
9371	Breakfast strips	2	item(s)	16	7	60	2	2	1	5	0.50	1.00	3.00	0
33705	Chik Nuggets	4	piece(s)	86	—	180	13	17	5	6	0.50	1.50	4.00	—
11587	Chik Patties	1	item(s)	71	36	150	9	16	2	6	1.00	1.50	2.50	—
2531	Garden veggie patties	1	item(s)	67	40	100	10	9	4	3	0.50	0.50	1.50	0
9412	Natural Touch low fat vegetarian chili, canned	1	cup(s)	230	173	170	18	21	11	1	—	—	—	0
33702	Spicy black bean veggie burger	1	item(s)	78	47	150	11	16	5	5	0.50	1.50	2.50	0
	Worthington													
9422	Chik Stiks	1	item(s)	47	27	110	10	4	2	6	1.00	1.00	3.00	0
9424	Chili, canned	1	cup(s)	230	167	290	19	21	9	15	2.50	3.50	9.00	0
9432	Crispychik patties	1	item(s)	71	37	150	9	16	2	6	1.00	1.50	3.50	0
9440	Dinner roast, frozen	1	slice(s)	85	53	180	12	5	3	12	1.50	5.00	5.00	0
9442	Fillets, frozen	2	piece(s)	85	48	180	16	8	4	9	1.00	3.50	4.50	0
9478	Meatless smoked beef, sliced	6	slice(s)	57	—	130	11	7	1	7	1.00	2.00	4.00	0
9480	Meatless smoked turkey, sliced	3	slice(s)	57	—	140	10	5	0	9	1.00	2.50	5.00	—
9462	Prosage links	2	item(s)	45	27	80	9	3	2	3	0.50	0.50	2.00	0
9486	Stripples bacon substitute	2	item(s)	16	7	60	2	2	1	5	0.50	1.00	2.50	0
9496	Vegetable Skallops	½	cup(s)	85	65	90	15	3	3	2	0.50	0.50	0.00	0
9434	Vegetarian cutlets	1	slice(s)	61	43	70	11	3	2	1	—	—	—	—
	DAIRY													
	Butter: *see* Fats & Oils													
	Cheese													
1433	Blue, crumbled	1	ounce(s)	28	12	100	6	1	0	8	5.29	2.21	0.23	—
884	Brick	1	ounce(s)	28	12	104	7	1	0	8	5.25	2.41	0.22	—
885	Brie	1	ounce(s)	28	14	94	6	<1	0	8	4.87	2.24	0.23	—
34821	Camembert	1	ounce(s)	29	15	87	6	<1	0	7	4.43	2.04	0.21	—
888	Cheddar or colby	1	ounce(s)	28	11	110	7	1	0	9	5.66	2.60	0.27	—
32096	Cheddar or colby, low fat	1	ounce(s)	28	18	49	7	1	0	2	1.23	0.59	0.06	—
5	Cheddar, shredded	¼	cup(s)	28	10	114	7	<1	0	9	5.96	2.65	0.27	—
889	Edam	1	ounce(s)	28	12	100	7	<1	0	8	4.92	2.28	0.19	—
890	Feta	1	ounce(s)	28	15	74	4	1	0	6	4.18	1.29	0.17	—
891	Fontina	1	ounce(s)	28	11	109	7	<1	0	9	5.37	2.43	0.46	—
8527	Goat, soft	1	ounce(s)	28	17	76	5	<1	0	6	4.14	1.37	0.14	—
893	Gouda	1	ounce(s)	28	12	100	7	1	0	8	4.93	2.17	0.18	—
894	Gruyere	1	ounce(s)	28	9	116	8	<1	0	9	5.30	2.81	0.49	—
895	Limburger	1	ounce(s)	28	14	92	6	<1	0	8	4.69	2.41	0.14	—
896	Monterey jack	1	ounce(s)	28	11	104	7	<1	0	8	5.34	2.45	0.25	—
13	Mozzarella, part skim milk	1	ounce(s)	28	15	71	7	1	0	4	2.83	1.26	0.13	—
12	Mozzarella, whole milk	1	ounce(s)	28	14	84	6	1	0	6	3.68	1.84	0.21	—
897	Muenster	1	ounce(s)	28	12	103	7	<1	0	8	5.35	2.44	0.19	—
898	Neufchatel	1	ounce(s)	28	17	73	3	1	0	7	4.14	1.90	0.18	—
14	Parmesan, grated	1	tablespoon(s)	5	1	22	2	<1	0	1	0.87	0.42	0.06	—
17	Provolone	1	ounce(s)	28	11	98	7	1	0	7	4.78	2.07	0.22	—
19	Ricotta, part skim milk	¼	cup(s)	62	46	85	7	3	0	5	3.03	1.42	0.16	—
18	Ricotta, whole milk	¼	cup(s)	62	44	107	7	2	0	8	5.10	2.23	0.24	—
20	Romano	1	tablespoon(s)	5	2	19	2	<1	0	1	0.86	0.39	0.03	—

PAGE KEY: A–58 = Breads/Baked Goods A–62 = Cereal/Rice/Pasta A–66 = Fruit A–70 = Vegetables/Legumes A–80 = Nuts/Seeds A–82 = Vegetarian A–84 = Dairy A–90 = Eggs A–90 = Seafood A–92 = Meats A–96 = Poultry A–96 = Processed Meats A–98 = Beverages A–102 = Fats/Oils A–104 = Sweets A–106 = Sauces/Condiments/Spices A–108 = Mixed Foods/Soups/Sandwiches A–114 = Fast Food A–130 = Convenience A–132 = Baby Foods

Chol (mg)	Calc (mg)	Iron (mg)	Magn (mg)	Pota (mg)	Sodi (mg)	Zinc (mg)	Vit A (RAE) (μg)	Thia (mg)	Vit E (mg α)	Ribo (mg)	Niac (mg)	Vit B_6 (mg)	Fola (μg)	Vit C (mg)	Vit B_{12} (μg)	Sele (μg)
0	20	0.72	—	—	120	—	—	—	—	—	—	—	—	0	—	—
0	60	1.80	—	—	400	—	—	—	—	—	—	—	—	0	—	—
0	60	1.80	—	—	720	—	—	—	—	—	—	—	—	4	—	—
24	72	0.00	37	232	672	1.07	0	0.12	—	0.18	1.30	0.10	12	0	<1	8
24	96	0.00	—	—	336	—	0	—	—	—	—	—	—	0	—	0
0	48	0.00	32	218	336	0.55	—	0.08	—	0.10	1.08	0.11	13	0	<.1	5
0	0	0.77	—	50	240	0.89	0	0.23	—	0.43	1.60	0.04	—	0	1	—
0	20	1.44	—	210	410	0.43	0	0.75	—	0.51	6.00	0.90	—	0	3	—
0	0	1.08	—	60	500	0.43	0	0.72	—	0.61	1.47	0.87	—	0	2	—
0	20	1.80	—	70	430	0.34	0	1.05	—	0.34	4.00	0.30	—	0	2	—
0	0	0.36	—	15	160	0.46	0	0.12	—	0.20	0.40	0.20	—	0	1	—
0	0	1.06	—	140	450	1.11	0	0.23	—	0.34	6.00	0.40	—	0	2	—
0	20	1.08	—	80	340	0.66	0	0.75	—	0.17	2.00	0.16	—	0	1	—
0	0	0.72	—	60	580	—	0	—	—	—	—	—	—	0	—	—
0	20	0.63	—	75	90	0.60	75	0.03	—	0.34	0.00	0.08	24	0	1	—
0	0	1.44	—	50	320	0.36	0	1.80	—	0.17	2.00	0.30	—	0	3	—
0	0	0.27	—	15	220	0.05	0	0.75	—	0.04	0.40	0.07	—	0	<1	—
0	40	3.60	—	330	590	—	0	1.20	—	0.26	5.00	0.40	—	0	3	—
0	0	1.80	—	210	540	0.31	0	1.80	—	0.17	2.00	0.20	—	0	1	—
0	40	0.72	—	180	350	0.58	—	6.47	—	0.10	0.00	0.00	—	0	0	—
0	40	1.80	—	480	870	1.36	—	0.60	—	0.21	0.00	0.30	—	0	0	—
0	40	1.80	44	320	470	0.93	0	—	—	0.14	0.00	0.21	—	0	<.1	—
0	20	1.80	—	100	300	0.31	0	0.60	—	0.17	6.00	0.40	—	0	2	—
0	40	3.60	—	420	1130	1.24	0	0.06	—	0.07	2.00	0.70	—	0	2	—
0	0	1.80	—	170	440	0.33	0	1.80	—	0.17	2.00	0.20	—	0	1	—
3	40	0.36	—	55	580	0.64	0	1.80	—	0.26	6.00	0.60	—	0	2	—
0	0	1.80	—	130	750	0.92	0	0.68	—	0.14	0.80	0.40	—	0	3	—
0	20	1.80	—	180	510	0.14	0	1.80	—	0.17	6.00	0.40	—	0	2	—
0	100	2.70	—	60	490	0.23	0	1.80	—	0.17	6.00	0.40	—	0	3	—
0	0	1.44	—	50	320	0.36	0	1.80	—	0.17	2.00	0.30	—	0	3	—
0	0	0.36	—	15	220	0.05	0	0.75	—	0.03	0.40	0.08	—	0	<1	—
0	0	0.72	—	10	410	0.67	0	0.03	—	0.03	0.00	0.01	—	0	0	—
0	0	0.00	—	30	340	0.43	0	0.03	—	0.04	0.00	0.04	—	0	0	—
21	150	0.09	7	73	395	0.75	56	0.01	0.07	0.11	0.29	0.05	10	0	<1	4
26	189	0.12	7	38	157	0.73	82	0.00	0.07	0.10	0.03	0.02	6	0	<1	4
28	52	0.14	6	43	176	0.67	49	0.02	0.07	0.15	0.11	0.07	18	0	<1	4
21	112	0.10	6	54	244	0.69	—	0.01	—	0.14	0.18	0.07	18	0	<1	4
27	192	0.21	7	36	169	0.86	74	0.00	0.08	0.11	0.03	0.02	5	0	<1	4
6	118	0.12	5	19	174	0.52	17	0.00	0.02	0.06	0.01	0.01	3	0	<1	4
30	204	0.19	8	28	175	0.88	75	0.01	0.08	0.11	0.02	0.02	5	0	<1	4
25	205	0.12	8	53	270	1.05	68	0.01	0.07	0.11	0.02	0.02	4	0	<1	4
25	138	0.18	5	17	312	0.81	35	0.04	0.05	0.24	0.28	0.12	9	0	<1	4
32	154	0.06	4	18	224	0.98	73	0.01	0.08	0.06	0.04	0.02	2	0	<1	4
13	40	0.54	5	7	105	0.26	82	0.02	0.05	0.11	0.12	0.07	3	0	<.1	1
32	196	0.07	8	34	229	1.09	46	0.01	0.07	0.09	0.02	0.02	6	0	<1	4
31	283	0.05	10	23	94	1.09	76	0.02	0.08	0.08	0.03	0.02	3	0	<1	4
25	139	0.04	6	36	224	0.59	95	0.02	0.06	0.14	0.04	0.02	16	0	<1	4
25	209	0.20	8	23	150	0.84	55	0.00	0.07	0.11	0.03	0.02	5	0	<1	4
18	219	0.06	6	24	173	0.77	36	0.01	0.04	0.08	0.03	0.02	3	0	<1	4
22	141	0.12	6	21	176	0.82	50	0.01	0.05	0.08	0.03	0.01	2	0	1	5
27	201	0.11	8	38	176	0.79	83	0.00	0.07	0.09	0.03	0.02	3	0	<1	4
21	21	0.08	2	32	112	0.15	83	0.00	—	0.05	0.04	0.01	3	0	<.1	1
4	55	0.05	2	6	76	0.19	6	0.00	0.01	0.02	0.01	0.00	1	0	<1	1
19	212	0.15	8	39	245	0.90	66	0.01	0.06	0.09	0.04	0.02	3	0	<1	4
19	167	0.27	9	77	77	0.82	66	0.01	0.04	0.11	0.05	0.01	8	0	<1	10
31	127	0.23	7	65	52	0.71	74	0.01	0.07	0.12	0.06	0.03	7	0	<1	9
5	53	0.04	2	4	60	0.13	5	0.00	0.01	0.02	0.00	0.00	<1	0	<.1	1

TABLE F–1

Food Composition (Computer code number is for Wadsworth Diet Analysis program) (For purposes of calculations, use "0" for t, <1, <.1, <.01, etc.)

DA + Code	Food Description	Quantity	Measure	Wt (g)	H_2O (g)	Ener (kcal)	Prot (g)	Carb (g)	Dietary Fiber (g)	Fat (g)	Fat Breakdown (g) Sat	Mono	Poly	Trans
	DAIRY—Continued													
900	Roquefort	1	ounce(s)	28	11	103	6	1	0	9	5.39	2.37	0.37	—
21	Swiss	1	ounce(s)	28	10	106	8	2	0	8	4.98	2.04	0.27	—
	Imitation cheese													
7998	Shredded imitation cheddar	¼	cup(s)	28	—	90	5	2	0	7	1.50	—	—	—
8028	Shredded imitation mozzarella	¼	cup(s)	28	—	80	6	1	0	6	1.00	—	—	—
	Cottage Cheese													
9	Low fat, 1% fat	½	cup(s)	113	93	81	14	3	0	1	0.73	0.33	0.04	—
8	Low fat, 2% fat	½	cup(s)	113	90	102	16	4	0	2	1.38	0.62	0.07	—
	Cream cheese													
11	Cream cheese	2	tablespoon(s)	29	16	101	2	1	0	10	6.37	2.85	0.37	—
17366	Fat free cream cheese	2	tablespoon(s)	30	23	29	4	2	0	<1	0.27	0.10	0.02	—
10438	Tofutti Better Than Cream Cheese	2	tablespoon(s)	30	—	80	1	1	0	8	2.00	—	6.00	—
	Processed cheese													
22	American cheese, processed	1	ounce(s)	28	11	106	6	<1	0	9	5.58	2.54	0.28	—
24	American cheese food, processed	1	ounce(s)	28	12	94	5	2	0	7	4.23	2.05	0.31	—
25	American cheese spread, processed	1	ounce(s)	28	14	82	5	2	0	6	3.78	1.77	0.18	—
9110	Kraft deluxe singles pasteurized process American cheese	1	ounce(s)	28	—	110	5	1	0	9	6.00	—	—	—
23	Swiss cheese, processed	1	ounce(s)	28	12	95	7	1	0	7	4.55	2.00	0.18	—
	Soy cheese													
10430	Nu Tofu cheddar flavored cheese alternative	1	ounce(s)	28	—	70	6	1	0	4	0.50	2.50	1.00	—
10435	Nu Tofu mozzarella flavored cheese alternative	1	ounce(s)	28	—	70	6	2	0	4	0.50	2.50	1.00	—
	Cream													
26	Half & half	1	tablespoon(s)	15	12	20	<1	1	0	2	1.07	0.50	0.06	—
28	Light coffee or table, liquid	1	tablespoon(s)	15	11	29	<1	1	0	3	1.80	0.84	0.11	—
30	Light whipping cream, liquid	1	tablespoon(s)	15	10	44	<1	<1	0	5	2.90	1.36	0.13	—
32	Heavy whipping cream, liquid	1	tablespoon(s)	15	9	52	<1	<1	0	6	3.45	1.60	0.21	—
34	Whipped cream topping, pressurized	1	tablespoon(s)	4	2	10	<1	<1	0	1	0.52	0.24	0.03	—
	Sour cream													
36	Sour cream	2	tablespoon(s)	24	17	51	1	1	0	5	3.13	1.45	0.19	—
30556	Fat free sour cream	2	tablespoon(s)	32	26	24	1	5	0	0	0.00	0.00	0.00	0
	Imitation cream													
3659	Coffeemate nondairy creamer, liquid	1	tablespoon(s)	16	—	20	0	2	0	1	0.00	0.50	0.00	—
40	Cream substitute, powder	1	teaspoon(s)	2	<.1	11	<.1	1	0	1	0.65	0.02	0.00	—
35972	Nondairy coffee whitener, liquid, frozen	1	tablespoon(s)	16	12	22	<1	2	0	2	0.31	1.20	0.00	—
35975	Nondairy dessert topping, pressurized	1	tablespoon(s)	5	3	12	<.1	1	0	1	0.88	0.09	0.01	—
35976	Nondairy dessert topping, frozen	1	tablespoon(s)	5	3	16	<.1	1	0	1	1.09	0.08	0.03	—
904	Imitation sour cream	2	tablespoon(s)	24	17	50	1	2	0	5	4.27	0.14	0.01	—
	Fluid milk													
57	Fat free, nonfat, or skim	1	cup(s)	245	223	83	8	12	0	<1	0.29	0.12	0.02	—
58	Fat free, nonfat, or skim, w/nonfat milk solids	1	cup(s)	245	221	91	9	12	0	1	0.40	0.16	0.02	—
54	Low fat, 1%	1	cup(s)	244	219	102	8	12	0	2	1.54	0.68	0.09	—
55	Low fat, 1%, w/nonfat milk solids	1	cup(s)	245	220	105	9	12	0	2	1.48	0.69	0.09	—
60	Low fat buttermilk	1	cup(s)	245	221	98	8	12	0	2	1.34	0.62	0.08	—
51	Reduced fat, 2%	1	cup(s)	244	218	122	8	11	0	5	2.35	2.04	0.17	—
52	Reduced fat, 2%, w/nonfat milk solids	1	cup(s)	245	218	125	9	12	0	5	2.93	1.36	0.17	—
50	Whole, 3.3%	1	cup(s)	244	216	146	8	11	0	8	4.55	1.98	0.48	—
	Canned													
61	Whole evaporated	2	tablespoon(s)	32	23	42	2	3	0	2	1.45	0.74	0.08	—
62	Fat free, nonfat, or skim evaporated	2	tablespoon(s)	32	25	25	2	4	0	<.1	0.04	0.02	0.00	—
63	Sweetened condensed	2	tablespoon(s)	38	10	123	3	21	0	3	2.10	0.93	0.13	—
	Dried Milk													
64	Dried buttermilk	¼	cup(s)	30	1	118	10	15	0	2	1.09	0.51	0.07	—
65	Instant nonfat dry milk w/added vitamin A	¼	cup(s)	17	1	63	6	9	0	<1	0.08	0.03	0.00	—
5234	Skim milk powder	¼	cup(s)	18	1	64	6	9	0	<1	0.08	0.03	0.01	—
907	Whole dry milk	¼	cup(s)	32	1	161	9	12	0	9	5.43	2.57	0.22	—
909	**Goat milk**	1	cup(s)	244	212	168	9	11	0	10	6.51	2.71	0.36	—
	Chocolate milk													
69	Low fat	1	cup(s)	250	211	158	8	26	1	3	1.54	0.75	0.09	—
68	Reduced fat	1	cup(s)	250	209	180	8	26	1	5	3.10	1.47	0.18	—
67	Whole milk	1	cup(s)	250	206	208	8	26	2	8	5.26	2.48	0.31	—
33156	Chocolate syrup, fortified, prepared w/milk	1	cup(s)	263	220	197	8	24	<1	8	5.22	2.44	0.31	—

PAGE KEY: A–58 = Breads/Baked Goods A–62 = Cereal/Rice/Pasta A–66 = Fruit A–70 = Vegetables/Legumes A–80 = Nuts/Seeds A–82 = Vegetarian A–84 = Dairy A–90 = Eggs A–90 = Seafood A–92 = Meats A–96 = Poultry A–96 = Processed Meats A–98 = Beverages A–102 = Fats/Oils A–104 = Sweets A–106 = Sauces/Condiments/Spices A–108 = Mixed Foods/Soups/Sandwiches A–114 = Fast Food A–130 = Convenience A–132 = Baby Foods

Chol (mg)	Calc (mg)	Iron (mg)	Magn (mg)	Pota (mg)	Sodi (mg)	Zinc (mg)	Vit A (RAE) (μg)	Thia (mg)	Vit E (mg α)	Ribo (mg)	Niac (mg)	Vit B_6 (mg)	Fola (μg)	Vit C (mg)	Vit B_{12} (μg)	Sele (μg)
25	185	0.16	8	25	507	0.58	82	0.01	—	0.16	0.21	0.03	14	0	<1	4
26	221	0.06	11	22	54	1.22	62	0.02	0.11	0.08	0.03	0.02	2	0	1	5
0	150	0.00	—	—	420	—	—	—	—	—	—	—	—	0	—	—
0	150	0.00	8	—	320	1.20	—	0.00	—	0.26	0.00	0.00	40	0	<1	—
5	69	0.16	6	97	459	0.43	12	0.02	0.01	0.19	0.14	0.08	14	0	1	10
9	78	0.18	7	108	459	0.47	24	0.03	0.02	0.21	0.16	0.09	15	0	1	12
32	23	0.35	2	35	86	0.16	106	0.00	0.09	0.06	0.03	0.01	4	0	<1	1
2	56	0.05	4	49	164	0.26	84	0.02	0.00	0.05	0.05	0.02	11	0	<1	1
0	0	0.00	—	—	135	—	0	—	—	—	—	—	—	0	—	—
27	156	0.05	8	48	422	0.81	72	0.01	0.08	0.10	0.02	0.02	2	0	<1	4
23	162	0.16	9	83	359	0.91	57	0.02	0.06	0.15	0.05	0.02	2	0	<1	5
16	160	0.09	8	69	382	0.74	49	0.01	0.05	0.12	0.04	0.03	2	0	<1	3
25	150	0.00	0	25	450	0.90	84	—	—	0.10	—	—	—	0	<1	—
24	219	0.17	8	61	388	1.02	56	0.00	0.10	0.08	0.01	0.01	2	0	<1	5
0	200	0.36	—	—	190	—	—	—	—	—	—	—	—	0	—	—
0	150	0.36	—	—	190	—	—	—	—	—	—	—	—	0	—	—
6	16	0.01	2	20	6	0.08	15	0.01	0.05	0.02	0.01	0.01	<1	<1	<.1	<1
10	14	0.01	1	18	6	0.04	27	0.00	0.08	0.02	0.01	0.00	<1	<1	<.1	<.1
17	10	0.00	1	15	5	0.04	42	0.00	0.13	0.02	0.01	0.00	1	<.1	<.1	<.1
21	10	0.00	1	11	6	0.03	62	0.00	0.16	0.02	0.01	0.00	1	<.1	<.1	<.1
3	4	0.00	<1	6	5	0.01	7	0.00	0.02	0.00	0.00	0.00	<1	0	<.1	<.1
11	28	0.01	3	35	13	0.06	42	0.01	0.14	0.04	0.02	0.00	3	<1	<.1	1
3	40	0.00	3	41	45	0.16	—	0.01	0.00	0.05	0.02	0.01	4	0	<.1	—
0	0	0.00	—	30	0	—	0	0.02	—	0.02	0.20	—	—	0	—	—
0	<1	0.02	<.1	16	4	0.01	<.1	0.00	0.01	0.00	0.00	0.00	0	0	0	<.1
0	1	0.00	<.1	30	13	0.00	—	0.00	—	0.00	0.00	0.00	0	0	0	<1
0	<1	0.00	<.1	1	3	0.00	—	0.00	—	0.00	0.00	0.00	0	0	0	<.1
0	<1	0.01	<.1	1	1	0.00	—	0.00	—	0.00	0.00	0.00	0	0	0	<1
0	1	0.09	1	39	24	0.28	0	0.00	0.18	0.00	0.00	0.00	0	0	0	1
5	223	1.23	22	238	108	2.08	149	0.11	0.02	0.45	0.23	0.09	12	0	1	8
5	316	0.12	37	419	130	1.00	149	0.10	0.00	0.43	0.22	0.11	12	2	1	5
12	264	0.85	27	290	122	2.12	142	0.05	0.02	0.45	0.23	0.09	12	0	1	8
10	314	0.12	34	397	127	0.98	145	0.10	—	0.42	0.22	0.11	12	2	1	6
10	284	0.12	27	370	257	1.03	17	0.08	0.12	0.38	0.14	0.08	12	2	1	5
20	271	0.24	27	342	115	1.17	134	0.10	0.07	0.45	0.22	0.09	12	<1	1	6
20	314	0.12	34	397	127	0.98	137	0.10	—	0.42	0.22	0.11	12	2	1	6
24	246	0.07	24	325	105	0.93	68	0.11	0.15	0.45	0.26	0.09	12	0	1	9
9	82	0.06	8	95	33	0.24	20	0.01	0.04	0.10	0.06	0.02	3	1	<.1	1
1	93	0.09	9	106	37	0.29	38	0.01	0.00	0.10	0.06	0.02	3	<1	<.1	1
13	109	0.07	10	142	49	0.36	28	0.03	0.06	0.16	0.08	0.02	4	1	<1	6
21	360	0.09	33	484	157	1.22	15	0.12	0.03	0.48	0.27	0.10	14	2	1	6
3	215	0.05	20	298	96	0.77	124	0.07	0.00	0.30	0.16	0.06	9	1	1	5
3	222	0.06	21	307	99	0.79	0	0.07	—	0.31	0.16	0.06	9	1	1	5
31	296	0.15	28	431	120	1.08	83	0.09	0.16	0.39	0.21	0.10	12	3	1	5
27	327	0.12	34	498	122	0.73	139	0.12	0.17	0.34	0.68	0.11	2	3	<1	3
8	288	0.60	33	425	153	1.03	145	0.10	0.05	0.42	0.32	0.10	13	2	1	5
18	285	0.60	33	423	150	1.03	138	0.09	0.10	0.41	0.32	0.10	13	2	1	5
30	280	0.60	33	418	150	1.03	65	0.09	0.15	0.41	0.31	0.10	13	2	1	5
34	292	2.68	32	460	147	0.92	—	0.09	—	0.55	6.53	0.11	13	2	1	5

TABLE F–1

Food Composition (Computer code number is for Wadsworth Diet Analysis program) (For purposes of calculations, use "0" for t, <1, <.1, <.01, etc.)

DA + Code	Food Description	Quantity	Measure	Wt (g)	H_2O (g)	Ener (kcal)	Prot (g)	Carb (g)	Dietary Fiber (g)	Fat (g)	Fat Breakdown (g)			
											Sat	Mono	Poly	Trans
	DAIRY—Continued													
908	Cocoa, hot, prepared w/milk	1	cup(s)	250	206	193	9	27	3	6	3.58	1.69	0.09	0.18
33184	Cocoa mix with aspartame, added sodium & vitamin A, no added calcium or phosphorus, prepared with water	1	cup(s)	192	177	56	2	10	1	<1	0.00	0.15	0.01	—
70	**Eggnog**	1	cup(s)	254	189	343	10	34	0	19	11.29	5.67	0.86	—
	Breakfast drinks													
10093	Carnation Instant Breakfast classic chocolate malt, prepared w/skim milk, no sugar added	1	cup(s)	243	—	142	11	21	<1	1	0.89	—	—	—
10091	Carnation Instant Breakfast strawberry creme, prepared w/skim milk	1	cup(s)	273	—	220	13	39	0	<1	0.40	—	—	—
10094	Carnation Instant Breakfast strawberry creme, prepared w/skim milk, no sugar added	1	cup(s)	243	—	134	12	21	0	<1	0.45	—	—	—
10092	Carnation Instant Breakfast vanilla creme, prepared w/skim milk, no sugar added	1	cup(s)	273	—	220	13	39	0	<1	0.40	—	—	—
1417	Ovaltine rich chocolate flavor, prepared w/skim milk	1	cup(s)	243	—	134	12	21	0	<1	0.45	—	—	—
8539	**Malted milk, chocolate mix, fortified, prepared w/milk**	1	cup(s)	265	216	223	9	29	1	9	4.95	2.17	0.54	—
	Milkshakes													
73	Chocolate	1	cup(s)	227	164	270	7	48	1	6	3.81	1.77	0.23	—
74	Vanilla	1	cup(s)	227	169	254	9	40	0	7	4.28	1.98	0.26	—
	Ice cream													
4776	Chocolate	½	cup(s)	66	37	143	3	19	1	7	4.49	2.12	0.27	—
16514	Chocolate, soft serve	½	cup(s)	87	50	177	3	24	1	8	5.17	2.43	0.31	—
12137	Chocolate fudge, fat free no sugar added	½	cup(s)	71	—	100	4	22	0	0	0.00	0.00	0.00	0
82	Light vanilla	½	cup(s)	66	42	109	4	18	<1	3	1.71	0.57	0.10	—
78	Light vanilla, soft serve	½	cup(s)	86	60	108	4	19	0	2	1.40	0.65	0.09	—
16523	Sherbet, all flavors	½	cup(s)	97	64	133	1	29	<1	2	1.12	0.51	0.08	—
4778	Strawberry	½	cup(s)	66	40	127	2	18	1	6	3.43	—	—	—
76	Vanilla	½	cup(s)	66	40	133	2	16	<1	7	4.48	1.96	0.30	—
12146	Vanilla chocolate swirl, fat free, no sugar added	½	cup(s)	71	—	100	4	20	0	0	0.00	0.00	0.00	0
	Soy desserts													
10694	Tofutti low fat vanilla fudge nondairy frozen dessert	½	cup(s)	70	—	120	2	24	0	2	1.00	—	—	—
15721	Tofutti premium chocolate supreme nondairy frozen dessert	½	cup(s)	60	—	180	3	18	0	11	2.00	—	—	—
15720	Tofutti premium vanilla nondairy frozen dessert	½	cup(s)	60	—	190	2	20	0	11	2.00	—	—	—
	Ice milk													
16516	Flavored, not chocolate	½	cup(s)	66	45	91	2	15	0	3	1.72	0.81	0.11	—
16517	Chocolate	½	cup(s)	66	43	95	3	17	<1	2	1.29	0.61	0.08	—
	Pudding													
25032	Chocolate	½	cup(s)	144	110	154	5	23	1	5	2.78	1.94	0.23	0
1923	Chocolate, sugar free, prepared w/2% milk	½	cup(s)	133	—	100	5	14	<1	3	1.50	—	—	—
1722	Rice	½	cup(s)	113	73	175	6	26	1	6	1.99	2.14	0.88	—
4747	Tapioca, ready to eat	1	item(s)	142	105	169	3	28	<1	5	0.85	2.24	1.93	—
25031	Vanilla	½	cup(s)	136	110	116	5	17	<.1	3	1.31	1.21	0.16	0
1924	Vanilla, sugar free, prepared w/2% milk	½	cup(s)	133	90	4	12	<1	2	1.50	10	150	0.00	—
	Frozen yogurt													
4785	Chocolate, soft serve	½	cup(s)	72	46	115	3	18	2	4	2.61	1.26	0.16	—
1747	Fruit varieties	½	cup(s)	113	80	144	3	24	0	4	2.63	1.11	0.11	—
4786	Vanilla, soft serve	½	cup(s)	72	47	117	3	17	0	4	2.46	1.14	0.15	—
	Milk substitutes													
	Lactose free													
16081	Fat free calcium fortified milk	1	cup(s)	240	—	90	9	13	0	0	0.00	—	—	0
36486	Low fat milk	1	cup(s)	240	—	110	8	13	0	3	1.50	—	—	—
36487	Reduced fat milk	1	cup(s)	240	—	130	8	13	0	5	3.00	—	—	—
36488	Whole milk	1	cup(s)	240	—	160	8	12	0	9	5.00	—	—	—
	Rice													
10083	Rice Dream carob rice beverage	1	cup(s)	240	—	150	1	32	0	3	0.00	—	—	—
10087	Rice Dream vanilla enriched rice beverage	1	cup(s)	240	—	130	1	28	0	2	0.00	—	—	—
17089	Rice Dream original rice beverage, enriched	1	cup(s)	240	—	120	1	25	0	2	0.00	—	—	—

PAGE KEY: A–58 = Breads/Baked Goods A–62 = Cereal/Rice/Pasta A–66 = Fruit A–70 = Vegetables/Legumes A–80 = Nuts/Seeds A–82 = Vegetarian A–84 = Dairy A–90 = Eggs A–90 = Seafood A–92 = Meats A–96 = Poultry A–96 = Processed Meats A–98 = Beverages A–102 = Fats/Oils A–104 = Sweets A–106 = Sauces/Condiments/Spices A–108 = Mixed Foods/Soups/Sandwiches A–114 = Fast Food A–130 = Convenience A–132 = Baby Foods

Chol (mg)	Calc (mg)	Iron (mg)	Magn (mg)	Pota (mg)	Sodi (mg)	Zinc (mg)	Vit A (RAE) (μg)	Thia (mg)	Vit E (mg α)	Ribo (mg)	Niac (mg)	Vit B_6 (mg)	Fola (μg)	Vit C (mg)	Vit B_{12} (μg)	Sele (μg)
20	263	1.20	58	493	110	1.58	128	0.10	0.08	0.46	0.33	0.10	13	1	1	7
<1	90	0.75	33	405	171	0.52	27	0.04	0.06	0.21	0.16	0.05	2	<1	<1	2
150	330	0.51	48	419	137	1.17	114	0.09	0.51	0.48	0.27	0.13	3	4	1	11
9	445	4.01	89	632	196	3.38	—	0.35	—	0.45	4.45	0.45	4	27	1	8
9	500	4.47	100	638	360	3.75	—	0.38	—	0.51	5.08	0.48	100	30	1	9
9	445	4.01	89	570	187	3.38	—	0.33	—	0.45	4.45	0.45	89	27	1	8
9	500	4.50	100	630	240	3.75	—	0.38	—	0.51	5.00	0.50	100	30	2	9
9	445	4.01	89	570	187	3.38	—	0.33	—	0.45	4.45	0.45	89	27	1	8
27	339	3.76	45	578	231	1.17	904	0.76	0.16	1.32	11.08	1.01	19	32	1	12
25	299	0.70	36	508	252	1.09	41	0.11	0.11	0.50	0.28	0.06	11	0	1	4
27	331	0.23	27	415	215	0.88	57	0.07	0.11	0.44	0.33	0.10	16	0	1	5
22	72	0.61	19	164	50	0.38	78	0.03	0.20	0.13	0.15	0.04	11	<1	<1	2
22	103	0.33	19	192	44	0.48	—	0.04	0.22	0.13	0.11	0.03	5	1	<1	—
0	80	0.36	—	—	60	—	—	—	—	—	—	—	—	0	—	—
17	77	0.05	9	137	49	0.48	91	0.02	0.08	0.11	0.06	0.02	3	<1	<1	1
10	135	0.05	12	190	60	0.46	25	0.04	0.05	0.17	0.10	0.04	5	1	<1	3
5	52	0.14	8	93	44	0.46	—	0.02	0.03	0.07	0.09	0.03	4	4	<1	—
19	79	0.14	9	124	40	0.22	63	0.03	—	0.17	0.11	0.03	8	5	<1	1
29	84	0.06	9	131	53	0.46	78	0.03	0.20	0.16	0.08	0.03	3	<1	<1	1
0	80	0.00	50	0					—							
0	0	0.00	—	8	90	—	0	—	—	—	—	—	—	0	—	—
0	0	0.00	—	7	180	—	0	—	—	—	—	—	—	0	—	—
0	0	0.00	—	2	210	—	0	—	—	—	—	—	—	0	—	—
9	91	0.07	10	138	56	0.29	—	0.04	0.06	0.17	0.06	0.04	4	1	<1	—
6	94	0.17	13	155	41	0.38	—	0.03	0.05	0.12	0.09	0.03	4	<1	<1	—
35	138	1.04	29	211	135	1.07	73	0.04	0.00	0.25	0.18	0.03	7	<1	<1	5
10	150	0.72	—	330	310	—	—	0.06	—	0.26	—	—	—	0	—	—
71	130	1.21	21	250	253	0.61	—	0.10	0.06	0.26	0.73	0.08	14	1	<1	—
1	119	0.33	11	136	226	0.38	0	0.03	0.43	0.14	0.44	0.03	4	1	<1	2
35	133	0.25	14	173	134	0.63	73	0.03	0.00	0.24	0.11	0.03	6	<1	<1	5
—	190	380	—	—	0.03	—	—	0.17	—	—	—	0	—	—		
4	106	0.90	19	188	71	0.35	32	0.03	—	0.15	0.22	0.05	8	<1	<1	2
15	113	0.52	11	176	71	0.32	—	0.05	0.10	0.20	0.08	0.05	5	1	<.1	—
1	103	0.22	10	152	63	0.30	42	0.03	0.08	0.16	0.21	0.06	4	1	<1	2
3	500	0.00	—	—	130	—	100	—	—	—	—	—	—	0	—	—
15	300	0.00	—	—	125	—	100	—	—	—	—	—	—	0	—	—
20	300	0.00	—	—	125	—	98	—	—	—	—	—	—	0	—	—
35	300	0.00	—	—	125	—	58	—	—	—	—	—	—	0	—	—
0	20	0.72	—	—	100	—	—	—	—	—	—	—	—	1	—	—
0	300	0.00	—	—	90	—	—	—	—	—	—	—	—	0	2	—
0	300	0.00	13	60	90	0.24	—	0.07	—	0.00	0.84	0.08	—	0	2	—

TABLE F–1
Food Composition (Computer code number is for Wadsworth Diet Analysis program) (For purposes of calculations, use "0" for t, <1, <.1, <.01, etc.)

DA + Code	Food Description	Quantity	Measure	Wt (g)	H_2O (g)	Ener (kcal)	Prot (g)	Carb (g)	Dietary Fiber (g)	Fat (g)	Fat Breakdown (g) Sat	Mono	Poly	Trans
	DAIRY—Continued													
	Soy													
34750	Soy Dream chocolate enriched soy beverage	1	cup(s)	240	—	210	7	37	1	4	0.50	—	—	—
34749	Soy Dream vanilla enriched soy beverage	1	cup(s)	240	—	150	7	22	0	4	0.50	—	—	—
13840	Vitasoy light chocolate soymilk	1	cup(s)	237	—	100	4	17	0	2	0.50	0.50	1.00	—
13839	Vitasoy light vanilla soymilk	1	cup(s)	237	—	70	4	10	0	2	0.50	0.50	1.00	—
13836	Vitasoy rich chocolate soymilk	1	cup(s)	237	—	160	7	24	1	4	0.50	1.00	2.50	—
13835	Vitasoy vanilla delite soymilk	1	cup(s)	237	—	120	8	13	1	4	0.50	1.00	2.50	—
	Yogurt													
3615	Custard style, fruit flavors	6	ounce(s)	170	127	190	7	32	0	4	2.00	—	—	—
3617	Custard style, vanilla	6	ounce(s)	170	134	190	7	32	0	4	2.00	0.94	0.10	—
32101	Fruit, low fat	1	cup(s)	245	184	243	10	46	0	3	1.82	0.77	0.08	—
29638	Fruit, nonfat, sweetened w/low calorie sweetener	1	cup(s)	241	208	122	11	19	1	<1	0.21	0.10	0.04	—
93	Plain, low fat	1	cup(s)	245	208	154	13	17	0	4	2.45	1.04	0.11	—
94	Plain, nonfat	1	cup(s)	245	209	137	14	19	0	<1	0.28	0.12	0.01	—
32100	Vanilla, low fat	1	cup(s)	245	194	208	12	34	0	3	1.97	0.84	0.09	—
5242	Yogurt beverage	1	cup(s)	245	200	172	6	33	0	2	1.39	0.59	0.06	—
38202	Yogurt smoothie, nonfat, all flavors	1	item(s)	325	—	290	10	60	6	0	0.00	0.00	0.00	0
	Soy yogurt													
10453	White Wave plain silk cultured	8	ounce(s)	227	—	120	5	22	1	3	0.00	—	—	0
34616	Stonyfield Farm Osoy chocolate-vanilla pack organic cultured	1	serving(s)	113	—	90	4	15	3	2	0.00	—	—	—
34617	Stonyfield Farm Osoy strawberry-peach pack organic cultured	1	serving(s)	113	—	90	4	15	3	2	0.00	—	—	—
	EGGS													
96	Raw, whole	1	item(s)	50	38	74	6	<1	0	5	1.55	1.91	0.68	—
97	Raw, white	1	item(s)	33	29	17	4	<1	0	<.1	0.00	0.00	0.00	—
98	Raw, yolk	1	item(s)	17	9	53	3	1	0	4	1.59	1.95	0.70	—
99	Fried	1	item(s)	46	32	92	6	<1	0	7	1.98	2.92	1.22	—
100	Hard boiled	1	item(s)	50	37	78	6	1	0	5	1.63	2.04	0.71	—
101	Poached	1	item(s)	50	38	74	6	<1	0	5	1.54	1.90	0.68	—
102	Scrambled, prepared w/milk & butter	2	item(s)	122	89	203	14	3	0	15	4.49	5.82	2.62	—
	Egg Substitute													
920	Frozen	¼	cup(s)	60	44	96	7	2	0	7	1.16	1.46	3.74	—
918	Liquid	¼	cup(s)	63	52	53	8	<1	0	2	0.41	0.56	1.01	—
4028	Egg Beaters	¼	cup(s)	61	—	30	6	1	0	0	0.00	0.00	0.00	0
	SEAFOOD													
	Fish													
	Cod													
6040	Atlantic cod or scrod, baked or broiled	3	ounce(s)	44	34	46	10	0	0	<1	0.07	0.05	0.13	—
1573	Atlantic cod, cooked, dry heat	3	ounce(s)	85	65	89	19	0	0	1	0.14	0.11	0.25	—
2905	**Eel, raw**	3	ounce(s)	85	58	156	16	0	0	10	2.01	6.12	0.81	—
	Fish fillets													
25079	Baked	3	ounce(s)	84	80	99	22	0	0	1	0.08	0.07	0.26	—
8615	Batter coated or breaded, fried	3	ounce(s)	85	0.04	197.19	12.46	14.42	0.42	10.44	2.39	2.19	5.32	—
25082	Broiled fish steaks	3	ounce(s)	86	69	129	24	0	0	3	0.37	0.87	0.84	—
25083	Poached fish steaks	3	ounce(s)	86	68	112	21	0	0	2	0.33	0.76	0.74	—
25084	Steamed fish fillets	3	ounce(s)	86	73	80	17	0	0	1	0.12	0.08	0.22	—
25089	**Flounder, baked**	3	ounce(s)	85	65	114	15	<1	<.1	6	1.15	2.17	1.44	0
1825	**Grouper, cooked, dry heat**	3	ounce(s)	85	62	100	21	0	0	1	0.25	0.23	0.34	—
	Haddock													
6049	Baked or broiled	3	ounce(s)	44	33	50	11	0	0	<1	0.07	0.07	0.14	—
1578	Cooked, dry heat	3	ounce(s)	85	63	95	21	0	0	1	0.14	0.13	0.26	—
1886	**Halibut, Atlantic & Pacific, cooked, dry heat**	3	ounce(s)	85	61	119	23	0	0	2	0.35	0.82	0.80	—
1582	**Herring, Atlantic, pickled**	4	piece(s)	60	33	157	9	6	0	11	1.43	7.17	1.01	—
1587	**Jack mackerel, solids, canned, drained**	2	ounce(s)	57	39	88	13	0	0	4	1.05	1.26	0.94	—
8580	**Octopus, common, cooked, moist heat**	3	ounce(s)	85	51	139	25	4	0	2	0.39	0.28	0.41	—
1831	**Perch, mixed species, cooked, dry heat**	3	ounce(s)	85	62	99	21	0	0	1	0.20	0.17	0.40	—
1592	**Pacific rockfish, cooked, dry heat**	3	ounce(s)	85	62	103	20	0	0	2	0.40	0.38	0.50	—
	Salmon													
29727	Smoked chinook (lox)	2	ounce(s)	57	<.1	66	10	0	0	2	0.52	1.14	0.56	—

PAGE KEY: A–58 = Breads/Baked Goods A–62 = Cereal/Rice/Pasta A–66 = Fruit A–70 = Vegetables/Legumes A–80 = Nuts/Seeds A–82 = Vegetarian A–84 = Dairy A–90 = Eggs A–90 = Seafood A–92 = Meats A–96 = Poultry A–96 = Processed Meats A–98 = Beverages A–102 = Fats/Oils A–104 = Sweets A–106 = Sauces/Condiments/Spices A–108 = Mixed Foods/Soups/Sandwiches A–114 = Fast Food A–130 = Convenience A–132 = Baby Foods

Chol (mg)	Calc (mg)	Iron (mg)	Magn (mg)	Pota (mg)	Sodi (mg)	Zinc (mg)	Vit A (RAE) (μg)	Thia (mg)	Vit E (mg α)	Ribo (mg)	Niac (mg)	Vit B_6 (mg)	Fola (μg)	Vit C (mg)	Vit B_{12} (μg)	Sele (μg)
0	300	1.80	60	350	160	0.60	33	0.15	—	0.07	0.80	0.12	60	0	3	—
0	300	1.80	40	260	140	0.60	33	0.15	—	0.07	0.80	0.12	60	0	3	—
0	300	0.72	24	200	140	0.90	0	0.09	—	0.34	—	—	24	0	1	—
0	300	0.72	24	200	110	0.90	0	0.09	—	0.34	—	—	24	0	1	—
0	300	1.08	40	320	150	0.90	0	0.15	—	0.34	—	—	60	0	1	—
0	40	0.72	—	320	115	—	0	—	—	—	—	—	—	0	—	—
15	200	0.00	16	310	90	—	0	—	—	0.26	—	—	—	0	—	—
15	200	0.00	16	300	90	—	0	—	—	0.26	—	—	—	0	—	—
12	338	0.15	32	434	130	1.64	27	0.08	0.05	0.40	0.21	0.09	22	1	1	7
3	370	0.62	41	550	139	1.83	0.10	0.17	0.17	0.50	0.11	0.09	26	1		
15	448	0.20	42	573	172	2.18	34	0.11	0.05	0.52	0.28	0.12	27	2	1	8
5	488	0.22	47	625	189	2.38	5	0.12	0.07	0.57	0.30	0.13	29	2	1	9
12	419	0.17	39	537	162	2.03	29	0.10	0.00	0.49	0.26	0.11	27	2	1	12
13	260	0.22	39	399	98	1.10	—	0.11	0.05	0.51	0.30	0.15	29	2	2	—
5	300	2.70	100	580	290	2.25	—	0.38	—	0.43	5.00	0.50	100	15	2	—
0	700	0.90	—	—	30	—	—	—	—	—	—	—	—	0	0	—
0	100	0.72	—	—	20	—	0	—	—	—	—	—	—	0	—	—
0	100	0.72	—	—	20	—	0	—	—	—	—	—	—	0	—	—
212	27	0.92	6	67	70	0.56	70	0.03	0.49	0.24	0.04	0.07	24	0	1	16
0	2	0.03	4	54	55	0.01	0	0.00	0.00	0.15	0.04	0.00	1	0	<.1	7
205	21	0.45	1	18	8	0.38	63	0.03	0.43	0.09	0.00	0.06	24	0	<1	9
210	27	0.91	6	68	94	0.55	91	0.03	0.56	0.24	0.04	0.07	23	0	1	16
212	25	0.60	5	63	62	0.53	85	0.03	0.51	0.26	0.03	0.06	22	0	1	15
211	27	0.92	6	67	147	0.55	70	0.03	0.48	0.24	0.04	0.07	24	0	1	16
429	87	1.46	15	168	342	1.22	174	0.06	1.04	0.53	0.10	0.14	37	<1	1	27
1	44	1.19	9	128	119	0.59	7	0.07	0.95	0.23	0.08	0.08	10	<1	<1	25
1	33	1.32	6	207	111	0.82	11	0.07	0.17	0.19	0.07	0.00	9	0	<1	16
0	20	1.08	4	85	115	0.60	113	0.15	—	0.85	0.20	0.08	60	0	1	—
24	6	0.22	19	108	35	0.26	—	0.04	—	0.03	1.11	0.13	5	<1	<1	17
47	12	0.42	36	207	66	0.49	12	0.07	0.69	0.07	2.14	0.24	7	1	1	32
107	17	0.43	17	231	43	1.38	887	0.13	3.40	0.03	2.98	0.06	13	2	3	6
44	8	0.32	29	489	86	0.49	10	0.03	—	0.05	2.48	0.46	8	3	1	44
28.89	15.3	1.79	20.39	272	452.2	0.37	10.19	0.09	—	0.09	1.78	0.08	17	0	0.94	7.73
37	55	0.99	98	529	64	0.49	55	0.06	—	0.08	6.88	0.36	13	0	1	43
33	48	0.86	85	460	55	0.43	48	0.06	—	0.08	5.97	0.33	12	0	1	37
42	13	0.30	25	323	42	0.35	12	0.07	—	0.06	1.92	0.22	6	1	1	32
44	19	0.35	47	225	281	0.21	39	0.06	0.41	0.08	2.03	0.19	7	3	2	34
40	18	0.97	31	404	45	0.43	43	0.07	—	0.01	0.32	0.30	9	0	1	40
33	19	0.60	22	177	39	0.21	—	0.02	—	0.02	2.05	0.15	4	0	1	18
63	36	1.15	43	339	74	0.41	16	0.03	—	0.04	3.94	0.29	11	0	1	34
35	51	0.91	91	490	59	0.45	46	0.06	—	0.08	6.05	0.34	12	0	1	40
8	46	0.73	5	41	522	0.32	155	0.02	1.03	0.08	1.98	0.10	1	0	3	35
45	137	1.16	21	110	215	0.58	74	0.02	0.58	0.12	3.50	0.12	3	1	4	21
82	90	8.11	51	536	391	2.86	77	0.05	1.02	0.06	3.21	0.55	20	7	31	76
98	87	0.99	32	292	67	1.22	9	0.07	—	0.10	1.62	0.12	5	1	2	14
37	10	0.45	29	442	65	0.45	60	0.04	1.33	0.07	3.33	0.23	9	0	1	40
13	6	0.48	10	99	1134	0.17	15	0.01	—	0.05	2.67	0.15	1	0	2	22

TABLE F-1

Food Composition (Computer code number is for Wadsworth Diet Analysis program) (For purposes of calculations, use "0" for t, <1, <.1, <.01, etc.)

DA + Code	Food Description	Quantity	Measure	Wt (g)	H_2O (g)	Ener (kcal)	Prot (g)	Carb (g)	Dietary Fiber (g)	Fat (g)	Fat Breakdown (g) Sat	Mono	Poly	Trans
	SEAFOOD—Continued													
1594	Broiled or baked w/butter	3	ounce(s)	85	54	155	23	0	0	6	1.16	2.29	2.33	—
2938	Coho, farmed, raw	3	ounce(s)	85	60	136	18	0	0	7	1.54	2.83	1.58	—
154	**Sardines, Atlantic, with bones, canned in oil**	2	item(s)	24	<.1	50	6	0	0	3	0.36	0.92	1.23	—
	Scallops													
155	Mixed species, breaded, fried	3	item(s)	47	<.1	100	8	5	0	5	1.24	2.09	1.32	—
1599	Steamed	3	ounce(s)	85	65	90	14	2	0	3	—	—	—	—
1839	**Snapper, mixed species, cooked, dry heat**	3	ounce(s)	85	60	109	22	0	0	1	0.31	0.27	0.50	—
	Squid													
1868	Mixed species, fried	3	ounce(s)	85	55	149	15	7	0	6	1.60	2.34	1.82	—
16617	Steamed or boiled	3	ounce(s)	85	63	90	15	3	0	1	0.35	0.11	0.51	—
1570	**Striped bass, cooked, dry heat**	3	ounce(s)	85	62	105	19	0	0	3	0.55	0.72	0.85	—
1601	**Sturgeon, steamed**	3	ounce(s)	85	59	111	17	0	0	4	0.97	2.04	0.73	—
1840	**Surimi, formed**	3	ounce(s)	85	65	84	13	6	0	1	0.16	0.13	0.38	—
1842	**Swordfish, cooked, dry heat**	3	ounce(s)	85	58	132	22	0	0	4	1.20	1.68	1.00	—
1846	**Tuna, yellowfin or ahi, raw**	3	ounce(s)	85	60	92	20	0	0	1	0.20	0.13	0.24	—
	Tuna, canned													
159	Light, canned in oil, drained	2	ounce(s)	57	34	113	17	0	0	5	0.87	1.68	1.64	—
355	Light, canned in water, drained	2	ounce(s)	57	42	66	14	0	0	<1	0.13	0.09	0.19	—
33211	Light, no salt, canned in oil, drained	2	ounce(s)	57	34	112	17	0	0	5	0.87	1.67	1.64	—
33212	Light, no salt, canned in water, drained	2	ounce(s)	57	43	66	14	0	0	<1	0.13	0.09	0.19	—
2961	White, canned in oil, drained	2	ounce(s)	57	36	105	15	0	0	5	0.73	1.85	1.69	—
351	White, canned in water, drained	2	ounce(s)	57	41	73	13	0	0	2	0.45	0.44	0.63	—
33213	White, no salt, canned in oil, drained	2	ounce(s)	57	36	105	15	0	0	5	0.94	1.41	1.92	—
33214	White, no salt, canned in water, drained	2	ounce(s)	57	42	73	13	0	0	2	0.45	0.44	0.63	—
	Yellowtail													
2970	Mixed species, raw	2	ounce(s)	57	42	83	13	0	0	3	0.73	1.13	0.81	—
8548	Mixed species, cooked, dry heat	3	ounce(s)	85	0.05	158.94	25.21	0	0	5.71	1.44	2.21	1.52	—
	Shellfish, meat only													
1857	Abalone, mixed species, fried	3	ounce(s)	85	51	161	17	9	0	6	1.40	2.33	1.42	—
16618	Abalone, steamed or poached	3	ounce(s)	85	41	177	29	10	0	1	0.25	0.18	0.18	—
	Crab													
1851	Blue crab, canned	2	ounce(s)	57	43	56	12	0	0	1	0.14	0.12	0.25	—
1852	Blue crab, cooked, moist heat	3	ounce(s)	85	66	87	17	0	0	2	0.19	0.24	0.58	—
8562	Dungeness crab, cooked, moist heat	3	ounce(s)	85	62	94	19	1	0	1	0.14	0.18	0.35	—
1860	**Clams, cooked, moist heat**	3	ounce(s)	85	54	126	22	4	0	2	0.16	0.15	0.47	—
1853	**Crayfish, farmed, cooked, moist heat**	3	ounce(s)	85	69	74	15	0	0	1	0.18	0.21	0.35	—
	Oysters													
8720	Baked or broiled	3	ounce(s)	85	69	90	6	3	0	6	1.38	2.18	1.88	—
152	Eastern, farmed, raw	3	ounce(s)	85	73	50	4	5	0	1	0.38	0.13	0.50	—
8715	Eastern, wild, cooked, moist heat	3	ounce(s)	85	60	116	12	7	0	4	1.31	0.53	1.65	—
8584	Pacific, cooked, moist heat	3	ounce(s)	85	55	139	16	8	0	4	0.87	0.66	1.52	—
1865	Pacific, raw	3	ounce(s)	85	70	69	8	4	0	2	0.43	0.30	0.76	—
1854	**Lobster, northern, cooked, moist heat**	3	ounce(s)	85	65	83	17	1	0	1	0.09	0.14	0.08	—
1862	**Mussels, blue, cooked, moist heat**	3	ounce(s)	85	52	146	20	6	0	4	0.72	0.86	1.03	—
	Shrimp													
1855	Mixed species, cooked, moist heat	3	ounce(s)	85	66	84	18	0	0	1	0.25	0.17	0.37	—
158	Mixed species, breaded, fried	3	ounce(s)	85	0.04	205.69	18.18	9.74	0.34	10.43	1.77	3.24	4.32	—
	BEEF, LAMB, PORK													
	Beef													
4450	Breakfast strips, cooked	2	slice(s)	23	0	101.47	7.07	0.31	0	7.77	3.24	3.8	0.35	—
174	Corned, canned	3	ounce(s)	85	49	213	23	0	0	13	5.25	5.07	0.54	—
33147	Cured, thin sliced	2	ounce(s)	57	31	87	18	2	0	1	0.54	0.48	0.04	—
4581	Jerky	1	ounce(s)	28	0	116.44	9.42	3.12	0.51	7.27	3.08	3.21	0.28	—
	Ground													
4411	Extra lean, broiled, well	3	ounce(s)	85	46	225	24	0	0	13	5.28	5.88	0.50	—
4417	Lean, broiled, medium	3	ounce(s)	85	47	231	21	0	0	16	6.16	6.87	0.59	—
4418	Lean, broiled, well	3	ounce(s)	85	45	238	24	0	0	15	5.89	6.56	0.56	—
4423	Regular, broiled, medium	3	ounce(s)	85	46	246	20	0	0	18	6.91	7.70	0.65	—
	Rib													
4183	Rib, whole, lean & fat, ¼" fat, roasted	3	ounce(s)	85	39	320	19	0	0	27	10.71	11.42	0.94	—
	Roast													
4264	Bottom round, lean & fat, ¼" fat, braised	3	ounce(s)	85	44	241	24	0	0	15	5.71	6.63	0.58	—

PAGE KEY: A–58 = Breads/Baked Goods A–62 = Cereal/Rice/Pasta A–66 = Fruit A–70 = Vegetables/Legumes A–80 = Nuts/Seeds A–82 = Vegetarian A–84 = Dairy A–90 = Eggs A–90 = Seafood A–92 = Meats A–96 = Poultry A–96 = Processed Meats A–98 = Beverages A–102 = Fats/Oils A–104 = Sweets A–106 = Sauces/Condiments/Spices A–108 = Mixed Foods/Soups/Sandwiches A–114 = Fast Food A–130 = Convenience A–132 = Baby Foods

Chol (mg)	Calc (mg)	Iron (mg)	Magn (mg)	Pota (mg)	Sodi (mg)	Zinc (mg)	Vit A (RAE) (µg)	Thia (mg)	Vit E (mg α)	Ribo (mg)	Niac (mg)	Vit B_6 (mg)	Fola (µg)	Vit C (mg)	Vit B_{12} (µg)	Sele (µg)
40	15	1.02	27	377	99	0.56	—	0.14	1.15	0.05	8.33	0.19	4	2	2	41
43	10	0.29	26	383	40	0.37	48	0.08	—	0.09	5.79	0.56	11	1	2	11
34	108	0.70	9	95	121	0.31	16	0.01	0.49	0.05	1.25	0.04	3	0	2	13
28	20	0.38	27	155	216	0.49	10	0.01	—	0.05	0.69	0.06	23	1	1	13
27	21	0.22	—	238	366	—	—	—	0.16	—	—	—	—	2	—	—
40	34	0.20	31	444	48	0.37	30	0.05	—	0.00	0.29	0.39	5	1	3	42
221	33	0.86	32	237	260	1.48	9	0.05	—	0.39	2.21	0.05	12	4	1	44
227	31	0.63	29	192	356	1.49	—	0.02	1.17	0.32	1.70	0.04	4	3	1	—
88	16	0.92	43	279	75	0.43	26	0.10	—	0.03	2.17	0.29	9	0	4	40
63	11	0.59	30	239	389	0.36	—	0.07	0.53	0.07	8.31	0.19	14	0	2	—
26	8	0.22	37	95	122	0.28	17	0.02	0.54	0.02	0.19	0.03	2	0	1	24
43	5	0.88	29	314	98	1.25	35	0.04	—	0.10	10.02	0.32	2	1	2	52
38	14	0.62	43	378	31	0.44	15	0.37	0.43	0.04	8.33	0.77	2	1	<1	31
10	7	0.79	18	118	202	0.51	13	0.02	0.50	0.07	7.06	0.06	3	0	1	43
17	6	0.87	15	134	192	0.44	10	0.02	0.19	0.04	7.53	0.20	2	0	2	46
10	7	0.79	18	117	28	0.51	13	0.02	—	0.07	7.03	0.06	3	0	1	43
17	6	0.87	15	134	28	0.44	10	0.02	—	0.04	7.53	0.20	2	0	2	46
18	2	0.37	19	189	225	0.27	3	0.01	1.30	0.04	6.63	0.24	3	0	1	34
24	8	0.55	19	134	214	0.27	3	0.00	0.48	0.02	3.29	0.12	1	0	1	37
18	2	0.37	19	189	28	0.27	14	0.01	—	0.04	6.63	0.24	3	0	1	34
24	8	0.55	19	134	28	0.27	3	0.00	—	0.02	3.29	0.12	1	0	1	37
31	13	0.28	17	238	22	0.29	16	0.08	—	0.02	3.86	0.09	2	2	1	21
60.34	24.64	0.53	32.29	457.29	42.5	0.56	26.35	0.14	—	0.04	7.41	0.15	3.4	2.46	1.06	39.77
80	31	3.23	48	241	502	0.81	2	0.19	—	0.11	1.62	0.13	12	2	1	44
143	50	4.85	69	295	980	1.38	—	0.29	6.74	0.13	1.90	0.22	6	3	1	—
50	57	0.48	22	212	189	2.28	1	0.05	1.04	0.05	0.78	0.09	24	2	<1	18
85	88	0.77	28	275	237	3.59	2	0.09	1.56	0.04	2.81	0.15	43	3	6	34
65	50	0.37	49	347	321	4.65	26	0.05	—	0.17	3.08	0.15	36	3	9	40
57	78	23.77	15	534	95	2.32	145	0.13	—	0.36	2.85	0.09	25	19	84	54
116	43	0.94	28	202	82	1.26	13	0.04	—	0.07	1.42	0.11	9	<1	3	29
42	37	5.30	38	126	418	72.22	60	0.07	0.99	0.06	1.04	0.05	8	3	15	—
21	37	4.91	28	105	151	32.23	7	0.09	—	0.06	1.08	0.05	15	4	14	54
89	77	10.19	81	239	359	154.37	46	0.16	—	0.15	2.11	0.10	12	5	30	61
85	14	7.82	37	257	180	28.25	124	0.11	0.72	0.38	3.08	0.08	13	11	24	131
43	7	4.35	19	143	90	14.14	69	0.06	—	0.20	1.71	0.04	9	7	14	65
61	52	0.33	30	299	323	2.48	22	0.01	0.85	0.06	0.91	0.07	9	0	3	36
48	28	5.71	31	228	314	2.27	77	0.26	—	0.36	2.55	0.09	65	12	20	76
166	33	2.63	29	155	190	1.33	58	0.03	1.17	0.03	2.20	0.11	3	2	1	34
150.44	56.95	1.07	34	191.25	292.39	1.17	47.59	0.1	—	0.11	2.6	0.08	20.39	1.27	1.58	35.44
26.89	2.03	0.7	6.1	93.11	509.17	1.43	0	0.02	0.07	0.05	1.46	0.07	1.8	0	0.77	6.05
73	10	1.77	12	116	855	3.03	0	0.02	0.13	0.12	2.07	0.11	8	0	1	36
45	3	1.58	11	140	1582	2.49	0	0.03	0.00	0.12	1.85	0.16	5	0	1	13
13.63	5.67	1.53	14.48	169.54	628.49	2.3	0	0.04	0.14	0.04	0.49	0.05	38.05	0	0.28	3.03
84	8	2.35	21	314	70	5.47	0	0.06	—	0.27	4.97	0.27	9	0	2	19
74	9	1.79	18	256	65	4.56	0	0.04	—	0.18	4.39	0.22	8	0	2	25
86	10	2.08	20	297	76	5.27	0	0.05	—	0.20	5.07	0.26	9	0	2	22
77	9	2.07	17	248	71	4.40	0	0.03	—	0.16	4.90	0.23	8	0	2	16
72	9	1.96	16	252	54	4.45	0	0.06	—	0.14	2.86	0.20	6	0	2	19
82	5	2.65	19	240	43	4.17	0	0.06	0.17	0.20	3.17	0.28	9	0	2	27

TABLE F–1

Food Composition (Computer code number is for Wadsworth Diet Analysis program) (For purposes of calculations, use "0" for t, <1, <.1, <.01, etc.)

DA + Code	Food Description	Quantity	Measure	Wt (g)	H_2O (g)	Ener (kcal)	Prot (g)	Carb (g)	Dietary Fiber (g)	Fat (g)	Fat Breakdown (g) Sat	Mono	Poly	Trans
	BEEF, LAMB, PORK—Continued													
169	Bottom round, separable lean, ¼" fat, roasted	3	ounce(s)	85	0.05	160.64	24.45	0	0	6.26	2.13	2.83	0.24	—
4147	Chuck, arm pot roast, lean & fat, ¼" fat, braised	3	ounce(s)	85	41	282	23	0	0	20	7.97	8.68	0.77	—
4161	Chuck, blade roast, lean & fat, ¼" fat, braised	3	ounce(s)	85	40	293	23	0	0	22	8.70	9.44	0.78	—
5853	Chuck, blade roast, separable lean, ¼" trim, pot roasted	3	ounce(s)	85	0.04	209.1	27.45	0	0	10.15	3.94	4.37	0.33	—
4295	Eye of round, lean, ¼" fat, roasted	3	ounce(s)	85	55	149	25	0	0	5	1.76	2.06	0.15	—
4285	Eye of round, lean & fat, ¼" fat, roasted	3	ounce(s)	85	51	195	23	0	0	11	4.23	4.66	0.39	—
	Steak													
1757	Rib, small end, lean, ¼" fat, broiled	3	ounce(s)	85	49	188	24	0	0	10	3.84	4.01	0.27	—
4349	Short loin, T-bone steak, lean, ¼" fat, broiled	3	ounce(s)	85	52	174	23	0	0	9	3.05	4.23	0.26	—
4348	Short loin, T-bone steak, lean & fat, ¼" fat, broiled	3	ounce(s)	85	43	274	19	0	0	21	8.29	9.58	0.75	—
4360	Top loin, prime, lean & fat, ¼" fat, broiled	3	ounce(s)	85	43	275	22	0	0	20	8.16	8.61	0.73	—
	Variety													
188	Liver, pan fried	3	ounce(s)	85	53	149	23	4	0	4	1.27	0.56	0.49	0.17
4447	Tongue, simmered	3	ounce(s)	85	49	236	16	0	0	19	6.91	8.59	0.56	0.71
	Lamb													
	Chop													
3275	Loin, domestic, lean & fat, ¼" fat, broiled	3	ounce(s)	85	44	269	21	0	0	20	8.36	8.25	1.43	—
3287	Shoulder, arm, domestic, lean & fat, ¼" fat, braised	3	ounce(s)	85	38	294	26	0	0	20	8.39	8.65	1.45	—
3290	Shoulder, arm, domestic, lean, ¼" fat, braised	3	ounce(s)	85	42	237	30	0	0	12	4.28	5.24	0.78	—
	Leg													
3264	Domestic, lean & fat, ¼" fat, cooked	3	ounce(s)	85	46	250	21	0	0	18	7.51	7.50	1.28	—
	Rib													
183	Domestic, lean, ¼" fat, broiled	3	ounce(s)	85	50	200	24	0	0	11	3.95	4.43	1.00	—
182	Domestic, lean & fat, ¼" fat, broiled	3	ounce(s)	85	40	307	19	0	0	25	10.80	10.30	2.01	—
	Shoulder													
187	Arm & blade, domestic, choice, lean, ¼" fat, roasted	3	ounce(s)	85	54	173	21	0	0	9	3.47	3.71	0.81	—
186	Arm & blade, domestic, choice, lean & fat, ¼" fat, roasted	3	ounce(s)	85	48	235	19	0	0	17	7.17	6.94	1.38	—
	Variety													
3375	Brain, pan fried	3	ounce(s)	85	52	232	14	0	0	19	4.82	3.42	1.94	—
3406	Tongue, braised	3	ounce(s)	85	49	234	18	0	0	17	6.66	8.50	1.06	—
	Pork													
	Cured													
161	Bacon, cured, broiled, pan fried or roasted	2	slice(s)	13	2	68	5	<1	0	5	1.73	2.33	0.57	0
29229	Bacon, Canadian style, cured	2	ounce(s)	57	38	89	12	1	0	4	1.26	1.79	0.36	—
35422	Breakfast strips, cured, cooked	3	slice(s)	34	9	156	10	<1	0	12	4.34	5.58	1.92	—
16561	Ham, smoked or cured, lean, cooked	1	slice(s)	42	28	66	11	0	0	2	0.77	1.06	0.27	—
189	Ham, cured, boneless, 11% fat, roasted	3	ounce(s)	85	55	151	19	0	0	8	2.65	3.77	1.20	—
1316	Ham, cured, extra lean, 5% fat, roasted	3	ounce(s)	85	58	123	18	1	0	5	1.54	2.23	0.46	—
29215	Ham, cured, extra lean, 4% fat, canned	2	ounce(s)	57	42	68	10	0	0	3	0.86	1.25	0.22	—
	Chop													
32671	Loin, blade, lean & fat, pan fried	3	ounce(s)	85	42	291	18	0	0	24	8.65	9.97	2.64	—
32672	Loin, center cut, lean & fat, pan fried	3	ounce(s)	85	45	236	25	0	0	14	5.11	6.00	1.62	—
32682	Loin, center rib, boneless, lean & fat, braised	3	ounce(s)	85	49	217	22	0	0	13	5.21	6.13	1.12	—
32603	Loin, center rib, lean, broiled	3	ounce(s)	85	48	186	26	0	0	8	2.94	3.78	0.53	—
32481	Loin, whole, lean, braised	3	ounce(s)	85	52	174	24	0	0	8	2.87	3.54	0.60	—
32478	Loin, whole, lean & fat, braised	3	ounce(s)	85	50	203	23	0	0	12	4.35	5.15	1.00	—
	Leg or ham													
32471	Rump portion, lean & fat, roasted	3	ounce(s)	85	48	214	25	0	0	12	4.47	5.42	1.17	—
32468	Whole, lean & fat, roasted	3	ounce(s)	85	47	232	23	0	0	15	5.50	6.70	1.43	—
	Ribs													
32696	Loin, country style, lean, roasted	3	ounce(s)	85	49	210	23	0	0	13	4.52	5.49	0.94	—
32693	Loin, country style, lean & fat, roasted	3	ounce(s)	85	43	279	20	0	0	22	7.83	9.36	1.71	—
	Shoulder													
32629	Arm picnic, lean, roasted	3	ounce(s)	85	51	194	23	0	0	11	3.66	5.09	1.02	—

PAGE KEY: A–58 = Breads/Baked Goods A–62 = Cereal/Rice/Pasta A–66 = Fruit A–70 = Vegetables/Legumes A–80 = Nuts/Seeds A–82 = Vegetarian A–84 = Dairy A–90 = Eggs A–90 = Seafood A–92 = Meats A–96 = Poultry A–96 = Processed Meats A–98 = Beverages A–102 = Fats/Oils A–104 = Sweets A–106 = Sauces/Condiments/Spices A–108 = Mixed Foods/Soups/Sandwiches A–114 = Fast Food A–130 = Convenience A–132 = Baby Foods

Chol (mg)	Calc (mg)	Iron (mg)	Magn (mg)	Pota (mg)	Sodi (mg)	Zinc (mg)	Vit A (RAE) (μg)	Thia (mg)	Vit E (mg α)	Ribo (mg)	Niac (mg)	Vit B_6 (mg)	Fola (μg)	Vit C (mg)	Vit B_{12} (μg)	Sele (μg)
66.3	4.25	2.66	23.79	332.35	56.09	3.92	0	0.06	—	0.2	3.45	0.31	10.19	0	2.29	23.29
84	9	2.64	16	209	51	5.81	0	0.06	0.19	0.20	2.70	0.24	8	0	3	21
88	11	2.64	16	196	54	7.07	0	0.06	0.15	0.20	2.06	0.22	4	0	2	21
73.94	11.05	3.12	19.54	223.55	60.34	8.72	0	0.06	—	0.23	0	0.24	—	0	2.09	22.69
59	4	1.66	23	336	53	4.03	0	0.08	—	0.14	3.19	0.32	6	0	2	23
61	5	1.56	20	308	50	3.69	0	0.07	0.15	0.14	2.97	0.30	6	0	2	22
68	11	2.18	23	335	59	5.94	0	0.09	0.12	0.19	4.08	0.34	7	0	3	19
50	5	3.11	22	278	65	4.34	0	0.09	0.12	0.21	3.94	0.33	7	0	2	9
58	7	2.56	18	234	58	3.56	0	0.08	0.19	0.18	3.29	0.28	6	0	2	10
67	8	1.89	20	294	54	3.85	0	0.07	—	0.15	3.96	0.31	6	0	2	19
324	5	5.24	19	298	65	4.45	6582	0.15	0.39	2.91	14.85	0.87	221	1	71	28
112	4	2.22	13	156	55	34.77	0	0.02	0.25	0.25	2.97	0.13	6	1	3	11
85	17	1.54	20	278	65	2.96	0	0.09	0.11	0.21	6.04	0.11	15	0	2	23
102	21	2.03	22	260	61	5.17	0	0.06	0.13	0.21	5.66	0.09	15	0	2	32
103	22	2.30	25	287	65	6.21	0	0.06	0.15	0.23	5.38	0.11	19	0	2	32
82	14	1.60	20	264	61	3.79	0	0.09	0.12	0.21	5.66	0.11	15	0	2	22
77	14	1.88	25	266	72	4.48	0	0.09	0.15	0.21	5.57	0.13	18	0	2	26
84	16	1.60	20	230	65	3.40	0	0.08	0.10	0.19	5.95	0.09	12	0	2	20
74	16	1.81	21	225	58	5.13	0	0.08	0.15	0.22	4.90	0.13	21	0	2	24
78	17	1.67	20	213	56	4.45	0	0.08	0.12	0.20	5.23	0.11	18	0	2	22
2128	18	1.73	19	304	133	1.70	0	0.14	—	0.31	3.87	0.20	6	20	20	10
161	9	2.24	14	134	57	2.54	0	0.07	—	0.36	3.14	0.14	3	6	5	24
14	1	0.18	4	71	291	0.44	1	0.05	0.04	0.03	1.40	0.04	<1	0	<1	8
28	5	0.39	10	195	799	0.79	0	0.43	0.12	0.10	3.53	0.22	2	0	<1	14
36	5	0.67	9	158	714	1.25	0	0.25	0.09	0.13	2.58	0.12	1	0	1	8
23	3	0.40	9	133	557	1.08	0	0.29	0.11	0.11	2.11	0.20	2	0	<1	—
50	7	1.14	19	348	1275	2.10	0	0.62	0.26	0.28	5.23	0.26	3	0	1	17
45	7	1.26	12	244	1023	2.45	0	0.64	0.21	0.17	3.42	0.34	3	0	1	17
22	3	0.53	10	206	712	1.09	0	0.47	0.10	0.13	3.01	0.26	3	0	<1	8
72	26	0.75	18	282	57	2.71	3	0.53	0.17	0.25	3.36	0.29	3	1	1	30
78	23	0.77	25	361	68	1.96	2	0.97	0.21	0.26	4.76	0.40	5	1	1	33
62	4	0.78	14	329	34	1.76	2	0.45	0.21	0.21	3.67	0.26	3	<1	<1	28
69	26	0.70	24	357	55	2.02	2	0.95	0.25	0.28	5.25	0.40	3	<1	1	40
67	15	0.96	17	329	43	2.11	2	0.56	0.18	0.23	3.90	0.33	3	1	<1	41
68	18	0.91	16	318	41	2.02	2	0.54	0.20	0.22	3.76	0.31	3	1	<1	39
82	10	0.89	23	318	53	2.40	3	0.64	0.19	0.28	3.96	0.27	3	<1	1	40
80	12	0.86	19	299	51	2.52	3	0.54	0.19	0.27	3.89	0.34	9	<1	1	39
79	25	1.10	20	297	25	3.24	2	0.49	—	0.29	3.97	0.37	4	<1	1	36
78	21	0.90	20	293	44	2.01	3	0.76	—	0.29	3.67	0.38	4	<1	1	32
81	8	1.21	17	299	68	3.46	2	0.49	—	0.30	3.67	0.35	4	<1	1	33

TABLE F-1
Food Composition (Computer code number is for Wadsworth Diet Analysis program) (For purposes of calculations, use "0" for t, <1, <.1, <.01, etc.)

DA + Code	Food Description	Quantity	Measure	Wt (g)	H_2O (g)	Ener (kcal)	Prot (g)	Carb (g)	Dietary Fiber (g)	Fat (g)	Fat Breakdown (g) Sat	Mono	Poly	Trans
	BEEF, LAMB, PORK—Continued													
32626	Arm picnic, lean & fat, roasted	3	ounce(s)	85	44	270	20	0	0	20	7.47	9.12	2.00	—
	Rabbit													
3366	Domesticated, roasted	3	ounce(s)	85	52	167	25	0	0	7	2.04	1.84	1.33	—
3367	Domesticated, stewed	3	ounce(s)	85	50	175	26	0	0	7	2.13	1.93	1.39	—
	Veal													
3391	Liver, braised	3	ounce(s)	85	51	163	24	3	0	5	1.69	0.97	0.88	0.26
3319	Rib, lean only, roasted	3	ounce(s)	85	55	150	22	0	0	6	1.77	2.26	0.57	—
1732	**Deer or venison, roasted**	3	ounce(s)	85	55	134	26	0	0	3	1.06	0.75	0.53	—
	POULTRY													
	Chicken													
29562	Flaked, canned	2	ounce(s)	57	0.03	97.47	10.37	0.05	0	5.87	1.62	2.32	1.29	—
	Fried													
29632	Breast, meat only, breaded, baked or fried	3	ounce(s)	85	44	193	25	7	<1	7	1.62	2.66	1.73	—
35327	Broiler breast, meat only, fried	3	ounce(s)	85	51	159	28	<1	0	4	1.10	1.46	0.91	—
36413	Broiler breast, meat & skin, flour coated, fried	3	ounce(s)	85	48	189	27	1	<.1	8	2.08	2.98	1.67	—
35389	Broiler drumstick, meat only, fried	3	ounce(s)	85	53	166	24	0	0	7	1.81	2.50	1.68	—
36414	Broiler drumstick, meat & skin, flour coated, fried	3	ounce(s)	85	48	208	23	1	<.1	12	3.11	4.61	2.75	—
35406	Broiler leg, meat only, fried	3	ounce(s)	85	52	177	24	1	0	8	2.12	2.92	1.89	—
35484	Broiler wing, meat only, fried	3	ounce(s)	85	51	179	26	0	0	8	2.13	2.62	1.76	—
29580	Patty, fillet, or tenders, breaded, cooked	3	ounce(s)	85	42	241	14	13	<1	15	4.62	7.25	1.87	—
	Roasted, meat only													
35409	Broiler chicken leg	3	ounce(s)	85	55	162	23	0	0	7	1.95	2.59	1.68	—
35486	Broiler chicken wing	3	ounce(s)	85	53	173	26	0	0	7	1.92	2.22	1.51	—
35138	Roasting chicken, dark meat	3	ounce(s)	85	57	151	20	0	0	7	2.07	2.82	1.70	—
35136	Roasting chicken, light meat	3	ounce(s)	85	58	130	23	0	0	3	0.92	1.29	0.79	—
35132	Roasting chicken	3	ounce(s)	85	57	142	21	0	0	6	1.54	2.13	1.28	—
	Stewed													
3174	Meat only, stewed	3	ounce(s)	85	0.05	150.44	23.19	0	0	5.7	1.56	2.03	1.3	—
1268	Gizzard, simmered	3	ounce(s)	85	58	124	26	0	0	2	0.57	0.45	0.30	0.11
1270	Liver, simmered	3	ounce(s)	85	57	142	21	1	0	6	1.75	1.20	1.08	0.08
	Duck													
1286	Domesticated, meat & skin, roasted	3	ounce(s)	85	44	286	16	0	0	24	8.22	10.97	3.10	—
1287	Domesticated, meat only, roasted	3	ounce(s)	85	55	171	20	0	0	10	3.54	3.15	1.22	—
	Goose													
35507	Domesticated, meat & skin, roasted	3	ounce(s)	85	44	259	21	0	0	19	5.84	8.72	2.14	—
35524	Domesticated, meat only, roasted	3	ounce(s)	85	49	202	25	0	0	11	3.88	3.69	1.31	—
1297	Liver pâté, smoked, canned	4	tablespoon(s)	52	19	240	6	2	0	23	7.51	13.32	0.44	—
	Turkey													
3256	Ground turkey, cooked	3	ounce(s)	85	51	200	23	0	0	11	2.88	4.16	2.75	—
222	Roasted, fryer roaster breast, meat only	3	ounce(s)	85	58	115	26	0	0	1	0.20	0.11	0.17	—
219	Roasted, dark meat, meat only	3	ounce(s)	85	54	159	24	0	0	6	2.06	1.39	1.84	—
220	Roasted, light meat, meat only	3	ounce(s)	85	56	133	25	0	0	3	0.88	0.48	0.73	—
3263	Patty, batter coated, breaded, fried	1	item(s)	94	47	266	13	15	<1	17	4.41	7.02	4.43	—
1302	Turkey roll, light meat	2	slice(s)	57	41	83	11	<1	0	4	1.15	1.42	0.99	—
1303	Turkey roll, light & dark meat	2	slice(s)	57	40	84	10	1	0	4	1.16	1.30	1.01	—
	PROCESSED MEATS													
	Beef													
1331	Corned beef loaf, jellied, sliced	2	slice(s)	57	39	87	13	0	0	3	1.47	1.52	0.18	—
	Bologna													
13458	Made w/chicken, pork, & beef	1	slice(s)	28	15	90	3	1	0	8	3.00	4.05	1.10	—
13461	Light, made w/pork, chicken, & beef	1	slice(s)	28	18	60	3	2	0	4	1.50	2.04	0.43	—
13459	Beef	1	slice(s)	28	15	90	3	1	0	8	3.50	4.26	0.31	—
13565	Turkey	1	slice(s)	28	19	50	3	1	0	4	1.00	1.09	0.98	—
	Chicken													
13562	Oven roasted white chicken	1	slice(s)	28	20	40	4	1	0	3	0.50	—	—	—
	Ham													
13581	Honey glazed, traditional carved	2	slice(s)	45	—	50	8	1	0	2	0.50	0.68	0.18	—
13777	Deli sliced cooked	1	slice(s)	28	—	30	5	1	0	1	0.50	0.39	0.11	—
13778	Deli sliced honey	1	slice(s)	28	—	35	5	1	0	1	0.50	0.39	0.11	—
8614	**Pork & beef mortadella, sliced**	2	slice(s)	46	24	143	8	1	0	12	4.37	5.23	1.44	—

PAGE KEY: A–58 = Breads/Baked Goods A–62 = Cereal/Rice/Pasta A–66 = Fruit A–70 = Vegetables/Legumes A–80 = Nuts/Seeds A–82 = Vegetarian A–84 = Dairy A–90 = Eggs A–90 = Seafood A–92 = Meats A–96 = Poultry A–96 = Processed Meats A–98 = Beverages A–102 = Fats/Oils A–104 = Sweets A–106 = Sauces/Condiments/Spices A–108 = Mixed Foods/Soups/Sandwiches A–114 = Fast Food A–130 = Convenience A–132 = Baby Foods

Chol (mg)	Calc (mg)	Iron (mg)	Magn (mg)	Pota (mg)	Sodi (mg)	Zinc (mg)	Vit A (RAE) (µg)	Thia (mg)	Vit E (mg α)	Ribo (mg)	Niac (mg)	Vit B_6 (mg)	Fola (µg)	Vit C (mg)	Vit B_{12} (µg)	Sele (µg)
80	16	1.00	14	276	60	2.93	2	0.44	—	0.26	3.33	0.30	3	<1	1	29
70	16	1.93	18	326	40	1.93	0	0.08	—	0.18	7.17	0.40	9	0	7	33
73	17	2.01	17	255	31	2.01	0	0.05	0.37	0.14	6.09	0.29	8	0	6	33
434	5	4.34	17	280	66	9.55	17973	0.15	0.58	2.43	11.18	0.78	281	1	72	16
98	10	0.82	20	264	82	3.82	0	0.05	0.31	0.25	6.38	0.23	12	0	1	9
95	6	3.80	20	285	46	2.34	0	0.15	—	0.51	5.70	—	—	0	—	11
35.34	7.98	0.9	6.84	148.19	410.39	0.8	19.37	0	—	0.07	3.6	0.19	—	0	0.16	—
67	19	1.05	25	223	450	0.84	—	0.08	—	0.10	10.98	0.47	4	0	<1	—
77	14	0.97	26	235	67	0.92	—	0.07	—	0.11	12.57	0.54	3	0	<1	22
76	14	1.01	26	220	65	0.94	—	0.07	—	0.11	11.69	0.49	5	0	<1	20
80	10	1.12	20	212	82	2.74	—	0.07	—	0.20	5.23	0.33	8	0	<1	17
77	10	1.14	20	195	76	2.46	—	0.07	—	0.19	5.13	0.30	9	0	<1	16
84	11	1.19	21	216	82	2.53	—	0.07	—	0.21	5.69	0.33	8	0	<1	16
71	13	0.97	18	177	77	1.80	—	0.04	—	0.11	6.16	0.50	3	0	<1	22
51	14	1.06	17	209	452	0.88	—	0.08	—	0.12	5.71	0.26	9	<1	<1	—
80	10	1.11	20	206	77	2.43	—	0.06	—	0.20	5.37	0.32	7	0	<1	19
72	14	0.99	18	179	78	1.82	—	0.04	—	0.11	6.22	0.50	3	0	<1	21
64	9	1.13	17	191	81	1.81	14	0.05	—	0.16	4.88	0.26	6	0	<1	17
64	11	0.92	20	201	43	0.66	7	0.05	0.23	0.08	8.90	0.46	3	0	<1	22
64	10	1.03	18	195	64	1.29	10	0.05	—	0.13	6.70	0.35	4	0	<1	21
70.55	11.89	0.99	17.85	153	59.5	1.69	12.75	0.04	0.23	0.13	5.19	0.22	5.09	0	0.18	17.76
315	14	2.71	3	152	48	3.76	0	0.02	0.17	0.18	2.65	0.06	4	0	1	35
479	9	9.89	21	224	65	3.38	3384	0.25	0.70	1.69	9.39	0.64	491	24	14	70
71	9	2.30	14	173	50	1.58	54	0.15	0.59	0.23	4.10	0.15	5	0	<1	17
76	10	2.30	17	214	55	2.21	20	0.22	0.59	0.40	4.34	0.21	9	0	<1	19
77	11	2.41	19	280	60	2.23	18	0.07	—	0.28	3.55	0.32	2	0	<1	19
82	12	2.44	21	330	65	2.70	10	0.08	—	0.33	3.47	0.40	10	0	<1	22
78	36	2.86	7	72	362	0.48	521	0.05	—	0.16	1.31	0.03	31	0	5	23
87	21	1.64	20	230	91	2.43	0	0.05	0.29	0.14	4.10	0.33	6	0	<1	32
71	10	1.30	25	248	44	1.48	0	0.04	0.08	0.11	6.37	0.48	5	0	<1	27
72	27	1.98	20	247	67	3.79	0	0.05	0.54	0.21	3.10	0.31	8	0	<1	35
59	16	1.15	24	259	54	1.73	0	0.05	0.08	0.11	5.81	0.46	5	0	<1	27
58	13	2.07	14	259	752	1.35	10	0.09	1.18	0.18	2.16	0.19	26	0	<1	19
24	23	0.73	9	142	277	0.88	0	0.05	0.07	0.13	3.97	0.18	2	0	<1	13
31	18	0.77	10	153	332	1.13	0	0.05	0.19	0.16	2.72	0.15	3	0	<1	17
27	6	1.16	6	57	540	2.32	0	0.00	—	0.06	1.00	0.07	5	0	1	10
30	0	0.36	6	43	290	0.40	0	—	—	—	—	—	—	0	—	—
15	0	0.36	6	46	310	0.45	0	—	—	—	—	—	—	0	—	—
20	0	0.36	4	47	310	0.57	0	0.01	—	0.03	0.68	0.05	4	0	<1	—
20	40	0.36	6	43	270	0.52	0	—	—	—	—	—	—	0	—	—
15	0	0.36	7	85	350	0.32	0	—	—	—	—	—	—	0	—	—
25	0	0.72	—	—	560	—	0	—	—	—	—	—	—	0	—	—
15	0	0.00	—	—	240	—	0	—	—	—	—	—	—	0	—	—
15	0	0.00	—	—	240	—	0	—	—	—	—	—	—	0	—	—
26	8	0.64	5	75	573	0.97	0	0.05	0.10	0.07	1.23	0.06	1	0	1	10

TABLE F–1

Food Composition (Computer code number is for Wadsworth Diet Analysis program) (For purposes of calculations, use "0" for t, <1, <.1, <.01, etc.)

DA + Code	Food Description	Quantity	Measure	Wt (g)	H_2O (g)	Ener (kcal)	Prot (g)	Carb (g)	Dietary Fiber (g)	Fat (g)	Fat Breakdown (g) Sat	Mono	Poly	Trans
	PROCESSED MEATS—Continued													
1323	**Pork olive loaf**	2	slice(s)	57	33	133	7	5	0	9	3.32	4.47	1.10	—
1324	**Pork pickle & pimento loaf**	2	slice(s)	57	32	149	7	3	0	12	4.45	5.45	1.47	—
	Sausages & frankfurters													
37296	Beerwurst beef beer salami (bierwurst)	1	slice(s)	29	17	74	4	1	0	6	2.50	2.69	0.21	—
37257	Beerwurst pork beer salami	1	slice(s)	21	13	50	3	<1	0	4	1.32	1.89	0.50	—
35338	Berliner, pork & beef	1	ounce(s)	28	17	65	4	1	0	5	1.72	2.27	0.45	—
37299	Braunschweiger pork liver sausage	1	slice(s)	15	0	51.34	1.97	0.34	0	4.48	1.52	2.08	0.52	—
37298	Bratwurst pork, cooked	1	piece(s)	74	42	181	10	2	0	14	5.15	6.73	1.51	—
1329	Cheesefurter or cheese smokie, beef & pork	1	item(s)	43	23	141	6	1	0	12	4.52	5.89	1.30	—
1330	Chorizo, beef & pork	2	ounce(s)	57	18	258	14	1	0	22	8.15	10.43	1.96	—
8600	Frankfurter, beef	1	item(s)	45	23	149	5	2	0	13	5.26	6.44	0.53	—
202	Frankfurter, beef & pork	1	item(s)	57	32	174	7	1	1	16	6.14	7.79	1.56	—
1293	Frankfurter, chicken	1	item(s)	45	26	116	6	3	0	9	2.49	3.82	1.82	—
3261	Frankfurter, turkey	1	item(s)	45	28	102	6	1	0	8	2.65	2.51	2.25	—
37275	Italian sausage, pork, cooked	1	item(s)	68	34	220	14	1	0	17	6.14	8.13	2.23	—
37307	Kielbasa, kolbassa, pork & beef	2⅛	ounce(s)	61	37	135	10	2	0	9	3.40	4.44	1.06	—
1333	Knockwurst or knackwurst, beef & pork	2	ounce(s)	57	31	174	6	2	0	16	5.79	7.26	1.66	—
37285	Pepperoni, beef & pork	1	slice(s)	11	3	55	2	<1	0	5	1.77	2.32	0.48	—
37313	Polish sausage, pork	2	slice(s)	57	31	163	8	2	—	14	4.91	6.42	1.46	—
206	Salami, beef, cooked, sliced	2	slice(s)	46	28	119	6	1	0	10	4.54	4.90	0.48	—
37272	Salami, pork, dry or hard	1	slice(s)	13	5	52	3	<1	0	4	1.52	2.05	0.48	—
3262	Salami, turkey	2	slice(s)	57	31	125	8	11	<.1	5	1.98	1.80	1.43	0
7162	Sausage, breakfast, turkey	2½	ounce(s)	100	67	190	17	<1	0	13	3.90	6.23	3.33	0
8620	Smoked sausage, beef & pork	2	ounce(s)	57	31	181	7	1	0	16	5.54	6.94	2.23	0
8619	Smoked, sausage, pork	2	ounce(s)	57	22	221	13	1	0	18	6.42	8.30	2.13	—
37273	Smoked, sausage, pork link	1	piece(s)	76	30	295	17	2	—	24	8.58	11.09	2.85	—
1336	Summer sausage, thuringer, or cervelat, beef & pork	2	ounce(s)	57	29	190	9	<1	0	17	6.82	7.35	0.68	—
37294	Vienna sausage, cocktail, beef & pork, canned	1	piece(s)	16	10	45	2	<1	0	4	1.49	2.01	0.27	—
	Spreads													
32419	Pork & beef sandwich spread	4	tablespoon(s)	60	36	141	5	7	<1	10	3.59	4.57	1.54	—
1318	Ham salad spread	¼	cup(s)	60	38	130	5	6	0	9	3.04	4.32	1.62	—
	Turkey													
16049	Breast, hickory smoked, slices	1	slice(s)	56	—	50	11	1	0	0	0.00	0.00	0.00	0
13606	Breast, hickory smoked fat free	1	slice(s)	28	—	25	4	1	0	0	0.00	0.00	0.00	0
16047	Breast, honey roasted, slices	1	slice(s)	56	—	60	11	2	0	0	0.00	0.00	0.00	0
16048	Breast, oven roasted, slices	1	slice(s)	56	—	50	11	1	0	0	0.00	0.00	0.00	0
13583	Breast, traditional carved	2	slice(s)	45	—	40	9	0	0	1	0.00	0.07	0.14	—
13604	Breast, oven roasted, fat free	1	slice(s)	28	—	25	4	1	0	0	0.00	0.00	0.00	0
13567	Turkey ham, 10% water added	1	slice(s)	28	20	35	5	0	0	1	0.00	0.22	0.31	—
13596	Turkey pastrami	2	ounce(s)	56	—	70	11	1	0	2	1.00	—	—	—
13597	Turkey salami	2	ounce(s)	56	—	120	8	1	0	9	2.50	2.92	2.30	—
	BEVERAGES													
	Alcoholic													
	Beer													
866	Ale, mild	12	fluid ounce(s)	360	332	148	1	13	1	0	0.00	0.00	0.00	—
686	Beer	12	fluid ounce(s)	356	336	118	1	6	<1	<1	0.00	0.00	0.00	0
869	Beer, light	12	fluid ounce(s)	354	337	99	1	5	0	0	0.00	0.00	0.00	0
16886	Beer, nonalcoholic	12	fluid ounce(s)	360	353	32	1	5	0	0	0.00	0.00	0.00	0
31608	Budweiser beer	12	fluid ounce(s)	355	328	143	1	11	0	0	0.00	0.00	0.00	0
31609	Bud Light beer	12	fluid ounce(s)	355	335	110	1	7	0	0	0.00	0.00	0.00	0
31613	Michelob Beer	12	fluid ounce(s)	355	323	155	1	13	0	0	0.00	0.00	0.00	0
31614	Michelob Light beer	12	fluid ounce(s)	355	330	134	1	12	0	0	0.00	0.00	0.00	0
	Gin, rum, vodka, whiskey													
687	Distilled alcohol, 80 proof	1	fluid ounce(s)	28	19	64	0	0	0	0	0.00	0.00	0.00	0
688	Distilled alcohol, 86 proof	1	fluid ounce(s)	28	18	70	0	<.1	0	0	0.00	0.00	0.00	0
689	Distilled alcohol, 90 proof	1	fluid ounce(s)	28	17	73	0	0	0	0	0.00	0.00	0.00	0
856	Distilled alcohol, 94 proof	1	fluid ounce(s)	28	17	76	0	0	0	0	0.00	0.00	0.00	0
857	Distilled alcohol, 100 proof	1	fluid ounce(s)	28	16	82	0	0	0	0	0.00	0.00	0.00	0
	Liqueurs													
3142	Coffee liqueur, 63 proof	1	fluid ounce(s)	35	14	107	<.1	11	0	<1	0.04	0.01	0.04	—
33187	Coffee liqueur, 53 proof	1	fluid ounce(s)	35	11	117	<.1	16	0	<1	0.04	0.01	0.04	—

PAGE KEY: A–58 = Breads/Baked Goods A–62 = Cereal/Rice/Pasta A–66 = Fruit A–70 = Vegetables/Legumes A–80 = Nuts/Seeds A–82 = Vegetarian A–84 = Dairy A–90 = Eggs A–90 = Seafood A–92 = Meats A–96 = Poultry A–96 = Processed Meats A–98 = Beverages A–102 = Fats/Oils A–104 = Sweets A–106 = Sauces/Condiments/Spices A–108 = Mixed Foods/Soups/Sandwiches A–114 = Fast Food A–130 = Convenience A–132 = Baby Foods

Chol (mg)	Calc (mg)	Iron (mg)	Magn (mg)	Pota (mg)	Sodi (mg)	Zinc (mg)	Vit A (RAE) (µg)	Thia (mg)	Vit E (mg α)	Ribo (mg)	Niac (mg)	Vit B_6 (mg)	Fola (µg)	Vit C (mg)	Vit B_{12} (µg)	Sele (µg)
22	62	0.31	11	169	843	0.78	34	0.17	0.14	0.15	1.04	0.13	1	0	1	9
21	54	0.58	10	193	789	0.80	12	0.17	0.24	0.14	1.17	0.11	3	0	1	8
18	3	0.44	4	67	265	0.71	0	0.02	—	0.04	0.99	0.05	1	0	1	5
12	2	0.16	3	53	261	0.36	0	0.12	—	0.04	0.69	0.07	1	0	<1	—
13	3	0.33	4	80	368	0.70	0	0.11	—	0.06	0.88	0.06	1	0	1	4
23.69	1.36	1.42	1.67	27.49	131.54	0.42	641.01	0.03	—	0.23	1.27	0.05	—	0	3.05	8.81
44	33	0.96	11	157	412	1.70	0	0.37	—	0.14	2.37	0.16	1	1	1	16
29	25	0.46	6	89	465	0.97	20	0.11	0.00	0.07	1.25	0.06	1	0	1	7
50	5	0.90	10	226	700	1.93	0	0.36	0.12	0.17	2.91	0.30	1	0	1	12
24	6	0.68	6	70	513	1.11	0	0.02	0.09	0.07	1.07	0.04	2	0	1	4
29	6	0.66	6	95	638	1.05	10	0.11	0.14	0.07	1.50	0.07	2	0	1	8
45	43	0.90	5	38	617	0.47	18	0.03	0.10	0.05	1.39	0.14	2	0	<1	8
48	48	0.83	6	81	642	1.40	0	0.02	0.28	0.08	1.86	0.10	4	0	<1	7
53	16	1.02	12	207	627	1.62	0	0.42	—	0.16	2.83	0.22	3	1	1	15
41	27	0.88	10	169	566	1.23	0	0.14	—	0.13	1.75	0.11	3	0	1	11
34	6	0.37	6	113	527	0.94	0	0.19	—	0.08	1.55	0.10	1	0	1	8
9	1	0.15	2	38	224	0.28	0	0.04	—	0.03	0.55	0.03	<1	0	<1	—
40	7	0.82	8	102	546	1.10	0	0.29	—	0.08	1.96	0.11	1	1	1	10
33	3	1.01	6	86	524	0.81	0	0.05	0.09	0.09	1.49	0.08	1	0	1	7
10	2	0.17	3	48	289	0.54	0	0.12	—	0.04	0.72	0.07	<1	0	<1	3
45	42	0.87	15	225	616	1.76	1	0.24	0.14	0.17	2.26	0.24	6	12	1	11
92	57	2.20	18	188	665	2.07	0	0.04	0.00	0.12	3.55	0.29	5	1	<1	—
33	7	0.43	7	101	517	0.71	7	0.11	0.07	0.06	1.67	0.09	1	0	<1	0
39	17	0.66	11	191	851	1.60	0	0.40	0.14	0.15	2.57	0.20	3	1	1	12
52	23	0.88	14	255	1137	2.14	0	0.53	—	0.20	3.43	0.27	4	0	1	16
43	7	1.44	8	154	704	1.45	0	0.09	0.12	0.19	2.44	0.15	1	0	3	12
8	2	0.14	1	16	152	0.26	0	0.01	—	0.02	0.26	0.02	1	0	<1	3
23	7	0.47	5	66	608	0.61	16	0.10	1.04	0.08	1.04	0.07	1	0	1	6
22	5	0.35	6	90	547	0.66	0	0.26	1.04	0.07	1.26	0.09	1	0	<1	11
25	0	0.72	—	—	730	—	0	—	—	—	—	—	—	0	—	—
10	0	0.00	—	—	300	—	0	—	—	—	—	—	—	0	—	—
20	0	0.72	—	—	640	—	0	—	—	—	—	—	—	0	—	—
20	0	0.72	—	—	620	—	0	—	—	—	—	—	—	0	—	—
20	0	0.72	—	—	540	—	0	—	—	—	—	—	—	0	—	—
10	0	0.00	—	—	330	—	0	—	—	—	—	—	—	0	—	—
20	0	0.36	6	81	310	0.73	0	—	—	—	—	—	—	0	—	—
40	0	0.72	—	—	590	—	0	—	—	—	—	—	—	0	—	—
50	40	0.72	—	—	500	—	0	—	—	—	—	—	—	0	—	—
0	18	0.11	—	—	18	—	0	0.02	0.00	0.10	1.63	—	—	0	<.1	—
0	18	0.07	21	89	14	0.04	0	0.02	0.00	0.09	1.61	0.18	21	0	<.1	2
0	18	0.14	18	64	11	0.11	0	0.03	0.00	0.11	1.39	0.12	14	0	<.1	2
0	25	0.04	32	90	18	0.04	—	0.02	0.00	0.10	1.63	0.18	22	0	<.1	—
0	18	0.11	21	89	9	0.07	0	0.02	0.00	0.09	1.61	0.18	21	0	<.1	4
0	18	0.14	18	64	9	0.11	0	0.03	0.00	0.11	1.39	0.12	15	0	<.1	4
0	18	0.11	21	89	9	0.07	0	0.02	0.00	0.09	1.61	0.18	21	0	<.1	4
0	18	0.14	18	64	9	0.11	0	0.03	0.00	0.11	1.39	0.12	15	0	<.1	4
0	0	0.01	0	1	<1	0.01	0	0.00	0.00	0.00	0.00	0.00	0	0	0	0
0	0	0.01	0	1	<1	0.01	0	0.00	0.00	0.00	0.00	0.00	0	0	0	0
0	0	0.01	0	1	<1	0.01	0	0.00	0.00	0.00	0.00	0.00	0	0	0	0
0	0	0.01	0	1	<1	0.01	0	0.00	0.00	0.00	0.00	0.00	0	0	0	0
0	0	0.01	0	1	<1	0.01	0	0.00	0.00	0.00	0.00	0.00	0	0	0	0
0	<1	0.02	1	10	3	0.01	0	0.00	—	0.00	0.05	0.00	0	0	0	<1
0	<1	0.02	1	10	3	0.01	0	0.00	0.00	0.00	0.05	0.00	0	0	0	<1

TABLE F-1

Food Composition (Computer code number is for Wadsworth Diet Analysis program) (For purposes of calculations, use "0" for t, <1, <.1, <.01, etc.)

DA + Code	Food Description	Quantity	Measure	Wt (g)	H_2O (g)	Ener (kcal)	Prot (g)	Carb (g)	Dietary Fiber (g)	Fat (g)	Fat Breakdown (g) Sat	Mono	Poly	Trans
	BEVERAGES—Continued													
736	Cordials, 54 proof	1	fluid ounce(s)	30	9	106	<.1	13	0	<.1	0.02	0.01	0.04	—
	Wine													
858	Champagne, domestic	5	fluid ounce(s)	150	—	105	<1	4	0	0	0.00	0.00	0.00	0
861	Red wine, California	5	fluid ounce(s)	150	133	125	<1	4	0	0	0.00	0.00	0.00	0
690	Sweet dessert wine	5	fluid ounce(s)	150	106	240	<1	21	0	0	0.00	0.00	0.00	0
1481	White wine	5	fluid ounce(s)	148	132	100	<1	1	0	0	0.00	0.00	0.00	0
1811	Wine cooler	10	fluid ounce(s)	300	270	150	<1	18	<.1	<.1	0.01	0.00	0.02	—
	Carbonated													
692	Club soda	12	fluid ounce(s)	355	355	0	0	0	0	0	0.00	0.00	0.00	0
12010	Coca-Cola Classic cola soda	12	fluid ounce(s)	360	—	146	0	41	0	0	0.00	0.00	0.00	0
12031	Coke diet cola soda	12	fluid ounce(s)	360	—	2	0	<1	0	0	0.00	0.00	0.00	0
693	Cola	12	fluid ounce(s)	426	380	179	<1	46	0	0	0.00	0.00	0.00	—
9522	Cola soda, decaffeinated	12	fluid ounce(s)	372	331	156	<1	40	0	0	0.00	0.00	0.00	0
1415	Cola, low calorie w/aspartame	12	fluid ounce(s)	355	354	4	<1	<1	0	0	0.00	0.00	0.00	0
9524	Cola, decaffeinated, low calorie w/aspartame	12	fluid ounce(s)	355	354	4	<1	<1	0	0	0.00	0.00	0.00	0
1412	Cream soda	12	fluid ounce(s)	371	321	189	0	49	0	0	0.00	0.00	0.00	0
31899	Diet 7 Up	12	fluid ounce(s)	360	—	0	0	0	0	0	0.00	0.00	0.00	0
695	Ginger ale	12	fluid ounce(s)	366	334	124	0	32	0	0	0.00	0.00	0.00	0
694	Grape soda	12	fluid ounce(s)	372	330	160	0	42	0	0	0.00	0.00	0.00	0
1876	Lemon lime soda	12	fluid ounce(s)	368	330	147	0	38	0	0	0.00	0.00	0.00	—
29392	Mountain Dew diet soda	12	fluid ounce(s)	360	—	0	0	0	0	0	0.00	0.00	0.00	0
29391	Mountain Dew soda	12	fluid ounce(s)	360	—	170	0	46	0	0	0.00	0.00	0.00	0
3145	Orange soda	12	fluid ounce(s)	372	326	179	0	46	0	0	0.00	0.00	0.00	0
1414	Pepper-type soda	12	fluid ounce(s)	368	329	151	0	38	0	<1	0.26	0.00	0.00	—
2391	Pepper-type or cola soda, low calorie w/saccharin	12	fluid ounce(s)	355	354	0	0	<1	0	0	0.00	0.00	0.00	0
29389	Pepsi diet cola soda	12	fluid ounce(s)	360	—	0	0	0	0	0	0.00	0.00	0.00	0
29388	Pepsi regular cola soda	12	fluid ounce(s)	360	—	150	0	41	0	0	0.00	0.00	0.00	0
696	Root beer	12	fluid ounce(s)	370	330	152	0	39	0	0	0.00	0.00	0.00	0
31898	7 Up	12	fluid ounce(s)	360	—	240	0	59	0	0	0.00	0.00	0.00	0
12034	Sprite diet soda	12	fluid ounce(s)	360	—	4	0	0	0	0	0.00	0.00	0.00	0
12044	Sprite soda	12	fluid ounce(s)	360	—	144	0	39	0	0	0.00	0.00	0.00	0
	Coffee													
731	Brewed	8	fluid ounce(s)	237	236	9	<1	0	0	0	0.00	0.00	0.00	0
9520	Brewed, decaffeinated	8	fluid ounce(s)	237	235	5	<1	1	0	0	0.00	0.00	0.00	0
16882	Cappuccino	8	fluid ounce(s)	240	224	78	4	6	<1	4	2.53	1.18	0.15	—
16883	Cappuccino, decaffeinated	8	fluid ounce(s)	240	224	78	4	6	<1	4	2.53	1.18	0.15	—
16880	Espresso	8	fluid ounce(s)	237	235	5	<1	1	0	0	0.00	0.00	0.00	0
16881	Espresso, decaffeinated	8	fluid ounce(s)	237	235	5	<1	1	0	0	0.00	0.00	0.00	0
732	Instant, prepared	8	fluid ounce(s)	239	237	5	<1	1	0	0	0.00	0.00	0.00	0
	Fruit drinks													
29357	Crystal Light low calorie lemonade drink	8	fluid ounce(s)	240	—	5	0	0	0	0	0.00	0.00	0.00	0
6012	Fruit punch drink w/added vitamin C, canned	8	fluid ounce(s)	276	242	129	0	33	<1	<.1	0.01	0.01	0.01	0
260	Grape drink, canned	8	fluid ounce(s)	250	221	113	<.1	29	0	0	0.00	0.00	0.00	0
266	Lemonade, from frozen concentrate	8	fluid ounce(s)	248	213	131	<1	34	<1	<1	0.02	0.00	0.04	—
268	Limeade, from frozen concentrate	8	fluid ounce(s)	247	220	104	<.1	26	0	<.1	0.00	0.00	0.00	—
31143	Gatorade Thirst Quencher, all flavors	8	fluid ounce(s)	240	—	50	0	14	0	0	0.00	0.00	0.00	0
17372	Kool-Aid (lemonade/punch/fruit drink)	8	fluid ounce(s)	248	220	108	<1	28	<1	<.1	0.01	0.01	0.02	—
17225	Kool-Aid sugar free, low calorie tropical punch mix, prepared	8	fluid ounce(s)	240	—	5	0	0	0	0	0.00	0.00	0.00	0
14266	Odwalla strawberry 'c' monster fruit drink	8	fluid ounce(s)	240	—	150	2	34	1	1	0.00	—	—	0
10080	Odwalla strawberry lemonade quencher	8	fluid ounce(s)	240	—	120	1	28	1	0	0.00	0.00	0.00	0
10099	Snapple fruit punch	8	fluid ounce(s)	240	—	110	0	29	0	0	0.00	0.00	0.00	0
10096	Snapple kiwi strawberry	8	fluid ounce(s)	240	211	110	0	28	0	0	0.00	0.00	0.00	0
	Slim Fast ready to drink shake													
16056	Dark chocolate fudge	11	fluid ounce(s)	325	—	220	10	42	5	3	1.00	1.50	0.50	—
16054	French vanilla	11	fluid ounce(s)	325	—	220	10	40	5	3	0.50	1.50	0.50	—
16055	Strawberries n cream	11	fluid ounce(s)	325	—	220	10	40	5	3	0.50	1.50	0.50	—
	Tea													
733	Tea, prepared	8	fluid ounce(s)	237	236	2	0	1	0	0	0.00	0.00	0.01	0
33179	Decaffeinated, prepared	8	fluid ounce(s)	237	236	2	0	1	0	0	0.00	0.00	0.01	0
1877	Herbal, prepared	8	fluid ounce(s)	237	236	2	0	<1	0	0	0.00	0.00	0.01	0
734	Instant tea mix, unsweetened, prepared	8	fluid ounce(s)	237	236	2	<.1	<1	0	0	0.00	0.00	0.00	0

PAGE KEY: A–58 = Breads/Baked Goods A–62 = Cereal/Rice/Pasta A–66 = Fruit A–70 = Vegetables/Legumes A–80 = Nuts/Seeds A–82 = Vegetarian A–84 = Dairy A–90 = Eggs A–90 = Seafood A–92 = Meats A–96 = Poultry A–96 = Processed Meats A–98 = Beverages A–102 = Fats/Oils A–104 = Sweets A–106 = Sauces/Condiments/Spices A–108 = Mixed Foods/Soups/Sandwiches A–114 = Fast Food A–130 = Convenience A–132 = Baby Foods

Chol (mg)	Calc (mg)	Iron (mg)	Magn (mg)	Pota (mg)	Sodi (mg)	Zinc (mg)	Vit A (RAE) (µg)	Thia (mg)	Vit E (mg α)	Ribo (mg)	Niac (mg)	Vit B_6 (mg)	Fola (µg)	Vit C (mg)	Vit B_{12} (µg)	Sele (µg)
0	<1	0.02	<1	5	2	0.01	0	0.00	0.00	0.00	0.02	0.00	0	0	0	—
0	—	—	—	—	—	—	—	—	—	—	0.00	—	—	—	0	—
0	12	1.43	16	171	15	0.15	0	0.02	0.00	0.04	0.12	0.05	1	0	<.1	—
0	12	0.36	14	138	14	0.11	0	0.03	0.00	0.03	0.32	0.00	0	0	0	1
0	13	0.47	15	118	7	0.10	0	0.01	—	0.01	0.10	0.02	0	0	0	<1
0	17	0.81	16	135	25	0.17		0.01	0.03	0.02	0.13	0.04	4	5	<.1	—
0	18	0.04	4	7	75	0.36	0	0.00	0.00	0.00	0.00	0.00	0	0	0	0
0	—	—	—	0	50	—	0	—	—	—	—	—	—	0	—	—
0	—	—	—	18	42	—	0	—	—	—	—	—	—	0	—	—
0	13	0.09	4	4	17	0.04	0	0.00	0.00	0.00	0.00	0.00	0	0	0	<1
0	11	0.07	4	4	15	0.04	0	0.00	0.00	0.00	0.00	0.00	0	0	0	<1
0	11	0.11	4	21	18	0.00	0	0.02	0.00	0.08	0.00	0.00	0	0	0	0
0	14	0.11	4	0	21	0.28	0	0.02	0.00	0.08	0.00	0.00	0	0	0	<1
0	19	0.19	4	4	44	0.26	0	0.00	0.00	0.00	0.00	0.00	0	0	0	0
0	—	—	—	116	53	—	—	—	0.00	—	—	—	—	—	—	—
0	11	0.66	4	4	26	0.18	0	0.00	0.00	0.00	0.00	0.00	0	0	0	<1
0	11	0.30	4	4	56	0.26	0	0.00	0.00	0.00	0.00	0.00	0	0	0	0
0	7	0.26	4	4	41	0.18	0	0.00	0.00	0.00	0.06	0.00	0	0	0	0
0	—	—	—	70	35	—	—	—	—	—	—	—	—	—	—	—
0	—	—	—	0	70	—	—	—	—	—	—	—	—	—	—	—
0	19	0.22	4	7	45	0.37	0	0.00	—	0.00	0.00	0.00	0	0	0	0
0	11	0.15	0	4	37	0.15	0	0.00	—	0.00	0.00	0	0	0	<1	
0	14	0.07	4	14	57	0.11	0	0.00	0.00	0.00	0.00	0.00	0	0	0	<1
0	—	—	—	30	35	—	—	—	—	—	—	—	—	—	—	—
0	—	—	—	0	35	—	—	—	—	—	—	—	—	—	—	—
0	18	0.18	4	4	48	0.26	0	0.00	0.00	0.00	0.00	0.00	0	0	0	<1
0	—	—	—	0	113	—	—	—	—	—	—	—	—	—	—	—
0	—	—	—	110	36	—	0	—	—	—	—	—	—	0	—	—
0	—	—	—	0	71	—	0	—	—	—	—	—	—	0	—	—
0	2	0.02	5	114	2	0.02	0	0.00	0.02	0.12	0.00	0.00	5	0	0	0
0	5	0.12	12	128	5	0.05	0	0.00	0.00	0.00	0.53	0.00	<1	0	0	0
17	152	0.26	22	250	62	0.50	—	0.04	0.10	0.20	0.37	0.05	5	1	<1	—
17	152	0.26	22	250	62	0.50	—	0.04	0.10	0.20	0.37	0.05	5	1	<1	—
0	5	0.12	12	128	5	0.05	0	0.00	0.05	0.00	0.53	0.00	<1	0	0	—
0	5	0.12	12	128	5	0.05	0	0.00	0.05	0.00	0.53	0.00	<1	0	0	—
0	10	0.10	7	72	5	0.02	0	0.00	0.00	0.00	0.56	0.00	0	0	0	<1
0	0	0.00	—	160	20	—	0	—	—	—	—	—	—	0	—	—
0	22	0.58	6	69	61	0.33	—	0.06	0.00	0.06	0.06	0.00	4	99	0	0
0	5	0.45	3	30	15	0.30	0	0.00	0.00	0.01	0.03	0.01	0	85	0	<1
0	10	0.52	5	50	7	0.07	0	0.02	0.02	0.07	0.05	0.02	2	13	0	<1
0	7	0.02	2	22	5	0.02	0	0.00	0.00	0.01	0.02	0.01	2	6	0	<1
0	10	0.18	—	30	110	—	—	—	—	—	—	—	—	1	—	—
0	14	0.46	5	50	31	0.20	—	0.04	—	0.05	0.05	0.01	4	42	0	1
0	0	0.00	—	10	10	—	0	—	—	—	—	—	—	6	—	—
0	20	1.44	—	330	40	—	—	—	—	—	—	—	—	600	0	—
0	20	0.00	—	70	30	—	0	—	—	—	—	—	—	60	0	—
0	0	0.00	—	20	10	—	0	—	—	—	—	—	—	0	0	—
0	0	0.00	—	40	10	—	0	—	—	—	—	—	—	0	0	—
5	400	2.70	140	600	220	2.25	—	0.53	—	0.60	7.00	0.70	120	60	2	18
5	400	2.70	140	600	220	2.25	—	0.53	—	0.60	7.00	0.70	120	60	2	18
5	400	2.70	140	600	220	2.25	—	0.53	—	0.60	7.00	0.70	120	60	2	18
0	0	0.05	7	88	7	0.05	0	0.00	0.00	0.03	0.00	0.00	12	0	0	0
0	0	0.05	7	88	7	0.05	0	0.00	0.00	0.03	0.00	0.00	12	0	0	0
0	5	0.19	2	21	2	0.09	0	0.02	0.00	0.01	0.00	0.00	2	0	0	0
0	7	0.05	5	47	7	0.02	0	0.00	0.00	0.00	0.09	0.00	0	0	0	0

TABLE F-1

Food Composition (Computer code number is for Wadsworth Diet Analysis program) (For purposes of calculations, use "0" for t, <1, <.1, <.01, etc.)

DA + Code	Food Description	Quantity	Measure	Wt (g)	H_2O (g)	Ener (kcal)	Prot (g)	Carb (g)	Dietary Fiber (g)	Fat (g)	Fat Breakdown (g)			
											Sat	Mono	Poly	Trans
	BEVERAGES—Continued													
735	Instant lemon flavored tea mix w/sugar, prepared	8	fluid ounce(s)	259	236	88	<1	22	0	<.1	0.01	0.00	0.02	—
	Water													
1413	Mineral water, carbonated	8	fluid ounce(s)	237	237	0	0	0	0	0	0.00	0.00	0.00	0
33183	Poland spring water, bottled	8	fluid ounce(s)	237	237	0	0	0	0	0	0.00	0.00	0.00	0
1	Tap water	8	fluid ounce(s)	237	237	0	0	0	0	0	0.00	0.00	0.00	—
1879	Tonic water	8	fluid ounce(s)	244	222	83	0	21	0	0	0.00	0.00	0.00	0
	FATS AND OILS													
	Butter													
104	Butter	1	tablespoon(s)	15	2	108	<1	<.1	0	12	6.13	5.00	0.43	—
921	Unsalted	1	tablespoon(s)	15	3	108	<1	<.1	0	12	7.71	3.15	0.46	—
107	Whipped	1	tablespoon(s)	11	2	82	<.1	<.1	0	9	5.76	2.67	0.34	—
944	Whipped, unsalted	1	tablespoon(s)	11	2	82	<.1	<.1	0	9	5.76	2.67	0.34	—
2522	Butter Buds, dry butter substitute	1	teaspoon(s)	2	—	8	0	2	0	0	0.00	0.00	0.00	0
	Fats, cooking													
2671	Beef tallow, semisolid	1	tablespoon(s)	13	0	115	0	0	0	13	6.37	5.35	0.51	—
922	Chicken fat	1	tablespoon(s)	13	<.1	115	0	0	0	13	3.81	5.72	2.68	—
5454	Household shortening w/vegetable oil	1	tablespoon(s)	13	0	115	0	0	0	13	3.39	5.56	2.75	2.20
111	Lard	1	tablespoon(s)	13	0	114	0	0	0	13	4.94	5.68	1.41	—
	Margarine													
114	Margarine	1	tablespoon(s)	14	2	101	<1	<1	0	11	2.23	5.05	3.58	—
116	Soft	1	tablespoon(s)	14	2	101	<1	<.1	0	11	1.95	4.02	4.88	—
117	Soft, unsalted	1	tablespoon(s)	14	3	101	<1	<1	0	11	1.95	5.26	3.62	—
928	Unsalted	1	tablespoon(s)	14	3	101	<.1	<.1	0	11	2.12	5.17	3.53	—
119	Whipped	1	tablespoon(s)	9	1	64	<.1	<.1	0	7	1.17	3.25	2.51	—
	Spreads													
16164	I Can't Believe It's Not Butter! whipped spread	1	tablespoon(s)	14	4	60	0	0	0	7	1.50	1.50	2.50	—
16157	Promise vegetable oil spread, stick	1	tablespoon(s)	14	4	90	0	0	0	10	2.50	2.00	4.00	—
	Oils													
2681	Canola	1	tablespoon(s)	14	0	120	0	0	0	14	0.97	8.01	4.03	—
120	Corn	1	tablespoon(s)	14	0	120	0	0	0	14	1.73	3.29	7.98	0.04
122	Olive	1	tablespoon(s)	14	0	119	0	0	0	14	1.82	9.98	1.35	—
124	Peanut	1	tablespoon(s)	14	0	119	0	0	0	14	2.28	6.24	4.32	—
2693	Safflower	1	tablespoon(s)	14	0	120	0	0	0	14	0.84	10.15	1.95	—
923	Sesame	1	tablespoon(s)	14	0	120	0	0	0	14	1.93	5.40	5.67	—
130	Soybean w/cottonseed oil	1	tablespoon(s)	14	0	120	0	0	0	14	2.45	4.01	6.54	—
128	Soybean, hydrogenated	1	tablespoon(s)	14	0	120	0	0	0	14	2.03	5.85	5.11	—
2700	Sunflower	1	tablespoon(s)	14	0	120	0	0	0	14	1.77	6.28	4.95	—
357	**Pam original no stick cooking spray**	1	serving(s)	0	—	0	0	0	0	0	0.00	0.00	0.00	—
	Salad dressing													
132	Blue cheese	2	tablespoon(s)	31	10	154	1	2	0	16	3.03	3.76	8.51	—
133	Blue cheese, low calorie	2	tablespoon(s)	32	25	32	2	1	0	2	0.82	0.57	0.78	—
1764	Caesar	2	tablespoon(s)	30	10	158	<1	1	<.1	17	2.64	4.05	9.86	—
29654	Creamy, reduced calorie, fat free, cholesterol free, sour cream and/or buttermilk & oil	2	tablespoon(s)	32	24	34	<1	6	0	1	0.16	0.21	0.46	—
29617	Creamy, reduced calorie, sour cream and/or buttermilk & oil	2	tablespoon(s)	30	22	48	<1	2	0	4	0.63	0.98	2.40	—
134	French	2	tablespoon(s)	31	11	143	<1	5	0	14	1.76	2.63	6.56	—
135	French, low fat	2	tablespoon(s)	33	18	76	<1	10	<1	4	0.36	1.92	1.64	—
136	Italian	2	tablespoon(s)	29	17	86	<1	3	0	8	1.32	1.86	3.80	—
137	Italian, diet	2	tablespoon(s)	30	25	23	<1	1	0	2	0.14	0.66	0.51	—
139	Mayonnaise type	2	tablespoon(s)	29	12	115	<1	7	0	10	1.44	2.65	5.29	—
942	Oil & vinegar	2	tablespoon(s)	31	15	140	0	1	0	16	2.84	4.62	7.52	—
1765	Ranch	2	tablespoon(s)	30	12	146	<1	2	<.1	16	2.32	3.85	8.92	—
3666	Ranch, reduced calorie	2	tablespoon(s)	30	21	62	<1	2	<.1	6	1.13	1.79	2.89	—
940	Russian	2	tablespoon(s)	31	11	151	<1	3	0	16	2.23	3.61	9.00	—
939	Russian, low calorie	2	tablespoon(s)	33	21	46	<1	9	<.1	1	0.20	0.29	0.75	—
941	Sesame seed	2	tablespoon(s)	31	12	136	1	3	<1	14	1.90	3.64	7.68	—
142	Thousand island	2	tablespoon(s)	31	15	115	<1	5	<1	11	1.59	2.46	5.68	—
143	Thousand island, low calorie	2	tablespoon(s)	31	19	62	<1	7	<1	4	0.23	1.98	0.82	—
	Sandwich spreads													
138	Mayonnaise w/soybean oil	1	tablespoon(s)	14	2	99	<1	1	0	11	1.64	2.70	5.89	0.04

PAGE KEY: A–58 = Breads/Baked Goods A–62 = Cereal/Rice/Pasta A–66 = Fruit A–70 = Vegetables/Legumes A–80 = Nuts/Seeds A–82 = Vegetarian A–84 = Dairy A–90 = Eggs A–90 = Seafood A–92 = Meats A–96 = Poultry A–96 = Processed Meats A–98 = Beverages A–102 = Fats/Oils A–104 = Sweets A–106 = Sauces/Condiments/Spices A–108 = Mixed Foods/Soups/Sandwiches A–114 = Fast Food A–130 = Convenience A–132 = Baby Foods

Chol (mg)	Calc (mg)	Iron (mg)	Magn (mg)	Pota (mg)	Sodi (mg)	Zinc (mg)	Vit A (RAE) (μg)	Thia (mg)	Vit E (mg α)	Ribo (mg)	Niac (mg)	Vit B_6 (mg)	Fola (μg)	Vit C (mg)	Vit B_{12} (μg)	Sele (μg)
0	5	0.05	5	49	8	0.03	0	0.00	0.00	0.04	0.09	0.01	0	<1	0	<1
0	33	0.00	0	0	2	0.00	0	0.00	—	0.00	0.00	0.00	0	0	0	0
0	2	0.02	2	0	2	0.00	0	0.00	—	0.00	0.00	0.00	0	0	0	0
0	4.74	0.00	2.37	0	4.74	0	0	0	0.57	0	0	0	0	0	0	0
0	2	0.02	0	0	10	0.24	0	0.00	0.00	0.00	0.00	0.00	0	0	0	0
32	4	0.00	<1	4	86	0.01	103	0.00	0.35	0.01	0.01	0.00	<1	0	<.1	<1
32	4	0.00	<1	4	2	0.01	103	0.00	0.35	0.01	0.01	0.00	<1	0	<.1	<1
25	3	0.02	<1	3	94	0.01	78	0.00	0.26	0.00	0.00	0.00	<1	0	<.1	<1
25	3	0.02	<1	3	1	0.01	—	0.00	0.26	0.00	0.01	0.00	<1	0	<.1	—
0	0	0.00	0	2	70	0.00	0	0.00	0.00	0.00	0.00	0.00	<1	0	0	—
14	0	0.00	0	0	0	0.00	0	0.00	0.35	0.00	0.00	0.00	0	0	0	<.1
11	0	0.00	0	0	0	0.00	0	0.00	0.35	0.00	0.00	0.00	0	0	0	<.1
0	0	0.00	0	0	0	0.00	0	0.00	—	0.00	0.00	0.00	0	0	0	—
12	0	0.00	0	0	0	0.01	0	0.00	0.08	0.00	0.00	0.00	0	0	0	<.1
0	4	0.01	<1	6	133	0.00	115	0.00	1.27	0.01	0.00	0.00	<1	<.1	<.1	0
0	4	0.00	<1	5	152	0.00	103	0.00	0.99	0.00	0.00	0.00	<1	<.1	<.1	0
0	4	0.00	<1	5	4	0.00	103	0.00	1.23	0.00	0.00	0.00	<1	<.1	<.1	0
0	2	0.00	<1	4	<1	0.00	115	0.00	1.80	0.00	0.00	0.00	<1	<.1	<.1	0
0	2	0.00	<1	3	97	0.00	—	0.00	0.45	0.00	0.00	0.00	<.1	<.1	<.1	—
0	10	0.18	—	4	70	—	—	1.65	0.00	0.00	0.00	—	—	1	—	—
0	10	0.18	—	9	90	—	—	0.00	—	0.00	0.00	—	—	1	—	—
0	0	0.00	0	0	0	0.00	0	0.00	2.33	0.00	0.00	0.00	0	0	0	0
0	0	0.00	0	0	0	0.00	0	0.00	1.94	0.00	0.00	0.00	0	0	0	0
0	<1	0.09	0	<1	<1	0.00	0	0.00	1.94	0.00	0.00	0.00	0	0	0	0
0	0	0.00	0	0	0	0.00	0	0.00	2.12	0.00	0.00	0.00	0	0	0	0
0	0	0.00	0	0	0	0.00	0	0.00	4.64	0.00	0.00	0.00	0	0	0	0
0	0	0.00	0	0	0	0.00	0	0.00	0.19	0.00	0.00	0.00	0	0	0	0
0	0	0.00	0	0	0	0.00	0	0.00	1.65	0.00	0.00	0.00	0	0	0	0
0	0	0.00	0	0	0	0.00	0	0.00	1.10	0.00	0.00	0.00	0	0	0	0
0	0	0.00	0	0	0	0.00	0	0.00	—	0.00	0.00	0.00	0	0	0	0
0	0	0.00	—	0	0	—	0	—	0.00	—	—	—	—	0	0	—
5	25	0.06	0	11	335	0.08	21	0.00	1.84	0.03	0.03	0.01	9	1	<.1	<1
<1	28	0.16	2	2	384	0.08	—	0.01	0.08	0.03	0.02	0.01	1	<.1	<.1	—
1	7	0.05	1	9	323	0.03	—	0.00	1.57	0.00	0.01	0.00	1	0	<.1	—
0	12	0.08	2	43	320	0.06	0	0.00	0.21	0.02	0.01	0.01	1	0	0	—
0	2	0.04	1	11	307	0.01	—	0.00	0.72	0.00	0.01	0.01	4	<1	<.1	—
0	7	0.25	2	21	261	0.09	7	0.01	1.56	0.02	0.06	0.00	0	0	<.1	0
0	4	0.28	3	35	262	0.07	9	0.01	0.10	0.02	0.15	0.02	1	0	0	1
0	2	0.19	1	14	486	0.04	1	0.00	1.47	0.01	0.00	0.02	0	0	0	1
2	3	0.20	1	26	410	0.06	<1	0.00	0.06	0.00	0.00	0.02	0	0	0	2
8	4	0.06	1	3	209	0.05	19	0.00	0.61	0.01	0.00	0.00	2	0	<.1	<1
0	0	0.00	0	2	<1	0.00	0	0.00	1.44	0.00	0.00	0.00	0	0	0	0
1	4	0.03	1	8	354	0.01	—	0.00	1.85	0.00	0.00	0.00	<1	<.1	<.1	—
<1	5	0.01	1	8	414	0.02	—	0.00	0.73	0.01	0.01	0.00	<1	<1	<.1	—
6	6	0.18	1	48	266	0.13	5	0.02	1.02	0.02	0.18	0.01	3	2	<.1	<1
2	6	0.20	0	51	283	0.03	1	0.00	0.13	0.00	0.00	0.00	1	2	<.1	1
0	6	0.18	0	48	306	0.03	1	0.00	1.53	0.00	0.00	0.00	0	0	0	<1
8	5	0.37	2	33	269	0.08	3	0.45	1.25	0.02	0.13	0.00	0	0	0	<1
<1	5	0.28	2	62	254	0.06	5	0.01	0.31	0.01	0.13	0.00	0	0	0	0
5	2	0.07	<1	5	78	0.02	12	0.00	0.72	0.00	0.00	0.08	1	0	<.1	<1

TABLE F–1
Food Composition
(Computer code number is for Wadsworth Diet Analysis program) (For purposes of calculations, use "0" for t, <1, <.1, <.01, etc.)

DA + Code	Food Description	Quantity	Measure	Wt (g)	H_2O (g)	Ener (kcal)	Prot (g)	Carb (g)	Dietary Fiber (g)	Fat (g)	Fat Breakdown (g) Sat	Mono	Poly	Trans
	FATS AND OILS—Continued													
2708	Mayonnaise w/soybean & safflower oils	1	tablespoon(s)	14	0	98.94	0.15	0.37	0	10.95	1.18	1.79	7.59	—
140	Mayonnaise, low calorie	1	tablespoon(s)	16	10	37	<.1	3	0	3	0.53	0.72	1.70	—
141	Tartar sauce	2	tablespoon(s)	28	9	144	<1	4	<.1	14	2.14	4.13	7.57	—
	SWEETS													
4799	**Butterscotch or caramel topping**	2	tablespoon(s)	41	13	103	1	27	<1	<.1	0.05	0.01	0.00	—
	Candy													
1786	Almond Joy candy bar	1	item(s)	49	5	240	2	29	2	13	9.00	3.63	0.74	0
1785	Bit-o-Honey candy	6	item(s)	40	2	170	1	34	0	3	2.00	0	20	—
33375	Butterscotch candy	2	piece(s)	12	1	47	<.1	11	0	<1	0.25	0.10	0.01	—
1701	Chewing gum, stick	1	item(s)	3	<.1	7	0	2	<.1	<.1	0.00	0.00	0.00	—
33378	Chocolate fudge w/nuts, prepared	2	piece(s)	38	3	175	2	26	1	7	2.29	1.41	2.81	—
1787	Jelly beans	15	item(s)	43	3	159	0	40	<.1	<.1	0.00	0.00	0.00	—
1784	Kit Kat wafer bar	1	item(s)	42	1	220	3	27	1	11	7.00	3.53	0.34	0
4674	Krackel candy bar	1	item(s)	41	1	220	3	26	1	11	6.00	3.94	0.37	0
4934	Licorice	4	piece(s)	44	7	147	1	34	1	1	0.18	0.07	0.00	—
1780	Life Savers candy	1	item(s)	2	—	8	0	2	0	<.1	0.00	—	—	0
1790	Lollipop	1	item(s)	28	—	108	0	28	0	0	0.00	0.00	0.00	0
4679	M & Ms peanut chocolate candy, small bag	1	item(s)	49	1	250	5	30	2	13	5.00	5.42	2.07	—
1781	M & Ms plain chocolate candy, small bag	1	item(s)	48	1	240	2	34	1	10	6.00	3.30	0.30	—
4673	Milk chocolate bar	1	item(s)	91	1	483	8	53	2	28	16.69	7.20	0.63	—
1783	Milky Way bar	1	item(s)	58	4	270	2	41	1	10	5.00	3.50	0.35	—
1788	Peanut brittle	1½	ounce(s)	43	<1	206	3	30	1	8	1.76	3.43	1.94	—
1789	Reese's peanut butter cups	2	piece(s)	45	1	250	5	25	1	14	5.00	6.17	2.34	0
4689	Reese's pieces candy, small bag	1	item(s)	46	1	230	6	26	1	11	7.00	0.97	0.46	0
33399	Semisweet chocolate candy, made w/butter	½	ounce(s)	14	<.1	68	1	9	1	4	2.49	1.41	0.13	—
1782	Snickers bar	1	item(s)	59	3	280	4	35	1	14	5.00	6.13	2.89	—
4694	Special Dark chocolate bar	1	item(s)	41	<1	220	2	24	3	13	8.00	4.59	0.41	0
4695	Starburst fruit chews, original fruits	1	package	59	4	240	0	48	0	5	1.00	2.10	1.83	—
4698	Taffy	3	piece(s)	45	2	169	<.1	41	0	1	0.92	0.43	0.05	—
4699	Three Musketeers bar	1	item(s)	60	4	260	2	46	1	8	4.50	2.59	0.27	—
4702	Twix caramel cookie bars	2	item(s)	58	2	280	3	37	1	14	5.00	7.75	0.49	—
4705	York peppermint pattie	1	item(s)	42	4	170	1	34	1	3	2.00	1.32	0.12	0
	Frosting, icing													
4760	Chocolate frosting, ready to eat	2	tablespoon(s)	28	5	112	<1	18	<1	5	1.55	2.54	0.60	—
4771	Creamy vanilla frosting, ready to eat	2	tablespoon(s)	28	4	118	0	19	<.1	5	0.84	1.37	2.24	0
17291	Dec-a-Cake variety pack candy decoration	1	teaspoon(s)	4	—	15	0	3	0	1	0.00	—	—	—
536	White icing	2	tablespoon(s)	40	3	163	<1	32	0	4	0.86	2.07	1.19	—
	Gelatin													
13697	Gelatin snack, all flavors	1	item(s)	99	97	70	1	17	0	0	0.00	0.00	0.00	0
2616	Mixed fruit gelatin mix, sugar free, low calorie, prepared	½	cup(s)	121	—	10	1	0	0	0	0.00	0.00	0.00	0
548	**Honey**	1	tablespoon(s)	21	4	64	<.1	17	<.1	0	0.00	0.00	0.00	0
	Jams, Jellies													
23054	Jams, jellies, preserves, all flavors	1	tablespoon(s)	20	<.1	56	<.1	14	<1	<.1	0.00	0.01	0.00	—
23278	Jams, jellies, preserves, all flavors, low sugar	1	tablespoon(s)	18	<.1	25	<.1	6	<1	<.1	0.00	0.01	0.02	—
545	**Marshmallows**	4	item(s)	29	5	92	1	23	<.1	<.1	0.02	0.02	0.01	—
4800	**Marshmallow cream topping**	2	tablespoon(s)	28	6	91	<1	22	<.1	<.1	0.02	0.02	0.01	—
555	**Molasses**	1	tablespoon(s)	20	4	58	0	15	0	<.1	0.00	0.01	0.01	—
4780	**Popsicle or ice pop**	1	item(s)	59	47	42	0	11	0	0	0.00	0.00	0.00	—
	Sugar													
559	Brown, packed	1	teaspoon(s)	5	<.1	17	0	4	0	0	0.00	0.00	0.00	0
563	Powdered, sifted	⅓	cup(s)	33	<.1	130	0	33	0	<.1	0.01	0.01	0.02	—
561	White granulated	1	teaspoon(s)	4	<.1	15	0	4	0	0	0.00	0.00	0.00	—
	Sugar Substitute													
1760	Equal sweetener, packet	1	item(s)	1	<.1	4	<.1	1	0	0	0.00	0.00	0.00	0
13029	Splenda granular no calorie sweetener	1	teaspoon(s)	1	—	2	0	1	0	0	0.00	0.00	0.00	0
1759	Sweet n Low sugar substitute, packet	1	item(s)	1	<.1	4	0	1	0	0	0.00	0.00	0.00	0
	Syrup													
3148	Chocolate	2	tablespoon(s)	38	12	105	1	24	1	<1	0.19	0.11	0.01	—
29676	Maple	¼	cup(s)	80	26	209	0	54	0	<1	0.03	0.05	0.08	—
4795	Pancake	¼	cup(s)	80	30	187	0	49	1	0	0.00	0.00	0.00	0

PAGE KEY: A–58 = Breads/Baked Goods A–62 = Cereal/Rice/Pasta A–66 = Fruit A–70 = Vegetables/Legumes A–80 = Nuts/Seeds A–82 = Vegetarian A–84 = Dairy A–90 = Eggs A–90 = Seafood A–92 = Meats A–96 = Poultry A–96 = Processed Meats A–98 = Beverages A–102 = Fats/Oils A–104 = Sweets A–106 = Sauces/Condiments/Spices A–108 = Mixed Foods/Soups/Sandwiches A–114 = Fast Food A–130 = Convenience A–132 = Baby Foods

Chol (mg)	Calc (mg)	Iron (mg)	Magn (mg)	Pota (mg)	Sodi (mg)	Zinc (mg)	Vit A (RAE) (μg)	Thia (mg)	Vit E (mg α)	Ribo (mg)	Niac (mg)	Vit B_6 (mg)	Fola (μg)	Vit C (mg)	Vit B_{12} (μg)	Sele (μg)
8.14	2.48	0.06	0.13	4.69	78.38	0.01	11.59	0	3.04	0	0	0.07	1.1	0	0.03	0.22
4	<.1	0.00	<.1	2	80	0.02	0	0.00	0.32	0.00	0.00	0.00	0	0	0	—
11	6	0.21	1	10	200	0.05	—	0.00	0.97	0.00	0.01	0.08	2	<1	<.1	—
<1	22	0.08	3	34	143	0.08	11	0.00	—	0.04	0.02	0.01	1	<1	<.1	0
3	20	0.36	33	138	70	0.40	0	0.02	—	0.08	0.24	—	—	0	—	—
0.00	—	—	85	—	0	—	—	—	—	—	—	0	—	—		
1	<1	0.00	<1	<1	47	0.00	3	0.00	0.01	0.00	0.00	0.00	0	0	0	<.1
0	0	0.00	0	<.1	<.1	0.00	0	0.00	0.00	0.00	0.00	0.00	0	0	0	<.1
5	21	0.75	21	68	16	0.54	14	0.03	0.10	0.04	0.12	0.03	6	<.1	<.1	1
0	1	0.06	1	16	21	0.02	0	0.00	0.00	0.00	0.00	0.00	0	0	0	<1
3	40	0.36	16	126	25	0.52	8	0.07	—	0.23	1.07	0.05	60	0	<.1	2
3	60	0.37	—	169	80	—	0	—	—	—	—	—	—	0	—	—
0	3	0.13	3	28	109	0.07	0	0.01	0.08	0.02	0.04	0.00	0	0	0	—
0	<1	0.04	—	0	1	—	0	0.00	—	0.00	0.00	—	—	0	—	0
0	0	0.00	—	—	11	—	0	0.00	—	0.00	0.00	—	—	0	—	1
5	40	0.36	36	171	25	1.13	15	0.03	—	0.07	1.60	0.04	17	1	<.1	2
5	40	0.36	20	127	30	0.46	15	0.03	—	0.07	0.11	0.01	3	1	<1	1
22	228	0.83	61	399	92	1.00	20	0.06	—	0.26	0.15	0.10	11	2	<1	—
5	60	0.18	20	140	95	0.41	15	0.02	—	0.07	0.20	0.03	6	1	<1	3
5	11	0.52	18	71	189	0.37	17	0.06	1.09	0.02	1.13	0.03	20	0	<.1	1
3	20	0.36	40	233	140	0.82	7	0.11	—	0.08	2.08	0.07	25	0	<.1	2
0	40	0.00	20	182	90	0.35	25	0.04	—	0.07	1.31	0.03	13	0	<.1	1
3	5	0.44	16	52	2	0.23	<1	0.01	—	0.01	0.06	0.01	<1	0	0	<1
5	40	0.36	42	—	140	1.38	15	0.03	—	0.07	1.60	0.05	23	1	<.1	3
3	0	0.72	46	136	0	0.60	0	0.01	—	0.03	0.16	0.01	1	0	0	1
0	10	0.18	1	1	0	0.00	—	0.00	—	0.00	0.00	0.00	0	30	0	<1
4	1	0.03	<1	2	40	0.02	—	0.00	—	0.01	0.01	0.00	0	0	<.1	—
5	20	0.36	18	80	110	0.33	14	0.02	—	0.03	0.20	0.01	0	1	<1	2
5	40	0.36	18	117	115	0.45	15	0.09	—	0.13	0.69	0.02	14	1	<1	1
0	0	0.36	25	71	10	0.31	0	0.01	—	0.04	0.34	0.01	2	0	<.1	—
0	2	0.40	6	55	51	0.08	0	0.00	0.44	0.00	0.03	0.00	<1	0	0	<1
0	1	0.04	<1	10	52	0.02	0	0.00	0.43	0.08	0.06	0.00	2	0	0	<.1
0	0	0.00	—	—	15	—	0	—	—	—	—	—	—	0	—	—
<1	5	0.02	—	7	92	—	—	0.00	0.33	0.01	0.00	—	—	<.1	—	—
0	0	0.00	—	0	40	—	0	—	—	—	—	—	—	0	—	—
0	0	0.00	0	0	50	0.00	0	0.00	0.00	0.00	0.00	0.00	0	0	0	—
0	1	0.09	<1	11	1	0.05	0	0.00	0.00	0.01	0.03	0.01	<1	<1	0	<1
0	4	0.10	1	15	6	0.01	0.00	0.00	0.00	0.02	0.01	0.00	2.20	1.76	0.00	—
0	2	0.05	1	19	<1	0.02	0.76	0.00	0.01	0.01	0.03	0.01	—	4.93	0.00	—
0	1	0.07	1	1	23	0.01	0	0.00	0.00	0.00	0.02	0.00	<1	0	0	<1
0	1	0.06	1	1	23	0.01	0	0.00	0.00	0.00	0.02	0.00	<1	0	0	1
0	41	0.94	48	293	7	0.06	0	0.01	0.00	0.00	0.19	0.13	0	0	0	4
0	0	0.00	1	2	7	0.01	0	0.00	0.00	0.00	0.00	0.00	0	0	0	0
0	4	0.09	1	16	2	0.01	0	0.00	0.00	0.00	0.00	0.00	<.1	0	0	<.1
0	<1	0.01	0	1	<1	0.00	0	0.00	0.00	0.01	0.00	0.00	0	0	0	<1
0	<.1	0.00	0	<.1	0	0.00	0	0.00	0.00	0.00	0.00	0.00	0	0	0	<.1
0	0	0.00	0	0	0	0.00	0	0.00	0.00	0.00	0.00	0.00	0	0	0	0
0	10	0.18	—	—	<1	—	—	0.02	0.00	0.02	0.20	—	—	1	0	—
0	0	0.00	0	—	0	0.00	0	0.00	0.00	0.00	0.00	0.00	0	0	0	0
0	5	0.79	24	84	27	0.27	0	0.00	0.00	0.02	0.12	0.00	1	<.1	0	1
0	54	0.96	11	163	7	3.33	0	0.00	0.00	0.01	0.02	0.00	0	0	0	<1
0	2	0.02	2	12	66	0.06	0	0.00	0.00	0.01	0.01	0.00	0	0	0	0

TABLE F–1

Food Composition (Computer code number is for Wadsworth Diet Analysis program) (For purposes of calculations, use "0" for t, <1, <.1, <.01, etc.)

DA + Code	Food Description	Quantity	Measure	Wt (g)	H_2O (g)	Ener (kcal)	Prot (g)	Carb (g)	Dietary Fiber (g)	Fat (g)	Fat Breakdown (g) Sat	Mono	Poly	Trans
	SAUCES, SPICES, CONDIMENTS													
	Spices													
807	Allspice, ground	1	teaspoon(s)	2	<1	5	<1	1	<1	<1	0.05	0.01	0.04	—
1171	Anise seeds	1	teaspoon(s)	2	<1	7	<1	1	<1	<1	0.01	0.21	0.07	—
729	Baker's yeast active	1	teaspoon(s)	4	<1	12	2	2	1	<1	0.02	0.10	0.00	—
683	Baking powder, double acting, w/phosphate	1	teaspoon(s)	5	<1	2	<.1	1	<.1	0	0.00	0.00	0.00	0
1611	Baking soda	1	teaspoon(s)	5	<.1	0	0	0	0	0	0.00	0.00	0.00	0
8552	Basil	1	teaspoon(s)	1	1	<1	<.1	<.1	<.1	<.1	0.00	0.00	0.00	—
34959	Basil, fresh	1	piece(s)	1	<1	<1	<.1	<.1	<.1	<.1	0.00	0.00	0.00	—
808	Basil, ground	1	teaspoon(s)	1	<.1	4	<1	1	1	<.1	0.00	0.01	0.03	—
809	Bay leaf	1	teaspoon(s)	1	<.1	2	<.1	<1	<1	<.1	0.01	0.01	0.01	—
11720	Betel leaves	1	ounce(s)	28	—	17	2	2	0	<.1	—	—	—	—
818	Black pepper	1	teaspoon(s)	2	<1	5	<1	1	1	<.1	0.02	0.02	0.02	—
730	Brewer's yeast	1	teaspoon(s)	3	<1	8	1	1	1	0	0.00	0.00	0.00	0
35417	Capers	1	teaspoon(s)	4	—	2	0	0	0	0	0.00	0.00	0.00	—
1172	Caraway seeds	1	teaspoon(s)	2	<1	7	<1	1	1	<1	0.01	0.15	0.07	—
819	Cayenne pepper	1	teaspoon(s)	2	<1	6	<1	1	<1	<1	0.06	0.05	0.15	—
1173	Celery seeds	1	teaspoon(s)	2	<1	8	<1	1	<1	1	0.04	0.32	0.07	—
1174	Chervil, dried	1	teaspoon(s)	1	<.1	1	<1	<1	<.1	<.1	0.00	0.01	0.01	—
810	Chili powder	1	teaspoon(s)	3	<1	8	<1	1	1	<1	0.08	0.09	0.19	—
8553	Chives, chopped	1	teaspoon(s)	1	1	<1	<.1	<.1	<.1	<.1	0.00	0.00	0.00	—
8556	Cilantro	1	teaspoon(s)	2	1	<1	<.1	<.1	<.1	<.1	0.00	0.00	0.00	—
811	Cinnamon, ground	1	teaspoon(s)	2	<1	6	<.1	2	1	<.1	0.01	0.01	0.01	—
812	Cloves, ground	1	teaspoon(s)	2	<1	7	<1	1	1	<1	0.11	0.03	0.15	—
1175	Coriander leaf, dried	1	teaspoon(s)	1	<.1	2	<1	<1	<.1	<.1	0.00	0.01	0.00	—
1176	Coriander seeds	1	teaspoon(s)	2	<1	5	<1	1	1	<1	0.02	0.24	0.03	—
1706	Cornstarch	1	tablespoon(s)	8	1	30	<.1	7	<.1	<.1	0.00	0.00	0.00	—
11729	Cumin, ground	1	teaspoon(s)	5	—	11	<1	1	1	<1	—	—	—	—
1177	Cumin seeds	1	teaspoon(s)	2	<1	8	<1	1	<1	<1	0.03	0.29	0.07	—
1178	Curry powder	1	teaspoon(s)	2	<1	7	<1	1	1	<1	0.04	0.11	0.05	—
1179	Dill seeds	1	teaspoon(s)	2	<1	6	<1	1	<1	<1	0.02	0.20	0.02	—
1180	Dill weed, dried	1	teaspoon(s)	1	<.1	3	<1	1	<1	<.1	0.00	0.01	0.00	—
34949	Dill weed, fresh	5	piece(s)	1	1	<1	<.1	<.1	<.1	<.1	0.00	0.01	0.00	—
4949	Fennel leaves, fresh	1	teaspoon(s)	1	1	<1	<.1	<.1	0	<.1	0.00	0.00	0.00	—
1181	Fennel seeds	1	teaspoon(s)	2	<1	7	<1	1	1	<1	0.01	0.20	0.03	—
1182	Fenugreek seeds	1	teaspoon(s)	4	<1	12	1	2	1	<1	0.05	—	—	—
11733	Garam masala, powder	1	ounce(s)	28	—	107	4	13	0	4	—	—	—	—
1067	Garlic clove	1	item(s)	3	2	4	<1	1	<.1	<.1	0.00	0.00	0.01	—
813	Garlic powder	1	teaspoon(s)	3	<1	9	<1	2	<1	<.1	0.00	0.00	0.01	—
1183	Ginger, ground	1	teaspoon(s)	2	<1	6	<1	1	<1	<1	0.03	0.02	0.02	—
1068	Ginger root	2	teaspoon(s)	4	3	3	<.1	1	<.1	<.1	0.01	0.01	0.01	—
35497	Leeks, bulb & lower leaf, freeze-dried	¼	cup(s)	1	<.1	3	<1	1	<.1	<.1	0.00	0.00	0.01	—
1184	Mace, ground	1	teaspoon(s)	2	<1	8	<1	1	<1	1	0.16	0.19	0.07	—
1185	Marjoram, dried	1	teaspoon(s)	1	<.1	2	<.1	<1	<1	<.1	0.00	0.01	0.03	—
1186	Mustard seeds, yellow	1	teaspoon(s)	3	<1	15	1	1	<1	1	0.05	0.65	0.18	—
814	Nutmeg, ground	1	teaspoon(s)	2	<1	12	<1	1	<1	1	0.57	0.07	0.01	—
2747	Onion flakes, dehydrated	1	teaspoon(s)	2	<.1	6	<1	1	<1	<.1	0.00	0.00	0.00	—
1187	Onion powder	1	teaspoon(s)	2	<1	7	<1	2	<1	<.1	0.00	0.00	0.01	—
815	Oregano, ground	1	teaspoon(s)	2	<1	5	<1	1	1	<1	0.04	0.01	0.08	—
816	Paprika	1	teaspoon(s)	2	<1	6	<1	1	1	<1	0.04	0.03	0.17	—
817	Parsley, dried	1	teaspoon(s)	0	<.1	1	<.1	<1	<.1	<.1	0.00	0.01	0.00	—
1189	Poppy seeds	1	teaspoon(s)	3	<1	15	1	1	<1	1	0.14	0.18	0.86	—
1190	Poultry seasoning	1	teaspoon(s)	2	<1	5	<1	1	<1	<1	0.05	0.02	0.03	—
1191	Pumpkin pie spice, powder	1	teaspoon(s)	2	<1	6	<.1	1	<1	<1	0.11	0.02	0.01	—
1192	Rosemary, dried	1	teaspoon(s)	1	<1	4	<.1	1	1	<1	0.09	0.04	0.03	—
11723	Rosemary, fresh	1	teaspoon(s)	1	<1	1	<.1	<1	<.1	<.1	0.02	0.01	0.01	—
2722	Saffron powder	1	teaspoon(s)	1	<.1	2	<.1	<1	<.1	<.1	0.01	0.00	0.01	—
11724	Sage	1	ounce(s)	28	—	34	1	4	0	1	—	—	—	—
1193	Sage, ground	1	teaspoon(s)	1	<.1	2	<.1	<1	<1	<.1	0.05	0.01	0.01	—
822	Salt, table	¼	teaspoon(s)	2	<.1	0	0	0	0	0	0.00	0.00	0.00	0
30189	Salt substitute	¼	teaspoon(s)	1	—	<.1	0	<.1	0	0	0.00	0.00	0.00	0
30190	Salt substitute, seasoned	¼	teaspoon(s)	1	—	1	<.1	<1	0	<.1	0.00	—	—	—
1194	Savory, ground	1	teaspoon(s)	1	<1	4	<.1	1	1	<.1	0.05	—	—	—
820	Sesame seed kernels, toasted	1	teaspoon(s)	3	<1	15	<1	1	<1	1	0.18	0.49	0.57	—
11725	Sorrel	1	tablespoon(s)	9	—	2	<1	<1	<.1	<.1	0.00	—	—	—

PAGE KEY: A–58 = Breads/Baked Goods A–62 = Cereal/Rice/Pasta A–66 = Fruit A–70 = Vegetables/Legumes A–80 = Nuts/Seeds A–82 = Vegetarian A–84 = Dairy A–90 = Eggs A–90 = Seafood A–92 = Meats A–96 = Poultry A–96 = Processed Meats A–98 = Beverages A–102 = Fats/Oils A–104 = Sweets A–106 = Sauces/Condiments/Spices A–108 = Mixed Foods/Soups/Sandwiches A–114 = Fast Food A–130 = Convenience A–132 = Baby Foods

Chol (mg)	Calc (mg)	Iron (mg)	Magn (mg)	Pota (mg)	Sodi (mg)	Zinc (mg)	Vit A (RAE) (μg)	Thia (mg)	Vit E (mg α)	Ribo (mg)	Niac (mg)	Vit B_6 (mg)	Fola (μg)	Vit C (mg)	Vit B_{12} (μg)	Sele (μg)
0	13	0.13	3	20	1	0.02	1	0.00	—	0.00	0.05	0.00	1	1	0	<.1
0	14	0.78	4	30	<1	0.11	<1	0.01	—	0.01	0.06	0.01	<1	<1	0	<1
0	3	0.66	4	80	2	0.26	0	0.09	0.00	0.22	1.59	0.06	94	<.1	<.1	1
0	339	0.52	2	<1	363	0.00	0	0.00	0.00	0.00	0.00	0.00	0	0	0	<.1
0	0	0.00	0	0	1259	0.00	0	0.00	0.00	0.00	0.00	0.00	0	0	0	<.1
0	1	0.03	1	4	<.1	0.01	2	0.00	—	0.00	0.01	0.00	1	<1	0	<.1
0	1	—	<1	2	<.1	0.00	—	0.00	—	0.00	0.01	0.00	<1	—	0	<.1
0	30	0.59	6	48	<1	0.08	7	0.00	0.10	0.00	0.10	0.03	4	1	0	<.1
0	5	0.26	1	3	<1	0.02	2	0.00	—	0.00	0.01	0.01	1	<1	0	<.1
0	110	2.29	—	156	2	—	—	0.04	—	0.07	0.20	—	—	1	0	—
0	9	0.61	4	26	1	0.03	<1	0.00	0.02	0.01	0.02	0.01	<1	<1	0	<.1
0	6	0.47	6	51	3	0.21	0	0.42	—	0.11	1.00	0.07	104	0	0	0
0	0	0.00	—	—	140	—	0	—	—	—	—	—	—	0	—	—
0	14	0.34	5	28	<1	0.12	<1	0.01	0.05	0.01	0.08	0.01	<1	<1	0	<1
0	3	0.14	3	36	1	0.04	37	0.01	0.54	0.02	0.16	0.04	2	1	0	<1
0	35	0.90	9	28	3	0.14	<.1	0.01	0.02	0.01	0.06	0.02	<1	<1	0	<1
0	8	0.19	1	28	<1	0.05	2	0.00	—	0.00	0.03	0.01	2	<1	0	<1
0	7	0.37	4	50	26	0.07	39	0.01	—	0.02	0.21	0.10	3	2	0	<1
0	1	0.02	<1	3	<.1	0.01	2	0.00	0.76	0.00	0.01	0.00	1	1	0	<.1
0	1	0.03	<1	8	1	0.00	—	0.00	—	0.00	0.02	0.00	1	1	0	<.1
0	28	0.88	1	12	1	0.05	<1	0.00	0.02	0.00	0.03	0.01	1	1	0	<.1
0	14	0.18	6	23	5	0.02	1	0.00	0.18	0.01	0.03	0.01	2	2	0	<1
0	7	0.25	4	27	1	0.03	2	0.01	—	0.01	0.06	0.00	2	3	0	<1
0	13	0.29	6	23	1	0.08	0	0.00	—	0.01	0.04	—	0	<1	0	<1
0	<1	0.04	<1	<1	1	0.00	0	0.00	0.00	0.00	0.00	0.00	0	0	0	<1
0	20	—	—	44	5	—	—	—	—	—	—	—	—	—	—	—
0	20	1.39	8	38	4	0.10	1	0.01	0.07	0.01	0.10	0.01	<1	<1	0	<1
0	10	0.59	5	31	1	0.08	1	0.01	0.44	0.01	0.07	0.02	3	<1	0	<1
0	32	0.34	5	25	<1	0.11	<.1	0.01	—	0.01	0.06	0.01	<1	<1	0	<1
0	18	0.49	5	33	2	0.03	3	0.00	—	0.00	0.03	0.02	2	1	0	0
0	2	—	1	7	1	0.01	—	0.00	—	0.00	0.02	0.00	2	—	0	—
0	1	0.03	—	4	<.1	—	—	0.00	—	0.00	0.01	0.00	—	<1	0	—
0	24	0.37	8	34	2	0.07	<1	0.01	—	0.01	0.12	0.01	—	<1	0	0
0	7	1.24	7	28	2	0.09	<1	0.01	—	0.01	0.06	0.02	2	<1	0	<1
0	215	9.25	94	411	28	1.07	—	0.10	—	0.09	0.71	—	0	0	0	—
0	5	0.05	1	12	1	0.03	0	0.01	0.00	0.00	0.02	0.04	<.1	1	0	<1
0	2	0.08	2	31	1	0.07	0	0.01	0.02	0.00	0.02	0.08	<.1	1	0	1
0	2	0.21	3	24	1	0.08	<1	0.00	0.32	0.00	0.09	0.02	1	<1	0	1
0	1	0.02	2	17	1	0.01	0	0.00	0.01	0.00	0.03	0.01	<1	<1	0	<.1
0	3	0.06	1	19	<1	0.01	<1	0.01	—	0.00	0.03	0.01	3	1	0	<.1
0	4	0.24	3	8	1	0.04	1	0.01	—	0.01	0.02	0.00	1	<1	0	<.1
0	12	0.50	2	9	<1	0.02	2	0.00	0.01	0.00	0.02	0.01	2	<1	0	<.1
0	17	0.33	10	23	<1	0.19	<.1	0.02	0.10	0.01	0.26	0.01	3	<.1	0	4
0	4	0.07	4	8	<1	0.05	<1	0.01	0.00	0.00	0.03	0.00	2	<.1	0	<.1
0	4	0.03	2	27	<1	0.03	<.1	0.01	0.00	0.00	0.02	0.03	3	1	0	<.1
0	8	0.05	3	20	1	0.05	0	0.01	0.01	0.00	0.01	0.03	3	<1	0	<.1
0	24	0.66	4	25	<1	0.07	5	0.01	0.28	0.00	0.09	0.02	4	1	0	<.1
0	4	0.50	4	49	1	0.09	55	0.01	0.63	0.04	0.32	0.08	2	1	0	<.1
0	4	0.29	1	11	1	0.01	2	0.00	0.02	0.00	0.02	0.00	1	<1	0	<.1
0	41	0.26	9	20	1	0.29	0	0.02	0.03	0.00	0.03	0.01	2	<.1	0	<.1
0	15	0.53	3	10	<1	0.05	2	0.00	0.03	0.00	0.04	0.02	2	<1	0	<1
0	12	0.34	2	11	1	0.04	<1	0.00	0.02	0.00	0.04	0.01	1	<1	0	<1
0	15	0.35	3	11	1	0.04	2	0.01	—	0.01	0.01	0.02	4	1	0	<.1
0	2	0.05	1	5	<1	0.01	1	0.00	—	0.00	0.01	0.00	1	<1	0	—
0	1	0.08	2	12	1	0.01	<1	0.00	—	0.00	0.01	0.01	1	1	0	<.1
0	170	—	45	110	1	0.48	—	0.03	—	—	—	—	—	—	0	—
0	12	0.20	3	7	<.1	0.03	2	0.01	0.05	0.00	0.04	0.02	2	<1	0	<.1
0	<1	0.00	<.1	<1	581	0.00	0	0.00	0.00	0.00	0.00	0.00	0	0	0	<.1
0	7	0.00	<.1	604	<.1	—	0	—	—	—	—	—	—	0	—	—
0	0	0	476	<1	—	0	—	—	—	—	—	—	0	—	—	
0	30	0.53	5	15	<1	0.06	4	0.01	—	—	0.06	0.03	—	1	0	<.1
0	4	0.21	9	11	1	0.28	<.1	0.03	0.01	0.01	0.15	0.00	3	0	0	<.1
0	—	—	—	—	<1	—	—	—	—	—	—	—	—	—	—	—

TABLE F–1
Food Composition (Computer code number is for Wadsworth Diet Analysis program) (For purposes of calculations, use "0" for t, <1, <.1, <.01, etc.)

DA + Code	Food Description	Quantity	Measure	Wt (g)	H_2O (g)	Ener (kcal)	Prot (g)	Carb (g)	Dietary Fiber (g)	Fat (g)	Fat Breakdown (g) Sat	Mono	Poly	Trans
	SAUCES, SPICES, CONDIMENTS—Continued													
11721	Spearmint	1	teaspoon(s)	2	2	1	<.1	<1	<1	<.1	0.00	0.00	0.01	—
35498	Sweet green peppers, freeze-dried	¼	cup(s)	2	<.1	5	<1	1	<1	<.1	0.01	0.00	0.03	—
11726	Tamarind leaves	1	ounce(s)	28	—	33	2	5	0	1	—	—	—	—
11727	Tarragon	1	ounce(s)	28	—	14	1	2	0	<1	—	—	—	—
1195	Tarragon, ground	1	teaspoon(s)	2	<1	5	<1	1	<1	<1	0.03	0.01	0.06	—
11728	Thyme, fresh	1	teaspoon(s)	1	1	1	<.1	<1	<1	<.1	0.00	0.00	0.00	—
821	Thyme, ground	1	teaspoon(s)	1	<1	4	<1	1	1	<1	0.04	0.01	0.02	—
1196	Turmeric, ground	1	teaspoon(s)	2	<1	8	<1	1	<1	<1	0.07	0.04	0.05	—
11995	Wasabi	1	tablespoon(s)	14	11	11	1	2	<1	<.1	—	—	—	—
1188	White pepper	1	teaspoon(s)	2	<1	7	<1	2	1	<.1	0.02	0.02	0.01	—
	Condiments													
674	Catsup or ketchup	1	tablespoon(s)	15	11	14	<1	4	<1	<.1	0.01	0.01	0.04	—
703	Dill pickle	1	ounce(s)	28	26	5	<1	1	<1	<.1	0.01	0.00	0.02	—
1641	Horseradish sauce, prepared	1	teaspoon(s)	5	3	10	<1	<1	<.1	1	0.59	0.28	0.04	—
140	Mayonnaise, low calorie	1	tablespoon(s)	16	10	37	<.1	3	0	3	0.53	0.72	1.70	—
138	Mayonnaise w/soybean oil	1	tablespoon(s)	14	2	99	<1	1	0	11	1.64	2.70	5.89	0.04
1682	Mustard, brown	1	teaspoon(s)	5	4	5	<1	<1	<.1	<1	—	—	—	—
700	Mustard, yellow	1	teaspoon(s)	5	4	3	<1	<1	<.1	<1	0.01	0.11	0.03	—
706	Sweet pickle relish	1	tablespoon(s)	15	9	20	<.1	5	<1	<.1	0.01	0.03	0.02	—
141	Tartar sauce	2	tablespoon(s)	28	9	144	<1	4	<.1	14	2.14	4.13	7.57	—
	Sauces													
685	Barbecue sauce	2	tablespoon(s)	31	25	23	1	4	<1	1	0.08	0.24	0.21	—
834	Cheese sauce	¼	cup(s)	70	49	121	5	5	<1	9	4.19	2.67	1.81	—
32123	Chili enchilada sauce, green	2	tablespoon(s)	57	53	15	1	3	1	<1	0.04	0.04	0.13	0
32122	Chili enchilada sauce, red	2	tablespoon(s)	32	24	27	1	5	2	1	0.08	0.05	0.43	0
29688	Hoisin sauce	1	tablespoon(s)	16	7	35	1	7	<1	1	0.09	0.15	0.27	—
16670	Mole poblano sauce	½	cup(s)	133	103	155	5	11	2	11	2.67	5.15	2.91	—
29689	Oyster sauce	1	tablespoon(s)	16	13	8	<1	2	<.1	<.1	0.01	0.01	0.01	—
1655	Pepper sauce or tabasco	1	teaspoon(s)	5	5	1	<.1	<.1	<.1	<.1	0.01	0.00	0.02	—
347	Salsa	2	tablespoon(s)	16	14	4	<1	1	<1	<.1	0.00	0.00	0.02	—
841	Soy sauce	1	tablespoon(s)	18	13	10	1	2	0	<.1	0.00	0.00	0.01	—
839	Sweet & sour sauce	2	tablespoon(s)	39	30	37	<.1	9	<.1	<.1	0.00	0.00	0.00	—
1613	Teriyaki sauce	1	tablespoon(s)	18	12	15	1	3	<.1	0	0.00	0.00	0.00	0
25294	Tomato sauce	½	cup(s)	112	100	46	2	8	2	1	0.18	0.29	0.72	0
728	White sauce, medium	¼	cup(s)	63	47	92	2	6	<1	7	1.78	2.78	1.79	—
1654	Worcestershire sauce	1	teaspoon(s)	6	4	4	0	1	0	0	0.00	0.00	0.00	0
	Vinegar													
30853	Balsamic	1	tablespoon(s)	15	—	10	0	2	0	0	0.00	0.00	0.00	0
727	Cider	1	tablespoon(s)	15	14	2	0	1	0	0	0.00	0.00	0.00	0
1673	Distilled	1	tablespoon(s)	15	14	2	0	1	0	0	0.00	0.00	0.00	0
15439	Tarragon	1	tablespoon(s)	16	—	0	0	0	0	0	0.00	0.00	0.00	0
	MIXED FOODS, SOUPS, SANDWICHES													
	Mixed Dishes													
16652	Almond chicken	1	cup(s)	242	186	280	22	16	3	15	1.91	6.07	5.62	—
25224	Barbecued chicken	2	piece(s)	177	100	325	27	15	<1	17	4.63	6.78	3.71	0
25227	Bean burrito	1	item(s)	149	82	327	17	33	6	15	8.30	4.73	0.85	0
9516	Beef & vegetable fajita	1	item(s)	223	144	397	23	35	3	18	5.50	7.53	3.45	—
16796	Beef or pork egg roll	2	item(s)	128	85	227	10	19	1	12	2.88	5.96	2.64	—
177	Beef stew w/vegetables, prepared	1	cup(s)	245	201	220	16	15	3	11	4.40	4.50	0.50	—
30233	Beef stroganoff w/noodles	1	cup(s)	256	190	343	20	23	2	19	7.37	5.62	4.47	—
16651	Cashew chicken	1	cup(s)	242	131	644	43	17	3	46	7.75	20.83	14.47	—
475	Cheese pizza	2	slice(s)	126	60	281	15	41	0	6	3.08	1.98	0.98	—
30330	Cheese quesadilla	1	item(s)	54	19	183	6	18	1	10	3.49	3.42	2.16	—
215	Chicken & noodles, prepared	1	cup(s)	240	170	365	22	26	1	18	5.10	7.10	3.90	—
30239	Chicken & vegetables w/broccoli, onion, bamboo shoots in soy based sauce	1	cup(s)	162	112	287	22	6	1	19	5.13	7.65	4.68	—
25093	Chicken cacciatore	1	cup(s)	230	166	266	28	5	1	14	3.98	5.78	3.11	0
28020	Chicken fried turkey steak	3	ounce(s)	85	48	122	13	12	1	2	0.59	0.37	0.78	—
218	Chicken pot pie	1	cup(s)	252	154	542	23	42	3	31	9.79	12.52	7.03	—
30240	Chicken teriyaki	1	cup(s)	244	163	339	51	13	1	7	1.78	2.03	1.71	—
25119	Chicken waldorf salad	½	cup(s)	100	68	178	14	6	1	11	1.76	3.18	5.05	0
25099	Chili con carne	¾	cup(s)	215	175	197	14	21	7	7	2.55	2.83	0.54	0
1062	Coleslaw	¾	cup(s)	90	73	62	1	11	1	2	0.35	0.64	1.22	—
1896	Combination pizza, w/meat & vegetables	2	slice(s)	158	75	368	26	43	5	11	3.07	5.09	1.83	—

PAGE KEY: A–58 = Breads/Baked Goods A–62 = Cereal/Rice/Pasta A–66 = Fruit A–70 = Vegetables/Legumes A–80 = Nuts/Seeds A–82 = Vegetarian A–84 = Dairy A–90 = Eggs A–90 = Seafood A–92 = Meats A–96 = Poultry A–96 = Processed Meats A–98 = Beverages A–102 = Fats/Oils A–104 = Sweets A–106 = Sauces/Condiments/Spices A–108 = Mixed Foods/Soups/Sandwiches A–114 = Fast Food A–130 = Convenience A–132 = Baby Foods

Chol (mg)	Calc (mg)	Iron (mg)	Magn (mg)	Pota (mg)	Sodi (mg)	Zinc (mg)	Vit A (RAE) (µg)	Thia (mg)	Vit E (mg α)	Ribo (mg)	Niac (mg)	Vit B_6 (mg)	Fola (µg)	Vit C (mg)	Vit B_{12} (µg)	Sele (µg)
0	4	0.23	1	9	1	0.02	4	0.00	—	0.00	0.02	0.00	2	<1	0	—
0	2	0.17	3	51	3	0.04	3	0.02	0.06	0.02	0.12	0.04	4	30	0	<.1
0	85	1.48	20	—	—	—	—	0.07	—	0.03	1.16	—	—	1	0	—
0	48	—	14	128	3	0.17	—	0.04	—	—	—	—	—	1	0	—
0	18	0.52	6	48	1	0.06	3	0.00	0.10	0.02	0.14	0.04	4	1	0	<.1
0	3	0.14	1	5	<.1	0.01	2	0.00	—	0.00	0.01	0.00	<1	1	0	—
0	26	1.73	3	11	1	0.09	3	0.01	—	0.01	0.07	0.01	4	1	0	<.1
0	4	0.91	4	56	1	0.10	0	0.00	—	0.01	0.11	0.04	1	1	0	<.1
0	13	0.11	—	—	—	—	—	0.02	—	0.01	0.07	—	—	11	0	—
0	6	0.34	2	2	<1	0.03	0	0.00	0.10	0.00	0.01	0.00	<1	1	0	<.1
0	3	0.08	3	57	167	0.04	7	0.00	0.22	0.07	0.23	0.02	2	2	0	<.1
0	3	0.15	3	33	363	0.04	3	0.00	0.03	0.01	0.02	0.00	<1	1	0	0
2	5	0.00	1	7	15	0.01	—	0.00	0.03	0.01	0.00	0.00	1	<.1	<.1	—
4	<.1	0.00	<.1	2	80	0.02	0	0.00	0.32	0.00	0.00	0.00	0	0	0	—
5	2	0.07	<1	5	78	0.02	12	0.00	0.72	0.00	0.00	0.08	1	0	<.1	<1
0	6	0.09	1	7	68	0.02	0	0.00	0.09	0.00	0.01	0.00	<1	<.1	0	—
0	4	0.09	2	8	56	0.03	<1	0.00	0.01	0.00	0.02	0.00	<1	<1	0	2
0	<1	0.13	1	4	122	0.02	1	0.00	0.06	0.00	0.03	0.00	<1	<1	0	0
11	6	0.21	1	10	200	0.05	—	0.00	0.97	0.00	0.01	0.08	2	<1	<.1	—
0	6	0.28	6	54	255	0.06	<1	0.01	0.01	0.01	0.28	0.02	1	2	0	<1
20	128	0.15	6	21	578	0.68	56	0.00	—	0.08	0.02	0.01	3	<1	<.1	2
0	5	0.36	9	126	62	0.11	—	0.03	0.00	0.02	0.63	0.06	6	44	0	0
0	7	1.05	11	231	114	0.15	—	0.02	0.00	0.22	0.61	0.34	7	<1	0	<1
<1	5	0.16	4	19	258	0.05	0	0.00	0.04	0.03	0.19	0.01	4	<.1	0	<1
1	37	1.51	57	283	305	0.95	—	0.07	1.72	0.09	1.82	0.09	14	5	<.1	—
0	5	0.03	1	9	437	0.01	0	0.00	0.00	0.02	0.24	0.00	2	<.1	<.1	1
0	1	0.06	1	6	32	0.01	4	0.00	—	0.00	0.01	0.01	<.1	<1	0	—
0	5	0.16	2	34	69	0.04	5	0.01	0.19	0.01	0.13	0.02	3	2	0	<.1
0	3	0.36	6	32	1029	0.07	0	0.01	0.00	0.02	0.61	0.03	3	0	0	—
0	5	0.20	1	8	98	0.01	0	0.00	—	0.01	0.12	0.04	<1	0	0	—
0	5	0.31	11	41	690	0.02	0	0.01	0.00	0.01	0.23	0.02	4	0	0	<1
0	21	1.08	19	431	199	0.30	48	0.05	0.39	0.05	1.18	0.13	15	15	0	1
4	74	0.21	9	98	221	0.26	—	0.04	—	0.12	0.25	0.03	3	1	<1	—
0	6	0.30	1	45	56	0.01	—	0.00	0.00	0.01	0.04	0.00	0	1	0	—
0	0	0.00	—	—	0	—	0	—	—	—	—	—	—	0	—	—
0	1	0.09	3	15	<1	0.00	0	0.00	0.00	0.00	0.00	0.00	0	0	0	<.1
0	1	0.09	0	2	<1	0.00	0	0.00	0.00	0.00	0.00	0.00	0	0	0	5
0	0	0.00	—	0	0	—	0	—	—	—	—	—	—	0	0	—
40	69	1.97	60	549	526	1.62	—	0.09	4.11	0.20	9.48	0.44	26	7	<1	—
120	26	1.64	31	387	477	2.69	69	0.07	0.01	0.37	6.92	0.39	15	5	<1	19
38	331	2.95	45	384	514	1.92	119	0.24	0.01	0.29	1.82	0.15	115	4	<1	18
45	84	3.74	37	476	757	3.51	—	0.39	0.80	0.30	5.37	0.38	23	27	2	—
74	30	1.66	20	248	547	0.91	—	0.32	1.28	0.25	2.55	0.19	20	4	<1	—
71	29	2.90	—	613	292	—	—	0.15	0.51	0.17	4.70	—	—	17	<.1	15
74	70	3.26	37	393	818	3.66	—	0.21	1.25	0.31	3.80	0.21	17	1	2	—
96	74	2.92	94	640	1355	2.24	—	0.23	4.11	0.22	19.76	0.88	64	11	<1	—
19	233	1.16	32	219	672	1.63	147	0.37	—	0.33	4.96	0.09	69	3	1	27
13	132	1.21	13	77	230	0.64	—	0.13	0.43	0.14	1.09	0.04	6	15	<.1	—
103	26	2.20	—	149	600	—	—	0.05	—	0.17	4.30	—	—	0	—	29
84	22	1.38	29	344	962	1.70	—	0.08	1.12	0.17	7.90	0.32	13	8	<1	—
103	45	2.21	37	444	451	2.01	53	0.10	0.00	0.21	9.20	0.54	15	8	<1	22
27	69	1.34	19	197	139	1.08	5	0.15	0.00	0.18	3.46	0.22	21	<1	<1	16
69	64	3.38	38	393	651	1.93	607	0.40	1.06	0.40	7.24	0.24	31	11	<1	—
157	52	3.27	67	589	3209	3.75	—	0.15	0.59	0.37	16.69	0.89	23	6	1	—
42	20	0.78	24	197	246	1.13	21	0.04	0.62	0.10	4.05	0.25	15	2	<1	11
27	43	3.16	50	646	865	2.44	25	0.13	0.02	0.23	3.01	0.18	56	10	1	10
7	41	0.53	9	163	21	0.18	48	0.06	—	0.06	0.24	0.11	24	29	0	1
41	202	3.07	36	357	765	2.23	117	0.43	—	0.35	3.92	0.19	65	3	1	22

TABLE F-1

Food Composition (Computer code number is for Wadsworth Diet Analysis program) (For purposes of calculations, use "0" for t, <1, <.1, <.01, etc.)

DA + Code	Food Description	Quantity	Measure	Wt (g)	H_2O (g)	Ener (kcal)	Prot (g)	Carb (g)	Dietary Fiber (g)	Fat (g)	Fat Breakdown (g) Sat	Mono	Poly	Trans
	MIXED FOODS, SOUPS, SANDWICHES—Continued													
1574	Crab cakes, from blue crab	1	item(s)	60	43	93	12	<1	0	5	0.89	1.69	1.36	—
32144	Enchiladas w/green chili sauce (enchiladas verdes)	1	item(s)	144	104	207	9	18	3	12	6.35	3.65	0.96	0
2793	Falafel patty	3	item(s)	51	18	170	7	16	0	9	1.22	5.19	2.12	—
28546	Fettuccine alfredo	1	cup(s)	222	81	247	11	42	1	3	1.61	0.79	0.43	0
32146	Flautas	3	item(s)	162	78	438	25	36	4	22	8.22	8.80	2.29	—
29629	Fried rice w/meat or poultry	1	cup(s)	198	129	329	12	41	1	12	2.27	3.53	5.69	—
16649	General tso chicken	1	cup(s)	146	91	293	19	16	1	17	3.98	6.27	5.27	—
1826	Green salad	¾	cup(s)	104	99	17	1	3	2	<.1	0.01	0.00	0.04	—
1814	Hummus	½	cup(s)	123	80	218	6	25	5	11	1.38	6.04	2.56	—
16650	Kung pao chicken	1	cup(s)	162	88	431	29	11	2	31	5.19	13.95	9.69	—
16622	Lamb curry	1	cup(s)	236	188	256	28	3	1	14	3.93	4.92	3.35	—
25253	Lasagna w/ground beef	1	cup(s)	237	157	288	18	22	2	15	7.47	4.84	0.84	0
442	Macaroni & cheese	1	cup(s)	200	122	393	15	40	1	19	8.18	6.72	2.66	—
25105	Meat loaf	1	slice(s)	115	85	244	17	7	<1	16	6.15	6.89	0.83	0
16646	Moo shi pork	1	cup(s)	151	77	512	19	5	1	46	6.84	14.80	22.07	—
16788	Nachos w/beef, beans, cheese, tomatoes, & onions	7	item(s)	551	284	1496	40	119	19	99	22.34	40.19	30.69	—
1668	Pepperoni pizza	2	slice(s)	142	66	362	20	40	1	14	4.47	6.28	2.33	—
655	Potato salad	½	cup(s)	125	95	179	3	14	2	10	1.79	3.10	4.67	—
29637	Ravioli, meat filled, w/tomato or meat sauce, canned	1	cup(s)	251	196	220	9	38	2	4	1.58	1.49	0.41	—
25109	Salisbury steaks w/mushroom sauce	1	serving(s)	135	102	251	17	9	1	15	5.98	6.67	0.76	0
16637	Shrimp creole w/rice	1	cup(s)	243	176	311	27	28	1	9	1.83	3.79	2.88	—
497	Spaghetti & meat balls w/tomato sauce, prepared	1	cup(s)	248	174	330	19	39	3	12	3.90	4.40	2.20	—
28585	Spicy thai noodles (pad thai)	8	ounce(s)	231	74	222	9	36	3	6	0.83	3.33	1.83	0
33073	Stir fried pork & vegetables w/rice	1	cup(s)	235	173	349	15	34	2	16	5.55	6.87	2.62	0
28588	Stuffed shells	2½	item(s)	299	189	292	18	33	3	10	3.81	3.57	1.62	0
16821	Sushi w/egg in seaweed	6	piece(s)	156	117	190	9	20	<1	8	2.09	3.02	1.55	—
16819	Sushi w/vegetables & fish	6	piece(s)	156	102	217	8	44	2	1	0.16	0.14	0.20	—
16820	Sushi w/vegetables in seaweed	6	piece(s)	156	110	182	3	41	1	<1	0.10	0.11	0.11	—
25266	Sweet & sour pork	¾	cup(s)	249	206	264	29	17	1	8	2.59	3.51	1.48	0
16824	Tabouli, tabbouleh, or tabuli	1	cup(s)	160	124	199	3	16	4	15	2.04	10.83	1.37	—
25276	Three bean salad	½	cup(s)	99	82	95	2	10	3	6	0.76	1.41	3.48	0
160	Tuna salad	½	cup(s)	103	65	192	16	10	0	9	1.58	2.96	4.23	0
25241	Turkey & noodles	1	cup(s)	319	228	271	24	21	1	9	2.39	3.48	2.27	0
16794	Vegetable egg roll	2	item(s)	128	90	202	5	20	2	12	2.46	5.71	2.65	—
16818	Vegetable sushi, no fish	6	piece(s)	156	99	225	5	50	2	<1	0.11	0.10	0.14	—
	Sandwiches													
1744	Bacon, lettuce & tomato w/mayonnaise	1	item(s)	164	97	349	11	34	2	19	4.54	7.22	6.07	—
30287	Bologna & cheese w/margarine	1	item(s)	111	46	350	13	28	1	20	8.55	8.40	2.28	—
30286	Bologna w/margarine	1	item(s)	83	34	256	7	26	1	13	4.08	6.31	2.07	—
16546	Cheese	1	item(s)	83	31	262	10	27	1	13	5.59	4.77	1.67	—
8789	Cheeseburger, large, plain	1	item(s)	185	72	609	30	47	0	33	14.84	12.74	2.44	—
8624	Cheeseburger, large, w/bacon, vegetables, & condiments	1	item(s)	195	85	608	32	37	2	37	16.24	14.49	2.71	—
1745	Club w/bacon, chicken, tomato, lettuce, & mayonnaise	1	item(s)	246	137	555	31	48	3	26	5.94	—	—	—
1908	Cold cut submarine w/cheese & vegetables	1	item(s)	228	132	456	22	51	2	19	6.81	8.23	2.28	—
30247	Corned beef	1	item(s)	130	75	268	19	25	2	10	3.75	3.96	0.80	—
25283	Egg salad	1	item(s)	126	72	278	10	29	1	13	2.96	3.97	4.79	—
16686	Fried egg	1	item(s)	96	50	226	10	26	1	9	2.29	3.51	1.64	—
16547	Grilled cheese	1	item(s)	83	27	292	10	27	1	16	6.22	6.29	2.54	—
16659	Gyro w/onion & tomato	1	item(s)	105	67	170	12	21	1	4	1.53	1.41	0.43	—
1906	Ham & cheese	1	item(s)	146	74	352	21	33	2	15	6.44	6.74	1.38	—
31890	Ham w/mayonnaise	1	item(s)	112	55	282	14	27	1	13	3.06	5.04	3.79	—
756	Hamburger, double patty, large, w/condiments & vegetables	1	item(s)	226	121	540	34	40	0	27	10.52	10.33	2.80	—
8793	Hamburger, large, plain	1	item(s)	137	58	426	23	32	2	23	8.38	9.88	2.14	—
8795	Hamburger, large, w/vegetables & condiments	1	item(s)	218	121	512	26	40	3	27	10.42	11.42	2.20	—
25134	Hot chicken salad	1	item(s)	98	49	239	16	23	1	9	2.83	2.61	2.76	0
1411	Hot dog w/bun, plain	1	item(s)	98	53	242	10	18	2	15	5.11	6.85	1.71	—
25133	Hot turkey salad	1	item(s)	98	50	221	16	23	1	7	2.23	1.76	2.28	0
30249	Pastrami	1	item(s)	134	71	331	14	27	2	18	6.18	8.74	1.02	—

PAGE KEY: A–58 = Breads/Baked Goods A–62 = Cereal/Rice/Pasta A–66 = Fruit A–70 = Vegetables/Legumes A–80 = Nuts/Seeds A–82 = Vegetarian A–84 = Dairy A–90 = Eggs A–90 = Seafood A–92 = Meats A–96 = Poultry A–96 = Processed Meats A–98 = Beverages A–102 = Fats/Oils A–104 = Sweets A–106 = Sauces/Condiments/Spices A–108 = Mixed Foods/Soups/Sandwiches A–114 = Fast Food A–130 = Convenience A–132 = Baby Foods

Chol (mg)	Calc (mg)	Iron (mg)	Magn (mg)	Pota (mg)	Sodi (mg)	Zinc (mg)	Vit A (RAE) (µg)	Thia (mg)	Vit E (mg α)	Ribo (mg)	Niac (mg)	Vit B_6 (mg)	Fola (µg)	Vit C (mg)	Vit B_{12} (µg)	Sele (µg)
90	63	0.65	20	194	198	2.45	34	0.05	—	0.05	1.74	0.10	32	2	4	24
27	266	1.08	38	251	276	1.27	—	0.07	0.03	0.16	1.28	0.18	45	59	<1	6
0	28	1.74	42	298	150	0.77	1	0.07	—	0.08	0.53	0.06	47	1	0	1
9	153	1.88	32	123	386	1.48	51	0.35	0.00	0.34	2.60	0.06	103	1	<1	35
73	146	2.66	61	223	886	3.44	0	0.10	0.10	0.17	3.00	0.27	96	0	1	37
102	36	2.66	31	182	821	1.42	—	0.30	1.60	0.19	3.51	0.24	24	3	<1	—
65	27	1.49	24	250	906	1.40	—	0.10	1.62	0.19	6.28	0.28	17	12	<1	—
0	13	0.65	11	178	27	0.22	59	0.03	—	0.05	0.57	0.08	38	24	0	<1
0	60	1.93	36	213	298	1.34	0	0.11	0.92	0.06	0.49	0.49	73	10	0	3
64	49	1.96	63	428	907	1.50	—	0.15	4.32	0.15	13.23	0.59	43	8	<1	—
89	36	2.97	40	495	495	6.62	—	0.09	1.30	0.28	8.05	0.20	27	1	3	—
68	222	2.33	40	437	493	2.81	108	0.19	0.22	0.29	3.02	0.20	50	10	1	22
30	323	2.26	42	263	800	1.95	327	0.25	0.72	0.40	2.18	0.10	12	<1	<1	—
85	54	2.09	21	278	423	3.55	27	0.08	0.00	0.29	3.77	0.13	20	<1	2	17
172	30	1.45	26	330	1078	1.83	—	0.50	5.39	0.38	2.90	0.31	22	8	1	—
82	699	6.71	205	1067	1611	7.55	—	0.31	7.71	0.50	5.62	0.85	59	14	1	—
28	129	1.87	17	305	534	1.04	105	0.27	—	0.47	6.09	0.11	74	3	<1	26
85	24	0.81	19	318	661	0.39	40	0.10	—	0.08	1.11	0.18	9	13	0	5
17	28	2.04	23	337	1354	1.19	—	0.22	0.70	0.20	2.88	0.14	17	22	<1	—
60	64	2.21	23	282	370	3.66	27	0.11	0.00	0.30	4.00	0.13	22	<1	2	17
181	101	4.44	64	439	381	1.73	—	0.29	2.07	0.10	4.77	0.22	12	18	1	—
89	124	3.70	—	665	1009	—	82	0.25	—	0.30	4.00	—	—	22	—	22
37	32	1.58	50	187	598	1.08	38	0.18	0.36	0.13	1.88	0.17	44	22	<.1	3
46	39	2.65	32	394	574	2.07	80	0.51	0.38	0.20	5.07	0.30	102	18	<1	23
35	241	3.18	63	462	543	1.68	280	0.32	0.00	0.36	4.64	0.30	109	15	<1	36
217	42	1.63	18	128	527	0.98	—	0.12	0.67	0.29	1.33	0.13	29	2	<1	—
11	24	2.18	25	204	340	0.79	—	0.26	0.25	0.07	2.77	0.15	14	4	<1	—
0	20	1.54	20	99	153	0.70	—	0.20	0.12	0.04	1.86	0.14	10	2	0	—
74	41	1.78	35	622	624	2.53	64	0.80	0.20	0.37	6.69	0.65	14	10	1	50
0	29	1.25	36	246	799	0.48	—	0.08	2.43	0.05	1.14	0.11	31	29	0	—
0	26	0.96	15	144	224	0.31	12	0.04	0.89	0.06	0.26	0.06	31	9	0	3
13	17	1.03	19	182	412	0.57	25	0.03	0.00	0.07	6.87	0.08	8	2	1	42
77	60	2.69	33	379	576	2.64	108	0.23	0.29	0.32	6.40	0.30	60	1	1	34
60	29	1.61	18	193	548	0.51	—	0.16	1.28	0.21	1.59	0.10	27	6	<1	—
0	23	2.40	23	158	369	0.84	—	0.28	0.16	0.06	2.44	0.13	15	4	0	—
20	76	2.54	27	328	837	0.98	—	0.39	1.16	0.27	3.81	0.20	31	15	<1	—
35	221	2.18	24	185	940	1.68	—	0.30	0.56	0.33	2.77	0.12	21	<.1	1	—
16	60	1.96	15	112	598	0.85	—	0.29	0.50	0.21	2.73	0.08	19	<.1	<1	—
19	216	1.75	20	135	655	1.14	—	0.25	0.47	0.29	2.04	0.07	19	<.1	<1	—
96	91	5.46	39	644	1589	5.55	185	0.48	—	0.57	11.17	0.28	74	0	3	39
111	162	4.74	45	332	1043	6.83	82	0.31	—	0.41	6.63	0.31	86	2	2	33
72	116	4.05	47	463	855	1.65	—	0.61	1.53	0.44	11.92	0.59	48	9	1	—
36	189	2.51	68	394	1651	2.58	71	1.00	—	0.80	5.49	0.14	87	12	1	31
46	67	2.67	20	187	1177	2.24	—	0.24	0.21	0.25	3.23	0.10	22	2	1	—
217	107	2.60	18	147	494	0.94	94	0.26	0.13	0.43	2.27	0.16	82	1	1	24
207	80	2.25	17	120	433	0.85	—	0.27	0.66	0.41	2.06	0.10	34	0	<1	—
19	219	1.76	21	137	696	1.15	—	0.19	0.72	0.28	1.86	0.06	13	<.1	<1	—
34	46	1.85	21	209	272	2.30	—	0.24	0.26	0.21	3.14	0.13	18	4	1	—
58	130	3.24	16	291	771	1.37	96	0.31	0.29	0.48	2.69	0.20	76	3	1	23
36	59	2.10	23	245	1033	1.50	—	0.71	0.50	0.31	4.89	0.26	19	0	<1	—
122	102	5.85	50	570	791	5.67	5	0.36	—	0.38	7.57	0.54	77	1	4	26
71	74	3.58	27	267	474	4.11	0	0.29	—	0.29	6.25	0.23	60	0	2	27
87	96	4.93	44	480	824	4.88	24	0.41	—	0.37	7.28	0.33	83	3	2	34
39	114	1.93	20	150	470	1.22	28	0.20	0.28	0.23	4.93	0.20	54	<1	<1	17
44	24	2.31	13	143	670	1.98	0	0.24	—	0.27	3.65	0.05	48	<.1	1	26
37	113	2.04	22	167	459	1.09	23	0.19	0.29	0.21	4.36	0.23	54	<1	<1	20
51	68	2.64	23	243	1335	2.69	—	0.29	0.27	0.27	4.77	0.13	21	2	1	—

TABLE F–1
Food Composition (Computer code number is for Wadsworth Diet Analysis program) (For purposes of calculations, use "0" for t, <1, <.1, <.01, etc.)

DA + Code	Food Description	Quantity	Measure	Wt (g)	H_2O (g)	Ener (kcal)	Prot (g)	Carb (g)	Dietary Fiber (g)	Fat (g)	Fat Breakdown (g) Sat	Mono	Poly	Trans
	MIXED FOODS, SOUPS, SANDWICHES—Continued													
16701	Peanut butter	1	item(s)	93	24	344	13	37	3	17	3.55	8.16	4.58	—
30306	Peanut butter & jelly	1	item(s)	93	24	330	11	42	3	15	3.00	6.87	3.82	—
1910	Roast beef, plain	1	item(s)	139	68	346	22	33	1	14	3.61	6.80	1.71	—
1909	Roast beef submarine w/mayonnaise & vegetables	1	item(s)	216	127	410	29	44	—	13	7.09	1.84	2.61	—
1907	Steak w/mayonnaise & vegetables	1	item(s)	204	104	459	30	52	2	14	3.81	5.34	3.35	—
25288	Tuna salad	1	item(s)	179	102	414	24	29	2	22	3.61	5.46	11.43	—
31891	Turkey w/mayonnaise	1	item(s)	143	75	330	29	26	1	11	2.61	3.25	4.40	—
30283	Turkey submarine w/cheese, lettuce, tomato, & mayonnaise	1	item(s)	277	156	583	37	51	3	25	7.15	8.03	7.81	—
	Soups													
25296	Bean	1	cup(s)	301	253	191	14	29	6	2	0.67	0.83	0.53	0
711	Bean with pork, condensed, prepared w/water	1	cup(s)	265	223	180	8	24	9	6	1.59	2.28	1.91	—
713	Beef noodle, condensed, prepared w/water	1	cup(s)	244	224	83	5	9	1	3	1.15	1.24	0.49	—
825	Cheese, condensed, prepared w/milk	1	cup(s)	251	207	231	9	16	1	15	9.11	4.09	0.45	—
826	Chicken broth, condensed, prepared w/water	1	cup(s)	244	234	39	5	1	0	1	0.39	0.59	0.27	—
25297	Chicken noodle	1	cup(s)	286	258	117	11	11	1	3	0.78	1.10	0.66	—
827	Chicken noodle, condensed, prepared w/water	1	cup(s)	241	222	75	4	9	1	2	0.65	1.11	0.55	—
724	Chicken noodle, dehydrated, prepared w/water	1	cup(s)	252	237	58	2	9	<1	1	0.31	0.52	0.39	—
823	Cream of asparagus, condensed, prepared w/milk	1	cup(s)	248	213	161	6	16	1	8	3.32	2.08	2.23	—
824	Cream of celery, condensed, prepared w/milk	1	cup(s)	248	214	164	6	15	1	10	3.94	2.46	2.65	—
708	Cream of chicken, condensed, prepared w/milk	1	cup(s)	248	210	191	7	15	<1	11	4.64	4.46	1.64	—
715	Cream of chicken, condensed, prepared w/water	1	cup(s)	244	221	117	3	9	<1	7	2.07	3.27	1.49	—
709	Cream of mushroom, condensed, prepared w/milk	1	cup(s)	248	210	203	6	15	<1	14	5.13	2.98	4.61	—
716	Cream of mushroom, condensed, prepared w/water	1	cup(s)	244	220	129	2	9	<1	9	2.44	1.71	4.22	—
25298	Cream of vegetable	1	cup(s)	285	251	165	7	15	2	9	1.56	4.62	1.92	—
16689	Egg drop	1	cup(s)	244	229	73	8	1	0	4	1.15	1.52	0.59	—
25138	Golden squash	1	cup(s)	258	224	144	8	21	2	4	0.84	2.18	0.88	0
16663	Hot & sour	1	cup(s)	244	210	161	15	5	1	8	2.72	3.40	1.20	—
28054	Lentil chowder	1	cup(s)	229	188	150	11	27	12	<1	0.09	0.08	0.22	0
28560	Macaroni & bean	1	cup(s)	229	129	136	6	21	5	3	0.48	2.06	0.59	0
714	Manhattan clam chowder, condensed, prepared w/water	1	cup(s)	244	224	78	2	12	1	2	0.38	0.38	1.29	—
28561	Minestrone	1	cup(s)	230	177	99	4	16	5	2	0.32	1.30	0.43	0
717	Minestrone, condensed, prepared w/water	1	cup(s)	241	220	82	4	11	1	3	0.55	0.70	1.11	—
28038	Mushroom & wild rice	1	cup(s)	230	188	81	4	12	2	<1	0.05	0.02	0.15	0
828	New England clam chowder, condensed, prepared w/milk	1	cup(s)	248	211	164	9	17	1	7	2.95	2.26	1.09	—
28036	New England style clam chowder	1	cup(s)	229	207	83	3	15	2	<1	0.08	0.03	0.05	0
28566	Old country pasta	1	cup(s)	228	164	135	6	20	3	3	1.17	1.60	0.63	0
725	Onion, dehydrated, prepared w/water	1	cup(s)	246	237	27	1	5	1	1	0.12	0.32	0.07	—
16667	Shrimp gumbo	1	cup(s)	244	206	171	10	19	3	7	1.34	3.02	2.05	—
28037	Southwestern corn chowder	1	cup(s)	229	202	102	5	18	2	<1	0.12	0.12	0.20	0
25140	Split pea	1	cup(s)	165	117	85	4	19	2	<1	0.07	0.03	0.18	0
718	Split pea with ham, condensed, prepared w/water	1	cup(s)	253	207	190	10	28	2	4	1.77	1.80	0.63	—
710	Tomato, condensed, prepared w/milk	1	cup(s)	248	210	161	6	22	3	6	2.90	1.61	1.12	—
719	Tomato, condensed, prepared w/water	1	cup(s)	244	220	85	2	17	<1	2	0.37	0.44	0.95	—
726	Tomato vegetable, dehydrated, prepared w/water	1	cup(s)	253	237	56	2	10	1	1	0.38	0.30	0.08	—
28595	Turkey noodle	1	cup(s)	228	203	106	8	14	2	2	0.27	1.06	0.67	0
28051	Turkey vegetable	1	cup(s)	227	203	98	11	8	2	1	0.32	0.17	0.30	0
720	Vegetable beef, condensed, prepared w/water	1	cup(s)	244	224	78	6	10	<1	2	0.85	0.81	0.12	—
28598	Vegetable gumbo	1	cup(s)	229	168	153	4	26	3	4	0.61	2.93	0.56	0
25141	Vegetable	1	cup(s)	252	225	96	5	20	4	—	0.06	0.04	0.16	0
721	Vegetarian vegetable, condensed, prepared w/water	1	cup(s)	241	223	72	2	12	—	2	0.29	0.82	0.72	—

PAGE KEY: A–58 = Breads/Baked Goods A–62 = Cereal/Rice/Pasta A–66 = Fruit A–70 = Vegetables/Legumes A–80 = Nuts/Seeds A–82 = Vegetarian A–84 = Dairy A–90 = Eggs A–90 = Seafood A–92 = Meats A–96 = Poultry A–96 = Processed Meats A–98 = Beverages A–102 = Fats/Oils A–104 = Sweets A–106 = Sauces/Condiments/Spices A–108 = Mixed Foods/Soups/Sandwiches A–114 = Fast Food A–130 = Convenience A–132 = Baby Foods

Chol (mg)	Calc (mg)	Iron (mg)	Magn (mg)	Pota (mg)	Sodi (mg)	Zinc (mg)	Vit A (RAE) (µg)	Thia (mg)	Vit E (mg α)	Ribo (mg)	Niac (mg)	Vit B_6 (mg)	Fola (µg)	Vit C (mg)	Vit B_{12} (µg)	Sele (µg)
1	80	2.47	62	272	479	1.25	0	0.33	2.39	0.25	6.46	0.17	43	0	<.1	—
1	68	2.11	53	239	409	1.06	—	0.27	2.02	0.21	5.45	0.15	37	<1	<.1	—
51	54	4.23	31	316	792	3.39	11	0.38	—	0.31	5.87	0.26	57	2	1	29
73	41	2.81	67	330	845	4.38	30	0.41	—	0.41	5.96	0.32	71	6	2	26
73	92	5.16	49	524	798	4.53	20	0.41	—	0.37	7.30	0.37	90	6	2	42
53	100	3.29	35	302	795	1.08	46	0.26	0.35	0.26	12.29	0.48	70	1	2	71
69	78	3.10	34	315	490	2.94	—	0.30	0.74	0.33	6.64	0.46	24	0	<1	—
70	324	3.88	51	552	2408	2.66	—	0.53	1.19	0.49	12.50	0.54	46	5	2	—
5	80	3.08	61	590	690	1.41	26	0.27	0.03	0.15	3.61	0.23	139	3	<1	8
3	85	2.15	48	421	996	1.09	48	0.09	0.80	0.03	0.59	0.04	34	2	<.1	8
5	15	1.10	5	100	952	1.54	7	0.07	0.68	0.06	1.07	0.04	20	<1	<1	7
48	289	0.80	20	341	1019	0.68	359	0.06	—	0.33	0.50	0.08	10	1	<1	7
0	10	0.51	2	210	776	0.24	0	0.01	0.05	0.07	3.35	0.02	5	0	<1	0
24	26	1.34	16	335	776	0.77	49	0.15	0.02	0.16	5.57	0.13	40	1	<1	10
7	17	0.77	5	55	1106	0.39	36	0.05	0.10	0.06	1.39	0.03	22	<1	<1	6
10	5	0.50	8	33	577	0.20	3	0.20	0.13	0.08	1.09	0.03	18	0	<.1	10
22	174	0.87	20	360	1042	0.92	62	0.10	—	0.28	0.88	0.06	30	4	<1	8
32	186	0.69	22	310	1009	0.20	114	0.07	—	0.25	0.44	0.06	7	1	<1	5
27	181	0.67	17	273	1047	0.67	179	0.07	—	0.26	0.92	0.07	7	1	1	8
10	34	0.61	2	88	986	0.63	163	0.03	—	0.06	0.82	0.02	2	<1	<.1	7
20	179	0.60	20	270	918	0.64	35	0.08	1.24	0.28	0.91	0.06	10	2	<1	4
2	46	0.51	5	100	881	0.59	15	0.05	0.95	0.09	0.72	0.01	5	1	<.1	1
1	68	1.38	17	312	784	0.74	100	0.12	1.06	0.20	3.27	0.12	37	10	<1	5
103	21	0.75	5	220	729	0.48	—	0.02	0.29	0.19	3.03	0.05	15	0	<1	—
4	203	1.63	39	412	500	1.72	454	0.17	0.53	0.38	1.15	0.15	32	10	1	8
34	29	1.89	29	382	1561	1.51	—	0.27	0.12	0.25	4.97	0.20	13	1	<1	—
<1	47	4.07	55	590	26	1.44	163	0.21	0.06	0.12	1.69	0.30	164	13	0	3
<1	64	1.86	32	254	489	0.46	174	0.15	0.35	0.13	1.36	0.09	59	7	0	9
2	27	1.63	12	188	578	0.98	56	0.03	0.34	0.04	0.82	0.10	10	4	4	9
0	68	1.76	31	273	423	0.38	138	0.10	0.23	0.10	0.69	0.07	47	12	0	4
2	34	0.92	7	313	911	0.75	118	0.05	—	0.04	0.94	0.10	36	1	0	8
0	27	1.08	26	332	267	0.87	4	0.06	0.07	0.21	2.97	0.14	18	4	<.1	4
22	186	1.49	22	300	992	0.79	57	0.07	0.45	0.24	1.03	0.13	10	3	10	13
2	69	1.29	26	430	236	0.66	34	0.07	0.02	0.12	1.02	0.20	17	12	3	4
6	51	2.32	47	434	319	0.69	114	0.20	0.01	0.15	2.42	0.23	65	17	<.1	9
0	12	0.15	5	64	849	0.05	0	0.03	0.00	0.06	0.48	0.00	2	<1	0	2
51	99	2.34	51	515	515	0.93	—	0.19	1.90	0.10	2.54	0.19	59	26	<1	—
1	65	1.10	24	374	200	0.73	46	0.08	0.09	0.14	1.65	0.22	27	37	<1	2
0	30	1.25	33	352	608	0.57	112	0.12	0.00	0.09	1.67	0.21	61	9	0	<1
8	23	2.28	48	400	1007	1.32	23	0.15	—	0.08	1.47	0.07	3	2	<1	8
17	159	1.81	22	449	744	0.30	64	0.13	1.24	0.25	1.52	0.16	17	68	<1	2
0	12	1.76	7	264	695	0.24	29	0.09	2.32	0.05	1.42	0.11	15	66	0	<1
0	8	0.63	20	104	1146	0.18	10	0.06	0.35	0.05	0.79	0.05	10	6	0	5
24	27	1.40	22	200	372	0.67	81	0.20	0.02	0.11	2.68	0.15	45	5	<1	13
20	36	1.30	22	383	328	0.90	110	0.08	0.01	0.09	3.33	0.27	21	10	<1	9
5	17	1.12	5	173	791	1.54	95	0.04	0.37	0.05	1.03	0.08	10	2	<1	4
0	52	1.90	35	313	471	0.56	15	0.17	0.58	0.07	1.59	0.16	51	18	0	4
0	41	2.45	38	688	674	0.78	118	0.12	0.00	0.13	2.37	0.27	33	23	0	5
0	22	1.08	7	210	822	0.46	116	0.05	—	0.05	0.92	0.06	10	1	0	4

TABLE F–1

Food Composition (Computer code number is for Wadsworth Diet Analysis program) (For purposes of calculations, use "0" for t, <1, <.1, <.01, etc.)

DA + Code	Food Description	Quantity	Measure	Wt (g)	H_2O (g)	Ener (kcal)	Prot (g)	Carb (g)	Dietary Fiber (g)	Fat (g)	Fat Breakdown (g) Sat	Mono	Poly	Trans
	FAST FOOD													
	Arby's													
36094	Au jus sauce	1	serving(s)	85	—	5	<1	1	<.1	<.1	0.02	—	—	—
751	Beef 'n cheddar sandwich	1	item(s)	198	—	480	23	43	2	24	8.00	—	—	—
9279	Cheddar curly fries	1	serving(s)	170	—	460	6	54	4	24	6.00	—	—	—
36131	Chocolate shake	1	serving(s)	397	—	480	10	84	0	16	8.00	—	—	—
36045	Curly fries, large	1	serving(s)	198	—	620	8	78	7	30	7.00	—	—	—
36044	Curly fries, medium	1	serving(s)	128	—	400	5	50	4	20	5.00	—	—	—
9265	Fish fillet sandwich	1	item(s)	220	—	529	23	50	2	27	7.00	9.20	10.60	—
752	Ham 'n cheese sandwich	1	item(s)	170	—	340	23	35	1	13	4.50	—	—	—
36048	Homestyle fries, large	1	serving(s)	213	—	560	6	79	6	24	6.00	—	—	—
36047	Homestyle fries, medium	1	serving(s)	142	—	370	4	53	4	16	4.00	—	—	—
33465	Homestyle fries, small	1	serving(s)	113	—	300	3	42	3	13	3.50	—	—	—
9267	Italian sub sandwich	1	item(s)	312	—	780	29	49	3	53	15.00	—	—	—
36041	Market Fresh grilled chicken caesar salad w/o dressing	1	serving(s)	338	—	230	33	8	3	8	3.50	—	—	—
9291	Roast beef deluxe sandwich, light	1	item(s)	182	—	296	18	33	6	10	3.00	5.00	2.00	—
9251	Roast beef sandwich, giant	1	item(s)	228	—	480	32	41	3	23	10.00	—	—	—
9249	Roast beef sandwich, junior	1	item(s)	129	—	310	16	34	2	13	4.50	—	—	—
750	Roast beef sandwich, regular	1	item(s)	157	—	350	21	34	2	16	6.00	—	—	—
2009	Roast beef sandwich, super	1	item(s)	245	—	470	22	47	3	23	7.00	—	—	—
9269	Roast beef sub sandwich	1	item(s)	334	—	760	35	47	3	48	16.00	—	—	—
9295	Roast chicken deluxe sandwich, light	1	item(s)	194	—	260	23	33	3	5	1.00	—	—	—
9293	Roast turkey deluxe sandwich, light	1	item(s)	194	—	260	23	33	3	5	0.50	—	—	—
36132	Strawberry shake	1	serving(s)	397	—	500	11	87	0	13	8.00	—	—	—
9273	Turkey sub sandwich	1	item(s)	306	—	630	26	51	2	37	9.00	—	—	—
36130	Vanilla shake	1	serving(s)	397	—	470	10	83	0	15	7.00	—	—	—
	Auntie Anne's													
35371	Cheese dipping sauce	1	serving(s)	35	—	100	3	4	0	8	4.00	—	—	—
35353	Cinnamon sugar soft pretzel	1	item(s)	120	—	350	9	74	2	2	0.00	—	—	—
35354	Cinnamon sugar soft pretzel w/butter	1	item(s)	120	—	450	8	83	3	9	5.00	—	—	—
35372	Marinara dipping sauce	1	serving(s)	35	—	10	0	4	0	0	0.00	0.00	0.00	0
35357	Original soft pretzel	1	item(s)	120	—	340	10	72	3	1	0.00	—	—	—
35358	Original soft pretzel w/butter	1	item(s)	120	—	370	10	72	3	4	2.00	—	—	—
35359	Parmesan herb soft pretzel	1	item(s)	120	—	390	11	74	4	5	2.50	—	—	—
35360	Parmesan herb soft pretzel w/butter	1	item(s)	120	—	440	10	72	9	13	7.00	—	—	—
35361	Sesame soft pretzel	1	item(s)	120	—	350	11	63	3	6	1.00	—	—	—
35362	Sesame soft pretzel w/butter	1	item(s)	120	—	410	12	64	7	12	4.00	—	—	—
35364	Sour cream & onion soft pretzel	1	item(s)	120	—	310	9	66	2	1	0.00	—	—	—
35366	Sour cream & onion soft pretzel w/butter	1	item(s)	120	—	340	9	66	2	5	3.00	—	—	—
35373	Sweet mustard dipping sauce	1	serving(s)	35	—	60	1	8	0	2	1.00	—	—	—
35367	Whole wheat soft pretzel	1	item(s)	120	—	350	11	72	7	2	0.00	—	—	—
35368	Whole wheat soft pretzel w/butter	1	item(s)	120	—	370	11	72	7	5	1.50	—	—	—
	Boston Market													
34975	Bbq baked beans	¾	cup(s)	201	—	270	8	48	12	5	2.00	—	—	—
34976	Black beans & rice	1	cup(s)	227	—	300	8	45	5	10	1.50	—	—	—
34978	Butternut squash	¾	cup(s)	193	—	150	2	25	6	6	4.00	—	—	—
35006	Caesar side salad	1	serving(s)	119	—	300	5	13	1	26	4.50	—	—	—
34979	Chicken gravy	1	ounce(s)	28	—	15	0	2	0	1	0.00	—	—	—
34973	Chicken pot pie	1	item(s)	425	—	750	26	57	2	46	14.00	—	—	—
35007	Cole slaw	¾	cup(s)	184	—	300	2	30	3	19	3.00	—	—	—
35057	Cornbread	1	item(s)	68	—	200	3	33	1	6	1.50	—	—	—
35008	Cranberry walnut relish	¾	cup(s)	210	—	350	3	75	3	5	0.00	—	—	—
34980	Creamed spinach	¾	cup(s)	181	—	260	9	11	2	20	13.00	—	—	—
34981	Glazed carrots	¾	cup(s)	153	—	280	1	35	4	15	3.00	—	—	—
34983	Green bean casserole	¾	cup(s)	170	—	80	1	9	2	5	1.50	—	—	—
34982	Green beans	¾	cup(s)	85	—	70	1	6	2	4	0.50	—	—	—
34967	Half chicken, w/skin	1	item(s)	277	—	590	70	4	0	33	10.00	—	—	—
34984	Homestyle mashed potatoes	¾	cup(s)	173	—	210	4	30	2	9	5.00	—	—	—
34985	Homestyle mashed potatoes & gravy	1	cup(s)	201	—	230	4	32	3	9	5.00	—	—	—
34969	Honey glazed ham	5	ounce(s)	142	—	210	24	10	0	8	3.00	—	—	—
34988	Hot cinnamon apples	¾	cup(s)	181	—	250	0	56	3	5	0.50	—	—	—
34989	Macaroni & cheese	¾	cup(s)	192	—	280	13	33	1	11	6.00	—	—	—
34970	Meatloaf	5	ounce(s)	142	—	282	20	15	1	17	7.28	—	—	—
35012	Old-fashioned potato salad	¾	cup(s)	150	—	200	3	22	2	12	2.00	—	—	—

PAGE KEY: A–58 = Breads/Baked Goods A–62 = Cereal/Rice/Pasta A–66 = Fruit A–70 = Vegetables/Legumes A–80 = Nuts/Seeds A–82 = Vegetarian A–84 = Dairy A–90 = Eggs A–90 = Seafood A–92 = Meats A–96 = Poultry A–96 = Processed Meats A–98 = Beverages A–102 = Fats/Oils A–104 = Sweets A–106 = Sauces/Condiments/Spices A–108 = Mixed Foods/Soups/Sandwiches A–114 = Fast Food A–130 = Convenience A–132 = Baby Foods

Chol (mg)	Calc (mg)	Iron (mg)	Magn (mg)	Pota (mg)	Sodi (mg)	Zinc (mg)	Vit A (RAE) (µg)	Thia (mg)	Vit E (mg α)	Ribo (mg)	Niac (mg)	Vit B_6 (mg)	Fola (µg)	Vit C (mg)	Vit B_{12} (µg)	Sele (µg)
0	0	0.00	—	—	386	—	0	—	—	—	—	—	—	0	—	—
90	100	3.60	—	—	1240	—	0	—	—	—	—	—	—	1	—	—
5	60	1.80	—	—	1290	—	0	—	—	—	—	—	—	15	—	—
45	500	0.72	—	—	370	—	38	—	—	—	—	—	—	2	—	—
0	0	2.70	—	—	1540	—	0	—	—	—	—	—	—	21	—	—
0	0	1.80	—	—	990	—	0	—	—	—	—	—	—	15	—	—
43	90	3.78	—	450	864	—	10	0.35	—	0.31	5.60	—	—	1	—	—
90	150	2.70	—	—	1450	—	20	—	—	—	—	—	—	1	—	—
0	0	1.80	—	—	1070	—	0	—	—	—	—	—	—	30	—	—
0	0	1.08	—	—	710	—	0	—	—	—	—	—	—	21	—	—
0	0	0.72	—	—	570	—	0	—	—	—	—	—	—	15	—	—
120	250	2.70	—	—	2440	—	—	—	—	—	—	—	—	2	—	—
80	200	1.80	—	—	920	—	—	—	—	—	—	—	—	42	—	—
42	130	4.50	—	392	826	—	40	0.27	—	0.49	8.40	—	—	8	—	—
110	60	5.40	—	—	1440	—	0	—	—	—	—	—	—	0	—	—
70	60	2.70	—	—	740	—	0	—	—	—	—	—	—	0	—	—
85	60	3.60	—	—	950	—	0	—	—	—	—	—	—	0	—	—
85	80	3.60	—	—	1130	—	40	—	—	—	—	—	—	1	—	—
130	300	4.50	—	—	2230	—	40	—	—	—	—	—	—	4	—	—
40	100	2.70	—	—	1010	—	—	—	—	—	—	—	—	2	—	—
40	80	1.80	—	—	980	—	—	—	—	—	—	—	—	1	—	—
15	350	0.36	—	—	340	—	36	—	—	—	—	—	—	1	—	—
100	200	0.36	—	—	2170	—	—	—	—	—	—	—	—	2	—	—
45	500	1.08	—	—	360	—	39	—	—	—	—	—	—	2	—	—
10	100	0.00	—	—	510	—	—	—	—	—	—	—	—	0	—	—
0	20	1.98	—	—	410	—	0	—	—	—	—	—	—	0	—	—
25	30	2.34	—	—	430	—	—	—	—	—	—	—	—	0	—	—
0	0	0.00	—	—	180	—	0	—	—	—	—	—	—	0	—	—
0	30	2.34	—	—	900	—	0	—	—	—	—	—	—	0	—	—
10	30	2.16	—	—	930	—	—	—	—	—	—	—	—	0	—	—
10	80	1.80	—	—	780	—	—	—	—	—	—	—	—	1	—	—
30	60	1.80	—	—	660	—	—	—	—	—	—	—	—	1	—	—
0	20	2.88	—	—	840	—	0	—	—	—	—	—	—	0	—	—
15	20	2.70	—	—	860	—	—	—	—	—	—	—	—	0	—	—
0	30	1.98	—	—	920	—	—	—	—	—	—	—	—	0	—	—
10	40	2.16	—	—	930	—	—	—	—	—	—	—	—	0	—	—
40	0	0.00	—	—	120	—	0	—	—	—	—	—	—	0	—	—
0	30	1.98	—	—	1100	—	0	—	—	—	—	—	—	0	—	—
10	30	2.34	—	—	1120	—	—	—	—	—	—	—	—	0	—	—
0	100	3.60	—	—	540	—	42	—	—	—	—	—	—	6	—	—
0	40	1.80	—	—	1050	—	0	—	—	—	—	—	—	4	—	—
20	80	1.08	—	—	560	—	1150	—	—	—	—	—	—	30	—	—
15	100	0.72	—	—	690	—	—	—	—	—	—	—	—	9	—	—
0	0	0.00	—	—	180	—	0	—	—	—	—	—	—	0	—	—
110	40	4.50	—	—	1530	—	—	—	—	—	—	—	—	1	—	—
20	60	0.72	—	—	540	—	108	—	—	—	—	—	—	36	—	—
25	0	1.08	—	—	390	—	0	—	—	—	—	—	—	0	—	—
0	0	5.40	—	—	0	—	0	—	—	—	—	—	—	0	—	—
55	250	2.70	—	—	740	—	—	—	—	—	—	—	—	9	—	—
0	40	1.08	—	—	80	—	1000	—	—	—	—	—	—	1	—	—
5	20	0.72	—	—	670	—	—	—	—	—	—	—	—	2	—	—
0	40	0.36	—	—	250	—	30	—	—	—	—	—	—	5	—	—
290	0	2.70	—	—	1010	—	0	—	—	—	—	—	—	0	—	—
25	40	0.36	—	—	590	—	53	—	—	—	—	—	—	15	—	—
25	60	0.36	—	—	780	—	—	—	—	—	—	—	—	15	—	—
75	0	1.08	—	—	1460	—	0	—	—	—	—	—	—	0	—	—
0	20	0.36	—	—	45	—	—	—	—	—	—	—	—	0	—	—
30	300	1.44	—	—	890	—	—	—	—	—	—	—	—	0	—	—
68	91	2.46	—	—	592	—	—	—	—	—	—	—	—	1	—	—
15	60	1.08	—	—	450	—	0	—	—	—	—	—	—	6	—	—

TABLE F–1

Food Composition (Computer code number is for Wadsworth Diet Analysis program) (For purposes of calculations, use "0" for t, <1, <.1, <.01, etc.)

DA + Code	Food Description	Quantity	Measure	Wt (g)	H_2O (g)	Ener (kcal)	Prot (g)	Carb (g)	Dietary Fiber (g)	Fat (g)	Fat Breakdown (g) Sat	Mono	Poly	Trans
	FAST FOOD—Continued													
34965	Quarter chicken, dark meat, no skin	1	item(s)	95	—	190	22	1	0	10	3.00	—	—	—
34966	Quarter chicken, dark meat, w/skin	1	item(s)	125	—	320	30	2	0	21	6.00	—	—	—
34963	Quarter chicken, white meat, no skin or wing	1	item(s)	140	—	170	33	2	0	4	1.00	—	—	—
34964	Quarter chicken, white meat, w/skin & wing	1	item(s)	152	—	280	40	2	0	12	3.50	—	—	—
34993	Rice pilaf	1	cup(s)	137	—	140	2	24	1	4	0.50	—	—	—
34968	Rotisserie turkey breast, skinless	5	ounce(s)	142	—	170	36	3	0	1	0.00	—	—	—
34998	Savory stuffing	1	cup(s)	132	—	190	4	27	2	8	1.50	—	—	—
34999	Squash casserole	¾	cup(s)	187	—	330	7	20	3	24	13.00	—	—	—
35003	Steamed vegetables	1	cup(s)	102	—	30	2	6	2	0	0.00	—	—	0
35004	Sweet potato casserole	¾	cup(s)	181	—	280	3	39	2	13	4.50	—	—	—
35005	Whole kernel corn	¾	cup(s)	146	—	180	5	30	2	4	0.50	—	—	—
	Burger King													
29731	Biscuit with sausage, egg, & cheese	1	item(s)	189	—	650	20	38	1	46	14.00	—	—	1
3739	BK Broiler chicken sandwich	1	item(s)	258	—	550	30	52	3	25	5.00	—	—	—
14249	Cheeseburger	1	item(s)	133	—	360	19	31	2	17	8.00	—	—	0.50
14251	Chicken sandwich	1	item(s)	224	—	660	25	53	3	39	8.00	—	—	2.20
3808	Chicken Tenders, 8 pieces	1	serving(s)	123	—	340	22	20	1	19	5.00	—	—	3.50
14259	Chocolate shake, small	1	item(s)	333	—	620	12	72	2	32	21.00	—	—	0
29732	Croissanwich w/sausage & cheese	1	item(s)	107	—	420	14	23	1	31	11.00	—	—	2
14261	Croissanwich w/sausage, egg, & cheese	1	item(s)	157	—	520	19	24	1	39	14.00	—	—	1.93
3809	Double cheeseburger	1	item(s)	189	—	540	32	32	2	31	15.00	—	—	1.50
14244	Double Whopper	1	item(s)	374	—	980	52	52	4	62	22.00	—	—	2
14245	Double Whopper w/cheese	1	item(s)	399	—	1070	57	53	4	70	27.00	—	—	2.50
14250	Fish Fillet sandwich	1	item(s)	185	—	520	18	44	2	30	8.00	—	—	1.12
14255	French fries, medium, salted	1	item(s)	117	—	360	4	46	4	18	5.00	—	—	4.50
14262	French toast sticks	1	serving(s)	112	—	390	6	46	2	20	4.50	—	—	4.50
14248	Hamburger	1	item(s)	121	—	310	17	31	2	13	5.00	—	—	0.50
14263	Hash brown rounds, small	1	serving(s)	75	—	230	2	23	2	15	4.00	—	—	5.0
14256	Onion rings, medium	1	serving(s)	91	—	320	4	40	3	16	4.00	—	—	3.50
39000	Tendercrisp chicken sandwich	1	item(s)	310	—	810	28	72	6	47	8.00	—	—	4.28
14258	Vanilla shake, small	1	item(s)	305	—	560	11	56	1	32	21.00	—	—	0
1736	Whopper	1	item(s)	291	—	710	31	52	4	43	13.00	—	—	1
14243	Whopper w/cheese	1	item(s)	316	—	800	36	53	4	50	18.00	—	—	2
	Carl's Jr													
10801	Carl's Catch fish sandwich	1	item(s)	201	—	530	18	55	2	28	7.00	—	1.89	—
10862	Carl's Famous Star hamburger	1	item(s)	254	—	590	24	50	3	32	9.00	—	—	—
10866	Charboiled chicken salad-to-go	1	item(s)	350	—	200	25	12	4	7	3.00	—	1.02	—
10855	Charboiled Sante Fe chicken sandwich	1	item(s)	220	—	540	28	37	2	31	8.00	—	—	—
10790	Chicken stars (6 pieces)	6	item(s)	90	—	260	13	14	1	16	4.50	—	1.71	—
34864	Chocolate shake, small	1	item(s)	595	—	530	14	96	0	10	7.00	—	—	—
10797	Crisscut fries	1	serving(s)	139	—	410	5	43	4	24	5.00	—	—	—
10799	Double western bacon cheeseburger	1	item(s)	308	—	920	51	65	3	50	21.00	—	6.55	—
34855	Famous bacon cheeseburger	1	item(s)	279	—	700	31	51	3	41	13.00	—	—	—
14238	French fries, small	1	serving(s)	92	—	290	5	37	3	14	3.00	—	—	—
10798	French toast dips w/o syrup	1	serving(s)	105	—	370	6	42	1	20	2.50	—	1.35	—
34856	Hamburger	1	item(s)	119	—	280	14	36	1	9	3.50	—	—	—
10802	Onion rings	1	serving(s)	127	—	430	7	53	3	22	5.00	—	0.84	—
38925	Six Dollar burger	1	item(s)	539	—	1000	39	72	6	82	25.00	—	—	—
34858	Spicy chicken sandwich	1	item(s)	198	—	480	14	47	2	26	5.00	—	—	—
34867	Strawberry shake, small	1	item(s)	595	—	510	14	91	0	10	7.00	—	—	—
10865	Super Star hamburger	1	item(s)	345	—	790	41	51	3	47	15.00	—	—	—
10818	Vanilla shake, small	1	item(s)	595	—	470	15	78	0	11	7.00	—	—	—
10770	Western bacon cheeseburger	1	item(s)	225	—	660	31	64	3	30	12.00	—	4.85	—
	Chick Fil-A													
38746	Biscuit w/bacon, egg, & cheese	1	item(s)	155	—	430	16	38	1	24	9.00	—	—	2.85
38747	Biscuit w/egg	1	item(s)	135	—	340	11	38	1	16	4.50	—	—	3
38748	Biscuit w/egg & cheese	1	item(s)	148	—	390	13	38	1	21	7.00	—	—	2.98
38753	Biscuit w/gravy	1	item(s)	191	—	310	5	44	1	13	3.50	—	—	3.98
38752	Biscuit w/sausage, egg, & cheese	1	item(s)	189	—	540	18	43	1	33	13.00	—	—	2.67
38741	Biscuit, plain	1	item(s)	78	—	260	4	38	1	11	2.50	—	—	2.97
38771	Carrot & raisin salad	1	item(s)	91	—	130	1	22	2	5	1.00	—	—	0
38761	Chargrilled chicken cool wrap	1	item(s)	245	—	380	29	54	3	6	3.00	—	—	0
38766	Chargrilled chicken garden salad	1	item(s)	275	—	180	22	9	3	6	3.00	—	—	0
38758	Chargrilled chicken sandwich	1	item(s)	157	—	280	26	30	1	7	1.50	—	—	0

PAGE KEY: A–58 = Breads/Baked Goods A–62 = Cereal/Rice/Pasta A–66 = Fruit A–70 = Vegetables/Legumes A–80 = Nuts/Seeds A–82 = Vegetarian A–84 = Dairy A–90 = Eggs A–90 = Seafood A–92 = Meats A–96 = Poultry A–96 = Processed Meats A–98 = Beverages A–102 = Fats/Oils A–104 = Sweets A–106 = Sauces/Condiments/Spices A–108 = Mixed Foods/Soups/Sandwiches A–114 = Fast Food A–130 = Convenience A–132 = Baby Foods

Chol (mg)	Calc (mg)	Iron (mg)	Magn (mg)	Pota (mg)	Sodi (mg)	Zinc (mg)	Vit A (RAE) (µg)	Thia (mg)	Vit E (mg α)	Ribo (mg)	Niac (mg)	Vit B_6 (mg)	Fola (µg)	Vit C (mg)	Vit B_{12} (µg)	Sele (µg)
115	0	1.08	—	—	440	—	0	—	—	—	—	—	—	0	—	—
155	0	1.80	—	—	500	—	0	—	—	—	—	—	—	0	—	—
85	0	0.72	—	—	480	—	0	—	—	—	—	—	—	0	—	—
135	0	1.08	—	—	510	—	0	—	—	—	—	—	—	0	—	—
0	20	1.08	—	—	520	—	—	—	—	—	—	—	—	4	—	—
100	20	1.80	—	—	850	—	0	—	—	—	—	—	—	0	—	—
5	40	1.44	—	—	620	—	—	—	—	—	—	—	—	2	—	—
70	200	0.72	—	—	1110	—	—	—	—	—	—	—	—	5	—	—
0	40	0.35	—	—	135	—	389	—	—	—	—	—	—	18	—	—
10	40	1.08	—	—	190	—	—	—	—	—	—	—	—	9	—	—
0	0	0.36	—	—	170	—	20	—	—	—	—	—	—	5	—	—
190	150	2.70	—	—	1600	—	90	—	—	—	—	—	—	0	—	—
105	60	3.60	—	—	1110	—	—	0.46	—	0.23	10.50	—	—	6	—	—
50	150	3.60	—	—	790	—	63	0.25	—	0.32	4.18	—	—	1	—	—
70	80	2.70	—	—	1330	—	—	0.47	—	0.30	9.59	—	—	0	—	—
50	20	0.72	—	—	840	—	—	0.14	—	0.12	10.93	—	—	0	—	—
95	350	1.08	—	—	310	—	42	0.11	—	0.56	0.24	—	—	0	—	—
45	100	3.60	—	—	840	—	—	—	—	—	—	—	—	0	—	—
210	300	4.50	—	—	1090	—	140	0.36	—	0.42	4.35	—	—	0	—	—
100	250	4.50	—	—	1050	—	100	0.26	—	0.45	6.37	—	—	1	—	—
160	150	9.00	—	—	1070	—	—	0.40	—	0.60	11.08	—	—	9	—	—
185	300	9.00	—	—	1500	—	—	0.40	—	0.67	11.07	—	—	9	—	—
55	150	2.70	—	—	840	—	14	—	—	—	—	—	—	1	—	—
0	20	0.72	—	—	640	—	0	0.16	—	0.48	2.32	—	—	9	—	—
0	60	1.80	—	—	440	—	0	0.19	—	0.22	2.86	—	—	0	—	—
40	76	3.60	—	—	580	—	9	0.25	—	0.29	4.26	—	—	1	—	—
0	0	0.36	—	—	450	—	0	0.11	—	0.07	2.11	—	—	1	—	—
0	97	0.00	—	—	460	—	0	0.14	—	0.09	2.33	—	—	0	—	—
60	80	4.50	—	—	1800	—	—	—	—	—	—	—	—	9	—	—
95	300	0.36	—	—	220	—	39	0.11	—	0.64	0.22	—	—	0	—	—
85	150	6.30	—	—	980	—	52	0.39	—	0.44	7.33	—	—	9	—	—
110	250	6.30	—	—	1420	—	157	0.39	—	0.51	7.31	—	—	9	—	—
80	150	1.80	—	—	1030	—	60	—	—	—	—	—	—	2	—	—
70	100	4.50	—	—	910	—	—	—	—	—	—	—	—	6	—	—
75	150	1.80	—	—	440	—	—	—	—	—	—	—	—	5	—	—
95	200	2.70	—	—	1210	—	—	—	—	—	—	—	—	6	—	—
40	20	1.08	—	—	480	—	0	—	—	—	—	—	—	0	—	—
45	600	1.08	—	—	350	—	0	—	—	—	—	—	—	0	—	—
0	20	1.80	—	—	950	—	0	—	—	—	—	—	—	12	—	—
155	300	7.20	—	—	1770	—	—	—	—	—	—	—	—	1	—	—
95	200	5.40	—	—	1310	—	102	—	—	—	—	—	—	6	—	—
0	0	1.08	—	—	180	—	0	—	—	—	—	—	—	21	—	—
0	40	1.08	—	—	430	—	0	0.26	—	0.24	2.00	—	—	0	—	—
35	80	2.70	—	—	480	—	0	—	—	—	—	—	—	1	—	—
0	20	0.72	—	—	700	—	0	—	—	—	—	—	—	4	—	—
135	350	5.40	—	—	1690	—	—	—	—	—	—	—	—	21	—	—
40	100	2.70	—	—	1220	—	—	—	—	—	—	—	—	6	—	—
45	600	0.00	—	—	330	—	0	—	—	—	—	—	—	0	—	—
130	100	7.20	—	—	980	—	—	—	—	—	—	—	—	9	—	—
50	600	0.00	—	—	350	—	0	—	—	—	—	—	—	0	—	—
85	200	5.40	—	—	1410	—	40	—	—	—	—	—	—	1	—	—
265	150	3.60	—	—	1070	—	—	—	—	—	—	—	—	0	—	—
245	80	2.70	—	—	740	—	—	—	—	—	—	—	—	0	—	—
260	150	2.70	—	—	960	—	—	—	—	—	—	—	—	0	—	—
5	60	1.80	—	—	930	—	0	—	—	—	—	—	—	0	—	—
280	150	3.60	—	—	1030	—	—	—	—	—	—	—	—	0	—	—
0	60	1.80	—	—	670	—	0	—	—	—	—	—	—	0	—	—
0	20	0.36	—	—	90	—	—	—	—	—	—	—	—	4	—	—
70	200	2.70	—	—	1060	—	—	—	—	—	—	—	—	6	—	—
70	150	0.72	—	—	660	—	—	—	—	—	—	—	—	30	—	—
70	80	1.80	—	—	980	—	0	—	—	—	—	—	—	2	—	—

TABLE F-1
Food Composition (Computer code number is for Wadsworth Diet Analysis program) (For purposes of calculations, use "0" for t, <1, <.1, <.01, etc.)

DA + Code	Food Description	Quantity	Measure	Wt (g)	H_2O (g)	Ener (kcal)	Prot (g)	Carb (g)	Dietary Fiber (g)	Fat (g)	Fat Breakdown (g) Sat	Mono	Poly	Trans
	FAST FOOD—Continued													
38759	Chargrilled deluxe chicken sandwich	1	item(s)	195	—	290	27	31	2	7	1.50	—	—	0
38742	Chicken biscuit	1	item(s)	137	—	400	16	43	2	18	4.50	—	—	2.83
38743	Chicken biscuit w/cheese	1	item(s)	151	—	450	19	43	2	23	7.00	—	—	2.85
38762	Chicken caesar wrap	1	item(s)	227	—	460	36	52	2	10	6.00	—	—	0
38757	Chicken deluxe sandwich	1	item(s)	208	—	420	28	39	2	16	3.50	—	—	0
38764	Chicken salad sandwich	1	item(s)	153	—	350	20	32	5	15	3.00	—	—	0
38756	Chicken sandwich	1	item(s)	170	—	410	28	38	1	15	3.50	—	—	0
38768	Chick-n-Strip salad	1	item(s)	331	—	390	34	22	4	18	5.00	—	—	0
38763	Chick-n-Strips	4	item(s)	127	—	290	29	14	1	13	2.50	—	—	0
38770	Coleslaw	1	item(s)	105	—	210	1	14	2	17	2.50	—	—	0
38755	Hash browns	1	serving(s)	84	—	170	2	20	2	9	4.50	—	—	1
38765	Hearty breast of soup	1	cup(s)	241	—	140	8	18	1	4	1.00	—	—	0
38778	Icedream, small cone	1	item(s)	135	—	160	4	28	0	4	2.00	—	—	0
38774	Icedream, small cup	1	serving(s)	213	—	230	5	38	0	6	3.50	—	—	0
38775	Lemonade	1	cup(s)	255	—	170	0	41	0	1	0.00	—	—	0
38776	Lemonade, diet	1	cup(s)	255	—	25	0	5	0	0	0.00	0.00	0.00	0
38777	Nuggets	8	item(s)	113	—	260	26	12	1	12	2.50	—	—	0
38769	Side salad	1	item(s)	108	—	60	3	4	2	3	1.50	—	—	0
38767	Southwest chargrilled salad	1	item(s)	303	—	240	22	17	5	8	3.50	—	—	0
38772	Waffle potato fries, small, salted	1	serving(s)	85	—	280	3	37	5	14	5.00	—	—	1.50
	Cinnabon													
39569	Caramel Pecanbon	1	item(s)	272	—	1100	16	141	8	56	10.00	—	—	5
39572	Caramellata Chill w/whipped cream	16	fluid ounce(s)	480	—	406	10	61	0	14	8.00	—	—	—
39571	Cinnapoppers	1	serving(s)	74	—	368	4	41	2	21	11.00	—	—	1
39567	Classic roll	1	item(s)	221	—	813	15	117	4	32	8.00	—	—	5
39568	Minibon	1	item(s)	92	—	339	6	49	2	13	3.00	—	—	2
39573	Mochalatta chill w/whipped cream	16	fluid ounce(s)	480	—	362	9	55	0	13	8.00	—	—	—
39570	Stix	5	item(s)	85	—	379	6	41	1	21	6.00	—	—	4
	Dairy Queen													
1466	Banana split	1	item(s)	369	—	510	8	96	3	12	8.00	3.00	0.50	0
38552	Brownie Earthquake	1	serving(s)	304	—	740	10	112	0	27	16.00	—	—	3
38561	Chocolate chip cookie dough blizzard, small	1	item(s)	319	—	720	12	105	0	28	14.00	—	—	2.50
1464	Chocolate malt, small	1	item(s)	418	—	650	15	111	0	16	10.00	—	—	0.50
38541	Chocolate shake, small	1	item(s)	397	—	560	13	93	1	15	10.00	—	—	0.50
17257	Chocolate soft serve	½	cup(s)	94	—	150	4	22	0	5	3.50	—	—	0
1463	Chocolate sundae, small	1	item(s)	163	—	280	5	49	0	7	4.50	1.00	1.00	0
1462	Dipped cone, small	1	item(s)	156	—	340	6	42	1	17	9.00	4.00	3.00	1
38555	Oreo cookies blizzard, small	1	item(s)	283	—	570	11	83	1	21	10.00	—	—	2.50
38547	Royal Treats Peanut Buster parfait	1	item(s)	305	—	730	16	99	2	31	17.00	—	—	0
17256	Vanilla soft serve	½	cup(s)	94	—	140	3	22	0	5	3.00	—	—	0
	Domino's													
31606	Barbeque wings	1	item(s)	25	—	50	6	2	<1	2	0.65	—	—	—
31604	Breadsticks	1	item(s)	37	—	116	3	18	1	4	0.79	—	—	—
37551	Buffalo chicken kickers	1	item(s)	24	14	47	4	3	<1	2	0.39	—	—	—
37548	Cinnastix	1	item(s)	32	8	122	2	15	1	6	1.15	—	—	—
	Classic hand tossed pizza													
31573	America's favorite feast, 12"	2	slice(s)	205	99	508	22	57	4	22	9.20	—	—	—
31574	America's favorite feast, 14"	2	slice(s)	283	138	697	30	79	5	30	12.70	—	—	—
37543	Bacon cheeseburger feast, 12"	2	slice(s)	198	60	549	25	55	3	26	11.62	—	—	—
37545	Bacon cheeseburger feast, 14"	2	slice(s)	275	121	762	35	75	4	36	16.10	—	—	—
37546	Barbeque feast, 12"	2	slice(s)	192	85	506	22	62	3	20	9.08	—	—	—
37547	Barbeque feast, 14"	2	slice(s)	262	115	691	30	85	4	27	12.24	—	—	—
31569	Cheese, 12"	2	slice(s)	159	—	375	15	55	3	11	4.81	—	—	—
31570	Cheese, 14"	2	slice(s)	219	—	516	21	75	4	15	6.72	—	—	—
37538	Deluxe feast, 12"	2	slice(s)	201	102	465	20	57	3	18	7.66	—	—	—
37540	Deluxe feast, 14"	2	slice(s)	273	138	627	26	78	5	24	10.20	—	—	—
31685	Deluxe, 12"	2	slice(s)	213	—	465	20	57	3	18	7.65	—	—	—
31694	Deluxe, 14"	2	slice(s)	273	—	627	26	78	5	24	10.20	—	—	—
31686	Extravaganzza, 12"	2	slice(s)	245	127	576	27	59	4	27	11.56	—	—	—
31695	Extravaganzza, 14"	2	slice(s)	329	171	773	36	88	5	36	15.42	—	—	—
31575	Hawaiian feast, 12"	2	slice(s)	204	105	450	21	58	3	16	7.20	—	—	—
31576	Hawaiian feast, 14"	2	slice(s)	283	147	623	29	80	5	22	10.09	—	—	—
31687	Meatzza, 12"	2	slice(s)	213	—	560	26	57	3	26	11.40	—	—	—
31696	Meatzza, 14"	2	slice(s)	293	139	753	35	78	5	34	15.24	—	—	—

PAGE KEY: A–58 = Breads/Baked Goods A–62 = Cereal/Rice/Pasta A–66 = Fruit A–70 = Vegetables/Legumes A–80 = Nuts/Seeds A–82 = Vegetarian A–84 = Dairy A–90 = Eggs A–90 = Seafood A–92 = Meats A–96 = Poultry A–96 = Processed Meats A–98 = Beverages A–102 = Fats/Oils A–104 = Sweets A–106 = Sauces/Condiments/Spices A–108 = Mixed Foods/Soups/Sandwiches A–114 = Fast Food A–130 = Convenience A–132 = Baby Foods

Chol (mg)	Calc (mg)	Iron (mg)	Magn (mg)	Pota (mg)	Sodi (mg)	Zinc (mg)	Vit A (RAE) (μg)	Thia (mg)	Vit E (mg α)	Ribo (mg)	Niac (mg)	Vit B_6 (mg)	Fola (μg)	Vit C (mg)	Vit B_{12} (μg)	Sele (μg)
70	80	1.80	—	—	990	—	—	—	—	—	—	—	—	5	—	—
30	60	2.70	—	—	1200	—	0	—	—	—	—	—	—	0	—	—
45	150	2.70	—	—	1430	—	—	—	—	—	—	—	—	0	—	—
80	500	2.70	—	—	1390	—	—	—	—	—	—	—	—	1	—	—
60	100	2.70	—	—	1300	—	—	—	—	—	—	—	—	2	—	—
65	150	1.80	—	—	880	—	—	—	—	—	—	—	—	0	—	—
60	100	2.70	—	—	1300	—	—	—	—	—	—	—	—	0	—	—
80	200	0.36	—	—	860	—	—	—	—	—	—	—	—	30	—	—
65	20	0.36	—	—	730	—	—	—	—	—	—	—	—	1	—	—
20	40	0.36	—	—	180	—	—	—	—	—	—	—	—	27	—	—
10	0	0.72	—	—	350	—	—	—	—	—	—	—	—	0	—	—
25	40	1.08	—	—	900	—	—	—	—	—	—	—	—	0	—	—
15	100	0.36	—	—	80	—	—	—	—	—	—	—	—	0	—	—
25	150	0.00	—	—	100	—	—	—	—	—	—	—	—	0	—	—
0	0	0.36	—	—	10	—	0	—	—	—	—	—	—	15	—	—
0	0	0.36	—	—	5	—	0	—	—	—	—	—	—	15	—	—
70	40	1.08	—	—	1090	—	0	—	—	—	—	—	—	0	—	—
10	100	0.00	—	—	75	—	—	—	—	—	—	—	—	15	—	—
60	200	1.08	—	—	770	—	—	—	—	—	—	—	—	24	—	—
15	20	0.00	—	—	105	—	0	—	—	—	—	—	—	21	—	—
63	—	—	—	—	600	—	—	—	—	—	—	—	—	—	—	—
46	—	—	—	—	187	—	—	—	—	—	—	—	—	—	—	—
62	—	—	—	—	104	—	—	—	—	—	—	—	—	—	—	—
67	—	—	—	—	801	—	—	—	—	—	—	—	—	—	—	—
27	—	—	—	—	337	—	—	—	—	—	—	—	—	—	—	—
46	100	0.00	—	—	252	—	—	—	—	—	—	—	—	0	—	—
16	—	—	—	—	413	—	—	—	—	—	—	—	—	—	—	—
30	250	1.80	—	860	180	—	—	0.15	—	0.60	0.20	—	—	15	—	—
50	250	1.80	—	—	350	—	—	—	—	—	—	—	—	0	—	—
50	350	2.70	—	—	370	—	—	—	—	—	—	—	—	1	—	—
55	450	1.80	—	—	370	—	—	—	—	—	—	—	—	2	—	—
50	450	1.44	—	—	280	—	—	0.12	—	—	—	—	—	2	—	—
15	100	0.72	—	—	75	—	—	—	—	—	—	—	—	0	—	—
20	200	1.08	—	278	140	—	—	0.06	—	0.24	0.20	—	—	0	—	—
20	200	1.08	—	290	130	—	—	0.06	—	0.26	0.20	—	—	1	—	—
40	350	2.70	—	—	430	—	—	—	—	—	—	—	—	1	—	—
35	300	1.80	—	—	400	—	—	—	—	—	—	—	—	1	—	—
15	150	0.72	—	—	70	—	150	—	—	—	—	—	—	0	—	—
26	6	0.32	—	—	175	—	—	—	—	—	—	—	—	<.1	—	—
0	<.1	0.87	—	—	152	—	—	—	—	—	—	—	—	6	—	—
9	3	0.00	—	—	163	—	—	—	—	—	—	—	—	0	—	—
0	6	0.70	—	—	110	—	—	—	—	—	—	—	—	<.1	—	—
49	202	3.70	—	—	1221	—	—	—	—	—	—	—	—	1	—	—
68	281	5.10	—	—	1685	—	—	—	—	—	—	—	—	1	—	—
60	293	3.56	—	—	1274	—	—	—	—	—	—	—	—	0	—	—
84	395	4.96	—	—	1809	—	—	—	—	—	—	—	—	0	—	—
46	—	—	—	—	1206	—	—	—	—	—	—	—	—	—	—	—
63	393	4.42	—	—	1672	—	—	—	—	—	—	—	—	2	—	—
23	187	2.99	—	—	776	—	131	—	—	—	—	—	—	0	—	—
32	261	4.13	—	—	1080	—	184	—	—	—	—	—	—	0	—	—
40	199	3.56	—	—	1063	—	—	—	—	—	—	—	—	1	—	—
53	276	4.84	—	—	1432	—	—	—	—	—	—	—	—	2	—	—
40	199	3.56	—	—	1063	—	—	—	—	—	—	—	—	1	—	—
53	276	4.85	—	—	1432	—	—	—	—	—	—	—	—	2	—	—
60	290	4.08	—	—	1348	—	—	—	—	—	—	—	—	1	—	—
89	403	5.48	—	—	1780	—	—	—	—	—	—	—	—	2	—	—
41	274	3.30	—	—	1102	—	—	—	—	—	—	—	—	2	—	—
57	384	4.57	—	—	1544	—	—	—	—	—	—	—	—	3	—	—
344	282	3.71	—	—	1463	—	—	—	—	—	—	—	—	<1	—	—
85	393	5.04	—	—	1947	—	—	—	—	—	—	—	—	<1	—	—

TABLE F–1

Food Composition (Computer code number is for Wadsworth Diet Analysis program) (For purposes of calculations, use "0" for t, <1, <.1, <.01, etc.)

DA + Code	Food Description	Quantity	Measure	Wt (g)	H_2O (g)	Ener (kcal)	Prot (g)	Carb (g)	Dietary Fiber (g)	Fat (g)	Fat Breakdown (g)			
											Sat	Mono	Poly	Trans
	FAST FOOD—Continued													
31571	Pepperoni feast, extra pepperoni & cheese, 12"	2	slice(s)	196	87	534	24	56	3	25	10.92	—	—	—
31572	Pepperoni feast, extra pepperoni & cheese, 14"	2	slice(s)	270	121	732	33	77	4	34	15.00	—	—	—
31577	Vegi feast, 12"	2	slice(s)	203	107	439	19	57	4	16	7.09	—	—	—
31578	Vegi feast, 14"	2	slice(s)	278	147	304	27	78	5	22	9.89	—	—	—
37549	Dot cinnamon	1	item(s)	28	8	99	2	15	1	4	0.68	—	—	—
31605	Double cheesy bread	1	item(s)	35	11	123	4	13	1	6	2.06	—	—	—
31607	Hot wings	1	item(s)	25	—	45	5	1	<1	2	0.65	—	—	—
	Thin crust pizza													
31583	America's favorite, 12"	¼	item(s)	159	—	408	19	34	2	23	9.77	—	—	—
31584	America's favorite, 14"	¼	item(s)	202	—	557	26	47	3	31	13.19	—	—	—
31579	Cheese, 12"	¼	item(s)	106	—	273	12	31	2	12	9.37	—	—	—
31580	Cheese, 14"	¼	item(s)	148	—	382	17	43	2	17	6.72	—	—	—
31688	Deluxe, 12"	¼	item(s)	159	—	363	16	34	2	19	7.64	—	—	—
31697	Deluxe, 14"	¼	item(s)	202	—	494	22	47	3	25	10.20	—	—	—
31689	Extravaganzza, 12"	¼	item(s)	159	—	425	20	34	3	24	9.41	—	—	—
31698	Extravaganzza, 14"	¼	item(s)	202	—	571	27	48	4	31	12.44	—	—	—
31585	Hawaiian, 12"	¼	item(s)	159	—	349	18	35	2	16	7.20	—	—	—
31586	Hawaiian, 14"	¼	item(s)	202	—	489	25	48	3	23	10.09	—	—	—
31690	Meatzza, 12"	¼	item(s)	159	—	458	23	33	2	27	11.39	—	—	—
31699	Meatzza, 14"	¼	item(s)	202	—	619	31	46	3	36	15.24	—	—	—
31581	Pepperoni, extra pepperoni & cheese 12"	¼	item(s)	159	—	420	20	32	2	24	10.46	—	—	—
31582	Pepperoni, extra pepperoni & cheese 14"	¼	item(s)	202	—	586	28	45	3	34	14.55	—	—	—
31587	Vegi, 12"	¼	item(s)	159	—	338	16	34	3	17	7.08	—	—	—
31588	Vegi, 14"	¼	item(s)	202	—	471	22	47	3	23	9.89	—	—	—
	Ultimate deep dish pizza													
31596	America's favorite, 12"	2	slice(s)	235	—	617	26	59	4	33	12.88	—	—	—
31702	America's favorite, 14"	2	slice(s)	311	—	851	36	84	5	44	17.35	—	—	—
31590	Cheese, 12"	2	slice(s)	181	—	482	19	56	3	22	7.91	—	—	—
31591	Cheese, 14"	2	slice(s)	257	—	677	26	80	5	30	10.88	—	—	—
31589	Cheese, 6"	1	item(s)	215	—	598	23	68	4	28	9.94	—	—	—
31691	Deluxe, 12"	2	slice(s)	235	—	527	23	59	4	29	10.75	—	—	—
31700	Deluxe, 14"	2	slice(s)	311	—	788	31	84	5	38	14.36	—	—	—
31692	Extravaganzza, 12"	2	slice(s)	235	—	635	27	59	4	34	12.52	—	—	—
31701	Extravaganzza, 14"	2	slice(s)	311	—	866	36	85	6	45	16.60	—	—	—
31599	Hawaiian, 12"	2	slice(s)	235	—	558	24	60	4	26	10.31	—	—	—
31600	Hawaiian, 14"	2	slice(s)	311	—	784	35	85	5	36	14.25	—	—	—
31693	Meatzza, 12"	2	slice(s)	235	—	667	30	58	4	37	14.50	—	—	—
31703	Meatzza, 14"	2	slice(s)	311	—	914	40	83	5	49	19.40	—	—	—
31593	Pepperoni, extra pepperoni & cheese 12"	2	slice(s)	235	—	629	26	57	4	34	13.57	—	—	—
31594	Pepperoni, extra pepperoni & cheese 14"	2	slice(s)	311	—	880	37	82	5	47	18.71	—	—	—
31602	Vegi, 12"	2	slice(s)	235	—	547	22	59	4	26	10.19	—	—	—
31603	Vegi, 14"	2	slice(s)	311	—	765	32	84	6	36	14.05	—	—	—
31598	With ham & pineapple tidbits, 6"	1	item(s)	430	—	619	25	70	4	28	10.19	—	—	—
31595	With Italian sausage, 6"	1	item(s)	430	—	642	25	70	4	31	11.33	—	—	—
31592	With pepperoni, 6"	1	item(s)	430	—	647	25	69	4	32	11.70	—	—	—
31601	With vegetables, 6"	1	item(s)	430	—	619	23	71	5	29	10.11	—	—	—
	In-n-Out Burger													
34374	Cheeseburger	1	item(s)	268	—	480	22	39	3	27	10.00	—	—	—
34391	Cheesburger w/mustard & ketchup	1	item(s)	268	—	400	22	41	3	18	9.00	—	—	—
34390	Cheeseburger, lettuce leaves instead of buns	1	item(s)	300	—	330	18	11	2	25	9.00	—	—	—
34377	Chocolate shake	1	item(s)	425	—	690	9	83	0	36	24.00	—	—	—
34375	Double-Double cheeseburger	1	item(s)	328	—	670	37	40	3	41	18.00	—	—	—
34393	Double-Double cheeseburger w/mustard & ketchup	1	item(s)	328	—	590	37	42	3	32	17.00	—	—	—
34392	Double-Double cheeseburger, lettuce leaves instead of buns	1	item(s)	361	—	520	33	11	2	39	17.00	—	—	—
34376	French fries	1	item(s)	125	—	400	7	54	2	18	5.00	—	—	—
34373	Hamburger	1	item(s)	243	—	390	16	39	3	19	5.00	—	—	—
34389	Hamburger w/mustard & ketchup	1	item(s)	243	—	310	16	41	3	10	4.00	—	—	—
34388	Hamburger, lettuce leaves instead of buns	1	item(s)	275	—	240	12	10	2	17	4.50	—	—	—

PAGE KEY: A–58 = Breads/Baked Goods A–62 = Cereal/Rice/Pasta A–66 = Fruit A–70 = Vegetables/Legumes A–80 = Nuts/Seeds A–82 = Vegetarian A–84 = Dairy A–90 = Eggs A–90 = Seafood A–92 = Meats A–96 = Poultry A–96 = Processed Meats A–98 = Beverages A–102 = Fats/Oils A–104 = Sweets A–106 = Sauces/Condiments/Spices A–108 = Mixed Foods/Soups/Sandwiches A–114 = Fast Food A–130 = Convenience A–132 = Baby Foods

Chol (mg)	Calc (mg)	Iron (mg)	Magn (mg)	Pota (mg)	Sodi (mg)	Zinc (mg)	Vit A (RAE) (µg)	Thia (mg)	Vit E (mg α)	Ribo (mg)	Niac (mg)	Vit B_6 (mg)	Fola (µg)	Vit C (mg)	Vit B_{12} (µg)	Sele (µg)
57	279	3.36	—	—	1349	—	155	—	—	—	—	—	—	<1	—	—
78	390	4.66	—	—	1855	—	233	—	—	—	—	—	—	<1	—	—
34	279	3.44	—	—	987	—	—	—	—	—	—	—	—	1	—	—
47	389	4.71	—	—	1369	—	—	—	—	—	—	—	—	2	—	—
0	6	0.59	—	—	86	—	—	—	—	—	—	—	—	<.1	—	—
6	47	0.66	—	—	164	—	—	—	—	—	—	—	—	<1	—	—
26	5	0.30	—	—	354	—	—	—	—	—	—	—	—	1	—	—
51	318	1.52	—	—	1285	—	—	—	—	—	—	—	—	<1	—	—
69	444	2.07	—	—	1751	—	—	—	—	—	—	—	—	1	—	—
23	225	0.97	—	—	835	—	125	—	—	—	—	—	—	0	—	—
32	315	1.36	—	—	1172	—	175	—	—	—	—	—	—	0	—	—
40	237	1.54	—	—	1123	—	—	—	—	—	—	—	—	1	—	—
53	330	2.08	—	—	1523	—	—	—	—	—	—	—	—	2	—	—
53	245	1.95	—	—	1408	—	—	—	—	—	—	—	—	1	—	—
69	340	2.59	—	—	1871	—	—	—	—	—	—	—	—	2	—	—
41	312	1.28	—	—	1162	—	—	—	—	—	—	—	—	2	—	—
57	437	1.80	—	—	1635	—	—	—	—	—	—	—	—	3	—	—
64	320	1.69	—	—	1523	—	—	—	—	—	—	—	—	<1	—	—
454	446	2.27	—	—	2039	—	—	—	—	—	—	—	—	<1	—	—
54	316	1.34	—	—	1362	—	162	—	—	—	—	—	—	<1	—	—
76	442	1.87	—	—	1900	—	227	—	—	—	—	—	—	<1	—	—
34	317	1.42	—	—	1047	—	—	—	—	—	—	—	—	1	—	—
47	442	1.94	—	—	1460	—	—	—	—	—	—	—	—	2	—	—
58	334	4.43	—	—	1573	—	—	—	—	—	—	—	—	1	—	—
78	464	6.24	—	—	2155	—	—	—	—	—	—	—	—	1	—	—
30	241	3.88	—	—	1123	—	151	—	—	—	—	—	—	<1	—	—
41	335	5.53	—	—	1575	—	210	—	—	—	—	—	—	1	—	—
36	295	4.67	—	—	1341	—	174	—	—	—	—	—	—	1	—	—
47	253	4.45	—	—	1410	—	—	—	—	—	—	—	—	2	—	—
62	349	6.25	—	—	1927	—	—	—	—	—	—	—	—	2	—	—
60	261	4.86	—	—	1696	—	—	—	—	—	—	—	—	2	—	—
78	359	6.76	—	—	2275	—	—	—	—	—	—	—	—	2	—	—
48	328	4.19	—	—	1449	—	—	—	—	—	—	—	—	2	—	—
67	457	5.97	—	—	2039	—	—	—	—	—	—	—	—	3	—	—
379	336	4.60	—	—	1810	—	—	—	—	—	—	—	—	1	—	—
501	466	6.44	—	—	2443	—	—	—	—	—	—	—	—	1	—	—
61	332	4.25	—	—	1650	—	187	—	—	—	—	—	—	1	—	—
85	462	6.04	—	—	2304	—	260	—	—	—	—	—	—	1	—	—
41	333	4.33	—	—	1334	—	—	—	—	—	—	—	—	2	—	—
57	462	6.11	—	—	1864	—	—	—	—	—	—	—	—	2	—	—
43	298	4.84	—	—	1498	—	—	—	—	—	—	—	—	1	—	—
45	302	4.89	—	—	1478	—	—	—	—	—	—	—	—	1	—	—
47	299	4.81	—	—	1524	—	168	—	—	—	—	—	—	1	—	—
36	307	5.10	—	—	1472	—	—	—	—	—	—	—	—	5	—	—
60	200	3.60	—	—	1000	—	188	—	—	—	—	—	—	15	—	—
55	200	3.60	—	—	1080	—	182	—	—	—	—	—	—	15	—	—
60	200	1.08	—	—	720	—	—	—	—	—	—	—	—	18	—	—
95	300	0.72	—	—	350	—	143	—	—	—	—	—	—	0	—	—
120	350	5.40	—	—	1430	—	184	—	—	—	—	—	—	15	—	—
115	350	5.40	—	—	1510	—	229	—	—	—	—	—	—	15	—	—
120	350	1.08	—	—	1160	—	275	—	—	—	—	—	—	18	—	—
0	20	1.80	—	—	245	—	0	—	—	—	—	—	—	0	—	—
40	40	3.60	—	—	640	—	50	—	—	—	—	—	—	15	—	—
35	40	3.60	—	—	720	—	75	—	—	—	—	—	—	15	—	—
40	40	1.08	—	—	370	—	—	—	—	—	—	—	—	18	—	—

TABLE F-1

Food Composition (Computer code number is for Wadsworth Diet Analysis program) (For purposes of calculations, use "0" for t, <1, <.1, <.01, etc.)

DA + Code	Food Description	Quantity	Measure	Wt (g)	H_2O (g)	Ener (kcal)	Prot (g)	Carb (g)	Dietary Fiber (g)	Fat (g)	Fat Breakdown (g) Sat	Mono	Poly	Trans
	FAST FOOD—Continued													
34379	Strawberry shake	1	item(s)	425	—	690	8	91	2	33	22.00	—	—	—
34378	Vanilla shake	1	item(s)	425	—	680	9	78	2	37	25.00	—	—	—
	Jack in the Box													
30392	Bacon ultimate cheeseburger	1	item(s)	353	—	1120	52	59	2	55	28.00	—	—	3.13
1740	Breakfast Jack	1	item(s)	133	—	310	14	34	1	14	5.00	—	—	0
14074	Cheeseburger	1	item(s)	116	—	300	14	31	2	13	6.00	—	—	0.89
14106	Chicken breast pieces	5	piece(s)	150	—	360	27	24	1	17	3.00	—	—	4.48
37241	Chicken club salad	1	item(s)	535	—	310	28	15	5	16	6.00	—	—	0
14111	Chocolate ice cream shake	1	item(s)	315	—	660	11	89	1	29	18.00	—	—	1
14075	Double cheeseburger	1	item(s)	155	—	410	20	32	1	22	11.00	—	—	—
14098	French fries, jumbo	1	serving(s)	142	—	410	4	55	4	20	4.50	—	—	5.34
14099	French fries, super scoop	1	serving(s)	198	—	580	6	77	6	28	6.00	—	—	7.07
14073	Hamburger	1	item(s)	104	—	250	12	30	2	9	3.50	—	—	0.88
14090	Hash browns	1	serving(s)	57	—	150	1	13	2	10	2.50	—	—	3
14072	Jack's Spicy Chicken sandwich	1	item(s)	253	—	580	24	53	3	31	6.00	—	—	2.81
1468	Jumbo Jack hamburger	1	item(s)	269	—	600	22	58	3	31	11.00	—	—	1.55
1469	Jumbo Jack hamburger w/cheese	1	item(s)	294	—	690	26	60	3	38	16.00	—	—	1.55
1470	Onion rings	1	serving(s)	119	—	500	6	51	3	30	5.00	—	—	10
33141	Sausage, egg, & cheese biscuit	1	item(s)	223	—	760	25	33	2	60	20.00	—	—	5.72
14095	Seasoned curly fries	1	serving(s)	125	—	400	6	45	5	23	5.00	—	—	7
14077	Sourdough Jack	1	item(s)	244	—	700	30	36	3	49	16.00	—	—	2.98
37249	Southwest chicken salad	1	serving(s)	598	—	340	28	31	9	13	6.00	—	—	0
14112	Strawberry ice cream shake	1	item(s)	313	—	640	10	84	0	28	18.00	—	—	1
14078	Ultimate cheeseburger	1	item(s)	328	—	990	41	59	2	66	28.00	—	—	3.05
14110	Vanilla ice cream shake	1	item(s)	285	—	570	12	65	0	29	18.00	—	—	1
	Jamba Juice													
31646	Banana berry smoothie	24	fluid ounce(s)	719	—	470	5	112	5	2	0.50	—	—	—
31647	Caribbean passion smoothie	24	fluid ounce(s)	730	—	440	4	102	4	2	1.00	—	—	—
38422	Carrot juice	16	fluid ounce(s)	472	—	100	3	23	0	1	0.00	—	—	—
31648	Chocolate mood smoothie	24	fluid ounce(s)	612	—	690	16	142	2	8	4.50	—	—	—
31649	Citrus squeeze smoothie	24	fluid ounce(s)	729	—	450	4	105	5	2	1.00	—	—	—
31650	Coffee mood smoothie	24	fluid ounce(s)	560	—	596	13	121	1	6	4.00	—	—	—
31651	Coldbuster smoothie	24	fluid ounce(s)	724	—	430	5	100	5	3	1.00	—	—	—
31652	Cranberry craze smoothie	24	fluid ounce(s)	731	—	420	6	97	4	2	1.00	—	—	—
31654	Jamba powerboost smoothie	24	fluid ounce(s)	730	—	440	6	103	7	2	0.00	—	—	—
38423	Lemonade	16	fluid ounce(s)	483	—	300	1	75	0	0	0.00	0.00	0.00	0
31656	Lime sublime smoothie	24	fluid ounce(s)	721	—	450	3	104	6	2	1.00	—	—	—
31657	Mango-a-go-go smoothie	24	fluid ounce(s)	739	—	500	4	117	4	2	1.00	—	—	—
38424	Orange juice, freshly squeezed	16	fluid ounce(s)	496	—	220	3	52	1	1	0.00	—	—	—
38426	Orange/carrot juice	16	fluid ounce(s)	484	—	160	3	37	0	1	0.00	—	—	—
31660	Orange-a-peel smoothie	24	fluid ounce(s)	726	—	440	9	102	5	1	0.00	—	—	—
31665	Protein berry pizzaz smoothie	24	fluid ounce(s)	710	—	440	20	92	6	2	0.00	—	—	—
31667	Raspberry refresher smoothie	24	fluid ounce(s)	636	—	442	3	101	8	3	0.90	—	—	—
31668	Razzmatazz smoothie	24	fluid ounce(s)	730	—	480	3	112	4	2	1.00	—	—	—
31669	Strawberries wild smoothie	24	fluid ounce(s)	725	—	450	6	105	4	0	0.00	—	—	—
38421	Strawberry tsunami smoothie	24	fluid ounce(s)	740	—	530	4	128	4	2	1.00	—	—	—
38427	Vibrant C juice	16	fluid ounce(s)	448	—	210	2	50	1	0	0.00	0.00	0.00	0
38428	Wheatgrass juice, freshly squeezed	1	ounce(s)	32	—	5	1	1	0	0	0.00	0.00	0.00	0
	Kentucky Fried Chicken (KFC)													
31850	BBQ baked beans	1	serving(s)	156	—	190	6	33	6	3	1.00	—	—	0.29
31853	Biscuit	1	item(s)	56	—	180	4	20	1	10	2.50	—	—	3.44
31851	Coleslaw	1	serving(s)	142	—	232	2	26	3	14	2.00	—	—	0.27
31842	Colonel's Crispy Strips	3	item(s)	150	—	340	28	20	0	16	4.50	—	—	4.47
31849	Corn on the cob	1	item(s)	162	—	150	5	35	2	2	0.00	—	—	0
3761	Extra Crispy chicken, breast	1	item(s)	162	—	470	34	19	0	28	8.00	—	—	4.50
3762	Extra Crispy chicken, drumstick	1	item(s)	60	—	160	12	5	0	10	2.50	—	—	1.50
3763	Extra Crispy chicken, thigh	1	item(s)	114	—	370	21	12	0	26	7.00	—	—	3
3764	Extra Crispy chicken, whole wing	1	item(s)	52	—	190	10	10	0	12	3.50	—	—	2
31833	Honey BBQ wing pieces	6	item(s)	189	—	607	33	33	1	38	10.00	—	—	5.42
10810	Hot & spicy chicken, breast	1	item(s)	179	—	450	33	20	0	27	8.00	—	—	0
10813	Hot & spicy chicken, drumstick	1	item(s)	60	—	140	13	4	0	9	2.50	—	—	0
10811	Hot & spicy chicken, thigh	1	item(s)	128	—	390	22	14	0	28	8.00	—	—	0
10812	Hot & spicy chicken, whole wing	1	item(s)	55	—	180	11	9	0	11	3.00	—	—	0
10859	Hot wings pieces	6	piece(s)	135	—	471	27	18	2	33	8.00	—	—	4.03
31848	Macaroni & cheese	1	serving(s)	153	—	180	7	21	2	8	3.00	—	—	2.81

PAGE KEY: A–58 = Breads/Baked Goods A–62 = Cereal/Rice/Pasta A–66 = Fruit A–70 = Vegetables/Legumes A–80 = Nuts/Seeds A–82 = Vegetarian A–84 = Dairy A–90 = Eggs A–90 = Seafood A–92 = Meats A–96 = Poultry A–96 = Processed Meats A–98 = Beverages A–102 = Fats/Oils A–104 = Sweets A–106 = Sauces/Condiments/Spices A–108 = Mixed Foods/Soups/Sandwiches A–114 = Fast Food A–130 = Convenience A–132 = Baby Foods

Chol (mg)	Calc (mg)	Iron (mg)	Magn (mg)	Pota (mg)	Sodi (mg)	Zinc (mg)	Vit A (RAE) (µg)	Thia (mg)	Vit E (mg α)	Ribo (mg)	Niac (mg)	Vit B_6 (mg)	Fola (µg)	Vit C (mg)	Vit B_{12} (µg)	Sele (µg)
85	250	0.00	—	—	280	—	134	—	—	—	—	—	—	0	—	—
90	300	0.00	—	—	390	—	145	—	—	—	—	—	—	0	—	—
160	300	7.20	—	600	2260	—	—	—	—	—	—	—	—	1	—	—
210	150	3.60	—	210	770	—	—	—	—	—	—	—	—	4	—	—
40	150	3.60	—	180	840	—	40	—	—	—	—	—	—	0	—	—
80	20	1.80	—	430	970	—	—	—	—	—	—	—	—	1	—	—
65	300	3.60	—	1010	890	—	—	—	—	—	—	—	—	54	—	—
110	350	0.36	—	720	270	—	215	—	—	—	—	—	—	0	—	—
70	250	4.50	—	280	920	—	—	—	—	—	—	—	—	1	—	—
0	20	1.08	—	550	690	—	0	—	—	—	—	—	—	6	—	—
0	20	1.44	—	770	960	—	0	—	—	—	—	—	—	9	—	—
30	100	3.60	—	155	610	—	0	—	—	—	—	—	—	0	—	—
0	10	0.18	—	190	230	—	0	—	—	—	—	—	—	0	—	—
60	150	1.80	—	470	950	—	—	—	—	—	—	—	—	9	—	—
45	164	4.92	—	390	980	—	—	—	—	—	—	—	—	10	—	—
75	250	4.50	—	420	1360	—	—	—	—	—	—	—	—	9	—	—
0	40	2.70	—	140	420	—	40	—	—	—	—	—	—	18	—	—
280	100	2.70	—	240	1390	—	—	—	—	—	—	—	—	0	—	—
0	40	1.80	—	580	890	—	—	—	—	—	—	—	—	0	—	—
80	200	4.50	—	450	1220	—	—	—	—	—	—	—	—	9	—	—
60	300	4.50	—	1020	920	—	—	—	—	—	—	—	—	48	—	—
110	350	0.00	—	610	220	—	202	—	—	—	—	—	—	0	—	—
130	300	7.20	—	480	1670	—	—	—	—	—	—	—	—	1	—	—
115	400	0.00	—	630	220	—	218	—	—	—	—	—	—	0	—	—
5	200	1.08	32	1000	85	0.30	—	0.06	0.32	0.26	1.20	0.40	33	15	0	0
5	100	1.80	24	810	60	0.30	—	0.09	0.64	0.26	5.00	0.50	100	78	0	1
0	150	2.70	80	1030	250	0.90	0	0.53	—	0.26	5.00	0.70	80	18	0	6
25	500	1.08	32	760	280	0.60	0	0.09	0.00	0.85	0.40	0.08	9	6	1	4
5	150	1.80	60	1150	50	0.30	—	0.30	0.40	0.26	1.90	0.40	100	168	0	1
28	455	0.30	49	634	429	1.50	—	0.10	0.16	0.60	0.30	0.10	18	7	1	3
5	100	1.08	60	1240	35	15.00	—	0.38	17.71	0.34	3.00	0.40	122	1302	0	1
5	250	1.44	16	500	90	0.30	—	0.03	0.64	0.26	5.00	0.50	100	54	0	1
0	1100	1.44	480	1110	40	15.00	—	5.25	17.71	5.78	66.00	6.80	640	294	10	70
0	20	0.00	8	200	10	0.00	0	0.03	0.00	0.17	14.00	1.80	320	36	0	0
5	150	1.80	32	660	75	0.60	—	0.12	0.32	0.26	7.00	0.80	160	66	<1	1
5	100	1.08	24	800	60	0.30	—	0.15	1.61	0.26	5.00	0.70	120	72	0	1
0	60	1.08	60	990	0	0.30	0	0.45	—	0.14	2.00	0.20	160	246	0	0
0	100	1.80	60	1010	125	0.60	0	0.45	—	0.26	3.00	0.50	120	132	0	3
0	250	1.80	60	1350	100	0.30	—	0.38	0.64	0.43	3.00	0.40	140	240	0	1
0	1100	2.62	39	650	240	0.58	0	0.09	0.31	0.10	1.55	0.40	58	60	0	4
3	104	2.20	56	806	47	0.80	—	0.10	0.40	0.30	1.60	0.40	43	35	<1	1
5	150	1.80	32	790	70	0.60	—	0.09	0.32	0.26	6.00	0.90	160	60	0	1
0	250	1.80	32	1020	115	0.30	—	0.03	0.32	0.34	1.20	0.20	32	60	0	1
5	100	1.08	24	480	10	0.30	0	0.06	—	0.34	14.00	1.80	320	90	0	1
0	20	1.08	40	720	0	0.30	0	0.30	—	0.10	1.60	0.40	80	678	0	0
0	0	1.80	8	80	0	0.00	0	0.03	—	0.03	0.40	0.04	16	4	0	3
5	80	1.80	—	—	760	—	—	—	—	—	—	—	—	1	—	—
0	20	1.08	—	—	560	—	—	—	—	—	—	—	—	1	—	—
8	30	0.18	—	—	284	—	65	—	—	—	—	—	—	34	—	—
70	10	0.72	—	—	1140	—	—	—	—	—	—	—	—	1	—	—
0	10	0.18	—	—	20	—	10	—	—	—	—	—	—	4	—	—
135	19	1.44	—	—	1230	—	—	—	—	—	—	—	—	1	—	—
70	9	0.65	—	—	415	—	—	—	—	—	—	—	—	1	—	—
120	19	1.04	—	—	710	—	—	—	—	—	—	—	—	1	—	—
55	9	0.34	—	—	390	—	—	—	—	—	—	—	—	1	—	—
193	40	1.44	—	—	1145	—	—	—	—	—	—	—	—	5	—	—
130	10	1.07	—	—	1450	—	—	—	—	—	—	—	—	1	—	—
65	20	0.68	—	—	380	—	—	—	—	—	—	—	—	1	—	—
125	10	1.44	—	—	1240	—	—	—	—	—	—	—	—	1	—	—
60	10	0.72	—	—	420	—	—	—	—	—	—	—	—	1	—	—
150	40	1.44	—	—	1230	—	—	—	—	—	—	—	—	1	—	—
10	150	0.18	—	—	860	—	350	—	—	—	—	—	—	1	—	—

TABLE F-1

Food Composition (Computer code number is for Wadsworth Diet Analysis program) (For purposes of calculations, use "0" for t, <1, <.1, <.01, etc.)

DA + Code	Food Description	Quantity	Measure	Wt (g)	H_2O (g)	Ener (kcal)	Prot (g)	Carb (g)	Dietary Fiber (g)	Fat (g)	Fat Breakdown (g) Sat	Mono	Poly	Trans
	FAST FOOD—Continued													
31847	Mashed potatoes with gravy	1	serving(s)	136	—	120	1	17	2	6	1.00	—	—	0.50
10825	Original Recipe chicken, breast	1	item(s)	161	—	370	40	11	0	19	6.00	—	—	2.50
10826	Original Recipe chicken, drumstick	1	item(s)	59	—	140	14	4	0	8	2.00	—	—	1
10827	Original Recipe chicken, thigh	1	item(s)	126	—	360	22	12	0	25	7.00	—	—	1.50
10828	Original Recipe chicken, whole wing	1	item(s)	47	—	145	11	5	0	9	2.50	—	—	1
3760	Original Recipe chicken sandwich w/sauce	1	item(s)	200	—	450	29	33	2	22	5.00	—	—	—
31834	Original Recipe chicken sandwich w/o sauce	1	item(s)	187	—	360	29	21	1	13	3.50	—	—	—
31852	Potato salad	1	serving(s)	160	—	230	4	23	3	14	2.00	—	—	0.31
10845	Potato wedges	1	serving(s)	156	—	376	6	53	5	15	4.20	—	—	6.12
10853	Rotisserie Gold chicken, breast & wing w/skin	4	ounce(s)	114	—	218	26	1	0	12	3.51	—	—	—
10851	Rotisserie Gold chicken, thigh & leg w/skin	4	ounce(s)	114	—	260	23	1	0	18	5.15	—	—	—
10852	Rotisserie Gold chicken, thigh & leg w/o skin	4	ounce(s)	117	—	217	27	0	0	12	3.50	—	—	—
31843	Spicy Crispy Strips	3	item(s)	115	—	335	25	23	1	15	4.00	—	—	—
10854	Tender Roast chicken, breast w/o skin	1	item(s)	118	—	169	31	1	0	4	1.20	—	—	—
	Long John Silver													
39392	Baked cod	1	serving(s)	101	—	120	22	1	0	5	1.00	—	—	—
3777	Batter dipped fish sandwich	1	item(s)	177	—	440	17	48	3	20	5.00	—	—	—
37568	Battered fish	1	item(s)	92	—	230	11	16	0	13	4.00	—	—	—
37569	Breaded clams	1	serving(s)	85	—	240	8	22	1	13	2.00	—	—	—
39404	Clam chowder	1	item(s)	227	—	220	9	23	0	10	4.00	—	—	—
39398	Cocktail sauce	1	ounce(s)	28	—	25	0	6	0	0	0.00	0.00	0.00	0
3770	Coleslaw	1	serving(s)	113	—	200	1	15	4	15	2.50	1.76	4.10	—
39394	Crunchy shrimp basket	21	item(s)	114	—	340	12	32	2	19	5.00	—	—	—
39400	French fries, large	1	item(s)	142	—	390	4	56	5	17	4.00	—	—	—
3774	Fries regular	1	serving(s)	85	—	230	3	34	3	10	2.50	7.40	5.10	—
3779	Hushpuppy	1	piece(s)	23	—	60	1	9	1	3	0.50	—	—	—
3781	Shrimp batter-dipped	1	piece(s)	14	—	45	2	3	0	3	1.00	—	—	—
39399	Tartar sauce	1	ounce(s)	28	—	100	0	4	0	9	1.50	—	—	—
39395	Ultimate fish sandwich	1	item(s)	199	—	500	20	48	3	25	8.00	—	—	—
	McDonald's													
2247	Barbecue sauce	1	serving(s)	28	—	45	0	10	0	0	0.00	0.00	0.00	0
737	Big Mac hamburger	1	item(s)	216	—	590	24	47	3	34	11.00	—	—	1.48
738	Cheeseburger	1	item(s)	121	—	330	15	36	2	14	6.00	—	—	1.02
29775	Chicken McGrill sandwich	1	item(s)	213	—	400	25	37	2	17	3.00	—	—	0
3792	Chicken McNuggets	4	item(s)	72	—	210	10	12	1	13	2.50	—	—	1.13
1873	Chicken McNuggets	6	item(s)	108	—	310	15	18	2	20	4.00	—	—	1.69
73	Chocolate milkshake	8	fluid ounce(s)	227	164	270	7	48	1	6	3.81	1.77	0.23	—
29774	Crispy chicken sandwich	1	item(s)	219	—	500	22	46	2	26	4.50	—	—	1.50
743	Egg McMuffin	1	item(s)	138	—	300	18	29	2	12	4.50	—	—	0.42
742	Filet-o-fish sandwich	1	item(s)	156	—	470	15	45	1	26	5.00	—	—	1.11
2257	French fries, large	1	serving(s)	176	—	540	8	68	6	26	4.50	—	—	6.18
1872	French fries, small	1	serving(s)	68	—	210	3	26	2	10	1.50	—	—	2.30
2244	French fries, super size	1	serving(s)	198	—	610	9	77	7	29	5.00	—	—	—
33822	Fruit n' yogurt parfait	1	item(s)	338	—	380	10	76	2	5	2.00	—	—	0.18
2251	Garden salad	1	item(s)	177	—	35	2	7	3	0	0.00	0.00	0.00	0
739	Hamburger	1	item(s)	107	—	280	12	35	2	10	4.00	—	—	0.51
2003	Hash browns	1	item(s)	53	—	130	1	14	1	8	1.50	—	—	2
2249	Honey sauce	1	item(s)	14	—	45	0	12	0	0	0.00	0.00	0.00	—
33816	McSalad Shaker chef salad	1	item(s)	206	—	150	17	5	2	8	3.50	—	—	—
33817	McSalad Shaker garden salad	1	item(s)	149	—	100	7	4	2	6	3.00	—	—	—
33818	McSalad Shaker grilled chicken caesar salad	1	item(s)	163	—	100	17	3	2	3	1.50	—	—	—
38396	Newman's Own cobb salad dressing	1	item(s)	59	—	120	1	9	0	9	1.50	—	—	0.01
38397	Newman's Own creamy caesar salad dressing	1	item(s)	59	—	190	2	4	0	18	3.50	—	—	0.29
38398	Newman's Own low fat balsamic vinaigrette salad dressing	1	item(s)	44	—	40	0	4	0	3	0.00	—	—	0.01
38399	Newman's Own ranch salad dressing	1	item(s)	59	—	290	1	4	0	30	4.50	—	—	0.22
1874	Plain hotcakes w/syrup & margarine	3	item(s)	228	—	600	9	104	0	17	3.00	—	—	4
740	Quarter Pounder hamburger	1	item(s)	172	—	430	23	37	2	21	8.00	—	—	1.01
741	Quarter Pounder hamburger w/cheese	1	item(s)	200	—	530	28	38	2	30	13.00	—	—	1.51
2005	Sausage McMuffin w/egg	1	item(s)	164	—	450	20	29	2	28	10.00	—	—	0.59

PAGE KEY: A–58 = Breads/Baked Goods A–62 = Cereal/Rice/Pasta A–66 = Fruit A–70 = Vegetables/Legumes A–80 = Nuts/Seeds A–82 = Vegetarian A–84 = Dairy A–90 = Eggs A–90 = Seafood A–92 = Meats A–96 = Poultry A–96 = Processed Meats A–98 = Beverages A–102 = Fats/Oils A–104 = Sweets A–106 = Sauces/Condiments/Spices A–108 = Mixed Foods/Soups/Sandwiches A–114 = Fast Food A–130 = Convenience A–132 = Baby Foods

Chol (mg)	Calc (mg)	Iron (mg)	Magn (mg)	Pota (mg)	Sodi (mg)	Zinc (mg)	Vit A (RAE) (μg)	Thia (mg)	Vit E (mg α)	Ribo (mg)	Niac (mg)	Vit B_6 (mg)	Fola (μg)	Vit C (mg)	Vit B_{12} (μg)	Sele (μg)
1	10	0.36	—	—	440	—	—	—	—	—	—	—	—	1	—	—
145	20	1.14	—	—	1145	—	—	—	—	—	—	—	—	1	—	—
75	10	0.70	—	—	440	—	—	—	—	—	—	—	—	1	—	—
165	10	1.00	—	—	1060	—	—	—	—	—	—	—	—	1	—	—
60	10	0.36	—	—	370	—	—	—	—	—	—	—	—	1	—	—
70	40	1.80	—	—	940	—	—	—	—	—	—	—	—	1	—	—
60	40	1.80	—	—	890	—	—	—	—	—	—	—	—	1	—	—
15	20	2.70	—	—	540	—	100	—	—	—	—	—	—	1	—	—
4	36	1.55	—	—	1323	—	—	—	—	—	—	—	—	8	—	—
102	7	0.12	—	—	718	—	—	—	—	—	—	—	—	1	—	—
127	8	0.14	—	—	764	—	—	—	—	—	—	—	—	1	—	—
128	10	0.18	—	—	772	—	—	—	—	—	—	—	—	1	—	—
70	20	0.90	—	—	1140	—	—	—	—	—	—	—	—	1	—	—
112	10	0.18	—	—	797	—	—	—	—	—	—	—	—	1	—	—
90	20	0.72	—	—	240	—	—	—	—	—	—	—	—	0	—	—
35	60	3.60	—	—	1120	—	—	—	—	—	—	—	—	9	—	—
30	20	1.80	—	—	700	—	—	—	—	—	—	—	—	5	—	—
10	20	1.08	—	—	1110	—	—	—	—	—	—	—	—	0	—	—
25	150	0.72	—	—	810	—	—	—	—	—	—	—	—	0	—	—
0	0	0.00	—	—	250	—	—	—	—	—	—	—	—	0	—	—
20	40	0.36	—	223	340	0.70	34	0.07	—	0.08	2.35	—	—	18	—	—
105	500	1.80	—	—	720	—	—	—	—	—	—	—	—	1	—	—
0	0	0.00	—	—	580	—	—	—	—	—	—	—	—	24	—	—
0	0	0.00	—	370	350	0.30	—	0.09	—	0.02	1.60	—	—	15	—	—
0	20	0.36	—	—	200	—	—	—	—	—	—	—	—	0	—	—
15	0	0.00	—	—	125	—	—	—	—	—	—	—	—	1	—	—
15	0	0.00	—	—	250	—	—	—	—	—	—	—	—	0	—	—
50	150	3.60	—	—	1310	—	—	—	—	—	—	—	—	9	—	—
0	10	0.18	—	45	250	—	3	—	—	—	—	—	—	4	—	—
85	300	4.50	—	430	1090	—	60	—	—	—	—	—	—	4	—	—
45	250	2.70	—	250	830	—	60	—	—	—	—	—	—	2	—	—
60	200	2.70	—	440	890	—	—	—	—	—	—	—	—	6	—	—
35	20	0.72	—	180	460	—	—	—	—	—	—	—	—	1	—	—
50	20	0.72	—	260	680	—	—	—	—	—	—	—	—	1	—	—
25	299	0.70	36	508	252	1.09	41	0.11	0.11	0.50	0.28	0.06	11	0	1	4
50	200	2.70	—	400	1100	—	—	—	—	—	—	—	—	6	—	—
235	300	2.70	—	210	830	—	—	—	0.72	—	—	—	—	1	—	—
50	200	1.80	—	280	890	—	40	—	—	—	—	—	—	1	—	—
0	20	1.44	—	1210	350	—	—	—	—	—	—	—	—	21	—	—
0	10	0.36	—	470	135	—	—	—	—	—	—	—	—	9	—	—
0	20	1.44	—	1370	390	—	—	—	—	—	—	—	—	24	—	—
15	300	1.80	—	550	240	—	—	—	—	—	—	—	—	24	—	—
0	40	1.09	—	410	20	—	—	—	—	—	—	—	—	24	—	—
30	200	2.70	—	230	590	—	5	—	—	—	—	—	—	2	—	—
0	10	0.36	—	210	330	—	—	—	—	—	—	—	—	2	—	—
0	10	0.18	—	7	0	—	—	—	—	—	—	—	—	1	—	—
95	150	1.44	—	360	740	—	323	—	—	—	—	—	—	15	—	—
75	150	1.08	—	290	120	—	273	—	—	—	—	—	—	15	—	—
40	100	1.08	—	420	240	—	—	—	—	—	—	—	—	12	—	—
10	40	0.18	—	13	440	—	—	—	0.00	—	—	—	—	1	—	—
20	60	0.18	—	16	500	—	—	—	15.40	—	—	—	—	1	—	—
0	10	0.18	—	9	730	—	—	—	0.00	—	—	—	—	2	—	—
20	40	0.18	—	64	530	—	—	—	—	—	—	—	—	1	—	—
20	100	4.50	—	280	770	—	—	—	—	—	—	—	—	1	—	—
70	200	4.50	—	370	840	—	10	—	—	—	—	—	—	2	—	—
95	350	4.50	—	420	1310	—	100	—	—	—	—	—	—	2	—	—
255	300	2.70	—	260	930	—	115	—	0.72	—	—	—	—	1	—	—

TABLE F–1

Food Composition (Computer code number is for Wadsworth Diet Analysis program) (For purposes of calculations, use "0" for t, <1, <.1, <.01, etc.)

DA + Code	Food Description	Quantity	Measure	Wt (g)	H_2O (g)	Ener (kcal)	Prot (g)	Carb (g)	Dietary Fiber (g)	Fat (g)	Fat Breakdown (g) Sat	Mono	Poly	Trans
	FAST FOOD—Continued													
3163	Strawberry milkshake	8	fluid ounce(s)	226	168	256	8	43	1	6	3.93	—	—	—
74	Vanilla milkshake	8	fluid ounce(s)	227	169	254	9	40	0	7	4.28	1.98	0.26	—
	Pizza Hut													
39009	Hot chicken wings	2	item(s)	57	—	110	11	1	0	6	2.00	—	—	0.25
14025	Meat Lovers hand tossed pizza	1	slice(s)	125	—	320	16	30	2	15	7.00	—	—	0.53
14026	Meat Lovers pan pizza	1	slice(s)	130	—	360	16	29	2	20	7.00	—	—	0.53
31009	Meat Lovers stuffed crust pizza	1	slice(s)	188	—	500	25	44	3	25	11.00	—	—	1.11
14024	Meat Lovers thin 'n crispy pizza	1	slice(s)	112	—	310	15	22	2	18	8.00	—	—	0.57
14031	Pepperoni Lovers hand tossed pizza	1	slice(s)	114	—	300	15	30	2	14	7.00	—	—	0.50
14032	Pepperoni Lovers pan pizza	1	slice(s)	119	—	350	15	29	2	19	8.00	—	—	0.50
31011	Pepperoni Lovers stuffed crust pizza	1	slice(s)	171	—	480	23	44	3	24	11.00	—	—	1.05
14030	Pepperoni Lovers thin 'n crispy pizza	1	slice(s)	94	—	270	13	22	2	14	7.00	—	—	0.51
10834	Personal Pan pepperoni pizza	1	slice(s)	59	—	150	7	18	—	6	2.50	—	—	0.97
10842	Personal Pan supreme pizza	1	slice(s)	73	—	170	8	19	1	7	3.00	—	—	0.95
39013	Personal Pan Veggie Lovers pizza	1	slice(s)	69	—	150	6	19	1	6	2.00	—	—	0.50
14028	Veggie Lovers hand tossed pizza	1	slice(s)	120	—	220	10	31	2	6	3.00	—	—	0.25
14029	Veggie Lovers pan pizza	1	slice(s)	125	—	260	10	31	2	12	4.00	—	—	0.26
31010	Veggie Lovers stuffed crust pizza	1	slice(s)	181	—	370	17	45	3	14	7.00	—	—	0.53
14027	Veggie Lovers thin 'n crispy pizza	1	slice(s)	110	—	190	8	23	2	7	3.00	—	—	0.54
39012	Wing blue cheese dipping sauce	1	item(s)	43	—	230	2	2	0	24	5.00	—	—	1
39011	Wing ranch dipping sauce	1	item(s)	43	—	210	1	4	0	22	3.50	—	—	0.50
	Starbucks													
38042	Apple cider, tall steamed	12	fluid ounce(s)	360	—	180	0	45	0	0	0.00	0.00	0.00	0
38052	Cappuccino, tall	12	fluid ounce(s)	360	—	120	7	10	0	6	4.00	—	—	—
38053	Cappuccino, tall nonfat	12	fluid ounce(s)	360	—	80	7	11	0	0	0.00	0.00	0.00	0
38054	Cappuccino, tall soy milk	12	fluid ounce(s)	360	—	100	5	13	1	3	0.00	—	—	—
38059	Cinnamon spice mocha, tall nonfat w/o whipped cream	12	fluid ounce(s)	360	—	170	11	32	0	0	0.50	0.00	0.00	0
38057	Cinnamon spice mocha, tall w/whipped cream	12	fluid ounce(s)	360	—	320	10	31	0	17	11.00	—	—	—
38051	Espresso, single shot	1	fluid ounce(s)	30	—	5	0	1	0	0	0.00	0.00	0.00	—
38088	Flavored syrup, 1 pump	1	serving(s)	10	—	20	0	5	0	0	0.00	0.00	0.00	0
32562	Frappuccino coffee drink, lite mocha	9½	fluid ounce(s)	281	—	100	7	12	3	3	2.00	—	—	0
38079	Frappuccino, grande chocolate malt	16	fluid ounce(s)	480	—	470	15	87	2	10	3.50	—	—	—
38075	Frappuccino, grande mocha malt	12	fluid ounce(s)	360	—	430	14	91	1	7	4.00	—	—	—
32561	Frappuccino low fat coffee drink, all flavors	9½	fluid ounce(s)	281	—	190	6	39	0	3	2.00	—	—	—
38067	Frappuccino, tall caramel	12	fluid ounce(s)	360	—	210	4	43	0	3	1.50	—	—	—
38078	Frappuccino, tall chocolate	12	fluid ounce(s)	360	—	290	13	52	1	5	1.00	—	—	—
38069	Frappuccino, tall chocolate brownie	12	fluid ounce(s)	360	—	270	5	51	1	7	4.50	—	—	—
38070	Frappuccino, tall coffee	12	fluid ounce(s)	360	—	190	4	38	0	3	1.50	—	—	—
38071	Frappuccino, tall espresso	12	fluid ounce(s)	360	—	160	4	33	0	2	1.50	—	—	—
38073	Frappuccino, mocha	12	fluid ounce(s)	360	—	220	5	44	0	3	1.50	—	—	—
38072	Frappuccino, tall mocha coconut	12	fluid ounce(s)	360	—	300	5	58	2	7	5.00	—	—	—
38080	Frappuccino, tall vanilla	12	fluid ounce(s)	360	—	260	11	47	0	4	1.00	—	—	—
38074	Frappuccino, tall white chocolate	12	fluid ounce(s)	360	—	240	5	48	0	4	2.50	—	—	—
33111	Latte, tall w/nonfat milk	12	fluid ounce(s)	360	335	123	12	17	0	1	0.40	0.16	0.02	0
33112	Latte, tall w/whole milk	12	fluid ounce(s)	360	325	212	11	17	0	11	6.90	3.24	0.42	—
33109	Macchiato, tall caramel w/nonfat milk	12	fluid ounce(s)	360	—	140	7	27	0	1	0.40	—	—	—
33110	Macchiato, tall caramel w/whole milk	12	fluid ounce(s)	360	—	190	6	27	0	7	4.00	—	—	—
33107	Mocha coffee drink, tall nonfat, w/o whipped cream	12	fluid ounce(s)	360	—	180	12	33	1	2	1.50	0.68	0.08	—
38089	Mocha syrup	1	serving(s)	17	—	25	1	6	0	1	0.00	—	—	—
33108	Mocha, tall w/whole milk	12	fluid ounce(s)	360	—	340	12	33	1	20	12.00	3.48	0.44	—
38084	Tazo chai black tea, tall	12	fluid ounce(s)	360	—	210	6	36	0	5	3.50	—	—	—
38083	Tazo chai black tea, tall nonfat	12	fluid ounce(s)	360	—	170	6	37	0	0	0.00	0.00	0.00	0
38087	Tazo chai black tea, tall soy milk	12	fluid ounce(s)	360	—	190	4	39	1	2	0.00	—	—	—
38063	Tazo chai creme frappuccino, tall	12	fluid ounce(s)	360	—	280	11	51	0	4	1.00	—	—	—
38076	Tazo iced tea, tall	12	fluid ounce(s)	360	—	60	0	16	0	0	0.00	0.00	0.00	0
38077	Tazo tea, grande lemonade	16	fluid ounce(s)	480	—	120	0	31	0	0	0.00	0.00	0.00	0
38065	Tazoberry creme frappuccino, tall	12	fluid ounce(s)	360	—	240	4	54	1	1	0.00	—	—	—
38066	Tazoberry frappuccino, tall	12	fluid ounce(s)	360	—	140	1	36	1	0	0.00	0.00	0.00	0
38045	Vanilla creme steamed nonfat milk, tall w/whipped cream	12	fluid ounce(s)	360	—	180	12	32	0	0	0.00	0.00	0.00	—
38046	Vanilla creme steamed soy milk, tall w/whipped cream	12	fluid ounce(s)	360	—	300	8	37	1	12	6.00	—	—	—

PAGE KEY: A–58 = Breads/Baked Goods A–62 = Cereal/Rice/Pasta A–66 = Fruit A–70 = Vegetables/Legumes A–80 = Nuts/Seeds A–82 = Vegetarian A–84 = Dairy A–90 = Eggs A–90 = Seafood A–92 = Meats A–96 = Poultry A–96 = Processed Meats A–98 = Beverages A–102 = Fats/Oils A–104 = Sweets A–106 = Sauces/Condiments/Spices A–108 = Mixed Foods/Soups/Sandwiches A–114 = Fast Food A–130 = Convenience A–132 = Baby Foods

Chol (mg)	Calc (mg)	Iron (mg)	Magn (mg)	Pota (mg)	Sodi (mg)	Zinc (mg)	Vit A (RAE) (µg)	Thia (mg)	Vit E (mg α)	Ribo (mg)	Niac (mg)	Vit B_6 (mg)	Fola (µg)	Vit C (mg)	Vit B_{12} (µg)	Sele (µg)
25	256	0.25	29	412	188	0.82	59	0.10	—	0.44	0.40	0.10	7	2	1	5
27	331	0.23	27	415	215	0.88	57	0.07	0.11	0.44	0.33	0.10	16	0	1	5
70	0	0.36	—	—	450	—	—	—	—	—	—	—	—	0	—	—
40	150	1.80	—	—	830	—	—	—	—	—	—	—	—	6	—	—
40	150	2.70	—	—	810	—	—	—	—	—	—	—	—	6	—	—
65	250	2.70	—	—	1450	—	—	—	—	—	—	—	—	9	—	—
45	150	1.80	—	—	880	—	—	—	—	—	—	—	—	9	—	—
40	200	1.80	—	—	730	—	58	—	—	—	—	—	—	2	—	—
40	200	2.70	—	—	710	—	58	—	—	—	—	—	—	2	—	—
65	300	2.70	—	—	1300	—	—	—	—	—	—	—	—	4	—	—
40	200	1.44	—	—	700	—	58	—	—	—	—	—	—	2	—	—
15	80	1.44	—	—	340	—	38	—	—	—	—	—	—	1	—	—
15	80	1.86	—	—	400	—	—	—	—	—	—	—	—	4	—	—
10	80	1.80	—	—	280	—	—	—	—	—	—	—	—	4	—	—
15	150	1.80	—	—	490	—	—	—	—	—	—	—	—	9	—	—
15	150	2.70	—	—	470	—	—	—	—	—	—	—	—	9	—	—
35	250	2.70	—	—	980	—	—	—	—	—	—	—	—	12	—	—
15	150	1.44	—	—	480	—	—	—	—	—	—	—	—	12	—	—
25	20	0.00	—	—	550	—	0	—	—	—	—	—	—	0	—	—
10	0	0.00	—	—	340	—	0	—	—	—	—	—	—	0	—	—
0	0	1.08	—	—	15	—	0	—	—	—	—	—	—	0	0	—
25	250	0.00	—	—	95	—	0	—	—	—	—	—	—	1	0	—
3	200	0.00	—	—	100	—	0	—	—	—	—	—	—	0	0	—
0	250	0.72	—	—	75	—	0	—	—	—	—	—	—	0	0	—
5	300	0.72	—	—	150	—	0	—	—	—	—	—	—	0	0	—
70	350	1.08	—	—	140	—	0	—	—	—	—	—	—	2	0	—
0	0	0.00	—	—	0	—	0	—	—	—	—	—	—	0	0	—
0	0	0.00	—	—	0	—	0	—	—	—	—	—	—	0	0	—
13	200	1.08	—	—	80	—	—	—	—	—	—	—	—	0	—	—
15	250	2.70	—	—	420	—	0	—	—	—	—	—	—	12	0	—
20	250	1.08	—	—	390	—	0	—	—	—	—	—	—	0	0	—
12	220	0.00	—	—	110	—	—	—	—	—	—	—	—	0	—	—
10	150	0.00	—	—	180	—	0	—	—	—	—	—	—	0	0	—
3	400	1.80	—	—	300	—	0	—	—	—	—	—	—	5	0	—
10	150	1.44	—	—	220	—	0	—	—	—	—	—	—	0	0	—
10	150	0.00	—	—	180	—	0	—	—	—	—	—	—	0	0	—
10	100	0.00	—	—	160	—	0	—	—	—	—	—	—	0	0	—
10	150	0.72	—	—	180	—	0	—	—	—	—	—	—	0	0	—
10	150	1.08	—	—	220	—	0	—	—	—	—	—	—	0	0	—
3	400	0.00	—	—	280	—	0	—	—	—	—	—	—	4	0	—
10	150	0.00	—	—	210	—	0	—	—	—	—	—	—	0	0	—
6	420	0.18	40	—	174	1.35	—	0.12	—	0.47	0.36	0.14	18	4	1	—
46	400	0.18	47	254	165	1.28	—	0.13	—	0.54	0.35	0.14	17	3	1	—
25	250	0.36	—	—	110	—	—	—	—	—	—	—	—	2	—	—
25	200	0.36	—	—	105	—	—	—	—	—	—	—	—	1	—	—
5	350	2.70	—	—	150	—	—	—	—	—	—	—	—	2	—	—
0	0	0.72	—	—	0	—	0	—	—	—	—	—	—	0	0	—
47	300	0.18	—	—	169	—	—	—	—	—	—	—	—	2	—	—
20	200	0.36	—	—	85	—	0	—	—	—	—	—	—	1	0	—
5	200	0.36	—	—	95	—	0	—	—	—	—	—	—	0	0	—
0	200	0.72	—	—	70	—	0	—	—	—	—	—	—	0	0	—
3	400	0.00	—	—	280	—	0	—	—	—	—	—	—	4	0	—
0	0	0.00	—	—	0	—	0	—	—	—	—	—	—	0	0	—
0	0	0.00	—	—	15	—	0	—	—	—	—	—	—	5	0	—
0	150	0.00	—	—	125	—	0	—	—	—	—	—	—	1	0	—
0	0	0.00	—	—	30	—	0	—	—	—	—	—	—	0	0	—
5	350	0.00	—	—	170	—	0	—	—	—	—	—	—	0	0	—
30	400	1.44	—	—	130	—	0	—	—	—	—	—	—	0	0	—

TABLE F-1

Food Composition (Computer code number is for Wadsworth Diet Analysis program) (For purposes of calculations, use "0" for t, <1, <.1, <.01, etc.)

DA + Code	Food Description	Quantity	Measure	Wt (g)	H_2O (g)	Ener (kcal)	Prot (g)	Carb (g)	Dietary Fiber (g)	Fat (g)	Fat Breakdown (g) Sat	Mono	Poly	Trans
	FAST FOOD—Continued													
38044	Vanilla creme steamed whole milk, tall w/whipped cream	12	fluid ounce(s)	360	—	340	10	31	0	18	12.00	—	—	—
38090	Whipped cream	1	serving(s)	27	—	100	0	2	0	9	6.00	—	—	—
38062	White chocolate mocha, tall nonfat w/o whipped cream	12	fluid ounce(s)	360	—	260	12	45	0	4	3.00	—	—	—
38061	White chocolate mocha, tall w/whipped cream	12	fluid ounce(s)	360	—	410	11	44	0	20	13.00	—	—	—
38048	White hot chocolate, tall w/o whipped cream	12	fluid ounce(s)	360	—	300	15	51	0	5	3.50	—	—	—
38047	White hot chocolate, tall w/whipped cream	12	fluid ounce(s)	360	—	460	13	50	0	22	15.00	—	—	—
38050	White hot chocolate soy milk, tall w/whipped cream	12	fluid ounce(s)	360	—	420	11	56	1	16	9.00	—	—	—
	Subway													
34023	Asiago caesar chicken wrap	1	item(s)	244	—	413	22	47	2	15	3.00	—	—	0
38622	Atkins-friendly chicken bacon ranch wrap	1	item(s)	213	—	480	40	19	11	27	9.00	—	—	0
38623	Atkins-friendly turkey bacon melt wrap	1	item(s)	199	—	430	32	22	12	25	9.00	—	—	0
34029	Bacon & egg breakfast sandwich	1	item(s)	127	—	302	14	29	1	15	4.00	—	—	0
32045	Chocolate chip cookie	1	item(s)	48	—	209	3	29	1	10	3.50	—	—	1.07
32048	Chocolate chip M&M cookie	1	item(s)	48	—	210	2	29	1	10	3.00	—	—	2.67
32049	Chocolate chunk cookie	1	item(s)	48	—	210	2	30	1	10	3.00	—	—	2.67
4024	Classic Italian B.M.T. sandwich, 6", white bread	1	item(s)	250	—	453	21	40	3	24	8.00	—	—	0
16397	Club salad	1	item(s)	323	—	145	17	12	3	4	1.00	—	—	0
3422	Club sandwich, 6", white bread	1	item(s)	253	—	294	22	40	3	5	1.50	—	—	0
4030	Cold cut trio sandwich, 6", white bread	1	item(s)	254	—	415	19	40	3	20	7.00	—	—	0
34030	Ham & egg breakfast sandwich	1	item(s)	147	—	291	15	30	1	12	3.00	—	—	0
3885	Ham sandwich, 6", white bread	1	item(s)	219	—	261	17	39	3	5	1.50	—	—	0
34026	Honey mustard melt sandwich, 6", Italian bread	1	item(s)	258	—	373	23	47	3	11	5.00	—	—	—
34027	Horseradish roast beef sandwich, 6", Italian bread	1	item(s)	230	—	401	18	42	3	17	3.00	—	—	—
4651	Meatball sandwich, 6", white bread	1	item(s)	284	—	501	23	46	4	25	10.00	—	—	0.75
15839	Melt sandwich, 6", white bread	1	item(s)	256	—	380	23	41	3	15	5.00	—	—	—
32046	Oatmeal raisin cookie	1	item(s)	48	—	197	3	29	1	8	2.00	—	—	2.67
32047	Peanut butter cookie	1	item(s)	48	—	220	3	26	1	12	3.00	—	—	1.07
3957	Roast beef sandwich, 6", white bread	1	item(s)	220	—	264	18	39	3	5	1.00	—	—	0
16403	Roasted chicken breast salad	1	item(s)	304	—	137	16	12	3	3	0.50	—	—	—
16378	Roasted chicken breast sandwich, 6", white bread	1	item(s)	234	—	311	25	40	3	6	1.50	—	—	0
34028	Southwest steak & cheese sandwich, 6", Italian bread	1	item(s)	255	—	412	23	42	4	18	6.00	—	—	—
4032	Spicy italian sandwich, 6", white bread	1	item(s)	213	—	458	19	42	2	24	9.00	—	—	0
4031	Steak & cheese sandwich, 6", white bread	1	item(s)	253	—	362	23	41	4	13	4.50	—	—	0
34024	Steak & cheese wrap	1	item(s)	245	—	353	22	46	3	9	4.00	—	—	—
32050	Sugar cookie	1	item(s)	48	—	222	2	28	1	12	3.00	—	—	3.73
16402	Tuna salad	1	item(s)	314	—	238	13	11	3	16	4.00	—	—	—
15844	Tuna sandwich, 6", white bread	1	item(s)	252	—	419	18	39	3	21	5.00	—	—	—
15834	Turkey breast & ham sandwich, 6", white bread	1	item(s)	229	—	267	18	40	3	5	1.00	—	—	0
34025	Turkey breast & bacon wrap	1	item(s)	228	—	318	19	45	2	7	2.50	—	—	—
16376	Turkey breast sandwich, 6", white bread	1	item(s)	220	—	254	16	39	3	4	1.00	—	—	0
16375	Veggie delite, 6", white bread	1	item(s)	163	—	200	7	37	3	3	0.50	—	—	0
32051	White macadamia nut cookie	1	item(s)	48	—	221	2	27	1	12	3.00	—	—	1.07
	Taco Bell													
29906	7-layer burrito	1	item(s)	283	—	530	18	67	10	22	8.00	—	—	3
744	Bean burrito	1	item(s)	198	—	370	14	55	8	10	3.50	—	—	2
749	Beef burrito supreme	1	item(s)	248	—	440	18	51	7	18	8.00	—	—	2
33417	Beef chalupa supreme	1	item(s)	153	—	390	14	31	3	24	10.00	—	—	3
29910	Beef gordita supreme	1	item(s)	153	—	310	14	30	3	16	7.00	—	—	0.50
2014	Beef soft taco	1	item(s)	99	—	210	10	21	2	10	4.50	—	—	1
10860	Beef soft taco supreme	1	item(s)	134	—	260	11	22	3	14	7.00	—	—	1
2018	Big beef burrito supreme	1	item(s)	291	—	510	23	52	11	23	9.00	6.55	1.61	—
14467	Big chicken burrito supreme	1	item(s)	255	—	460	27	50	3	17	6.00	—	—	—
34472	Chicken burrito supreme	1	item(s)	248	—	410	21	50	5	14	6.00	—	—	2
33418	Chicken chalupa supreme	1	item(s)	153	—	370	17	30	1	20	8.00	—	—	3

PAGE KEY: A–58 = Breads/Baked Goods A–62 = Cereal/Rice/Pasta A–66 = Fruit A–70 = Vegetables/Legumes A–80 = Nuts/Seeds A–82 = Vegetarian A–84 = Dairy A–90 = Eggs A–90 = Seafood A–92 = Meats A–96 = Poultry A–96 = Processed Meats A–98 = Beverages A–102 = Fats/Oils A–104 = Sweets A–106 = Sauces/Condiments/Spices A–108 = Mixed Foods/Soups/Sandwiches A–114 = Fast Food A–130 = Convenience A–132 = Baby Foods

Chol (mg)	Calc (mg)	Iron (mg)	Magn (mg)	Pota (mg)	Sodi (mg)	Zinc (mg)	Vit A (RAE) (µg)	Thia (mg)	Vit E (mg α)	Ribo (mg)	Niac (mg)	Vit B_6 (mg)	Fola (µg)	Vit C (mg)	Vit B_{12} (µg)	Sele (µg)
75	40	0.00	—	—	160	—	0	—	—	—	—	—	—	2	0	—
40	0	0.00	—	—	10	—	0	—	—	—	—	—	—	0	0	—
5	400	0.00	—	—	210	—	0	—	—	—	—	—	—	0	0	—
70	400	0.00	—	—	210	—	0	—	—	—	—	—	—	2	0	—
10	450	0.00	—	—	250	—	0	—	—	—	—	—	—	0	0	—
75	500	0.00	—	—	250	—	0	—	—	—	—	—	—	4	0	—
35	500	1.44	—	—	210	—	0	—	—	—	—	—	—	0	0	—
46	40	2.70	—	—	1320	—	—	—	—	—	—	—	—	15	—	—
90	350	2.70	—	—	1340	—	—	—	—	—	—	—	—	7	—	—
65	300	2.70	—	—	1650	—	—	—	—	—	—	—	—	5	—	—
185	60	1.80	—	—	480	—	—	—	—	—	—	—	—	15	—	—
12	0	1.00	—	—	135	—	0	—	—	—	—	—	—	0	—	—
13	0	1.00	—	—	135	—	0	—	—	—	—	—	—	0	—	—
12	0	1.00	—	—	150	—	0	—	—	—	—	—	—	0	—	—
56	100	2.70	—	—	1740	—	—	—	—	—	—	—	—	24	—	—
30	40	1.80	—	—	1070	—	—	—	—	—	—	—	—	30	—	—
30	40	3.60	—	—	1250	—	60	—	—	—	—	—	—	24	—	—
57	150	3.60	—	—	1670	—	100	—	—	—	—	—	—	24	—	—
189	60	2.70	—	—	700	—	67	—	—	—	—	—	—	15	—	—
25	40	2.70	—	—	1260	—	—	—	—	—	—	—	—	24	—	—
41	100	2.70	—	—	1570	—	—	—	—	—	—	—	—	24	—	—
27	40	3.60	—	—	880	—	—	—	—	—	—	—	—	24	—	—
56	100	3.60	—	—	1350	—	—	—	—	—	—	—	—	24	—	—
41	100	2.70	—	—	1690	—	—	—	—	—	—	—	—	24	—	—
14	0	1.00	—	—	180	—	0	—	—	—	—	—	—	0	—	—
0	0	1.00	—	—	200	—	0	—	—	—	—	—	—	0	—	—
20	40	3.60	—	—	840	—	60	—	—	—	—	—	—	24	—	—
36	40	1.08	—	—	730	—	—	—	—	—	—	—	—	30	—	—
48	60	3.60	—	—	880	—	—	—	—	—	—	—	—	24	—	—
44	100	6.30	—	—	1120	—	—	—	—	—	—	—	—	24	—	—
57	30	3.00	—	—	1498	—	—	—	—	—	—	—	—	13	—	—
37	100	6.30	—	—	1200	—	—	—	—	—	—	—	—	24	—	—
37	150	7.20	—	—	1400	—	—	—	—	—	—	—	—	15	—	—
18	0	1.00	—	—	170	—	0	—	—	—	—	—	—	0	—	—
42	100	1.08	—	—	880	—	177	—	—	—	—	—	—	30	—	—
42	100	2.70	—	—	1180	—	100	—	—	—	—	—	—	24	—	—
23	40	2.70	—	—	1210	—	—	—	—	—	—	—	—	24	—	—
24	60	2.70	—	—	1490	—	—	—	—	—	—	—	—	15	—	—
15	40	2.70	—	—	1000	—	—	—	—	—	—	—	—	24	—	—
0	40	1.80	—	—	500	—	—	—	—	—	—	—	—	24	—	—
13	0	1.00	—	—	140	—	0	—	—	—	—	—	—	0	—	—
25	300	3.59	—	—	1360	—	—	—	—	—	—	—	—	5	—	—
10	200	2.69	—	—	1200	—	53	—	—	—	—	—	—	5	—	—
40	200	2.70	—	—	1330	—	351	—	—	—	—	—	—	9	—	—
40	150	1.80	—	—	600	—	—	—	—	—	—	—	—	5	—	—
35	150	2.70	—	—	590	—	—	—	—	—	—	—	—	5	—	—
25	100	1.80	—	—	620	—	44	—	—	—	—	—	—	2	—	—
40	150	1.80	—	—	630	—	73	—	—	—	—	—	—	5	—	—
60	150	2.70	—	493	1500	—	877	—	—	0.07	—	—	—	5	—	—
70	101	1.46	—	—	1200	—	—	—	—	—	—	—	—	2	—	—
45	200	2.70	—	—	1270	—	—	—	—	—	—	—	—	9	—	—
45	100	1.08	—	—	530	—	—	—	—	—	—	—	—	5	—	—

TABLE F-1

Food Composition (Computer code number is for Wadsworth Diet Analysis program) (For purposes of calculations, use "0" for t, <1, <.1, <.01, etc.)

DA + Code	Food Description	Quantity	Measure	Wt (g)	H_2O (g)	Ener (kcal)	Prot (g)	Carb (g)	Dietary Fiber (g)	Fat (g)	Fat Breakdown (g) Sat	Mono	Poly	Trans
	FAST FOOD—Continued													
29900	Chicken fajita wrap supreme	1	item(s)	255	—	510	20	53	3	24	7.76	—	—	—
29895	Choco taco ice cream dessert	1	item(s)	113	—	310	3	37	1	17	10.00	—	—	—
10794	Cinnamon twists	1	serving(s)	35	—	160	0	28	0	5	1.00	—	—	1.50
14465	Grilled chicken burrito	1	item(s)	198	—	390	19	49	3	13	4.00	—	—	—
29911	Grilled chicken gordita supreme	1	item(s)	153	—	290	17	28	2	12	5.00	—	—	0
14463	Grilled chicken soft taco	1	item(s)	99	—	190	14	19	0	6	2.50	—	—	—
29912	Grilled steak gordita supreme	1	item(s)	153	—	290	16	28	2	13	6.00	—	—	0.50
29904	Grilled steak soft taco	1	item(s)	127	—	280	12	21	1	17	4.50	—	—	1
29905	Grilled steak soft taco supreme	1	item(s)	135	—	240	15	20	2	11	5.00	—	—	—
2021	Mexican pizza	1	serving(s)	216	—	550	21	46	7	31	11.00	—	—	5
2011	Nachos	1	serving(s)	99	—	320	5	33	2	19	4.50	—	—	5
2012	Nachos bellgrande	1	serving(s)	308	—	780	20	80	12	43	13.00	—	—	10
34473	Steak burrito supreme	1	item(s)	248	—	420	19	50	6	16	7.00	—	—	2
33419	Steak chalupa supreme	1	item(s)	153	—	370	15	29	2	22	8.00	—	—	3
29899	Steak fajita wrap supreme	1	item(s)	255	—	510	21	52	3	25	8.00	—	—	—
747	Taco	1	item(s)	78	—	170	8	13	3	10	4.00	—	—	0.50
2015	Taco salad w/salsa, with shell	1	serving(s)	533	—	790	31	73	13	42	15.00	—	—	8.75
14459	Taco supreme	1	item(s)	113	—	220	9	14	3	14	7.00	—	—	1
748	Tostada	1	item(s)	170	—	250	11	29	7	10	4.00	—	—	1.50
29901	Veggie fajita wrap supreme	1	item(s)	255	—	470	11	55	3	22	7.00	—	—	—
	CONVENIENCE MEALS													
	Banquet													
29961	Barbeque chicken meal	1	item(s)	281	—	330	16	37	2	13	3.00	—	—	—
14788	Boneless white fried chicken meal	1	item(s)	234	—	490	14	49	2	27	7.00	—	—	—
29960	Fish sticks meal	1	item(s)	187	—	270	13	31	3	10	3.00	—	—	—
29957	Lasagna with meat sauce meal	1	item(s)	312	—	320	15	46	7	9	4.00	—	—	—
14777	Macaroni & cheese meal	1	item(s)	340	—	420	15	57	5	14	8.00	—	—	—
1741	Meatloaf meal	1	item(s)	269	—	240	14	20	4	11	4.00	—	—	—
39418	Pepperoni pizza meal	1	item(s)	191	—	480	11	56	5	23	8.00	—	—	—
33759	Roasted white turkey meal	1	item(s)	255	—	230	14	30	5	6	2.00	—	—	—
1743	Salisbury steak meal	1	item(s)	269	197	380	12	28	3	24	12.00	—	—	—
	Budget Gourmet													
1914	Cheese manicotti w/meat sauce	1	item(s)	284	194	420	18	38	4	22	11.00	6.00	1.34	—
1915	Chicken w/fettucini	1	item(s)	284	—	380	20	33	3	19	10.00	—	—	—
3986	Light beef stroganoff	1	item(s)	248	177	290	20	32	3	7	4.00	—	—	—
3996	Light sirloin of beef in herb sauce	1	item(s)	269	214	260	19	30	5	7	4.00	2.30	0.31	—
3987	Light vegetable lasagna	1	item(s)	298	227	290	15	36	5	9	1.79	0.89	0.60	—
	Healthy Choice													
36979	Bowls chicken teriyaki with rice	1	item(s)	298	—	330	19	50	5	6	2.00	2.00	2.00	—
9425	Cheese French bread pizza	1	item(s)	170	—	360	20	57	5	5	1.50	—	—	—
9306	Chicken enchilada suprema meal	1	item(s)	320	252	360	13	59	8	7	3.00	2.00	2.00	—
9316	Lemon pepper fish meal	1	item(s)	303	—	280	11	49	5	5	2.00	1.00	2.00	—
9322	Traditional salisbury steak meal	1	item(s)	354	250	360	23	45	5	9	3.50	4.00	1.00	—
9359	Traditional turkey breasts meal	1	item(s)	298	—	330	21	50	4	5	2.00	1.50	1.50	—
9451	Zucchini lasagna	1	item(s)	383	—	280	13	47	5	4	2.50	—	—	—
	Stouffers													
2363	Cheese enchiladas with mexican rice	1	serving(s)	276	—	370	12	48	5	14	5.00	—	—	—
2313	Cheese French bread pizza	1	serving(s)	294	—	370	14	43	3	16	6.00	—	—	—
11138	Cheese manicotti w/tomato sauce	1	item(s)	255	—	330	17	35	3	13	8.00	—	—	—
2366	Chicken pot pie	1	item(s)	284	—	740	23	56	4	47	18.00	12.41	10.48	—
11116	Homestyle baked chicken breast w/mashed potatoes & gravy	1	item(s)	252	—	260	19	21	1	11	3.00	—	—	—
11146	Homestyle beef pot roast & potatoes	1	item(s)	252	—	270	16	25	3	12	4.50	—	—	—
11152	Homestyle roast turkey breast w/stuffing & mashed potatoes	1	item(s)	273	—	300	16	34	2	11	3.00	—	—	—
11043	Lean Cuisine Cafe Classics baked chicken & whipped potatoes w/stuffing	1	item(s)	227	—	240	17	33	3	5	1.50	1.50	1.00	0
11046	Lean Cuisine Cafe Classics honey mustard chicken	1	item(s)	213	—	260	18	37	1	4	1.50	1.00	1.00	0
360	Lean Cuisine Everyday Favorites chicken chow mein w/rice	1	item(s)	255	—	210	12	33	2	3	1.00	1.00	0.50	0
9467	Lean Cuisine Everyday Favorites fettucini alfredo	1	item(s)	262	—	280	13	40	2	7	3.50	2.00	1.00	0
11055	Lean Cuisine Everyday Favorites lasagna w/meat sauce	1	item(s)	291	—	300	19	41	3	8	4.00	2.00	0.50	0
9479	Lean Cuisine French bread deluxe pizza	1	item(s)	174	—	330	18	44	3	9	3.50	1.50	1.00	0

PAGE KEY: A–58 = Breads/Baked Goods A–62 = Cereal/Rice/Pasta A–66 = Fruit A–70 = Vegetables/Legumes A–80 = Nuts/Seeds A–82 = Vegetarian A–84 = Dairy A–90 = Eggs A–90 = Seafood A–92 = Meats A–96 = Poultry A–96 = Processed Meats A–98 = Beverages A–102 = Fats/Oils A–104 = Sweets A–106 = Sauces/Condiments/Spices A–108 = Mixed Foods/Soups/Sandwiches A–114 = Fast Food A–130 = Convenience A–132 = Baby Foods

Chol (mg)	Calc (mg)	Iron (mg)	Magn (mg)	Pota (mg)	Sodi (mg)	Zinc (mg)	Vit A (RAE) (µg)	Thia (mg)	Vit E (mg α)	Ribo (mg)	Niac (mg)	Vit B_6 (mg)	Fola (µg)	Vit C (mg)	Vit B_{12} (µg)	Sele (µg)
57	165	1.52	—	—	1182	—	—	—	—	—	—	—	—	7	—	—
20	60	0.72	—	—	100	—	—	—	—	—	—	—	—	0	—	—
0	0	0.37	—	—	150	—	0	—	—	—	—	—	—	0	—	—
40	151	1.44	—	—	1240	—	—	—	—	—	—	—	—	2	—	—
45	100	1.80	—	—	530	—	—	—	—	—	—	—	—	5	—	—
30	100	1.08	—	—	550	—	15	—	—	—	—	—	—	1	—	—
35	100	2.70	—	—	520	—	—	—	—	—	—	—	—	4	—	—
30	100	1.44	—	—	650	—	29	—	—	—	—	—	—	4	—	—
35	100	1.08	—	—	510	—	29	—	—	—	—	—	—	4	—	—
45	350	3.60	—	—	1030	—	—	—	—	—	—	—	—	6	—	—
4	80	0.72	—	—	530	—	0	—	—	—	—	—	—	0	—	—
35	200	2.70	—	—	1300	—	162	—	—	—	—	—	—	6	—	—
35	200	2.70	—	—	1260	—	789	—	—	—	—	—	—	9	—	—
35	100	1.44	—	—	520	—	—	—	—	—	—	—	—	4	—	—
50	150	1.80	—	—	1200	—	—	—	—	—	—	—	—	6	—	—
25	60	1.08	—	—	350	—	44	—	—	—	—	—	—	2	—	—
65	400	6.23	—	—	1670	—	—	—	—	—	—	—	—	21	—	—
40	80	1.44	—	—	360	—	73	—	—	—	—	—	—	5	—	—
15	150	1.44	—	—	710	—	281	—	—	—	—	—	—	5	—	—
30	150	1.44	—	—	990	—	—	—	—	—	—	—	—	6	—	—
50	40	1.08	—	—	1210	—	0	—	—	—	—	—	—	5	—	—
65	60	1.08	—	—	1150	—	—	—	—	—	—	—	—	0	—	—
30	60	1.44	—	—	690	—	—	—	—	—	—	—	—	2	—	—
20	100	2.70	—	—	1170	—	—	—	—	—	—	—	—	0	—	—
20	150	1.44	—	—	1330	—	0	—	—	—	—	—	—	0	—	—
30	0	1.80	—	—	1040	—	0	—	—	—	—	—	—	0	—	—
35	150	1.80	—	—	870	—	0	—	—	—	—	—	—	0	—	—
25	60	1.80	—	—	1070	—	—	—	—	—	—	—	—	4	—	—
60	40	1.44	—	—	1140	—	0	—	—	—	—	—	—	0	—	—
85	300	2.70	45	484	810	2.29	—	0.45	—	0.51	4.00	0.23	31	0	1	—
85	100	2.70	—	—	810	—	—	0.15	—	0.43	6.00	—	—	0	—	—
35	40	1.80	39	280	580	4.71	—	0.17	—	0.37	4.28	0.27	19	2	3	—
30	40	1.80	58	540	850	4.81	—	0.16	—	0.29	5.53	0.37	38	6	2	—
15	283	3.03	79	420	780	1.39	—	0.22	—	0.45	3.13	0.32	75	59	<1	—
40	20	0.72	—	—	600	—	—	—	—	—	—	—	—	15	—	—
10	350	3.60	—	—	600	—	—	—	—	—	—	—	—	12	—	—
30	40	1.44	—	—	580	—	—	—	—	—	—	—	—	4	—	—
30	40	0.36	—	—	580	—	—	—	—	—	—	—	—	30	—	—
45	80	2.70	—	—	580	—	—	—	—	—	—	—	—	21	—	—
35	40	1.44	—	—	600	—	—	—	—	—	—	—	—	0	—	—
10	200	1.80	—	—	310	—	—	—	—	—	—	—	—	0	—	—
25	200	1.44	—	360	890	—	—	—	—	—	—	—	—	12	—	—
15	200	1.80	—	240	880	—	—	—	—	—	—	—	—	0	—	—
40	350	1.08	—	430	810	—	—	—	—	—	—	—	—	1	—	—
65	150	2.70	—	—	1170	—	—	—	—	—	—	—	—	2	—	—
50	20	0.72	—	500	760	—	0	—	—	—	—	—	—	0	—	—
35	20	1.80	—	790	820	—	—	—	—	—	—	—	—	6	—	—
35	40	0.72	—	450	1190	—	0	—	—	—	—	—	—	0	—	—
30	80	0.72	—	480	690	—	—	—	—	—	—	—	—	0	—	—
35	60	0.36	—	370	640	—	—	—	—	—	—	—	—	0	—	—
30	20	0.36	—	310	620	—	—	—	—	—	—	—	—	0	—	—
20	200	0.36	—	260	670	—	0	—	—	—	—	—	—	0	—	—
30	200	1.08	—	590	650	—	—	—	—	—	—	—	—	5	—	—
20	100	1.80	—	390	630	—	—	—	—	—	—	—	—	9	—	—

TABLE F–1
Food Composition (Computer code number is for Wadsworth Diet Analysis program) (For purposes of calculations, use "0" for t, <1, <.1, <.01, etc.)

DA + Code	Food Description	Quantity	Measure	Wt (g)	H_2O (g)	Ener (kcal)	Prot (g)	Carb (g)	Dietary Fiber (g)	Fat (g)	Fat Breakdown (g) Sat	Mono	Poly	Trans
	CONVENIENCE MEALS—Continued													
	Weight Watchers													
11164	Smart Ones chicken enchiladas suiza entree	1	serving(s)	255	—	270	15	33	2	9	3.50	—	—	—
11155	Smart Ones garden lasagna entree	1	item(s)	312	—	270	14	36	5	7	3.50	—	—	—
11187	Smart Ones pepperoni pizza	1	item(s)	158	—	390	23	46	4	12	4.00	—	—	—
31514	Smart Ones spicy penne pasta & ricotta	1	item(s)	289	—	280	11	45	4	6	2.00	—	—	—
31512	Smart Ones spicy szechuan style vegetables & chicken	1	item(s)	255	—	220	11	39	3	2	0.50	—	—	—
	BABY FOODS													
787	Apple juice	4	fluid ounce(s)	127	112	60	0	15	<1	<1	0.02	0.00	0.04	—
778	Applesauce, strained	4	tablespoon(s)	64	55	31	<1	8	1	<1	0.02	0.01	0.04	—
779	Bananas w/tapioca, strained	4	tablespoon(s)	60	50	34	<1	9	1	<.1	0.02	0.01	0.01	—
604	Carrots, strained	4	tablespoon(s)	56	52	15	<1	3	1	<.1	0.01	0.00	0.03	—
770	Chicken noodle dinner, strained	4	tablespoon(s)	64	55	42	2	6	1	1	0.38	0.55	0.30	—
801	Green beans, strained	4	tablespoon(s)	60	0.05	15	0.77	3.53	1.13	0.05	0.01	0	0.03	—
910	Human milk, mature	2	fluid ounce(s)	62	54	43	1	4	0	3	1.24	1.02	0.31	—
760	Mixed cereal, prepared w/whole milk	4	ounce(s)	114	85	128	5	18	1	4	2.19	1.25	0.43	—
772	Mixed vegetable dinner, strained	2	ounce(s)	57	50	23	1	5	1	<.1	0.00	0.00	0.06	—
762	Rice cereal, prepared w/whole milk	4	ounce(s)	114	85	131	4	19	<1	4	2.64	1.02	0.16	—
758	Teething biscuits	1	item(s)	11	1	43	1	8	<1	<1	0.17	0.16	0.09	—

PAGE KEY: A–58 = Breads/Baked Goods A–62 = Cereal/Rice/Pasta A–66 = Fruit A–70 = Vegetables/Legumes A–80 = Nuts/Seeds A–82 = Vegetarian A–84 = Dairy A–90 = Eggs A–90 = Seafood A–92 = Meats A–96 = Poultry A–96 = Processed Meats A–98 = Beverages A–102 = Fats/Oils A–104 = Sweets A–106 = Sauces/Condiments/Spices A–108 = Mixed Foods/Soups/Sandwiches A–114 = Fast Food A–130 = Convenience A–132 = Baby Foods

Chol (mg)	Calc (mg)	Iron (mg)	Magn (mg)	Pota (mg)	Sodi (mg)	Zinc (mg)	Vit A (RAE) (μg)	Thia (mg)	Vit E (mg α)	Ribo (mg)	Niac (mg)	Vit B_6 (mg)	Fola (μg)	Vit C (mg)	Vit B_{12} (μg)	Sele (μg)
50	250	1.08	—	—	660	—	—	—	—	—	—	—	—	4	—	—
30	350	1.80	—	—	610	—	—	—	—	—	—	—	—	6	—	—
45	450	1.80	—	320	650	—	55	—	—	—	—	—	—	5	—	—
5	150	2.70	—	250	400	—	—	—	—	—	—	—	—	6	—	—
10	150	1.80	—	—	730	—	—	—	—	—	—	—	—	2	—	—
0	5	0.72	4	115	4	0.04	1	0.01	0.76	0.02	0.11	0.04	0	73	0	<1
0	3	0.14	2	45	1	0.01	1	0.01	0.38	0.02	0.04	0.02	1	25	0	<1
0	3	0.12	6	53	5	0.04	1	0.01	0.36	0.02	0.11	0.07	4	10	0	<1
0	12	0.21	5	110	21	0.08	321	0.01	0.29	0.02	0.26	0.04	8	3	0	<1
10	17	0.41	9	89	15	0.35	70	0.03	0.13	0.04	0.46	0.04	7	<.1	<.1	2
0	23.39	0.44	14.39	94.8	1.2	0.12	27	0.01	0.31	0.05	0.2	0.02	21	3.11	0	0.18
9	20	0.02	2	31	10	0.10	38	0.01	0.05	0.02	0.11	0.01	3	3	<.1	1
12	250	11.85	31	226	53	0.81	28	0.49	—	0.66	6.56	0.07	12	1	<.1	—
0	12	0.19	6	69	5	0.09	77	0.01	—	0.02	0.29	0.04	5	2	0	<1
12	272	13.85	51	216	52	0.73	25	0.53	—	0.57	5.91	0.13	9	1	<1	4
0	29	0.39	4	36	40	0.10	3	0.03	0.03	0.06	0.48	0.01	5	1	<.1	3

Glossary

absorption the passage of nutrients or substances into cells or tissues; nutrients pass into intestinal cells after digestion and then into the circulatory system (for example, into the bloodstream).

Acceptable Macronutrient Distribution Range (AMDR) a range of intakes for a particular energy source (carbohydrates, fat, protein) that is associated with a reduced risk of chronic disease while providing adequate intakes of essential nutrients.

accreditation approval; in the case of hospitals or university departments, approval by a professional organization of the educational program offered. There are phony accrediting agencies; the genuine ones are listed in a directory called *Accredited Institutions of Postsecondary Education*.

acesulfame-K (AY-see-sul-fame) a derivative of acetoacetic acid approved for use in the United States in 1988; also called acesulfame potassium. Because it is not metabolized by the body, acesulfame-K does not contribute calories and is excreted from the body unchanged. It is currently approved for use in more than 70 countries and is found in more than 100 international products, including chewing gum, gelatins, nondairy creamers, powdered drink mixes, and puddings.

acetaldehyde (ass-et-AL-duh-hide) a substance into which drinking alcohol (ethanol) is metabolized.

acid–base balance equilibrium between acid and base concentrations in the body fluids.

acidosis (a-sih-DOSE-sis) blood acidity above normal, indicating excess acid.

acids compounds that release hydrogens in a watery solution; acids have a low pH.

acquired immune deficiency syndrome (AIDS) an immune system disorder caused by the human immunodeficiency virus (HIV).

activities of daily living (ADL) include bathing, dressing, grooming, transferring from bed to chair, going to the bathroom, and feeding one's self.

added sugars sugars and other caloric sweeteners that are added to foods during processing or preparation. Added sugars do not include naturally occurring sugars such as those that occur in milk and fruits.

adequacy characterizes a diet that provides all of the essential nutrients, fiber, and energy (calories) in amounts sufficient to maintain health.

Adequate Intake (AI) the average amount of a nutrient that appears to be adequate for individuals when there is not sufficient scientific research to calculate an RDA. The AI exceeds the EAR and possibly the RDA.

adverse reaction an unusual response to food, including food allergies and food intolerances.

aerobic requiring oxygen.

aflatoxin a poisonous toxin produced by molds.

agave a plant with spiny-margined leaves and flowers.

age-related macular degeneration oxidative damage to the central portion of the eye—called the macula—that allows you to focus and see details clearly (peripheral vision remains unimpaired). The carotenoids lutein and zeaxanthin—found in broccoli, Brussels sprouts, kale, spinach, corn, lettuce, and peas—may protect normal macular function.

alcohol clear, colorless volatile liquid; the most commonly ingested form is ethyl alcohol or ethanol (EtOH).

alcohol abuse continued use of alcohol in spite of negative psychological, social, family, employment, or school problems because of alcohol.

alcohol dehydrogenase a liver enzyme that mediates the metabolism of alcohol.

alcohol dependency (alcoholism) a dependency on alcohol marked by compulsive uncontrollable drinking with negative effects on physical health, family relationships, and social health.

alcoholic hepatitis inflammation and injury to the liver due to excess alcohol consumption.

alkalosis (al-kah-LOH-sis) blood alkalinity above normal.

alternative sweeteners nutritive (calorie-containing) sweeteners such as fructose, sorbitol, mannitol, and xylitol.

amaranth a golden-colored grain.

amenorrhea cessation of menstruation; can be associated with strenuous athletic training.

amine (a-MEEN) **group** the nitrogen-containing portion of an amino acid.

amino (a-MEEN-o) **acids** building blocks of protein; each is a compound with an amine group at one end, an acid group at the other, and a distinctive side chain.

anabolic steroids synthetic male hormones with a chemical structure similar to that of cholesterol; such hormones have wide-ranging effects on body functioning.

anaerobic not requiring oxygen.

anaphylaxis (an-ah-fa-LAX-is) a potentially fatal reaction to a food allergen causing reduced oxygen supply to the heart and other body tissues. Symptoms include difficulty breathing, low blood pressure, pale skin, a weak, rapid pulse, and loss of consciousness.

anemia any condition in which the blood is unable to deliver oxygen to the cells of the body. Examples include a shortage or abnormality of the red blood cells. Many nutrient deficiencies and diseases can cause anemia.

anorexia nervosa literally "nervous lack of appetite," a disorder (usually seen in teenage girls) involving self-starvation to the extreme.

antibodies large proteins of the blood and body fluids that are produced by one type of immune cell in response to invasion of the body by unfamiliar molecules (mostly foreign proteins). Antibodies inactivate the foreign substances and so protect the body. The foreign substances are called *antigens*.

antioxidant (anti-OX-ih-dant) a compound that protects other compounds from oxygen by itself reacting with oxygen; helps to prevent damage done to the body as a result of chemical reactions that involve the use of oxygen.

antioxidant nutrients vitamins and minerals that protect other compounds from damaging reactions involving oxygen by themselves reacting with oxygen. The antioxidant nutrients are vitamin C, vitamin E, and beta-carotene. The mineral selenium also has a role in antioxidant reactions in the body.

appendicitis inflammation and/or infection of the appendix, a sac protruding from the large intestine.

appetite the psychological desire to find and eat food, experienced as a pleasant sensation, often in the absence of hunger.

appropriate technology a technology that utilizes locally abundant resources in preference to locally scarce resources. For developing countries, which usually have a large labor force and little capital, the appropriate technology would therefore be labor intensive.

arousal heightened activity of certain brain centers, may be associated with excitement and anxiety.

artesian water or **artesian well water** water drawn from a well that taps a confined water-bearing rock or rock formation.

artificial sweeteners nonnutritive sugar replacements such as acesulfame-K, aspartame, neotame, saccharin, and sucralose.

aspartame a dipeptide containing the amino acids aspartic acid and phenylalanine and used in the United States and Canada since 1981. Although it is digested as protein and supplies calories, it is so sweet that only small amounts, which contribute negligible calories, are needed to sweeten foods. Thus, it is classified as a nonnutritive sweetener. Aspartame is often sold under the trade name NutraSweet, and it is also blended with lactose and an anticaking agent and sold commercially as Equal.

atherosclerosis (ATH-er-oh-scler-OH-sis) a type of cardiovascular disease characterized by the formation of fatty deposits, or plaques, in the inner walls of the arteries.

atrophic gastritis an age-related condition characterized by the stomach's inability to produce enough acid, which in turn leads to vitamin B_{12} deficiencies.

atrophy a decrease in size in response to disuse.

balance a feature of a diet that provides a number of types of foods in balance with one another, such that foods rich in one nutrient do not crowd out of the diet foods that are rich in another nutrient.

basal metabolic rate (BMR) the rate at which the body spends energy to support its basal metabolism. The BMR accounts for the largest component of a person's daily energy (calorie) needs.

basal metabolism the sum total of all the chemical activities of the cells necessary to sustain life, exclusive of voluntary activities—that is, the ongoing activities of the cells when the body is at rest.

bases compounds that accept hydrogens from solutions; bases have a high pH.

behavior modification a process developed by psychologists for helping people make lasting behavior changes.

beriberi the thiamin deficiency disease, characterized by irregular heartbeat, paralysis, and extreme wasting of muscle tissue.

beta-carotene an orange pigment found in plants that is converted into vitamin A inside the body. Beta-carotene is also an antioxidant.

bialy a flat breakfast roll that is softer than a bagel.

bifidus factor (BIFF-id-us) a factor in colostrum and breast milk that favors the growth in the infant's intestinal tract of the "friendly" bacteria *Lactobacillus bifidus* so that other, less desirable intestinal inhabitants will not flourish.

bile a mixture of compounds, including cholesterol, made by the liver, stored in the gallbladder, and secreted into the small intestine. Bile emulsifies lipids to ready them for enzymatic digestion and helps transport them into the intestinal wall cells.

binders in foods, chemical compounds that can combine with nutrients (especially minerals) to form complexes the body cannot absorb. Examples of such binders are **phytic** (FIGHT-ic) **acid** and **oxalic** (ox-AL-ic) **acid.**

bing thin pancakes.

binge-eating disorder an eating disorder characterized by uncontrolled chronic episodes of overeating (compulsive overeating) without other symptoms of eating disorders. Typically, the episodes of binge eating occur at least twice a week for a period of six months or more.

bioelectrical impedance estimation of body fat content made by measuring how quickly electrical current is conducted through the body.

biological value (BV) a measure of protein quality, assessed by determining how well a given food or food mixture supports nitrogen retention.

biotechnology the use of biological systems or living organisms to make or modify products. Includes traditional methods used in making products such as wine, beer, yogurt, and cheese; cross breeding to enhance crop production; and modification of living plants, animals, and fish through the manipulation of genes (genetic engineering).

bisphosphonates drugs that decrease the risk of fractures by acting on the bone-dismantling cells (osteoclasts) and inhibiting their resorption of bone tissue; examples are Fosamax, Actonel, and Boniva.

black, cuban, or **turtle beans** These medium-size, black-skinned ovals have a rich, sweet taste. They are best served in Mexican and Latin American dishes or thick soups and stews.

black-eyed peas These peas are small, oval shaped, and creamy white with a black spot. They have a vegetable flavor with mealy texture. Use in salads with rice and greens.

body mass index an index of a person's weight in relation to height that correlates with total body fat content.

bok choy a vegetable with broad, white or greenish-white stalks and dark green leaves; also called *Chinese chard*.

bolillo a roll-like bread often used instead of tortillas or to make sandwiches.

bone density a measure of bone strength that reflects the degree of bone mineralization. The DEXA (dual-energy X-ray absorptiometry) bone density test compares your bone density to that of a healthy young adult. Bone density tests—using a dual beam of low-level X rays—take a snapshot of bone density in the spine, wrist, and hip.

bran the fibrous protective covering of a whole grain and source of fiber, B vitamins, and trace minerals.

buffers compounds that help keep a solution's acidity (amount of acid) or alkalinity (amount of base) constant.

bulimia nervosa, bulimarexia (byoo-LEE-me-uh, byoo-lee-ma-REX-ee-uh) binge eating (literally, "eating like an ox"), combined with an intense fear of becoming fat and usually followed by self-induced vomiting or taking laxatives.

burritos warm flour tortillas stuffed with a mixture of egg, meat, beans, and/or avocado.

caffeine a type of compound, called a *methylxanthine*, found in coffee beans, cola nuts, cocoa beans, and tea leaves. A central nervous system stimulant, caffeine's effects include increasing the heart rate, boosting urine production, and raising the metabolic rate.

caffeine dependence syndrome dependence on caffeine characterized by at least three of the four following criteria: withdrawal symptoms such as headache and fatigue; caffeine consumption despite knowledge that it may be causing harm; repeated, unsuccessful attempts to cut back on caffeine; and tolerance to caffeine.

calcitonin a hormone used as a drug to decrease the rate of bone loss in osteoporosis. Administered as a nasal spray (Miacalcin) or by injection, calcitonin works by inhibiting the bone resorption activity of osteoclasts.

calorie the unit used to measure energy.

calorie control control of consumption of energy (calories); a feature of a sound diet plan.

carbohydrates compounds made of single sugars or multiple sugars and composed of carbon, hydrogen, and oxygen atoms.

carcinogen (car-SIN-oh-jen) a cancer-causing substance.

cardiovascular conditioning or **training effect** the effect of regular exercise on the cardiovascular system—including improvements in heart, lung, and muscle function and increased blood volume.

carotenoids (kah-ROT-eh-noyds) a group of pigments (yellow, orange, and red) found in plant foods. Examples include beta-carotene (a precursor of vitamin A), alpha-carotene, lycopene, lutein, and zeaxanthin. Carotenoids have a variety of effects in the body including antioxidant activity and enhancement of immune function.

cassava a starchy root that is never eaten raw because it must be cooked to eliminate its bitter smell.

cellophane noodles thin, translucent noodles made from mung beans.

central obesity excess fat on the abdomen and around the trunk.

challah an egg-containing yeast bread, often braided, typically served on the Sabbath and holidays.

chapatti a flat unleavened bread made with finely milled wholemeal flour.

cherimoya a fruit with a rough green outer skin and sherbetlike flesh.

chilaquiles tortilla casserole often made with eggs or meat.

chiles relleños roasted mild green chili pepper stuffed with cheese, dipped in egg batter, and fried.

Chinese broccoli a green leafy vegetable often stir fried; also called *Chinese kale*.

chitterlings (chitlins) pig intestine.

chlorophyll the green pigment of plants that traps energy from sunlight and uses this energy in photosynthesis (the synthesis of carbohydrate by green plants).

cholesterol (koh-LESS-ter-all) one of the sterols, manufactured in the body for a variety of purposes, and also found in animal-derived foods.

choline a nonessential nutrient used by the body to synthesize various compounds, including the phospholipid lecithin; the body can make choline from the amino acid methionine.

chorizo spicy beef or pork sausage.

choy sum a bright green vegetable commonly stir fried; also called *field mustard* or *flowering Chinese cabbage*.

chylomicron (KIGH-loh-MY-cron) a type of lipoprotein that transports newly digested fat—mostly triglyceride—from the intestine through lymph and blood.

cirrhosis a chronic, degenerative disease of the liver in which the liver cells become infiltrated with fibrous tissues; blood flow through the liver is obstructed, causing back pressure and eventually leading to coma and death unless the cause of the disease is removed; the most common cause of cirrhosis is chronic alcohol abuse.

club soda artificially carbonated water containing added salts and minerals.

coenzymes enzyme helpers; small molecules that interact with enzymes and enable them to do their work. Many coenzymes are made from water-soluble vitamins.

cofactor a mineral element that, like a coenzyme, works with an enzyme to facilitate a chemical reaction.

collagen (COLL-a-jen) the chief protein of most connective tissue, including scars, ligaments, tendons, and the underlying matrix on which bones and teeth are built.

colon cancer cancer of the large intestine (colon), the terminal portion of the digestive tract.

colostrum (co-LAHS-trum) a milk-like secretion from the breast, rich in protective factors, present during the first day or so after delivery and before milk appears.

complementary proteins two or more food proteins whose amino acid assortments complement each other in such a way that the essential amino acids limited in or missing from each are supplied by the others.

complete proteins proteins containing all the essential amino acids in the right proportion relative to need. The *quality* of a food protein is judged by the proportions of essential amino acids that it contains relative to our needs. Animal and soy proteins are the highest in quality.

complex carbohydrates long chains of sugars (glucose) arranged as starch or fiber; also called polysaccharides.

constipation hardness and dryness of bowel movements associated with discomfort in passing them.

contaminants potentially dangerous substances, such as lead, that can accidentally get into foods.

contamination iron iron found in foods as the result of contamination by inorganic iron salts from iron cookware, iron-containing soils, and the like.

control group a group of individuals with characteristics that match the group being treated in an intervention study but who receive a sham treatment or no treatment at all.

correlation a simultaneous change in two factors, such as a decrease in blood pressure with regular aerobic activity (a direct or positive correlation) or the decrease in incidence of bone fractures with increasing calcium intakes (an inverse or negative correlation).

correspondence school a school from which courses can be taken and degrees granted by mail. Schools that are accredited offer respectable courses and degrees.

cortical bone the dense outer ivory-like layer of bone that provides an exterior shell over trabecular bone.

cretinism (CREE-tin-ism) severe mental and physical retardation of an infant caused by iodine deficiency during pregnancy.

cross-contamination the inadvertent transfer of bacteria from one food to another that occurs, for instance, by chopping vegetables on the same cutting board that was used to skin poultry.

cross-reaction the reaction of one antigen with antibodies developed against another antigen.

culture knowledge, beliefs, customs, laws, morals, art, and literature acquired by members of a society and passed along to succeeding generations.

Daily Value the amount of fat, sodium, fiber, and other nutrients health experts say should make up a healthful diet. The % Daily Values that appear on food labels tell you the percentage of a nutrient that a serving of the food contributes to a healthful diet.

degenerative disease chronic disease characterized by deterioration of body organs as a result of misuse and neglect. Poor eating habits, smoking, lack of exercise, and other lifestyle habits often contribute to degenerative diseases, including heart disease, cancer, osteoporosis, and diabetes.

Delaney clause a provision in the 1958 Food Additives Amendment that prohibited manufacturers from using any substance that was known to cause cancer in animals or humans at any dose level.

denaturation the change in shape of a protein brought about by heat, alcohol, acids, bases, salts of heavy metals, or other agents.

dental caries decay of the teeth, or cavities.

dental plaque a colorless film, consisting of bacteria and their by-products, that is constantly forming on the teeth.

designer estrogens (Selective Estrogen Receptor Modulators—SERMs) drugs that act on *estrogen receptors* in osteoblasts to promote an increase in bone mass; an example is raloxifene (Evista).

diabetes (dye-uh-BEET-eez) a disorder (technically termed *diabetes mellitus*) characterized by insufficiency or relative ineffectiveness of insulin, which renders a person unable to regulate the blood glucose level normally.

Dietary Reference Intakes (DRI) a set of reference values for energy and nutrients that can be used for planning and assessing diets for healthy people.

digestion the process by which foods are broken down into smaller absorbable products.

digestive system the body system composed of organs and glands associated with the ingestion and processing of food for absorption of nutrients into the body.

dim sum steamed or fried dumplings stuffed with pork, shrimp, beef, sweet paste, or preserves.

dipeptides (dye-PEP-tides) protein fragments two amino acids long. A peptide is a strand of amino acids.

diploma mill a correspondence school that grinds out degrees—sometimes worth no more than the cost of the paper they are printed on—the way a grain mill grinds out flour.

disaccharides pairs of single sugars linked together (*di* means two).

discretionary calorie allowance the balance of calories remaining in a person's energy allowance, after accounting for the number of calories needed to meet recommended nutrient intakes through consumption of nutrient-dense foods in low-fat or no added sugar forms. The calories assigned to discretionary calories may be used to increase intake from the basic food groups; to select foods from these groups that are higher in fat or with added sugars; to add oils, solid fats, or sugars to foods or beverages; or to consume alcohol.

disordered eating eating food as an outlet for emotional stress rather than in response to internal physiological cues.

diuretics (dye-you-RET-ics) medications and other substances causing increased water excretion.

diverticulosis (dye-ver-tic-you-LOCE-iss) outpocketings of weakened areas of the intestinal wall, like blowouts in a tire, that can rupture, causing dangerous infections.

drugs substances that can modify one or more of the body's functions.

dysentery (DISS-en-terry) an infection of the digestive tract that causes diarrhea.

eating disorder general term for several conditions (anorexia nervosa, bulimia nervosa, binge-eating disorder) that exhibit an excessive preoccupation with body weight, a fear of body fatness, and a distorted body image.

eclampsia a severe extension of preeclampsia characterized by convulsions; may lead to coma.

edema (eh-DEEM-uh) swelling of body tissue caused by leakage of fluid from the blood vessels, seen in (among other conditions) protein deficiency.

electrolytes compounds that partially dissociate in water to form ions; examples are sodium, potassium, and chloride.

empty-calorie foods a phrase used to indicate that a food supplies calories but negligible nutrients.

emulsifier a substance that mixes with both fat and water and can break fat globules into small droplets, thereby suspending fat in water.

endosperm the soft, white inside portion of a grain or kernel that contains starch and protein and provides energy.

endurance the ability to sustain an effort for a long time. One type, **muscle endurance,** is the ability of a muscle to contract repeatedly within a given time without becoming exhausted. Another type, **cardiovascular endurance,** is the ability of the cardiovascular system to sustain effort over a period of time.

energy the capacity to do work, such as moving or heating something.

enriched refers to the process by which the B vitamins thiamin, riboflavin, niacin, folic acid, and the mineral iron are added to refined grains and grain products at levels specified by law.

enterotoxin a toxic compound, produced by microorganisms, that harms mucous membranes, as in the gastrointestinal tract.

enzymes protein catalysts. A catalyst facilitates a chemical reaction without itself being altered in the process.

EPA, DHA eicosapentaenoic (EYE-cossa-PENTA-ee-NO-ic) acid, docosahexaenoic (DOE-cossa-HEXA-ee-NO-ic) acid; omega-3 fatty acids made from linolenic acid in the tissues of fish.

epidemiological study a study of a population that searches for possible correlations between nutrition factors and health patterns over time.

epithelial (ep-ih-THEE-lee-ul) **tissue** the cells that form the outer surface of the body and line the body cavities and the principal passageways leading to the exterior. Examples include the cornea, digestive tract lining, respiratory tract lining, and skin. The epithelial cells produce mucus to protect these tissues from bacteria and other potentially harmful substances. Without this mucus, infections become more likely.

ergogenic aids anything that helps to increase the capacity to work or exercise.

essential amino acids amino acids that cannot be synthesized by the body or that cannot be synthesized in amounts sufficient to meet physiological need.

essential fatty acid a fatty acid that cannot be synthesized in the body in amounts sufficient to meet physiological need.

essential nutrients nutrients that must be obtained from food because the body cannot make them for itself.

Estimated Average Requirement (EAR) the amount of a nutrient that is estimated to meet the requirement for the nutrient in half of the people of a specific age and gender. The EAR is used in setting the RDA.

Estimated Energy Requirement (EER) the average calorie intake that is predicted to maintain energy balance in a healthy adult of a defined age, gender, weight, height, and level of physical activity, consistent with good health.

estrogen a major female hormone; important in connection with nutrition because it maintains calcium balance and because its secretion abruptly declines at menopause.

estrogen replacement therapy (ERT) administration of estrogen to replace the natural hormone that declines with menopause.

ethnic cuisine the traditional foods eaten by the people of a particular culture.

exchange lists lists of foods with portion sizes specified. The foods on a single list are similar with respect to nutrient and calorie content and thus can be mixed and matched in the diet (see the food photos in Appendix B).

exercise stress test a test that monitors heart function during exercise to detect abnormalities that may not show up under ordinary conditions; exercise physiologists and trained physicians or health care professionals can administer the test.

experimental group the participants in a study who receive the real treatment or intervention under investigation.

external cue theory the theory that some people eat in response to such external factors as the presence of food or the time of day rather than to such internal factors as hunger.

famine widespread lack of access to food caused by natural disasters, political factors, or war; characterized by a large number of deaths due to starvation and malnutrition.

fat cell theory states that during the growing years, fat cells respond to overfeeding by producing additional fat cells; the number of fat cells eventually becomes relatively fixed, and overfeeding from this point on causes the body to enlarge existing fat cells.

fats lipids that are solid at normal room temperature.

fatty acids basic units of fat composed of chains of carbon atoms with an acid group at one end and hydrogen atoms attached all along their length.

female athlete triad a condition characterized by disordered eating, lack of menstrual periods, and low bone density.

fetal alcohol syndrome (FAS) the cluster of symptoms seen in an infant or child whose mother consumed excess alcohol during pregnancy, including retarded growth, impaired development of the central nervous system, and facial malformations. A lesser condition—called *fetal alcohol effect (FAE)*—causes learning impairment and other more subtle abnormalities in infants exposed to alcohol during pregnancy.

fibers the indigestible residues of food, composed mostly of polysaccharides. The best known of the fibers are **cellulose, hemicellulose, pectin,** and **gums.**

First Amendment the amendment to the U.S. Constitution that guarantees freedom of the press.

fitness the body's ability to meet physical demands, composed of four components:

flexibility, strength, muscle endurance, and cardiovascular endurance.

flexibility the ability to bend or extend without injury; depends on the elasticity of the muscles, tendons, and ligaments and on the condition of the joints.

fluid balance distribution of fluid among body compartments.

fluorosis (floor-OH-sis) discoloration of the teeth from ingestion of too much fluoride during tooth development.

foam cells cells from the immune system containing scavenged oxidized LDL cholesterol that are thought to initiate arterial plaque formation.

food additive any substance added to food, including substances used in the production, processing, treatment, packaging, transportation, or storage.

food allergen a substance in food—usually a protein—that is seen by the body as harmful and causes the immune system to mount an allergic reaction.

food allergy an adverse reaction to an otherwise harmless substance that involves the body's immune system.

food aversion a strong desire to avoid a particular food.

food banks nonprofit community organizations that collect surplus commodities from the government and edible but often unmarketable foods from private industry for use by nonprofit charities, institutions, and feeding programs at nominal cost.

food composition tables tables that list the nutrient profile of commonly eaten foods.

food group plan a diet-planning tool, such as MyPyramid, that groups foods according to similar origin and nutrient content and then specifies the amount of food a person should eat from each group.

food insecurity the inability to acquire or consume an adequate quality or sufficient quantity of food in socially acceptable ways, or the uncertainty that one will be able to do so.

food intolerance a general term for any adverse reaction to a food or food component that does not involve the body's immune system.

food intoxication illness caused by eating food that contains a harmful toxin or chemical.

food recovery such activities as salvaging perishable produce from grocery stores and wholesale food markets; rescuing surplus prepared food from restaurants, corporate cafeterias, and caterers; and collecting nonperishable, canned or boxed processed food from manufacturers, supermarkets, or people's homes. The items recovered are donated to hungry people.

food security access by all people at all times to enough food for an active and healthy life. Food security has two aspects: ensuring that adequate food supplies are available and ensuring that households whose members suffer from undernutrition have the ability to acquire food, either by producing it themselves or by being able to purchase it.

foodborne illness illness occurring as a result of ingesting food or water contaminated with a poisonous substance, such as a toxin or chemical (*food intoxication*) or an infectious agent, such as bacteria, viruses, or parasites (*foodborne infection*); commonly called *food poisoning*.

foodborne infection illness caused by eating a food containing bacteria or other microorganisms capable of growing and thriving in a person's tissues.

fortified foods foods to which nutrients have been added. Typically, commonly eaten foods are chosen for fortification with added nutrients to help prevent a deficiency (iodized salt, milk with vitamin D) or to reduce the risk of chronic disease (juices with added calcium).

free radicals atoms or molecules with one or more unpaired electrons that make the atom or molecule unstable and highly reactive.

fructose (FROOK-toce) fruit sugar—the sweetest of the single sugars.

functional food a general term for foods that provide an *additional* physiological or psychological benefit beyond that of meeting basic nutritional needs. Also called *medical foods* or *designer foods*—foods "fortified" with phytochemicals or plants bred to contain high levels of phytochemicals.

functional tolerance actual change in sensitivity to a drug resulting in hallucinations and convulsions when alcohol is removed.

fusion cuisine a term used to describe food that combines the elements of two or more cuisines—say, European and Oriental—to create a new one.

galactose (ga-LACK-toce), a monosaccharide; occurs bonded to glucose in the sugar of milk.

garbanzo beans or **chickpeas** These legumes are large, round, and tan colored. They have a nutty flavor and crunchy texture. Use in soups and stews and puréed for dips.

gefilte fish a chopped fish mixture often made with pike and whitefish as well as matzoh crumbs, eggs, and seasonings.

genetic engineering the use of biotechnology to alter the genes of a plant in an effort to create a new plant with different traits; also called genetic modification. In some cases, a plant's gene(s) may be deleted or altered, or a gene(s) may be introduced from different organisms or species. The foods or crops produced are called genetically engineered (GE) or genetically modified (GM) foods or crops.

germ the nutrient-rich and fat-dense inner part of a whole grain.

gestational diabetes the appearance of abnormal glucose tolerance during pregnancy.

ghee a clarified butter without any milk solids or water.

ghrelin (GREH-lin) a hormone released by the stomach that signals the hypothalamus in the brain to stimulate eating.

gleaning the harvesting of excess food from farms, orchards, and packing houses to feed the hungry.

glucagon (GLUE-cuh-gon) a hormone released by the pancreas that signals the liver to release glucose into the bloodstream.

glucose (GLOO-koce) the building block of carbohydrate; a single sugar used in both plant and animal tissues as quick energy.

glutinous rice short-grained, opaque white rice that turns sticky when cooked.

glycemic effect (also called *glycemic response*) the effect of food on a person's blood glucose and insulin response—how fast and how high the blood glucose rises and how quickly the body responds by bringing it back to normal.

glycemic index (GI) a ranking of foods based on their potential to raise blood glucose levels.

glycerol (GLISS-er-all) an organic compound that serves as the backbone for triglycerides.

glycogen (GLY-co-gen) a polysaccharide composed of chains of glucose, manufactured in the body and stored in liver and muscle. As a storage form of glucose, liver glycogen can be broken down by the liver to maintain a constant blood glucose level when carbohydrate intake is inadequate.

GOBI an acronym formed from the elements of UNICEF's Child Survival campaign—Growth charts, Oral rehydration therapy, Breast milk, and Immunization.

goiter (GOY-ter) enlargement of the thyroid gland caused by iodine deficiency.

GRAS (Generally Recognized As Safe) list a list of ingredients, established by the FDA, that had long been in use and were believed safe. The list is subject to revision as new facts become known.

grazing eating small amounts of food at intervals throughout the day rather than—or in addition to—eating regular meals.

great northern beans This variety is medium white and kidney shaped. Enjoy the delicate flavor and firm texture in salads, soups, and main dishes.

green soybeans (Edamame) These large soybeans are harvested when the beans are still green and sweet and can be served as a snack or a main vegetable dish, after boiling in water for 15 to 20 minutes. They are high in protein and fiber and contain no cholesterol.

grits coarsely ground cornmeal.

ground water water that comes from an underground body of water that does not come into contact with any surface water.

guava a sweet, juicy fruit with green or yellow skin and red or yellow flesh.

hard water water with a high concentration of minerals such as calcium and magnesium.

hazard state of danger; used to refer to any circumstance in which harm is possible.

HDL (high-density lipoprotein) carries cholesterol in the blood back to the liver for recycling or disposal.

health claim a statement on the food label linking the nutritional profile of a food to a reduced risk of a particular disease, such as osteoporosis or cancer. Manufacturers must adhere to strict government guidelines when making such claims.

health fraud conscious deceit practiced for profit, such as the promotion of a false or an unproven product or therapy.

health promotion helping people achieve their maximum potential for good health.

Healthy Eating Index (HEI) a summary measure of the quality of one's diet. The HEI provides an overall picture of how well one's diet conforms to the nutrition recommendations contained in the Dietary Guidelines for Americans and the USDA Food Guide. The Index factors in such dietary practices as consumption of total fat, saturated fat, cholesterol, and sodium, and the variety of foods in the diet.

heat stroke an acute and dangerous reaction to heat buildup in the body, requiring emergency medical attention; also called *sun stroke*.

heavy metals any of a number of mineral ions, such as mercury and lead, so named because of their relatively high atomic weight. Many heavy metals are poisonous.

heme (HEEM) **iron** the iron-holding part of the hemoglobin protein, found in meat, fish, and poultry. About 40 percent of the iron in meat, fish, and poultry is bound into heme. Meat, fish, and poultry also contain a factor (MFP factor) other than heme that promotes the absorption of iron, even of the iron from other foods eaten at the same time as the meat.

hemoglobin (HEEM-oh-globe-in) the oxygen-carrying protein of the blood; found in the red blood cells.

hemorrhoids (HEM-or-oids) swollen, hardened (varicose) veins in the rectum, usually caused by the pressure resulting from constipation.

histamine a substance released by cells of the immune system during an allergic reaction to an antigen, causing inflammation, itching, hives, dilation of blood vessels, and a drop in blood pressure.

hominy hulled, dried corn kernels.

homocysteine (ho-mo-SIS-teen) a chemical that appears to be toxic to the blood vessels of the heart. High blood levels of homocysteine have been associated with low blood levels of vitamin B_{12}, vitamin B_6, and folate.

hormones chemical messengers. Hormones are secreted by a variety of glands in the body in response to altered conditions. Each affects one or more target tissues or organs and elicits specific responses to restore normal conditions.

hunger the physiological drive to find and eat food, experienced as an unpleasant sensation.

husk the outer, inedible covering of a grain.

hydrogenation (high-droh-gen-AY-shun) the process of adding hydrogen to unsaturated fat to make it more solid and more resistant to chemical change.

hydrolyzed vegetable protein (HVP) Hydrolyzed vegetable protein (HVP) is a protein obtained from any vegetable, including soybeans. The protein is broken down into amino acids by a chemical process called acid hydrolysis. HVP is a flavor enhancer that can be used in soups, broths, sauces, gravies, flavoring and spice blends, canned and frozen vegetables, and meats and poultry.

hyperglycemia an abnormally high blood glucose concentration, often a symptom of diabetes.

hypertension sustained high blood pressure.

hypertrophy an increase in size in response to use.

hypoglycemia (HIGH-po-gligh-SEEM-eeuh) an abnormally low blood glucose concentration—below 60 to 70 mg/100 ml.

hypothalamus (high-poh-THALL-ah-mus) a part of the brain that senses a variety of conditions in the blood, such as temperature, salt content, and glucose content, and then signals other parts of the brain or body to change those conditions when necessary.

immunity specific disease resistance derived from the immune system's memory of prior exposure to specific disease agents and its ability to mount a swift response against them.

incidental food additives (or indirect additives) substances that accidentally get into food as a result of contact with it during growing, processing, packaging, storing, or some other stage before the food is consumed.

incomplete protein a protein lacking or low in one or more of the essential amino acids.

ingredients list a listing of the ingredients in a food, with items listed in descending order of predominance by weight. All food labels are required to bear an ingredients list.

inorganic being or composed of matter other than plant or animal.

insoluble fiber includes the fiber types called cellulose, hemicellulose, and lignin. Insoluble fibers do not dissolve in water.

insulin a hormone secreted by the pancreas in response to high blood glucose levels; it assists cells in drawing glucose from the blood.

integrated pest management the use of biological controls, crop rotation, genetic engineering, and other tactics to reduce chemical use in the growing of crops.

intentional food additives substances intentionally added to food; examples include nutrients, colors, spices, and herbs.

intervention study a population study examining the effects of a treatment on experimental subjects compared to a control group.

intestinal flora the normal bacterial inhabitants of the digestive tract.

intrinsic factor a compound made in the stomach that is necessary for the body's absorption of vitamin B_{12}.

ions (EYE-ons) electrically charged particles, such as sodium (positively charged) and chloride (negatively charged).

iron-deficiency anemia a reduction of the number and size of red blood cells and a loss of their color because of iron deficiency.

iron overload a condition in which the body contains more iron than it needs or can handle; excess iron is toxic and can damage the liver. The most common cause of iron overload is the genetic disorder hemochromatosis.

irradiation the process of exposing a substance to low doses of radiation, using gamma rays, X-rays, or electricity (electron beams) to kill insects, bacteria, and other potentially harmful microorganisms.

isoflavones compounds found in many fruits, vegetables, and soy-based foods that are thought to play a role in fighting breast cancer by blocking the action of the hormone estrogen.

jicama a crisp, bean root vegetable that is tan outside and white inside and is always eaten raw; as popular in Mexico as the potato is in the United States.

jujube Chinese date.

kasha cracked buckwheat, barley, millet, or wheat that is served as a cooked cereal or potato substitute.

ketosis (kee-TOE-sis) abnormal amounts of ketone bodies in the blood and urine; ketone bodies are produced from the incomplete breakdown of fat when glucose is unavailable for the brain and nerve cells.

kidney beans These familiar beans are large, red, and kidney shaped (the white variety is called *cannellini*). They have a bland taste and soft texture but tough skins. Use in chili, bean stews, and Mexican dishes (for red) or Italian dishes (for white).

knish a potato pastry filled with ground meat, potato, or kasha.

kosher fit, proper, or in accordance with religious law.

kwashiorkor (kwash-ee-OR-core) a deficiency disease caused by inadequate protein in the presence of adequate food energy.

lactoferrin (lak-toe-FERR-in) a factor in breast milk that binds and helps absorb iron and keeps it from supporting the growth of the infant's intestinal bacteria.

lacto-ovovegetarian Milk and milk products and eggs are included in this diet, but meat, poultry, fish, and seafood are excluded.

lactose a double sugar composed of glucose and galactose; commonly known as milk sugar.

lactose intolerance inability to digest lactose as a result of a lack of the necessary enzyme lactase. Symptoms include nausea, abdominal pain, diarrhea, or excessive gas that occurs anywhere from 15 minutes to a couple of hours after consuming milk or milk products.

lactovegetarian Milk and milk products are included in this diet, but meat, poultry, fish, seafood, and eggs are excluded.

lassi a milkshake type of drink made from yogurt.

LDL (low-density lipoprotein) carries cholesterol (much of it synthesized in the liver) to body cells. A high blood cholesterol level usually reflects high LDL.

lecithin (LESS-ih-thin) a phospholipid, a major constituent of cell membranes, manufactured by the liver and also found in many foods.

legumes (leg-GYOOMS) plants of the bean and pea family having roots with nodules that contain bacteria that can trap nitrogen from the air in the soil and make it into compounds that become part of the seed. The seeds are rich in high-quality protein compared with those of most other plant foods.

lentils These legumes are small, flat, and round. Usually brown colored, lentils also can be green, pink, or red. They have a mild taste with firm texture. Best used when combined with grains or vegetables in salads, soups, or stews.

leptin an appetite-suppressing hormone produced in the fat cells that communicates information about the body's fat content to the brain (*leptos* means slender).

lifestyle diseases conditions that may be aggravated by modern lifestyles that include too little exercise, poor diets, and excessive drinking and smoking. Lifestyle diseases are also referred to as *diseases of affluence*.

lima or **butter beans** Limas are soft and mealy in texture. They are flat, oval shaped, and white tinged with green. The smaller variety has a milder taste. Use in soups and stews.

limiting amino acid a term given to the essential amino acid in shortest supply (relative to the body's need) in a food protein; it therefore *limits* the body's ability to make its own proteins.

linoleic (lin-oh-LAY-ic) **acid, linolenic** (lin-oh-LEN-ic) **acid** polyunsaturated fatty acids, essential for human beings.

lipids a family of compounds that includes triglycerides (fats and oils), phospholipids (lecithin), and sterols (cholesterol).

lipoprotein lipase (LPL) an enzyme located on the surfaces of fat cells that enables the cell to convert blood triglycerides into fatty acids and glycerol to be pulled into the cell for reassembly and storage as body fat.

lipoproteins (LIP-oh-PRO-teens) clusters of lipids associated with protein that serve as transport vehicles for lipids in blood and lymph. The four main types of lipoproteins are chylomicrons, VLDL, LDL, and HDL.

liposuction a type of surgery (also called *lipectomy*) that vacuums out fat cells that have accumulated, typically in the buttocks and thighs. If the person continues to eat more calories than are expended through physical activity, fat will return to the fat cells that remain in those regions.

litchi small, round fruits with orange-red skin and opaque white flesh; also called *litchee* or *lychee*.

longan a small round fruit with smooth brown skin and clear pulp.

low birthweight (LBW) a birthweight of 5½ lb (2,500 g) or less, used as a predictor of poor health in the newborn and as a probable indicator of poor nutrition status of the mother during and/or before pregnancy.

lox smoked salmon.

lymph (LIMF) the body fluid that transports the products of fat digestion toward the heart and eventually drains back into the bloodstream; lymph consists of the same components as blood with the exception of red blood cells.

macrobiotic diet a vegan diet composed mostly of whole grains, beans, and certain vegetables; taken to extremes, a macrobiotic diet can result in malnutrition and death.

mad cow disease (bovine spongiform encephalopathy or BSE) a rare and fatal degenerative disease first diagnosed in 1986 in cattle in the United Kingdom. The bovine disease may be passed to humans who eat the meat of infected animals and may lead to death due to brain and nerve damage.

macular degeneration a progressive loss of function of part of the retina (macula) that is necessary for focused vision; often leads to blindness.

major mineral an essential mineral nutrient found in the human body in amounts greater than 5 grams.

malnutrition the impairment of health resulting from a relative deficiency or excess of food energy and specific nutrients necessary for health.

maltose a double sugar composed of two glucose units.

mantou steamed bread.

marasmus (ma-RAZ-mus) an energy deficiency disease; starvation.

margin of safety from a food safety standpoint, the margin is a zone between the maximum amount of a substance that appears to be safe and the amount allowed in the food supply.

matzoh a cracker-like bread eaten most often at Passover.

meat alternatives Meat alternatives made from soybeans contain soy protein or tofu and other ingredients mixed together to simulate various kinds of meat. These meat alternatives are sold as frozen, canned, or dried foods.

meat replacements textured vegetable protein products formulated to look and taste like meat, fish, or poultry. Many of these are designed to match the known nutrient contents of animal protein foods.

megadose a dose of ten or more times the amount normally recommended. An overdose is an amount high enough to cause toxicity symptoms. Megadoses taken over a long period often result in an overdose.

menopause the time of life at which a woman's menstrual cycle ceases, usually at about 45 to 50 years of age.

metabolic syndrome a cluster of interrelated symptoms, including obesity, high blood pressure, abnormal blood lipids, and insulin resistance; highly associated with development of type 2 diabetes and heart disease.

metabolic tolerance increased efficiency of removing high levels of alcohol from the blood due to long-term exposure leading to more drinking and possible addiction.

metabolism the sum of all physical and chemical reactions taking place in living cells; includes reactions in which the cells derive energy (ATP) from the energy nutrients and reactions in which the cells synthesize body compounds (glycogen, body fat, body proteins).

milk allergy the most common food allergy; caused by the protein in raw milk.

mineral water water that is drawn from an underground source and contains at least 250 parts per million of dissolved solids.

minerals small, naturally occurring, inorganic, chemical elements; the minerals serve as structural components of the body and in many vital body processes.

miso a rich, salty condiment that characterizes the essence of Japanese cooking. The Japanese make miso soup and use miso to flavor a variety of foods. A smooth paste, miso is made from soybeans and a grain such as rice, plus salt and a mold culture, and then aged in cedar vats for one to three years.

moderation the attribute of a diet that provides no unwanted constituent in excess.

monoglyceride (mon-oh-GLISS-er-ide) a glycerol molecule with one fatty acid attached to it. A *diglyceride* is a glycerol molecule with two fatty acids attached to it.

monosaccharide a single sugar (*mono* means one).

monounsaturated fatty acid a fatty acid containing one point of unsaturation, found mostly in vegetable oils such as olive, canola, and peanut.

multinational corporations international companies with direct investments and/or operative facilities in more than one country. U.S. oil and food companies are examples.

neotame a zero-calorie, heat-stable sweetener that is a derivative of the dipeptide composed of aspartic acid and phenylalanine. It is 7,000 to 13,000 times as sweet as sugar depending on how it is used. Unlike aspartame, it is not metabolized to phenylalanine, and thus no special labeling for people with PKU is required.

neural tube defects birth defects involving abnormalities of the brain and spinal cord related to a woman's intake of folic acid before and during pregnancy. The two main types are *spina bifida* (incomplete closure of the bony casing around the spinal cord) and *anencephaly* (a partially or completely missing brain).

neurotoxin a poisonous compound that disrupts the nervous system.

neurotransmitters chemicals that are released at the end of a nerve cell when a nerve impulse arrives there; they diffuse across the gap to the next nerve cell and alter the membrane of that second cell to either excite it or inhibit it.

niacin equivalents (NEs) the amount of niacin present in food, including the niacin that can theoretically be made from tryptophan contained in the food.

night blindness slow recovery of vision following flashes of bright light at night; an early symptom of vitamin A deficiency.

nitrogen balance the amount of nitrogen consumed compared with the amount of nitrogen excreted in a given time period.

nondairy soy frozen dessert Nondairy frozen desserts are made from soy milk or soy yogurt.

nonheme iron the iron found in plant foods.

nursing bottle syndrome (also called *baby bottle tooth decay*) decay of all the upper and sometimes the back lower teeth that occurs in infants given carbohydrate-containing liquids when they sleep.

nutrient content claims claims such as "low-fat" and "low-calorie" used on food labels to help consumers who don't want to scrutinize the Nutrition Facts panel get an idea of a food's nutritional profile. These claims must adhere to specific definitions set forth by the Food and Drug Administration.

nutrient dense refers to a food that supplies large amounts of nutrients relative to the number of calories it contains. The higher the level of nutrients and the fewer the number of calories, the more nutrient dense the food is.

nutrients substances obtained from food and used in the body to promote growth, maintenance, and repair. The nutrients include carbohydrate, fat, protein, vitamins, minerals, and water.

nutrition the study of foods, their nutrients and other chemical components, their actions and interactions in the body, and their influence on health and disease.

Nutrition Facts panel a detailed breakdown of the nutritional content of a serving of a food that must appear on virtually all packaged foods sold in the United States.

nutritional yeast a fortified food supplement containing B vitamins, iron, and protein that can be used to improve the quality of a vegetarian diet.

nutritionist a person who claims to be capable of advising people about their diets. Some nutritionists are registered dietitians, whereas others are self-described experts whose training may be questionable.

obesity body weight that is high enough above normal weight to constitute a health hazard; conventionally defined as weight 20 percent or more above the desirable weight for height, or a BMI of 30 or greater.

oils lipids that are liquid at normal room temperature.

olestra an artificial fat derived from vegetable oils and sugar combined in such a way that the body cannot break them down. Sold under the brand name Olean®, olestra does not contribute calories to food.

omega-3 fatty acids polyunsaturated fatty acids that have their end-most double bonds after the third carbon in the chain.

omega-6 fatty acids polyunsaturated fatty acids that have their end-most double bonds after the sixth carbon in the chain.

oral rehydration therapy (ORT) the treatment of dehydration (usually due to diarrhea caused by infectious disease) with an oral solution; ORT as developed by UNICEF is intended to enable a mother to mix a simple solution for her child from substances that she has at home.

organic of, related to, or containing carbon compounds.

organic halogens compounds that contain one or more of a class of atoms called halogens, including fluorine, chlorine, iodine, or bromine.

organically grown foods crops or livestock grown and processed according to USDA regulations concerning use of pesticides, herbicides, fungicides, fertilizers, preservatives, other synthetic chemicals, growth hormones, antibiotics, or other drugs.

oriental radish a large, cylindrically shaped vegetable with smooth skin; also called *daikon*.

osteoblast a bone-building cell; responsible for formation of bone.

osteoclast a bone-destroying cell; responsible for resorption and removal of bone.

osteomalacia (os-tee-o-mal-AY-shuh) the disease resulting from vitamin D deficiency in adults and characterized by softening of the bones. (Its counterpart in children is called rickets.) Symptoms include bowed legs and a curved spine.

osteoporosis (OSS-tee-oh-pore-OH-sis) also known as *adult bone loss;* a disease in which the bones become porous and fragile.

overload an extra physical demand placed on the body. A principle of training is that for a body system to improve, its workload must be increased by increments over time.

overnutrition calorie or nutrient overconsumption severe enough to cause disease or increased risk of disease; a form of malnutrition.

overweight conventionally defined as weight between 10 and 20 percent above the desirable weight for height, or a body mass index (BMI) of 25.0 through 29.9.

ovovegetarian Eggs are included in this diet, but milk and milk products, meat, poultry, fish, and seafood are excluded.

oxidation interaction of compound with oxygen; a damaging effect by a chemically reactive and unstable form of oxygen.

oxidized LDL-cholesterol (o-LDL) the cholesterol in LDLs that is attacked by reactive oxygen molecules inside the walls of the arteries; o-LDL is taken up by scavenger cells and deposited in plaque.

parathyroid hormone a hormone used as a drug (Teriparatide) to stimulate new bone formation; administered by injection once a day in the thigh or abdomen.

pasteurization the process of sterilizing food via heat treatment.

peak bone mass the highest bone density achieved for an individual; accumulated over the first three decades of life; typically occurs by 30 years of age. After age 30, bone resorption slowly begins to exceed bone formation.

pellagra (pell-AY-gra) niacin deficiency characterized by the "4 Ds": diarrhea, dermatitis (inflammation of the skin), dementia, and death.

peptide bond a bond that connects one amino acid with another.

periodontal disease inflammation or degeneration of the tissues that surround and support the teeth.

pesticides chemicals applied intentionally to plants, including foods, to prevent or eliminate pest damage. Pests include all living organisms that destroy or spoil foods: bacteria, molds and fungi, insects, and rats and other rodents, to name a few.

pH the concentration of hydrogen ions. The lower the pH, the stronger the acid: pH 2 is a strong acid, pH 7 is neutral, and a pH above 7 is alkaline (base).

phenylketonuria (FEN-il-KEY-toe-NU-ree-ah) or **PKU** an inborn error of metabolism, detectable at birth, in which the body lacks the enzyme needed to convert the amino acid phenylalanine to the amino acid tyrosine. If not detected and treated, derivatives of phenylalanine accumulate in the blood and tissues, where they can cause severe damage, including mental retardation.

phospholipid (FOSS-foh-LIP-id) a lipid similar to a triglyceride but containing phosphorus; one of the three main classes of lipids.

photosynthesis a process in which plants use the green pigment chlorophyll to trap the energy of the sun and produce glucose from carbon dioxide and water.

phytochemicals (FIGH-toe-CHEM-icals) physiologically active compounds found in plants that are not essential nutrients but appear to help promote health and reduce risk for cancer, heart disease, and other conditions. Also called *phytonutrients*.

pica the craving of nonfood items such as clay, ice, and laundry starch; does not appear to be limited to any particular geographic area, race, sex, culture, or social status.

pigment a molecule capable of absorbing certain wavelengths of light. Pigments in the eye permit us to perceive different colors.

pinto beans These medium ovals are mottled beige and brown with an earthy flavor. They are most often used in Mexican dishes, such as refried beans, stews, or dips.

placebo (plah-SEE-bo) a sham treatment given to a control group; an inert, harmless "treatment" that the group's members cannot recognize as different from the real thing. Using a placebo minimizes the chance that an effect of the treatment will appear to have occurred due to the healing effect of the *belief* in the treatment (known as the *placebo effect*), rather than the treatment itself.

placebo effect an improvement in a person's sense of well-being or physical health in response to the use of a placebo (a substance having no medicinal properties or medicinal effects).

placenta (pla-SEN-tuh) the organ inside the uterus in which the mother's and fetus's circulatory systems intertwine and in which exchange of materials between maternal and fetal blood takes place. The fetus receives nutrients and oxygen across the placenta; the mother's blood picks up carbon dioxide and other waste materials to be excreted via her lungs and kidneys.

plantain a greenish, starchy banana; because it is starchy even when ripe, it is never eaten raw and is usually pan fried.

point of unsaturation a site in a molecule where the bonding is such that additional hydrogen atoms can easily be added.

polysaccharide a long chain of 10 or more glucose molecules linked together in straight or branched chains; another term for complex carbohydrates.

polyunsaturated fatty acid (sometimes abbreviated PUFA) a fatty acid in which two or more points of unsaturation occur, found in nuts and vegetable oils such as safflower, sunflower, and soybean, and in fatty fish.

postnatal after birth.

poverty the state of having too little money to meet minimum needs for food, clothing, and shelter. The U.S. Department of Agriculture defines the poverty level in the United States as an annual income of $20,000 for a family of four.

precursor a compound that can be converted into another compound. For example, beta-carotene is a precursor of vitamin A.

preeclampsia a condition characterized by hypertension, fluid retention, and protein in the urine.

preformed vitamin A vitamin A in its active form.

pregnancy-induced hypertension (PIH) high blood pressure that develops during the second half of pregnancy.

premenstrual syndrome (PMS) a cluster of physical, emotional, and psychological symptoms that some women experience seven to ten days before menstruation. Symptoms can include acne, anxiety, food cravings (especially for sweets), back pain, breast tenderness, cramps, depression, fatigue, headaches, irritability, moodiness, water retention, and weight gain.

prenatal prior to birth.

protein digestibility–corrected amino acid score (PDCAAS) a measuring tool for determining protein quality. PDCAAS reflects both a protein's digestibility and its proportion of amino acids relative to human needs.

protein quality a measure of the essential amino acid content of a protein relative to the essential amino acid needs of the body.

protein sparing a description of the effect of carbohydrate and fat, which, by being available to yield energy, allow amino acids to be used to build body proteins.

protein synthesis the process by which cells assemble amino acids into proteins. All individuals are unique because of minute differences in the ways their body proteins are made. The instructions for making all the proteins in our bodies are transmitted in the genetic information we receive at conception.

protein-energy malnutrition (PEM), also called **protein-calorie malnutrition (PCM),**

the world's most widespread malnutrition problem, includes both kwashiorkor and marasmus as well as the states in which they overlap.

proteins compounds composed of atoms of carbon, hydrogen, oxygen, and nitrogen and arranged as strands of amino acids. Some amino acids also contain atoms of sulfur.

psyllium seed husk, an ingredient in certain cereals and bulk-forming laxatives; contains both soluble and insoluble properties.

purified water (also known as **demineralized water, distilled water, deionized water,** or **reverse osmosis water**) water from which all the minerals have been removed, thereby eliminating the possibility that the minerals might corrode, say, a steam iron.

purslane a leafy vegetable that can be used in salads or cooked like spinach.

quackery fraud. A quack is a person who practices health fraud.

queso blanco, fresco, or **Mexicano** soft white cheese made of part-skim milk.

Recommended Dietary Allowance (RDA) the average daily amount of a nutrient that is sufficient to meet the nutrient needs of nearly all (97–98 percent) healthy individuals of a specific age and gender.

red beans This versatile bean is a medium-size, dark-red oval. The taste and texture are similar to kidney beans. Use in soups and stews, and serve with rice.

reference dose the estimated amount of a chemical that could be consumed daily without causing harmful effects.

reference protein egg white protein, the standard with which other proteins are compared to determine protein quality.

refined refers to the process by which the coarse parts of food products are removed. For example, refining wheat into flour involves removing three of the four parts of the kernel—the chaff, the bran, and the germ—leaving only the endosperm.

registered dietitian (RD) a professional who has graduated from a program of dietetics accredited by the Commission on Accreditation for Dietetics Education (CADE) of the American Dietetic Association (ADA), has completed an internship program or the equivalent to gain practical skills, has passed a registration examination, and maintains competencies through continuing education. Some states require licensing for dietitians, thereby requiring anyone who wants to use the title "dietitian" to receive permission by passing a state examination.

regulation a legal mandate that must be obeyed. Failure to follow a regulation brings about legal consequences.

requirement the minimum amount of a nutrient that will prevent the development of deficiency symptoms. Requirements differ from the RDA and AI, which include a substantial margin of safety to cover the requirements of different individuals.

retina (RET-in-uh) the paper-thin layer of light-sensitive cells lining the back of the inside of the eye.

retinal (RET-in-al) one of the active forms of vitamin A that functions in the pigments of the eye. Other active forms of vitamin A include retinol and retinoic acid.

retinol one of the active forms of vitamin A.

retinol activity equivalents (RAE) a measure of the amount of retinol the body will derive from a food containing preformed vitamin A or beta-carotene and other vitamin A precursors. Note that some tables list vitamin A in terms of *International Units (IU)*. See Appendix B for methods of converting from one measure to another.

rice sticks flat, opaque, wide noodles made from rice flour.

rice vermicelli thin, white noodles made from rice flour.

rickets a disease that occurs in children as a result of vitamin D deficiency and that is characterized by abnormal growth of bone, which in turn leads to bowed legs and an outward-bowed chest.

risk for pesticide residues, the harm a substance may confer. Scientists estimate risk by assessing the amount of a chemical that each person in a population might consume over time (also called *exposure*) and by considering how toxic the substance might be (*toxicity*).

roti an unleavened flatbread, resembling a tortilla, usually made from whole-wheat flour.

saccharin a zero-calorie sweetener discovered in 1879 and used in the United States since the turn of the century. Saccharin is the sweetening agent in Sweet 'N Low and Sugar Twin.

salt a pair of charged mineral particles, such as sodium ($Na+$) and chloride ($Cl–$), that associate together. In water, they dissociate and help to carry electric current—that is, they become electrolytes.

satiety (sah-TIE-eh-tee) the feeling of fullness and satisfaction that occurs after a meal.

saturated fatty acid a fatty acid carrying the maximum possible number of hydrogen atoms (having no points of unsaturation). Saturated fats are found in animal foods like meat, poultry, and full-fat dairy products, and in tropical oils such as palm and coconut.

schmaltz chicken fat.

scurvy the vitamin C deficiency disease characterized by bleeding gums, tooth loss, and, in severe cases, death.

Second Harvest a national food-banking network to which the majority of food banks belong.

seltzer tap water injected with carbon dioxide and containing no added salts.

serving the standard amount of food used as a reference to give advice regarding how much to eat (such as a 1 cup serving of milk).

set-point theory the theory that the body tends to maintain a certain weight by adjusting hunger, appetite, and food energy intake on the one hand and metabolism (energy output) on the other so that a person's conscious efforts to alter weight may be foiled.

simple carbohydrates (sugars) the single sugars (monosaccharides) and the pairs of sugars (disaccharides) linked together.

Simplesse® the trade name for a protein-based, low-calorie artificial fat, approved by the FDA for use in foods such as frozen desserts; cannot be used for frying or baking.

skinfold test a method in which the thickness of a fold of skin on the back of the arm (the triceps), below the shoulder blade (subscapular), or in other areas is measured with an instrument called a caliper. Obesity is defined by triceps skinfold thickness equal to or greater than 18–19 mm in adult men or 25–26 mm in women.

social group a group of people, such as a family, who depend on one another and share a set of norms, beliefs, values, and behaviors.

soft water water containing a high sodium concentration.

solid fats fats that are solid at room temperature, such as butter, lard, and shortening. These fats may be visible or may be a constituent of foods such as milk, cheese, meats, or baked products.

soluble fiber includes the fiber types called pectin, gums, mucilages, some hemicelluloses, and algal substances (for example, carrageenan). Soluble fibers either dissolve or swell when placed in water.

sopa rice or pasta that is fried and cooked in consommé.

soy cheese Soy cheese is made from soy milk. Its creamy texture makes it an easy substitute for sour cream or cream cheese.

soy flour Soy flour is made from roasted soybeans ground into a fine powder. Soy flour gives a protein boost to recipes. Soy flour is gluten-free so yeast-raised breads made with soy flour are more dense in texture.

soy milk, soy beverages Soybeans that are soaked, ground fine, and strained, produce a fluid called soybean milk. Soy milk is an excellent source of high-quality protein and B vitamins.

soy nuts Roasted soy nuts are whole soybeans that have been soaked in water and then baked until browned. Soy nuts can be found in a variety of flavors, including chocolate covered. High in protein and isoflavones, soy nuts are similar in texture and flavor to peanuts.

soy-nut butter Made from roasted, whole soy nuts, which are then crushed and blended with soy oil and other ingredients, soy-nut butter has a slightly nutty taste, significantly less fat than peanut butter, and provides many other nutritional benefits.

soy protein, texturized Texturized soy protein usually refers to products made from texturized soy flour. Texturized soy flour is made by running defatted soy flour through an extrusion cooker, which allows for many different forms and sizes. When hydrated, it has a chewy texture. It is widely used as a meat extender.

soy yogurt Soy yogurt is made from soy milk. Its creamy texture makes it an easy substitute for sour cream or cream cheese.

soybeans As soybeans mature in the pod they ripen into a hard, dry bean. Most soybeans are yellow. However, brown and black varieties are also available. Whole soybeans can be cooked and used in sauces, stews, and soups.

sparkling bottled water water whose carbon dioxide (the ingredient that makes soda pop bubbly) is naturally present. That is, carbonation is not added from an outside source.

split peas Green or yellow, these small, halved peas supply an earthy flavor with a mealy texture. They are best used in soups and with rice or grains.

sports anemia a temporary condition of low blood hemoglobin level, associated with the early stage of athletic training.

spring water water derived from an underground formation from which water flows naturally to the surface of the earth and to which minerals have not been added or taken away. It may be collected either at the spring itself or through a hole tapping the underground formation feeding the spring.

stanol esters members of the sterol class of lipids, derived from plants, and capable of reducing blood cholesterol levels when eaten as part of a low-fat diet.

staple food a food used frequently or daily in the diet.

staple grain a grain used frequently or daily in the diet—for example, corn (in Mexico) or rice (in Asia).

starch a plant polysaccharide composed of hundreds of glucose molecules, digestible by human beings.

static stretches stretches that lengthen tissues without injury; characterized by long-lasting, painless, pleasurable stretches.

sterols (STEER-alls) lipids with a structure similar to that of cholesterol; one of the three main classes of lipids.

strength the ability of muscles to work against resistance.

stress fracture bone damage or breakage caused by stress on bone surfaces during exercise.

sucralose a nonnutritive sweetener used as a tabletop sweetener and for use in a variety of desserts, confections, and nonalcoholic beverages. Sucralose is a noncaloric, heat-stable sweetener derived from a chlorinated form of sugar. Although sucralose is made from sugar, the body does not recognize it as sugar, and the sucralose molecule is excreted in the urine essentially unchanged.

sucrose (SOO-crose) a double sugar composed of glucose and fructose.

sugar alcohols (mannitol, sorbitol, isomalt, xylitol) can be derived from fruits or commercially produced from dextrose and are absorbed more slowly and metabolized differently than other sugars in the human body. The sugar alcohols are not readily used by ordinary mouth bacteria and therefore are associated with less cavity formation.

target heart rate the heartbeat rate that will achieve a cardiovascular conditioning effect for a given person—fast enough to push the heart but not so fast as to strain it.

taro a starchy vegetable with brown hairy skin and a pink-purple interior.

tempeh a traditional Indonesian food; a chunky, tender soybean cake. Whole soybeans, sometimes mixed with another grain such as rice or millet, are fermented into a rich cake of soybeans with a smoky nutty flavor. Tempeh can be marinated and grilled and added to soups, casseroles, or chili.

tofu Tofu, also known as soybean curd, is a soft cheese-like food made by curdling fresh hot soy milk with a coagulant. Tofu is a bland product that easily absorbs the flavors of other ingredients with which it is cooked. Tofu is rich in high-quality protein and B vitamins and is low in sodium.

Tolerable Upper Intake Level (UL) the maximum amount of a nutrient that is unlikely to pose any risk of adverse health effects to most healthy people. The UL is not intended to be a recommended level of intake.

tolerance decrease of effectiveness of a drug after a period of prolonged or heavy use.

tolerance level the maximum amount of a particular substance allowed in food.

tonic water artificially carbonated water with added sugar and/or high-fructose corn syrup, sodium, and quinine.

toxicants poisons, that is, agents that cause physical harm or death when present in large amounts.

toxicity the ability of a substance to harm living organisms. All substances are toxic if present in high enough concentrations.

trabecular (tra-BECK-you-lar) **bone** the lacy inner network of calcium-containing crystals; spongelike in appearance, it supports the bone's structure.

trace mineral an essential mineral nutrient found in the human body in amounts less than 5 grams.

***trans* fatty acid** a type of fatty acid created when an unsaturated fat is hydrogenated. Found primarily in margarines, shortenings, commercial frying fats, and baked goods, *trans* fatty acids have been implicated in research as culprits in heart disease.

transgenetics the process of transferring genes from one species to another unrelated species.

transport proteins proteins that carry nutrients and other molecules in body fluids. Some transport proteins reside in cell membranes and act as "pumps" where they pick up compounds on one side of the membrane and release them on the other side as needed.

triglycerides (try-GLISS-er-ides) the major class of dietary lipids, including fats and oils. A triglyceride is made up of three units known as *fatty acids* and one unit called *glycerol*.

trimester one-third of the normal duration of pregnancy; the first trimester is 0 to 13 weeks, the second is 13 to 26 weeks, and the third trimester is 26 to 40 weeks.

tripeptides (try-PEP-tides) protein fragments three amino acids long.

under-5-mortality rate (U5MR) the number of children who die before the age of five for every 1,000 live births.

undernutrition severe underconsumption of calories or nutrients leading to disease or increased susceptibility to disease; a form of malnutrition.

underwater weighing (hydrostatic weighing) a measure of density and volume; the less a person weighs underwater compared to the person's out-of-water weight, the greater the proportion of body fat (fat is less dense or more buoyant than lean tissue).

underweight weight 10 percent or more below the desirable weight for height, or a BMI of less than 18.5.

UNICEF the United Nations International Children's Emergency Fund, now referred to as the United Nations Children's Fund.

unique radiolytic products substances unique to irradiated food and apparently created during the process of irradiation.

unsaturated fatty acid a fatty acid with one or more points of unsaturation. Unsaturated fats are found in foods from both plant and animal sources. Unsaturated fatty acids are further divided into monounsaturated fatty acids and polyunsaturated fatty acids.

unspecified eating disorders some people suffer from unspecified eating disorders; that is, they exhibit some but not all of the criteria for specific eating disorders.

urea (yoo-REE-uh) the principal nitrogen excretion product of metabolism, generated mostly by the removal of amine groups from unneeded amino acids or from those amino acids being sacrificed to a need for energy.

variety a feature of a diet in which different foods are used for the same purposes on different occasions—the opposite of *monotony*.

vegan (VEE-gun) a person who eats only plant foods; also called *strict vegetarian*.

vegetarian a person who excludes animal flesh from their diet and in some cases other animal products, such as milk, cheese, and eggs.

vitamins organic, or carbon-containing, essential nutrients that are vital to life and needed in minute amounts.

VLDL (very-low-density lipoprotein) carries fats packaged or made by the liver to various tissues in the body.

waist circumference a measure used to assess abdominal (visceral) fat; excess fat in the abdomen increases a person's risk for health problems.

water provides the medium for life processes.

well water ground water derived from a rock formation by way of a hole bored, drilled, or otherwise constructed in the ground.

white navy beans These beans are small, white ovals and are best used in soups and stews and as baked beans.

whole food a food that is altered as little as possible from the plant or animal tissue from which it was taken—such as milk, oats, potatoes, or apples.

whole grain refers to a grain that is milled in its entirety (all but the husk), not refined. Whole grains include wheat, corn, rice, rye, oats, amaranth, barley, buckwheat, sorghum, and millet. Two others—bulgur and couscous—are processed from wheat grains.

yard-long beans thin, tender string beans that grow to as long as 18 inches.

xeropthalmia (ZEER-ahf-THALL-me-uh) a severe form of eye disease in which the cornea hardens and may cause blindness. The problem results from vitamin A deficiency.

yo-yo dieting the practice of losing weight and then regaining it only to lose it and regain it again.

zapote an apple-size fruit with green skin and black flesh.

Index

D

E

Q

R

W

X

Y

Z

How to Read a Food Label

You can use the food label to help you make informed food choices for healthy eating practices. The food label allows you to compare similar products, determine the nutritional value of the foods you choose, and can increase your awareness of the links between good nutrition and reduced risk of chronic diet-related diseases.

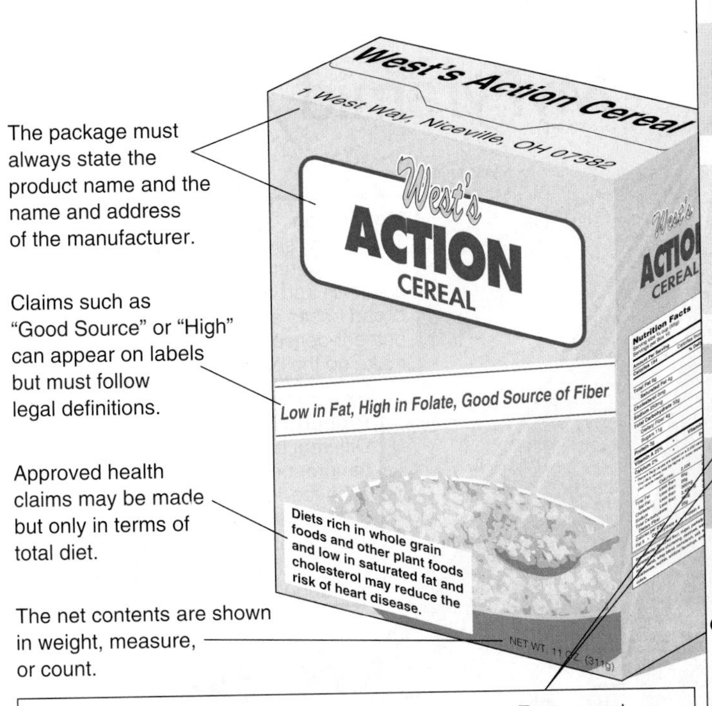

West's Action Cereal
1 West Way, Niceville, OH 07582
West's
ACTION
CEREAL
Low in Fat, High in Folate, Good Source of Fiber
Diets rich in whole grain foods and other plant foods and low in saturated fat and cholesterol may reduce the risk of heart disease.
NET WT. 11 OZ (311g)

The package must always state the product name and the name and address of the manufacturer.

Claims such as "Good Source" or "High" can appear on labels but must follow legal definitions.

Approved health claims may be made but only in terms of total diet.

The net contents are shown in weight, measure, or count.

The only vitamins required to appear on the Nutrition Facts panel are vitamins A and C. If a manufacturer makes a nutrition claim about another vitamin, however, the amount of that nutrient in a serving of the product must also be stated on the panel. For instance, the cereal shown is touted as "High in Folate," so the percentage of the recommended intake for folate in a serving of the cereal (25%) is stated on the label. The manufacturer has the option of listing any other nutrients as well.

Nutrition Facts
Serving size 3/4 cup (55g)
Servings per Box 5

Amount Per Serving	
Calories 167	Calories from Fat 27
	% Daily Value*
Total Fat 3g	5%
Saturated Fat 1g	5%
***Trans* Fat** 0g	
Cholesterol 0mg	0%
Sodium 250mg	10%
Total Carbohydrate 32g	11%
Dietary Fiber 4g	16%
Sugars 11g	
Protein 3g	
Vitamin A	0%
Vitamin C	15%
Calcium	10%
Iron	25%
Vitamin D	0%
Thiamin	25%
Riboflavin	25%
Niacin	25%
Folate	25%
Phosphorus	15%
Magnesium	10%
Zinc	6%
Copper	8%

*Percent Daily values are based on a 2,000 calorie diet. Your daily values may be higher or lower depending on your calorie needs:

	Calories:	2,000	2,500
Total Fat	Less than	65g	80g
Sat Fat	Less than	20g	25g
Cholesterol	Less than	300mg	300mg
Sodium	Less than	2,400mg	2,400mg
Total Carbohydrate		300g	375g
Dietary Fiber		25g	30g

Calories per gram
Fat 9 • Carbohydrate 4 • Protein 4

Calorie/gram reminder

Ingredients: Whole oats, milled corn, enriched wheat flour (contains niacin, reduced iron, thiamin mononitrate, riboflavin, folic acid), dextrose, maltose, high fructose corn syrup, brown sugar, partially hydrogenated cottonseed oil, coconut oil, walnuts, vitamin C (sodium ascorbate), vitamin A (palmitate), iron.

Ingredients in descending order of predominance by weight

Start here

Serving size, number of servings per container, and calorie information

Limit these nutrients.

Information on sodium is required on food labels.

Get enough of these nutrients.

Guide to the % Daily Value: Quantities of nutrients per serving and percentage of Daily Value for nutrients based on a 2,000 calorie energy intake.
5% or less is low.
10% or more is good.
20% or more is high.

Calcium and iron are required on the label. Look for foods that provide 10% or more of these minerals.

Reference values

This allows comparison of some values for nutrients in a serving of the food with the needs of a person requiring 2,000 or 2,500 calories per day.

Daily Values (DV) Used on Food Labels[a]

Daily Reference Values (DRVs)[b]

Food Component	Amount
protein[c]	50 g
fat	65 g[d]
saturated fat	20 g
cholesterol	300 mg[e]
total carbohydrate	300 g
fiber	25 g
sodium	2,400 mg
potassium	3,500 mg

Reference Daily Intakes (RDI)

Nutrient	Amount	Nutrient	Amount
Thiamin	1.5 mg	Calcium	1,000 mg
Riboflavin	1.7 mg	Iron	18 mg
Niacin	20 mg	Zinc	15 mg
Biotin	300 μg	Iodine	150 μg
Pantothenic Acid	10 mg	Copper	2 mg
Vitamin B_6	2 mg	Chromium	120 μg
Folate	400 μg[f]	Selenium	70 μg
Vitamin B_{12}	6 μg	Molybdenum	75 μg
Vitamin C	60 mg	Manganese	2 mg
Vitamin A	5,000 IU[g]	Chloride	3,400 mg
Vitamin D	400 IU[g]	Magnesium	400 mg
Vitamin E	30 IU[g]	Phosphorus	1 g
Vitamin K	80 μg		

[a]Based on 2,000 calories a day for adults and children over 4 years old.
[b]Formerly the U.S. RDA, based on National Academy of Sciences' 1968 Recommended Dietary Allowances.
[c]DRV for protein does not apply to certain populations; Reference Daily Intake (RDI) for protein has been established for these groups: children 1 to 4 years: 16 g; infants under 1 year: 14 g; pregnant women: 60 g; nursing mothers: 65 g.
[d](g) grams
e(mg) milligrams
[f](μg) micrograms
[g]Equivalent values for the three RDI nutrients expressed as IU are: vitamin A, 900 RE (assumes a mixture of 40% retinol and 60% beta-carotene); vitamin D, 10 μg; vitamin E, 20 mg.

MyPyramid: Steps to a Healthier You

Variety
The colors of the pyramid illustrate variety: each color represents one of the five food groups, plus one for oils.

Gradual Improvement
Gradual Improvement is encouraged by the slogan. It suggests that individuals can benefit from taking small steps to improve their diet and lifestyle each day.

MyPyramid
STEPS TO A HEALTHIER YOU
MyPyramid.gov

Activity
A person climbing steps reminds consumers to be physically active each day.

Moderation
The narrow slivers of color at the top imply moderation in foods rich in solid fats and added sugars. The broad bases at the bottom represent nutrient-dense foods that should make up the bulk of the diet.

Personalization
Personalization is shown by the person on the steps, the slogan, and the URL. Find the kinds and amounts of food to eat each day at www.MyPyramid.gov.

Proportionality
Different band widths suggest the proportional contribution of each food group to a healthy diet. Greater intakes of grains, vegetables, fruit, and milk are encouraged by the broad bases of orange, green, red, and blue.

GRAINS Make half your grains whole	VEGETABLES Vary your veggies	FRUITS Focus on fruits	OILS*	MILK Get your calcium-rich foods	MEAT & BEANS Go lean with protein
Eat at least 3 oz of whole grain cereals, breads, crackers, rice, or pasta every day. 1 oz is about 1 slice of bread, about 1 cup of breakfast cereal or 1/2 cup of cooked rice, cereal, or pasta.	Eat more dark-green veggies like broccoli, spinach, and other dark leafy greens. Eat more orange vegetables like carrots and sweet potatoes. Eat more dry beans and peas like pinto beans, kidney beans, and lentils.	Eat a variety of fruit. Choose fresh, frozen, canned, or dried fruit. Go easy on fruit juices.		Go low fat or fat free when you choose milk, yogurt, and other milk products. If you don't or can't consume milk, choose lactose-free products or other calcium sources such as fortified foods and beverages. 1 cup = 1 1/2 oz natural cheese, or 2 oz processed cheese.	Choose low fat or lean meats and poultry. Bake it, broil it, or grill it. Vary your protein routine–choose more fish, beans, peas, nuts, and seeds. 1 oz = 1 oz meat, poultry, or fish; 1/4 cup cooked dry beans; 1 egg; 1 tbsp peanut butter; 1/2 oz nuts/seeds.
For a 2,000 calorie diet, you need the amounts below from each food group. To find the amounts that are right for you, go to www.MyPyramid.gov.					
Eat 6 oz every day.	Eat 2 1/2 cups every day.	Eat 2 cups every day.		Get 3 cups every day; for kids aged 2 to 8, It's 2.	Eat 5 1/2 oz every day.

*Make most of your fat choices from fish, nuts, and vegetable oils. Find your allowance for oils at www.MyPyramid.gov.

SOURCE: U.S. Department of Agriculture, 2005.

The Best Of *The Mailbox*
Intermediate Edition

A collection of ideas
from the first ten years of *The Mailbox* magazine

Editor In Chief: Margaret Michel

Editorial Manager: Charlotte Perkins

Editors:
Lynn Bemer
Diane Badden
Karen Shelton
Becky Simpson
Kathy Wolf

Artists:
Teresa Davidson
Teresa Fogleman
Jennifer Tipton
Irene Maag Wareham

Contributing Artist:
Marilynn G. Barr

The Best Of *The Mailbox*

About This Book

It's been ten busy, rewarding years since we published the first issue of *The Mailbox* magazine. To celebrate, we've compiled in this volume many of the best, teacher-tested ideas published in the intermediate edition of *The Mailbox* since 1979. These practical ideas were sent to us by teachers across the United States. Our staff of editors selected ideas from over 50 issues to provide you with this creative and useful volume.

With *The Mailbox,* our commitment to you has always been to provide the most valuable teacher resource on the market. We take pride in bringing to you the brightest ideas and most timely teaching units. We've included many of our regularly featured sections of the magazine plus many special units.

- **Bulletin Boards:** We hope you'll involve your students in making colorful and motivating bulletin boards. Reproducible patterns make the task even easier!
- **File Folder Ideas:** These easy-to-store learning activities make basic skills fun.
- **Rainy Day Activities:** Students will love these creative art activities for every season of the year. You'll appreciate the complete instructions and inexpensive materials.
- **Lifesavers:** These management tips will help you create a positive classroom, encourage student discipline, and organize your materials.
- **Spotlight On Centers:** Adapt these easy-to-make manipulatives and learning centers to fit the needs of your students.
- **Game Plans:** Add these terrific ideas to your grab bag of individual and group games.
- **Book Specials:** To motivate reading and literature appreciation, try these activities and reproducible worksheets based on popular children's books.
- **Special Units:** We've featured many of our most popular teaching units plus reproducible worksheets. Research skills, solar system, punctuation, parent conferences, and dinosaurs are just some of the helpful topics.
- **Science Specials:** Make science something special with these motivating activities.
- **Our Readers Write:** We love hearing from all of you, and one thing we often hear is that this section of teaching tips is a favorite!
- **Fun With Numbers:** Getting kids excited about math is as easy as one, two, three with these fun activities.
- **Reproducible Patterns and Worksheets:** You'll find practical, creative reproducibles that you'll use again and again.
- **Answer Keys:** All the answers to this volume's reproducible worksheets at your fingertips.